Y0-AFI-963

INSTRUMENTATION AND PROCESS CONTROL

SIXTH EDITION

Franklyn W. Kirk
Thomas A. Weedon
Philip Kirk

atp AMERICAN TECHNICAL PUBLISHERS
Orland Park, Illinois 60467-5756

American Technical Publishers, Inc., Editorial Staff

Editor in Chief:
Jonathan F. Gosse
Vice President—Production:
Peter A. Zurlis
Multimedia Manager:
Carl R. Hansen
Technical Editor:
Eric F. Borreson
Copy Editor:
Talia J. Lambarki
Cover Design:
Bethany J. Fisher

Illustration/Layout:
Melanie G. Doornbos
Nick W. Basham
Thomas E. Zabinski
Bethany J. Fisher
Digital Resources Development:
Robert E. Stickley
Kathleen A. Moster
Daniel Kundrat
Adam T. Schuldt
Cory S. Butler

6 7 8 9 – 14 – 9 8 7 6 5 4 3 2

Printed in the United States of America

ISBN 978-0-8269-3442-0

 This book is printed on recycled paper.

ACKNOWLEDGMENTS

Technical information and assistance was provided by the following companies, organizations, and individuals:

Allen-Bradley
American Society of Heating, Refrigeration and Air-Conditioning Engineers, Inc. (ASHRAE)
ASI Robicon
ASCO Valve, Inc.
Badger Meter, Inc.
Baldor Motors and Drives
Banner Engineering Corp.
Bell & Gossett
BERK-TEK
Brooks Instrument
Cleaver-Brooks
Dwyer Instruments, Inc.
Endress + Hauser
Fireye, Inc.
Fisher Controls International, LLC
Fluid Components International
Fluke Corporation
The Foxboro Company
GE Panametrics
Gerald Liu, P. Eng
GE Thermometrics
Hach
Hart Scientific
Hedland
Honeywell, Inc.
Honeywell's MicroSwitch Division
Honeywell Sensing & Control
Industrial Scientific Corporation
Intec Controls Corporation
Ircon, Inc.
Jackson Systems, LLC
K-TEK, LLC
Kay-Ray/Sensall, Inc.
McDonnell & Miller
NASA/JPL/Caltech
NDC Infrared Engineering
Norcross Corporation
Oseco, Inc.
Pepperl + Fuchs, Inc.
Pyromation
Rockwell Automation, Allen-Bradley Company, Inc.
Rosemount Analytical Instruments
Southern Forest Products Association
Sprecher + Schuh
Thermo Electron Corporation
Trerice, H.O., Co.
TSI Incorporated
Vishay BLH
Watts Regulator Company
Weiss Instruments, Inc.
Worcester Controls Corporation
Xycom, Inc.

CONTENTS

CONTENTS

DIGITAL RESOURCES

- Quick Quizzes®
- Illustrated Glossary
- Flash Cards
- Review Questions
- Instrumentation Resources
  - Pneumatic Transmission and Control
  - Principles of Electricity
  - Instrumentation Tables
- Media Library
- ATPeResources.com

INTRODUCTION

Instrumentation and Process Control presents all aspects of instrumentation and builds on the quality of previous editions as an industry-leading instructional tool. This comprehensive textbook is a revision of the classic textbook by Franklyn W. Kirk, supplemented by the expertise of Thomas A. Weedon and Philip Kirk.

Instrumentation and Process Control has been used for many years to teach the principles of industrial instruments and control systems. New content has been added to explain the operation of battery-powered instruments and new instruments when taking difficult airflow measurements. In addition, the section on Ethernet communication has been completely updated to include the most recent innovations in the field of network communication, and the section on personal protective equipment has been updated to highlight NFPA 70E® requirements. Detailed illustrations, photographs from leading companies in the field, and concise text enhance learning. Informative tech facts and illustrative vignettes provide supplemental content throughout the textbook.

Instrumentation and Process Control offers the following features:

- Provides a technician-level approach to the field of instrumentation.
- Contains twelve sections covering instruments, digital communications, controls, and applications.
- Takes a systems approach to integrating instruments into a complex control system.
- Includes comprehensive applications of instruments and controllers in typical industrial control systems. Examples show how to install instruments and protect them from damaging environmental conditions.

The book is organized in a logical sequence beginning with an introduction to the field of instrumentation and continuing through all the other elements of a control system. Throughout the text, emphasis is placed on the fundamental scientific principles that underlie the operation of the instruments. Applications are thoroughly illustrated throughout the text as well as in a separate chapter that shows how to put instruments together into systems that can be used to control a process.

Section 1, *Introduction to Instrumentation,* provides an overview of industrial instrumentation and the principles of instruments, instrumentation diagrams, and control.

Section 2, *Temperature Measurement,* covers the scientific principles of temperature, heat transfer, and temperature measurement. Extensive coverage is provided of the details necessary to install temperature instruments in systems that will work in practice. The use of thermowells, break protection, bridge circuits, and calibration procedures is included.

Section 3, *Pressure Measurement,* covers the scientific fundamentals of pressure, hydrostatics, Pascal's law, and pressure measurement. Details of instrument protection, such as chemical seals, wet legs, valve manifolds, snubbers, and siphons, are included.

Section 4, *Level Measurement,* covers the scientific rules of point and continuous level measurement, level switches, ultrasonic and radar principles, and weigh systems. Coverage of bulk solids measurement, water columns and try cocks, diaphragm seals, and compensation for elevation changes is included.

Section 5, *Flow Measurement,* covers the scientific principles of flow, Reynolds number, compressible gases, and all types of flow measurement. Additional detail is added on the location of pressure taps, differential pressure measurement, blocking valves, flow integrators and switches, and bulk solids flow.

Section 6, *Analyzers,* covers the scientific essentials used in typical industrial analysis measurements. Extensive coverage is given on liquid, gas, oxygen, humidity, moisture, viscosity, refractive index, conductivity, pH, and ORP analyzers. Sampling systems, sampling lags, and calibration are also covered.

Section 7, *Position Measurement,* covers the operation of mechanical and proximity switches as well as their application in industrial operations. Information is provided on applications including sensor installation, rotary speed sensing, continuous web handling, and safety light curtains. The switches and sensors provide position information for operational status, alarm, or interlocking systems.

Section 8, *Transmission and Communication,* covers the operation of modern wireless and industrial digital communication systems as well as provides information on electrical and pneumatic communication systems. Information is included on digital and hex numbering systems, network addressing and configuration, and the OSI reference model protocol. Detail on ground loops, transmitter calibration, wiring formats, and connectors is included to help the technician understand the operation of communication systems.

Section 9, *Automatic Control,* covers the operation of modern automatic control systems as well as provides coverage of electrical controllers. Additional information is provided on how process dynamics, such as load changes, gain, lag, and dead time, affect control. Extensive coverage of proportional, integral, and derivative control strategies, and controller tuning is provided. Advanced control strategies, such as artificial intelligence, fuzzy logic, and neural networks, are also discussed.

Section 10, *Final Elements,* covers the operation of final control elements including control valves, regulators, dampers, actuators and positioners, solenoid valves, variable-speed drives, and electric power controllers. Extensive detail is provided on how to install valves in systems and how to calibrate actuators.

Section 11, *Safety Systems,* covers the detailed operation of individual safety devices such as safety and relief valves, rupture discs, burner control systems, shutoff valves, and alarm systems. An overview of hazardous location classifications, enclosures, risk evaluation, and safety-instrumented systems is included.

Section 12, *Instrumentation and Control Applications,* covers how the instruments and control systems described in the previous chapters are assembled into complex control systems. Information is included on split range valves, high and low selectors, and gap action, cascade, and ratio controllers. Examples are included to show how to control temperature using heat exchangers, pressure using control valves, level using pump-up level control, and flow using ratio controllers. In addition, a detailed example is provided of a lead-lag combustion control system with oxygen trim. A boiler drum level and feedwater flow control system is described in detail.

The *Answers* include answers to odd-numbered questions from the Review Questions.

The *Appendix* includes instrumentation symbols, tables, and other information used by technicians.

The *Glossary* provides definitions of instrumentation terms introduced in the text.

The *Instrumentation and Process Control* online Digital Resources is a self-study aid designed to add to the content included in the textbook. The online Digital Resources includes Quick Quizzes® that provide an interactive review of key topics, an Illustrated Glossary, Flash Cards, Media Clips, Instrumentation resources, and related instrumentation reference material. Two chapters, *Pneumatic Transmission and Control* and *Principles of Electricity* are included.

Instrumentation and Process Control is one of several high-quality training products available from ATP. To obtain information about related training products, visit the ATP website at atplearning.com.

BOOK FEATURES

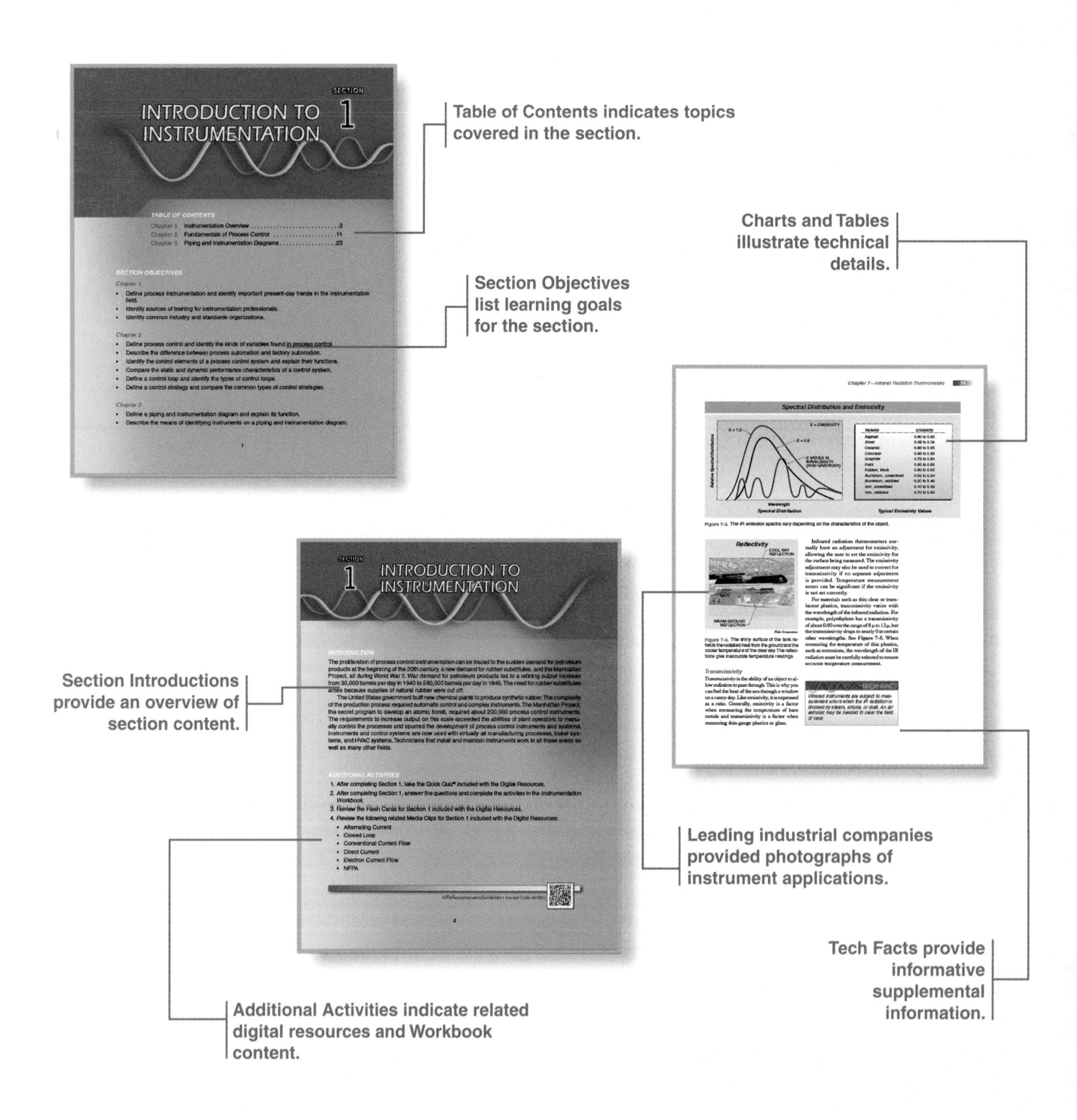

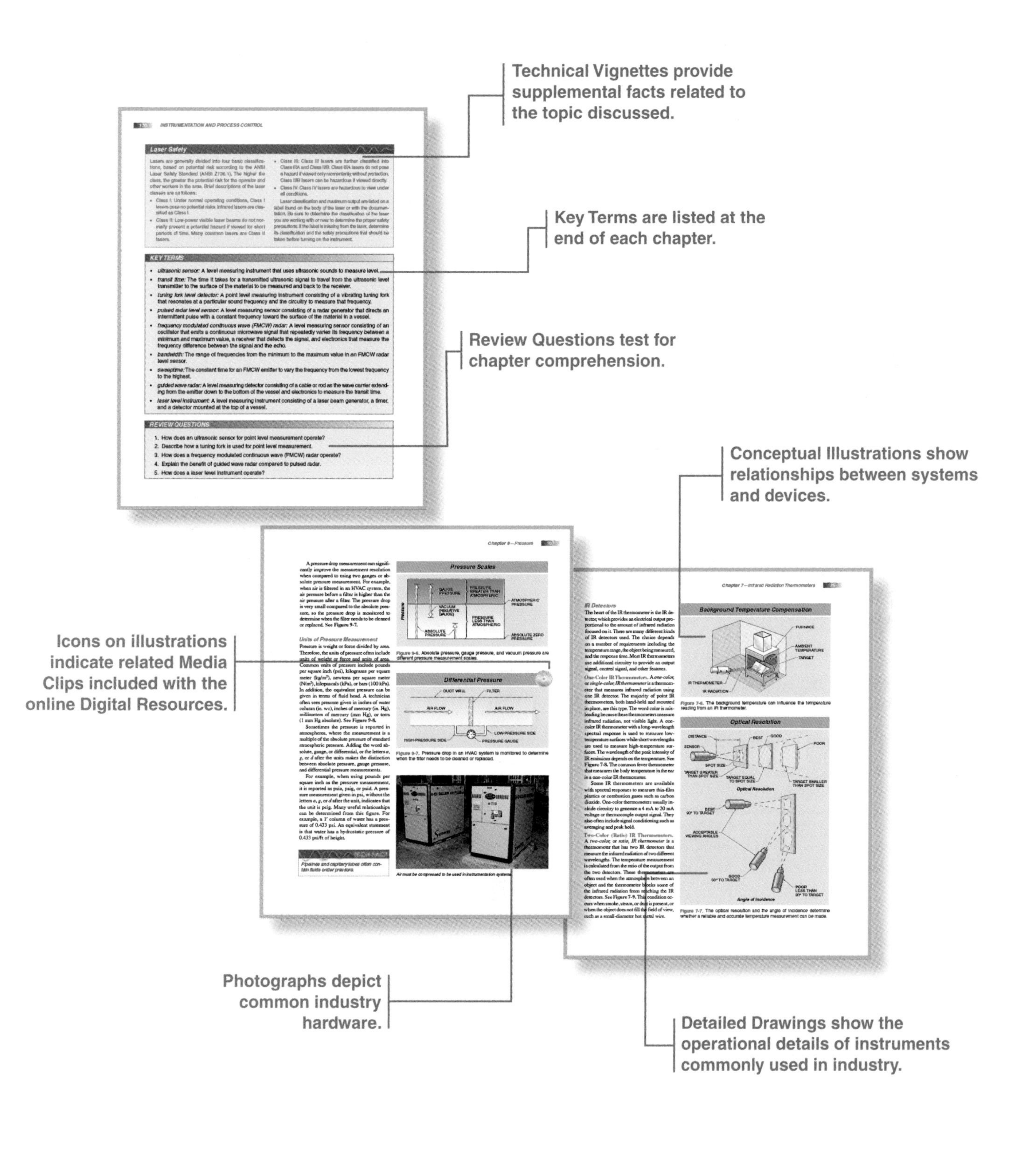

Technical Vignettes provide supplemental facts related to the topic discussed.
Key Terms are listed at the end of each chapter.
Review Questions test for chapter comprehension.
Conceptual Illustrations show relationships between systems and devices.
Icons on illustrations indicate related Media Clips included with the online Digital Resources.
Photographs depict common industry hardware.
Detailed Drawings show the operational details of instruments commonly used in industry.

DIGITAL FEATURES

Instrumentation and Process Control includes digital resources that enhance chapter concepts and promote learning. QuickLinks™ and Quick Response (QR) codes, included with each section, offer easy access to digital resources.

The Digital Resources include the following:

- **Quick Quizzes®** that provide 20 interactive questions per section, with embedded links to highlighted content within the book and to the Illustrated Glossary
- An **Illustrated Glossary** that provides a helpful reference to commonly used terms, with selected terms linked to illustrations and media clips
- **Flash Cards** that provide a self-study tool for learning the terms and definitions used in the book
- **Review Questions** that allow learners to demonstrate comprehension of chapter content
- **Instrumentation Resources** that provide data tables and supplemental information for instrumentation-related topics
- A **Media Library** that includes animations and video clips that enhance book content
- **ATPeResources.com**, which links to online resources that support continued learning

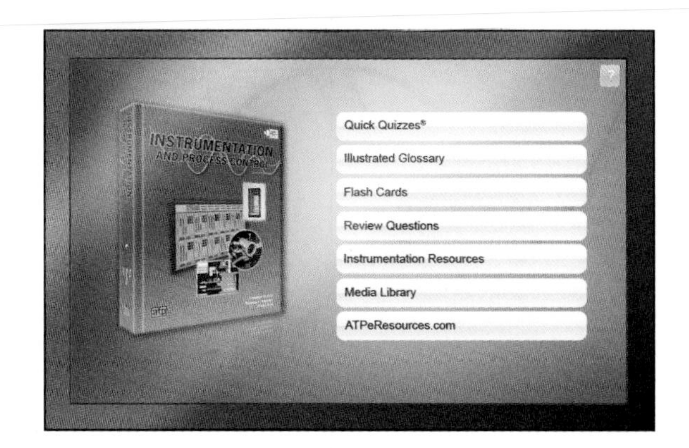

USING QUICKLINKS

To access a QuickLink using an access code, follow these simple steps:

1. Key ATPeResources.com/QuickLinks into a browser.
2. Open the page and type in the access code provided on the last page of the chapter.
3. Instantly access the digital resources.

USING QUICK RESPONSE (QR) CODES

To access a QR code, follow these simple steps:

1. Download a QR code reader app to a mobile device. (Visit atplearning.com/QR for more information.)
2. Open the app and scan the QR code on the book page.
3. Instantly access the digital resources.

RESOURCES

WORKBOOK

The *Instrumentation and Process Control Workbook* includes review questions and activities that reinforce and expand upon the information presented in the textbook. Activities help learners reinforce and apply chapter concepts by introducing typical real-world applications and having the student apply knowledge gained from the textbook.

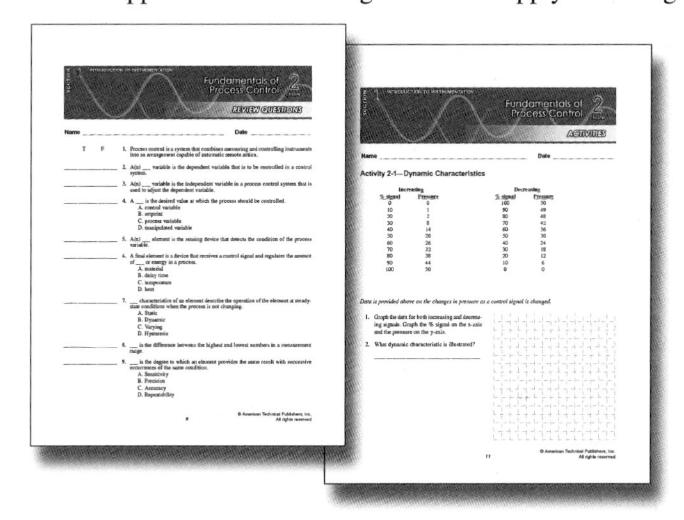
Fundamentals of Process Control
REVIEW QUESTIONS
Name Date
Fundamentals of Process Control
ACTIVITIES
Name Date
Activity 2-1—Dynamic Characteristics

INSTRUCTOR'S RESOURCE GUIDE

The *Instrumentation and Process Control Instructor's Resource Guide* provides a comprehensive teaching resource that includes a detailed Instructional Guide, editable PowerPoint® Presentations, an interactive Image Library, multiple Assessments, and Answer Keys in addition to access to the digital resources used with the textbook.

- The **Instructional Guide** explains how to best use the learning resources provided and includes Instructional Plans for each chapter that identify images that enhance the learning experience.
- **PowerPoint® Presentations** provide a review of key concepts and illustrations from the textbook.
- The **Image Library** provides all the numbered figures in a format that can be manipulated for instructional use.
- **Assessments** include sets of questions based on objectives and key concepts from each chapter and consist of a pretest, posttest, and test banks. The test banks can be used with most test development software packages and learning management systems.
- **Answer Keys** list answers to pretest, posttest, textbook, and workbook questions.

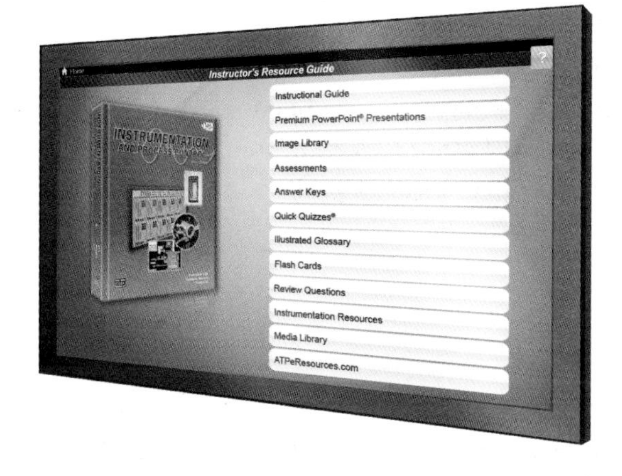

To obtain information on related products, visit the American Technical Publishers website at atplearning.com.

The Publisher

ABOUT THE AUTHORS

Mr. Franklyn Kirk was the founder of the Franklyn W. Kirk Company, an industrial instrument distributorship. Mr. Kirk developed and taught a four-semester course in Basic Instrumentation at the Technical Institute of Fenn College (now Cleveland State University) in Cleveland. He co-authored the first edition of the text, *Instrumentation*. Technical schools worldwide have used this text and the subsequent editions. Mr. Kirk continued to teach at Fenn and later for the Continuing Education Department of Cleveland State University. Mr. Kirk received the Technical Educator Award from the Cleveland Technical Societies Council. This award goes to an individual who has demonstrated outstanding performance, creative abilities, technical competence, and integrity in the practice of his or her discipline.

Mr. Thomas Weedon is an engineering graduate of the Case Institute of Technology and a licensed Professional Engineer in Ohio. Mr. Weedon has many years of industrial experience in instrument engineering. He developed a course in instrumentation for a training program for new engineers. The course was taught for many years. Mr. Weedon was involved in the control system design of many chemical plants around the country, including one of the first computer-controlled chemical plants in his field. He was one of the founders of Instrumentation Technology, Inc., a control systems consulting company providing services to chemical equipment makers and plants worldwide. Consulting provided him with extensive experience in all types of instrument applications including large-scale DCS and PLC installations with modern digital and wireless instruments. He is the author of the Instrumentation chapter in the five-volume book, *Handbook of Chlor-Alkali Technology*.

Mr. Philip Kirk is President of the Franklyn W. Kirk Company. Mr. Kirk has many years of experience in the instrumentation industry. His instrumentation experience includes specifying and designing instrumentation system solutions for industrial and commercial users. In this capacity, Mr. Kirk has acquired a practical knowledge of process sensors, controllers, software, and industrial instrumentation.

SECTION 1

INTRODUCTION TO INSTRUMENTATION

TABLE OF CONTENTS

SECTION OBJECTIVES

Chapter 1

- Define process control instrumentation and identify important present-day trends in the instrumentation field.
- Identify sources of training for instrumentation professionals.
- Identify common industry and standards organizations.

Chapter 2

- Define process control and identify the kinds of variables found in process control.
- Describe the difference between process automation and factory automation.
- Identify the control elements of a process control system and explain their functions.
- Compare the static and dynamic performance characteristics of a control system.
- Define a control loop and identify the types of control loops.
- Define a control strategy and compare the common types of control strategies.

Chapter 3

- Define a piping and instrumentation diagram and explain its function.
- Describe the means of identifying instruments on a piping and instrumentation diagram.

SECTION 1

INTRODUCTION TO INSTRUMENTATION

INTRODUCTION

The proliferation of process control instrumentation can be traced to the sudden demand for petroleum products at the beginning of the 20th century, a new demand for rubber substitutes, and the Manhattan Project, all during World War II. War demand for petroleum products led to a refining output increase from 30,000 barrels per day in 1940 to 580,000 barrels per day in 1945. The need for rubber substitutes arose because supplies of natural rubber were cut off.

The United States government built new chemical plants to produce synthetic rubber. The complexity of the production process required automatic control and complex instruments. The Manhattan Project, the secret program to develop an atomic bomb, required about 200,000 process control instruments. The requirements to increase output on this scale exceeded the abilities of plant operators to manually control the processes and spurred the development of process control instruments and systems. Instruments and control systems are now used with virtually all manufacturing processes, boiler systems, and HVAC systems. Technicians that install and maintain instruments work in all these areas as well as many other fields.

ADDITIONAL ACTIVITIES

1. After completing Section 1, take the Quick Quiz® included with the Digital Resources.
2. After completing each chapter, answer the questions and complete the activities in the *Instrumentation and Process Control Workbook.*
3. Review the Flash Cards for Section 1 included with the Digital Resources.
4. Review the following related Media Clips for Section 1 included with the Digital Resources:
   - Alternating Current
   - Closed Loop
   - Conventional Current Flow
   - Direct Current
   - Electron Current Flow
   - NFPA
5. After completing each chapter, refer to the Digital Resources for the Review Questions in PDF format.

ATPeResources.com/QuickLinks • Access Code: 467690

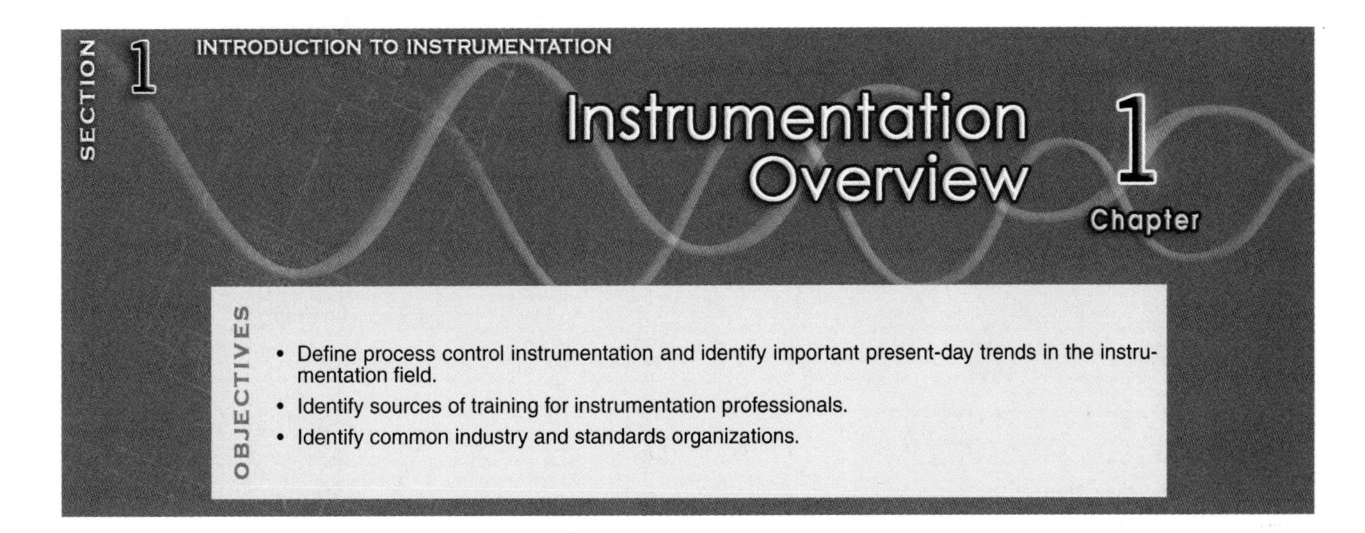

Instrumentation Overview

OBJECTIVES

- Define process control instrumentation and identify important present-day trends in the instrumentation field.
- Identify sources of training for instrumentation professionals.
- Identify common industry and standards organizations.

INSTRUMENTATION

Much has been written about the advances of science and technology in the last century and the advances yet to come. What is easily overlooked is the vital role process control instrumentation plays in the manufacture and implementation of new technologies.

Process control instrumentation is the technology of using instruments to measure and control manufacturing, conversion, or treating processes to create the desired physical, electrical, and chemical properties of materials. Process control instrumentation provides the means for manufacturing everything from snack foods to automobile interiors; the capability for environmental control of municipal wastewater and fossil fuel power generation stack emissions; the method for sophisticated air conditioning in the workplace, office buildings, and homes; and much more.

Another vital aspect of process control instrumentation is its relationship to other disciplines that support industrial processes and manufacturing. Process control instrumentation measures, controls, and interacts with computer, electrical, hydraulic, and mechanical systems. Individuals involved with process control instrumentation need to have some knowledge of the other disciplines as well as an understanding of the fundamentals of process control systems.

Instrumentation and Industry

There are two significant trends that challenge those working in the field of process control instrumentation. The first trend is the replacement of simple, repairable mechanical devices by sophisticated electronic or digital systems. Therefore, the opportunity to actually see the internal working parts of an instrument is disappearing. **See Figure 1-1.** Technicians need to gain an understanding of the fundamentals of instruments from sources other than the instruments themselves to help them to work with process control instrumentation and systems.

Technicians often read instruments as part of the job of managing complex industrial processes.

Technology Changes

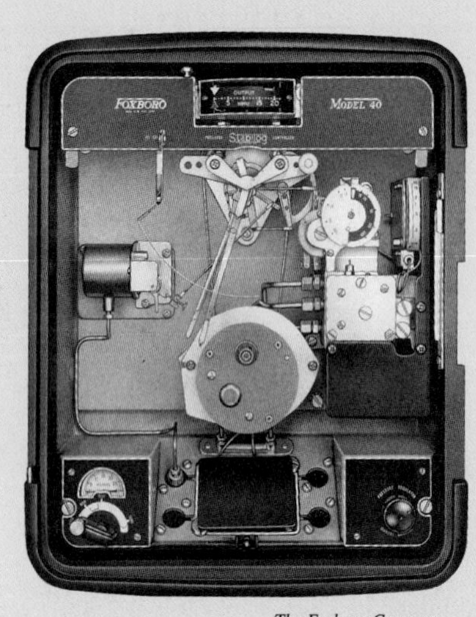

The Foxboro Company

Pneumatic Controller

Rockwell Automation, Allen-Bradley Company, Inc.

Digital Controller Modules

Figure 1-1. Pneumatic controllers have visible internal components that make it easy to see how they work. Modern digital controllers are more versatile and reliable, but the internal working parts are not as easy to see and understand.

A very early digital computer designed specifically for process control systems was developed by Westinghouse in 1959 for use in the steel industry to control rolling mill drives.

The second trend is the increasing rate of change in the complexity and sophistication of process control instrumentation. The technology of process control instrumentation is linked to the technology of computers and computer systems, which continue to evolve at an increasing rate. Advances in computer technology are quickly incorporated into process control instrumentation. Continuing education about process control instrumentation continues to be critically important to those individuals involved in the design, implementation, and utilization of the process control systems of today and tomorrow.

Digital communication systems are becoming more complex as technology evolves. If this trend continues, the work of instrumentation and control personnel may be divided into two categories. The first category is instrumentation and control applications related to the process, and the second category is the application and support of digital communications. These require different skill sets. The true experts of the future will be those who are able to master both areas of knowledge.

As a result of these trends, technicians are increasingly being asked to perform many tasks. Multiskilled maintenance technicians are becoming more popular in industry because they offer flexibility and cost effectiveness in the maintenance organization. Many technicians are directly responsible for the design, installation, and operation of process control instrumentation. For example, boiler operators or stationary engineers, HVAC technicians, electricians, computer programmers, and many others in related fields may all be responsible for process control instrumentation to some degree.

For example, a boiler operator or stationary engineer is responsible for maintaining and operating complex industrial equipment. Boiler control systems typically measure the temperature, pressure, and water level in the boiler; steam flow rate out of the boiler; conductivity of returned condensate; and the composition of the stack gas. **See Figure 1-2.** All of these are measured with instruments. The boiler operator is often responsible for the instruments used to measure these variables.

Technicians working in the HVAC field are often responsible for installing and troubleshooting systems that include pressure, temperature, flow, and humidity measurements along with the controllers that keep the systems working as designed. Digital building automation systems are being installed in new buildings and as upgrades in older buildings. Technicians need to understand the instruments employed as well as the digital communication systems that are used. Technicians often take measurements in the field to find the source of problems and to verify correct operation of the entire system. **See Figure 1-3.**

Computer programmers or engineers are typically responsible for the selection and programming of programmable controllers that are used to control industrial processes. Engineers are required to work with vendors to select the proper size of instruments, valves, and other fittings. An understanding of how the instruments operate is essential for this kind of work.

Electricians may be called upon to install and configure instruments in the field. This typically includes installing smart transmitters and controllers. Modern digital communications systems require different wiring standards and procedures than older analog systems. Many control systems use fiber-optic cables as the network communications layer. Electricians are often responsible for troubleshooting electrical systems that are part of control systems. **See Figure 1-4.**

Stack Gas Analysis

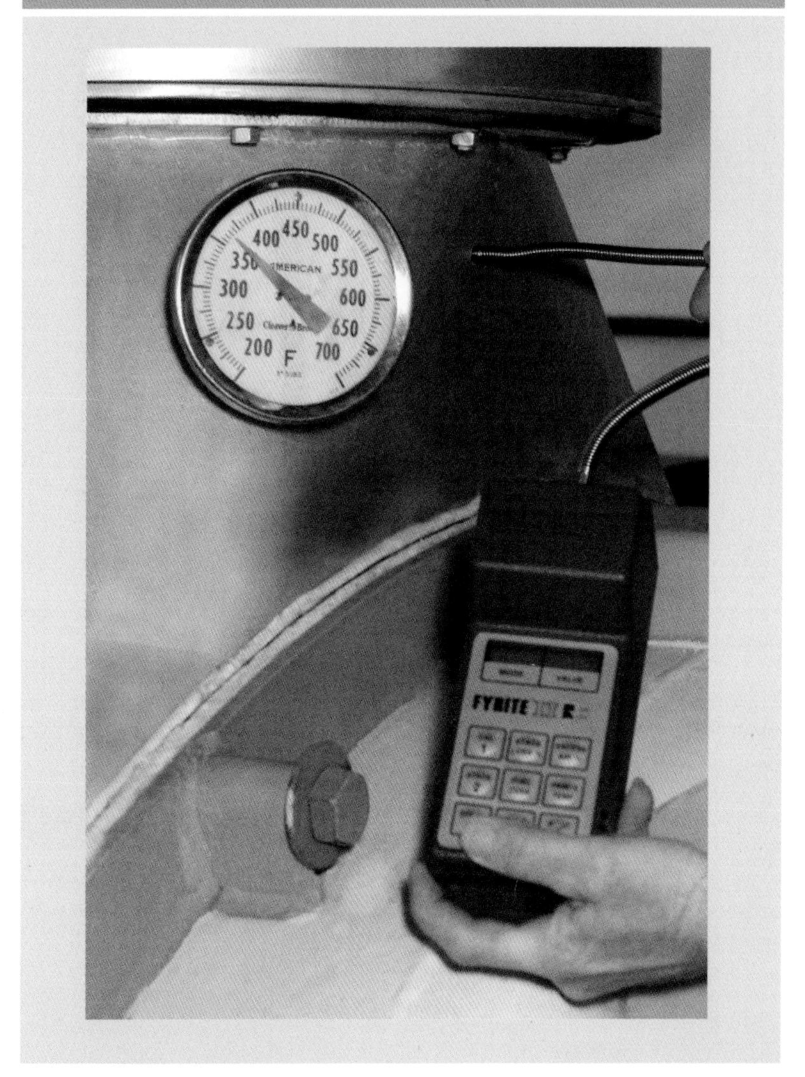

Figure 1-2. A boiler operator is responsible for calibrating the instruments used to control a boiler.

NASA/JPL/Caltech

Miniature diode lasers are used as part of the communication systems used in digital control systems.

Airflow Measurement

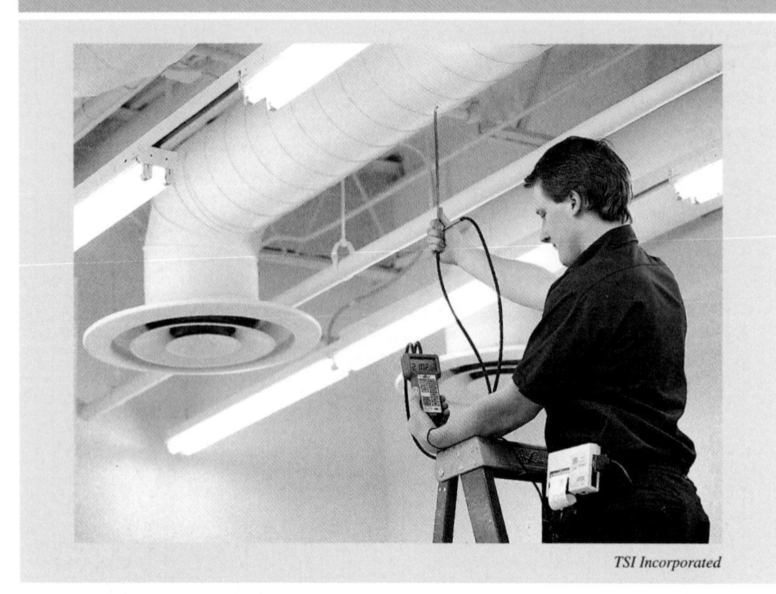

TSI Incorporated

Figure 1-3. An HVAC technician measures airflow to troubleshoot an air-handling system.

Electrical Troubleshooting

Fluke Corporation

Figure 1-4. An electrician is often required to troubleshoot electrical systems related to instrument systems.

Training

Successful technicians practice lifelong learning. Once basic skills are mastered, higher levels of skill mastery are required. Technicians need to use each task to refine knowledge and skills. In addition, new process control instrumentation equipment is being developed regularly. For example, modern control loops contain instruments that can use inputs from many types of devices and combine digital and analog output signals to a controller. **See Figure 1-5.** The ability to adapt as the industry changes is crucial for continued success. New skills are required to remain current with advancing technology and to grow in the profession.

There are only a few places where technicians and engineers can get formal training in process control instrumentation. Some four-year engineering schools and two-year technical colleges provide training in process control instrumentation as part of another program. A few technical colleges have complete programs in instrumentation as well as individual courses within other programs. The majority of technicians working with instrumentation and control systems have learned on the job. They have obtained their knowledge from coworkers, from any books that they could find that were not too theoretical, and by exposure to situations in the workplace.

Many process control instrument manufacturers provide training for their products. For example, many manufacturers of controllers provide training for using their products as well as training in general process control topics. In addition, third-party training companies provide training in many related topics. Many short courses are provided at industry trade shows. The Internet makes tutorials, instruction manuals, product specifications, and research papers available to everyone.

There are many trade union training programs that help technicians gain and develop the skills needed to be successful. For example, the International Union of Operating Engineers (IUOE) is a trade union that offers apprenticeship programs for stationary engineers. The IUOE represents stationary engineers who work in operations and maintenance in building and industrial complexes. A stationary engineer apprenticeship program is a four-year program that combines work experience and training.

Union electrical workers receive training through the National Joint Apprenticeship and Training Committee (NJATC) of the National Electrical Contractors Association (NECA) and the International Brotherhood of Electrical Workers (IBEW). Electrical workers can take courses in instrumentation-related applications.

Training Simulators. A new trend in training is using computer-based process simulators. It is expensive and time consuming to develop a simulator for a single process installation. The most frequent uses for simulators are for complex systems where there are many systems in use or where a great number of people must be trained. Simulators are also used where it would be dangerous to train on actual equipment.

Typical examples of simulator use include pilot training with flight simulators; marine captain training with simulated ships; and engineer training with simulated locomotives. All of these applications require highly trained operators, and mistakes on the actual equipment could be catastrophic.

Some chemical plants have found it worthwhile to simulate portions of their chemical processes if the processes are complex and normal learning mistakes could result in significant financial loss. In the future, more computer programs will be available that will make it easier and more economical to implement simulations for more processes.

INDUSTRY AND STANDARDS ORGANIZATIONS

Manufacturing processes have been improved over the years through the efforts of many organizations. These organizations have established safety, quality, and consistency standards and provide a vehicle for product and process improvement. These organizations can be broadly classified as government agencies, standards organizations, technical societies, private organizations, and trade associations. **See Figure 1-6.**

Figure 1-5. A modern transmitter can receive inputs from several types of instruments and sends signals in both digital and analog formats.

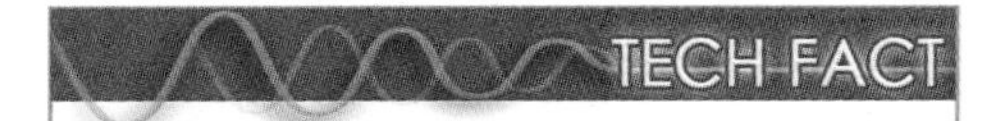

The National Joint Apprenticeship and Training Committee (NJATC) is a committee of the National Electrical Contractors Association and the International Brotherhood of Electrical Workers that provides training programs for electrical workers.

Industry and Standards Organizations

Government Agencies

EPA
Environmental Protection Agency
Ariel Rios Building
1200 Pennsylvania Avenue, NW
Washington, DC 20460
www.epa.gov

OSHA
Occupational Safety and Health Administration
200 Constitution Avenue, NW
Washington, DC 20210
www.osha.gov

NIOSH
National Institute for Occupational Safety and Health
Hubert H. Humphrey Building
200 Independence Avenue, SW
Washington, DC 20201
www.cdc.gov/niosh

NIST
National Institute of Standards and Technology
100 Bureau Drive
Gaithersburg, MD 20899-3460
www.nist.gov

Standards Organizations

ANSI
American National Standards Institute
1819 L Street, NW
Washington, DC 20036
www.ansi.org

CSA®
Canadian Standards Association
5060 Spectrum Way
Mississauga, Ontario L4W 5N6
www.csa.ca

ISO
International Organization for Standardization
1, Ch. de la Voie-Creuse, Case postale 56
CH-1211 Geneva 20, Switzerland
www.iso.ch

Technical Societies

ISA
International Society of Automation
67 Alexander Drive
P.O. Box 12277
Research Triangle Park, NC 27709
www.isa.org

IEEE
Institute of Electrical and Electronics Engineers, Inc.
IEEE Corporate Office
3 Park Avenue, 17th Floor
New York, NY 10016-5997
www.ieee.org

ASHRAE
American Society of Heating, Refrigerating and Air-Conditioning Engineers
1791 Tullie Circle, N.E.
Atlanta, GA 30329
www.ashrae.org

ASTM International
100 Barr Harbor Drive
West Conshohocken, PA 19428-2959
www.astm.org

ASME International
3 Park Avenue
New York, NY 10016-5990
www.asme.org

AIChE
American Institute of Chemical Engineers
3 Park Avenue
New York, NY 10016-5991
www.aiche.org

Private Organizations

FM Global
1301 Atwood Avenue
P.O. Box 7500
Johnston, RI 02919
www.fmglobal.com

NFPA
National Fire Protection Association
1 Batterymarch Park
Quincy, MA 02169-7471
www.nfpa.org

UL
Underwriters Laboratories, Inc.
333 Pfingsten Road
Northbrook, IL 60062-2096
www.ul.com

Trade Associations

NEMA®
National Electrical Manufacturers Association
1300 N. 17th Street, Suite 1752
Rosslyn, VA 22209
www.nema.org

API
American Petroleum Institute
1220 L Street, NW
Washington, DC 20005-4070
www.api.org

Fieldbus Foundation
9005 Mountain Ridge Drive
Bowie Building-Suite 200
Austin, TX 78759
www.fieldbus.org

Profibus International
16101 N. 82nd Street
Suite 3B
Scottsdale, AZ
www.profibus.com

Hart Communication Foundation
9390 Research Boulevard
Suite I-350
Austin, TX 78759 USA
www.hartcomm.org

Figure 1-6. There are many industry and standards organizations that influence production operations.

Government Agencies

Government agencies are federal, state, and local government organizations and departments that establish rules and regulations related to safety, health, environmental protection, and equipment installation and operation. The Occupational Safety and Health Administration (OSHA) is a federal agency under the U.S. Department of Labor that develops and enforces workplace safety and health regulations. The Environmental Protection Agency (EPA) is a federal agency established in 1970 that issues and enforces regulations affecting public health and the environment.

The National Institute of Standards and Technology (NIST), formerly the National Bureau of Standards (NBS), is a nonregulatory government agency within the Department of Commerce that develops and promotes measurement standards and technology to enhance productivity and facilitate trade. The NIST Laboratories develop and periodically update tables of thermocouple voltages.

Standards Organizations

Standards organizations are organizations affiliated with governmental agencies and organizations. Standards organizations coordinate the development of codes and standards among member organizations.

The American National Standards Institute (ANSI) is a national membership organization that helps identify industrial and public needs for national standards. The Canadian Standards Association (CSA®) is a Canadian national organization that develops standards and provides facilities for certification testing to national and international standards.

The International Organization for Standardization (ISO) is a nongovernmental international organization comprised of national standards institutions of about 150 countries (one per country). ISO provides a worldwide forum for the standards development process. ANSI is the United States representative to ISO. The International Electrotechnical Commission (IEC) is an international organization that prepares and publishes standards for electrical, electronic, and related technologies.

Technical Societies

Technical societies are organizations composed of groups of engineers and technical personnel united by professional interest. Technical societies develop and issue standards through the national standards organizations.

The International Society of Automation (ISA), formerly known as the Instrumentation, Systems, and Automation Society, is a technical society that develops standards related to instrumentation. The Institute of Electrical and Electronics Engineers (IEEE) is a technical society that publishes standards related to electrical power, computer engineering, telecommunications, and other areas. The American Society of Heating, Refrigerating and Air-Conditioning Engineers (ASHRAE®) is a technical society dedicated to advancing the art and science of heating, ventilation, air conditioning, and refrigeration, and related human factors.

ASTM International, formerly the American Society for Testing and Materials, is a technical society devoted to developing and publishing voluntary, full-consensus standards. ASME International, formerly the American Society of Mechanical Engineers, is an education and technical society dedicated to the advancement of the mechanical engineering profession. The American Institute of Chemical Engineers (AIChE) is a technical society that promotes the safe use of chemicals.

TECH FACT

Many technical societies and trade associations work with ANSI to jointly produce and publish standards.

Private Organizations

Private organizations are organizations that develop standards from an accumulation of knowledge and experience with materials, methods, and practices. The National Fire Protection Association (NFPA) is an international organization that provides scientifically based guidance in assessing the hazards of the products of combustion.

The NFPA publishes the National Electric Code® (NEC®). The NEC® requires that anyone working on electrical systems be a qualified person. OSHA defines a qualified person as "one knowledgeable in the construction and operation of the electric power generation, transmission, and distribution equipment involved, along with the associated hazards."

FM Global is a worldwide property insurance and risk management company that was formerly known as Factory Mutual. The Underwriters Laboratories Inc. (UL) is an independent organization that tests equipment and products to verify conformance to national codes and standards. Profibus International is an independent organization that develops standards for using Profibus and Profinet communications protocols. The Hart Communications Foundation is an independent organization that develops standards for using Hart communications protocols.

Trade Associations

Trade associations are organizations that represent producers and distributors of specific products. Trade associations often develop standards that are specific to their industry. The National Electrical Manufacturers Association® (NEMA®) is a trade association representing manufacturers of electrical equipment. The American Petroleum Institute (API) is the primary trade association of the United States petroleum industry that provides public policy services for advancing petroleum technology, industry equipment, and performance standards. The Fieldbus Foundation is a trade association dedicated to developing a single international, interoperable fieldbus standard.

REVIEW QUESTIONS

1. Define process control instrumentation.
2. What are two trends related to working in the field of process control instrumentation?
3. What are at least three sources of training available to instrumentation technicians?
4. Compare the roles of government agencies and standards organizations in the instrumentation field.
5. What are the roles of technical societies, private organizations, and trade associations in the instrumentation field?

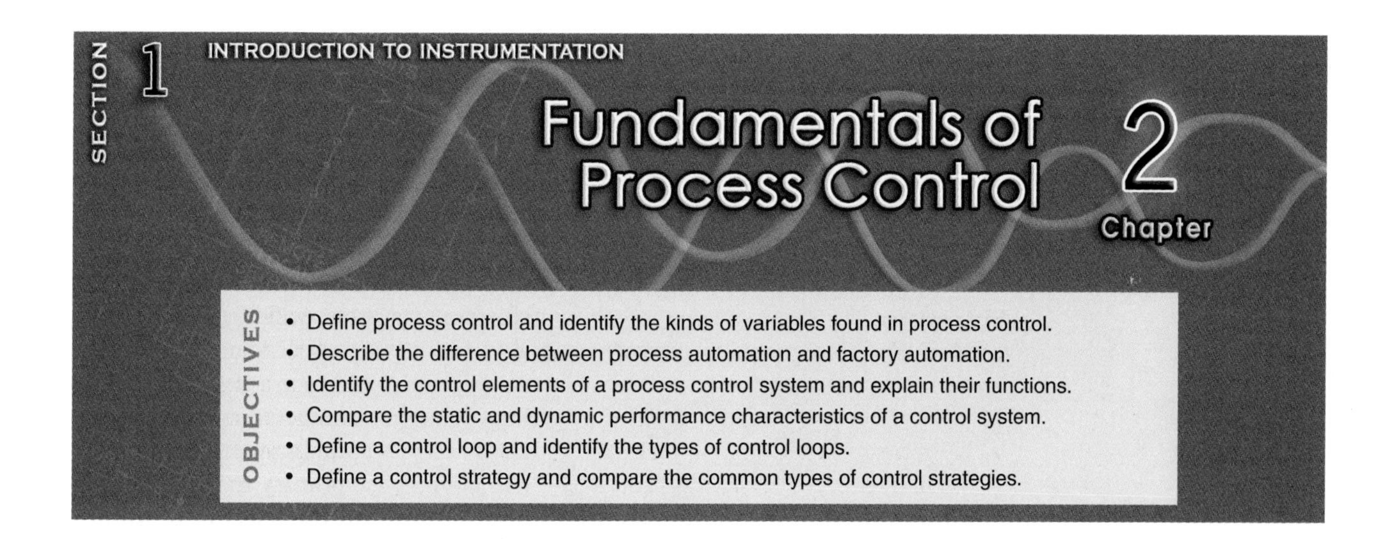

PROCESS CONTROL

Process control is a system that combines measuring materials and controlling instruments into an arrangement capable of automatic action. A process control system usually requires more than measuring and controlling the physical and chemical characteristics of a process material. It may be necessary to measure and control the characteristics of secondary materials to properly control a process. For example, the temperature and flow rate of the steam used in a heat exchanger to heat a process liquid must be controlled to control the process fluid to a desired temperature.

Variables

A *variable* is a value measured by an instrument. A *process variable (PV)* is the dependent variable that is to be controlled in a control system. Typical examples of process variables are temperature, pressure, level, and flow. A *control variable,* or *manipulated variable,* is the independent variable in a process control system that is used to adjust the dependent variable, the process variable.

Typical examples of control variables are the flow rate of steam used to heat a process, the flow rate of a refrigerant used to cool a process, and the amount of reagent required to neutralize a wastewater stream. A *setpoint (SP)* is the desired value at which the process should be controlled and is used by the controller for comparison with the process variable (PV). An operator or engineer often determines the setpoint value, but in some cases the setpoint is calculated by the process control system.

For example, the temperature of a process liquid at the exit of a heat exchanger is the process variable. The desired temperature of the process liquid is the setpoint and the steam flow to the heat exchanger is the control variable. A boiler requires feedwater to replace water boiled off as steam. The amount of feedwater is determined by measuring the level of water in the boiler. The process variable is the level of water in the boiler. The control variable is the flow rate of the feedwater into the boiler. The desired water level in the boiler is the setpoint. Boilers include auxiliary level-measuring equipment to shut down a boiler in the event of a low-water condition. **See Figure 2-1.**

> **TECH-FACT**
> *In 1927, the ITS-27 was introduced as the first International Temperature Scale for scientific and industrial measurements.*

Low-Water Fuel Cutoff

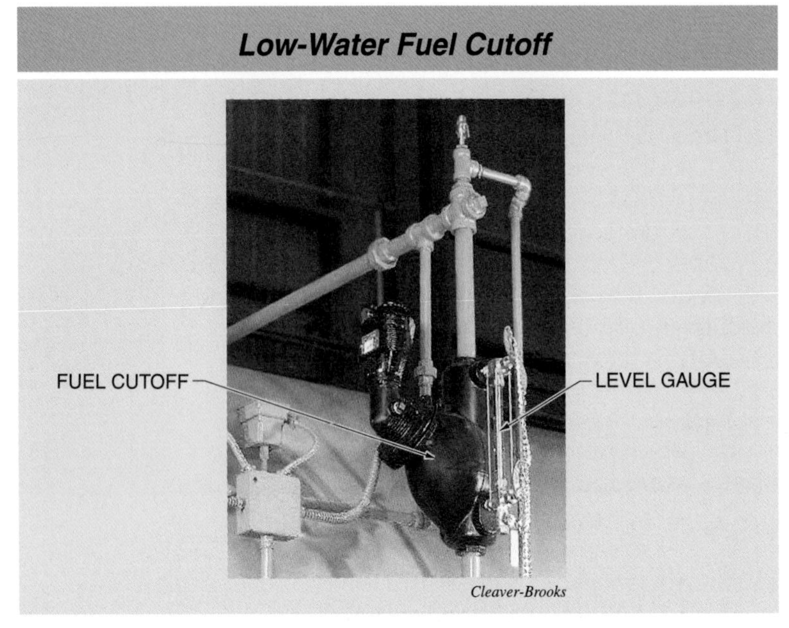

Figure 2-1. A low-water fuel cutoff is a level-measuring device that shuts down a boiler when the water level drops below the lowest allowed level.

Automation

Automation can be broadly categorized as process automation and factory automation. While there is considerable overlap between these two types of automation, there are many differences in the types of control and the order of management priorities.

Boilers use primary elements to measure the steam pressure and temperature, feedwater flow and conductivity, boiler water level, fuel flow and pressure, and composition of the stack gas.

Process Automation. Process automation refers to processes involving batch and continuous flow of liquids, gases, and bulk solids. **See Figure 2-2.** Common operations include chemical reactions, cooling and heating, and mixing and separating products. Control of process automation includes continuous process control and measuring many analog variables. Process automation places priority on process safety, process uptime, high capital investment, long plant lifetimes, and data management and product traceability. Common instrumentation systems take analog measurements of temperature, pressure, levels, and flow. Emphasis is placed on managing processes in hazardous locations, fieldbus communications at moderate rates, and continuous control.

Factory Automation. Factory automation refers to processes usually involving the piece flow of product. **See Figure 2-3.** Factory automation is typically used in the manufacturing of discrete products. Common operations include mechanical processing, assembly, moving, aligning, and transporting. Control of factory automation includes machine control and measuring many binary values. Factory automation places priority on production speed, low cost, real-time processing, and flexibility. Common instrumentation systems take binary measurements of part location, position, and state. Emphasis is placed on position sensors, barcode ID and tracking systems, high speed fieldbus communication, and motion control.

Control Elements

Variables such as temperature, pressure, flow, level, humidity, density, viscosity, and others frequently affect the outcome of a manufacturing process. In many cases, a process control system is used to control these variables. A process control system uses a primary element, a control element, and a final element. Collectively the individual elements are referred to as control elements. **See Figure 2-4.**

The *primary element* is the sensing device that detects the condition of the process variable. Examples of primary elements are thermocouples, pressure gauges, level gauges, and flowmeters.

The *control element,* or *controller,* is a device that compares a process measurement to a setpoint and changes the control variable (CV) to bring the process variable (PV) back to the setpoint. The controller accomplishes this by sending a signal to the final element. This signal may be air pressure, electric current, voltage, or a digital signal combined with at least one other signal.

A *final element* is a device that receives a control signal and regulates the amount of material or energy in a process. Common types of final elements are control valves, hydraulic pistons, variable-speed drives, relays, pumps, and dampers.

In some cases, a fourth element is added to the description of a process control system. A *measuring element* is a device that establishes a scaled value for the measured process variable. For example, a temperature transmitter is a measuring element that converts a thermocouple voltage to a scaled temperature value.

Performance Characteristics

The elements of a control system, such as instruments, controllers, and final elements, are subject to performance limitations. For example, a temperature-sensor primary element takes time to respond to changes. A sensor in a protective cover responds more slowly to a temperature change than an unprotected one. These characteristics can be classified as static or dynamic characteristics.

TECH FACT

Fieldbus networks are digital communication protocols that are optimized for use in the process automation industries. Device networks are digital protocols optimized for use in factory automation industries.

Process Automation

Processes including liquids, gases, and bulk solids:
- Chemical reactions
- Heating, cooling, mixing, and separating
- Measuring properties

Outdoor facilities and high ambient temperatures
Significant environmental protection requirements
Continuous process control
Many analog variables

Order of Priorities
- Process safety
- Process uptime
- Long facility lifetime (20+ years)
- Protection of investment and cost of ownership
- Process security
- Data authenticity and security
- Product traceability

Figure 2-2. Process automation refers to processes involving batch and continuous flow of liquids, gases, and bulk solids.

Factory Automation

Synchronized manufacturing processes:
- Moving, aligning, transporting parts
- Mechanical processing
- Measuring dimensions

Compact plants and limited ambient temperature range
Limited environmental protection requirements
State detection and motion control
Binary values dominate measurements

Order of Priorities
- Productivity (speed, high clock frequencies, etc.)
- Low costs
- Real-time processing
- High positioning accuracy (part location)
- Scalability and flexibility
- Predictable/remote maintenance
- Product traceability
- High-speed communication

Figure 2-3. Factory automation refers to processes usually involving piece flow of product.

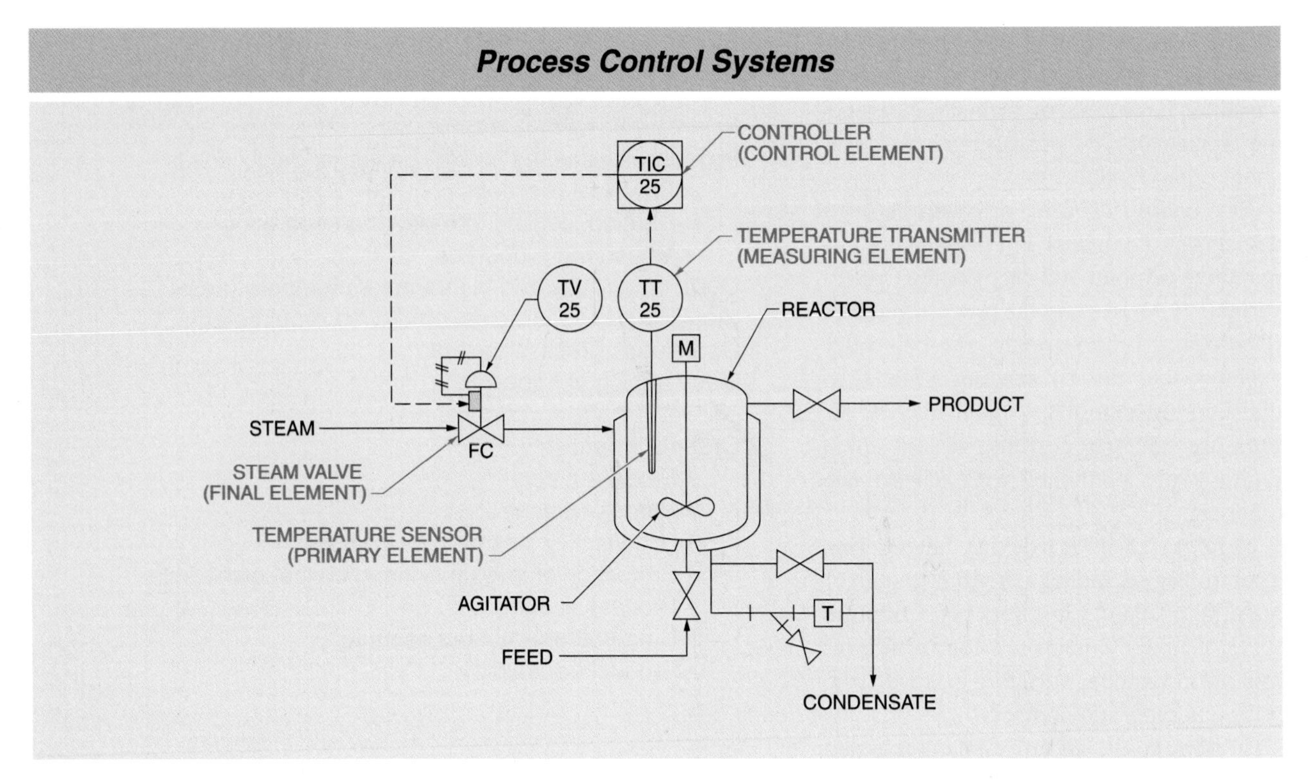

Figure 2-4. Control elements are part of a control loop used to maintain a chemical reactor at a desired temperature.

Static Characteristics. Static characteristics are the characteristics of an element that describe the operation of the element at steady-state conditions when the process is not changing. **See Figure 2-5.**

Range is the boundary of the values that identify the minimum and maximum limits of an element. For example, a temperature sensor may have a range of –50°F to 200°F. Likewise, control valves are available that operate over a variety of ranges. The valve is sized to be able to regulate the fluid flow as required by the process. An instrument or controller may be calibrated to only use part of the maximum range. An operating range is a part of the total range. Range is specified with two numbers representing the lowest and the highest values. *Span* is the difference between the highest and lowest numbers in the range.

Bias is a systematic error or offset introduced into a measurement system. Bias typically shows up as an error in measurement where the measurements are all on one side of the true value. For example, a temperature sensor may read 2° high under all conditions. A bias may be intentionally introduced into control strategy.

Accuracy is the degree to which an observed value matches the actual value of a measurement over a specified range. Manufacturers usually specify control element accuracy as the worst-case accuracy over the entire range. Accuracy is often stated as a percentage of the full-scale range or as a percentage of the reading. However, there is no standard definition of this word. Manufacturers may have different meanings when they specify accuracy.

Precision is the closeness to which elements provide agreement among measured values. Precision does not describe the same thing as accuracy. Precision only measures agreement among the measured values. It does not compare the measured values to a standard or true value.

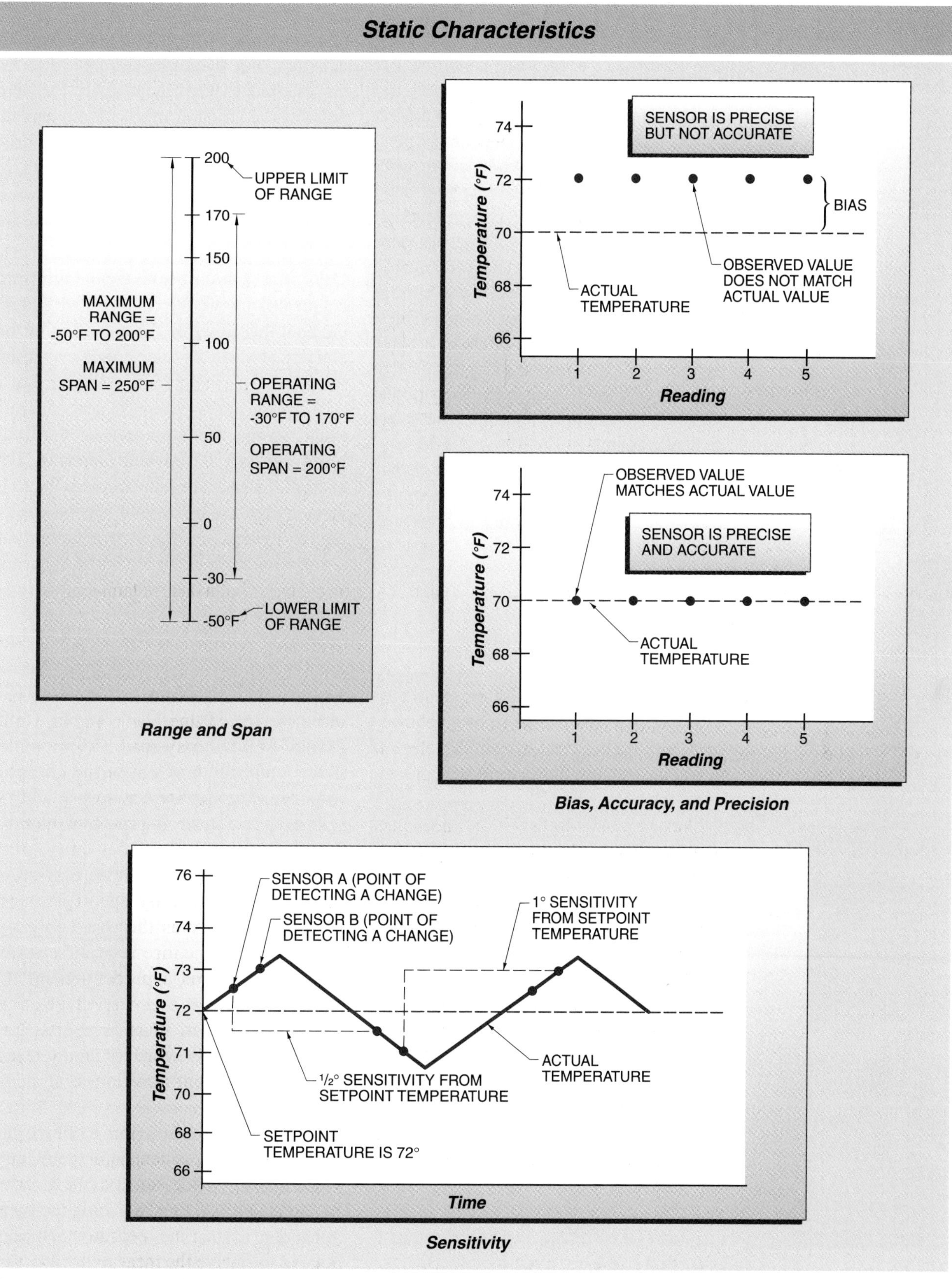

Figure 2-5. The static characteristics of control elements are the properties during steady-state operations.

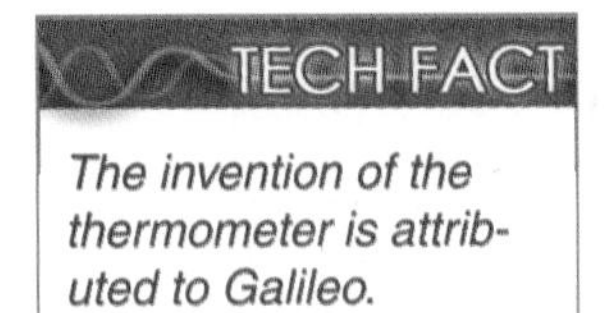

The invention of the thermometer is attributed to Galileo.

Repeatability is the degree to which an element provides the same result with successive occurrences of the same condition. For example, an air pressure switch is expected to actuate at the same air pressure every time. A lack of repeatability means that the air pressure switch actuates at different air pressures at different times. Repeatability is usually reported as a percent of the average reading.

Sensitivity is the smallest change of a value a primary element can detect, the smallest change in input that can cause a control element to change its output, or the smallest change a final element can produce. In some applications, it is important to be able to measure very small changes in the controlled variable. For example, Sensor A can detect a change as small as ½°F from the set temperature. Sensor B can only detect a change of 1°F from the set temperature. Therefore, Sensor A is twice as sensitive as Sensor B. *Dead zone* is a condition where there is no response to a change because the change is less than the sensitivity of an element.

Alarm systems are used to monitor air for the presence of toxic or suffocating atmospheres near refrigerant systems.

Drift is a gradual change in a variable over time when the process conditions are constant. For example, the operation of valves change over time as the valve stem gets dirty and friction affects the movement. Instruments with solid-state circuitry drift as the circuitry ages or with changes in ambient conditions. Microprocessor-based instruments have the ability to compensate for drift.

Dynamic Characteristics. Dynamic characteristics are the characteristics of an element that describe the operation of the element at unsteady-state conditions when the process is changing. **See Figure 2-6.**

Reproducibility is the closeness of agreement among repeated measured values when approached from both directions. For example, when a thermocouple voltage is measured, a control signal can be generated to adjust a process. Reproducibility is usually reported as a percent of the reading. Reproducibility includes hysteresis and drift.

Response time is the time it takes an element to respond to a change in the value of the measured variable or to produce a 100% change in the output signal due to a 100% change in the input signal. For example, if the temperature of a material changes, response time determines how quickly a temperature sensor indicates or records that change.

Fidelity is the ability of an element to follow a change in the value of an input. *Dynamic error* is the difference between a changing value and the momentary instrument reading or the controller action.

Hysteresis is a property of physical systems that do not react immediately to the forces applied to them or do not return completely to their original state. Systems that exhibit hysteresis are systems whose condition depends on their immediate history. Frictional or magnetic forces may cause hysteresis. Hysteresis affects valve actuators by slowing down the response to a changing control signal because of a worn linkage or overtightened packing nut. **See Figure 2-7.**

Stability is the ability of a measurement to exhibit only natural, random variation where there are no known identifiable external effects causing the variation. For example, a pressure measurement has stability when external effects, such as valves closing and pumps starting, do not affect the pressure being measured.

Linearity is the closeness to which multiple measurements approximate a straight line on a graph. Linearity is usually measured as nonlinearity and expressed as linearity. *Nonlinearity* is the degree to which multiple measurements do not approximate a straight line on a graph.

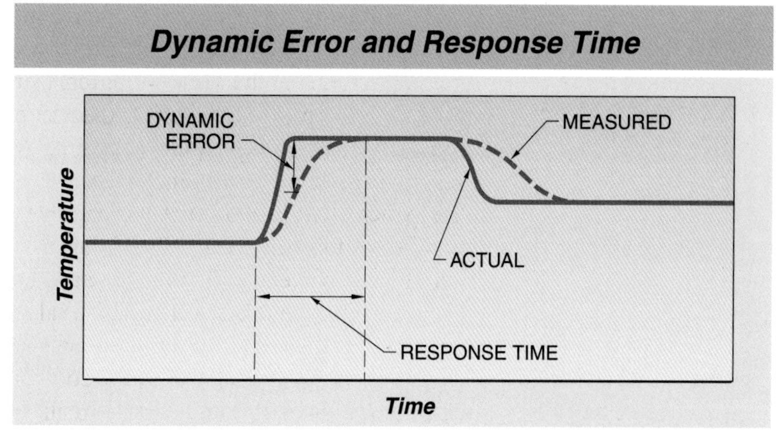

Figure 2-6. The dynamic characteristics of control elements are the properties during changing conditions.

Control Loops

An *instrument loop* is a control system in which one or more instruments are connected together to perform a task. A *control loop* is a control system in which information is transferred from a primary element to the controller, from the controller to the final element, and from the final element to the process. A process control system may measure and control many variables in a process at the same time. **See Figure 2-8.**

Closed Loops. A *closed loop* is a control system that provides feedback to the controller on the state of the process variable due to changes made by the final control element. The primary element measures the process variable and sends a signal to the controller. The controller compares the value of the variable to the setpoint and sends a signal to a final element. The primary element detects the value of the process variable after the final element has made changes. This closes the loop.

For example, a thermostat consists of a primary element and a controller. If the room temperature is below the setpoint, the thermostat controller sends a signal to the final element (boiler, furnace, or heat pump) that increases the amount of heat added to the room. When the air temperature is warm enough, the thermostat controller turns off the final element.

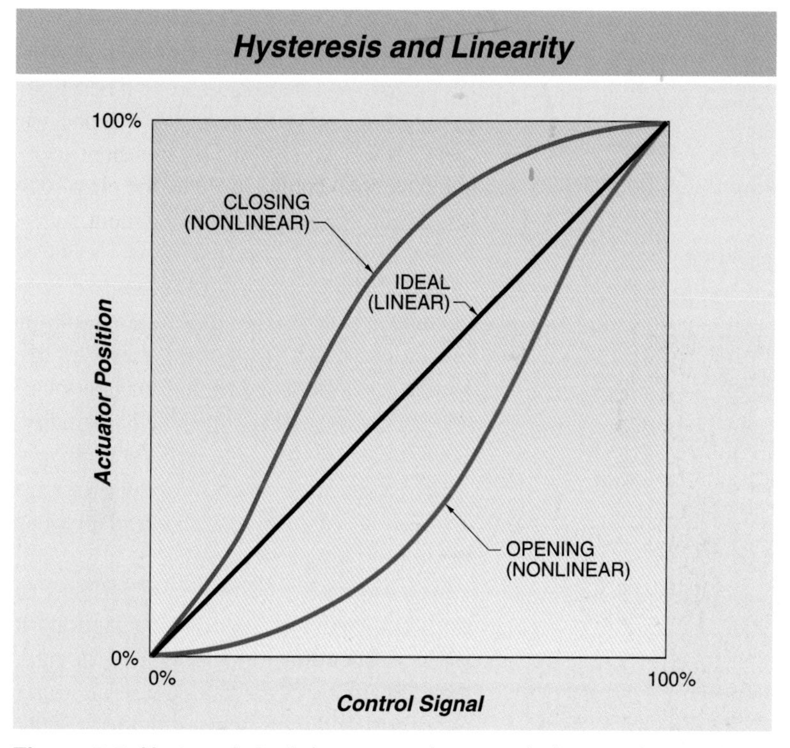

Figure 2-7. Hysteresis is the property of a control element that results in different performance when a measurement is increasing than when the measurement is decreasing.

TECH FACT

An instrument technician should understand the operation of each instrument used in a control loop as well as the operation of the entire control loop to verify that proper control is achieved.

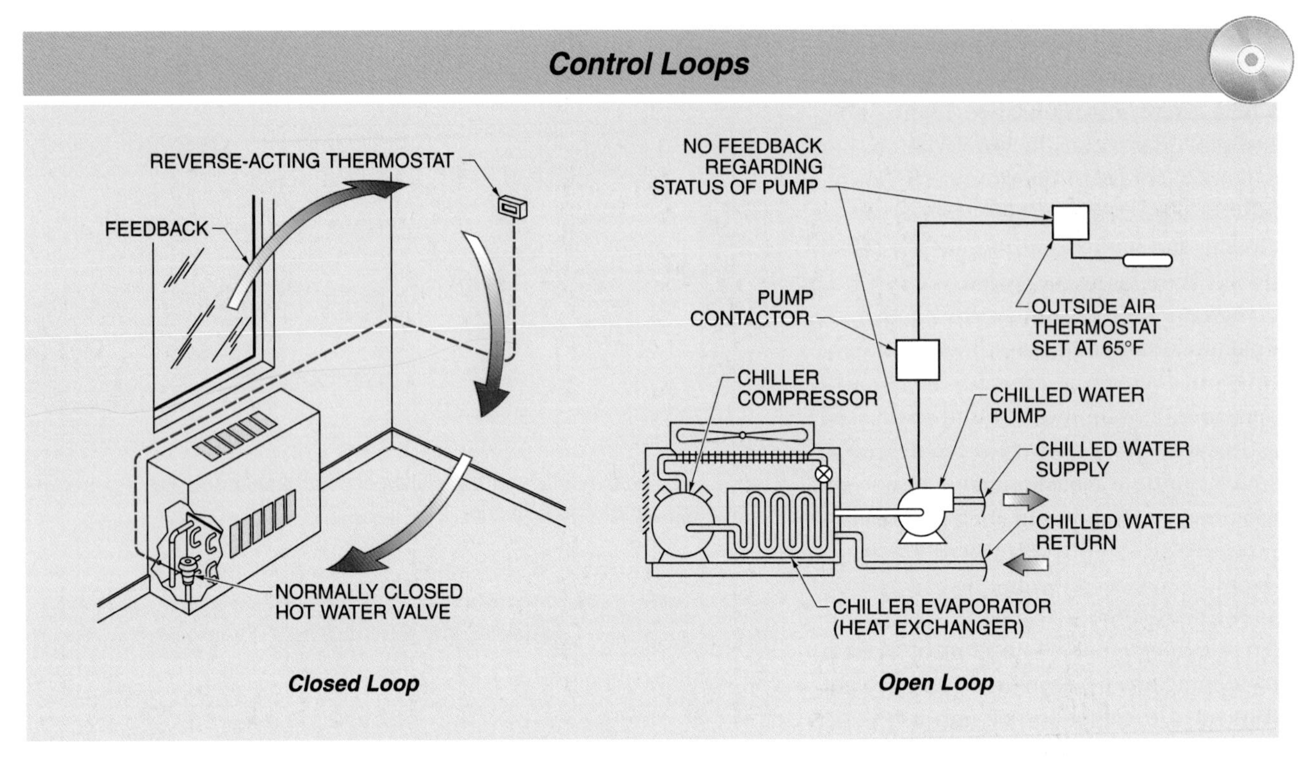

Figure 2-8. Both closed loop and open loop control systems are commonly used in industry.

Open Loops. An *open loop* is a control system that sends a control signal to a final element but does not verify the results of that control. An open loop exists if the feedback is absent or disabled, such as when the control element is operated manually. Open-loop control can be used for relatively simple processes where the relationship between the control variable and process variable is well understood.

For example, a pump can be used to send chilled water supply to refrigerate a room. Depending on the system design, the temperature of the chilled water coming out of the chiller evaporator can vary if the temperature of the outside air varies. Furthermore, if the temperature variation can be tolerated, eliminating the feedback and using simpler open-loop control can reduce the cost of the refrigeration system.

Common Control Strategies

A control strategy is a method of determining the output of a controller based on the input to the controller and the setpoint of the process. In other words, a signal is sent to a controller. The method of determining the proper controller output to send to the final element is a control strategy. There are many types of strategies that are used to control processes. Three of the most common are ON/OFF control, proportional control, and time proportional control.

ON/OFF Control. ON/OFF control is the simplest and most common control strategy. A controller activates or deactivates the final element depending on whether the measured variable is above or below the setpoint. For example, when the temperature in a room drops low enough, a thermostat activates the circuit that warms the air in the room. **See Figure 2-9.** When the temperature reaches the desired value, the thermostat deactivates the circuit. A sump pump level control system activates a pump to remove the water that has collected in a sump and deactivates the pump when the sump is empty.

In an ON/OFF control system, there is a tendency for the process variable to go above (overshoot) or below (undershoot) the setpoint. *Deadband* is the range of values where a change in measurement value does not result in a change in controller output. Deadband is the difference between the value at which the controller activates and the value at which the controller deactivates. Many process control devices permit the deadband to be adjusted.

Proportional Control. Proportional control is a control strategy that uses the difference between the setpoint and the process variable to determine a control output that is sent to a final element. Optionally, the cumulative amount of difference and the rate of change of the process variable can also be used to determine a control output. The predominant type of controller used in proportional control is a microprocessor-based electronic controller that can accept analog or digital input signals and provide an analog or digital output signal. PLCs are being used increasingly in control applications in place of single-loop controllers.

For example, a level controller for a tank of water opens and closes a valve based on the position of a float. **See Figure 2-10.** When the float is up all the way, the controller sends a signal that closes the water source valve. When the float is down all the way, the controller sends a signal that opens the valve. When the float is in between the limits, the controller sends a signal that opens the valve proportionally to where the float is located between the limits.

Time Proportional ON/OFF Control. A time proportional ON/OFF controller is a control strategy that has a predetermined output period during which the output contact is held closed (or power is ON) for a variable portion of the output period. Time proportional ON/OFF control allows a process to be regulated more precisely than is possible with simple ON/OFF control. A modulated response to the output signal is accomplished by varying the percentage of time the final element is ON and OFF over a defined time interval, called cycle time. Shorter cycle time increases the precision of the control over the process. Longer cycle time reduces wear and tear on the operating components. **See Figure 2-11.**

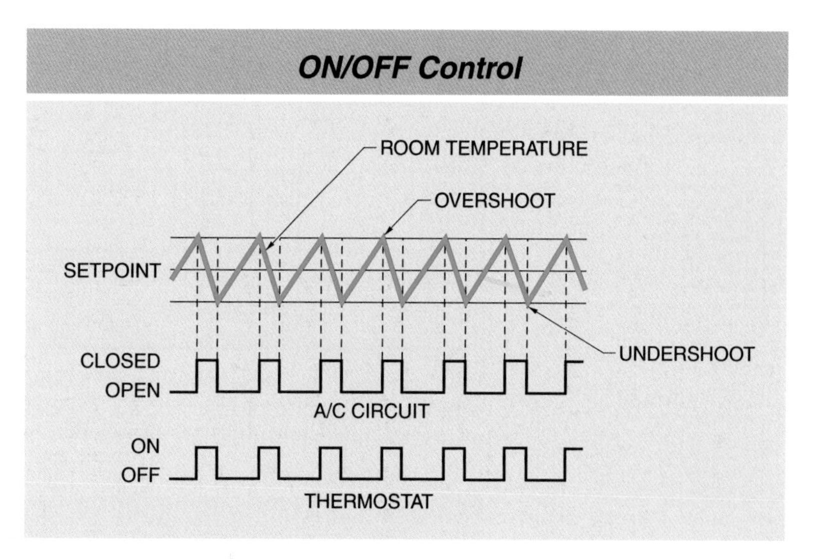

Figure 2-9. ON/OFF control is the simplest type of control.

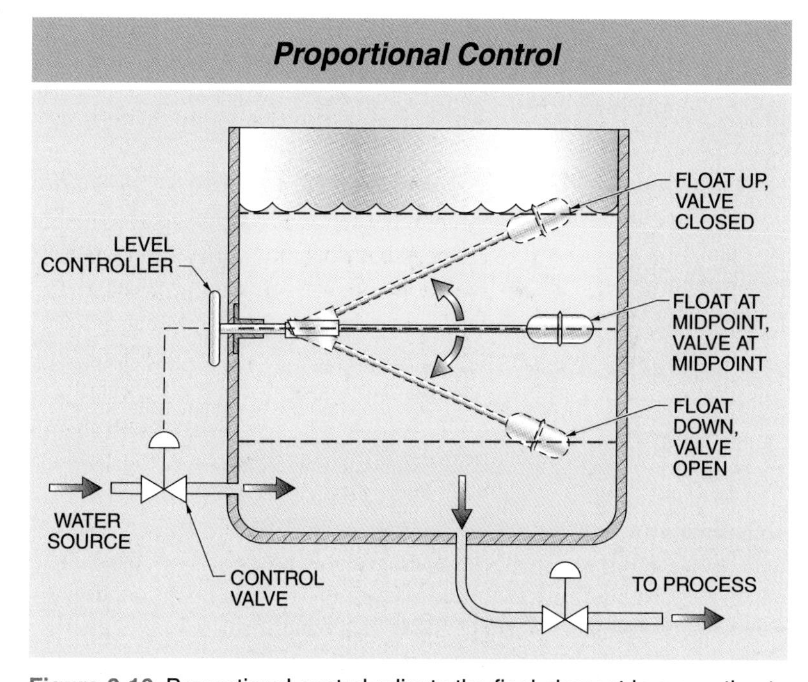

Figure 2-10. Proportional control adjusts the final element in proportion to where the process variable is within the measurement range.

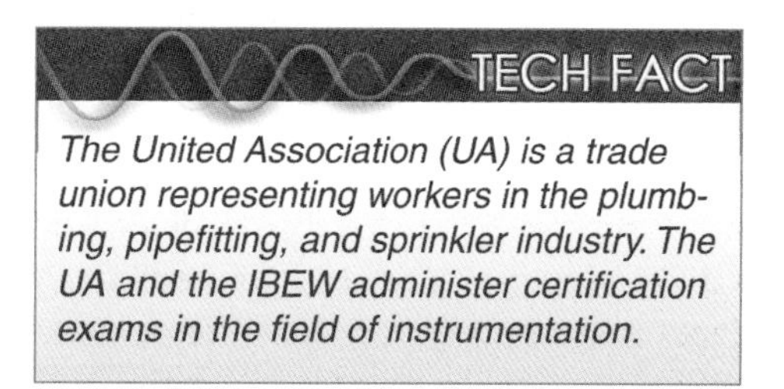

The United Association (UA) is a trade union representing workers in the plumbing, pipefitting, and sprinkler industry. The UA and the IBEW administer certification exams in the field of instrumentation.

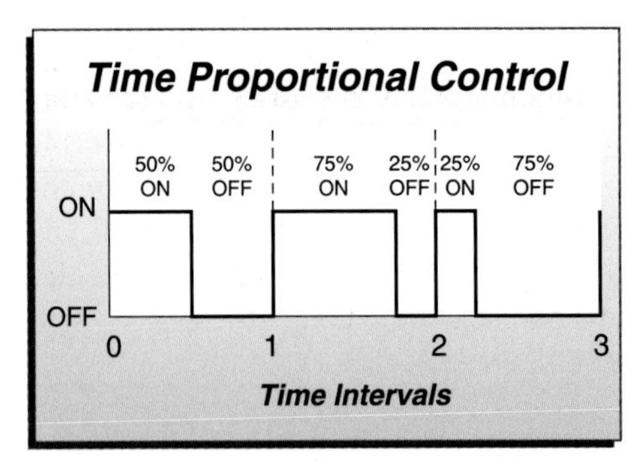

Figure 2-11. Time proportional control modulates ON/OFF control by adjusting the amount of time the final element is ON and OFF.

The Mohs mineral hardness scale was created by the mineralogist Friedrich Mohs to measure hardness. Two minerals are compared to decide which is harder.

Instrument Calibration

Calibration or recertification of process control instrumentation is often required as part of a maintenance program or to meet the requirements of a standard, such as ISO 9000. Calibration is used to certify that an instrument meets performance requirements when tested with calibrated and certified test equipment. It can also be used to adjust an instrument to correct detected inaccuracies.

When calibrating an instrument, it is common practice to document the as-found and as-calibrated performance of the instrument. Certification normally includes data of the true value and the value from the instrument. Certification may be performed by in-house staff or by an outside contractor. Calibration management software can be used to keep track of calibration data.

Measurement Scales

A useful basic principle of measurement is the concept of measurement scales. The four types of measurement scales are categorical or nominal, ordinal, interval, and ratio scales. Measurement scales are used to help determine what types of calculations and statistics are usable on a data set. For example, an average cannot be calculated for a nominal or ordinal measurement scale, but it can be calculated for an interval or ratio measurement scale.

A *categorical scale, or nominal scale,* is a measurement scale that uses unique identifiers such as numbers or labels to represent a variable. For example, computers commonly use 0 and 1 or OFF and ON. A tag number on an instrument or transmitter gives the number of the loop. The loop number is a categorical variable. There is no ranking of variables because there is no "greater than" or "less than" when discussing the variables.

An *ordinal scale* is a measurement scale that establishes rank by "more or less" or "larger or smaller." For example, the Mohs mineral hardness scale ranks diamonds as 10, corundum as 9, topaz as 8, and so on. This means that diamonds are harder than corundum, but it does not establish how much harder. Wood and lumber are graded on ordinal scales where A-Select is higher quality than B-Select. The grades are defined clearly, but there is no quantitative determination of how much one grade is better than the other.

An *interval scale* is a measurement scale where there are defined intervals between the values on the scale but the zero value is arbitrary. For example, it can be said that 80°F is higher than 40°F, however, 80°F is not twice as hot as 40°F because the choice of the zero value of the Fahrenheit scale is arbitrary.

A *ratio scale* is a measurement scale where there are defined intervals between the values on the scale and the zero value corresponds to none, or zero, of the object. Length is a ratio scale because it has a defined interval, typically inches or feet; the zero value of the scale corresponds to an object of no length; and we can say that a length of 4″ is twice as long as a length of 2″. Other common examples of ratio scales are voltage, weight, density, absolute temperature, and absolute pressure.

Measurement Scales

Scale	Description	Example
Categorical (Nominal)	Uses numbers to code data; no ranking; represents qualities instead of quantities	ON, OFF; male, female; loop number
Ordinal	Quantifies differences by order with larger or smaller ranking	Mineral hardness; lumber grades
Interval	Defined intervals; quantifies differences by "how much" larger	Fahrenheit temperatures; time of day
Ratio	Defined intervals and true zero	Length, weight, density, age, and frequency

Instrument Management Software

In addition to calibration management, software systems are available to keep track of all instruments in a whole plant. The software system maintains a complete database of information and historical issues. It can even help with the design of new loops, their numbering, documentation, wiring drawings, installation details, etc. These are very powerful systems, but it takes a large investment of money and time to document all the existing instrument systems before the new ones can be added.

KEY TERMS

- *process control:* A system that combines measuring materials and controlling instruments into an arrangement capable of automatic action.
- *variable:* A value measured by an instrument.
- *process variable:* The dependent variable that is to be controlled in a control system.
- *control variable, or manipulated variable:* The independent variable in a process control system that is used to adjust the dependent variable, the process variable.
- *setpoint (SP):* The desired value at which the process should be controlled and is used by the controller for comparison with the process variable (PV).
- *primary element:* The sensing device that detects the condition of the process variable.
- *control element, or controller:* A device that compares a process measurement to a setpoint and changes the control variable (CV) to bring the process variable (PV) back to the setpoint.
- *final element:* A device that receives a control signal and regulates the amount of material or energy in a process.
- *measuring element:* A device that establishes a scaled value for the measured process variable.
- *range:* The boundary of the values that identify the minimum and maximum limits of an element.
- *span:* The difference between the highest and lowest numbers in the range.
- *bias:* A systematic error or offset introduced into a measurement system.
- *accuracy:* The degree to which an observed value matches the actual value of a measurement over a specified range.
- *precision:* The closeness to which elements provide agreement among measured values.
- *repeatability:* The degree to which an element provides the same result with successive occurrences of the same condition.
- *sensitivity:* The smallest change of a value a primary element can detect, the smallest change in input that can cause a control element to change its output, or the smallest change a final element can produce.
- *dead zone:* A condition where there is no response to a change because the change is less than the sensitivity of an element.
- *drift:* A gradual change in a variable over time when the process conditions are constant.
- *reproducibility:* The closeness of agreement among repeated measured values when approached from both directions.

KEY TERMS *(continued)*

- *response time:* The time it takes an element to respond to a change in the value of the measured variable or to produce a 100% change in the output signal due to a 100% change in the input signal.
- *fidelity:* The ability of an element to follow a change in the value of an input.
- *dynamic error:* The difference between a changing value and the momentary instrument reading or the controller action.
- *hysteresis:* A property of physical systems that do not react immediately to the forces applied to them or do not return completely to their original state.
- *stability:* The ability of a measurement to exhibit only natural, random variation where there are no known identifiable external effects causing the variation.
- *linearity:* The closeness to which multiple measurements approximate a straight line on a graph.
- *nonlinearity:* The degree to which multiple measurements do not approximate a straight line on a graph.
- *instrument loop:* A control system in which one or more instruments are connected together to perform a task.
- *control loop:* A control system in which information is transferred from a primary element to the controller, from the controller to the final element, and from the final element to the process.
- *closed loop:* A control system that provides feedback to the controller on the state of the process variable due to changes made by the final control element.
- *open loop:* A control system that sends a control signal to a final element but does not verify the results of that control.
- *deadband:* The range of values where a change in measurement value does not result in a change in controller output.
- *categorical scale, or nominal scale:* A measurement scale that uses unique identifiers such as numbers or labels to represent a variable.
- *ordinal scale:* A measurement scale that establishes rank by “more or less” or “larger or smaller”.
- *interval scale:* A measurement scale where there are defined intervals between the values on the scale but the zero value is arbitrary.
- *ratio scale:* A measurement scale where there are defined intervals between the values on the scale and the zero value corresponds to none, or zero, of the object.

REVIEW QUESTIONS

1. List and define the kinds of variables found in process control.
2. List and define the static performance characteristics of a control system.
3. List and define the dynamic performance characteristics of a control system.
4. What is the difference between an open loop and a closed loop?
5. What are the similarities between the common types of control strategies?

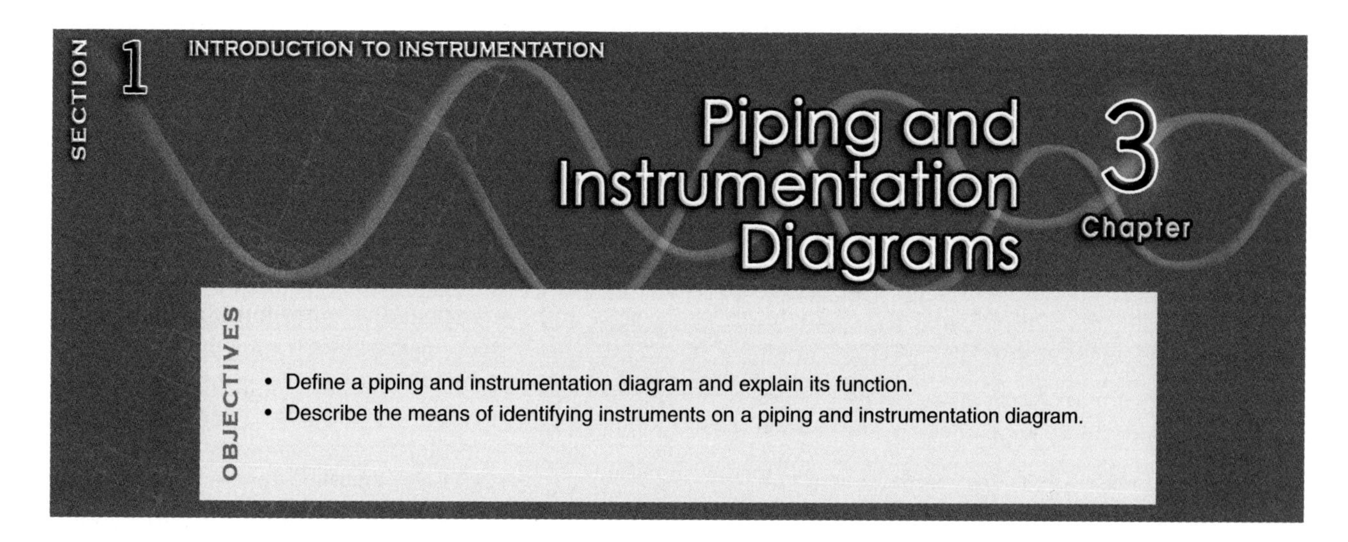

Piping and Instrumentation Diagrams

Chapter 3

OBJECTIVES

- Define a piping and instrumentation diagram and explain its function.
- Describe the means of identifying instruments on a piping and instrumentation diagram.

PIPING AND INSTRUMENTATION DIAGRAMS

Instrumentation, like many other fields, has a common language of symbols used to describe instruments. In addition, there are many government agencies and organizations that issue regulations, standards, and codes that need to be followed.

A commonly accepted way to communicate details of instrumentation systems is through piping and instrumentation diagrams. A *piping and instrumentation diagram (P&ID)* is a schematic diagram of the relationship between instruments, controllers, piping, and system equipment. A P&ID typically includes the instrumentation and designations; all valves and their designations; mechanical equipment; flow direction; control inputs and outputs and interlocks; and details of the pipe connections, sample lines, and other fittings. Standard symbols and abbreviations have been established for many common components. **See Appendix.**

The International Society of Automation (ISA) has developed a standard for instrumentation symbols that is very comprehensive. Nearly all companies use this standard for their internal P&ID documents, but modify some of the details and the instrument numbering procedures to better suit their own needs. Fortunately, the majority of the symbol usage is recognizable and it only takes a short time to learn any particular system.

INSTRUMENTATION SYMBOLS AND NUMBERING

Instruments and modes of communication are identified on P&IDs by standard designations and symbols. The letters of the alphabet are assigned common meanings, such as "T" for temperature. Letters are combined to show multiple pieces of information. The first letter identifies the function of an instrument and the variable that is being measured. Succeeding letters are used to modify the functional description. **See Figure 3-1. See Appendix.**

Variable speed drives are a common part of many control systems.

Typical Instrument Symbols and Numbers

Tags

L	Measured Variable (Level)
LT	Functional Identification (Level Transmitter)
08	Loop Number
LT 08	Complete Instrument Tag

Note: Hyphens may be used as separators

Typical Combinations

LC	Level Controller (Blind)
LIC	Level Indicating Controller
LRC	Level Recording Controller
LI	Level Indicator
LT	Level Transmitter
LV	Level Valve

Figure 3-1. The ISA has developed standard symbols and nomenclature used in instrumentation diagrams.

A P&ID typically includes symbols for the equipment and the instruments. Each instrument and related piece of equipment is identified with a balloon that contains identifying information. A *balloon,* or *bubble,* is a circular symbol used to identify the function and loop number of an instrument or device on a P&ID. A complex industrial process may have many control loops and many interconnected drawings. All the instruments and control elements in a single loop have a common number designation. The loop numbering convention is at the option of the user. Coded information may be added to the loop numbers, such as building numbers. However, simplicity is recommended.

In the past, when all instruments were individual pieces of equipment and control strategies were much simpler, a P&ID was able to convey almost all the knowledge about the control systems that was necessary. However, a P&ID cannot show the complete control strategy implemented in distributed control systems, programmable logic controllers, computer control, and more complex control systems. The most complex systems use additional types of documents to describe how the system is to function. Even with this limitation, P&IDs are still the best documents for understanding the instruments and controls and how they relate to the process equipment.

One of the best ways of learning how to read and use instrument symbols is by studying actual P&ID drawings. A drawing includes many standard symbols and labels. A review of the symbols and their use will help clarify the standards. The primary P&ID of a plating process is included as an example. **See Figure 3-2.** This example includes several loops that are used to control the composition and flow of electrolyte to the plating process. The drawing shows links to other sheets and drawings that are not included here.

Recirculation Tanks

The drawing labeled 0833MF05 is one drawing of a series that makes up a complete P&ID. This drawing has a recirculation tank, M-3-RT-1, and the circulation piping system as the main subject. **See Figure 3-3.** The lower left corner of the sheet has the title that identifies this sheet as "Plating Module No. 3 Flowsheet" and "Sheet 1 of 2," and gives the drawing number as "0833MF05."

Level Measurement. This tank's level is measured with a level gauge. A balloon is used to identify the level gauge as LG 362. The first letter represents level. The second letter is a modifier representing a gauge. The level gauge is in control loop 362. The details of the level gauge are not shown on this sheet.

TECH FACT

During high-temperature operation, steel piping expands as it is heated. An engineering thermal analysis must be performed to determine the proper supports, expansion loops, and expansion joints.

Piping and Instrumentation Diagrams (P&IDs)

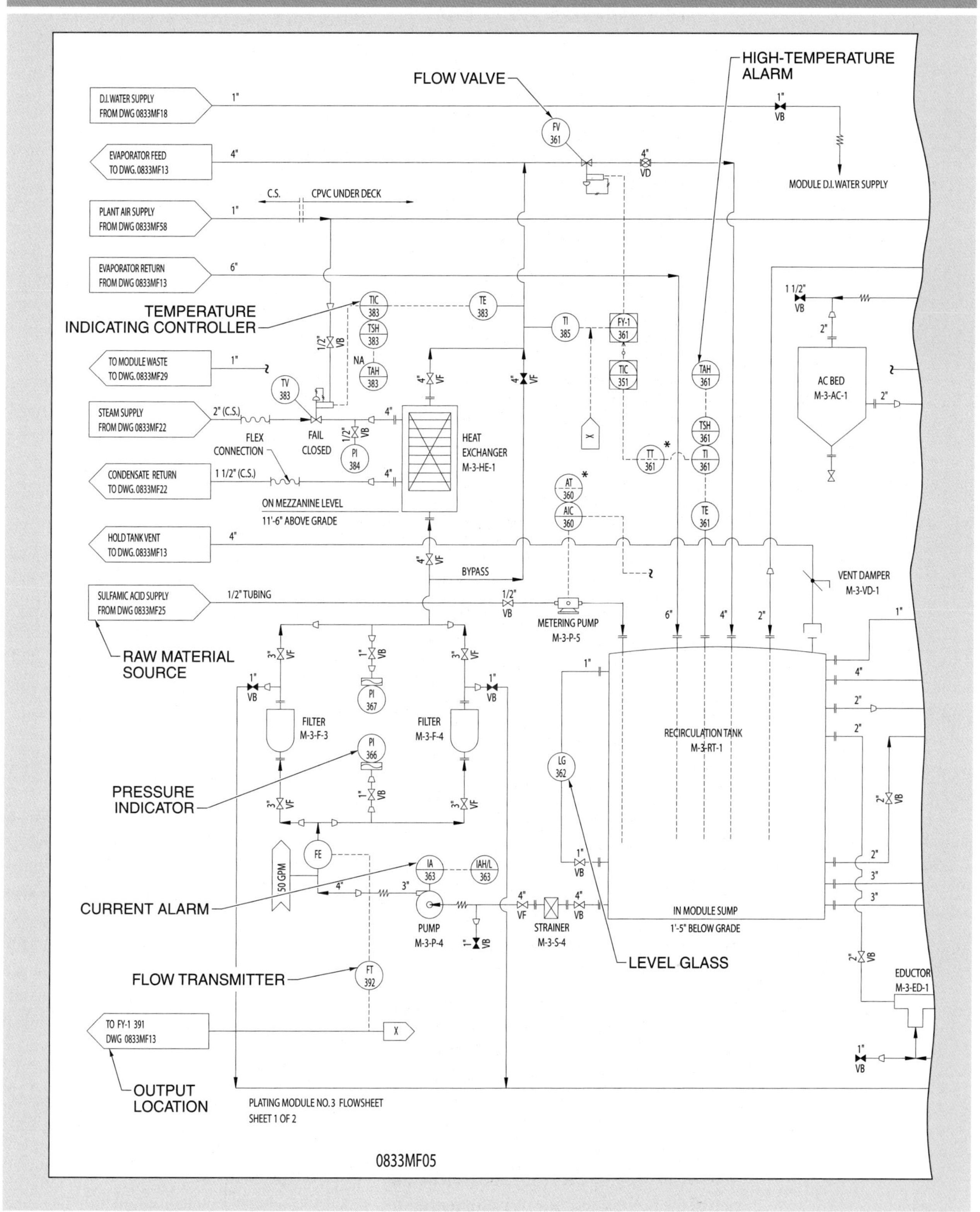

Figure 3-2. Piping and instrumentation diagrams (P&IDs) contain information about the instruments being used as well as the equipment employed.

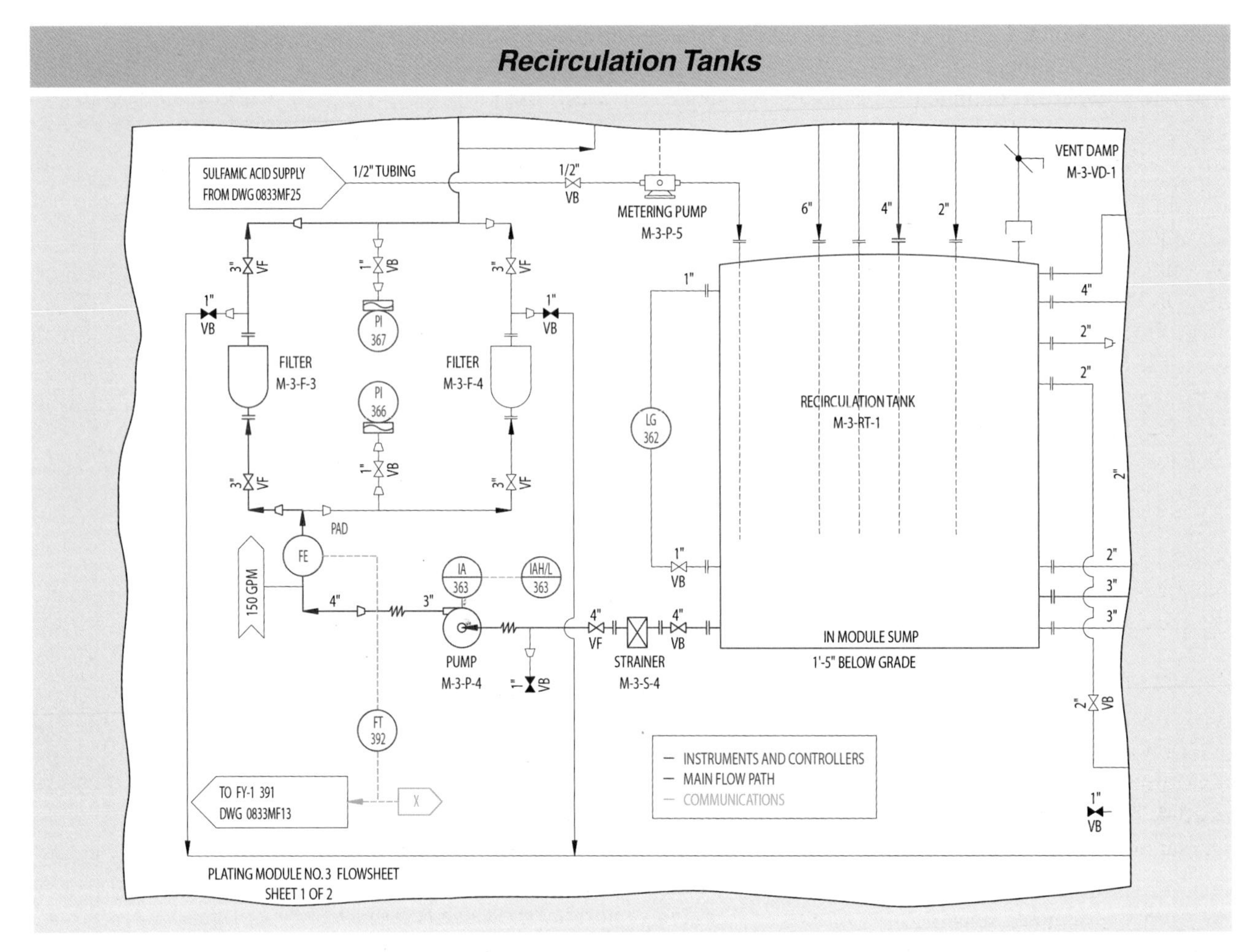

Figure 3-3. A detail view of the recirculation tank drawing shows how the instruments are related to the pumping system.

Pump Alarms. On the left side of the tank is a centrifugal pump, M-3-P-4. The M-3 represents plating module no. 3 and the P-4 represents pump no. 4. An alarm device monitors the current drawn by the pump. A balloon is used to show the alarm as IA 363. The first letter represents current. The second letter is a modifier representing an alarm. The alarm is in loop 363. There is a solid line through the center of the balloon. This line means that the alarm is mounted in an instrument panel. The alarm is wired to an annunciator alarm window, IAH/L 363, that is tripped on either high or low motor current. The dashed line connecting the alarm and the annunciator represents electrical communication between the devices.

Flow Measurement. A flow element, FE 392, is in the discharge line from the pump. The flow element is wired to a flow transmitter, FT 392. The dashed line represents electrical communication between the devices. The flow element and the flow transmitter are both in control loop 392. There are many different types of flowmeters, but in this case the letters "PAD" represent a paddle-type flow sensor.

The output of the flow transmitter is a 4 mA to 20 mA electrical current signal. The signal splits below the transmitter. One signal goes off the sheet to the left to FY-1 391 on Drawing 0833MF13. This means that the user needs to look at this drawing to identify what happens to the signal at that point. The other signal goes to the right to a transfer arrow "X." Transfer arrows are used to show a connection between two instruments where a continuous signal line would make the P&ID look too cluttered and hard to read.

Filter Pressure. There are two filters after the pump and flow element. There are two pressure indicators, PI 366 and PI 367, that measure the pressure at the inlets and outlets of the filters. The first letter represents pressure. The second letter is the modifier that represents an indicator. The small rectangle with the wavy line through it represents a chemical seal used to protect the pressure gauge from the corrosive electrolyte solution. The pressure gauges with a chemical seal are ordered as a single unit so there is no separate instrument tag number for the chemical seal.

Temperature Control

After passing through the filters of a recirculation tank, electrolyte passes through a steam-heated heat exchanger. **See Figure 3-4.** The temperature of the electrolyte is measured after exiting the heat exchanger.

Heat Exchanger Temperature Measurement. A temperature sensor element, TE 383, is used to measure the temperature for the control loop. The first letter represents temperature. The second letter represents a sensor element. This sensor element could be a thermocouple, a resistance element, or a thermistor.

A local temperature indicator, TI 385, is also shown. This instrument is used as a local backup indicator for the temperature control loop. This local indicator can be one of several different types of instruments. The selected type would be identified in the "Temperature Indicator" data sheets or specifications.

In almost all cases, a thermometer with a thermowell is used. A thermowell is only given a separate tag number on the rare occasions where it has to be purchased separately from the dial thermometer. A small circle centered on the TI connection line and touching the process line is sometimes used to represent the thermowell. In actual practice, temperature measurements are almost never used without a thermowell, so the thermowell symbol is not always used.

Temperature Controller. The temperature sensor element, TE 383, is connected electrically to the controller, TIC 383, as shown by the dashed horizontal line. The solid horizontal line through the balloon indicates that the instrument is panel mounted. The controller has a built-in high-temperature alarm switch, TSH 383. The first letter represents temperature. The second letter is a modifier that represents a switch. The third letter is another modifier that represents high.

The alarm switch is connected to an annunciator window, TAH 383. The first letter represents temperature again. The second letter represents alarm and the third letter represents high. The balloons labeled TSH 383 and TIC 383 are touching. This indicates that the two instruments are contained within one device.

The output of the temperature controller is a 4 mA to 20 mA electrical current signal connected to the positioner on the control valve, TV 383. The dashed line represents the electrical connection. The control valve symbol indicates that the valve is operated with air pressure that moves a diaphragm that throttles the steam going to the heat exchanger. The valve has a fail closed action (air to open). The temperature-sensor element, the controller, the alarm switch, the annunciator, and the control valve are all in loop 383.

The heated electrolyte is split into two streams. One stream can be seen going off the drawing to the left as the evaporator feed to Drawing 0833MF13. The other stream is returned to the recirculation tank through valve TV 361. Note that the valve is labeled a temperature valve even though the valve controls the electrolyte flow. This is because the valve is part of the temperature control loop 361.

A pressure gauge, PI 384, is used to measure the pressure on the steam side of the heat exchanger and is used for troubleshooting purposes. The pressure can be used to determine the steam temperature.

> **TECH FACT**
>
> *Fieldbus wiring standards require terminators at each end of a network to balance the impedance.*

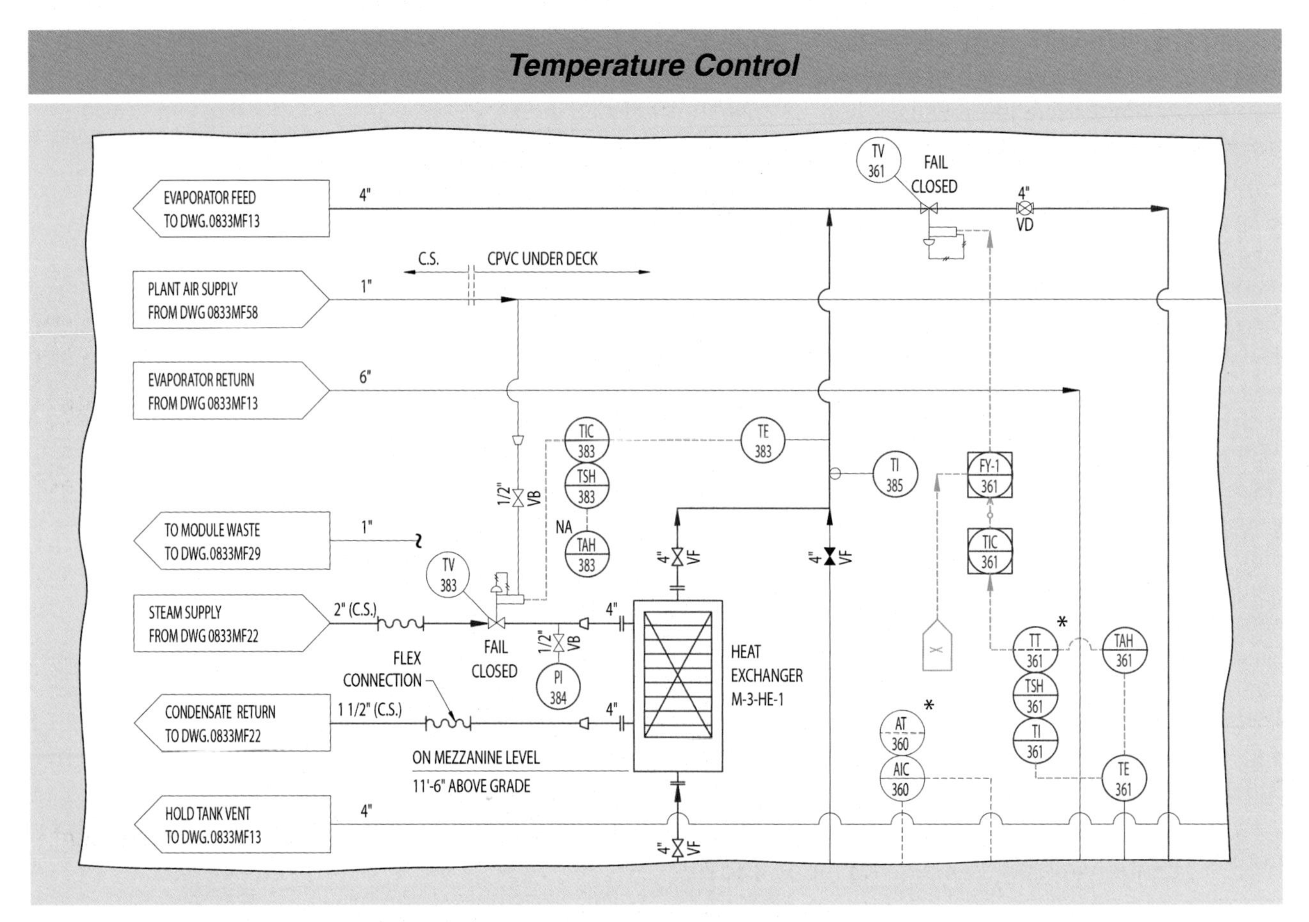

Figure 3-4. A detail view of the controls shows how information is transferred from instruments to final control elements.

Recirculation Tank Temperature Measurement. The recirculation tank temperature is measured with temperature element TE 361, which is electrically connected to temperature indicator TI 361. The temperature element can be either a thermocouple or a resistance element.

This temperature indicator has a built-in high-temperature alarm switch, TSH 361, and a built-in temperature transmitter, TT 361. Both of these instrument balloons are touching the TI 361 balloon, since all of these functions are contained in the TI 361 case. The asterisk (*) indicates that the TT 361 current output is wired through a 250 Ω resistor to produce a differential voltage for a future data acquisition system. The temperature switch is electrically connected to a high-temperature alarm, TAH 361.

The transmitter output is wired to a separate temperature controller, TIC 361, which is a microprocessor-type controller. The temperature controller's digital output goes to a signal high limit device, FY-1 361. The dashed line with a circle represents a digital signal. The total electrolyte flow measurement is sent to the high limit device through the transfer arrow and is used in a special function to limit the control signal to the recycle control valve, TV 361. The output signal from FY-1 361 is a 4 mA to 20 mA electrical current and goes to the positioner of control valve TV 361. This control valve has a fail closed action (air to open).

This completes a very brief description of the instruments in the plating P&ID. This description shows how instruments are labeled and tagged, and how instruments are put together in control loops. There are a number of control functions that cannot be fully explained using just the instrument symbols. Other forms of documents such as written descriptions, equations, and diagrams are used to provide the details needed.

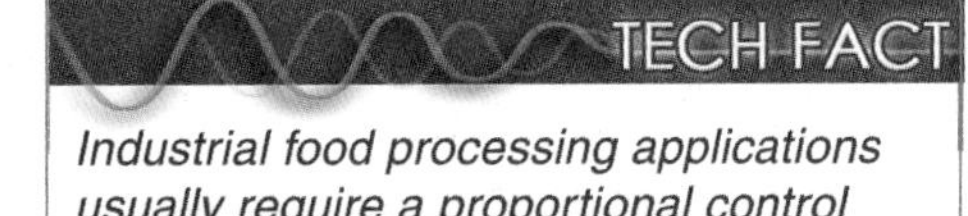

Industrial food processing applications usually require a proportional control circuit to maintain a very narrow temperature range to ensure food safety.

KEY TERMS

- *piping and instrumentation diagram (P&ID):* A schematic diagram of the relationship between instruments, controllers, piping, and system equipment.
- *balloon,* or *bubble:* A circular symbol used to identify the function and loop number of an instrument or device on a P&ID.

REVIEW QUESTIONS

1. Define piping and instrumentation diagram (P&ID).
2. How are instruments identified on piping and instrumentation diagrams?
3. What is the difference between the first letter and succeeding letters for describing an instrument and the variable that is being measured?
4. What is the purpose of a balloon on a piping and instrumentation diagram?
5. What are the limitations of a piping and instrumentation diagram in describing complex control systems?

SECTION

TEMPERATURE MEASUREMENT 2

TABLE OF CONTENTS

SECTION OBJECTIVES

Chapter 4

- Define temperature and identify the most common temperature scales.
- Define and compare the three types of heat transfer.
- Identify the common units of heat energy.
- Define specific heat and heat capacity.

Chapter 5

- Describe the principles of thermal expansion.
- Compare the types of thermometers that use the principles of thermal expansion.
- Explain how pressure-spring bulb location affects operation.

Chapter 6

- Define thermocouple, and identify the phenomena that govern the behavior of thermocouples.
- Describe the purpose of a cold junction, and explain how cold junction compensation is used.
- Describe the construction of a thermocouple.
- List several factors that affect the choice of thermocouple wires.
- List and define typical thermocouple measurement circuits.
- Define resistance temperature detector, describe its construction, and explain how it is used.
- Define thermistor, describe its construction, and explain how it is used.

SECTION

2 TEMPERATURE MEASUREMENT

Chapter 7

- Define IR thermometers and explain how they are used.
- Compare blackbodies, graybodies, and non-graybodies.
- Compare one-color and two-color IR thermometers.
- Explain how fiber-optic connections and infrared windows are used in harsh environments.
- Explain the use of disappearing filament pyrometers.
- Explain the use of thermal imagers.

Chapter 8

- Describe the response time and time constant of thermometers.
- Describe the function and design of thermowells.
- List the considerations for using thermocouples in various environments.
- Describe the function and use of resistance bridge circuits.
- Describe the calibration of thermometers.

INTRODUCTION

The first practical thermometer was the human eye. Prehistoric craftsmen made metal tools and weapons by observing the color of the metal to determine when to form the objects. This method was used until fairly modern times. From experience, the craftsmen learned to approximate the temperature in an oven or furnace by the color of the interior of the heating chamber or the color of the object being heated.

In the early 18th century, Daniel Fahrenheit invented the mercury thermometer and a corresponding temperature scale. In the early 19th century, Frederick William Herschel conducted experiments with mercury thermometers and prisms and proved that infrared radiation warmed up the thermometers. In the late 19th century, Josef Stefan extended these ideas and used infrared radiation to measure the temperature of the sun.

ADDITIONAL ACTIVITIES

1. After completing Section 2, take the Quick Quiz® included with the Digital Resources.
2. After completing each chapter, answer the questions and complete the activities in the *Instrumentation and Process Control Workbook.*
3. Review the Flash Cards for Section 2 included with the Digital Resources.
4. Review the following related Media Clips for Section 2 included with the Digital Resources:
   - Forced Convection
   - Temperature Measurement
5. After completing each chapter, refer to the Digital Resources for the Review Questions in PDF format.

ATPeResources.com/QuickLinks • Access Code: 467690

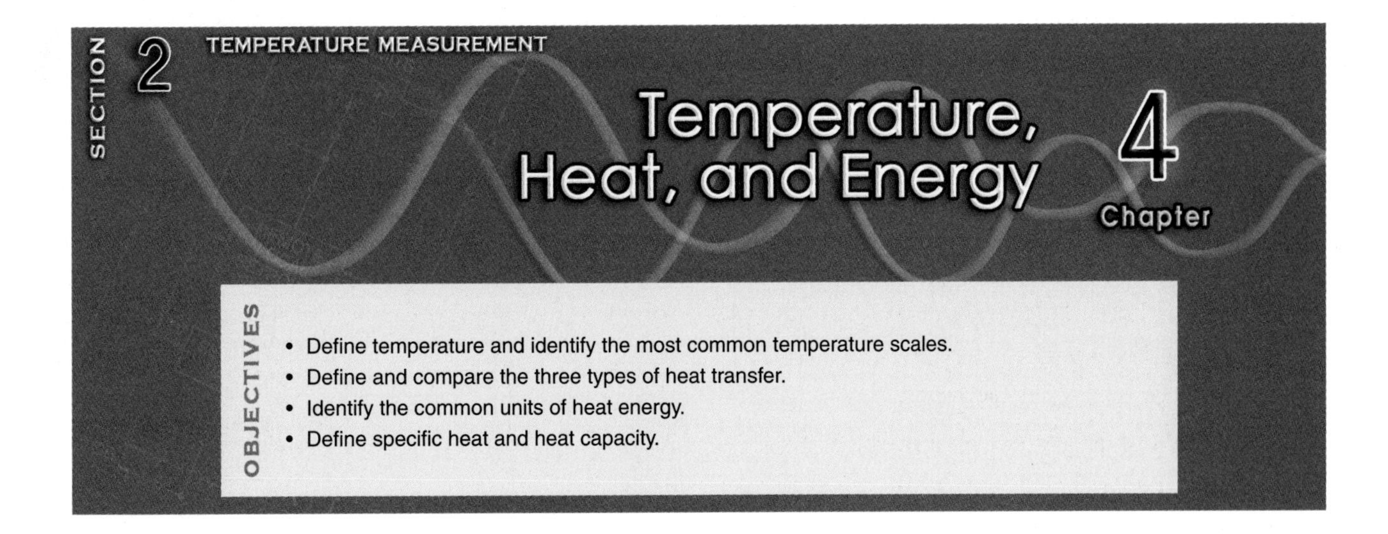

TEMPERATURE SCALES

There are changes in the physical or chemical state of most substances when they are heated or cooled. This is why temperature is one of the most important of the measured variables encountered in the industrial environment and the most often measured of the variables described in this text. Many temperature measurements are involved in heat transfer, boiler operation, HVAC systems, welding, and many other industrial processes.

Temperature is the degree or intensity of heat measured on a definite scale. Temperature is an indirect measurement of the heat energy contained in molecules. When molecules have a low level of energy they are cold, and as energy increases they get warmer. The energy is in the form of molecular movement or vibration of the molecules. A *thermometer* is an instrument that is used to indicate temperature. Many technologies are used to measure temperature and they are also called thermometers.

Absolute zero is the lowest temperature possible, where there is no molecular movement and the energy is at a minimum. Absolute zero is a theoretical state and cannot actually be reached. This condition is the zero point for the absolute temperature scales. The four common temperature scales are the Fahrenheit, Rankine, Celsius, and Kelvin scales. **See Figure 4-1.**

Fahrenheit and Rankine Scales

The Fahrenheit (°F) temperature scale is the most common temperature scale in use in the United States. On the Fahrenheit scale, the freezing point of water is 32°F, the boiling point of water is 212°F at standard atmospheric pressure, and there are 180 degrees between the fixed points. The Rankine (°R) scale is the absolute equivalent of the Fahrenheit scale. The Rankine scale has its zero point at absolute zero, the scale divisions are the same as the Fahrenheit scale, and the scales are offset by 459.67°. To convert between Fahrenheit and Rankine temperatures, add or subtract 459.67 degrees as follows:

$$°R = °F + 459.67$$

or

$$°F = °R - 459.67$$

where

$°R$ = degrees Rankine

$°F$ = degrees Fahrenheit

Sprecher + Schuh

A temperature of nearly 2800°F is obtained during the continuous casting process by which iron rounds are manufactured.

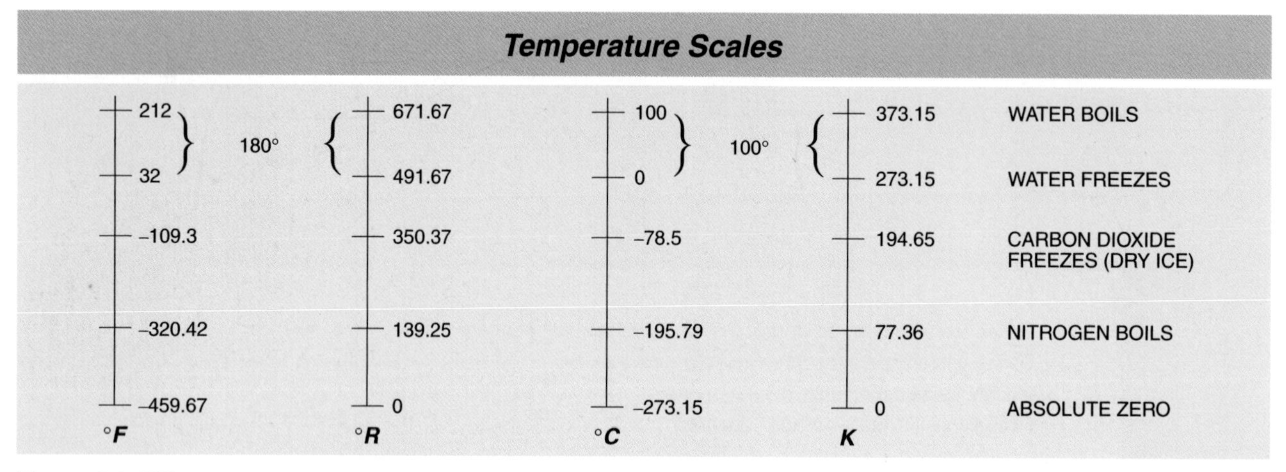

Figure 4-1. Different temperature scales are used in different applications.

For example, to convert 32°F to degrees Rankine, use the following:

$$°R = °F + 459.67$$
$$°R = 32 + 459.67$$
$$°R = \mathbf{491.67}$$

To convert 212°F to degrees Rankine, use the following:

$$°R = °F + 459.67$$
$$°R = 212 + 459.67$$
$$°R = \mathbf{671.67}$$

The difference between 671.67° and 491.67° on the Rankine scale is 180 degrees, the same as the difference between 212° and 32° on the Fahrenheit scale.

Celsius and Kelvin Scales

The Celsius (°C) temperature scale is another temperature scale that is sometimes used in the United States, but it is primarily used in other countries. The Celsius scale, formerly known as the Centigrade scale, is universally used for scientific measurements. On the Celsius scale, the freezing point of water is 0°C, the boiling point of water is 100°C at standard atmospheric pressure, and there are 100 degrees between the fixed points.

The Kelvin (K) scale is the absolute equivalent of the Celsius scale. The Kelvin scale has its zero point at absolute zero, the scale divisions are the same as the Celsius scale, and the scales are offset by 273.15 degrees. When using the Kelvin scale, the word "degree" and the degree symbol (°) are not used. To convert between Celsius and Kelvin temperatures, simply add or subtract 273.15 degrees as follows:

$$K = °C + 273.15$$

or

$$°C = K - 273.15$$

where

K = Kelvin

$°C$ = degrees Celsius

For example, to convert 0°C to Kelvin, use the following:

$$K = °C + 273.15$$
$$K = 0 + 273.15$$
$$K = \mathbf{273.15}$$

To convert 100°C to Kelvin, use the following:

$$K = °C + 273.15$$
$$K = 100 + 273.15$$
$$K = \mathbf{373.15}$$

The difference between 373.15 and 273.15 on the Kelvin scale is 100 the same as the difference between 100° and 0° on the Celsius scale.

Fahrenheit and Celsius Conversions

It is often necessary to convert temperature readings from Fahrenheit to Celsius or Celsius to Fahrenheit. **See Figure 4-2.** Make these conversions as follows:

$$°C = \frac{5}{9} \times (°F - 32)$$

or

$$°F = \left(\frac{9}{5} \times °C\right) + 32$$

For example, to convert 32°F to degrees Celsius, use the following:

$$°C = \frac{5}{9} \times (°F - 32)$$

$$°C = \frac{5}{9} \times (32 - 32)$$

$$°C = \frac{5}{9} \times (0)$$

$$°C = \mathbf{0}$$

HEAT TRANSFER

Heat transfer is the movement of thermal energy from one place to another. When objects are at the same temperature, they are in thermal equilibrium. *Thermal equilibrium* is the state where objects are at the same temperature and there is no heat transfer between them. When two substances are at different temperatures, there is heat transfer from the substance at the higher temperature to the substance at the lower temperature. The transfer of heat stops when both substances are in thermal equilibrium. Heat transfer occurs by conduction, convection, and radiation. **See Figure 4-3.** When fluid moves, it carries along any present heat.

Conduction

Conduction is heat transfer that occurs when molecules in a material are heated and the heat is passed from molecule to molecule through the material. For example, conduction occurs when one end of a metal rod is heated in a flame or when metals are welded. The molecules are heated and move faster. The faster-moving molecules transfer energy through collisions from molecule to molecule across the metal until they reach the opposite end of the workpiece. When heat is transferred through conduction, there is no flow of material.

Conduction also occurs between two different materials that are in direct contact. The process of heat transfer is the same, but the rate of heat transfer differs depending on the substances. Gases and liquids are generally poor heat conductors. The thermal conductivity of iron is much greater than the thermal conductivity of water.

Temperature Conversions

	°F	°R	°C	K
°F	—	°R = °F + 459.67	°C = 5/9 × (°F − 32)	K = 5/9 × (°F + 459.67)
°R	°F = °R − 459.67	—	°C = (5/9 × °R) − 273.15	K = 5/9 × °R
°C	°F = (9/5 × °C) + 32	°R = 9/5 × (°C + 273.15)	—	K = °C + 273.15
K	°F = (9/5 × K) − 459.67	°R = 9/5 × K	°C = K − 273.15	—

Figure 4-2. Temperature scale conversions are easy to perform.

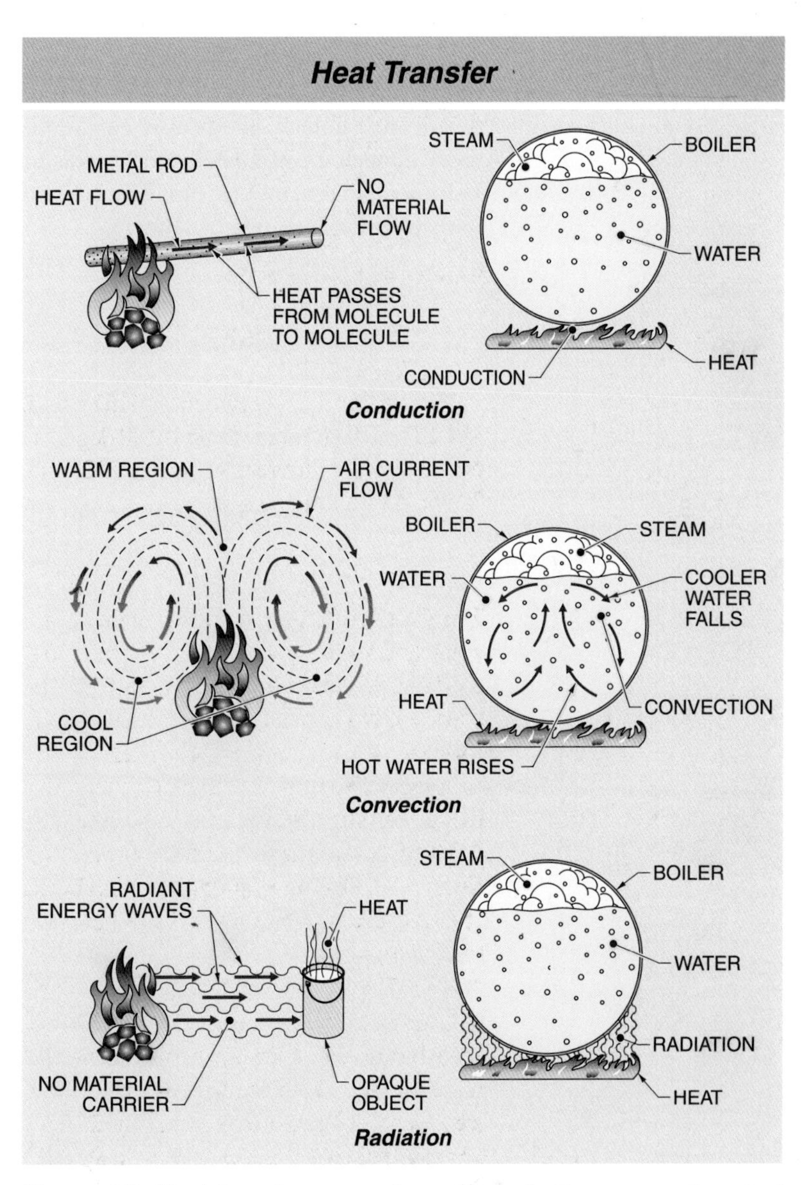

Figure 4-3. Heat transfer occurs through conduction, convection, and radiation.

For example, heat is transferred from a boiler's heating surface to the boiler water to produce steam. Heat is conducted from the hot combustion gases through the metal wall to the water. The hot side of the metal (next to the fire) is nearly the temperature of the fire. The cool side of the metal (next to the water) is nearly the temperature of the water. The temperature within the metal wall varies between these two extremes.

Another example of conduction is a heat sink. **See Figure 4-4.** A *heat sink* is a heat conductor used to remove heat from sensitive electronic parts. For example, a transistor conducts current during its normal operation. Some of the power is lost as heat. That heat must be removed from the transistor to prevent heat-related failure. The heat is conducted away through a metal bracket or chassis to radiator fins. From the fins, heat is transferred away by convection or radiation.

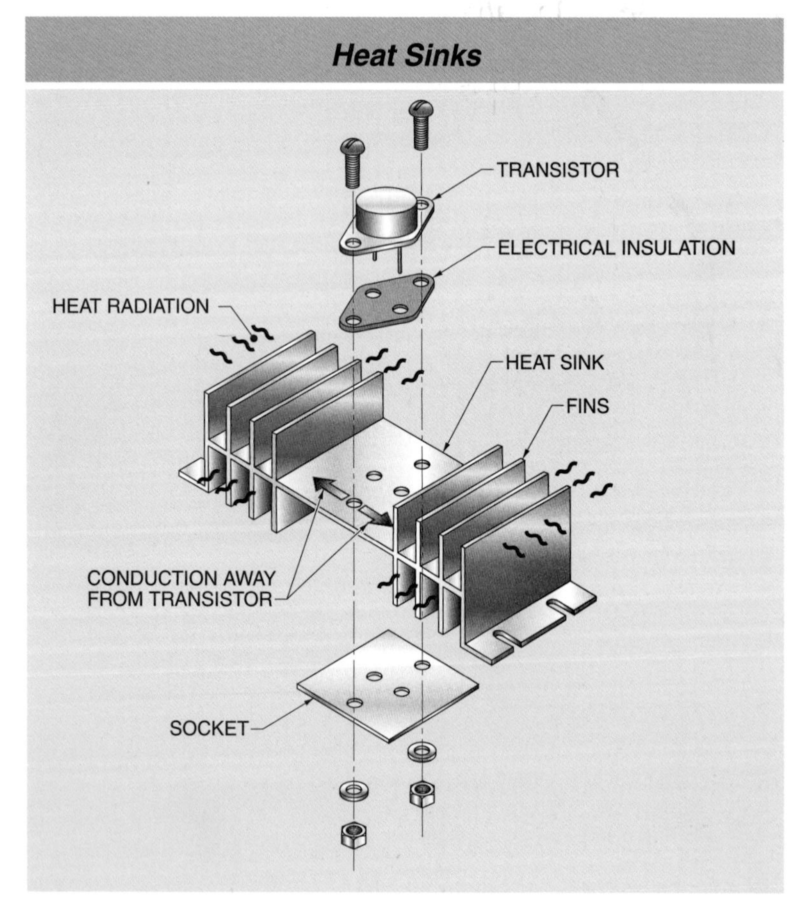

Figure 4-4. A heat sink conducts heat away from a temperature-sensitive object.

TECH FACT

Two important early ideas about heat are the suggestion that heat is conserved and the distinction between the quantity and quality of heat. The quality of heat is now called temperature and the study of temperature is called thermometry. The study of the quantity of heat is called calorimetry.

Convection

Convection is heat transfer by the movement of gas or liquid from one place to another caused by a pressure difference. The two types of convection are natural convection and forced convection.

Natural Convection. *Natural convection* is the unaided movement of a gas or liquid caused by a pressure difference due to a difference in density within the gas or liquid. Heat is transferred by currents that circulate between warm and cool regions in a fluid. For example, a flame heats the boiler's heating surface. The hot boiler surface heats the water. The water is heated so quickly that conduction within the water cannot transfer the heat away fast enough. The hot water is less dense and begins to rise to the surface and is replaced by cooler, denser water that moves to the bottom near the heat.

Forced Convection. *Forced convection* is the movement of a gas or liquid due to a pressure difference caused by the mechanical action of a fan or pump. For example, warm air in a forced-air heating system travels through the ducts because of the pressure difference created by the blower fan. Once the air is in a room and mixed, natural sources of heating and cooling cause air in parts of the room to cool down or warm up. This may be due to heat generated in a room, air infiltration, or heat transfer through walls and windows. Cool air is denser than warm air and natural convection moves the cooler, denser air to the floor and into the cool air return. From there, forced convection moves the air back through the heating system. **See Figure 4-5.**

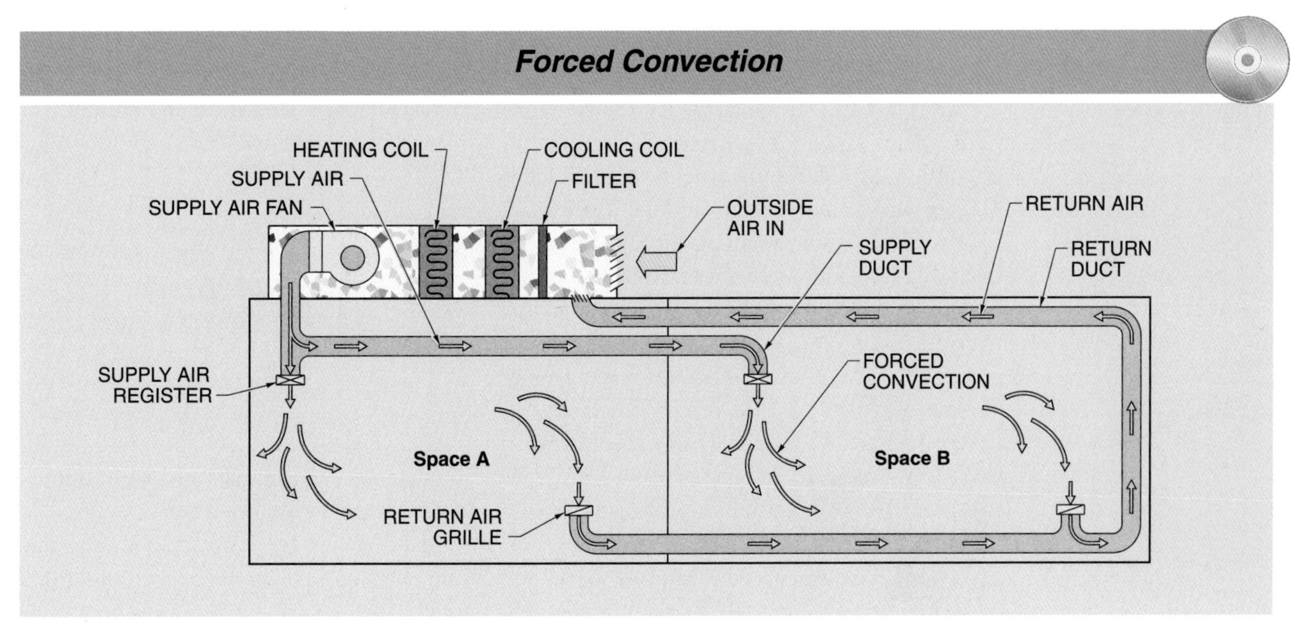

Figure 4-5. An HVAC system uses forced convection to transfer heat throughout a building space.

Radiation

Radiation is heat transfer by electromagnetic waves emitted by a higher-temperature object and absorbed by a lower-temperature object. All objects emit radiant energy. The amount of emitted energy depends on the temperature and nature of the surface of the object. Radiant energy waves move through air or space without producing heat. Heat is only produced when the radiant energy waves contact an object that absorbs the energy waves. The energy is transferred to the surface molecules of the object, which are warmed by the energy from the electromagnetic waves. When heat is transferred by radiation, there is no flow of material.

For example, when a metal is heated to glowing red, a person that is standing a distance away from the metal can feel the radiant energy. When a boiler furnace door is opened, the heat can immediately be felt even though the air temperature between the fire and the person does not change very much. In both cases, the electromagnetic waves emitted by the hot object travel through the air and warm the person. Another example of radiation is solar heating where radiation from the sun is used to heat water. The heated water is then moved by convection to the area where it is needed. **See Figure 4-6.**

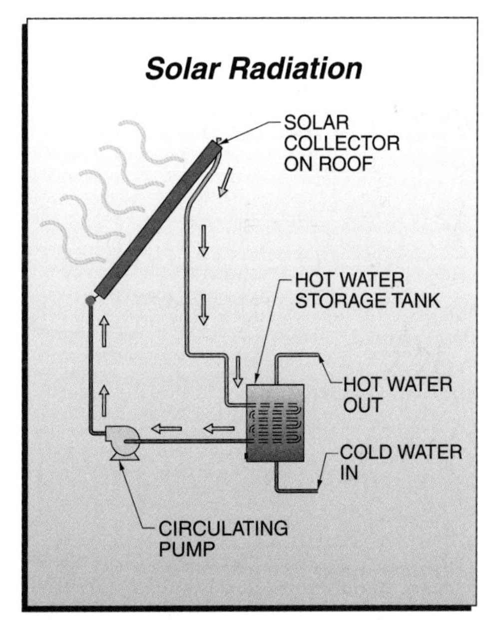

Figure 4-6. A solar heater collects radiant energy from the sun to warm water.

Heat Capacity

Heat is a form of energy. *Heat capacity* is the amount of energy needed to change the temperature of a material by a certain amount. Heat energy is commonly measured in units of the British thermal unit and the calorie.

Reference Temperatures

The freezing and boiling temperatures of water are inadequate to define a temperature scale for industrial processes that may involve much lower or higher temperatures, as for example with cryogenic gases or molten metals. Many more known fixed points are needed to define a temperature scale. The International Temperature Scale of 1990 (ITS-90), adopted by the International Committee of Weights and Measures in 1989, uses 17 points to define the Kelvin temperature scale and thus the Celsius temperature scale.

The ITS-90 also defines the Kelvin as 1/273.16 of the thermodynamic temperature of the triple point of water. A *triple point* is the condition where all three phases of a substance—gas, liquid, and solid—can coexist in equilibrium. The triple point of water occurs at 32.018°F (273.16 K) and 0.08865 psi (611.657 pascals) pressure.

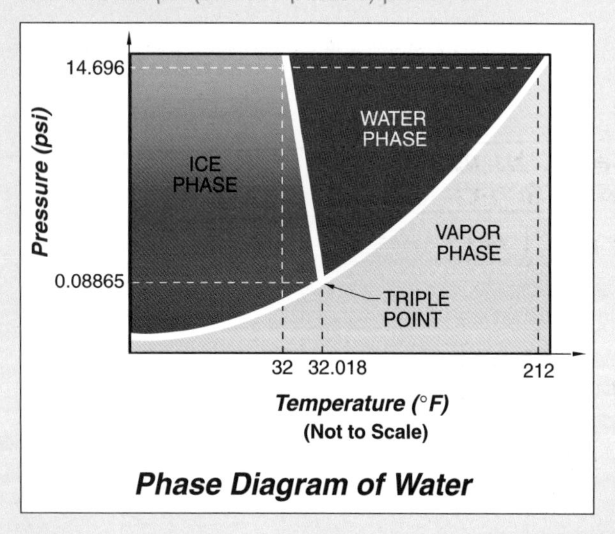

Phase Diagram of Water

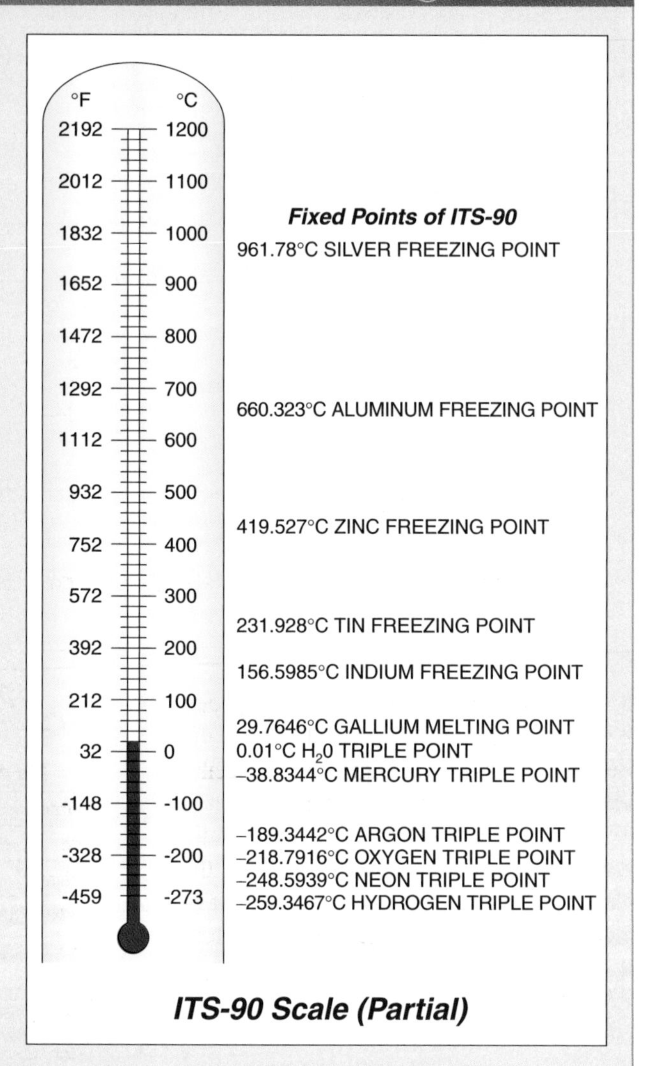

ITS-90 Scale (Partial)

High-temperature ducts, pipes, and valves are covered with insulation to protect workers and to conserve energy.

A *British thermal unit (Btu)* is the amount of energy necessary to change the temperature of 1 lb of water by 1°F from 59°F to 60°F. A *calorie (cal)* is the amount of energy necessary to change the temperature of 1 g of water by 1°C from 14.5°C to 15.5°C. There are other definitions of a calorie that are slightly different. **See Figure 4-7.** Because the heat capacity of most substances varies slightly with changes in temperature, the Btu and calorie are defined at specific temperatures. However, any difference in heat capacity with the associated temperature change is usually small.

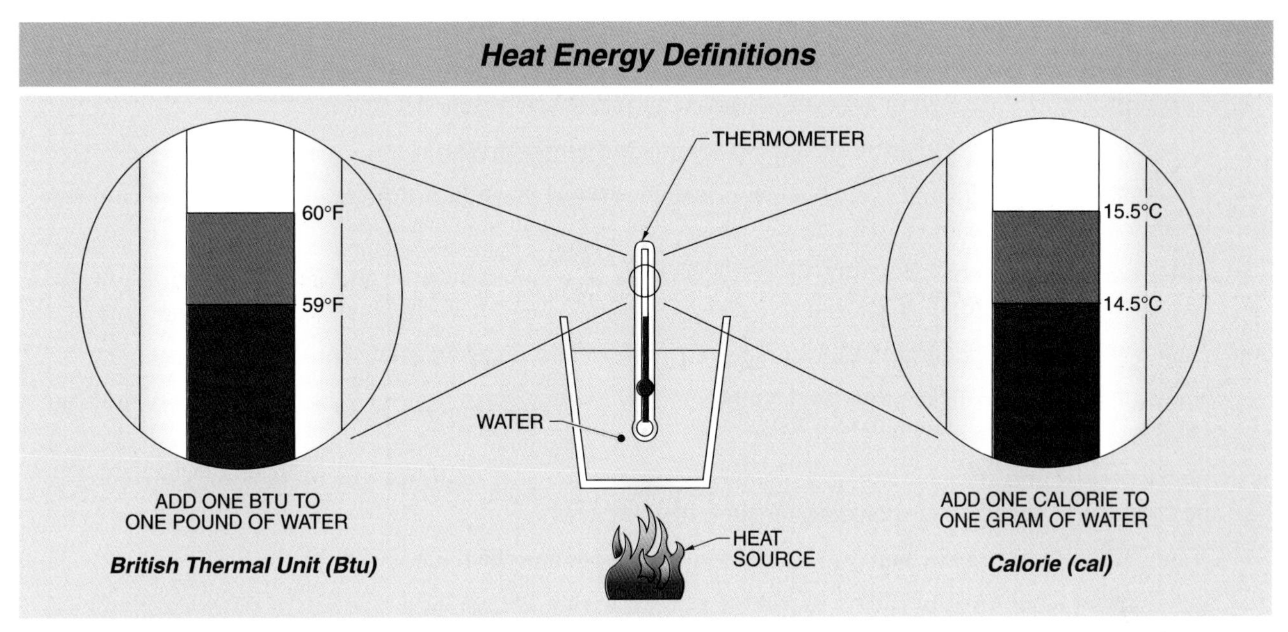

Figure 4-7. The Btu and the calorie are defined based on a specific temperature change of a specific quantity of water.

Specific Heat

Specific heat is the ratio of the heat capacity of a liquid to the heat capacity of water at the same temperature. **See Figure 4-8.** Specific heat has no unit of measurement since it is a ratio, but is almost identical numerically to the heat capacity because of the way the Btu and calorie are defined. The amount of energy required to change the temperature of a substance is expressed by the following:

$$Q = M \times C \times (T_2 - T_1)$$

where

Q = amount of energy (in Btu)

M = mass of the substance (in lb)

C = heat capacity (in Btu/lb°F)

T_2 = final temperature (in °F)

T_1 = initial temperature (in °F)

For example, water has an average heat capacity of 1.0 Btu/lb°F. How much energy is required to heat 1 lb of water from 50°F to 212°F in a boiler?

$$Q = M \times C \times (T_2 - T_1)$$

$$Q = 1 \times 1.0 \times (212 - 50)$$

$$Q = 1 \times 1.0 \times (162)$$

$$\mathbf{Q = 162\ Btu}$$

This shows that it takes 162 Btu to increase the temperature of 1 lb of water from 50°F to 212°F. In other words, it takes 162 Btu/lb to increase the temperature of any amount of water from 50°F to 212°F.

Specific Heats of Common Substances

Material	Specific Heat
Water	1.0
Air	0.24
Alumel®	0.13
Aluminum	0.22
Ammonia (gas)	0.54
Ammonia (liquid)	1.1
Chromel®	0.11
Constantan	0.094
Copper	0.092
Ethyl alcohol	0.60
Glass, Pyrex®	0.20
10% Hydrochloric Acid	0.75
Iron	0.11
Mercury	0.033
Nicrosil	0.11
Nisil	0.12
Platinum	0.032
25% Sodium Chloride Brine	0.79
Steam	0.48
98% Sulfuric Acid	0.35

Figure 4-8. The specific heats of common substances vary considerably.

KEY TERMS

- *temperature:* The degree or intensity of heat measured on a definite scale.
- *thermometer:* An instrument that is used to indicate temperature.
- *absolute zero:* The lowest temperature possible, where there is no molecular movement and the energy is at a minimum.
- *triple point:* The condition where all three phases of a substance — gas, liquid, and solid— can coexist in equilibrium.
- *heat transfer:* The movement of thermal energy from one place to another.
- *thermal equilibrium:* The state where objects are at the same temperature and there is no heat transfer between them.
- *conduction:* Heat transfer that occurs when molecules in a material are heated and the heat is passed from molecule to molecule through the material.
- *heat sink:* A heat conductor used to remove heat from sensitive electronic parts.
- *convection:* Heat transfer by the movement of gas or liquid from one place to another caused by a pressure difference.
- *natural convection:* The unaided movement of a gas or liquid caused by a pressure difference due to a difference in density within the gas or liquid.
- *forced convection:* The movement of a gas or liquid due to a pressure difference caused by the mechanical action of a fan or pump.
- *radiation:* Heat transfer by electromagnetic waves emitted by a higher-temperature object and absorbed by a lower-temperature object.
- *heat:* A form of energy.
- *heat capacity:* The amount of energy needed to change the temperature of a material by a certain amount.
- *triple point:* The condition where all three phases of a substance — gas, liquid, and solid — can coexist in equilibrium.
- *British thermal unit (Btu):* The amount of energy necessary to change the temperature of 1 lb of water by 1°F from 59°F to 60°F.
- *calorie (cal):* The amount of energy necessary to change the temperature of 1 g of water by 1°C from 14.5°C to 15.5°C.
- *specific heat:* The ratio of the heat capacity of a liquid to the heat capacity of water at the same temperature.

REVIEW QUESTIONS

1. Define temperature.
2. What are the four most common temperature scales?
3. Define and compare the three types of heat transfer.
4. What are the common units of heat energy?
5. Define heat capacity and specific heat.

SECTION 2 TEMPERATURE MEASUREMENT

Chapter 5 Thermal Expansion Thermometers

OBJECTIVES

- Describe the principles of thermal expansion.
- Compare the types of thermometers that use the principles of thermal expansion.
- Explain how pressure-spring bulb location affects operation.

THERMAL EXPANSION

Materials usually expand when heated and contract when cooled. For example, a metal rod that is heated or cooled changes in length and volume. The *coefficient of linear expansion* is the amount a unit length of a material lengthens or contracts with temperature changes.

The *coefficient of volumetric expansion* is the amount a unit volume of a material expands or contracts with temperature changes. The coefficient of volumetric expansion is three times the value of the coefficient of linear expansion. Volumetric expansion is usually used only for liquids.

These coefficients are very small and are often reported as μ in./in.°F (millionths of an inch of change per inch of original length per degree Fahrenheit of temperature change). **See Figure 5-1.** The expansion of materials is usually approximately linear with changes in temperature. This allows the expansion to be used to indicate temperature. The expansion is calculated as follows:

$$L_2 = L_1 \times [1 + \alpha \times (T_2 - T_1)]$$

or

$$V_2 = V_1 \times [1 + \beta \times (T_2 - T_1)]$$

where

L_2 = length at the final temperature (in in.)

L_1 = length at the initial temperature (in in.)

T_2 = final temperature (in °F)

T_1 = initial temperature (in °F)

V_2 = volume at the final temperature (in cu in.)

V_1 = volume at the initial temperature (in cu in.)

α = coefficient of linear expansion (in in./in.°F)

β = coefficient of volumetric expansion (in cu in./cu in.°F)

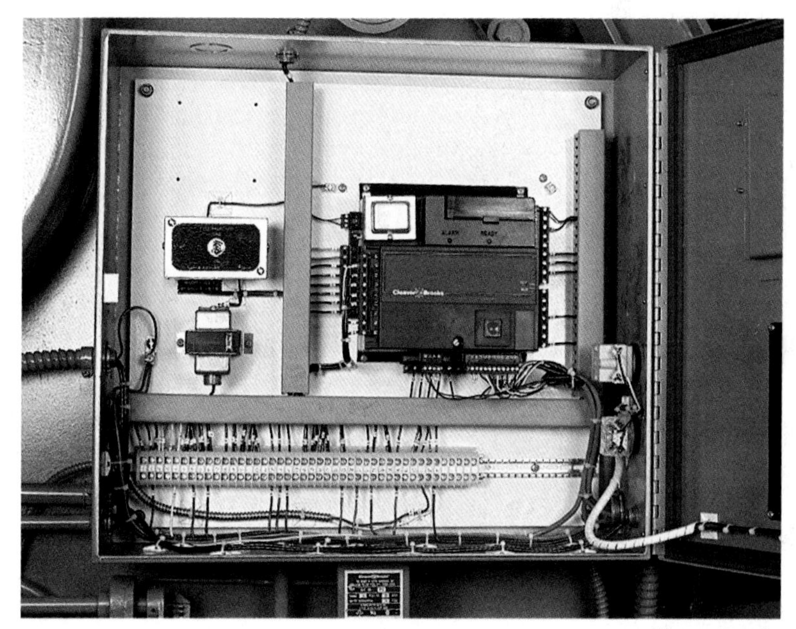

Cleaver-Brooks

Temperature measurements are essential to the operation of a boiler flame safeguard system. The flame safeguard system controls sequenced operations required for burner startup, firing rate, and shutdowns.

Coefficent of Thermal Expansion of Common Substances		
Material	α, ***Linear****	β, ***Volumetric***†
Alcohol	—	620.0
Alumel®	6.7	20.1
Aluminum, alloy 3003	12.9	38.7
Brass, yellow	10.5	31.5
Chromel®	7.3	21.9
Constantan	8.2	24.6
Copper	9.2	27.6
Glass	5.0	15.0
Glass, borosilicate	1.8	5.4
Invar®	0.68	2.4
Iron, ductile	6.5	19.5
Mercury	—	101.0
Platinum	5.0	15.0
Steel, austenitic	10.0	30.0
Steel, carbon	6.7	20.1
Water	—	115.0

* in millionths of an in. per in. per °F at room temperature
† in millionths of a cu in. per cu in. per °F at room temperature
Note: To convert to Celsius, multiply value in table by 1.8.

Figure 5-1. The coefficient of thermal expansion determines how much the size of an object changes when heated or cooled.

For example, copper has a coefficient of linear expansion of 9.2 μ in./in.°F (9.2 millionths of an inch per inch per °F), which is equivalent to 0.0000092 in./in.°F. What is the final length of a 6″ copper strip heated from 68°F to 72°F?

$$L_2 = L_1 \times \left[1 + \alpha \times (T_2 - T_1)\right]$$
$$L_2 = 6 \times \left[1 + 0.0000092 \times (72 - 68)\right]$$
$$L_2 = 6 \times \left[1 + 0.0000092 \times (4)\right]$$
$$L_2 = 6 \times \left[1 + 0.0000368\right]$$
$$L_2 = 6 \times \left[1.0000368\right]$$
$$L_2 = \mathbf{6.00022''}$$

In comparison, the metal Invar® has a coefficient of linear expansion of 0.68 μ in./in.°F (0.68 millionths of an inch per inch per °F). A 6″ strip of Invar expands much less than the copper strip under the same conditions.

$$L_2 = L_1 \times \left[1 + \alpha \times (T_2 - T_1)\right]$$
$$L_2 = 6 \times \left[1 + 0.00000068 \times (72 - 68)\right]$$
$$L_2 = 6 \times \left[1 + 0.00000068 \times (4)\right]$$
$$L_2 = 6 \times \left[1 + 0.00000272\right]$$
$$L_2 = 6 \times \left[1.00000272\right]$$
$$L_2 = \mathbf{6.0000163''}$$

This shows that copper expands much more than Invar when both metals are exposed to the same temperature change. The concept of thermal expansion is used in liquid-in-glass, bimetallic, and pressure-spring thermometers.

LIQUID-IN-GLASS THERMOMETERS

The volume of a liquid changes when the temperature changes. When liquid is placed in a glass tube, the top of the liquid moves with a change in temperature. This is the basic operation of liquid-in-glass thermometers. A *liquid-in-glass thermometer* is a thermal expansion thermometer consisting of a sealed, narrow-bore glass tube with a bulb at the bottom filled with a liquid. Liquid-in-glass thermometers are commonly called glass-stem thermometers.

The volumetric expansion of liquids is typically many times greater than that of glass. Since the volume of the liquid changes more than the change in the glass, the liquid moves up or down in the tube with changes in temperature. **See Figure 5-2.** Liquid-in-glass thermometers can typically be used for temperatures from –112°F to 760°F.

Reference marks are made on the thermometer using a regulated water bath that can establish and maintain temperatures with very little variation. The spaces between the reference marks are evenly subdivided. A larger number of reference marks enables the thermometer to be read more accurately. Some liquid-in-glass thermometers are calibrated to be completely immersed and others are calibrated for partial immersion. The thermometer should be immersed as recommended by the manufacturer to obtain the most accurate readings.

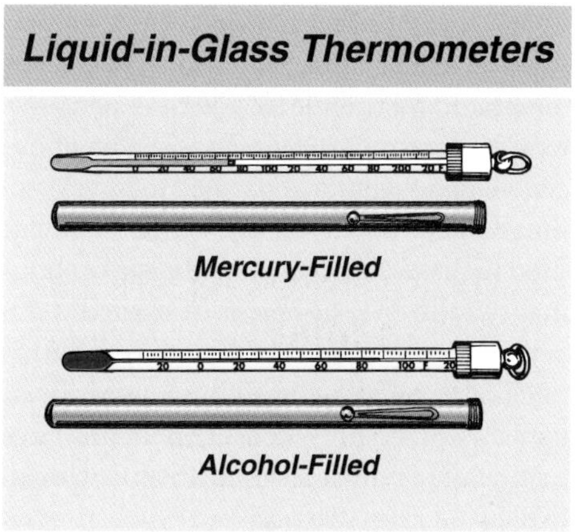

Liquid-in-Glass Thermometers

Mercury-Filled

Alcohol-Filled

Figure 5-2. The simplest liquid-in-glass expansion device is the common thermometer.

The liquid used in thermometers is usually alcohol or some other organic liquid with a red dye added to improve visibility. Mercury was commonly used in the past, but the use of mercury is being discouraged because of the hazards associated with mercury spills if the glass thermometer breaks. Mercury is classified as a hazardous material and spills must be handled in accordance with federal, state, and local environmental statutes and regulations.

The most common type of liquid-in-glass thermometer used for process measurements is the industrial thermometer. With this type of thermometer, the glass tube is not marked or scaled. Instead, the graduations are engraved on metal plates. Both the tube and the scales are enclosed in a metal case. The lower portion of the glass tube extends out of the bottom of the case into a metal bulb chamber. **See Figure 5-3.** The chamber contains a liquid with excellent heat transfer characteristics that improves thermal conductivity between the metal bulb chamber and the glass thermometer.

Industrial thermometers are available in vertical, horizontal, or oblique designs. Industrial thermometers are fitted with an external pipe thread that enables them to be screwed into a pipe.

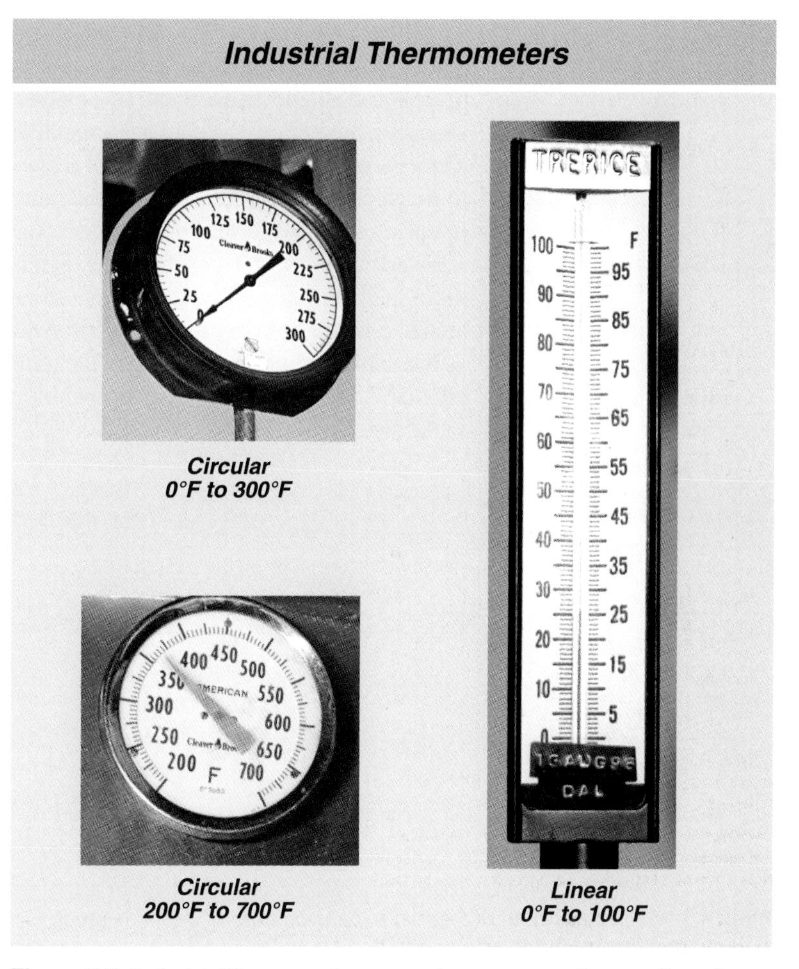

Figure 5-3. Industrial thermometers come in many mounting configurations.

BIMETALLIC THERMOMETERS

The principle of differential thermal expansion is the basis of operation for some thermometers such as bimetallic expansion thermometers. When one material has a greater coefficient of thermal expansion than another material, the difference in expansion can be used as a measure of temperature by direct reading or by connection to a mechanical linkage. This same differential expansion can produce a force that actuates devices in direct relation to the temperature.

A *bimetallic thermometer* is a thermal expansion thermometer that uses a strip consisting of two metal alloys with different coefficients of thermal expansion that are fused together and formed into a single strip and a pointer or indicating mechanism calibrated for temperature reading.

A *bimetallic element* is a bimetallic strip that is usually wound into a spiral, helix, or coil and allows movement for a given change in temperature. Industrial bimetallic elements can be wound in a variety of shapes to fit special requirements. When the helix or spiral is heated, it unwinds because the alloy with the greater coefficient expands more than the other alloy. A pointer can be attached to the helix by a shaft that moves as the helix unwinds and indicates the temperature on a calibrated circular temperature scale. **See Figure 5-4.**

Alloys whose coefficients of thermal expansion can be closely controlled make the bimetallic thermometer a very dependable temperature-measuring device. The alloys are welded together and rolled to tight specifications to make the bimetallic elements used in these thermometers. In applications that require measurement of a small temperature range, alloys with widely differing rates of thermal expansion are used. If the application requires measurement of a large temperature range, alloys with more similar expansion rates are used.

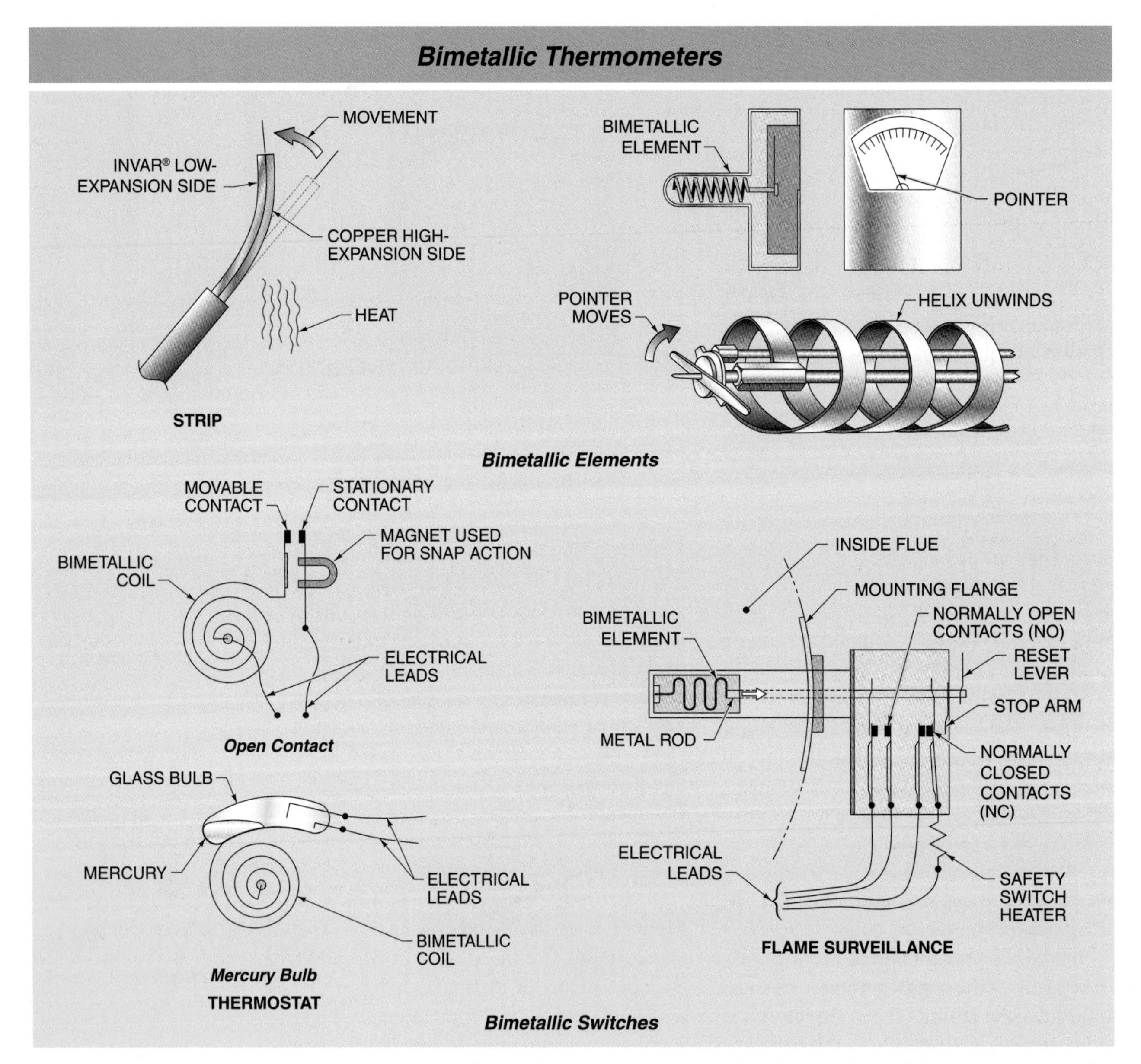

Figure 5-4. Bimetallic thermometers work through differential thermal expansion and can be used as switches to activate circuits.

Bimetallic thermometers are used to measure temperatures from –300°F to 800°F. They can indicate temperatures as high as 1000°F, but not on a continuous basis because the bimetallic element (helix) tends to overstretch at this temperature, causing permanent inaccuracy.

A bimetallic temperature element can easily be used as a temperature switch. For example, a bimetallic element can be used in a thermostat that operates a furnace or air conditioning system. The movement caused by the expansion or contraction of the element can be used to operate an ON/OFF switch.

Similarly, a bimetallic element can be mounted in a boiler or furnace flue and used with a timer as a flame surveillance control. The burner control circuit attempts to start a flame. If the temperature in the flue increases, the bimetallic element expands and forces two electrical contacts together, indicating the presence of a flame. If the temperature of the flue gas does not increase, the electrical contacts do not touch, the timer runs out, and the burner control circuit shuts the furnace down.

PRESSURE-SPRING THERMOMETERS

A *pressure-spring thermometer* is a thermal expansion thermometer consisting of a filled, hollow spring attached to a capillary tube and bulb where the fluid in the bulb expands or contracts with temperature changes. **See Figure 5-5.** A *capillary tube* is a small diameter tube with an interior hole 10 thousandths of an inch (0.010″) in diameter and is designed to contain very little liquid. A *bulb* is a cylinder larger in diameter than a capillary tube that contains the vast majority of the fluid in a thermometer. The volume of fluid contained by the capillary tube and the pressure spring is designed to be small in comparison to the volume in the bulb to minimize errors due to changes in ambient temperatures.

One end of the capillary tube is attached to the pressure spring. The spring can be in the same shape as the C-shaped Bourdon tube, but it is often in the shape of a spiral or helix. The other end of the tubing is joined to a sensing bulb. The bulb is the part of the thermometer that makes contact with the substance to be measured.

Bourdon Tube Pressure-Spring Thermometers

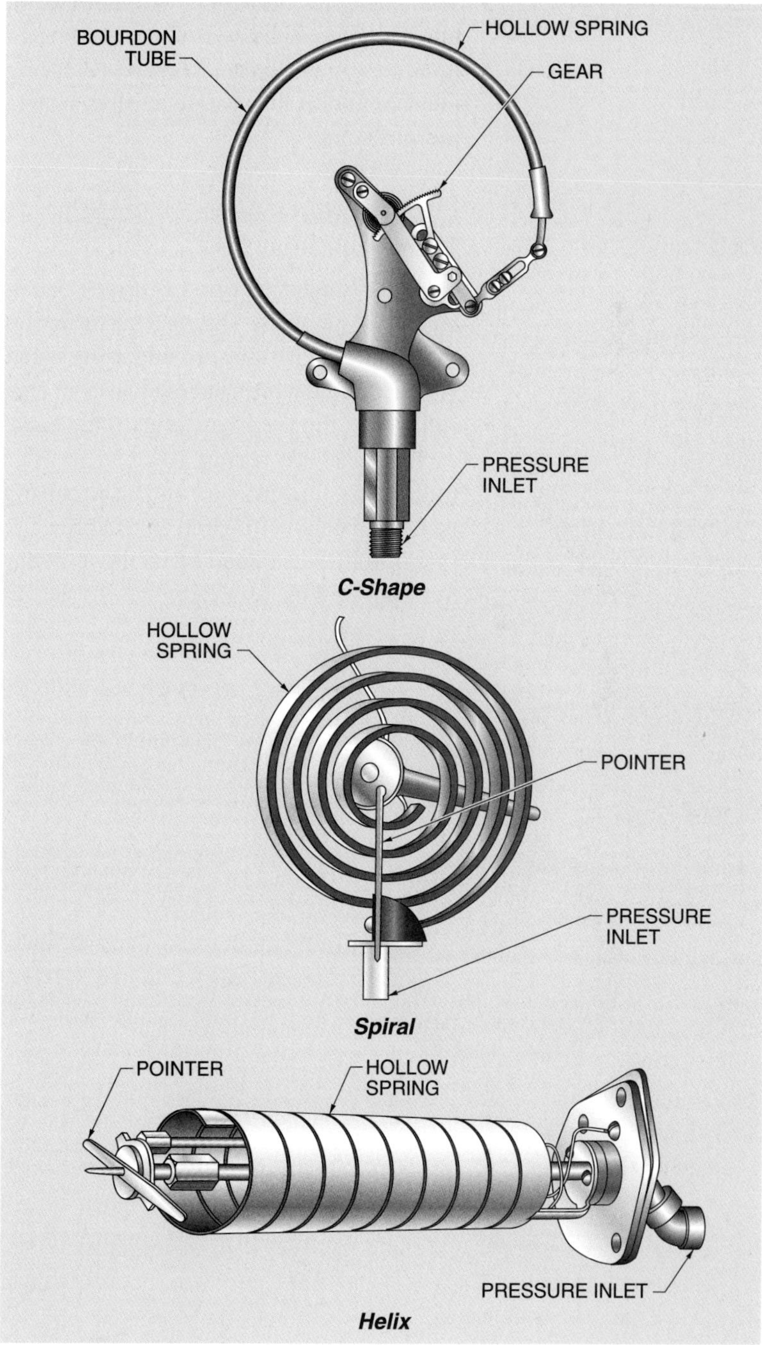

Figure 5-5. Bourdon tubes can be manufactured in C-shape, spiral, and helix configurations.

When the spring, tubing, and bulb contain a suitable fluid, they form a measuring unit just as a liquid-in-glass thermometer does with its glass tube, liquid, and bulb. As the temperature of the fluid in the bulb changes, the fluid expands or contracts to change the pressure in the system. This causes the spring to move. The spring is attached to an indicator or other device calibrated to show the temperature of the bulb. A pressure-spring thermometer is also called a filled-system thermometer.

Liquid-Filled Pressure-Spring Thermometers

A *liquid-filled pressure-spring thermometer* is a pressure-spring thermometer that is filled with a liquid under pressure. When the bulb is immersed in a heated substance, the liquid expands. This causes the pressure spring to unwind. The indicating, recording, and controlling mechanisms are attached to the pressure spring and are actuated by its movements. **See Figure 5-6.**

The volume of liquid in the system determines its measuring range. The wider the range desired, the greater the required volume. The bulb expands with temperature, but since this expansion is small compared to the expansion of the liquid, the effect is negligible. If the thermometer is calibrated properly, allowances can be made for expansion of the bulb.

Some liquid-filled thermometers are designed to work with the indicator at a normal ambient temperature. The more expensive models incorporate a bimetallic compensation in the indicator to correct for changes in ambient temperatures. The capillaries are about ⅛″ in outside diameter with a small bore about 0.01″ in diameter. The capillary can be supplied bare, but it is usual practice to protect the capillary with flexible spiral stainless steel armor. The armor can also be plastic covered to protect from corrosion.

TECH FACT

A Bourdon tube is a hollow tube with an elliptical cross section. The tube expands when pressure is applied to the open end.

Liquid-Filled Pressure-Spring Thermometer

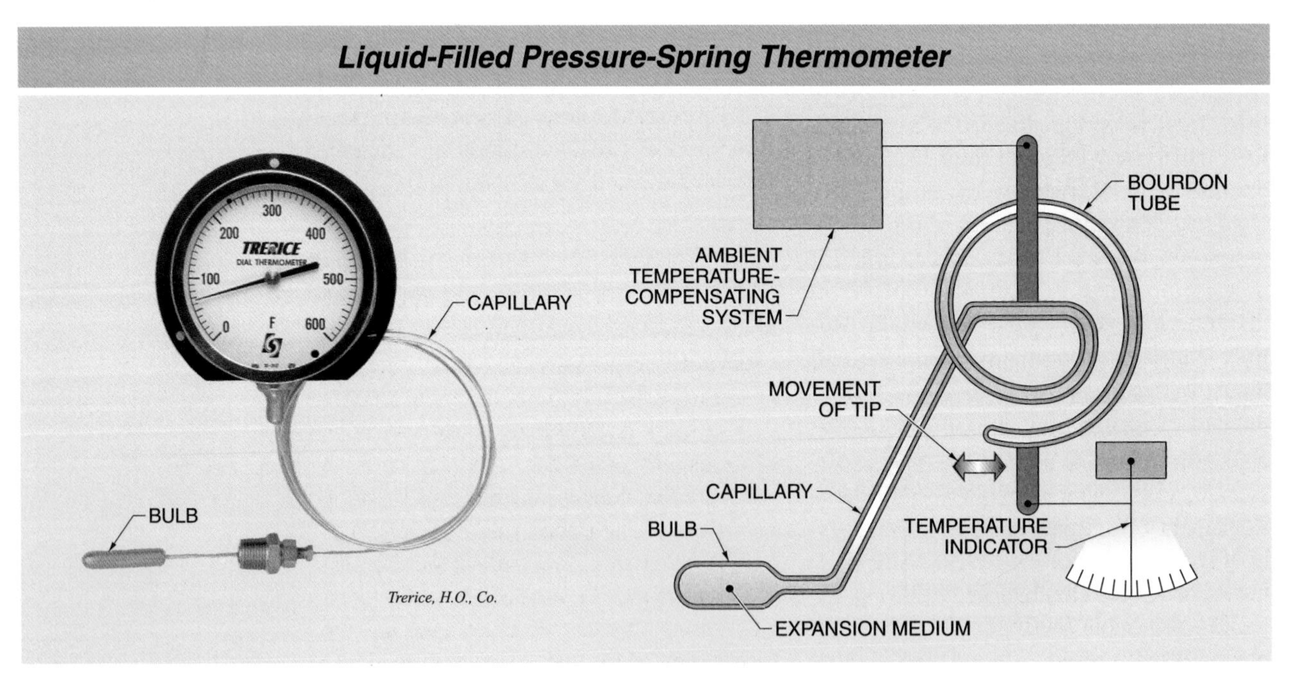

Figure 5-6. A liquid-filled pressure-spring thermometer uses the thermal expansion of a liquid to pressurize a Bourdon tube that is calibrated in temperature units.

Capillaries should never be kinked and should be run where the capillary can be fully supported and not near any heat sources. Excess capillary should be coiled uncut and secured in a protected location. Small differences in elevation, 5′ or 10′, between the bulb and the indicator should not cause any error in reading. Greater differences in elevation can cause errors, and where such an elevation difference exists, the manufacturer should be notified so that the system can be compensated for the elevation difference.

Mercury has long been used as the liquid fill because of its uniform thermal expansion, but has been discontinued due to the potential hazards posed by mercury. Stainless steel bulbs and capillary tubing were used in mercury-filled systems because mercury is corrosive to copper and brass. Filled-system thermometers are available with other liquids and have a copper or stainless steel bulb and capillary tubing.

Gas-Filled Pressure-Spring Thermometers

A *gas-filled pressure-spring thermometer* is a pressure-spring thermometer that measures the increase in pressure of a confined gas (kept at constant volume) due to a temperature increase. Nitrogen is the gas most often used for such systems because it is chemically inert and possesses a favorable coefficient of thermal expansion. Except for the size of the bulb, the gas-filled system is identical to the liquid-filled types.

Vapor-Pressure Pressure-Spring Thermometers

A *vapor-pressure pressure-spring thermometer* is a pressure-spring thermometer that uses the change in vapor pressure due to temperature change of an organic liquid to determine the temperature. A liquid partially fills the system, with vapor above the liquid. The movement of the pressure spring is caused by a change in vapor pressure as the bulb is heated or cooled. The change in volume of the liquid is negligible. **See Figure 5-7.** Water in a pressure cooker responds in a similar manner. The pressure increases as the water is heated and changed to steam (water vapor).

Vapor-Pressure Pressure-Spring Thermometers

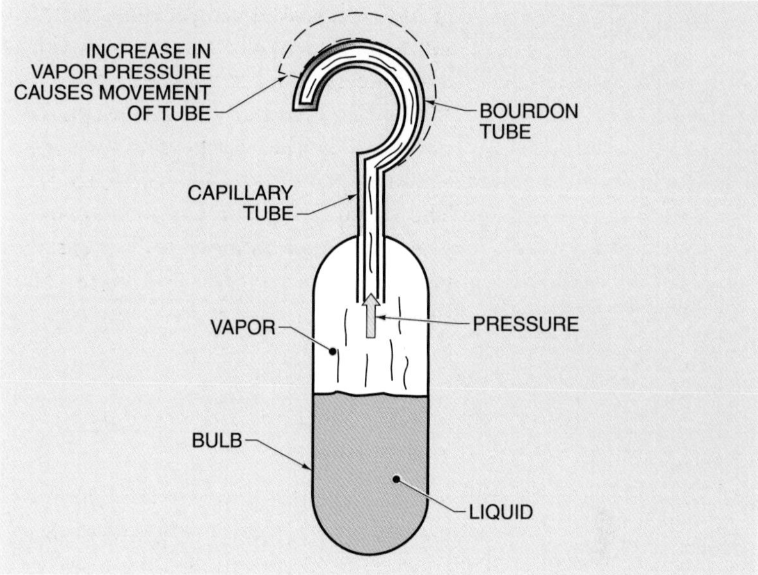

Figure 5-7. A vapor-pressure pressure-spring thermometer uses the vapor pressure from liquid in the bulb to pressurize a Bourdon tube.

Vapor pressure changes nonlinearly with temperature (unit increase in pressure for each unit of temperature rise). The vapor pressure changes exponentially. At lower temperatures, the increase of vapor pressure for each unit of temperature change is small. At higher temperatures, the change of vapor pressure is much greater. **See Figure 5-8.** Therefore, graduations on the scales of these thermometers are more widely spaced at the higher readings than at the lower readings.

Bulb Location

The difference in height between the bulb and the pressure spring can introduce error, especially in a partially filled vapor-pressure system. Since this system is not filled under pressure, as are the liquid- and gas-filled systems, any column of fluid can create a pressure that causes an erroneous reading. In advance of purchase, the manufacturer should be advised if the bulb in the required application is to be located above the pressure spring, at what specific height, and the orientation of the bulb. The system should then be installed per manufacturer specifications. **See Figure 5-9.**

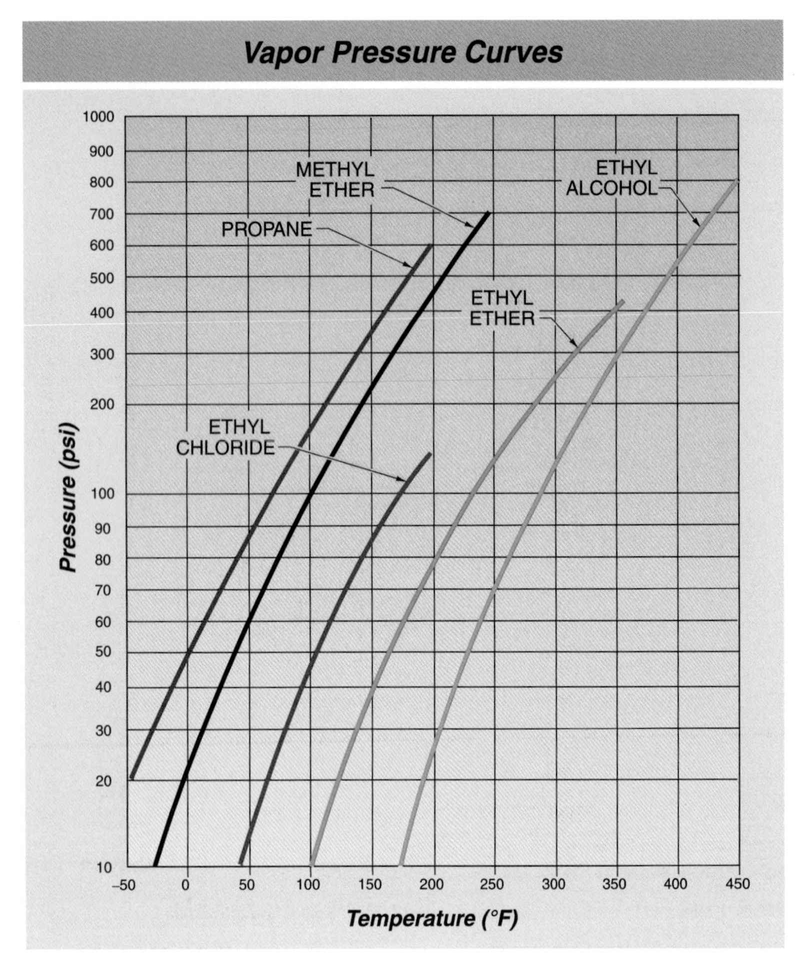

Figure 5-8. Vapor pressure increases in a nonlinear fashion when temperature increases.

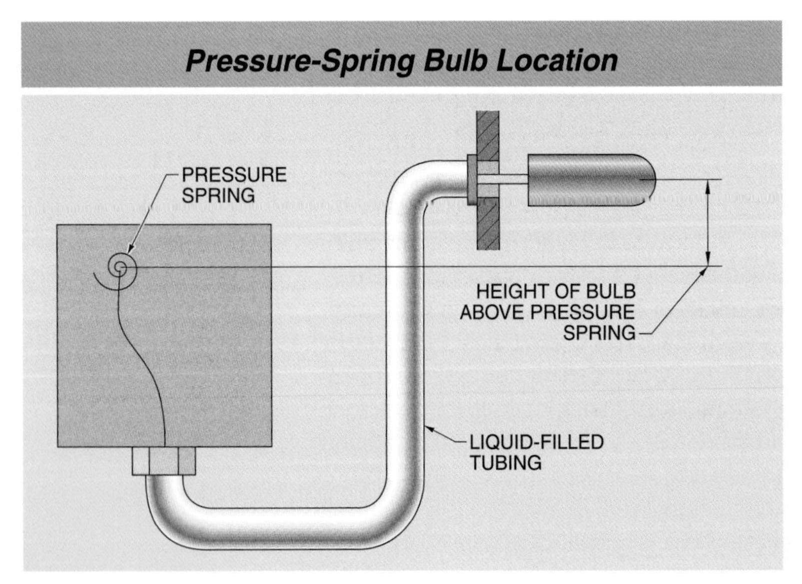

Figure 5-9. When a pressure-spring bulb is mounted above the pressure spring, the instrument must be calibrated to account for the hydrostatic pressure of the liquid in the line.

Liquid-filled systems may be filled under a little more pressure than a vapor-pressure system, but it is not so high that elevation differences between the bulb and the indicator can be completely ignored. If there is 50′ or more of difference, the head of fluid affects the measurement.

Vapor-pressure thermometers provide a more accurate temperature measurement if the desired temperature falls into the right range. In addition, vapor-pressure thermometers can generate a greater amount of power to make it easier to operate switch mechanisms, so they are frequently used for driving temperature switches.

In using a vapor-pressure thermometer, it is necessary to be aware of the maximum and minimum temperatures at the bulb and at the indicator. In most applications, the bulb is always warmer than the indicator. This results in the capillary tube and the pressure-sensing element in the indicator being filled with liquid. This is acceptable as long as this situation is constant.

If the condition reverses, with the capillary and pressure-sensing element being warmer than the bulb, all the liquid is now in the bulb. This could result in an error in the temperature measurement due to a change in head sensed by the element. In most temperature-measurement applications, it is preferable to use a liquid-filled thermometer because it is simpler to select and install.

TECH FACT

The boiling point of a liquid is the temperature at which vapor pressure equals the pressure above the liquid. Industrial processes are used to manipulate the pressure to force liquids to boil at a desired temperature. Food products can be dried without damage by lowering the pressure so that the water boils at a low temperature.

HEAT-SENSITIVE MATERIALS

There are many materials that melt or change state at specific temperatures. These materials can be used as temperature indicators. Liquid crystal indicators are heat-sensitive temperature-indicating substances, available as sheets and strips, where the color of the strip changes progressively on the strip with changes in temperature. Liquid crystals respond to temperatures over the range of –22°F to 250°F. A temperature scale is typically printed on the strip so the position of the color indicates the temperature of the strip.

Heat-sensitive dyes are colorants available in the form of strips or circles, with multiple dots of heat-sensitive dye having different critical temperatures. Most heat-sensitive color dyes permanently change color when the dye reaches a specific temperature. Overall temperature ranges are from 100°F to 500°F. A temperature scale is printed beside the dyes on the indicator. They are used to indicate the maximum temperature encountered. The response of the dye is sensitive to the rate of temperature change and length of time the temperature is maintained.

Disposable temperature-melt materials change phase at a specific temperature and are available in the form of crayons, pellets, and liquid lacquers for temperatures from 100°F to 2500°F. The use of a temperature-indicating crayon is a simple method of measuring temperature. Crayons are commonly used in the welding industry to indicate temperature while a workpiece is being preheated. They are available for various temperatures. In use, a workpiece is marked with one or more crayons. The crayons melt during preheat, indicating that the required temperature range has been attained.

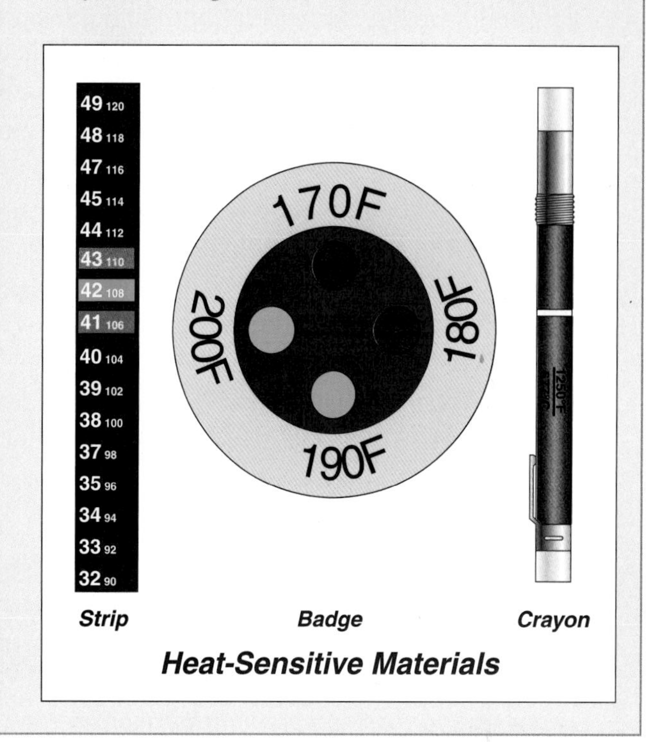

Heat-Sensitive Materials

Response Time

An important consideration in using pressure-spring thermometers is the time it takes to respond to temperature change. The system installation can influence the speed of response. The bulb must be installed so that it senses only the temperature of the process or material into which it is immersed. It should be shielded from reflected or radiant heat. At no point should the bulb be in contact with cold metal, which lowers the temperature reading.

Of the various types of pressure-spring thermometers, the gas-filled systems have the fastest response, followed by the vapor pressure systems and the liquid-filled systems. The response of all systems is faster if the substance whose temperature is to be measured is a liquid rather than a gas because the heat transfer is better. Response is also affected by the velocity of the measured material as it moves past the bulb. A faster flow provides a faster thermometer response.

KEY TERMS

- *coefficient of linear expansion:* The amount a unit length of a material lengthens or contracts with temperature changes.
- *coefficient of volumetric expansion:* The amount a unit volume of a material expands or contracts with temperature changes.
- *liquid-in-glass thermometer:* A thermal expansion thermometer consisting of a sealed, narrow-bore glass tube with a bulb at the bottom filled with a liquid.
- *bimetallic thermometer:* A thermal expansion thermometer that uses a strip consisting of two metal alloys with different coefficients of thermal expansion that are fused together and formed into a single strip, and a pointer or indicating mechanism calibrated for temperature reading.
- *bimetallic element:* A bimetallic strip that is usually wound into a spiral, helix, or coil and allows movement for a given change in temperature.
- *pressure-spring thermometer:* A thermal expansion thermometer consisting of a filled, hollow spring attached to a capillary tube and bulb where the fluid in the bulb expands or contracts with temperature changes.
- *capillary tube:* A small diameter tube with an interior hole 10 thousandths of an inch (0.010″) in diameter and is designed to contain very little liquid.
- *bulb:* A cylinder larger in diameter than a capillary tube that contains the vast majority of the fluid in a thermometer.
- *liquid-filled pressure-spring thermometer:* A pressure-spring thermometer that is filled with a liquid under pressure.
- *gas-filled pressure-spring thermometer:* A pressure-spring thermometer that measures the increase in pressure of a confined gas (kept at constant volume) due to a temperature increase.
- *vapor-pressure pressure-spring thermometer:* A pressure-spring thermometer that uses the change in vapor pressure due to temperature change of an organic liquid to determine the temperature.

REVIEW QUESTIONS

1. Describe the principles of thermal expansion.
2. How can thermal expansion be used in a liquid-in-glass thermometer?
3. How can thermal expansion be used in a bimetallic thermometer?
4. How can thermal expansion be used in a pressure-spring thermometer?
5. How does bulb location affect pressure-spring operation?

SECTION 2 TEMPERATURE MEASUREMENT

Chapter 6 Electrical Thermometers

OBJECTIVES

- Define thermocouple, and identify the phenomena that govern the behavior of thermocouples.
- Describe the purpose of a cold junction, and explain how cold junction compensation is used.
- Describe the construction of a thermocouple.
- List several factors that affect the choice of thermocouple wires.
- List and define typical thermocouple measurement circuits.
- Define resistance temperature detector, describe its construction, and explain how it is used.
- Define thermistor, describe its construction, and explain how it is used.

THERMOCOUPLES

A *thermocouple* is an electrical thermometer consisting of two dissimilar metal wires joined at one end and a voltmeter to measure the voltage at the other end of the two wires. **See Figure 6-1.** A *thermocouple junction* is the point where the two dissimilar wires are joined. The *hot junction,* or *measuring junction,* is the joined end of the thermocouple that is exposed to the process where the temperature measurement is desired.

When the hot junction is at a different temperature than the cold junction, a measurable voltage is generated across the cold junction. The *cold junction,* or *reference junction,* is the end of a thermocouple used to provide a reference point. Controlling or accurately measuring the temperature at the cold junction is essential for accurate temperature measurement with a thermocouple.

Thermocouple Operating Principles

Phenomena that govern the behavior of a thermocouple are the Seebeck effect, the Peltier effect, the Thomson effect, the law of intermediate temperatures, and the law of intermediate metals. Thermocouple measurements can be made with a cold junction or with cold junction compensation.

Seebeck Effect. The *Seebeck effect* is a thermoelectric effect where continuous current is generated in a circuit where the junctions of two dissimilar conductive materials are kept at different temperatures. When the circuit is opened at the cold junction, an electrical potential difference (the Seebeck voltage) exists across the two dissimilar wires at that junction. The voltage produced by exposing the measuring junction to heat depends on the composition of the two wires and the temperature difference between the hot junction and the cold junction. **See Figure 6-2.**

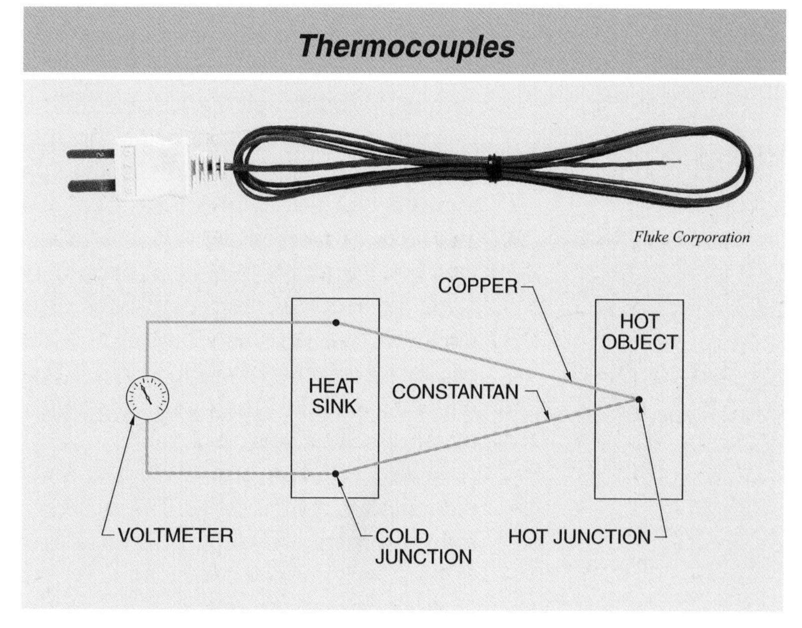

Figure 6-1. A thermocouple creates an electrical potential when the junction is at an elevated temperature.

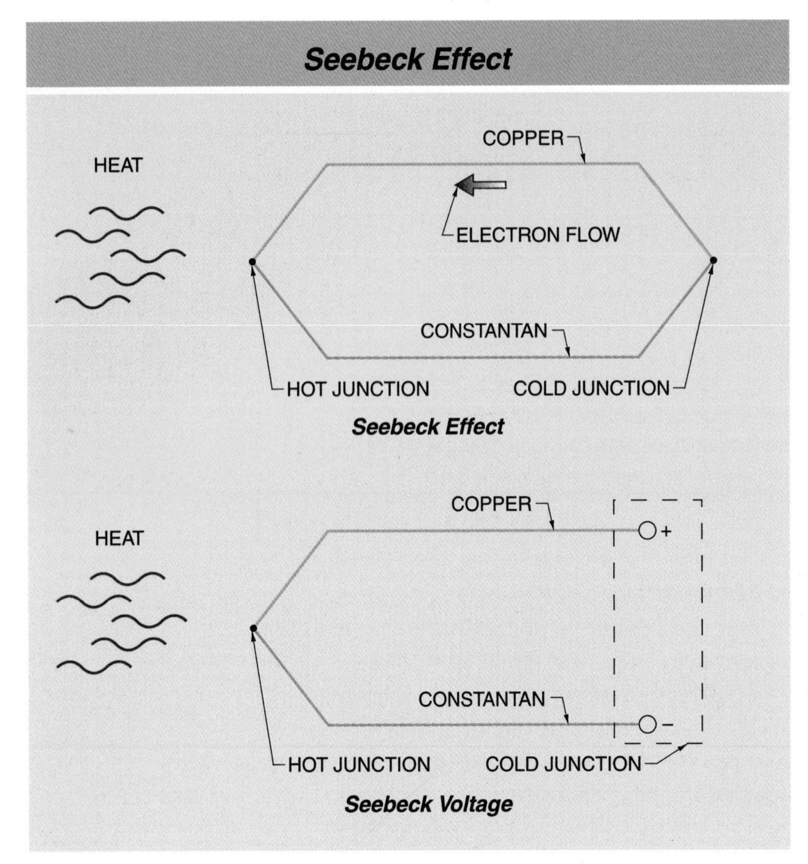

Figure 6-2. The Seebeck effect causes an electrical potential when two dissimilar wires are joined and the end is heated.

Peltier Effect. The *Peltier effect* is a thermoelectric effect where heating and cooling occurs at the junctions of two dissimilar conductive materials when a current flows through the junctions. Heat is either given off or absorbed at the dissimilar material junctions, depending on the direction of electron flow. This can also be stated to say that a voltage is generated in a thermocouple circuit due solely to the presence of dissimilar wires. It is not the same as resistance heating of wires caused by current flow. The Peltier effect only occurs at the junctions of dissimilar materials. **See Figure 6-3.**

With the Peltier effect, there is a net motion of charged particles. These charges transport energy. The energy levels differ in the two types of conductors. As the charges move from one material to the other, they either give up this energy or absorb more energy from the junction. A material cools off as it gives up energy to the moving charges and it warms up as it absorbs energy. This is the basis of thermoelectric cooling. In many thermoelectric coolers, the dissimilar conductors can be P-type and N-type semiconductors. This allows modern chip manufacturing techniques to be used to make small and inexpensive coolers. Thermoelectric cooling is commonly used to cool computer chips, integrated circuits and power supplies, spectroscopy instruments, and many types of detectors.

Thomson Effect. The *Thomson effect* is a thermoelectric effect where heat is generated or absorbed when an electric current passes through a conductor in which there is a temperature gradient. The Thomson effect is not the same as resistance heating of wires caused by current flow. It only occurs when current is flowing through a conductor that has a temperature gradient.

There are three types of the Thomson effect. The first type is the positive Thomson effect, where heat is generated as current flows through the object from hot to cold and heat is absorbed as current flows from cold to hot. This occurs in copper and zinc. The second type is the negative Thomson effect, where heat is absorbed as current flows from hot to cold and heat is generated as current flows from cold to hot. This occurs in iron, nickel, and cobalt. The third type is the zero Thomson effect, where there is no heat generated or absorbed as current flows in the circuit. This occurs only in lead. **See Figure 6-4.**

The voltage that produces the Seebeck current is the sum of the Peltier-effect voltage at the junctions and the two Thomson-effect voltages along the dissimilar wires. This is the basis of thermoelectric thermometry. The National Institute of Standards and Technology (NIST) has published tables that give the voltage produced for given temperature differences for many pairs of dissimilar wires. These tables use a reference temperature of 32°F (0°C).

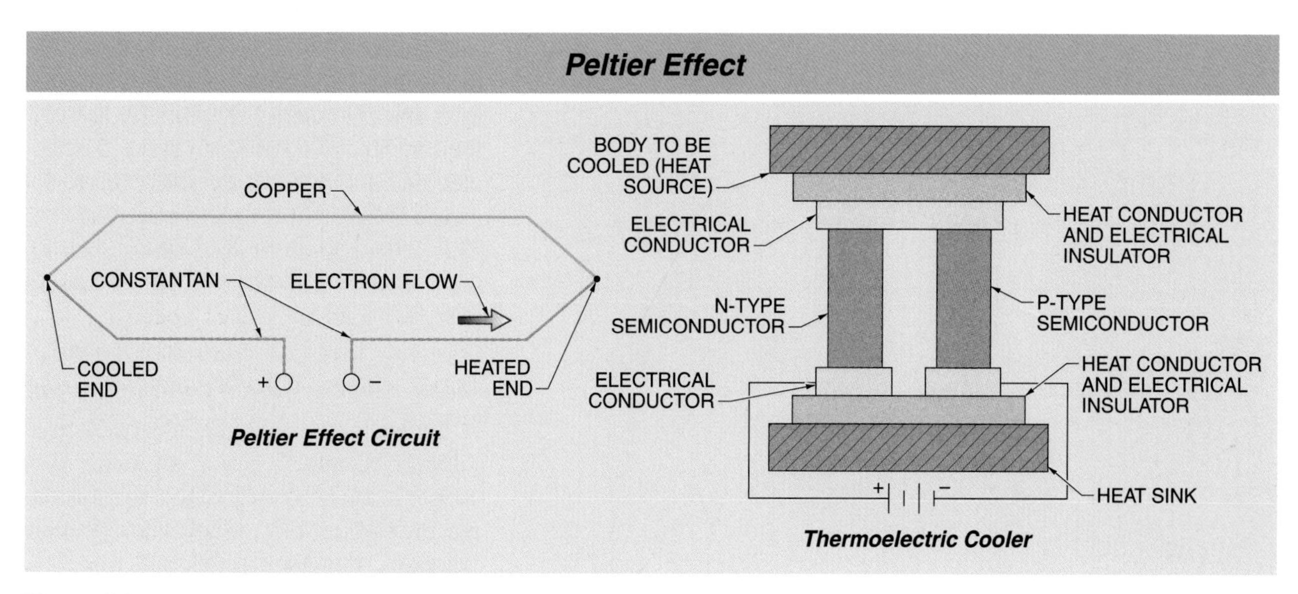

Figure 6-3. The Peltier effect can be used to build thermoelectric coolers.

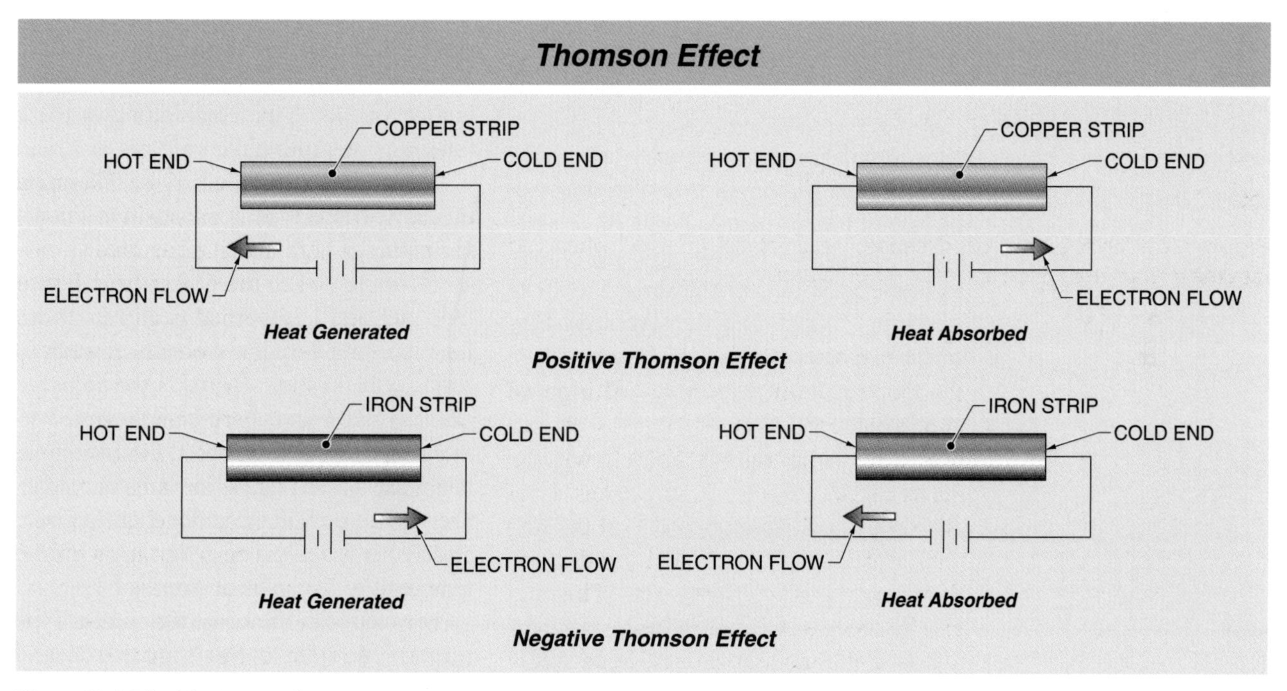

Figure 6-4. The Thomson effect causes heating or cooling when there is current flow through a temperature gradient in a wire.

Law of Intermediate Temperatures. The *law of intermediate temperatures* is a law stating that in a thermocouple circuit, if a voltage is developed between two temperatures T_1 and T_2, and another voltage is developed between temperatures T_2 and T_3, the thermocouple circuit generates a voltage that is the sum of those two voltages when operating between temperatures T_1 and T_3. **See Figure 6-5.** Therefore, it is possible to use a reference junction with any fixed temperature T_2 that is lower than T_3. This is the basis of cold junction temperature compensation in thermocouples. A temperature-sensitive resistor, or thermistor, is used to measure the reference temperature and an adjustment is made to the measured voltage to determine the temperature at the measured junction.

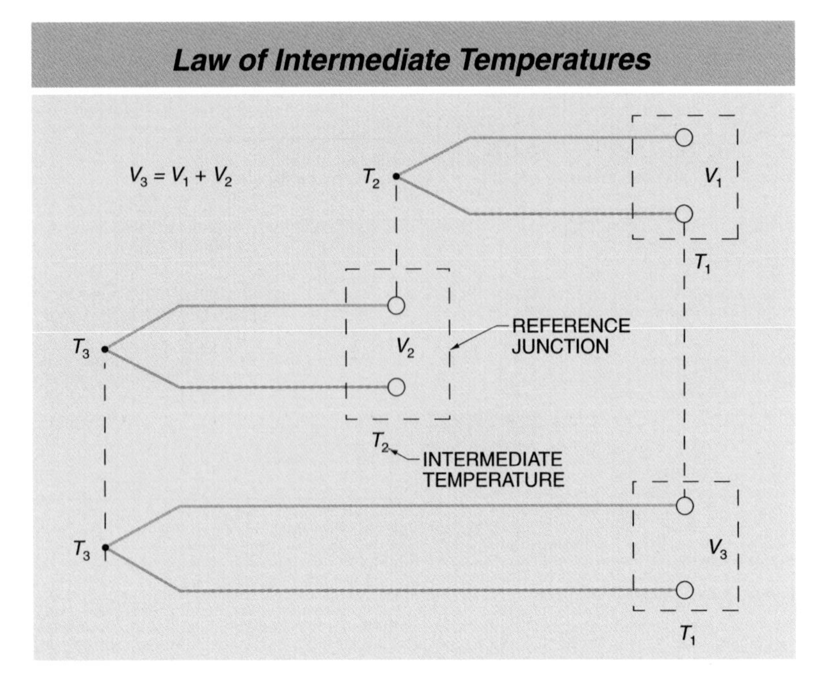

Figure 6-5. The law of intermediate temperatures states that the temperature at the end of the wires determines the electrical potential regardless of the intermediate temperatures.

Law of Intermediate Metals. The *law of intermediate metals* is a law stating that the use of a third metal in a thermocouple circuit does not affect the voltage, as long as the temperature of the three metals at the point of junction is the same. **See Figure 6-6.** Therefore, metals different from the thermocouple materials can be used as extension wires in the circuit. This is a common practice in industry.

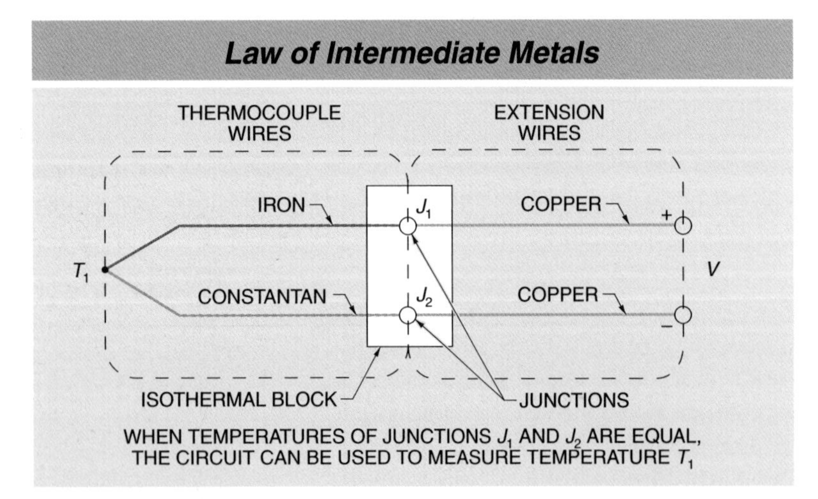

Figure 6-6. The law of intermediate metals states that other metals may be used in a thermocouple circuit as long as the junctions are at the same temperature.

For example, since platinum is very expensive, the extension lead wires used with platinum-platinum/rhodium thermocouples are often made of copper. Another application might use one copper wire and a second made from an alloy. The extension lead wires for Chromel®/Alumel® thermocouples can be copper and constantan, or iron and an alloy.

Measurement with a Cold Junction

A Seebeck voltage cannot be measured directly because when a voltmeter with copper leads is connected to a thermocouple, the connections are new thermocouple junctions. **See Figure 6-7.** Junction J_1 is the desired thermocouple junction. J_2 and J_3 are junctions between the thermocouple wires and the copper leads of the voltmeter. Both of these new junctions generate Seebeck voltages that oppose the thermocouple voltage. The temperatures at these junctions determine the voltages at J_2 and J_3. Therefore, it is necessary to determine the temperatures at J_2 and J_3 before the temperature at J_1 can be measured.

A simple way to determine the temperatures at J_2 and J_3 is to hermetically seal those junctions and immerse them in ice water to form a cold junction. Then both junctions are at exactly 32°F and no temperature measurement is needed. The NIST thermocouple tables use the ice point as the reference value. The measured voltage in the thermocouple circuit can be looked up in the tables and the temperature T_1 can be determined.

For example, the hot junction of a Type J thermocouple, consisting of iron and constantan wires, is placed in a process stream. The thermocouple wires are connected to the copper voltmeter leads. The thermocouple-voltmeter junctions are kept at 32°F and a voltage of 17.710 mV is measured. **See Figure 6-8.**

In order to determine the temperature T_1, look up the voltage in the thermocouple table. A voltage of 17.710 mV for a Type J thermocouple corresponds to a temperature of 325°C (617°F). This method of measuring voltage produces a very accurate temperature measurement. However, it is very inconvenient to use an ice water bath to measure process temperatures.

Cold Junction Compensation. *Cold junction compensation* is the process of using automatic compensation to calculate temperatures when the reference junction is not at the ice point and is often achieved by measuring the temperature of the cold junction with a thermistor. The voltmeter junctions are wired to an isothermal (constant temperature) block. Cold junction compensation measures the temperature of the isothermal block and calculates the equivalent reference voltage. **See Figure 6-9.**

However, since the junctions J_2 and J_3 are no longer at 32°F, the thermocouple tables cannot be used directly. The law of intermediate temperatures can be used to solve this problem. First, measure the temperature of the isothermal block. Use the thermocouple table to look up the equivalent reference voltage (V_{Ref}). Measure the voltage across the wires at the cold junction (V) and add it to V_{Ref} to find the thermocouple voltage (V_{TC}). Use the thermocouple voltage table to convert the voltage to temperature.

For example, with a Type J thermocouple, an isothermal block at 122°F (50°C), and a measured voltage of 8.471 mV across the wires at the cold junction, calculate the process temperature. Look up 50°C in the thermocouple table. From the table, the reference voltage is 2.585 mV. Add the reference voltage to the measured voltage to get 11.056 (8.471 + 2.585 = 11.056). Next, look up 11.056 in the thermocouple table. The voltage corresponds to 205°C (401°F).

TECH FACT

Thermocouple wires must not be allowed to be in electrical contact anywhere except at the hot junction. Many measurement errors are caused by unintended thermocouple junctions. Whenever wires of two different metals are connected together, a thermocouple junction is formed. Care must be taken to ensure that the proper extension wires are used.

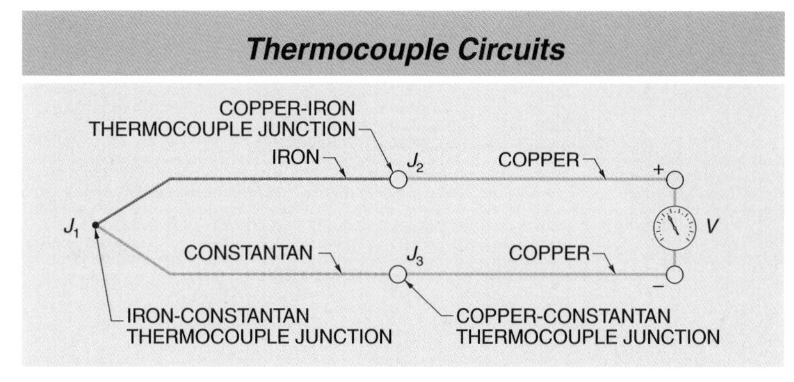

Figure 6-7. A complete thermocouple circuit includes extra junctions from the copper wire in the leads to the voltmeter.

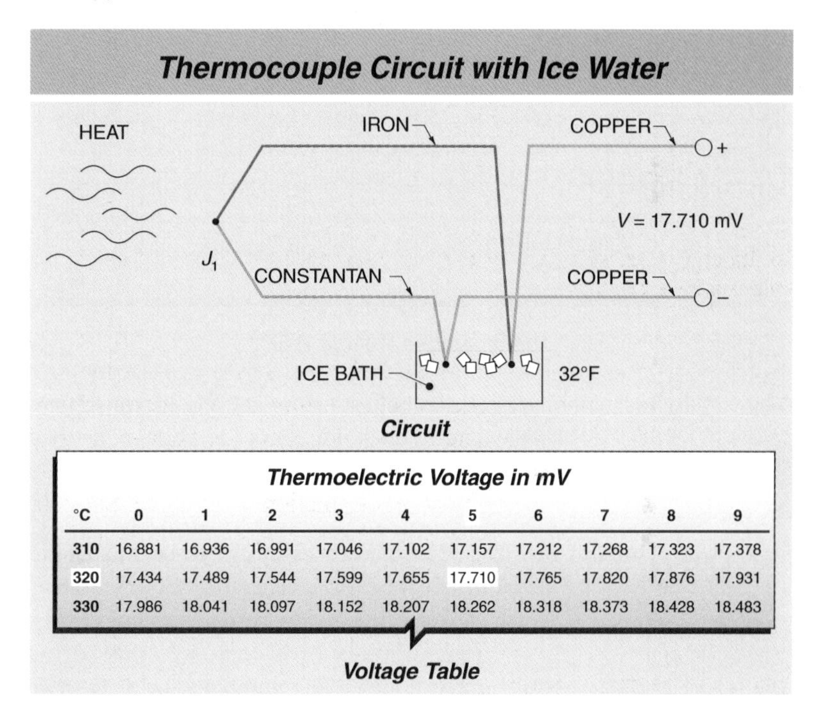

°C	0	1	2	3	4	5	6	7	8	9
310	16.881	16.936	16.991	17.046	17.102	17.157	17.212	17.268	17.323	17.378
320	17.434	17.489	17.544	17.599	17.655	17.710	17.765	17.820	17.876	17.931
330	17.986	18.041	18.097	18.152	18.207	18.262	18.318	18.373	18.428	18.483

Figure 6-8. A 32°F ice bath is the reference temperature for thermocouple tables.

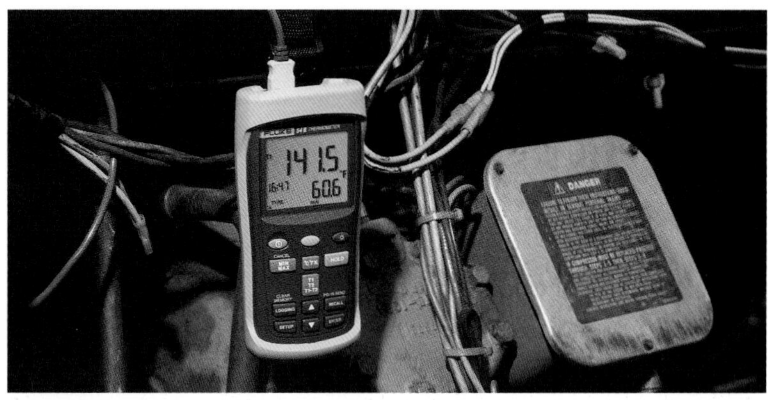

Fluke Corporation

A digital thermocouple can be used to troubleshoot an HVAC system.

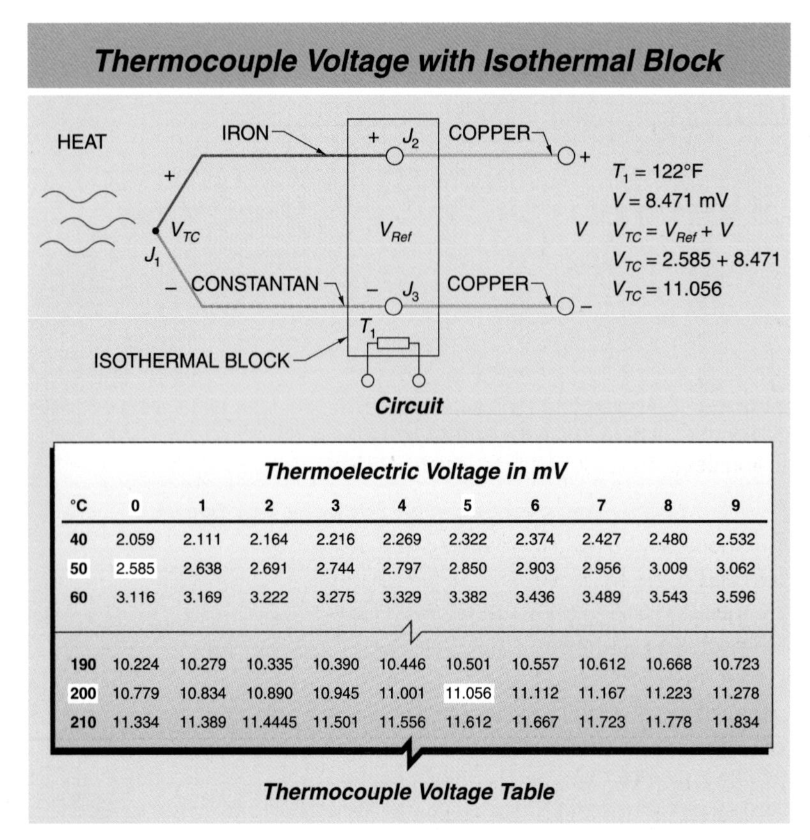

°C	0	1	2	3	4	5	6	7	8	9
40	2.059	2.111	2.164	2.216	2.269	2.322	2.374	2.427	2.480	2.532
50	2.585	2.638	2.691	2.744	2.797	2.850	2.903	2.956	3.009	3.062
60	3.116	3.169	3.222	3.275	3.329	3.382	3.436	3.489	3.543	3.596
190	10.224	10.279	10.335	10.390	10.446	10.501	10.557	10.612	10.668	10.723
200	10.779	10.834	10.890	10.945	11.001	11.056	11.112	11.167	11.223	11.278
210	11.334	11.389	11.4445	11.501	11.556	11.612	11.667	11.723	11.778	11.834

Figure 6-9. An isothermal block can be used to establish a reference temperature for the cold junction.

Figure 6-10. A modern digital thermocouple includes a voltage to temperature conversion, cold junction compensation, and a digital readout of the temperature.

Thermocouples are often used in situations where the resistance temperature-measuring devices are not appropriate because of high temperatures or a corrosive measuring environment. The combination of the two temperature-measuring devices allows accurate measurement of temperature over a very broad range. Most modern thermocouple systems include a direct readout along with automatic compensation for the temperature of the isothermal block. **See Figure 6-10.**

TECH-FACT

Tiny impurities in thermocouple wires can change the thermocouple voltage. The impurities can be introduced in the initial manufacture of the wires. In addition, they can be caused by changes in the wires when they are heated to elevated temperatures and react with the process material or atmosphere.

Thermocouple Construction

The original and still commonly used method of constructing a thermocouple consists of welding the two thermocouple wires together and then slipping ceramic beads down the open ends of the wires. The ceramic beads provide separation of the two thermocouple wires and electrical insulation from the thermowell. The length of the thermocouple wires is selected to match the length of the thermowell and the associated additional pieces. The open ends of the wires terminate at a special insulated termination block. Each wire is screw-clamped to a block made of the same materials as the wire. The thermocouple assembly is designed to slip into the thermowell with the terminal block fastened into the thermocouple head, pressing the welded tip against the bottom of the thermowell. **See Figure 6-11.**

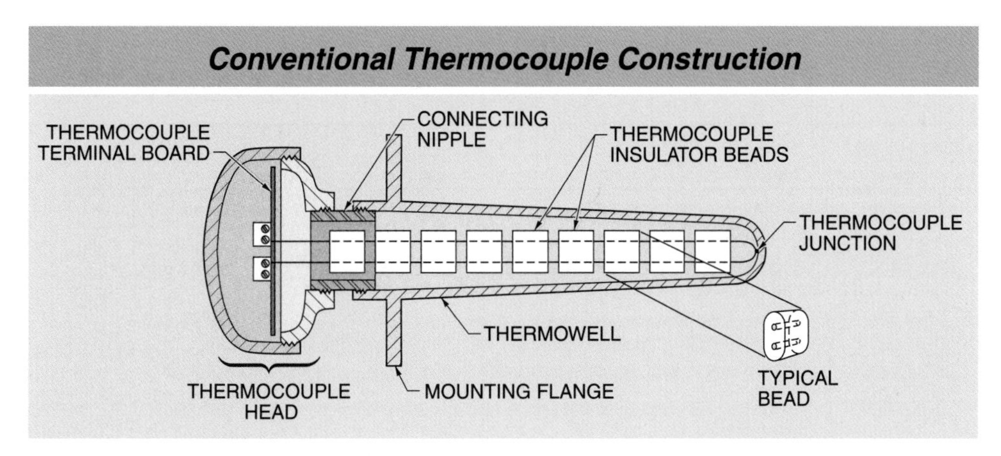

Figure 6-11. Conventional thermocouple construction uses insulator beads to isolate the two thermocouple wires.

Seebeck Coefficient. One of the factors in evaluating a pair of materials for use as a thermocouple is the thermoelectric difference (Seebeck coefficient) between them. The voltage is proportional to the temperature difference between the two thermocouple junctions. A large thermoelectric difference is needed to measure low temperatures. The output voltage is small because of the small temperature difference between the measured temperature and the temperature of the cold junction. A larger thermoelectric difference increases the output voltage. The Seebeck coefficient gives an indication of the output of the thermocouple pair, but it is not the only factor that should be used for the selection. Other factors in evaluating thermocouples are the useful temperature range and resistance to extreme environments. **See Figure 6-12.**

Thermocouple Wires

A factor in evaluating a pair of materials for use as a thermocouple is the required precision. Thermocouple wires are available as precision, standard, and extension grade wires. Precision grade wires have an error rating of ±0.5% of reading or 1°C, whichever is greater. Standard grade wires have an error rating of ±0.75% of reading or 2°C, whichever is greater. Extension grade wires have an error rating of ±1% of reading or 4°C, whichever is greater. Wire diameters are available from 0.001″ and up. Standard diameters are 0.01″, 0.02″, 0.032″, 0.040″, 1⁄16″, 1⁄8″, 3⁄16″, and 1⁄4″. Smaller wires have a faster response time but are more fragile.

One wire of a thermocouple pair is positive and one wire is negative. The name of the metal appearing before the slash indicates the positive wire when the measuring junction is higher than the cold junction. The name of the metal after the slash indicates the negative wire. When one of the wires is an alloy, the different metals of the alloy are separated by a hyphen. For example, in a platinum-rhodium/platinum thermocouple, the platinum-rhodium alloy is the positive wire and platinum is the negative wire.

Fluke Corporation

Digital thermocouples allow a quick temperature measurement during testing or maintenance of industrial equipment.

Thermocouple Characteristics

Designation	Useful Temperature Range °F	°C	Uses
B	122 to 3100	50 to 1700	Resistant to oxidation and corrosion; hydrogen, carbon, and some metals can contaminate the Pt/Rh wires; do not insert into metal tubes
R	32 to 2650	1 to 1450	Resistant to oxidation and corrosion; hydrogen, carbon, and some metals can contaminate the Pt/Rh wires; do not insert into metal tubes
S	32 to 2650	1 to 1450	Resistant to oxidation and corrosion; hydrogen, carbon, and some metals can contaminate the Pt/Rh wires; do not insert into metal tubes
E	–325 to 1650	–200 to 900	Highest emf output, acceptable in mildly oxidizing or reducing atmospheres
J	32 to 1340	1 to 750	Acceptable in vacuum or reducing atmospheres; not recommended for low temperatures
T	–325 to 700	–200 to 350	Acceptable in mildly oxidizing and reducing atmospheres, inert atmospheres, and vacuum and where moisture is present; low temperature and cryogenic applications
K	–325 to 2300	–200 to 1250	Acceptable in clean oxidizing and inert atmospheres; limited use in vacuum and reducing atmospheres
N	–450 to 2400	–270 to 1300	Alternative for Type K, more stable
C	32 to 4200	0 to 2315	Acceptable in vacuum, inert, or reducing atmospheres; poor oxidation resistance, subject to embrittlement
D	32 to 4200	0 to 2315	Acceptable in vacuum, inert, or reducing atmospheres; poor oxidation resistance, subject to embrittlement
G	32 to 4200	0 to 2315	Acceptable in vacuum, inert, or reducing atmospheres; poor oxidation resistance, subject to embrittlement

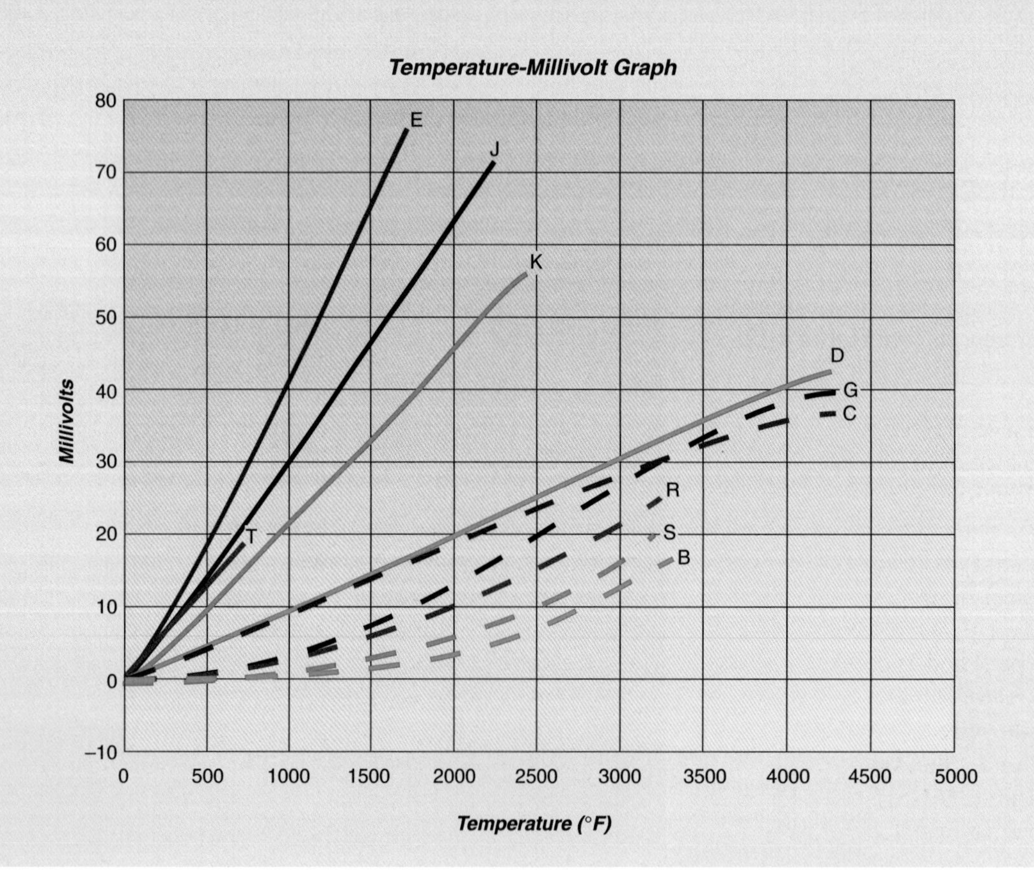

Figure 6-12. Thermocouple designations require that the wires follow a particular voltage-temperature curve.

Interpolation

There are times when measured voltage values fall between two numbers in a table. The user has two options. The first and simplest option is to use the temperature that falls closest to the measured voltage. This is acceptable when the desired accuracy does not require a more specific value.

The second option is to use interpolation to calculate the temperature based upon the difference between the adjacent numbers and where the measured voltage falls in that difference. Interpolation assumes that the voltage changes evenly (linearly) between the lower and upper numbers. The procedure is to calculate where the voltage falls as a fraction of the total difference and then multiply that by the temperature difference and add it to the lower temperature. The formula to use is as follows:

$$T = T_l + \frac{(V - V_l)}{(V_h - V_l)} \times (T_h - T_l)$$

where

T = object temperature to be determined

T_l = temperature corresponding to low voltage in table

T_h = temperature corresponding to high voltage in table

V = measured voltage

V_h = voltage above measured voltage in table

V_l = voltage below measured voltage in table

For example, if the measured voltage of a Type J thermocouple with a cold junction temperature of 32°F is 10.975 mV, what is the temperature at the hot junction? Find the two numbers above and below the measured voltage. These are the low and high voltage for the equation.

V_l = 10.945

V_h = 11.001

Find the temperatures that correspond to those voltages. These are the low and high temperatures for the equation.

T_l = 203°C

T_h = 204°C

The measured voltage is

V = 10.975

Thermoelectric Voltage in mV

°C	0	1	2	3	4	5	6	7	8	9
190	10.224	10.279	10.335	10.390	10.446	10.501	10.557	10.612	10.668	10.723
200	10.779	10.834	10.890	10.945	11.001	11.056	11.112	11.167	11.223	11.278
210	11.334	11.389	11.445	11.501	11.556	11.612	11.667	11.723	11.778	11.834

Substitute the numbers into the equation as follows:

$$T = T_l + \frac{(V - V_l)}{(V_h - V_l)} \times (T_h - T_l)$$

$$T = 203 + \frac{(10.975 - 10.945)}{(11.001 - 10.945)} \times (204 - 203)$$

$$T = 203 + \frac{(0.030)}{(0.056)} \times (1)$$

$$T = 203 + 0.54 \times 1$$

T = **203.54°C**

For a Type J thermocouple, the temperature that corresponds to 10.975 V is 203.54°C. If the round-off method is used, the temperature is read as 204°C.

There are several types of thermocouple wires. The choice of the wire pair depends on the application. The most common thermocouple wires are combinations of iron/constantan, Type J; copper/constantan, Type T; Chromel/Alumel, Type K; Chromel/constantan, Type E; and platinum-rhodium/platinum, Type R and Type S. These letter designations and color codes for different thermocouple wire combinations have been agreed on by the American National Standards Institute (ANSI), ASTM International, the International Society of Automation (ISA), and many other organizations. **See Figure 6-13.**

The letter designation specifies a voltage-temperature relationship, not a particular wire composition. Thermocouples of a given type may have variations in composition as long as their resultant temperature-voltage relationships remain within specified tolerances. All materials manufactured in compliance with the established thermoelectric voltage standards are equally acceptable.

TECH FACT

Thermocouple tables are published by the National Institute of Standards and Technology (NIST).

International Thermocouple Color Codes

ANSI Code	Alloy Combination		US & Canadian		International	Czech/ British	Dutch/ German	French	Maximum Temperature Range
	Positive Lead	Negative Lead	Thermocouple Grade	Extension Grade					
J	IRON Fe (magnetic)	COPPER-NICKEL Cu-Ni (CONSTANTAN)							Thermocouple Grade –210°C to 1200°C –346°F to 2193°F Extension Grade 0°C to 200°C 32°F to 392°F
K	NICKEL-CHROMIUM Ni-Cr (CHROMEL)	NICKEL-ALUMINUM Ni-Al (magnetic) (ALUMEL)							Thermocouple Grade –270°C to 1372°C –454°F to 2501°F Extension Grade 0°C to 200°C 32°F to 212°F
T	COPPER Cu	COPPER-NICKEL Cu-Ni (CONSTANTAN)							Thermocouple Grade –270°C to 400°C –454°F to 752°F Extension Grade –60°C to 100°C –76°F to 212°F
E	NICKEL-CHROMIUM Ni-Cr (CHROMEL)	COPPER-NICKEL Cu-Ni (CONSTANTAN)							Thermocouple Grade –270°C to 1000°C –454°F to 1832°F Extension Grade 0°C to 200°C 32°F to 392°F
N	NICKEL-CHROMIUM-SILICON Ni-Cr-Si	NICKEL-SILICON-MAGNESIUM Ni-Si-Mg					NO STANDARD, USE US COLOR CODES		Thermocouple Grade –270°C to 1300°C –454°F to 1832°F Extension Grade 0°C to 200°C 32°F to 392°F
R	PLATINUM-13% RHODIUM Pt-13% Rh	PLATINUM Pt	NOT ESTABLISHED						Thermocouple Grade –50°C to 1768°C –58°F to 3214°F Extension Grade 0°C to 150°C 32°F to 392°F
S	PLATINUM-10% RHODIUM Pt-10% Rh	PLATINUM Pt	NOT ESTABLISHED						Thermocouple Grade –50°C to 1768°C –58°F to 3214°F Extension Grade 0°C to 150°C 32°F to 300°F
B	PLATINUM-30% RHODIUM Pt-30% Rh	PLATINUM-6% RHODIUM Pt-6% Rh	NOT ESTABLISHED		—	NO STANDARD, USE COPPER WIRE		NO STANDARD, USE COPPER WIRE	Thermocouple Grade 50°C to 1820°C 122°F to 3308°F Extension Grade 0°C to 100°C 32°F to 212°F

Figure 6-13. Thermocouple color codes have been standardized in many countries.

Platinum Thermocouples. Thermocouple types B, R, and S are all constructed with platinum or a platinum alloy as the negative wire. These thermocouples share many characteristics such as stability.

Type S has been specified as the standard for temperature calibration between the antimony point, 630.74°C (1167.33°F), and the gold point, 1064.43°C (1947.97°F). These thermocouples are very susceptible to metal vapor diffusion at high temperatures. Therefore, they should be used inside a nonmetallic sheath. The Type B thermocouple passes through a minimum between 32°F (0°C) and 108°F (42°C). This means that any voltage in that range could correspond to two different temperatures. Therefore, the Type B thermocouple cannot be used below about 122°F (50°C).

Constantan Thermocouples. Thermocouple types E, J, and T are all constructed with constantan as the negative wire. Constantan is not a specific alloy. It consists of a range of copper-nickel alloys of varying construction. In fact, the constantan used in a Type J thermocouple is not the same constantan used in a Type T thermocouple.

The Type E thermocouple is widely used for low-temperature measurements because it has the highest Seebeck coefficient of any common thermocouple pair. The Type J thermocouple is widely used because of its low cost. However, Type J thermocouples are often subject to inaccuracies because of difficulties in making and keeping the iron wire pure. Type J thermocouples cannot be used above 1400°F (760°C) because of a phase change that occurs in the iron. Type T thermocouples have a copper wire as one of the leads. Since copper is usually the type of wire used in the leads to a voltmeter, there is one less junction to be concerned about with a Type T thermocouple.

Nickel-Chromium Thermocouples. Thermocouple types K and N both use nickel-chromium alloys for the positive wire. Type K thermocouples are somewhat unstable when exposed to high temperatures for extended periods due to oxidation of the Chromel wire. They also exhibit hysteresis of several degrees between the temperatures of 392°F and 1112°F (200°C and 600°C).

Type K thermocouples are subject to "green rot" when a reducing atmosphere such as hydrogen comes in contact with the wires in the presence of a small amount of oxygen. The chromium in the Chromel wire oxidizes and becomes magnetic. The alloy often turns green and the voltage changes. At high temperatures, hydrogen can even diffuse through metal thermowells and destroy the thermocouple. A common solution is to use an air-purged thermowell.

Thermocouple Type N has improved alloys for better stability above 932°F (500°C) and better oxidation resistance. The positive wire, Nicrosil, is constructed of a nickel-chromium alloy with a small amount of silicon. The negative wire, Nisil, is constructed of a nickel-silicon alloy with a small amount of magnesium.

Tungsten Thermocouples. Thermocouple types C, D, and G are constructed with tungsten alloys in both the positive and negative wires. These thermocouples are used in vacuum or reducing atmospheres. The thermocouples break down quickly in oxidizing atmospheres. Pure tungsten becomes very brittle above about 2200°F (1200°C). Rhenium alloys are used to allow higher temperature applications.

Other Thermocouples. Many other materials are used in thermocouples in specialized applications. Platinel® thermocouples were designed for high-temperature applications in turboprop engines. Platinel can be used continuously up to 2372°F (1300°C) and intermittently up to 2462°F (1350°C) in oxidizing and neutral atmospheres. Platinel II® is a palladium-platinum-gold/gold-palladium thermocouple. Nickel-nickel/molybdenum sometimes replaces Chromel/Alumel. Molybdenum wires can be used up to about 3000°F (1650°C) when protected with a silica coating. Other combinations are Chromel/white gold, molybdenum/tungsten, tungsten/iridium, and iridium/iridium-rhodium.

Sheathed Thermocouple Construction. The sheathed thermocouple was developed to eliminate the shortcomings of the conventional thermocouple assembly. The thermocouple wires are welded together, as in the conventional thermocouple, but the wires are then inserted into a thin-walled tube and packed with a magnesium oxide powder. The thin metal tube can be 304 or 316 stainless steel, Inconel, or other material. The whole assembly is then drawn through various dies to reduce the diameter.

The drawing process stretches out the tube and the thermocouple wires. Sheathed thermocouples have been drawn down to a diameter of 0.010″, but common diameters are ¼″, ⅛″, and 1⁄16″. **See Figure 6-14.** This construction method protects the thermocouple wires from exposure to air or other gases, thus providing a longer life, and creates a thermocouple which is a smaller size and a size that fits closely inside the thermowell bore.

TECH FACT

Thinner thermocouple wires improve the response time but increase the resistance of the circuit and make it more susceptible to noise.

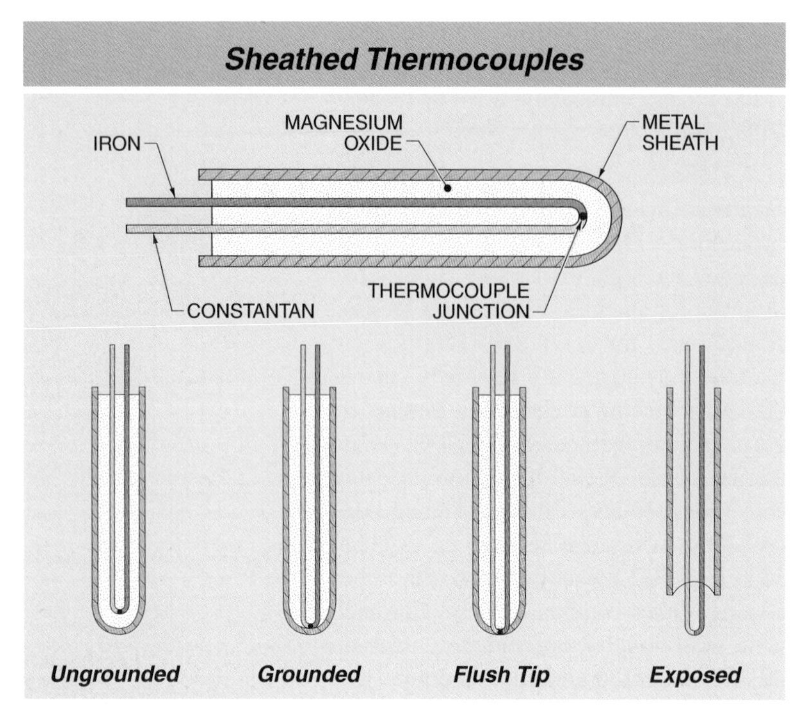

Figure 6-14. Sheathed thermocouples can be wired in several ways for different applications.

The temperature-measuring tip of the thermocouple can be made as an ungrounded, grounded, flush, or exposed tip. The ungrounded form has the thermocouple junction separated from the sheath by magnesium oxide. The grounded form has the thermocouple junction in contact with the sheath, but still protected by the sheath. Grounded thermocouples have a faster temperature response than ungrounded forms because the heat is conducted through the metal wall.

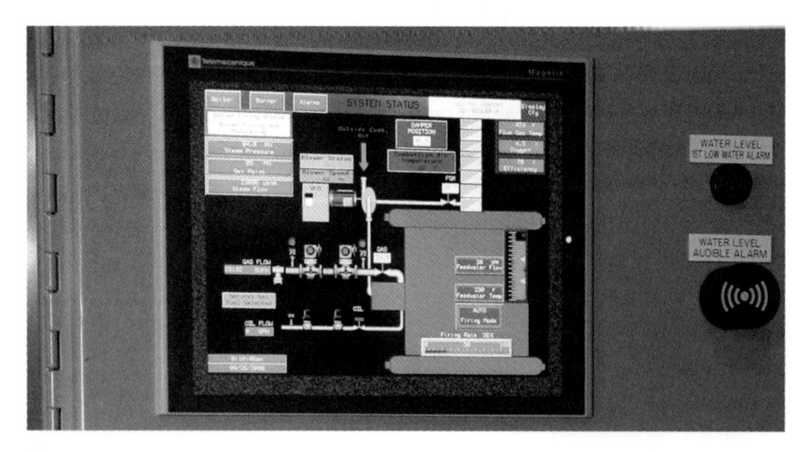

Temperature measurements are sent to a control panel and used to control an industrial process.

The flush form has the junction and end of the tube cut off and exposed to the outside environment. The exposed-tip form has the sheath cut back from the thermocouple junction so that the junction and a short length of wires are exposed. The flush and exposed forms have very fast temperature response but are subject to corrosion and damage. Bare sheathed thermocouples can be inserted through a thermocouple compression fitting into a process stream. This can be used for air ducts where thermowells are not needed. The sheaths can be bent if care is taken and the radius is not too small for the sheath diameter.

Thermocouple Measurement Circuits

Thermocouples can be wired in other types of circuits to accomplish specialized measurements. These circuits include difference thermocouples, thermopiles, averaging thermocouples, pyrometers, null-current thermocouples, and high-input-impedance circuits.

Difference Thermocouples. A *difference thermocouple* is a pair of thermocouples connected together to measure a temperature difference between two objects. The lower temperature thermocouple is wired so that the polarity is reversed from the high-temperature thermocouple. Therefore, the voltage output of the two thermocouples is equivalent to the temperature difference of the two measurements. **See Figure 6-15.** A difference thermocouple typically can measure differences of about 50°F or more.

Thermopiles. A *thermopile* is an electrical thermometer consisting of several thermocouples connected in series to provide a higher voltage output. In a thermopile, the individual voltages of each thermocouple are added together. A thermopile can be used to measure extremely small temperature differences or it can be used to increase the voltage of a circuit to be able to trip a contact. Thermopiles have been designed that are capable of measuring temperature differences as small as a few millionths of a degree. **See Figure 6-16.**

Thermopiles are commonly used to make optical temperature measurements of furnaces or hot materials where the temperature is high enough to emit visible light. A lens system in the optical temperature sensor is aimed at either the glowing brickwork of the furnace or the material in the furnace. The radiation is focused on the hot junctions of the thermopile, raising their temperature. The cold junctions are maintained at a cooler temperature by either air or water cooling. The output of the thermopile is sufficient to be used by a transmitter, recorder, or controller. This type of temperature sensing is commonly used in glass furnaces, kilns, and steel mills.

Averaging Thermocouples. An *averaging thermocouple* is an electrical thermometer consisting of a set of parallel-connected thermocouples that is commonly used to measure an average temperature of an object or area. For example, in a large tank or reactor, a set of thermocouples is inserted in a protective tube or thermowell in the top of the vessel. The different thermocouples are positioned at different depths in the tube and the circuit averages the voltage readings. In an averaging thermocouple, the resistance of the different thermocouple circuits must be very similar. Since the wires are all different lengths, the best way to ensure that each circuit has equivalent resistance is to put a relatively large resistor, called a swamping resistor, in each circuit. **See Figure 6-17.**

Thermocouple Pyrometers. A *thermocouple pyrometer* is an electrical thermometer consisting of a plain electrical meter with a measurement range of 20 mV to 50 mV, a thermocouple, and a balancing resistor. The electrical meters require a certain resistance range and the meter is selected to match the resistance of the thermocouple circuit. The meter is kept at ambient temperature and requires a constant loop resistance. Variations in loop resistance cause errors in measurement. Thus the size of the thermocouple wires used, the thermocouple extension wire size, and the distance from the thermocouple to the meter must be carefully selected to be within the allowable resistance. **See Figure 6-18.**

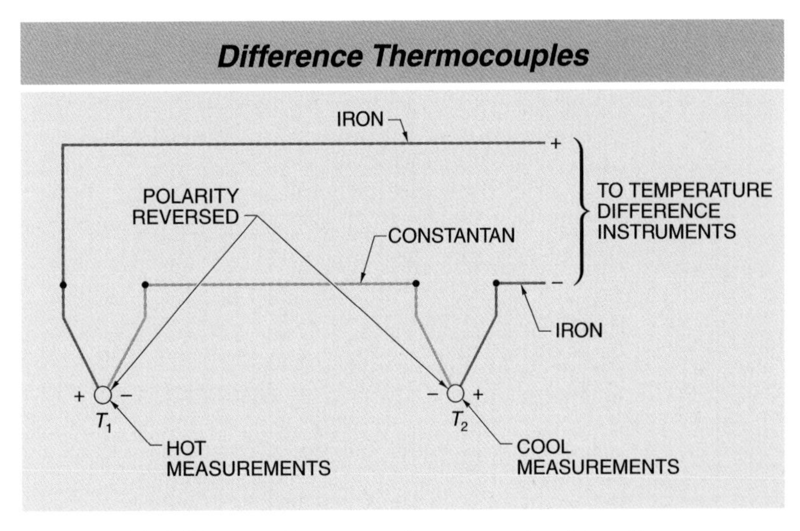

Figure 6-15. Difference thermocouples are made of two thermocouples wired in series with reversed polarity.

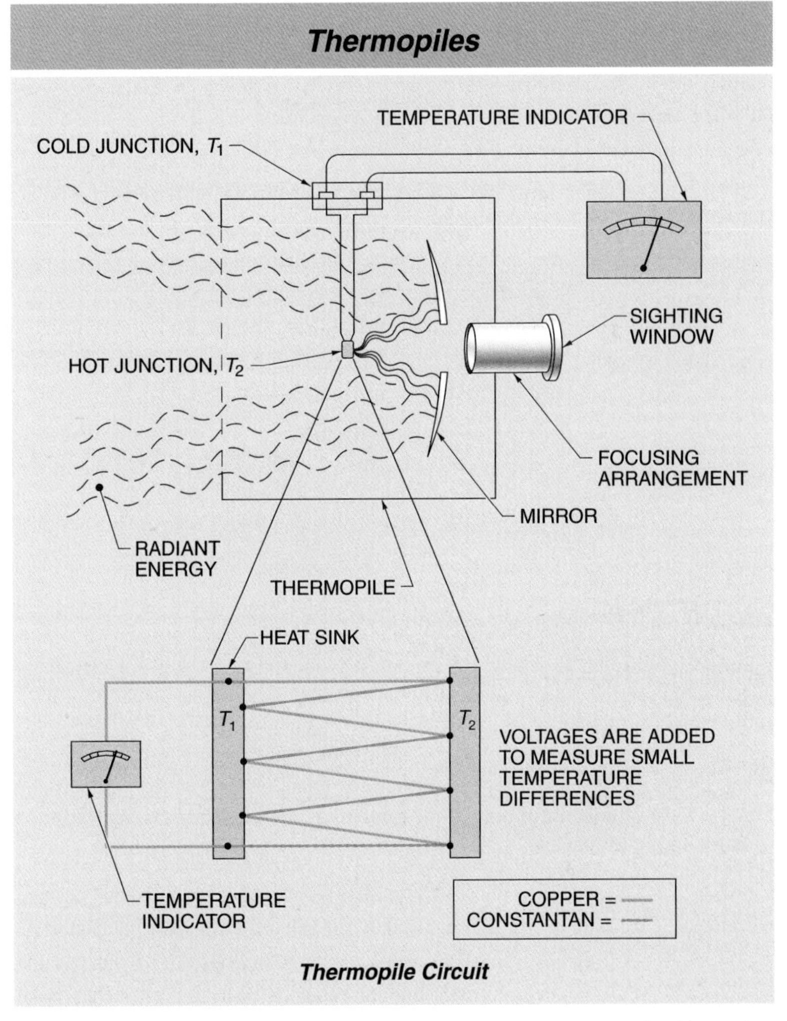

Figure 6-16. A thermopile consists of several thermocouples wired in series in order to amplify the signal.

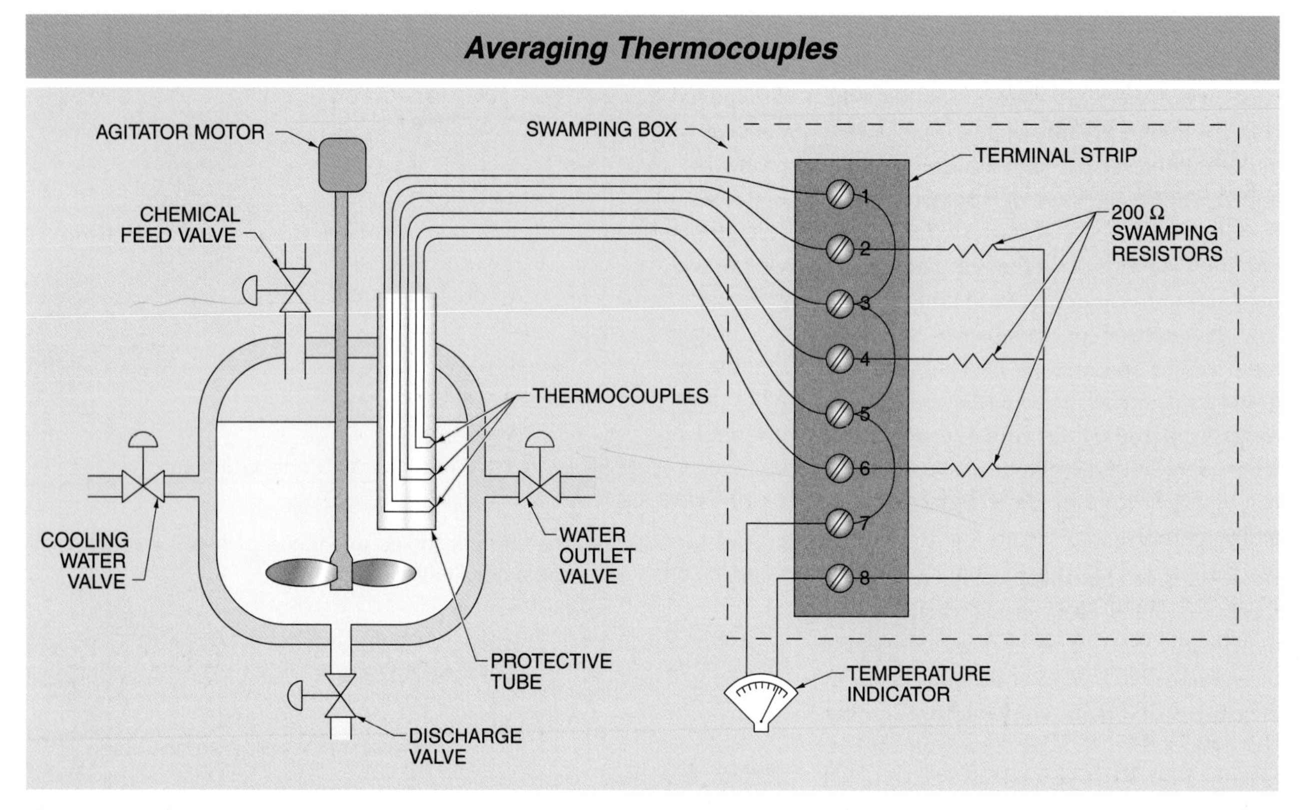

Figure 6-17. A swamping box uses resistors in each thermocouple circuit to eliminate errors when measuring an average reading of a set of thermocouples.

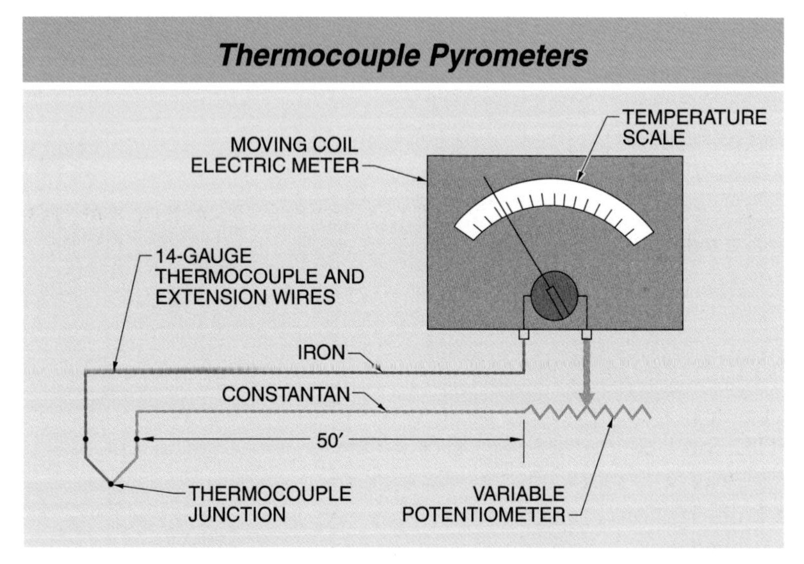

Figure 6-18. A thermocouple pyrometer uses a variable potentiometer to balance loop resistance.

Typically, use of 14-gauge thermocouple wire and extension wire allows a 50′ distance. The balancing resistor is usually a potentiometer that can be adjusted to make the total loop resistance equal to the required resistance. The meter scale is calibrated in degrees for the type of thermocouple that must be used. The system is entirely self-contained, requiring no external power, and is ideal for a local thermocouple measurement installation. It does not have a high degree of accuracy, but is acceptable for a noncritical measurement.

Null-Current Thermocouples. A *null-current thermocouple* is a circuit and a voltage generator that can be adjusted to exactly balance the voltage output of a thermocouple. The current through the measurement circuit is measured and causes a counter voltage source to be changed until the current is zero. A precision variable resistor called a slide wire is usually used to generate the feedback volts. The wiper, or movable contact, of the slide wire is directly connected to the temperature-indicating pointer or the recording pen.

If the potentials are in balance there is no current flow through the connecting wires to cause any voltage losses and subsequent errors in measurement. The measurement instrument can now be located hundreds of feet from the thermocouples with no worry about measurement errors. Zero reference junctions are used in these instruments as well as open measurement failure actions. This technology was in common use prior to the development of high-input-impedance solid-state circuits. It was used in portable instruments, single- and multipoint recorders, and controllers. There are many still in service.

High-Input-Impedance Thermocouple Circuits. The development of the field-effect transistor (FET) and other high-input-impedance solid-state devices revolutionized electronic measurement technology. With high input impedance, the current flow through the connecting wires to the thermocouple is so small as to be insignificant. This allows the elimination of the high-maintenance slide wire and the mechanical driving mechanisms.

Thermocouple instruments automatically correct for the thermocouple nonlinearity and can be field selected to work with different types of thermocouples. They all contain zero junction references and open circuit protection actions. Instrument sizes could be as small as 1″ square up to full-size recorders. Multipoint recorders can handle many different types of thermocouples at the same time.

RESISTANCE TEMPERATURE DETECTORS (RTDs)

A *resistance temperature detector (RTD)* is an electrical thermometer consisting of a high-precision resistor with resistance that varies with temperature, a voltage or current source, and a measuring circuit. RTDs are accurate and reliable temperature sensors especially for low temperatures and small ranges. They are generally more expensive than thermocouples and are not used for high temperatures or corrosive measuring environments. Thermocouples are often used in situations where the use of an RTD is not appropriate because of high temperatures or a corrosive measuring environment. The combination of the two temperature measuring devices allows accurate measurement of temperature over a very broad range.

RTD Operating Principles

An RTD increases its resistance when it is exposed to heat. This gives the RTD a positive temperature coefficient (PTC). A protective sheath material covers the RTD wires, which are coiled around an insulator that serves as a support. **See Figure 6-19.** Unlike a thermocouple, an RTD does not generate its own voltage. An external source of voltage or current must be incorporated into the circuit. The voltage drop across an RTD provides a much larger output than the Seebeck voltage of a thermocouple, allowing an RTD to be more precise over a small temperature range.

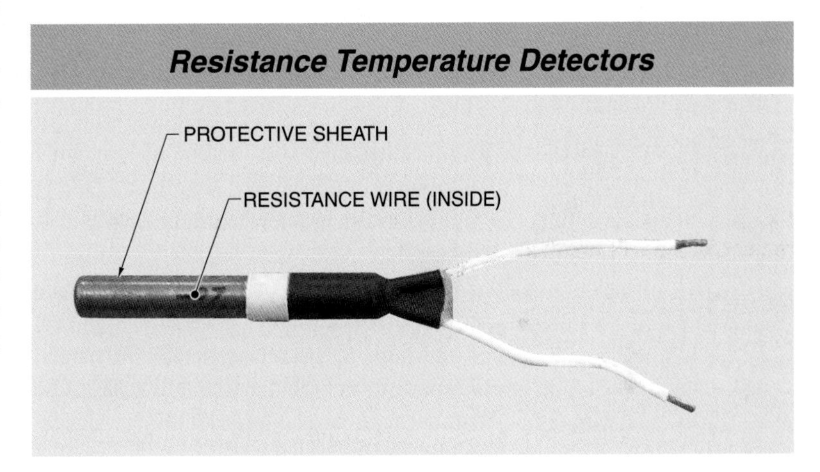

Figure 6-19. A resistance temperature detector (RTD) contains a resistor with a resistance that varies with temperature.

RTD Construction

The heat-sensitive element of a wound RTD consists of a carefully made electrical resistor manufactured in the form of a bulb. Platinum, nickel, or copper wire wrapped around an insulator is most often used for the resistance wire of the element. The bulb consists of a fine resistance wire wrapped around an insulator and enclosed in glass. The most common form for an RTD is very similar to a sheathed thermocouple with an outside diameter of ¼″ so that it can fit into standard thermowells. **See Figure 6-20.**

Resistance Temperature Detector Construction

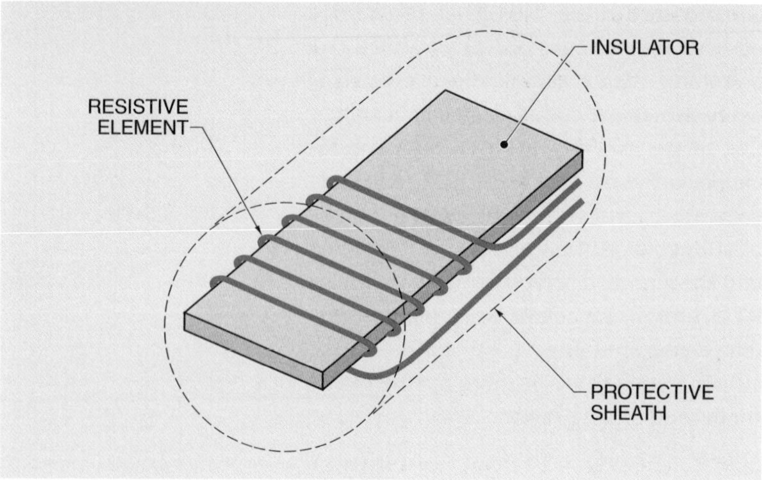

Figure 6-20. RTDs consist of precision wires wrapped around an insulator and encapsulated in a protective sheath.

Platinum wire is the best material for an RTD because it is useful over a wide temperature range from –400°F to 1200°F. Nickel is frequently used because it is economical and, over its useful range of –250°F to 600°F, its resistance per degree of temperature change is greater than that of platinum. Copper is generally restricted to temperatures below that of nickel. The useful range of a copper RTD is –325°F to 250°F.

Every RTD made from the same material should be interchangeable without requiring recalibration of the instrument being used. For this reason, RTDs are manufactured to have a fixed resistance at a certain temperature. Platinum RTDs generally have a resistance of 100 Ω at 32°F (0°C). There are two different alpha (α) curve numbers for platinum RTDs, 0.00385 and 0.00392. The α curve number is the change in resistance in ohms per °C. Some nickel RTDs have a resistance of 120 Ω at 0°C and an α number of 0.00672 Ω per °C. There may be other nickel RTDs based on other conditions.

In newer styles of RTDs, the conductor is deposited as a thin film on a ceramic substrate. This style of RTD has lower cost and better resistance to shock and environmental effects but tends to be less stable.

The response time of an RTD is about the same or somewhat slower than the response of a thermocouple if it is used under similar conditions and in similar enclosures. However, both RTDs and thermocouples are usually enclosed in some type of thermowell that slows down response. The lag in temperature measurements caused by the thermowell is usually small compared to the lag of the entire temperature-measurement process.

THERMISTORS

A *thermistor* is a temperature-sensitive resistor consisting of solid-state semiconductors made from sintered metal oxides and lead wires, hermetically sealed in glass. They are available in several shapes such as rods, disks, beads, washers, and flakes. **See Figure 6-21.** The electrical resistance of most thermistors decreases with an increase in temperature. Therefore, most thermistors have a negative temperature coefficient (NTC). However, there are some applications where a PTC thermistor is used. PTC thermistors are made from strontium and barium titanate mixtures.

Thermistors have much higher resistance than RTDs. Thermistors are typically available with resistances ranging from 100 Ω to 100 MΩ. Therefore, lead wire resistance is not a concern and two-wire devices are adequate. Thermistors can typically be used over a temperature measurement range of –22°F to 212°F. However, prolonged exposure to elevated temperatures, even below the maximum limit specified by the manufacturer, can cause permanent damage. A typical application uses a thermistor for only a fraction of that range.

NTC Thermistors

A typical NTC thermistor has a sensitivity range of –50 Ω/°F to –500 Ω/°F at room temperature. This means that the resistance decreases by about 50 Ω to 500 Ω for a temperature increase of 1°F, which makes the thermistor much more sensitive to small temperature changes than a thermocouple or RTD. However, the change in resistance with temperature is very nonlinear, which limits the temperature measurement range and accuracy of thermistors. The temperature is calculated from a lookup table of resistance and temperature or from an equation that relates the two variables.

Thermistors

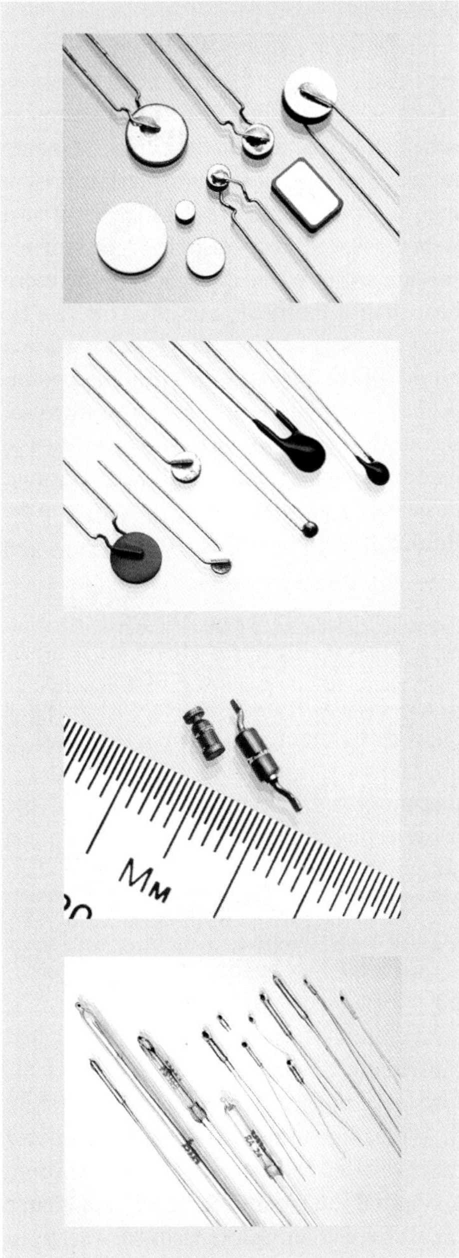

GE Thermometrics

Figure 6-21. Thermistors are available in a variety of shapes and sizes.

NTC thermistors are well suited for many applications that require a large change in resistance when a small change of temperature occurs. For example, a thermistor can be used to sound an alarm if the temperature increases above a setpoint. **See Figure 6-22.** As the temperature increases, the resistance of the thermistor decreases. As the resistance of the thermistor decreases, the current flow increases and there is a larger voltage drop across the alarm. The alarm sounds as long as the temperature is high.

PTC Thermistors

A PTC thermistor is characterized by an extremely large resistance change for a small temperature span. The resistance of a thermistor begins to change rapidly at the switch temperature. This point can be changed from below zero to above 300°F. Even though there is a very small current flow at higher resistances, a PTC thermistor can be used as a switch in many applications.

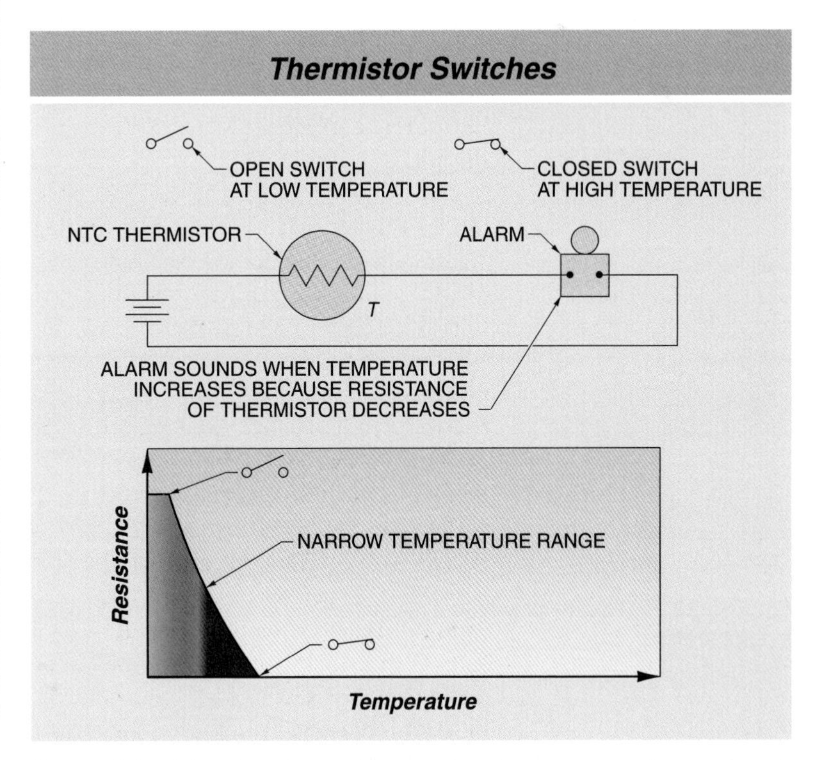

Figure 6-22. The changing resistance of a thermistor can be used as a temperature switch.

The increase in resistance of a PTC thermistor at the switch temperature makes it suitable for current-limiting applications. **See Figure 6-23.** For currents lower than the limiting current, the power generated in the unit is insufficient to heat the PTC thermistor to its switch temperature. However, as the current increases to the critical level, the resistance of the PTC thermistor increases at a rapid rate so that any further increase in power dissipation results in a current reduction. The time required for the PTC thermistor to get into the current-limiting mode is controlled by the heat capacity of the PTC thermistor, its dissipation constant, and the ambient temperature.

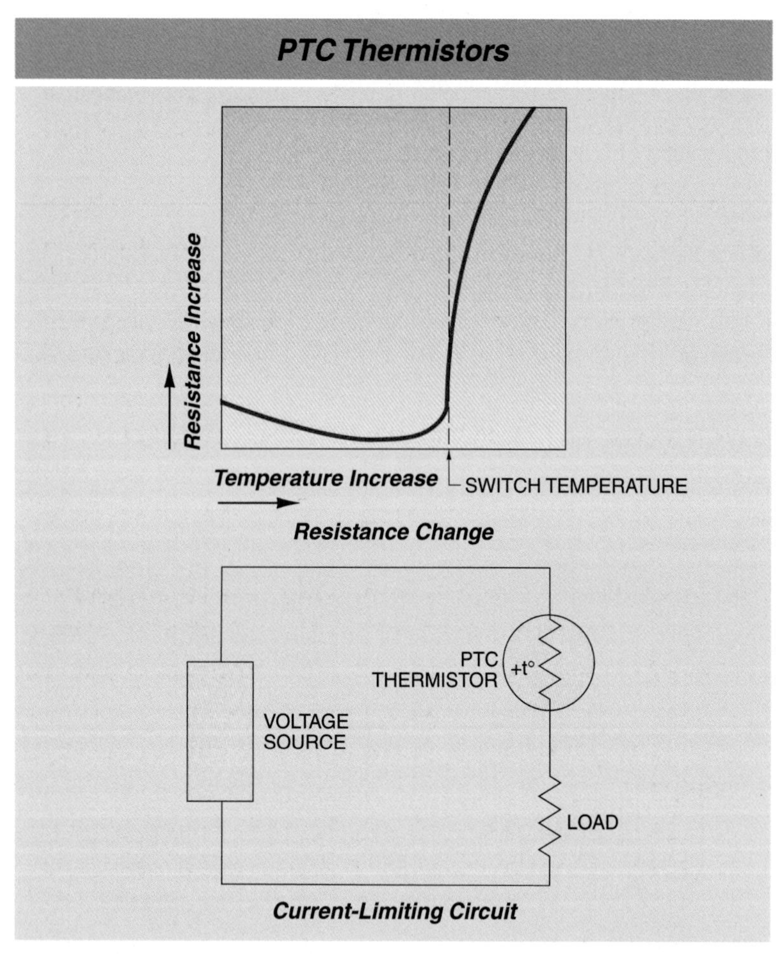

Figure 6-23. PTC thermistors have a resistance that increases with increases in temperature.

Zero Power Sensing

Zero power sensing is a thermistor circuit with the current kept low enough that power is dissipated without causing the thermistor to self-heat enough to cause erroneous readings. For example, the power dissipation constant of a chip thermistor is about 4.0 mW/°F in still air. In order to keep the accuracy within 0.1°F, the current must be low enough that the power dissipation is 0.4 mW or less. The accuracy of a thermistor is slightly inferior to that of the conventional RTD.

Self-Heat

A thermistor can be used in a self-heat application where the rate of heat loss is the variable of interest. When a thermistor is heated by increasing the current, the rate of heat loss to the environment depends on the environment in which the thermistor is being used. For example, a thermistor submerged in water dissipates much more heat than a thermistor in air. This can be used as a level switch that alarms when a liquid level reaches a predetermined point. A thermistor can also be used to measure the flow rate of fluids by measuring the heat loss to the flowing medium. The rate of heat loss depends on the velocity of the fluid past the thermistor.

SEMICONDUCTOR THERMOMETERS

Many semiconductor devices have a change in electrical properties with a change in temperature. Semiconductor thermometers are typically produced in the form of integrated circuits. For example, a silicon P-N diode junction with a constant current bias has about a –2 mV/°C temperature coefficient.

Since the P-N junction is the basic building block of diodes, transistors, and ICs, temperature sensing can be incorporated into many devices at low cost. This allows easy measurement of the temperature of electrical circuits. A semiconductor thermometer can be fairly accurate when calibrated, but is usually used as a coarse measurement in thermal shutdown applications. Semiconductor thermometers are typically available to measure over a range of –10°F to 400°F.

KEY TERMS

- *thermocouple:* An electrical thermometer consisting of two dissimilar metal wires joined at one end and a voltmeter to measure the voltage at the other end of the two wires.
- *thermocouple junction:* The point where the two dissimilar wires are joined.
- *hot junction,* or *measuring junction:* The joined end of the thermocouple that is exposed to the process where the temperature measurement is desired.
- *cold junction,* or *reference junction:* The end of a thermocouple used to provide a reference point.
- *Seebeck effect:* A thermoelectric effect where continuous current is generated in a circuit where the junctions of two dissimilar conductive materials are kept at different temperatures.
- *Peltier effect:* A thermoelectric effect where heating and cooling occurs at the junctions of two dissimilar conductive materials when a current flows through the junctions.
- *Thomson effect:* A thermoelectric effect where heat is generated or absorbed when an electric current passes through a conductor in which there is a temperature gradient.
- *law of intermediate temperatures:* A law stating that in a thermocouple circuit, if a voltage is developed between two temperatures T_1 and T_2, and another voltage is developed between temperatures T_2 and T_3, the thermocouple circuit generates a voltage that is the sum of those two voltages when operating between temperatures T_1 and T_3.
- *law of intermediate metals:* A law stating that the use of a third metal in a thermocouple circuit does not affect the voltage, as long as the temperature of the three metals at the point of junction is the same.
- *cold junction compensation:* The process of using automatic compensation to calculate temperatures when the reference junction is not at the ice point and is often achieved by measuring the temperature of the cold junction with a thermistor.
- *difference thermocouple:* A pair of thermocouples connected together to measure a temperature difference between two objects.
- *thermopile:* An electrical thermometer consisting of several thermocouples connected in series to provide a higher voltage output.
- *averaging thermocouple:* An electrical thermometer consisting of a set of parallel-connected thermocouples that is commonly used to measure an average temperature of an object or area.
- *thermocouple pyrometer:* An electrical thermometer consisting of a plain electrical meter with a measurement range of 20 mV to 50 mV, a thermocouple, and a balancing resistor.
- *null-current thermocouple:* A circuit and a voltage generator that can be adjusted to exactly balance the voltage output of a thermocouple.
- *resistance temperature detector (RTD):* An electrical thermometer consisting of a high-precision resistor with resistance that varies with temperature, a voltage or current source, and a measuring circuit.
- *thermistor:* A temperature-sensitive resistor consisting of solid-state semiconductors made from sintered metal oxides and lead wires, hermetically sealed in glass.
- *zero power sensing:* A thermistor circuit with the current kept low enough that power is dissipated without causing the thermistor to self-heat enough to cause erroneous readings.

REVIEW QUESTIONS

1. Define a thermocouple.
2. List and describe the phenomena that govern the behavior of thermocouples.
3. Describe the purpose of a cold junction and explain how cold junction compensation is used.
4. List several factors that affect the choice of thermocouple wires.
5. List and define several thermocouple measurement circuits.

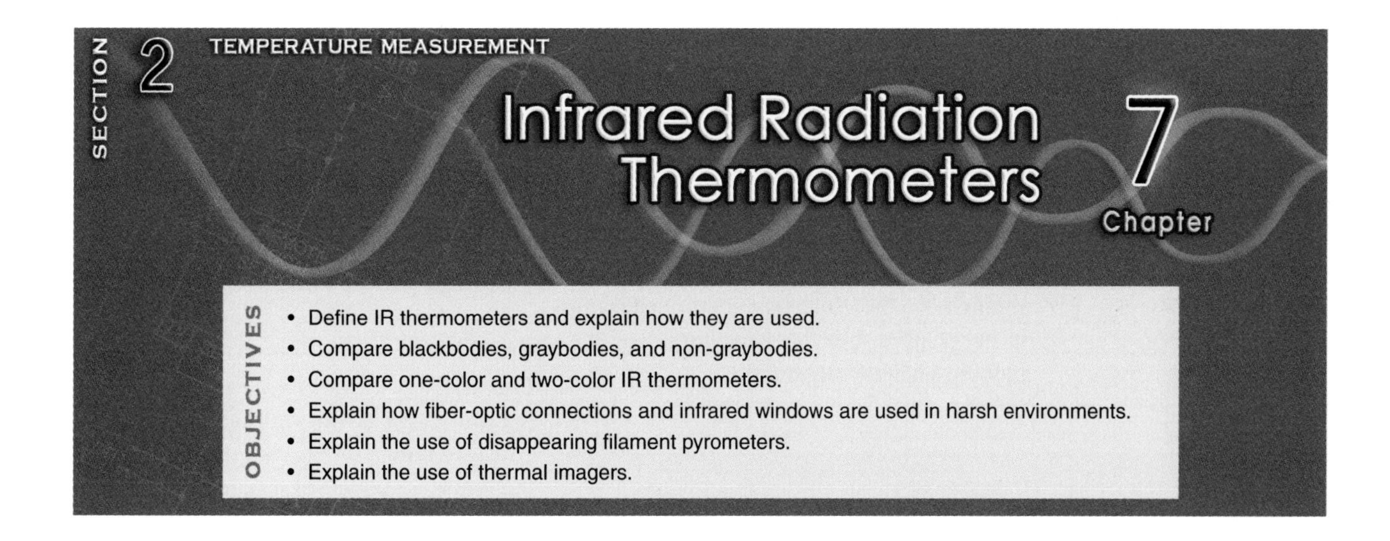

IR THERMOMETER OPERATING PRINCIPLES

Bodies that are at thermal equilibrium must balance the energy entering that object, such as heat or light, with the energy leaving that object. The energy leaving the surface of an object is often emitted as electromagnetic radiation. An *IR thermometer* is a thermometer that measures the infrared radiation (IR) emitted by an object to determine its temperature. Infrared radiation is that part of the electromagnetic spectrum with longer wavelengths than visible light. **See Figure 7-1.**

Infrared radiation thermometers generally have very quick response times. They can typically make many measurements per second. An IR thermometer can be used in areas where it is very difficult to use a contact thermometer.

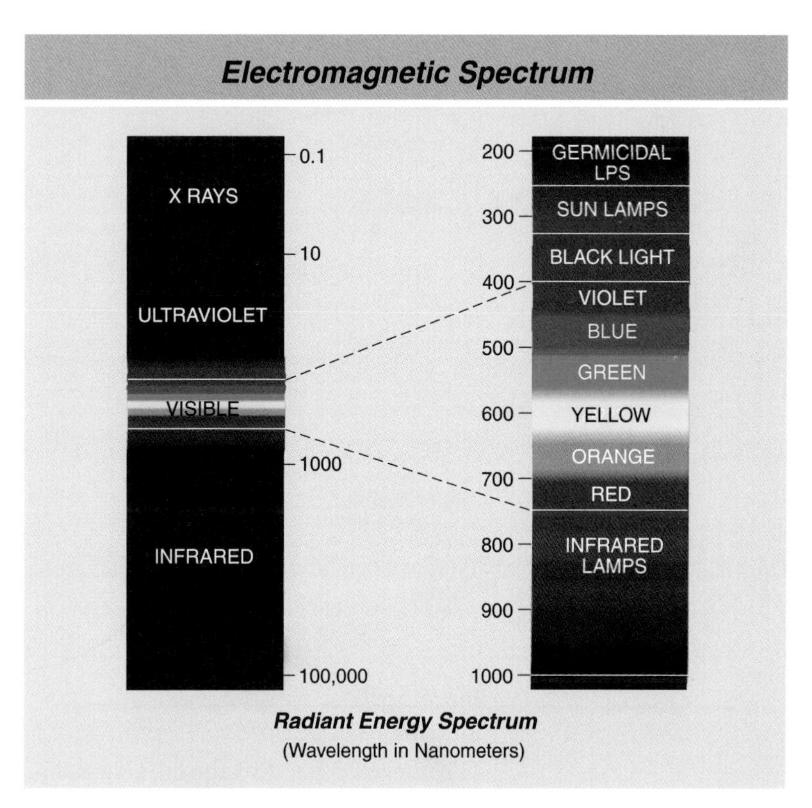

Figure 7-1. Infrared radiation is that part of the electromagnetic spectrum at longer wavelengths than visible light.

Spectral Response

Spectral response is the range of infrared wavelengths measured by an IR thermometer. Common units of measurement for infrared wavelengths are nanometers (nm, 10^{-9} meters) and microns (μ, 10^{-6} meters). The IR spectrum is between 0.79 μ and 1000 μ, just longer than the visible light spectrum. The spectral response of typical IR thermometers is 8 μ to 14 μ. The wavelengths measured are generally those that are not absorbed by atmospheric gases such as water vapor, carbon dioxide, and nitrogen oxide.

Bodies at low temperature emit infrared radiation only at long wavelengths. As the temperature increases, the amount of emitted infrared radiation from the surface increases dramatically and the wavelength of peak emittance becomes shorter. **See Figure 7-2.** If the body is hot enough, some of the emitted radiation is in the visible light part of the electromagnetic spectrum and the wavelength is seen as a color. For example, as a steel billet is heated, its color changes from red at about 1112°F (600°C), to orange at about 1292°F (700°C), to yellow at about 2012°F (1100°C), and to white at about 2732°F (1500°C). The infrared segment of the electromagnetic spectrum is not visible, but the wavelength of the radiation can be measured.

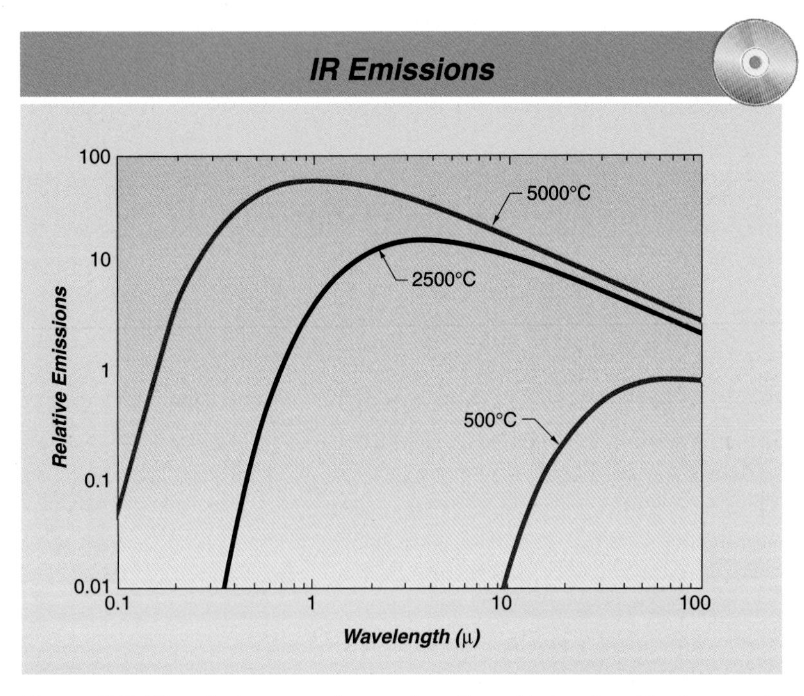

Figure 7-2. IR emissions vary in wavelength and intensity with changes in temperature.

Blackbodies, Graybodies, and Non-Graybodies

The reference standard for infrared radiation is called a blackbody. A *blackbody* is an ideal body that completely absorbs all radiant energy of any wavelength falling on it and reflects none of this energy from the surface. Therefore, the blackbody absorbs 100% of the radiant energy that falls on it and the blackbody reflectivity is zero. It is a complete radiator, emitting a greater amount of thermal energy at a faster rate than any other body at the same temperature and under the same conditions.

Any body other than a blackbody reflects a portion of infrared radiation. A *graybody* is a body that emits and reflects a portion of all wavelengths of radiation equally. A *non-graybody* is a body that emits and reflects radiation to a varying degree depending on the wavelength of the infrared radiation.

Emissivity

Emissivity is the ability of a body to emit radiation and is the ratio of the relative emissive power of any radiating surface to the emissive power of a blackbody radiator at the same temperature. **See Figure 7-3.** For example, polished silver has an emissivity of 0.02 while dark asphalt has an emissivity of 0.95. Any material that is or was once alive generally has an emissivity near 0.95, such as wood, skin, cloth, and rubber. Emissivity is expressed as follows:

$$\text{Emissivity} = \frac{\text{Total radiation from a non-blackbody}}{\text{Total radiation from a blackbody}}$$

An IR thermometer measures the total emissions from a body. Any emissivity value less than 1 means that an inaccuracy can be introduced into the measurement because the actual emissions are less than that of a blackbody. An IR thermometer normally has a setting for the emissivity of the body being measured.

Reflectivity

Reflectivity is the ability of an object to reflect radiation. Bodies with a high emissivity reflect very little. Bodies with a low emissivity are very reflective. Materials with high reflectivity can be very hard to measure. For example, when heat radiates from the ground onto the bottom of a tanker truck, the shiny surface of the tank reflects the radiated heat and gives an inaccurate high temperature reading. The shiny surface can also reflect the cooler temperature of a clear sky onto the top of the tank, which gives an inaccurate low temperature reading. **See Figure 7-4.**

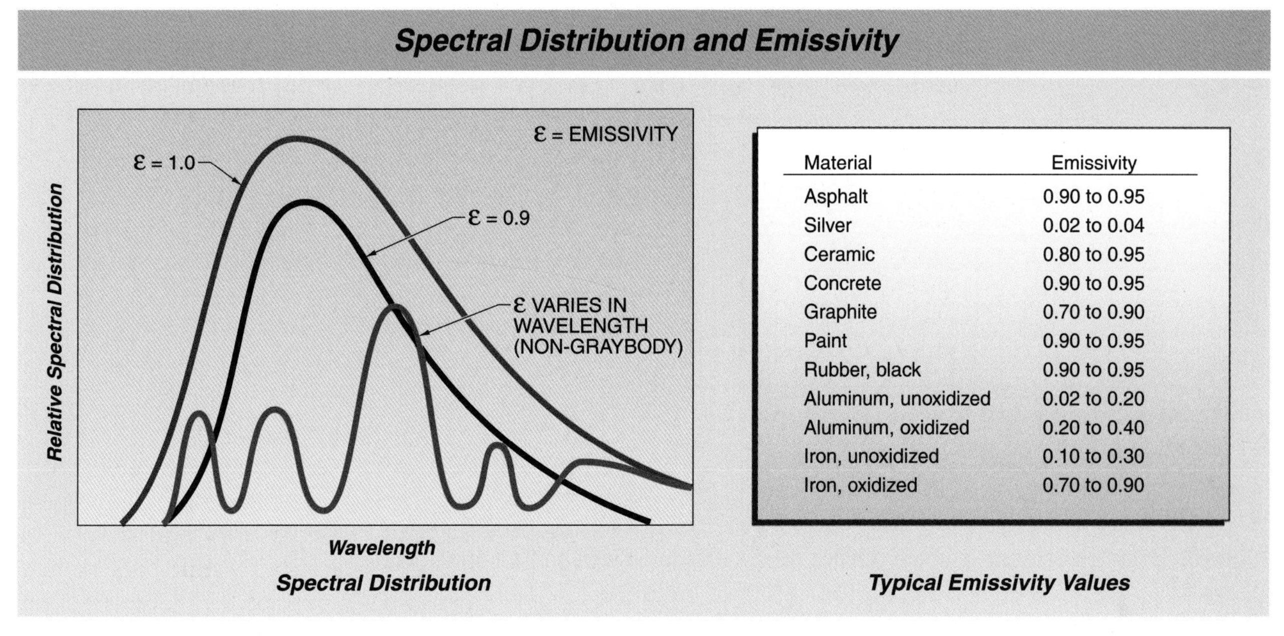

Material	Emissivity
Asphalt	0.90 to 0.95
Silver	0.02 to 0.04
Ceramic	0.80 to 0.95
Concrete	0.90 to 0.95
Graphite	0.70 to 0.90
Paint	0.90 to 0.95
Rubber, black	0.90 to 0.95
Aluminum, unoxidized	0.02 to 0.20
Aluminum, oxidized	0.20 to 0.40
Iron, unoxidized	0.10 to 0.30
Iron, oxidized	0.70 to 0.90

Figure 7-3. The IR emission spectra vary depending on the characteristics of the object.

Fluke Corporation

Figure 7-4. The shiny surface of the tank reflects the radiated heat from the ground and the cooler temperature of the clear sky. The reflections give inaccurate temperature readings.

Transmissivity

Transmissivity is the ability of an object to allow radiation to pass through. This is why you can feel the heat of the sun through a window on a sunny day. Like emissivity, it is expressed as a ratio. Generally, emissivity is a factor when measuring the temperature of bare metals and transmissivity is a factor when measuring thin-gauge plastics or glass.

Infrared radiation thermometers normally have an adjustment for emissivity, allowing the user to set the emissivity for the surface being measured. The emissivity adjustment may also be used to correct for transmissivity if no separate adjustment is provided. Temperature measurement errors can be significant if the emissivity is not set correctly.

For materials such as thin clear or translucent plastics, transmissivity varies with the wavelength of the infrared radiation. For example, polyethylene has a transmissivity of about 0.90 over the range of 8 μ to 12 μ, but the transmissivity drops to nearly 0 at certain other wavelengths. **See Figure 7-5.** When measuring the temperature of thin plastics, such as extrusions, the wavelength of the IR radiation must be carefully selected to ensure accurate temperature measurement.

Infrared instruments are subject to measurement errors when the IR radiation is blocked by steam, smoke, or dust. An air exhaust may be needed to clear the field of view.

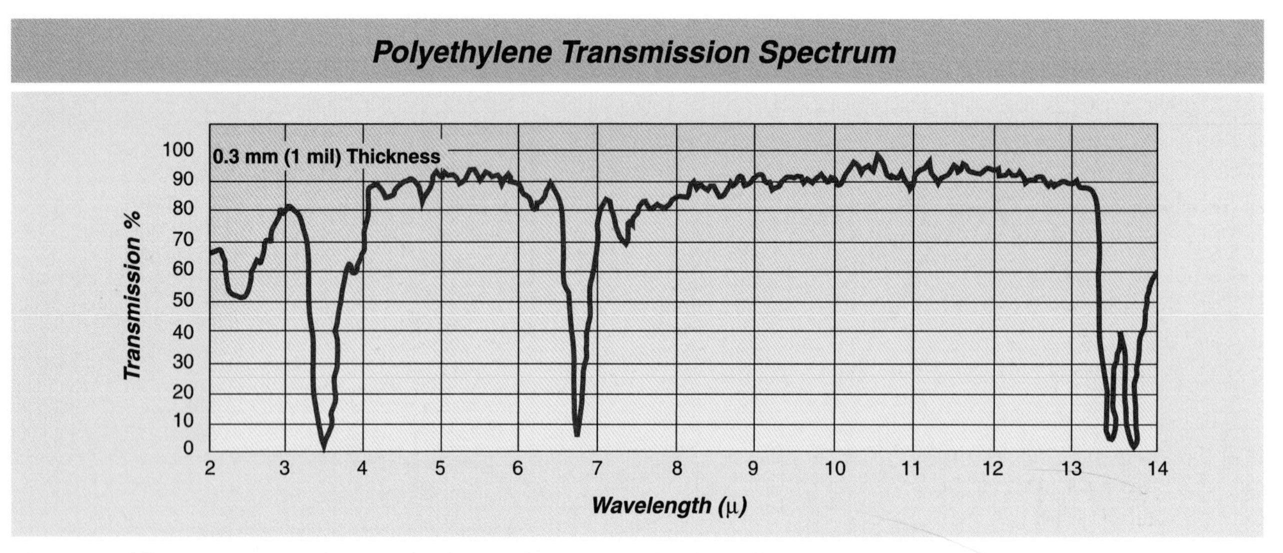

Figure 7-5. The temperature of some plastics can be measured with an IR thermometer.

Background Temperature Compensation

The emissivity correction performed by IR thermometers generally assumes that the hot body is in an ambient temperature environment. For example, if a painted automobile body exits a paint-drying oven at 200°F and has an emissivity of 0.95, then 5% of the infrared radiation detected by the IR thermometer is reflected from the ambient factory surroundings. If the 200°F automobile body is measured while inside the paint-drying oven and the inside walls of the paint-drying oven are 400°F, then the 5% is from a much hotter source. The assumption of ambient temperature reflection is no longer true and the temperature measurement by the IR thermometer is less accurate. **See Figure 7-6.**

This error is generally small when the emissivity setting is high, such as 0.95, and the source of reflected infrared radiation is near ambient temperature. As the temperature of the source of reflected infrared radiation increases and as the emissivity value of the surface decreases (the reflectivity increases), the error increases. *Background temperature compensation* is a process by which some IR thermometers allow the temperature of a reflected source to be measured or specified.

Optical Resolution. An IR thermometer focuses IR energy onto the IR detector by means of a lens and aperture. The ratio D:S is often used to describe the optical resolution. D is the distance to the object and S is the size of the spot that is measured at the focus distance. For example, if an IR thermometer has an optical resolution (D:S) of 10:1, the size of the measured spot at 10′ is 1′ in diameter when the thermometer is perpendicular to the surface of a body.

At distances other than the focus distance, the measured spot is different from that specified by the D:S ratio. For best accuracy, the body to be measured must fill the field of view and the angle of incidence must be as close to 90° as is practical. **See Figure 7-7.** If the angle of incidence is not close to 90°, erroneous readings can occur because the spot size gets larger and the amount of incident radiation changes.

IR THERMOMETER CONSTRUCTION

The three main components of an IR thermometer are the infrared (IR) detector, electrical circuitry, and the lens or fiber-optic assembly. The electrical circuitry analyzes the infrared radiation and may use one or more inputs from the detector to best measure the temperature.

IR Detectors

The heart of the IR thermometer is the IR detector, which provides an electrical output proportional to the amount of infrared radiation focused on it. There are many different kinds of IR detectors used. The choice depends on a number of requirements including the temperature range, the object being measured, and the response time. Most IR thermometers use additional circuitry to provide an output signal, control signal, and other features.

One-Color IR Thermometers. A *one-color,* or *single-color, IR thermometer* is a thermometer that measures infrared radiation using one IR detector. The majority of point IR thermometers, both hand-held and mounted in place, are this type. The word color is misleading because these thermometers measure infrared radiation, not visible light. A one-color IR thermometer with a long-wavelength spectral response is used to measure low-temperature surfaces while short wavelengths are used to measure high-temperature surfaces. The wavelength of the peak intensity of IR emissions depends on the temperature. **See Figure 7-8.** The common fever thermometer that measures the body temperature in the ear is a one-color IR thermometer.

Some IR thermometers are available with spectral responses to measure thin-film plastics or combustion gases such as carbon dioxide. One-color thermometers usually include circuitry to generate a 4 mA to 20 mA voltage or thermocouple output signal. They also often include signal conditioning such as averaging and peak hold.

Two-Color (Ratio) IR Thermometers. A *two-color,* or *ratio, IR thermometer* is a thermometer that has two IR detectors that measure the infrared radiation of two different wavelengths. The temperature measurement is calculated from the ratio of the output from the two detectors. These thermometers are often used when the atmosphere between an object and the thermometer blocks some of the infrared radiation from reaching the IR detectors. **See Figure 7-9.** This condition occurs when smoke, steam, or dust is present, or when the object does not fill the field of view, such as a small-diameter hot metal wire.

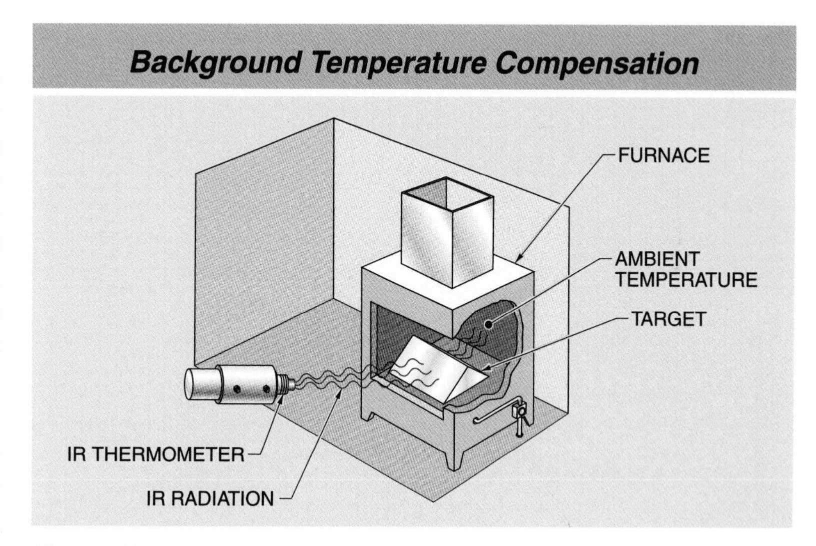

Figure 7-6. The background temperature can influence the temperature reading from an IR thermometer.

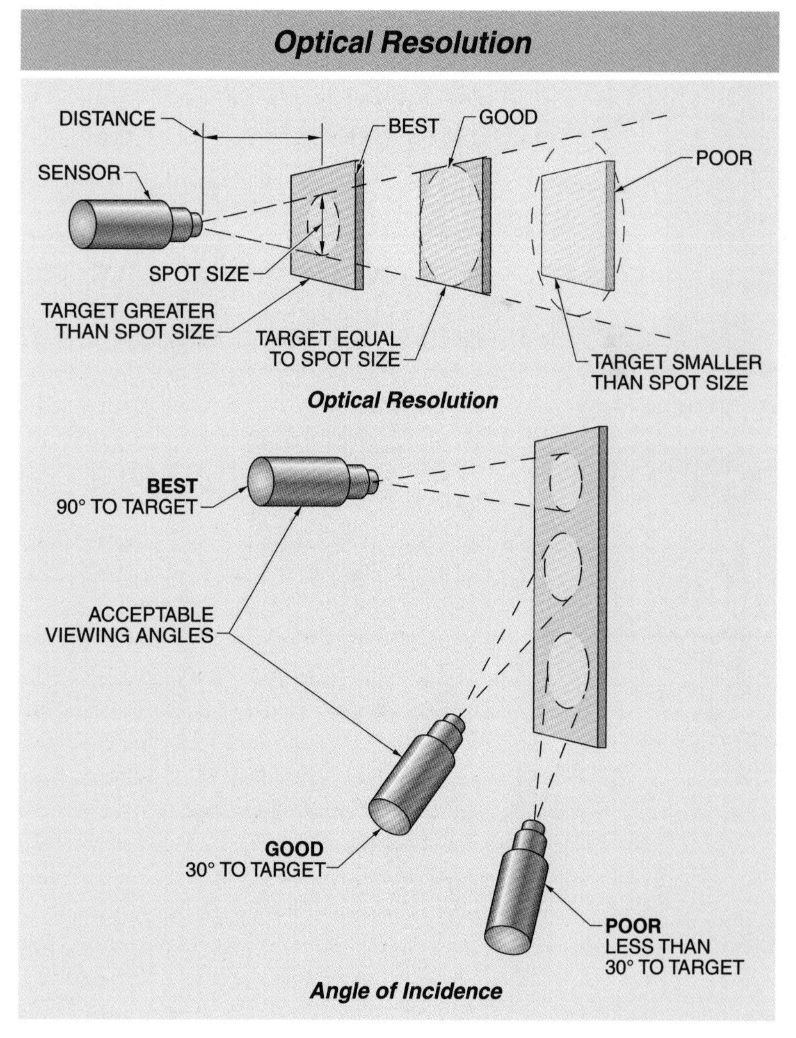

Figure 7-7. The optical resolution and the angle of incidence determine whether a reliable and accurate temperature measurement can be made.

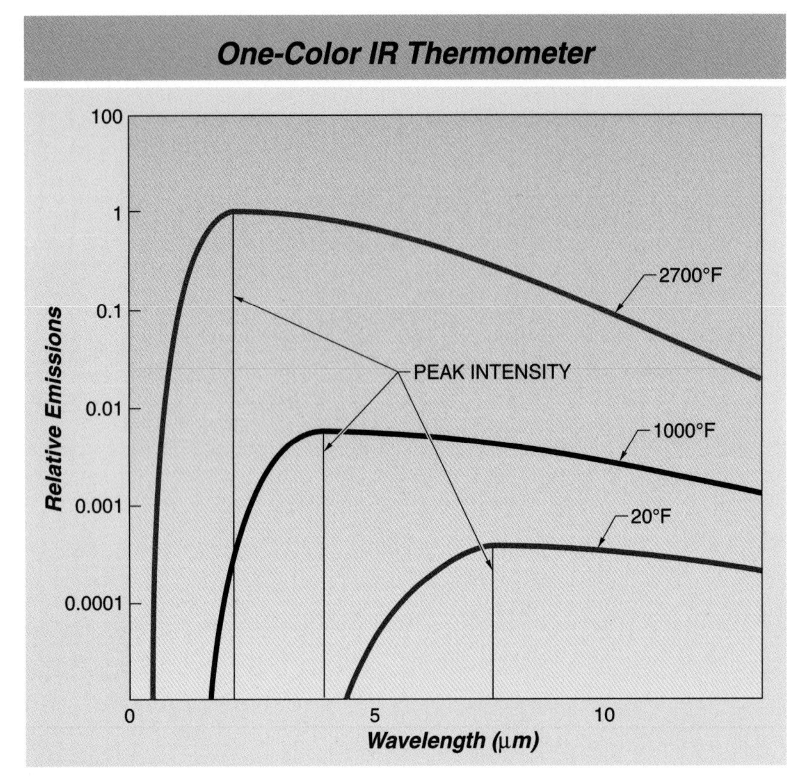

Figure 7-8. A one-color IR thermometer measures the intensity of emissions and calculates the temperature from that measurement.

In these cases, the reduction or attenuation in infrared radiation at the different wavelengths that reach the detectors is usually equal. Therefore, the ratio of the two IR detectors is equal, although the total infrared radiation to both detectors has been reduced. Therefore, the measurement of the object's temperature by a two-color thermometer is unaffected by these conditions. The temperature of the smoke, steam, or dust and the temperature of the item that fills up the balance of the spot size must be much colder than the object. Wavelengths of sensitivity for both detectors are selected so both are insensitive to the colder material.

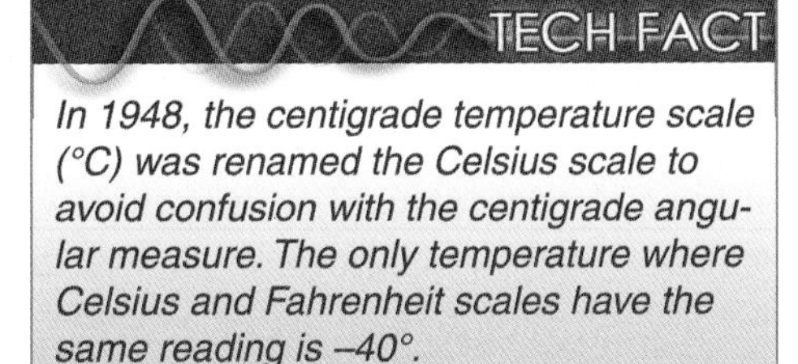

In 1948, the centigrade temperature scale (°C) was renamed the Celsius scale to avoid confusion with the centigrade angular measure. The only temperature where Celsius and Fahrenheit scales have the same reading is –40°.

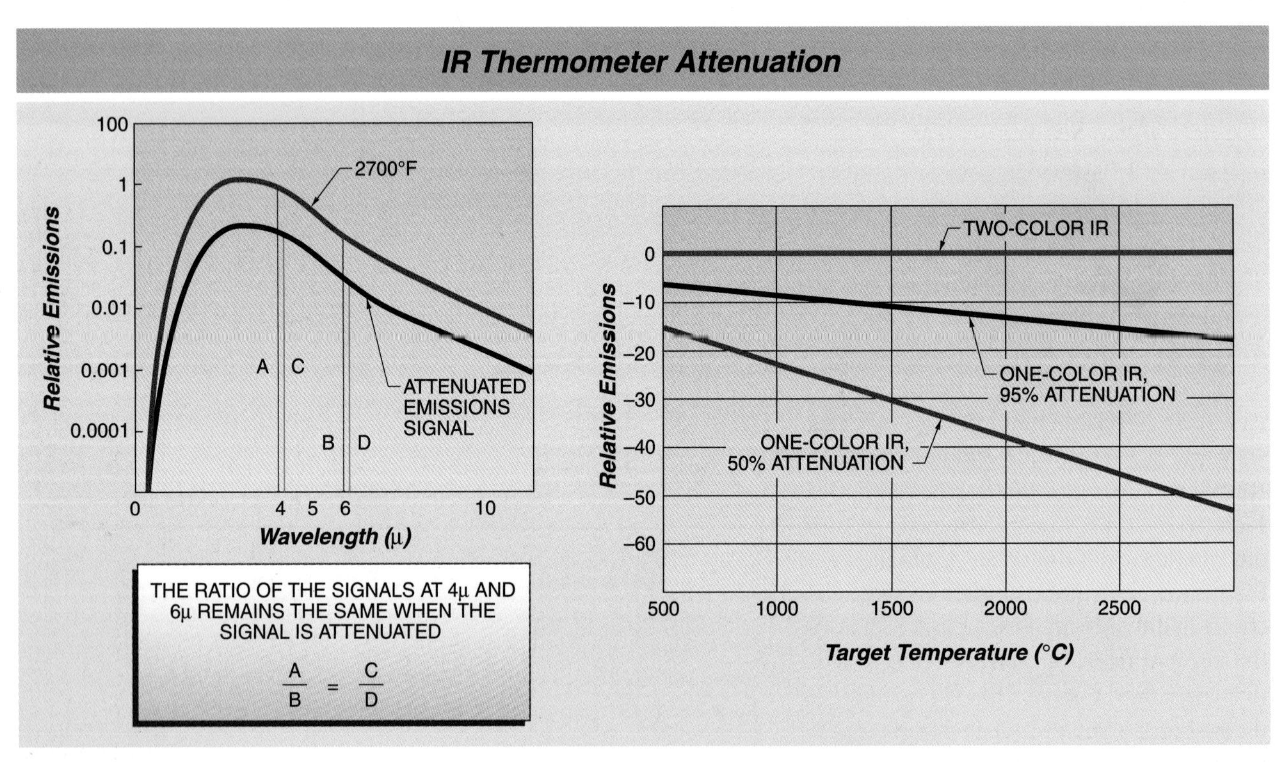

Figure 7-9. A two-color IR thermometer measures the ratio of emissions at two different wavelengths so it is less subject to errors from attenuation.

As long as the emissivity of the measured surface is the same for the two wavelengths of the two detectors, the two-color IR thermometer is independent of the emissivity of the surface, even if the emissivity changes at different temperatures. If the emissivity of the surface is different at the two wavelengths, then one detector receives disproportionately more infrared radiation than the other. The two detectors have a slope adjustment to compensate for a difference in the infrared radiation detected.

Two-color thermometers are generally more affected by background infrared radiation than one-color thermometers. Therefore, one-color thermometers are typically used to measure the temperature of a surface in a very hot environment such as a furnace.

Fiber-Optic IR Connections

Sometimes it is not possible to place an IR thermometer in a suitable location. For example, it is very difficult to mount an IR thermometer in tight spaces. In such situations, a small lens assembly connected to a fiber-optic cable may be used to transmit the infrared radiation. The other end of the fiber-optic cable is connected to the detector or detectors of an IR thermometer. **See Figure 7-10.** Because a glass fiber is used, only short wavelengths of infrared radiation can be measured, which limits the temperature measurement range of these thermometers.

Infrared Windows

Sometimes it is necessary to physically isolate an IR thermometer from a process, such as a hydrogen atmosphere furnace. Because longer wavelengths of infrared radiation do not transmit through glass, other materials must be used as windows for temperature measurements below about 1000°F. **See Figure 7-11.** When the temperature measurement permits, two-color thermometers have an advantage in that contaminants on the window do not affect the measurement, as is the case with one-color IR thermometers.

Fiber-Optic IR Thermometers

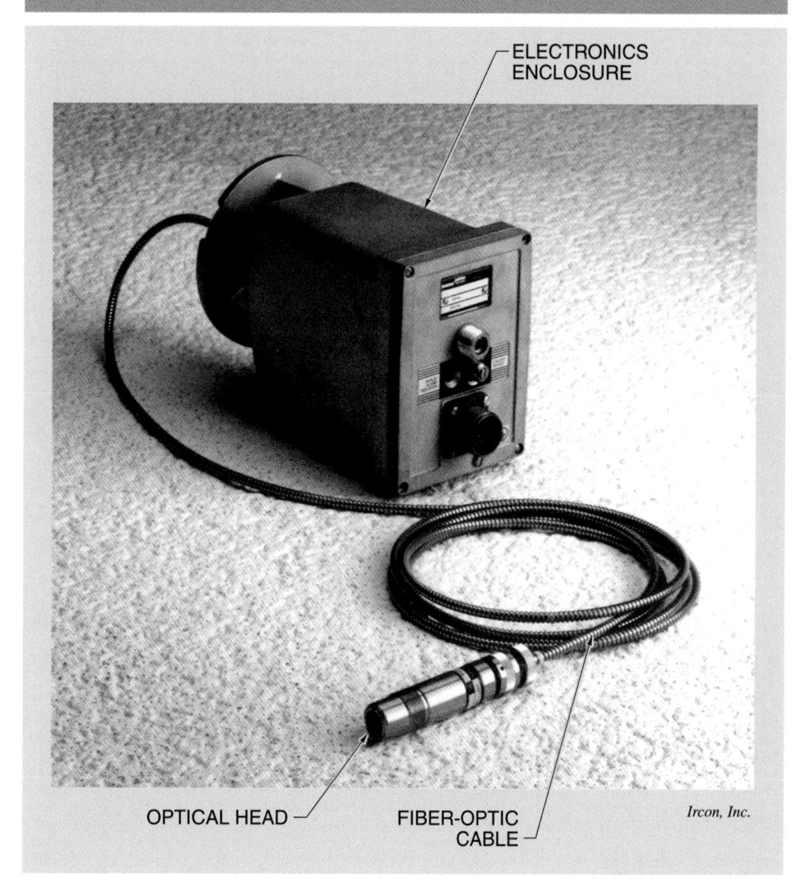

Figure 7-10. A fiber-optic IR thermometer can be used in harsh environments or tight spaces.

Fluke Corporation

A handheld IR imager can be used to measure the temperature in hard-to-reach spaces to help with troubleshooting a control system.

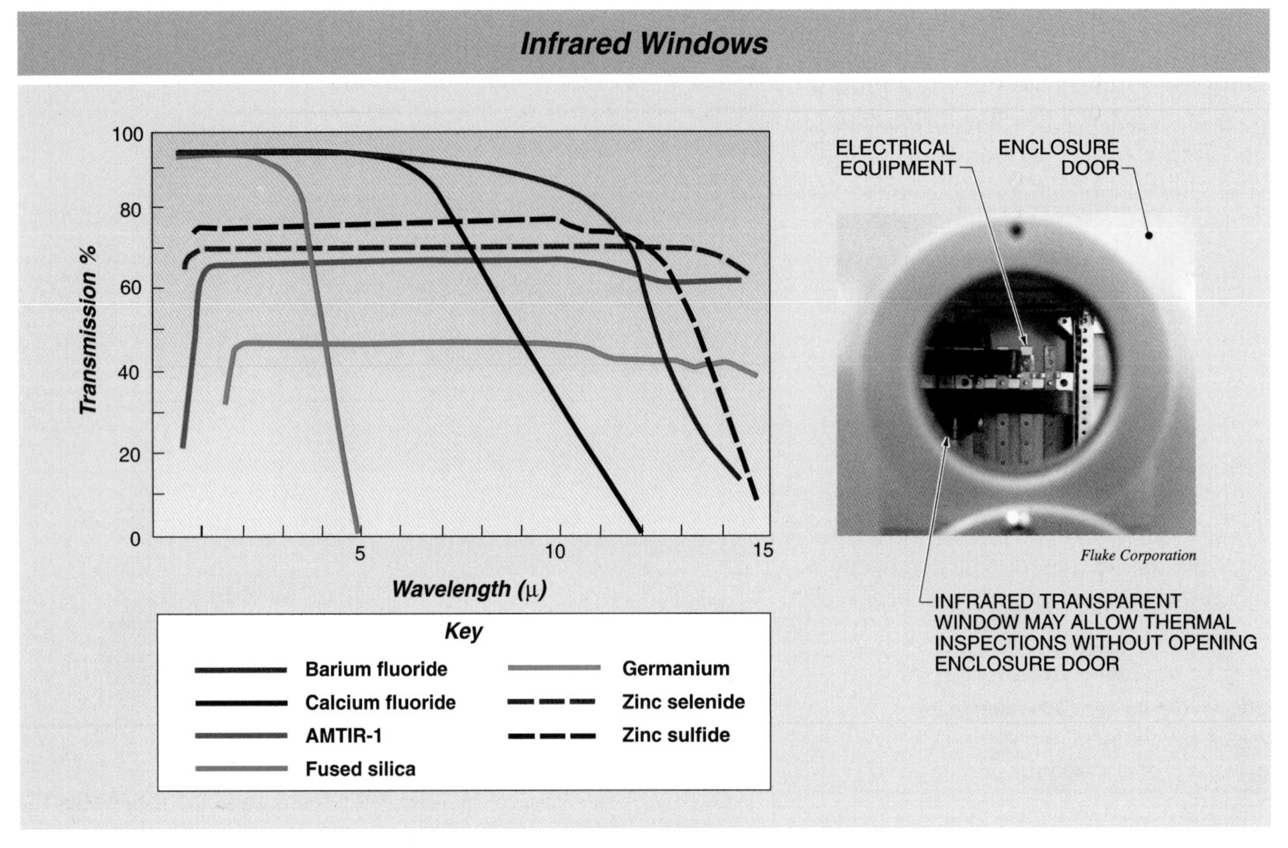

Figure 7-11. Different window glasses have different transmissivities. The correct window must be chosen to be able to use an IR thermometer through the window.

Line Scanner IR Thermometers

Dozens of individual one-color IR thermometers could be mounted side by side so the fields of view are successive circles in a line, but this approach is expensive and difficult to install in practice. A *line scanner IR thermometer* is a thermometer that uses a rotating mirror with a single, very fast response detector or a linear array of IR detectors to measure successive areas with a single stationary device.

Line scanners are capable of hundreds of individual measurements per scan at speeds up to several hundred scans per second. **See Figure 7-12.** With companion software, a thermal image can be generated of wide moving material or a rotating object such as extruded plastic sheet, hot steel strip, paper or film web, or a rotary cement kiln. Line scanners can also be coupled with process control systems.

DISAPPEARING FILAMENT PYROMETERS

A *pyrometer* is an instrument used to measure the temperature of an object that is hot enough to emit visible light. A *disappearing filament pyrometer* is a high-temperature thermometer that has an electrically heated, calibrated tungsten filament contained within a telescope tube. **See Figure 7-13.**

The telescope tube also contains a red glass filter that restricts the brightness of the hot source to one specific wavelength of red light. The current to the filament is manually adjusted until the apparent brightness matches that of the target source. When the brightness of the filament matches the brightness of the hot source, the filament disappears from view. Measuring circuitry in the pyrometer converts the filament current value to a temperature reading on a temperature indicator.

A disappearing filament pyrometer can focus on, and measure temperatures of, small areas or moving parts while not being directly exposed to the high temperatures. The accuracy of the temperature depends partly on the emissivity. The temperature measurement is lower than the actual temperature if the emissivity is significantly different from 1.

Another possible source of error is when there is smoke or gas between the hot source and the pyrometer that absorbs radiation and introduces errors. In addition, no two people see color the same and the accuracy of the measurement depends on human judgment. Therefore, the temperature of a surface as measured by two users with the same disappearing filament pyrometer may not be exactly the same.

A disappearing filament pyrometer can only be used with surfaces that have a visible color change above about 1000°F. Disappearing filament pyrometers are typically used over the measuring range of 1000°F to 10,000°F.

THERMAL IMAGERS

A *thermal imager* is an infrared device that uses a two-dimensional array of IR detectors to generate an image showing the temperature of an object. **See Figure 7-14.** Thermal imagers are either nonradiometric or radiometric. A *nonradiometric thermal imager* is an imager in which a surface-temperature image is generated but the actual temperature at a specific position is unknown.

A *radiometric thermal imager* is an imager in which the temperature measurement at all positions in the image is known. For industrial applications it is often important to know the actual temperatures. For example, an image of the three fuses in a power system shows where the equipment is the hottest. The image also shows the actual temperature.

Thermal imagers are commonly used in industry to identify equipment that is running hotter than normal. They can also be used to provide visual images of the temperature distribution within an object. Thermal imagers can be used to conduct a thermal survey of a building to determine whether insulation is needed and to show the exact locations of heat loss.

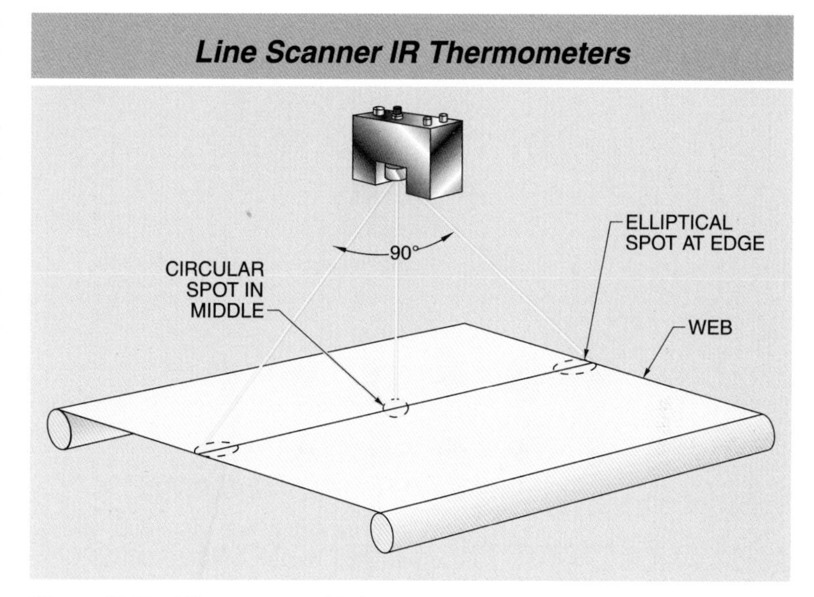

Figure 7-12. A line scanner IR thermometer measures the temperature at many places across a moving object.

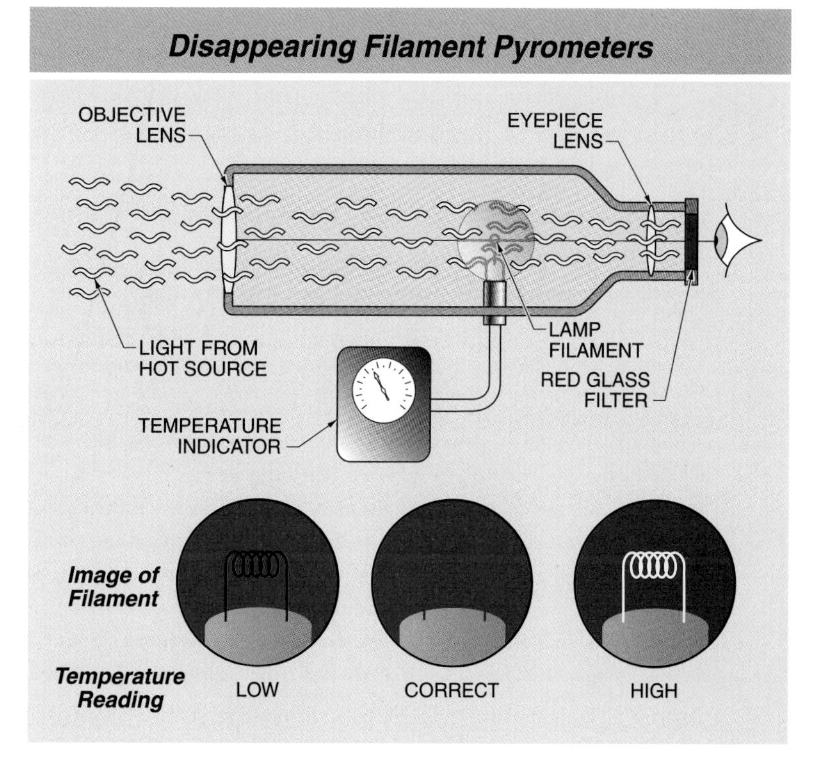

Figure 7-13. A disappearing filament pyrometer is used to measure the temperature of a glowing object.

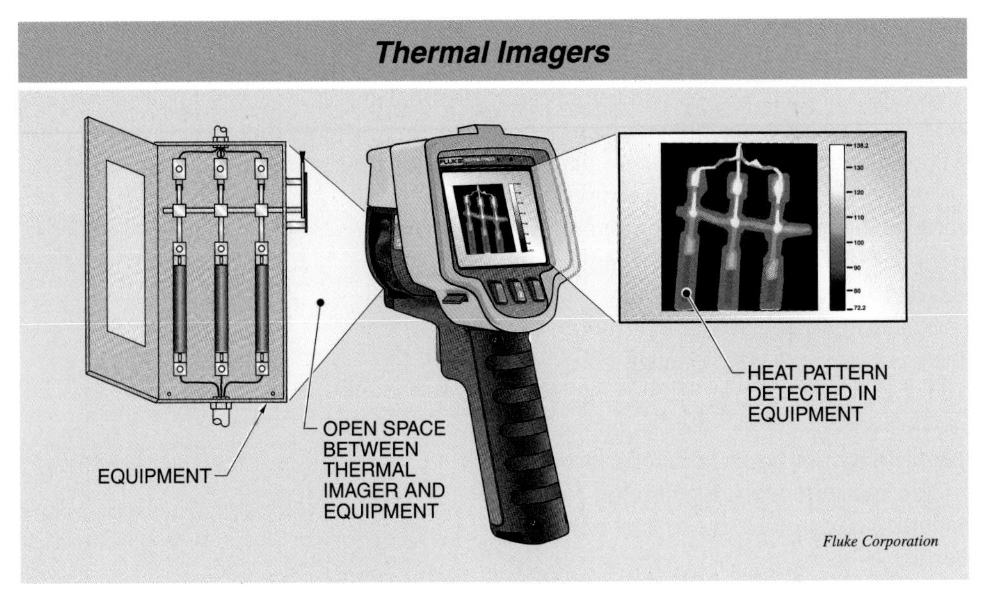

Figure 7-14. A thermal imager generates an image showing the temperature of an object.

Night-Viewing Equipment

Night-viewing goggles, scopes, and viewers are often used by the military but are also available to the public. They all use the principles of thermal imaging to convert IR radiation into visible wavelengths. This generally produces a green image. Night-viewing equipment is widely used by police departments in night work and by fire fighters to find victims in smoke-filled buildings.

KEY TERMS

- *IR thermometer:* A thermometer that measures the infrared radiation (IR) emitted by an object to determine its temperature.
- *spectral response:* The range of infrared wavelengths measured by an IR thermometer.
- *blackbody:* An ideal body that completely absorbs all radiant energy of any wavelength falling on it and reflects none of this energy from the surface.
- *graybody:* A body that emits and reflects a portion of all wavelengths of radiation equally.
- *non-graybody:* A body that emits and reflects radiation to a varying degree depending on the wavelength of the infrared radiation.
- *emissivity:* The ability of a body to emit radiation and is the ratio of the relative emissive power of any radiating surface to the emissive power of a blackbody radiator at the same temperature.
- *reflectivity:* The ability of an object to reflect radiation.
- *transmissivity:* The ability of objects to allow radiation to pass through.
- *background temperature compensation:* A process by which some IR thermometers allow the temperature of a reflected source to be measured or specified.
- *one-color,* or *single-color, IR thermometer:* A thermometer that measures infrared radiation using one IR detector.

KEY TERMS *(continued)*

- *two-color,* or *ratio, IR thermometer:* A thermometer that has two IR detectors that measure the infrared radiation of two different wavelengths.
- *line scanner IR thermometer:* A thermometer that uses a rotating mirror with a single, very fast response detector or a linear array of IR detectors to measure successive areas with a single stationary device.
- *pyrometer:* An instrument used to measure the temperature of an object that is hot enough to emit visible light.
- *disappearing filament pyrometer:* A high-temperature thermometer that has an electrically heated, calibrated tungsten filament contained within a telescope tube.
- *thermal imager:* An infrared device that uses a two-dimensional array of IR detectors to generate an image showing the temperature of an object.
- *nonradiometric thermal imager:* An imager in which a surface-temperature image is generated but the actual temperature at a specific position is unknown.
- *radiometric thermal imager:* An imager in which the temperature measurement at all positions in the image is known.

REVIEW QUESTIONS

1. What is an IR thermometer and how is it used?
2. Define blackbody, graybody, and non-graybody.
3. Define emissivity, reflectivity, transmissivity, and background temperature compensation.
4. What is the difference between one-color and two-color IR thermometers, and when are they used?
5. How are disappearing filament pyrometers used?

Practical Temperature Measurement and Calibration

Chapter 8

OBJECTIVES

- Describe the response time and time constant of thermometers.
- Describe the function and design of thermowells.
- List the considerations for using thermocouples in various environments.
- Describe the function and use of resistance bridge circuits.
- Describe the calibration of thermometers.

RESPONSE TIME

All temperature-measuring instruments take time to respond to changes in temperature. The rate of heat transfer from a hot object to a cold object is very complex and depends on many factors including the heat transfer coefficient of the materials, the mass and heat capacity of the materials, and the temperature difference. These factors and others affect the length of time it takes a temperature-measuring instrument to respond to temperature changes.

A simple process can often be described as a first order system. For a first order system, a time constant can be used to develop a model that describes the process. A *time constant (t)* is the time required for a process to change by 63.2% of its total change when an input to the process is changed. For example, the time constant of a thermometer is the time required for the thermometer to change by 63.2% of the total change when the process temperature changes or when the thermometer is inserted into the process. **See Figure 8-1.**

If the ambient temperature is 70°F and the process temperature is 170°F, the total temperature change is 100°F when the thermometer is inserted into the process. The time constant is the time it takes for the thermometer to change by 63.2% of 100°F, or 63.2°F. In other words, the time constant is the time it takes for the thermometer reading to change from 70°F to 133.2°F.

It is often more advantageous to know the time required to reach 90%, 95%, or 99% of final value, instead of the 63.2%. Therefore, manufacturers frequently state the time required to reach these higher percentages when they list the response speed of their thermometers. However, the time constant at 63.2% is used in modeling. A thermowell slows down the response to temperature changes.

THERMOWELLS

In most cases, temperature instruments cannot be used without protection from the environment in which they are used. A *thermowell* is a closed tube used to protect a temperature instrument from process conditions and to allow instrument maintenance to be performed without draining the process fluid. **See Figure 8-2.** Thermowells are also called thermocouple wells, sheaths, and protecting tubes. Even if the measuring environment is not harmful to the sensing element, a thermowell provides the ability to remove the sensing element for servicing and calibration without having to shut down the process and drain a pipeline or reactor.

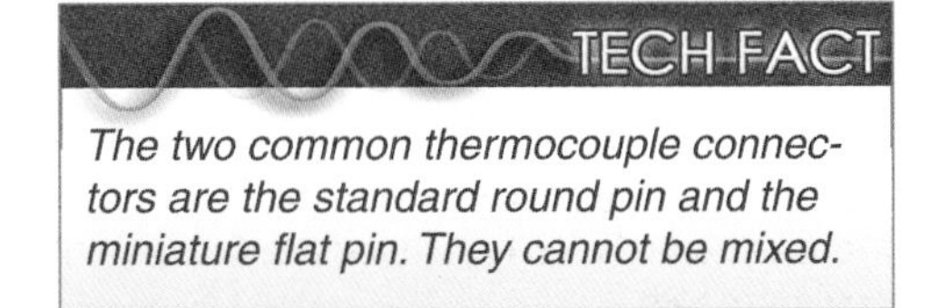

The two common thermocouple connectors are the standard round pin and the miniature flat pin. They cannot be mixed.

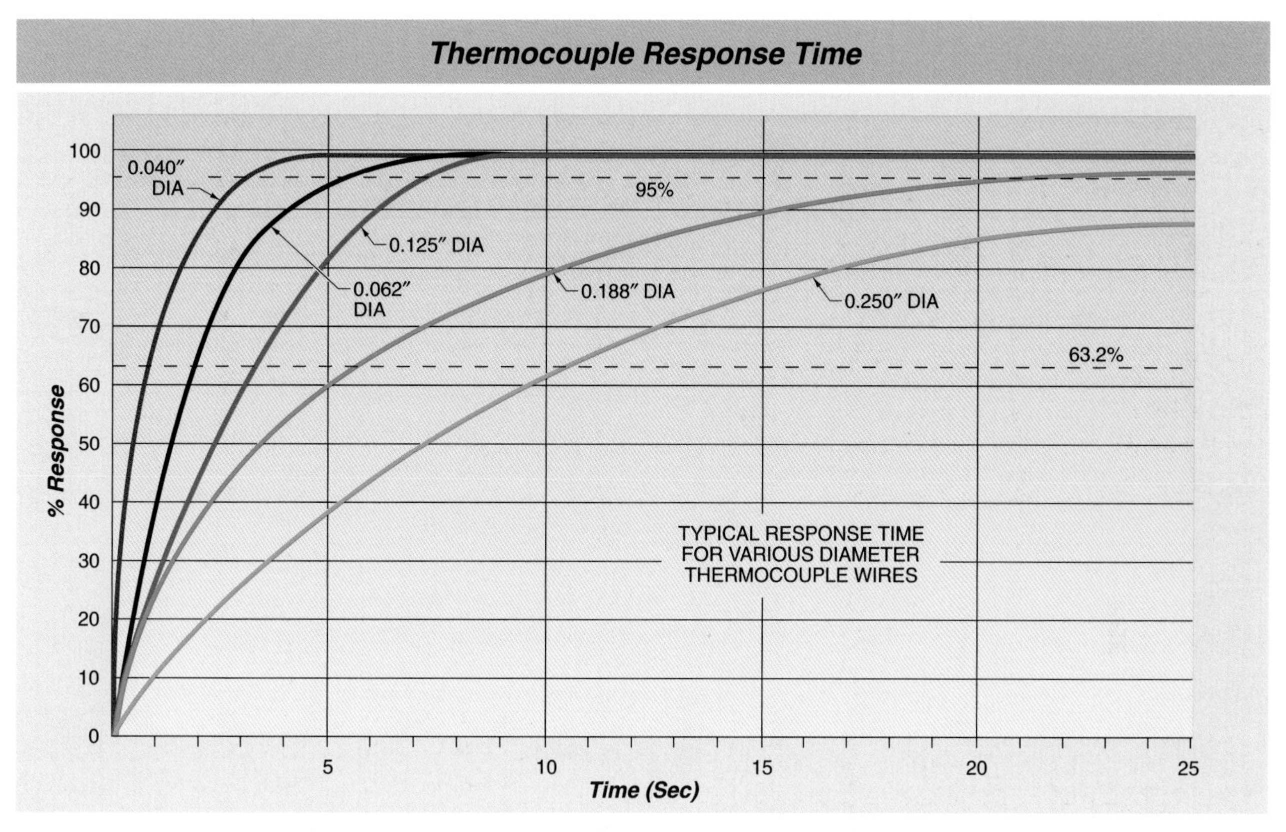

Figure 8-1. The response time of a thermocouple to a sudden temperature change depends on the size of the thermocouple wire.

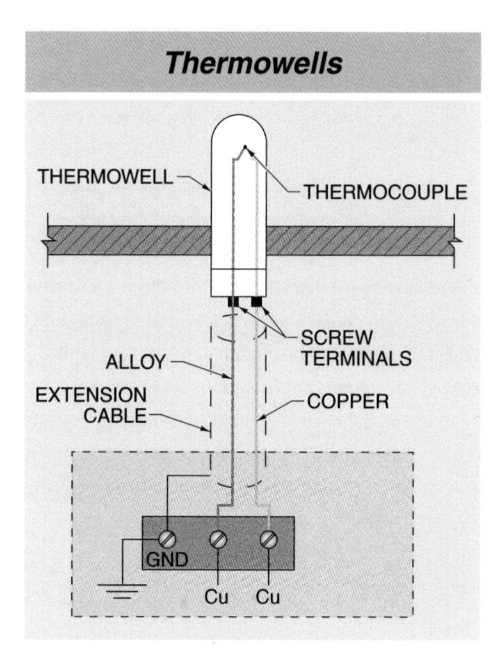

Figure 8-2. Thermowells are used to protect thermocouple wires from harsh and corrosive environments.

Thermowell Construction

Thermowells may be made of metals such as iron, steel, nickel, or Inconel®; silica compounds such as corundum or Carbofrax®; or metal-ceramic compounds such as chromium oxide and aluminum oxide. Sometimes a thermocouple must be enclosed in a primary metal protector as well as a secondary silica protector. The selection of the complete thermocouple assembly should be made carefully. Manufacturers provide a considerable amount of data for this purpose. **See Figure 8-3.**

Thermowells are available in built-up and bar stock construction. Built-up thermowells are fabricated from various-size tubes by welding the various parts together. This fabrication method leaves the welds exposed to the process fluids. The welds can be more susceptible to corrosion than the base metals.

Thermowell Materials

Type	Recommended Maximum Temperature*	Applications
Alsint: 99.7% alumina	3100	Used with platinum thermocouples; high resistance to alkaline and other fluxes; good resistance to thermal shock
Carbon steel	1000	Low cost; little corrosion resistance
Chrome-moly steels	1100	High strength steels used in boilers
Corundum: 60%-80% silica	3000	For steel industry where thermal shock may be high
Hastelloy® C	2000	Excellent corrosion resistance in chlorinated environments; widely used in chemical industry
Incoloy® 800	2000	Superior resistance to sulfur atmospheres
Inconel® 600	2200	Excellent mechanical strength at high temperatures; poor resistance to sulfur atmospheres above 900°F
Monel® 400	1000	Use with seawater, hydrofluoric acid, sulfuric acid, and most alkalis
Nickel	2000	For special chemical applications
70%-90% Silicon carbide	2550	Resistant to thermal shock; not suitable for strongly oxidizing atmospheres
304 Stainless steel	1800	Low cost; for acid and alkaline solutions; used extensively in food industry
310 Stainless steel	2100	Good protection against oxidation; suitable for use in sulfur-bearing atmospheres
316 Stainless steel	1650	Most common thermowell material; resistant to hydrogen sulfide
446 Stainless steel	2100	Poor mechanical strength at high temperatures; used for nitric acid, sulfuric acid, and most alkalis

* in °F

Figure 8-3. The choice of thermowell material is determined by the exposure to the application conditions.

Bar stock thermowells are machined from solid bars of the selected material. The only welding needed is the attachment of a flange if one is used. The insertion portion of the thermowell can be machined in three different shapes: straight, tapered, and stepped. Straight thermowells have the same diameter over the whole insertion length. Straight thermowells are used for low-pressure and low-stress installations because the straight shape cannot resist the forces applied by a flowing stream as well as the other styles.

Tapered thermowells start with a large diameter at the process connection and have an even taper down to the tip. Tapered thermowells are used for high-pressure and high-stress applications because the wider base of the thermowell helps the thermowell resist the stresses of flowing fluids. The most common thermowell is the stepped form, which has a larger insertion diameter over most of the length but steps down in diameter over the temperature-sensing end. **See Figure 8-4.**

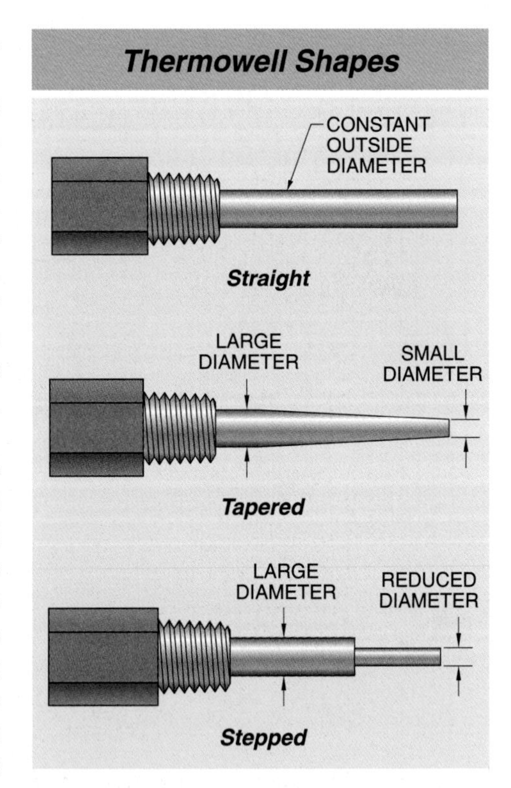

Figure 8-4. Thermowells can be constructed in straight, tapered, or stepped form.

Process Connections. Process connections can be threaded or flanged fittings. Threaded connections used in the U.S. are usually ¾″ NPT, but are available in 1″ NPT and other sizes on special order. Flanged connections are available in 1″, 1½″, and 2″ in either 150 lb ASA or 300 lb ASA flanges. Available materials of construction are carbon steel, 304 or 316 stainless steel, Hastelloy® C, Monel®, nickel, Inconel®, and titanium. **See Figure 8-5.** Essentially any bar stock material can be made into a thermowell. Tantalum is also available as a thin metal sheath slipped over a straight-flanged thermowell.

There is a connection for the instrument on the outside of the process connection. This instrument connection is usually ½″ NPT. It can be directly on the outside of the process connection or there can be a lagging extension of either 2″ or 3″. The purpose of a lagging extension is to extend the instrument connection outside of any thermal insulation. Any thermowell that is used for any remote-mounted temperature measuring device, whether it is a filled system or electrical, should be supplied with a union connection. The union is necessary so that the sensing element can be removed from the well without twisting the sensor connections. **See Figure 8-6.**

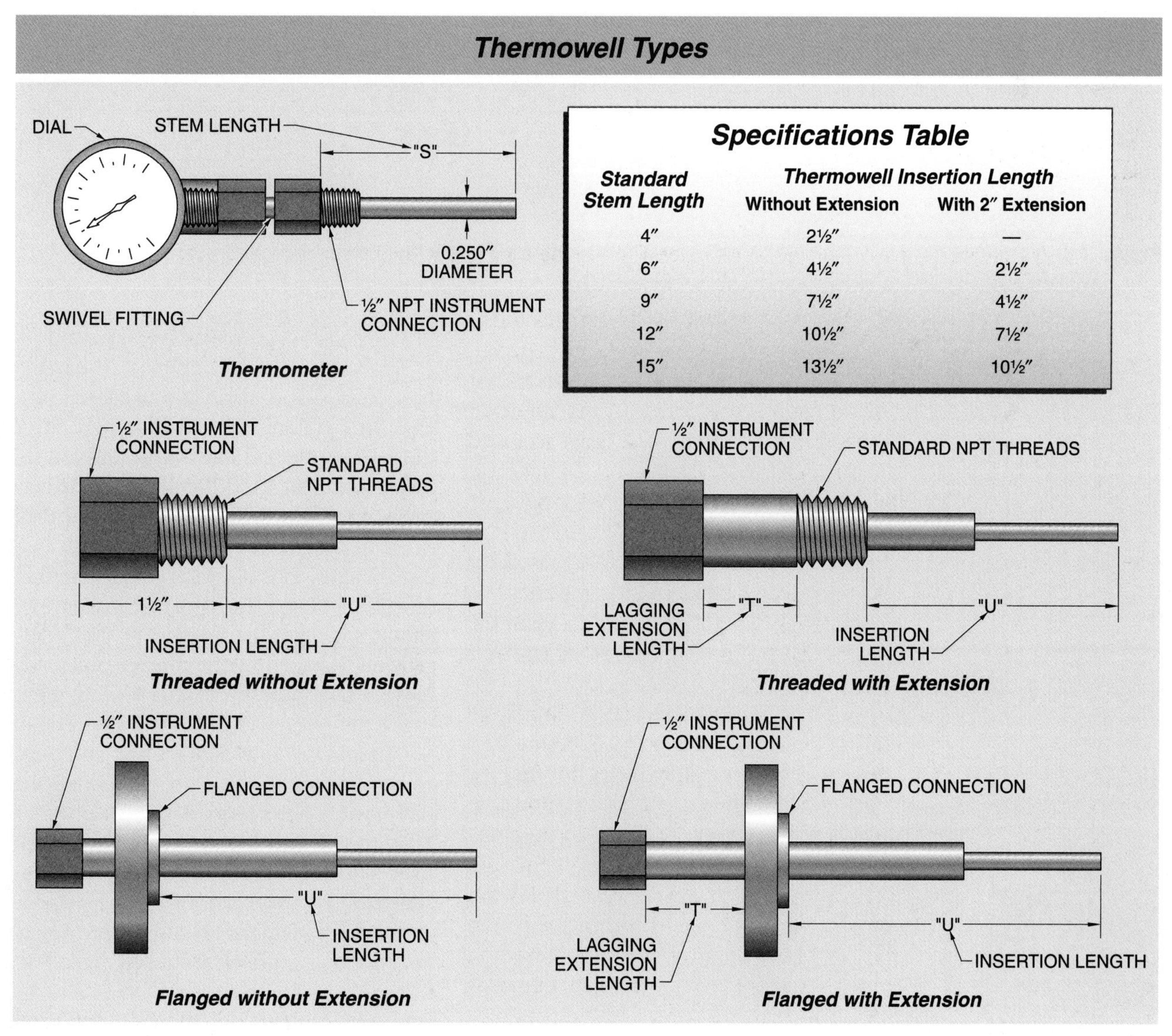

Specifications Table

Standard Stem Length	Thermowell Insertion Length — Without Extension	Thermowell Insertion Length — With 2″ Extension
4″	2½″	—
6″	4½″	2½″
9″	7½″	4½″
12″	10½″	7½″
15″	13½″	10½″

Figure 8-5. Thermowell fittings and extensions determine the thermometer length that can be used.

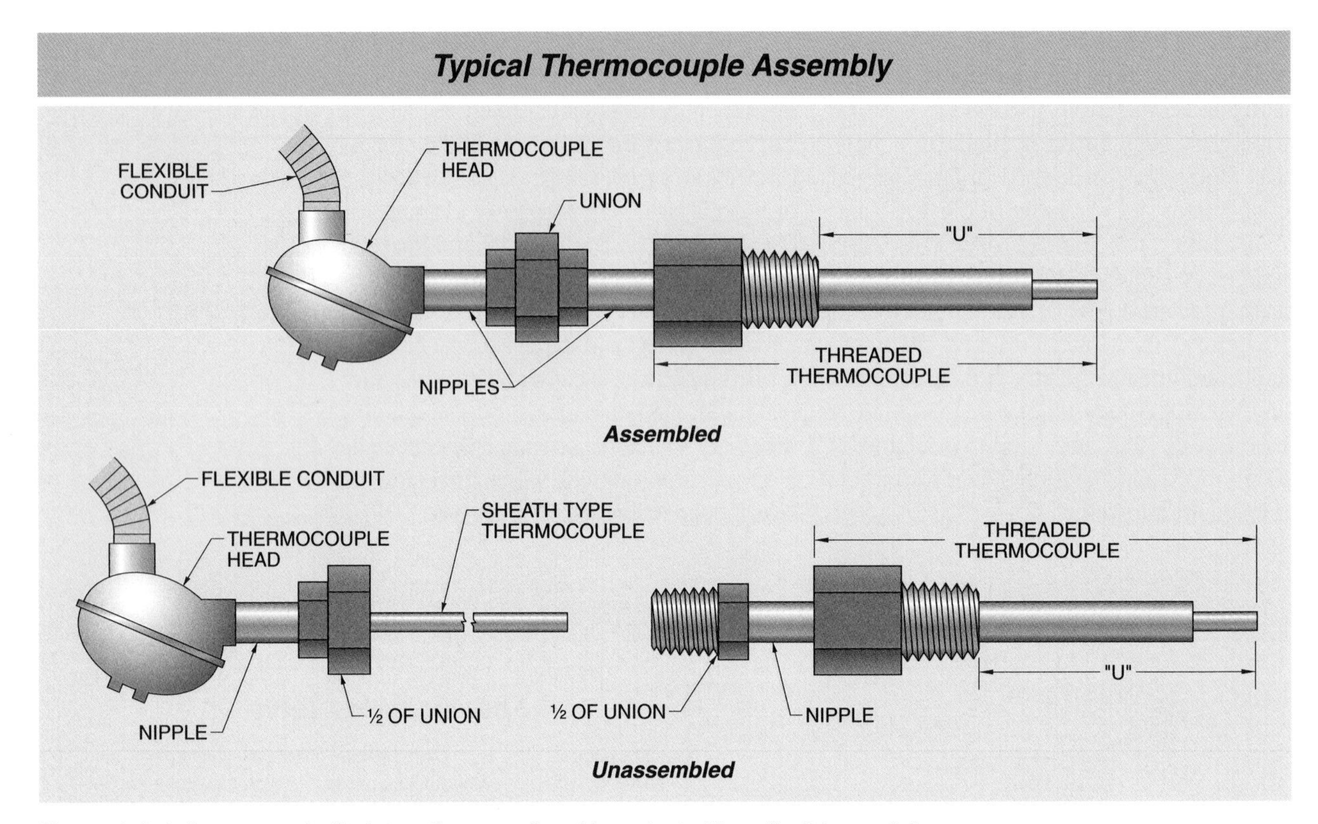

Figure 8-6. A thermocouple fits into a thermowell and is protected by a flexible conduit.

ASI Robicon

Thermowells can be designed for high temperatures and hostile environments.

To ensure that the thermowell has the best chance of accurately sensing the process temperature, the sensitive portion of the insertion length must be in the actively flowing stream. It is difficult to insert a thermowell into the side of pipes smaller than 4″. It is far better to install thermowells into pipe elbows. This allows a longer insertion length to be used. The tip of the thermowell should face into the flowing stream. Thermowell installations in smaller pipes can cause a serious restriction of flow. In those cases, the piping is increased in size around the thermowell to ensure there is sufficient free space for the flow to get around the thermowell. **See Figure 8-7.**

A thermowell must often be considered to be a pressure fitting. If a thermowell is inserted into a pressurized vessel, all applicable rules and regulations about pressure-vessel construction must be followed. The wall thickness of the thermowell must also be sufficient to be able to withstand the pressure in the vessel.

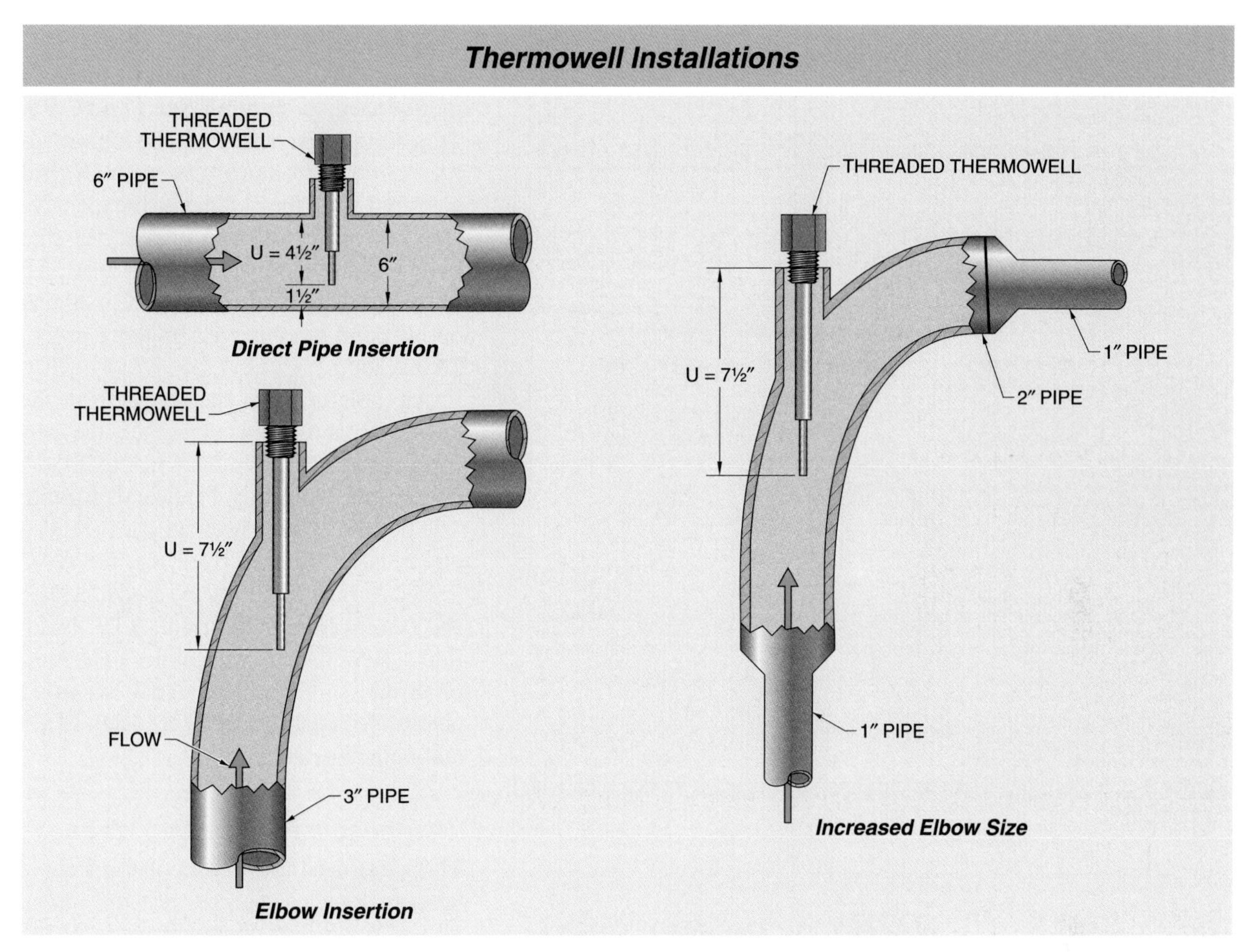

Figure 8-7. A thermowell can be installed directly into a pipe or duct or into a pipe elbow.

Insertion Length. Thermowells are primarily selected by the insertion length (U). The insertion length is the length of the thermowell that is in contact with the process fluid. Thermowells are available from most manufacturers in the standard U lengths of 2½″, 4½″, 7½″, 10½″, 13½″, and 17½″, and in custom lengths. Standard bimetallic and filled dial thermometers have stem lengths that match these insertion lengths and screw directly into the thermowell.

The bore in the thermowell is the diameter of the hole for the sensing element. The simplest type of thermowell is designed for test instruments such as glass-stem thermometers. The bore size can be 0.260″ or 0.375″. Dial thermometers, either bimetallic or filled types, usually use a 0.260″ bore. Thermocouple sheath types and RTDs usually use bores of 0.260″.

Thermowell Vibration. On rare occasions, consideration must be given to resonance and vortex shedding when a thermowell is inserted into moving fluid. When an object is inserted into a flowing liquid or gas stream, turbulent wakes are generated as vortices break away from the object. The frequency of the wakes depends on the geometry of the object and the velocity of the fluid flow. The thermowell has its own natural vibration frequency based on the insertion depth and the mass and stiffness of the thermowell.

The thermowell must be designed so that the two frequencies do not match and cause resonance that can break the thermowell. **See Figure 8-8.** This is usually a concern only when the fluid velocity is very high. Information on this topic can be obtained from the thermowell manufacturer.

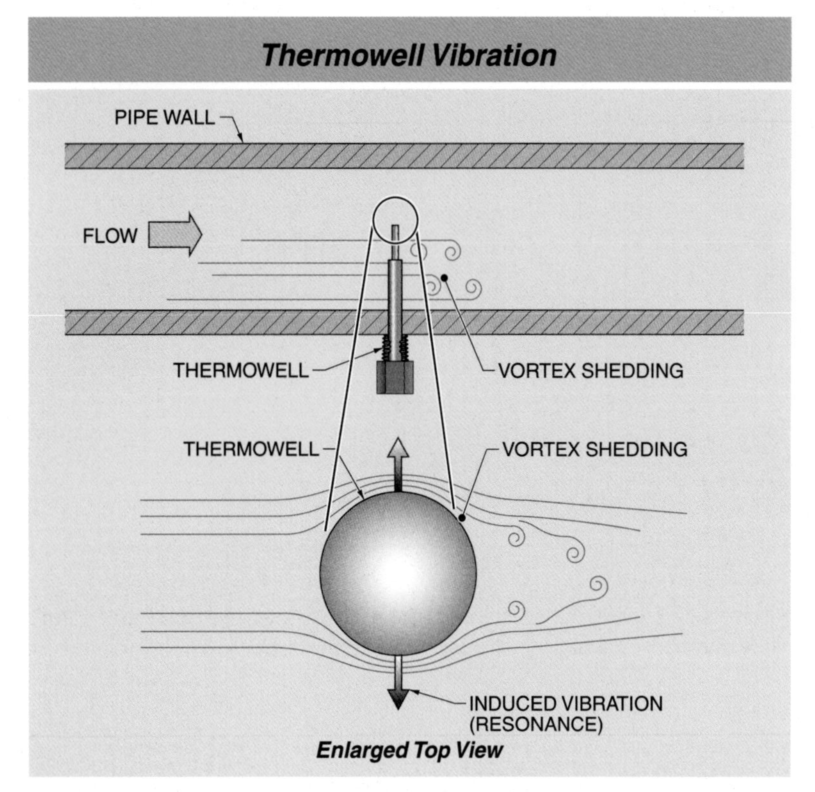

Figure 8-8. A thermowell can be damaged by resonant vibration caused by flowing liquid.

PRACTICAL THERMOCOUPLE MEASUREMENT

The size of the wire and the protection of the thermocouple are important factors to consider when evaluating thermocouples. Although thermocouples are protected from contact with the process by a thermowell, they are exposed to the air inside the thermowell. Because this exposure can shorten the life of a thermocouple, the selected thermocouple wire size is usually a relatively large 14-gauge wire.

The wire size affects both the sensitivity and the maximum operating temperature of the thermocouple. Thermocouple wire sizes range from fine, 40 AWG, to heavy, 8 AWG. Thinner wires respond faster to temperature changes, but break down faster in extreme environments. For example, an 8-gauge wire in a grounded thermocouple responds in about 20 seconds to a change in temperature while a 16-gauge wire in the same situation responds in about 8 seconds.

Transmitters

Locally mounted transmitters are commonly used for control. **See Figure 8-9.** Transmitters can vary from the high-quality "smart" transmitters with digital electronics to the relatively inexpensive "hockey puck," disk-shaped styles that mount inside the thermocouple head. All of these transmitters have 4 mA to 20 mA outputs and are powered by the 4 mA to 20 mA loop.

An isolator takes the signal from the transmitter and electrically isolates the thermocouple signal from the 4 mA to 20 mA signal and from any electrical ground. The best 24 VDC power supplies are also electrically isolated from the 115 VAC power. A transformer can be used to isolate AC current from DC current. A DC isolation system usually uses optics to accomplish the isolation. The signal is carried across the isolation barrier by using LEDs and optical detectors.

Thermocouple Extension Wires

The wires used to carry the thermocouple temperature signals are constructed of special alloys. Extension wires are made of the same type of materials as the thermocouple but of a lower purity. In most cases, the wire can be 20 gauge or smaller. Even though thermocouple wire is relatively expensive, copper wires should never be used as a replacement. For either single pair or multipair cables, the wires should be protected with cable shields to minimize electromagnetic interference (noise) pickup.

Multiconductor cables are run from the centralized location to a field junction box. From there, the wires are continued as single pairs. Thermocouple extension wire shields should only be grounded at one end, and all segments of the shields for a temperature signal are connected together. Field connections should be kept to a minimum. Thermocouple extension wires should always be run separately from any power sources or 4 mA to 20 mA control signals. If these design practices are followed, there should be no noise problems.

Ground Loops

The combination of the isolator, transmitter, and power supply allows the use of grounded thermocouples without any danger of ground loop electrical currents between thermocouples since there is only one ground point in the sensing portion of the circuit. A *ground loop* is current flow from one grounded point to a second grounded point in the same powered loop due to differences in the actual ground potential.

The use of a nonisolated transmitter with a grounded thermocouple allows two electrically common points through which a ground loop current can flow. One point is the grounded thermocouple and the other point is the common 24 VDC power supply used for both 4 mA to 20 mA loops. **See Figure 8-10.** Even a very small ground loop current can cause very large errors in measurement. If there is any question about whether a thermocouple transmitter is isolated, an ungrounded thermocouple should be used.

Noise Reduction

There are many possible sources of noise in a thermocouple circuit. The best solution to noise problems is to install the circuit in a way that prevents the noise from entering the circuit. In those rare cases where this cannot be done, filtering is the simplest method of reducing thermocouple noise. An analog low-pass filter can be placed directly at the input of the voltmeter. However, a filter reduces crosstalk at the expense of response time.

Another method of reducing noise is integration. Integration is an analog-to-digital (A/D) conversion technique that averages noise out over a full line cycle. Power line noise and its harmonics are eliminated. Noise can also be reduced somewhat by increasing the wire size on the thermocouple. The increased wire size reduces the resistance of the wires and therefore the coupling capacitance is also reduced.

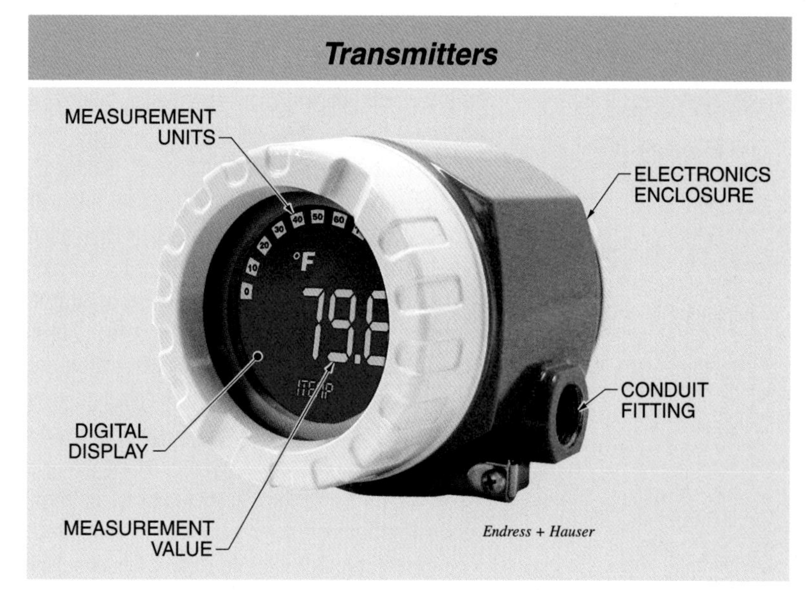

Figure 8-9. Locally mounted transmitters are commonly used for control. They can have digital or 4 mA to 20 mA analog outputs.

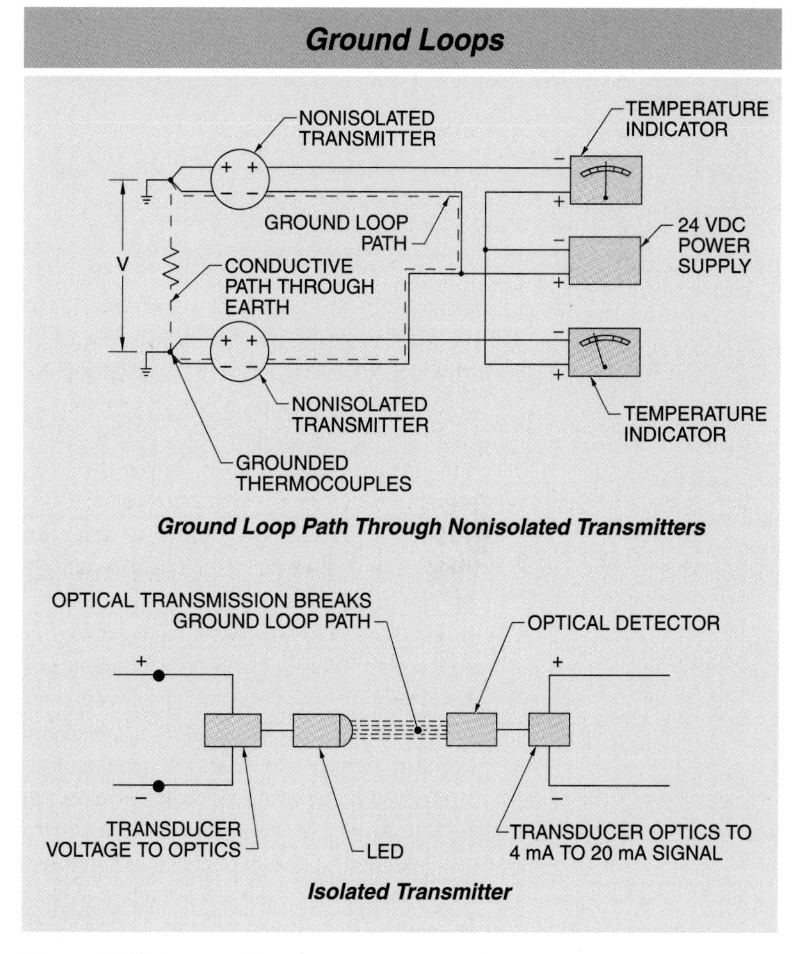

Figure 8-10. A ground loop can cause errors in measurement.

Break Protection

If one of the two wires of a thermocouple breaks, the voltage at the cold junction becomes zero. This also occurs when the temperatures of the hot and cold junctions are the same. This condition can be hazardous, especially when a thermocouple is used to control heating. Many electronic devices that use thermocouples as the input signal offer thermocouple break protection. *Thermocouple break protection* is a circuit where an electronic device sends a low-level current across the thermocouple. **See Figure 8-11.** As long as there is current flow, the thermocouple wires are intact. If the circuit is open, the thermocouple is broken or no longer connected. The electronic device can detect the absence of the current and signal an alarm.

Thermocouple Break Protection Circuits

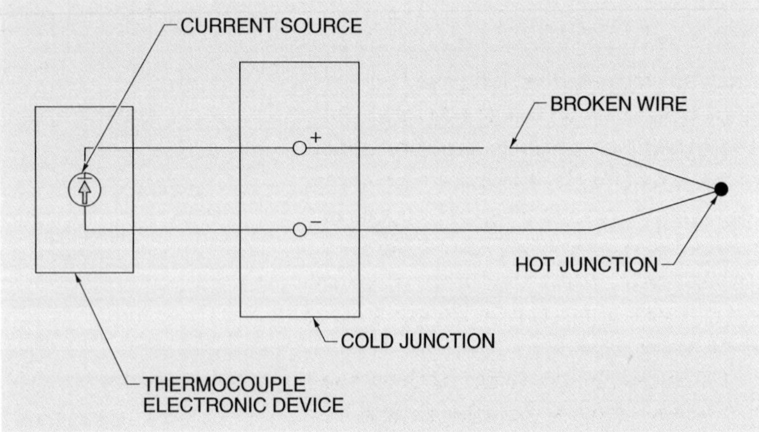

Figure 8-11. Thermocouple break protection can detect breaks in the thermocouple wires.

Typically, the break protection device causes the indication of the measured temperature to become the maximum value (upscale break protection) or minimum value (downscale break protection). Upscale break protection is selected as a fail-safe circuit for heating applications because a broken thermocouple wire has a voltage reading of zero, indicating that more heat is needed. This turns on the heater and causes overheating. Downscale break protection is commonly used for cooling applications to shut off the cooling source to prevent damage. The resistance of a thermocouple is very low, so the voltage drop from the current flow across the thermocouple is low and usually has a negligible effect on the voltage at the cold junction. Alternatively, the voltage drop due to the current flow can be subtracted from the measured voltage.

Decalibration

Decalibration is the process of unintentionally altering the physical makeup of a thermocouple wire so that it no longer conforms to the limits of the voltage-temperature curves. Decalibration can result from diffusion of atmospheric particles into the wires caused by high temperatures. Decalibration can also be caused by high-temperature annealing caused by high temperatures or temperature gradients. Decalibration can also be caused by cold-working the metal by rough handling or vibration. If a thermocouple circuit is found to be drifting, it may be necessary to replace any wires that have a temperature gradient, not just the hot junction. The best solution is to install a thermocouple circuit in such a way that there are no steep temperature gradients, for example by using metallic sleeving or by careful placement of the thermocouple wire.

Shunt Impedance

Shunt impedance is an unintended circuit caused by damaged thermocouple insulation. If the leakage resistance is smaller than the resistance of the thermocouple wire, the hot junction appears to be at or near the point of leakage. **See Figure 8-12.** To minimize this problem, insulation can be chosen that is appropriate to the temperatures and conditions to which the thermocouple is exposed. Choosing insulation appropriate to the temperatures and conditions that the thermocouple is exposed to can minimize this problem. Larger wire sizes also help minimize this problem by reducing the resistance of the thermocouple circuit. This increases response time.

Galvanic Reaction

When electrolytes such as water come in contact with thermocouples, there is a galvanic reaction between the two thermocouple wires that is much greater than the Seebeck voltage.

Thermocouple wire insulation may also contain dyes that dissolve in water to increase the effect. If thermocouples become wet, they must be dried out or replaced.

Thermocouple Aging

Thermocouple aging is the process by which thermocouples gradually change their voltage-temperature curve due to extended time in extreme environments. In Type J and Type T thermocouples, the pure metal wire tends to oxidize. In Type K thermocouples, the nickel wire often reacts with sulfur, a common constituent of flue gas and refinery streams. Type R and Type S thermocouples are highly sensitive to impurities.

Platinum easily adsorbs other metals and the changed alloy no longer follows the specified voltage-temperature curve. At high temperatures, platinum also reacts quickly with phosphorus and silicon. Silicon can be released from insulating ceramics in reducing atmospheres. Great care must be taken to ensure that thermocouples are used as specified by the manufacturer.

Some temperature measurement devices generate a pseudothermocouple signal. If these devices have a high resistance across the thermocouple signal wires, the voltage drop due to thermocouple break protection can significantly affect the voltage at the cold junction and thus introduce an error in the temperature measurement.

Battery-Powered Temperature Indicators

Local temperature indicators are used to provide temperature measurements in the field. These indicators can be used to check for a transmitted signal, as a temperature measurement for use by the operators, or to provide additional temperature information regarding the operation of a piece of equipment. The vast majority of these applications are handled by locally installed bimetallic dial thermometers.

There are occasions when a dial thermometer is not suitable because the temperature is too high or too low or because the piping is too small for a thermowell. A temperature indicator powered by a lithium battery, with thermocouple or RTD inputs, can provide the local temperature readings needed. The indicator housing typically has a NEMA 4X rating. Lithium battery life is normally over 2 years.

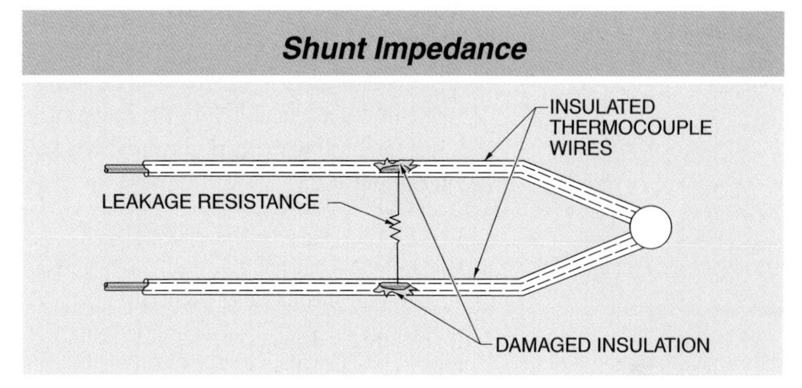

Figure 8-12. Shunt impedance is caused by damaged thermocouple insulation.

PRACTICAL RTD MEASUREMENTS

An RTD can be used in a thermowell in the same manner as a sheathed thermocouple. The installation practices are very similar to those required by thermocouples except the signal wires are 20-gauge copper and grouped into three wires per measurement. Two of the wires have the same color. Individual triads and triad bundles need to have shields.

RTDs can also be wired with four wires. Two wires of the same color are used for the resistance measurement. Two wires of a different color supply voltage to the sensor.

As with thermocouple transmitters, RTD transmitters are available that range from sophisticated "smart" transmitters to simple "hockey puck," disk-shaped forms. Since RTDs are never grounded, the transmitter does not have to be electrically isolated, although isolation is generally a good idea.

An RTD can fail in service, but the failure modes are more complex than with thermocouples because there are three or four wires and the failure of each wire causes a different response. Overall, RTDs are subject to fewer failure problems than thermocouples.

RESISTANCE BRIDGE CIRCUITS

A *resistance bridge* is a circuit used to precisely measure an unknown resistance and consists of the unknown electrical resistance, several known resistances, and a voltage meter. A Wheatstone bridge, with variations, is often used as the resistance bridge. **See Figure 8-13.** Resistance bridges are often used to measure the resistance of RTDs or thermistors as they change with temperature.

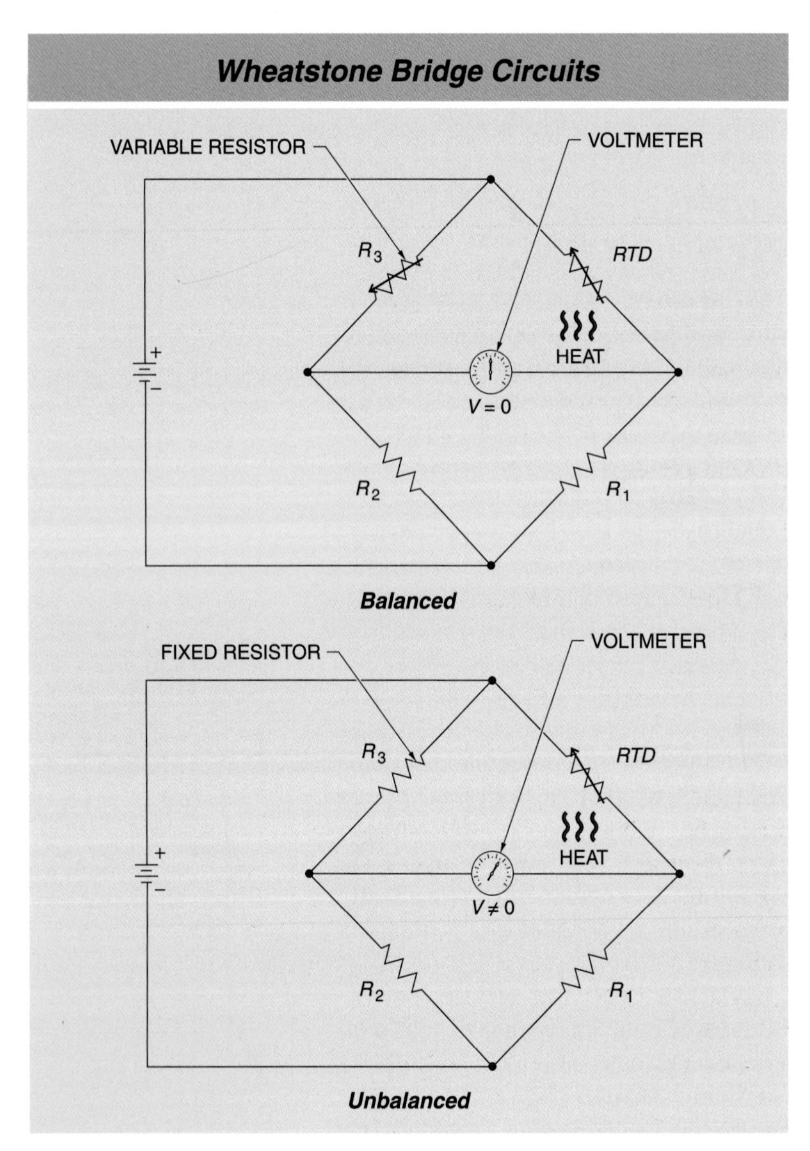

Figure 8-13. A Wheatstone bridge circuit is used to measure the resistance change of an RTD.

A balanced resistance bridge is a resistance bridge with the resistances adjusted so there is equal current flow through the legs of the bridge and zero potential across the bridge. The fixed resistors R_1 and R_2 are matched to each other to have the same resistance.The variable resistor R_3 is adjusted to match the resistance of the RTD in order to balance the bridge. The resistance of the variable resistor is proportional to the temperature. An unbalanced resistance bridge is a resistance bridge with fixed resistances and the voltage across the bridge is proportional to the temperature of the RTD.

When the temperature of the RTD changes, the resistance changes and the bridge is no longer balanced. The voltmeter now registers a potential across the bridge. The variable resistor is adjusted to balance the circuit again. The change in resistance is proportional to the change in temperature. Caution must be exercised to minimize resistance heating of the RTD and circuit. This is done by specifying a low current flow through the circuit.

As with thermocouples, it is possible to have temperature averaging and difference measurements, but these are harder to implement with RTDs in analog systems. If the RTD signals go to a digital control system, the averaging or the subtracting can be done directly with the digital temperature values.

Two-Wire RTD Bridge

A difficulty with a Wheatstone bridge as shown is that the leads to the RTD in a 2-wire circuit have resistance of their own that is not accounted for in the circuit. This can add significant error to the measurement. For example, a typical 100′ lead wire can add 1 Ω to 5 Ω to the circuit, depending on the wire. The RTD itself may only have 100 Ω at 32°F. A solution to this source of error is to use a 3-wire RTD circuit. **See Figure 8-14.**

Three-Wire RTD Bridge

A 3-wire RTD modifies the Wheatstone bridge circuit by adding another lead wire from the RTD. The extra wire is used to compensate for the resistance of the lead wires.

The RTD leads are matched so that they are all of equal resistance. The resistance of the leads L_1 and L_2 are in separate legs of the circuit and their effect balances out. When the bridge is balanced, there is no current flow through L_3 and across the bridge, and the potential at the voltmeter is zero. **See Figure 8-15.**

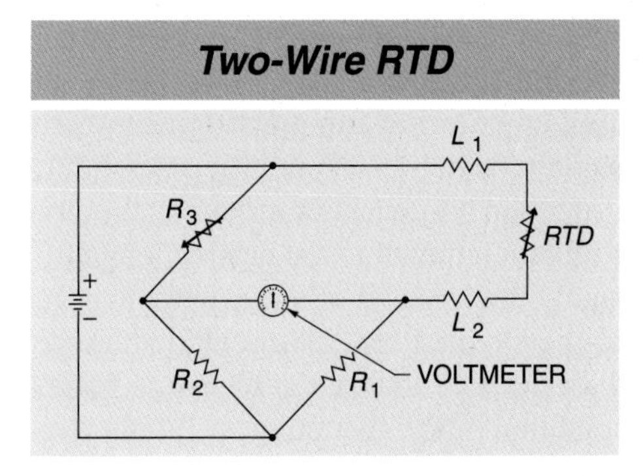

Figure 8-14. Lead wire resistance in an RTD bridge circuit can cause significant errors.

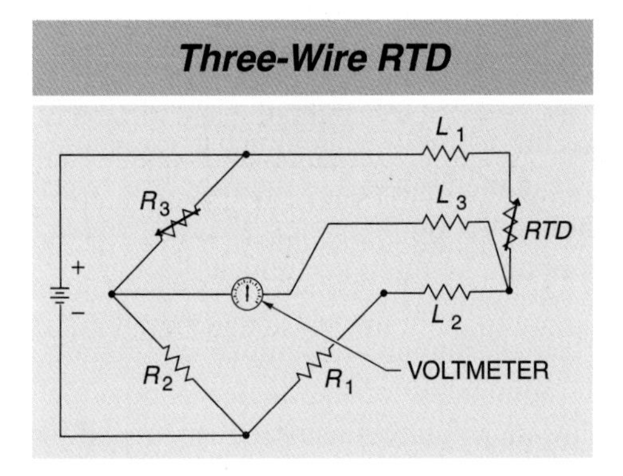

Figure 8-15. A 3-wire RTD adds another lead wire to compensate for the resistance of the lead wires.

Four-Wire RTD Bridge

In addition to 2-wire and 3-wire RTDs, 4-wire RTDs are also available. In a 4-wire RTD, the sensing RTD takes the place of two resistors in the Wheatstone bridge. The voltage that normally connects to the top and bottom of the bridge is carried directly to the sensing RTD. Therefore, no current is routed through the RTD sensing wires. There is no need to compensate for wire resistance as with the 3-wire RTD circuit. **See Figure 8-16.**

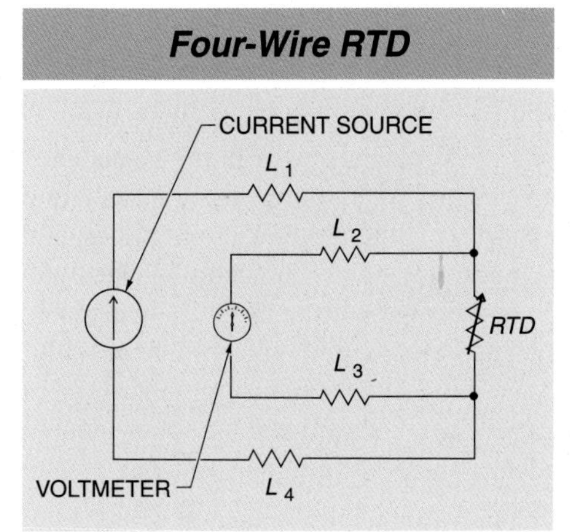

Figure 8-16. In a 4-wire RTD, the sensing RTD takes the place of two resistors in the Wheatstone bridge.

Thermistor Bridge Circuits

For temperature measurement, thermistors are commonly used in bridge circuits. Because of the very significant nonlinearity of the thermistor response, the bridge is usually modified by placing a resistor in parallel with the thermistor. The resistor must be chosen to be at the midpoint of the thermistor resistance over the temperature change of interest. For example, if the temperature range of interest is 55°F to 95°F and a thermistor has a resistance of 2252 Ω at the midpoint, 75°F, the resistor needs to be close to 2252 Ω. **See Figure 8-17.** The other resistors in the bridge are chosen to give the desired voltage output range.

CALIBRATION

Temperature-measuring instruments may need to be calibrated in the field or maintenance shop. Thermal expansion and electrical thermometers are commonly calibrated by inserting the thermometer in a box or bath at a known temperature. A comparison is made between the actual temperature and the reading from the thermometer. Infrared radiation thermometers are calibrated with a blackbody calibrator. In addition, transmitters need to be calibrated to ensure that the signal from an instrument is properly sent on to an indicator or controller.

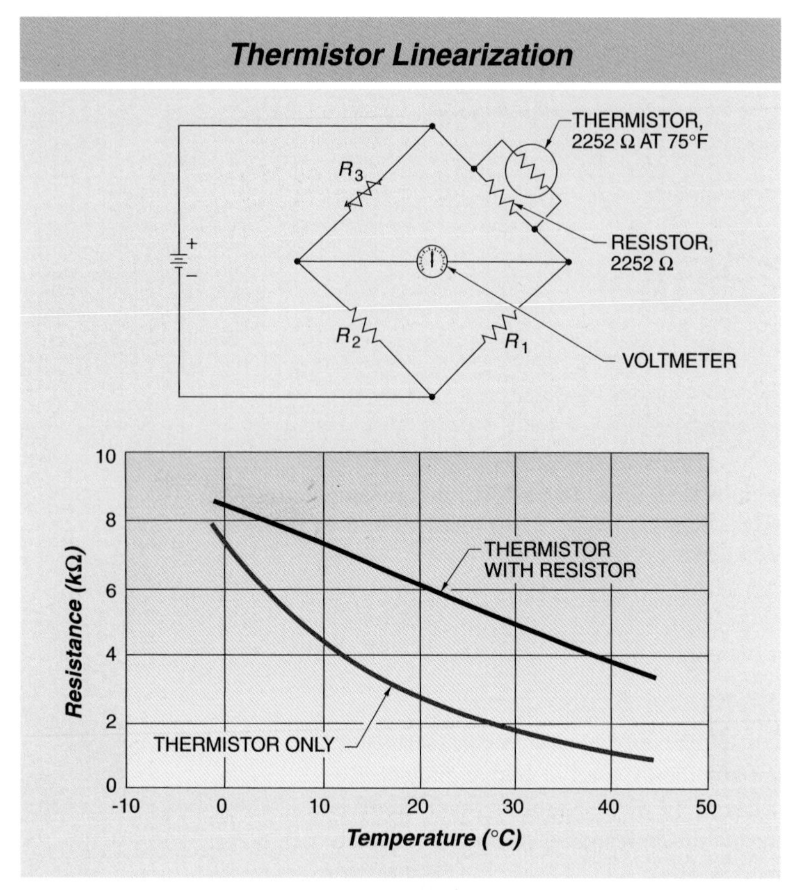

Figure 8-17. The nonlinear characteristics of a thermistor can be compensated for with a modified bridge circuit.

Dry Well and Microbath Calibrators

A *dry well calibrator* is a temperature-controlled well or box where a thermometer can be inserted and the output compared to the known dry well temperature. **See Figure 8-18.** Dry wells are constructed of high-stability metal blocks with holes drilled in them to accept a reference and a thermometer under test. Dry wells are available to calibrate a temperature range of about –50°F to almost 2200°F.

A *microbath* is a small tank containing a stirred liquid used to calibrate thermometers. The use of a thermal bath eliminates problems resulting from poor fit in a block, so microbaths are especially suited for calibrating odd-shaped probes. Microbaths are available to calibrate a temperature range of about –20°F to about 400°F.

Dry wells and microbaths have a temperature controller that maintains the calibrator at a constant temperature. The actual temperature of a dry well block or microbath is measured with a reference thermometer. An external reference thermometer is usually made of a platinum RTD for maximum precision and accuracy. An internal reference thermometer allows the calibrator to be more portable, but generally the reference is less accurate than an external reference temperature measurement.

To perform a calibration, set the dry well or microbath to the desired temperature at the low end of the expected range of the desired measurements. Measure the temperature to determine the as-found condition of the thermometer. Verify that the measured temperature matches the reference temperature within the desired tolerances. Set the dry well or microbath to a temperature at the midpoint of the expected range of the desired measurements. Again, verify that the measured temperature matches the reference temperature. Set the dry well or microbath to a temperature at the high end of the expected range and verify that the measured temperature matches the reference temperature.

If a calibration shows that a liquid-in-glass thermometer is out of calibration, no adjustment of the device is possible. The thermometer must be discarded or used as is with an offset. If an electrical thermometer is out of calibration, the thermometer can be replaced or the conversion coefficients can be adjusted in the readout device.

TECH-FACT

Lasers used with fiber-optic lines are at maximum efficiency at a certain wavelength and must be controlled within 0.1 nm. The laser wavelength changes with temperature. A thermistor and a Peltier effect cooler can be used to keep the laser temperature within tolerance.

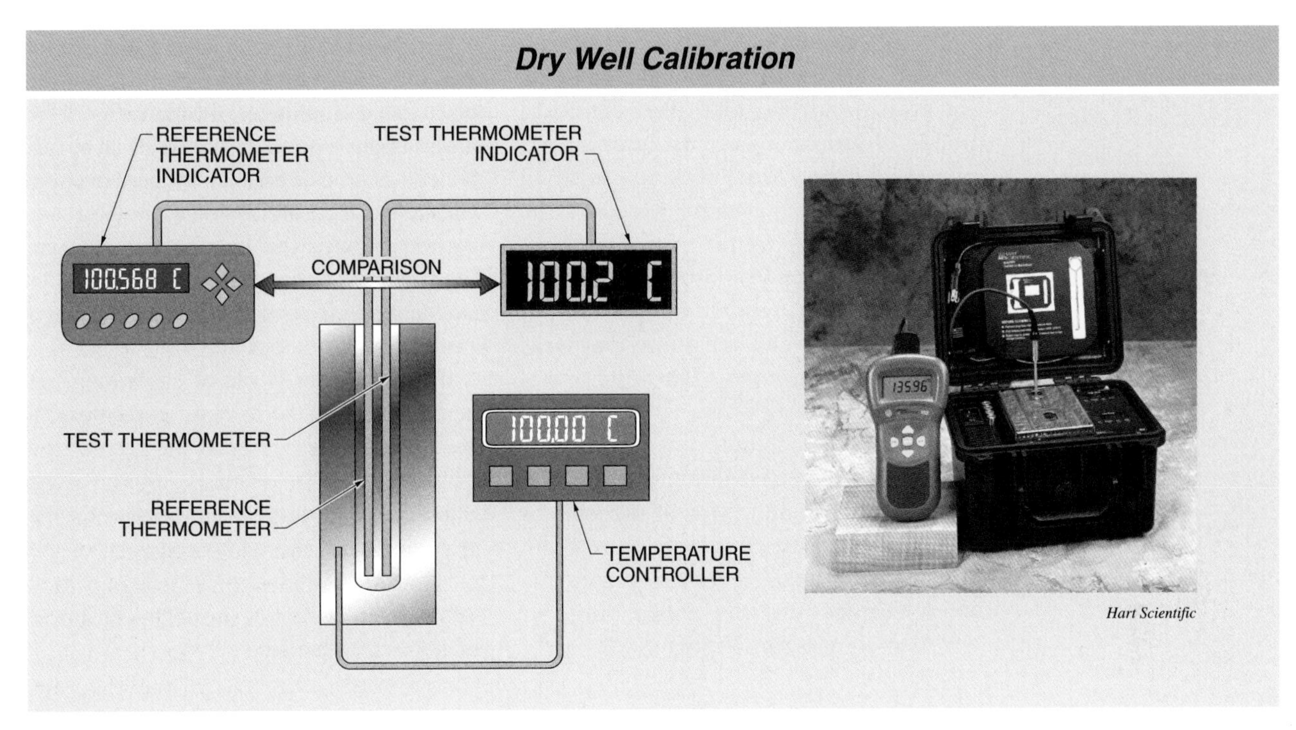

Figure 8-18. A dry well calibrator is used to calibrate thermometers.

Blackbody Calibrators

A *blackbody calibrator* is a device used to calibrate infrared thermometers. Blackbody calibrators have unheated or heated surfaces whose emissivity is nearly 1.0. The temperature of the surface is often measured with a certified RTD. In operation, the surface is heated to a specified temperature and the IR thermometer is aimed at the spot on the calibrator. The manufacturer provides the actual emissivity value for the calibrator. **See Figure 8-19.**

Transmitter Calibration

A standard temperature transmitter can be calibrated by removing the input wires and replacing them with an electronic calibrator. The calibrator generates an electrical signal that replicates the signal from an electrical thermometer. Connect the test leads from the calibrator to the transmitter. The output from the thermocouple jacks simulates a temperature input to the transmitter. The red and black test leads provide loop power to the transmitter and measure the current resulting from temperature changes into the transmitter. Select the low end of the transmitter range as a zero point and enter a span. Verify that the output from the transmitter matches the specified instrument tolerance at various input levels.

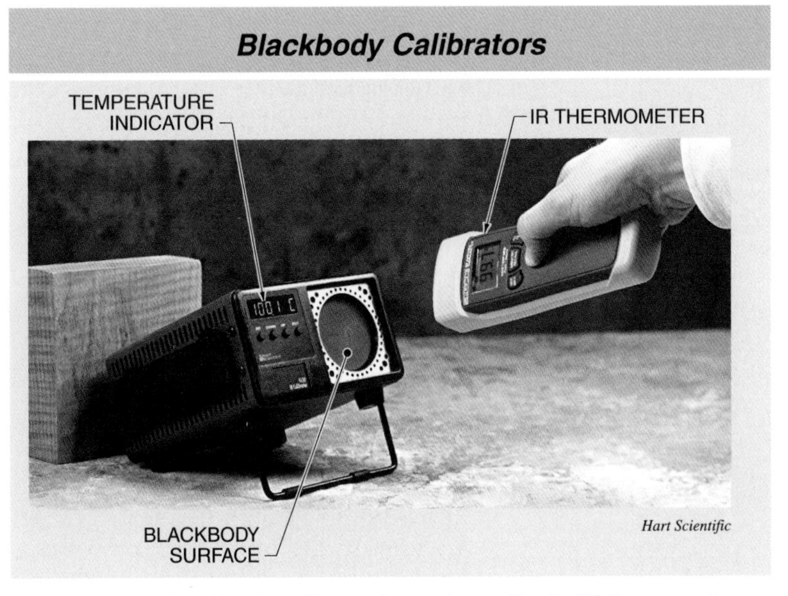

Figure 8-19. A blackbody calibrator is used to calibrate IR thermometers.

Many transmitters include an optical isolator to eliminate an electrical path for ground currents and other electrical noise. A loop-powered isolator draws operating power from the input side of the isolator on the same wires as the 4 mA to 20 mA signal from the transmitter. This type of isolator can easily be calibrated. First, remove the transmitter from the circuit and substitute a process meter or loop calibrator. The process meter can be set to generate any current in the range of 4 mA to 20 mA to simulate signals from the transmitter. Connect an ammeter to the output side of the isolator and verify that the isolator is putting out the correct signal current to match the input current over the entire range of 4 mA to 20 mA. **See Figure 8-20.**

A two-wire isolating transmitter draws operating power from the output side of the transmitter and modulates the current from the power supply to provide a signal. It can be calibrated by removing the input thermocouple wires and using a process meter or loop calibrator to simulate the thermocouple voltage. The simulated voltage enters the transmitter and the signal is conditioned and output as a 4 mA to 20 mA output.

If a transmitter is out of calibration, set the calibrator to the lowest temperature in the expected range. Adjust the zero of the transmitter until the output is 4.00 A. Set the calibrator to the highest temperature in the expected range and adjust the span of the transmitter until the output is 20.00 A. Repeat both steps to verify that the span adjustment did not change the zero.

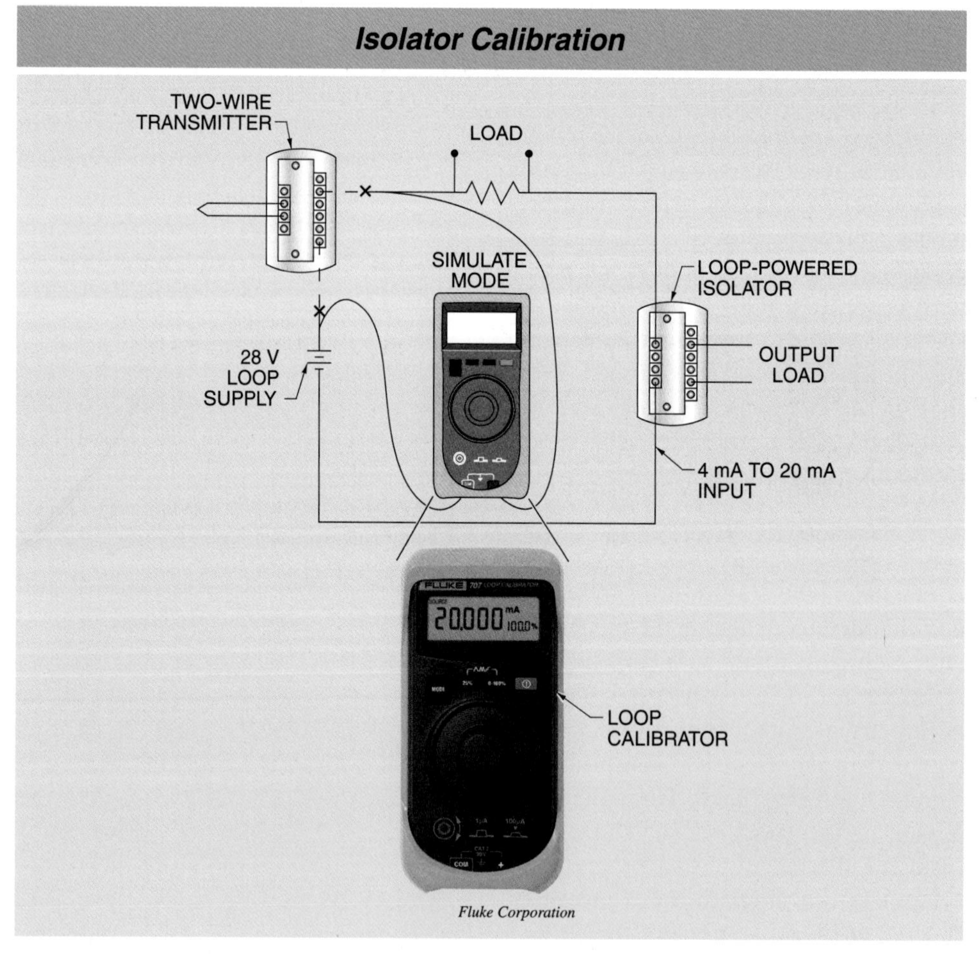

Figure 8-20. An isolator can be easily calibrated with a loop meter or process meter.

KEY TERMS

- *time constant:* The time required for a process to change by 63.2% of its total change when an input to the process is changed.
- *thermowell:* A closed tube used to protect a temperature instrument from process conditions and to allow instrument maintenance to be performed without draining the process fluid.
- *ground loop:* Current flow from one grounded point to a second grounded point in the same powered loop due to differences in the actual ground potential.
- *thermocouple break protection:* A circuit where an electronic device sends a low-level current across the thermocouple.
- *decalibration:* The process of unintentionally altering the physical makeup of a thermocouple wire so that it no longer conforms to the limits of the voltage-temperature curves.
- *shunt impedance:* An unintended circuit caused by damaged thermocouple insulation.
- *thermocouple aging:* The process by which thermocouples gradually change their voltage-temperature curve due to extended time in extreme environments.
- *resistance bridge:* A circuit used to precisely measure an unknown resistance and consists of the unknown electrical resistance, several known resistances, and a voltage meter.
- *dry well calibrator:* A temperature-controlled well or box where a thermometer can be inserted and the output compared to the known dry well temperature.
- *microbath:* A small tank containing a stirred liquid used to calibrate thermometers.
- *blackbody calibrator:* A device used to calibrate infrared thermometers

REVIEW QUESTIONS

1. Describe the response time and time constant of thermometers.
2. What is the purpose of a thermowell?
3. What is a ground loop, and how can problems associated with it be prevented?
4. What is a resistance bridge?
5. Define dry well calibrator, microbath, and blackbody calibrator.

SECTION 3

PRESSURE MEASUREMENT

TABLE OF CONTENTS

SECTION OBJECTIVES

Chapter 9

- Define pressure, atmospheric pressure, head, and hydrostatic pressure.
- Identify and define the types of mechanical pressure sources, and identify a common application of Pascal's law.
- Identify four common pressure scales and common units of pressure measurement.

Chapter 10

- Identify types of manometers and their working principles.
- Identify types of mechanical pressure sensors and their working principles.

Chapter 11

- Describe the operation of a resistance pressure transducer.
- Compare the operation of a capacitance pressure transducer and a differential pressure (d/p) cell.
- Compare the operation of inductance and reluctance pressure transducers.
- Describe the operation of a piezoelectric pressure transducer.

Chapter 12

- List important considerations in pressure measurement applications.
- List important considerations in pressure measurement with manometers.
- Identify means of protecting pressure sensors from hazardous environments.
- Describe devices for calibrating pressure sensors.

SECTION

3 PRESSURE MEASUREMENT

INTRODUCTION

Evangelista Torricelli invented the mercury barometer in 1643. He became the first person to create a vacuum, something that had been thought impossible. The earliest pressure measurements used liquid column instruments based on the same principles. Blaise Pascal was the first to explain barometric pressure as the weight of the atmosphere. Pascal also demonstrated that when pressure is applied to a confined fluid, the pressure is transmitted equally in all directions. This is the basis of all modern hydraulic systems.

Robert Boyle, Jacques-Alexandre Charles, and Joseph Gay-Lussac discovered the relationships between pressure, volume, temperature, and mass of gases that led to the development of the combined gas law.

The development of steam engines in the early 19th century led to a demand for instruments to measure higher pressures than had been used previously. The first pressure gauge with a flexible measuring element was the Bourdon tube pressure gauge. In 1848, Eugène Bourdon received a patent on the device and it was named after him. In 1852, Edward Ashcroft purchased the United States rights to the Bourdon patent. The Bourdon tube gauge is sometimes called an Ashcroft gauge because of this.

ADDITIONAL ACTIVITIES

1. After completing Section 3, take the Quick Quiz® included with the Digital Resources.
2. After completing each chapter, answer the questions and complete the activities in the *Instrumentation and Process Control Workbook.*
3. Review the Flash Cards for Section 3 included with the Digital Resources.
4. Review the following related Media Clips for Section 3 included with the Digital Resources:
   - Filters
   - Pigtail Siphon
5. After completing each chapter, refer to the Digital Resources for the Review Questions in PDF format.

ATPeResources.com/QuickLinks • Access Code: 467690

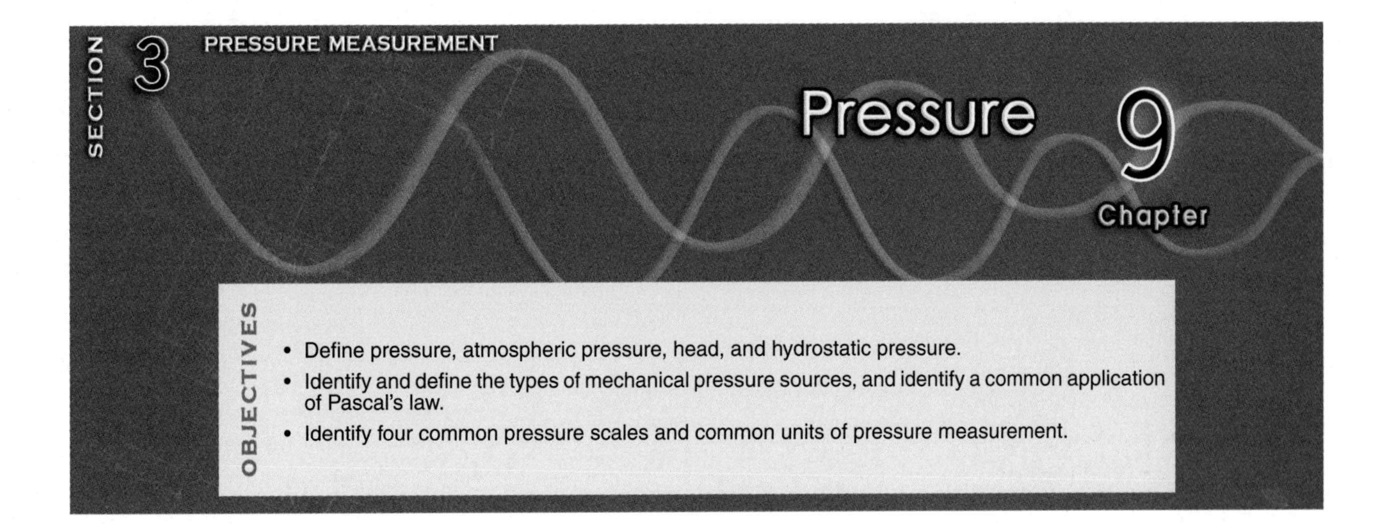

PRESSURE AND FORCE

The pressure of almost any liquid or gas that is stored or moved must be known to ensure safe and reliable operations. *Pressure* is force divided by the area over which that force is applied. *Force* is anything that changes or tends to change the state of rest or motion of a body. *Area* is the number of unit squares equal to the surface of an object. The formulas relating pressure, force, and area are:

$$P = \frac{F}{A}$$

$$A = \frac{F}{P}$$

$$F = P \times A$$

where

P = pressure (in lb per sq in., or psi)

F = force (in lb)

A = area of contact with the fluid (in sq in.)

Pressure increases as force increases or area decreases. For instance, a hydraulic actuator piston is used to convert pressure to linear motion. Force is transmitted through the piston rod in a straight line to move a load. The piston moves the load if the force generated by the piston is greater than the load. The amount of force available depends on the pressure and the size of the piston. **See Figure 9-1.** For example, if a cylinder is pressurized to 1600 pounds per square inch (psi) in a 3-inch diameter cylinder that has a piston surface area of 7.07 sq in., the force is calculated as follows:

$$F = P \times A$$

$$F = 1600 \times 7.07$$

$$F = \mathbf{11,312\ lb}$$

Force increases as pressure or area increases. The pressure can be increased to amplify the force. For example, if the pressure is doubled from 1600 psi to 3200 psi, then the force is also doubled as follows:

$$F = P \times A$$

$$F = 3200 \times 7.07$$

$$F = \mathbf{22,624\ lb}$$

The area of the piston can be increased to amplify the force. For example, if the area of the piston is doubled from 7.07 sq in. to 14.14 sq in. while maintaining the original pressure, the force is doubled as follows:

$$F = P \times A$$

$$F = 1600 \times 14.14$$

$$F = \mathbf{22,624\ lb}$$

Fluid Pressure

A *fluid* is any material that flows and takes the shape of its container. Gases and liquids are both fluids. Fluid pressure may be due to the weight of a fluid column, or due to applied mechanical energy. Mechanical energy is provided by such devices as a pump or blower and stored in the form of a fluid under pressure, at an elevated height, or both.

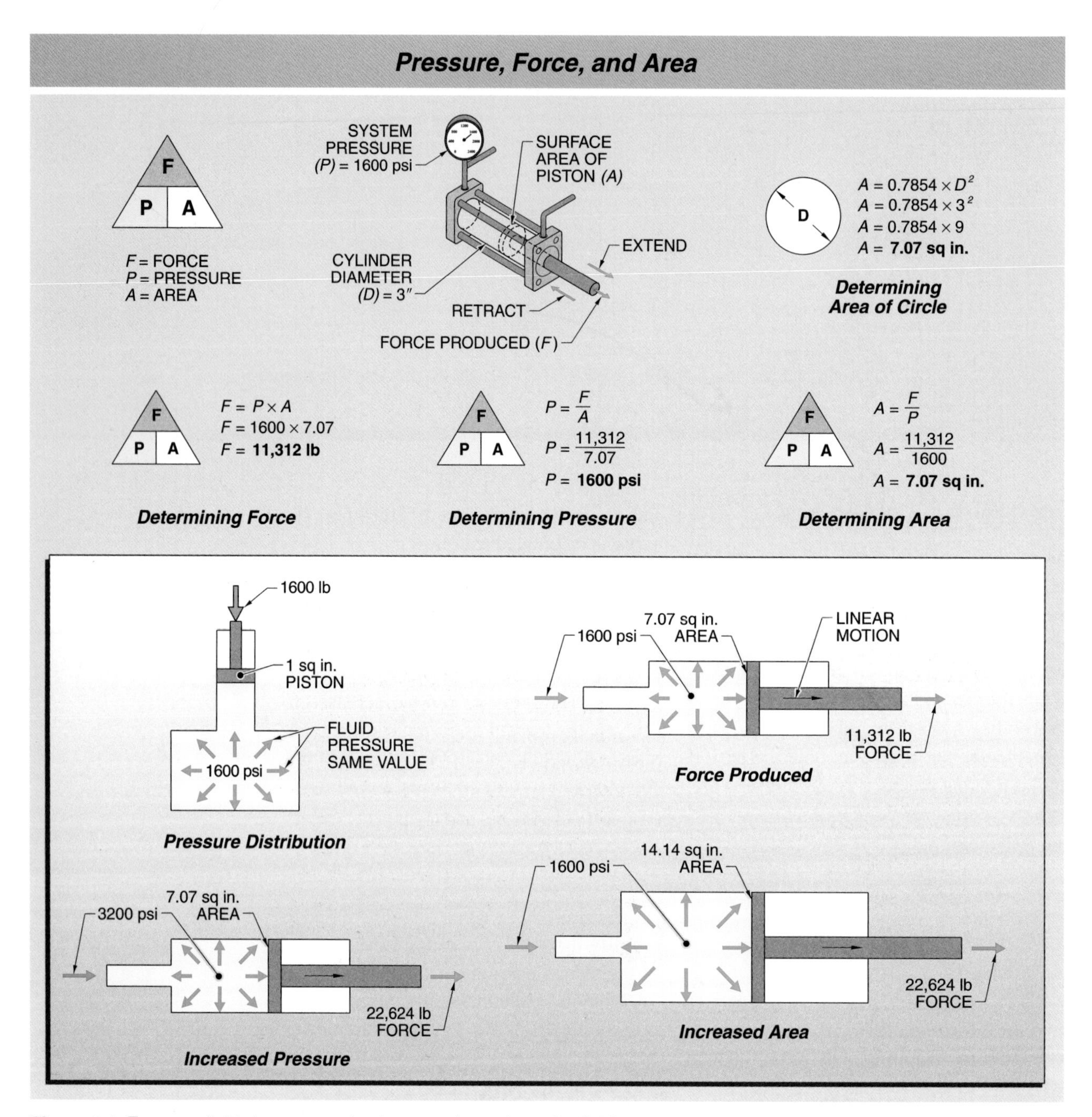

Figure 9-1. Force available in a pressurized system depends on the fluid pressure and the area over which the force is applied.

Atmospheric Pressure. *Atmospheric pressure* is the pressure due to the weight of the atmosphere above the point where it is measured. **See Figure 9-2.** Atmospheric pressure changes at different elevations because at higher elevations there is less weight of air above that elevation than at lower elevations. At mean sea level, the standard pressure of air is 14.696 psi, usually rounded to 14.7 psi. This value is often expressed as 1 atmosphere of pressure. In Denver, Colorado, at 5280′ of elevation, the pressure of air is 12.2 psi. Atmospheric pressure also changes from day to day with changes in the weather.

Atmospheric Pressure

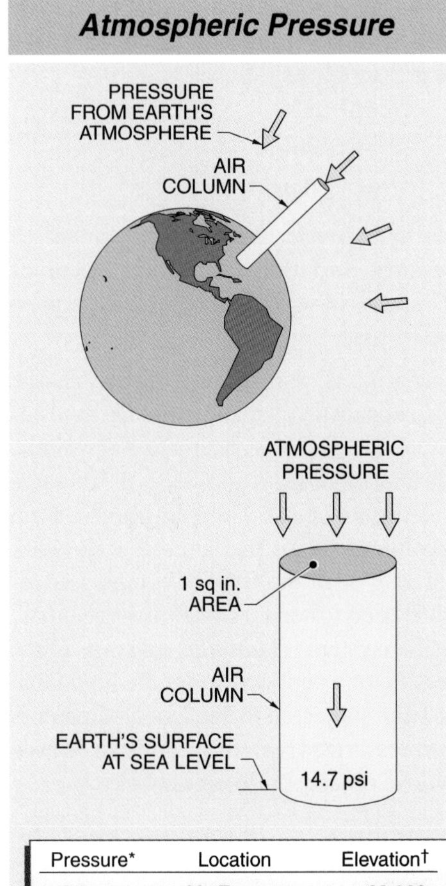

Pressure*	Location	Elevation†
5.0	Mt. Everest	29,002
11.1	Mexico City	7556
12.2	Denver, CO	5280
14.7	Sea level	0

Atmospheric Pressure

* in psi
† in ft

Figure 9-2. Atmospheric pressure is the pressure of air due to the weight of the air from the surface of the earth to the top of the atmosphere.

Head. *Head* is the actual height of a column of liquid. A container or vessel can be any shape, but head is only determined by the height of the liquid. For example, the head of water in water towers of different shapes depends only on the height of the water. **See Figure 9-3.** Head is expressed in units of length such as inches or feet, and includes a statement of which liquid is being used. For example, head may be expressed as feet of water or inches of mercury.

Fluid Head

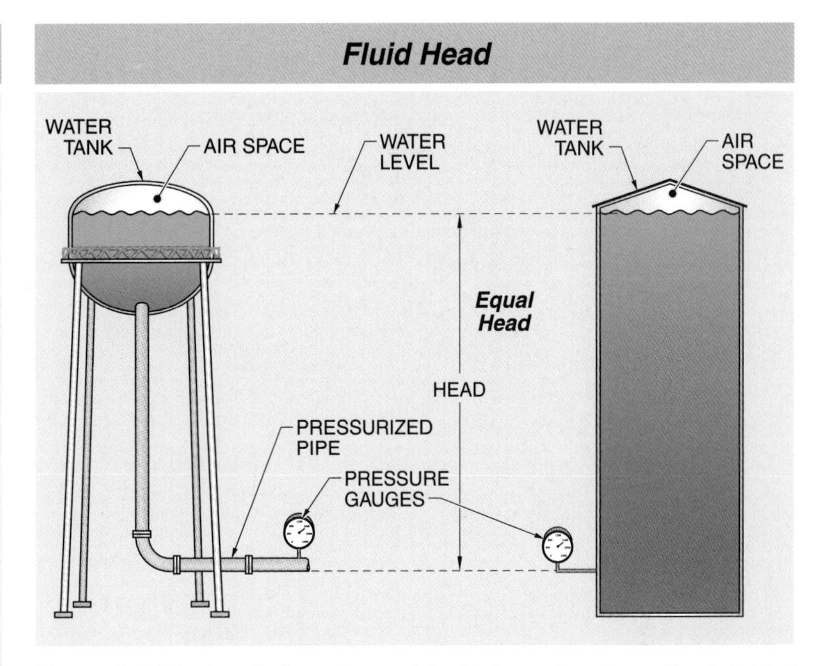

Figure 9-3. The head of a column of liquid depends only on the height of the column, not the shape of the container.

Hydrostatic Pressure. *Hydrostatic pressure* is the pressure due to the head of a liquid column. Frequently, this is referred to as pressure head. Pressure is independent of the shape of the container and depends only on the properties of the fluid and the height. For example, mercury and water have very different densities. Since mercury is much denser than water, a shorter column of mercury produces a hydrostatic pressure equivalent to a much taller column of water. **See Figure 9-4.**

A variable air volume controller uses air pressure measurements to determine the adjustments needed to deliver the required amount of air to a building space.

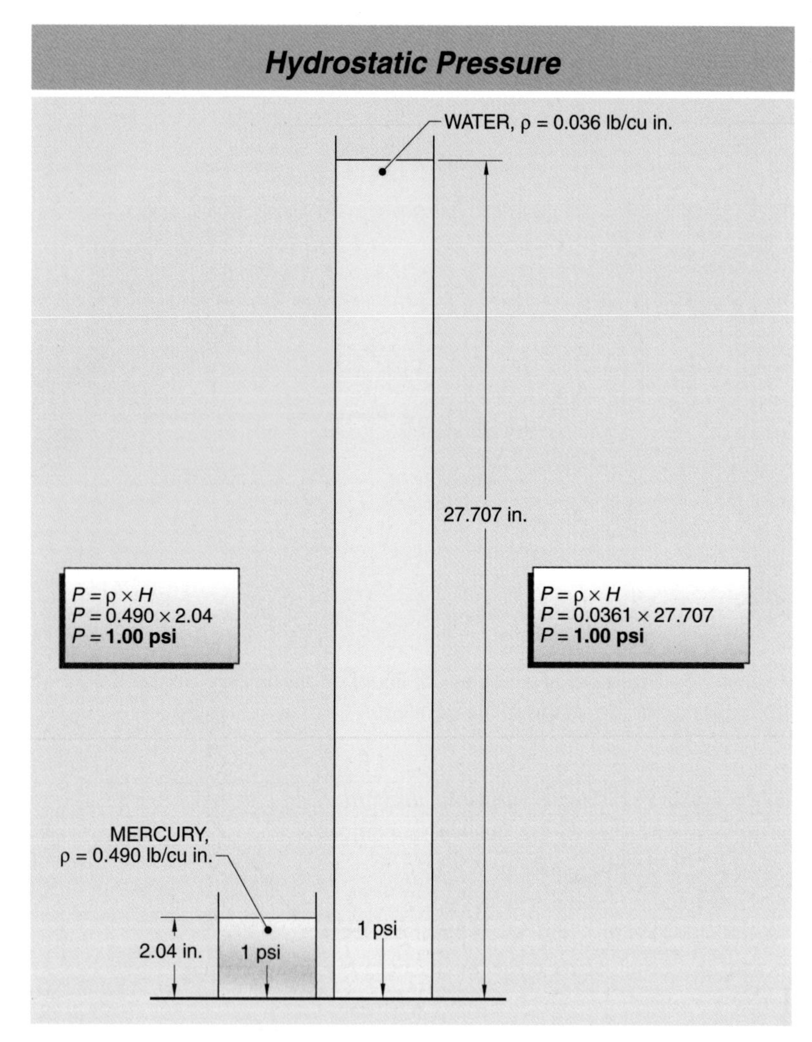

Figure 9-4. Equivalent pressure is obtained by different heights of columns of mercury and water.

The head of a column of liquid can be related to the actual pressure based on the height and density of the liquid by the following formula:

$$P = \rho \times h$$

where

P = pressure (in psi)

ρ = density (in lb per cu in., or lb/cu in.)

h = height (in in.)

For example, the density of water is 0.0361 lb/cu in. If a column of water is 27.7″ high, the pressure is as follows:

$$P = \rho \times h$$

$$P = 0.0361 \times 27.7$$

$$P = \mathbf{1.00\ psi}$$

The density of mercury is 0.490 lb/cu in. If the head of a column of mercury is 2.04″, the pressure is as follows:

$$P = \rho \times h$$

$$P = 0.490 \times 2.04$$

$$P = \mathbf{1.00\ psi}$$

The hydrostatic head of 2.04 inches of mercury and the hydrostatic head of 27.7 inches of water are both equal to a pressure of 1 psi.

Mechanical Pressure. Pressure may also be mechanical energy in the form of a fluid under pressure, such as pneumatic or hydraulic pressure. *Pneumatic pressure* is the pressure of air or another gas. This may be in the form of compressed air that is used to do work or it may be in the form of compressed air that is used to send a signal in a pneumatic control system. *Hydraulic pressure* is the pressure of a confined hydraulic liquid that has been subjected to the action of a pump. The pressurized hydraulic fluid can be used to move objects or do other work.

Pascal's Law

Pascal's law is a law stating that the pressure applied to a confined static fluid is transmitted with equal intensity throughout the fluid. A hydraulic press shows the application of Pascal's law.

For example, two cylinders of different diameters are filled with hydraulic fluid and fitted with pistons. **See Figure 9-5.** The cylinders are connected in a way that when the pressure is increased in the small pump cylinder, the pressure also increases in the large load cylinder. The force on the larger piston is much larger than the force on the small piston because the increased pressure is exerted over the large surface area of the load piston. This magnifies the force applied to the small cylinder. However, the large cylinder moves proportionally less.

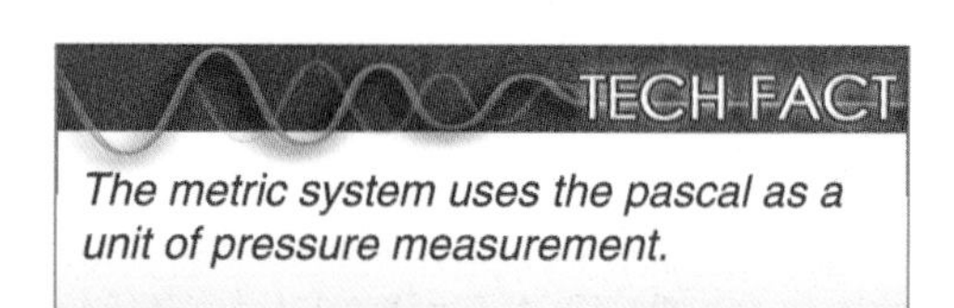

The metric system uses the pascal as a unit of pressure measurement.

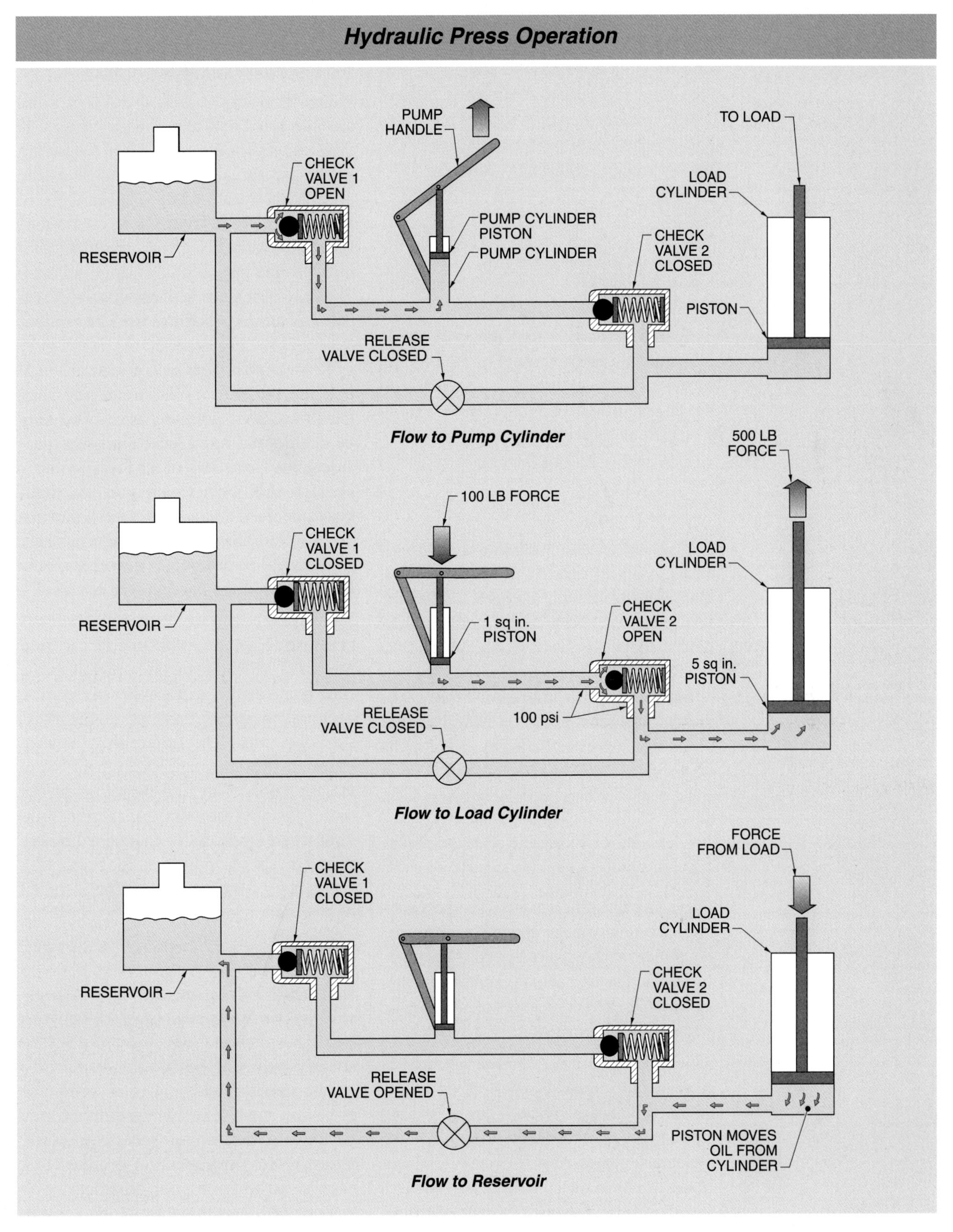

Figure 9-5. Hydraulic presses amplify force through the application of Pascal's law.

For the first step, the pump handle is raised, opening check valve 1 and withdrawing fluid from the reservoir. In the next step, the arm is pushed down with a force of 100 lb onto an area of 1 sq in. This closes check valve 1 and opens check valve 2. The press pressure created is as follows:

$$P = \frac{F}{A}$$

$$P = \frac{100}{1}$$

$$P = \mathbf{100\ psi}$$

Pascal's law says that this pressure is transmitted equally throughout the piping to the larger load cylinder. The force on the larger piston having an area of 5 sq in. would be as follows:

$$F = P \times A$$

$$F = 100 \times 5$$

$$F = \mathbf{500\ lb}$$

These steps are repeated as many times as necessary to raise the load piston more than is possible with one cycle. After the load piston has been raised to the desired height, the release valve is opened and the fluid returns to the reservoir.

However, when the pistons move, the distance the smaller piston travels is proportionately greater than the distance the larger piston travels, satisfying the law of conservation of energy. If the smaller piston moves 5″, the larger one moves 1.0″.

Pressure Scales

There are many different ways to report pressure, depending on the application. Pressure is reported in many units as well as on different scales. The four common pressure scales are absolute, gauge, vacuum, and differential pressure. **See Figure 9-6.**

Absolute Pressure. *Absolute pressure* is pressure measured with a perfect vacuum as the zero point of the scale. When measuring absolute pressure, the units increase as the pressure increases. Absolute pressure cannot be less than zero and is unaffected by changes in atmospheric pressure. Certain equations that relate pressure to other variables call for the use of absolute pressure. *Absolute zero pressure* is a perfect vacuum. Absolute zero pressure cannot be reached in practice.

Gauge Pressure. *Gauge pressure* is pressure measured with atmospheric pressure as the zero point of the scale. When measuring gauge pressure, the units increase as the pressure increases. *Negative gauge pressure* is gauge pressure that is less than atmospheric pressure. Negative gauge pressure indicates the presence of a partial vacuum. The only difference between absolute pressure and gauge pressure is the zero point of the scale.

If a vessel is kept at constant absolute pressure, the gauge pressure can vary when the atmospheric pressure varies. This may be significant if very accurate pressure measurements are needed or the measurements are made at different locations or elevations. For example, if a process requires a particular absolute pressure, the gauge pressure reading will be different if the process is in Denver than if the process is at sea level.

Vacuum Pressure. *Vacuum pressure* is pressure less than atmospheric pressure measured with atmospheric pressure as the zero point of the scale. When measuring vacuum, the units increase as the pressure decreases. The differences between absolute pressure and vacuum pressure are the zero point and direction of the scale. Vacuum pressure measurement is used when a process is maintained at less than atmospheric pressure. For example, a vacuum pressure gauge may be installed on the suction side of a pump to check for a clogged suction line, a dirty strainer, or a closed suction valve.

Differential Pressure. *Differential pressure* is the difference in pressure between two measurement points in a process. The actual pressure at the different points may not be known and there is no reference pressure used. The two pressures may be above or below atmospheric pressure. *Pressure drop* is a pressure decrease that occurs due to friction or obstructions as an enclosed fluid flows from one point in a process to another.

A pressure drop measurement can significantly improve the measurement resolution when compared to using two gauges or absolute pressure measurement. For example, when air is filtered in an HVAC system, the air pressure before a filter is higher than the air pressure after a filter. The pressure drop is very small compared to the absolute pressure, so the pressure drop is monitored to determine when the filter needs to be cleaned or replaced. **See Figure 9-7.**

Units of Pressure Measurement

Pressure is weight or force divided by area. Therefore, the units of pressure often include units of weight or force and units of area. Common units of pressure include pounds per square inch (psi), kilograms per square meter (kg/m^2), newtons per square meter (N/m^2), kilopascals (kPa), or bars (100 kPa). In addition, the equivalent pressure can be given in terms of fluid head. A technician often sees pressure given in inches of water column (in. WC), inches of mercury (in. Hg), millimeters of mercury (mm Hg), or torrs (1 mm Hg absolute). **See Figure 9-8.**

Sometimes the pressure is reported in atmospheres, where the measurement is a multiple of the absolute pressure of standard atmospheric pressure. Adding the word absolute, gauge, or differential, or the letters *a*, *g*, or *d* after the units makes the distinction between absolute pressure, gauge pressure, and differential pressure measurements.

For example, when using pounds per square inch as the pressure measurement, it is reported as psia, psig, or psid. A pressure measurement given in psi, without the letters *a*, *g*, or *d* after the unit, indicates that the unit is psig. Many useful relationships can be determined from this figure. For example, a 1′ column of water has a pressure of 0.433 psi. An equivalent statement is that water has a hydrostatic pressure of 0.433 psi/ft of height.

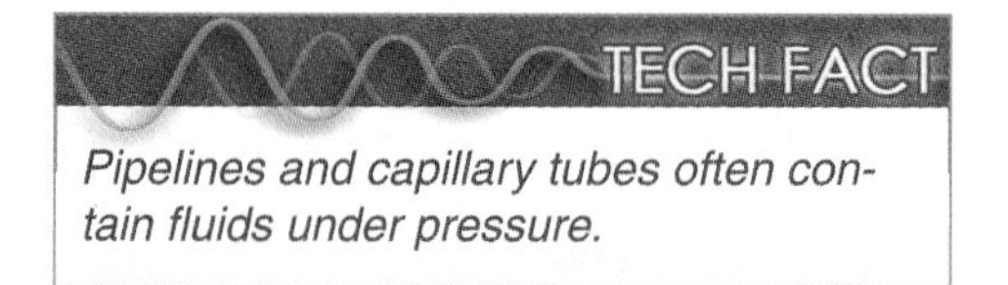

Pipelines and capillary tubes often contain fluids under pressure.

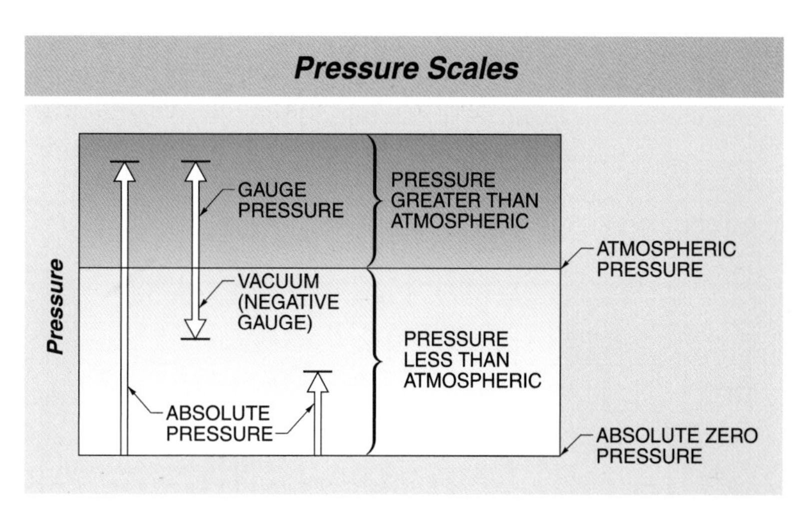

Figure 9-6. Absolute pressure, gauge pressure, and vacuum pressure are different pressure measurement scales.

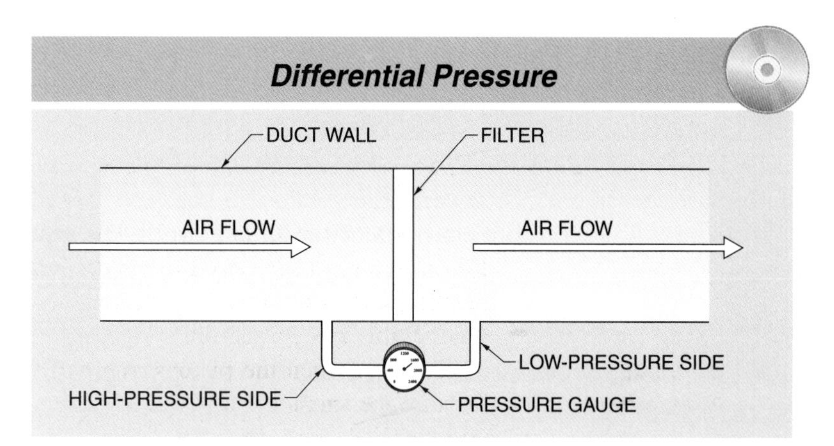

Figure 9-7. Pressure drop in an HVAC system is monitored to determine when the filter needs to be cleaned or replaced.

Air must be compressed to be used in instrumentation systems.

Pressure Equivalents

	kg per sq cm	lb per sq in.	atm	bar	in. of Hg	kilopascals	in. of water	ft of water
kg per sq cm	1	14.22	0.9678	0.98067	28.96	98.067	394.05	32.84
lb per sq in.	0.07031	1	0.06804	0.06895	2.036	6.895	27.7	2.309
atm	1.0332	14.696	1	1.01325	29.92	101.325	407.14	33.93
bar	1.01972	14.5038	0.98692	1	29.53	100	402.156	33.513
in. of Hg	0.03453	0.4912	0.03342	0.033864	1	3.3864	13.61	11.134
kilopascals	0.0101972	0.145038	0.0098696	0.01	0.2953	1	4.02156	0.33513
in. of water	0.002538	0.0361	0.002456	0.00249	0.07349	0.249	1	0.0833
ft of water	0.03045	0.4332	0.02947	0.029839	0.8819	2.9839	12	1

1 ounce/sq in. = 0.0625 lb/sq in.

Note: use multiplier at convergence of row and column

Figure 9-8. Pressure is measured in many units. Conversions from one to another are sometimes necessary.

KEY TERMS

- *pressure:* Force divided by the area over which that force is applied.
- *force:* Anything that changes or tends to change the state of rest or motion of a body.
- *area:* The number of unit squares equal to the surface of an object.
- *fluid:* Any material that flows and takes the shape of its container.
- *atmospheric pressure:* The pressure due to the weight of the atmosphere above the point where it is measured.
- *head:* The actual height of a column of liquid.
- *hydrostatic pressure:* The pressure due to the head of a liquid column.
- *pneumatic pressure:* The pressure of air or another gas.
- *hydraulic pressure:* The pressure of a confined hydraulic liquid that has been subjected to the action of a pump.
- *Pascal's law:* A law stating that the pressure applied to a confined static fluid is transmitted with equal intensity throughout the fluid.
- *absolute pressure:* Pressure measured with a perfect vacuum as the zero point of the scale.
- *absolute zero pressure:* A perfect vacuum.
- *gauge pressure:* Pressure measured with atmospheric pressure as the zero point of the scale.
- *negative gauge pressure:* Gauge pressure that is less than atmospheric pressure.
- *vacuum pressure:* Pressure less than atmospheric pressure measured with atmospheric pressure as the zero point of the scale.
- *differential pressure:* The difference in pressure between two measurement points in a process.
- *pressure drop:* A pressure decrease that occurs due to friction or obstructions as an enclosed fluid flows from one point in a process to another.

REVIEW QUESTIONS

1. Define pressure and describe the relationship between pressure, force, and area.
2. Define atmospheric pressure, head, and hydrostatic pressure.
3. What are the types of mechanical pressure sources, and how can they be used?
4. What is Pascal's law, and how can it be applied?
5. List and describe four common pressure scales.

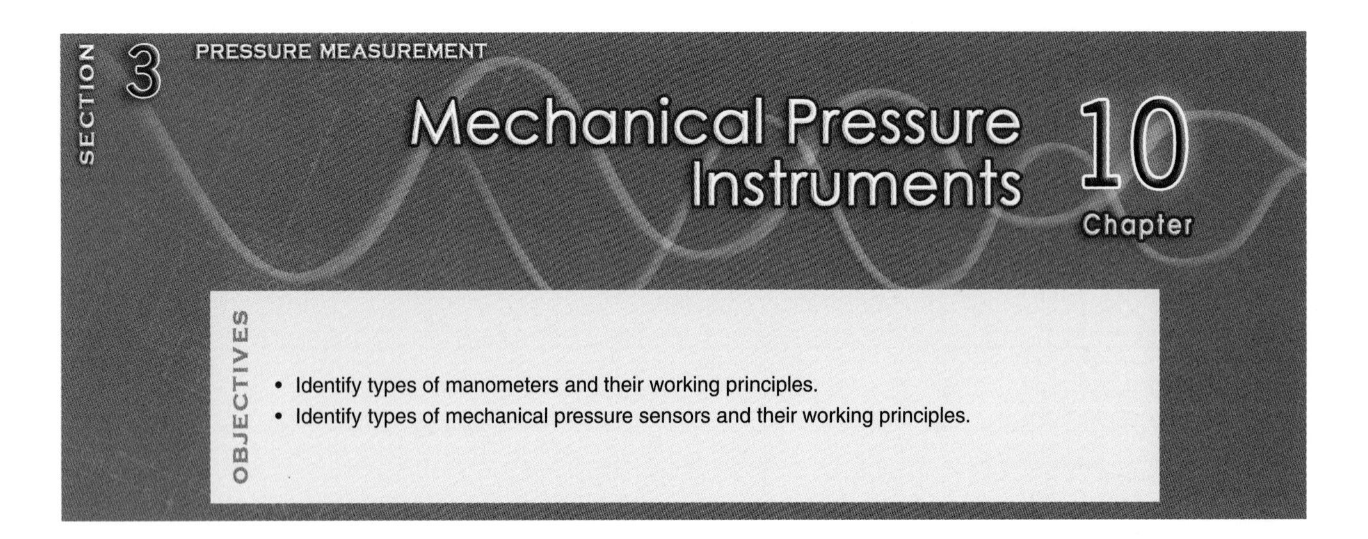

MANOMETERS

A *manometer* is a device for measuring pressure with a liquid-filled tube. A manometer is the simplest device for measuring pressure. In a manometer, a fluid under pressure is allowed to push against a liquid in a tube. The movement of the liquid is proportional to the pressure. Water is the most common liquid used in manometers. Other liquids may be used as long as the densities are known. However, the EPA discourages the use of mercury. The common types of manometers are the U-tube manometer, inclined tube manometer, well-type manometer, and the barometer.

U-Tube Manometers

A *U-tube manometer* is a clear tube bent into the shape of an elongated letter U. Many U-tube manometers have a graduated scale placed in the center between the vertical columns, or legs, with the scale markings increasing above and below the zero point. Other manometers have a scale that can be adjusted up or down to make it easier to set the zero point on the scale at the manometer liquid level.

Liquid is poured into the tube of the manometer until the level in both vertical columns is at the middle of the scale or zero. In operation, a pressure is applied to one of the columns and the other column is left open to atmospheric pressure. The level in the higher-pressure side decreases and the level in the lower-pressure side increases. The difference in the height of the two liquid columns represents the applied pressure. **See Figure 10-1.**

Manometer Operation. In operation, different pressures are applied to the two halves of the U-tube. One column may have atmospheric pressure applied while the other column is connected to the process. This is used to measure gauge pressure. For example, to measure the draft of a boiler, the manometer has one end of the tube connected to the breeching or the flue and the other column open to air to measure the pressure difference. If the manometer liquid is water and the applied pressure causes the water to rise 1½″ on one side and to fall 1½″ on the other side, the total height, or head, of the water is 3″. Care must be taken when using U-tube manometers. Overpressure can cause the liquid to spill out the end of the tube.

U-Tube Manometer Sizing. U-tube manometers are available in lengths from 6″ to 60″, with the most common sizes being 15″ and 30″. In addition to the various manometer lengths that are available, manufacturers offer manometer fluids with a choice of densities. The manufacturers identify these different fluids by their specific gravity. *Specific gravity* is the ratio of the density of a fluid to the density of a reference fluid. Water is the usual reference fluid for liquids.

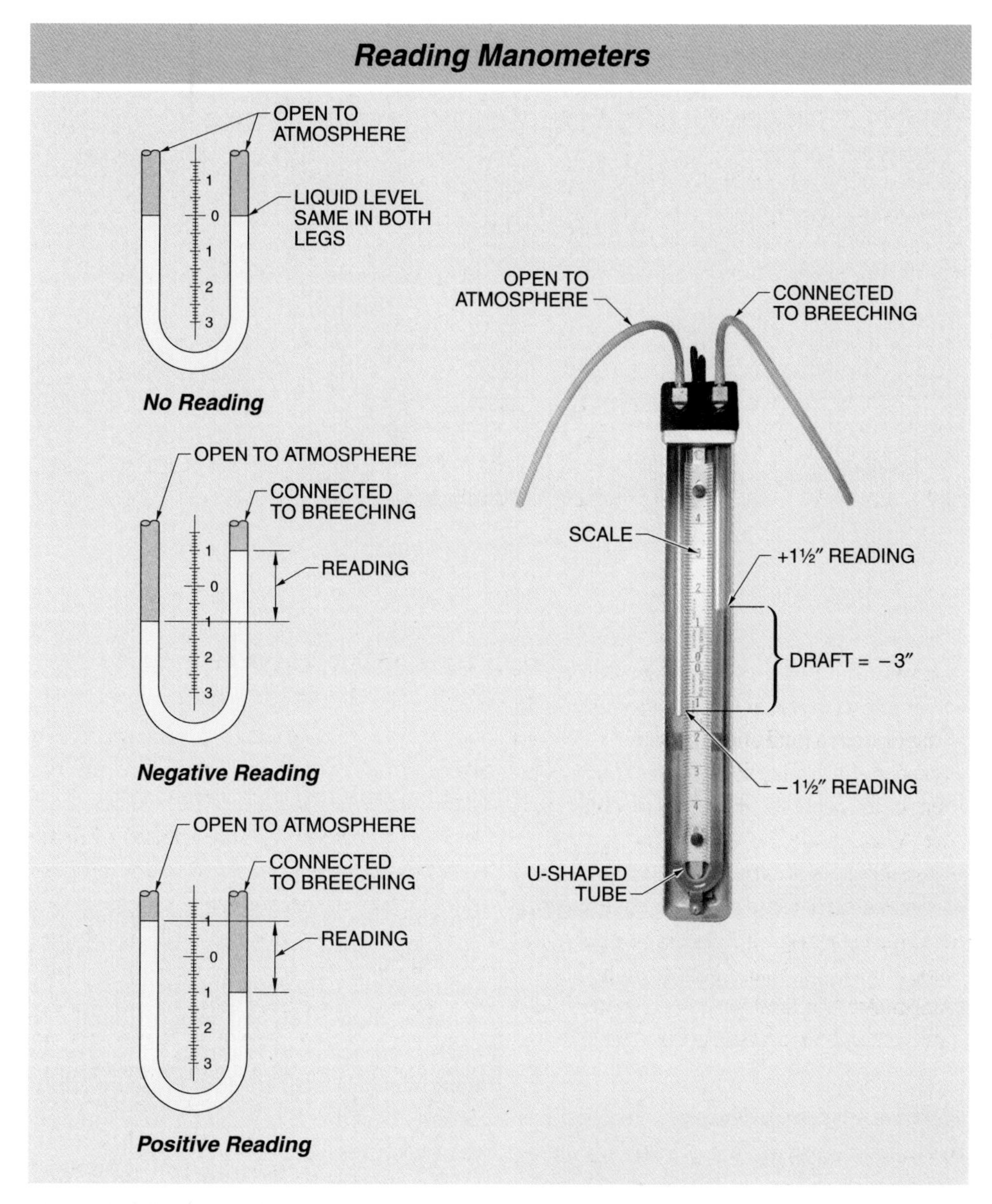

Figure 10-1. A glass tube is bent into a U-tube manometer. The applied pressure is determined by measuring the difference between the liquid level in the two legs.

Typical choices of specific gravities for manometer fluids are 0.826, 1.000, 1.750, and 2.950. All of these manometer fluids are organic chemicals and immiscible with water. A dye is often added to improve readability. The different combinations of manometer length and manometer fluid make it easier to choose a combination that will satisfy the required pressure measurement range application.

Inclined-Tube Manometers

An *inclined-tube manometer* is a manometer with a reservoir serving as one end and the measuring column at an angle to the horizontal to reduce the vertical height. The fill liquid is usually water. **See Figure 10-2.** Since an inclined-tube manometer must be mounted level to the ground, a bubble level or a screw-type leveling adjustment is often included with an inclined-tube manometer.

Inclined-Tube Manometers

LEVELING BUBBLE
PRESSURE APPLIED
OPEN TO ATMOSPHERE

© Dwyer Instruments, Inc.

Figure 10-2. An inclined-tube manometer has one column at an angle to lengthen the scale and improve readability.

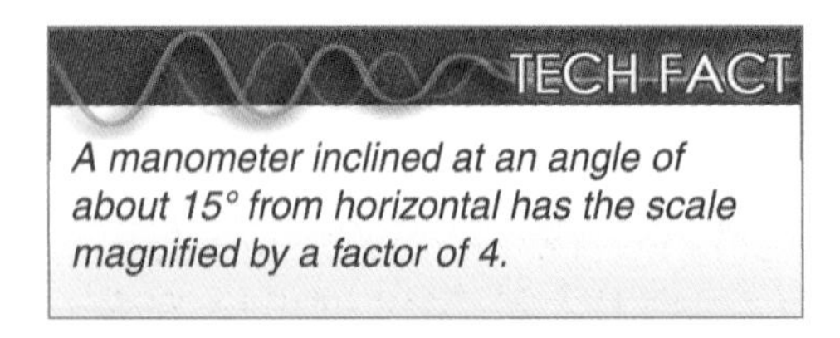

TECH FACT

A manometer inclined at an angle of about 15° from horizontal has the scale magnified by a factor of 4.

Many pressure measurements are used during boiler operation.

The purpose of the angled tube is to lengthen the scale for easier reading. For example, when the angle is 30°, 1 vertical inch becomes 2″ on the inclined scale. The scale is then stretched proportionally to enable an accurate reading. This type of manometer is used for low-pressure applications because it is difficult to accurately read low pressures in a vertical tube.

For example, an HVAC system may have a static pressure drop of only 0.1 to 0.2 inches of water or a boiler furnace draft may have a pressure differential of about 4 to 6 inches of water. Under these circumstances, it is easier to get an accurate reading with an inclined-tube manometer than with a standard U-tube manometer.

Even the smallest collection of condensed water in an inclined-tube manometer can generate very significant measurement errors. This occurs by either increasing the total manometer fluid volume and changing the zero point, or by applying an additional head to the reservoir and changing the differential measurement.

Well-Type Manometers

A *well-type manometer* is a manometer with a vertical glass tube connected to a metal well, with the measuring liquid in the well at the same level as the zero point on the tube scale. **See Figure 10-3.** The pressure to be measured is applied to the well side and the reading taken on the tube side. A well-type manometer is simpler to read than a U-type manometer since only one measurement is required.

An applied pressure will push down the liquid level in the well and the liquid level in the tube will rise just as in a U-tube manometer. In a well-type manometer, the well has a much larger surface area than the area of the tube. Therefore, the level in the tube changes proportionally more than the level in the well. The actual applied pressure is equal to the difference between the increased level of the liquid in the column and the decreased level of the liquid in the well. Since it is difficult to read a change in level of the liquid in the well, all measurements are taken on the tube side of the manometer and a compensated measurement scale is used.

Well-type manometers are affected by condensed water, but the errors are usually small due to the larger reservoir area. The larger reservoir area distributes the collected condensate so that the head is less. Continual collection of condensate would eventually cause serious errors in the readings. An additional problem is that the collected condensate cannot be seen in the reservoir.

Well Drop. *Well drop* is the ratio of the area of a well-type manometer tube to the area of a well. The measurement scale is decreased in proportion to the well drop. A reading of 1″ on the scale corresponds to 1″ of head, but the scale measurement is actually less than 1″ to compensate for the movement of the level in the well.

For example, the inside area of the well may be 7.00 sq in. and the inside area (bore) of the tube may be 0.25 sq in. This means that an applied pressure to the well that moves the liquid down 0.25″ raises the column by 7.00″ above the new level in the well. The height of the column above the original zero point is 6.75″.

In this case, the well drop is 0.25 ÷ 7.0 = 0.036. The measurement scale is reduced by a factor of 0.036. On this compensated scale, the scale is reduced by 0.036 inch for every inch of height to compensate for the well drop.

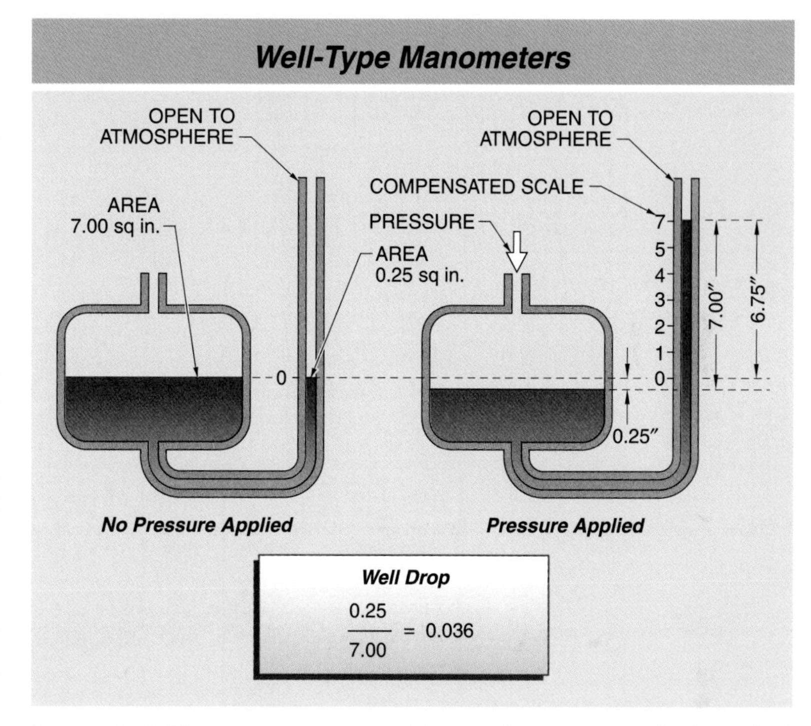

Figure 10-3. The pressure measured by a well-type manometer is applied to the well and the movement of the liquid in the column is magnified.

Barometers

A *barometer* is a manometer used to measure atmospheric pressure. *Barometric pressure* is a pressure reading made with a barometer. The earliest barometer was a long vertical glass tube that had been sealed at the bottom and filled with mercury. The open end was then turned upside down into a container of mercury without allowing any air into the tube. **See Figure 10-4.** The mercury in the tube falls to a level where the head of the mercury is equal to the atmospheric pressure. When the atmospheric pressure changes, the level of the mercury changes. A scale along the tube indicates the height of the column of mercury, equal to the barometric pressure, measured in inches or millimeters.

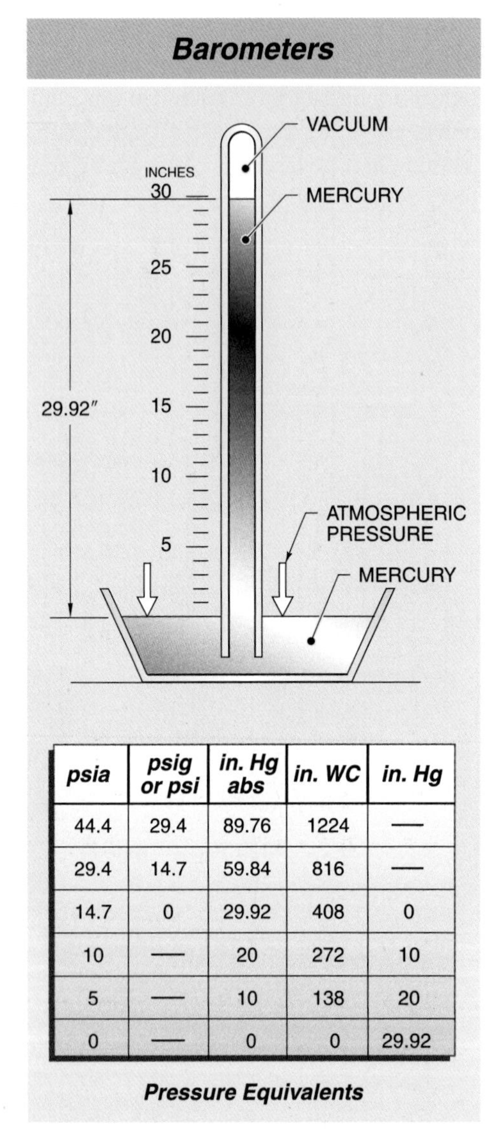

psia	psig or psi	in. Hg abs	in. WC	in. Hg
44.4	29.4	89.76	1224	—
29.4	14.7	59.84	816	—
14.7	0	29.92	408	0
10	—	20	272	10
5	—	10	138	20
0	—	0	0	29.92

Figure 10-4. A barometer is used to measure atmospheric pressure.

Barometric instruments are primarily used in weather stations, airports, or any other facility interested in weather observations. A thermometer is often attached to the tube to allow for temperature compensation.

An *aneroid barometer* is an instrument consisting of a mechanical pressure sensor with a linkage to a pointer that is used to measure atmospheric pressure without the use of a liquid manometer. Such instruments are often provided with electronic circuitry that can produce digital readouts and output signals for remote reading.

MECHANICAL PRESSURE SENSORS

The most common type of instrument uses, in one form or another, an elastic deformation element. An *elastic deformation element* is a device consisting of metal, rubber, or plastic components such as diaphragms, capsules, pressure springs, or bellows that flex, expand, or contract in proportion to the pressure applied within them or against them. Some elements are attached to mechanical linkages used in gauges, switches, or controllers. Other elements are combined with electrical components that convert the deformation into signals for indication, transmission, and control.

There are several mechanical devices used for continuous pressure measurement. Most of the mechanical sensors are elastic deformation pressure elements. In mechanical designs, the movement of the pressure element is transferred through a mechanical linkage that moves a pointer in a gauge. In electrical systems, a transducer converts the movement of the pressure element to an electrical signal that can be used for control. A *transducer* is a device that converts one form of energy to another, such as converting pressure to voltage.

Diaphragms

A *diaphragm* is a mechanical pressure sensor consisting of a thin, flexible disc that flexes in response to a change in pressure. Pressure-sensing diaphragms are commonly made of steel, stainless steel, titanium, beryllium copper, bronze, rubber, or other materials. A diaphragm is usually secured at its outer edges between matching base plates resembling flanges. A spring acting on the center of the diaphragm may be used to provide the counteracting force to the applied pressure. Diaphragms vary in thickness and area and may be flat or convoluted to provide greater displacement. The larger ones are made of thinner materials for lower pressures and the smaller ones are made of thicker materials for higher pressures. **See Figure 10-5.**

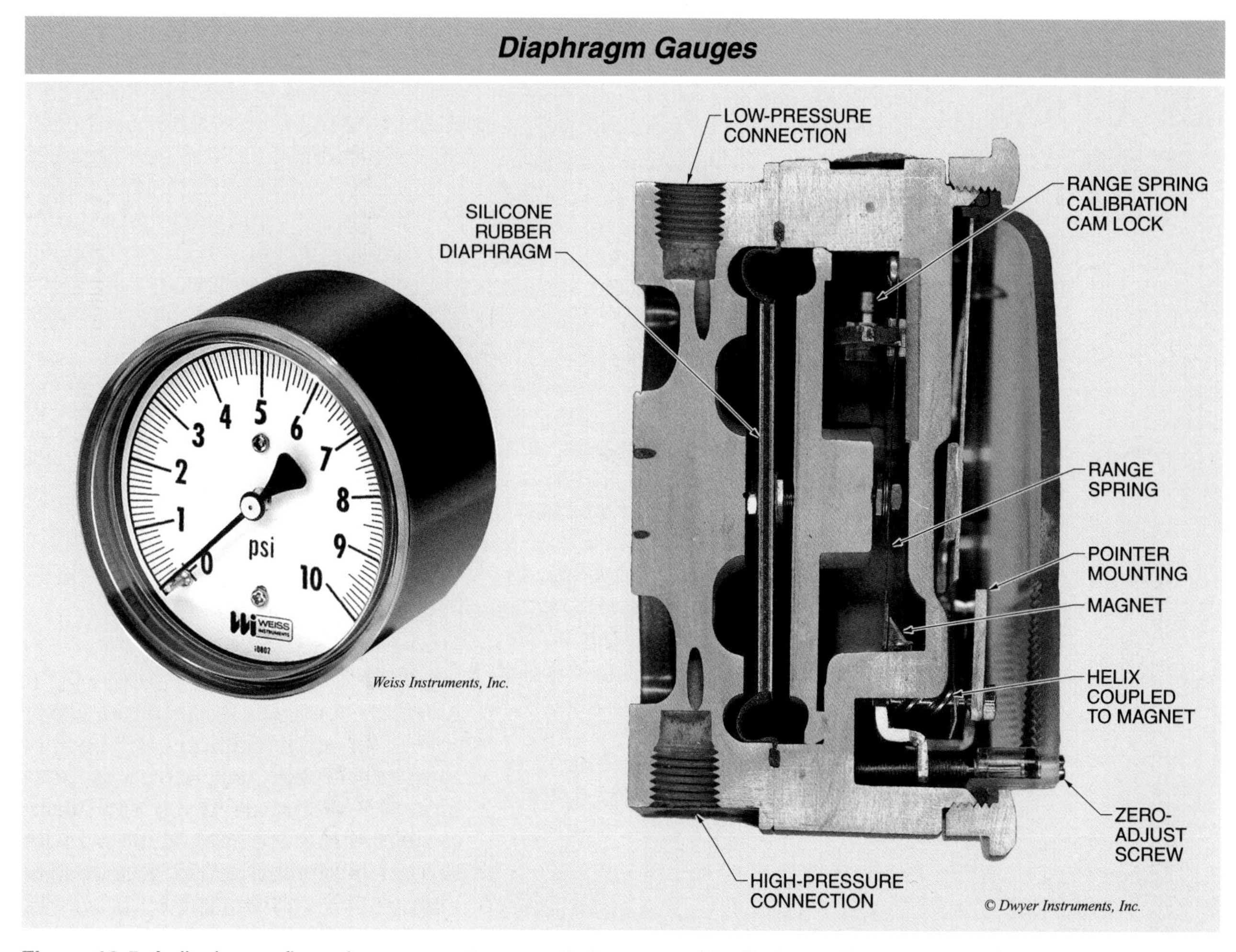

Figure 10-5. A diaphragm flexes in response to an applied pressure. The flexing motion moves a pointer.

Capsules

A *capsule* is a mechanical pressure sensor consisting of two convoluted metal diaphragms with their outer edges welded, brazed, or soldered together to provide an empty chamber between them. One of the diaphragms is connected at its center to metal tubing that admits fluid to the chamber. The other diaphragm is fitted with a mechanical connection to the indicator, or fitted with a transducer. Capsules may be used singly or stacked to provide greater travel. **See Figure 10-6.**

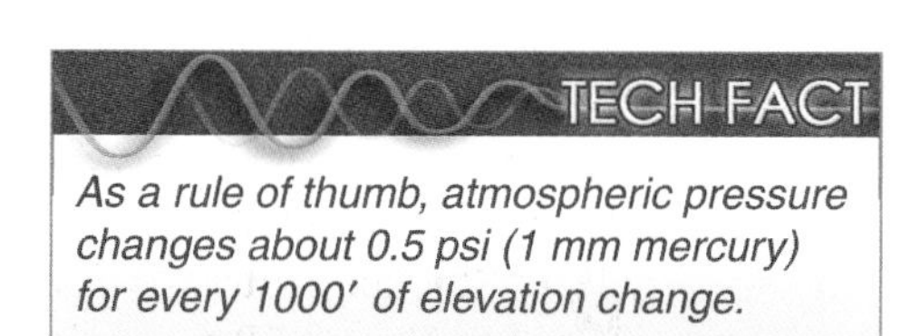

As a rule of thumb, atmospheric pressure changes about 0.5 psi (1 mm mercury) for every 1000′ of elevation change.

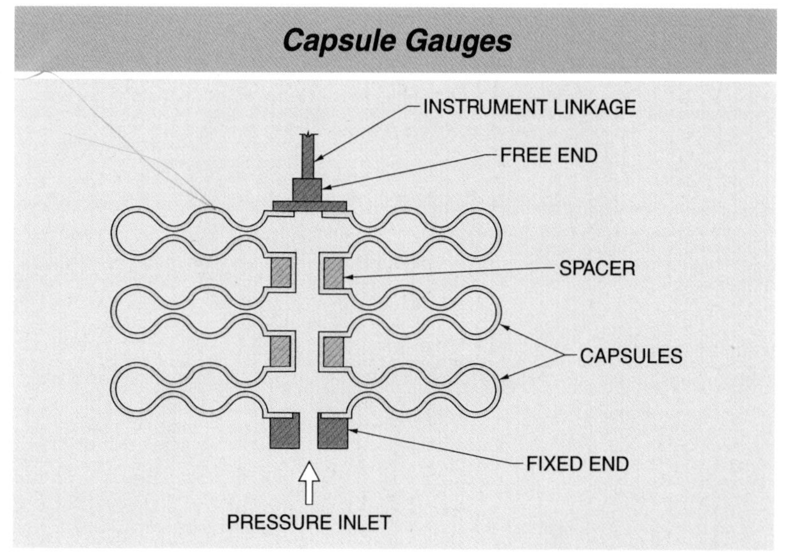

Figure 10-6. A capsule gauge has two diaphragms attached together to provide an empty chamber between them. Capsules may be stacked to amplify the movement.

Pressure Springs

A *pressure spring* is a mechanical pressure sensor consisting of a hollow tube formed into a helical, spiral, or C shape. To construct a pressure spring, a tube made from flattened, seamless metal tubing is formed into the desired shape. Beryllium copper, steel, and stainless steel are commonly used materials. The spring is welded, brazed, or flanged to metal tubing that admits the pressurized fluid from the process. The wall thickness determines the maximum pressure for a pressure spring.

The Bourdon tube is the original pressure spring and is named after its inventor. A Bourdon tube is a C-shaped tube that is flattened into an elliptical cross section. As the tube is filled with pressurized fluid, the increasing pressure tends to return the spring to its original circular cross section. This straightens the tube and causes movement of the end of the spring. Bourdon tubes are available to measure pressures up to 20,000 psi. A pressure spring can easily be used as a pressure switch by adding a linkage to a switching device. **See Figure 10-7.**

Bourdon Tube Pressure Switches

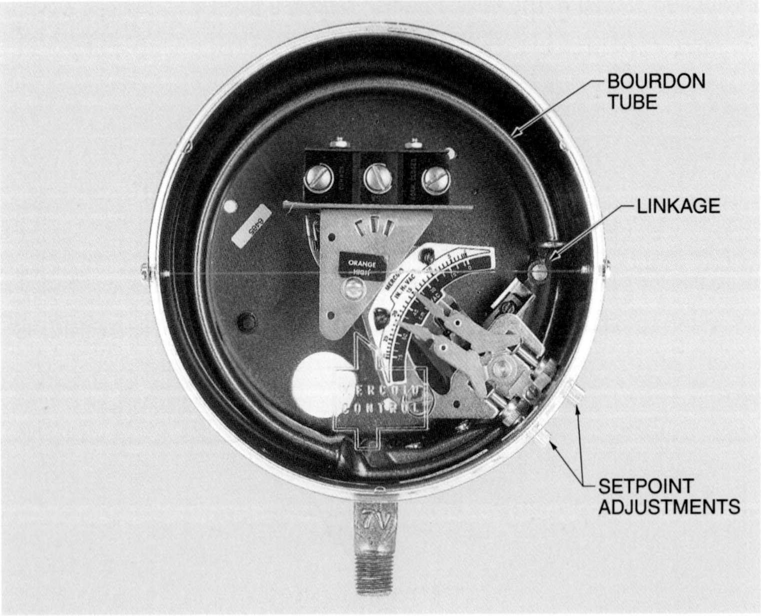

© Dwyer Instruments, Inc.

Figure 10-7. A Bourdon tube pressure switch activates a circuit at a pressure setpoint.

Another model of the Bourdon tube encloses the tube in a chamber that admits a second pressurized fluid. This model is used to make differential pressure measurements. The low-pressure fitting may be connected to atmospheric pressure to give gauge pressure, or to another process pressure to give differential pressure. The surrounding pressure in the chamber opposes the pressure inside the Bourdon tube.

This arrangement makes it possible for the movement of the Bourdon tube to be the only mechanical action necessary to provide the indication of differential pressure. The rotational motion provided by the mechanism attached to the Bourdon tube is magnetically linked to the meter pointer. **See Figure 10-8.**

Bellows

A *bellows* is a mechanical pressure sensor consisting of a one-piece, collapsible, seamless metallic unit with deep folds formed from thin-wall tubing with an enclosed spring to provide stability, or with an assembled unit of welded sections. **See Figure 10-9.** The deflection of a bellows is dependent on its diameter, its thickness, and the material. The free end of the bellows is attached to a linkage that moves a pointer, or it is linked to a transducer.

A design for differential pressure measurement has one bellows as part of a low-pressure chamber and the other bellows as part of a high-pressure chamber. The differential movement of the bellows is linked to a pointer through a rack-and-pinion system.

TECH FACT

Pressure gauges used for steam service can be damaged if live steam comes in contact with the gauge's internal components. A pigtail or U-tube siphon that is filled with water will force the steam to condense and cool. The pressure is transmitted to the gauge without the damaging effects of the live steam.

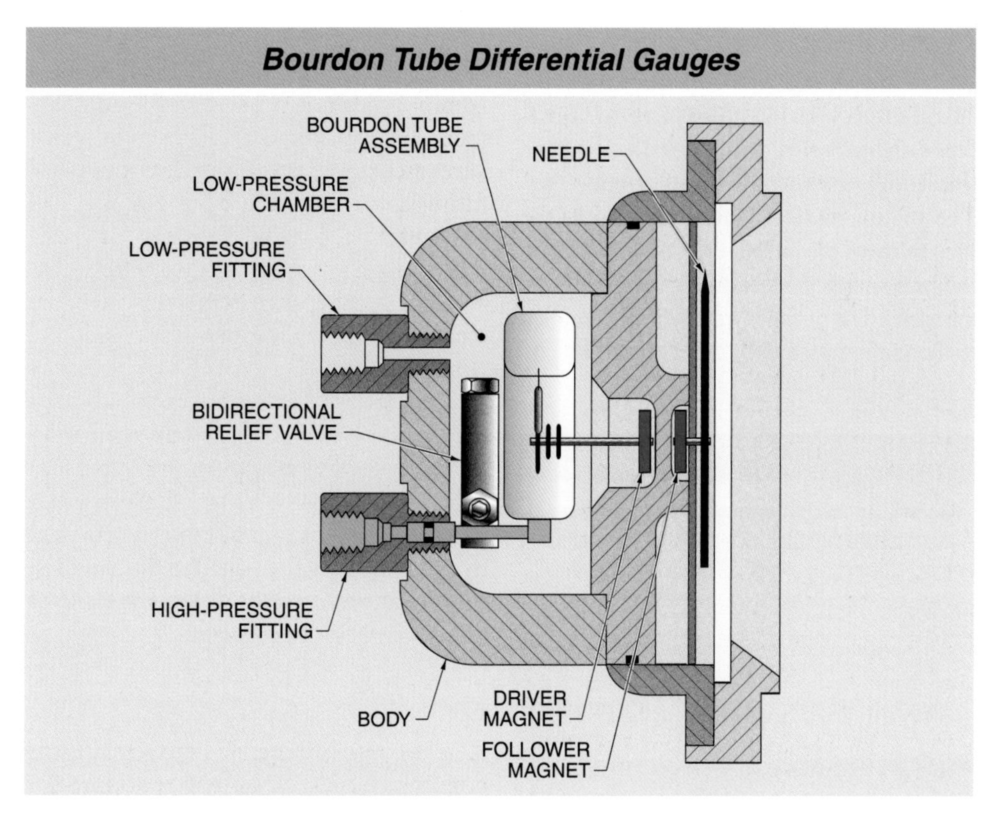

Figure 10-8. A Bourdon tube may be used to measure differential pressure.

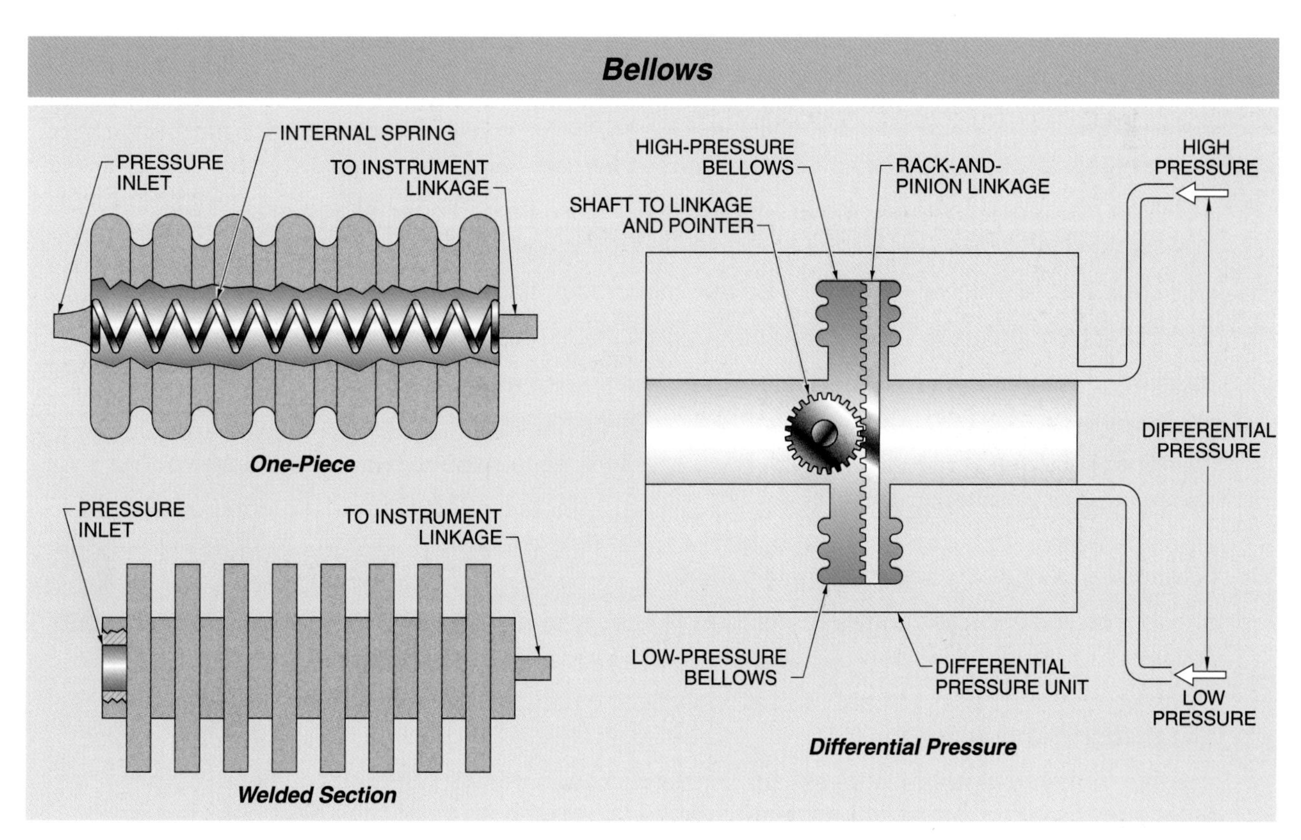

Figure 10-9. A bellows is an elastic deformation element that flexes with changes in pressure.

Double-Ended Pistons

A *double-ended piston* is a mechanical pressure sensor consisting of a differential pressure gauge with a piston that admits pressurized fluid at each end. The piston motion that results from the inequality of the pressures is opposed by an internal spring that establishes the range of the meter. The piston is magnetically coupled to the pointer assembly. **See Figure 10-10.**

A mercury-filled manometer that is used to measure atmospheric pressure needs to be about 30″ high, while a water-filled manometer needs to be about 34′ high.

Double-Ended Piston

HIGH-PRESSURE INLET
MAGNET
RANGE SPRING
LOW-PRESSURE INLET
MAGNET
POINTER ASSEMBLY

Figure 10-10. A double-ended piston is used to measure pressure by balancing the force from the pressure on the piston with the force needed to compress a spring.

KEY TERMS

- *manometer:* A device for measuring pressure with a liquid-filled tube.
- *U-tube manometer:* A clear tube bent into the shape of an elongated letter U.
- *specific gravity:* The ratio of the density of a fluid to the density of a reference fluid.
- *inclined-tube manometer:* A manometer with a reservoir serving as one end and the measuring column at an angle to the horizontal to reduce the vertical height.
- *well-type manometer:* A manometer with a vertical glass tube connected to a metal well, with the measuring liquid in the well at the same level as the zero point on the tube scale.
- *well drop:* The ratio of the area of a well-type manometer tube to the area of a well.
- *barometer:* A manometer used to measure atmospheric pressure.
- *barometric pressure:* A pressure reading made with a barometer.
- *aneroid barometer:* An instrument consisting of a mechanical pressure sensor with a linkage to a pointer that is used to measure atmospheric pressure without the use of a liquid manometer.
- *elastic deformation element:* A device consisting of metal, rubber, or plastic components such as diaphragms, capsules, springs, or bellows that flex, expand, or contract in proportion to the pressure applied within them or against them.
- *transducer:* A device that converts one form of energy to another, such as converting pressure to voltage.
- *diaphragm:* A mechanical pressure sensor consisting of a thin, flexible disc that flexes in response to a change in pressure.
- *capsule:* A mechanical pressure sensor consisting of two convoluted metal diaphragms with their outer edges welded, brazed, or soldered together to provide an empty chamber between them.

KEY TERMS *(continued)*

- *pressure spring:* A mechanical pressure sensor consisting of a hollow tube formed into a helical, spiral, or C shape.
- *bellows:* A mechanical pressure sensor consisting of a one-piece, collapsible, seamless metallic unit with deep folds formed from thin-wall tubing with an enclosed spring to provide stability, or with an assembled unit of welded sections.
- *double-ended piston:* A mechanical pressure sensor consisting of a differential pressure gauge with a piston that admits pressurized fluid at each end.

REVIEW QUESTIONS

1. How does a manometer operate?
2. List and describe the common types of manometers.
3. Why is an inclined-tube manometer more suitable than a standard U-tube manometer for measuring the small pressure differences in an HVAC system?
4. Define elastic deformation element.
5. Describe the common types of mechanical pressure sensors.

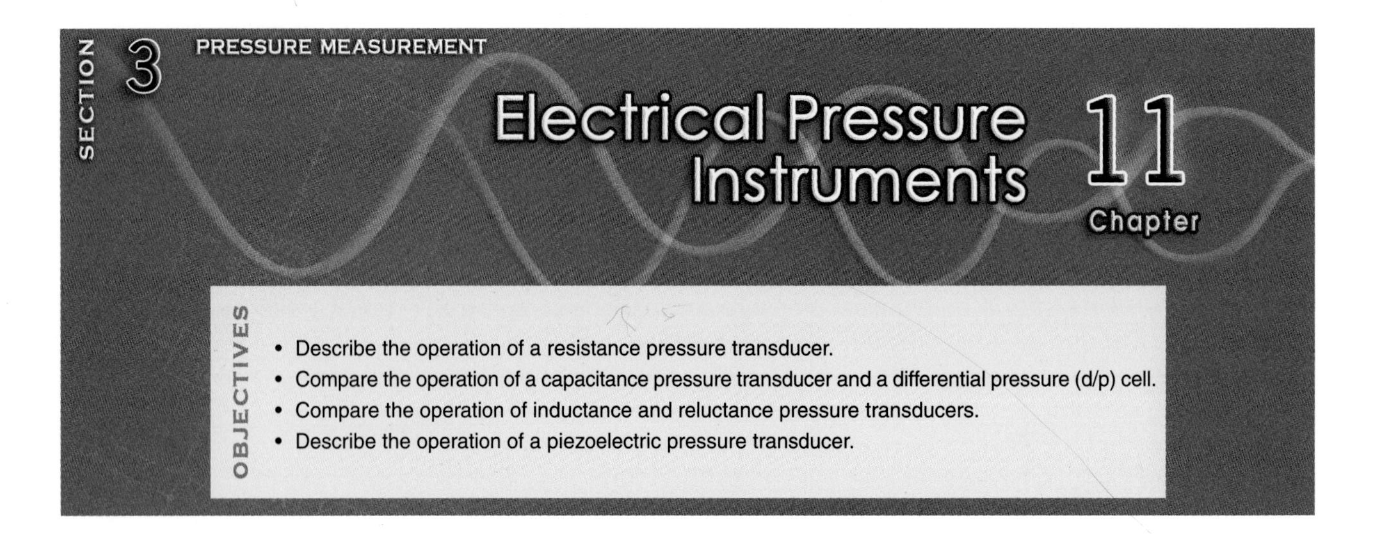

ELECTRICAL TRANSDUCERS

An *electrical transducer* is a device that converts input energy into output electrical energy. A pressure transducer is often used to convert the mechanical displacement of a diaphragm caused by a change in applied external force into an electrical signal. A secondary converter, such as an analog to digital converter, converts the analog electrical signal to a digital signal. However, all secondary converters require a power supply to produce a signal of sufficient strength for conditioning or transmission.

A *pressure transmitter* is a pressure transducer with a power supply and a device that conditions and converts the transducer output into a standard analog or digital output. A *pressure switch* is a pressure-sensing device that provides a discrete output (contact make or break) when applied pressure reaches a preset level within the switch.

Resistance

A *resistance pressure transducer* is a diaphragm pressure sensor with a strain gauge as the electrical output element. Resistance pressure transducers are the most widely used electrical pressure transducers. A *strain gauge* is a transducer that measures the deformation, or strain, of a rigid body as a result of the force applied to the body. **See Figure 11-1.**

A bonded strain gauge consists of a fine wire laid out in the form of a grid on a substrate. The electrical resistance of the wire changes when the grid is stretched or distorted. The amount of resistance change depends on the wire material and diameter and the configuration of the strain gauge.

The grid is attached to a diaphragm and is stretched or distorted when a pressure is applied to the diaphragm. An unbonded strain gauge has a part of the gauge that moves with the force applied, with wires connecting the moving parts. The movement changes the resistance of the wires by stretching them, as in a bonded strain gauge, creating a voltage differential proportional to the applied force. The measuring range of the transducer is determined by the combined characteristics of the diaphragm and the strain gauge.

The two main methods of combining a bonded strain gauge and a diaphragm are with a strain foil gauge and with a thin-film strain gauge. A *strain foil gauge* is a strain gauge that has the wire grid impressed on nonmetallic foil and then the assembly is mechanically bonded to the metal diaphragm. A *thin-film strain gauge* is a strain gauge that has the wire grid sputter-deposited on the diaphragm surface. Other types of strain gauges are available.

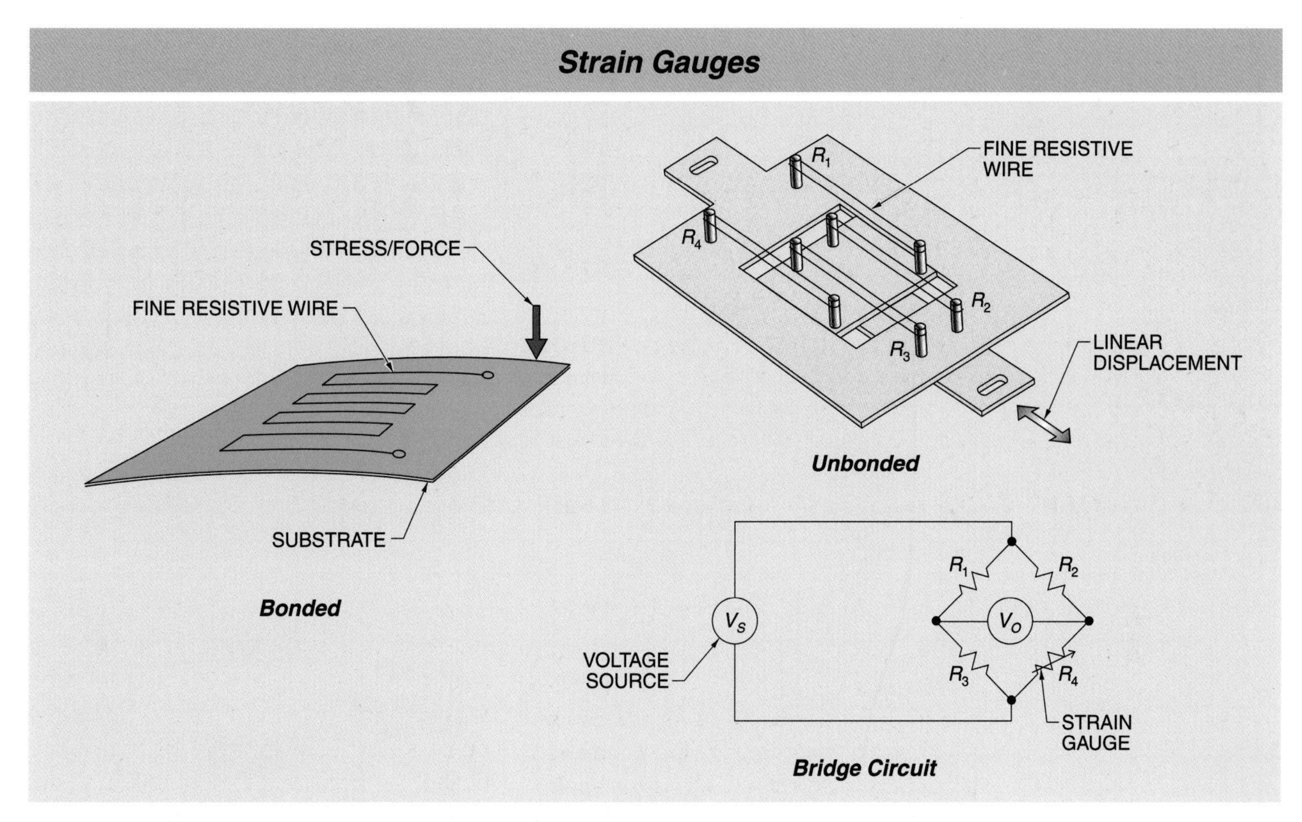

Figure 11-1. A strain gauge is used to measure the deformation of a diaphragm.

It is important that the strain gauge be affected only by the distortion due to applied pressure and not by other effects like the heat of the wire. Therefore, the circuits in which strain gauges are used are low-energy circuits requiring careful installation and maintenance. Strain gauges by themselves are usually rather delicate electrical devices. When a strain gauge is properly packaged for industrial use, the overall device is rugged and reliable.

Bridge Circuits. In order to measure a change in resistance, it must occur in a circuit. A bridge circuit is commonly used, with the strain gauge resistance occupying one or more arms of the bridge. When the balance of the bridge is upset due to a change in the resistance of one arm (the strain gauge), the imbalance can be detected and amplified.

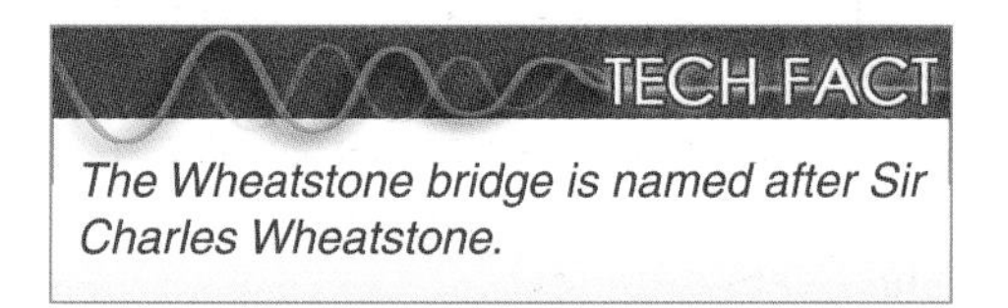

The Wheatstone bridge is named after Sir Charles Wheatstone.

The circuit is typically designed so that the amount of imbalance is proportional to the applied pressure. The strain gauge is supplied with a DC excitation voltage, typically 10 V. The output from the strain gauge is stated as millivolts per volt (mV/V) of excitation at full-scale pressure. Most strain gauge pressure transducers have integral electronic circuitry that converts the mV/V output of the strain gauge to a common current, voltage, or digital signal. The strain gauge and diaphragm system must be calibrated so that the relationship between applied pressure and output voltage is known with sufficient accuracy.

Temperature Compensation. Changes in ambient temperature affect strain gauge circuits, so these circuits require temperature compensation. A dummy gauge, added to the bridge-conditioning circuit, is often used to provide this compensation. **See Figure 11-2.** As the temperature increases, the resistance in the strain gauge changes as a result of both temperature and pressure.

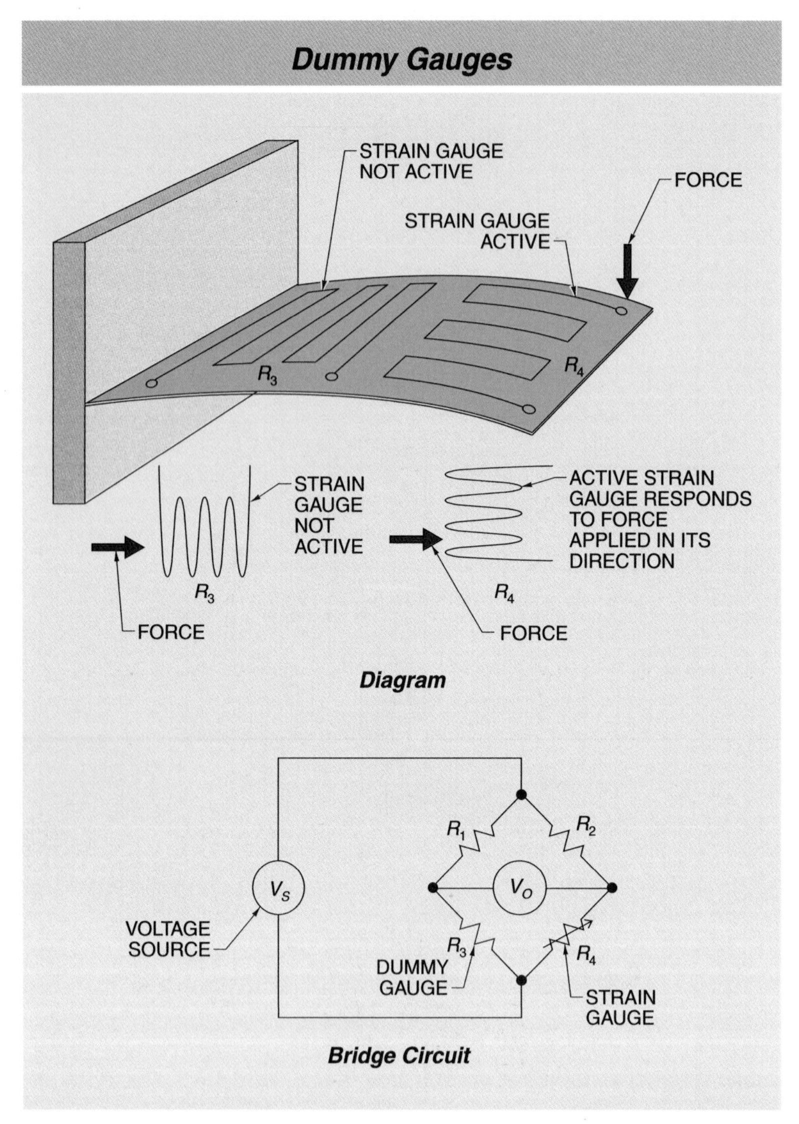

Figure 11-2. A dummy gauge can be used to compensate for resistance changes due to changes in temperature.

The resistance in the dummy gauge also changes due to temperature, since it is made of the same material as the measuring gauge. However, it does not change due to pressure because stress is measured in only one direction. Therefore, the dummy gauge provides a value for resistance change based solely on temperature change. The bridge circuit uses the information from the dummy circuit to compensate for temperature resistance change in the measuring gauge value, resulting in a resistance change based only on force.

Capacitance

A *capacitance pressure transducer* is a diaphragm pressure sensor with a capacitor as the electrical element. A basic capacitor consists of two small, thin electrically conductive plates of equal area that are parallel to one another and separated by a dielectric material. The capacitive plates are typically constructed of stainless steel, but other alloys are used in highly corrosive service. **See Figure 11-3.**

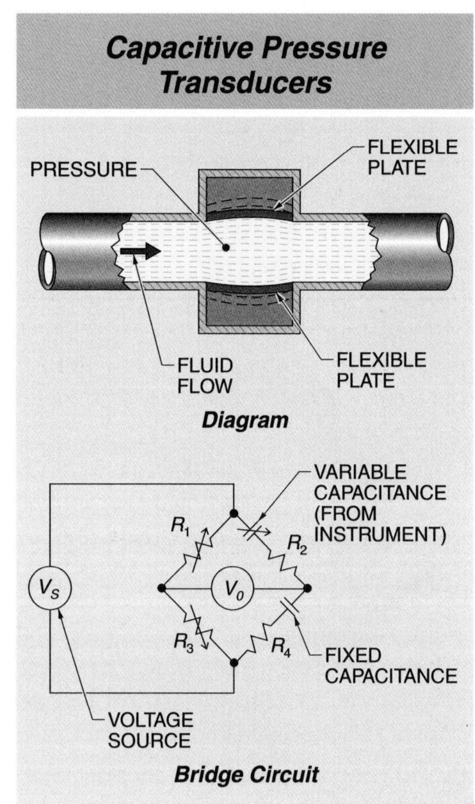

Figure 11-3. A capacitance pressure transducer is a diaphragm pressure sensor with a capacitor as the electrical element.

In a typical industrial capacitive pressure sensor, part of the diaphragm is one plate and the mounting surface is the other. When pressure distorts the diaphragm and alters the distance between the plates, the capacitance of the sensor changes. Capacitive transducers are used in high-frequency bridge circuits. The change in the capacitance of the sensor causes a variation in the impedance that varies with the applied pressure. The circuits must be constructed with great care to avoid the presence of stray capacitance that can cause erroneous readings.

Differential Pressure Cells. A differential pressure (d/p) cell converts a differential pressure to an electrical output signal. A differential pressure transmitter sends the output of a d/p cell to another location where the signal is used for recording, indicating, or control. Many electronic d/p cells consist of a diaphragm surrounded by silicone oil where the process pressure is applied to one side of the diaphragm. Another process pressure or atmospheric pressure is applied to the other side of the diaphragm. **See Figure 11-4.**

The center diaphragm is a stretched spring element that deflects in response to a differential pressure applied across it. Two capacitor plates measure the position of the center diaphragm. The differential capacitance between the center diaphragm and the plates is converted to a digital or 4 mA to 20 mA analog signal.

An absolute pressure cell converts an absolute pressure measurement into an electrical output signal. It is similar to a differential pressure cell except that one side is a sealed vacuum instead of a second process pressure.

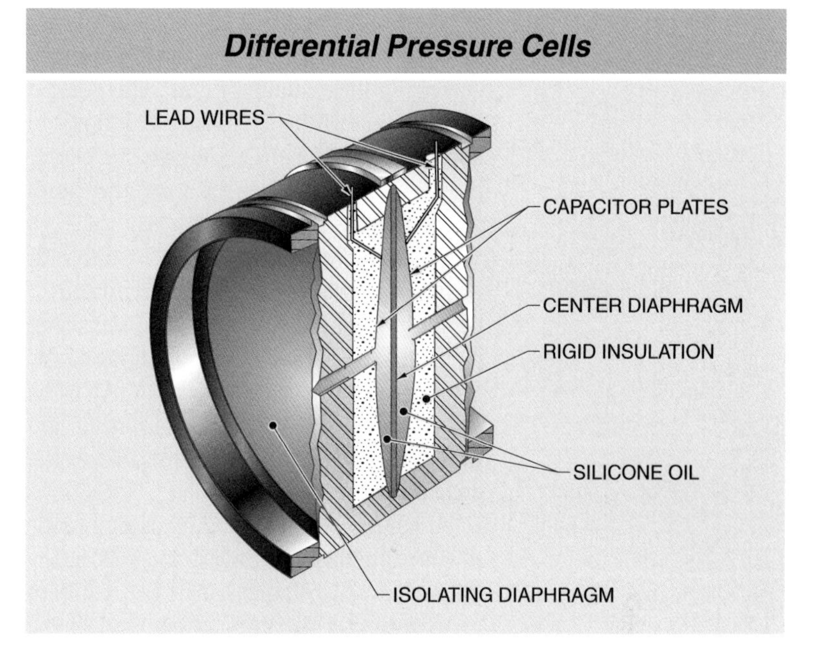

Figure 11-4. A differential pressure cell uses a diaphragm with a capacitance circuit to measure differential pressure.

Inductance

An *inductance pressure transducer* is a diaphragm or bellows pressure sensor with electrical coils and a movable ferrite core as the electrical element. The core moves inside the coils as pressure is applied. *Inductance* is the property of an electric circuit that opposes a changing current flow. **See Figure 11-5.**

A *linear-voltage differential transformer (LVDT)* is an inductance transducer consisting of two coils wound on a single nonconductive tube. The primary coil is wrapped around the middle of the tube. The secondary coil is divided with one half wrapped around each end of the tube.

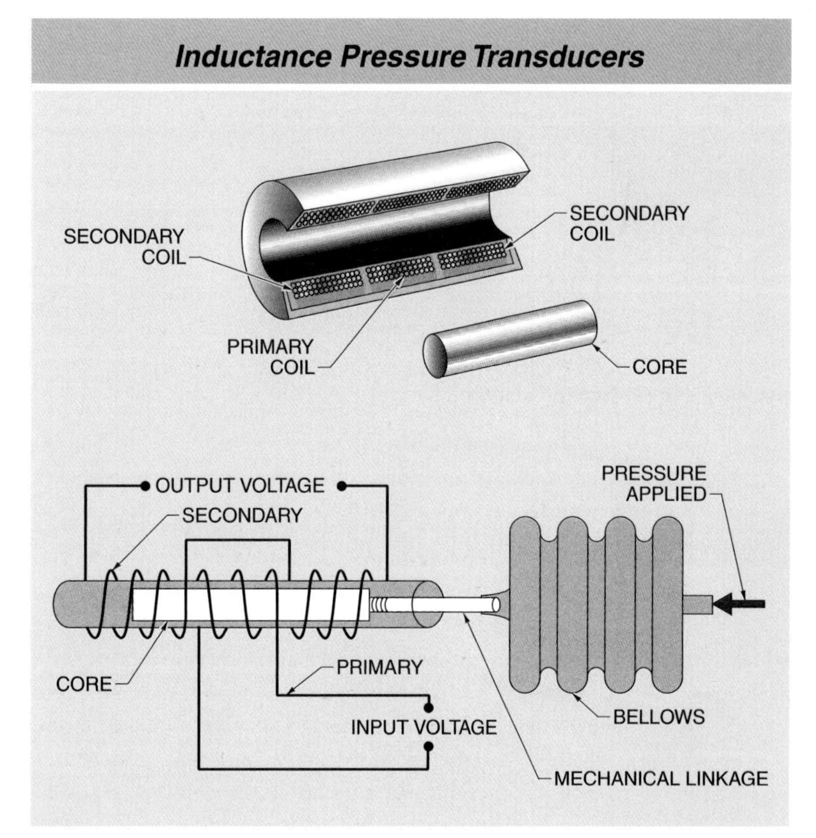

Figure 11-5. An inductance pressure transducer is a bellows or diaphragm pressure sensor with electrical coils and a movable iron core as the electrical element. The core moves inside the coils to change the electrical output.

A supplied alternating current passes through the primary coil and produces current in the secondary coil. The current in the halves of the secondary coil flows in opposite directions. This allows the center position of the core to be a neutral point where there is no output.

When more of the core is in one half of the secondary than the other, more of the output of the transducer comes from that portion of the core. The output of the secondary coil is stated as mV/V AC. The AC output is rectified and amplified and becomes a DC current proportional to the pressure.

A modern LVDT often provides rectification circuits to convert the secondary output to a DC voltage signal. The resultant voltage ranges from –V at one end of the core movement, through 0 at the center position, and +V at the other end of the core movement. **See Figure 11-6.**

A chiller is used to compress a refrigerant. The chiller control system operates a variable-speed drive in response to pressure measurement.

Reluctance

A *reluctance pressure transducer* is a diaphragm pressure sensor with a metal diaphragm mounted between two stainless steel blocks. *Reluctance* is the property of an electric circuit that opposes a magnetic flux. Embedded in each block is a magnetic core and coil assembly with a gap between the diaphragm and the core. The blocks have pressure ports and passages for the fluid media to exert pressure against the diaphragm. The movement of the diaphragm increases the gap on one side of the diaphragm and decreases the gap on the other side.

The magnetic reluctance varies with the gap. This variation is proportional to the change in applied pressure and produces a signal used in a bridge circuit. **See Figure 11-7.** The change in reluctance is measured in a bridge circuit. The applied voltage is typically about 5 VAC with a frequency of approximately 5 kHz.

Reluctance transducers have a relatively high output for small changes in pressure. This makes them a good choice for measuring a small range of pressure. In addition, a change in range can be accomplished by replacing the diaphragm.

TECH FACT

Piezoelectric crystals come in many configurations and compositions. They have good linearity over a wide range. Manufacturers supply specialized crystal transducers for high temperature and corrosive conditions. New designs can be used continuously at 1000°F and intermittently at 1200°F.

Piezoelectric

A *piezoelectric pressure transducer* is a diaphragm pressure sensor combined with a crystalline material that is sensitive to mechanical stress in the form of pressure. This type of transducer produces an electrical output proportional to the pressure on the diaphragm. No external power is needed.

Linear-Voltage Differential Transformer (LVDT)

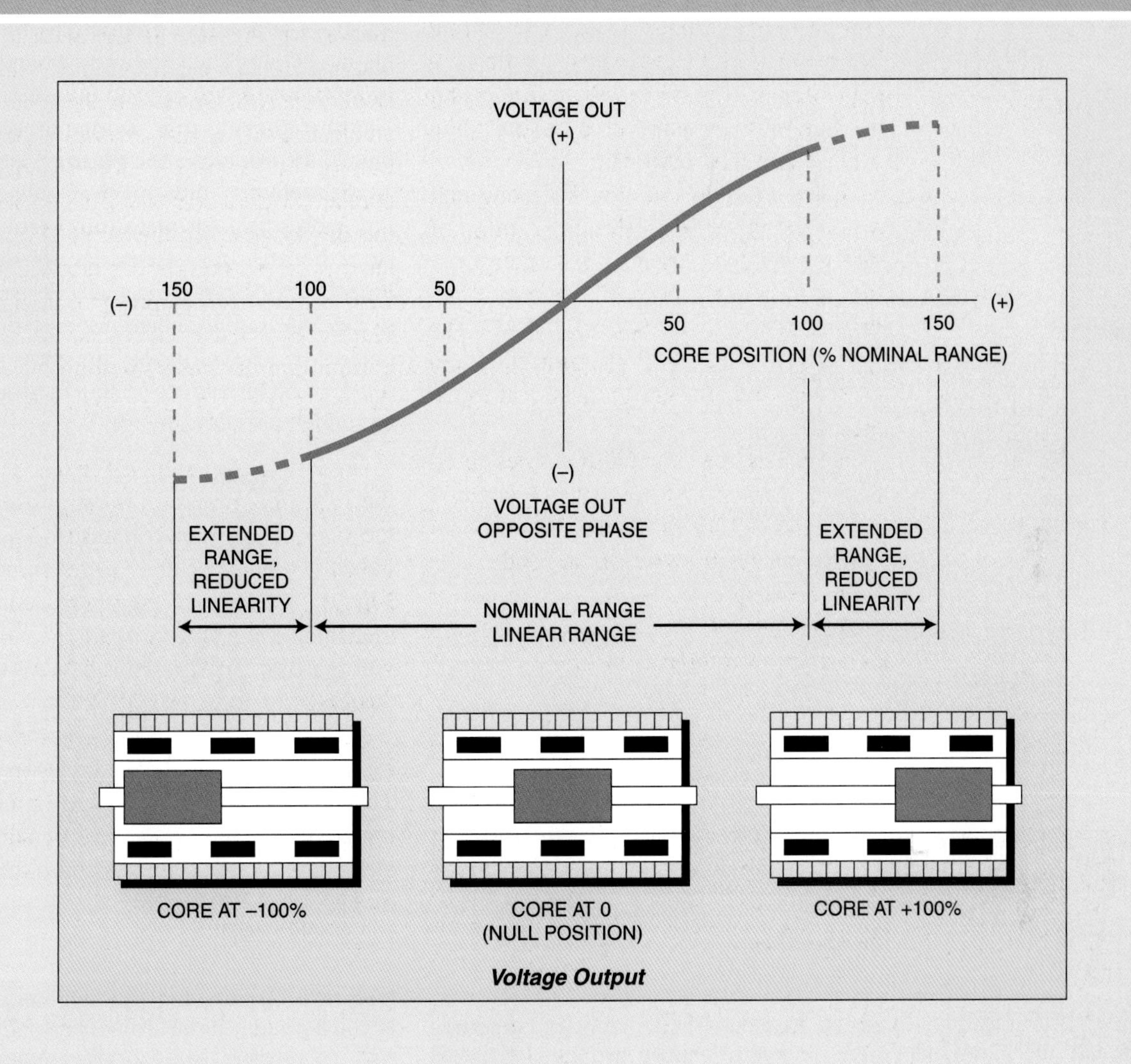

Voltage Output

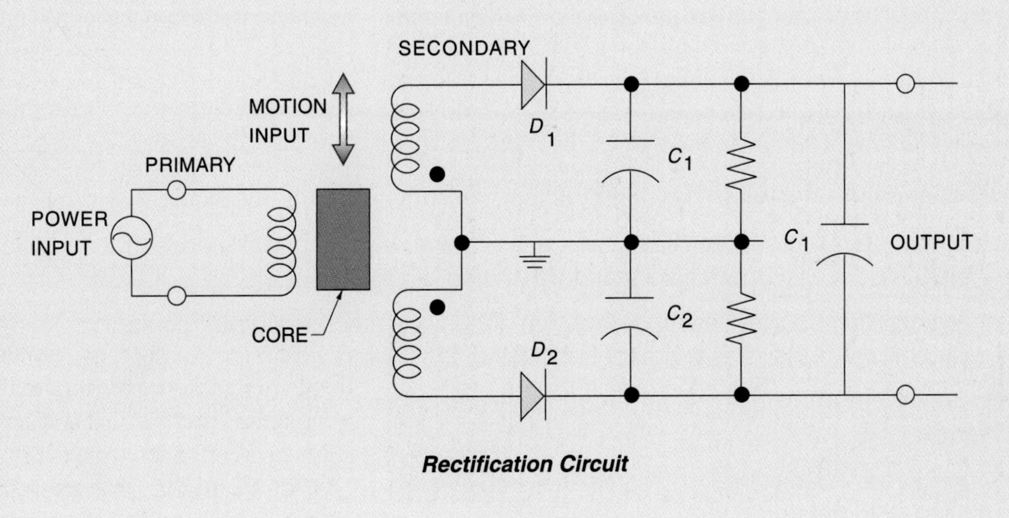

Rectification Circuit

Figure 11-6. The AC voltage output of an LVDT is rectified to produce a DC voltage proportional to position.

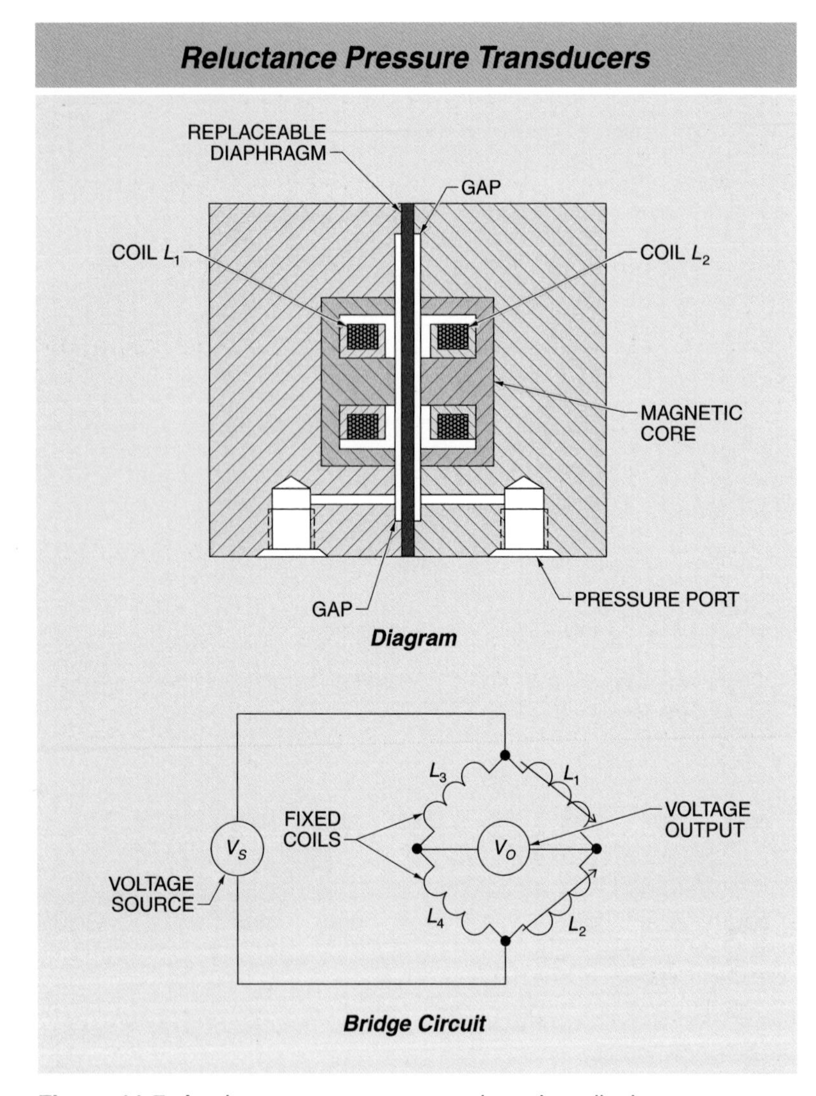

Figure 11-7. A reluctance pressure transducer is a diaphragm pressure sensor with the metal diaphragm mounted between two stainless steel blocks. The electrical reluctance changes with movement of the diaphragm.

As the crystal is compressed, a small electric potential is developed across the crystal. The potential produced by the crystal can be amplified and conditioned to be proportional to the applied pressure. Temperature compensation is often included as part of the circuitry. **See Figure 11-8.**

Piezoelectric pressure transducers are not appropriate for measuring static pressures because the signal decays rapidly. Piezoelectric pressure transducers are used to measure rapidly changing pressure that results from explosions, pressure pulsations, or other sources of shock, vibration, or sudden pressure change.

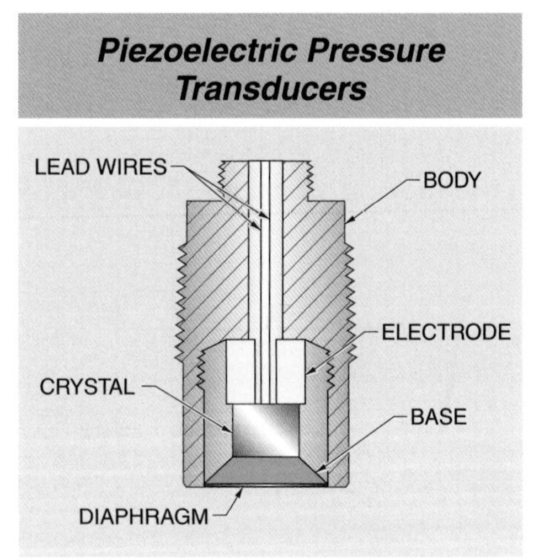

Figure 11-8. A piezoelectric pressure transducer is a diaphragm pressure sensor combined with a crystalline material that is sensitive to mechanical stress in the form of pressure.

KEY TERMS

- *electrical transducer:* A device that converts input energy into output electrical energy.
- *pressure transmitter:* A pressure transducer with a power supply and a device that conditions and converts the transducer output into a standard analog or digital output.
- *pressure switch:* A pressure-sensing device that provides a discrete output (contact make or break) when applied pressure reaches a preset level within the switch.
- *resistance pressure transducer:* A diaphragm pressure sensor with a strain gauge as the electrical output element.
- *strain gauge:* A transducer that measures the deformation, or strain, of a rigid body as a result of the force applied to the body.

KEY TERMS (continued)

- *strain foil gauge:* A strain gauge that has the wire grid impressed on nonmetallic foil and then the assembly is mechanically bonded to the metal diaphragm.
- *thin-film strain gauge:* A strain gauge that has the wire grid sputter-deposited on the diaphragm surface.
- *capacitance pressure transducer:* A diaphragm pressure sensor with a capacitor as the electrical element.
- *inductance pressure transducer:* A diaphragm or bellows pressure sensor with electrical coils and a movable ferrite core as the electrical element.
- *inductance:* The property of an electric circuit that opposes a changing current flow.
- *linear-voltage differential transformer (LVDT):* An inductance transducer consisting of two coils wound on a single nonconductive tube.
- *reluctance pressure transducer:* A diaphragm pressure sensor with a metal diaphragm mounted between two stainless steel blocks.
- *reluctance:* The property of an electric circuit that opposes a magnetic flux.
- *piezoelectric pressure transducer:* A diaphragm pressure sensor combined with a crystalline material that is sensitive to mechanical stress in the form of pressure.

REVIEW QUESTIONS

1. How is strain gauge resistance measured?
2. How is temperature compensation used with a strain gauge?
3. Describe the operation of a capacitance pressure transducer, and explain how it can be used in a differential pressure transmitter.
4. Compare the operation of inductance and reluctance pressure transducers.
5. How does a piezoelectric pressure transducer operate?

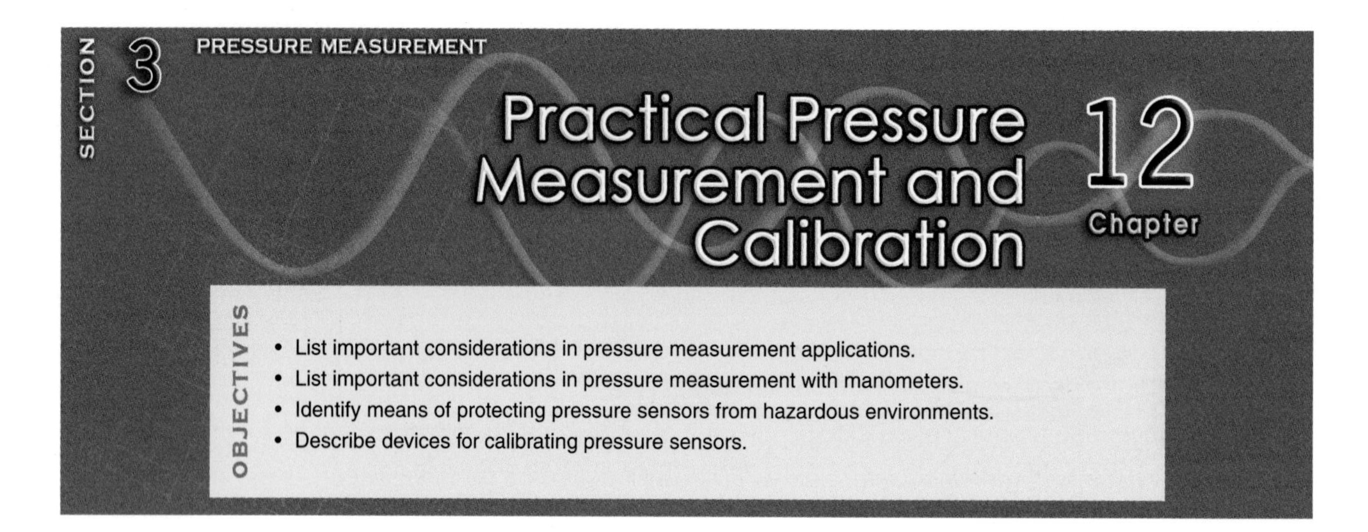

PRESSURE MEASUREMENT APPLICATIONS

There are many practical problems that must be overcome in measuring pressure. When measuring pressure in tanks and pressure vessels, gauges must be mounted in the correct way to account for differences in pressure head. Many processes can damage pressure gauges, and gauge protection is required to successfully measure pressure. In addition, pressure sensors require maintenance and calibration.

Pressure Measurement in Tanks and Pressure Vessels

Measuring the pressure of liquids stored in tanks and pressure vessels can be difficult because the liquid could damage the wetted materials of the pressure-measuring device. Common methods of protecting pressure instruments include the use of dry legs and wet legs. **See Figure 12-1.**

Dry Legs. A *dry leg* is an impulse line that is filled with a noncondensing gas. A dry leg can be used when process vapors are noncorrosive and nonplugging, and when their condensation rates, at normal operating temperatures, are very low. A dry leg enables the d/p cell to compensate for the pressure pushing down on the liquid's surface, in the same way as the effect of barometric pressure is canceled out in open tanks. It is important to keep this reference leg dry because accumulation of condensate or other liquids would cause error in the level measurement. A drain reservoir is commonly included as part of the dry leg.

Wet Legs. A *wet leg* is an impulse line filled with fluid that is compatible with the pressure-measuring device. The fluid may or may not be the same as the fluid in the tank. Care must be taken when installing a differential pressure-sensing unit. If the pressurized fluid is a gas and a wet leg is used, the gauge must be placed at the same elevation as the top level of the liquid in the wet leg. If the gauge is placed below the top of the liquid in the wet leg, a correction must be made for the head of the pressurized liquid above the instrument position.

A *seal pot* is a surge tank that may be installed in a wet leg to prevent volume changes from forcing the fluid into the process as well as protecting the sensing element from high temperatures as with steam applications. If two seal pots are used in the wet legs of a differential pressure measurement, the seal pots must be at the same elevation. If an instrument is inserted through a fitting into a pressurized vessel, all applicable codes and regulations governing pressure vessel construction must be followed.

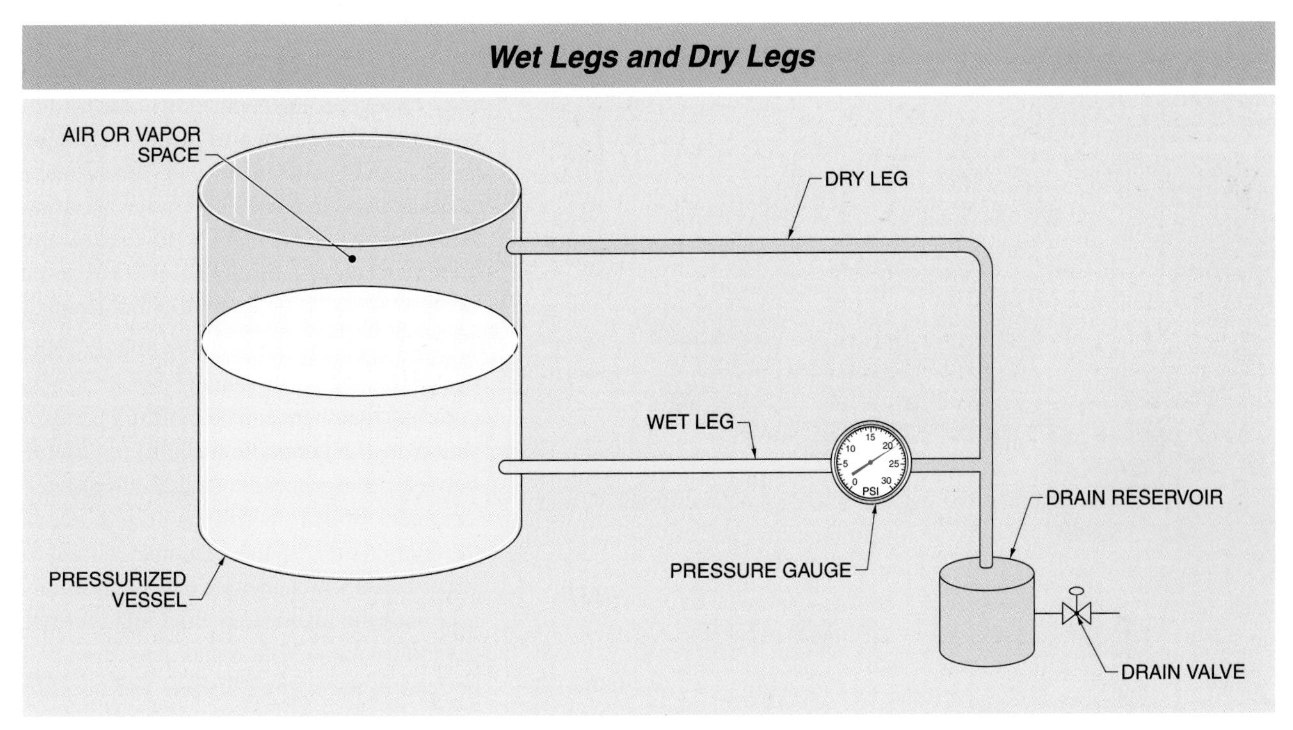

Figure 12-1. Wet legs and dry legs are components of differential pressure measurement of liquid in a vessel.

Wet Leg Pressure Adjustment. Wet leg adjustments must be made if the pressure gauge is not mounted at the same elevation as the liquid level. If the pressure gauge on a boiler is mounted below the top of the liquid, the gauge reads high because of the head of the liquid in the line. **See Figure 12-2.** The pressure is calculated as follows:

$$P = \rho \times h$$

where

P = pressure (in psi)

ρ = density (in lb/cu in.)

h = height (in in.)

The density of water is 0.0361 lb/cu in. If the gauge is mounted 10′ (120″) below the top of the water level of the wet leg, the pressure adjustment is:

$$P = \rho \times h$$

$$P = 0.0361 \times 120$$

$$P = \mathbf{4.33\ psi}$$

If the steam pressure is 6 psig measured with a dry leg above the boiler, the pressure at a gauge located 10′ below the liquid level of the wet leg reads 10.33 psig (6 + 4.33 = 10.33).

Fluke Corporation

A hand-held electronic calibrator can be used to calibrate pressure transmitters in the field.

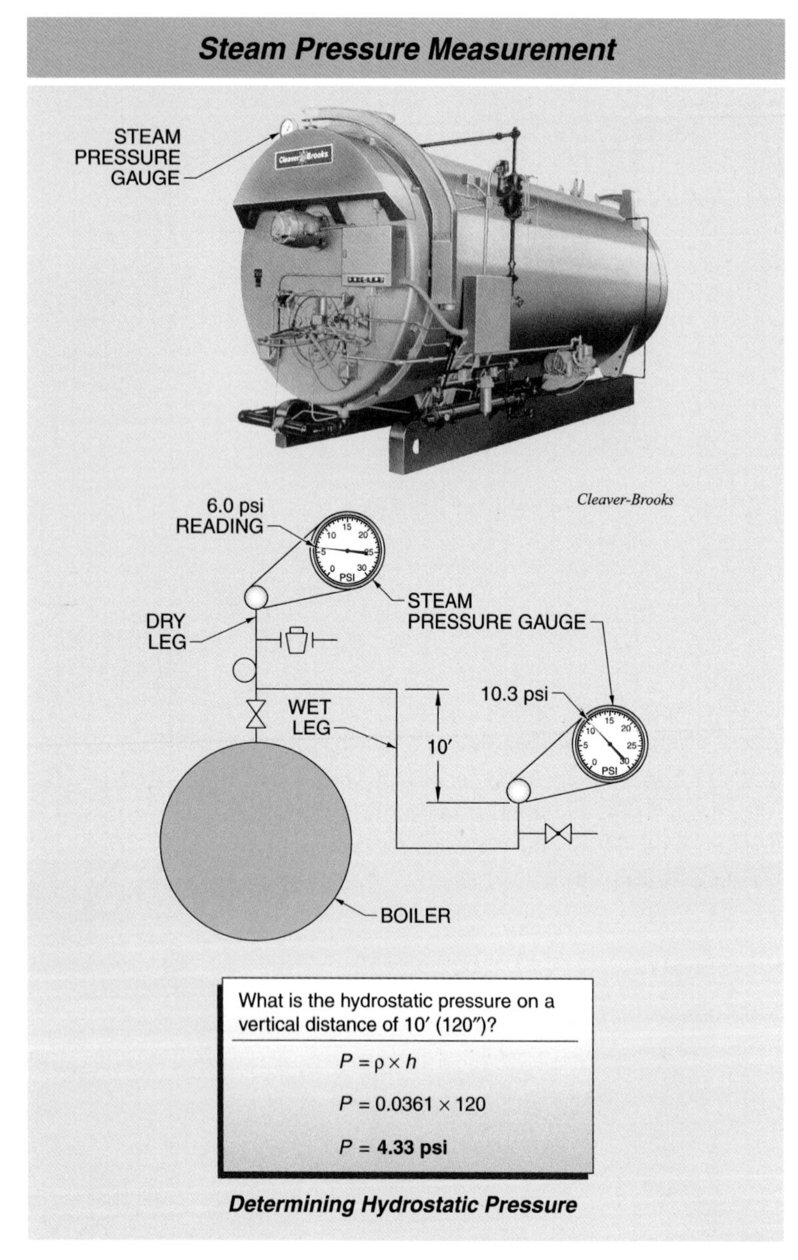

Figure 12-2. Hydrostatic pressure differences must be taken into account when measuring pressure with a wet leg.

Manometer Measurements

There are many unique difficulties in using manometers to measure pressure. These difficulties can cause errors in measurement if they are not considered during installation and maintenance. These difficulties include moisture condensation, measuring liquid pressure, and installing valve manifolds.

Temperature Correction. A manometer normally requires a temperature correction for maximum accuracy because the density of materials changes with temperature. This means that the height of a column of a fluid may change at different temperatures. For example, the density of water varies by 0.5% from 70°F to 100°F. This means that a pressure measurement with a water-filled manometer can vary by 0.5% when the ambient conditions change from 70°F to 100°F, even though the actual pressure does not change.

Moisture Condensation. When measuring gas pressures with a manometer, it is quite common that there can be moisture condensation in the connection piping or tubing between the process and the manometer. This can result in a collection of water in the manometer. If the manometer fluid is water-based, the condensed water will simply mix with the manometer fluid and increase the total volume. The manometer fluid will be read in the normal manner and no error will be generated as long as the scale can be zeroed at the new level.

If the manometer fluid has a gravity of 1.000 but is immiscible with water, the condensed water will collect on top of the manometer fluid on the process side. Since the gravity of the manometer fluid and the water are the same, measuring the differential between the tops of the two liquids will provide the correct reading.

If a manometer is using a fluid with a gravity greater than 1.0, the condensed water will be collected in the manometer on top of the manometer fluid on the process side. **See Figure 12-3.** This will cause an error in the measured differential. The height of the condensed water will increase the differential of the manometer fluid. A correct reading can be obtained by measuring the differential of the manometer fluid only, converting this to inches of water column, and then subtracting the height of the condensed water.

If there is a possibility of condensed water collecting in a manometer, the connecting piping or tubing design should include a condensate collection pot to intercept the condensed water before it reaches the manometer. It must be remembered that any manometer installation using a water-based fluid or subject to the collection of condensate can be damaged if exposed to freezing conditions.

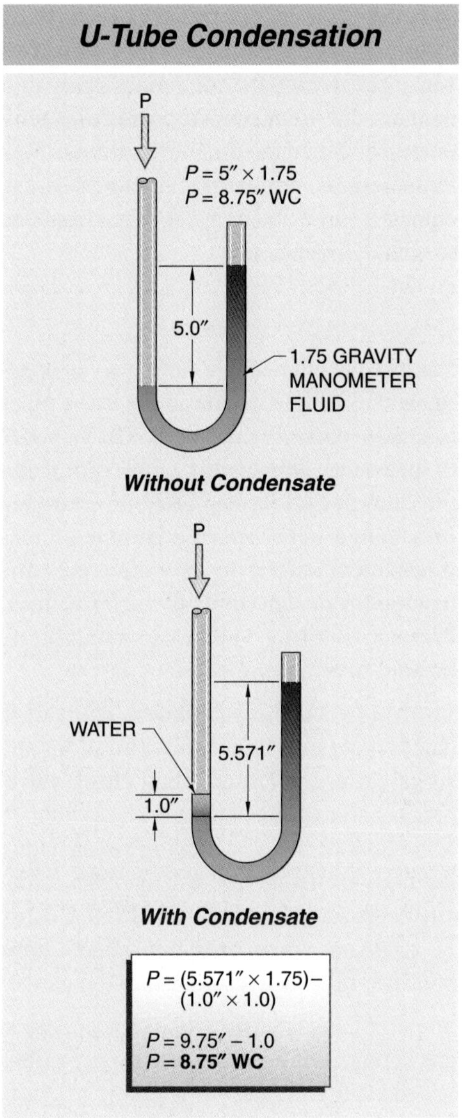

U-Tube Condensation

P

P = 5″ × 1.75
P = 8.75″ WC

5.0″

1.75 GRAVITY MANOMETER FLUID

Without Condensate

P

WATER

5.571″

1.0″

With Condensate

P = (5.571″ × 1.75) − (1.0″ × 1.0)
P = 9.75″ − 1.0
P = 8.75″ WC

Figure 12-3. Measurement compensation is required when there is liquid condensation in a U-tube manometer.

Measuring Liquids. Manometers can also be used to measure the pressure of liquids in pipelines. When this is to be done, the process side of the manometer and the connecting piping or tubing must be filled with the process fluid. This is done so that a consistent liquid head is applied to the manometer at all times. For this type of installation, the process fluid, the material of construction of the manometer, and the manometer fluid must all be compatible. For this reason, measuring liquid pressures is usually only done for water services.

Process liquids in the manometer change the effective gravity of the manometer fluid. A portion of the difference in elevation of the manometer fluid is due to the equal elevation of the process fluid. The effective manometer fluid specific gravity, when used to measure liquid pressures, is equal to the normal manometer fluid gravity minus the process fluid gravity. **See Figure 12-4.** The process pressure is calculated as follows:

$$P = (h_m \times SG_m) - (h_m \times SG_p)$$

where

h_m = height of manometer liquid (in in.)

SG_m = specific gravity of manometer liquid

SG_p = specific gravity of process liquid

Assume that a 15″ manometer is to measure the pressure in a pipe filled with water with a gravity of 1.0, using mercury as a manometer fluid with a specific gravity of 13.6. The manometer is located at the same elevation as the piping. If the mercury differential is 10.0″, calculate the process pressure as follows:

$$P = (h_m \times SG_m) - (h_m \times SG_p)$$

$$P = (10.0 \times 13.6) - (10.0 \times 1.0)$$

$$P = 136.0 - 10.0$$

$$P = \mathbf{126\ inches\ water\ column}$$

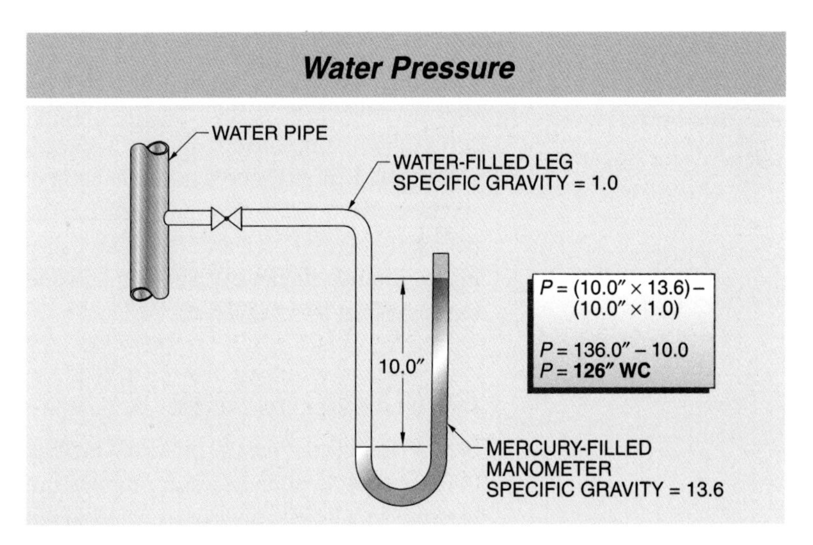

Figure 12-4. A manometer with water in one leg must have a pressure adjustment.

Valve Manifolds. The simplest instrument for measuring differential pressure produced by a differential pressure flow element is a U-tube manometer. When a U-tube manometer is used to measure flow, a three-valve manifold is used with the manometer for easy shutoff and equalizing. The three-valve manifold is necessary so that the manometer can be opened up to the process stream without blowing out the manometer fluid. There is a specific valve operating procedure for the three-way valve manifold that slowly equalizes the pressure and prevents the loss of the manometer fluid.

Cutting in the manometer is the procedure for operating the manometer three-way manifold valves to connect a manometer to a process without blowing out the manometer fluid. **See Figure 12-5.** The procedure starts with the high- and low-pressure shut-off valves closed and the equalizer valve open. Slowly open the high-pressure shut-off valve. This pressurizes both sides of the manometer to the static pressure in the process piping through the equalizer. Next, close the equalizer valve and slowly open the low-pressure shut-off valve.

Cutting out the manometer is the procedure for operating the manometer three-way manifold valves to disconnect the manometer from the process. This procedure is exactly the opposite of cutting in the manometer. Close the low-pressure shut-off valve and slowly open the equalizer valve. This allows equal pressures to be placed on both sides of the manometer. Close the high-pressure shut-off valve. Note that the manometer still contains pressurized fluids. This can be dangerous if gases or vapors can escape. As pressurized gases or vapors escape, the potential energy is released and can cause injury. Liquids do not expand as much and are somewhat less dangerous.

GAUGE PROTECTION

Sensors frequently must be protected in order to function. It is often necessary to measure pressure in a hostile environment. Isolation or block valves are typically installed between the process and the pressure-measuring instrument to facilitate removal and maintenance but do not provide protection against damage due to excessive pressure or temperature. Since pressure pulsations, high temperatures, and corrosive materials can damage sensors, other methods must be used to protect instruments.

Overpressure Protection

Pumps starting and stopping, valves opening and closing, and vibrations in the pipes can cause pressure pulsations. The response of pressure sensors is quite rapid and pressure pulsations can cause overranging in a gauge. *Overranging* is subjecting a mechanical sensor to excessive pressure beyond the design limits of the instrument. Pressure-limiting valves are available to prevent overranging.

Snubbers. Adding snubbers (pulsation dampers) to inlet lines limits pulsations and surges. A porous filter, a ball check valve, or a variable orifice may serve as a snubber. Some snubbers have a moving piston within the body of the device that also cleans out any scale and sediment. **See Figure 12-6.** In addition, some pressure gauges have pressure-relief valves built into the gauge.

Sensing Lines. Sensing lines, especially in gas service, may affect pressure pulsations because they may act as a flow restriction. Long runs of narrow-bore tubing tend to slow response and dampen pressure surges. Large-bore tubing may also slow response due to the absorption of pressure variations by the large amount of fluid. Snubbers and other flow restrictors provide overpressure protection at the expense of response time. Pressure gauges are slower to respond to true pressure changes when equipped with overpressure protection.

TECH FACT

Deadweight testers are very accurate calibrators for pressure instruments but are not very portable. Hand-held electronic testers can be carried where needed.

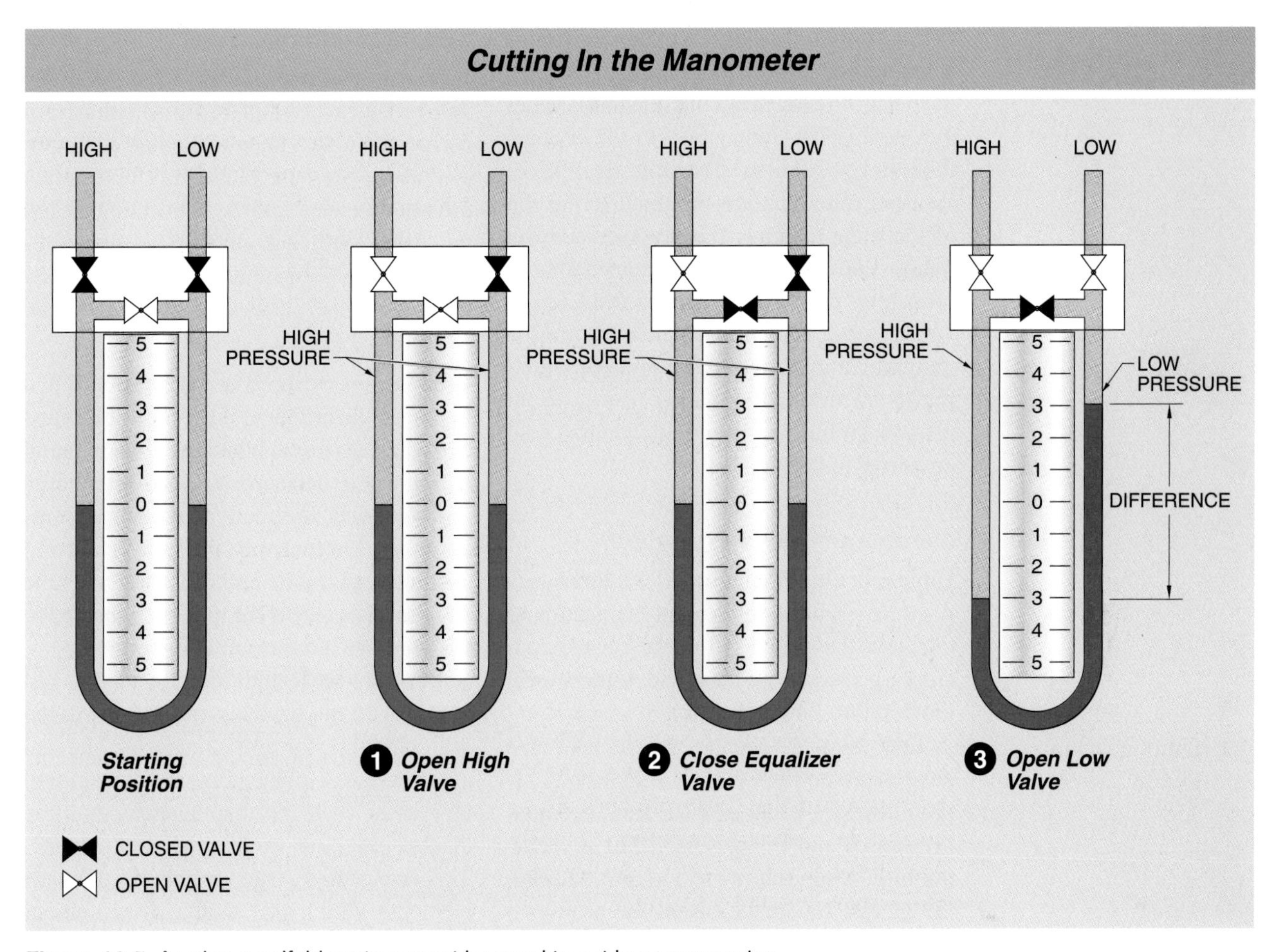

Figure 12-5. A valve manifold system must be used to cut in a manometer.

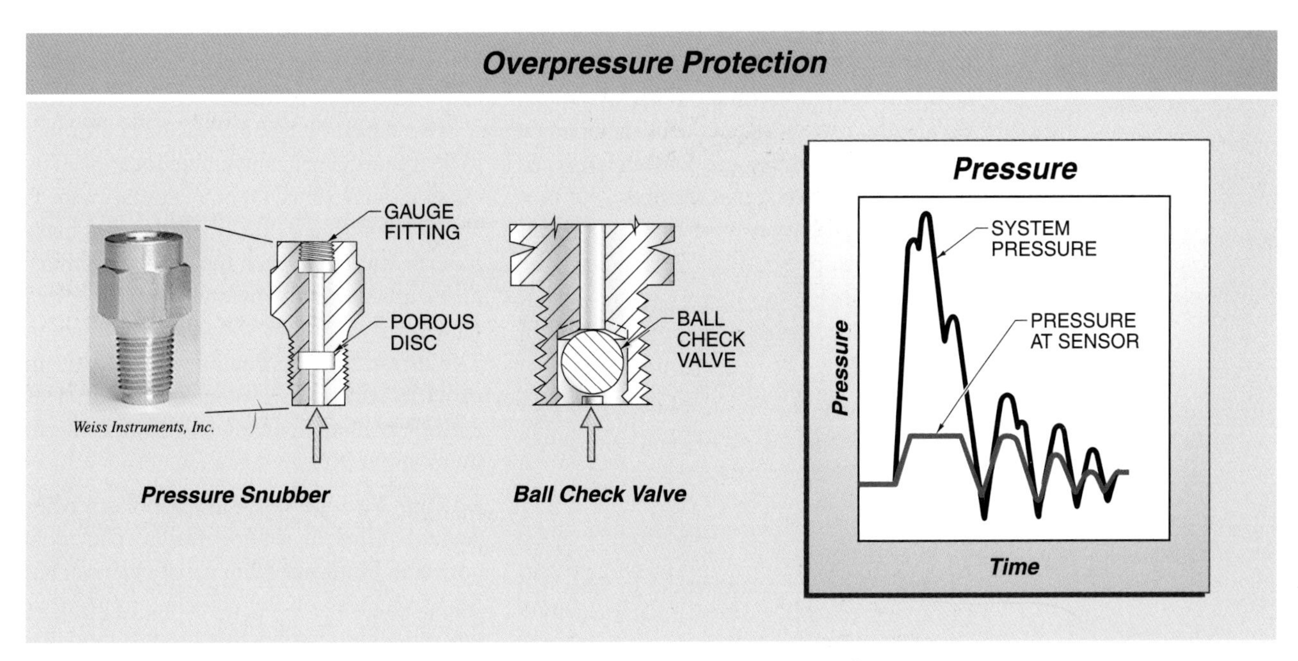

Figure 12-6. Pressure snubbers and ball check valves protect pressure sensors from overpressure surges.

Manometers. In addition to damage to instruments from overpressure, it has always been easy to blow fluid out of a manometer by opening the connection to the process too quickly. Manometer manufacturers offer fluid traps that are mounted to the top of the manometer in the pressure-sensing line or lines to catch the manometer fluid, or at least the majority of the fluid, if the wrong actions are taken when putting the manometer into service. Manometers used for flow measurement services should always include a three-way shutoff and equalizer manifold.

Overtemperature Protection

High process temperatures can damage a pressure sensor. Adding enough inlet tubing to allow the process fluid to cool before entering the sensor may widen the useful temperature range of pressure sensors.

For example, boiler steam pressure gauges are equipped with a siphon. A siphon forms a water seal, or trap, between the sensor and the process, thus preventing high-temperature steam from entering the gauge. As steam enters the piping to the gauge, it begins to condense in the lower part of the siphon, keeping the high-temperature steam from entering the gauge while allowing pressure to be transmitted normally. **See Figure 12-7.** Two common types of siphons are the pigtail siphon and the U-tube siphon. A lever cock installed on the pigtail allows the gauge to be changed out for maintenance and service.

TECH-FACT

Chemical isolating sealing systems are widely used to extend the life of gauges and even to allow gauges to be used that would not survive without the sealing system. Wetted parts are typically made of stainless steel but are available in many other materials, such as brass, for specialized chemical applications. Manufacturers provide chemical compatibility charts to help in selection.

Sealing Systems

Isolating sealing systems have been developed to protect pressure sensors from corrosive fluids. The process fluid acts upon a separate pressure element. The separate pressure element and the connecting tubing are filled with a noncorrosive pressure-transmitting liquid. Frequently the seal assembly is closely coupled to the instrument. **See Figure 12-8.**

Diaphragm Seals. A *diaphragm seal* is a metal or elastomeric diaphragm clamped between two metal housings. One housing is connected to the process. The other housing is connected directly to the pressure instrument. Alternatively, the second housing is connected to one end of a capillary tube and the other end of the tube is connected to the pressure measurement instrument.

Pressure seals with integral sealing liquids should not be disassembled due to the difficulty of restoring that integrity. The seal must have a displacement volume equal to or greater than the displacement volume of the pressure-measuring instrument. This is especially true for bellows pressure elements, which have a large displacement volume. Seals with elastomeric diaphragms have greater displacement than seals with metal diaphragms. Additionally, the spring force of the seal diaphragm may be too great for very low pressure measurements, such as furnace draft pressure.

The capillary tube allows the instrument to be mounted in a convenient location. The outside diameter (OD) of a capillary tube is about 1/8″ with a 0.005″ to 0.010″ diameter hole in the center for the fluid. Capillary tubes are usually protected from physical damage by flexible stainless steel armor. The armor can be coated with plastic to provide corrosion protection. Capillaries can be obtained in nearly any desired length up to about 20′.

Sealing Fluids. Seal manufacturers offer several different sealing fluids. The most common fluids are glycerin or silicone, but for services involving chlorine, oxygen, or any other highly oxidizing process fluid, the fluid is halocarbon oil.

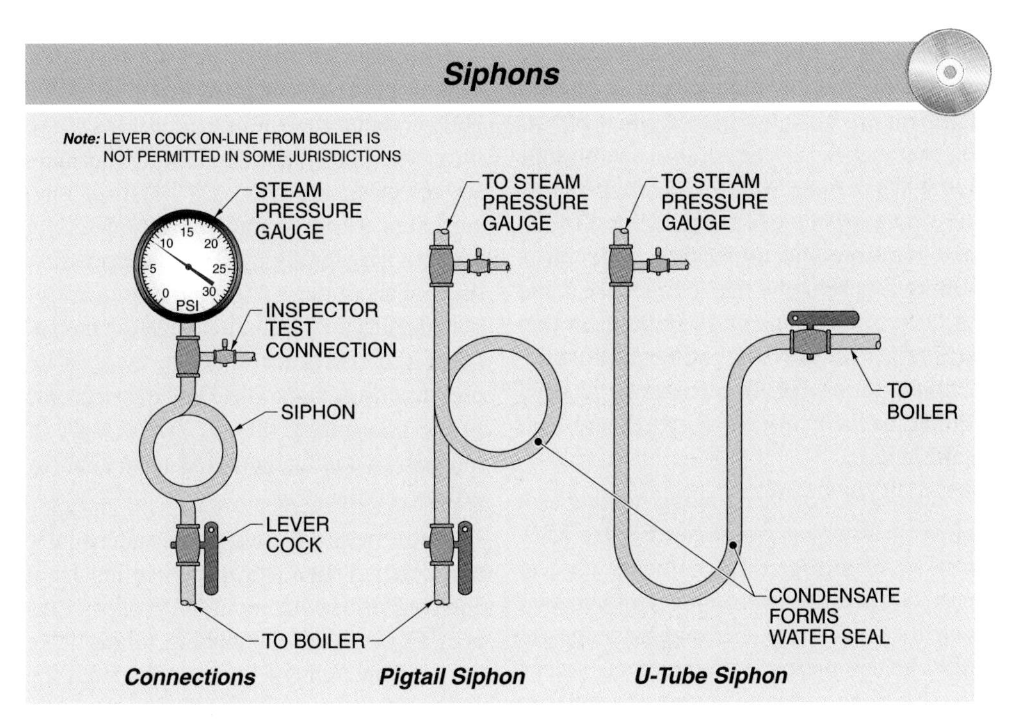

Figure 12-7. A siphon forms a water seal, or trap, preventing high-temperature steam from entering a pressure gauge.

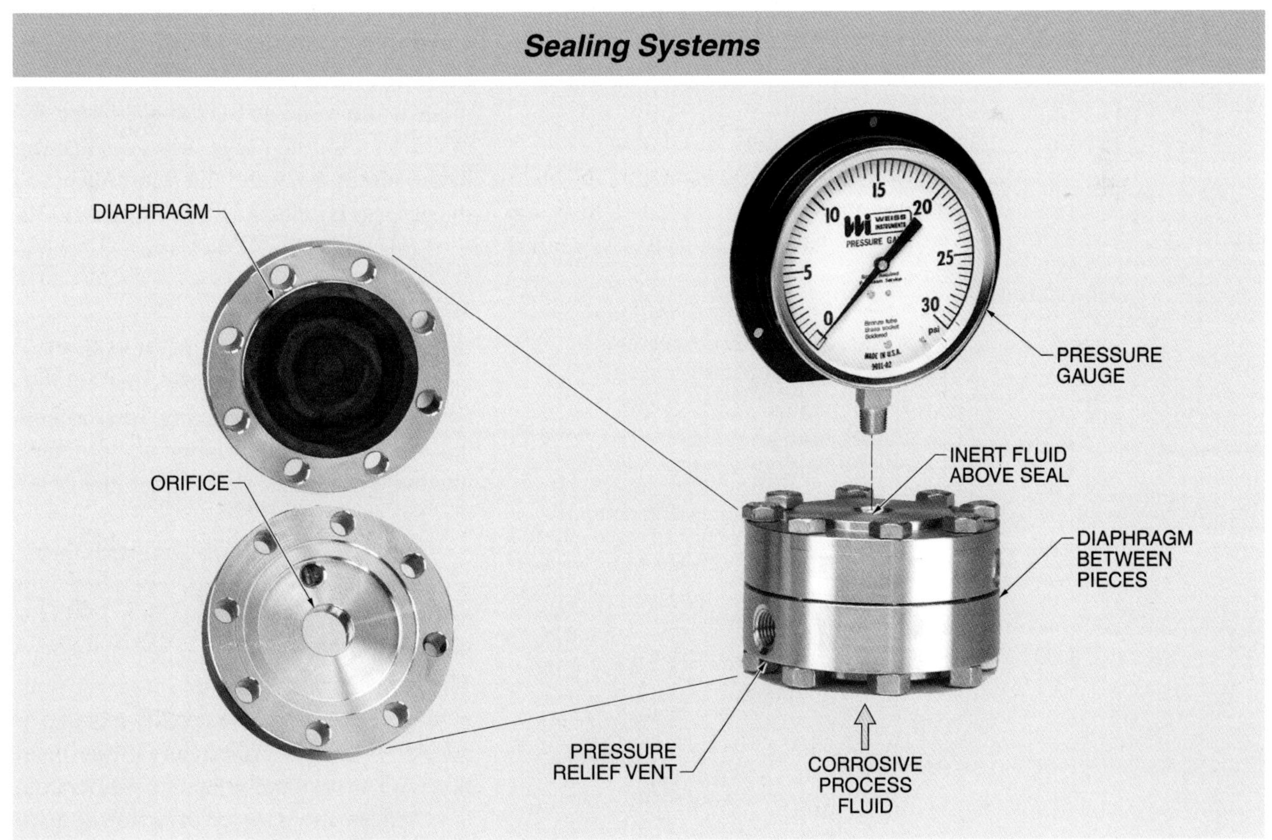

Figure 12-8. A sealing system protects a pressure sensor from corrosive fluids by isolating the gauge.

Thermal expansion or contraction of the sealing fluid due to ambient or process temperature changes has minimal effects on accuracy when the measurement range is 15 psig or more. Measuring smaller pressure ranges becomes much more difficult and requires special designs. Elevation differences between the chemical seal and the instrument generate a static head that will affect the accuracy of the instrument. Compensation for the difference in head should be incorporated into the instrument calibration.

Gas Purges. Another sealing method employs an air or gas purge. **See Figure 12-9.** Dry air or nitrogen is commonly used to protect a pressure instrument from contact with the process gas. The selected purge gas must be compatible with the process since there will be a flow of purge gas into the process. A suitable purge gas is fed through a pressure regulator with a setting slightly higher than the maximum process pressure to be measured. This is done to prevent the purge gas from overpressuring the instrument in case of a blocked connection tube to the process.

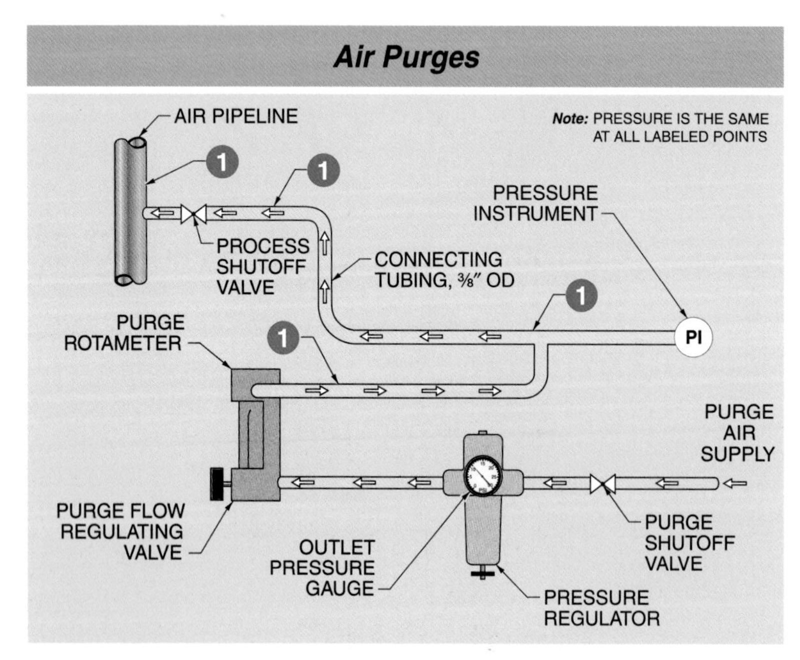

Figure 12-9. An air purge is used to protect an instrument from corrosive process fluids.

Downstream of the pressure regulator, a small purge rotameter is installed with a built-in regulating valve to adjust the purge flow rate. Gas purge rotameters commonly are selected with relatively low flow rate, sufficient to purge the connection tubing without developing a measurable pressure drop. The outlet of the purge rotameter connects to the tubing between the instrument and the process. Since there is no pressure drop due to the flow of purge gas in the connecting tubing, the pressure at the instrument is exactly the same as at the process connection.

Liquid Purges. Liquid purges are usually done with water to prevent solids from entering or depositing at the process connection or in the connecting tubing. The purge water must be compatible with the process fluid. In the case of a water purge, a pressure regulator is usually not used. The water purge rotameter is commonly selected with a range of 0 cm^3/min to 200 cm^3/min and is set at 100 cm^3/min. A valve built into the purge rotameter adjusts the purge water flow. The piping arrangement is the same as for the gas purge application with the purge water connecting to the tubing between the instrument and the process connection. **See Figure 12-10.**

GAUGE CALIBRATION

Pressure sensors require maintenance and calibration. For many years, the calibration of pressure-measuring instruments has been performed using manometers, deadweight testers, and digital electronic test gauges.

There are calibration standards problems with some units of pressure. For example, there is not a single standard definition of the unit "inches of water." The international standard reference temperature is based on the density of water at 39.2°F (4°C). The American Gas Association (AGA) standard reference temperature is based on the density of water at 60°F (15.6°C). The technician should ensure that the proper calibration standard is used.

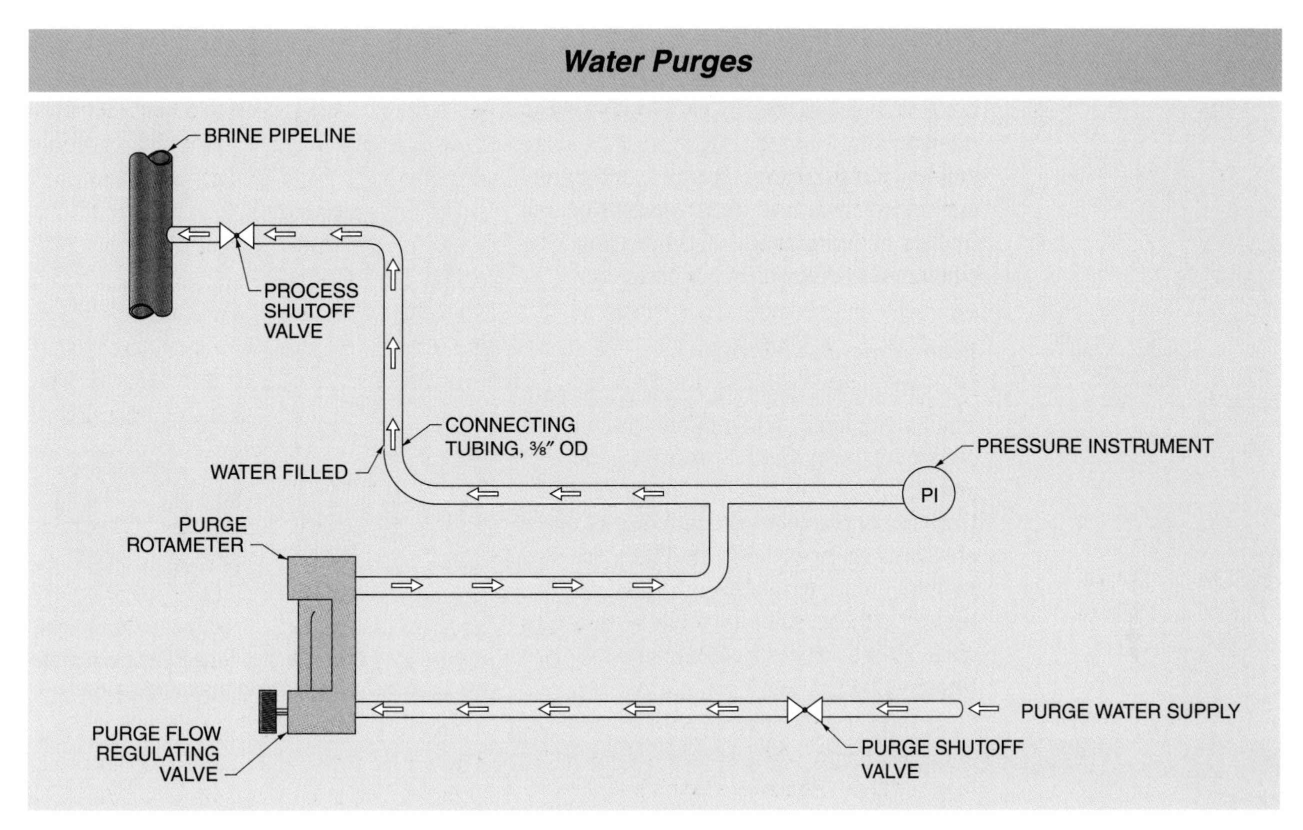

Figure 12-10. A water purge is used to protect an instrument when water can be added to the process fluid.

Deadweight Testers

A *deadweight tester* is a hydraulic pressure-calibrating device that includes a manually operated screw press, a weight platform supported by a piston, a set of weights, and a fitting to connect the tester to a gauge. **See Figure 12-11.** A deadweight tester is another application of Pascal's law. Pressure is created in the hydraulic system by turning the screw press. The pressure acts on the piston supporting the weights.

When the hydraulic pressure is sufficient to begin to lift the weights off the rest position, the pressure is equal to the total of the weights divided by the area of the piston. The weight platform should be gently rotated to overcome the break torque of the piston. The pressurized fluid from the deadweight tester is allowed to enter the pressure gauge. By changing the weights as needed, a gauge can be calibrated over any desired range. For example, if the weights total 1 lb and the piston has an area of 0.2 sq in., the pressure is calculated as follows:

$$P = \frac{F}{A}$$

$$P = \frac{10}{0.2}$$

$$P = \mathbf{50\ psi}$$

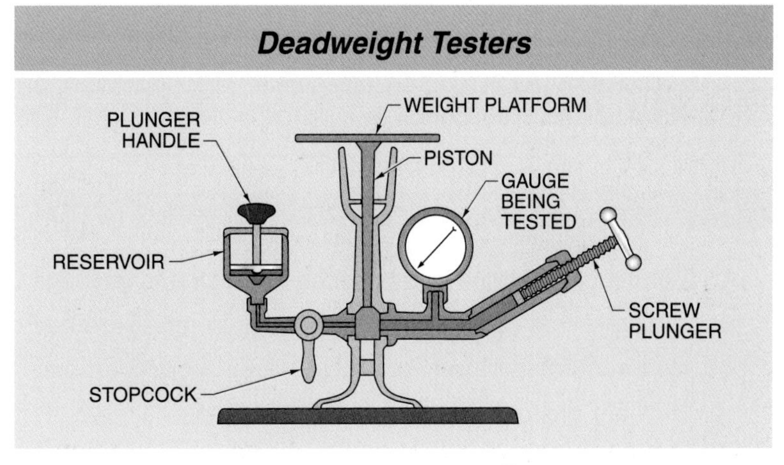

Figure 12-11. A deadweight tester is used to calibrate pressure sensors.

Manometers

Manometers are sometimes used to calibrate other pressure sensors. However, manometers are limited to low pressures and ambient temperatures. The measuring units are scaled in actual linear measurements (inches or centimeters) and must often be converted to other pressure units.

Electronic Calibrators

Portable digital calibration test gauges are available as hand-held models with manual pneumatic or hydraulic pressure generator and electronic transmission for certification. Units are available for vacuum service or for pressures up to 10,000 psi. There are also portable tabletop deadweight models. The sensors in these calibrators may be thin-film strain gauges or piezoelectric crystals. **See Figure 12-12.**

Electronic Calibrators

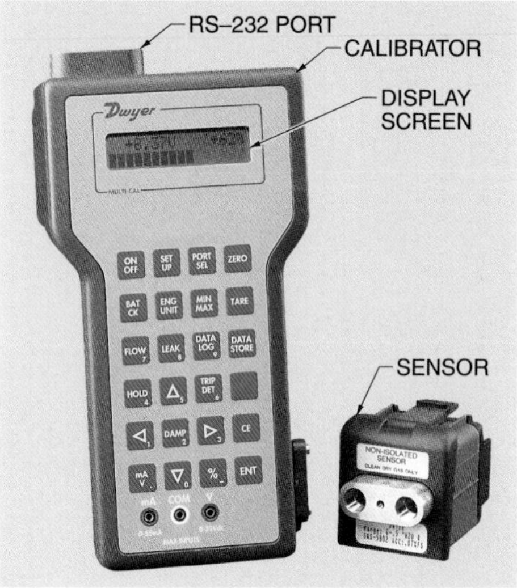

RS–232 PORT
CALIBRATOR
DISPLAY SCREEN
SENSOR

Figure 12-12. An electronic calibrator simplifies the process of calibrating pressure gauges.

KEY TERMS

- *dry leg:* An impulse line that is filled with a noncondensing gas.
- *wet leg:* An impulse line filled with fluid that is compatible with the pressure-measuring device.
- *seal pot:* A surge tank that may be installed in a wet leg to prevent volume changes from forcing the fluid into the process as well as protecting the sensing element from high temperatures as with steam applications.
- *cutting in the manometer:* The procedure for operating the manometer three-way manifold valves to connect a manometer to a process without blowing out the manometer fluid.
- *cutting out the manometer:* The procedure for operating the manometer three-way manifold valves to disconnect the manometer from the process.
- *overranging:* Subjecting a mechanical sensor to excessive pressure beyond the design limits of the instrument.
- *diaphragm seal:* A metal or elastomeric diaphragm clamped between two metal housings.
- *deadweight tester:* A hydraulic pressure-calibrating device that includes a manually operated screw press, a weight platform supported by a piston, a set of weights, and a fitting to connect the tester to a gauge.

REVIEW QUESTIONS

1. What are the similarities between dry legs and wet legs used in pressure measurement?
2. What are four considerations for pressure measurement with manometers?
3. How can pressure sensors be protected from overpressure surges?
4. How are gas or liquid purges used in pressure measurement?
5. What are the common types of pressure calibrators?

SECTION 4

LEVEL MEASUREMENT

TABLE OF CONTENTS

SECTION OBJECTIVES

Chapter 13

- Explain how the shape of a tank or vessel affects the relationship between level and volume.
- Define and compare point level measurement and continuous level measurement.
- Identify and describe level measurement with visual inspection.
- Explain the use of indirect pressure measurements for measuring level.
- Define float, displacer, and paddle wheel switch.

Chapter 14

- Describe the operation of capacitance level instruments.
- Compare conductivity and inductive probes.
- Identify and describe the operation of photometric sensors.
- Describe the operation of magnetostrictive and thermal dispersion sensors.

Chapter 15

- List and describe the operation of ultrasonic level instruments.
- List and describe the operation of radar level instruments.
- Describe the operation of laser level instruments.

Chapter 16

- Describe nuclear level detectors.
- Explain the use of weigh systems.

SECTION

4 LEVEL MEASUREMENT

Chapter 17

- Describe the flow properties of bulk solids and explain how these properties can make it difficult to measure level.
- Describe methods of calibrating load cells.
- List considerations in measuring the water level in a boiler.
- List considerations in measuring the level of corrosive fluids.
- Describe devices for calibrating pressure sensors.

INTRODUCTION

Level measurement has been performed for thousands of years. The earliest and simplest method of measuring level was inserting a pole into the water of a river or canal. The pole was pulled out and the wetted part of the pole was measured to determine the depth. Early sailors measured the depth of the ocean by placing a weight on the end of a rope and dropping the weight overboard. Markers or special knots on the rope indicated the ocean depth. A modern variation of this idea is a tape level gauge where a float connected to a tape moves up and down with changes in level and the tape is read outside the tank.

Archimedes, an ancient Greek scientist, discovered that the apparent weight of a floating object is reduced by the weight of the liquid displaced. This idea was developed into the modern displacer where the buoyant force on an object is measured to determine the level in a tank. In the late 18th century, several boiler engineers developed the first float-type regulators for boilers. There are many variations on this design including mechanical switches, floating mercury switches, and float-actuated reed switches.

ADDITIONAL ACTIVITIES

1. After completing Section 4, take the Quick Quiz® included with the Digital Resources.
2. After completing each chapter, answer the questions and complete the activities in the *Instrumentation and Process Control Workbook.*
3. Review the Flash Cards for Section 4 included with the Digital Resources.
4. Review the following related Media Clips for Section 4 included with the Digital Resources:
   - Gauge Glass
   - Level Gauge
   - Low Water Fuel Cutoff
   - Water Column
5. After completing each chapter, refer to the Digital Resources for the Review Questions in PDF format.

ATPeResources.com/QuickLinks • Access Code: 467690

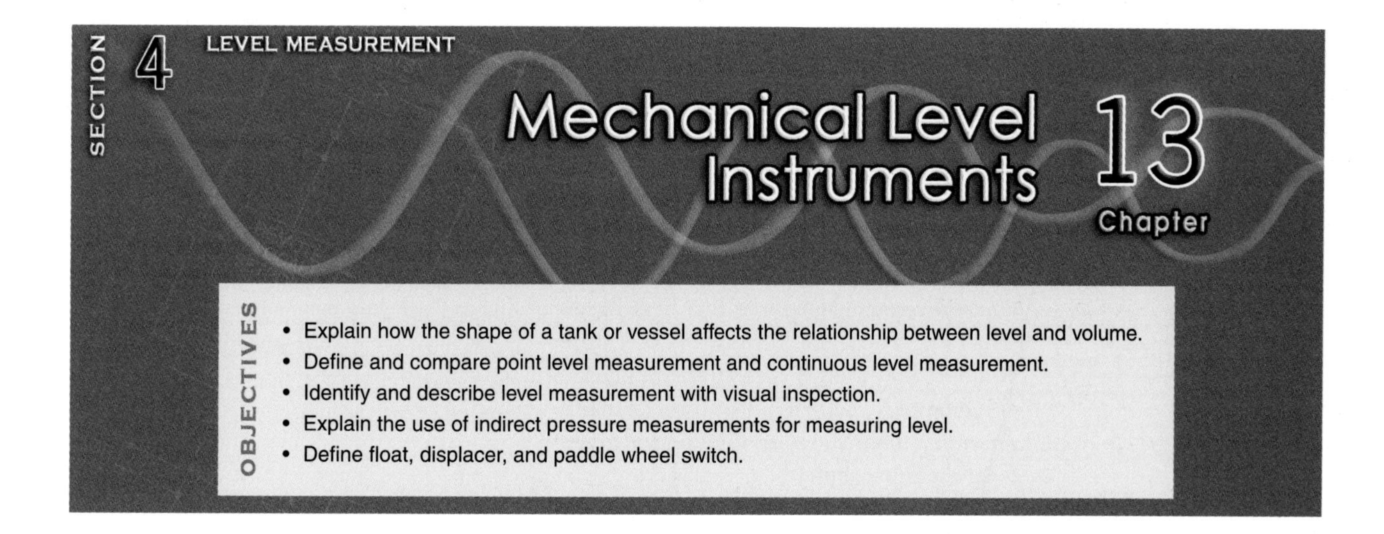

LEVEL MEASUREMENT

The amount of water, fuel, solvent, bulk solids, or other material stored in tanks, pressure vessels, silos, and other containers is important when operating a manufacturing process. Level measurement and control instruments are essential for maintaining adequate quantities of materials for a process. In addition, level measurement is essential for safety systems in boilers and for overflow and spill prevention systems in tanks and silos.

There are some processes that require continuous measurement and others that need to know only that the level is within the desired limits. Frequently, level switches are installed to prevent overflow of a vessel or to supply material to a vessel when it runs low. There are also occasions when it is necessary to keep track of the interface of two liquids or, in some cases, the interface level of liquid and foam.

Level and Volume

Level measurement is often used to measure the volume of material in a tank or vessel. The shape and position of a tank affect the relationship between level and volume. For a vertical cylindrical tank with a flat bottom, the relationship is uniform and each unit of level represents an equal unit of volume. **See Figure 13-1.** Many vertical cylindrical tanks have dished bottoms. A dished bottom is a special convex shape when viewed from the outside that is designed to handle internal pressures. The relationship between level and volume for a vertical dished-bottom tank is uniform for all levels except for the dished end.

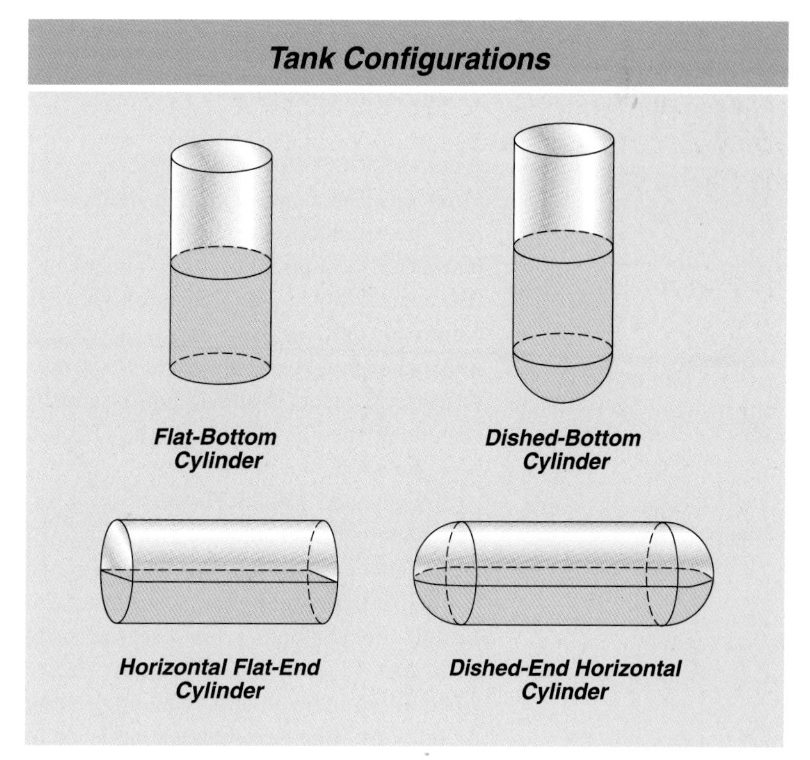

Figure 13-1. The shape of a storage tank determines the relationship between level and volume.

For a horizontal cylindrical tank, this is not true. One unit of level in the middle of the tank represents more volume than one at the bottom or top. In addition, horizontal tanks add more complexity to the level-volume relationship because the ends are dished or hemispherical. Flat ends are rarely used for pressurized horizontal tanks because a flat plate is a very weak structure. The calculations to determine volume from a level measurement can be very complex. Tank manufacturers usually provide a table showing volume for given level measurement.

Level is normally measured in linear units of length or translated into units of volume or weight. Typical linear units of length measure are feet and inches. Typical volume units are gallons or cubic feet. Typical weight units are pounds or tons.

An oil/water separator is a settling tank where denser water sinks to the bottom and oil floats on top. The level measurement of the interface is used to determine when to pump out clean water for disposal.

Point Level Measurements

Point level measurement is a method of level measurement where the only concern is whether the amount of material is within the desired limits. The measured value is commonly used to sound alarms or to determine when to activate or deactivate pumps or other material handling equipment. A sensing element is installed at the selected level position.

If high- and low-level operation is required, one sensor is required for each level point. This ensures that the level remains within the limits set by the positions of the sensors. For example, point level measurement can be used to prevent overflows when filling a tank or silo, to avoid running a pump dry when emptying a tank, or to sound an alarm when a surge tank is above or below a normal level.

Continuous Level Measurements

Continuous level measurement is a method of tracking the change of level over a range of values. This is commonly used for inventory tracking and for determining when to add or remove material from containers. For example, the level of a material in a tank must be known in order to maintain a safe level when transferring material, or the water level in a boiler must be known at all times to prevent a low-water condition that can lead to boiler damage or an explosion.

VISUAL INSPECTION

The simplest way to determine the level of a material in a container or tank is to visually inspect the container. While it is rare to open containers and observe the contents, other methods of visual observation are still used. Some plastic tanks are made of a material that is translucent so that the liquid level can be seen from the outside. Fiberglass tanks can sometimes be purchased with translucent vertical strips through which the liquid level can be seen. Common visual level measuring instruments are gauge glasses, magnetically coupled level gauges, and cable and weight systems.

Gauge Glasses

A *gauge glass* is a continuous level measuring instrument that consists of a glass tube connected above and below the liquid level in a tank and that allows the liquid level to be observed visually. **See Figure 13-2.** The liquid level in the glass tube is the same as the level in the tank. The gauge glass occupies the vertical space between the gauge cocks. The gauge cocks include ball check valves to prevent the loss of process fluid if the gauge glass should break. The gauge glass is a thick-walled glass tube fastened to the gauge cocks with compression fittings.

The assembly is attached to the vessel using upper and lower pipe flanges or fittings. Guard rods or an overall plastic protection tube can be obtained to protect the gauge glass. Gauge glasses can also be enclosed in metal enclosures to protect against breakage when used in boilers or high-pressure vessels, or for other safety requirements. A limited choice of materials is available for the gauge cocks; thus, gauge glasses are usually used for noncorrosive services. There are also flat glass gauges that typically are used at higher pressures.

The most corrosive applications can be handled with a double-walled glass pipe gauge glass. The inner wall is a heavy glass pipe and the outer pipe is plastic used as a shield. All wetted parts are either glass or Teflon®.

Armored Gauge Glasses. An *armored gauge glass* is a gauge glass that uses flat glasses enclosed in metal bodies and covers to protect against breakage when used in boilers or high-pressure vessels, or for other safety requirements. Armored gauges are typically used for high-pressure applications or for more dangerous materials and are available as a transparent type that uses two windows on each side of the gauge, or a single-window reflex type.

Armored gauge glasses can be obtained with or without gauge cocks or valves. Some chemical plants choose to use the same type of valves for their armored gauges as is used in the process piping. Armored gauges are available in various section lengths, which are then combined into the required length. If a very tall tank needs to have armored level gauges that cannot be handled with one gauge, several are used at overlapping locations so that all levels are visible.

Reflex Gauge Glasses. A *reflex gauge glass* is a flat gauge glass with a special vertical sawtooth surface that acts as a prism to improve readability. The light entering the portion of the prism in contact with the liquid is refracted into the tank and the glass appears dark. The light entering the portion of the prism above the liquid is reflected back out of the gauge and the glass appears silvery white. This feature is useful with clear or translucent liquids that are hard to see in a conventional gauge glass. **See Figure 13-3.**

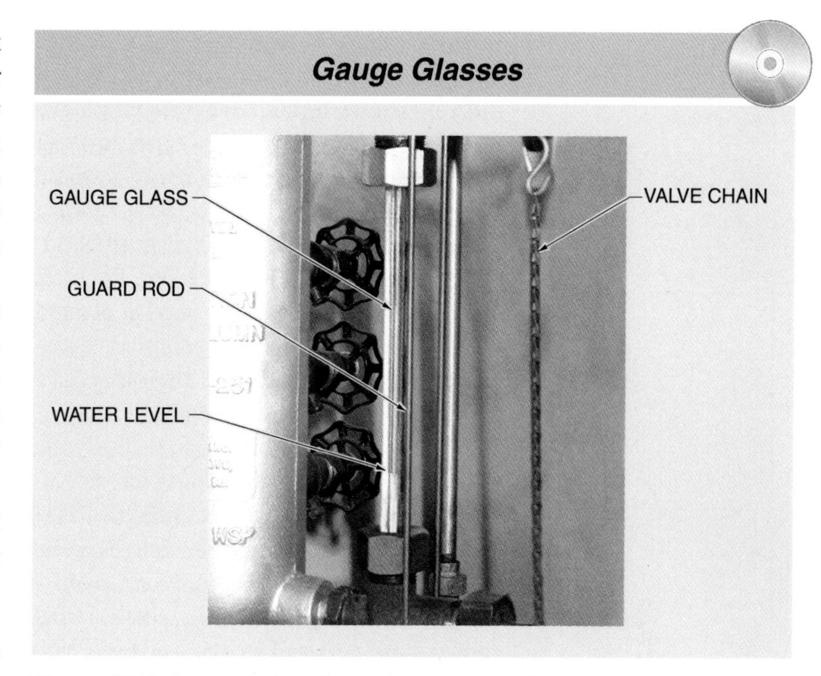

Figure 13-2. A gauge glass is used to measure the water level in a boiler.

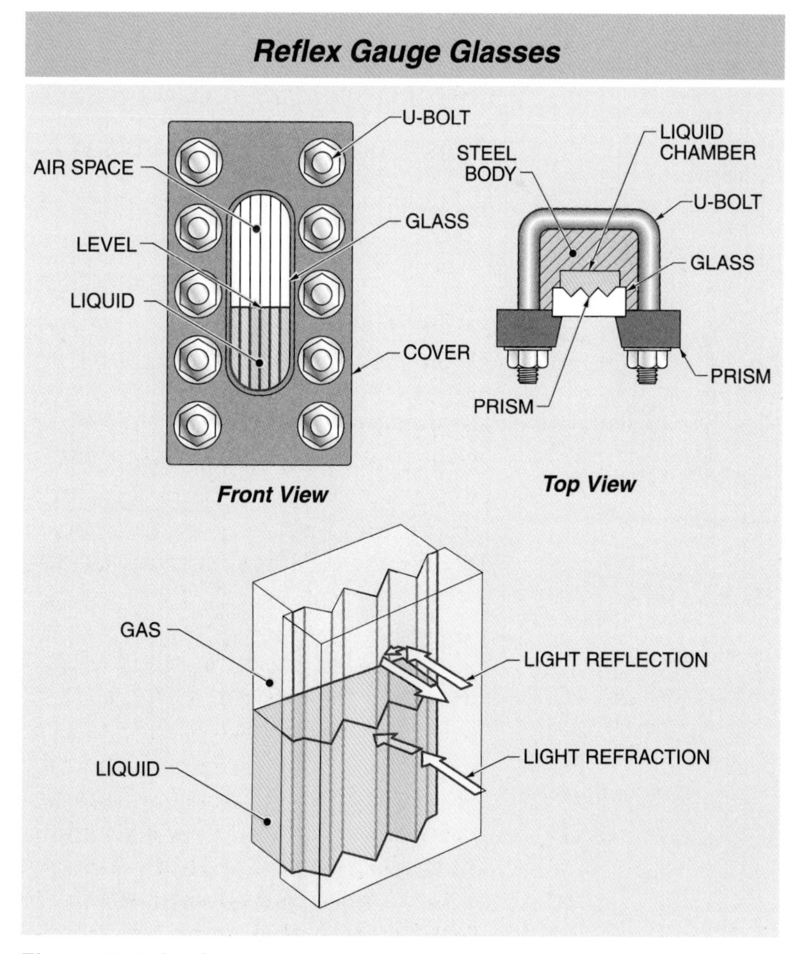

Figure 13-3. A reflex gauge glass uses light refraction to show level.

Magnetically Coupled Level Gauges

To determine liquid levels of corrosive applications, magnetically coupled level gauges are used. A *magnetically coupled level gauge* is a gauge that consists of a stainless steel float containing a magnet riding in a stainless steel tube where the level indicator consists of horizontally pivoted magnetized vanes painted yellow or white on one side and black on the other in a housing bolted to the level tube.

As the liquid level raises the float, the vanes flip from showing the black side to showing the yellow or white side. **See Figure 13-4.** The magnetic level gauge can be mounted to the side of a tank for easy visibility or to the top of a tank with the float down in the tank and an extension rod with the magnet in the gauge tube.

The magnetic field that connects the magnet in the float to the indicators on the outside of the tube is relatively weak. The magnetic field can be disrupted if the instrument is located close to an area where very high electrical currents are used such as near electric furnaces or electrolyzers.

Cable and Weight Systems

A *cable and weight system* is an intermittent full-range level measuring assembly consisting of a manual or remotely operated switch, a relay and a servomotor, a plumb bob for a weight, and a cable. The relay and servomotor are mounted at the top of the silo or tank. The servomotor lowers the weight until the weight touches the surface of the stored material and the tension on the cable is relieved. This causes the servomotor to stop momentarily.

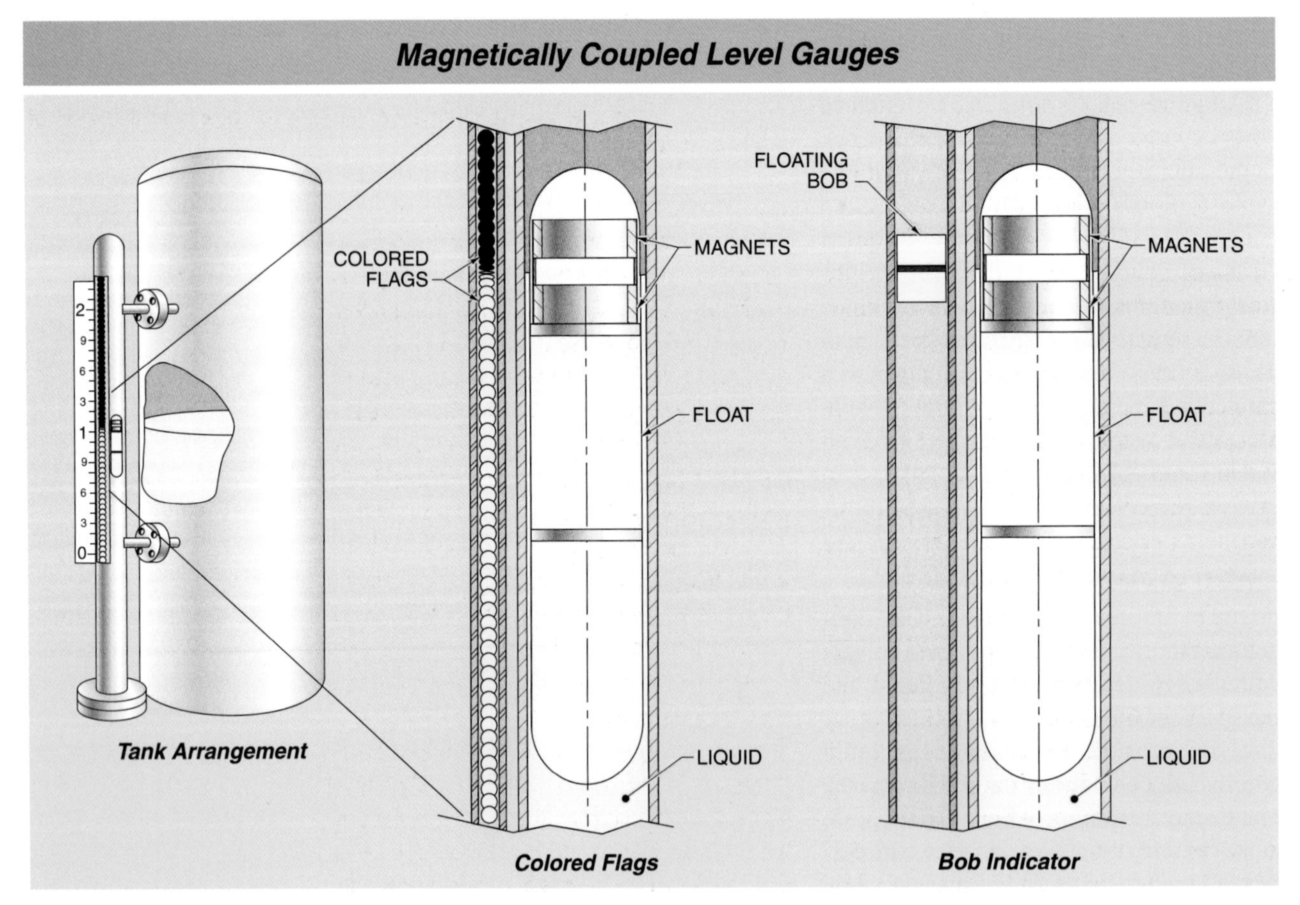

Figure 13-4. Magnetically coupled level gauges use a magnetic float to flip colored flags or move a floating bob to indicate level.

The cable length is read on an indicator at the tank or used as an input to a transmitter. The weight then returns to a rest position above the maximum level. **See Figure 13-5.** A cable and weight system is commonly used to measure granular materials in bins and silos. Dusty material can cause problems with the drive mechanism.

PRESSURE

There are many applications where it is more convenient to measure the pressure at the bottom of a tank than to measure the actual location of the top of the liquid. For example, a tank may be sealed to prevent the escape of volatile or toxic fluids. Common methods of using pressure to measure level are hydrostatic pressure and bubbler systems.

Hydrostatic Pressure

Hydrostatic pressure variations present at the base of a liquid column provide the means of determining liquid level in a storage vessel. As long as the liquid in the tank has a constant density, variations in pressure are caused only by variations in level. Some types of instruments use a large diaphragm that presses against a mechanical switch, and are suitable for use with granular bulk solids in tanks and silos.

For an open vessel, the simplest way to measure pressure is to connect a pressure gauge, switch, or transmitter to the side of a vessel at the lowest practical level so that any rise in level creates an increase in hydrostatic pressure. Therefore, a gauge or transmitter connected to the bottom of the tank can be calibrated in units of level.

Pressure Connections. If a pressure connection is higher than the tank bottom, any liquid below the connection is below the zero point of the gauge. This makes it necessary to correct the scale of the pressure sensor if the true level is to be measured. The distance from the measurement point to the bottom of the tank must be added to all the measurements on the scale to allow the true level to be indicated.

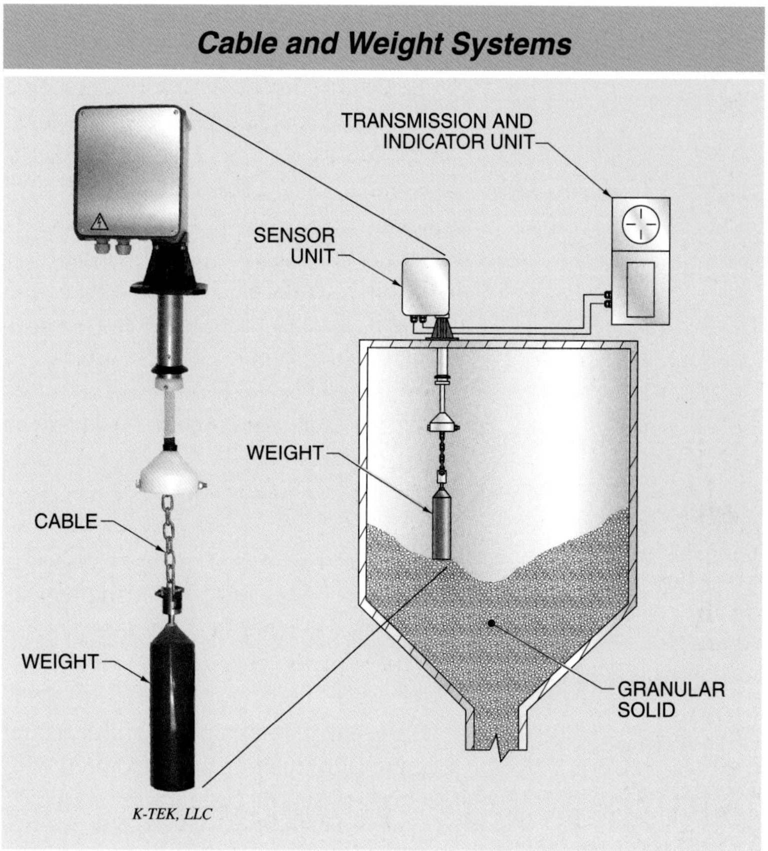

Cable and Weight Systems

Figure 13-5. A cable and weight system is used to determine the level of granular solids in a tank or silo.

For example, an open vertical cylindrical water tank 20′ high is filled with water and has a pressure connection 2′ from the bottom. **See Figure 13-6.** To measure the actual tank level, the measurement instrument is calibrated to measure a range of 0′ to 20′, but the zero calibration is suppressed 2′. When the level is at the connection point, the indicated level is 2′. With the level at 20′ the indicated level is 20′. This arrangement results in a situation where the indicated level can never read less than the height of the measurement connection, even if the tank is completely empty.

In general practice, the level of the connection is established as the zero level point. The measuring instrument is calibrated for a range of 0′ to 18′, the distance from the measurement connection (2′) to the maximum level (20′). The gauge is calibrated with a 0% to 100% output range, corresponding to an actual level range of 2′ to 20′.

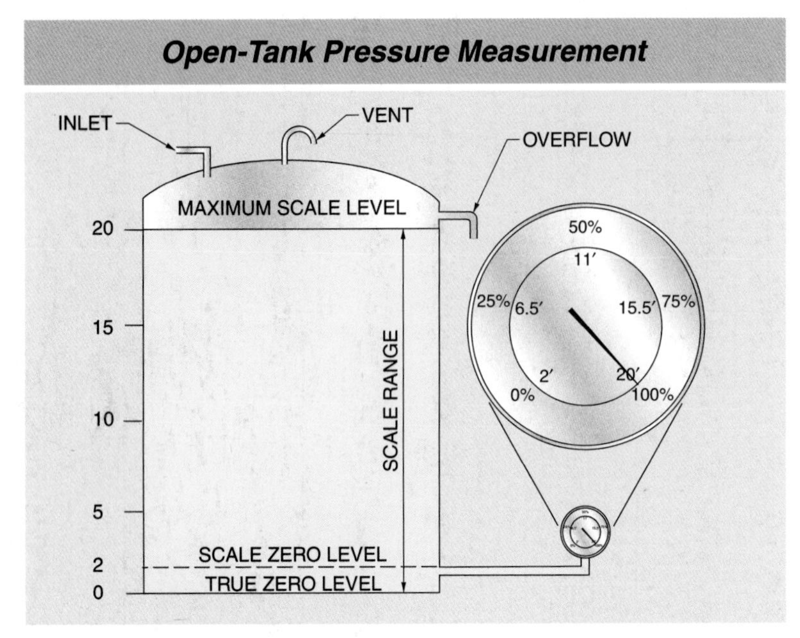

Figure 13-6. Pressure measurement can be used to determine level when the density is known.

If the tank is pressurized, a differential pressure (d/p) instrument is required with the high-pressure connection at the 2′ level and the low-pressure connection in the vapor space above the liquid. **See Figure 13-7.** The pressure in the high-pressure line is the sum of the tank pressure and the hydrostatic pressure due to the liquid height. The pressure in the low-pressure line is the tank pressure only.

For example, the same tank is pressurized to 50 psi. The low-pressure measurement above the water is 50 psi. The high-pressure measurement is 50 psi plus the hydrostatic pressure in the tank. The differential pressure is due to the height of the liquid. In this case, the differential pressure is 18′ of water. The correction for suppression gives a level of 20′. This method is useful when the vessel is sealed to prevent the escape of volatile or toxic liquids or vapors.

Modified Pressure Cells. Standard d/p cells used to measure level are susceptible to attack by corrosive fluids or blockage by fluids with solids. Several special d/p cells have been developed for the measurement of level. Flush diaphragm d/p cells are available with either 2″ or 3″ flanged connections. **See Figure 13-8.** The 3″ size is the most common and most accurate. The flanged connection, containing the high-pressure diaphragm, is designed to connect directly to a 3″ connection on the tank. The low-pressure connection is a ½″ threaded connection in the body of the d/p cell. The choices of materials of construction for the wetted parts of the diaphragm connection are quite varied, and available materials can be specified to handle most corrosive fluids.

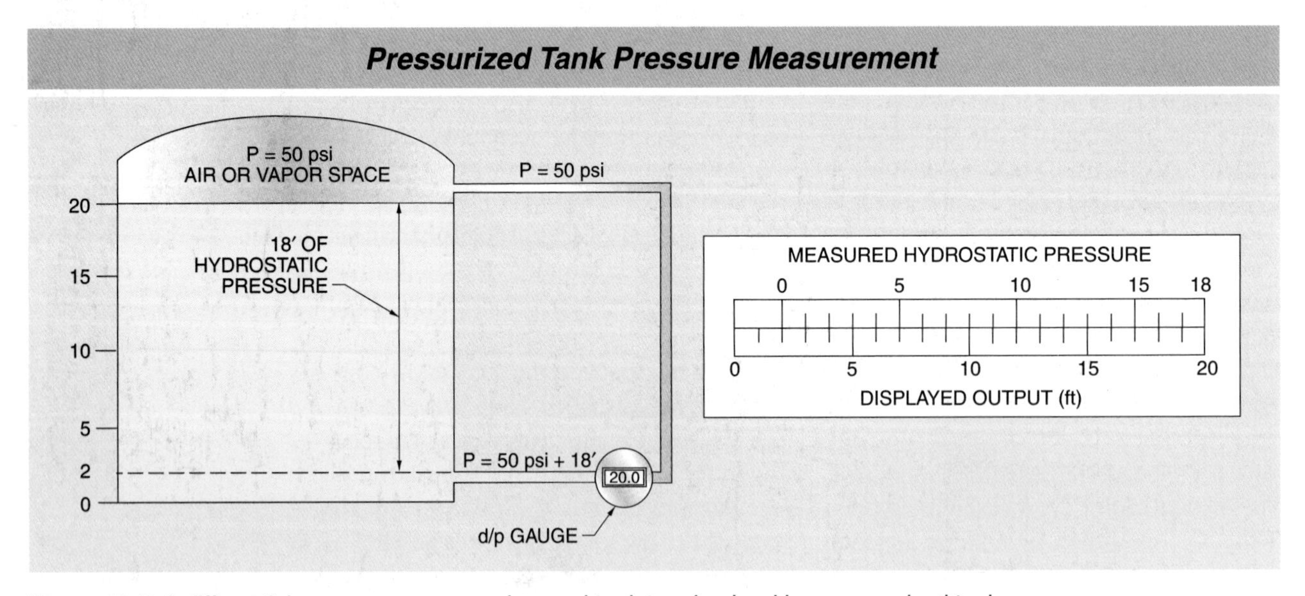

Figure 13-7. A differential pressure gauge can be used to determine level in a pressurized tank.

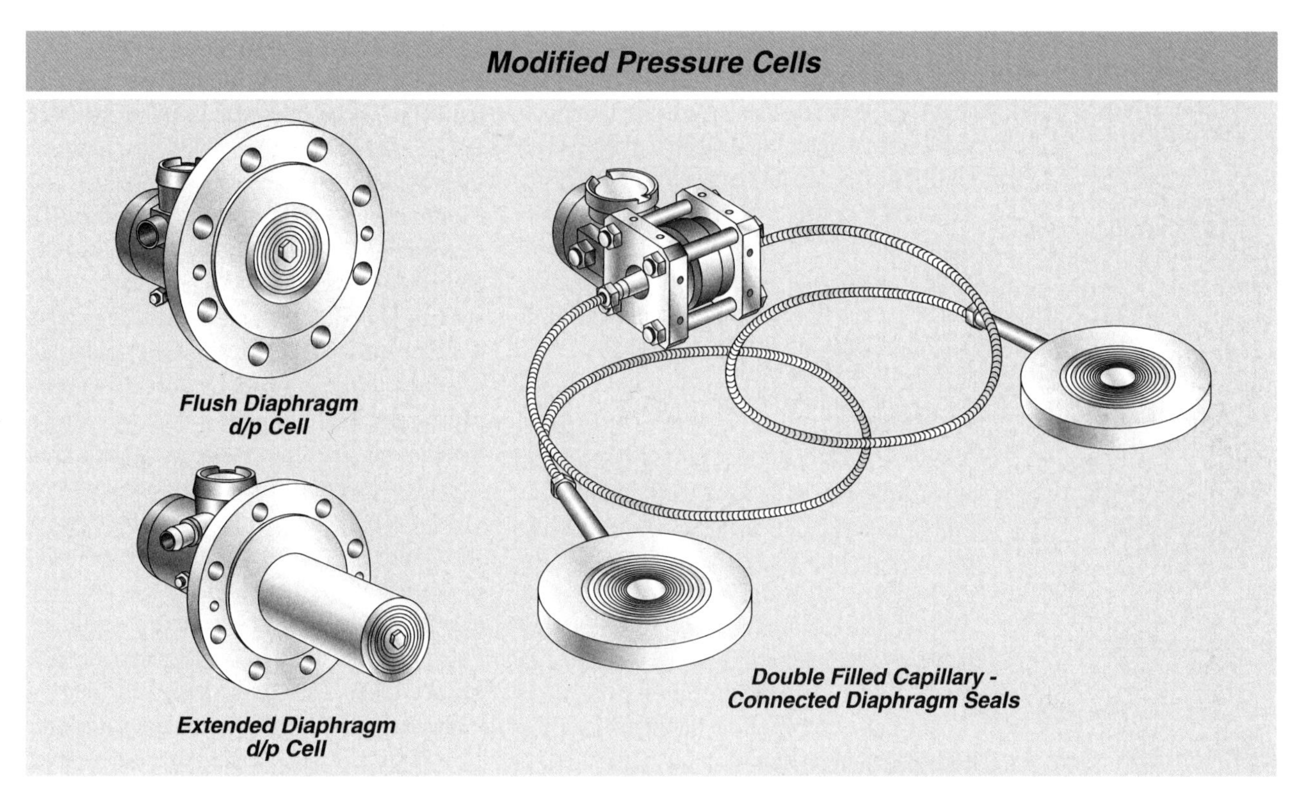

Figure 13-8. Level differential pressure cells come in a variety of configurations for different applications.

A second type of flush diaphragm d/p cell is the extended diaphragm d/p cell for use with fluids containing high quantities of solids. The extended diaphragm d/p cell uses a 4″ flanged connection with an extension of the diaphragm into the tank flanged nozzle so that the pressure-sensing surface is flush with the inside of the tank. This is the high-pressure connection. The low-pressure connection is a ½″ threaded connection in the body. The choice of materials of construction is not as extensive as for the flush diaphragm d/p cell.

A third type of level measuring d/p cell is the double filled system using flanged diaphragm seals and capillary systems. The diaphragm seal flange connections should be 3″. To minimize ambient temperature effects, it is important that the volume of the filling fluid be exactly equal for the high- and low-pressure measurement connections. It is also best to use measurement ranges of 0″ to 30″ differential or greater. A full range of materials of construction is available to handle all corrosive fluids.

Fuel oil tanks can be manufactured with a flat end because the oil is not stored under high pressure.

In all d/p cell level measurement applications of pressurized tanks, except the double filled capillary and diaphragm seal system, it is possible to have condensed vapors collect as liquids in the low-pressure connection. This can seriously affect the level measurement accuracy. To prevent this error in level measurement, the low-pressure connection can be purged with air or nitrogen, whichever is compatible with the process vapor.

Bubblers

A *bubbler* is a level measuring instrument consisting of a tube extending to the bottom of a vessel; a pressure gauge, single-leg manometer, transmitter, or recorder; a purge flowmeter or sight feed bubble; and a pressure regulator. A gas is slowly fed into the bubbler system until the pressure is equal to the hydrostatic head of the liquid in the tank. At that point, the gas flow bubbles out of the end of the bubble tube.

The gas flow rates are low to eliminate any pressure loss in the bubbler system tubing. Sight feed bubblers are commonly used to regulate the gas flow through the tube because they show the flow rate as a series of bubbles. Small gas flow rotameters can also be used. The hydrostatic head is easily converted to level. As the level in the vessel changes, the pressure in the tube and the measuring instrument changes proportionally. **See Figure 13-9.**

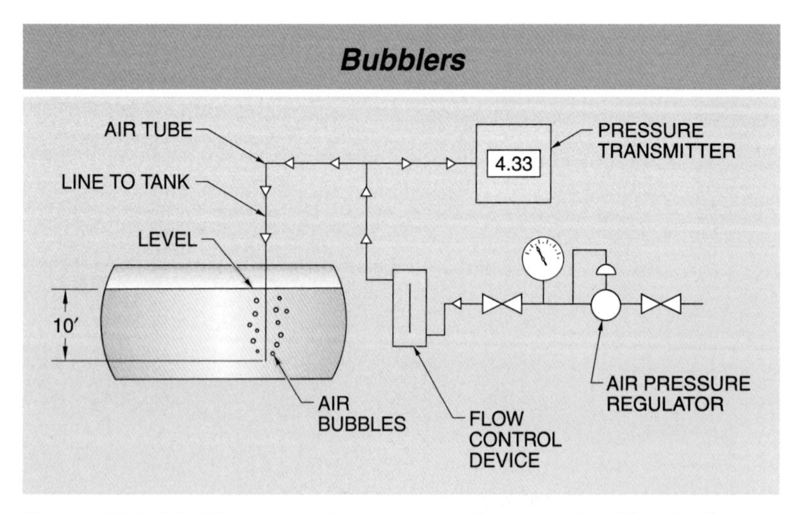

Figure 13-9. A bubbler uses air pressure to determine level in a tank.

For example, water has a hydrostatic pressure of 0.433 psi per foot of height. If a tank has 10′ of water in it, the equivalent pressure is 4.33 psi ($10 \times 0.433 = 4.33$). The gas tube must have a pressure of at least 4.33 psi to overcome the hydrostatic pressure of the 10′ of water. The pressure regulator can be set to any value higher than 4.33 psi and the pressure measured in the air tube is 4.33 psi.

Bubblers are typically used with tanks that are open to the atmosphere, but the system can be adapted to a closed tank by using a differential pressure gauge or transmitter. The high-pressure side of the measurement instrument is connected to the bubbler pressure and the low-pressure side connected to the top of the vessel. The low-pressure side is usually arranged to have a small gas flow the same as the high-pressure side. **See Figure 13-10.**

It is also possible to use single-leg manometers for the local indication of the tank level when using a bubbler system. The vertical column of manometer fluid makes a convenient representation of the liquid level in the tank. The manometer scale is calibrated to represent feet of the tank liquid. Single-leg manometers can also be used with pressurized tanks.

The gas sent through the tube must be compatible with the liquid in the tank. Air is the most common gas used, but nitrogen is used on occasion. A bubbler is susceptible to freezing or plugging and must include means to clean out the tube. Solid deposits tend to form at the interface of the bubbler gas and the tank liquid. If the solids are water soluble, a small trickle of water added to the top of the bubble tube can considerably extend the time between bubbler cleanouts.

A bubbler is also subject to error if the liquid density changes. A change in the density of the process liquid affects the measurement and requires compensation in the measurement. A liquid with a density that is lower than expected indicates a lower level than the actual level in the tank. This can cause the tank to be overfilled, resulting in an overflow. When calibrating the instrument, the scales must be set up so that spills cannot occur even at the lowest possible liquid density expected.

FLOATS AND DISPLACERS

Some types of level instruments depend on the buoyancy of an object to measure level. A floating object determines the surface of a liquid. A solid object lowered down to the top of the material in a silo determines the level of the solids. Various types of floats and displacers are commonly used to measure level.

Floats

A *float* is a point level measuring instrument consisting of a hollow ball that floats on top of a liquid in a tank and is attached to the instrument. The float is connected by a lever to an ON/OFF switch actuated by the movement of the float. **See Figure 13-11.** Floats can be used to indicate tank levels, actuate alarms or shutdown switches, or even mechanically control valves.

The switch can start a pump when the float is at one position and stop the pump at the other position. The float may be inside the tank or enclosed in an attached cage or in a stilling well to minimize turbulence. Alarm contacts may also be included.

Tape Floats. A *tape float* is a continuous level measuring instrument consisting of a floating object connected by a chain, rope, or wire to a counterweight, which is the level pointer. The float rides up and down on two guide wires that keep the float in a specified position. A scale fastened to the outside of the tank shows the reversed tank level with 100% being at the bottom and 0% being at the top. When the float is at the bottom of the tank, the tank is empty and the counterweight is at the top. When the float is at the top of the tank, the indicator pointer is at the bottom of the scale.

As the liquid level changes, the float moves up or down with the liquid. The counterweight keeps tension on the chain and the pointer moves up or down in the reverse direction to indicate the level of the liquid surface. A tape float is subject to possible corrosion and mechanical problems with buildup on the chain causing hang-ups on the pulleys.

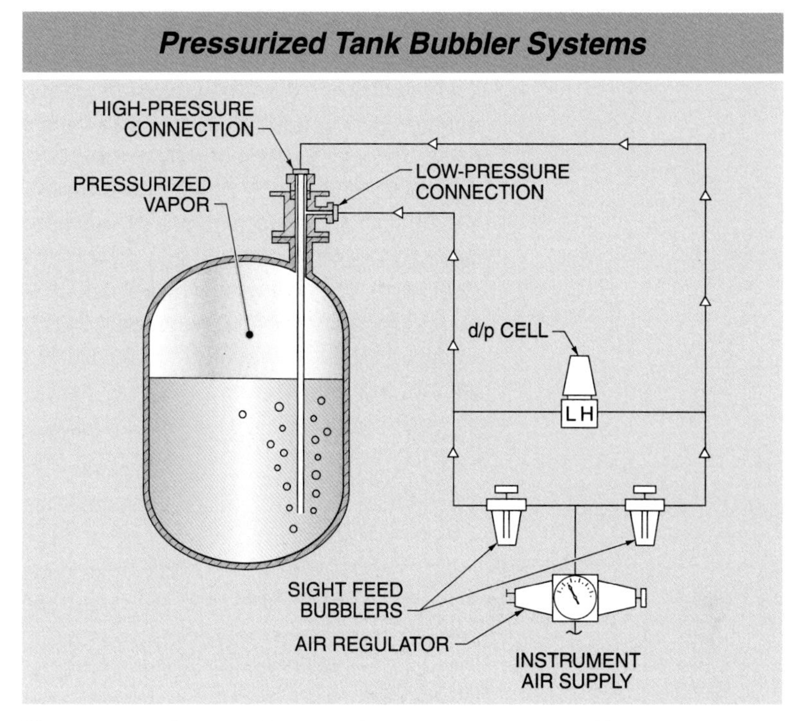

Figure 13-10. A pressurized tank bubbler system uses a differential pressure transmitter to determine the pressure due to the height of the liquid.

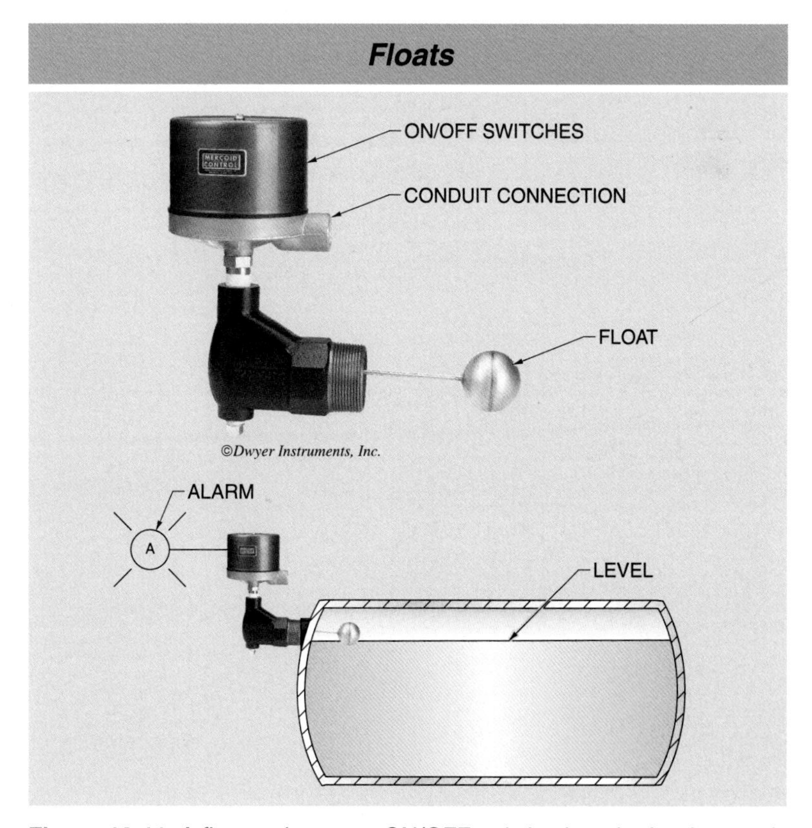

Figure 13-11. A float activates an ON/OFF switch when the level exceeds the alarm level.

Tanks containing flammable liquids are often padded with nitrogen to a pressure of a few inches of water. In these applications the tape between the float and the pointer passes through a group of pulleys, forming a pressure seal that is filled with a liquid. The float should be located in a stilling tank or in a part of the vessel that minimizes turbulence. These devices typically are used for local display only but may be used with a transmitter for continuous level measuring instruments. **See Figure 13-12.** An alternative readout consists of a housing containing a spring-powered take-up reel and a dial readout of the level.

Float-and-Dial Instruments. Special float-and-dial level instruments are available for horizontal tanks. A float is attached to a long arm, long enough for the float to reach the top and bottom of the tank, and is coupled through a seal to a dial level indicator. The dial scale is calibrated for the volume in the tank. **See Figure 13-13.** These level gauges are used for measuring clean, noncorrosive liquids stored under pressure, such as ammonia or methyl chloride, and must be ordered for each specific tank.

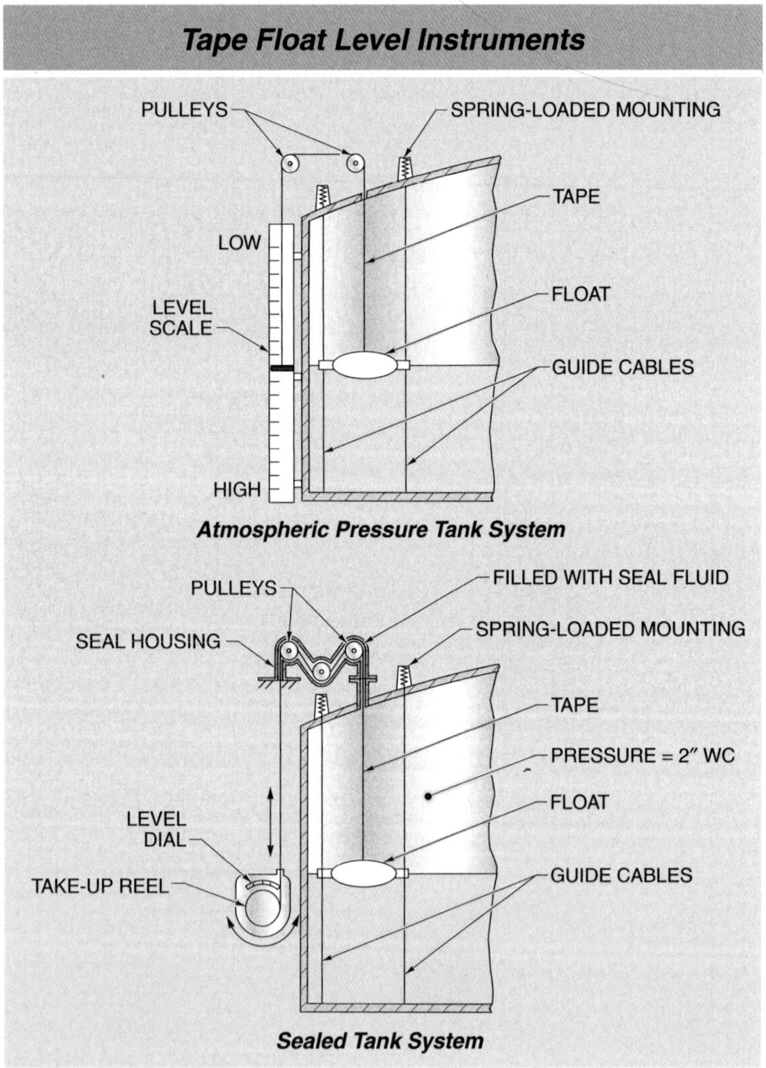

Tape Float Level Instruments

PULLEYS
SPRING-LOADED MOUNTING
TAPE
LOW
FLOAT
LEVEL SCALE
GUIDE CABLES
HIGH

Atmospheric Pressure Tank System

PULLEYS
FILLED WITH SEAL FLUID
SEAL HOUSING
SPRING-LOADED MOUNTING
TAPE
PRESSURE = 2″ WC
FLOAT
LEVEL DIAL
TAKE-UP REEL
GUIDE CABLES

Sealed Tank System

Figure 13-12. A tape float uses a floating object connected to a tape or transmitter to measure level.

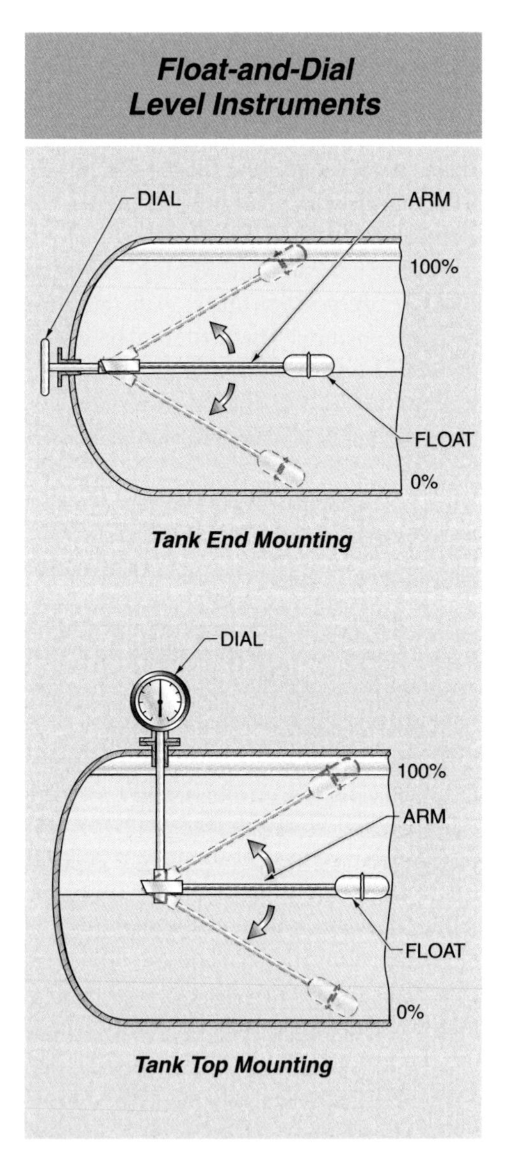

Figure 13-13. A float-and-dial level instrument is used in horizontal tanks to measure level.

Float Control Valves. There are float systems that are mounted in a housing and connected to a control valve mechanism. The housing is mounted at the desired control level. A drop in level opens the valve, allowing water to be fed into the tank. As the tank fills, the float rises and shuts off the valve. **See Figure 13-14.** These float control systems are used to control feedwater to small boilers.

TECH FACT

Paddle wheel switches are commonly used in pneumatic conveying systems to determine if any material has dropped out or if a conveying tube is plugged. When a tube is plugged, material backs up and blocks the free rotation of the paddle wheel.

Float Control Valves

TANK
FEEDER CONTROL VALVE
BLOWDOWN VALVE
MAKEUP WATER FLOW
McDonnell & Miller

Figure 13-14. A float control valve prevents a low-water condition in a boiler.

Displacers

A *displacer* is a liquid level measuring instrument consisting of a buoyant cylindrical object, heavier than the liquid, that is immersed in the liquid and connected to a spring or torsion device that measures the buoyancy of the cylinder. Cylinders are available in various lengths from 14″ to 60″. The measured level ranges match the displacer lengths.

The displacer is usually mounted in a housing called a cage that eliminates fluid turbulence. When the displacer is supported by an arm connected to the sealed end of a torsion tube, the sensing system becomes separated from the output devices. A rod running down the center of the torsion tube, fastened at the sealed end, transfers the amount of rotation at the sealed end of the torsion tube to the various output devices. **See Figure 13-15.**

An output device can be a pneumatic transmitter, a pneumatic controller, or a 4 mA to 20 mA transmitter. The displacer, torsion tube, wetted parts, and mounting cage are available in various materials. This type of level instrument has been used in many applications including liquid chlorine. Displacers supported by spring systems are more susceptible to corrosion and contamination.

A reflex gauge glass is used for applications where the liquid is hard to see in a standard gauge glass.

Displacers can be used to operate several level switches because of the greater amount of travel of the displacer system compared to float-type level switches. A magnet connected to the displacer support mechanism is used to actuate the level switches.

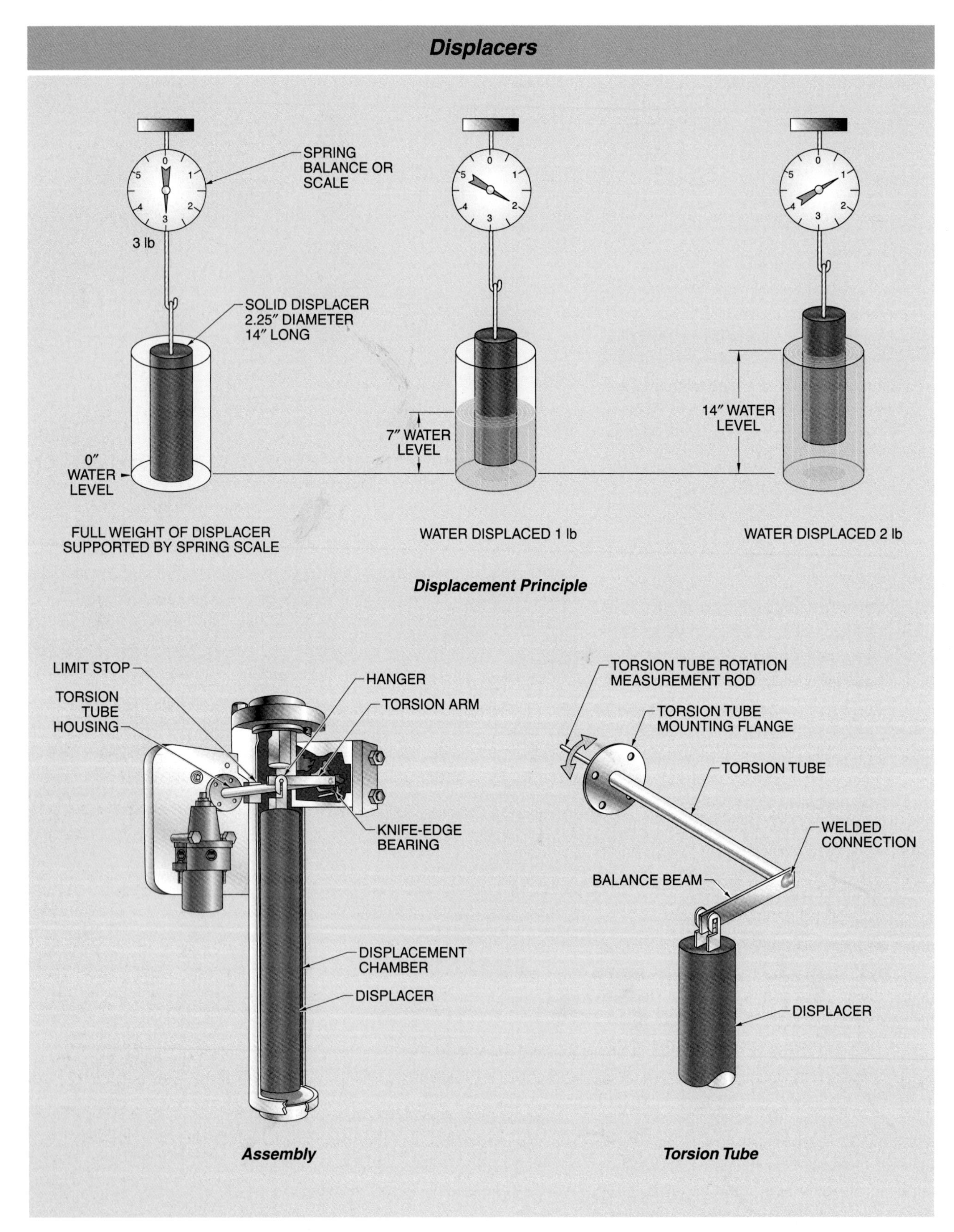

Figure 13-15. A displacer measures the buoyancy of an object to determine the level of a liquid.

PADDLE WHEEL SWITCHES

A *paddle wheel switch* is a point level measuring device consisting of a drive motor and a rotating paddle wheel mounted inside a tank. As the level in the tank rises and touches the paddle wheel, the torque required to turn the paddle wheel increases. The increased torque activates a switch that can be used to stop or start equipment or signal an alarm. Damage to the paddle wheel drive motor and mechanism is prevented by a clutch that slips when torque becomes too high.

A paddle wheel switch is commonly used to measure the level of granular solids in pneumatic conveying equipment and in bins, tanks, and silos. Paddle wheel switches are subject to problems from vibration and damage from material being added to a tank or silo. **See Figure 13-16.**

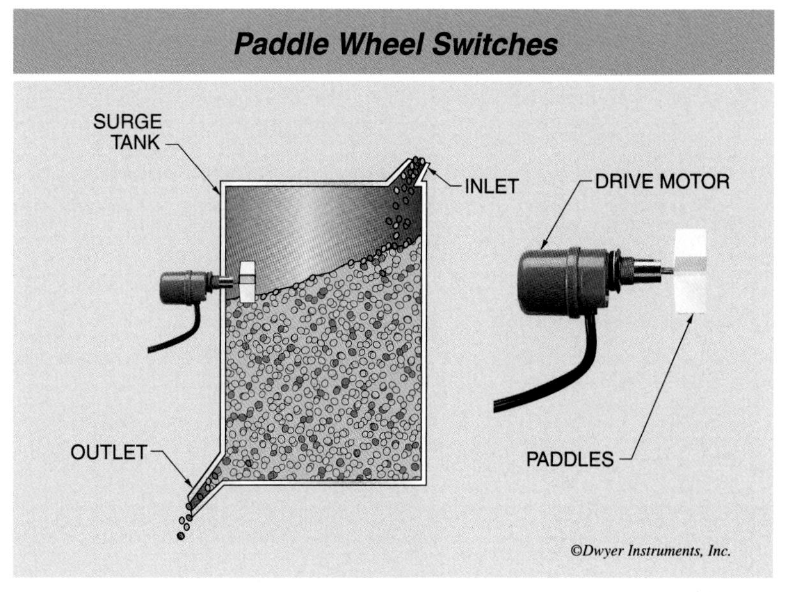

Figure 13-16. A paddle wheel switch stops rotating when it comes in contact with granular solids in a tank or hopper.

KEY TERMS

- *point level measurement:* A method of level measurement where the only concern is whether the amount of material is within the desired limits.
- *continuous level measurement:* A method of tracking the change of level over a range of values.
- *gauge glass:* A continuous level measuring instrument that consists of a glass tube connected above and below the liquid level in a tank and that allows the liquid level to be observed visually.
- *armored gauge glass:* A gauge glass that uses flat glasses enclosed in metal bodies and covers to protect against breakage when used in boilers or high-pressure vessels, or for other safety requirements.
- *reflex gauge glass:* A flat gauge glass with a special vertical sawtooth surface that acts as a prism to improve readability.
- *magnetically coupled level gauge:* A gauge that consists of a stainless steel float containing a magnet riding in a stainless steel tube where the level indicator consists of horizontally pivoted magnetized vanes painted yellow or white on one side and black on the other in a housing bolted to the level tube.
- *cable and weight system:* An intermittent full-range level measuring assembly consisting of a manual or remotely operated switch, a relay and a servomotor, a plumb bob for a weight, and a cable.
- *bubbler:* A level measuring instrument consisting of a tube extending to the bottom of a vessel; a pressure gauge, single-leg manometer, transmitter, or recorder; a purge flowmeter or sight feed bubble; and a pressure regulator.
- *float:* A point level measuring instrument consisting of a hollow ball that floats on top of a liquid in a tank and is attached to the instrument.

KEY TERMS *(continued)*

- *tape float:* A continuous level measuring instrument consisting of a floating object connected by a chain, rope, or wire to a counterweight, which is the level pointer.
- *displacer:* A liquid level measuring instrument consisting of a buoyant cylindrical object, heavier than the liquid, that is immersed in the liquid and connected to a spring or torsion device that measures the buoyancy of the cylinder.
- *paddle wheel switch:* A point level measuring device consisting of a drive motor and a rotating paddle wheel mounted inside a tank.

REVIEW QUESTIONS

1. What is the difference between point and continuous level measurements?
2. Define gauge glass and describe the different types of gauge glasses.
3. Why may pressure measurement be used as an indirect measurement of level?
4. Describe the operation of a tape float.
5. Describe the operation of a paddle wheel switch.

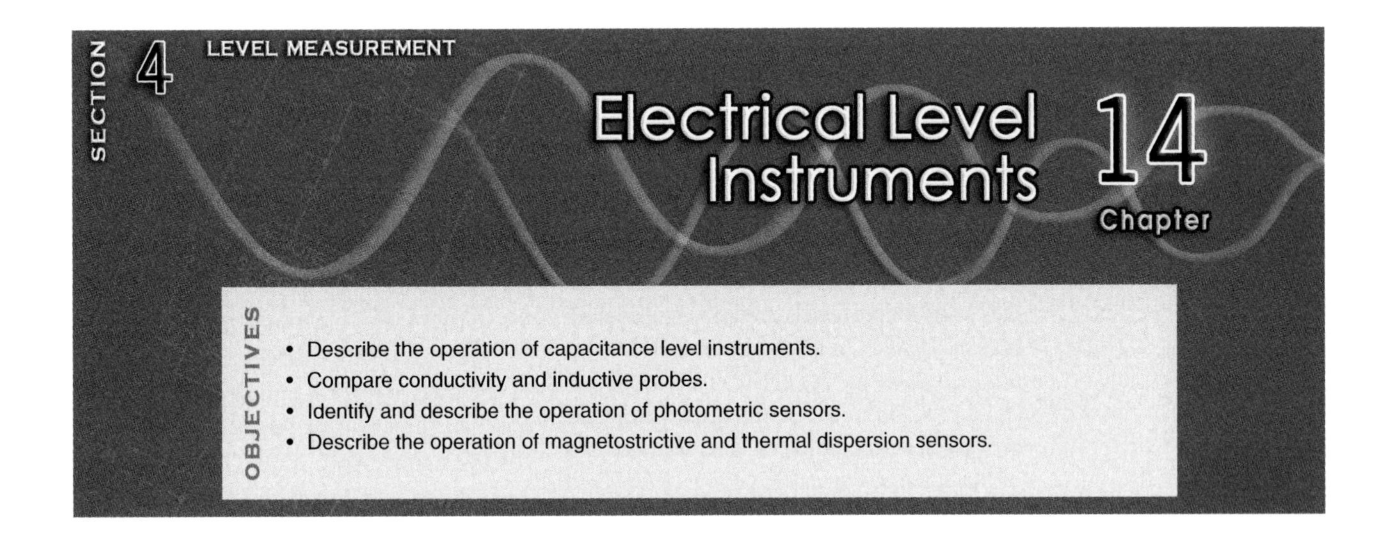

CAPACITANCE SENSORS

The use of capacitance to measure level is based on the electrical relationships between capacitance and frequency. Reactance is the term used to describe the resistance of a circuit to the flow of alternating current. In practical applications there is usually a small amount of resistance in addition to the capacitance, but the resistance must be larger than the impedance of the capacitor for the level measurement to be effective. As long as the resistance is high compared to the capacitance reactance, the resistance has almost no effect on the capacitance level measurement. This means that capacitance level measurement does not work well with very conductive liquids.

Capacitance

Capacitance is the ability of an electrical device to store charge as the result of the separation of charge. *Admittance* is the ability of a circuit to conduct alternating current and is the reciprocal of impedance. A *capacitor* is an electrical device that stores electrical energy by means of an electrostatic field. A *dielectric* is the insulating material between the conductors of a capacitor.

The effectiveness of a dielectric is compared to that of air or a vacuum. The *dielectric constant* is the ratio of the insulating ability of a material to the insulating ability of a vacuum. See **Figure 14-1.** For air, the dielectric constant is 1.0 and for water it is about 80. This means that water is 80 times more effective as a dielectric than air. Since dielectric constants are ratios, there are no units.

Dielectric Comparison

Material	Dielectric Constant*
Vacuum	1
Acetone	21
Ammonia (-27°F)	22
Calcium Carbonate	9.1
Ethanol	24
Freon 12	2.4
Kerosene	1.8
Methanol	33
Naphthalene	2.5
Sand	3.5
Sugar	3.0
Toluene	2.4
Water	80
Water (212°F)	48

*at 68°F unless noted

Figure 14-1. The dielectric constant determines the effectiveness of a capacitor.

TECH FACT

Level measurements for liquid calculated from a differential pressure measurement are subject to errors if the process liquid can plug the capillary tubes.

Capacitance Probes

A *capacitance probe* is a part of a level measuring instrument and consists of a metal rod inserted into a tank or vessel, with a high-frequency alternating voltage applied to it and a means to measure the current that flows between the rod and a second conductor. The metal rod is electrically insulated from the tank. A bare metal rod can only be used with nonconductive liquids. Liquids that are conductive require the use of a plastic-coated rod.

The amount of capacitance depends on the dielectric constant, the surface area of the conductors, and the distance between the conductors. If the capacitance probe is fixed in place, the surface area and distance between the conductors cannot change. As the level in the vessel rises, the capacitance increases because the material in the vessel replaces the air or vapor between the conductors. The granular solid or liquid material has a higher dielectric constant than air, and the changed capacitance is measured and used to determine the level.

The current is proportional to the admittance or capacitance from the metallic rod to the second conductor. For many configurations, the most convenient second conductor is the tank wall. Therefore, a capacitance probe measures current flow from the metal rod, through the material in a tank, to ground.

© *Dwyer Instruments, Inc.*

A float-actuated level switch can be used to activate a circuit.

In order for a current to flow in a capacitive circuit, alternating current must be used. The power supply for the circuit is generally standard 120 VAC although other voltages are available. This is converted by an oscillator to a 100 kHz radio frequency (RF) input to a bridge circuit. For this reason, these instruments are often called RF capacitance probes.

The capacitance probe is in one arm of the bridge with a tuning capacitor and another capacitor is used to compensate for the capacitance of the coaxial cable. The bridge is balanced when the probe is not in contact with the material in the tank. When the level of material in the vessel covers the probe, the capacitance changes, the bridge becomes unbalanced, and a current starts to flow through the bridge. The current can be converted to DC by a demodulator and then amplified to activate a relay or other electrical device. **See Figure 14-2.**

Single-Point Capacitance Probes

A single-point level-control capacitance probe is used as a switch to signal an alarm or actuate a circuit when the level in a tank or vessel reaches a specified level. The probe is one conductor of an electrical capacitor. When the probe is covered or uncovered by the process material, the capacitance suddenly changes and the amount of current that can flow changes. The change is detected in a bridge circuit.

If the material in the vessel is conductive, an insulated probe is used. The probe is one conductor, the material is the other conductor, and the insulated shield is the dielectric. The material must be connected to ground either through the metal walls of a tank or through a separate grounding probe. If the material in the vessel is nonconductive, an uninsulated probe can be used. The probe is one conductor, the vessel wall is the other conductor, and the material in the vessel is the dielectric. **See Figure 14-3.**

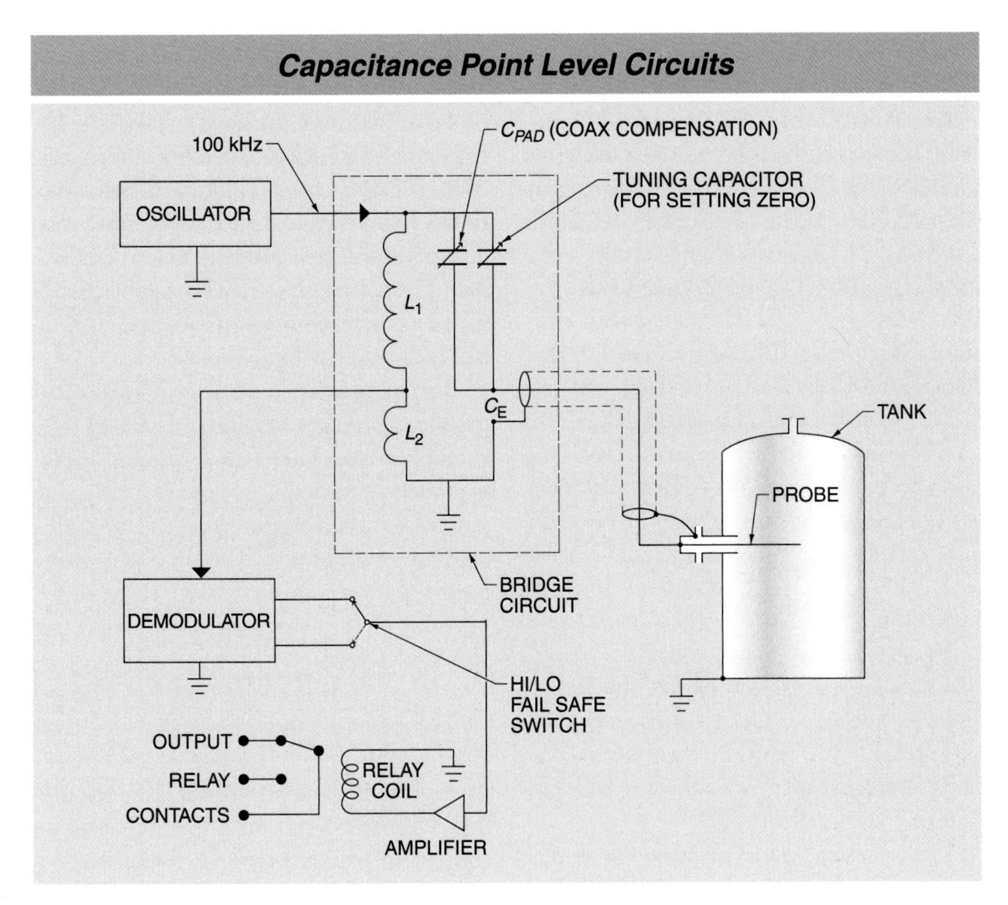

Figure 14-2. A bridge circuit is used to measure the capacitance of a sensor.

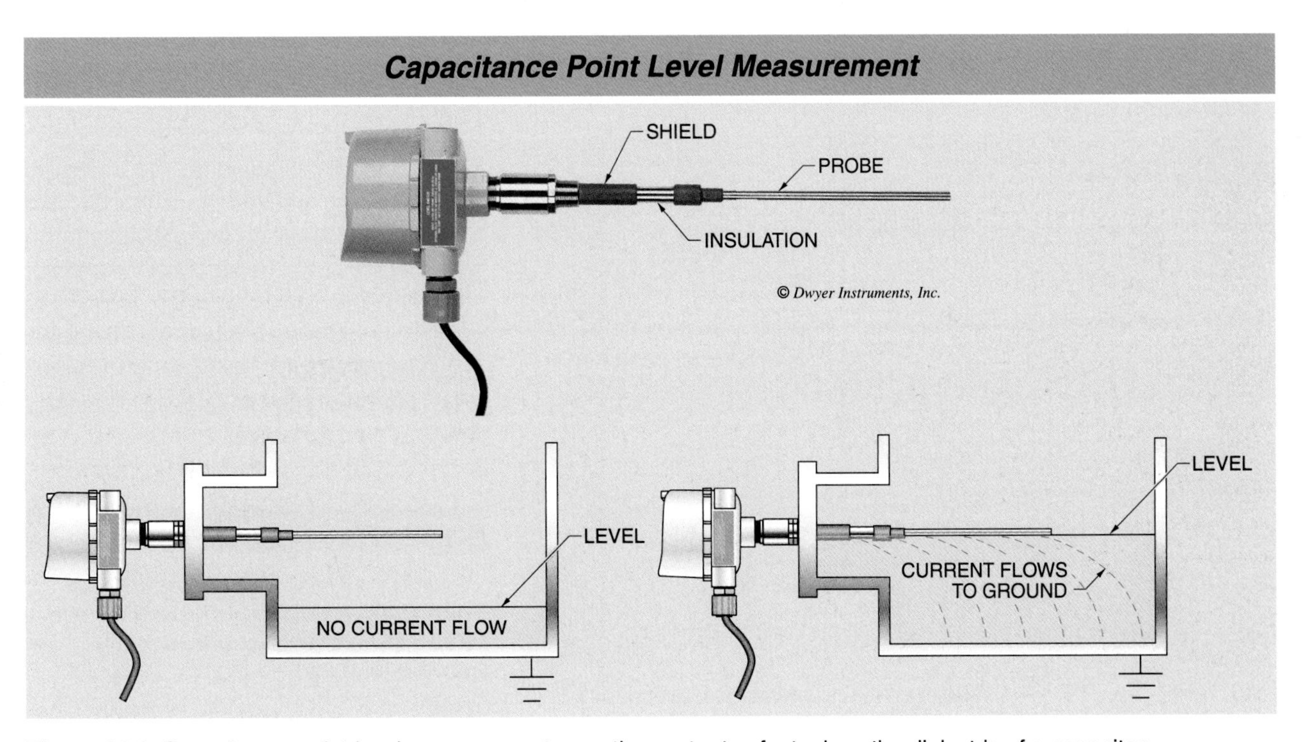

Figure 14-3. Capacitance point level measurement uses the contents of a tank as the dielectric of a capacitor.

Capacitance Continuous Probes

For capacitance continuous level measuring instruments, a vertically suspended capacitance probe can be used for measuring the level in a tank. As the level in the tank rises up the probe, the amount of current that can flow increases. The increased current is directly proportional to level. **See Figure 14-4.**

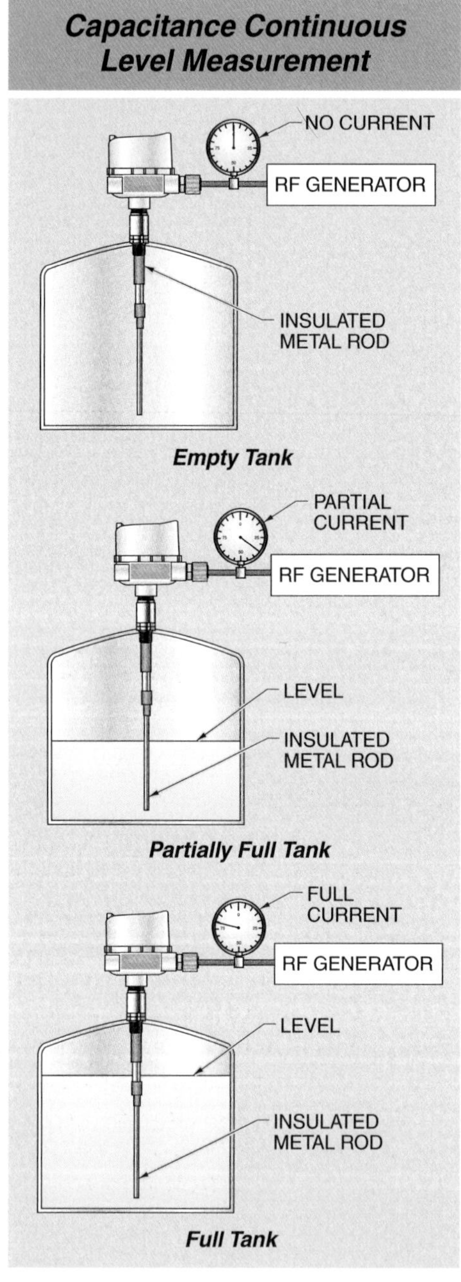

Figure 14-4. Capacitance continuous level measurement determines the changing capacitance of a tank as it fills to determine the level.

Capacitance Level Measurement

The buildup of process material on the probe sometimes presents a problem because the presence of a coating can act as a dielectric even when the level is below the probe. Manufacturers have developed probe designs and added adjustments to diminish the effect of coating as well as changes in the physical properties of the process fluid like density or composition.

For example, a conductive shield enclosing the rod and its insulation is added. The shield is connected to the oscillator so that the rod and the shield are always at the same potential and no current can flow through a coating on the shield. The tip of the probe is still able to conduct current through the material to the tank so that the level can still be detected.

A concern is the length of the shielded cable connecting the probe to the electrical circuit. If the cable is longer than 3′ or 4′, its own capacitance becomes large enough that it must be neutralized by including an equivalent capacitance and balancing the two so that the current to the demodulator is unaffected by the cable capacitance. **See Figure 14-5.**

The calibration of RF continuous level measuring instruments requires the use of special equipment normally provided by the manufacturer. Capacitance probes are affected by changes in the dielectric constant of the material in the vessel. Care must be taken to use these units only in situations where the dielectric constant does not change. If this cannot be done, the system must be recalibrated when the dielectric constant does change. RF capacitance probe level detectors can operate with granular solids as well as liquids.

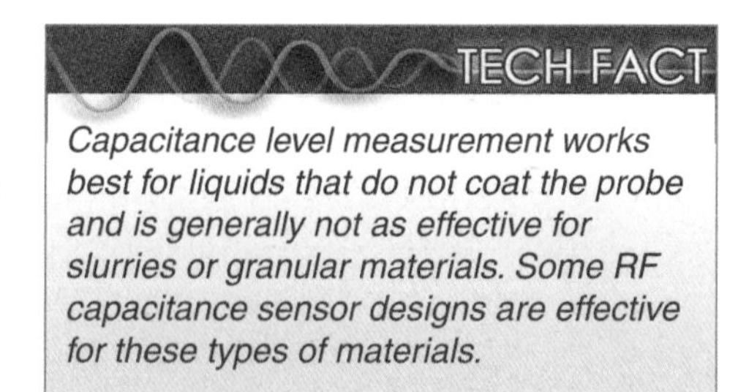

Capacitance level measurement works best for liquids that do not coat the probe and is generally not as effective for slurries or granular materials. Some RF capacitance sensor designs are effective for these types of materials.

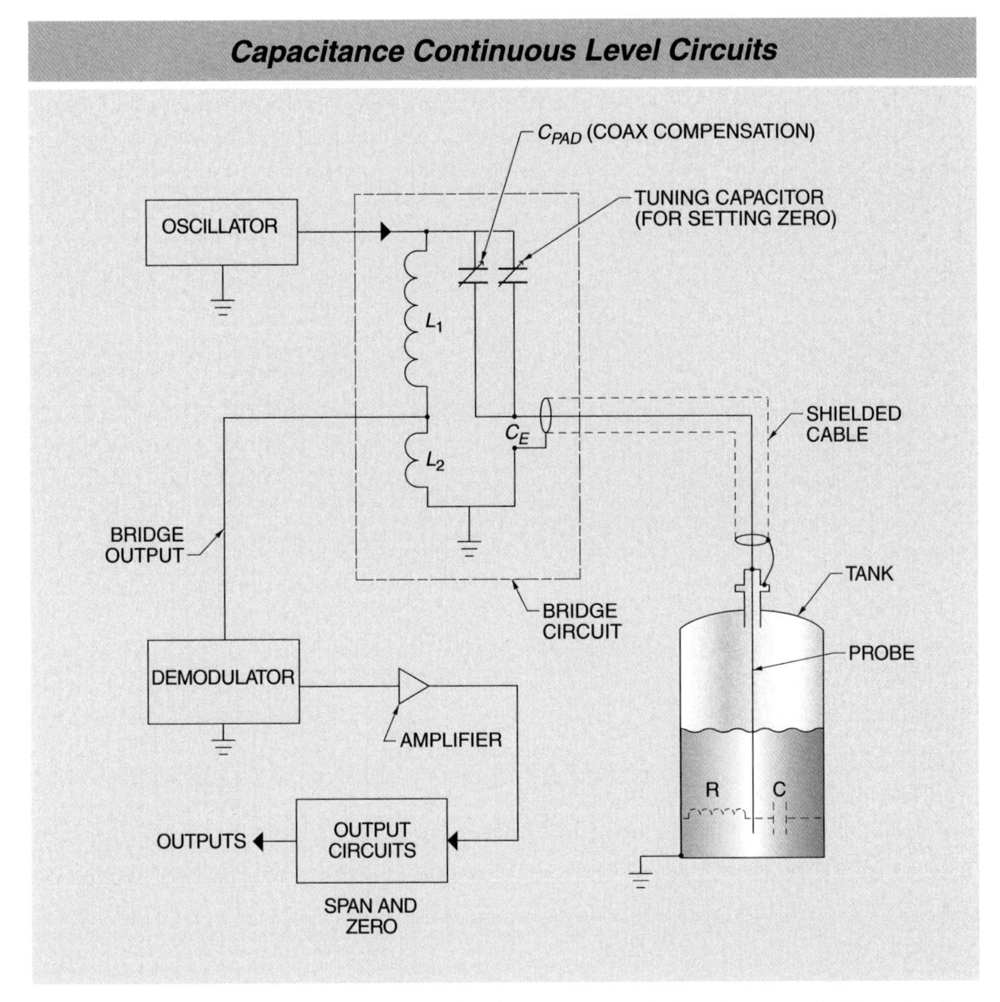

Figure 14-5. A capacitance continuous level measurement circuit uses a bridge circuit to measure the capacitance and a demodulator and amplifier to send a level signal.

CONDUCTIVITY AND INDUCTIVE PROBES

A *conductivity probe* is a point level measuring system consisting of a circuit of two or more probes or electrodes, or an electrode and the vessel wall where the material in the vessel completes the circuit as the level rises in the vessel. Conductivity level probes can only be used with conductive liquids. The body of the probe, often called an electrode holder, resembles an automotive spark plug and is threaded into a receptacle or bolted to a flange at the top of the tank to act as an electrical ground. The center element of the probe is the conductor. The length of the center element establishes the level control point.

Probes may be supplied with high-level AC or low-level DC voltage. The nature of the process liquid is the determining factor in the choice of power supply. The sensing circuits can be either inductive or electronic circuits. The inductive circuits use a high-voltage AC power supply to force enough current through the liquid and the electrodes to keep a relay energized. **See Figure 14-6.** For example, boiler water requires over 200 volts. This can be a potential shock hazard for operators or technicians. The electronic circuits use 6 VDC power on the electrodes and a transistorized sensing circuit that eliminates any danger of electrical shocks.

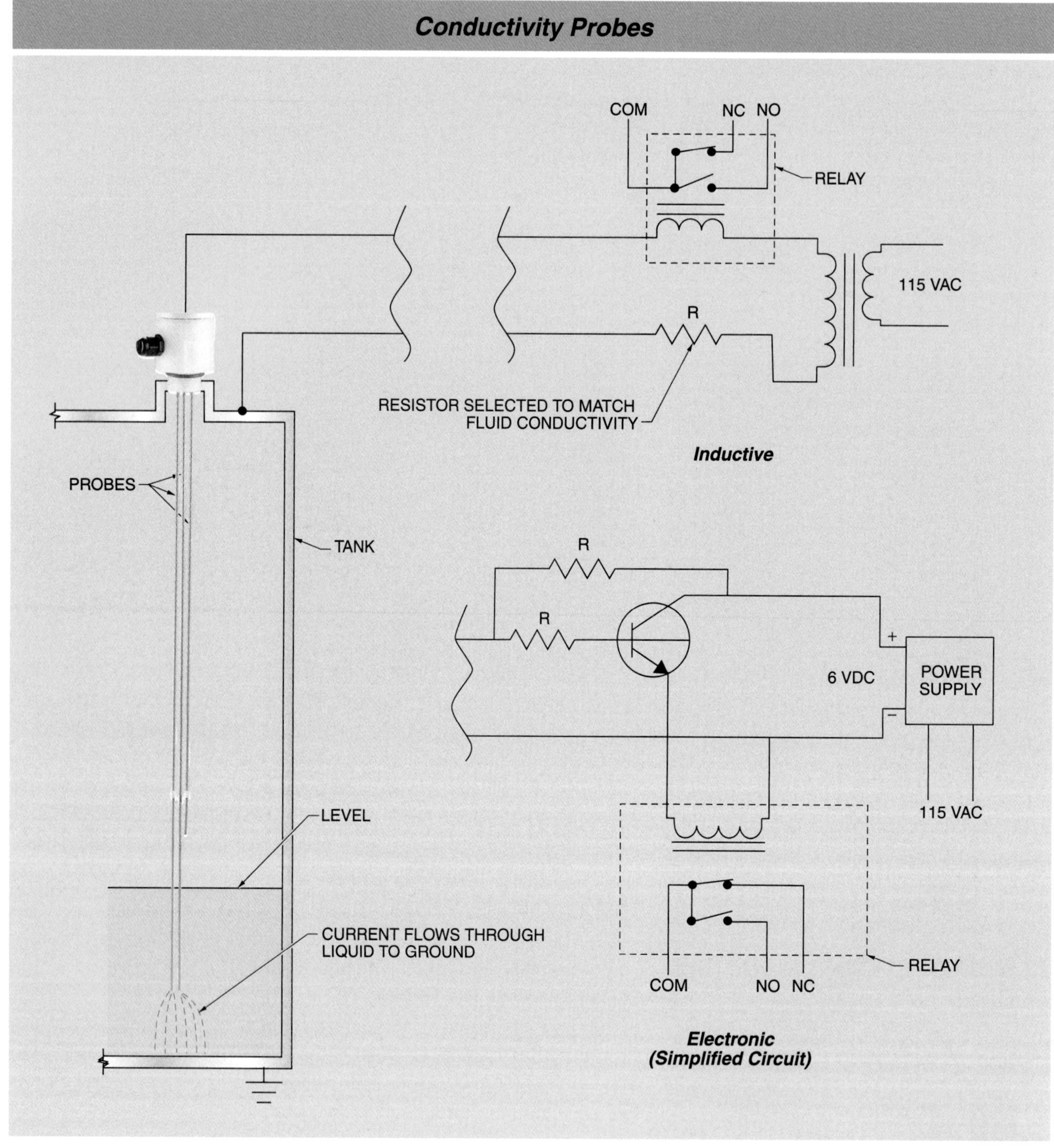

©Dwyer Instruments, Inc.

Figure 14-6. A conductivity probe measures electrical conductivity to determine the level of a tank.

Electrode holders are available for handling multiple electrodes. This allows low, high, and alarm level operating points to be configured with a single instrument. If the vessel is made of a nonconductive material, part of the probe can serve as the ground. In this case, the wires to the longest electrode are grounded. Using the longest electrode as the ground ensures that the circuit is grounded whenever any part of the circuit is in contact with the conductive liquid.

An *inductive probe* is a point level measuring instrument consisting of a sealed probe containing a coil, an electrical source that generates an alternating magnetic field, and circuitry to detect changes in inductance. The sealed probe is inserted into a vessel containing conductive material. As the level changes, the magnetic field of the probe interacts with the conductive material. The interaction is detected by measuring the inductive reactance. **See Figure 14-7.**

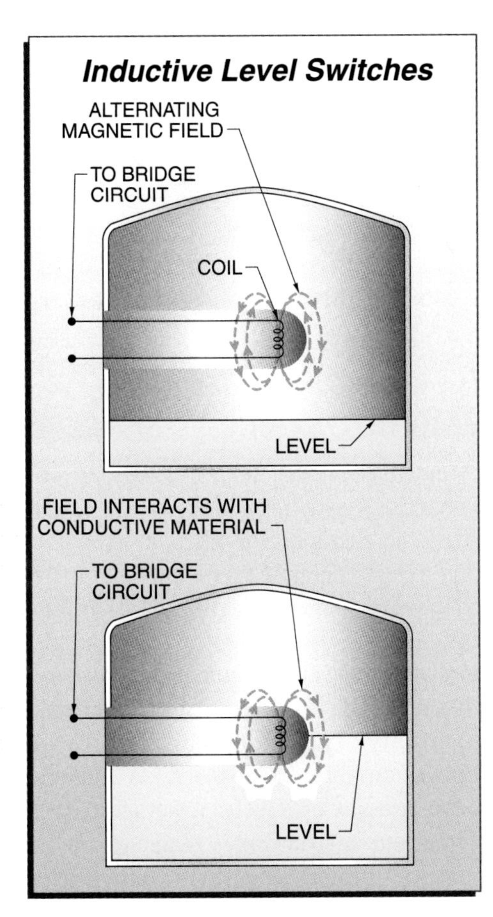

Figure 14-7. An inductive level switch uses a bridge circuit to measure changing inductance to determine when the level of material in a tank reaches the level of the switch.

PHOTOMETRIC SENSORS

There are situations where a light beam can be used as a point level measurement. The two common types of optical probes are beam-breaking photometric sensors and optical liquid-level sensors.

Beam-Breaking Photometric Sensors

A *beam-breaking photometric sensor* is a point level measuring instrument consisting of a light source and a detector that indicates a level of the contents of a vessel when the beam is broken. These units are subject to false signals from outside light sources and from dust or splashes of liquid. The lenses must be kept clean to maintain the strength of the light beam and the ability of the detector to see the beam. **See Figure 14-8.**

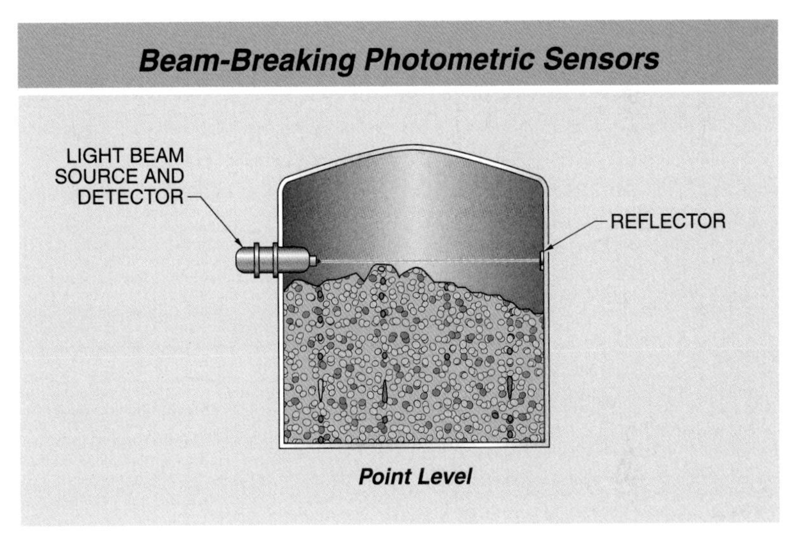

Figure 14-8. A beam-breaking photometric sensor sends a light beam across a tank to a reflector. When the beam is blocked, the tank is full.

Optical Liquid-Level Sensors

An *optical liquid-level sensor* is a liquid point level measuring instrument where a light source and a light detector, shielded from each other, are mounted in a housing and the light source is directed against the inside of a glass or plastic cone-shaped prism. If the cone is above the liquid, the light is reflected from the cone back to the detector. If the cone is submerged in the liquid, the light is refracted into the liquid and is not sensed by the detector. **See Figure 14-9.** The light detector circuits control relay contacts that can be used for alarms or control.

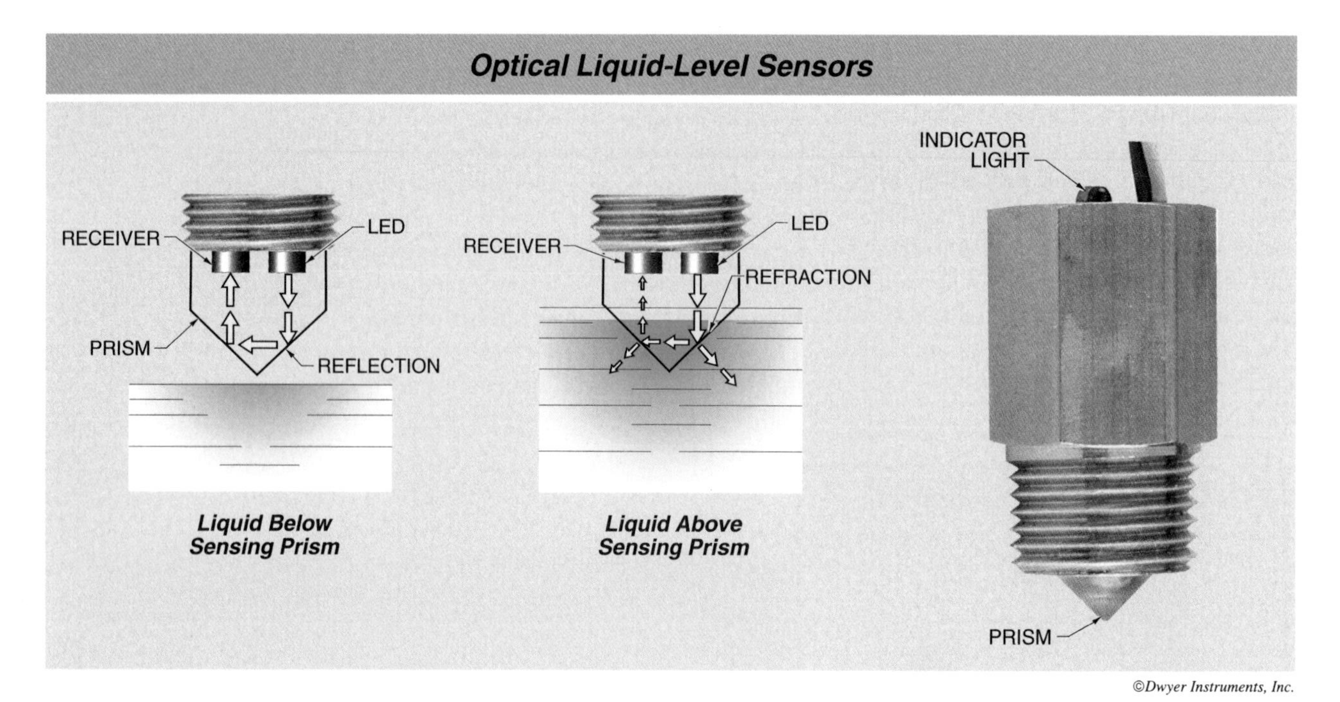

©Dwyer Instruments, Inc.

Figure 14-9. An optical liquid-level sensor uses the reflectance of a prism to determine when the prism is covered with a liquid.

TECH-FACT

Beam-breaking photometric sensors are very similar to the optical switches used to measure the position of objects in assembly and conveying operations.

Intec Controls Corporation

Reactor control systems often include tank level monitoring.

MAGNETOSTRICTIVE SENSORS

A *magnetostrictive sensor* is the part of a continuous level measuring system consisting of an electronics module, a waveguide, and a float containing a magnet that is free to move up and down a pipe that is inserted into a vessel from the top. Within the pipe is a waveguide constructed of magnetostrictive material. "Magnetostrictive" refers to a property of certain ferrous alloys having dimensions that change in response to magnetic stress. Conversely, when an external force puts a strain on a magnetostrictive material, the internal magnetic flux changes. **See Figure 14-10.**

The electronics module at the top end of the waveguide generates a current pulse that creates a magnetic field in the waveguide. The interaction of the magnetic field with the magnets in the float results in the generation of a second pulse in the waveguide that reflects back to the top. The time between the generated pulse from the electronics module and the return pulse is a function of the distance between the magnets within the float and the waveguide.

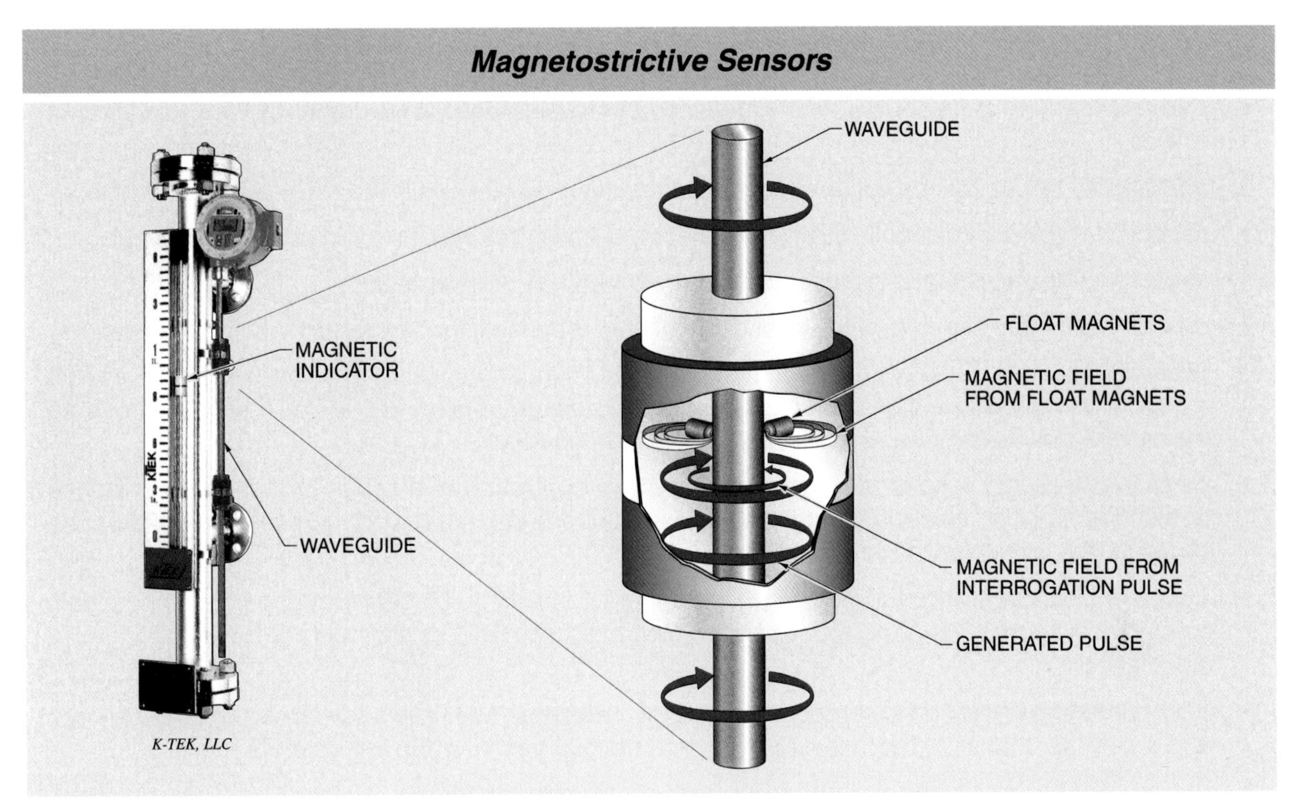

Figure 14-10. A magnetostrictive sensor uses a magnetic pulse to determine the location of a moving float.

THERMAL DISPERSION SENSORS

A *thermal dispersion sensor* is a point level measuring instrument consisting of two probes that extend from the detector into the vessel, with one of the probe tips being heated. The detector monitors the difference in temperature between the heated probe tip and the unheated probe. **See Figure 14-11.** When the liquid covers the probe tips, the temperature of the heated probe drops because the heat is removed by the liquid. The decreased differential temperature is detected and a switch is activated.

An alternative design uses a constant current flow through a thermistor in one of the probes. As the probe is immersed in the liquid, heat is conducted away from the heated probe. The resistance of the thermistor changes with changes in temperature and the change in resistance is detected in a bridge circuit. When the resistance changes beyond a setpoint, a contact is closed and a circuit is energized.

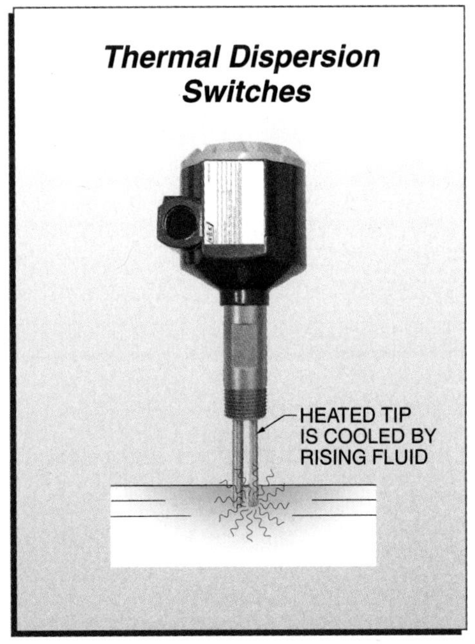

Figure 14-11. A thermal dispersion sensor measures the temperature difference between two sensor tips as the heat is carried away by a fluid.

KEY TERMS

- *capacitance:* The ability of an electrical device to store charge as the result of the separation of charge.
- *admittance:* The ability of a circuit to conduct alternating current and is the reciprocal of impedance.
- *capacitor:* An electrical device that stores electrical energy by means of an electrostatic field.
- *dielectric:* The insulating material between the conductors of a capacitor.
- *dielectric constant:* The ratio of the insulating ability of a material to the insulating ability of a vacuum.
- *capacitance probe:* A part of a level measuring instrument and consists of a metal rod inserted into a tank or vessel, with a high-frequency alternating voltage applied to it and a means to measure the current that flows between the rod and a second conductor.
- *conductivity probe:* A point level measuring system consisting of a circuit of two or more probes or electrodes, or an electrode and the vessel wall where the material in the vessel completes the circuit as the level rises in the vessel.
- *inductive probe:* A point level measuring instrument consisting of a sealed probe containing a coil, an electrical source that generates an alternating magnetic field, and circuitry to detect changes in inductance.
- *beam-breaking photometric sensor:* A point level measuring instrument consisting of a light source and a detector that indicates a level of the contents of a vessel when the beam is broken.
- *optical liquid-level sensor:* A liquid point level measuring instrument where a light source and a light detector, shielded from each other, are mounted in a housing and the light source is directed against the inside of a glass or plastic cone-shaped prism.
- *magnetostrictive sensor:* The part of a continuous level measuring system consisting of an electronics module, a waveguide, and a float containing a magnet that is free to move up and down a pipe that is inserted into a vessel from the top.
- *thermal dispersion sensor:* A point level measuring instrument consisting of two probes that extend from the detector into the vessel, with one of the probe tips being heated.

REVIEW QUESTIONS

1. How do capacitance probes operate?
2. What is the difference between single-point and continuous capacitance probes?
3. Compare conductivity and inductive probes.
4. How do photometric sensors operate?
5. How do thermal dispersion sensors operate?

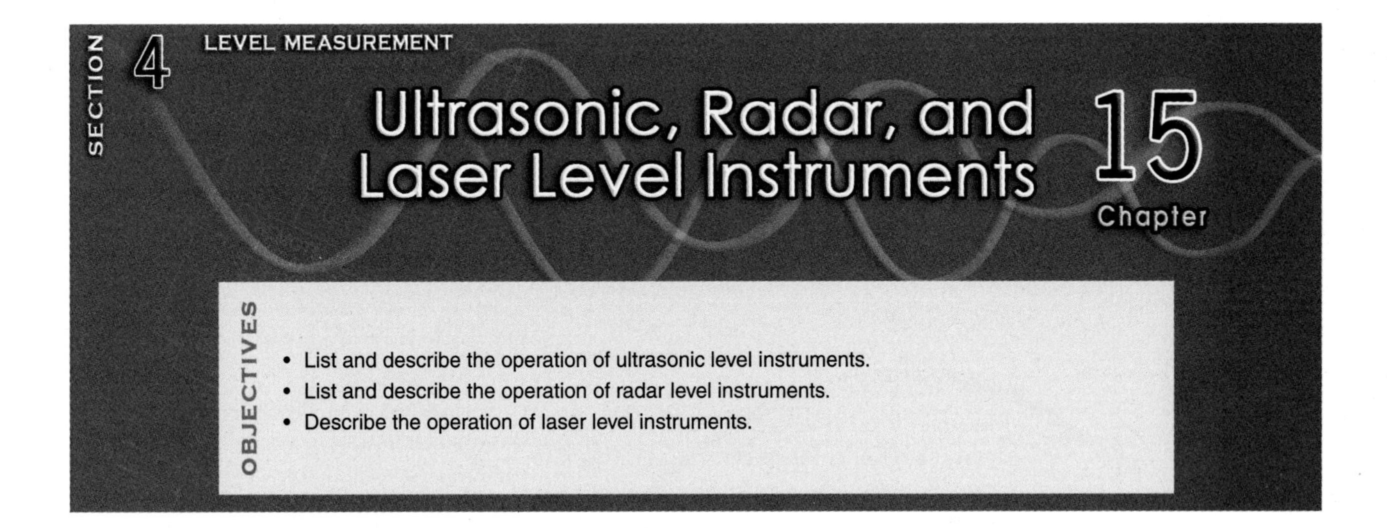

ULTRASONIC SENSORS

An *ultrasonic sensor* is a level measuring instrument that uses ultrasonic sounds to measure level. The transmitter generates a high-frequency sound directed at the surface of the material in the vessel.

Ultrasonic Point Level Measurement

A design of an ultrasonic level instrument for point level measurement uses two similar crystals. One of the crystals is used for transmission of the ultrasonic signal and the other for its reception. Both are enclosed in a probe but are separated by a small integral air gap. **See Figure 15-1.**

When the gap is exposed to air or vapor, the ultrasonic signal is not able to pass through the gap in sufficient strength to be received. When the level rises and material fills the gap, the ultrasonic signal from the transmitter passes through the gap because liquids and granular solids carry sound waves more efficiently than air or vapor. Reception of the transmitted signal causes the electrical circuitry to operate an ON/OFF relay.

This type of ultrasonic level measuring instrument is used primarily for granular solids, but is also suitable for use with noncorrosive liquids and slurries. If the liquid is a slurry or is sticky, a wider gap permits it to drain more readily from the gap. There is a limited selection of materials of construction for the wetted parts so these devices cannot be used for highly corrosive liquids.

Strong industrial noise or vibration at or near the signal frequency can cause interference and false signals. Dust may cause false signals or may build up on the probe, causing signal attenuation. Ultrasonic level measurement cannot be used in a vacuum or with liquids that are covered with foam.

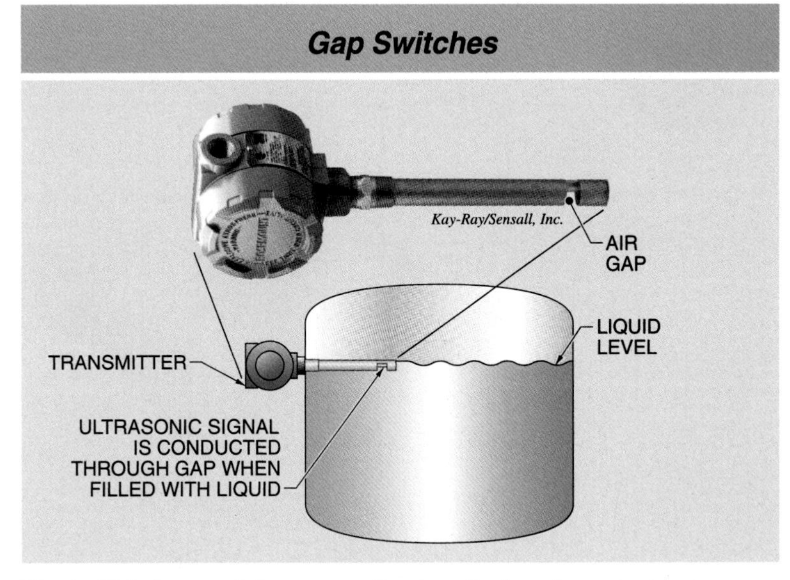

Figure 15-1. A gap switch measures the strength of an ultrasonic signal across a small gap to determine when the material in the tank has reached the switch.

The presence of internal vessel bracing, mixer shafts, or other interior objects can cause interference with the signal path and should be avoided. It is best to locate the sensor away from the walls of the vessel to minimize the effect of the beam hitting the side walls.

Transit Time

Transit time is the time it takes for a transmitted ultrasonic signal to travel from the ultrasonic level transmitter to the surface of the material to be measured and back to the receiver. The electronic circuitry in the receiver measures the transit time and calculates the distance. **See Figure 15-2.**

Ultrasonic Sensors

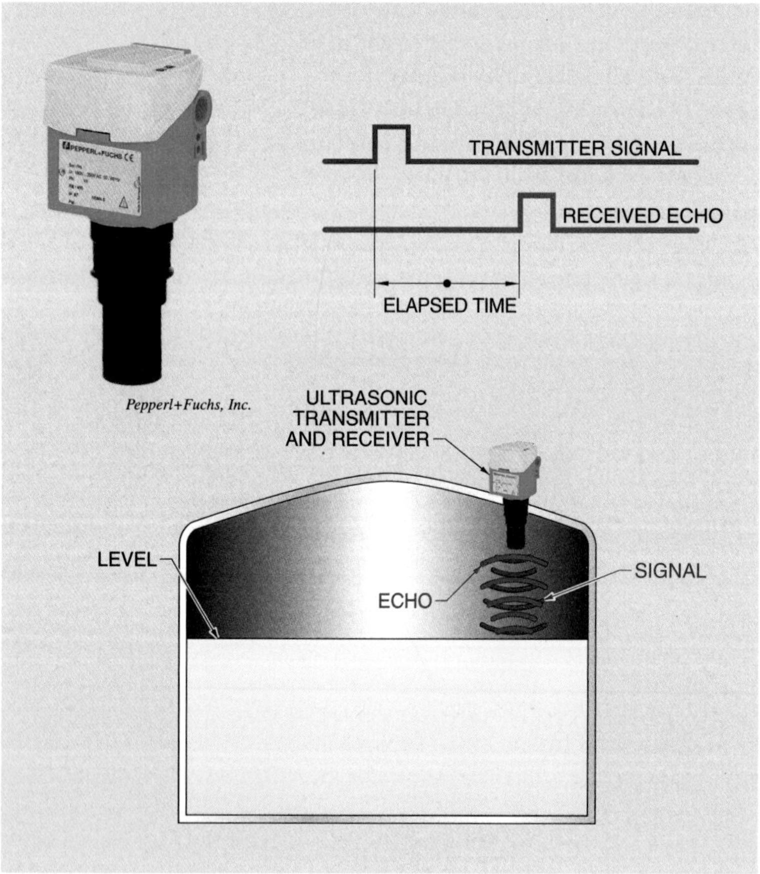

Figure 15-2. Ultrasonic sensors measure the transit time from the transmitter to the surface of the material to determine level.

As the level in the vessel changes, the transit time changes. The distance from the receiver down to the surface is calculated from the transit time. The level is then calculated by subtracting this distance from the height of the vessel. For example, if the distance from the receiver to the surface of the material is determined to be 5′ and the location of the receiver is known to be 30′ above the bottom of the vessel, the level is 25′ (30 – 5 = 25).

The transit time depends on the speed of the signal in the air or vapor space above the material. Several factors affect the speed of the signal. Any changes in the temperature, pressure, humidity, and composition of gas above the material affect the speed of the ultrasonic signal. A change in the speed of the ultrasonic signal introduces an error in the level measurement unless the instrument compensates for these factors.

Tuning Forks

A *tuning fork level detector* is a point level measuring instrument consisting of a vibrating tuning fork that resonates at a particular sound frequency and the circuitry to measure that frequency. **See Figure 15-3.** The tuning fork is mounted in a fitting so that it is exposed to the process material. When the fork is uncovered, it vibrates at a fixed frequency. When material in the tank covers the fork, the frequency changes enough to actuate an output relay. The relay contacts can be used for feeder control or as an alarm.

Tuning fork level sensors are commonly used for single-point level detection of liquids or granular solids, although there may be differences in design details depending on whether the application is for liquids or solids. The materials of construction for the wetted parts are very limited and should not be used for corrosive fluids.

Tuning forks should not be used in vibrating bins where the frequency of the bin would be close to the harmonic of the tuning fork.

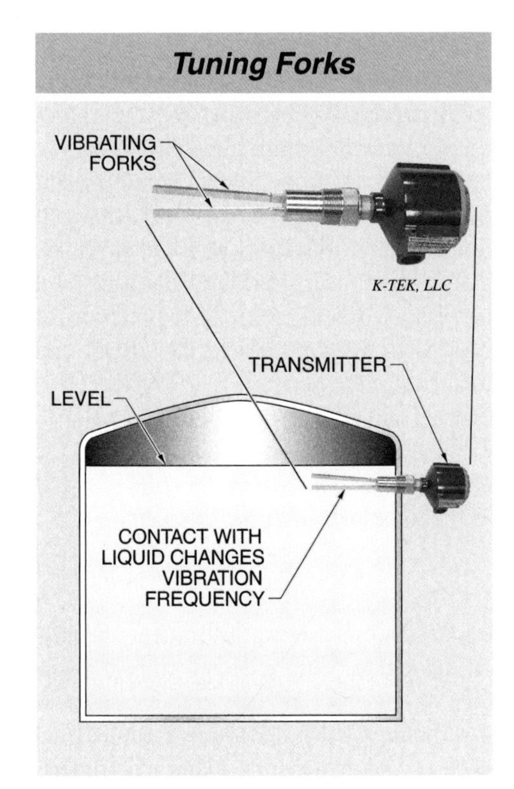

Figure 15-3. The vibration frequency of a tuning fork changes when it is in contact with material in a tank.

RADAR

Radar level sensors use very high frequency (about 10 GHz) radio waves. These radio waves are aimed at the surface of the material in the storage vessel. The radio waves are reflected off the material in the vessel and returned to the emitting source. Common types of radar systems are pulsed, frequency modulated continuous wave, and guided wave radar.

Pulsed Radar

A *pulsed radar level sensor* is a level measuring sensor consisting of a radar generator that directs an intermittent pulse with a constant frequency toward the surface of the material in a vessel. The time it takes for the pulse to travel to the surface of the material, reflect off the material, and return to the source is a function of the distance from the sensor to the material surface. **See Figure 15-4.**

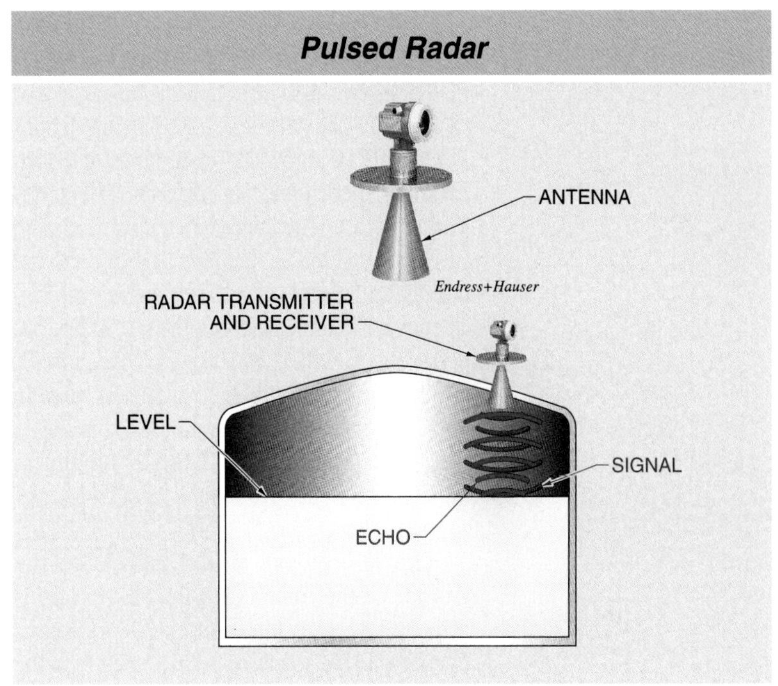

Figure 15-4. Pulsed radar measures the transit time from the transmitter to the surface of the material to determine level.

The Foxboro Company

Level measurements are often sent to a control room where they are used to ensure that other parts of a process have sufficient raw materials.

The two common types of antennae used to emit either pulses or continuous waves are in the form of a cone or a rod. A cone antenna is larger and sturdier than a rod and less subject to material buildup or condensation. Rod antennae are smaller and less expensive, making them more suitable for use in smaller vessels.

Frequency Modulated Continuous Wave (FMCW) Radar

A *frequency modulated continuous wave (FMCW) radar* is a level measuring sensor consisting of an oscillator that emits a continuous microwave signal that repeatedly varies its frequency between a minimum and maximum value, a receiver that detects the signal, and electronics that measure the frequency difference between the signal and the echo. *Bandwidth* is the range of frequencies from the minimum to the maximum value in an FMCW radar level sensor. *Sweeptime* is the constant time for an FMCW emitter to vary the frequency from the lowest frequency to the highest.

The microwave signal travels to the surface of the material in the vessel and is reflected back to the emitter. The reflected echo signal has a different frequency than the emitted signal that is being generated at that instant. **See Figure 15-5.** These differences vary directly with the distance between the emitter and the surface of the material in the vessel. This permits the radar detector to be calibrated in units of level.

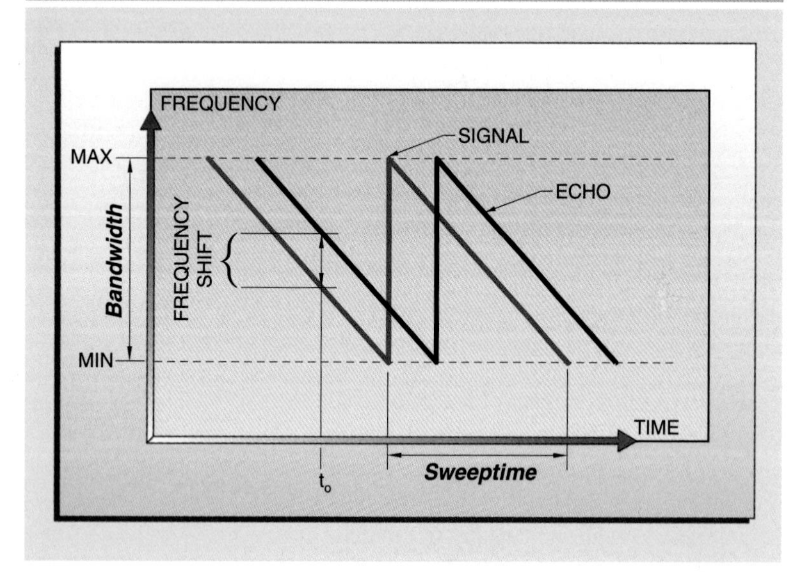

Figure 15-5. Frequency modulated radar measures the frequency difference of the radar signal and the echo to determine level.

Guided Wave Radar

A *guided wave radar* is a level measuring detector consisting of a cable or rod as the wave carrier extending from the emitter down to the bottom of the vessel and electronics to measure the transit time. A time domain reflectometer (TDR) is another name for guided wave radar. The material in the vessel reflects, or echoes, some of the microwave energy at the point where the carrier and material make contact. **See Figure 15-6.** The transit time is measured and used to calculate the level. This guiding reduces the effect of dust above granular solids as well as the turbulence in some liquids.

Radar Applications

Radar level measuring instruments depend on the signal being reflected from the surface of the material back to the receiver. It is difficult to use radar to measure the level of materials that absorb, rather than reflect, a radar signal. Water and most other liquids are good reflectors. They do not absorb much microwave energy. Many other materials do absorb this energy and are harder to measure.

The dielectric constant of a material determines the degree of absorption and therefore the strength of the reflected wave. For materials with a dielectric constant less than 1.5, it is difficult to measure level with radar. For materials with a dielectric constant between 1.5 and 2.0, it is usually possible to measure level with radar. Some applications with materials with low dielectric constant require a stilling well to help focus the beam.

Radar level sensors need to be calibrated on installation and periodically after that. The calibration is fairly simple since the sensors measure the time of flight or a frequency shift of the signal. The distance and the velocity of the radar signal determine the time of flight. Since radar is an electromagnetic pulse, the velocity of the radar signal is fixed at the speed of light. Any changes in the time of flight are related to the distance the signal has to travel.

Since radar units are mounted at the top of a vessel or tank, there are few problems with tank integrity and product leakage, buildup on the sensors, or errors due to changes in the product. Mixer blades and other objects in the vessel can cause false echoes, while dust in the air above a granular solids surface may absorb the radar signal. Radar level instruments should be avoided in vessels handling liquids where vapor space is a condensing environment. Condensation on the radar emitter disrupts the transmission signal. It is important to install and aim the radar continuous level sensor correctly to eliminate or minimize these sources of error.

LASERS

A *laser level instrument* is a level measuring instrument consisting of a laser beam generator, a timer, and a detector mounted at the top of a vessel. Laser beams are intense, narrow light beams that can travel long distances. A crystal-emitted pulsing laser beam with a wavelength of about 900 nm is directed at the surface of the process material. The laser beam is reflected back to the emitter where a very accurate timing device measures the out-and-back interval. The travel time varies with the level. **See Figure 15-7.** Because laser beams are light beams, dust and vapor can interfere with their transmission and reception. Their intense, narrowly focused light can have a destructive effect on one's eyesight.

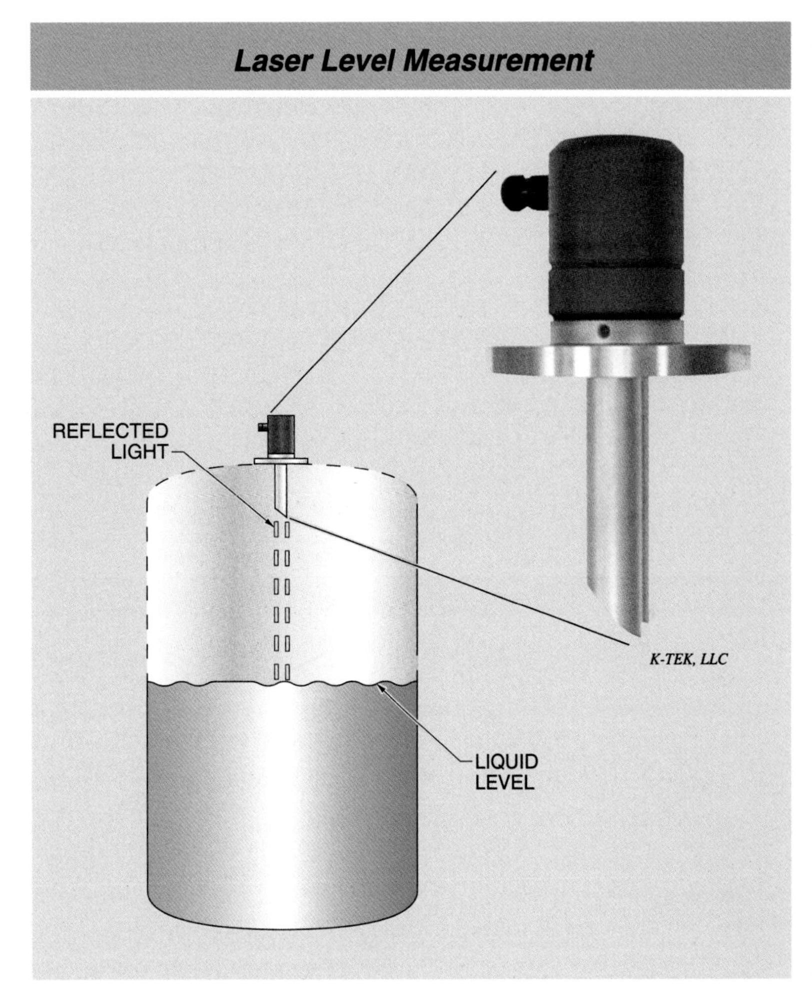

Figure 15-7. A laser measures the transit time of reflected light to determine level.

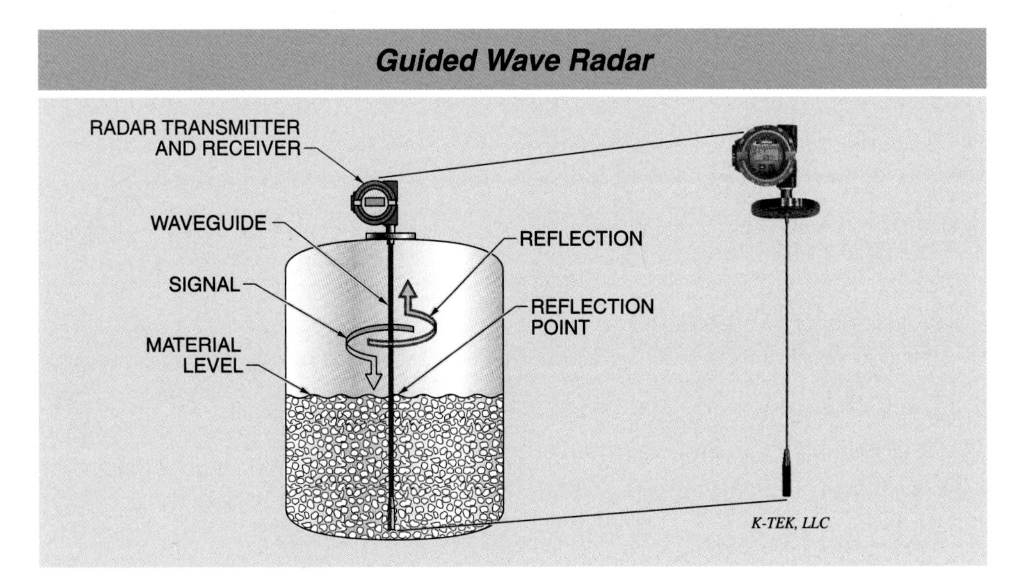

Figure 15-6. Guided wave radar uses a waveguide to direct a radar signal down a wire.

Laser Safety

Lasers are generally divided into four basic classifications, based on potential risk according to the ANSI Laser Safety Standard (ANSI Z136.1). The higher the class, the greater the potential risk for the operator and other workers in the area. Brief descriptions of the laser classes are as follows:

- Class I: Under normal operating conditions, Class I lasers pose no potential risks. Infrared lasers are classified as Class I.
- Class II: Low-power visible laser beams do not normally present a potential hazard if viewed for short periods of time. Many common lasers are Class II lasers.
- Class III: Class III lasers are further classified into Class IIIA and Class IIIB. Class IIIA lasers do not pose a hazard if viewed only momentarily without protection. Class IIIB lasers can be hazardous if viewed directly.
- Class IV: Class IV lasers are hazardous to view under all conditions.

Laser classification and maximum output are listed on a label found on the body of the laser or with the documentation. Be sure to determine the classification of the laser you are working with or near to determine the proper safety precautions. If the label is missing from the laser, determine its classification and the safety precautions that should be taken before turning on the instrument.

KEY TERMS

- *ultrasonic sensor:* A level measuring instrument that uses ultrasonic sounds to measure level.
- *transit time:* The time it takes for a transmitted ultrasonic signal to travel from the ultrasonic level transmitter to the surface of the material to be measured and back to the receiver.
- *tuning fork level detector:* A point level measuring instrument consisting of a vibrating tuning fork that resonates at a particular sound frequency and the circuitry to measure that frequency.
- *pulsed radar level sensor:* A level measuring sensor consisting of a radar generator that directs an intermittent pulse with a constant frequency toward the surface of the material in a vessel.
- *frequency modulated continuous wave (FMCW) radar:* A level measuring sensor consisting of an oscillator that emits a continuous microwave signal that repeatedly varies its frequency between a minimum and maximum value, a receiver that detects the signal, and electronics that measure the frequency difference between the signal and the echo.
- *bandwidth:* The range of frequencies from the minimum to the maximum value in an FMCW radar level sensor.
- *sweeptime:* The constant time for an FMCW emitter to vary the frequency from the lowest frequency to the highest.
- *guided wave radar:* A level measuring detector consisting of a cable or rod as the wave carrier extending from the emitter down to the bottom of the vessel and electronics to measure the transit time.
- *laser level instrument:* A level measuring instrument consisting of a laser beam generator, a timer, and a detector mounted at the top of a vessel.

REVIEW QUESTIONS

1. How does an ultrasonic sensor for point level measurement operate?
2. Describe how a tuning fork is used for point level measurement.
3. How does a frequency modulated continuous wave (FMCW) radar operate?
4. Explain the benefit of guided wave radar compared to pulsed radar.
5. How does a laser level instrument operate?

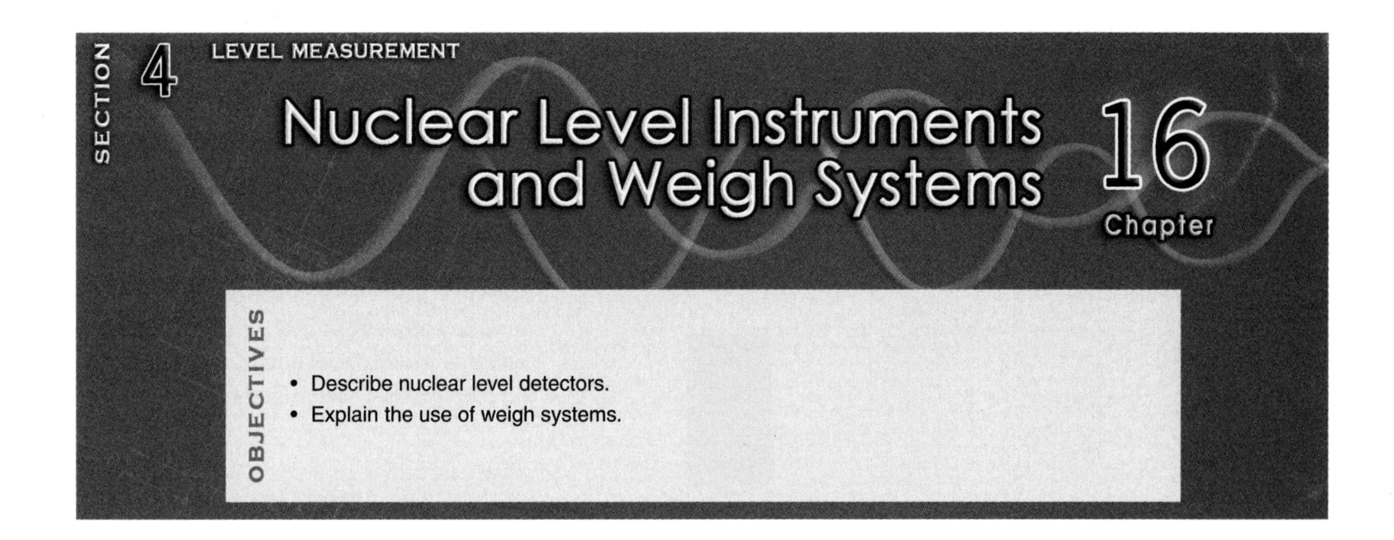

NUCLEAR LEVEL INSTRUMENTS

A *nuclear level detector* is a level measuring system consisting of a radioactive source that directs radiation through a vessel to a detector, such as a Geiger counter, on the other side. Nuclear level sensors are used for process materials that are extremely hot, corrosive, toxic, or under very high pressure, and so are not suitable for intrusive level detectors.

Radioactive elements such as cesium 137 or cobalt 60 provide the radioactive source in the form of gamma rays. The amount of radioactive energy required is calculated based upon vessel wall thickness and distance between source and detector. Cobalt 60 provides a stronger radioactive signal than cesium 137 but has a shorter life.

Federal, state, and local authorities closely regulate the use of nuclear energy sources. Nuclear level sensors are relatively expensive to purchase, install, and operate. However, they are sometimes the only way to measure level under extreme conditions.

Point Level Measurement

For point level measurement, a radioactive source is mounted externally on one side of a vessel at the selected level. The source is enclosed in a protective housing with a window allowing the radiation to be directed toward the detector. The nuclear energy detector is mounted externally on the opposite side at the same level.

The nuclear energy source produces a beam of radiation whose frequency is proportional to the strength of the radiation. When the level of the material is high enough to block the beam path, the amount of radioactive energy that reaches the detector is lower than when the level is lower. This causes a controller to send a signal to stop or start a feeder, light a lamp, or sound an alarm. **See Figure 16-1.**

Continuous Level Measurement

For continuous level measurement, the same sources described for point level detection are used, but a scintillation counter is used as the nuclear sensor. A *scintillation counter* is a device that detects and measures nuclear radiation as it strikes a sensitive material, known as a phosphor, producing tiny flashes of visible light. Phosphors include zinc sulfide, sodium iodide, and some liquids and organic substances. The attenuation of the source is used to determine the level.

> **TECH FACT**
> *Nuclear instruments are rarely used because of the high cost of nuclear materials and the licensing requirements for handling radioactive materials.*

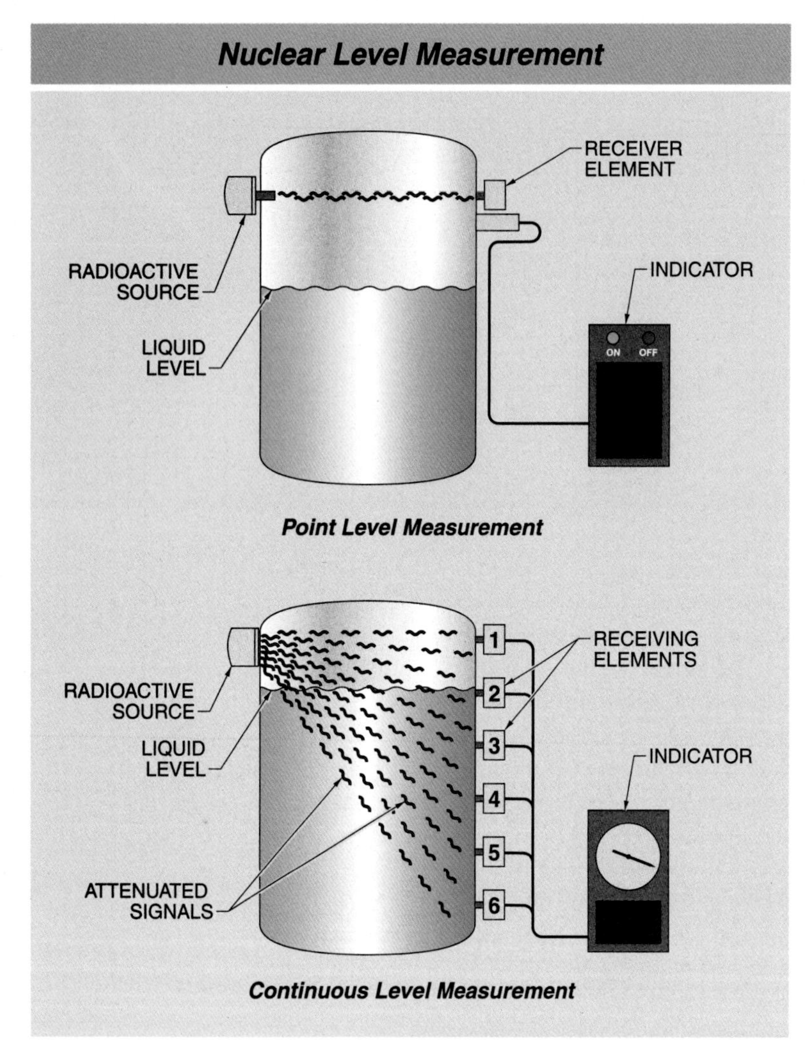

Figure 16-1. Nuclear radiation is attenuated by the material in a tank.

The scintillation counter provides a longer detection span than the Geiger counter. The mounting and positioning follows the pattern of the single-point installation except that the nuclear source window allows the radiation to be directed over the whole height of the vessel. The source is externally mounted on one side of the vessel and the detector is mounted externally on the opposite side. Scintillation counters allow the use of multiple detectors and circuit linkage for increased measuring range. In very high-pressure applications where the vessel wall thickness is too great for a nuclear source, the nuclear source can be mounted inside the vessel.

LOAD CELLS

Weighing a vessel containing either liquids or solids is a very accurate method of determining level. Weighing a vessel and its contents calls for the installation of load cells. A *load cell* is a device used to weigh large items and typically consists of either piston-cylinder devices that produce hydraulic output pressure or strain gauge assemblies that provide electrical output proportional to the applied load. **See Figure 16-2.**

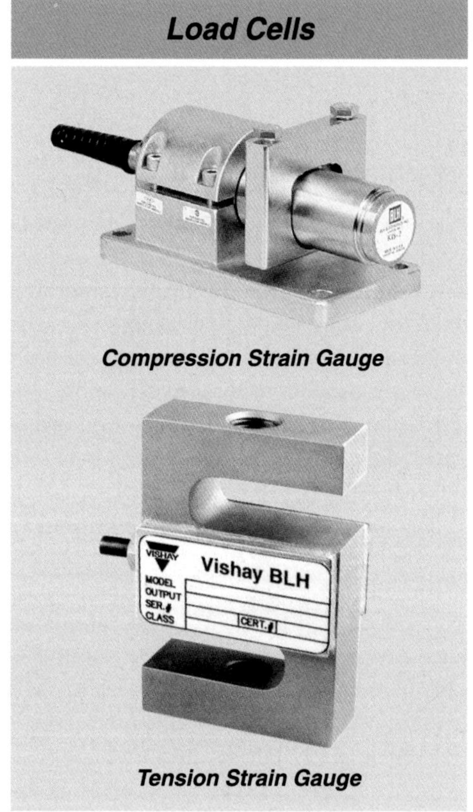

Figure 16-2. The most common configurations for load cells are compression and tension.

Strain Gauge Load Cells

A strain gauge load cell typically has a beam, column, or other stress member with strain gauges bonded to it. When a weight load is impressed against a member, the strain gauge is deformed and its electrical resistance changes. The balance of the bridge provides an output that is proportional to the force acting upon the load cell.

The three types of strain gauge load cells are shear, compression, and tension. Shear-type strain gauge load cells are used to measure the weight in vertical vessels. **See Figure 16-3.** This is because the load cell provides a secure mounting for the vessel without the need for stay rods. Compression-type strain gauge load cells are used for long, horizontal vessels where one end of the vessel needs to be free floating to allow for dimensional changes with temperature changes. Tension-type strain gauge load cells are used for tanks that hang from beams or from a ceiling.

Compression load cell applications require the use of stay rods to stabilize the vessel. **See Figure 16-4.** Compression load cells are not restrained at the top or bottom ends and thus cannot keep the vessel from moving laterally. Horizontal, lateral, and lengthwise stay rods are used to prevent the vessel fixed end from moving in any direction. Lateral stay rods are used at the movable end of the vessel so that only lengthwise movement is allowed.

Individual load cells are sized for different maximum applied loads. If four load cells are to be used, the maximum total weight of the vessel is divided by four. The load cells are selected with a safety margin that is 50% to 100% greater than the maximum calculated load that is expected to be applied to each load cell.

Hydraulic Load Cells

Hydraulic load cells are part of a closed hydraulic pressure system in which the load cell transfers the pressure acting on the cell from the weight of the vessel and its contents to a piston. The piston pressurizes the system hydraulic fluid in a diaphragm chamber. This change in pressure varies with the load acting on the load cell. The pressure measurement can be converted to a level measurement when the density of the material and the configuration of the tank are known.

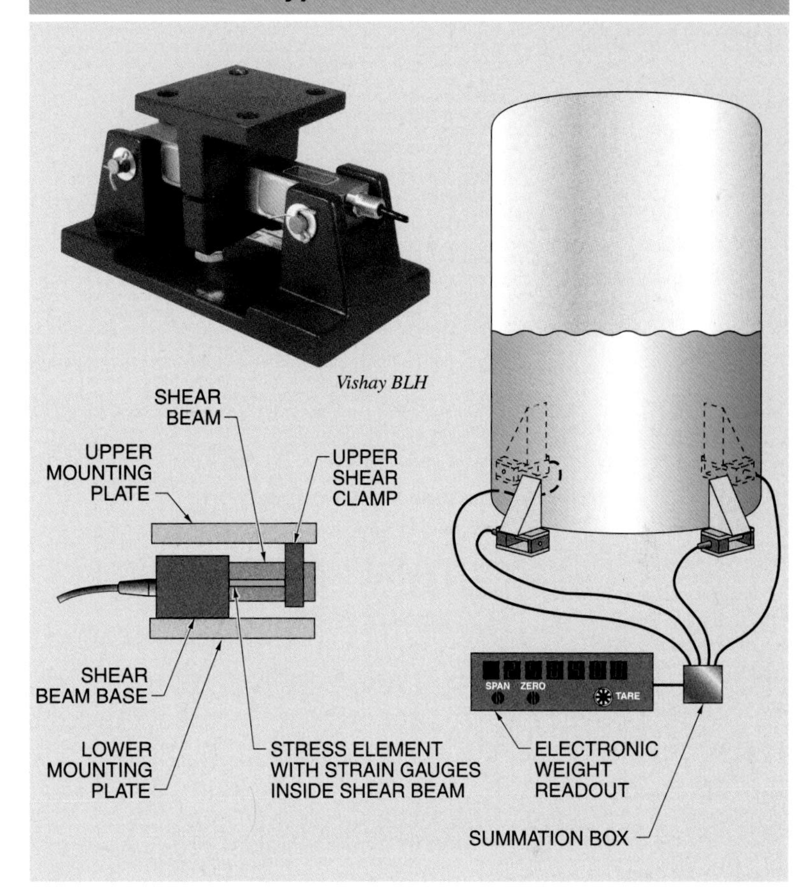

Figure 16-3. Shear-type electronic load cells are placed under the feet of a tank to measure the weight.

Bulk raw materials are often weighed to determine the quantity used.

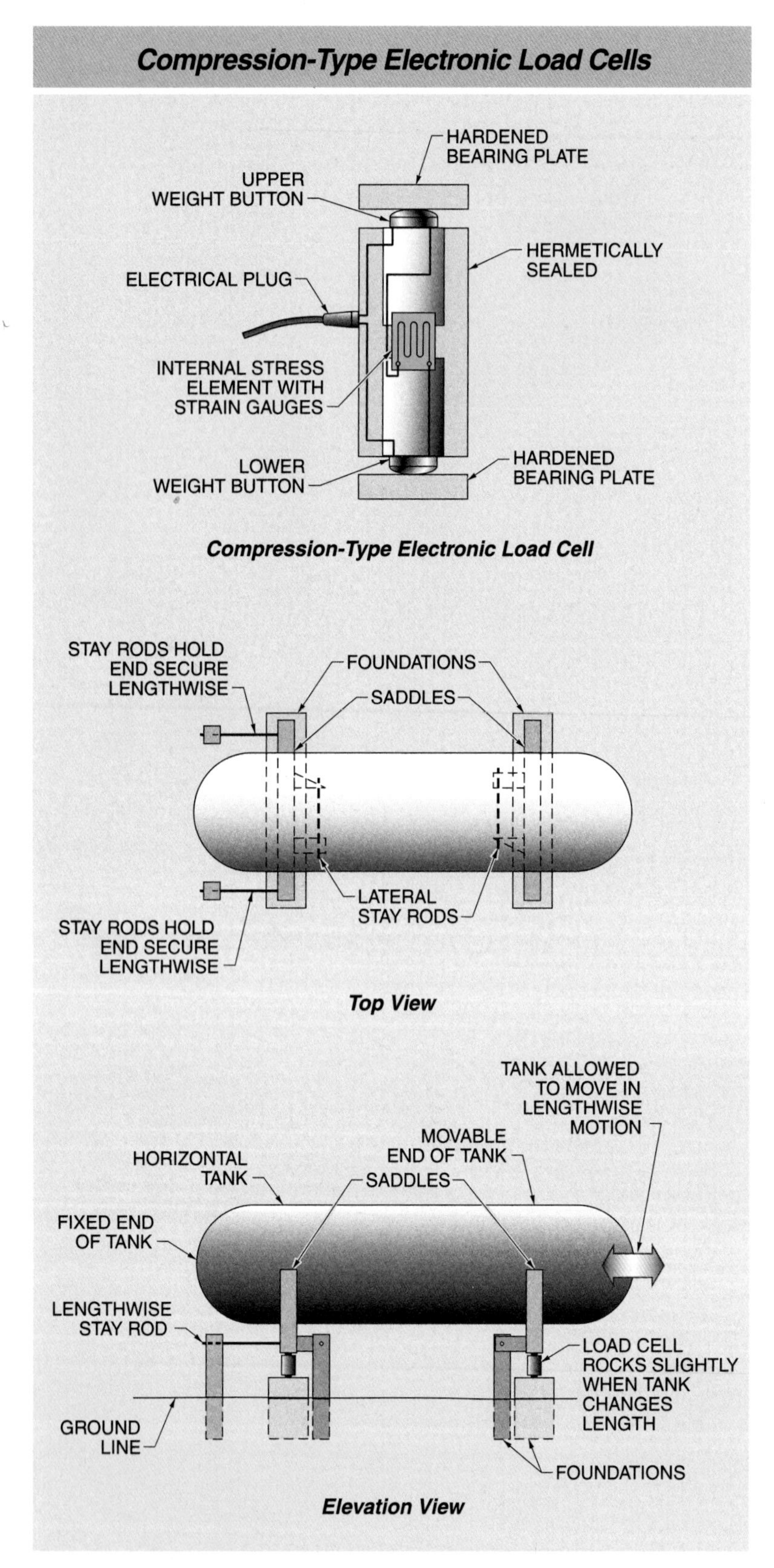

Figure 16-4. A compression-type electronic load cell measures the applied stress to a compressive strain gauge to determine weight.

In practice, a single hydraulic load cell can only be used to measure a vessel and its contents if there are additional pivoted supports to provide a stable base for the vessel. The most common arrangement is with a vessel that has four equally spaced supports. Two supports are placed on pivots and a beam supports the remaining two, with the hydraulic load cell located in the middle of the beam. This is a simple and inexpensive arrangement that allows half of the weight of the vessel and its contents to be supported by the load cell. This configuration can only be used when the vessel contains a liquid. The weight of a granular solid in a vessel may not be evenly distributed because of uneven flow. **See Figure 16-5.**

The pressure readout instrument is set to indicate zero weight when the vessel is empty. This eliminates the tare weight of the vessel and piping. The *tare weight* is the weight of the vessel, piping, and equipment that is supported by the load cell(s).

Since the hydraulic load cell can only sense one-half of the vessel weight, a special readout scale should be made that doubles the actual measured weight so that the total weight will be indicated. To improve the accuracy, three or four hydraulic load cells can be used. The hydraulic pressures are combined in a hydraulic summation unit. The problem with using more than one hydraulic load cell is ensuring that the weight is equally distributed among all the cells so that the load cells cannot bottom out.

Piping Restraints

A problem with the use of load cells is that rigid piping restraints cause errors in the weight readings. Pipes are usually flanged to a vessel and restraints are used to support the pipes, which keep the pipes from moving. The greater the vertical movement, the greater the effect of the piping restraint. However, all load cell systems require some vertical movement with increased load.

Piping restraints to the vessel also tend to support some of the weight of the vessel and act as springs helping to hold up the vessel. Strain gauge load cells only compress about

0.005″ when fully loaded. Hydraulic load cell installations are more susceptible to errors caused by piping stresses than strain gauge load cell systems because the hydraulic load cells have greater vertical displacement when a load is applied. Ambient temperature changes can also create measurement errors.

Piping restraints on the vessel are reduced by having long, unrestrained horizontal runs or by using flexible joints. The supporting steel structure of the vessel must also be very stiff. Deformations of the supporting structure under loaded conditions also contribute to the piping restraint.

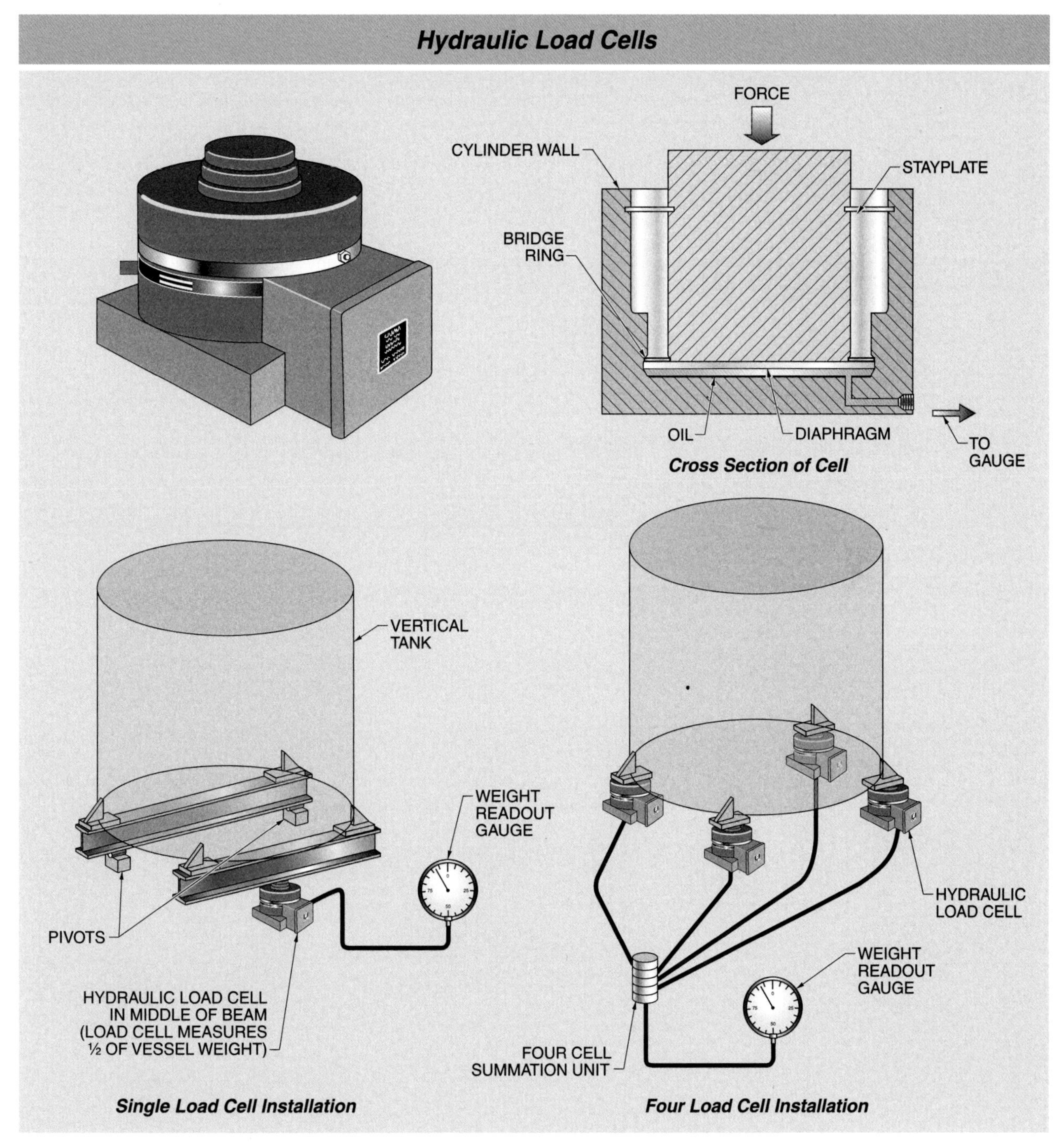

Figure 16-5. Single or multiple hydraulic load cells can be used to weigh the contents of a tank.

KEY TERMS

- *nuclear level detector:* A level measuring system consisting of a radioactive source that directs radiation through a vessel to a detector, such as a Geiger counter, on the other side.
- *scintillation counter:* A device that detects and measures nuclear radiation as it strikes a sensitive material, known as a phosphor, producing tiny flashes of visible light.
- *load cell:* A device used to weigh large items and typically consists of either piston-cylinder devices that produce hydraulic output pressure or strain gauge assemblies that provide electrical output proportional to the applied load.
- *tare weight:* The weight of the vessel, piping, and equipment that is supported by the load cell(s).

REVIEW QUESTIONS

1. Explain why nuclear level detectors are used.
2. How are nuclear level detectors used for point or continuous level measurements?
3. Define load cell.
4. How does a strain gauge load cell operate?
5. How does a hydraulic load cell operate?

SECTION 4 LEVEL MEASUREMENT

Practical Level Measurement and Calibration

Chapter 17

OBJECTIVES

- Describe the flow properties of bulk solids and explain how these properties can make it difficult to measure level.
- Describe methods of calibrating load cells.
- List considerations in measuring the water level in a boiler.
- List considerations in measuring the level of corrosive fluids.
- Describe devices for calibrating pressure sensors.

BULK SOLIDS IN SILOS AND TANKS

The flow and handling of bulk solids is extremely complex. A *bulk solid* is a granular solid, such as gravel, sand, sugar, grain, cement, or other solid material that can be made to flow. The top surface of the bulk solid in a silo may not be even across the top. The surface may be heaped up in the middle as the silo is being filled or it may be lowered in the middle as the bulk solid flows out the bottom of the silo. Measurement of the level of bulk solids in a silo depends on the flow properties of the bulk solid.

Flow Properties

The flow properties of bulk solids make level measurement difficult. The two types of flow of bulk solids in silos are funnel flow and mass flow. *Funnel flow* is the flow of a bulk solid where the material empties out of the bottom of a silo and the main material flow is down the center of the silo, with stagnant areas at the sides and bottom of the silo. *Mass flow* is the flow of a bulk solid where all material in a silo flows down toward the bottom at the same rate. Mass flow is the most desired flow regime, but it rarely exists in practice. **See Figure 17-1.**

The undesirable effects of funnel flow include ratholing and bridging. *Ratholing* is a condition arising in a silo when material in the center has flowed out the feeder at the bottom, leaving large areas of stagnant material on the sides. *Bridging* is a condition arising in a silo when material has built up over the feeder, blocking all flow out of the silo. The top surface of a bulk solid in a silo can be very uneven. Care must be taken when measuring level in a silo to measure it at a point that represents a level at typical flow conditions.

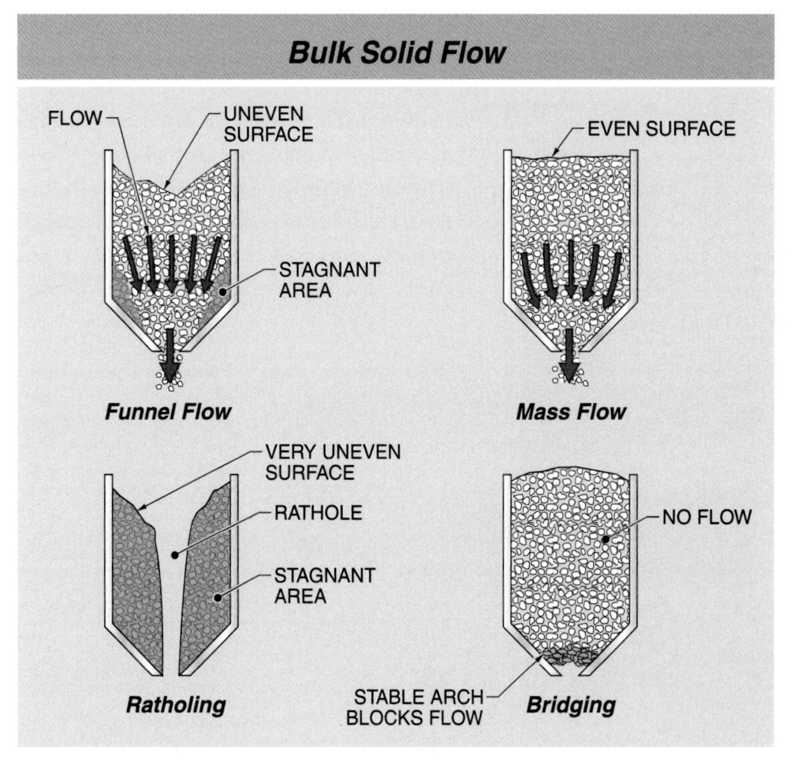

Figure 17-1. The flow of bulk solids is very complex and can affect a level determination.

Silo design significantly affects the flow of bulk solids. Free-flowing materials will flow out of bins with flat or nearly flat bottoms while cohesive materials need bins with steep sides to be able to flow.

Dust

Many bulk solids are very dusty when dumped into a silo. The presence of dust in a silo can interfere with most types of level measuring instruments. The dust can block most light and laser measurement instruments. Some radar instruments can penetrate the dust and get a reflection from the surface.

LOAD CELL CALIBRATION

Load cells and the electronic readouts need to be calibrated. It is necessary to select a maximum measured weight range of the material in the vessel. The electronic readout is zeroed with no weight in the vessel. Newer digital indicating transmitters have the ability to provide calibration for the load cells. In many cases, this may be all that is needed to provide a reasonable calibration.

It is also possible to use a load cell simulator to set the range of a weigh cell transmitter. The simulator provides a signal that simulates the signal from the load cells at various weights. The problem with this type of calibration is that it does not account for piping stresses that might add or subtract forces to the load cells. **See Figure 17-2.**

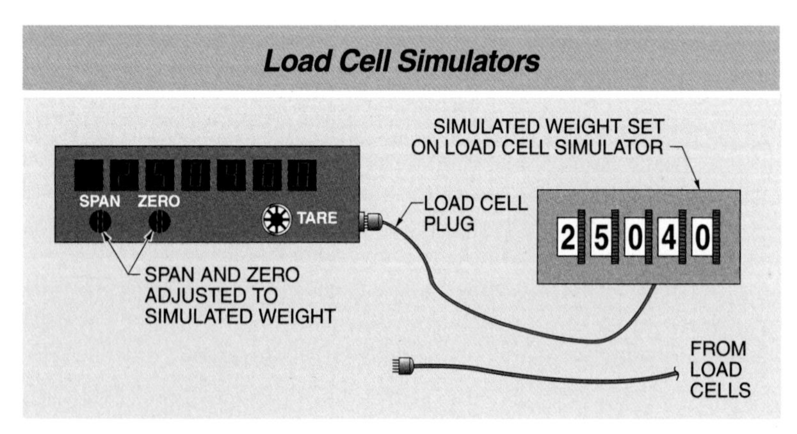

Figure 17-2. Electronic load cells can be calibrated with a load cell simulator.

A simple calibration method uses physical weights piled onto the vessel a little at a time, checking the indicated weight against the actual applied weight. This is an accurate method, but it is usually very difficult to apply the full weight to the vessel. In addition, this method requires hard work and takes a considerable amount of time. This method indicates that there are piping restraints if the indicated weight has significant errors from the actual applied weight that cannot be removed by recalibration. **See Figure 17-3.**

Another calibration method uses known amounts of liquid added to the vessel as the calibration weight. This method allows adding sufficient weight to cover the whole measurement range, but requires accurately weighing small containers of the liquid or the addition of a mass flowmeter to a pipe to the vessel. This method is also labor intensive and takes a considerable amount of time. **See Figure 17-4.**

The final method uses a portable load cell calibration system, a readout instrument, and hydraulic jacks. The jacks are positioned in line with the calibration load cells to either simulate loads on the vessel or shift weight from the application weigh vessel to the portable load cell calibration system. This method is quick and accurate, but expensive. In addition, the supporting structure must be designed for use with this system. **See Figure 17-5.**

A system is calibrated when the amount of applied weight matches the scale readout, within the error specified by the manufacturer. In addition, the calibration needs to be checked while adding the weights as well as while removing the weights to verify that there is no hysteresis.

The calibration method should be more accurate than the weigh system itself, which means that it should have less than 0.1% error. The simplest calibration method is to use a load cell simulator to provide an accurate electronic input signal, equivalent to the normal load cell outputs, to the electronic readout instrument. This method does not work well when there are piping restraints that could shift the calibration from the ideal condition.

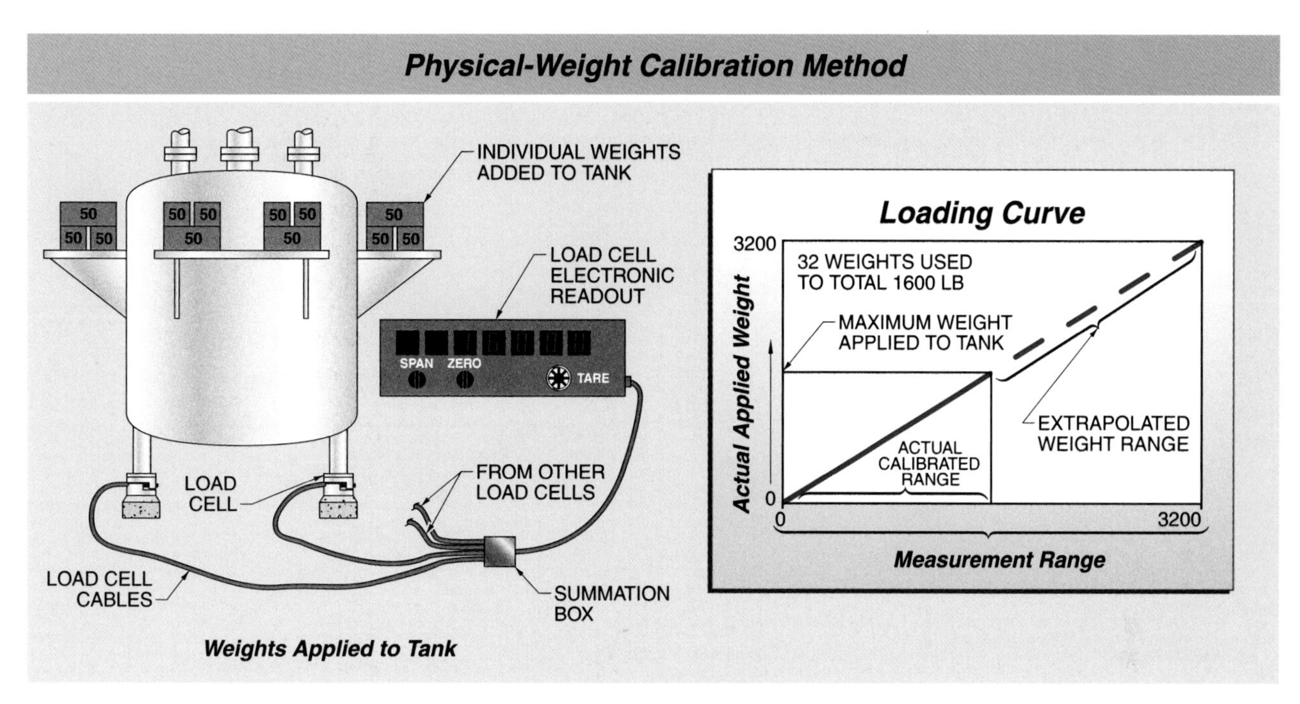

Figure 17-3. A load cell system can be calibrated by adding known weights to a tank.

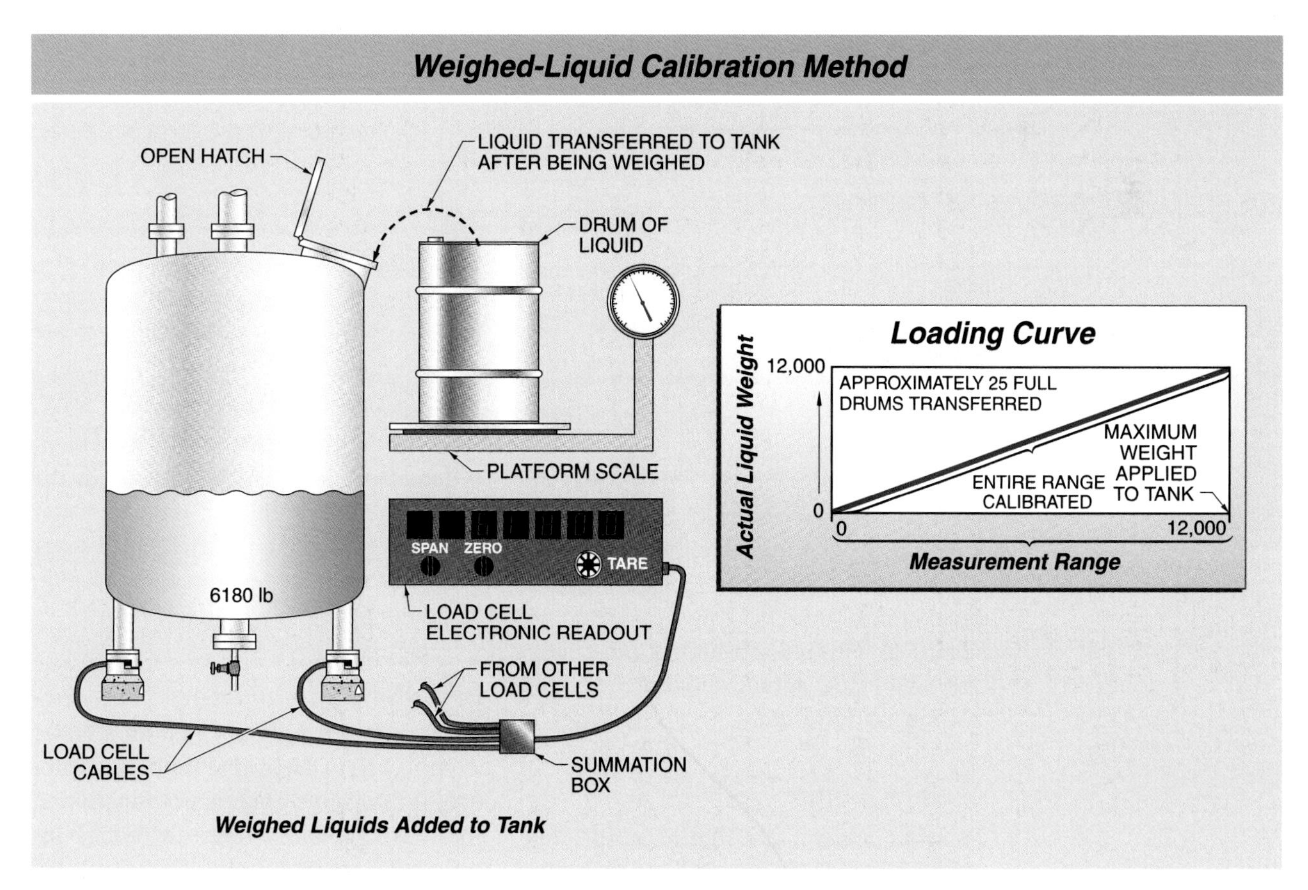

Figure 17-4. Drums of a weighed liquid can be used to calibrate a tank weighing system.

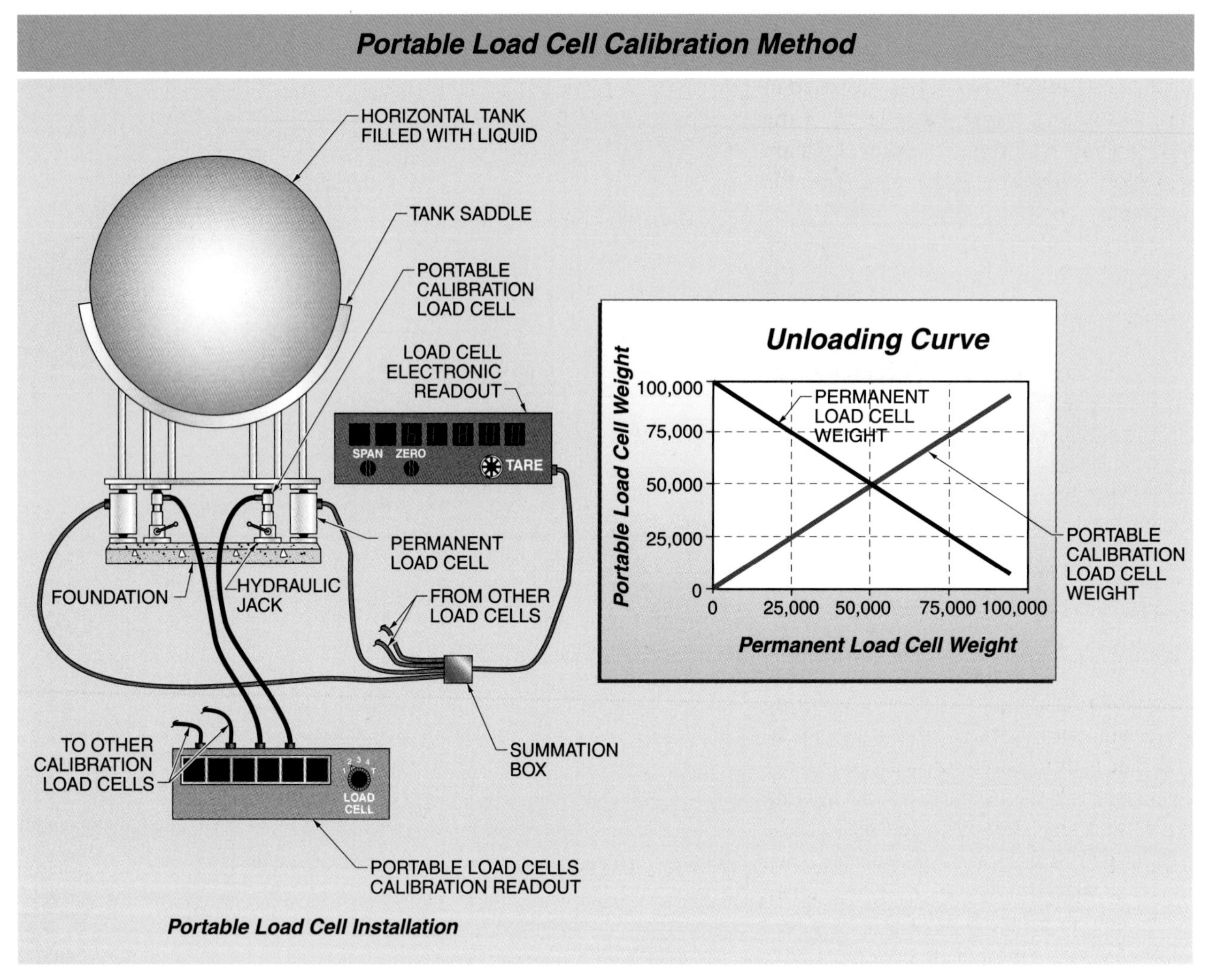

Figure 17-5. A portable calibrator can be used with hydraulic jacks to calibrate load cells.

WATER LEVEL IN A BOILER

There are several special procedures required when measuring water level in a boiler. Regulations require that all boilers have two means of measuring the water level. This can be either two gauge glasses, or one gauge glass and one remote level indicator, or one gauge glass and try cocks. In addition, all boilers must have two automatic burner shutdown devices for low water level.

Water Column

A *water column* is a boiler fitting that reduces the turbulence of boiler water to provide an accurate water level in the gauge glass. When the boiler is producing steam, the water inside the boiler is constantly in motion. This makes it difficult to determine how much water is in the boiler. The water column reduces water turbulence, allowing the true boiler water level to be indicated by the water level in the gauge glass. **See Figure 17-6.**

Try Cocks

A *try cock* is a valve located on a water column used to determine the boiler water level if the gauge glass is not functional. There are typically three try cocks installed on a water column. The middle try cock is installed at the normal operating water level (NOWL). The top try cock is mounted at the highest acceptable water level. The bottom try cock is mounted at the lowest acceptable water level. If the boiler water is at the proper level, steam and water should be discharged from the middle try cock.

Water discharged from the top try cock indicates a high water condition in the boiler. Steam discharged from the bottom try cock indicates a low water condition in the boiler. **See Figure 17-7.** Try cocks are typically used for pressures up to 250 psi. Above that pressure, it is difficult to distinguish between the water and the flash steam that blows out of a try cock.

Gauge Glass Blowdown

A water column and gauge glass accumulate sludge and/or sediment which must be periodically removed by blowdown. *Blowdown* is the process of discharging water and undesirable accumulated material. The water column blowdown valve is opened for about 5 sec to 10 sec, allowing water and any sludge or sediment to be discharged. The gauge glass blowdown valve is opened to perform a blowdown of the gauge glass. Both blowdowns are typically performed each shift.

Free flow of boiler water in the water column and gauge glass is crucial for providing an accurate boiler water level reading. For example, if the top line to the gauge glass is closed or clogged, the gauge glass fills with water. This gives a false indication that the boiler is full of water when the boiler water level may actually be dangerously low. If the bottom line to the gauge glass is closed or clogged, the water level in the gauge glass remains stationary instead of slightly fluctuating up and down during normal boiler operation.

After a period of time, the water level begins to rise in the gauge glass from steam condensing on top, which also gives a false reading. Any time the accuracy of a gauge glass water level is suspect, the try cocks should be used to determine the actual boiler water level.

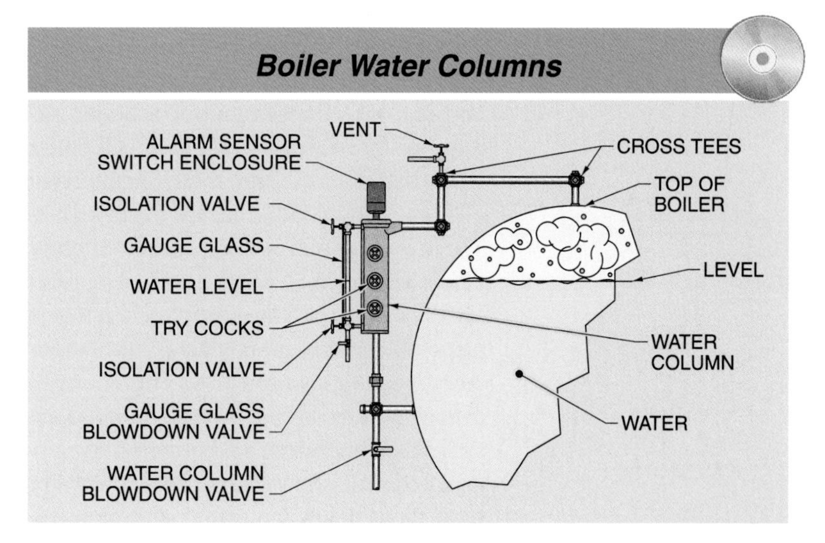

Figure 17-6. A water column is part of the instrument used to measure water level in a boiler.

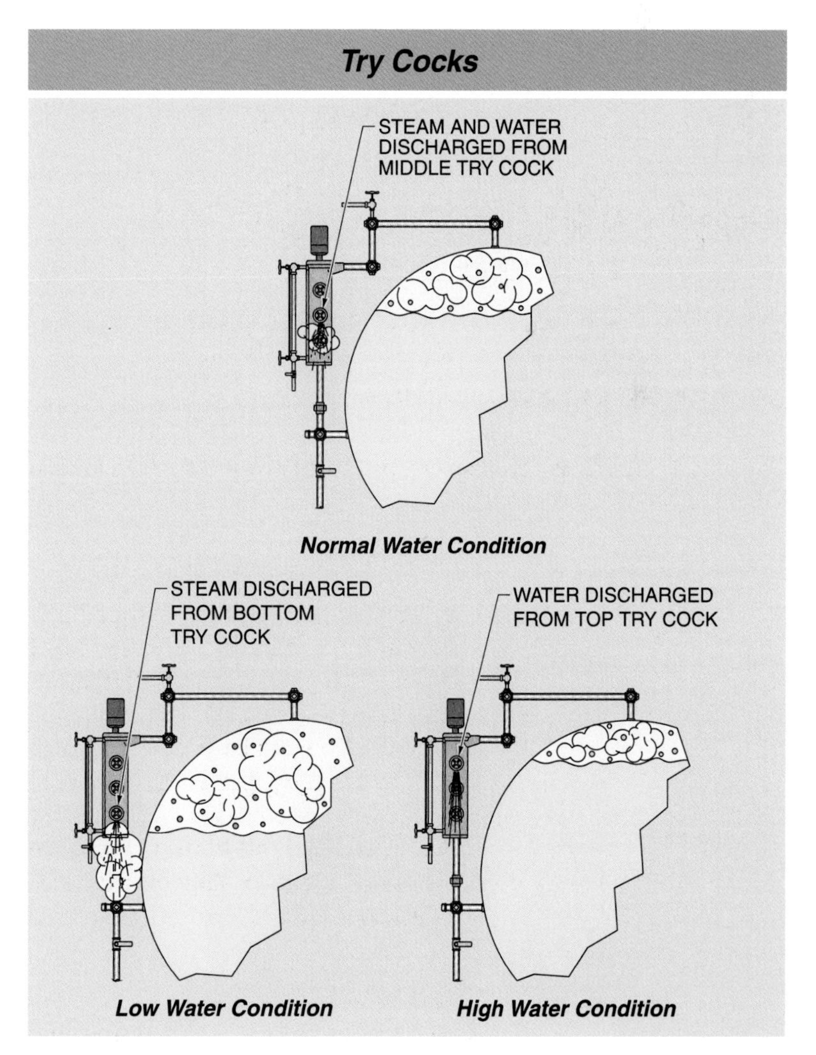

Figure 17-7. Try cocks are a backup system used to determine the water level in a boiler.

Low Water Fuel Cutoff

A loss of water in a boiler can lead to the burning out of tubes and/or a boiler explosion. A *low water fuel cutoff* is a boiler fitting that shuts the burner OFF in the event of a low water condition. **See Figure 17-8.** The low water fuel cutoff is located slightly below the normal water level. A typical cutoff consists of a level float switch that is part of the permissive contacts of the burner control safety shutdown system. An open contact fails the burner permissive interlock system, de-energizes the burner safety system, and shuts down the burner if the water level drops below the safe operating level.

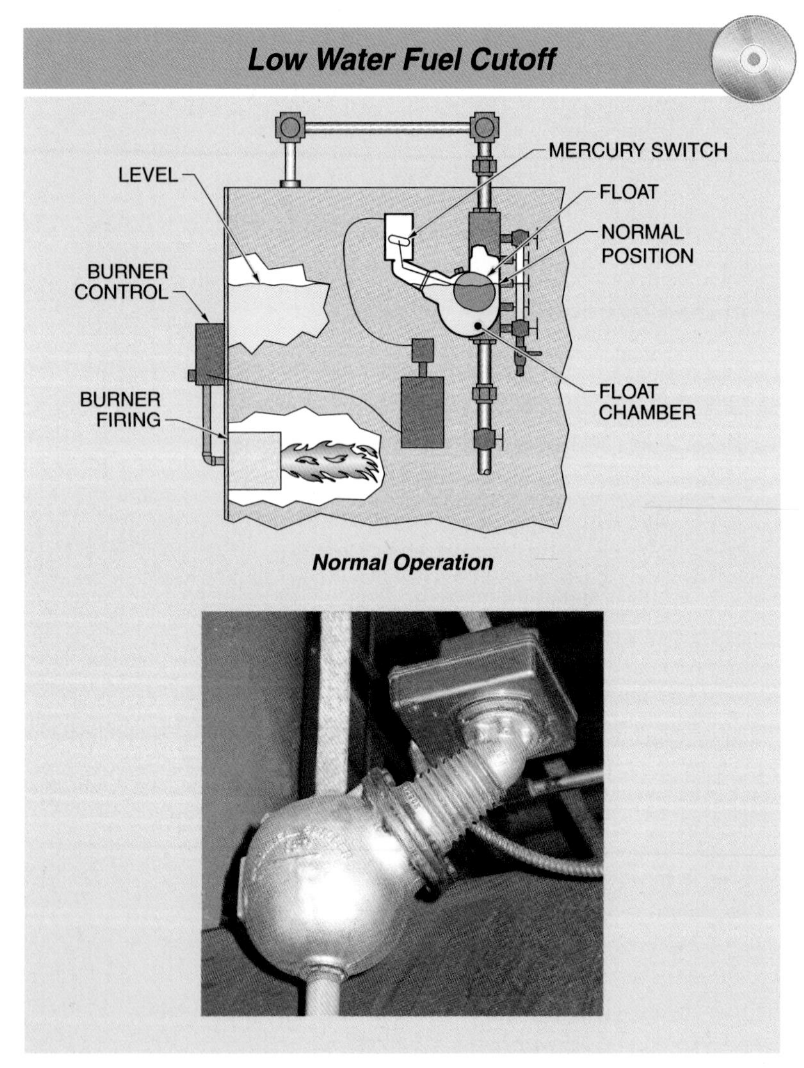

Figure 17-8. A low water fuel cutoff shuts off the fuel flow to a boiler when the water level gets too low.

CORROSIVE FLUIDS

There are times when a pressure measurement is used to determine level and the process liquid is incompatible with the pressure gauge. The pressure gauge must be protected from the corrosive fluids. A common way to respond to this problem is to use diaphragm seals with d/p cells.

Diaphragm Seals

A transmitter can be protected from a corrosive process fluid by connecting capillary tubing, with diaphragm seals, that is filled with liquid of constant specific gravity. The diaphragm seals protect the integrity of the liquid in the capillary tubing and the liquid protects the transmitter. Common diaphragm fill fluids are glycerin and silicone. Other fill fluids are available with specific gravities ranging from 0.85 to 1.85.

Seal Connections. The bottom connection is near the bottom of the tank and the top connection is near the top of the liquid level or in the vapor space above the liquid in the tank. By convention, the top connection on the tank is connected to the low-pressure connection on the transmitter. However, because the density of seal fluid in the capillary may be higher than the density of the process fluid, the actual pressure on the low-pressure connection may be higher than at the high-pressure connection of the transmitter. The differential pressure at the transmitter is determined by calculating the pressure due to the head of liquid on each side of the transmitter as follows:

$$P = \rho \times H$$

where

P = pressure (in in. WC)

ρ = specific gravity of capillary fluid

H = head of capillary fluid (in in.)

and

$$dP = P_{HC} - P_{LC}$$

where

dP = differential pressure

P_{HC} = pressure at high-pressure side of transmitter

P_{LC} = pressure at low-pressure side of transmitter

For example, a sealed tank containing 20% sulfuric acid with a specific gravity of 1.151 is stored under pressure of 10 psig. The level-sensing pressure connections are 96″ (8′) apart, and the liquid used to fill the capillary tubes has a specific gravity of 1.20. The low-pressure side of the transmitter is connected at an elevation of 96″. The high-pressure side of the transmitter is connected at an elevation of 0″. **See Figure 17-9.**

Transmitter Calibration. Transmitter calibration requires the determination of the zero and span values. The level transmitter calibration values for measuring the acid level from the elevation of the bottom connection to the elevation of the top connection are calculated from the density and height of the liquid.

When the tank is empty, the differential pressure reading across the transmitter is due entirely to the head of the fluid in the capillary tubing because the pressure at the high-pressure connection is zero. The pressure applied to the tank can be ignored here because a d/p transmitter measures differential pressure and the applied pressure is equal on both sides of the transmitter and cancels out.

$P_{LC} = \rho \times H_C$

$P_{LC} = 1.20 \times 96$

$P_{LC} = \mathbf{115.2\ in.\ WC}$

$P_{HC} = 0$ in. WC

$dP = P_{HC} - P_{LC}$

$dP = 0 - 115.2$

$dP = \mathbf{-115.2\ in.\ WC}$

Since the transmitter measures a pressure difference of –115.2 in. WC when the tank is empty, the zero point of the transmitter must be suppressed by 115.2 in. WC.

The span of the transmitter is the difference between the pressures at the high-pressure connection when the tank is empty and when it is full. When the tank is empty, the pressure is zero. Therefore, the span is equivalent to the pressure due to the head of the acid. This is calculated as follows:

$P_{HC} = \rho \times H$

$P_{HC} = 1.151 \times 96$

$P_{HC} = \mathbf{110.5\ in.\ WC}$

The transmitter calibration values are a span of 110.5 in. WC and a zero suppression of 115.2 in. WC.

Level Transmitter Application Requiring Suppression

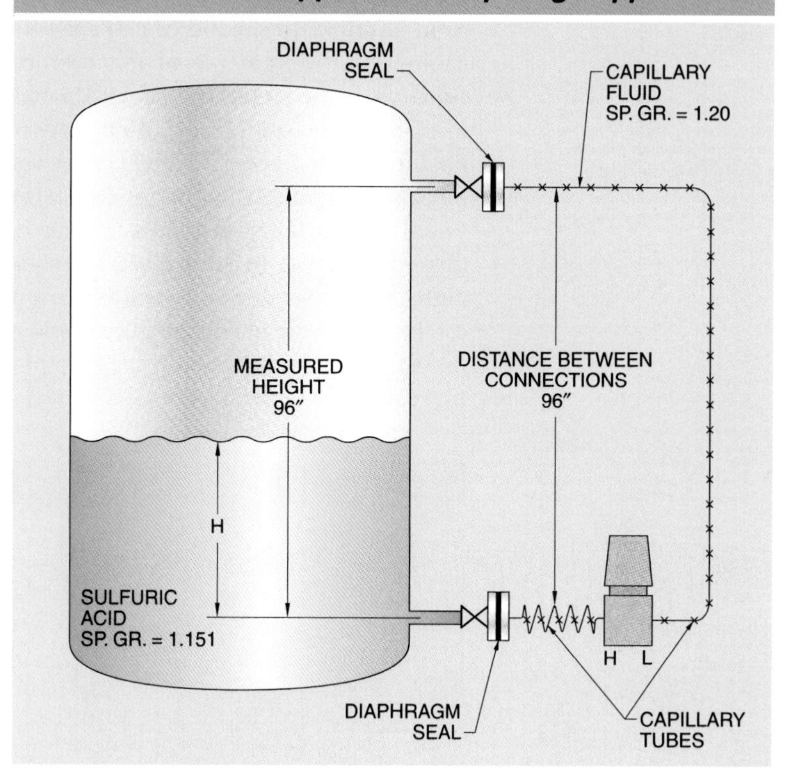

Figure 17-9. A double diaphragm seal system may require the use of transmitter suppression.

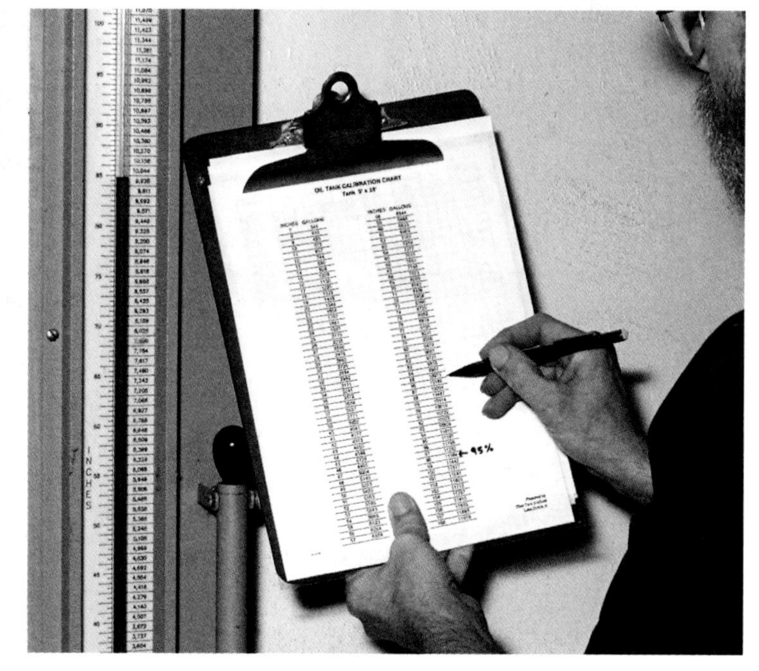

Remote-reading fuel oil tank level gauges convert a pressure reading into the movement of a liquid in a glass tube.

Some level measurement applications may require that the level transmitter have an elevation incorporated into its calibration. An example of this type of situation is the use of a simple d/p cell to directly measure the head of process fluid in a vented vessel, but with the level transmitter mounted below the low-level vessel connection. **See Figure 17-10.** In this arrangement there is always an additional head pressure applied to the d/p cell high-pressure connection. Even when the vessel is empty, the d/p cell output represents a positive level in the vessel. An elevation value is added to the level transmitter calibration to compensate for this constant head, which is added to the normal measurement level.

TECH FACT

Tank foundations are typically made of reinforced concrete with foundation footings below the frost line or that rest on bedrock. A steel saddle may be welded to the tank to make it into a portable tank system.

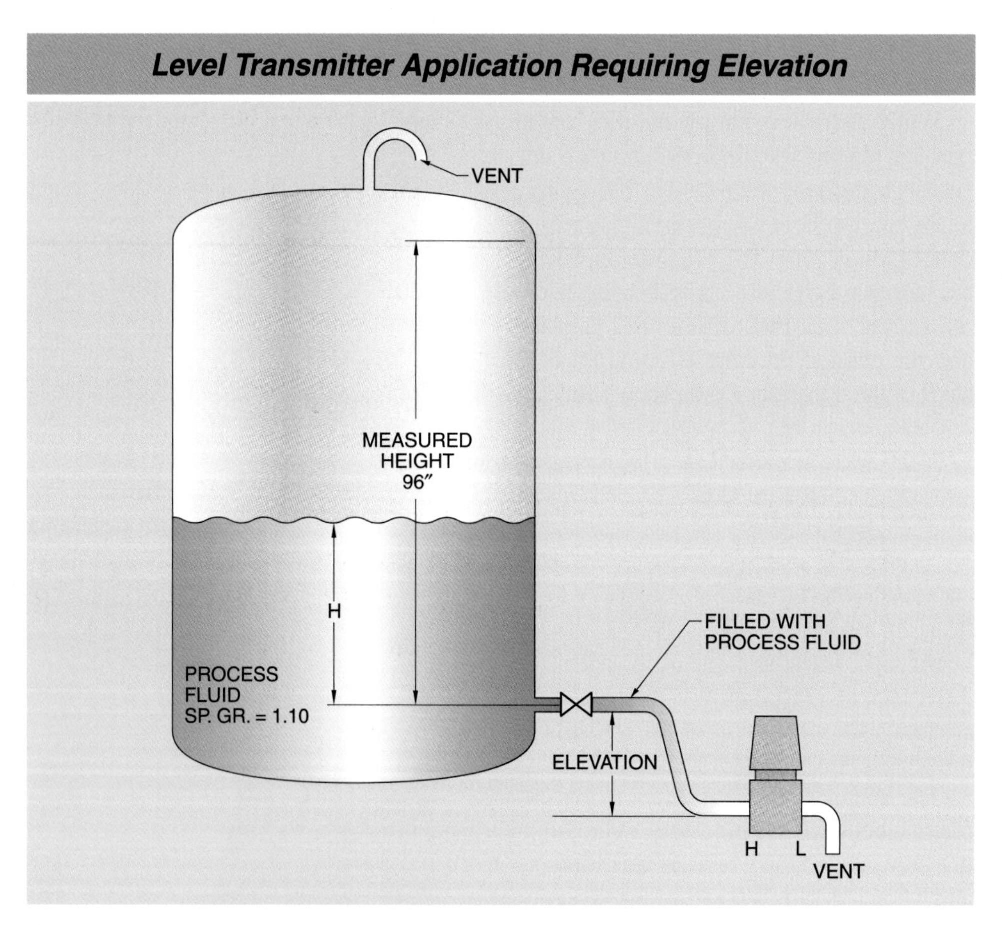

Figure 17-10. Transmitter elevation may be required when the transmitter is below the tank bottom and a filled tube connects the tank and transmitter.

KEY TERMS

- *bulk solid:* A granular solid, such as gravel, sand, sugar, grain, cement, or other solid material that can be made to flow.
- *funnel flow:* The flow of a bulk solid where the material empties out of the bottom of a silo and the main material flow is down the center of the silo, with stagnant areas at the sides and bottom of the silo.
- *mass flow:* The flow of a bulk solid where all material in a silo flows down toward the bottom at the same rate.
- *ratholing:* A condition arising in a silo when material in the center has flowed out the feeder at the bottom, leaving large areas of stagnant material on the sides.
- *bridging:* A condition arising in a silo when material has built up over the feeder, blocking all flow out of the silo.
- *water column:* A boiler fitting that reduces the turbulence of boiler water to provide an accurate water level in the gauge glass.
- *try cock:* A valve located on a water column used to determine the boiler water level if the gauge glass is not functional.
- *blowdown:* The process of discharging water and undesirable accumulated material.
- *low water fuel cutoff:* A boiler fitting that shuts the burner OFF in the event of a low water condition.

REVIEW QUESTIONS

1. What are the flow properties of bulk solids?
2. How does funnel flow make it difficult to measure level?
3. List and describe common types of load cell calibration.
4. Define water column, and explain how it can be used to measure the water level in a boiler.
5. What is the purpose of diaphragm seals?

SECTION 5

FLOW MEASUREMENT

TABLE OF CONTENTS

SECTION OBJECTIVES

Chapter 18

- Compare flow rate and total flow.
- Identify the physical properties of fluids and other factors that affect flow and the selection of a flow measurement method.
- Define and identify the three gas laws and the combined gas law.

Chapter 19

- List and describe the types of primary flow elements used with differential pressure measurements.
- Explain the operating principles of differential pressure flowmeters.

Chapter 20

- Explain the use and operation of variable-area flowmeters.
- Describe positive-displacement flowmeters and explain their use.
- Compare turbine meters and paddle wheel meters and explain their use.
- Explain the use of open-channel flow measurements.

Chapter 21

- Describe magnetic flowmeters and explain their use.
- List and describe ultrasonic flowmeters.
- Explain the operation of mass flowmeters.

SECTION

5 FLOW MEASUREMENT

Chapter 22

- Describe the requirements for making differential instrument connections and instrument locations.
- Explain the purpose of blocking valves and manifolds.
- List and describe accessory flow devices.

INTRODUCTION

Interest in measurement of water and air flow goes back to the earliest recorded history. Early civilizations needed to measure the flow of irrigation water through the aqueducts to ensure fair distribution. The early Greek philosophers Aristotle and Archimedes studied fluid dynamics, pneumatics, and aerodynamics. Archimedes invented a water pump that uses a screw to move water. In his honor, this is now called an Archimedes screw.

The first modern studies of fluids began in the late 18th century when Swiss physicist Daniel Bernoulli introduced the idea of conservation of energy in fluid flows. His work is the basis for many of the fluid flow measuring instruments in use today. The Bernoulli principle is studied by engineers today. In the middle of the 19th century, Austrian physicist Christian Doppler developed the ideas behind what is now known as the Doppler effect, used in Doppler flowmeters. In the late 19th century, British engineer Osborne Reynolds developed the idea of a single number that can be used to describe fluid flow. The Reynolds number is a fundamental measure of the type of flow.

ADDITIONAL ACTIVITIES

1. After completing Section 5, take the Quick Quiz® included with the Digital Resources.
2. After completing each chapter, answer the questions and complete the activities in the *Instrumentation and Process Control Workbook*.
3. Review the Flash Cards for Section 5 included with the Digital Resources.
4. Review the following related Media Clip for Section 5 included with the Digital Resources:
   - Orifice Plate
5. After completing each chapter, refer to the Digital Resources for the Review Questions in PDF format.

ATPeResources.com/QuickLinks • Access Code: 467690

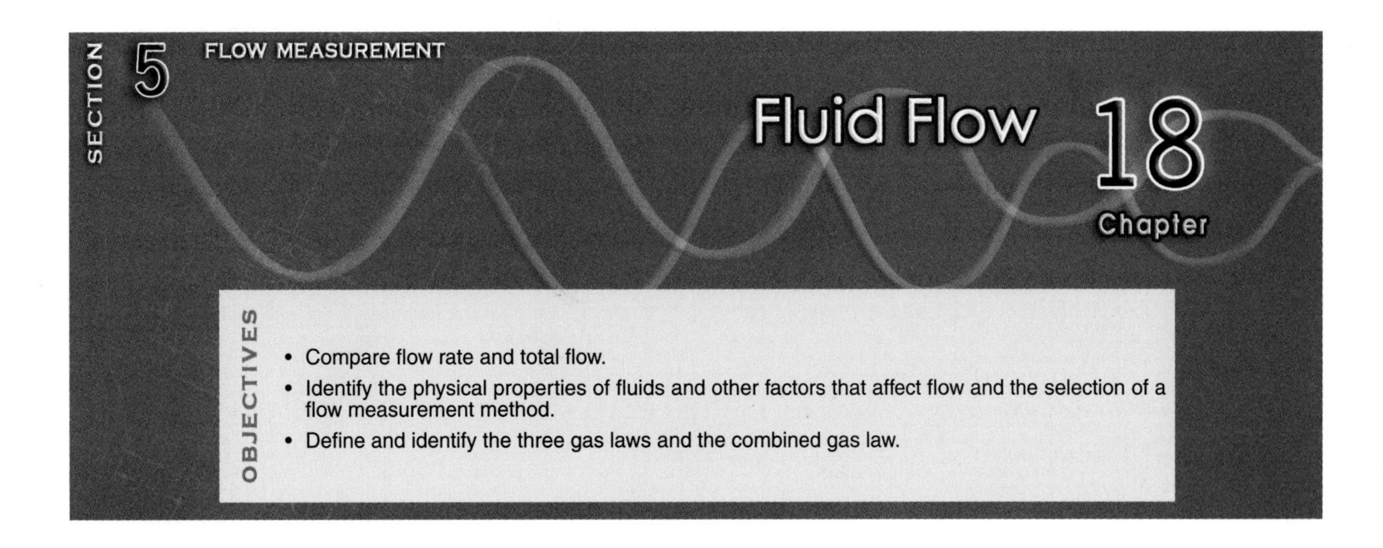

CHARACTERISTICS OF FLUID FLOW

Fluid flow is the movement of liquids in pipes or channels, and gases or vapors in pipes or ducts. The need to measure fluid flow has generated a variety of instruments based on various scientific principles. It is often more convenient to measure the flow of a fluid by measuring some other characteristic that varies in a predictable and reliable way with the rate of flow, such as a drop in pressure caused by a restriction in a pipeline.

The most important characteristic of a fluid that affects flow is whether the fluid is a liquid, gas, or vapor. At certain temperatures and pressures, most fluids can change phase between vapor, liquid, or solid. For example, water can be heated to make steam or cooled to become ice. Gases can be condensed to liquids like liquid nitrogen or liquid oxygen, or to a solid like dry ice.

> **TECH FACT**
> *Turbulence is the state of agitated motion of a fluid. The unpredictability of turbulent flow has long confounded scientists. Supercomputer modeling and new experiments have shown that the onset of turbulence occurs abruptly and is described by the strange attractors of the chaos theory.*

Flow Rate and Total Flow

The two types of flow measurements are flow rate and total flow. *Flow rate* is the quantity of fluid passing a point at a particular moment. *Total flow* is the quantity of fluid that passes a point during a specific time interval. For example, the flow rate of pumping a fluid may be given in gallons per hour and the total flow is the total gallons pumped.

The flow rate of liquids is expressed in volumetric or mass units. The common volumetric units used in the United States are gallons per minute (gpm) and gallons per hour (gph). In other areas, the Imperial gallon is used. Commonly used metric units are liters per minute (l/min), cubic meters per hour (cu m/hr or m^3/hr), and cubic centimeters per minute (cm^3/min or cc/min). The unit of mass in the United States is pounds per hour (lb/hr). The metric unit of mass is kilograms per hour (kg/hr). **See Figure 18-1.**

The flow rate of gases and vapors is usually expressed in volumetric units. In the United States, gas flows are commonly stated as standard cubic feet per minute (scfm) or standard cubic feet per hour (scfh). The metric units are standard cubic centimeters per minute (sccm) or cubic meters per hour (m^3/hr). Steam flow is generally expressed in mass units of pounds per hour (lb/hr) or kilograms per hour (kg/hr).

Flow Rate Conversions

Given Value	To Convert to					
	gpm	**gph**	**l/min**	**m^3/hr**	**cm^3/min**	**ft^3/min**
gpm	1	gpm × 60	gpm × 3.785	gpm × 0.2271	gpm × 3785	gpm × 0.1337
gph	gph × 0.01667	1	gph × 0.06309	gph × 0.003785	gph × 63.09	gph × 8.022
l/min	l/min × 0.2642	l/min × 15.85	1	l/min × 0.06	l/min × 1000	l/min × 0.0353
m^3/hr	m^3/hr × 4.403	m^3/hr × 264.2	m^3/hr × 16.67	1	m^3/hr × 16,667	m^3/hr × 0.5886
cm^3/min	cm^3/min × 0.0002642	cm^3/min × 0.01585	cm^3/min × 0.001	cm^3/min × 0.00006	1	cm^3/min × 0.0000353
ft^3/min	ft^3/min × 7.479	ft^3/min × 0.1247	ft^3/min × 28.31	ft^3/min × 1.699	ft^3/min × 28,312	1

Figure 18-1. Flow is measured in many units. Conversion tables are used to convert from one unit to another.

PHYSICAL PROPERTIES

There are a number of physical properties common to most fluids that influence the selection of the method of measuring fluid flow. These include pressure, velocity, density, viscosity, compressibility, electrical capacitance and conductivity, thermal conductivity, and the response to sonic impulses, light, or mechanical vibration. Many of these properties can be measured to determine fluid flow rate. The fact that so many properties and characteristics can be measured accounts for the wide variety of flowmeters.

Photo used by permission of GE Panametrics

A liquid flow transmitter can easily be used with many types of sensors in various types of applications.

In addition to the physical properties of fluids, there are other factors that affect the flow. These factors include the configuration of the pipes or ducts; the location, style, and number of valves; and changes in elevation of the fluid. The most important factors affecting fluid flow are the properties of the fluid (density, specific gravity, and viscosity), the Reynolds number describing the type of flow, and the compressibility of the fluid.

Density and Specific Gravity

Density is mass per unit volume. Common units of density are pounds per cubic foot (lb/ft^3 or lb/cu ft) and grams per cubic centimeter (g/cm^3 or g/cu cm). Density varies with changes in temperature. *Specific gravity* is the ratio of the density of a fluid to the density of a reference fluid. For liquids, the reference fluid is usually water. For gases, the reference fluid is usually dry air.

When two liquids that do not mix are in a container, the one with lower specific gravity will float on top of the one with higher specific gravity. For example, many oils have specific gravities between 0.75 and 0.85 at ambient temperatures. Water has a specific gravity of 0.998 at 68°F. The oils will remain on top of the water unless agitated. This is why motor oil and gasoline float on water.

Viscosity

Absolute viscosity is the resistance to flow of a fluid and has units of centipoise (cP). *Kinematic viscosity* is the ratio of absolute viscosity to fluid density and has units of centistokes (cS). The viscosity of many commercial fluids, like oils, is commonly specified as an allowable range at a certain temperature.

For example, No. 6 fuel oil and motor oil specifications allow a broad range of viscosity at a specified temperature. At 68°F, water has a viscosity of 1.0 cP, SAE 30 oil has a viscosity of approximately 100 cP, and No. 6 fuel oil has a viscosity of approximately 850 cP. **See Figure 18-2.** Viscosity is affected by temperature and other factors and normally decreases with increasing temperature. Many fluids must be preheated before being pumped.

Typical Liquid Viscosities*

	Specific Gravity, SG	Absolute Viscosity, Centipoise	Kinematic Viscosity, Centistokes
Water	1.0	0.98	0.97
Gasoline	0.71	0.48	0.67
Ethylene Glycol	1.1	20	18
SAE 30 Motor Oil	0.91	96 (100°F)	106 (100°F)
No. 6 Fuel Oil	0.88	850 (68°F) 335 (122°F)	966 (68°F) 379 (122°F)
95% Sulfuric Acid	1.84	26.6	14.5
25% Sodium Chloride	1.2	2.9	2.4
Acetic Acid	1.05	1.2	1.15
Glycerin	1.3	1000	770
Acetone	0.79	0.33	0.42
n-Propyl Alcohol	0.80	2.2	2.75
Corn Oil	0.93	26.5 (130°F)	28.7 (130°F)
Molasses	1.43	430 to 7000 (100°F)	300 to 5000 (100°F)
Freon	1.33	0.20	0.15

*at room temperature unless otherwise noted

Figure 18-2. Liquid viscosities vary over a wide range.

Reynolds Number

A *Reynolds number* is the ratio between the inertial forces moving a fluid and viscous forces resisting that movement. It describes the nature of the fluid flow. The Reynolds number has no units of measure and is calculated from velocity or flow rate, density, viscosity, and the inside diameter of a pipe. Reynolds numbers commonly range from 100 to 1,000,000. However, they can be higher or lower than these values.

Velocity is the speed of fluid in the direction of flow and is often expressed in ft/sec. A streamline can be drawn to show the direction and magnitude of smooth flow at every point across a pipe profile. A flow profile can be drawn to represent the velocity of a fluid at different points across a pipe or duct. **See Figure 18-3.**

Laminar and Turbulent Flow. *Laminar flow* is smooth fluid flow that has a flow profile that is parabolic in shape with no mixing between the streamlines. Laminar flow in pipes occurs at Reynolds numbers below about 2100. A cross section of a laminar flow stream is a parabolic flow profile, with the maximum velocity in the center and the minimum velocity at the pipe walls.

Turbulent flow is fluid flow in which the flow profile is a flattened parabola, the streamlines are not present, and the fluid is freely intermixing. Turbulent flow in pipes typically occurs at Reynolds numbers above about 4000. The exact shape of the flattened profile depends on the Reynolds number.

There is a sudden transition between laminar flow and turbulent flow as the flow rate increases. The exact transition point cannot be easily predicted, but it normally occurs at Reynolds numbers between 2100 and 4000. The flow is often termed transitional at flow rates between these Reynolds numbers, even though the actual flow is laminar or turbulent. Many flowmeters require turbulent flow and specify Reynolds numbers above 10,000 to ensure that turbulent flow is the prevailing condition.

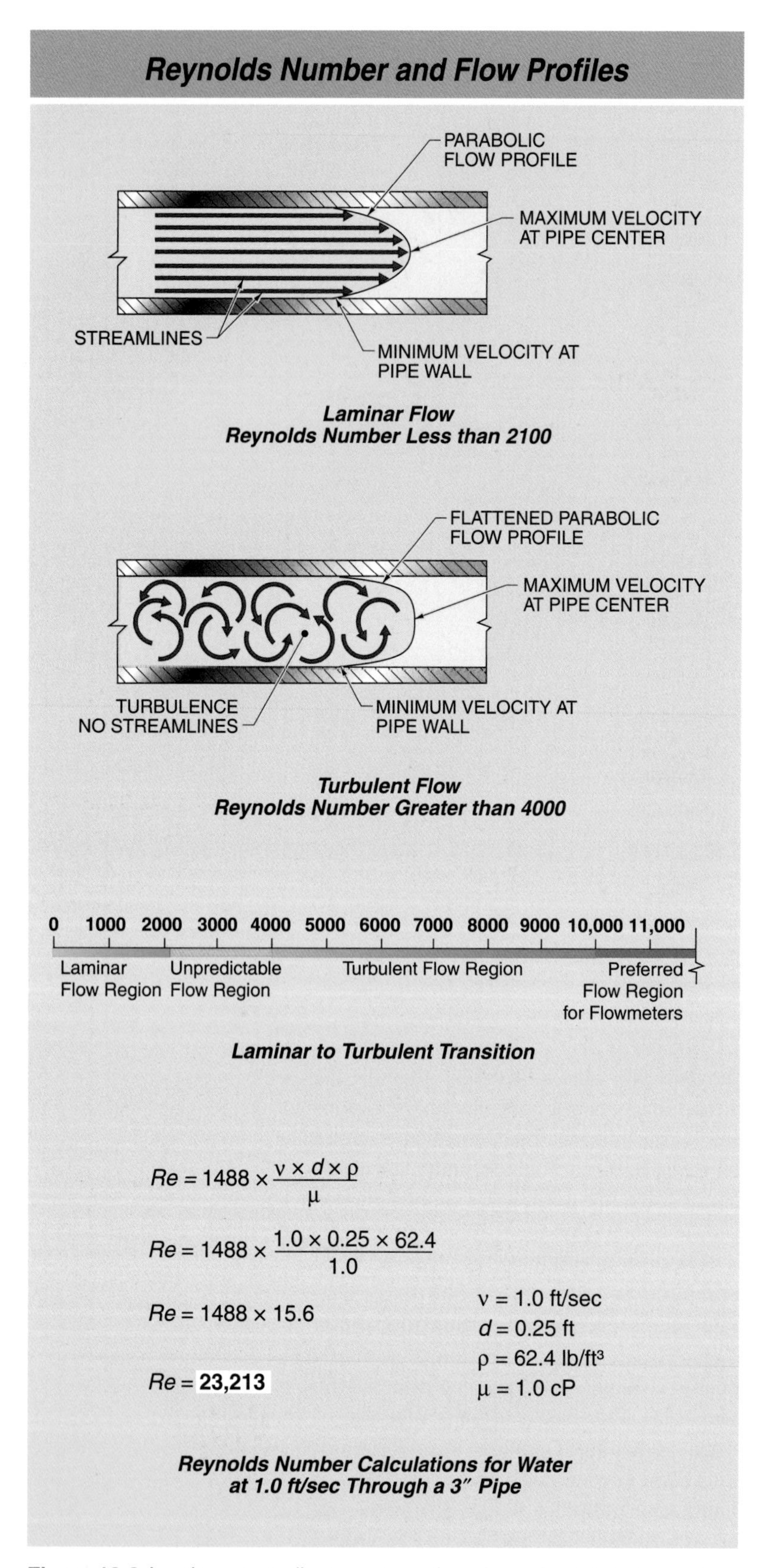

Figure 18-3. Laminar streamlines are smooth and steady while turbulent flow is chaotic. Laminar flow occurs at Reynolds numbers below 2100.

Calculating the Reynolds Number. There are many forms of the Reynolds number equation, depending on the measurement units used. One common form of this equation is as follows:

$$Re = 1488 \times \frac{\nu \times d \times \rho}{\mu}$$

where

Re = Reynolds number

ν = velocity (in ft/sec)

d = inside diameter of the pipe (in ft)

ρ = density (in lb/cu ft)

μ = viscosity (in cP)

1488 = constant conversion factor

For example, water has a viscosity of 1.0 cP (μ) at 68°F and a density of 62.4 lb/ft^3 (ρ). When the water is pumped at 1.0 ft/sec (ν) through a 3″ pipe (d, 0.25′), the Reynolds number is as follows:

$$Re = 1488 \times \frac{\nu \times d \times \rho}{\mu}$$

$$Re = 1488 \times \frac{1.0 \times 0.25 \times 62.4}{1.0}$$

$$Re = \mathbf{23{,}213}$$

A Reynolds number of 23,213 is in the turbulent flow region.

Compressibility

An *incompressible fluid* is a fluid where there is very little change in volume when subjected to a change in pressure. Liquids are essentially incompressible. For example, fluid power systems transmit power through an incompressible hydraulic fluid.

A *compressible fluid* is a fluid where the volume and density change when subjected to a change in pressure. Gases and vapors are compressible fluids. The specific gravity of a gas is usually calculated as the ratio of the density of that gas to the density of air at standard conditions. The density of air at standard atmospheric pressure is 0.808 lb/ft^3 (1.294 g/l) at 32°F and 0.0753 lb/ft^3 (1.205 g/l) at 68°F. Since gases and vapors are compressible, the pressure and temperature at the point of measurement affect their volume.

TECH FACT

A Newtonian liquid has a constant viscosity when energy, such as pumping, is applied. A non-Newtonian liquid has a viscosity that changes (usually decreases) when energy is applied. The flow properties of non-Newtonian fluids are difficult to predict.

Standard Conditions. A *standard condition* is an accepted set of temperature and pressure conditions used as a basis for measurement. Standard conditions compensate for the compressibility of fluids and may be different in different industries. For example, natural gas usage is based on standard conditions of 60°F and 14.73 psia. Standard conditions used in HVAC systems when sizing air-handling equipment are 70°F and 14.696 psia (usually rounded to 14.7 psia). Metric standard conditions are typically 0°C and 1.033 kg/cm^2 or 15°C and 1.0125 kg/cm^2. Gases measured at nonstandard conditions can be converted to standard conditions through the application of the gas laws. The use of standard conditions does not imply that the gas exists at those conditions. Standard conditions are used as units of measure.

Flowing Conditions. A *flowing condition* is the pressure and temperature of the gas or vapor at the point of measurement. For example, the flowing condition at the point of measurement of compressed air may be 100 psig and 80°F in a pipe line. Actual cubic feet per hour (acfh) is a typical unit of flowing volume at the temperature and pressure at the point of measurement. This information is useful when selecting pipe sizes, but is not usually used for flowmeters.

GAS LAWS

The three gas laws, Boyle's law, Charles' law, and Gay-Lussac's law, provide the basis for determining the volume of a gas at one set of pressure and temperature conditions when data from another set of conditions are known. **See Figure 18-4.**

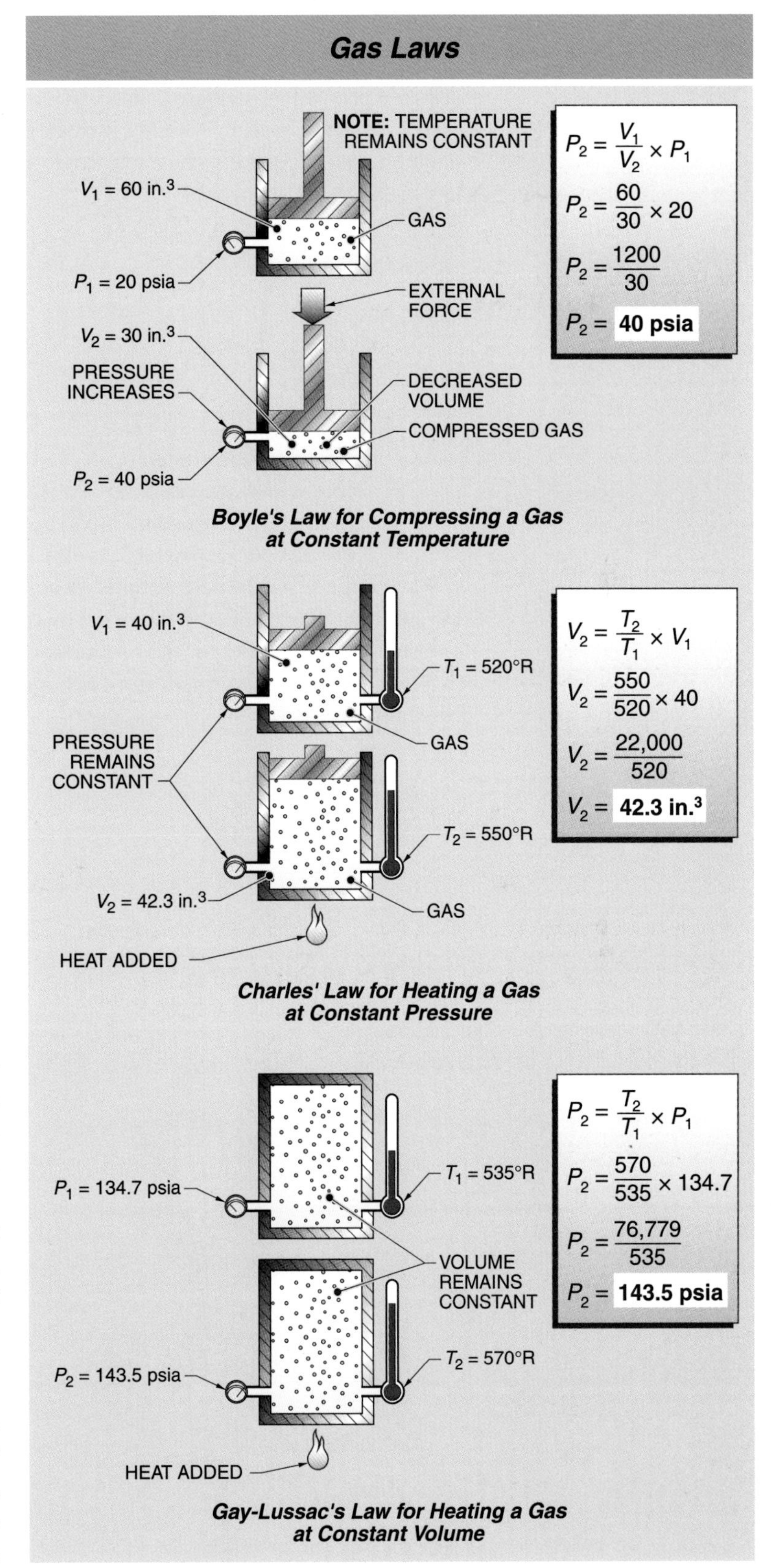

Figure 18-4. Gas laws show how gases behave with changes in temperature, pressure, and volume.

Boyle's Law

Boyle's law is a gas law that states that the absolute pressure of a given quantity of gas varies inversely with its volume provided the temperature remains constant. Boyle's law is expressed as follows:

$$\frac{P_2}{P_1} = \frac{V_1}{V_2}$$

or

$$P_2 = \frac{V_1}{V_2} \times P_1$$

where
P_2 = final pressure (in psia)
P_1 = initial pressure (in psia)
V_2 = final volume (in cubic units)
V_1 = initial volume (in cubic units)

For example, a cylinder contains 60 in.^{3} (V_1) of air at 20 psia (P_1). What is the final pressure (P_2) when the air is compressed to 30 in.^{3} (V_2) and the temperature is kept constant?

$$P_2 = \frac{V_1}{V_2} \times P_1$$

$$P_2 = \frac{60}{30} \times 20$$

$$P_2 = \frac{1200}{30}$$

P_2 = **40 psia at 30 in.^{3}**

Charles' Law

Charles' law is a gas law that states that the volume of a given quantity of gas varies directly with its absolute temperature provided the pressure remains constant. Charles' law is expressed as follows:

$$\frac{V_2}{V_1} = \frac{T_2}{T_1}$$

or

$$V_2 = \frac{T_2}{T_1} \times V_1$$

where
T_2 = final temperature (in °R)
T_1 = initial temperature (in °R)
V_2 = final volume (in cubic units)
V_1 = initial volume (in cubic units)

For example, a cylinder contains 40 in.^{3} (V_1) of air at 60°F ($T_1 = 60 + 460 = 520$°R). What is the final volume (V_2) when the temperature is increased to 90°F ($T_2 = 90 + 460 = 550$°R) and the pressure is kept constant?

$$V_2 = \frac{T_2}{T_1} \times V_1$$

$$V_2 = \frac{550}{520} \times 40$$

$$V_2 = \frac{22{,}000}{520}$$

V_2 = **42.3 in.^{3} at 90°F**

The Foxboro Company
An integral orifice differential pressure transmitter with an integral three-valve process manifold is used to measure and transmit a flow rate.

Gay-Lussac's Law

Gay-Lussac's law is a gas law that states that the absolute pressure of a given quantity of a gas varies directly with its absolute temperature provided the volume remains constant. Gay-Lussac's law is expressed as follows:

$$\frac{P_2}{P_1} = \frac{T_2}{T_1}$$

or

$$P_2 = \frac{T_2}{T_1} \times P_1$$

where
T_2 = final temperature (in °R)
T_1 = initial temperature (in °R)
P_2 = final volume (in psia)
P_1 = initial volume (in psia)

For example, a compressor tank holds air at 75°F (T_1 = 75 + 460 = 535°R) and 134.7 psia (P_1). What is the final pressure (P_2) when the temperature is increased to 110°F (T_2 = 110 + 460 = 570°R) and the volume is kept constant?

$$P_2 = \frac{T_2}{T_1} \times P_1$$

$$P_2 = \frac{570}{535} \times 134.7$$

$$P_2 = \frac{76,779}{535}$$

$$P_2 = \mathbf{143.5\ psia\ at\ 110°F}$$

Combined Gas Law

The three gas laws can be combined into one equation that shows the relationship between volume, pressure, and temperature. The combined gas law is expressed as follows:

$$\frac{V_1 \times P_1}{T_1} = \frac{V_2 \times P_2}{T_2}$$

or

$$V_2 = V_1 \times \frac{P_1}{P_2} \times \frac{T_2}{T_1}$$

where

V = volume (in ft³ or other volumetric terms)

P = pressure (in psia or other absolute pressure terms)

T = temperature (in °R or K)

and the subscripts refer to different sets of conditions.

For example, 100 ft³ (V_1) of compressed gas is measured in a meter at 75 psig (P_1 = 75 + 14.7 = 89.7 psia) and 150°F (T_1 = 150 + 460 = 610°R). What is the volume (V_2) at standard conditions of 14.7 psia (P_2) and 70°F (T_2 = 70 + 460 = 530°R)? **See Figure 18-5.**

$$V_2 = V_1 \times \frac{P_1}{P_2} \times \frac{T_2}{T_1}$$

$$V_2 = 100 \times \frac{89.7}{14.7} \times \frac{530}{610}$$

$$V_2 = 100 \times 6.10 \times 0.869$$

$$V_2 = \mathbf{530\ ft^3\ at\ 14.7\ psia\ and\ 70°F}$$

Since flow is volume per unit time, flow rates can be substituted for the volumetric terms in the equations above to allow flow rates at different conditions to be compared as follows:

$$F_2 = F_1 \times \frac{P_1}{P_2} \times \frac{T_2}{T_1}$$

where F_1 and F_2 are the volumetric flow rate in ft³/hr (or ft³/min)

The calculation is exactly the same except that the units of volumetric flow are used instead of units of volume.

> **TECH FACT**
> *Oxygen is a very reactive gas. Gases can be compressed enough that they liquefy. If organic materials, such as grease, are brought into contact with liquid oxygen, an explosion can occur.*

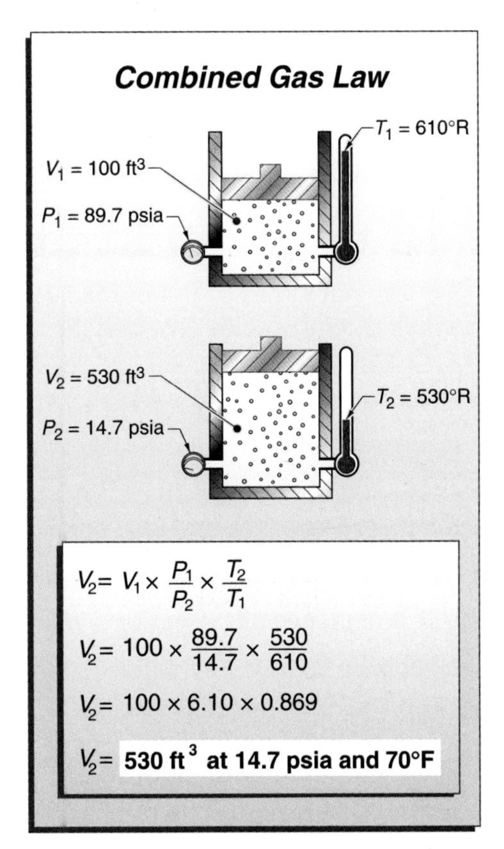

Figure 18-5. The combined gas law takes into account changes in temperature, pressure, and volume.

KEY TERMS

- *flow rate:* The quantity of fluid passing a point at a particular moment.
- *total flow:* The quantity of fluid that passes a point during a specific time interval.
- *density:* Mass per unit volume.
- *specific gravity:* The ratio of the density of a fluid to the density of a reference fluid.
- *absolute viscosity:* The resistance to flow of a fluid and has units of centipoise (cP).
- *kinematic viscosity:* The ratio of absolute viscosity to fluid density and has units of centistokes (cS).
- *Reynolds number:* The ratio between the inertial forces moving a fluid and viscous forces resisting that movement.
- *laminar flow:* Smooth fluid flow that has a flow profile that is parabolic in shape with no mixing between the streamlines.
- *turbulent flow:* Fluid flow in which the flow profile is a flattened parabola, the streamlines are not present, and the fluid is freely intermixing.
- *incompressible fluid:* A fluid where there is very little change in volume when subjected to a change in pressure.
- *compressible fluid:* A fluid where the volume and density change when subjected to a change in pressure.
- *standard condition:* An accepted set of temperature and pressure conditions used as a basis for measurement.
- *flowing condition:* The pressure and temperature of the gas or vapor at the point of measurement.
- *Boyle's law:* A gas law that states that the absolute pressure of a given quantity of gas varies inversely with its volume provided the temperature remains constant.
- *Charles' law:* A gas law that states that the volume of a given quantity of gas varies directly with its absolute temperature provided the pressure remains constant.
- *Gay-Lussac's law:* A gas law that states that the absolute pressure of a given quantity of a gas varies directly with its absolute temperature provided the volume remains constant.

REVIEW QUESTIONS

1. What is the difference between flow rate and total flow?
2. What are the properties of fluids that affect flow and influence the selection of flow instruments?
3. Explain the difference between absolute viscosity and kinematic viscosity.
4. Explain the difference between laminar and turbulent flow.
5. Identify the three gas laws and the combined gas law.

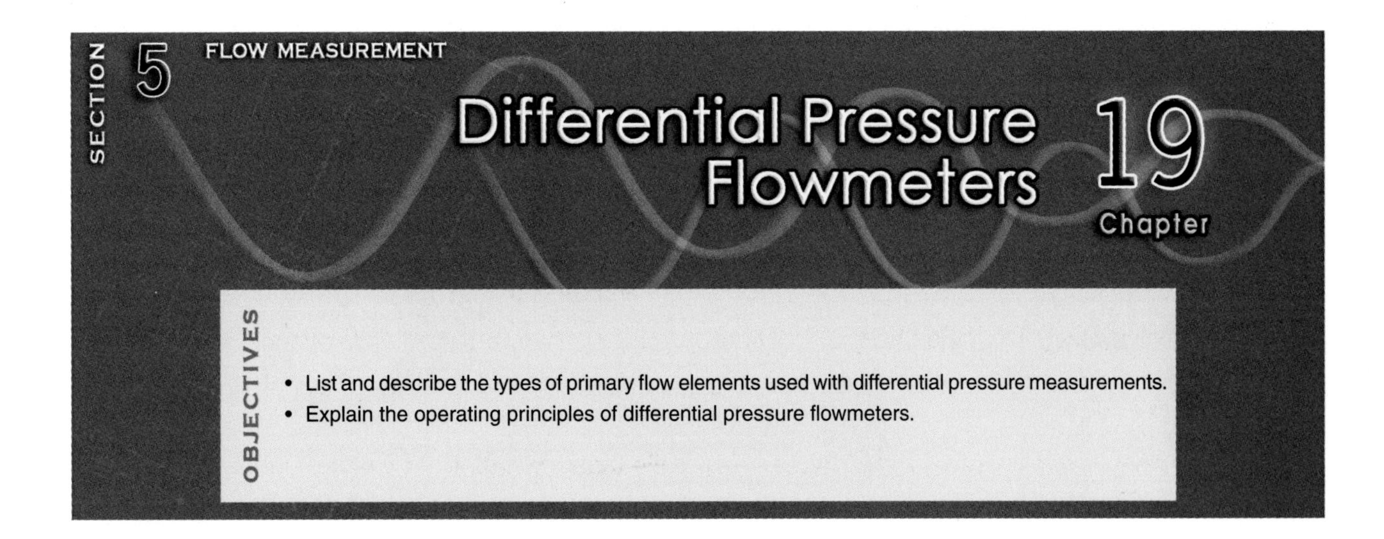

PRIMARY FLOW ELEMENTS

A pressure difference is created when a fluid passes through a restriction in a pipe. The point of maximum developed differential pressure is between the pressure upstream of the restriction and the pressure downstream of the restriction, at the point of highest velocity. The shape and configuration of the restriction affects the magnitude of the differential pressure and how much of the differential is recoverable. A low differential pressure recovery means that the flowing fluid permanently loses much of its pressure. A high differential pressure recovery is more energy efficient. A restriction in piping used for flow measurement is a primary flow element.

A *primary flow element* is a pipeline restriction that causes a pressure drop used to measure flow. Primary flow elements are designed to provide accuracy, low cost, ease of use, and pressure recovery, but not necessarily all in the same element. For example, the venturi tube has a high pressure recovery but it is relatively expensive and not easy to use. The most common primary element is the orifice plate. Other primary elements include flow nozzles, venturi tubes, low-loss flow tubes, and pitot tubes.

Orifice Plates

An *orifice plate* is a primary flow element consisting of a thin circular metal plate with a sharp-edged round hole in it and a tab that protrudes from the flanges. The tab has orifice plate information stamped onto it. The information usually includes pipe size, bore size, material, and type of orifice. Orifice plates are not always reversible so the stamping is on the upstream face. The orifice is held in place between two special pipe flanges called orifice flanges. **See Figure 19-1.**

Orifice plates are simple, inexpensive, and replaceable. The hole in the plate is generally in the center (concentric) but may be off-center (eccentric). Eccentric plates are usually used to prevent excessive buildup of foreign material or gases on the inlet side of the orifice.

The use of orifice plates is backed by considerable test data and experience. For example, the American Gas Association provides specifications and guidelines for the sharpness of the leading edge of the orifice bore and the maximum differential pressure allowed across an orifice. In addition, studies show measurement errors of nearly 20% when orifice plates are bent, warped, or installed backward.

TECH FACT

Water is considered incompressible even though its volume decreases by 1.8% at a pressure of 600 psi.

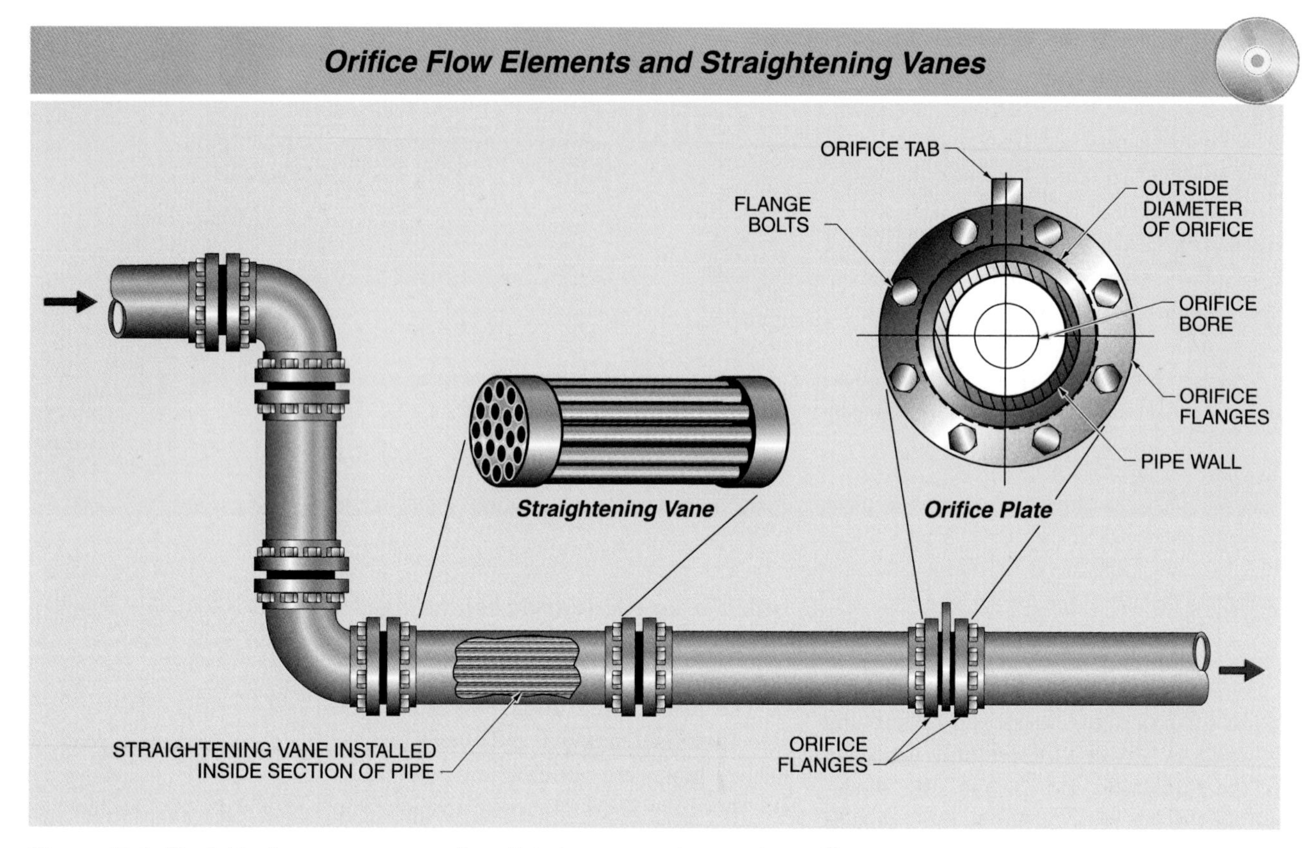

Figure 19-1. Straightening vanes remove flow disturbances upstream of an orifice.

To ensure measurement accuracy, a consistent flow pattern before and after an orifice must be created. Pipe fittings and valves introduce undesirable disturbances such as swirls and eddies that affect accuracy. Straight runs of about 20 times the pipe diameter before and 6 times the pipe diameter after the orifice plate are recommended to allow the flow disturbances to die out. For example, for a 4″ pipe, a straight run of 80″ before and 24″ after the orifice is required. When an insufficient straight run of upstream pipe is available, the use of a straightening vane upstream of the orifice plate reduces or eliminates the disturbances.

Orifice plates have the poorest recovery of differential pressure of any of the primary flow elements. The recovery is rarely higher than 50%. This means that 50% of the measurement differential is permanent pressure loss. A similar type of primary flow element is a flow nozzle.

Flow Nozzles

A *flow nozzle* is a primary flow element consisting of a restriction shaped like a curved funnel that allows a little more flow than an orifice plate and reduces the straight run pipe requirements. The nozzle is mounted between a pair of standard flanges. The pressure-sensing taps are located in the piping a fixed distance upstream and downstream of the flow nozzle. **See Figure 19-2.** The differential pressure recovery is slightly better than an orifice plate. When higher accuracy and energy efficiency are desired, a venturi tube can be used.

Venturi Tubes

A *venturi tube* is a primary flow element consisting of a fabricated pipe section with a converging inlet section, a straight throat, and a diverging outlet section. The static pressure connection is located at the entrance to the inlet section. The reduced pressure connection is in the throat. Venturi tubes are much more expensive than orifice plates but are more accurate and recover 90% or more of the differential pressure. This reduces the burden on pumps and the cost of power to run them. Venturi tubes are frequently used to measure large flows of water. For even higher energy efficiency, a low-loss flow tube can be used.

Pipe taps measure permanent drop. This gives accurate results at the expense of space. Flange taps save space and offer convenient installation. Vena contracta taps provide the largest pressure drop.

Low-Loss Flow Tubes

A *low-loss flow tube* is a primary flow element consisting of an aerodynamic internal cross section with the low-pressure connection at the throat. A low-loss tube is very expensive but has the highest recovered differential pressure of any primary flow element at 97%. Low-loss tubes are often used in applications where the line pressure is low and therefore the pressure recovery must be high. Low-loss flow tubes can often pay for themselves in energy savings in a short time. **See Figure 19-3.**

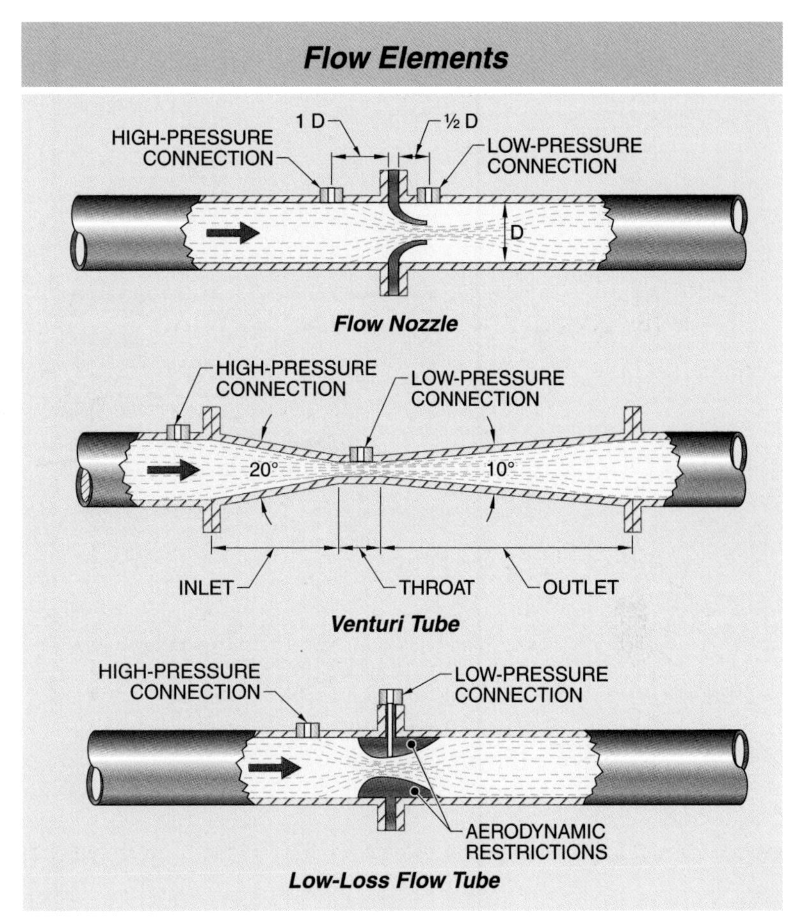

Figure 19-2. Flow elements have different internal designs to achieve different objectives.

Pitot Tubes

A *pitot tube* is a flow element consisting of a small bent tube with a nozzle opening facing into the flow. The nozzle is called the impact opening and senses the velocity pressure plus the static pressure. The static pressure is sensed at the pipe wall perpendicular to the fluid stream. Pitot tubes are commonly used to measure air velocity. For example, pitot tubes are used for measuring air velocity in ducts and for measuring the airspeed of planes. **See Figure 19-4.** A simple pitot tube senses the impact pressure at only one point even though the velocity varies across the whole stream. To overcome this disadvantage, the averaging pitot tube was developed.

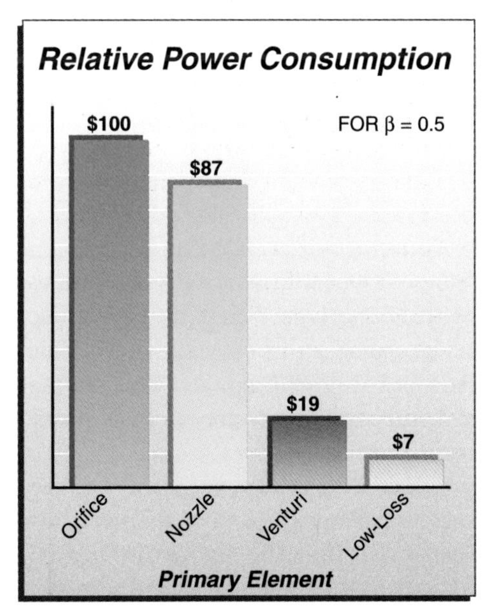

Figure 19-3. Low-loss flow elements have the lowest operating costs of any primary flow element.

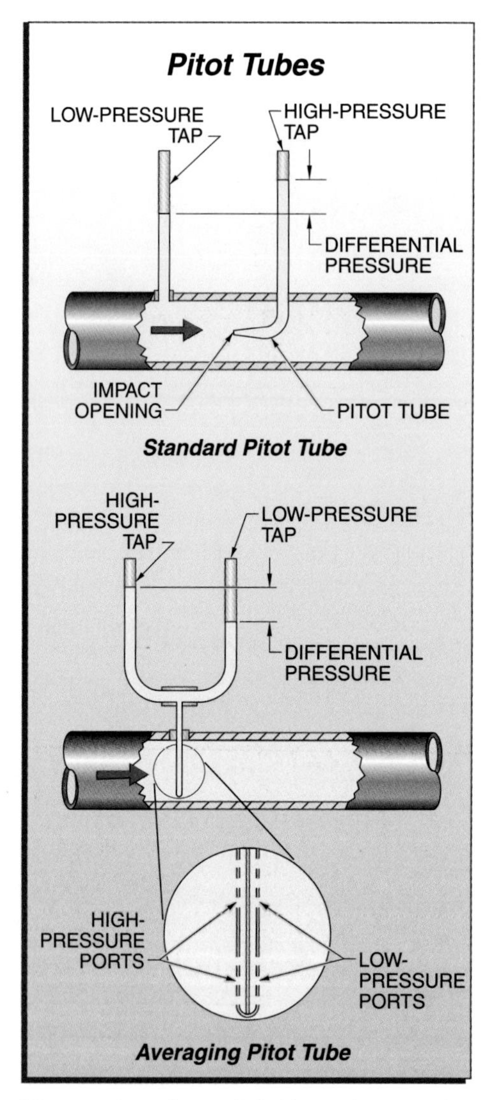

Figure 19-4. Averaging pitot tubes are improved devices for measuring flow.

An *averaging pitot tube* is a pitot tube consisting of a tube with several impact openings inserted through the wall of the pipe or duct and extending across the entire flow profile. The tube has several high-pressure ports facing upstream. The ports are spaced at specific locations across the flow stream and therefore provide an average impact pressure. The low-pressure ports are located on the same tube downstream from the high-pressure ports. The low-pressure ports are internally separate from the high-pressure ports.

A great amount of testing and documentation accompanied the acceptance of the averaging pitot tube as a dependable industrial flow measurement sensor. The averaging pitot tube is used to measure fluid flow in small or large pipes and air flow in ducts. The differential pressure developed by an averaging pitot tube is generally less than that produced by an orifice plate and the pressure recovery is usually very good. This combination of characteristics provides energy saving advantages.

OPERATING PRINCIPLES

The operating principles of all differential pressure flowmeters are based on the equations developed by Daniel Bernoulli, a late 18th century Swiss scientist. Bernoulli's experiments related the pressure and velocity of flowing water. He determined that at any point in a closed pipe there were three types of pressure head present, static head due to elevation, static head due to applied pressure, and velocity head.

The *Bernoulli equation* is an equation stating that the sum of the pressure heads of an enclosed flowing fluid is the same at any two locations. A flow element creates a restriction that forces fluid to flow faster. This increases the velocity head. According to the Bernoulli equation, if the velocity head increases, the static head due to applied pressure must decrease. This pressure decrease is used to calculate flow. The different types of pressure head can be converted to each other by changes in flow.

A primary flow element has the pressure measured upstream and downstream of the flow element. The flow stream contracts slightly before it passes through the flow element and continues to do so until it reaches maximum contraction and then slowly expands until it again fills the pipe. The *vena contracta* is the point of lowest pressure and the highest velocity downstream from a primary flow element. The actual location of the vena contracta point varies with flow rate and design of the flow element. **See Figure 19-5.** According to the Bernoulli equation, the velocity increases and the pressure decreases as the fluid flows through the restriction.

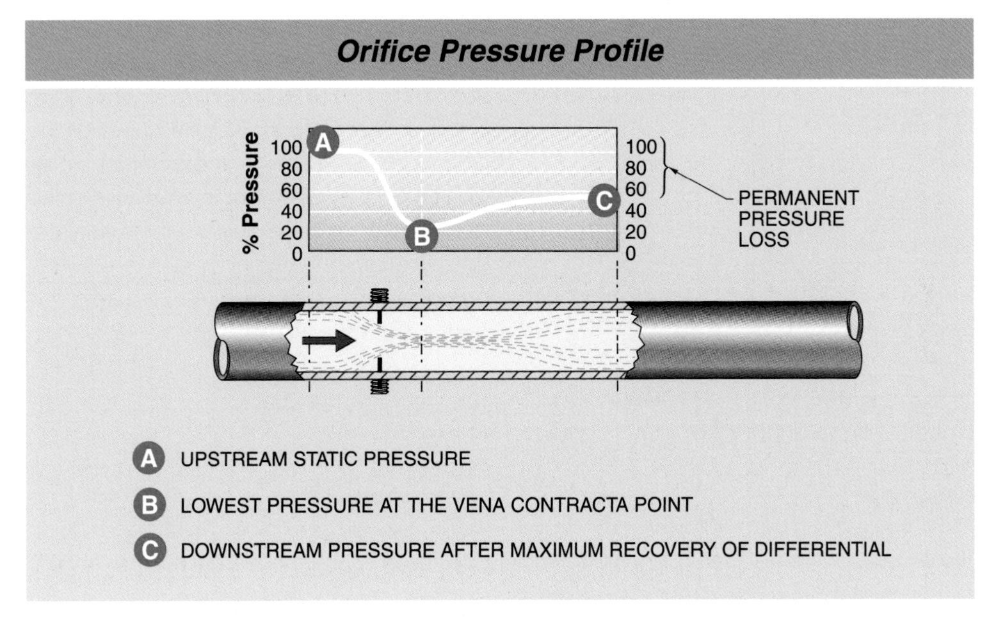

Figure 19-5. The vena contracta occurs after an orifice at the point of lowest pressure.

Differential Pressure Measurements

For turbulent flow, the flow rate is proportional to the square root of the differential pressure. The square root relationship affects the rangeability of the flow metering system. *Rangeability,* or *turndown ratio,* is the ratio of the maximum flow to the minimum measurable flow at the desired measurement accuracy. This is a characteristic of the instrument and is not adjustable. For example, if the maximum measurable flow rate of a flowmeter is 100 gpm of water and the minimum measurable flow rate is 20 gpm of water, the rangeability is 5 to 1. **See Figure 19-6.**

For differential pressure flow measurement, when the flow is at 20% of the maximum flow, the differential pressure is only 4% of the maximum. At 4% of the maximum differential pressure, measurement errors, transmitter calibration errors, and transmission errors, which are all based on percent of the maximum values, start to significantly affect the flow measurement accuracy. Thus the flow measurement range is usually limited to a rangeability of 5:1, or 100% down to about 20% of the maximum flow.

An orifice plate can be inserted into a gas line to measure flow.

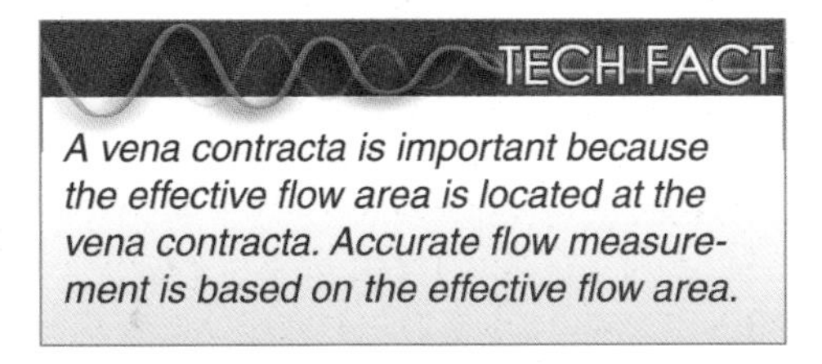

TECH-FACT

A vena contracta is important because the effective flow area is located at the vena contracta. Accurate flow measurement is based on the effective flow area.

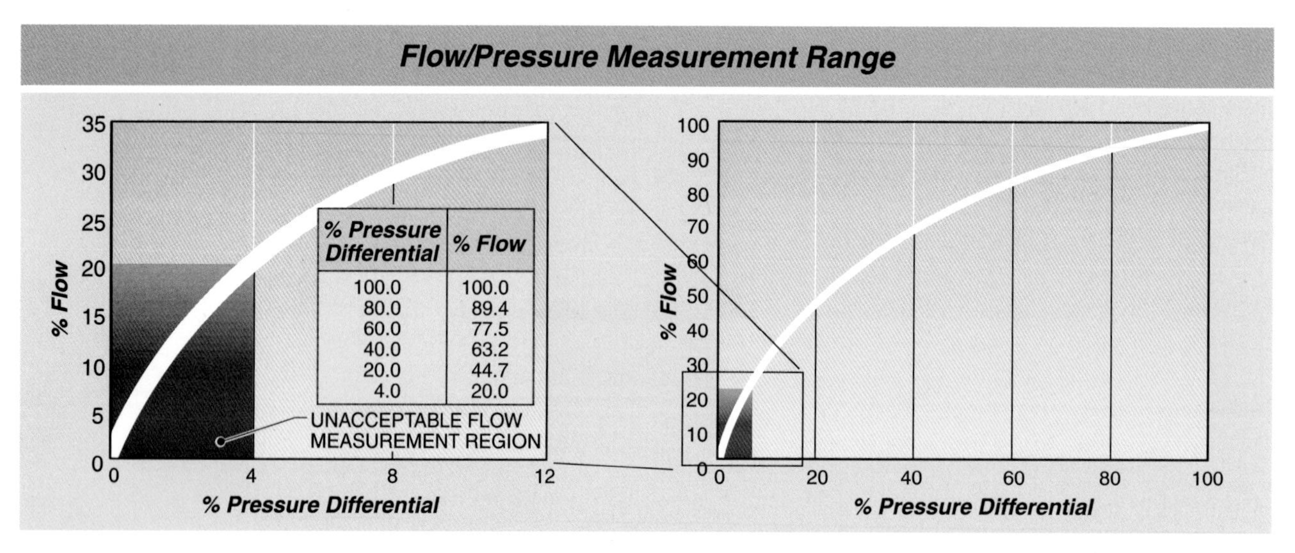

Figure 19-6. Flow varies with the square root of the pressure drop, restricting the meter to the range from full flow to 20% of full flow.

Differential Pressure Flow Equations

Differential pressure calculations for flow through an orifice plate require that the fluid flows be turbulent. Fluids exiting an orifice plate have a theoretical velocity that is proportional to the square root of the hydrostatic pressure causing the flow. In practice, the actual velocity is less than the theoretical.

The *coefficient of discharge* is the ratio of the actual velocity to the theoretical velocity of flow through an orifice. When applied to an orifice plate, the Reynolds number, the beta ratio, the differential pressure across the plate, the sharpness of the plate edge, and other minor factors affect the coefficient of discharge. The *beta ratio (d/D)* is the ratio between the diameter of the orifice plate (*d*) and the internal diameter of the pipe (*D*). In practice, beta ratios may vary from 0.2 to 0.75. Values for the coefficient of discharge have been determined experimentally and are available from the manufacturer of each type of flow measuring device.

The basic flow equation for all fluid flow in pipes relates the flow rate to the area of flow (or orifice diameter) and the square root of the pressure drop (head). The equation is customized to adjust for the different types of measurement units for each type of flow. The following equations show the relationship between flow rate and the various factors that affect that flow rate. The equations are approximations and should not be used for calculation of actual flow. Many manufacturers provide computer software that contains the formulas for precisely determining orifice plate sizes or flow rates.

Volumetric liquid flow:

$$Q = N \times d^2 \times C \times \sqrt{h \times SG}$$

where

Q = flow rate (in gpm)
N = constant conversion factor, 5.67
d = orifice diameter (in in.)
C = coefficient of discharge
h = differential pressure (in in. water)
SG = specific gravity of the liquid relative to water

Volumetric gas flow:

$$Q = N \times d^2 \times C \times Y \times \sqrt{\frac{h \times P_f}{SG \times T_f}}$$

where

Q = flow rate (in scfh)
N = constant conversion factor, 27,171
d = orifice diameter (in in.)
C = coefficient of discharge
Y = gas expansion factor
h = differential pressure (in in. water)
P_f = pressure at flowing conditions (in psia)
SG = specific gravity of gas relative to air
T_f = temperature at flowing conditions (in °R)

Steam flow:

$$W = N \times d^2 \times C \times Y \times \sqrt{\frac{h}{V}}$$

where

W = flow rate (in lb/hr)
N = constant conversion factor, 1244
d = orifice diameter (in in.)
C = coefficient of discharge
Y = gas expansion factor
h = differential pressure (in in. water)
V = specific volume (in cu ft per lb, or cu ft/lb)

Operating Ranges. A typical d/p cell has an adjustable operating range from a minimum of 0″ WC to 20″ WC, up to a maximum of 0″ WC to 450″ WC. A d/p cell has built-in zero and span adjustments as well as optional square root extraction. The zero and span adjustments permit the user to select the differential operating range. The standard analog electrical output signal range is 4 mA DC to 20 mA DC and is associated with the selected differential operating range. Cells with digital output conforming to different standards are readily available. The square root extraction feature alters the differential pressure signal so that the output is linear with flow.

For example, a typical d/p cell has built-in zero and span adjustments. The span adjustment permits the user to select a differential pressure less than the maximum that still produces full-scale output. The turndown ratio depends on the valve and actuator selected to operate under existing process conditions.

Pressure and Temperature Correction. Flow measurement is only accurate as long as the flowing conditions remain the same as when the system was designed. Changes in pressures and temperatures are common in gas and vapor flow measurements. Liquid flow measurements are usually more consistent. Flowing conditions that differ from the original flowmeter design calculation can result in significant flow measurement errors. When the original design conditions and the actual flowing conditions are known, the flowmeter's displayed flow rate can be changed to the correct value, where the flow is expressed at standard conditions, with the corrections as follows:

Pressure correction

$$PC = \sqrt{\frac{P_f}{P_d}}$$

where

PC = pressure correction

P_f = flowing pressure (in psia)

P_d = design pressure (in psia)

Temperature correction

$$TC = \sqrt{\frac{T_d}{T_f}}$$

where

TC = temperature correction

T_f = flowing temperature (in °R)

T_d = design temperature (in °R)

To obtain the correct flow, multiply the correction factors PC and TC by the displayed flow. **See Figure 19-7.**

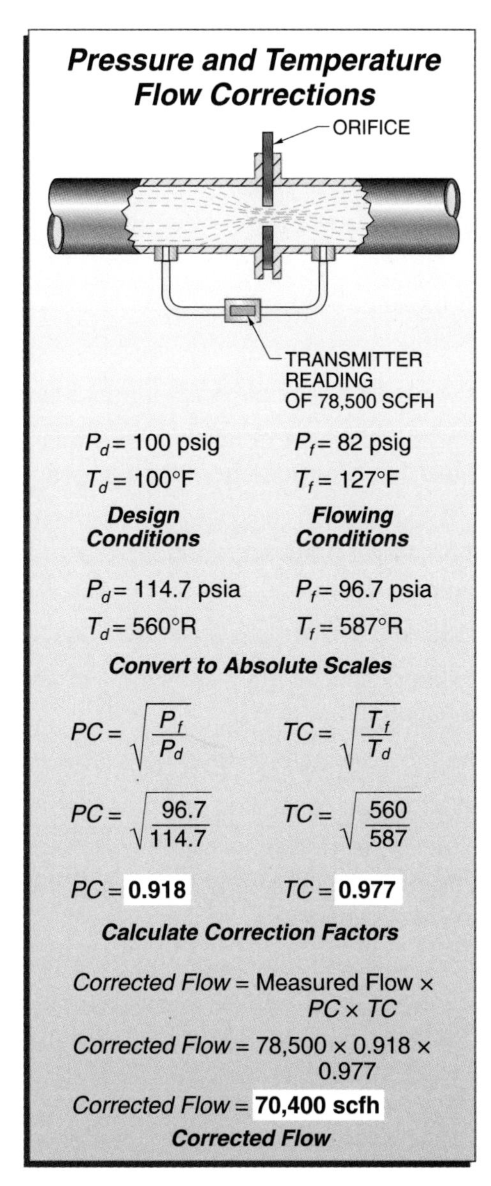

Figure 19-7. Gas and vapor flows must be corrected from measured flow conditions to design flow conditions.

For example, a meter reads 78,500 scfh under actual flowing conditions of 82 psig (P_f = 82 + 14.7 = **96.7 psia**) and 127°F (T_f = 127 + 460 = **587°R**). The orifice plate design conditions are 100 psig (P_d = 100 + 14.7 = **114.7 psia**) and 100°F (T_d = 100 + 460 = **560°R**).

The pressure correction is as follows:

$$PC = \sqrt{\frac{P_f}{P_d}}$$

$$PC = \sqrt{\frac{96.7}{114.7}}$$

$$PC = \mathbf{0.918}$$

An old fuel oil tank is the leading source of a fuel oil leak or spill.

The temperature correction is as follows:

$$TC = \sqrt{\frac{T_d}{T_f}}$$

$$TC = \sqrt{\frac{560}{587}}$$

The corrected flow is:

Corrected flow = uncorrected flow × $PC \times TC$

Corrected flow = 78,500 scfh × 0.918 × 0.977

Corrected flow = **70,400 scfh**

Automatic pressure and temperature compensation may be used in critical applications, such as boiler fuel flow or when flow is used for billing purposes. Instrument accessories are available that automatically calculate the correct flow using measured pressures and temperatures.

KEY TERMS

- *primary flow element:* A pipeline restriction that causes a pressure drop used to measure flow.
- *orifice plate:* A primary flow element consisting of a thin circular metal plate with a sharp-edged round hole in it and a tab that protrudes from the flanges.
- *flow nozzle:* A primary flow element consisting of a restriction shaped like a curved funnel that allows a little more flow than an orifice plate and reduces the straight run pipe requirements.
- *venturi tube:* A primary flow element consisting of a fabricated pipe section with a converging inlet section, a straight throat, and a diverging outlet section.
- *low-loss flow tube:* A primary flow element consisting of an aerodynamic internal cross section with the low-pressure connection at the throat.
- *pitot tube:* A flow element consisting of a small bent tube with a nozzle opening facing into the flow.
- *averaging pitot tube:* A pitot tube consisting of a tube with several impact openings inserted through the wall of the pipe or duct and extending across the entire flow profile.
- *Bernoulli equation:* An equation stating that the sum of the pressure heads of an enclosed flowing fluid is the same at any two locations.
- *vena contracta:* The point of lowest pressure and the highest velocity downstream from a primary flow element.
- *coefficient of discharge:* The ratio of the actual velocity to the theoretical velocity of flow through an orifice.
- *beta ratio (d/D):* The ratio between the diameter of the orifice plate (*d*) and the internal diameter of the pipe (*D*).
- *rangeability,* or *turndown ratio:* The ratio of the maximum flow to the minimum measurable flow at the desired measurement accuracy.

REVIEW QUESTIONS

1. What are the common types of primary flow elements?
2. Compare the cost and energy efficiency of orifice plates, venturi tubes, and low-loss tubes.
3. Define the Bernoulli equation.
4. Why is the rangeability of differential pressure measurement limited to about 5:1?
5. Why are temperature and pressure corrections needed when measuring the flow of gas or vapor?

Mechanical Flowmeters

20 Chapter

OBJECTIVES

- Explain the use and operation of variable-area flowmeters.
- Describe positive-displacement flowmeters and explain their use.
- Compare turbine meters and paddle wheel meters and explain their use.
- Explain the use of open-channel flow measurements.

VARIABLE-AREA FLOWMETERS

A *variable-area flowmeter* is a meter that maintains a constant differential pressure and allows the flow area to change with flow rate. The flow area changes through the movement of some type of flow restriction that repositions with changes in flow. This may be a fixed-size plug that moves in a tapered tube, a shaped plug that partially blocks an orifice, or a restriction that moves up and down on a cone. The most common type of variable-area flowmeter is the rotameter.

The operating principles of a variable-area flowmeter are different from the operating principles of a differential pressure flowmeter, although the basic flow equation is very similar. A differential pressure flowmeter maintains a constant flow area and measures the differential pressure. A variable-area flowmeter maintains a constant differential pressure and allows the area to change with flow rate. The differential pressure meter has a constant orifice area and the differential pressure varies with the flow rate.

A rotameter can only provide correct flow rates for compressible gases and vapors when the flowing conditions are the same as the design conditions. When the flowing conditions have changed, the pressure and temperature correction factors described for orifices are valid for rotameters.

Rotameters

A *rotameter* is a variable-area flowmeter consisting of a tapered tube and a float with a fixed diameter. The float of the rotameter changes its position in the tube to keep the forces acting on the float in equilibrium. One of the forces is gravity acting on the float. The other force is produced by the velocity of the process fluid. When the float changes position, the open area between the float diameter and the tube varies. Under most circumstances, the rotameter tube must be vertical.

Floats are available in different configurations and thus have different reference points. **See Figure 20-1.** Floats have a sharp edge at the point where the reading should be made on the scale. The scale can be calibrated for direct reading of flow rate, or for percent of full scale. The spool-shaped floats have a lower guide disk serving as a stabilizer and a top disk edge providing the point of measurement.

TECH FACT

Diaphragm seals or sealing fluids are often used to isolate a process fluid from direct contact with a sensing element. In this case, the location of the measuring element is almost always below the elevation of the flow element.

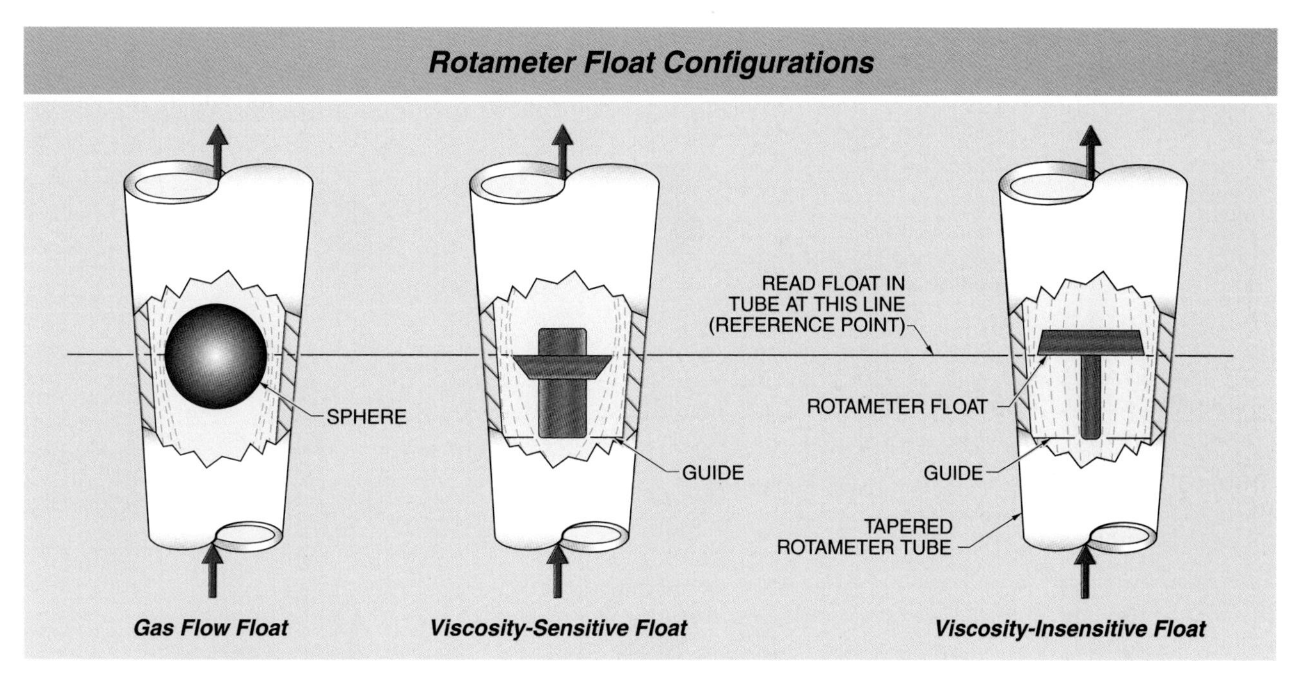

Figure 20-1. Rotameter floats have different configurations for different fluids and applications.

One early flowmeter design had slots which caused the float to spin to stabilize and center the float in the tube. Because of the spinning float, the term rotameter was used. In most flowmeters, the inside of the glass tube has three slight projecting ridges, called beads, which keep the float centered while it rides up and down. Other glass-tube rotameters use floats that are rod-guided.

Rotameter Tube Construction. A *clear-tube rotameter* is a rotameter consisting of a clear tube to allow visual observation of the flow rate. Clear-tube rotameters are available with 5″ or 10″ tubes and ½″ NPT to 2″ NPT pipe fittings. The main parts of a clear-tube rotameter are the tube, float, frame, inlet and outlet fittings, and safety shield. The tube scale may be etched on the tube or printed on a plastic or metal strip alongside the tube. **See Figure 20-2.**

Glass is the most common material used to make clear-tube rotameters. Plastic tubes are used for measuring fluids that are incompatible with glass. Typical applications of plastic tubes are high-temperature water with a high pH that softens glass, wet steam, caustic soda, and hydrofluoric acid. They are available for ¼″ to 3″ plastic pipe or tubing.

A *metal-tube tapered rotameter* is a rotameter consisting of a tapered metal tube and a rod-guided float. A rod that is attached to the float passes through top and bottom guides in the tube. A magnet in the float is coupled to a matching magnet and indicator located outside the tube. Magnetic-coupled rotameters cannot be used in areas where strong magnetic fields are generated, such as in electrolytic cell rooms, near large motors or generators, or in areas where high currents are present.

Variations on metal-tube meters are available with PVC flanged bodies and floats or with PTFE-lined stainless steel bodies and matching floats. An indicating electrical transmitter is often substituted for the indicator. Metal or opaque plastic tubes are used for applications involving fluids which obscure the float or those too hot, too corrosive, or involving high pressures. Excessive temperatures can diminish the magnetic coupling.

Modified Rotameters

Modified rotameters combine a standard rotameter with another device or modify the rotameter itself to achieve a specific function. Common types of modified rotameters are purge meters, bypass meters, shaped-float and orifice meters, and metering-cone meters.

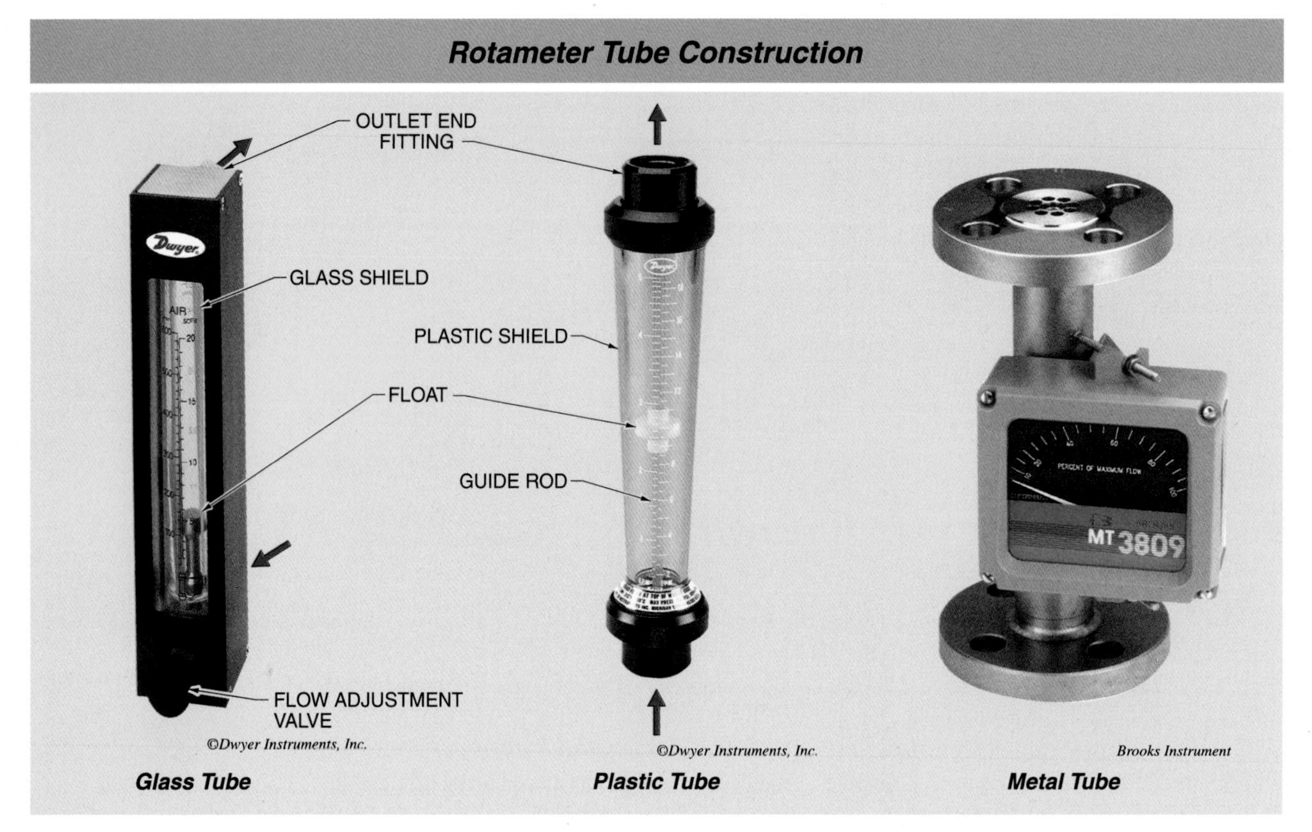

Figure 20-2. Rotameters can have clear tubes for direct reading or opaque tubes with a magnetic linkage to a meter.

Purge Meters. A *purge meter* is a small metal or plastic rotameter with an adjustable valve at the inlet or outlet of the meter to control the flow rate of the purge fluid. A purge meter is used for purging applications such as regulating a small flow of air or nitrogen into an enclosure to prevent the buildup of hazardous or noxious gases. A purge meter may also be used in a bubbler level measuring system. A small bead or ball moves in the fluid stream to indicate the flow rate. **See Figure 20-3.**

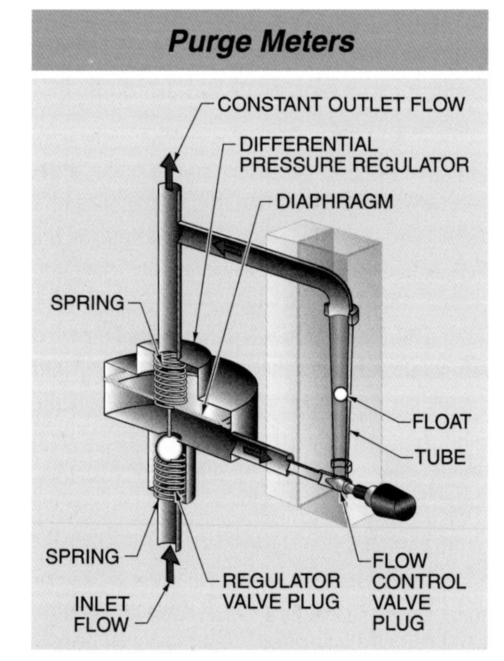

Figure 20-3. A purge meter can be used to regulate a small flow of air or nitrogen into an enclosure to prevent the buildup of hazardous or noxious gases.

Bypass Meters. A *bypass meter* is a combination of a rotameter with an orifice plate used to measure flow rates through large pipes. The advantage of a bypass meter is the linear scale of the rotameter and accompanying better rangeability. The differential pressure across the mainline orifice plate must be matched to the differential pressure across the rotameter at the maximum flow rate.

The rotameter manufacturer must provide a metering orifice plate at the inlet to the rotameter to accomplish this matching of differential pressures. This can be made easier by selecting one of the standard differential pressure ranges and using a main line orifice plate to match. Standard ranges may be 0″ to 50″, 0″ to 100″, 0″ to 200″, and 0″ to 400″ water column. **See Figure 20-4.**

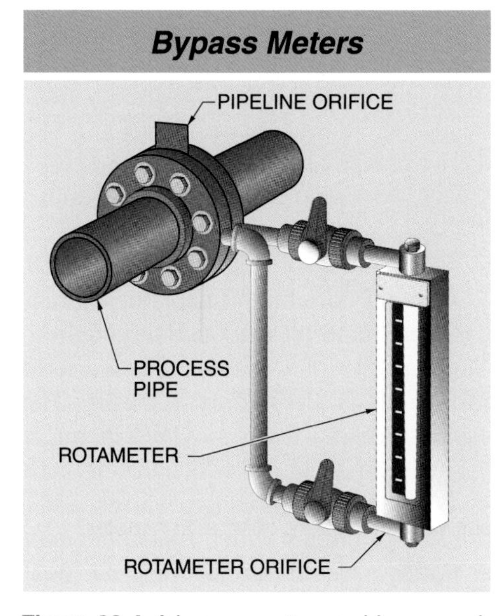

Figure 20-4. A bypass meter combines an orifice plate with a rotameter to obtain better rangeability from the linear scale of the rotameter.

Metering-Cone Meters. A *metering-cone meter* is a flowmeter consisting of a straight tube and a tapered cone, instead of a tapered tube, with an indicator that moves up and down the cone with changes in flow. The variable area is the annular space between the float and the tapered cone. The indicator is often spring-loaded to allow the meter to be mounted at any angle. **See Figure 20-5.**

Shaped-Float and Orifice Meters. A *shaped-float and orifice meter* is a flowmeter consisting of an orifice as part of the float assembly that acts as a guide. Instead of a tapered tube, the float has a shaped profile that provides more open flowing area as the float rises. The variable area is the annular space between the float and the disk. The external readouts are available as indicators or transmitters with or without alarms. **See Figure 20-6.**

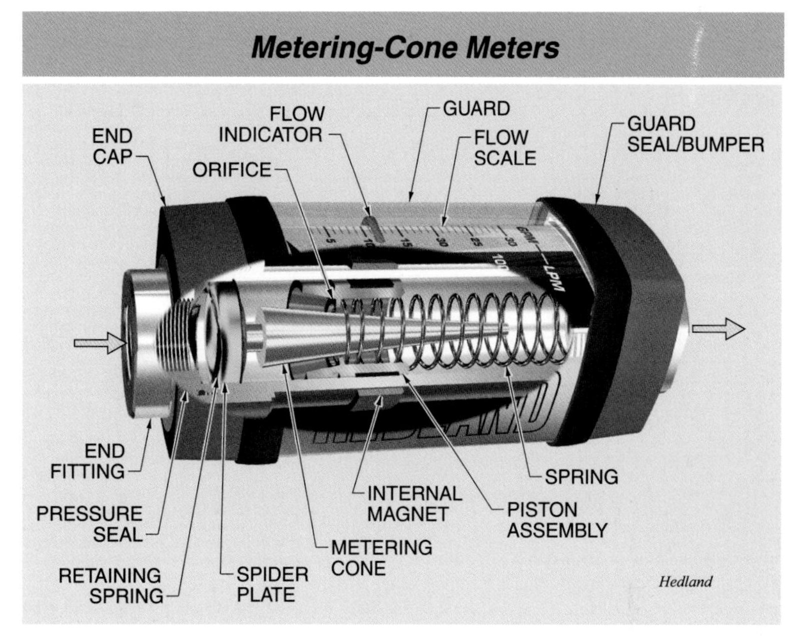

Figure 20-5. A metering-cone meter uses a straight tube and a tapered cone instead of the tapered tube of a standard rotameter.

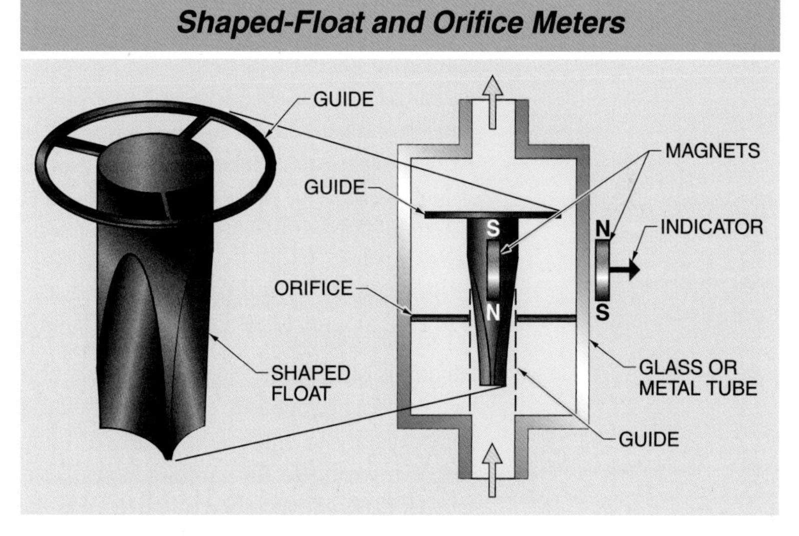

Figure 20-6. A shaped-float and orifice meter uses an orifice as part of the float assembly to act as a guide. The float has a shaped profile that provides more open flowing area as the float rises.

Rotameter Sizing Tables

Manufacturers of rotameters provide capacity tables listing the float and tube for specified flow. The listed capacity for liquids is based on water. The listed capacity for gases is based on air, with a specific gravity of 1.0 at standard conditions (usually 0 psig and 70°F).

Rotameter Flow Equations

The basic flow equation for all fluid flow in pipes relates the flow rate to the area of flow (or pipe diameter) and the square root of the pressure drop. The same fundamental equation that governs the orifice meter also applies to the variable-area meter, but with a constant pressure drop and a variable flow area. This basic equation has been modified to make it easier for rotameter manufacturers to determine the rotameter maximum flow for changes in the factors that affect the flow.

$$W = F \times K \times D_f \times \sqrt{\frac{W_f \times (\rho_f - \rho) \times \rho}{\rho_f}}$$

where

W = flow rate (in lb/min)

F = float position (0% – 100%)

K = flow coefficient based on taper and size of tube

D_f = diameter of float at indicating edge (in in.)

W_f = weight of float (in g, dry)

ρ_f = density of float (g/cm³ or specific gravity for liquids)

ρ = density of flowing fluid (g/cm³ or specific gravity for liquids)

This equation relates the maximum flow to the other factors that affect that flow. The format of this equation allows the development of factors that can provide change in maximum flow for a change in fluid density, or a change in float weight.

This equation shows that the flow rates measured by a variable-area meter are linear. Thus the rangeability of rotameters is 10 to 1.

There are two variables that establish the flow range, the tube taper and the float weight. There are usually one or two tapers for each meter size and several float weights available for each taper selection. The choice of meter is from tables listing the capacity for each combination. The meter scale may be customized to suit the application. The flow rates measured by a variable-area meter are linear. Thus the rangeability of in-line rotameters is 10 to 1. Bypass meter rangeability may not be as high because the meter is combined with a differential pressure element.

Rotameter Correction Factors

When the substance flowing through a rotameter is other than water or air, correction factors must be applied to convert the flowing fluid capacity to equivalent standard fluids before referring to the tables for available sizes. If the liquid to be measured has a higher specific gravity than water, the additional buoyancy on the float will result in a decrease of the effective weight of the float. As a result, it takes less flow to lift the float and the readings are in error.

For liquids, multiply the desired flow by the factor, where *SG* is the specific gravity of the process fluid, to obtain the capacity of water. For gases, factors for specific gravity, temperature, and pressure must be applied. Many manufacturers provide software that performs these calculations.

Another correction factor for alternative float material may also be required. The capacity of a rotameter being used for liquids is affected by viscosity. The capacity tables usually provide a listing of viscosity limits above which the effect of viscosity is great enough to warrant additional consultation with the manufacturer. A highly viscous liquid drags the float to a different position than a nonviscous liquid. Float shape influences the effect of viscosity so manufacturers have developed shapes that are less sensitive to variations in viscosity. These factors and the procedures for making the corrections are supplied with the manufacturer's capacity tables.

POSITIVE-DISPLACEMENT FLOWMETERS

A *positive-displacement flowmeter* is a flowmeter that admits fluid into a chamber of known volume and then discharges it. The number of times the chamber is filled during a given interval is counted. This type of meter is commonly used for measuring total flow in homes and factories. The chambers are arranged so that as one is filling, the other is being emptied. This action is registered by a counting mechanism. The total flow for a given interval is determined by reading the counters. **See Figure 20-7.**

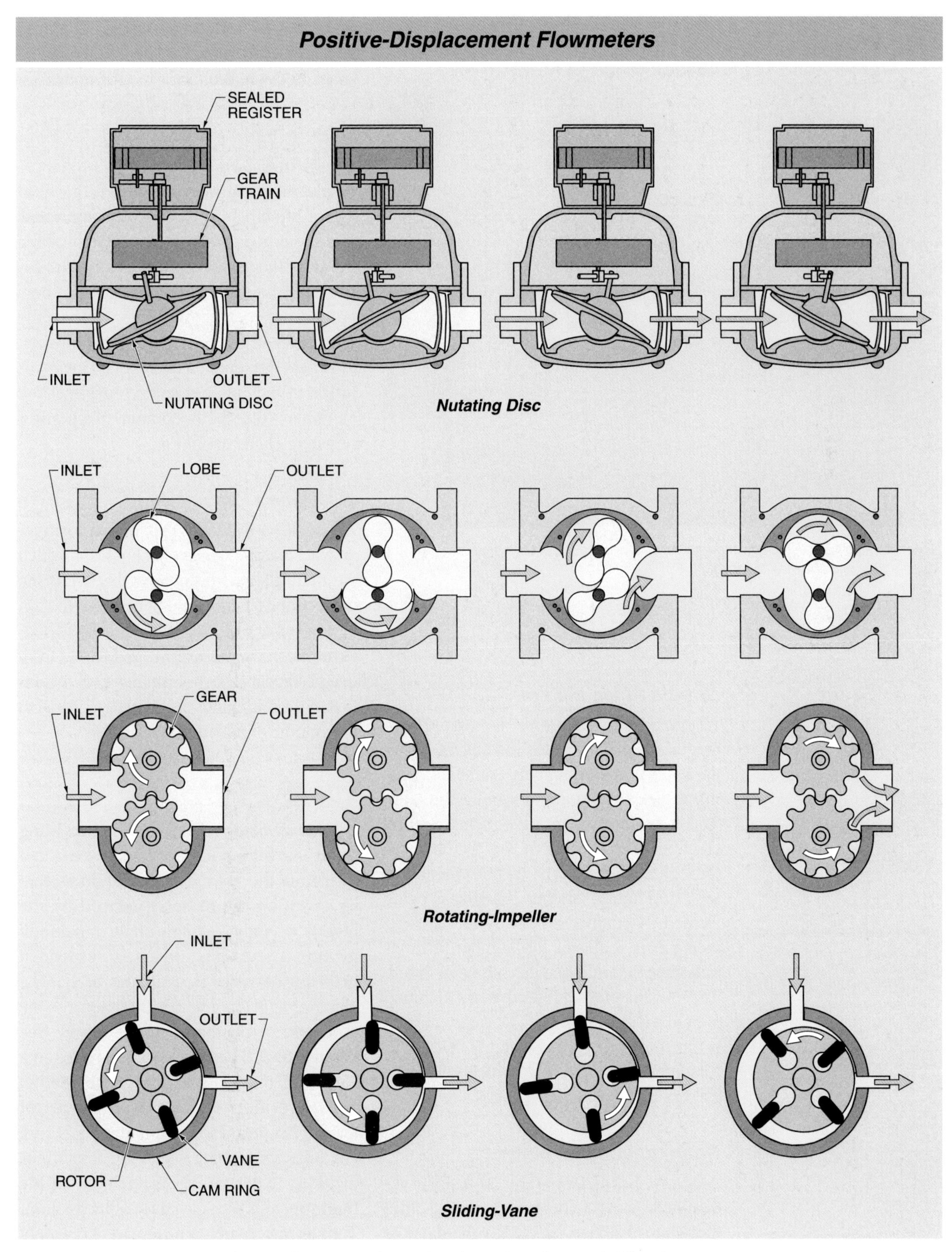

Figure 20-7. Positive-displacement flowmeters have chambers that alternately fill and empty.

Gas meters use positive displacement flowmeters because of their high accuracy and repeatability and low maintenance.

Nutating Disc Meters

A *nutating disc meter* is a positive-displacement flowmeter for liquids where the liquid flows through the chambers, causing a disk to rotate and wobble (nutate). The wobble momentarily forms a filled chamber. The rotation moves the chamber through the meter body. As this chamber is releasing liquid, another chamber is being formed. A counter driven by the rotating mechanism indicates the number of times a chamber has released its volume of liquid. The movement resembles that of a spinning coin just before it stops. The rotary motion is transferred to a gear assembly that drives a counter. This is the type of meter typically used as a domestic water meter.

Rotating-Impeller Meters

A *rotating-impeller meter* is a positive-displacement flowmeter where the fluid flows into chambers defined by the shape of the impellers. The impellers rotate, allowing fluid to flow into the chambers. The fluid measurement chambers are created in the space between the lobes and the housing.

Sliding-Vane Meters

A *sliding-vane meter* is a positive-displacement flowmeter where the fluid fills a chamber formed by sliding vanes mounted on a common hub rotated by the fluid. As the first chamber fills, the hub rotates on a fixed cam, moving the chambers around the meter. The vanes slide in the hub, maintaining contact with the housing, to isolate the separate chambers. One revolution of the hub is equal to four times the chamber volume. A counter mechanism registers each revolution.

Oscillating-Piston Meters

An *oscillating-piston meter* is a positive-displacement flowmeter for liquids in which the fluid fills one piston chamber while the other piston chamber is emptied. The motion of the oscillating piston automatically switches the incoming and exiting liquid flows from one piston chamber to the other. One revolution of the oscillating-piston meter is equal to the volume of both piston chambers.

TURBINE METERS

A *turbine meter* is a flowmeter consisting of turbine blades mounted on a wheel that measures the velocity of a liquid stream by counting the pulses produced by the blades as they pass an electromagnetic pickup. The turbine wheel is suspended between bearings in a tubular body. The electromagnetic pickup is threaded into the tube wall perpendicular to the wheel. **See Figure 20-8.** The bearings are offered in a choice of materials to satisfy the friction and corrosion-resistance requirements of the application. Flow straighteners before and after the turbine wheel are sometimes included to ensure that the velocity of the fluid stream is the sole cause of its rotation. Turbine meters are very accurate and widely used in blending applications.

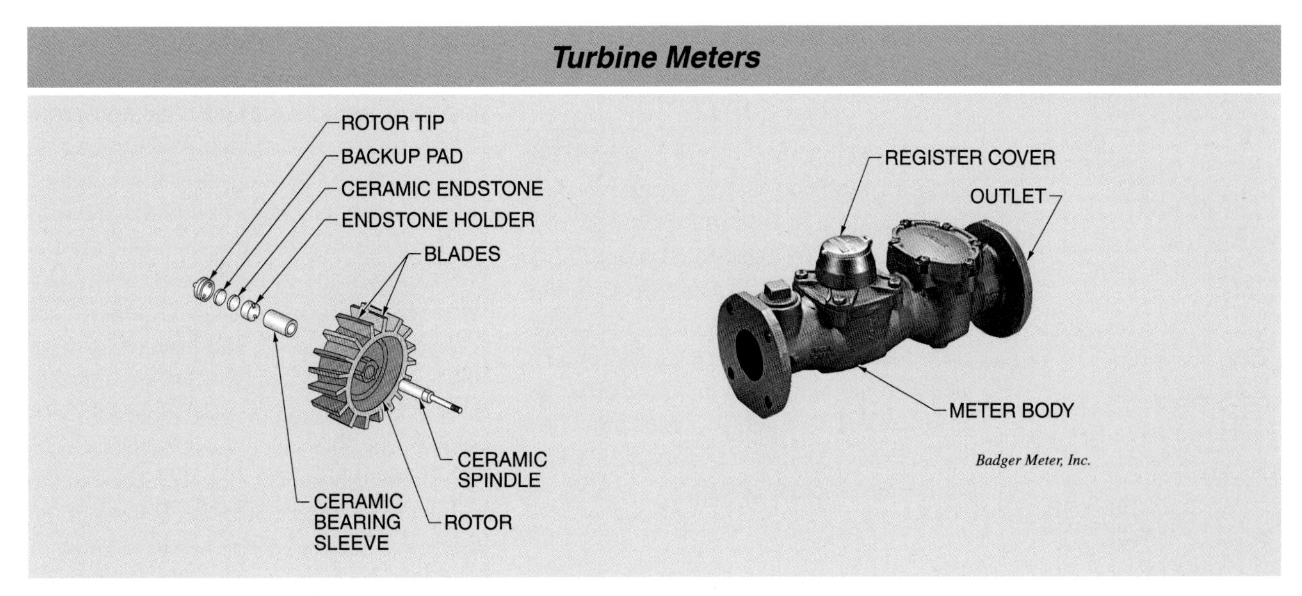

Figure 20-8. A turbine meter uses turbine blades and counts the pulses produced by the blades as they pass an electromagnetic pickup.

PADDLE WHEEL METERS

A *paddle wheel meter* is a flowmeter consisting of a number of paddles mounted on a shaft fastened in a housing, which can be inserted into a straight section of pipe. **See Figure 20-9.** The housing is inserted so that only half of the paddles are exposed to the liquid velocity. The paddles rotate in proportion to the liquid velocity like an old-fashioned water wheel.

The rotation can be detected by two methods. In one version, magnets are imbedded in the tips of the plastic paddles and are sensed by an electromagnetic coil in the housing. The other version has a coil mounted in the housing that creates a magnetic field. The passage of the metal paddle tips disrupts the magnetic field. In both versions, the frequency generated by the moving paddles is linearly related to the liquid flow rate. The frequency signal can be converted to a standard 4 mA to 20 mA analog signal or used directly to serve as the input to a digital controller.

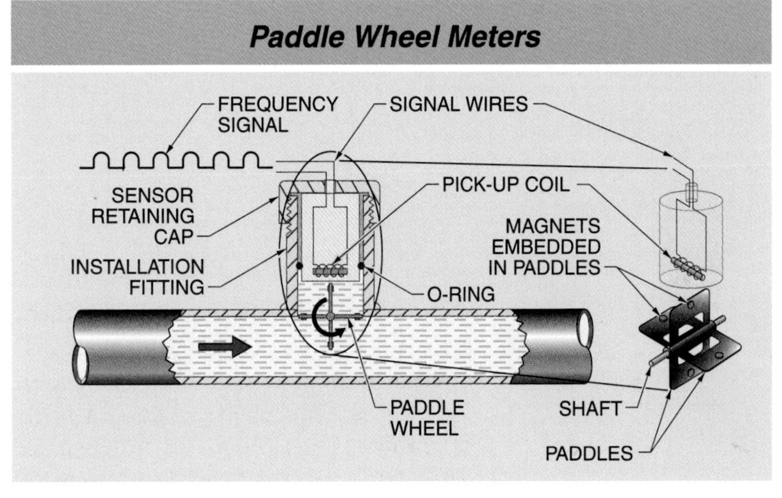

Figure 20-9. A paddle wheel uses a number of paddles mounted on a shaft fastened in a housing.

TECH FACT

Turbine meters need to be filled with liquid. Air pockets can cause the rotor to turn too fast and damage the bearings.

Plastic models are available in materials that are suitable for almost all corrosive fluids. The metal paddle types do not have the corrosion resistance offered by the plastic models, but they work in areas of strong magnetic environments. The insertable paddle meters are available in sizes from 1″ to 12″. There are also some special models designed for very small flows. Paddle wheel meters are only used for measuring liquid flows and then only for the less critical applications.

OPEN-CHANNEL FLOW MEASUREMENTS

Measurement of flow in open channels uses a restriction to create a head of liquid. The height of the head that is created is measured and used to determine the flow rate. The flow has a very restricted rangeability, just like differential measurements in piping. Detailed information on how to design open-channel flow systems can be found in special reference books or in information from vendors. Common open-channel flow measurement devices are weirs and Parshall flumes. **See Figure 20-10.**

Open-Channel Flow Measurements

Weir

Parshall Flume

Figure 20-10. Open-channel flow is measured with weirs or flumes.

Weirs

A *weir* is an open-channel flow measurement device consisting of a flat plate that has a notch cut into the top edge and is placed vertically in a flow channel. The rate of flow is determined by measuring the height of liquid in the stilling basin upstream of the weir. The crest is the bottom of the weir. The weir notch may be rectangular, trapezoidal (Cipolletti), or triangular (V-shaped), and all have a sharp upstream edge. V-shaped weirs are typically 30°, 45°, or 90° and provide rangeability of up to 15 to 1. The weir is installed in the outlet of a stilling basin. The flow is related to the height of the water above the bottom of the weir notch measured at a point upstream of the weir where the water has no draw-down.

The level can be measured with special instruments for measuring weir heads. The weir flow equation is only valid when the weir overflow is free-falling. Free fall is when there is an air space under the overflow and the sides of the overflow are not in contact with the channel sides. The requirement of having a free falling overflow means that a significant change in elevation is needed to be able to use a weir in an open channel.

Parshall Flumes

A *Parshall flume* is a special form of open-channel flow element that has a horizontal configuration similar to a venturi tube, with converging inlet walls, a parallel throat, and diverging outlet walls. A Parshall flume requires much less channel elevation change than a weir. The bottom profile is specially designed to generate a hydraulic jump in the throat. Flow can be calculated from a measurement of the elevation of the inlet water at a specific point. The measurement instruments are the same as are used for weirs. Parshall flumes are much less subject to problems from dirt or other fouling factors than a weir and have the ability to measure much larger flows.

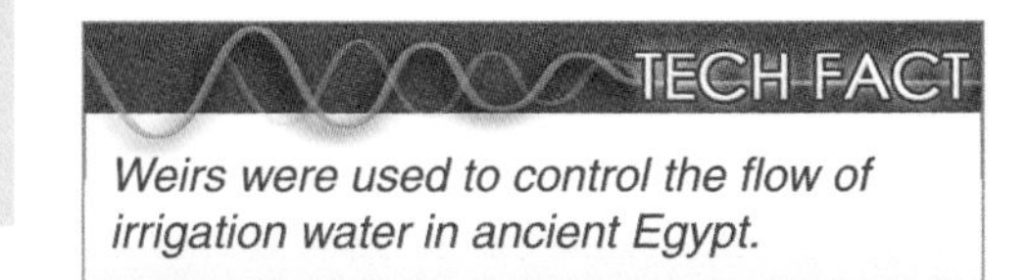

Weirs were used to control the flow of irrigation water in ancient Egypt.

Open-Channel Flow Equations

The measurement of flow in open channels is very complex. The equations to use depend on the configuration and size of the flow measuring device. For weirs, the equations are determined by the shape of the notch that is cut into the flat plate. For Parshall flumes, the flow equations are determined by the height of the flow for standard size devices and by the height of the flow and the width of the flume for nonstandard sizes.

Weir Flow Equations. Weirs can be either free-falling or submerged, but nearly all applications are free-falling. Therefore, only equations for free-falling weirs will be covered. In addition, only those weirs which have full end contractions will be listed. The most common types are rectangular, trapezoidal (Cipoletti), or triangular (V-notch).

For rectangular weirs with complete contractions and negligible velocity of approach:

$$Q_{sf} = 3.33 \times \left[L - (0.2 \times H_a)\right] \times H_a^{1.5}$$

For Cipoletti weirs with full end and bottom contractions:

$$Q_{sf} = 3.367 \times L \times H_a^{1.5}$$

For triangular or V-notch weirs:

$$Q_{sf} = \left[0.025 + (2.462 \times U)\right] \times H_a^{\left(2.5 - \frac{0.0195}{U^{0.75}}\right)}$$

where

Q_{sf} = flow rate (in ft³/sec)

L = crest width of weir (in ft)

H_a = upper head measured from crest (in ft)

U = slope of sides of notch measured from vertical, or tan(θ) in which θ is half of the total included angle of the notch

Parshall Flume Flow Equations. The Parshall flume flow equations vary with the width of the crest. The smaller sizes are available in a series of standard-width crests in pre-molded fiberglass inserts. The larger sizes are usually formed in place using concrete.

$$Q_{sf} = 0.992 \times H_a^{1.547} \text{ (for } L = 3'')$$

$$Q_{sf} = 2.06 \times H_a^{1.58} \text{ (for } L = 6'')$$

$$Q_{sf} = 3.07 \times H_a^{1.53} \text{ (for } L = 9'')$$

$$Q_{sf} = 4 \times L \times H_a^{1.52 \times L^{0.026}} \text{ (for } L = 1' \text{ to } 8')$$

$$Q_{sf} = \left[(3.6875 \times L) + 2.5\right] \times H_a^{1.6} \text{ (for } L = 8' \text{ and larger)}$$

where

Q_{sf} = flow rate (in cu ft per sec, or cu ft/sec)

L = width of Parshall flume throat (in ft)

H_a = upper head measured from crest (in ft)

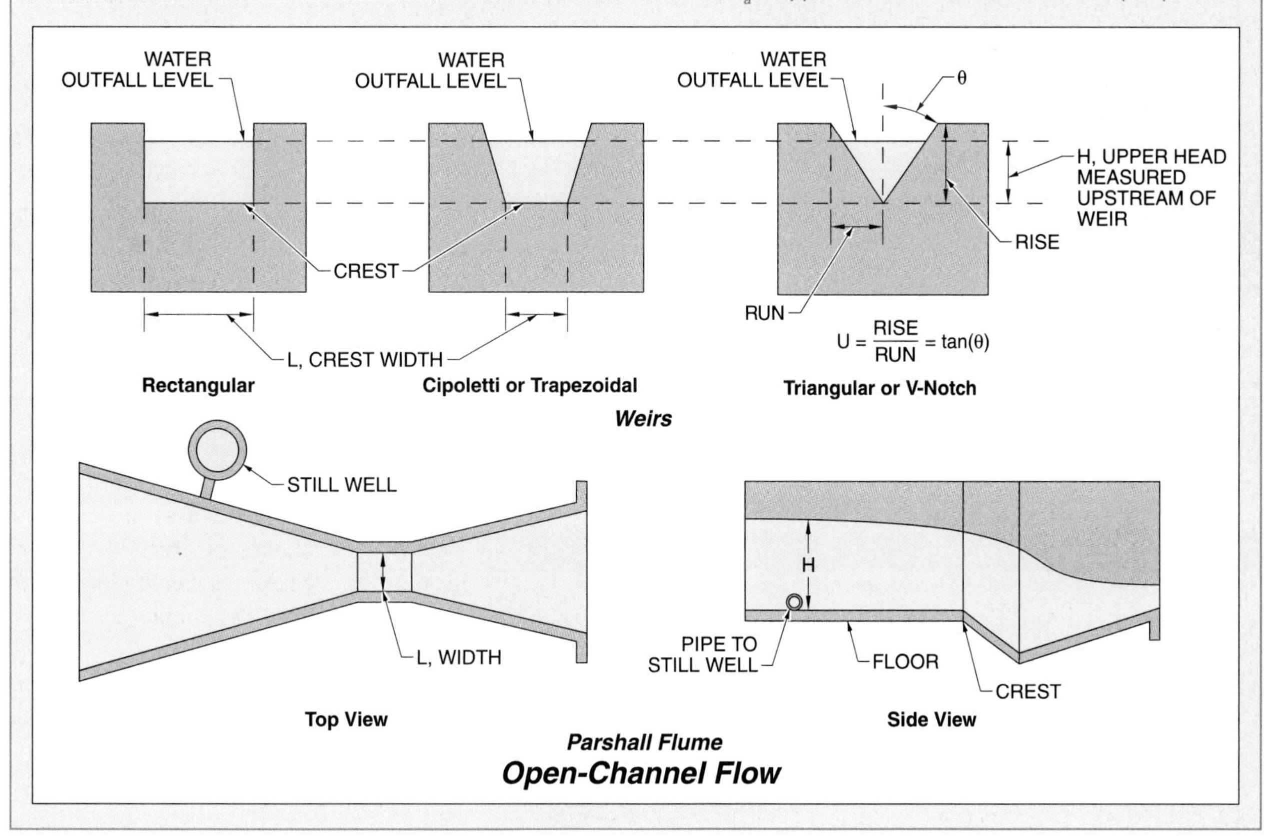

Open-Channel Flow

KEY TERMS

- *variable-area flowmeter:* A meter that maintains a constant differential pressure and allows the flow area to change with flow rate.
- *rotameter:* A variable-area flowmeter consisting of a tapered tube and a float with a fixed diameter.
- *clear-tube rotameter:* A rotameter consisting of a clear tube to allow visual observation of the flow rate.
- *metal-tube tapered rotameter:* A rotameter consisting of a tapered metal tube and a rod-guided float.
- *purge meter:* A small metal or plastic rotameter with an adjustable valve at the inlet or outlet of the meter to control the flow rate of the purge fluid.
- *bypass meter:* A combination of a rotameter with an orifice plate used to measure flow rates through large pipes.
- *metering-cone meter:* A flowmeter consisting of a straight tube and a tapered cone, instead of a tapered tube, with an indicator that moves up and down the cone with changes in flow.
- *shaped-float and orifice meter:* A flowmeter consisting of an orifice as part of the float assembly that acts as a guide.
- *positive-displacement flowmeter:* A flowmeter that admits fluid into a chamber of known volume and then discharges it.
- *nutating disc meter:* A positive-displacement flowmeter for liquids where the liquid flows through the chambers, causing a disk to rotate and wobble (nutate).
- *rotating-impeller meter:* A positive-displacement flowmeter where the fluid flows into chambers defined by the shape of the impellers.
- *sliding-vane meter:* A positive-displacement flowmeter where the fluid fills a chamber formed by sliding vanes mounted on a common hub rotated by the fluid.
- *oscillating-piston meter:* A positive-displacement flowmeter for liquids in which the fluid fills one piston chamber while the other piston chamber is emptied.
- *turbine meter:* A flowmeter consisting of turbine blades mounted on a wheel that measures the velocity of a liquid stream by counting the pulses produced by the blades as they pass an electromagnetic pickup.
- *paddle wheel meter:* A flowmeter consisting of a number of paddles mounted on a shaft fastened in a housing, which can be inserted into a straight section of pipe.
- *weir:* An open-channel flow measurement device consisting of a flat plate that has a notch cut into the top edge and is placed vertically in a flow channel.
- *Parshall flume:* A special form of open-channel flow element that has a horizontal configuration similar to a venturi tube, with converging inlet walls, a parallel throat, and diverging outlet walls.

REVIEW QUESTIONS

1. Compare the operation of a differential pressure flowmeter to a variable-area flowmeter.
2. What are the common types of variable-area flowmeters?
3. What are the common types of positive-displacement flowmeters?
4. Define turbine meter and paddle wheel meter.
5. List and define the common types of open-channel flow measurement devices.

SECTION 5 FLOW MEASUREMENT

Magnetic, Ultrasonic, and Mass Flowmeters

Chapter 21

OBJECTIVES

- Describe magnetic flowmeters and explain their use.
- List and describe ultrasonic flowmeters.
- Explain the operation of mass flowmeters.

MAGNETIC FLOWMETERS

A *magnetic meter,* or *magmeter,* is a flowmeter consisting of a stainless steel tube lined with nonconductive material, with two electrical coils mounted on the tube like a saddle. Two electrodes are in contact with the electrically conductive fluid but insulated from the metal tubes. The electrodes are located opposite one another and at right angles to the flow and the magnetic field. **See Figure 21-1.**

A fundamental principle of electromagnetism states that a voltage is generated when a conductor moves relative to a magnetic field. As a flowing conductive liquid moves within the nonconductive tube and passes through the magnetic field of the coils, a voltage is induced and detected by the electrodes. The amount of voltage depends on the strength of the magnetic field, the distance between the electrodes, and the conductivity and velocity of the liquid. When the magnetic field strength, the position of the electrodes, and the conductivity of the liquid remain constant, the voltage generated at the electrodes is linearly related to the velocity of the liquid stream.

A magmeter has no moving parts other than the flowing liquid. Magmeters have no flow-restricting components, they are available with wettable components that are designed for many corrosive fluids, and they are not adversely affected by complicated piping configurations. They have been applied in many water supply and wastewater facilities and are somewhat immune to internal buildups.

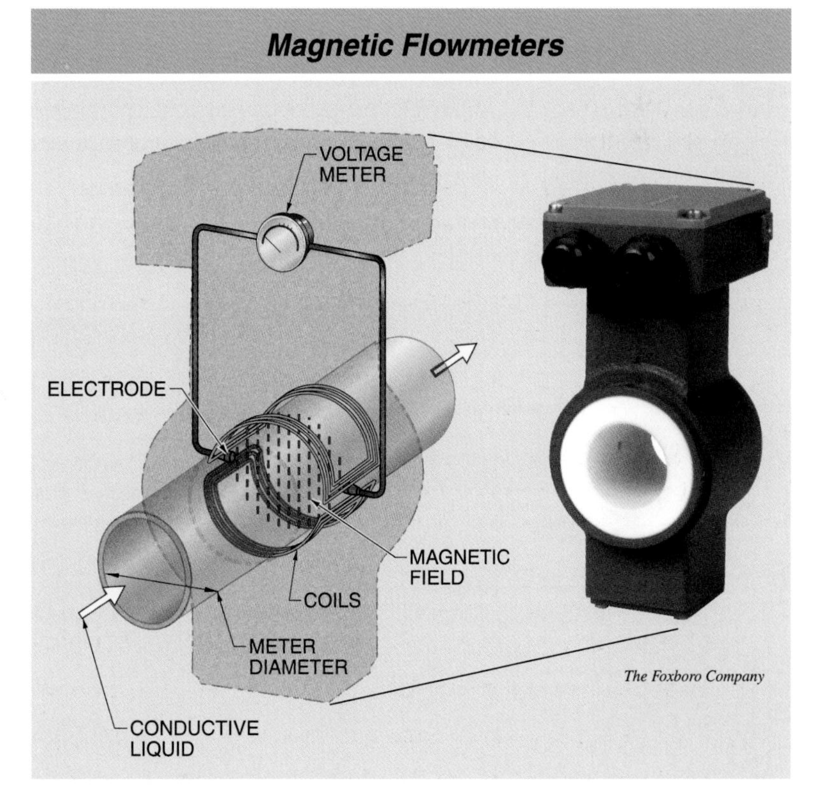

Figure 21-1. Magnetic flowmeters use a magnetic field to measure the flow of a conducting fluid.

ULTRASONIC FLOWMETERS

An *ultrasonic flowmeter* is a flowmeter that uses the principles of sound transmission in liquids to measure flow. They use either the change in frequency of sound reflected from moving elements or measure the change in the speed of sound in a moving liquid. One of the major advantages of ultrasonic flowmeters is that nothing protrudes into the flowing liquid.

Doppler Ultrasonic Meters

A *Doppler ultrasonic meter* is a flowmeter that transmits an ultrasonic pulse diagonally across a flow stream, which reflects off turbulence, bubbles, or suspended particles and is detected by a receiving crystal. **See Figure 21-2.** The frequency of the reflected pulses, when compared to the transmitted pulses, results in a Doppler frequency shift that is proportional to the velocity of the flowing stream.

This is the same principle as radar used to measure the speed of vehicles on the highway, but with different frequencies. Knowing the pipe size and velocity is sufficient to determine the volumetric flow rate. The success of this meter is dependent on the presence of particles or bubbles in the flowing liquid. Clear liquids or liquids with high solids entrainment cannot be measured with a Doppler meter.

Transit Time Ultrasonic Meters

A *transit time ultrasonic meter* is a flowmeter consisting of two sets of transmitting and receiving crystals, one set aimed diagonally upstream and the other aimed diagonally downstream. **See Figure 21-3.** The liquid velocity slows the upstream signal and increases the received frequency while speeding up the downstream signal and decreasing the received frequency.

The difference in the measured frequencies is used to calculate the transit time of the ultrasonic beams and thus the liquid velocity. The flowmeter circuitry is able to convert this information to flow rate by multiplying the velocity by the pipe area. This measurement method has been applied successfully to very large pipes carrying clean, noncorrosive, bubble-free liquids.

TECH FACT

Magmeters have limits on the flow rates that can be measured. It is often necessary to select a meter with a smaller diameter than the piping size. Piping transitions can be placed between the larger piping and the smaller meter. The transitions should be carefully designed in order to minimize the turbulence and head loss in the pipe.

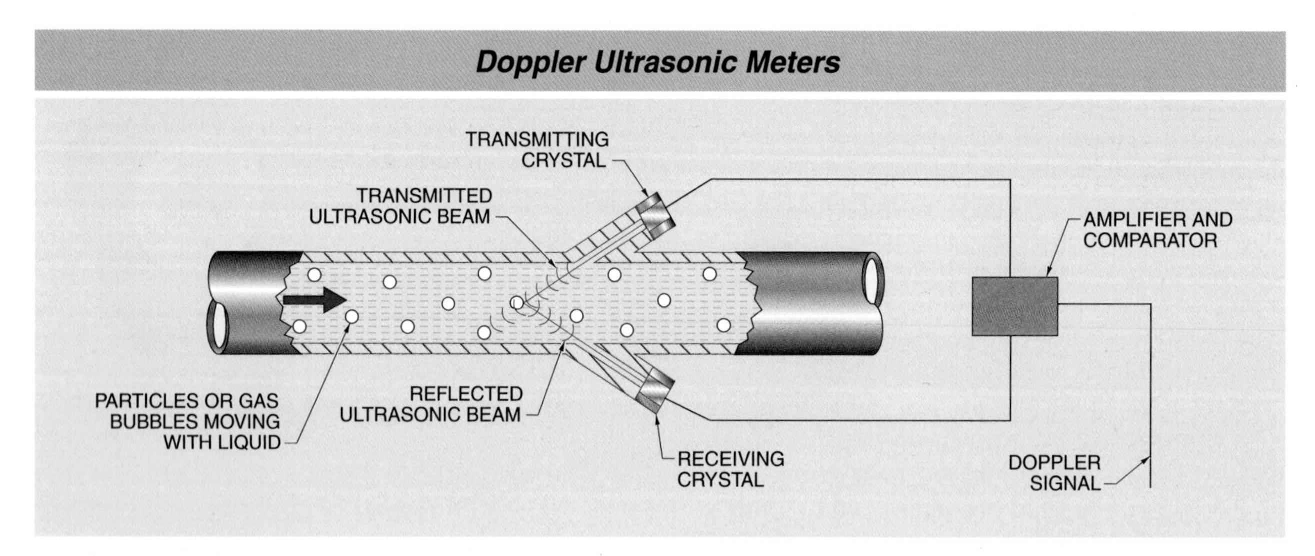

Figure 21-2. Doppler ultrasonic meters transmit an ultrasonic signal across the flow stream to a receiving crystal.

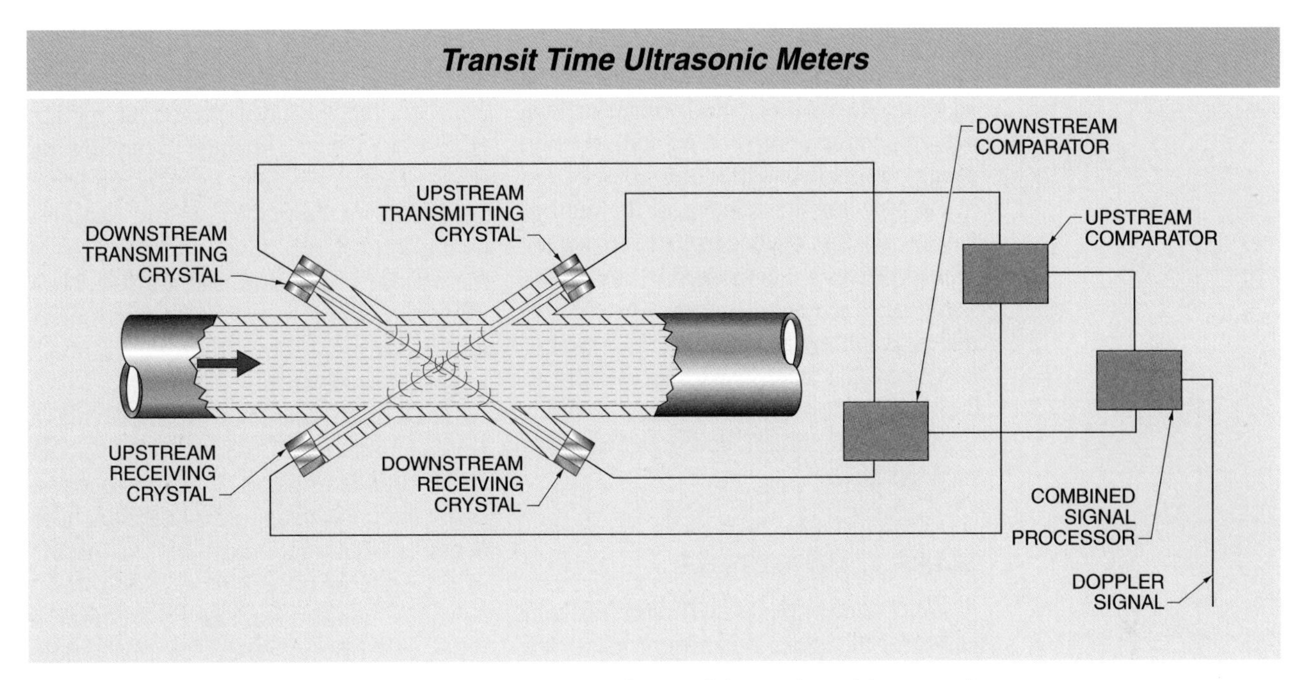

Figure 21-3. Transit time ultrasonic meters use two sets of transmitting and receiving crystals.

Vortex Shedding Meters

A *vortex shedding meter* is an electrical flowmeter consisting of a pipe section with a symmetrical vertical bluff body (a partial dam) across the flowing stream. A vortex shedding meter uses the formation of vortices as its principle of operation. A vortex is a fluid moving in a whirlpool or whirlwind motion. **See Figure 21-4.** A common way to explain a vortex is to look at a flag blowing in the wind. The flag ripples faster when the wind is blowing faster because of the increase in the vortices formed along the flag.

A common bluff body shape is a triangular block with the broad surface facing upstream. As the fluid is impeded by the bluff body, a vortex forms on one side of the body. It increases in size until it becomes too large to remain attached to the bluff body and breaks away. The formation of a vortex on one side of the bluff body alters the flowing stream so that another vortex is created on the other side and acts similarly. Alternating vortices are formed and travel downstream at a frequency that is linearly proportional to the speed of the flowing fluid and inversely linearly proportional to the width of the body.

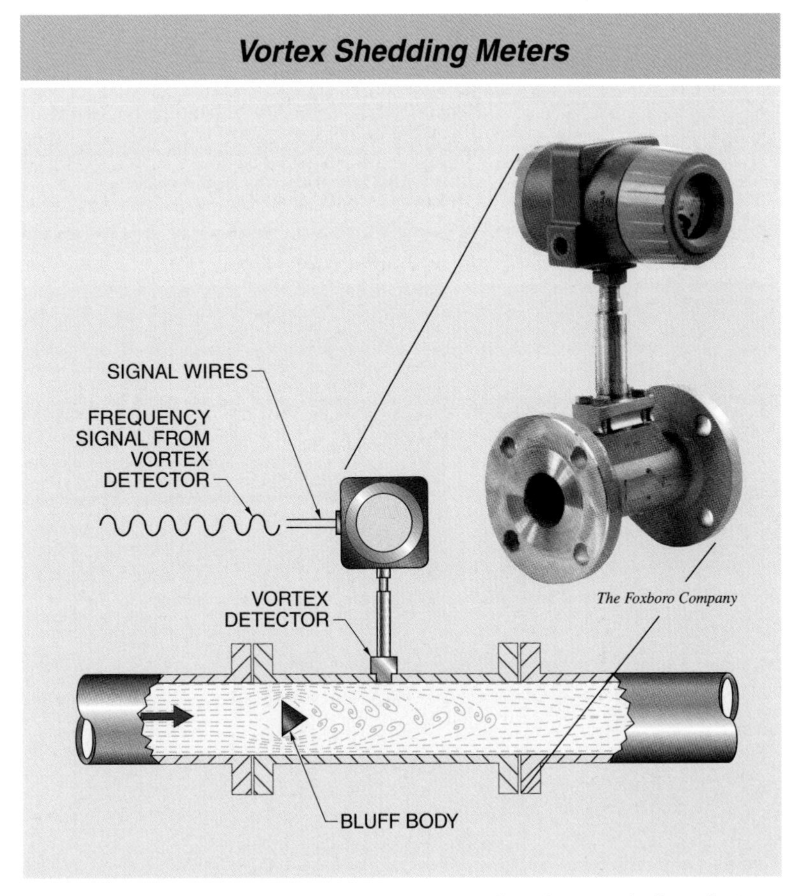

Figure 21-4. Vortex shedding meters create disturbances in flow that are measured to calculate flow.

The frequency of release of the vortices can be measured with ultrasonic sensors. In addition, the frequency can be measured using temperature, pressure, crystal, or strain gauge sensors. The area of the cross section of the pipe and the velocity of the fluid of the flow stream obtained from the frequency of the vortices are combined by the vortex shedding flowmeter to continuously determine the volumetric flow rate. This type of flowmeter has been used successfully to measure the flow of a wide variety of fluids such as steam, hot oil, and liquefied gases such as chlorine.

MASS FLOWMETERS

A *mass flowmeter* is a flowmeter that measures the actual quantity of mass of a flowing fluid. Mass flow measurement is a better way to determine the quantity of material than volumetric flow measurement. Changes in pressure and temperature can affect density, which then introduces errors into calculations that convert volumetric flow to actual quantity of material. Common types of mass flowmeters are the Coriolis meter and the thermal mass meter.

The airflow in a duct can be measured with a differential pressure meter and a transmitter sends a signal to the controller.

Coriolis Meters

A *Coriolis meter* is a mass flowmeter consisting of specially formed tubing that is oscillated at a right angle to the flowing mass of fluid. Coriolis force is the force generated by the inertia of fluid particles as the fluid moves toward or away from the axis of oscillation. **See Figure 21-5.** There are some newer mass flowmeters that use straight tubes to provide a much lower pressure drop.

The Coriolis mass flowmeter provides a very accurate measurement of the flow of either liquids or gases. It is an expensive meter but it has become more widely used because of its accuracy. It is available in a variety of materials to handle most corrosive conditions. The most common types of Coriolis meters cannot handle two-phase flows (mixtures of liquid and vapor). This is because turbulent two-phase flow upsets the resonant vibration balance of the meter. Some new specialty meters are designed for two-phase flows in which the liquid and vapor are evenly distributed and well mixed.

Coriolis Meter Operation. With Coriolis meters, the flow is divided and then passes through two tubes of equal length and shape. The tubes are firmly attached to the meter body (a pipe section). The two sections of tubing are made to oscillate at their natural frequency in opposite directions from each other. The fluid accelerates as it is vibrated and causes the tubing to twist back and forth while the tube oscillates.

Two detectors, one on the inlet and the other on the outlet, consist of a magnet and a coil mounted on each tubing section at the points of maximum motion. Each of these detectors develops a sine wave current due to the opposite oscillations of the two sections of tubing. The sine waves are in phase when there is no flow through the tubes. When there is flow, the tubes twist in opposite directions, resulting in the sine waves being out of phase. The degree of phase shift varies with the mass flow through the meter.

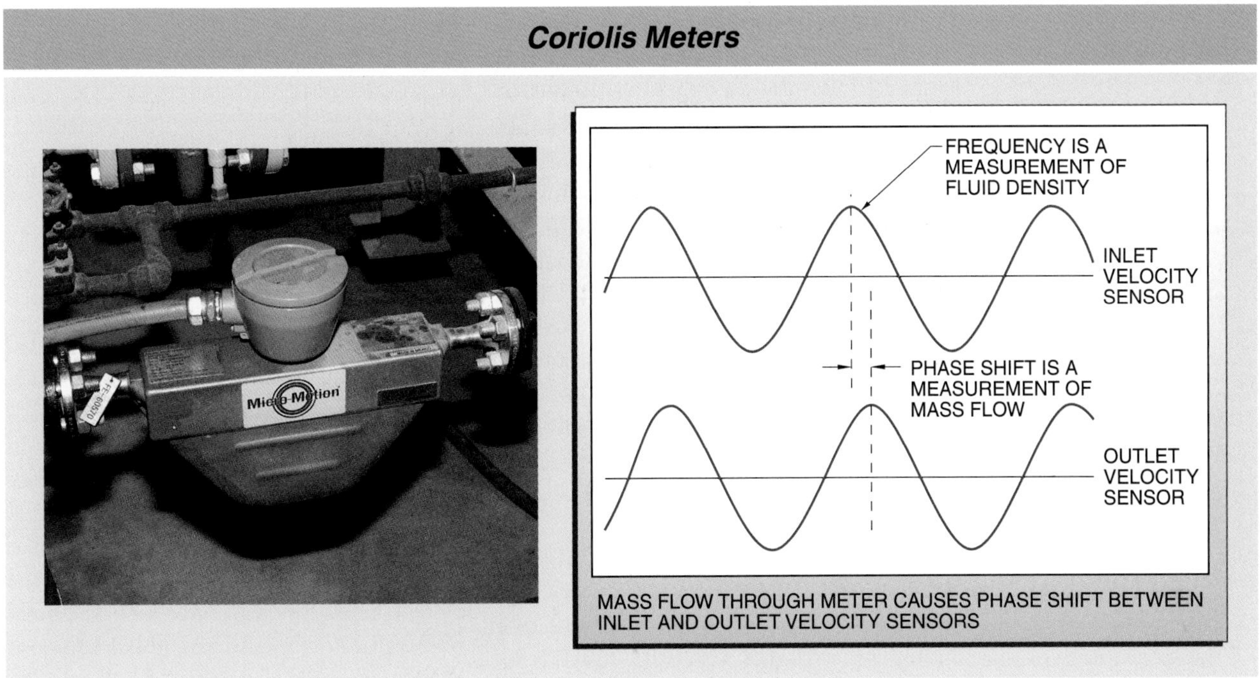

Coriolis Meters

FREQUENCY IS A MEASUREMENT OF FLUID DENSITY
INLET VELOCITY SENSOR
PHASE SHIFT IS A MEASUREMENT OF MASS FLOW
OUTLET VELOCITY SENSOR
MASS FLOW THROUGH METER CAUSES PHASE SHIFT BETWEEN INLET AND OUTLET VELOCITY SENSORS

Figure 21-5. Coriolis mass meters use the vibrations and twist of a tube to measure flow.

Density Measurement. An added feature of the Coriolis meter is that it can also measure fluid density. The natural frequency of the oscillating tubes can be related to the mass in the tubes. The tubes have known internal volume, and since mass divided by volume is density, the meter can be used to calculate the density. The temperature of the fluid is also measured so that the mass flow and density measurement are temperature-compensated. The measurements of mass flow, density, temperature, and oscillating frequency are available as output signals from the sensor.

Thermal Mass Meters

A *thermal mass meter* is a mass flowmeter consisting of two RTD temperature probes and a heating element that measure the heat loss to the fluid mass. Thermal mass meters are predominantly used for measuring gas flows. The two RTD probes are immersed in the flow stream. One probe is in an assembly that includes an adjacent heating element that is measured by the RTD probe. The other probe is separate and it measures the temperature of the flowing fluid. **See Figure 21-6.**

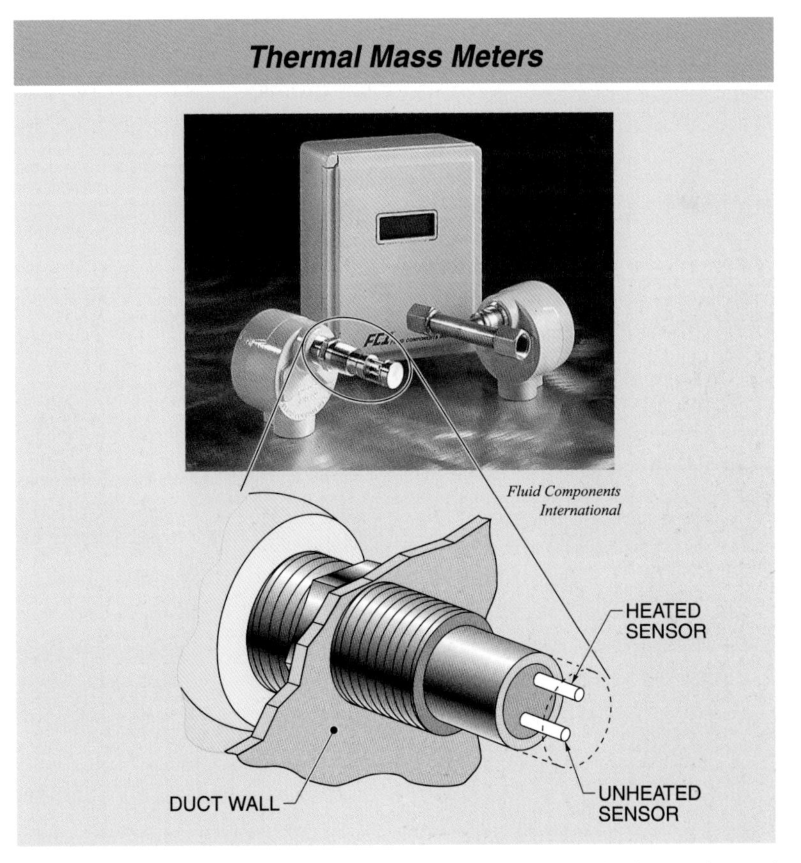

Figure 21-6. Thermal mass meters use the energy transfer from a heated element to measure flow.

The heated probe loses heat to the stream by convection. The electric circuitry is designed to maintain a constant difference in temperature between the two probes by varying the power to the heating element. The power becomes the measured variable of the system. The variations in power are proportional to the variations in mass flow rate. An alternative design uses constant power input to the heating element and measures the temperature difference between the RTD probes. The circuitry includes corrections for thermal conductivity, viscosity, and density as well as corrections for the design and construction features.

Thermal mass meters are relatively inexpensive and can be used to measure some low-pressure gases that are not dense enough for Coriolis meters to measure. Thermal mass meters are less accurate than many other types of meters.

Rotameters can be used to indicate the flow of water treatment chemicals.

The flow velocity for flammable fluids must be engineered to avoid the build-up of static charge. One-half the density, in lb/ft^3, multiplied by the square of the fluid velocity, in ft/s, and divided by 144 must be less than 30.

Belt Weighing Systems

Measuring the flow of granular (bulk) solids is a difficult task because of the basic properties of solids. Bulk solids vary greatly in flow properties. Some bulk solids are sticky and do not flow well. Other bulk solids are so fine and slippery that they flow like liquids. Bulk solids are usually transported by belt, screw, or drag conveyors. The most successful flow measurement is by the use of belt weighing systems.

A *belt weighing system* is a mass flow measuring system consisting of a specially constructed belt conveyor and a section that is supported by electronic weigh cells. **See Figure 21-7.** The conveyor belt is designed to minimize the transfer of the weight of the unmeasured section of the conveyor. Solids are deposited on the conveyor and carried onto the weighing section. The weight of solids on the measured section divided by the length of the measured section multiplied by the conveyor speed results in the flow in pounds per unit time (flow = weight ÷ length × speed).

The manufacturer should always be consulted regarding any new installation. Factors that affect the continued accuracy of the system change over time. These factors include belt stiffness, belt tension, and weigh and speed system calibration. Frequent recalibrations are required. This is usually done with known weight chains on the conveyor while it is running. There are many companies that specialize in the calibration of belt weighing systems.

There are some bulk solids that are not suitable for a belt weighing system. They are typically weighed as a batch and then dumped onto a conveyor for transport through the manufacturing process.

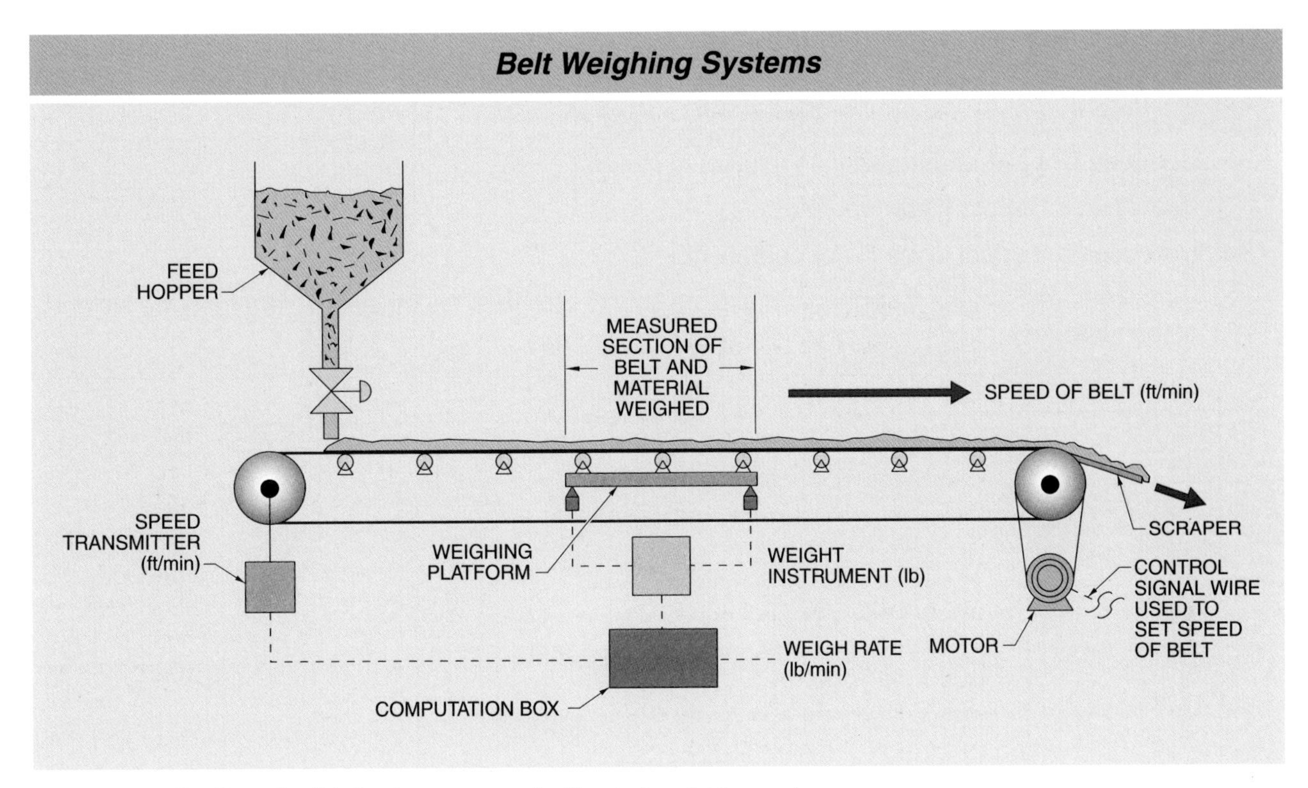

Figure 21-7. The flow of solids is often measured with a belt weighing system.

KEY TERMS

- *magnetic meter,* or *magmeter:* A flowmeter consisting of a stainless steel tube lined with nonconductive material, with two electrical coils mounted on the tube like a saddle.
- *ultrasonic flowmeter:* A flowmeter that uses the principles of sound transmission in liquids to measure flow.
- *Doppler ultrasonic meter:* A flowmeter that transmits an ultrasonic pulse diagonally across a flow stream, which reflects off turbulence, bubbles, or suspended particles and is detected by a receiving crystal.
- *transit time ultrasonic meter:* A flowmeter consisting of two sets of transmitting and receiving crystals, one set aimed diagonally upstream and the other aimed diagonally downstream.
- *vortex shedding meter:* An electrical flowmeter consisting of a pipe section with a symmetrical vertical bluff body (a partial dam) across the flowing stream.
- *mass flowmeter:* A flowmeter that measures the actual quantity of mass of a flowing fluid.
- *Coriolis meter:* A mass flowmeter consisting of specially formed tubing that is oscillated at a right angle to the flowing mass of fluid.
- *thermal mass meter:* A mass flowmeter consisting of two RTD temperature probes and a heating element that measure the heat loss to the fluid mass.
- *belt weighing system:* A mass flow measuring system consisting of a specially constructed belt conveyor and a section that is supported by electronic weigh cells.

REVIEW QUESTIONS

1. Describe magnetic meters and explain their operation.
2. Compare Doppler and transit time ultrasonic meters.
3. How does a Coriolis meter operate?
4. How does a thermal mass meter operate?
5. Describe a belt weighing system and explain how to calculate the mass flow rate from the weight of the material.

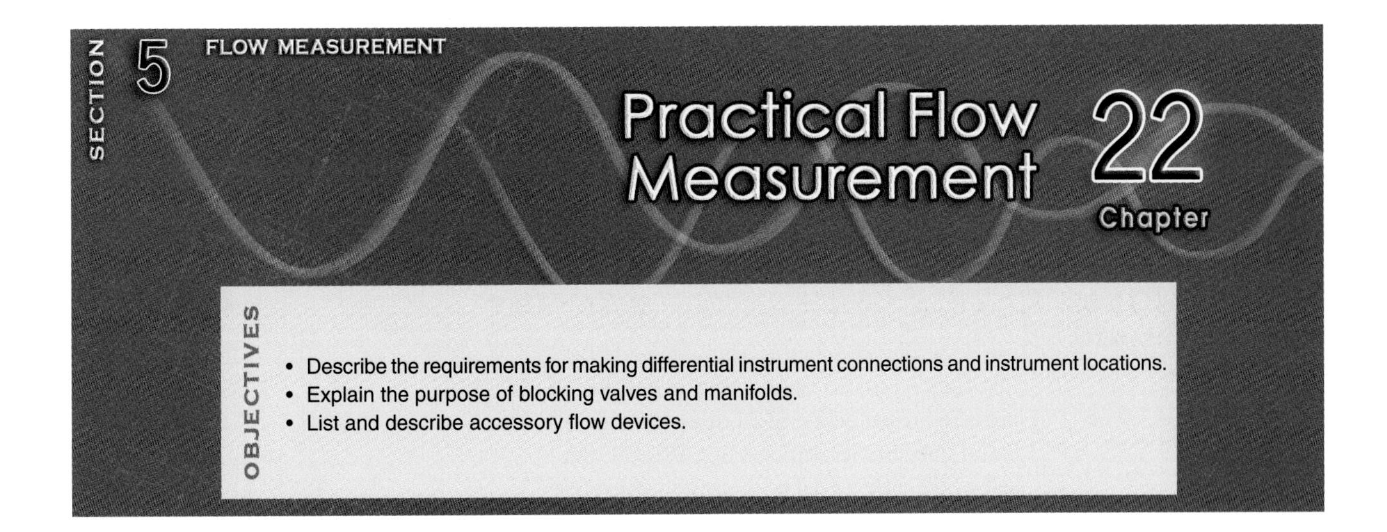

DIFFERENTIAL PRESSURE MEASUREMENT

There are several concerns with using differential pressure measurement to measure flow. The pressure instruments must be connected to the process to be useful. The location of the connections and the instruments depends on the type of process. Blocking valves and manifolds are often used to provide a convenient location to isolate the instrument from the process.

Differential Instrument Connections

The location of the differential connections for flowmeters varies. Two connections are needed to measure the differential. The connection upstream of the flow element (the high-pressure connection) measures the static pressure in the pipe. The connection downstream of the flow element (the low-pressure connection) measures the reduced pressure developed by the flow through the flow element. The difference in pressure, measured in inches of water, is the pressure differential. An averaging pitot has both connections located within a single tube inserted into the pipe or duct.

Taps. A *tap* is a pressure connection. For orifice plates, the three common variations are flange taps, vena contracta taps, and pipe taps. The names are based on the location of the taps. Flange taps are located in the orifice plate flanges 1″ upstream and downstream of the orifice plate faces. Vena contracta taps are located 1 pipe diameter upstream and about ½ pipe diameter downstream of the orifice plate faces. Pipe taps are located 2½ pipe diameters upstream and 8 pipe diameters downstream of the orifice plate faces. **See Figure 22-1.**

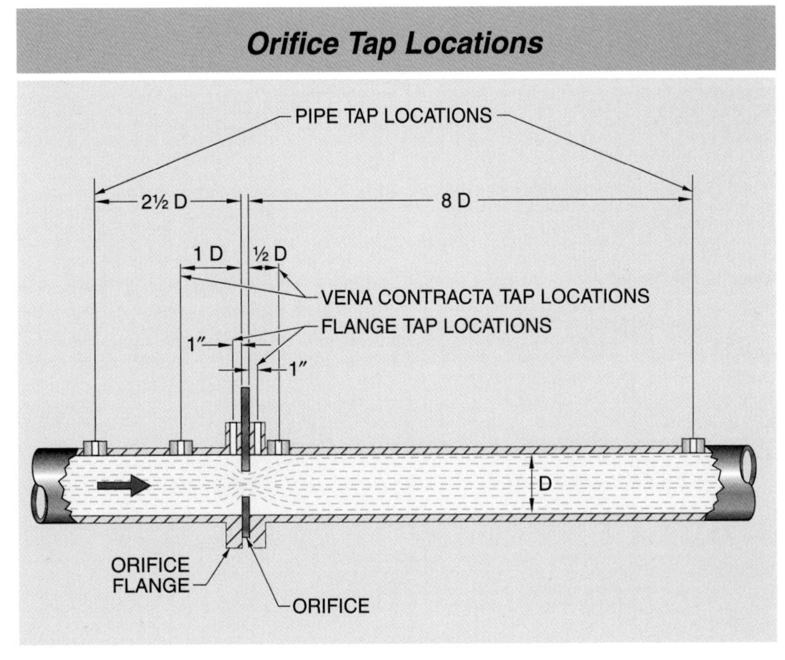

Figure 22-1. Flange taps, vena contracta taps, and pipe taps are located at different positions in a pipe for measuring pressure drop.

The pressure taps used with venturi tubes are part of the assembly and are supplied with the tubes. For flow nozzles and low-loss tubes, tap placements are specified by the manufacturer. The interconnections between the taps and the differential pressure (d/p) sensor may be made using tubing or piping.

Impulse Lines. An *impulse line* is the tubing or piping connection that connects the flowmeter taps to any of the differential pressure instruments. Most often the differential pressure instrument is a d/p cell that senses the differential pressure and converts the pressure to an electrical signal.

Differential Instrument Locations

When liquid flow is being measured, the measuring instrument must be mounted below the elevation of the flow element and the impulse lines must be filled with the liquid. Care must be exercised to ensure that no bubbles of air are trapped in the instrument or the impulse lines. The length of the impulse lines has no effect on the measurement accuracy as long as the two impulse lines start and end at equal elevations. **See Figure 22-2.**

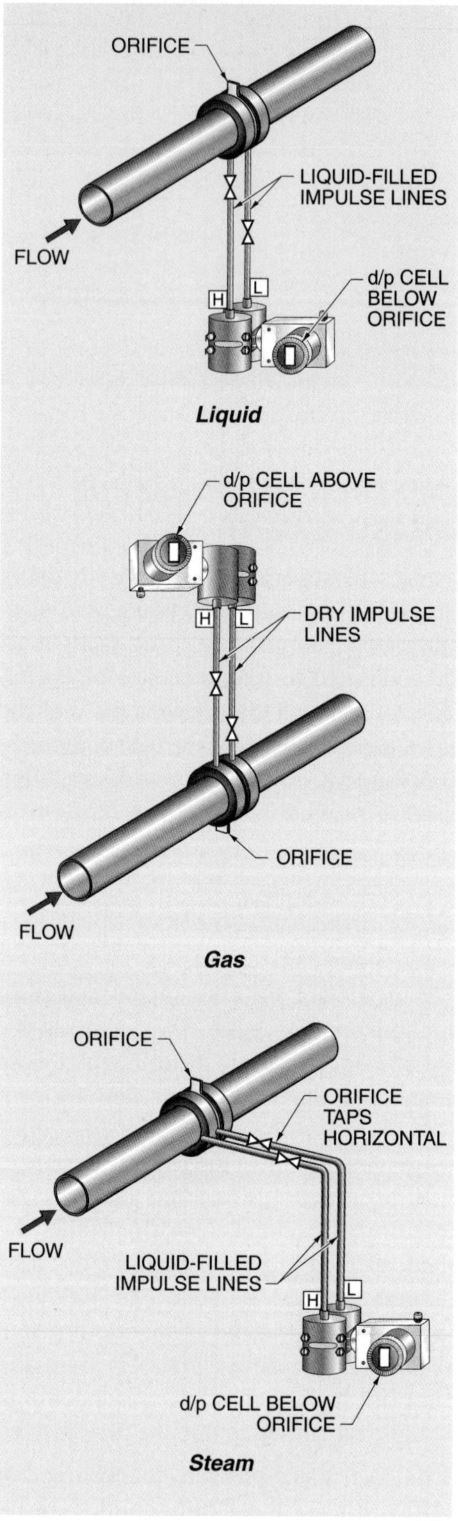

Figure 22-2. The location of a flow transmitter varies with the type of flowing fluid.

Flow switches are used in a boiler gas train.

When gas flow is being measured, the measuring instrument must be mounted above the elevation of the flow element and the diameter of the impulse lines must be large enough and routed so that any liquids that may condense in the impulse lines drain freely into the main piping.

When steam flow is being measured, the measuring instrument must be mounted below the elevation of the flow element even though the fluid is a vapor. Steam condenses to water very easily, and mounting the measuring instrument below the flow element allows the instrument and impulse lines to fill with condensate. The steam condensate protects the instrument from contact with the hot steam.

Mounting the instrument above the flow element can create drainage problems and can subject the instrument to prohibitively high temperatures. When installing the instrument and impulse lines, trapped air can be removed by initially blowing steam through the lines. The steam condensing process takes some time, but can be speeded up by cooling the outsides of the impulse lines with water or by manually backfilling the impulse lines with water.

Compact Orifice Flowmeters

A recent development in orifice flowmeters allows for a much simpler installation. The orifice holder is designed for the d/p cell and valve manifold to be directly bolted to the holder. The two available designs are the standard form and the conditioning form. **See Figure 22-3.**

With the standard form, the orifice has the conventional single bore in the center of the orifice plate and requires the normal lengths of upstream and downstream straight pipe diameters. The assembly offers a much easier installation than traditional orifice plates. They work well as long as the directly bolted d/p cell is not subjected to temperatures that are either too high or too low. In addition, the materials of construction may be limited.

The conditioning form has a unique design with four bores in the orifice in place of the conventional single bore. The four bores only require two straight pipe diameters upstream and downstream of the orifice. The stated accuracy of ±0.5% is provided even when there are severe upstream flow pattern disturbances, such as reducers, single and multiple 90° bends, a 10° swirl, or a butterfly valve. Special equations and conditioning factors are used to obtain this special feature.

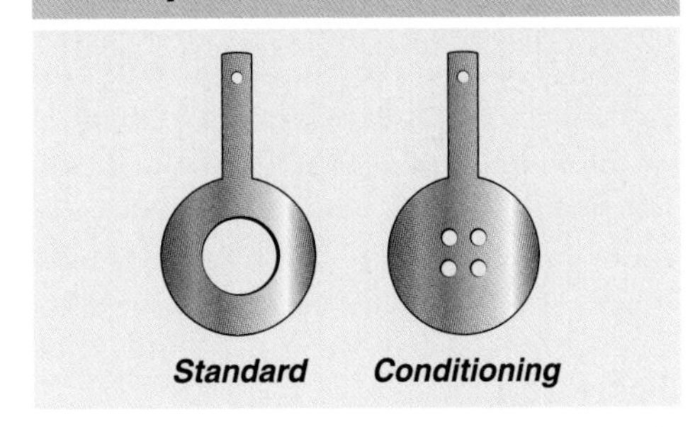

Figure 22-3. Compact orifice flowmeters can use standard and conditioning orifices.

Boiler Airflow Measurement

One of the most difficult flow measurements to take is of the combustion air for a boiler. It is difficult because it must have a 10:1 flow turndown ratio so that the proper air-fuel ratio can be maintained at all firing conditions. The flow is proportional to the square root of the differential pressure. Therefore, the differential pressure measurement device must be able to accurately measure a differential pressure turndown ratio of 100:1.

Boiler combustion air is usually supplied by combustion air fans with a very low discharge pressure. The low discharge pressure requires that the maximum differential also be very small. If the maximum differential were 1.00″WC, the minimum differential would be only 0.01″WC because of the 100:1 turndown ratio. In addition, the flow measurement has to be accurate over the entire range.

These are very difficult objectives to achieve. Custom-design flow elements are available with special differential pressure measurement transmitters. Temperature and pressure compensations are included along with special linearization techniques. These flow measurement assemblies provide a linear flow signal proportional to the airflow with ±3% accuracy and can be used with the various burner control strategies.

Blocking Valves and Manifolds

A *blocking valve* is a valve used at the differential measuring instrument to provide a convenient location to isolate the instrument from the impulse, equalizing, or venting lines and to provide a way to equalize the high- and low-pressure sides of the differential instrument. Equalizing the instrument is necessary so that the instrument can be periodically calibrated and zeroed. The blocking valves may be connected with individual pipe fittings and valves or can be part of a manifold assembly in one block. The manifold block can be bolted directly to the d/p cell. Typical blocking valves are single-valve equalizers, or three-valve, four-valve, and five-valve manifolds. **See Figure 22-4.**

ACCESSORY FLOW DEVICES

Other flow devices are instruments that are flow-metering accessory devices or other devices that do not actually measure flow but use flow principles to obtain information. Common examples of other flow devices are flow integrators that measure total flow and flow switches that can be configured to trigger an alarm or a switch.

Integrators

An *integrator* is a calculating device that totalizes the amount of flow during a specified time period. Integrators are available for use with pulse outputs as produced by turbine flowmeters or analog outputs (linear or square root) from orifice meters equipped

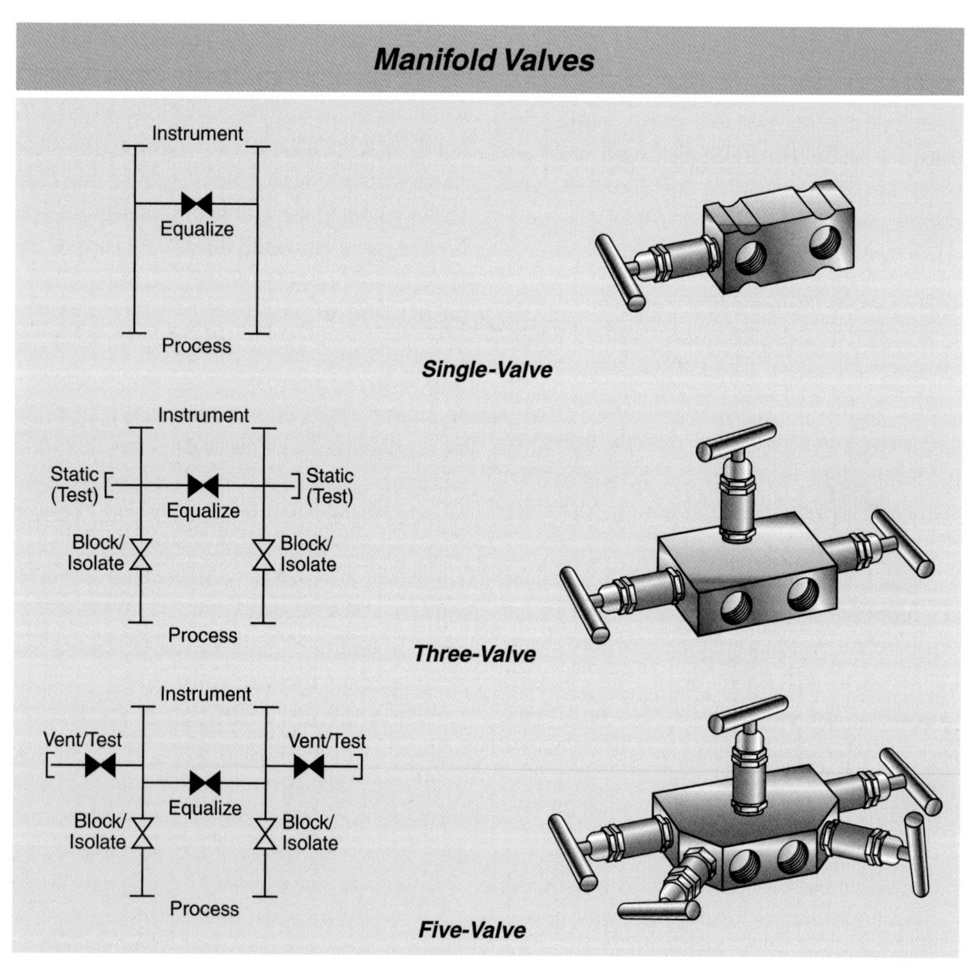

Figure 22-4. Manifold valves are used to isolate an instrument from a process and to allow in-line testing and calibration.

with analog-to-digital converters. Integrators are also available as mechanically connected units in large-case circular chart recorders, or as stand-alone models using pneumatic or electrically transmitted signals. An integrator follows a sequence of steps as the differential pressure increases. **See Figure 22-5.**

Except for their source of power and the method of converting differential pressure measurement to flow rate, all integrators resemble one another and perform the same function. The principal requirements are an accurate measurement of the differential pressure, a conversion to flow rate, a constant time input, and an easy-to-read counter.

Pneumatic integrators that receive 3 psig to 15 psig signals were developed to satisfy a demand for nonelectrical devices in applications where explosives or other hazards are present.

Flow Switches

A *flow switch* is a device used to monitor flowing streams and to provide a discrete electrical or pneumatic output action at a predetermined flow rate. Flow rate switches are used to generate alarm or shutdown signals on either high or low flows. A typical example is the monitoring of the flow from a pump that actuates an alarm if the flow stops. The flow switch function is usually composed of several different measurement principles such as an orifice plate and a differential pressure switch. **See Figure 22-6.** Common types of flow switches include differential pressure, blade, thermal, and rotameter switches.

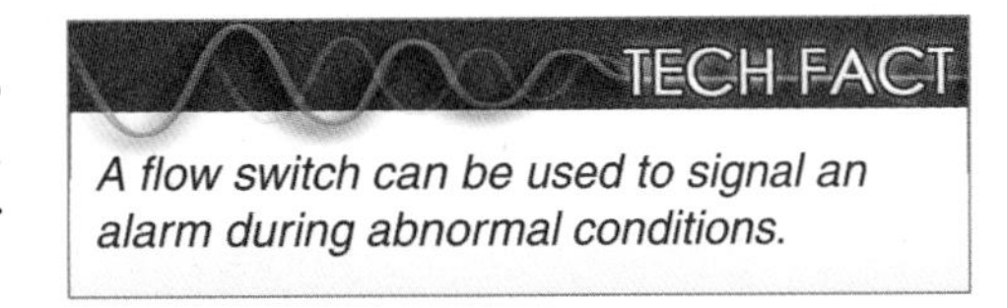

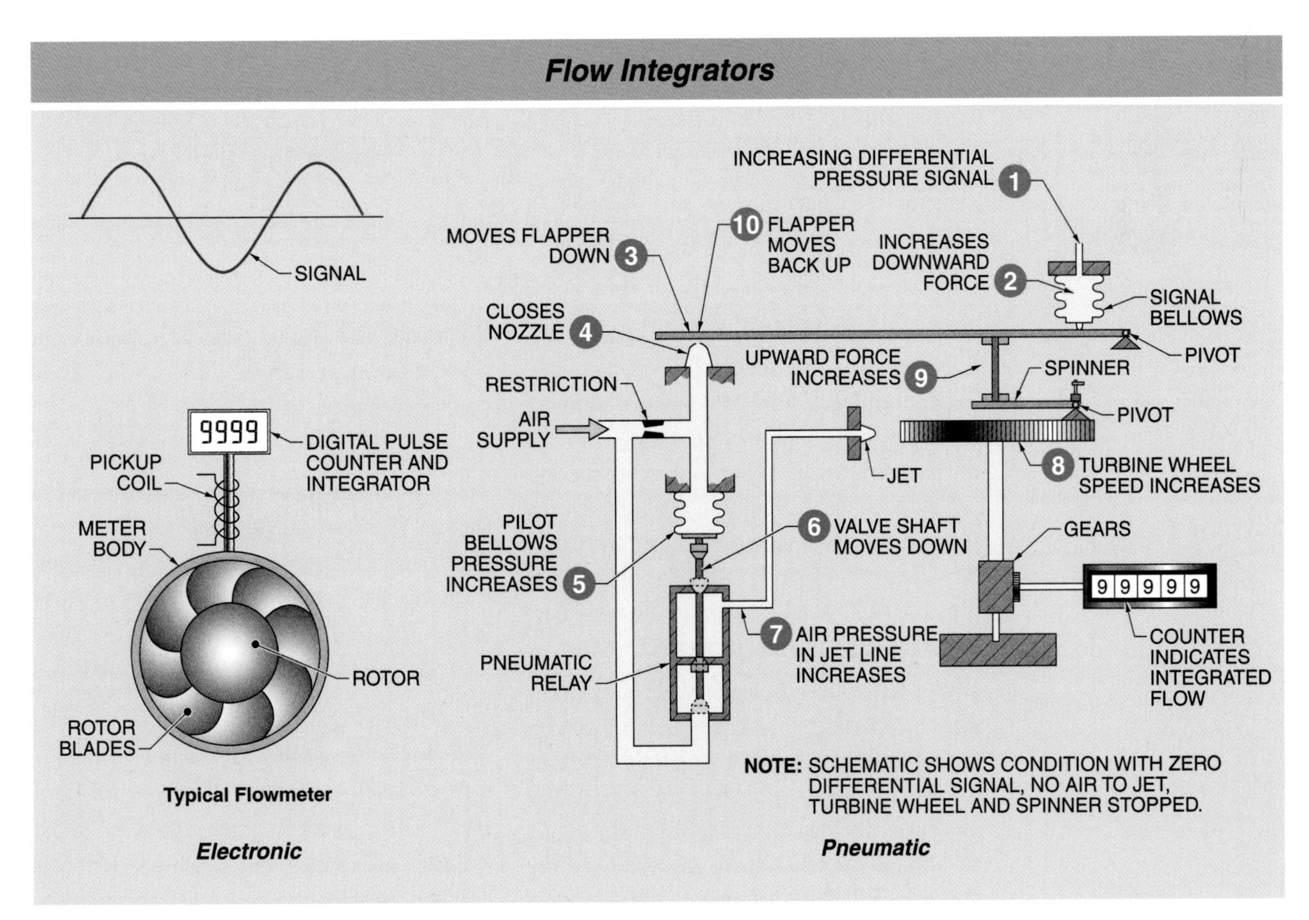

Figure 22-5. Pneumatic integrators are used in hazardous atmospheres where electrical devices are not used.

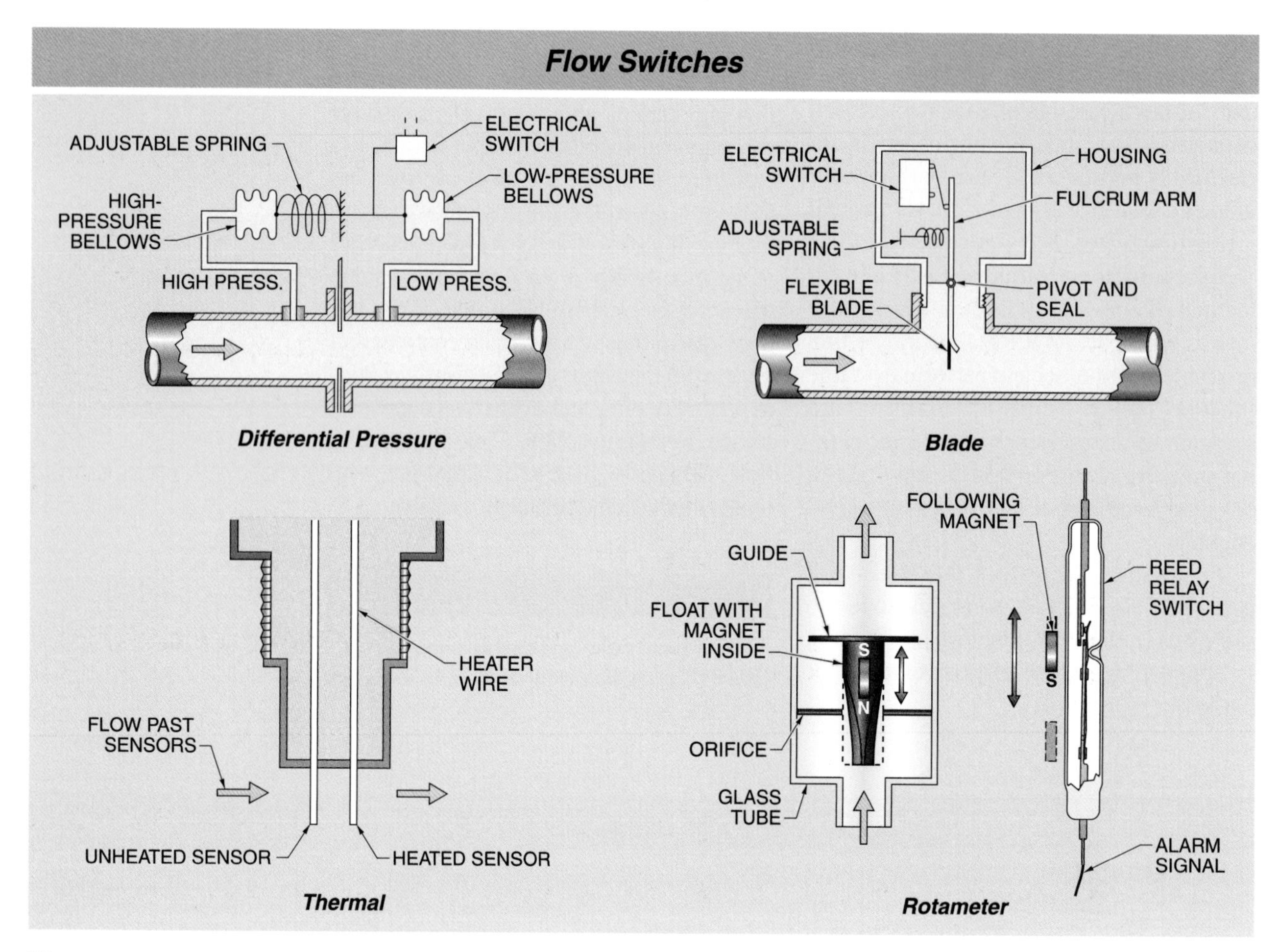

Figure 22-6. Flow switches are used to activate a signal or an alarm, or to actuate a control function, when flow reaches a predetermined level.

Differential Pressure Switches. A *differential pressure switch* is a flow switch consisting of a pair of pressure-sensing elements and an adjustable spring that can be set at a specific value to operate an output switch. The differential pressure switch measures the pressure drop across a primary flow element. In small sizes, the differential pressure switch and the primary flow element are combined into one device.

Blade Switches. A *blade switch* is a flow switch consisting of a thin, flexible blade inserted into a pipeline. Flowing fluid develops a force that presses against the blade. The motion of the blade is transferred through a seal and is opposed by an adjustable spring that establishes the trip point. An electrical or pneumatic switch senses the motion of the blade. The wetted parts are available in a variety of materials including PTFE. They can be used in ¾″ to 24″ pipes.

Thermal Switches. A *thermal switch* is a flow switch consisting of a heated temperature sensor. The flowing fluid carries away heat from the heated temperature sensor. The electronic circuits in the switch can be set to trip at some predetermined flow rate. Thermal flow switches are available in limited materials and in 1″ to 12″ sizes.

Rotameter Switches. A *rotameter switch* is a flow switch consisting of a shaped float, a fixed orifice, and a magnet embedded in the float that trips a magnetic sensing switch outside the tube. Rotameter-style flow switches are available in ½″ to 2″ sizes.

KEY TERMS

- *tap:* A pressure connection.
- *impulse line:* The tubing or piping connection that connects the flowmeter taps to any of the differential pressure instruments.
- *blocking valve:* A valve used at the differential measuring instrument to provide a convenient location to isolate the instrument from the impulse, equalizing, or venting lines and to provide a way to equalize the high- and low-pressure sides of the differential instrument.
- *integrator:* A calculating device that totalizes the amount of flow during a specified time period.
- *flow switch:* A device used to monitor flowing streams and to provide a discrete electrical or pneumatic output action at a predetermined flow rate.
- *differential pressure switch:* A flow switch consisting of a pair of pressure-sensing elements and an adjustable spring that can be set at a specific value to operate an output switch.
- *blade switch:* A flow switch consisting of a thin, flexible blade inserted into a pipeline.
- *thermal switch:* A flow switch consisting of a heated temperature sensor.
- *rotameter switch:* A flow switch consisting of a shaped float, a fixed orifice, and a magnet embedded in the float that trips a magnetic sensing switch outside the tube.

REVIEW QUESTIONS

1. List three common types of taps and describe typical locations for the taps when using an orifice plate.
2. What are the differential instrument locations for gas, liquid, and vapor flow measurement?
3. What is a blocking valve and why is it used?
4. What is an integrator and why is a pneumatic integrator used?
5. Define flow switch and list several common types of flow switches.

SECTION

ANALYZERS

TABLE OF CONTENTS

SECTION OBJECTIVES

Chapter 23

- Define the terms analysis, analyzer, and analyzer sampling system.
- Define thermal conductivity analyzer and explain its use.
- Identify the types of radiant-energy absorption analyzers and explain their use.
- Identify the types of oxygen analyzers and explain their use.
- Define opacity analyzer.

Chapter 24

- Define the common terms associated with humidity.
- Identify the types of humidity analyzers.
- Identify the types of solids moisture analyzers.

Chapter 25

- Identify common types of liquid analyzers and explain their use.
- Describe a hydrostatic pressure density analysis.
- Describe a slurry density analysis.
- Identify the common types of viscosity analyzers.
- Describe the function of turbidity analyzers and refractive index analyzers.

SECTION 6

ANALYZERS

Chapter 26

- Identify the analyzers that use conductivity and explain their use.
- Compare pH analyzers and oxidation-reduction potential (ORP) analyzers.
- Define composition analyzer and explain its use.
- Describe the function of automatic titration analyzers.

INTRODUCTION

The field of analysis is very broad and relatively new to industry. The developments of the modern industrial age led to the need for instruments to measure the products of the new factories.

In the early 19th century, William Herschel's discovery of infrared radiation led to further discoveries about the nature of light and energy. Chemists were able to combine Herschel's discoveries of infrared radiation with other discoveries of the nature of spectral lines and develop the science of spectroscopy. The development of spectroscopy eventually led to the development of infrared and ultraviolet analyzers.

In the late 19th century, Walther Hermann Nernst devised the first theories of electric potential and conduction in electrolytic solutions. These theories related the electrical energy produced by a chemical reaction to the chemical potential of the reaction. The theories of electric potential eventually led to the development of conductivity meters, pH meters, and ORP analyzers.

ADDITIONAL ACTIVITIES

1. After completing Section 6, take the Quick Quiz® included with the Digital Resources.
2. After completing each chapter, answer the questions and complete the activities in the *Instrumentation and Process Control Workbook.*
3. Review the Flash Cards for Section 6 included with the Digital Resources.
4. After completing each chapter, refer to the Digital Resources for the Review Questions in PDF format.

ATPeResources.com/QuickLinks • Access Code: 467690

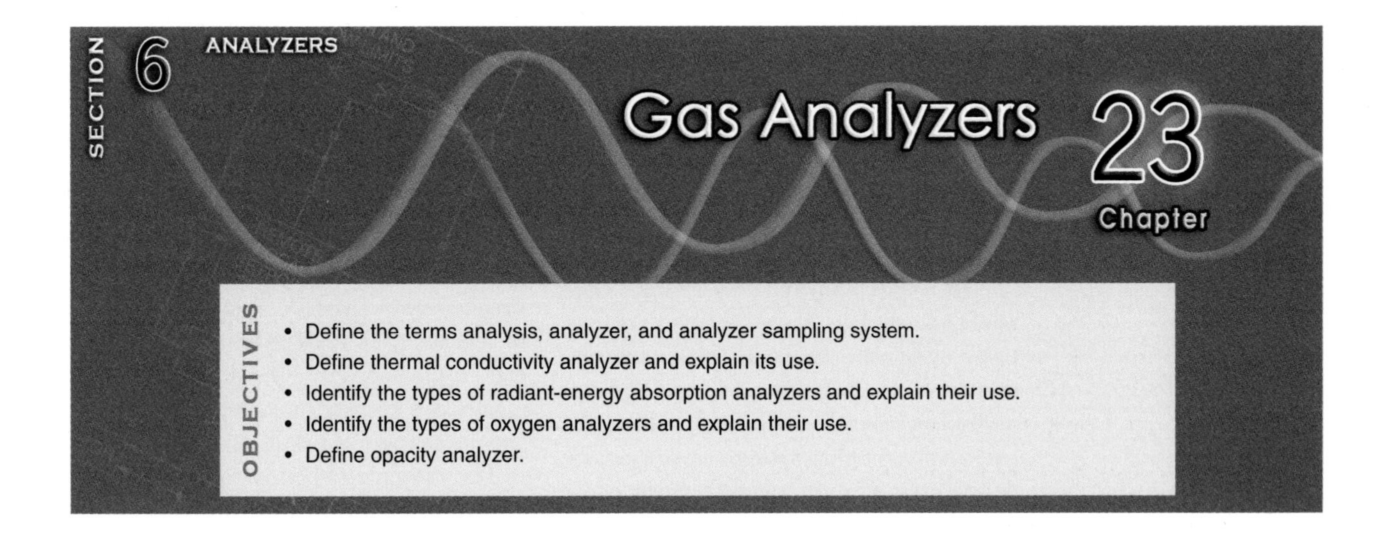

SECTION 6 ANALYZERS

Gas Analyzers

Chapter 23

OBJECTIVES

- Define the terms analysis, analyzer, and analyzer sampling system.
- Define thermal conductivity analyzer and explain its use.
- Identify the types of radiant-energy absorption analyzers and explain their use.
- Identify the types of oxygen analyzers and explain their use.
- Define opacity analyzer.

ANALYZERS

In many industries, it is vital to know the composition of raw materials, intermediate chemical compounds, and the final product. *Analysis* is the process of measuring the physical, chemical, or electrical properties of chemical compounds so that the composition and quantities of the components can be determined. Some of this analysis is done in centralized, inplant laboratories and some is done at or near the process.

An *analyzer* is an instrument used to provide an analysis of a sample from a process. The sample may be in the form of a gas, liquid, or solid. The sampling apparatus may be operated manually or automatically. Analyses performed in plant laboratories are very sophisticated and can provide accurate results, but the delay in completing the analysis can have very significant effects on the process.

In some processes, the area operators in the local control rooms perform simple analysis procedures. While this reduces the time between sampling the process and making corrections, it is not an ideal solution. Therefore, analyzers have been developed that are applied directly to the process stream to reduce the time needed for analysis.

An *on-line analyzer* is an instrument that is located in the process area and obtains frequent or continuous samples from the process. On-line analyzers can provide very rapid results, enabling controllers and operators to react quickly to changes in the process or to provide a continuous record of emissions. **See Figure 23-1.** On-line analyzers usually do not provide the accuracy that is obtainable by laboratory analyzers, but their timesaving benefits have made them valuable industrial tools. On-line analyzers are by far the most complex instruments in a plant. Analyzers require careful maintenance and the most attention to calibration of all instruments.

Analyzers

Figure 23-1. On-line oxygen analyzers are used to optimize combustion processes.

ANALYZER SAMPLING SYSTEMS

An *analyzer sampling system* is a system of piping, valves, and other equipment that is used to extract a sample from a process stream, condition it if necessary, and convey it to an analyzer. An in-line analyzer must be directly connected to a process stream in order to obtain a sample. Some analyzers continuously analyze the process samples, while others require time to analyze a sample and thus a sample is only fed intermittently. Although a great many sampling systems are of the same design, each one must be considered individually to ensure that it provides a representative sample of the process to the analyzer and at conditions suitable for the analyzer.

Representative Samples

A *representative sample* is a sample from a process in which the composition of the sample is the same as in the process piping. If the sample is not representative of the process, the analyzer results will not be accurate. Samples should be taken from the center of the pipe, duct, or vessel. Care should be taken to avoid entrainment of liquid or dirt in the sample lines. In addition, the process of transporting a sample from the process to the analyzer can change the composition of the sample.

For example, if the process stream is hot and there are condensable vapors present, the cooling of the sample could result in some of the component gases being absorbed in the condensate and removed from the sample prior to reaching the analyzer. Thus the analyzer would not measure a representative sample. In many cases, water vapor in the sample stream is not wanted. A drying process must be selected to ensure a representative sample gets to the analyzer. Filtering is needed in many cases to ensure that the sample is clean and the analyzer is not damaged.

For example, when sampling hot, wet stack gas, a filter must be installed at the entrance to the sample lines to remove any particulate contamination. Then a coalescing filter is used to remove any suspended liquids before the sample goes to the analyzer. **See Figure 23-2.**

Sample Transportation Lags

A *sample transportation lag* is the time that it takes for a sample to get from the process through the final analysis. Analyzers that are in direct control of the process require the sample to be processed very quickly to ensure control stability. Long lag times significantly degrade controller performance. Analyzers that process intermittent samples have lag times too long for conventional control methods. Special control strategies need to be used for these conditions.

Sampling systems that are too long or that require extensive conditioning can also cause problems when used in control loops. An additional problem with using analyzers in direct control of the process is that analyzers are usually the least reliable of all instrumentation. It is usually better to use the analyzer to provide a record of the analysis and use a secondary measurement to control the composition.

Industrial processes require the storage and analysis of raw materials.

TECH FACT

Before working with industrial chemicals, always refer to specific procedures detailed on the Material Safety Data Sheet (MSDS).

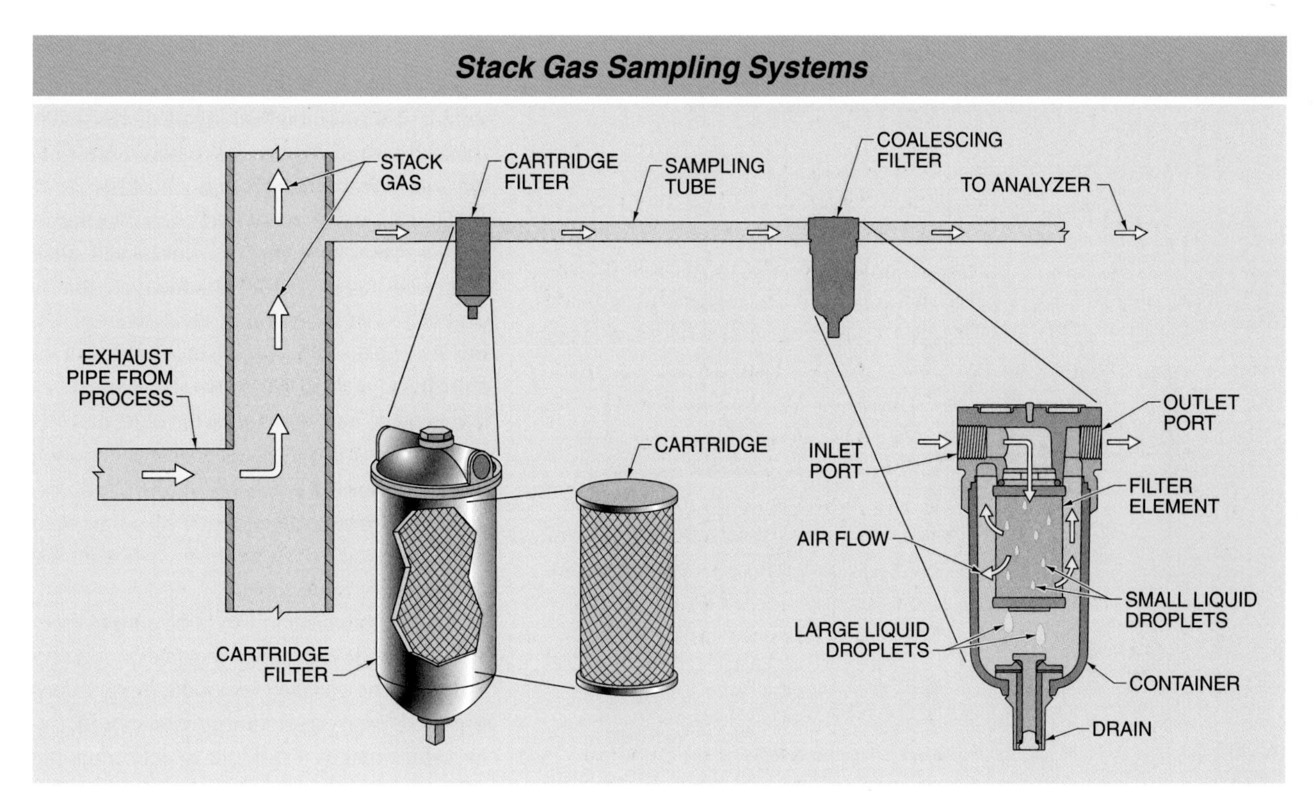

Figure 23-2. A stack gas sampling system may consist of filters and other conditioning aids to prepare the sample for analysis.

Analyzer Calibrations

An *analyzer calibration* is the process of substituting known sample compositions for the normal process sample so that the analyzer can be adjusted to read the correct values. Calibration samples are usually prepared for the upper and lower ends of the measuring range of the analyzer and sometimes for a concentration near the normal operating point. The calibration process can be made simpler by having pre-piped connections and valves for the calibration samples. This is important since critical analyzers may need to be calibrated frequently. In many cases, the accuracy of the analyzer is no better than the accuracy of the calibration process.

GAS ANALYZERS

A *gas analyzer* is an instrument that measures the concentration of an individual gas in a gaseous sample. Gas analyzers use many different technologies and scientific principles to measure various gas properties that can be used to distinguish one gas from another. One thing that the analyzers all have in common is that the gas samples for the analyzers must be clean and be at constant temperature and pressure. The most common types of gas analyzers are thermal conductivity, radiant-energy absorption, oxygen, and opacity analyzers.

Thermal Conductivity Analyzers

A *thermal conductivity analyzer* is a gas analyzer that measures the concentration of a single gas in a sample by comparing its ability to conduct heat to that of a reference gas. The thermal conductivity detector can be a two-chamber or four-chamber design. In a typical two-chamber design, the thermal conductivity of an unknown gas is compared to a sample gas with air or nitrogen used as a reference gas. **See Figure 23-3.** In a four-chamber design, the sample gas passes through two of the chambers and the reference gas passes through the remaining two. Each chamber contains identical heated resistance heating elements.

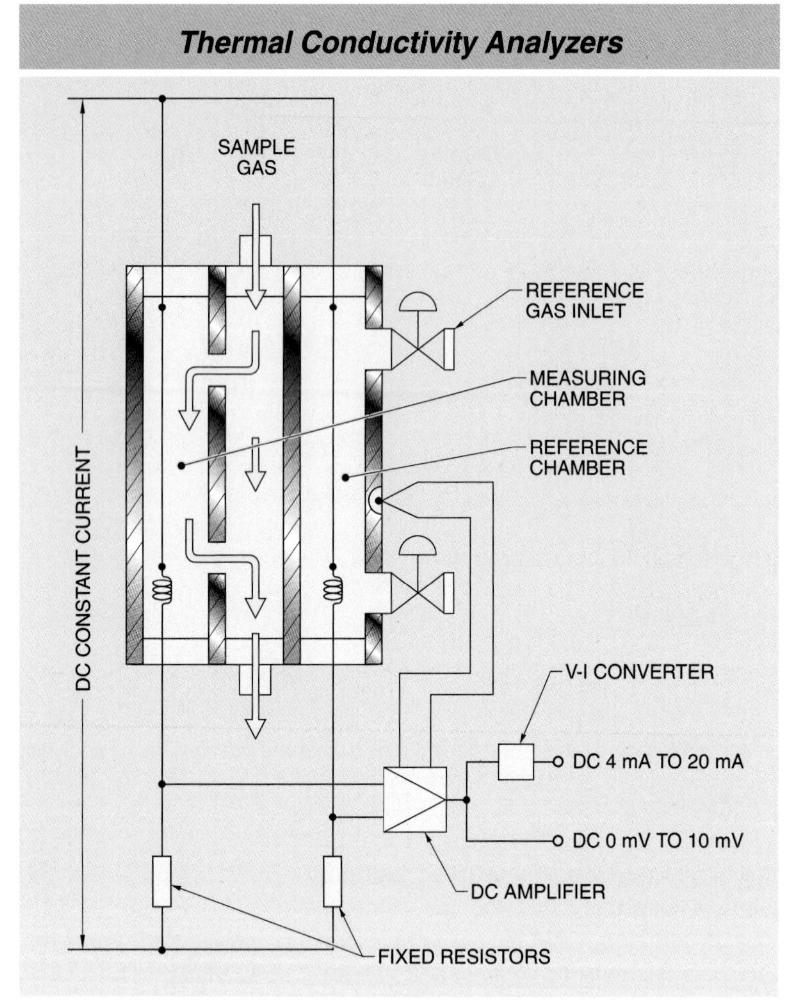

Figure 23-3. Two-chamber designs for a thermal conductivity analyzer allow a sample gas to be compared to a reference gas.

All gases conduct heat. In both thermal conductivity detector designs, the flow of gases across the heating elements conducts heat out of the heating element and lowers the temperature of a thermistor or thermocouple. The amount of the temperature change depends on the thermal conductivity and flow rate of the gas. This temperature change is measured, amplified, and converted to a digital or an analog output that is proportional to the concentration of one of the constituents of the mixture.

The thermal conductivity analysis method can only be used for binary gases with widely different thermal conductivities or in pseudo-binary gas mixtures where all but two of the gases have a fixed concentration. A binary gas is a gas mixture with only two gases in the sample. Thermal conductivity methods cannot be used to analyze the gas when more than two components vary in concentration.

The measurement of hydrogen concentration in almost any other gas works well since hydrogen has a thermal conductivity that is very different from that of most other gases. For example, some processes that generate chlorine can also allow small amounts of hydrogen to be contained in the chlorine. It is important to know the hydrogen concentration in the chlorine because in subsequent processing the hydrogen concentration can increase to the explosive range as chlorine is condensed and removed from the gas.

Some gases have low thermal conductivity and other gases have high thermal conductivity, but many gases have conductivity values similar to each other. Similarity in conductivity values makes it difficult to determine the difference between gases by thermal conductivity measurements. **See Figure 23-4.** For example, the amount of oxygen in air cannot be determined by the thermal conductivity method because the thermal conductivity of oxygen is not sufficiently different from the thermal conductivity of nitrogen.

Radiant-Energy Absorption Analyzers

A *radiant-energy absorption analyzer* is a gas analyzer that uses the principle that different gases absorb different, very specific, wavelengths of electromagnetic radiation in the infrared (IR) or ultraviolet (UV) regions of the electromagnetic spectrum. The drop in intensity of a specific wavelength after it has passed through a sample is directly related to the concentration of the gas in that sample.

The general arrangement of IR and UV analyzers is very similar. Both have optical systems and separate detection chambers. The beam is filtered or split, chopped, and then directed through two measurement chambers. One chamber contains a reference gas and the other the sample material. The most common types of radiant-energy absorption analyzers are nondispersive infrared analyzers and ultraviolet analyzers.

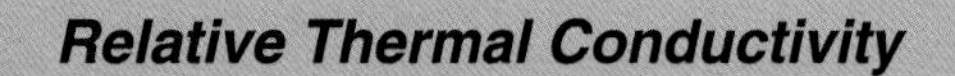

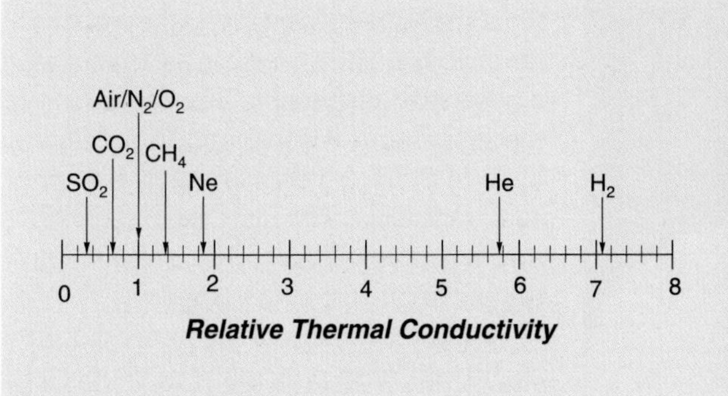

Gas	80°F microcal/ sec-cm-C	80°F milliwatt/ m-C	Relative Conductivity Air = 1
Air	62.2	26.2	1.00
Carbon Dioxide (CO_2)	39.7	16.8	0.64
Helium (He)	360	150	5.79
Hydrogen (H_2)	446	187	7.17
Methane (CH_4)	81.8	34.1	1.32
Neon (Ne)	116	48.3	1.86
Nitrogen (N_2)	62.4	26.0	1.00
Oxygen (O_2)	63.6	26.3	1.02
Sulfur Dioxide (SO_2)	23.0	9.6	0.37

Thermal Conductivity

Figure 23-4. Relative thermal conductivity determines the effectiveness of thermal conductivity analysis.

Electromagnetic Spectrum

Many important parts of the electromagnetic spectrum have their wavelengths measured in nanometers (nm, or 10^{-9} meters) or microns (µm, or 10^{-6} meters). Visible light is that part of the electromagnetic spectrum that we can see. Visible light has wavelengths of approximately 400 nm (violet) to 750 nm (red). Ultraviolet electromagnetic radiation is that part of the spectrum consisting of the wavelengths of electromagnetic radiation beyond the violet from approximately 10 nm to 400 nm. These wavelengths are shorter than visible light.

Infrared radiation is that part of the electromagnetic spectrum consisting of the wavelengths of electromagnetic radiation beyond the red from approximately 750 nm to 1,000,000 nm (0.750 µm to 1000 µm). These wavelengths are longer than visible light. The IR spectrum may be further divided into near, mid, and far infrared. The near and mid infrared regions of the spectrum are most commonly used in analysis.

There are no specific definitions of the boundaries between the different regions of the IR spectrum, but near IR is typically described as the range of wavelengths from 1000 nm to 2500 nm and mid IR is typically described as the range of wavelengths from 2500 nm to 5000 nm. The distinctions between the different regions of the IR spectrum are based on the use of different types of detectors to measure the electromagnetic radiation. The IR and UV parts of the spectrum are invisible to the eye. On occasion, the term "light" is incorrectly used to describe IR and UV radiation.

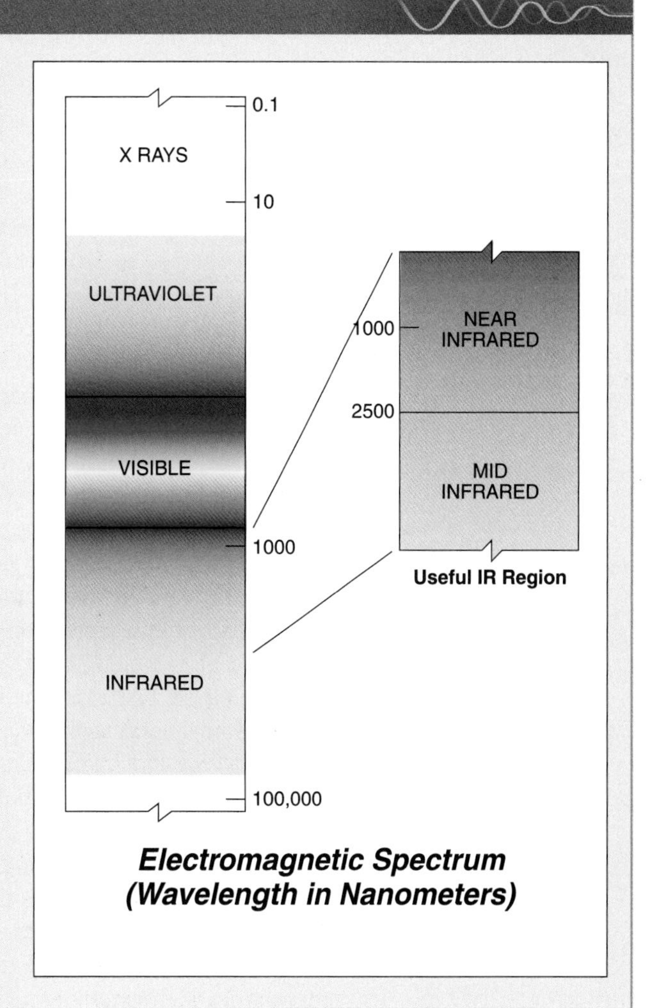

Electromagnetic Spectrum (Wavelength in Nanometers)

TSI Incorporated

A combustion analyzer is used to measure the composition of combustion gases.

Nondispersive Infrared (NDIR) Analyzers. A *nondispersive infrared (NDIR) analyzer* is a radiant-energy absorption analyzer consisting of an IR electromagnetic radiation source, an IR detector, and two IR absorption chambers. One IR absorption chamber contains a reference gas and the other contains the sample and uses a specific wavelength of IR radiation that is absorbed by a specific gaseous compound. In an NDIR analyzer, a single-wavelength IR source is split and often chopped and sent through the two chambers. By comparison, dispersive infrared analysis separates the radiation from the IR source into individual components with a diffraction grating or a prism and all wavelengths are used.

In an NDIR analyzer, the unabsorbed IR electromagnetic radiation is collected in a Luft-type pneumatic sensor or measured with a photoacoustic detector. Both detectors measure a pressure change when a sample gas absorbs IR radiation. The pressure of the reference gas is monitored to ensure that any pressure change is due to the absorption of the IR radiation and not environmental effects.

In a Luft-type sensor, the sample gas absorbs the IR radiation and heats up, while the reference gas does not absorb the IR radiation. The heated gas increases the pressure in the chamber. The increase in pressure moves a capacitive diaphragm. The movement of the diaphragm is measured in an electric circuit just like an ordinary capacitive pressure transducer. The increase in pressure is inversely proportional to the concentration of the gas of interest in the sample gas.

A photoacoustic detector works on a slightly different principle. The chopped IR radiation source heats the gas while the gas is absorbing the IR radiant energy and the pressure goes up in the detection chamber. While the chopper blocks the IR radiation source, the gas cools off and the pressure goes down. The alternating increase and decrease in pressure generates pressure pulsations. When the frequency of the pressure pulsations is within the range of human hearing it is referred to as sound.

The amplitude of the sound is measured with a pressure sensor, typically a microphone or a piezoelectric pressure transducer. The amplitude of the sound is inversely proportional to the concentration of the gas of interest in the sample gas. Photoacoustic detectors are desirable because they are suitable for low-cost silicon micromachining and they can easily be configured for different gases.

An NDIR analyzer is very specific for a single suitable gas regardless of the composition of the background gases. This is because the absorption of infrared radiation is directly related to the structure of the molecule. **See Figure 23-5.** Gases that are excellent candidates for NDIR analysis are water vapor, carbon monoxide (CO), carbon dioxide (CO_2), sulfur dioxide (SO_2), nitrous oxides (NO_x), and methane (CH_4).

In addition, any gas containing a carbon-hydrogen bond absorbs IR radiation. NDIR analyzers may measure ranges of up to 100% and down to about 1 part per million. For example, NDIR analyzers are commonly used to monitor the flue gases from combustion processes since individual analyzers can be selected for CO_2, CO, SO_2, or NO_x.

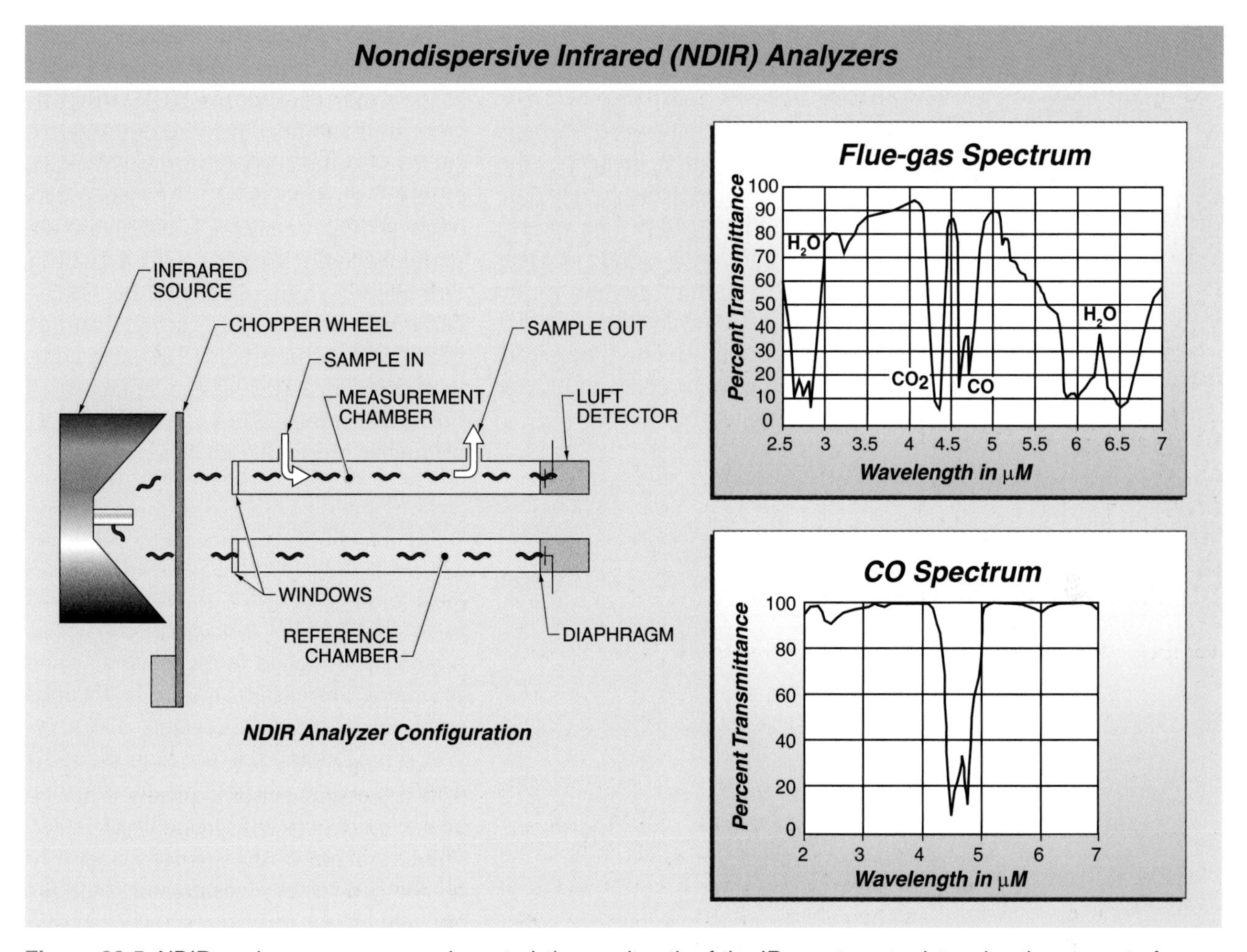

Figure 23-5. NDIR analyzers measure a characteristic wavelength of the IR spectrum to determine the amount of a gas present in a sample.

Tunable Diode Laser Absorption Spectroscopy

A diode can be used to convert electrical current into coherent (in phase), monochromatic (single frequency), laser light. This allows a laser system to monitor sample absorption at a precise frequency. For most purposes, this frequency does not have to be at a very specific value. Tunable diode laser absorption spectroscopy (TDLAS) technology has revolutionized the field of infrared (IR) absorption gas analyzers.

The great advantage of having a tunable diode as a light source for use in an IR gas analyzer is that the light source can be tuned to match the specific absorption peak of the gas to be analyzed. This allows gas streams to be analyzed that contain trace amounts of the gas of interest. In some cases, the systems are sensitive to concentrations as low as a few parts per billion (ppb).

The main frequency generated by a diode is determined by the composition of the diode. Diodes can be internally tuned by varying the temperature or the injection current density of the diode. Varying the temperature changes the frequency more than varying the injection current, but the temperature can only be changed slowly. The injection current density can be changed at a frequency of up to 10 GHz. Externally, the frequency can be further refined using special mirrors and/or filters. Typical tunable lasers contain InGaAsP-InP diodes, which are tunable over 900 nm to 1.6 μm, and InGaAsP-InAsP diodes, which are tunable over 1.6 μm to 2.2 μm.

Ultraviolet (UV) Analyzers. An *ultraviolet (UV) analyzer* is a radiant-energy absorption analyzer consisting of a UV electromagnetic radiation source, a sample cell, and a detector that measures the absorption of UV radiation by specific molecules. **See Figure 23-6.** The various photodetectors that can be used to measure the UV radiation exiting the absorption chambers are silicon diodes, vacuum phototubes, photomultiplier tubes, barrier layer (photovoltaic) cells, and linear photodiode array detectors.

Groups of gases that effectively absorb UV radiation include the halogens, such as fluorine (F), chlorine (Cl), bromine (Br), and iodine (I); oxidizing agents, such as sodium hypochlorite (NaOCl), chlorine dioxide (ClO_2), hydrogen peroxide (H_2O_2), ozone (O_3), and potassium permanganate ($KMnO_4$); and the sulfur compounds hydrogen sulfide (H_2S), carbonyl sulfide (COS), carbon disulfide (CS_2), and sulfur dioxide (SO_2). In addition, mercury in the air can be measured in concentrations as low as a few parts per billion by ultraviolet analyzers.

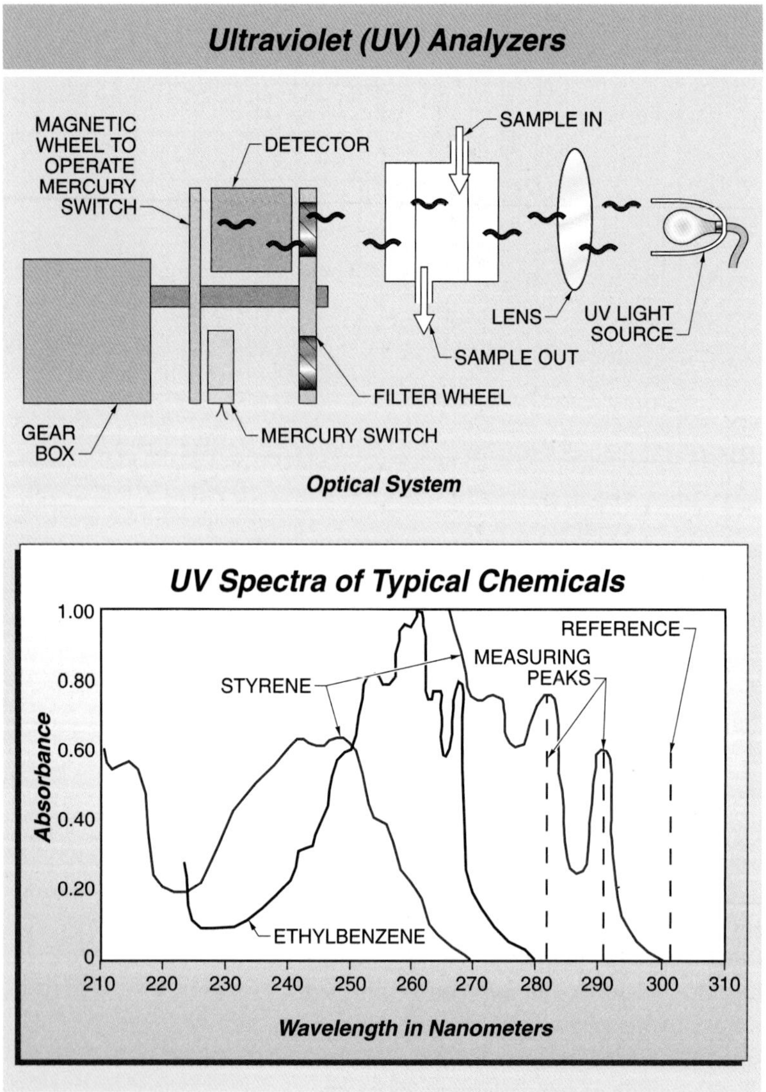

Figure 23-6. UV analyzers measure a characteristic wavelength of the UV spectrum to determine the amount of a gas present in a sample.

Oxygen Analyzers

The measurement of oxygen concentration is very important since oxygen is involved in many chemical processes. Oxygen is necessary to support nearly all combustion reactions. Since most oxygen is obtained from air, which only contains 20% oxygen, it is economical to use as much of the available oxygen as possible. Combustion processes require a precise ratio of fuel to oxygen. Measurement of the oxygen in the flue gas ensures the efficient use of the fuel and air.

In other applications, the presence of certain amounts of oxygen in a flammable gas can result in a hazardous or explosive mixture. The monitoring of the oxygen content ensures that a hazardous or explosive mixture is not present. Common types of oxygen analyzers are paramagnetic, thermoparamagnetic, zirconium oxide, and electrochemical oxygen analyzers.

Paramagnetic Oxygen Analyzers. A *paramagnetic oxygen analyzer* is an analyzer consisting of two diamagnetic spheres filled with nitrogen that are connected with a bar to form a "dumbbell" shaped assembly. This assembly is suspended horizontally by a vertical torsion system. The dumbbell assembly is mounted between a pair of magnetic poles in a stainless steel chamber with a glass window. This chamber is the sample cell. **See Figure 23-7.**

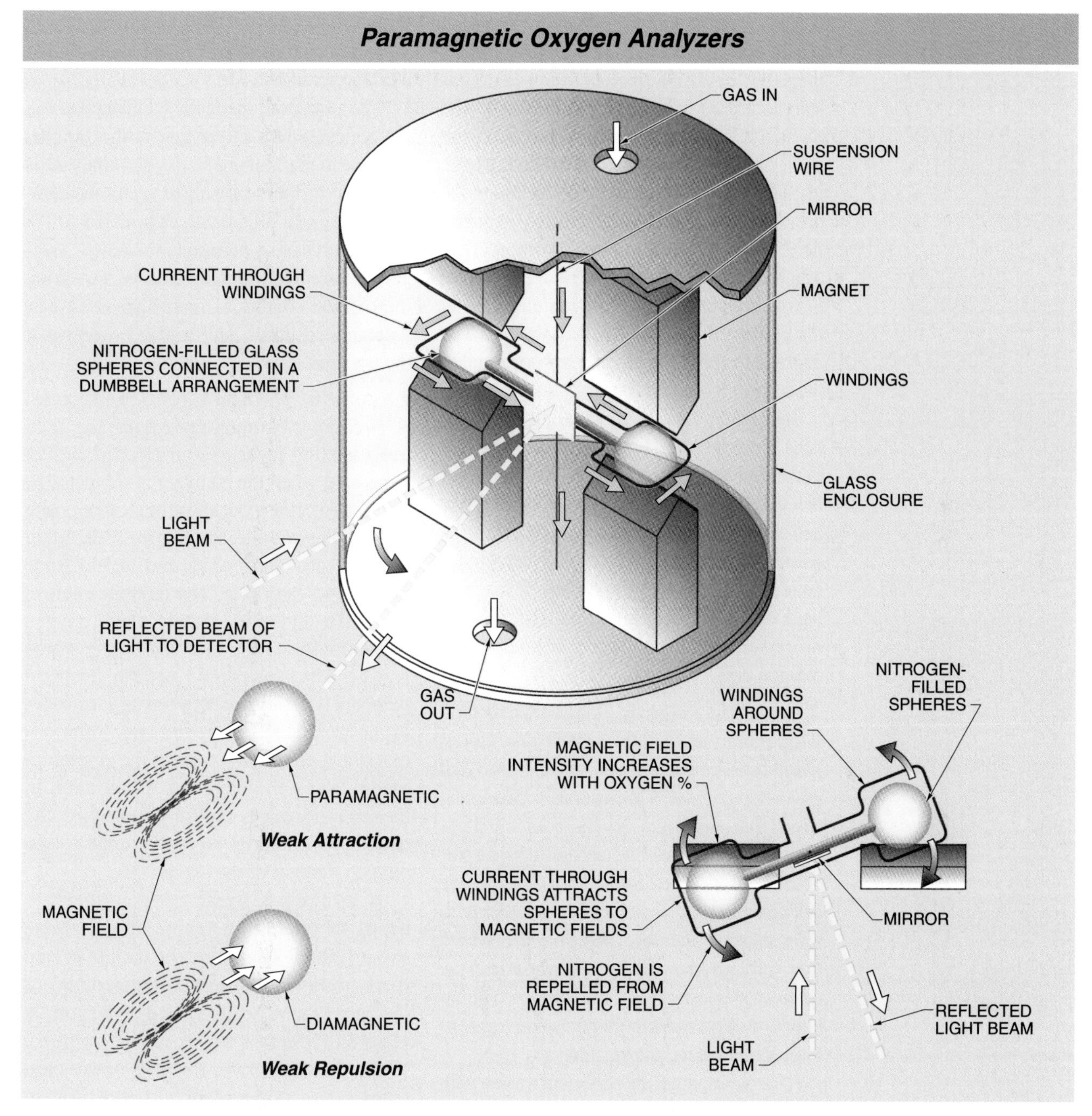

Figure 23-7. Paramagnetic oxygen analyzers measure the displacement of nitrogen-filled spheres in a magnetic field.

The oxygen molecule has a unique property in that it is strongly paramagnetic. Paramagnetic materials are attracted by a magnetic field and strengthen the field as they enter the field. Since this is a property of only a few molecules, magnetic fields can be used as a very specific analysis method. Nitrogen is diamagnetic. Diamagnetic materials are repelled by a magnetic field.

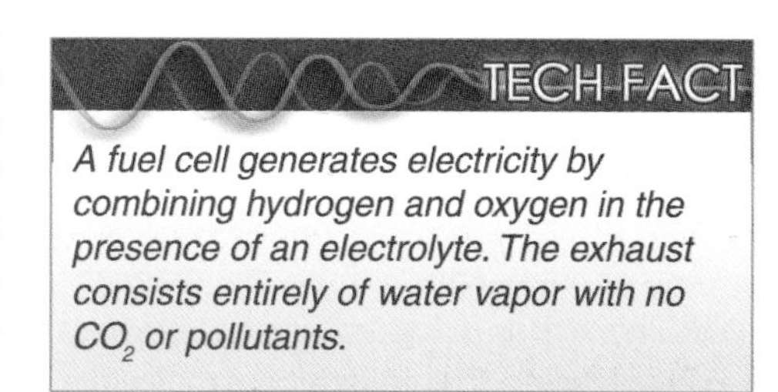

A fuel cell generates electricity by combining hydrogen and oxygen in the presence of an electrolyte. The exhaust consists entirely of water vapor with no CO_2 or pollutants.

The spheres that comprise the dumbbell are filled with nitrogen and placed within the magnetic field. The oxygen-containing gas sample passes through the sample cell. The oxygen is drawn into the magnetic field and strengthens the magnetic field. The nitrogen inside the spheres has the opposite magnetic property and is forced out of the magnetic field, causing the dumbbell to rotate. The dumbbell's position is sensed by a light beam reflected from a mirror mounted at the center of the dumbbell assembly. The degree of rotation is directly proportional to the oxygen concentration.

In another arrangement, a pair of photocells detect the reflected light. A current is generated at the photocells when the dumbbell begins to rotate. The current is amplified and conducted around the dumbbell through windings, which generates an electromagnetic field that causes the dumbbell to rotate back to its original position. The amount of current is directly proportional to the oxygen concentration.

Thermoparamagnetic Oxygen Analyzers. A *thermoparamagnetic oxygen analyzer* is an analyzer consisting of a sensing head that uses magnetic fields to generate a "magnetic wind" that carries the oxygen-containing gas sample through a sample cell and across a pair of thermistors. **See Figure 23-8.** The sample cell consists of a two-chamber cylinder with a measuring chamber positioned above a flow-through chamber. The temperature of the cell is held constant. The measuring chamber houses two permanent magnets and a pair of electrically heated thermistors.

The paramagnetic coefficient varies with temperature. This means that as the oxygen molecules are warmed by the first thermistor, the paramagnetic force on those molecules is reduced. When the paramagnetic force is reduced, there is less force attracting the molecules into the magnetic field. Therefore, colder oxygen molecules, which are more strongly attracted into the magnetic field, displace the heated oxygen molecules, which are less strongly attracted into the magnetic field. This sets up a flow of oxygen known as a magnetic wind. The cold oxygen molecules remove heat from the thermistor. The change in resistance of the thermistor is measured in a bridge circuit.

The amount of magnetic wind depends on the concentration of oxygen in the sample. The thermoparamagnetic oxygen analyzer is very specific for measuring oxygen concentrations and can measure ranges from 0% to 1% or 0% to 100%. A thermoparamagnetic oxygen analyzer is commonly used for measuring the oxygen content of inert blanketing gas, reactor feed gases, and sewage wastewater digester gas.

Zirconium Oxide Oxygen Analyzers. A *zirconium oxide oxygen analyzer* is an analyzer that measures an electric current generated when the analyzer is subjected to different oxygen concentrations on opposite sides of an electrode. **See Figure 23-9.** A thin layer of zirconium oxide has an electrode fastened to each side. The reference side is exposed to air at normal atmospheric oxygen concentration while the measurement side is exposed to the sample gas. An electrical potential is developed between the electrodes in the presence of different concentrations of oxygen on the two sides of the zirconium oxide. When the oxygen concentration in the reference gas is carefully controlled, the developed potential is proportional to the oxygen content of the sample gas.

A zirconium oxide oxygen analyzer needs to operate at temperatures above about 1000°F. Therefore, the sample stream needs to be at an elevated temperature or the instrument needs a sample heater. A zirconium oxide analyzer is fairly rugged, simple, and accurate. Because of the high temperature requirement, it is often used to measure the excess oxygen in burner flue gas to measure burner efficiency. The zirconium oxide oxygen analyzer can be selected to measure low concentrations of 0% to 2% or high concentrations of 0% to 25%.

TECH-FACT

Michael Faraday, an English chemist, discovered paramagnetism by studying how materials respond to magnetic fields.

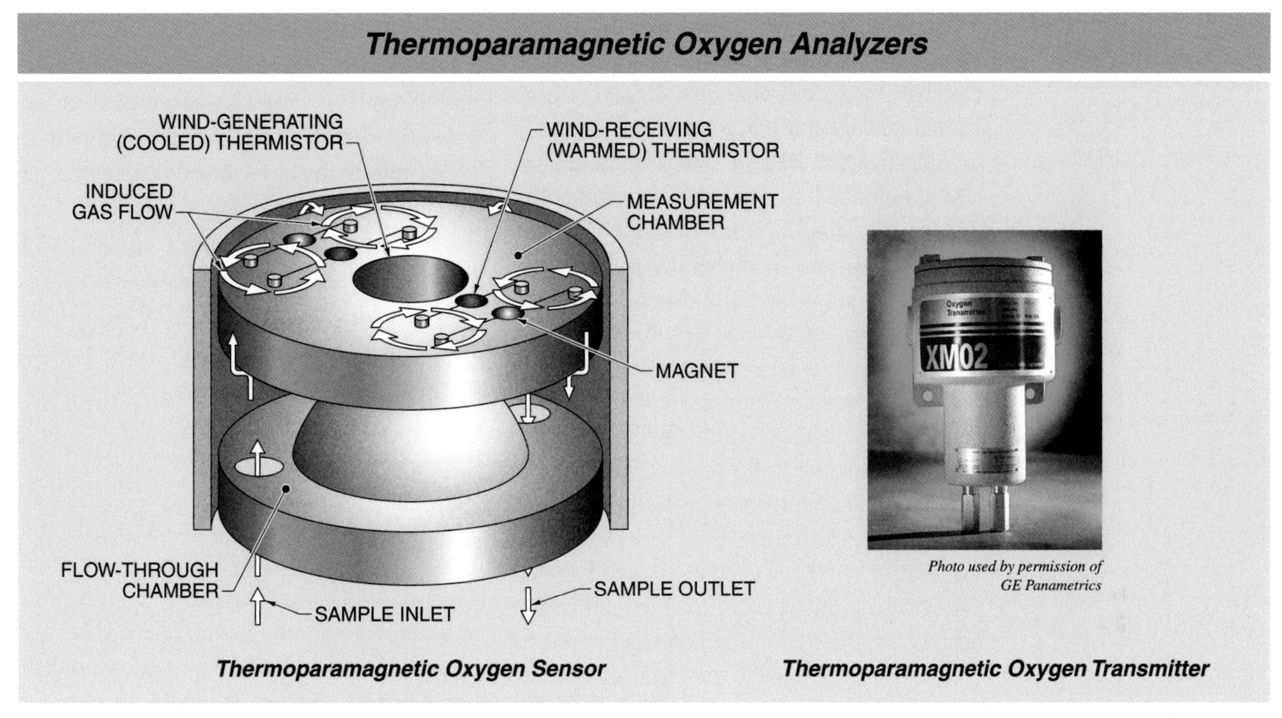

Figure 23-8. Thermoparamagnetic oxygen analyzers measure the flow rate of oxygen caused by a magnetic field to determine the oxygen concentration in a sample.

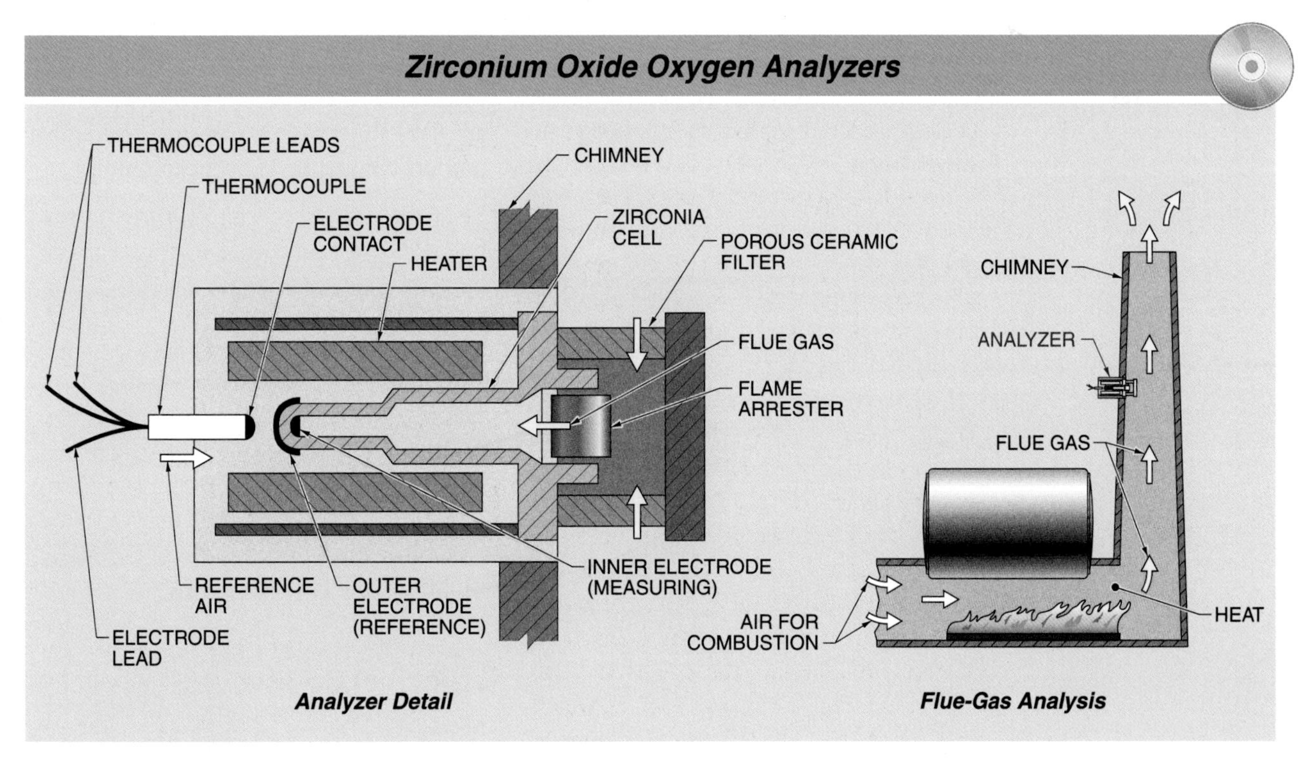

Figure 23-9. Zirconium oxide oxygen analyzers measure the oxygen content of a high-temperature gas stream. In a boiler application, the analyzer output is used to open or close dampers to control the combustion.

Electrochemical Oxygen Analyzers. An *electrochemical oxygen analyzer* is an analyzer that measures an electric current generated by the reaction of oxygen with an electrolytic reagent. **See Figure 23-10.** An oxygen cell is exposed to a gas sample where the sample passes through a permeable membrane to the cell cathode. The cathode is perforated and coated with gold, silver, platinum, or rhodium, depending on the application. Under the cathode are the electrolyte material and the lead anode.

Oxygen in the sample reacts with the electrolyte, generating an electrical current. The amount of current is proportional to the oxygen concentration. This process is very similar to a fuel cell used to generate power. A wide range of oxygen concentrations can be measured, but high concentrations of oxygen rapidly deplete the electrolyte reservoir. Other oxidizing gases that can react with the electrolyte could give false readings and poison the cell.

The gas sample can be obtained from a pressurized sampling system, from direct exposure to the atmosphere, or from the atmosphere with an internal sample pump. For example, an electrochemical oxygen analyzer can be used to measure the oxygen content in enclosed spaces such as an operator control room.

Hydrogen-Specific Gas Analyzers

Hydrogen (H_2) has always been a difficult gas to analyze. The classic method in the past has been the use of thermal conductivity because of the large difference in thermal conductivity between hydrogen and other gases. Unfortunately, thermal conductivity can only be used with binary gas samples.

A new technology has been developed that can measure hydrogen even with multiple inert background gases. The technology uses the property of hydrogen absorption by a palladium-nickel (Pd-Ni) alloy with a special proprietary coating. The palladium catalyzes the H_2 molecule into 2 H+ (atomic) hydrogen atoms so it can be absorbed. There is an equilibrium maintained between the H_2 and 2 H+.

The absorbed hydrogen is measured in two separate portions of the Pd-Ni alloy. In one portion, the absorbed hydrogen changes the bulk resistivity of the Pd-Ni alloy. This is used to measure hydrogen concentrations of 0.5% to multiple atmospheres. The other portion uses a capacitor circuit to measure the charge density caused by the hydrogen absorption, which covers a range from 15 ppm to 0.5%. These two sensing principles are used to provide a measurement signal proportional to the H2 concentration.

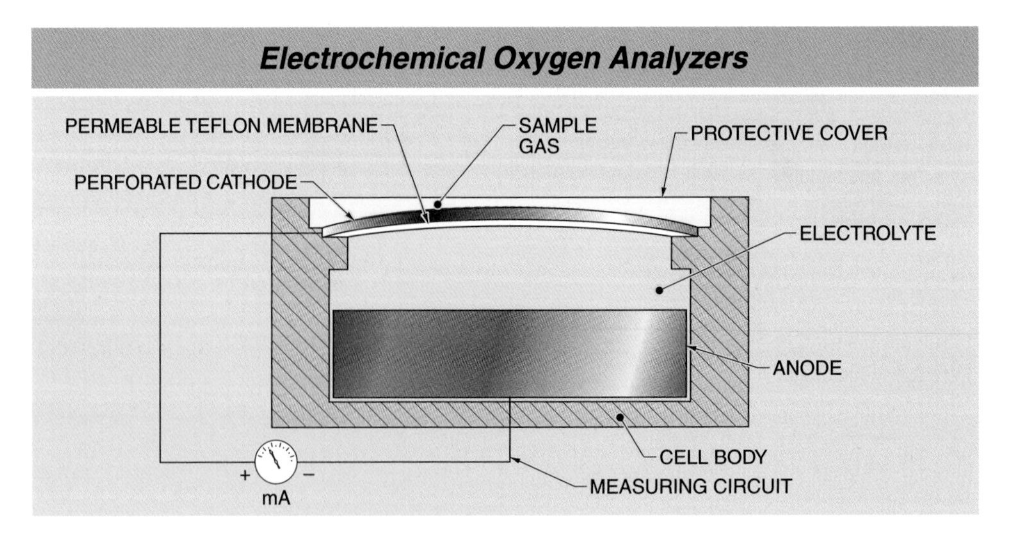

Figure 23-10. Electrochemical oxygen analyzers measure the oxygen content of a sample from the output of an electrochemical reaction between oxygen and a chemical solution.

Opacity Analyzers

An *opacity analyzer* is a gas analyzer consisting of a collimated (focused beam) light source and an analyzer to measure the received light intensity. The received light is measured by a silicon photodiode and converted to an output signal. An increase in particulate matter absorbs and scatters the transmitted light, reducing the light received at the detector. Thus, the decrease in light intensity is proportional to the amount of particulate matter.

An opacity analyzer can be either a single- or dual-pass design. A single-pass design has the light source and receiver on opposite sides of the stack. A dual-pass design has the light source and receiver in the same housing on one side of the stack. The light beam from the source is reflected by a mirror mounted on the opposite side of the stack. Both designs require that the connections into the stack be purged with clean air to keep the optics clean. The light source and receiver need to be periodically calibrated.

A common application for an opacity analyzer is measuring the amount of particulate matter in a flue stack. Coal-fired combustion processes are the most likely to have particulate matter in the flue stack. Burning coal produces fly ash which is usually removed from the flue gases with an electrostatic precipitator. The opacity analyzer monitors for any particulate breakthroughs. Natural gas and oil-fired combustion processes very rarely produce particulate matter in their flue gases. **See Figure 23-11.**

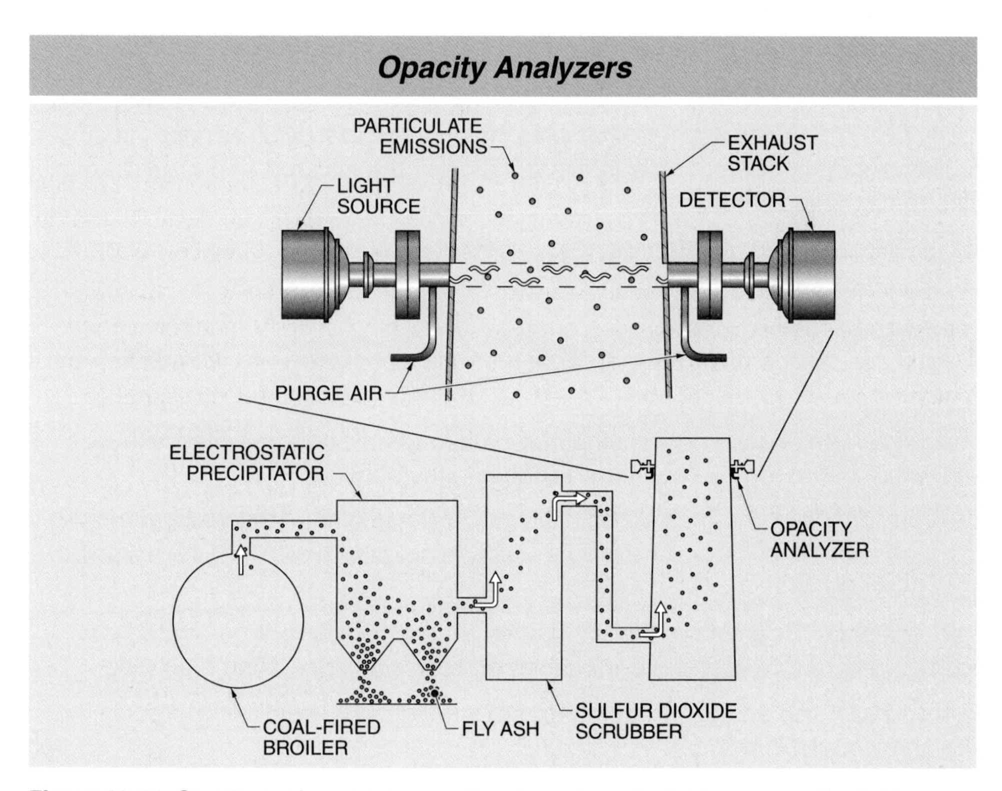

Figure 23-11. Opacity analyzers measure the attenuation of a light source as the light passes through a sample.

KEY TERMS

- *analysis:* The process of measuring the physical, chemical, or electrical properties of chemical compounds so that the composition and quantities of the components can be determined.
- *analyzer:* An instrument used to provide an analysis of a sample from a process.
- *on-line analyzer:* An instrument that is located in the process area and obtains frequent or continuous samples from the process.
- *analyzer sampling system:* A system of piping, valves, and other equipment that is used to extract a sample from a process stream, condition it if necessary, and convey it to an analyzer.
- *representative sample:* A sample from a process in which the composition of the sample is the same as in the process piping.
- *sample transportation lag:* The time that it takes for a sample to get from the process through the final analysis.
- *analyzer calibration:* The process of substituting known sample compositions for the normal process sample so that the analyzer can be adjusted to read the correct values.
- *gas analyzer:* An instrument that measures the concentration of an individual gas in a gaseous sample.
- *thermal conductivity analyzer:* A gas analyzer that measures the concentration of a single gas in a sample by comparing its ability to conduct heat to that of a reference gas.
- *radiant-energy absorption analyzer:* A gas analyzer that uses the principle that different gases absorb different, very specific, wavelengths of electromagnetic radiation in the infrared (IR) or ultraviolet (UV) regions of the electromagnetic spectrum.
- *nondispersive infrared (NDIR) analyzer:* A radiant-energy absorption analyzer consisting of an IR electromagnetic radiation source, an IR detector, and two IR absorption chambers.
- *ultraviolet (UV) analyzer:* A radiant-energy absorption analyzer consisting of a UV electromagnetic radiation source, a sample cell, and a detector that measures the absorption of UV radiation by specific molecules.
- *paramagnetic oxygen analyzer:* An analyzer consisting of two diamagnetic spheres filled with nitrogen that are connected with a bar to form a "dumbbell" shaped assembly.
- *thermoparamagnetic oxygen analyzer:* An analyzer consisting of a sensing head that uses magnetic fields to generate a "magnetic wind" that carries the oxygen-containing gas sample through a sample cell and across a pair of thermistors.
- *zirconium oxide oxygen analyzer:* An analyzer that measures an electric current generated when the analyzer is subjected to different oxygen concentrations on opposite sides of an electrode.
- *electrochemical oxygen analyzer:* An analyzer that measures an electric current generated by the reaction of oxygen with an electrolytic reagent.
- *opacity analyzer:* A gas analyzer consisting of a collimated (focused beam) light source and an analyzer to measure the received light intensity.

REVIEW QUESTIONS

1. Define representative sample and explain why a sample from a process needs to be representative of the material in the process.
2. What is a thermal conductivity analyzer, and how does it operate?
3. List and define the different types of radiant-energy absorption analyzers.
4. List and define the different types of oxygen analyzers.
5. What is an opacity analyzer, and how is it used?

HUMIDITY ANALYZERS

A *humidity analyzer* is an instrument that measures the amount of humidity in air. *Humidity* is the amount of water vapor in a given volume of air or other gases. Since most humidity measurements are made in air, only the measurement of water vapor in air will be used hereafter when discussing humidity measurement. The most common terms used in humidity measurement are humidity ratio, relative humidity, specific humidity, dry bulb temperature, wet bulb temperature, and dewpoint.

Absolute humidity, or *humidity ratio,* is the ratio of the mass of water vapor to the mass of dry air. It is typically expressed in pounds of water per pound of dry air, grains of water per cubic foot of dry air (there are 7000 grains to a pound), or grams of water per cubic centimeter of dry air.

Relative humidity (rh) is the ratio of the actual amount of water vapor in the air to the maximum amount of water vapor possible at the same temperature. Relative humidity is equivalent to the ratio of the actual water vapor pressure in the air to the maximum water vapor pressure at the same temperature. A saturated steam table referenced to temperature can provide the saturated water vapor pressure at different temperatures. *Saturated air* is a mixture of water and air where the relative humidity is 100%.

Specific humidity is the ratio of the mass of water vapor to the mass of dry air plus moisture. It is typically expressed as grains of water vapor per pound of the mixture of dry air and water vapor.

Dry bulb temperature is the ambient air temperature measured by a thermometer that is freely exposed to the air but shielded from other heating or cooling effects. *Wet bulb temperature* is the lowest temperature that can be obtained through the cooling effect of water evaporating into the atmosphere.

When air blows past a wetted thermometer bulb, water evaporates and cools the bulb. The amount of water evaporation and cooling depends on the relative humidity. When the relative humidity is high, very little water evaporates, the cooling effect is small, and the wet bulb temperature is only a little bit lower than the dry bulb temperature. When the relative humidity is low, more water evaporates, the cooling effect is larger, and the wet bulb temperature is quite a bit lower than the dry bulb temperature. The wet bulb temperature is an indirect measure of the amount of water vapor present in the air.

TECH FACT

Buildings often require the use of humidifiers and dehumidifiers to maintain humidity within the comfort zone.

Cooling towers use evaporation to cool water close to the wet bulb temperature.

Dewpoint is the temperature to which air must be cooled for the air to be saturated with water and any further cooling results in water condensing out. The dewpoint is another indirect measure of the amount of water vapor present in the air. The wet bulb temperature and the dewpoint are the same value for saturated air.

Psychrometric Charts

A *psychrometric chart* is a graph that graphically combines the properties of moist air at standard atmospheric pressure (14.7 psi). The graph combines dry bulb temperature, wet bulb temperature, dewpoint temperature, relative humidity, and absolute humidity so that all the properties of moist air can be obtained when any two of them are known. **See Figure 24-1.**

Humidity readings are commonly used in heating, ventilating, and air conditioning (HVAC) applications. In HVAC systems, the comfort area for conditioned air is between 73°F to 75°F and 40% to 50% relative humidity.

On a typical psychrometric chart, the dry bulb temperature lines are vertical and the dry bulb temperatures are indicated at the bottom of the chart. The saturation line is the curved line at the upper left of a psychrometric chart that shows the conditions of air that contains the maximum amount of water vapor at that temperature. The wet bulb temperature lines run at an angle downward toward the right from the saturation line. The dewpoint temperature lines run horizontally across to the right from the saturation line. The dewpoint and wet bulb temperature values are read at the point where these lines intersect with the saturation line.

The lines indicating the percentage of relative humidity curve upward from left to right and the percentage values are shown directly on the lines themselves. The absolute humidity is read on the vertical scale on the right of the chart, using the horizontal line that leads from the intersection of a relative humidity line with a dry bulb line. Many psychrometric charts include other information such as specific volume and enthalpy data.

Several examples show how this somewhat complex chart is used. Example A: Given a dry bulb temperature of 75°F and a wet bulb temperature of 65°F, determine the relative humidity, absolute humidity, and dewpoint.

Locate the 75°F dry bulb temperature reading at the bottom of the chart. Then locate the 65°F wet bulb reading along the saturation line. Follow the angled line from the saturation line down to the right until it intersects with the vertical line up from the dry bulb temperature. The relative humidity can be read at the point where these two lines intersect. The nearest curved line for the relative humidity is 60%. The relative humidity is just below that line at approximately 59%. A horizontal line drawn from the point of the intersection to the right-hand scale of the chart shows the absolute humidity. In this case, it is 0.011 lb of water per lb of dry air. A horizontal line drawn from the point of the intersection to the left to the saturation line shows the dewpoint of approximately 59°F.

Example B: Given a dry bulb temperature of 105°F and a dewpoint temperature of 75°F, determine the relative humidity, absolute humidity, and wet bulb temperature.

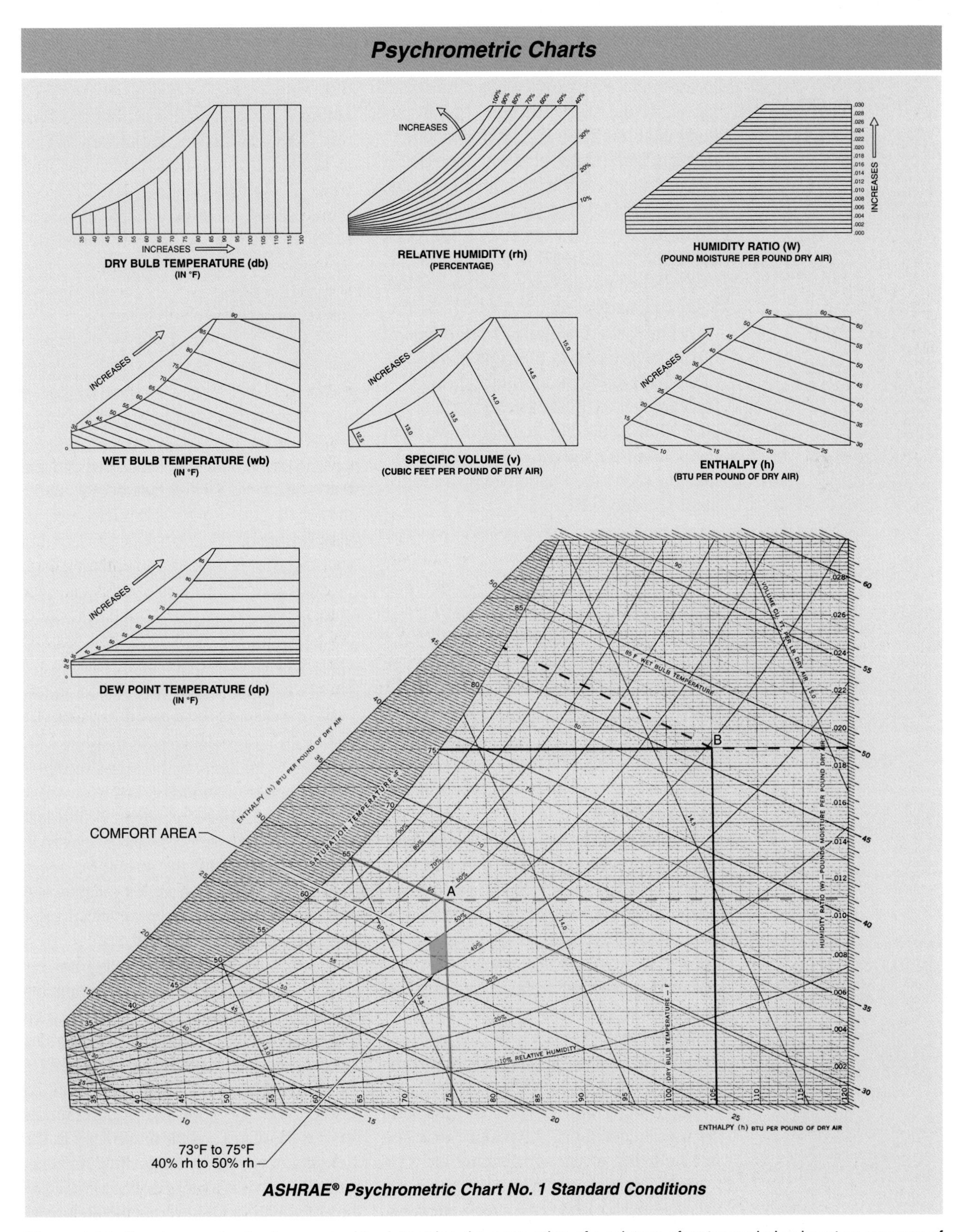

Figure 24-1. Psychrometric charts are used to determine the properties of a mixture of water and air when two or more of the properties are known.

Locate the 105°F dry bulb temperature reading at the bottom of the chart. Then locate the 75°F dewpoint temperature reading along the saturation line. Follow the horizontal line from the saturation line to the right until it intersects with the vertical line up from the dry bulb temperature. The relative humidity can be read at the point where these two lines intersect. The nearest curved line for the relative humidity is 40%. The relative humidity is just below that line at approximately 39%.

A horizontal line drawn from the point of the intersection to the right-hand scale of the chart shows the absolute humidity. In this case, it is 0.019 lb of water per lb of dry air. Following the diagonal line drawn from the point of the intersection to the saturation line shows the wet bulb temperature of approximately 82°F.

Psychrometers

A *psychrometer* is a humidity analyzer that uses two thermometers with the bulb of one thermometer kept moist (wet bulb) and the other bulb kept dry (dry bulb). Several types of psychrometers are available. Two of the most common are the recording psychrometer and the sling psychrometer. A *recording psychrometer* is a psychrometer consisting of wet and dry bulb thermometers connected to a recorder.

A *sling psychrometer* is a psychrometer consisting of two glass thermometers attached to an assembly that permits the two thermometers to be rotated through the air. One thermometer has a moistened cloth cover (or wick) over the sensing bulb to maintain a moist condition. The thermometers are rotated rapidly through the air.

The wet bulb shows a lower temperature than the dry bulb because of the evaporation of water from the wick. These temperature readings, along with a psychrometric chart, are used to determine the relative or absolute humidity of the surrounding air. The hygrometer and thermohygrometer family of instruments is replacing the sling psychrometer because they provide direct readouts in relative humidity values. **See Figure 24-2.**

Psychrometers

Fluke Corporation

Figure 24-2. The hygrometer and thermohygrometer family of instruments is replacing the sling psychrometer because they provide direct readouts in relative humidity values.

Hygrometers

A *hygrometer* is a humidity analyzer that measures the physical or electrical changes that occur in various materials as they absorb or release moisture. Some hygrometers employ human hair, animal membrane, or other materials that lengthen or stretch when they absorb water. A *hygroscopic material* is a material that readily absorbs and retains moisture with increases in humidity and releases moisture with decreases in humidity. The hygroscopic material undergoes changes in physical properties as the amount of moisture changes. In some applications, the material changes physical dimensions and a strain gauge is used to measure the change in tension of a fiber of the material. **See Figure 24-3.**

Another type of electrical hygrometer uses a capacitance probe as its primary element. In a common type of capacitance probe, an anodized aluminum strip provides a porous aluminum oxide layer as a dielectric. A thin film of gold is deposited on the oxide layer. The gold and the aluminum serve as the plates and the oxide layer as the dielectric of an electrical capacitor. An alternative design uses a polymer layer instead of the aluminum oxide layer as the dielectric. The output of the cell is a voltage that varies with the relative humidity.

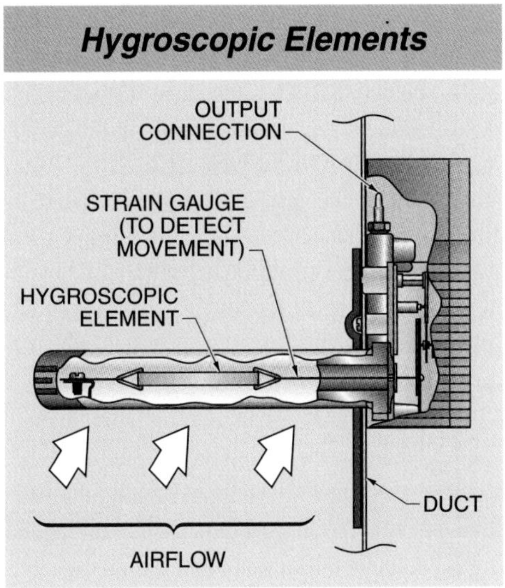

Hygroscopic Elements

OUTPUT CONNECTION
STRAIN GAUGE (TO DETECT MOVEMENT)
HYGROSCOPIC ELEMENT
DUCT
AIRFLOW

Figure 24-3. Hygroscopic elements change size with changes in humidity. The change in size is detected with a strain gauge.

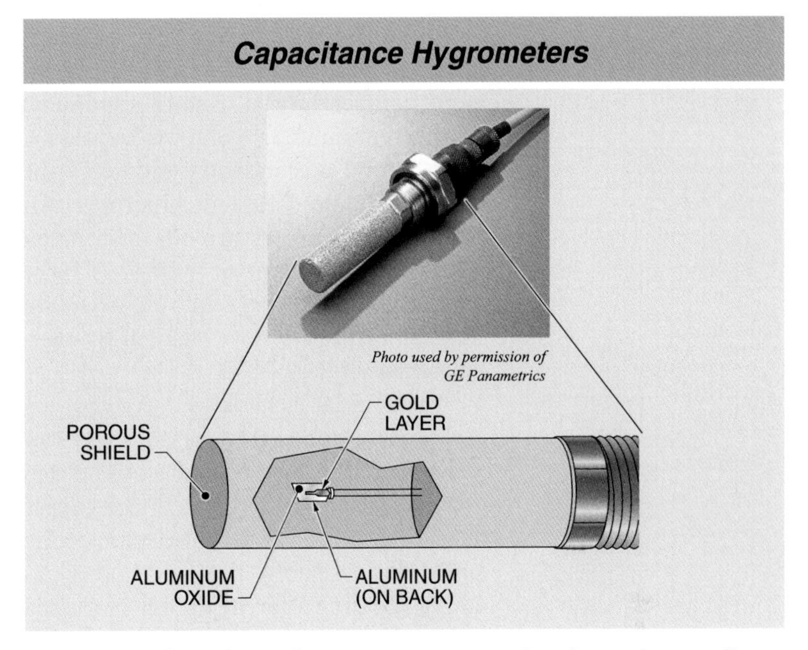

Figure 24-4. Capacitance hygrometers measure the change in capacitance as the sensor absorbs moisture.

When the probe is installed within the environment of a process containing water vapor, the dielectric constant of the capacitor changes with changes in humidity. This alters the electrical capacitance of the probe. The instrument can be calibrated for digital readout in relative humidity units. This type of hygrometer is commonly used because of its small size, rapid response, and very wide range of humidity and dewpoints that can be measured. **See Figure 24-4.**

Thermohygrometers. A *thermohygrometer* is a combination of a hygrometer, pressure sensor, and temperature-sensing instrument with digital processing to calculate relative humidity, absolute humidity, dry bulb temperature, dewpoint reading, and other properties for local display or transmission. The three actual measurements of temperature, pressure, and relative humidity allow the digital calculation of the other values. **See Figure 24-5.**

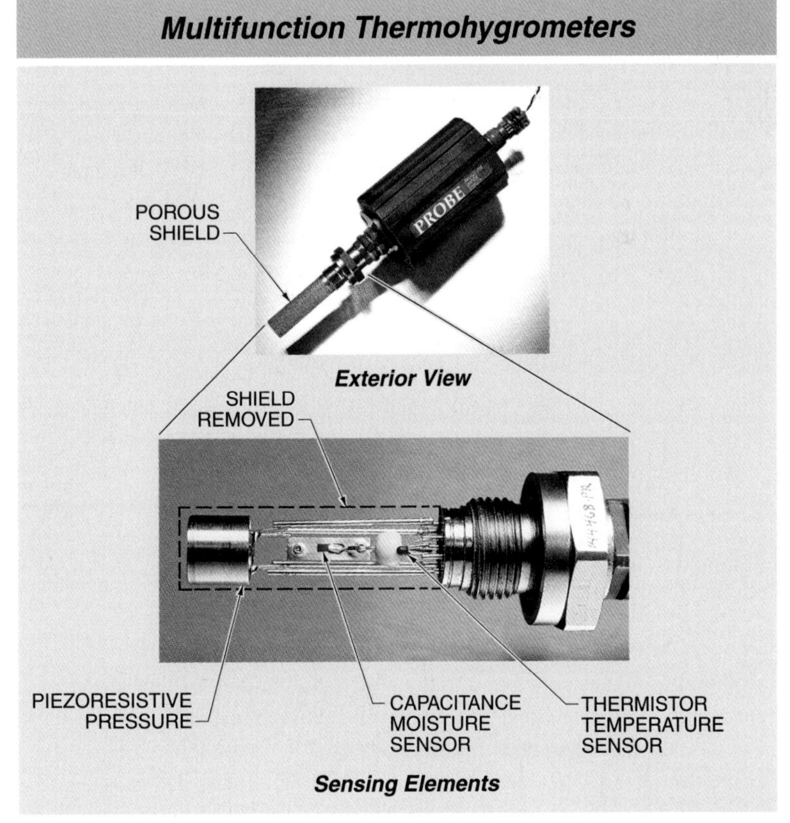

Photos used by permission of GE Panametrics

Figure 24-5. Thermohygrometers combine a thermistor with a hygrometer to provide temperature and humidity levels to a controller. This combination enables automatic calculation of the thermodynamic properties of moist air.

TECH FACT

Molecules only absorb radiation of certain wavelengths. The particular wavelengths depend on the chemical bonds.

Dewpoint Analyzers

In many industrial processes, the dewpoint temperature is a more significant measurement than relative humidity. The natural gas industry monitors the presence of moisture in pipelines. Air dryer manufacturers provide instrument air with a dewpoint of −40°F. A large majority of processes requiring dewpoint measurements are now handled by thermohygrometers.

The original technology for measuring dewpoint is still used to measure very low dewpoints. This method uses a gold-plated mirror surface that is bonded to a copper thermistor holder. A thermoelectric cooler chills this assembly. The air or other gas being measured for dewpoint is passed across the mirror. A light beam is directed at the mirror, which reflects the beam toward a photoelectric detector. **See Figure 24-6.**

As dew or frost forms on the mirror and clouds it, there is a change in the amount of light reflected to the detector. This change results in a proportional change in the power supplied to the thermoelectric chiller, thus diminishing the amount of dew or frost on the mirror. This feedback continues until equilibrium is reached. A temperature sensor in the cooling assembly measures the temperature of the mirror at equilibrium to determine the dewpoint.

> **TECH FACT**
>
> *Clouds form when air cools below the dewpoint. This can be seen in industrial processes where humid air is cooled. The condensing water vapor can produce fog inside a vessel that can obscure the view of the inside.*

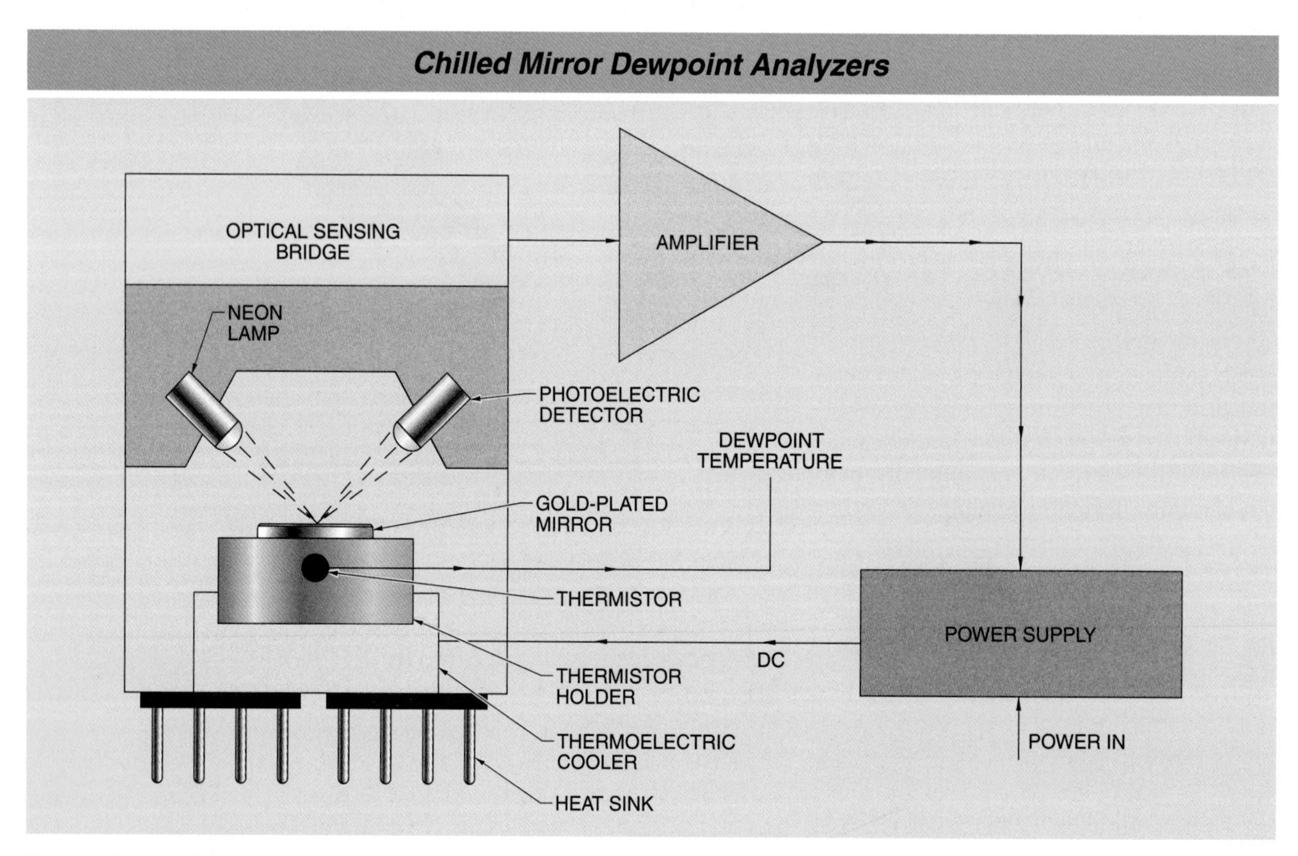

Figure 24-6. A chilled mirror is used to measure very low dewpoints. At the dewpoint, dew or frost condenses out on the glass and changes the reflectance.

Another method of determining dewpoint is based upon the behavior of a hygroscopic chemical, lithium chloride, in the presence of moisture. The electrical conductivity increases as the chemical absorbs water and forms a lithium chloride/water ionic solution. A *dew cell* is a fiberglass cylindrical core impregnated with lithium chloride and wrapped with a winding of gold wire around the core that serves as a heating element. Current must flow through the lithium chloride in order to complete the circuit. As the core absorbs moisture from the air, the conductance through the lithium chloride/water solution increases, allowing more current flow through the heating element, which increases the temperature of the dew cell.

Vapor pressure of water increases at higher temperatures. As the core is heated, it reaches the condition where the vapor pressure of the water in the dew cell increases above the vapor pressure of water in the air. At this point, water starts to evaporate from the dew cell, lowering the conductivity. The lowered conductivity reduces the heating effect until the dew cell cools off again and it starts absorbing moisture again. At constant humidity, the circuit will quickly reach equilibrium. The dewpoint is the temperature at which moisture is no longer absorbed. An integral thermistor or RTD measures the equilibrium temperature. **See Figure 24-7.**

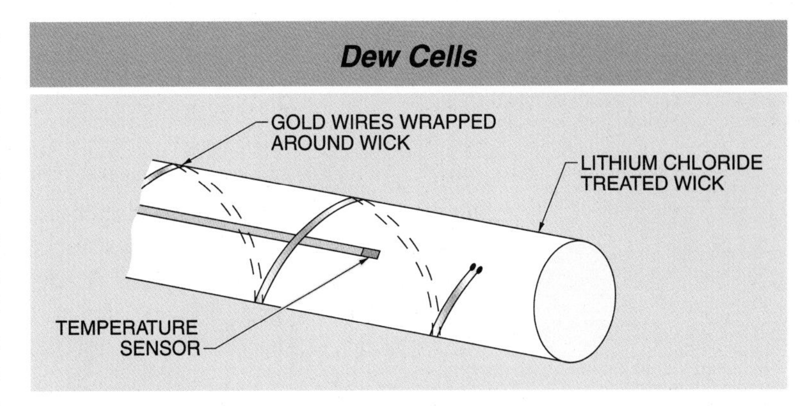

Figure 24-7. Dew cells measure the change in conductance of a wire wrapped around a lithium chloride-containing wick as moisture is absorbed by the chemical.

SOLIDS MOISTURE ANALYZERS

A *solids moisture analyzer* is an instrument used to measure the amount of moisture in a solid. The continuous measurement of moisture content in solids is crucial to the mass production of granular materials such as chemicals, plastics, grains, powders, and minerals. This is also true for continuous feeds of materials, such as paper and textiles, which are processed as continuous webs. Granular materials are carried through the process in pipes, on conveyor belts, or through chutes and hoppers.

Determining the moisture content of materials of these types requires a non-contact detection method. The two methods most suited to continuous high-speed processing are based on the use of near infrared radiation (spectroscopic) and microwaves. Lesser-known or rarely used methods are gravimetric analysis, electrical impedance analyzers, and nuclear analyzers. **See Figure 24-8.**

Near Infrared (NIR) Moisture Analyzers

A *near infrared (NIR) moisture analyzer* is a solids moisture analyzer that measures the reflectance of the process material and calculates moisture content. Near IR wavelengths of 1430 nm and 1940 nm are absorbed very readily by water and are not absorbed by other common materials. *Reflectance* is the ability of a material to reflect light or radiant energy. Reflectance depends on the chemical composition, moisture content, and surface texture.

An analyzer must compensate for factors other than moisture. Two wavelengths of infrared radiation are selected, a measuring wavelength and a reference wavelength. The amount of energy reflected by the process material is compared at each wavelength. A filter wheel, driven by a synchronous motor, carries two optical filters. One of the filters transmits light at the measuring wavelength, and the other at the reference wavelength.

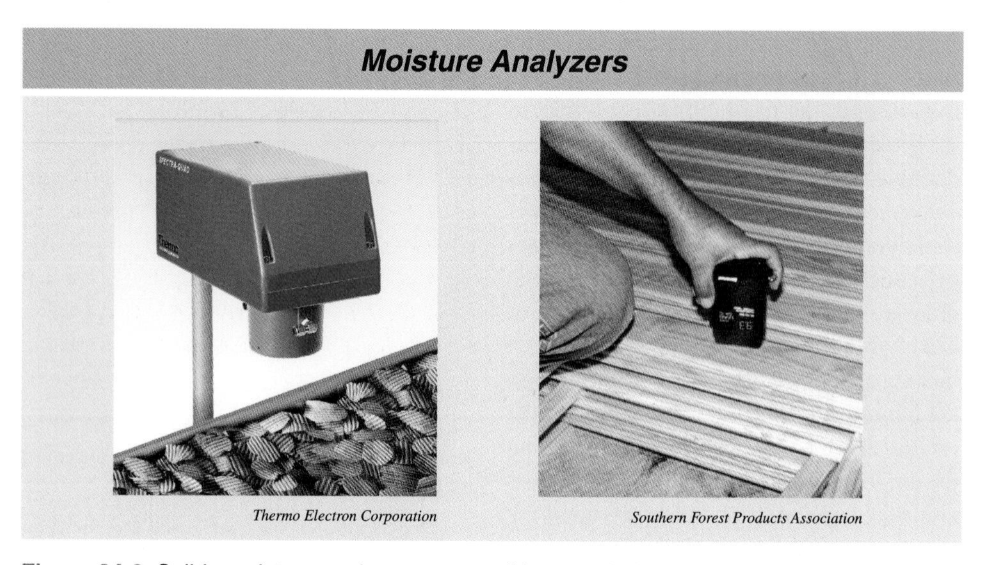

Figure 24-8. Solids moisture analyzers are used in many industries.

The changes due to factors other than moisture are cancelled by using the ratio of the two reflected signals. Both the reference and measuring wavelengths are subject to extraneous factors, but the moisture additionally absorbs only the measuring wavelength. The analyzer responds only to the changes in moisture content because the wavelengths selected are unique to the presence or absence of water. **See Figure 24-9.**

Figure 24-9. Near infrared analyzers use two IR beams to compensate for variations in chemical composition and surface texture.

The IR beam is projected onto the surface of the process material indirectly, using a system of lenses and mirrors. The function of the lenses is to provide a uniform distribution of the infrared waves on the process material to be measured. A concave mirror to the NIR detector cell directs the energy reflected from the process material. The cell converts the reflected radiant energy into an electrical signal.

A signal is produced in the detector cell by alternating the measuring and reference wavelengths. The amplitude of one indicates the reflectance at the measuring wavelength, and the amplitude of the other indicates the reflectance at the reference wavelength. These pulses are separated and converted into two DC signals, each corresponding to the amplitude of one of the wavelengths. The ratio of the two signals is proportional to the moisture content. Moisture in solids can be measured as low as 0.01% (100 ppm).

Microwave Moisture Analyzers

A *microwave* is the band of electromagnetic radiation between infrared and VHF broadcast frequencies, covering the range of approximately 3 mm to 3 m wavelengths. Water absorbs microwaves at a wavelength of approximately 122 mm. A *microwave moisture analyzer* is a solids moisture analyzer consisting of a transmitter that directs a microwave beam onto the material whose moisture is to be measured and a receiver.

Molecules of free water are made to rotate or spin when subjected to high-frequency microwave radiation. The rotation or spin of the water molecules absorbs energy from the microwave source beam and causes a phase shift in the microwave source beam. This is the same method that is used in a microwave oven to cook food. Microwave moisture analyzers use these principles to measure the water content in solids.

Some microwave moisture analyzers measure only the attenuation of the microwave beam. Other microwave moisture analyzers measure both the attenuation and the phase shift to determine the quantity of water. This provides more accurate results than just measuring the attenuation by itself. Analyzers that use these two methods use microwave frequencies that are tuned to the natural frequency of the water molecule. The most accurate microwave moisture analyzers utilize multiple frequencies to measure the water content.

Microwave moisture analyzers are available as either reflection type or a transmission type. The choice of technique depends on the nature of the materials to be measured, the physical form of the material, the location within the process where the measurement has to be made, and cost. **See Figure 24-10.**

The first technique is to direct microwaves at the surface of the material. The unabsorbed, reflected microwaves are altered in amplitude and phase and are measured by the receiver. In some cases, multiple microwave frequencies are used simultaneously, with each frequency being analyzed for amplitude and phase shift. This eliminates the effect of variables such as temperature, particle size, and volatile material.

The second technique is to transmit the microwaves through the process material. Sensors are mounted directly across webs, pipes, chutes, belts, or tanks, as needed. The method of mounting and the specific type of unit selected depends on the requirements of the application. Continuous web materials usually require that the microwaves scan the width of the strip to determine the overall moisture content. Since this method scans the entire product stream, it automatically measures the average moisture level. Usually, samples taken manually and dried in a laboratory are used to calibrate the microwave results.

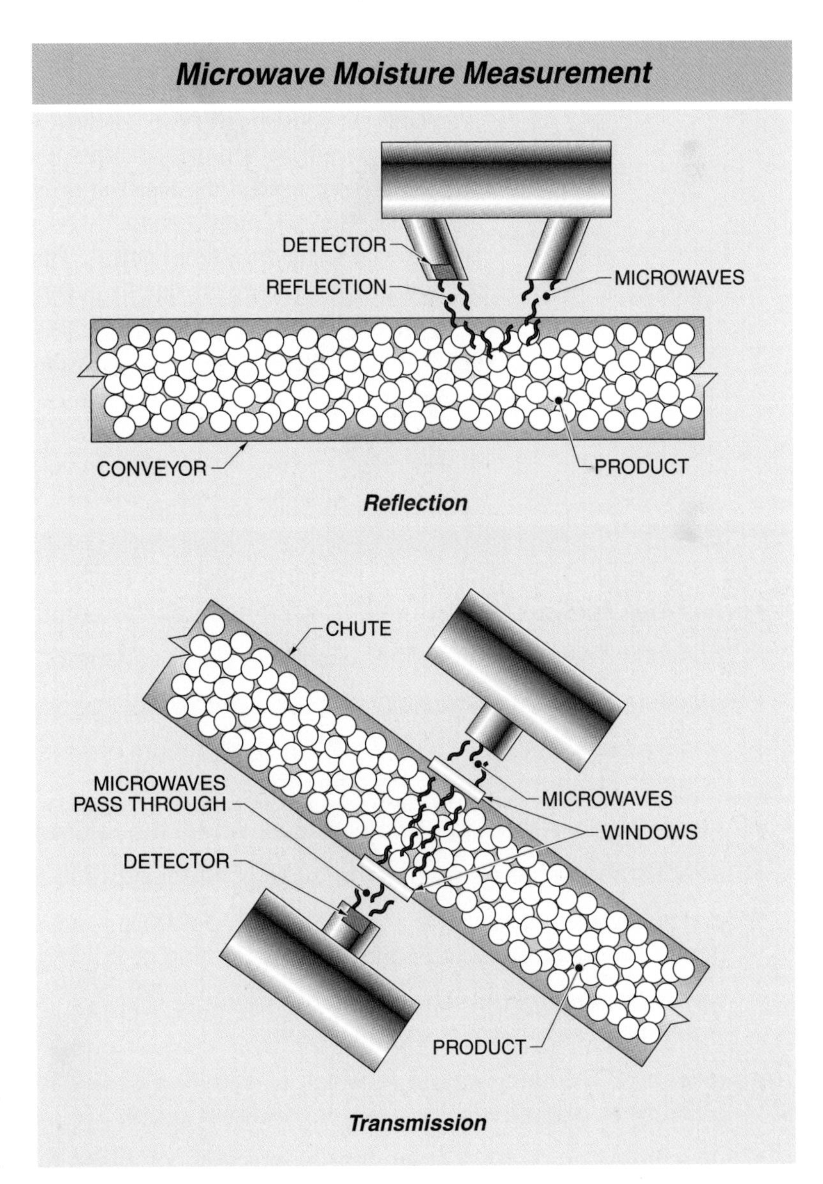

Figure 24-10. Microwave moisture analyzers are used in either reflection or transmission configurations.

Gravimetric Moisture Analysis

Gravimetric moisture analysis is a method of measuring moisture levels in solids by determining the loss of weight from evaporation when the sample is dried. The sample is weighed, dried in an oven, and weighed again. The difference in weight is the amount of water present. This method is seldom used for on-line analysis, mostly due to the time needed to dry the material. It is sometimes used as a calibration method for other moisture measurement methods.

Electrical Impedance Moisture Analysis

An *electrical impedance moisture analyzer* is an analyzer consisting of two conductive plates, separated by a nonconductive rib, that are pressed against the material to be measured. The measured conductivity is proportional to the moisture content. This method is subject to errors due to material density, material cross-sectional area, temperature, contact resistance, etc. When newer, more accurate methods were developed, the electrical impedance method was mostly phased out. However, this method is still used as a quick check in the lumber industry to determine the moisture of wood after drying.

Nuclear Solids Moisture Analyzers

A *nuclear solids moisture analyzer* is an analyzer that measures the amount of moisture in solids by measuring the speed of neutrons that strike the object. Fast neutrons are slowed down by scattering from the nuclei of hydrogen atoms, but are hardly slowed down by higher atomic weight atoms. Thus the number of slow neutrons is directly proportional to the number of hydrogen atoms. If water is the only source of hydrogen atoms, then the number of slow neutrons is proportional to the amount of moisture. This method is very rarely used because of the need to use nuclear sources and the associated license requirements, especially when other good methods are available.

KEY TERMS

- *humidity analyzer:* An instrument that measures the amount of humidity in air.
- *humidity:* The amount of water vapor in a given volume of air or other gases.
- *absolute humidity,* or *humidity ratio:* The ratio of the mass of water vapor to the mass of dry air.
- *relative humidity (rh):* The ratio of the actual amount of water vapor in the air to the maximum amount of water vapor possible at the same temperature.
- *saturated air:* A mixture of water and air where the relative humidity is 100%.
- *specific humidity:* The ratio of the mass of water vapor to the mass of dry air plus moisture.
- *dry bulb temperature:* The ambient air temperature measured by a thermometer that is freely exposed to the air but shielded from other heating or cooling effects.
- *wet bulb temperature:* The lowest temperature that can be obtained through the cooling effect of water evaporating into the atmosphere.
- *dewpoint:* The temperature to which air must be cooled for the air to be saturated with water and any further cooling results in water condensing out.
- *psychrometric chart:* A graph that graphically combines the properties of moist air at standard atmospheric pressure (14.7 psi).

KEY TERMS (continued)

- *psychrometer:* A humidity analyzer that uses two thermometers with the bulb of one thermometer kept moist (wet bulb) and the other bulb kept dry (dry bulb).
- *recording psychrometer:* A psychrometer consisting of wet and dry bulb thermometers connected to a recorder.
- *sling psychrometer:* A psychrometer consisting of two glass thermometers attached to an assembly that permits the two thermometers to be rotated through the air.
- *hygrometer:* A humidity analyzer that measures the physical or electrical changes that occur in various materials as they absorb or release moisture.
- *hygroscopic material:* A material that readily absorbs and retains moisture with increases in humidity and releases moisture with decreases in humidity.
- *thermohygrometer:* A combination of a hygrometer, pressure sensor, and temperature-sensing instrument with digital processing to calculate relative humidity, absolute humidity, dry bulb temperature, dewpoint reading, and other properties for local display or transmission.
- *dew cell:* A fiberglass cylindrical core impregnated with lithium chloride and wrapped with a winding of gold wire around the core that serves as a heating element.
- *solids moisture analyzer:* An instrument used to measure the amount of moisture in a solid.
- *near infrared (NIR) moisture analyzer:* A solids moisture analyzer that measures the reflectance of the process material and calculates moisture content.
- *reflectance:* The ability of a material to reflect light or radiant energy.
- *microwave:* The band of electromagnetic radiation between infrared and VHF broadcast frequencies, covering the range of approximately 3 mm to 3 m wavelengths.
- *microwave moisture analyzer:* A solids moisture analyzer consisting of a transmitter that directs a microwave beam onto the material whose moisture is to be measured and a receiver.
- *gravimetric moisture analysis:* A method of measuring moisture levels in solids by determining the loss of weight from evaporation when the sample is dried.
- *electrical impedance moisture analyzer:* An analyzer consisting of two conductive plates, separated by a nonconductive rib, that are pressed against the material to be measured.
- *nuclear solids moisture analyzer:* An analyzer that measures the amount of moisture in solids by measuring the speed of neutrons that strike the object.

REVIEW QUESTIONS

1. What is the difference between absolute humidity and relative humidity?
2. What is the difference between dry bulb temperature and wet bulb temperature?
3. What is the difference between a hygrometer and a thermohygrometer?
4. List and define the common types of solids moisture analyzers.
5. How does a near infrared analyzer use two frequencies to compensate for factors other than moisture?

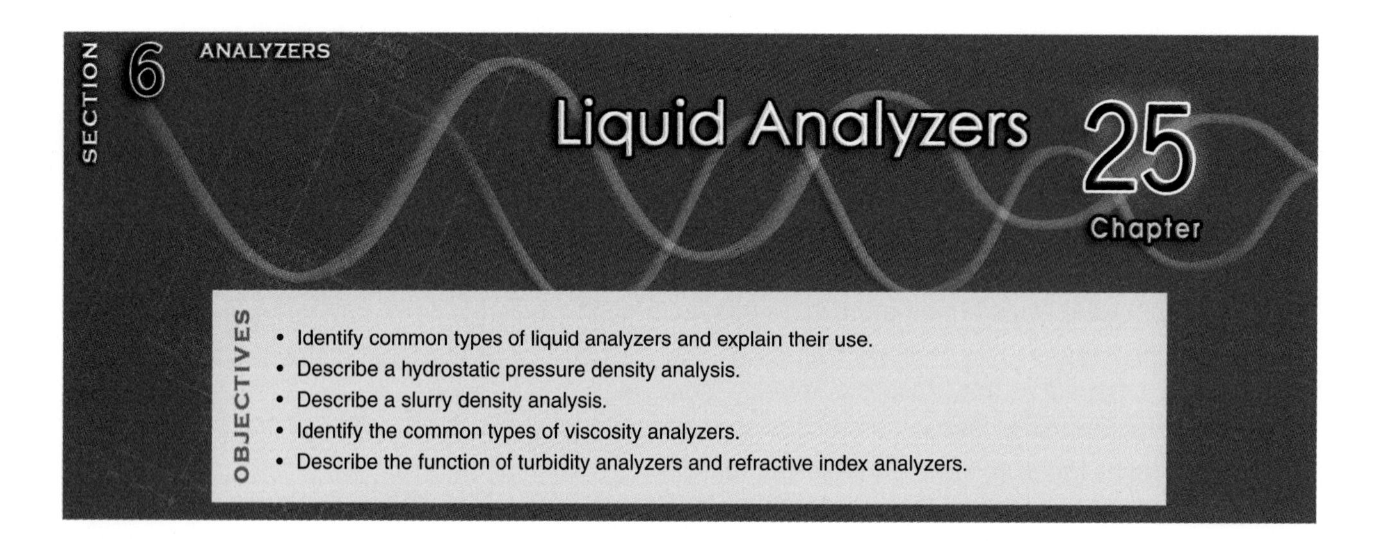

LIQUID DENSITY ANALYZERS

A *liquid analyzer* is an instrument that measures the properties of a liquid. Liquid analyzers may be used to determine a specific property of a liquid, such as density, viscosity, turbidity, or refractive index. A *liquid density analyzer* is an analyzer that measures the density of a liquid by measuring related variables such as buoyancy, the pressure developed by a column of liquid, or the natural frequency of a vibrating mass of liquid; or by nuclear radiation absorption. At 4°C (39.6°F), a cubic foot of water weighs 62.4 lb.

Many density-measuring instruments are calibrated in specific gravity units rather than density values. Common types of liquid density analyzers are hydrometers, differential pressure density analyzers, vibrating U-tube density analyzers, nuclear radiation density analyzers, and slurry density analyzers.

Hydrometers

A *hydrometer* is a liquid density analyzer with a sealed float consisting of a hollow, tubular glass cylinder with the upper portion much smaller in diameter, a scale on the small-diameter portion, and weights at the lower end to make it float upright, with the upper portion partially above the surface of the liquid. The hydrometer is the simplest device for measuring liquid density. **See Figure 25-1.**

The smaller diameter stem provides greater accuracy in reading the density. The scales can be specific gravity, density in g/cm^3, or a number of other specialized density scales particular to different industries. Hydrometers are available to cover different ranges of specific gravity.

The position of the hydrometer in the liquid depends on the density of the liquid. A less dense liquid causes the hydrometer to be positioned lower in the liquid because a greater volume of the liquid has to be displaced to equal the weight of the hydrometer. The temperature of the fluid must be measured so that a correction can be applied to the measured specific gravity to obtain a correct value.

The reading is taken by noting the point on the scale to which the hydrometer sinks. The meniscus is the surface of the liquid in the manometer. The surface is curved because the liquid adheres to the glass stem and side walls of the tube. The reading should be taken by looking horizontally at the meniscus across the flat part of the liquid surface. A reading taken on the curved part of the liquid surface will be in error.

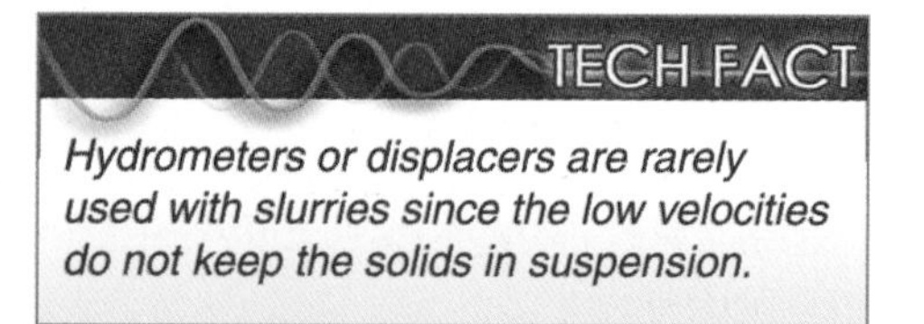

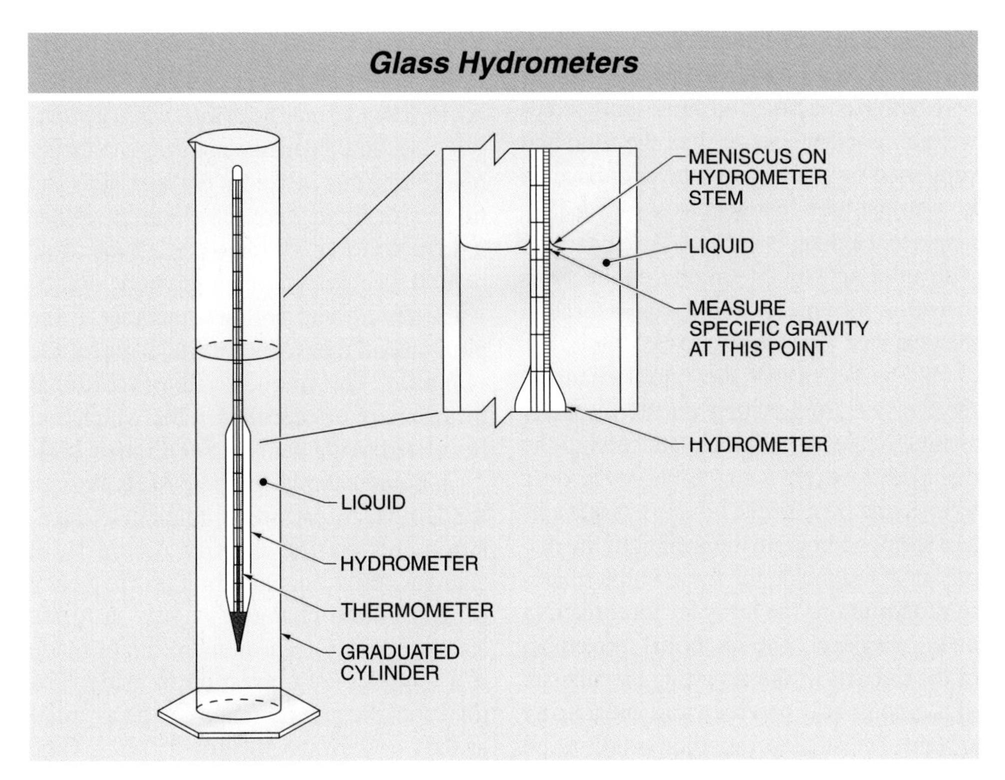

Figure 25-1. Hydrometers are used to measure the density of a liquid.

Continuous Hydrometers. A *continuous hydrometer* is a liquid density analyzer consisting of a float that is completely submerged in a liquid and has chains attached to the bottom to change the effective float weight to match changes in density of the liquid. A metal rod inside the float, which is also used as a weight, acts as a variable coupling to a set of external electrical coils.

An upward movement of the hydrometer float due to an increase in liquid density is stopped when the float picks up more of the chains. **See Figure 25-2.** Stability is then obtained at a new position. This hydrometer design can be used for continuous measurement but the sample flow must be restricted to limit the liquid velocity. The temperature of the liquid is measured continuously so that density corrections due to temperature can be made automatically. Very accurate density measurements can be made with this type of instrument.

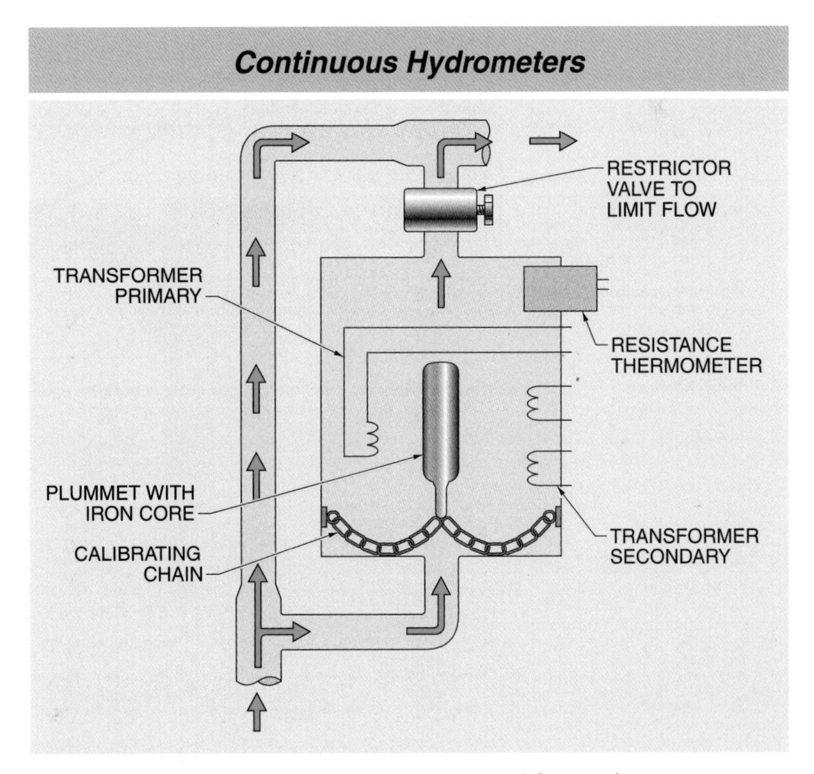

Figure 25-2. Continuous hydrometers are used for continuous measurement of density.

Displacement Hydrometers. A *displacement hydrometer* is a liquid density analyzer consisting of a fixed-volume and fixed-weight cylinder heavier than the fluid and supported by a spring or a lever connected to a torque tube that acts as a spring. The displacer element is enclosed in a chamber with inlet and outlet sample connections that allow the process liquid to pass through the chamber. **See Figure 25-3.**

As the density of the liquid changes, the buoyant force acting on the displacer changes, causing a linkage attached to the displacer to exert a rotational force on a torque tube. The torque tube twists and acts as a spring to hold up the weight of the displacer. A rod inside the torque tube rotates in proportion to the buoyant forces acting on the displacer. The rotational movement of the rod acts as the actuating mechanism of a pneumatic or electrical measuring system. The displacer method is only good for measuring a relatively wide range of specific gravities.

Vibrating U-Tube (Coriolis) Density Analyzers

A *vibrating U-tube (Coriolis) density analyzer* is a liquid density analyzer consisting of a U-tube that is fixed at the open ends and filled with liquid. An electric current excites a drive coil that vibrates the U-tube at its natural frequency around the node points. An armature and coil arrangement at the pick-up end measures the frequency of the vibration. The frequency is proportional to the mass of the filled tube, which can be related to the density. **See Figure 25-4.**

A liquid sample can be passed through the U-tube to provide a continuous measurement. As the density of the fluid changes, the resonant vibration frequency of the U-tube changes. A very accurate specific gravity measurement can be made if a temperature correction is made. The total specific gravity range can be as small as 0.01 or 0.02 specific gravity units. Care needs to be taken to prevent solids accumulation in the U-tube.

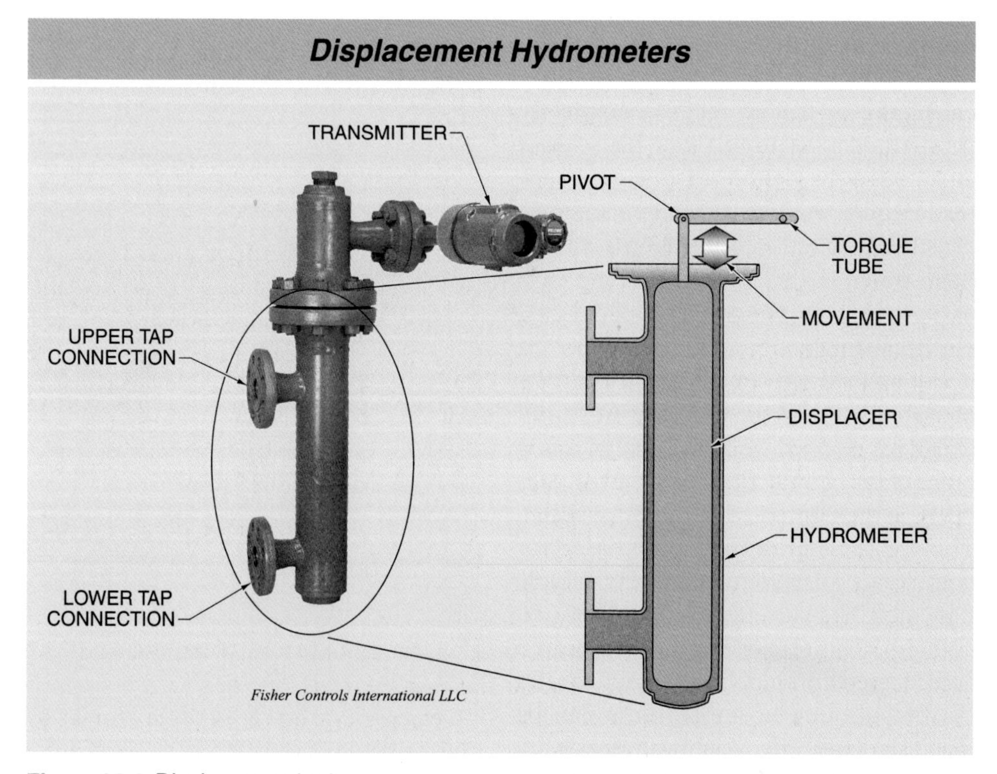

Figure 25-3. Displacement hydrometers measure the change in buoyancy force due to changes in the density of the liquid sample.

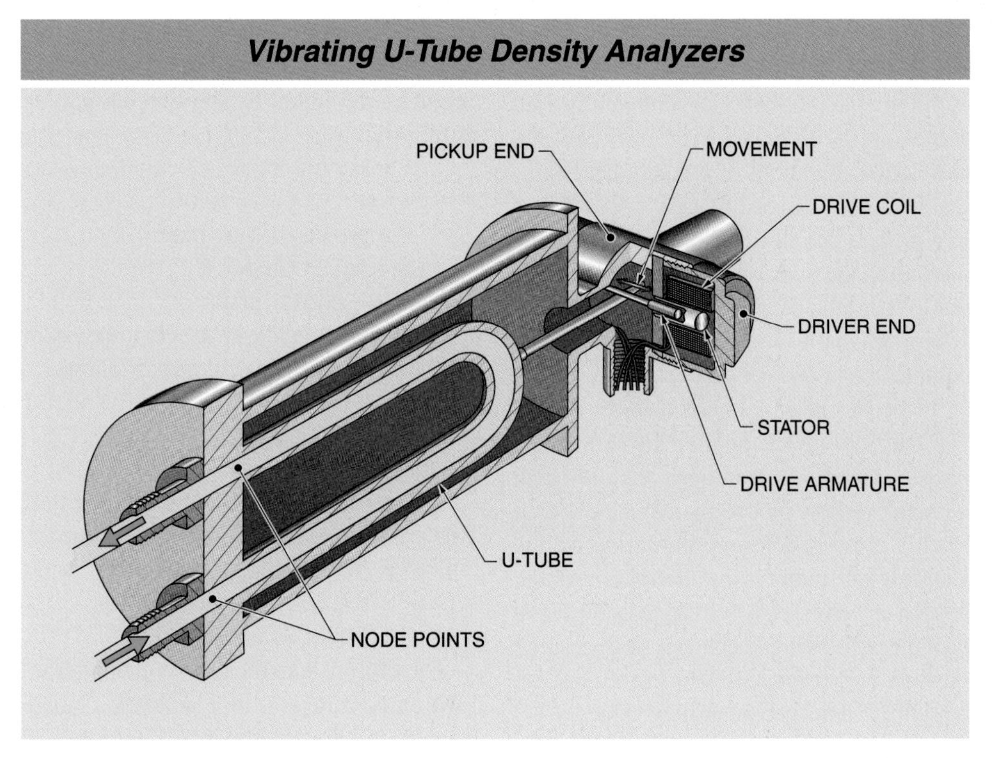

Figure 25-4. Vibrating U-tube density analyzers determine density by measuring the natural vibration frequency of a tube filled with liquid.

Nuclear Radiation Density Analyzers

A *nuclear radiation density analyzer* is a liquid density analyzer consisting of a suitable radioactive isotope source producing gamma rays directed through a chamber containing the liquid to be measured. A radioactive sensing cell detects the unabsorbed radiation. **See Figure 25-5.** The amount of radiation absorbed by any material varies directly with its density. The analyzer is calibrated with no liquid in the chamber and by using thin metal sheets placed in the radiation beam to simulate the normal absorption range of the liquid.

Typical radioactive sources are cesium 137 and cobalt 60. The radioactive source gradually loses strength due to decay, but this effect is known and is compensated in the electronics or by regular calibrations. The use of nuclear materials is closely regulated by federal, state, and local regulations. Care must be exercised in the handling and storage of these materials.

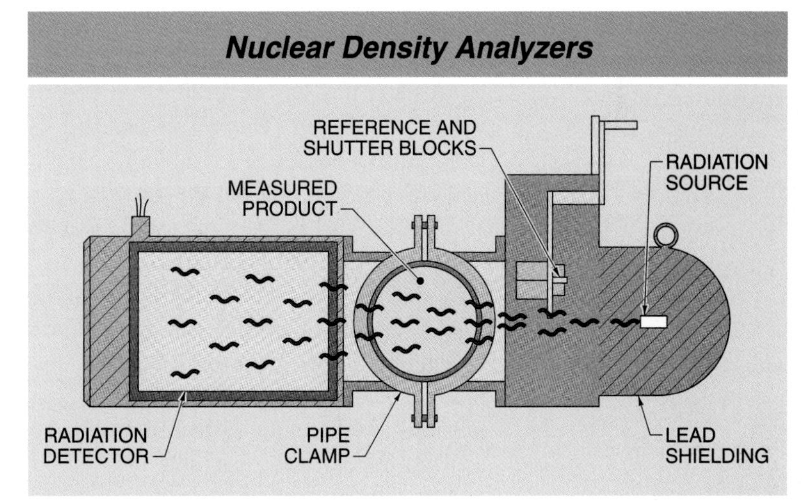

Figure 25-5. Nuclear density analyzers determine density by measuring the attenuation of a beam of radiation.

Nuclear radiation absorption is a very robust technique. For example, density measurements have been successfully applied to the measurement of polypropylene solids in propylene monomer at 500 psig through 5″ thick steel-walled vessels.

Hydrostatic Pressure Density Analyzers

A differential pressure measurement is used to measure the difference in pressure between two known elevations of a liquid. Determining the density or specific gravity of a liquid in a process vessel can be accomplished by measuring the hydrostatic pressure variation due to change in density. Hydrostatic pressure is proportional to the product of the liquid column height and density.

If the height of a liquid column is held constant, the only factor affecting hydrostatic pressure is density. In an open or closed tank with an overflow pipe to maintain a constant level, a pressure-sensing connection for the gauge or transmitter is installed at or near the bottom of the vessel. Any change in the density of the process liquid changes the measured pressure proportionately. The instrument can be calibrated in specific gravity values.

For a vessel under pressure, a differential pressure measuring system is required with two connections a known vertical distance apart. The upper connection is attached to the differential pressure sensor by a pipe or tubing that is filled with a sealing liquid of known density. This liquid introduces a hydrostatic pressure separate from that of the process liquid. The vessel pressure is cancelled out because it acts upon both high- and low-pressure inlets to the instrument.

Bubbler and purge systems similar to those used for level measurement can also be applied to density measurement with the modifications required to fix the column height. It may be necessary to install diaphragm seals at the upper and lower connections to protect the differential pressure sensor from corrosive liquids.

For example, in the simplest situation, a constant-level open tank has a pressure transmitter installed at a measured distance below the surface of the liquid. The height is constant and the pressure varies directly with the density. **See Figure 25-6.** Calculate the pressure differential for a process fluid with a specific gravity range of 1.00 to 1.10 and a vertical distance of 100″ between the measurement tap and the top of the fluid. To convert the process fluid head pressure to the equivalent head of water, multiply the height of the liquid by the specific gravity as follows:

$$P = H \times SG$$

where

P = pressure (in in. water)

H = height (in in.)

SG = specific gravity

For a process fluid with a specific gravity of 1.00, the pressure is calculated as follows:

$$P = H \times SG$$

$$P = 100 \times 1.00$$

$$P = \textbf{100 in. of water}$$

For a process fluid with a specific gravity of 1.10, the pressure is calculated as follows:

$$P = H \times SG$$

$$P = 100 \times 1.10$$

$$P = \textbf{110 in. of water}$$

Therefore, a specific gravity change from 1.00 to 1.10 results in a pressure change from 100 inches of water to 110 inches of water. The specific gravity of the liquid, up to the lower end of the measurement range, produces 100 inches of pressure, which provides no useful information. The transmitter calibration can be set up with an elevation of 100″ and a measurement span of 0 inches of water to 10 inches of water, corresponding to a specific gravity change of 1.00 to 1.10.

This calculation shows that a small differential pressure range must be used to measure relatively large ranges of specific gravity. Even then, large suppression or elevation values must be used. This shows that differential pressure measurements can only be used to measure a relatively wide range of specific gravity.

TECH FACT

Hydrostatic pressure density measurements calculated from a differential pressure measurement are subject to errors if the process liquid can plug the capillary tubes. Diaphragm seals can be used to protect the capillary tubes. The choice of a diaphragm seal should be made with help from the manufacturer to ensure chemical compatibility.

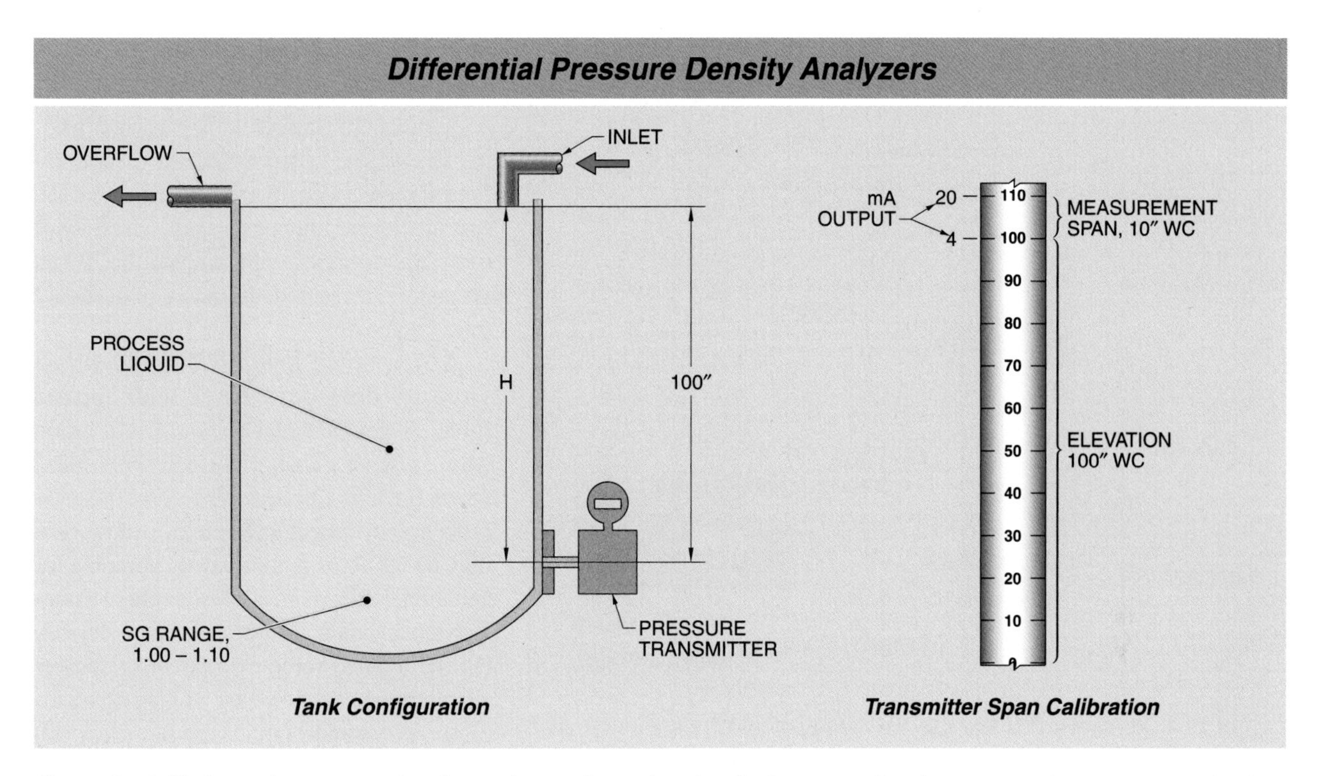

Figure 25-6. Hydrostatic pressure density analyzers determine density by measuring the pressure in a process tank when the height is fixed.

The maximum suppression or elevation value for a differential pressure transmitter cannot be greater than the maximum available differential range. Three typical differential pressure cells are designed to cover the ranges of a minimum of 0–1″ WC up to a maximum of 0–25″ WC; a minimum of 0–10″ WC up to a maximum of 0–277″ WC; and a minimum of 0–80″ WC up to a maximum of 0–1000″ WC. A differential pressure cell can be calibrated to any differential within its range, from the minimum to the maximum.

Slurry Density Analyzers

A *slurry density analyzer* is an analyzer that is used to measure the density of a slurry. A *slurry* is a liquid that contains suspended solids that are heavier than the liquid. In most cases the solids precipitate out of the liquid due to decreased agitation or decreased temperature, or due to a chemical reaction.

Pseudo density is the density of a slurry, determined by the total weight of a slurry including the solids, divided by the volume of the slurry. Many processes produce slurries that need to be measured as percent solids, not as a pseudo density. *Percent solids* is the volume of solids suspended in a slurry divided by the total volume of the slurry. Percent solids is not a term from which a density instrument's range can be selected, but a slurry's pseudo density can be used to determine the range of the density instrument.

The percent solids and pseudo density are related and can be easily determined from a slurry sample. **See Figure 25-7.** If the slurry is put into a container and allowed to settle, the solids collect in the bottom. When completely settled, the volume of solids divided by the volume of sample multiplied by 100 is the percent solids by volume. The total weight of the slurry sample divided by sample volume is the pseudo density.

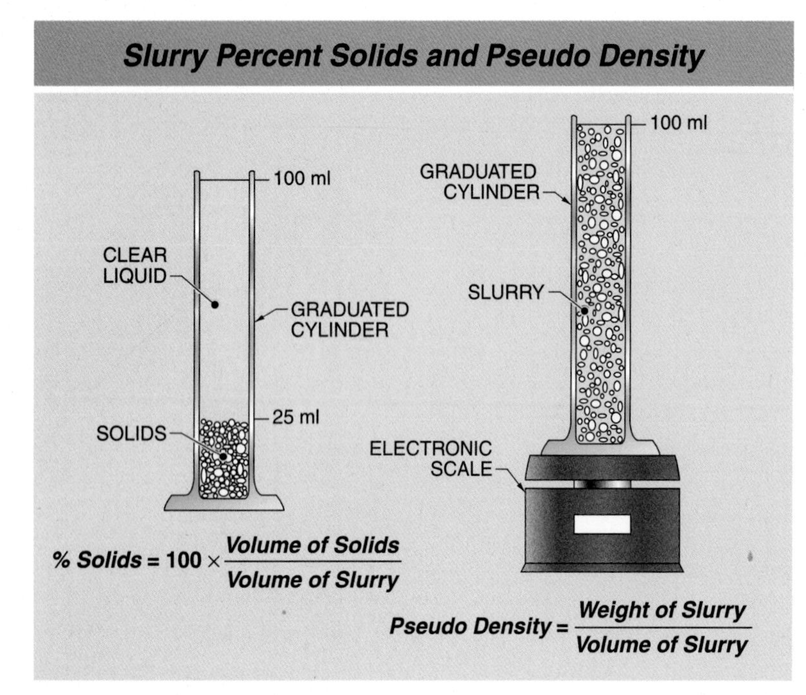

Figure 25-7. Percent solids and pseudo density of a slurry are determined by measuring the volume of the solids, the volume of the slurry, and the weight of the slurry sample.

Slurries must be agitated to keep the solids suspended so that they do not plug up processing equipment. All density instruments need to measure slurries when the solids are suspended. The differential pressure method of measuring density is dependent on agitation to measure a pseudo density. If the slurry is not agitated, the differential pressure measures the density of the clear liquid. Excess agitation can cause errors in differential pressure measurements. **See Figure 25-8.** The slurry suspension can also be measured by liquid density methods such as vibrating U-tubes and nuclear radiation absorption.

For example, one of the methods used in the production of caustic soda requires the evaporation of a caustic and salt solution. As water is removed during the evaporation, the salt solution exceeds its saturation point and salt crystals are formed. The crystals are mechanically removed and continuously deposited into a vessel. The amount of water or brine added to the salt is determined by the slurry density measurement in the slurry vessel. The density is controlled to produce a pumpable slurry.

VISCOSITY ANALYZERS

Viscosity is resistance to flow. A *viscosity analyzer* is a liquid analyzer that measures a liquid's resistance to flow at specific conditions. For example, viscosity must be measured at a specific temperature because the viscosity of most liquids varies with temperature. When viscosity is measured, the nature of the liquid must be known.

A *Newtonian liquid* is a liquid whose viscosity does not change with applied force. A *non-Newtonian liquid* is a liquid whose viscosity changes (usually decreases) when force is applied. The force may be from gravity causing the fluid to flow or it may be the force from mixing, pumping, or agitation. Latex paint is an example of a non-Newtonian liquid. It is very viscous when still, but the applied force of a brush or sprayer allows the paint to flow easily. Other examples of non-Newtonian liquids are tar, silica gel, and many foods such as ketchup, molasses, and peanut butter.

Newtonian fluids permit viscosity measurement by any of the instruments that are described in this section. Non-Newtonian fluids must be tested for their particular characteristics to determine their suitability for viscosity measurement. Common instruments that have been developed for continuous measurement of viscosity include the falling piston, rotating spindle, and orifice viscosity analyzers. A rheometer is a common laboratory instrument used to measure the flow properties of plastics and polymers.

Absolute and Kinematic Viscosity

Absolute viscosity is the resistance to flow of a fluid and has units of centipoise (cP). *Kinematic viscosity* is the ratio of absolute viscosity to fluid density and has units of centistokes (cS). Kinematic viscosity can be calculated from absolute viscosity as follows:

$$\upsilon_k = \frac{\upsilon_a}{\rho}$$

where

υ_k = kinematic viscosity (in cS)

υ_a = absolute viscosity (in cP)

ρ = liquid density (in g/cm^3)

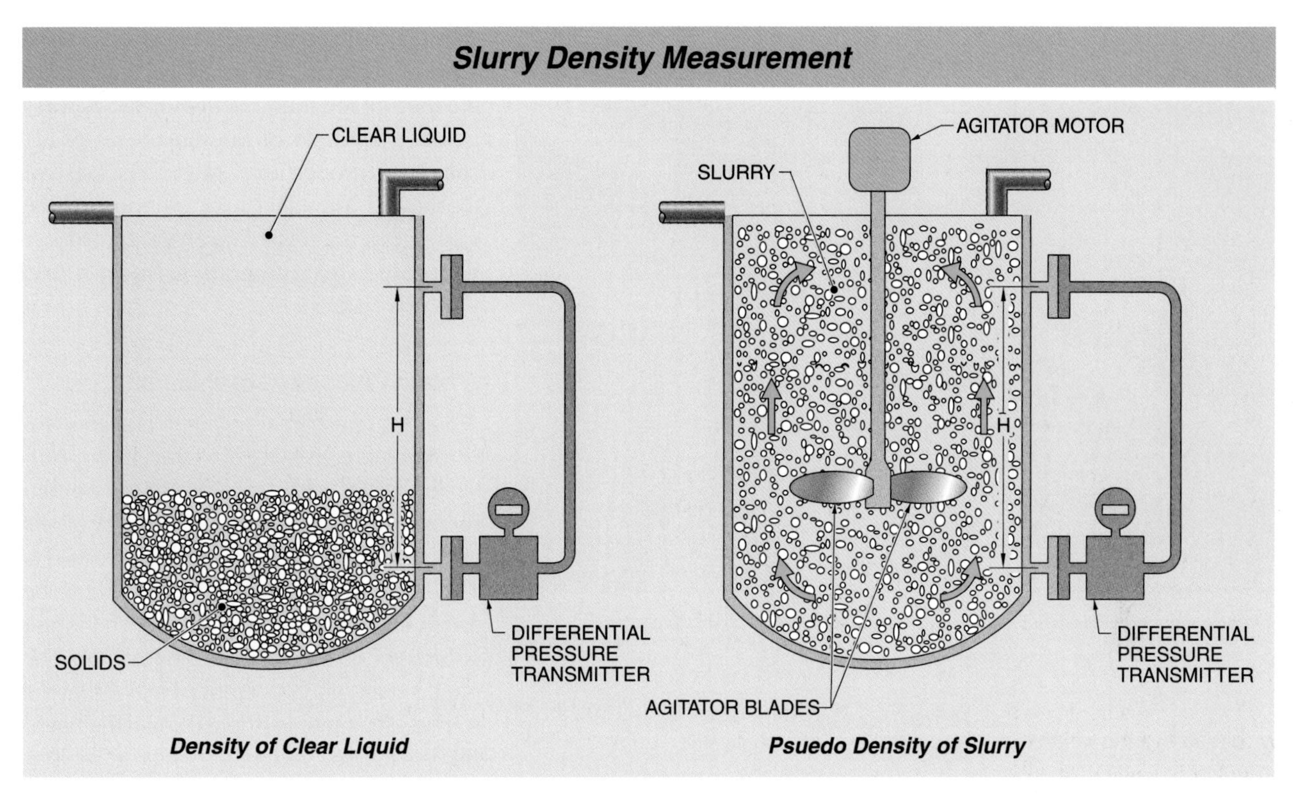

Figure 25-8. Slurries must be agitated to measure the pseudo density with differential pressure.

For example, if the absolute viscosity is 35 cP and the liquid has a density of 0.93 g/cm³, the kinematic viscosity is calculated as follows:

$$\upsilon_k = \frac{\upsilon_a}{\rho}$$

$$\upsilon_k = \frac{35}{0.93}$$

$$\upsilon_k = \mathbf{37.6\ cS}$$

Falling Piston Viscosity Analyzers

A *falling piston viscosity analyzer* is a viscosity analyzer consisting of a precision tube and a piston with a timed fall through a constant-temperature liquid. The time required for the piston to fall a measured distance through the liquid is proportional to the viscosity. A continuous viscosity-measuring instrument based upon this principle uses a mechanical assembly with a release to allow the piston to fall and a push rod to lift the piston back to the starting position. This operation occurs repeatedly to obtain regular readings. **See Figure 25-9.**

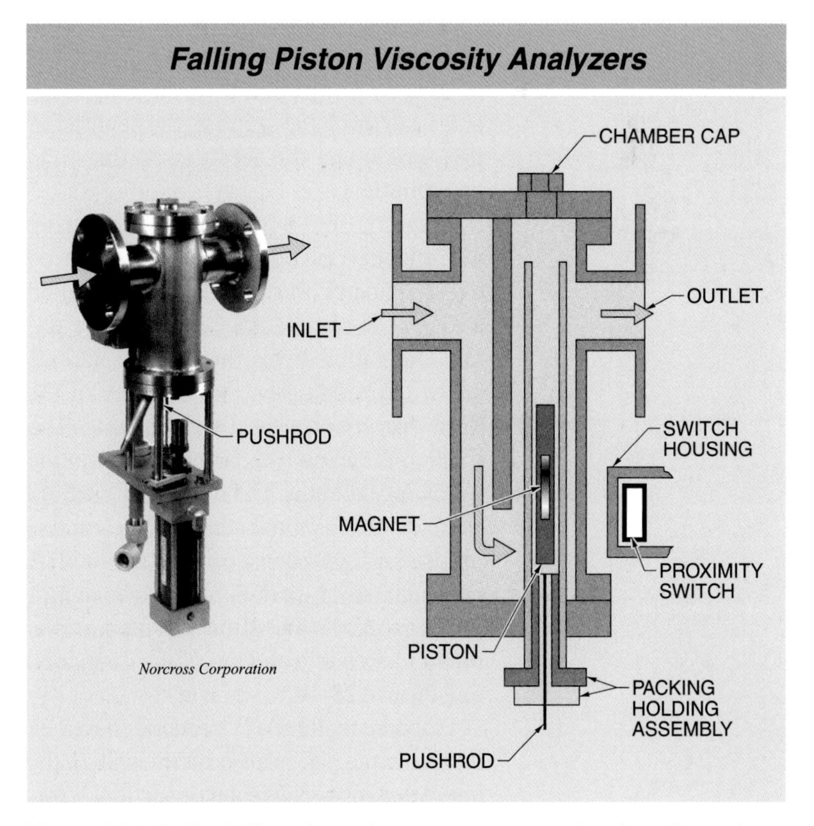

Figure 25-9. In-line falling-piston viscometers use a pushrod to raise a piston that is then released to fall through the liquid.

The viscosity of a liquid affects its flow through valves and piping systems.

Rotating Spindle Viscosity Analyzers

A *rotating spindle viscosity analyzer* is a viscosity analyzer consisting of a rotating spindle in a container of the sample liquid at a controlled temperature. The viscosity is directly proportional to the torque required to drive the spindle. Some units of this type rotate the container rather than the spindle.

One continuous viscosity-measuring instrument based on this principle replaced the container with a driven cylinder called a rotor. The spindle is replaced with a stator. The motor drives the rotor at various speeds, depending on the viscosity of the fluid that couples the two elements. The rotation of the rotor starts the fluid in motion following the rotor. The motion of the fluid exerts a torque on the stator. The torque is proportional to the viscosity of the fluid. A housing encases these elements and has piping connections that allow a sample of the liquid to flow into and out of the container. **See Figure 25-10.**

Another continuous viscometer based on this principle places a disc in the path of the liquid that is flowing in a pipe section. A torsion element links the disc with the shaft of a synchronous electric motor. An inductive transducer that surrounds the spindle shaft senses the braking torque that is imparted to the disc by the fluid friction of the process liquid. The angle of the movement of the spindle is proportional to the viscosity of the liquid. The electronic circuitry of the viscometer is able to convert the transducer voltage to a signal proportional to kinematic viscosity values.

Orifice Viscosity Analyzers

A *Saybolt universal viscometer* is a laboratory apparatus consisting of a temperature-controlled vessel with an orifice in the bottom for measuring the kinematic viscosity of oils and other viscous liquids. **See Figure 25-11.** The orifice is plugged and the liquid to be tested is poured into the vessel. The liquid is brought to the desired temperature. The plug is then quickly removed and a timer is started. The timer is stopped when the liquid fills the desired volume.

The units of viscosity measurement with this apparatus are Saybolt universal seconds. For more viscous liquids the orifice is changed, and the resultant measurement is expressed as Saybolt furol seconds. Saybolt viscometers are commonly used to measure the viscosity of lubricating oils. When measuring oils with a kinematic viscosity greater than about 20 centistokes, viscosity units are converted as follows:

$$\text{Saybolt Universal Seconds} = \text{centistokes} \times 4.64$$

$$\text{Saybolt Furol Seconds} = \text{centistokes} \times 0.470$$

There are several other orifice viscometers consisting of variations of a funnel-shaped cup with different size openings. The viscosity is reported as the time for the fluid to empty from the cup. Examples of this are the Zahn cup, the Shell cup, and the Norcross cup. In a similar technique, the Ubbelohde method uses gravity to force a fluid to flow through a capillary in a U-tube and the time to flow between two calibrated marks is recorded. These methods are generally used as sample viscosity analyzers. They are not suitable for continuous measurements.

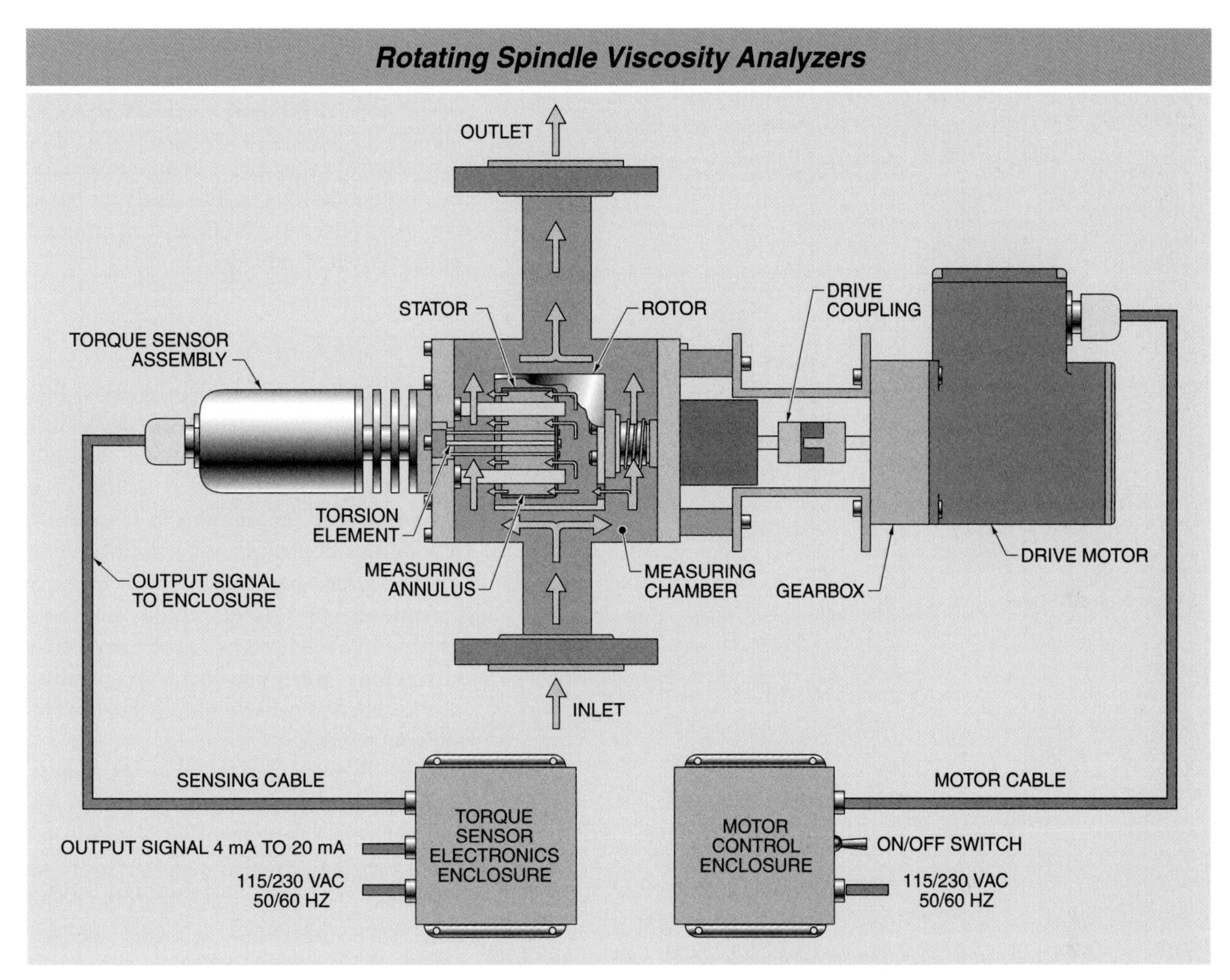

Figure 25-10. Rotating spindle viscometers continuously measure viscosity by measuring the torque generated on a stator by a rotor immersed in a flow of fluid.

Rheometers

A *rheometer* is a viscosity analyzer consisting of a heated, constant-temperature cylinder where a polymer is melted and forced through an orifice by a piston moving at a constant rate. **See Figure 25-12.** The force required to push the polymer through the orifice is measured. The displacement and the force can be graphed. This graph is a curve that provides information relating to the physical properties of the polymer.

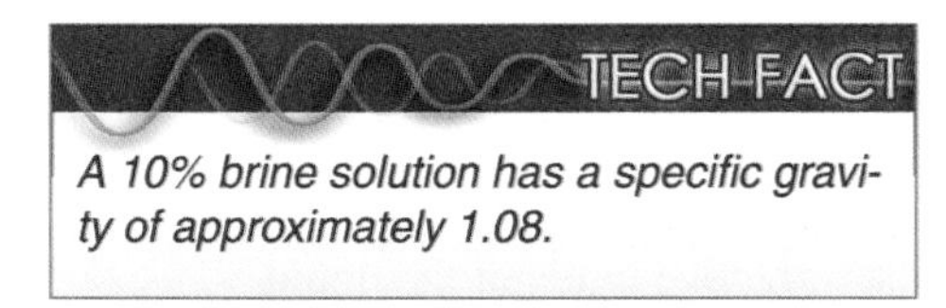

A 10% brine solution has a specific gravity of approximately 1.08.

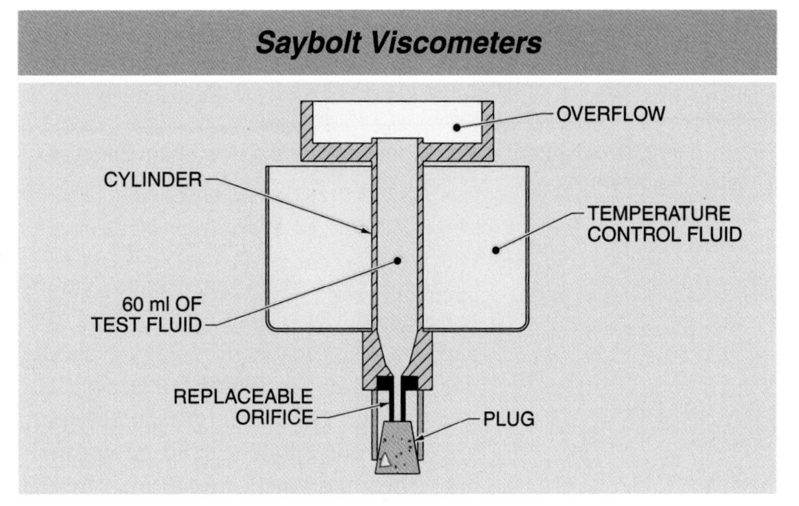

Figure 25-11. Saybolt viscometers are used to measure the time it takes an oil to drain through a fixed orifice after the plug is removed.

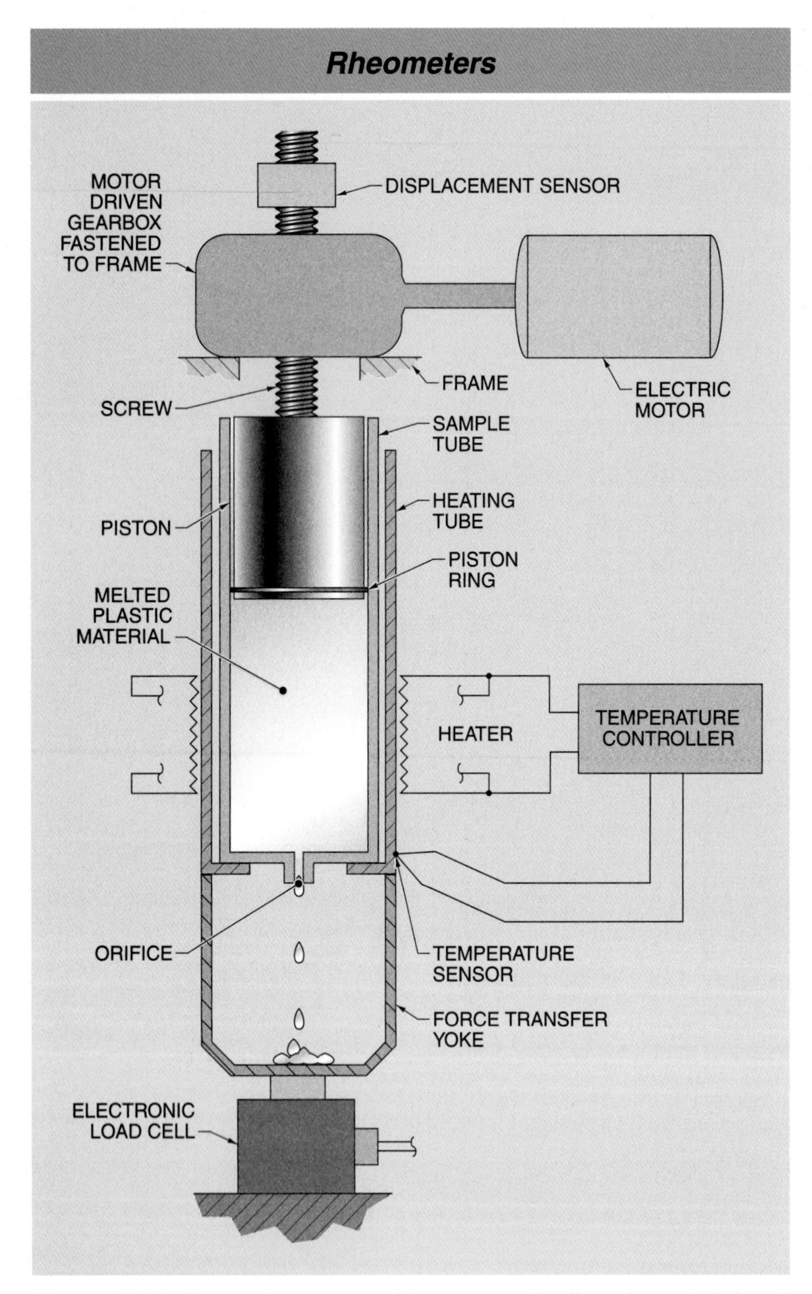

Figure 25-12. Rheometers are used to measure the flow characteristics of plastics and polymers.

Rheology is the science of the deformation and flow of matter. Polymers melt and flow when heated. Rheology generally deals with non-Newtonian fluids, especially polymers. The most common polymers are PVC, polyethylene, and polypropylene. These polymers are formed by the polymerization of monomers, such as vinyl chloride, ethylene, and propylene, using catalysts, temperature, and pressure. The raw plastics are then modified by the addition of plasticizers and other chemicals to change the viscosity, flexibility, and other properties. A rheometer is not capable of being a continuous on-line analyzer but is commonly used as a laboratory instrument in the polymer industry.

TURBIDITY ANALYZERS

A *turbidity analyzer* is a liquid analyzer that measures the amount of suspended solids in a liquid by the measurement of light scattering from the suspended particles. **See Figure 25-13.** The turbidity is commonly reported as nephelometric turbidity units (NTU), grams per liter, percent, or parts per million (ppm). Some turbidity analyzers use multiple light beams and detectors with a switching strategy to obtain comparative results that compensate for coatings on the viewing windows.

Turbidity is an important measurement used to detect solids breakthrough from clarifiers and filters or to determine when sufficient precoat material has been deposited on precoat style filters. In addition, NTUs are now the accepted standard used by EPA Method 180.1 for water and wastewater analysis.

REFRACTIVE INDEX ANALYZERS

The *refractive index* is the amount of bending of a light beam as it moves between fluids with different refractive index values. For example, an object placed partially in a glass of water looks like it bends at the interface between the water and the air. **See Figure 25-14.** A *refractive index analyzer* is a liquid analyzer consisting of a light source directed into a prism that has a flat surface in contact with the liquid to be measured. Sapphire and spinel are the materials most often used for the prism. A refractive index analyzer determines the concentration of a solution or dissolved solid by measuring the refractive index of the liquid.

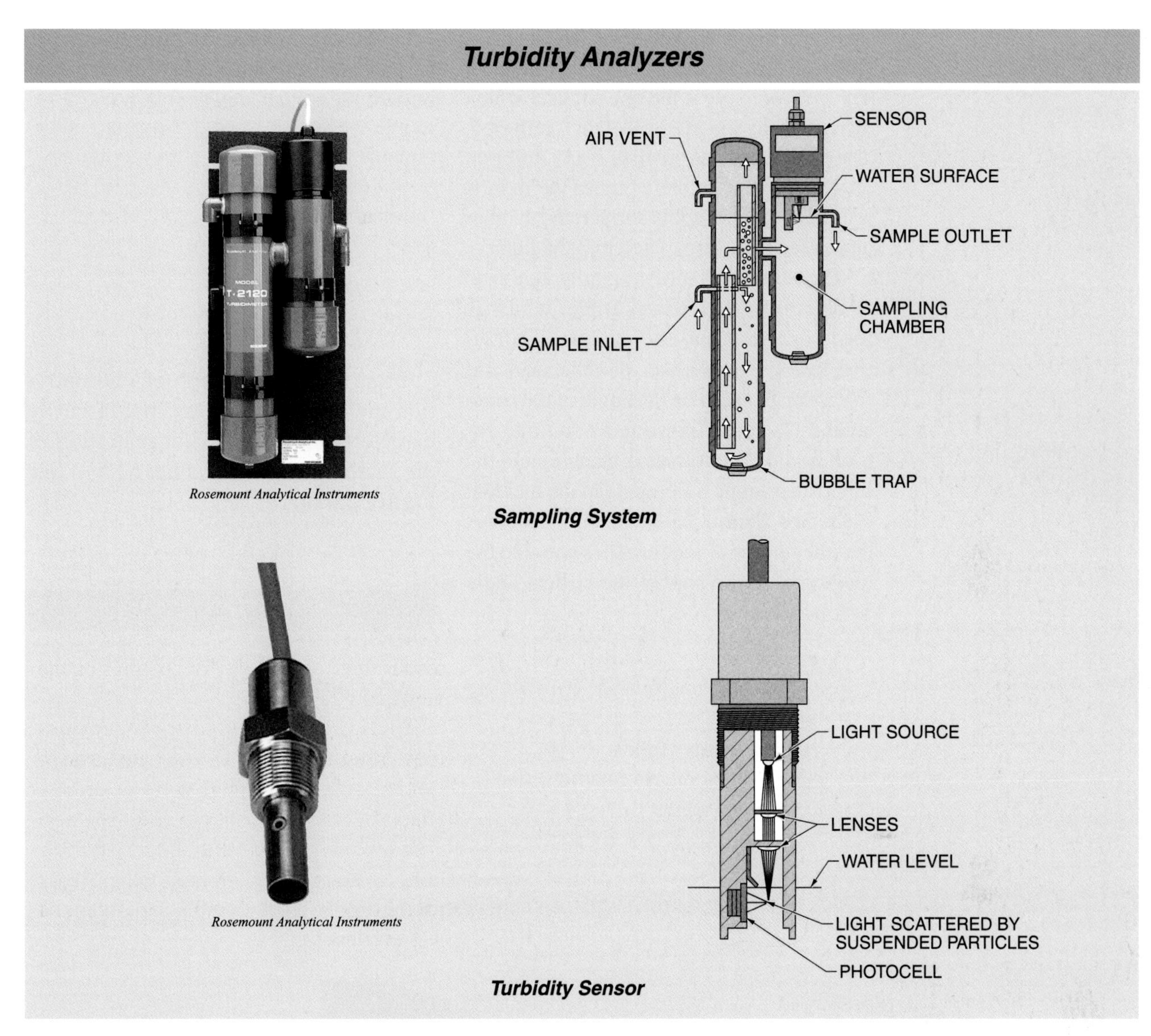

Figure 25-13. Turbidity measuring systems are used to measure the scattering of light in a liquid sample.

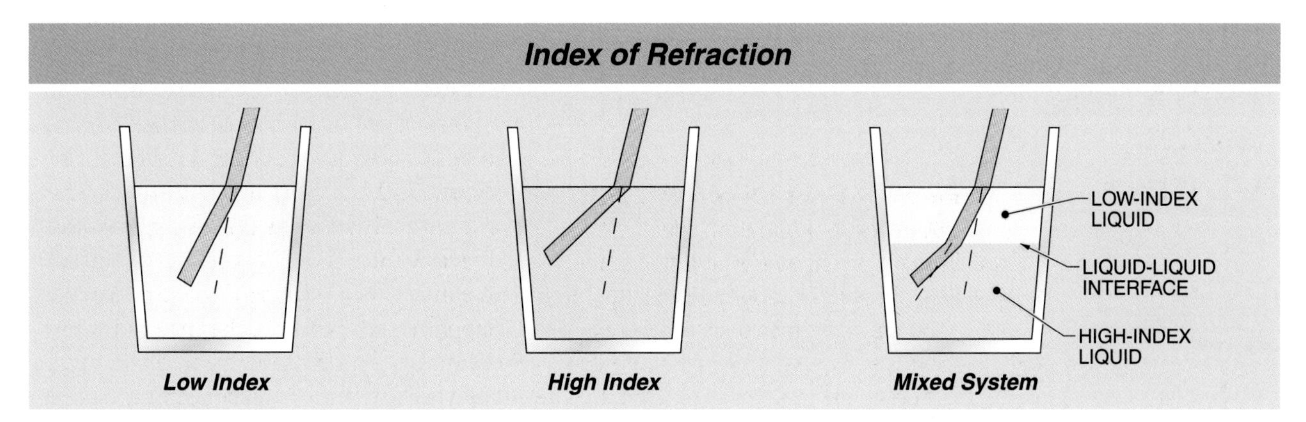

Figure 25-14. The index of refraction is the amount of bending of light as it passes through an interface between two different fluids.

A light beam directed toward a liquid surface may be reflected or it may be refracted. The *critical angle* is the one angle at which there is no refraction and all light is reflected. The light enters the prism at many different angles. At some of these angles the light is reflected from the liquid surface and at other angles the light is refracted into the liquid.

The reflected light is digitally analyzed to determine the critical angle, which is then converted into the measured refractive index. The amount of reflection and refraction depends on the angle of the beam source (incident light) and is unique for each liquid. The refracted beam enters the liquid at an angle different than the incident light. **See Figure 25-15.** Snell's law says that the refractive index, *nD,* is equal to the reciprocal of the sine of the critical angle as follows:

$$nD = \frac{1}{\sin(c)}$$

where
nD = refractive index
sin = trigonometric function, sine
c = critical angle

For example, when the critical angle, *c,* of a sucrose solution is 43.6°, the index of refraction is calculated as follows:

$$nD = \frac{1}{\sin(c)}$$

$$nD = \frac{1}{\sin(43.6)}$$

$$nD = \frac{1}{0.690}$$

$$nD = \mathbf{1.45}$$

From a table or graph of concentration and refractive index, it can easily be determined that the concentration of the sucrose solution is about 63.5%. **See Figure 25-16.**

TECH FACT

Andre Marie Ampere was born near Lyon, France in 1775. In 1815 Ampere published one of the first papers on light refraction. Later, Ampere identified a law for electrical circuits. The unit of electrical current is named after him.

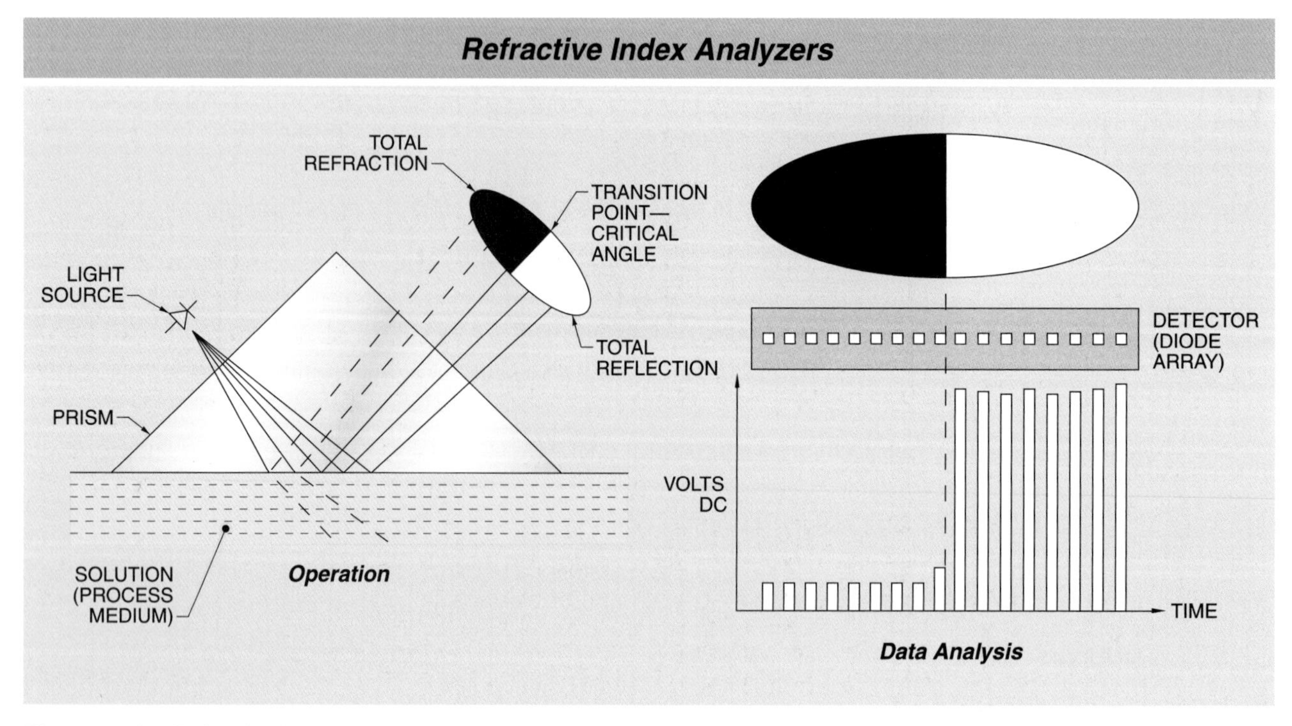

Figure 25-15. Refractive index analyzers measure the critical angle of reflection.

Sucrose Solution Percent by Weight Concentration

	0	0.1	0.2	0.3	0.4	0.5	0.6	0.7	0.8	0.9
62	1.44650	1.44673	1.44696	1.44719	1.44742	1.44765	1.44788	1.44811	1.44834	1.44858
63	1.44881	1.44904	1.44927	1.44950	1.44974	1.44997	1.45020	1.45043	1.45067	1.45090
64	1.45113	1.45137	1.45160	1.45184	1.45207	1.45230	1.45254	1.45277	1.45301	1.45324

Figure 25-16. A lookup table is used to relate the refractive index to the concentration of sucrose in water.

All liquids have a refractive index value and many liquids have similar refractive index vs. concentration ranges. Thus a refractive index value should not be used to identify the presence of a chemical, but only to determine the concentration of a known solution. A solution containing several dissolved materials presents problems because all contribute to the measured refractive index. A refractive index analyzer can only be used when there is one dissolved solid or only one of a group of dissolved solids that varies during the measured portion of the process.

To improve the measurement accuracy, the sample temperature is measured and used to correct the refractive index reading. Solids and gas bubbles entrained in the liquid reflect some of the refracted light and can cause some inaccuracies in the measurement if an analog detector is used. A digital detector eliminates this type of error. Typical ranges of refractive index vary between 1.3 and 1.6 with a resolution of 0.0002, which is equivalent to a resolution of 0.1% of concentration.

Refractive index analyzers have been used in many industries including numerous steps in the manufacturing of sugar; measuring the liquor concentrations in the pulp paper industry; and in the food industry including beer blending and manufacturing of fruit juices, soft drinks, and other beverages. Semiconductor fabrication plants use refractive index analyzers to measure the concentration of process liquids. The analyzer can be designed to sound an alarm if the composition is not within programmed specifications.

KEY TERMS

- *liquid analyzer:* An instrument that measures the properties of a liquid.
- *liquid density analyzer:* An analyzer that measures the density of a liquid by measuring related variables such as buoyancy, the pressure developed by a column of liquid, or the natural frequency of a vibrating mass of liquid; or by nuclear radiation absorption.
- *hydrometer:* A liquid density analyzer with a sealed float consisting of a hollow, tubular glass cylinder with the upper portion much smaller in diameter, a scale on the small-diameter portion, and weights at the lower end to make it float upright, with the upper portion partially above the surface of the liquid.
- *continuous hydrometer:* A liquid density analyzer consisting of a float that is completely submerged in a liquid and has chains attached to the bottom to change the effective float weight to match changes in density of the liquid.
- *displacement hydrometer:* A liquid density analyzer consisting of a fixed-volume and fixed-weight cylinder heavier than the fluid and supported by a spring or a lever connected to a torque tube that acts as a spring.
- *vibrating U-tube (Coriolis) density analyzer:* A liquid density analyzer consisting of a U-tube that is fixed at the open ends and filled with liquid.

KEY TERMS *(continued)*

- *nuclear radiation density analyzer:* A liquid density analyzer consisting of a suitable radioactive isotope source producing gamma rays directed through a chamber containing the liquid to be measured.
- *slurry density analyzer:* An analyzer that is used to measure the density of a slurry.
- *slurry:* A liquid that contains suspended solids that are heavier than the liquid.
- *pseudo density:* The density of a slurry, determined by the total weight of a slurry including the solids, divided by that volume of the slurry.
- *percent solids:* The volume of solids suspended in a slurry divided by the total volume of the slurry.
- *viscosity:* Resistance to flow.
- *viscosity analyzer:* A liquid analyzer that measures a liquid's resistance to flow at specific conditions.
- *Newtonian liquid:* A liquid whose viscosity does not change with applied force.
- *non-Newtonian liquid:* A liquid whose viscosity changes (usually decreases) when force is applied.
- *absolute viscosity:* The resistance to flow of a fluid and has units of centipoise (cP).
- *kinematic viscosity:* The ratio of absolute viscosity to fluid density and has units of centistokes (cS).
- *falling piston viscosity analyzer:* A viscosity analyzer consisting of a precision tube and a piston with a timed fall through a constant-temperature liquid.
- *rotating spindle viscosity analyzer:* A viscosity analyzer consisting of a rotating spindle in a container of the sample liquid at a controlled temperature.
- *Saybolt universal viscometer:* A laboratory apparatus consisting of a temperature-controlled vessel with an orifice in the bottom for measuring the kinematic viscosity of oils and other viscous liquids.
- *rheometer:* A viscosity analyzer consisting of a heated, constant-temperature cylinder where a polymer is melted and forced through an orifice by a piston moving at a constant rate.
- *rheology:* The science of the deformation and flow of matter.
- *turbidity analyzer:* A liquid analyzer that measures the amount of suspended solids in a liquid by the measurement of light scattering from the suspended particles.
- *refractive index:* The amount of bending of a light beam as it moves between fluids with different refractive index values.
- *refractive index analyzer:* A liquid analyzer consisting of a light source directed into a prism that has a flat surface in contact with the liquid to be measured.
- *critical angle:* The one angle at which there is no refraction and all light is reflected.

REVIEW QUESTIONS

1. Define hydrometer and describe its operation.
2. Define vibrating U-tube density analyzer and describe its operation.
3. What is the difference between a Newtonian fluid and a non-Newtonian fluid?
4. What are the common instruments used to measure viscosity?
5. What are a refractive index and a refractive index analyzer?

SECTION 6 ANALYZERS

Electrochemical and Composition Analyzers

Chapter 26

OBJECTIVES

- Identify the analyzers that use conductivity and explain their use.
- Compare pH analyzers and oxidation-reduction potential (ORP) analyzers.
- Define composition analyzer and explain its use.
- Describe the function of automatic titration analyzers.

ELECTROCHEMICAL ANALYZERS

Electrochemical analyzers use the measurement of the electrical properties of conductivity, pH, and oxidation-reduction potential (ORP) of liquids. These properties provide measurements relating to the amount of impurities in water, concentrations of a chemical in water, the acidity or alkalinity of a liquid, and the determination of the completion of chemical reactions. The sensing probes are small in size, are available in materials which are impervious to attack, and can be installed directly in the process stream or in a sample chamber. The measurements respond quickly to changes and work in most non-coating liquids.

Conductivity Analyzers

A *conductivity analyzer* is an electrochemical analyzer that measures the electrical conductivity of liquids and consists of two electrodes immersed in a solution. The electrical conductivity is proportional to the concentration of ions in a solution. The old measurement term for conductivity was "mho," the inverse of ohm, but this is seldom used today. A *siemens* is the modern unit of electrical conductivity and is the reciprocal of resistance. Because of the size of the measurement unit, liquid conductivity is usually measured in microsiemens per centimeter (μS/cm). Conductivity changes significantly with temperature so the measurement is usually compensated for the measured temperature.

The conductivity electrodes used in industry often have areas larger or smaller than 1 sq cm and are placed closer or farther apart than 1 cm. A *conductivity cell constant* is the ratio of the size of the actual electrodes to those of the standard conductivity cell. These constants range from 0.01 to 100. The larger cell constants are used to measure the more conductive solutions.

Conductivity measurements are most suitable at lower concentrations. Conductivity measurement is much more complex at higher concentrations because the response may become very nonlinear. Conductivity can even pass through a maximum with increasing concentration. This means that two different concentrations can have the same conductivity. A chart of conductivity for various materials and concentrations is useful to help understand the relationships. **See Figure 26-1.**

TECH FACT

Conductivity analyzers depend on dissolving molecules in water. When the molecules dissolve completely, the conductivity depends on the concentration. When molecules do not dissolve, the conductivity is not a good measure of concentration.

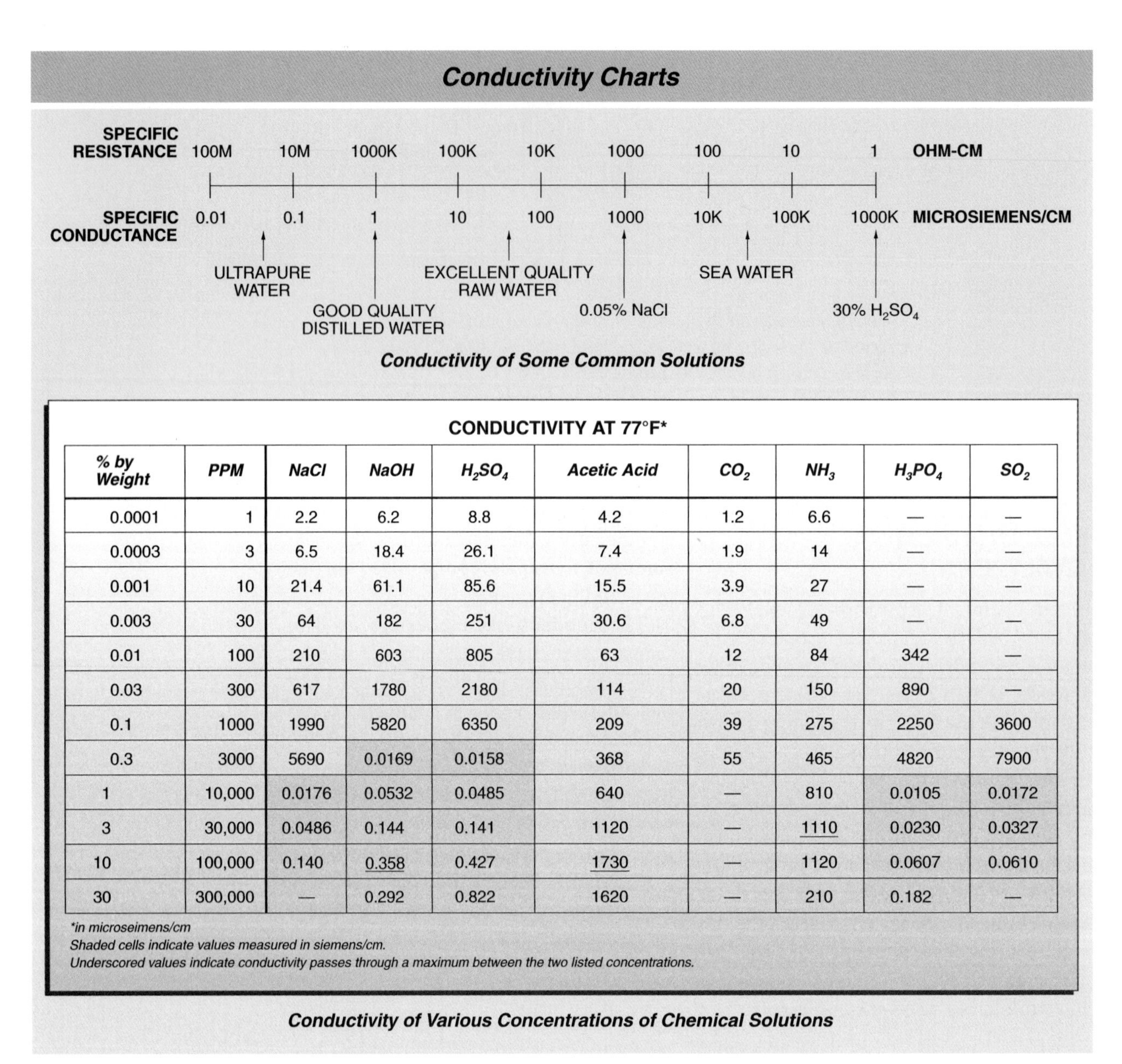

CONDUCTIVITY AT 77°F*

% by Weight	PPM	NaCl	NaOH	H_2SO_4	Acetic Acid	CO_2	NH_3	H_3PO_4	SO_2
0.0001	1	2.2	6.2	8.8	4.2	1.2	6.6	—	—
0.0003	3	6.5	18.4	26.1	7.4	1.9	14	—	—
0.001	10	21.4	61.1	85.6	15.5	3.9	27	—	—
0.003	30	64	182	251	30.6	6.8	49	—	—
0.01	100	210	603	805	63	12	84	342	—
0.03	300	617	1780	2180	114	20	150	890	—
0.1	1000	1990	5820	6350	209	39	275	2250	3600
0.3	3000	5690	0.0169	0.0158	368	55	465	4820	7900
1	10,000	0.0176	0.0532	0.0485	640	—	810	0.0105	0.0172
3	30,000	0.0486	0.144	0.141	1120	—	1110	0.0230	0.0327
10	100,000	0.140	0.358	0.427	1730	—	1120	0.0607	0.0610
30	300,000	—	0.292	0.822	1620	—	210	0.182	—

*in microsiemens/cm
Shaded cells indicate values measured in siemens/cm.
Underscored values indicate conductivity passes through a maximum between the two listed concentrations.

Conductivity of Various Concentrations of Chemical Solutions

Figure 26-1. The conductivity of a solution depends on the concentration of the electrolyte.

The measurement of conductivity is useful in many industrial processes for the determination of ionic concentrations. Since the measurement of conductivity is an indication of the total ions present in a solution, the scale of the instrument can be calibrated in ppm or percent concentration of electrolyte (conductive material in solution). This is frequently an important consideration in determining the purity of water or the completeness of a chemical reaction.

For example, conductivity is used to measure the total dissolved solids in boiler water. If the conductivity of boiler water is over a specified limit, the boiler must be blown down. Conductivity measurement is also used to detect the presence of liquid carryover in a steam line. A common use for a conductivity analyzer is for checking the quality of the recovered steam condensate being used for feedwater. **See Figure 26-2.**

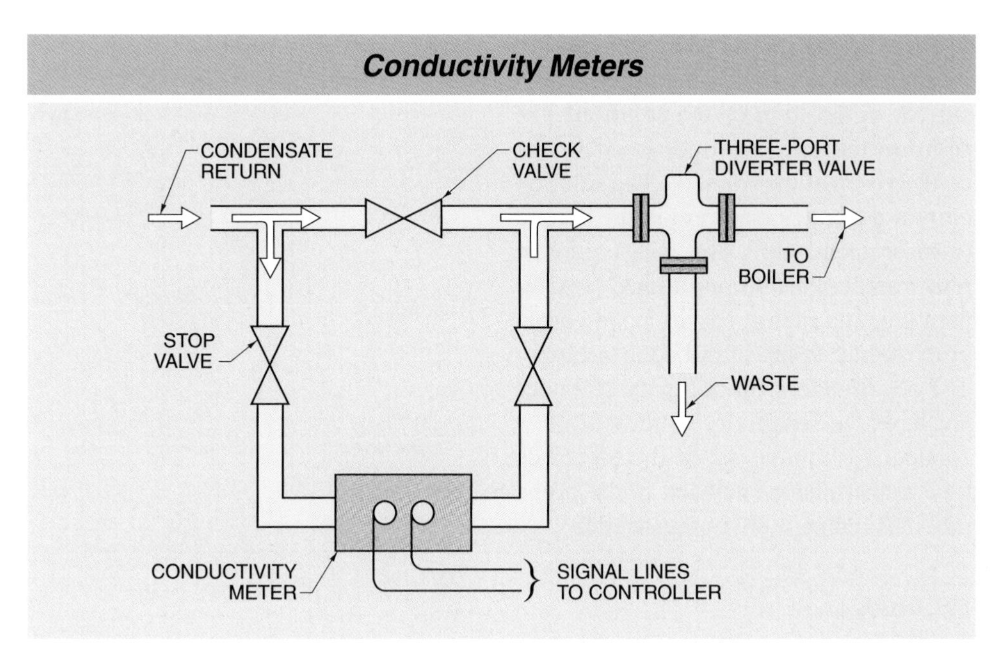

Figure 26-2. Conductivity meters are used to measure dissolved solids in condensate return. When the amount of solids exceeds a predetermined level, the condensate cannot be reused.

Conventional Conductivity Cells. Conductivity cells are available in probe form. They contain titanium-palladium or graphite electrodes with plastic insulation. Conventional conductivity cells are available in various cell constants. A thermistor may be enclosed in the cell for temperature compensation. Some conductivity cells have four electrodes with two used to measure the solution and the other two as reference electrodes. **See Figure 26-3.**

Electrodeless Conductivity Cells. A conductivity measuring system without electrodes is also available for measuring liquids that are very corrosive. In this system, two toroidally wound coils (wire-wrapped metal rings) are encapsulated in a flow-through protective plastic cover and immersed in the liquid solution. The coils surround the bore of the probe so that a loop of the solution couples them.

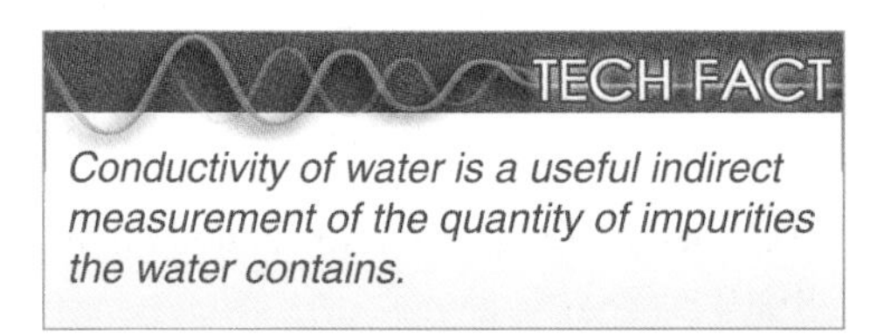

Conductivity of water is a useful indirect measurement of the quantity of impurities the water contains.

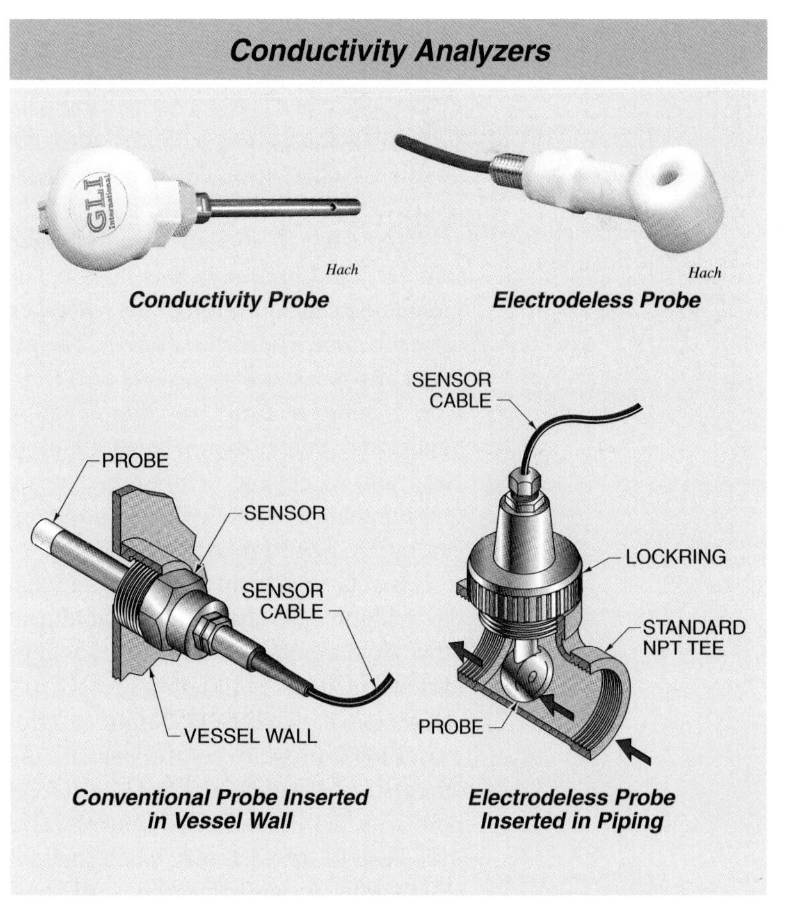

Figure 26-3. Conductivity analyzer probes come in a variety of configurations.

An AC signal produced by an oscillator is applied to one coil, which generates a current in the loop of the solution. The solution current is proportional to the conductivity of the solution. The solution current generates a current in the second (measurement) coil. The measured current is amplified and conditioned for temperature, and then it serves as an output proportional to the liquid concentration. The electrodeless conductivity cell does not have the sensitivity to measure low conductive liquids. As in the electrode type, a thermistor enclosed in the probe provides temperature compensation.

pH Analyzers

A *pH analyzer* is an electrochemical analyzer consisting of a cell that generates an electric potential when immersed in a sample. A *pH* is the measurement of the acidity or alkalinity of a solution caused by the dissociation of chemical compounds in water. As these compounds dissociate, electrically charged ions are formed in the solution. Some ions are positively charged and some are negatively charged.

Water itself dissociates into hydrogen ions (H^+) and hydroxyl ions (OH^-). The greater the concentration of the hydrogen ions, the more acidic the solution. The pH is usually measured on a scale of 0 to 14 with 7 being neutral, less than 7 being acidic, and greater than 7 being alkaline. Each unit of change of pH represents a tenfold change in the hydrogen ion concentration. **See Figure 26-4.**

It is very important to understand that the changes in pH are extremely nonlinear when determining the amount of acid or alkaline reagents required to neutralize a sample or change the pH. Nonlinear processes are difficult to control because the amount of control is different at different points on the pH scale. A control valve may have to meter a very small amount of solution at one pH and a very large amount of solution at another pH.

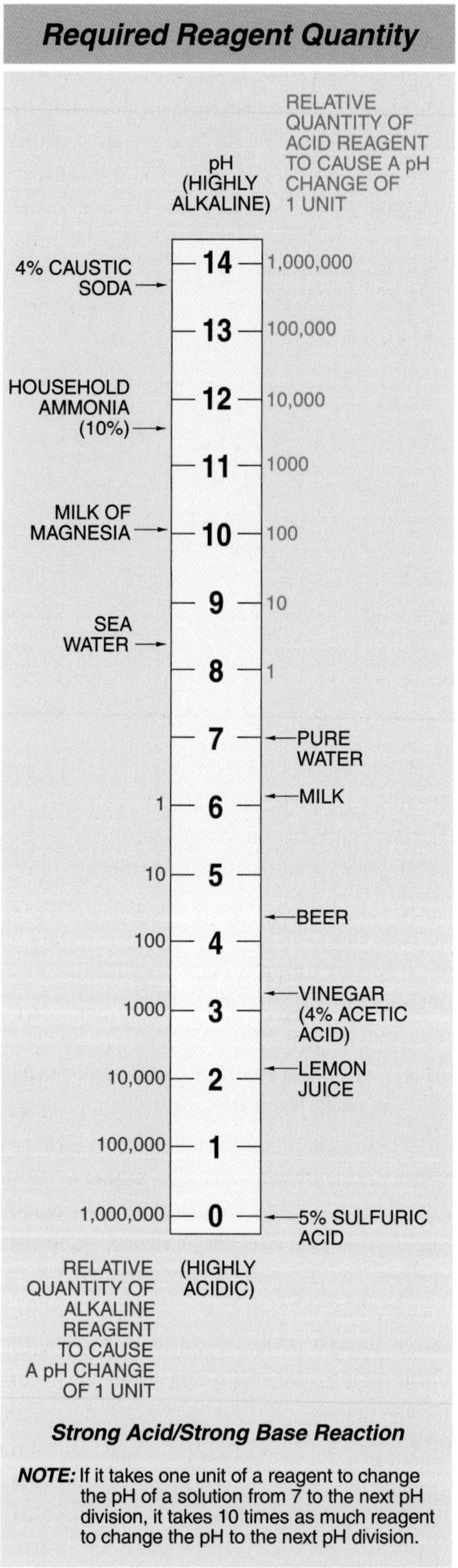

Required Reagent Quantity

Substance	Relative quantity of alkaline reagent to cause a pH change of 1 unit	pH (highly alkaline at top, highly acidic at bottom)	Relative quantity of acid reagent to cause a pH change of 1 unit	Substance
4% CAUSTIC SODA		14	1,000,000	
		13	100,000	
HOUSEHOLD AMMONIA (10%)		12	10,000	
		11	1000	
MILK OF MAGNESIA		10	100	
SEA WATER		9	10	
		8	1	
		7		PURE WATER
	1	6		MILK
	10	5		BEER
	100	4		
	1000	3		VINEGAR (4% ACETIC ACID)
	10,000	2		LEMON JUICE
	100,000	1		
	1,000,000	0		5% SULFURIC ACID

Strong Acid/Strong Base Reaction

NOTE: If it takes one unit of a reagent to change the pH of a solution from 7 to the next pH division, it takes 10 times as much reagent to change the pH to the next pH division.

Figure 26-4. The amount of reagent required to change the pH of a solution depends on the original pH and the amount of desired change.

A *buffered solution* is a solution of an acid or a base and another chemical compound in water where one of the parts of the compound is more reactive with the acid or base than the other. For example, sodium carbonate in water is a buffered solution. The sodium part of the compound reacts with an acid, but readily dissolves again. The carbonate part of the compound reacts with the acid, but further reactions proceed and consume some of the acid.

An *unbuffered solution* is a solution of a strong acid or strong base without any other chemicals that react with the acid or base. For an unbuffered solution, if it takes 1 unit of a reagent to change the pH of the solution from 8 to 7, it takes 10 units of reagent to change the pH from 9 to 8. For each further 1-unit change of pH, it takes 10 times as much reagent to neutralize the sample as the previous step. The same principle applies when starting with an acidic solution.

A *titration curve* is a graph that shows the quantities of reagent required to change the pH of a solution. A titration curve for a strong acid, hydrochloric acid (HCl), and a strong base, sodium hydroxide (NaOH), is a curve that is very steep near the neutral point. A strong acid, HCl, mixed with a weak base, sodium carbonate (Na_2CO_3), is a buffered solution. **See Figure 26-5.**

For a buffered Na_2CO_3 solution, the pH starts dropping when HCl is added. The acid then starts reacting with the Na_2CO_3 to form sodium chloride (NaCl), water (H_2O), and carbon dioxide (CO_2). These reactions consume some of the acid that is then not available to change the pH.

Therefore, in a buffered solution, more acid is required to change the pH than with an unbuffered solution. This has the effect of making the pH curve less steep at 7.0 and steeper at other portions of the titration curve. When dealing with pH systems it is vital to understand the titration curve. If the pH curve is steep at the desired control point, it is very difficult to control the pH of the process because a very small amount of reagent causes a very large change in pH.

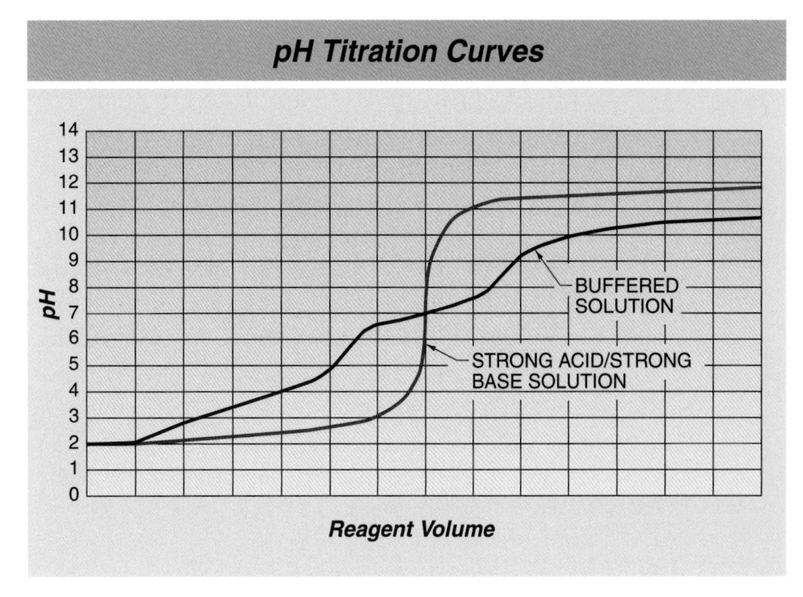

Figure 26-5. A pH titration curve is used to determine the amount of reagent required to change the pH of a solution.

TECH FACT

Buffers help control pH by preventing wide swings in chemical concentrations. The alkaline papermaking process uses a buffered calcium carbonate solution that maintains a stable pH value without continuous monitoring.

Common applications of pH measurement are the control of boiler water pH and the pretreatment of wastewater. Boiler water must be maintained in an alkaline condition to prevent corrosion in the boiler. Wastewater must be kept within a certain pH range before being discharged. The operator needs to understand the relationship between pH and the amount of chemical to add to the water in order to establish and maintain the desired pH.

pH Electrodes. The measurement of pH requires the use of specially designed electrodes. **See Figure 26-6.** Two electrodes must be used to obtain a measurement. One electrode produces a change in voltage with changes of the pH of the solution in which it is immersed. This is due to the special H^+ ion sensitive tip.

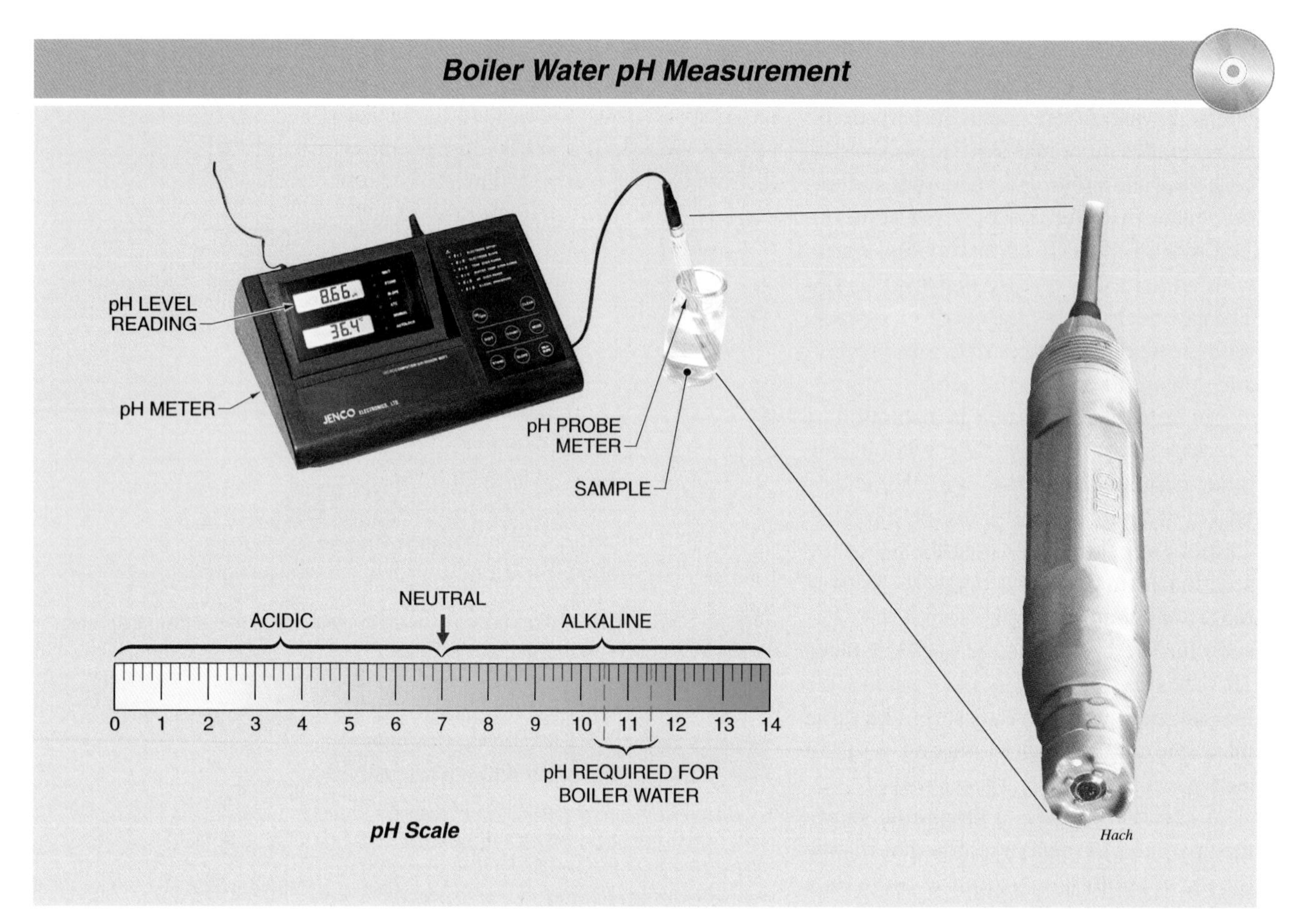

Figure 26-6. Boiler water must be maintained in an alkaline condition.

The most common pH sensitive electrode is the glass electrode. The other electrode maintains a constant voltage and is called the reference electrode. The most common types of reference electrodes are the solid-state electrode containing a field effect transistor, the silver chloride electrode, and the calomel electrode. In modern pH meters, the measuring electrode and the reference electrode are in the same tube that is immersed into the process fluid.

Together, these electrodes form an electrolytic cell whose output equals the difference in the voltage produced by the two electrodes. The voltage difference is measured and amplified with electronic circuits. A temperature-compensating resistor is frequently included to provide temperature compensation. The common styles of electrode holders are flow-through, where a sample flows through the holder; submersible, where both the electrodes and the holder are sealed so that they can be placed beneath a liquid surface; and insertable, where the electrodes are inserted through seals into a process pipeline.

Each pH electrode holder has its advantages and disadvantages depending on the process configuration. The insertable type is rarely used due to the additional problems that are encountered with maintaining the seals.

Oxidation-Reduction Potential (ORP) Analyzers

An *oxidation-reduction potential (ORP) analyzer* is an electrochemical analyzer consisting of a metal measuring electrode and a standard reference electrode that measure the voltage produced by an electrochemical reaction between the metals of the electrodes and the chemicals in solution. **See Figure 26-7.** The voltage is proportional to the chemical concentration.

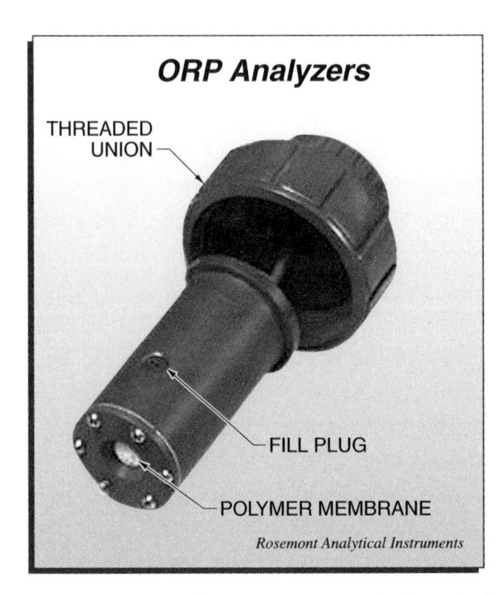

Figure 26-7. An ORP analyzer can be installed directly in a process line.

An ORP cell is very similar to a pH electrode and acts like the plates of a battery. Oxidation means a molecule or element loses electrons. Reduction means electrons are added to a molecule. Electrons move from one electrode to the other. This movement produces the ORP voltage.

Not all chemical activity results in a usable ORP output, but those that do can provide information on the chemical reaction unattainable by any other method. The most successful approach is to use ORP to measure the completeness of a process when there is an oxidation-reduction reaction occurring. The ORP probe's output voltage is a clear indication of the completeness of the reaction.

There are several metal combinations used for special conditions. A conventional pH measurement instrument can also be used for ORP measurements, but with a plain millivolt range. Manufacturers offer a selection of metal combinations for particular applications.

For example, in bleach manufacture where a diluted sodium hydroxide solution is chlorinated, a strong voltage is generated as the process approaches the end point of the reaction. This uses a silver/platinum ORP probe. In water treatment, where residual free chlorine in water is measured, a gold/silver ORP probe is used. In waste treatment, where anaerobic and aerobic microorganisms must be controlled, a standard ORP probe is used. The microorganisms are dependent on oxidation-reduction reactions and the reactions are easily measured with a standard ORP probe.

In chromium waste treatment, where toxic hexavalent chromium (Cr^{+6}) is reduced to trivalent chromium (Cr^{+3}) through a reaction with sulfur dioxide (SO_2), ferric sulfate ($FeSO_4$), or sodium bicarbonate ($NaHCO_3$), the output voltage of a standard ORP probe is a clear indication of the degree of treatment. Then the Cr^{+3} is further reduced to chromium hydroxide through the addition of an alkaline solution. A pH measurement is used to determine the degree of treatment in this reaction.

COMPOSITION ANALYZERS

A *composition analyzer* is an analyzer used to measure the quantity of multiple components in a single sample from a process. Composition analyzers are very complex and use special methods to isolate the individual components so that they can be analyzed.

Chromatography

Chromatography is the process of separating components of a sample transported by an inert carrier stream through a variety of media. Different components of the sample attach themselves to the media with different strengths. The carrier stream sweeps the weakly attached components from the material before the stronger attachments are broken. Therefore, the different components of the sample have different travel times through the column.

A *chromatograph analyzer* is an instrument consisting of a small stainless steel tube packed with a porous inert material such as silica gel or alumina, an injection valve assembly, and a detector. A *chromatographic column* is a stainless steel tube or length of capillary tubing of a chromatography instrument after the tube is filled with packing and ready to use.

Columns can be packed with many different materials and can be of different lengths, depending on the separation needed between components. The packing is porous and has spacing between the individual pieces of the packing. This allows the sample to flow through the column. The different components of the sample interact with the packing and travel through the column at different rates. **See Figure 26-8.**

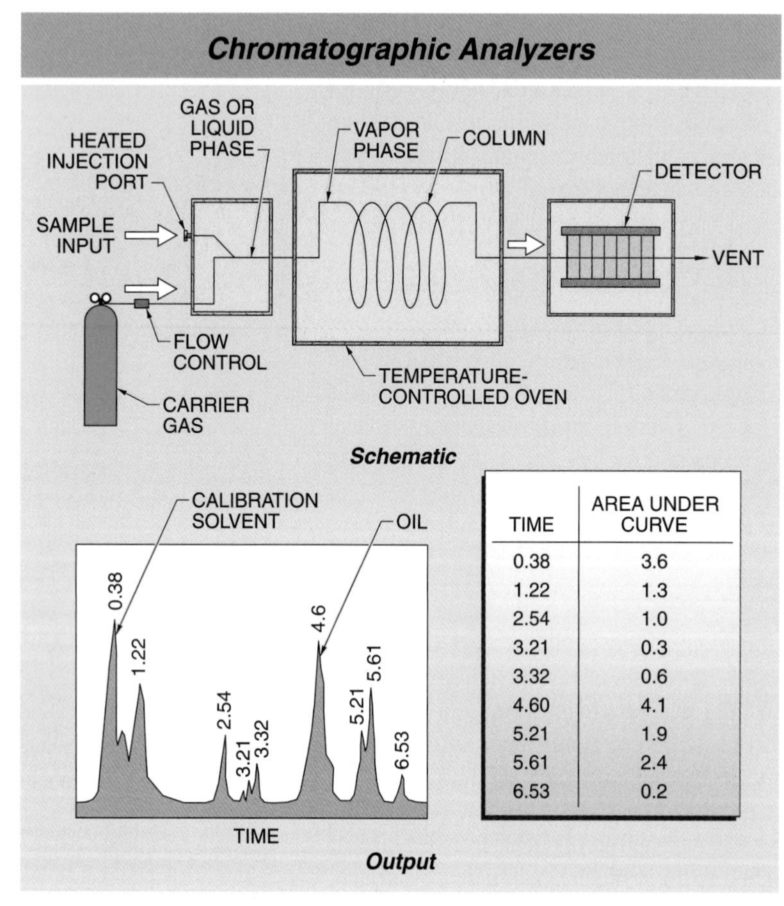

TIME	AREA UNDER CURVE
0.38	3.6
1.22	1.3
2.54	1.0
3.21	0.3
3.32	0.6
4.60	4.1
5.21	1.9
5.61	2.4
6.53	0.2

Figure 26-8. A chromatograph is used to determine the identity and quantity of components of a solution.

The injection port assembly combines the sample stream with the carrier gas or liquid and forces the combined fluid through the column and to the detector. The injection port also controls the carrier gas, such as hydrogen, nitrogen, helium, or argon, so that it is purging the column or carrying the sample into the column. During the analysis period, a constant flow of a carrier gas is purged through the separation column. Some components of the sample pass through the column more quickly than others; thus the components become separated.

The detector, which can be a thermal conductivity analyzer, flame ionization analyzer, or one of a number of other types, senses the differences between the carrier fluid and the individual components. The time between when the sample is injected into the column and when the peak is measured is unique to each compound.

A graph of the resulting analysis is a horizontal baseline with a series of peaks. The area or height of the peak represents the quantity of the specific component. The time and amplitude of each peak is established by specially prepared samples where the composition and quantity of all the components are known. Modern computerized gas chromatographs can provide digital or analog outputs for each of the desired components.

Gas Chromatographic (GC) Analyzers. A gas chromatographic analyzer is used to separate individual components of a gas sample so that the quantity of each component can be measured. The gas sample is injected into the column and carried along with the carrier gas stream. Gas chromatography is relatively quick and can be used as an in-line analyzer.

Liquid Chromatographic Analyzers. A liquid chromatographic analyzer is used to test many types of samples where a sample is sent through the column as a liquid. The sample can also be dissolved in a solvent or oil before testing. The solvent or oil can then be used as a calibration standard. Liquid chromatography is very slow compared to gas chromatography. *High-pressure liquid chromatography (HPLC)* is a type of liquid chromatography that uses high pressure to force the liquid sample through a column at a faster rate than the liquid would normally travel. As an alternative, a heated injection valve is used to vaporize the liquid to allow the sample to quickly move through the column as a gas.

Infrared Spectroscopy

Several types of composition analyzers use infrared radiation to identify the composition of a sample. Almost all molecules absorb infrared radiation. Each type of molecule absorbs infrared radiation at certain frequencies, depending on the type of chemical bond. This property of molecules provides a unique characteristic for each molecule. It provides a way to identify the molecule type (qualitative analysis) and the amount or quantity of the molecule (quantitative analysis).

The goal of any absorption spectroscopy is to measure how well a sample absorbs, transmits, or reflects infrared radiation at different wavelengths. The simplest way to do this is to shine a monochromatic beam through a sample, measure how much of the light is absorbed, and repeat for each different frequency. **See Figure 26-9.** Many infrared instruments use a wavenumber (cm^{-1}) instead of frequency. The wavenumber is related to the inverse of the frequency.

Chemical feed pumps are used to add treatment chemicals to process water to maintain proper pH and conductivity.

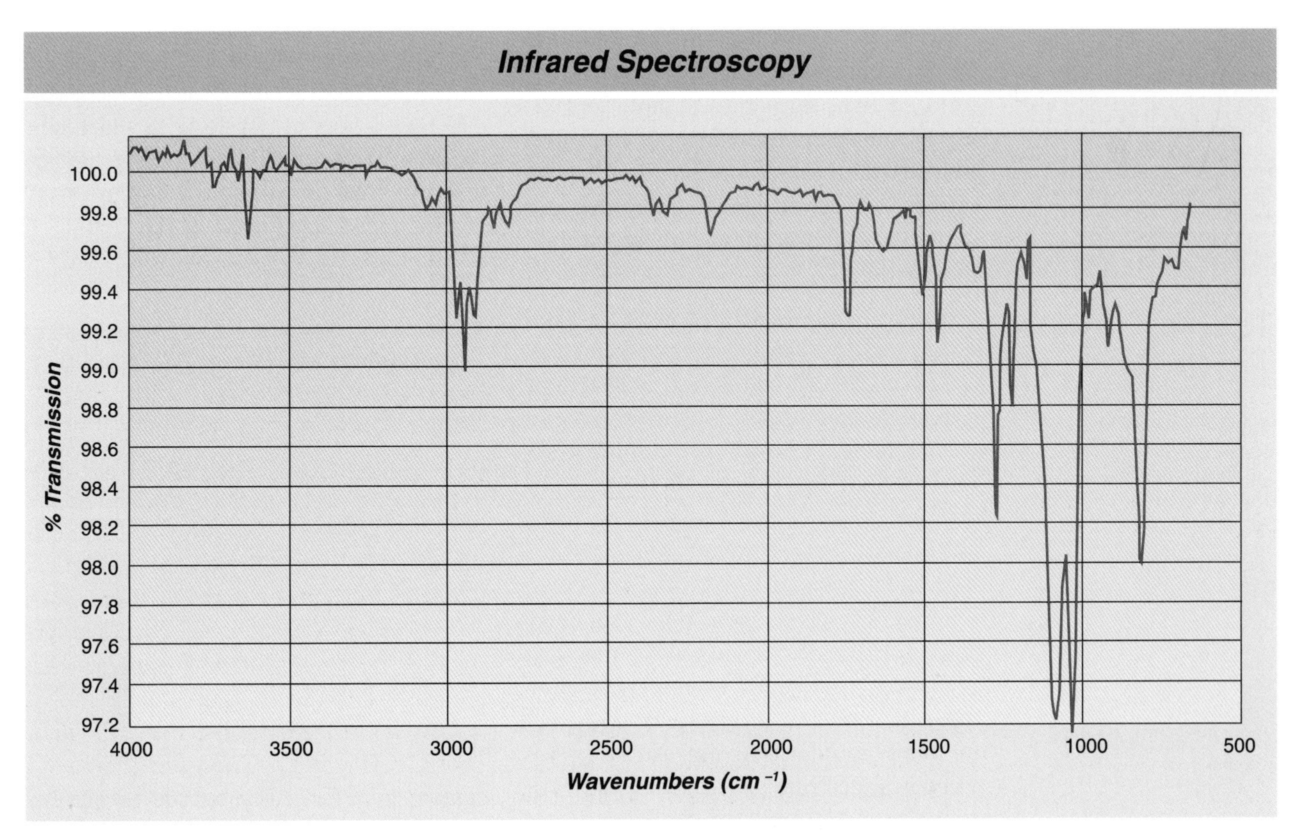

Figure 26-9. Infrared spectroscopy measures the transmission of infrared radiation through a sample at different frequencies.

Each measurement gives a peak in the spectrum that indicates the amount of infrared radiation absorbed. The height of the peak depends on the absorptivity and concentration of the molecule. The concentration can be determined by comparing the height of an absorption peak of the molecule in a reference gas spectrum to the height of the corresponding peak in a sample gas.

Near Infrared (NIR) Analyzers. A *near infrared (NIR) liquid analyzer* is an analyzer that uses infrared radiation to measure the organic molecules in a sample. The analyzer design is similar to the analyzer used to measure moisture in solids except that NIR light normally passes through the sample. To measure the moisture content of a product, two wavelengths of IR radiation are used in the analyzer. One wavelength is used for measurement and the other for reference. Small concentrations of many organic compounds in water can be accurately measured.

In addition, moisture as low as a few ppm can be measured in organic liquids. Samples are analyzed by passing the infrared light through a sample and using IR transparent windows or by reflection off the surface of the liquid. Near infrared measurement accuracy depends strongly on the accuracy of the calibration method.

Fourier Transform Infrared (FTIR) Analyzers. A *Fourier transform infrared (FTIR) analyzer* is an infrared analyzer that uses a Michelson interferometer to examine a sample with a broad spectrum of infrared radiation. **See Figure 26-10.** An FTIR analyzer passes a beam containing many different frequencies of light and measures how much of that beam is absorbed or reflected by the sample. Next, the beam is modified to use a different combination of frequencies to make a second measurement. This process is repeated many times to create an interferogram. An *interferogram* is a spectrum developed by an FTIR analyzer and is similar to the interference pattern developed by a hologram.

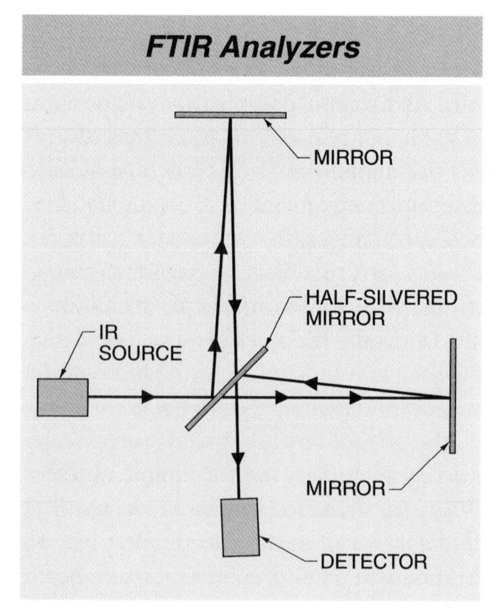

Figure 26-10. An FTIR analyzer uses a half-silvered mirror to split an IR beam.

In operation, the broadband light source shines into a certain configuration of mirrors that allows some frequencies to pass through but blocks others due to wave interference. For each new measurement, moving one of the mirrors modifies the beam. A computer then uses a Fourier transform to generate a spectrum of absorbance at each wavenumber from which the composition of the sample can be determined. **See Figure 26-11.**

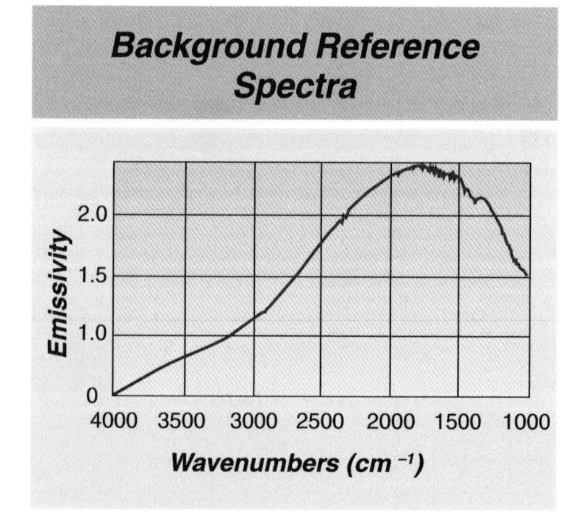

Figure 26-11. An FTIR should be run with the sample chamber empty in order to establish a background scan.

To compensate for interference from the sample chamber, a single beam spectrum is run on the sample chamber without any sample to establish a background reference. This background represents the conditions inside the spectrometer at the moment the background reference was made. It is important to maintain a stable environment inside the spectrometer, meaning the conditions must have minimal fluctuations.

TECH FACT

Boilers can operate at dangerously high temperatures and pressures. Precise control of boiler water pH, alkalinity, hardness, and the concentration of contaminants is essential to prevent corrosion and pitting, scaling, or other damage to the boiler that can cause a fatal explosion.

TITRATION ANALYZERS

A *titration analyzer* is a liquid analyzer consisting of an instrument where a measured quantity of a process sample is mixed with precise quantities of reagents and then quantitatively measured with a pH meter or by a colorimetric or other type of detector. **See Figure 26-12.**

In a typical system, a sample flows into the analyzer and the overflow siphon controls the volume of the liquid. The titration pump is activated and the titrating solution is added to the analyzer until the endpoint is reached. The sensors detect the endpoint and send a signal to shut off the titration pump. The concentration of the sample is proportional to the amount of titrating solution used.

An automatic titration analyzer is specially designed for the analysis of a specific component and is usually quite complex. It is used to measure quantities and chemicals that cannot be measured on-line in any other manner. Maintenance and calibration require a considerable amount of effort by skilled personnel.

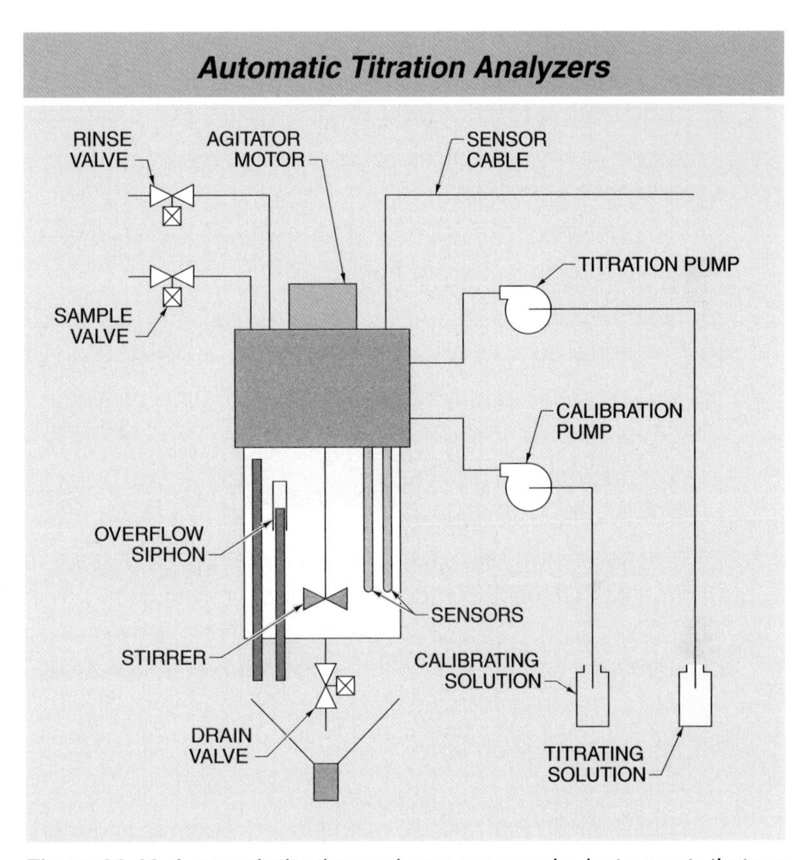

Figure 26-12. Automatic titration analyzers are complex instruments that are only used when no other measurement methods are available.

KEY TERMS

- *conductivity analyzer:* An electrochemical analyzer that measures the electrical conductivity of liquids and consists of two electrodes immersed in a solution.
- *siemens:* The modern unit of electrical conductivity and is the reciprocal of resistance.
- *conductivity cell constant:* The ratio of the size of the actual electrodes to those of the standard conductivity cell.
- *pH analyzer:* An electrochemical analyzer consisting of a cell that generates an electric potential when immersed in a sample.

KEY TERMS *(continued)*

- *pH:* The measurement of the acidity or alkalinity of a solution caused by the dissociation of chemical compounds in water.
- *buffered solution:* A solution of an acid or a base and another chemical compound in water where one of the parts of the compound is more reactive with the acid or base than the other.
- *unbuffered solution:* A solution of a strong acid or strong base without any other chemicals that react with the acid or base.
- *titration curve:* A graph that shows the quantities of reagent required to change the pH of a solution.
- *oxidation-reduction potential (ORP) analyzer:* An electrochemical analyzer consisting of a metal measuring electrode and a standard reference electrode that measure the voltage produced by an electrochemical reaction between the metals of the electrodes and the chemicals in solution.
- *composition analyzer:* An analyzer used to measure the quantity of multiple components in a single sample from a process.
- *chromatography:* The process of separating components of a sample transported by an inert carrier stream through a variety of media.
- *chromatograph analyzer:* An instrument consisting of a small stainless steel tube packed with a porous inert material such as silica gel or alumina, an injection valve assembly, and a detector.
- *chromatographic column:* A stainless steel tube or length of capillary tubing of a chromatography instrument after the tube is filled with packing and ready to use.
- *high-pressure liquid chromatography (HPLC):* A type of liquid chromatography that uses high pressure to force the liquid sample through a column at a faster rate than the liquid would normally travel.
- *near infrared (NIR) liquid analyzer:* An analyzer that uses infrared radiation to measure the organic molecules in a sample.
- *Fourier transform infrared (FTIR) analyzer:* An infrared analyzer that uses a Michelson interferometer to examine a sample with a broad spectrum of infrared radiation.
- *interferogram:* The spectrum developed by an FTIR analyzer and is similar to the interference pattern developed by a hologram.
- *titration analyzer:* A liquid analyzer consisting of an instrument where a measured quantity of a process sample is mixed with precise quantities of reagents and then quantitatively measured with a pH meter or by a colorimetric or other type of detector.

REVIEW QUESTIONS

1. What is the purpose of a conductivity analyzer?
2. Define pH and explain why pH is difficult to control in an industrial process.
3. How does an oxidation-reduction potential (ORP) analyzer function?
4. How can chromatography be used to identify the components of a sample?
5. Compare the operation of a near infrared (NIR) liquid analyzer and a Fourier transform infrared (FTIR) analyzer.

SECTION

7 POSITION MEASUREMENT

TABLE OF CONTENTS

SECTION OBJECTIVES

Chapter 27

- Describe the use of mechanical switches.
- List the types of proximity sensors, and describe their applications.

Chapter 28

- Describe some of the factors that affect sensor installation.
- Describe how sensors are used for rotary speed sensing.
- Describe how sensors are used for continuous web handling.
- Explain how an optical sensor is used as a light curtain.

SECTION

7 POSITION MEASUREMENT

INTRODUCTION

The fields of process automation and factory automation have been considered to be separate areas. However, these fields have been converging in recent years as control systems, networks, instruments, and sensors have become more capable. The Instrumentation Society of America (ISA) was founded in 1945 to promote good practices in instrumentation and process automation. Over the years, the organization changed its name several times to reflect the growing influence of factory automation. The new International Society of Automation (ISA) develops standards for many types of instruments and control in all types of automation.

Industrial switches and sensors are used in many automated industries for detecting position and motion. Some of the many applications include equipment safety shutoff systems, materials conveying, item counting, elevator control, packaging machinery stops, air or hydraulic positioning, valve positioning, and rotary speed sensing. The switches and sensors provide position information for operational status, alarm, or interlocking systems. Switches and sensors use a number of methods to detect the position of an object including direct mechanical contact and noncontact devices such as inductance, capacitance, photoelectric, and ultrasonic systems.

ADDITIONAL ACTIVITIES

1. After completing Section 7, take the Quick Quiz® included with the Digital Resources.
2. After completing each chapter, answer the questions and complete the activities in the *Instrumentation and Process Control Workbook.*
3. Review the Flash Cards for Section 7 included with the Digital Resources.
4. After completing each chapter, refer to the Digital Resources for the Review Questions in PDF format.

ATPeResources.com/QuickLinks • Access Code: 467690

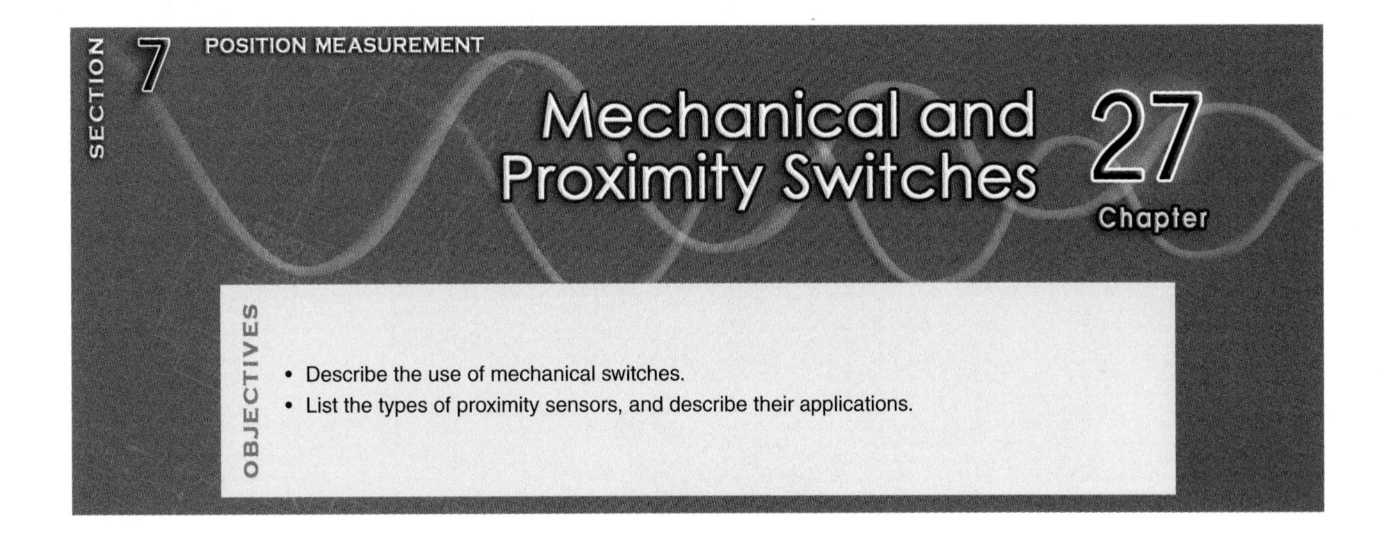

MECHANICAL SWITCHES

A *mechanical switch* is a switch that requires physical contact with an object to actuate a switch mechanism. Mechanical switches are widely used in industry because the trip points can be set very accurately. However, the switches can become damaged and fail if the switches are exposed to outdoor weather or corrosive service conditions. This failure is commonly due to corrosion of the moving parts, especially if the switch is not frequently actuated.

Mechanical Switch Designs

The switch-actuating mechanism typically is spring loaded so that the switch returns to its original position when no longer in contact with the object. The switch-actuating mechanism can have many forms. **See Figure 27-1.** One of the simplest forms is a simple button type that requires a direct pressure against the button.

Mechanical switches are also designed with a pivoting arm that actuates the internal switch mechanism. A pivoting arm reduces the amount of force required to actuate the switch mechanism, but increases the travel distance required to activate the mechanism. A major advantage of using a pivoting arm is that it allows an object to be detected in an enclosed area while the switch is mounted in a more accessible position. The arm can be straight or bent at an angle and of almost any length.

TECH FACT

Mechanical switches are often used to detect the presence of products on a conveyor line. The controller can detect the products and send them on to packaging or to another part of the process.

One of the most common arm designs uses a roller to contact the object. A roller is used where the arm has to rub on a moving object and is actuated by a change in the surface. The roller reduces both the wear and tear on the sensor arm tip and the frictional resistance against the moving object. An arm with a roller on the end provides a very flexible arrangement. The surface to be monitored can be linear with occasional changes in the profile that move the arm and actuate the switch.

Mechanical Switch Output

Mechanical switches come in a variety of sizes and can actuate either electrical or pneumatic signals. Electrical signals can be either AC or DC since the switches are direct-contact types. What is important is the electrical current that the switch can handle. The form of the switch can be normally open (NO), normally closed (NC), or Form C. A *Form C electrical contact* is a single-pole double-throw switch that allows both NO and NC contacts.

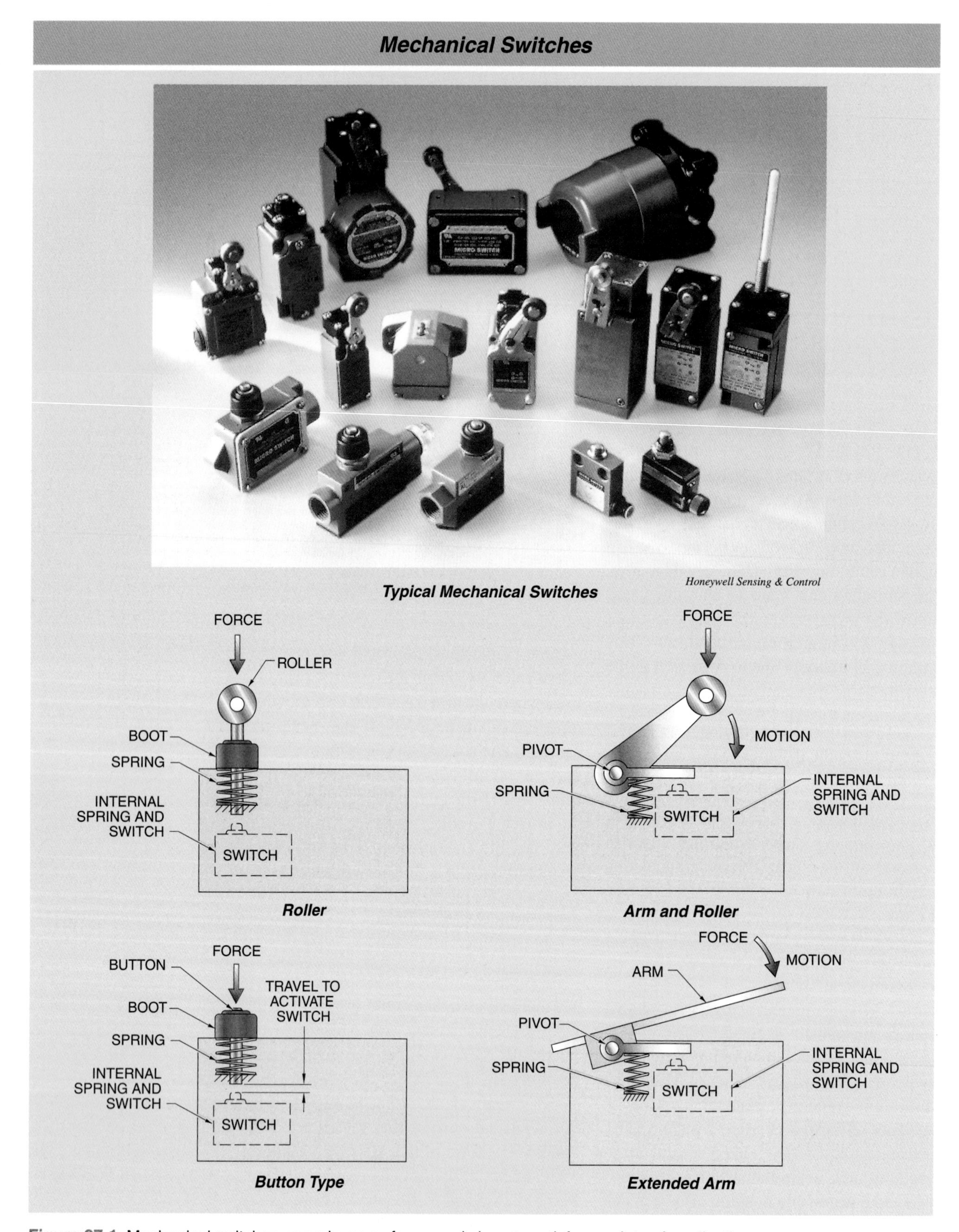

Figure 27-1. Mechanical switches come in many forms and sizes to satisfy a variety of applications.

The best choice is usually a Form C because this provides the opportunity to change the action of the switch in the field if necessary. If the switch housing is large enough, it may be possible to use a double-pole double-throw switch. This allows a single switch to actuate two separate electrical circuits. The electrical housings are typically available as general purpose (NEMA 12), waterproof (NEMA 4), or explosionproof (NEMA 7).

PROXIMITY SWITCHES

A proximity switch uses a proximity sensor to detect the presence of an object. A *proximity sensor* is a sensor that detects the presence of an object without requiring contact with the object. Proximity switches are typically used when the target object is light and easily damaged by physical contact, when the sensing rate needs to be very quick, when the target object must be sensed through plastic or glass, or when the sealed construction of proximity switches is suited to the environmental conditions.

Electrical Proximity Sensors

An *electrical proximity sensor* is a proximity sensor that uses inductance and capacitance properties to detect the presence of an object. The basic principles of the inductance and capacitance sensors are different, but the housings, sensing range principles, and interference due to the type of mounting are very similar for both types. **See Figure 27-2.** Both inductance and capacitance sensors have an output circuit with a transistor for DC sensors and a thyristor for AC sensors.

TECH FACT

A Zener diode allows current to flow in the forward direction and in the reverse direction if the voltage is higher than the breakdown voltage of the diode. Zener diodes are often used in electronic controls in hazardous locations because the diodes shunt any arc current to ground before the arc can occur and cause an explosion.

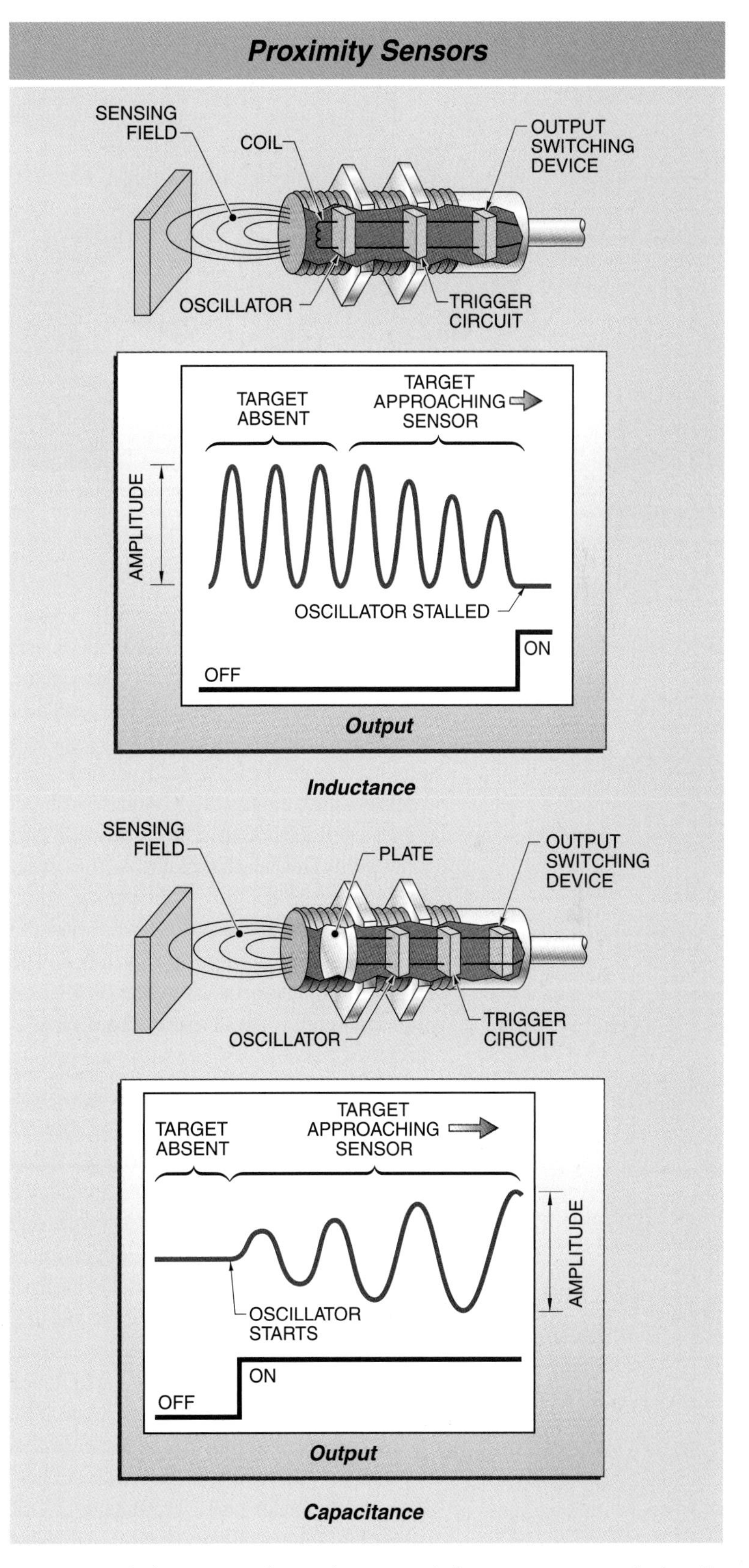

Figure 27-2. Inductance and capacitance proximity sensors are used primarily to sense metal objects.

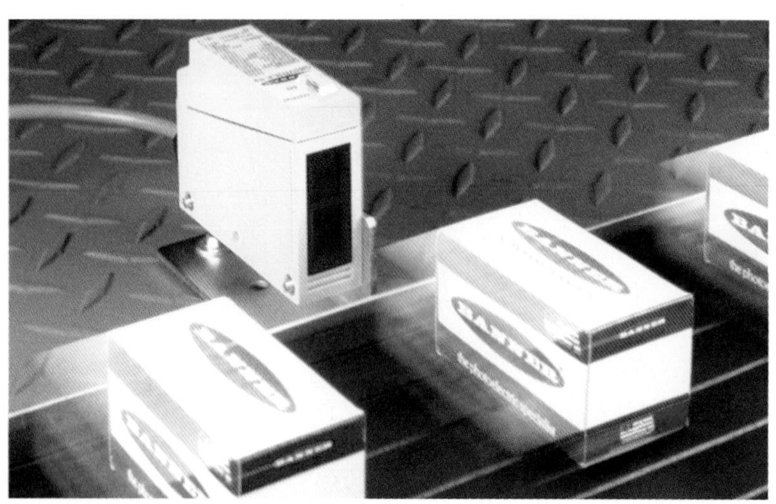

Banner Engineering Corp.

Retroreflective sensors are commonly used with conveyor systems.

Inductance Sensors. An *inductance proximity sensor* is a proximity sensor that consists of a sensor coil, an oscillator, a trigger circuit, and an output switching circuit. The oscillator generates a radio frequency (RF) signal that is emitted from the coil. Eddy currents are generated in an electrically conductive object as it enters the RF field. These eddy currents draw energy from the oscillator. As the object moves closer to the coil in the sensor, more energy is drawn from the oscillator and the amplitude of the oscillation is reduced. The trigger circuit detects when the oscillation stops and sends a signal to the output switching circuit that changes the state of the output.

Capacitance Sensors. A *capacitance proximity sensor* is a proximity sensor that consists of a sensor plate, an oscillator, a trigger circuit, and an output switching circuit. A capacitance sensor acts as a simple capacitor with one plate built into the sensor and the target object being the other plate. The oscillator is inactive when the target object is outside the sensing range. As the target object nears the sensor, the circuit detects the increasing capacitance between the object and the sensor plate. The circuit then starts oscillating and increases the amplitude of the oscillation as the object gets closer to the sensor plate. The trigger circuit detects when the oscillation begins and sends a signal to the output switching circuit that changes the state of the output.

Sensor Output Circuits. Sensor power and output circuits for both induction and capacitance sensors consist of 2-wire, 3-wire, and 4-wire DC, and 2-wire and 4-wire AC designs. Each form requires a different wiring design and different considerations for the output device attached to the sensor. **See Figure 27-3.** Transistor outputs are either NPN (sinking) or PNP (sourcing).

A 2-wire DC sensor can be normally open or normally closed and may or may not be polarity-sensitive. Since there are only two wires, the load is always in the circuit. The load is the output device being actuated. That means that in the low output state there is a small leakage current passing through the load. This leakage into the discrete inputs of a distributed control system (DCS) or a programmable logic controller (PLC) system may be sufficient to actuate the input channel. In the high output state, there is a voltage drop across the sensor so the load device does not receive the full voltage.

A 3-wire DC sensor can be normally open or normally closed. The choice of an NPN or PNP transistor determines whether the circuit is normally open or normally closed. Power is provided to the sensor through two polarized wires. The output load receives its power from the third wire. The load circuit is completed to the positive (+) power wire for NPN transistors and to the negative (–) power wire for PNP transistors. The choice between an NPN and a PNP transistor is based on the PLC or DCS input module that is being used. There is essentially zero current passing through the load during the low output state so there is no worry about inadvertently actuating the loads.

A 4-wire DC sensor can be normally open or normally closed. The choice of an NPN or PNP transistor determines whether the circuit is normally open or normally closed. Power is provided to the sensor through two polarized wires. The output loads receive their power from the third and fourth wires. The load circuits are completed to the positive (+) power wire for NPN transistors and to the negative (–) power wire for PNP transistors.

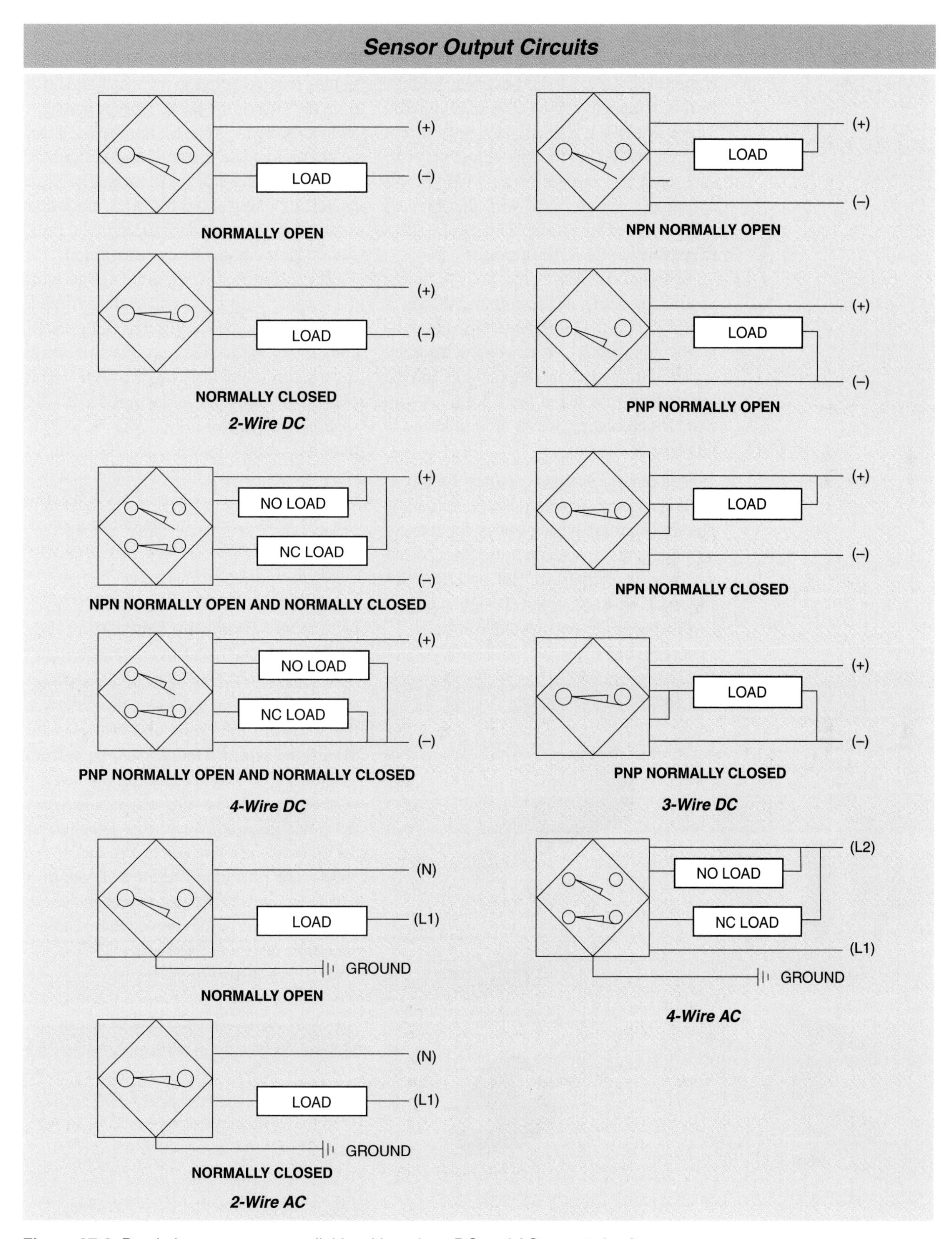

Figure 27-3. Proximity sensors are available with various DC and AC output circuits.

A 2-wire AC sensor has either normally open or normally closed outputs. AC power is provided to the L1 (hot) terminal and the load is in the line to the L2 (neutral) circuit. There is always a small leakage current through the load in the low output state, so care must be exercised as to what type of output devices are used with this type of sensor. A third wire, for grounding, needs to be attached to this type of sensor.

A 4-wire AC sensor has both normally open and normally closed outputs. AC power is provided to the L1 (hot) terminal and the L2 (neutral) circuit. The two output loads are powered from separate output wires and the wiring is terminated at the L2 line. A fifth wire, for grounding, needs to be attached to this type of sensor.

Sensor Styles. A *sensor style* is the physical housing into which a sensor element is placed. Common types of physical housing arrangements are surface-mounted, limit switch style, cylindrical, slot, and ring. **See Figure 27-4.** Each style is designed to best solve a particular sensing application.

Proximity Sensor Styles

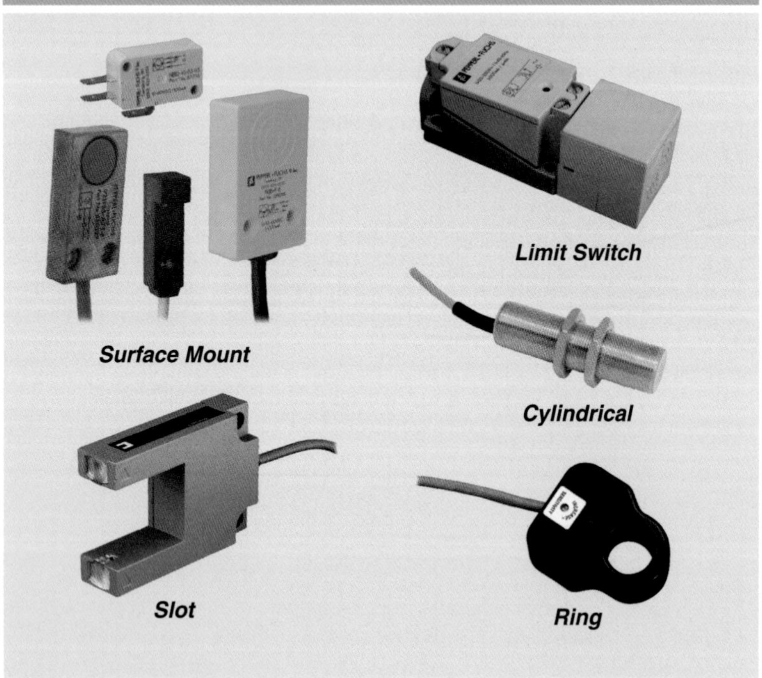

Pepperl+Fuchs, Inc.

Figure 27-4. Proximity sensors are available in a variety of forms.

The surface-mounted style is designed to be bolted to a flat surface. The sensing field is located either in the top of the head or in the side of the head. The limit switch style is surface-mounted, but the head can be positioned in any one of five directions: front, top, either side, or bottom. The limit switch style sensor has the same mounting dimensions as a standard mechanical limit switch so it can be easily substituted.

The cylindrical style is designed with the sensing field at the end of the cylinder. The outside of the cylinder is threaded and is supplied with nuts for mounting into a threaded hole or through a flat plate. The slot style is designed with the sensing field between two coils on a common axis, one on each side of the slot. It is activated when a metallic object passes between the coils. The ring style has the sensing field located inside the ring and is used to sense metallic objects when they enter the ring.

Ultrasonic Proximity Sensors

An *ultrasonic proximity sensor* is a proximity sensor that uses a pulse of sound waves to detect the presence of an object. Sound waves travel toward the target and are reflected toward the sensor. Frequencies range from 65 kHz to 400 kHz, depending on the type of sensor used. The elapsed time between the pulse generation and the detection of the reflection is related to the distance to the target. The ultrasound beam diverges as it leaves the sensor. Ultrasonic sensors have a dead zone close to the sensor where a target cannot be detected. **See Figure 27-5.**

TECH-FACT

Sensors used for position measurement can be connected in series to act as a safety system to prevent an unsafe startup or for shutdown if a dangerous condition exists. This uses AND logic to ensure that all conditions are met in order to operate safely. All sensors must detect normal conditions for the process to operate.

Ambient temperature has an effect on the accuracy of the distance measurement. The sensors are available in cylindrical or rectangular form in an assortment of materials. Sensors are available as single- and double-point outputs as well as analog outputs. The analog output ultrasonic sensors can be used to measure liquid levels. These have the longest ranges, from about 500 mm to about 6 m (24″ to 20′). The power for the sensors is either DC or AC.

Ultrasonic Sensor Outputs. The outputs from the ultrasonic sensor are available in many forms. The choice of the type of output depends on the application and the inputs needed by the controller. Common outputs are as follows:

- single-point transistor outputs using either an NPN (sinking) or PNP (sourcing) circuit that are programmable as either normally open or normally closed forms
- dual-point transistor outputs using two PNP (sourcing) circuits that are programmable as either normally open or normally closed
- single-point thyristor, normally open output
- analog outputs available as either a 4 mA to 20 mA current or a 0 VDC to 10 VDC voltage transmission signal
- 8 bit serial analog outputs in either binary or BCD format with a PNP transistor (sourcing) circuit

Photoelectric Proximity Sensors

A *photoelectric sensor* is a proximity sensor that uses visible light and infrared radiation sources to detect target objects. An LED is usually used as the light or infrared source. Visible light sources are commonly available in red, green, and yellow. **See Figure 27-6.** The sources can be pulsed, which increases the range and life of the source and makes the sensor less susceptible to interference by external light sources. The housings can be cylindrical for the smaller sizes and rectangular for the larger sizes. The wiring for the photoelectric sensors is 3-wire DC, 2-wire AC/DC (relay output), or 4-wire DC. The three common modes of photoelectric detection are diffused (proximity), retro-reflective, and through-beam detection.

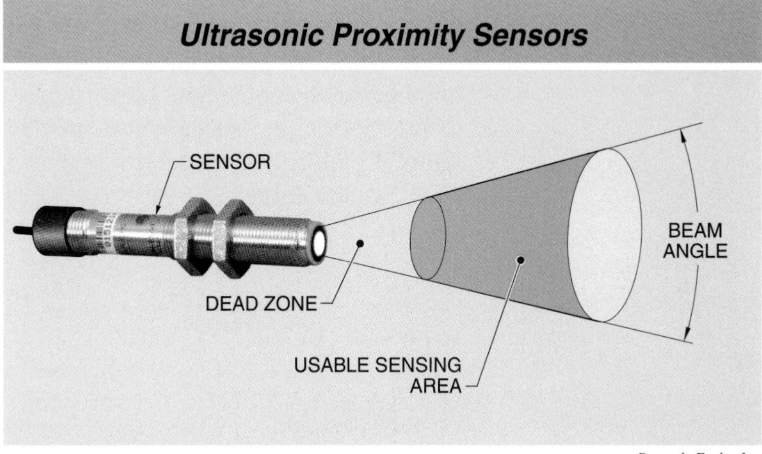

Pepperl+Fuchs, Inc.

Figure 27-5. Ultrasonic proximity sensors have a dead zone close to the sensor where an object cannot be detected.

Photoelectric Sensors

LIGHT SOURCE (LED)

DETECTOR

Figure 27-6. Photoelectric sensors use an LED to generate visible light or an IR beam.

Diffused Mode. A *diffused mode photoelectric sensor* is a photoelectric sensor that directs its source against a target object and detects a reflection from the target object. The reflection is diffused and is thus not very strong. An infrared light source is much stronger than a visible light source and is thus better suited to this type of sensor. **See Figure 27-7.** The color, finish, and size of the target object have a significant impact on whether this photoelectric mode is suitable for an application. Shiny targets reflect more light, but only at a specific angle. Therefore the sensor must be aimed directly at the target.

Photoelectric sensors can have problems in dusty or dirty environments where the light beam can be blocked.

Diffused Mode

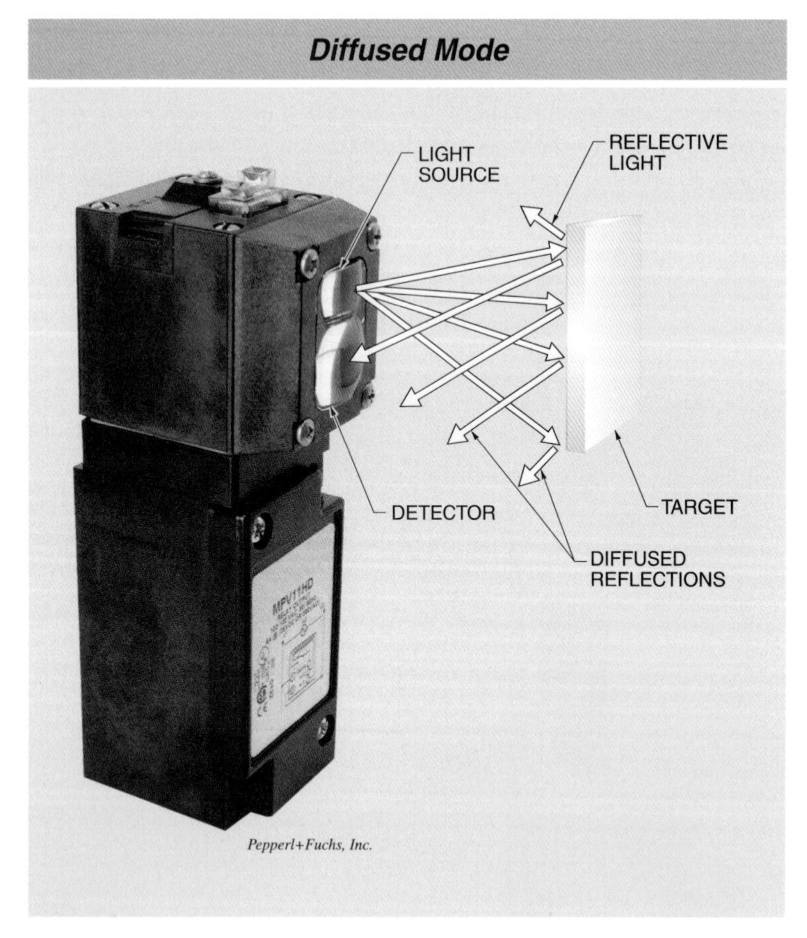

Figure 27-7. Diffused mode sensors measure the reflected light from an object.

Retro-Reflective Mode. A *retro-reflective mode photoelectric sensor* is a photoelectric sensor that uses a focused beam directed across the path of a target object and reflected back to the sensor. The sensor is actuated when there is no object in the path of the beam. The sensor is deactivated when the target object blocks the beam. This requires the reflector to be placed farther away than the object itself, such as across a conveyor. The longer distance that the beam has to travel normally suggests that the strongest source be used. The strongest source is an invisible infrared source, but a visible light source makes it easier to adjust the mirror.

A special corner cube reflector is typically used to reflect the light beam back to the detector. **See Figure 27-8.** The corner cube reflector has a triangular grooved surface that returns the light beam on a parallel axis. A polarizing filter in the light source can help eliminate the possibility of reflection off the target object, keeping the sensor actuated.

Through-Beam Mode. A *through-beam mode photoelectric sensor* is a photoelectric sensor that uses a beam aimed directly at a target object with a separate receiver to sense the beam. The presence of an object interrupts the beam and actuates the circuit. **See Figure 27-9.** The through-beam mode of sensing provides the greatest range. Because of the tightly focused source, the through-beam mode is less susceptible to atmospheric contamination.

An advantage of a separate emitter and receiver is the ability to use convergent sensing, or fixed-focus sensing. Since the beam source is tightly focused, it can be directed to reflect off the object. The receiver needs to be located to sense the reflected beam. This arrangement allows a through-beam photoelectric sensor to monitor an exact point, such as a bottle cap or a liquid level.

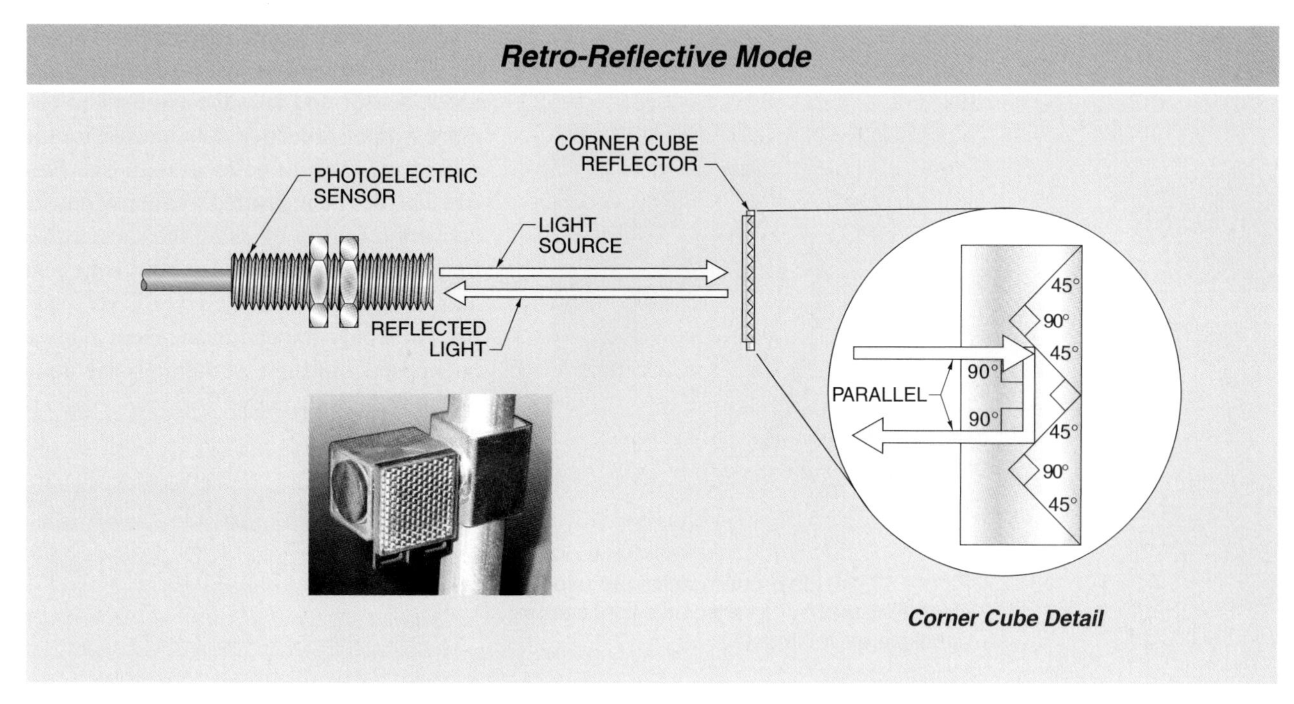

Figure 27-8. Retro-reflective sensors measure the reflected light from a corner cube reflector.

Fiber-Optic Cables. Fiber-optic cables can be used with either diffused-mode or through-beam mode sensors. A diffused-mode sensor uses one cable, but half of the fibers carry the source and the other half of the fibers carry the receiving signal. A through-beam mode sensor uses two fiber-optic cables, one for the source and the other for the receiver. The cables have protective coverings for physical protection. **See Figure 27-10.** The fiber-optic cables are used when the object to be sensed is in a position that would be impossible with standard photoelectric sensors. The small diameters of the fiber-optic cables also allow smaller objects to be detected.

Photoelectric Operating States. The operating states for photoelectric sensors can be confusing since each mode of the sensor can have normally ON or normally OFF outputs when there is no light present (dark on). At the same time, depending on the mode of sensing, the sensor is detecting light when the object is present or when the object is not present. **See Figure 27-11.**

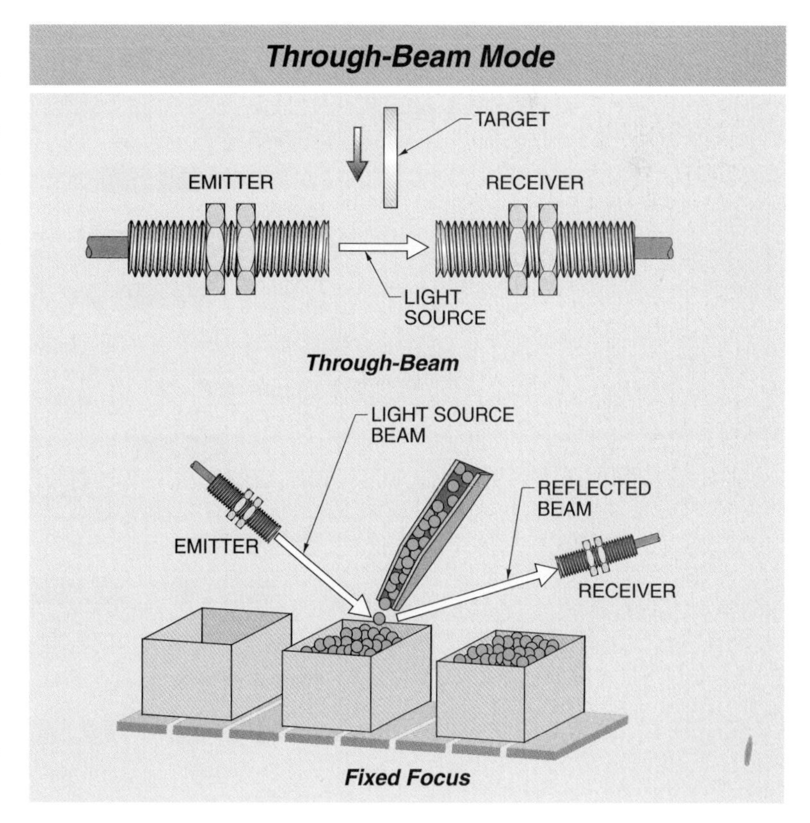

Figure 27-9. Through-beam sensors detect when an object interrupts the light beam.

Fiber-Optic Cables

Honeywell Sensing & Control

Figure 27-10. Fiber-optic cables are used to extend the reach of a sensor into tight spaces or hostile environments.

Optical Speed Measuring. Optical speed measuring can be done with a retro-reflective mode sensor and an adhesive-backed circular ring of reflective dots located on the end of the shaft to be measured. **See Figure 27-12.** The measured frequency can be converted to rpm by using the same equation that was used with the split ring gear and proximity sensor. This type is very useful when only the end of the shaft surface is exposed and there is nothing else upon which to fasten a target.

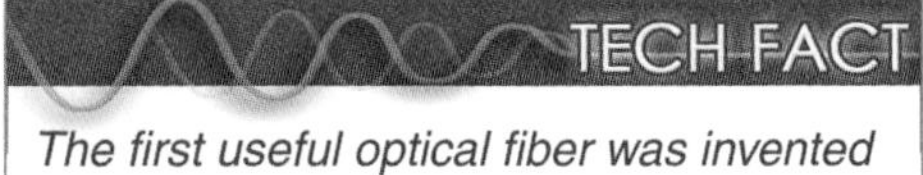

The first useful optical fiber was invented in 1970 by researchers at Corning Glass Works.

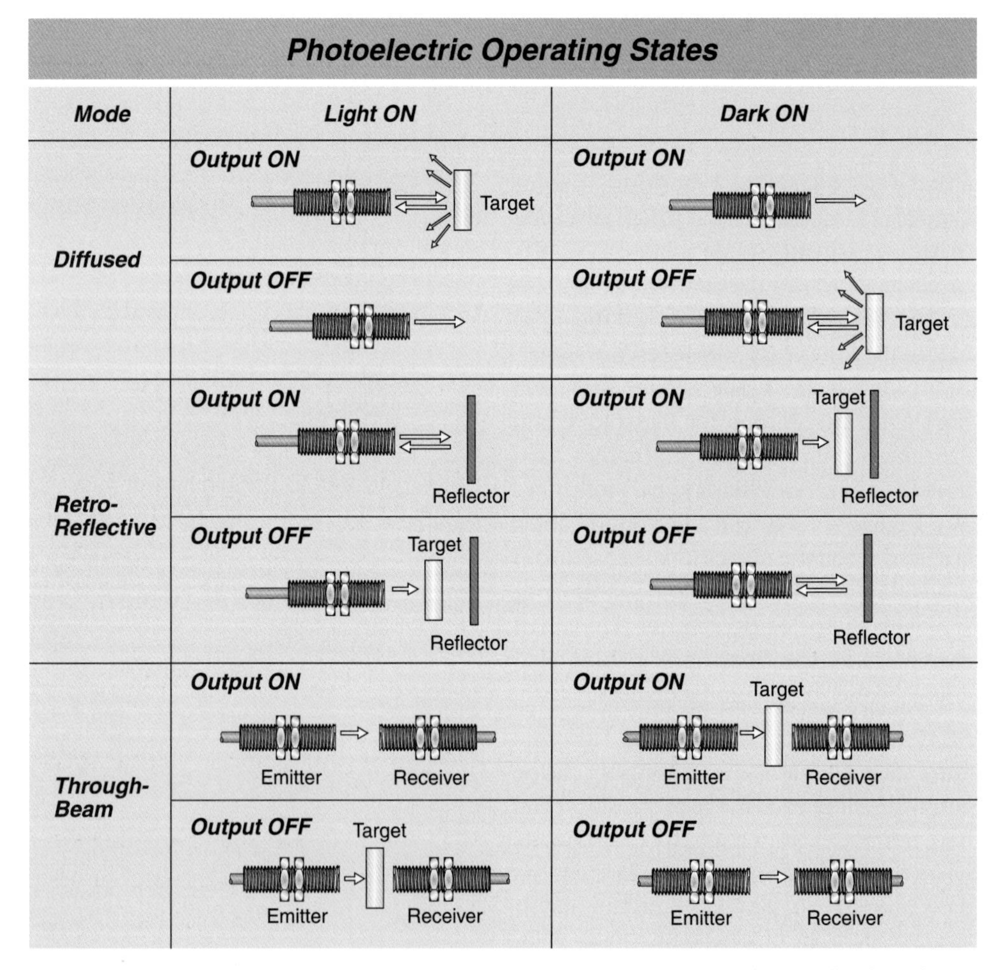

Figure 27-11. Photoelectric sensors are available in normally open and normally closed configurations. Objects are detected by reflected light or a blocked beam of light.

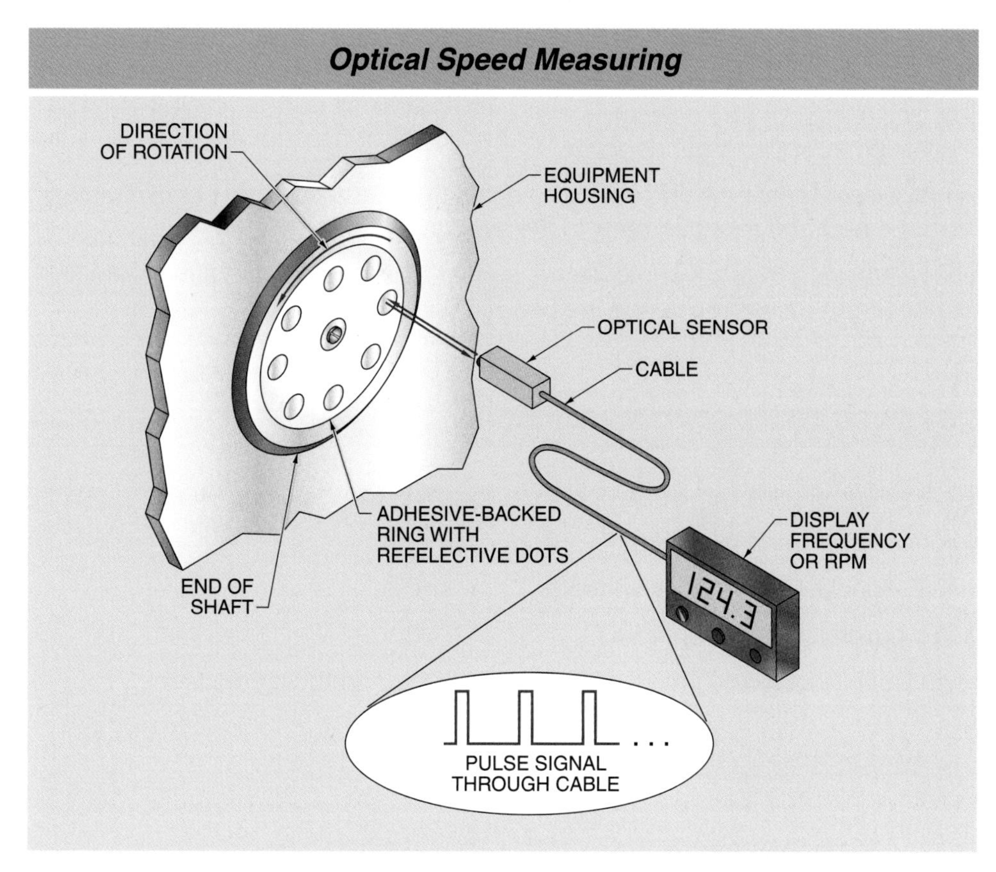

Figure 27-12. Optical speed measuring using reflective discs is easy to install.

KEY TERMS

- *mechanical switch:* A switch that requires physical contact with an object to actuate a switch mechanism.
- *Form C electrical contact:* A single-pole double-throw switch that allows both NO and NC contacts.
- *proximity sensor:* A sensor that detects the presence of an object without requiring contact with the object.
- *electrical proximity sensor:* A proximity sensor that uses inductance and capacitance properties to detect the presence of an object.
- *inductance proximity sensor:* A proximity sensor that consists of a sensor coil, an oscillator, a trigger circuit, and an output switching circuit.
- *capacitance proximity sensor:* A proximity sensor that consists of a sensor plate, an oscillator, a trigger circuit, and an output switching circuit.
- *sensor style:* The physical housing into which a sensor element is placed.
- *ultrasonic proximity sensor:* A proximity sensor that uses a pulse of sound waves to detect the presence of an object.
- *photoelectric sensor:* A proximity sensor that uses visible light and infrared radiation sources to detect target objects.

KEY TERMS (continued)

- *diffused mode photoelectric sensor:* A photoelectric sensor that directs its source against a target object and detects a reflection from the target object.
- *retro-reflective mode photoelectric sensor:* A photoelectric sensor that uses a focused beam directed across the path of a target object and reflected back to the sensor.
- *through-beam mode photoelectric sensor:* A photoelectric sensor that uses a beam aimed directly at a target object with a separate receiver to sense the beam.

REVIEW QUESTIONS

1. Describe a mechanical switch and list two types of switch-actuating mechanisms.
2. What is a proximity sensor, and when is it used?
3. Compare inductive and capacitance proximity sensors.
4. How does an ultrasonic proximity sensor operate?
5. List and define the types of photoelectric proximity sensors and their operating modes.

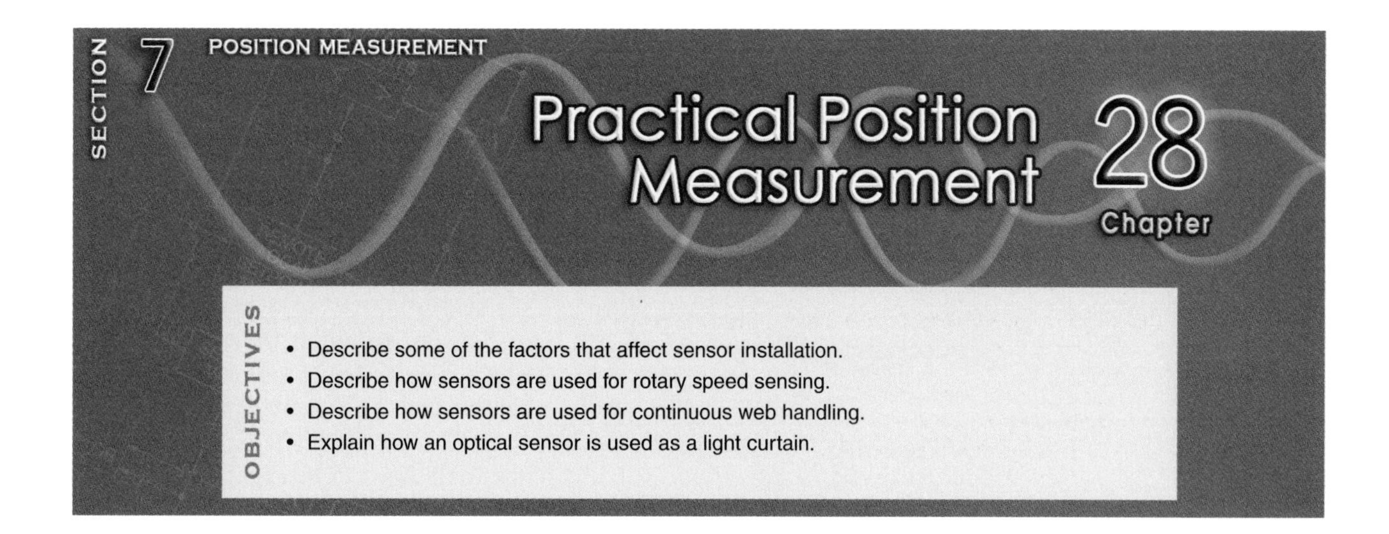

SENSOR INSTALLATION

There are several factors to take into account when installing sensors. Proximity sensors have a limited sensing range that depends on the sensor design and the material being sensed. Sensors need to be mounted and installed to prevent interferences with detection.

Sensing Ranges

A *sensing range* is the distance from the end of a proximity sensor to where an object can be detected. The sensing range is related to the size of the coil. A larger sensing range requires a larger coil and therefore the sensor is larger. The sensor diameter can vary from 3 mm up to 65 mm (0.118″ up to 2.56″) with equivalent ranges of 0.6 mm and 50 mm (0.024″ and 1.97″). The ideal target object should have a side dimension equal to the diameter of the sensor or three times the nominal sensing range, whichever is greater. The target object thickness should be a minimum of about 1 mm (0.04″) thick.

The nominal sensing range is based on the target object being constructed of mild steel. The sensing range is reduced if the target object is constructed of material other than steel. The reduction factors are 0.85 for stainless steel, 0.4 for aluminum and brass, and 0.3 for copper. The nominal sensing range multiplied by the reduction factor is the actual sensing range.

The nominal range, S_n, is a theoretical range and does not take into consideration actual production tolerances and environmental factors. The actual range, S_r, is the range based on 68°F and a standard voltage. The effective range, S_u, is the range after adjustment for environmental factors, temperature, and voltage. The working range, S_w, is the range after taking into consideration all other factors. This is the range at which the target always actuates the sensor.

Actuation Directions

An *actuation direction* is the direction an object moves relative to a proximity sensor. The actuation directions that can be used to bring the target object to the sensor are lateral and axial. The actuation of the sensor occurs when the edge of the target reaches the solid line of the lateral actuation diagram. The axial actuation diagram also shows the effect of hysteresis. Hysteresis is the difference in range between the actuation point and the release point. **See Figure 28-1.**

> **TECH-FACT**
>
> *Photoelectric sensors are very similar to optical level sensors. However, the presence of an object is detected instead of the level in a tank or container.*

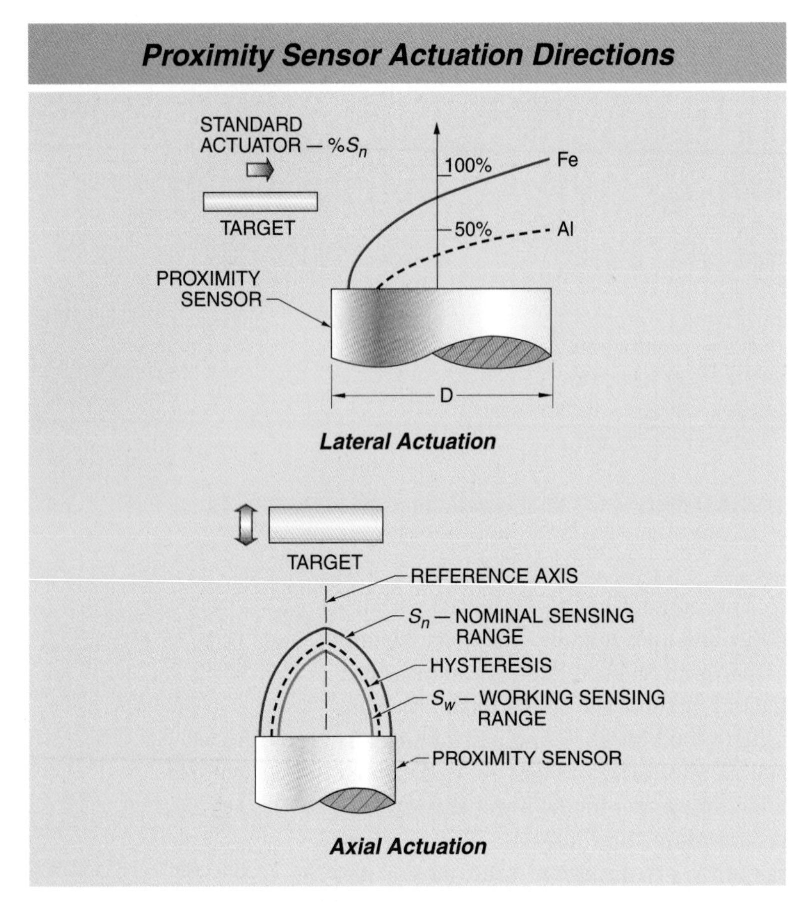

Figure 28-1. Proximity sensors can be actuated by laterally or axially moving targets.

Mounting and Interferences

Depending on the style of the sensor, the mounting method can cause problems with the ability of the sensor to detect the target object. Most of the problems are associated with the cylindrical style sensor. The basic design of the cylindrical sensor has a threaded exterior surface except for a distance a little more than the sensing range from the sensing end. The unenclosed end of the sensor contains the inductance coil or the capacitance plate. Although the basic design focuses the detection field axially from the end of the sensor, there is some lateral field generated. If this type of sensor is flush mounted in a metal plate, the metal around the end of the sensor is sensed by the lateral field and continuously activates the sensor. This prevents the sensor from sensing the target object. **See Figure 28-2.**

If a sensor must be flush mounted, shielded versions are typically available that restrict the lateral field at the expense of reducing the sensing range. Unshielded sensors must have their ends exposed a distance of twice the sensor diameter. Lateral clearance must be no less than the sensor diameter around the end of the sensor.

Sensors mounted too close to each other can also be a problem. Unshielded sensors must be separated from each other by about three diameters. Shielded sensors only need to be separated by about one diameter. Target objects spaced too close to each other may continuously activate the sensor. Thus the spacing between target objects that are the size of the sensor must not be less than two sensor diameters.

Surface-mounted sensors also have mounting restrictions that are similar to those for cylindrical sensors, but surface-mounted sensors, due to their physical design, are rarely flush mounted. As long as the sensing end of the sensor is extended beyond the mounting surface, there should be no interference.

ROTARY SPEED SENSING

Rotary speed sensing is the use of sensors to measure the speed of a rotating object. Common applications of rotary speed sensing are sensing the speed of rotating equipment such as conveyor drives, grinding equipment, or mixers. In many cases, rotating equipment must rotate at a specific speed to ensure the proper operation of a production process. In addition, a sudden change in the speed of rotation can indicate a problem in the process and the sensor can be installed to sound an alarm.

Frequency Measurement

Speed sensors measure how often an object moves past the sensor. The sensor detects a target on the rotating object and determines the rate at which the target moves past the sensor. The rate is expressed as a frequency. The frequency can be calculated as follows:

$$f = \frac{\omega \times N}{60}$$

where

f = frequency (in Hz)

ω = rotational speed (in rpm)

N = number of targets in one revolution

For example, a rotating shaft with one target that is rotating at 100 rpm has the frequency calculated as follows:

$$f = \frac{\omega \times N}{60}$$

$$f = \frac{100 \times 1}{60}$$

$$f = \frac{100}{60}$$

$$f = \mathbf{1.67\ Hz}$$

Another example is the speed sensor for a vessel agitator using a ring gear with 72 teeth rotating at 40 rpm. With the ring gear, each tooth is a target. The frequency is calculated as follows:

$$f = \frac{\omega \times N}{60}$$

$$f = \frac{40 \times 72}{60}$$

$$f = \frac{2880}{60}$$

$$f = \mathbf{48\ Hz}$$

Common ranges for proximity speed sensors are 0.1 to 1.0 Hz, 1.0 to 10.0 Hz, and 10.0 to 100.0 Hz. Targets can be fastened to the equipment shaft in a number of ways using pieces of steel angle or channel and a clamp. **See Figure 28-3.** Moving targets, like all rotating objects, should have physical guards around them.

> **TECH FACT**
>
> *Antoine Henri Becquerel discovered radioactivity when he placed unexposed photographic plates in a drawer along with some crystals containing uranium.*

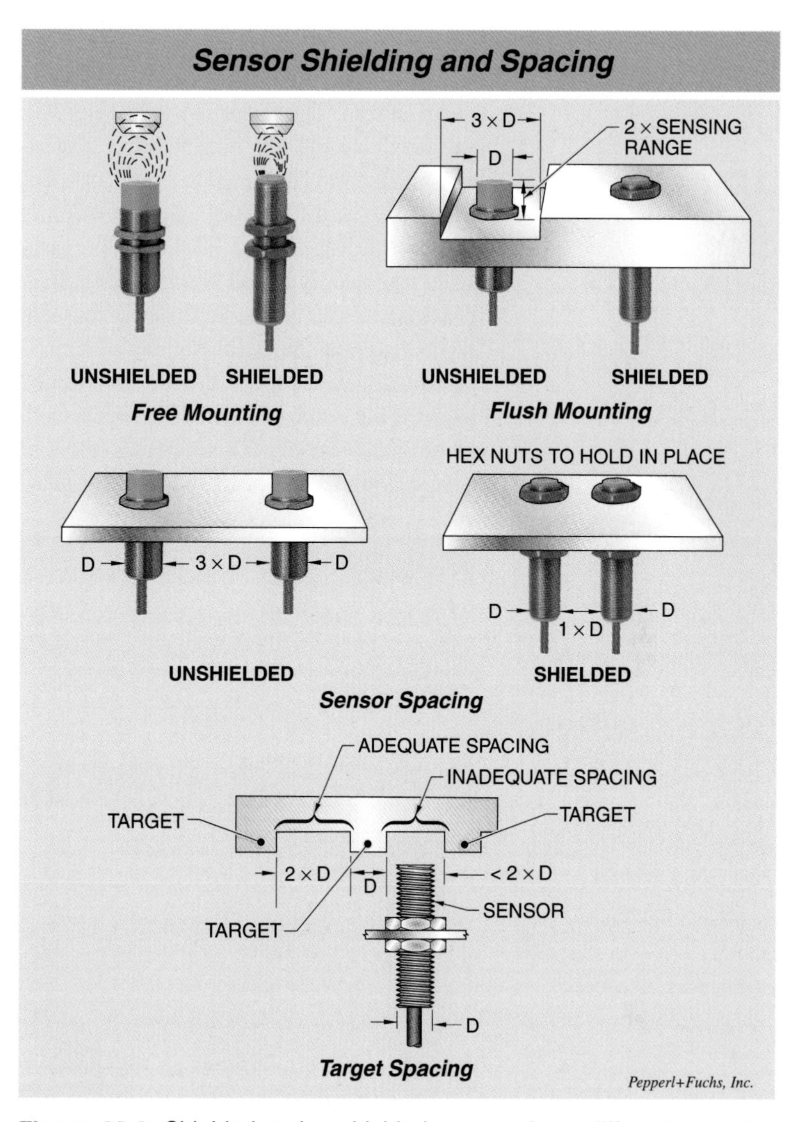

Figure 28-2. Shielded and unshielded sensors have different mounting requirements. Sensor spacing and target spacing must also be taken into account when installing sensor systems.

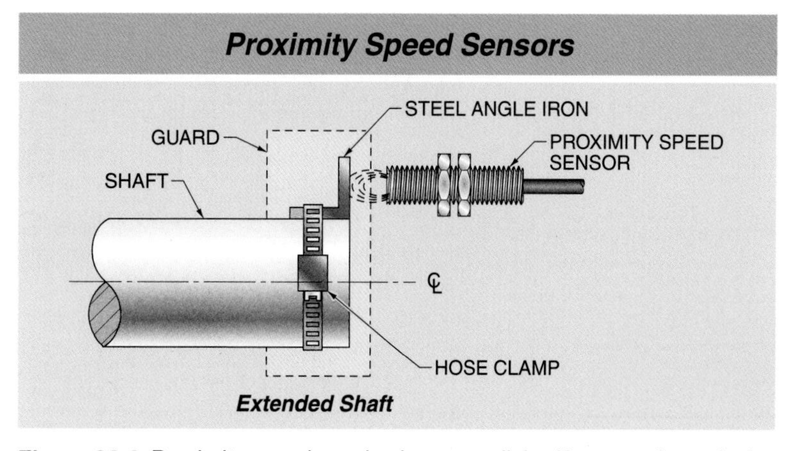

Figure 28-3. Proximity speed sensing is accomplished by mounting a device to a shaft and measuring the frequency of the signal from the sensor.

A proximity speed measurement is designed to accurately measure the rotational speed in rpm. The target is a split ring with gear teeth cut into the outside edge. The interior diameter is designed to match standard shaft sizes used for vessel agitators or other equipment. **See Figure 28-4.** The properly sized ring gear is bolted around the shaft. A proximity sensor is mounted to sense the teeth on the ring gear. Since the number of teeth is known, the exact speed can be determined from the measured frequency of the gear teeth passing the sensor. The electronic readout device contains the speed and alarm points.

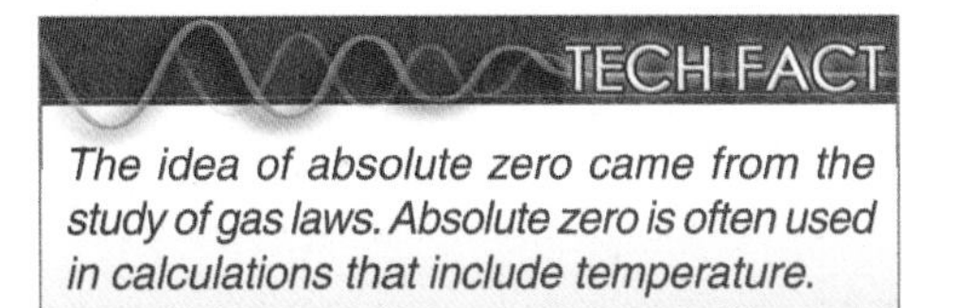

The idea of absolute zero came from the study of gas laws. Absolute zero is often used in calculations that include temperature.

Speed Sensing Alarms

A proximity speed sensor can be designed to provide a discrete output signal when the speed of the equipment being monitored is above or below a preset value. **See Figure 28-5.** A drop in the speed of the equipment away from the setpoint actuates the output of the proximity speed sensor. This type of sensor is usually used for equipment that is driven by belts, chains, or couplings. A typical application is to start a shutdown sequence if the driving linkage breaks.

Without modification, the action of the speed sensor and the associated logic circuit prevents the equipment from starting up because the speed is below the setpoint. There is a short override built into the sensor that provides an output to the logic circuit to allow the equipment to start and get up to operating speed.

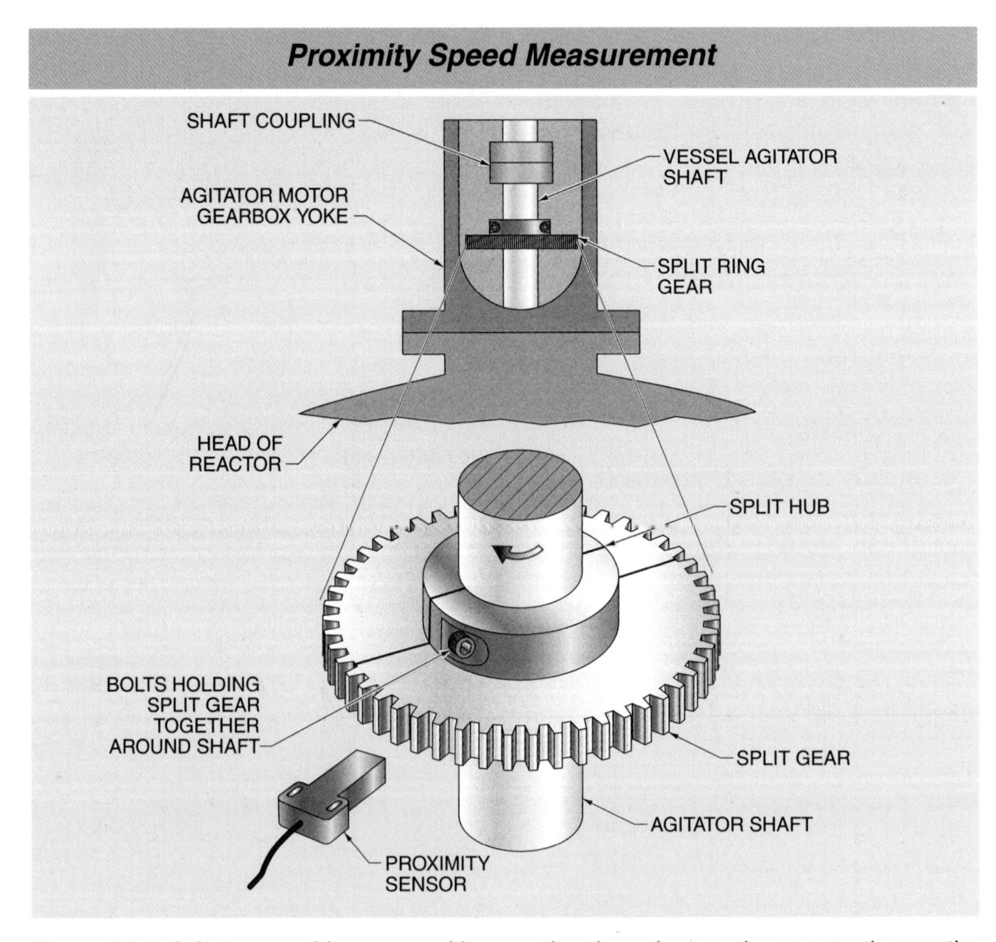

Figure 28-4. Agitator speed is measured by counting the pulses as the gear teeth pass the sensor.

CONTINUOUS WEB HANDLING

Many production operations use a web, or strip, of material. A *web* is a continuous length of material that is fed through a process. A common web handling procedure involves unwinding a web from a roll, guiding it through a process, measuring the thickness, and winding it on another roll. Common applications include printing and producing paper, metal foils, and plastic films. **See Figure 28-6.**

Webs are handled with a series of rolls that control the material through the process. The rolls can be used to squeeze the material to a desired thickness, chemically treat the material, print onto the web, or pull or push the web through the process. All web-handling systems require similar sensors and controls. Common applications of continuous web handling include web guiding, web thickness measurement, and continuous web loop control.

Web Guiding

A web must be centered while going through rolls. The frame of the machinery can damage the web if it moves to the edge of the handling equipment. It is difficult to guide a web through a process without continuous guidance. It is common for the rolls to be slightly out-of-round and for conveyors to track slightly off-center. When the web moves off-center, sensors detect the location of the edge, and a controller adjusts a pair of rolls to guide the web back to the center of the machine. **See Figure 28-7.**

There are a number of devices used to detect the edge of a web. One of the oldest types of sensors consists of a pneumatic jet and pressure detector. The web begins to block a jet of air when it moves toward the edge. The pressure detector senses the change in airflow, and the controller adjusts the web back toward the center. Modern edge sensors use proximity or optical sensors. When the edge of the web blocks the sensor, the controller signals the guide rolls to bring the web back to the center.

Speed Sensing Alarms

Figure 28-5. A proximity speed sensor can be designed to provide a discrete output signal when the speed of the equipment being monitored is above or below a preset value.

Continuous Web Handling

Figure 28-6. A common web handling procedure involves unwinding a web from a roll, guiding it through a process, measuring the thickness, and winding it on another roll.

Figure 28-7. When a web moves off-center, sensors detect the location of the edge and a controller adjusts a pair of rolls to guide the web back to the center of the machine.

Web Thickness Measurement

Many applications require that a base material be a precise thickness. There are a number of noncontact methods used to measure the thickness of webs. With all of these web thickness measurements, it is preferable to scan the measurement across the width of the web so that a thickness measurement can be made as representative of the whole web. **See Figure 28-8.**

Web Thickness Measurement

WEB THICKNESS PROFILE

SCANNER SWEEPS AROSS WEB

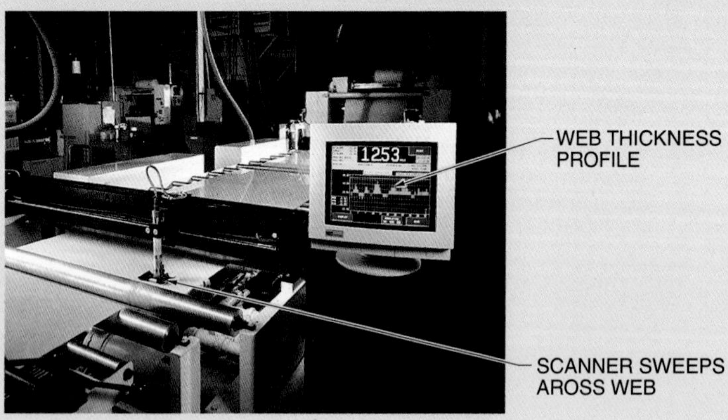

NDC Infrared Engineering

Figure 28-8. It is best to scan a measurement across the width of a web so that a thickness measurement can be made as representative of the whole web.

In some applications, the web can be passed over a fixed support to provide a reference point for the measurement. This means that the bottom of the web is always at a known position and only the position of the top surface needs to be measured. There are a number of devices that are used to detect the top of web material.

One of the oldest types of sensors consists of an air nozzle placed just above the top of the web. This type of sensor can measure thickness to within 0.001″. Depending on the composition of the web material, a capacitance proximity sensor may be used to measure the height of the web surface. The capacitance sensor area, the frequency of the sensor circuit, and the material of the web determine the signal sensitivity and the required measurement distance. The latest technology uses a laser triangulation dimension measuring system to measure from a fixed position to the top surface of the web.

There are other applications where the web cannot be passed over a fixed support. This means that there is no fixed reference point for the measurement. In this type of application, an X-ray thickness measurement can be made of thin films with thicknesses of 1 μm to 1000 μm. This type of sensor can measure the thickness to within a resolution of 0.1 μm. This is good for measuring the thickness of foam, plastics, and rubber.

Continuous Web Loop Control

A web is normally driven by sets of rolls. The various roll speeds must be closely matched to prevent damage to the web. It is very difficult to keep all sets of rolls running at exactly the same speed. Any difference in speed between a set of upstream and downstream driving rolls tends to either stretch the strip or cause a loop. To prevent the strip from being broken or damaged, it is common for take-up loops to be maintained between each driving roll.

The set of rolls at the beginning or the end of the strip processing is the primary speed controller. All the other roll speeds are set to match the primary speed controller. The loop before or after the primary and second drive rolls is monitored for size. If the web material is flexible, a weighted idler roll can be used to maintain the loop.

An ultrasonic sensor can be used to measure the vertical position of the top of the web or the idler roll. A physical linear position measurement can also be used. This serves as an indication of whether the second drive roll is slower or faster than the primary drive rolls. If the idler roll is rising, the following drive rolls are going too fast and have to be slowed. The opposite is true if the idler roll is dropping. A similar arrangement is set up between any other drive rolls.

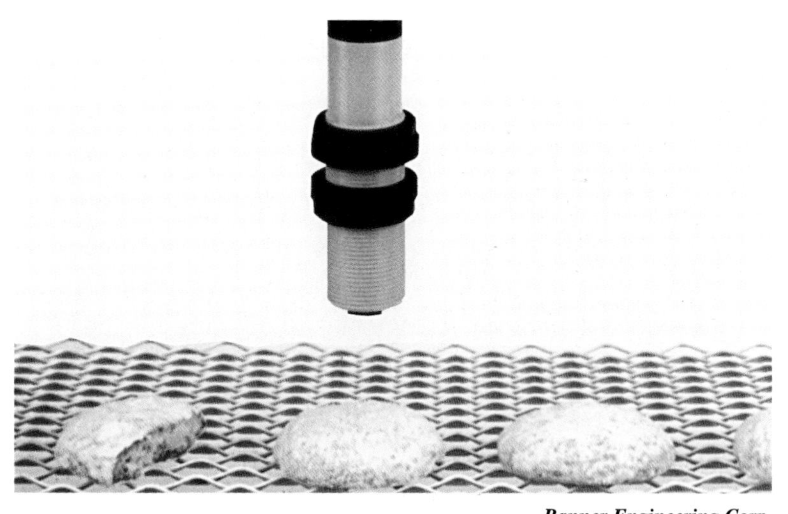

Banner Engineering Corp.

Position sensors can be used to detect damage to objects on a conveyor belt.

SAFETY LIGHT CURTAINS

A *safety light curtain* is a series of closely spaced light sources mounted on a rail and used as a safety device to shut down equipment if an operator reaches into a protected space. Optical sensors used as receivers are also mounted on another rail using the same spacing as the light sources. The two rails are mounted directly opposite each other so that an area of space has a curtain of light. **See Figure 28-9.** An object that enters the curtain breaks some of the light beams and generates an output. Light curtains can be used to protect operators from hazardous areas and moving parts.

TECH FACT

A low-range pressure switch can be used as a break detector for continuous web or sheet materials. A constant stream of low-pressure air is applied to the web. A pressure switch is on the other side of the web. If the web breaks, the pressure switch detects the air stream and sounds an alarm or shuts down the machine.

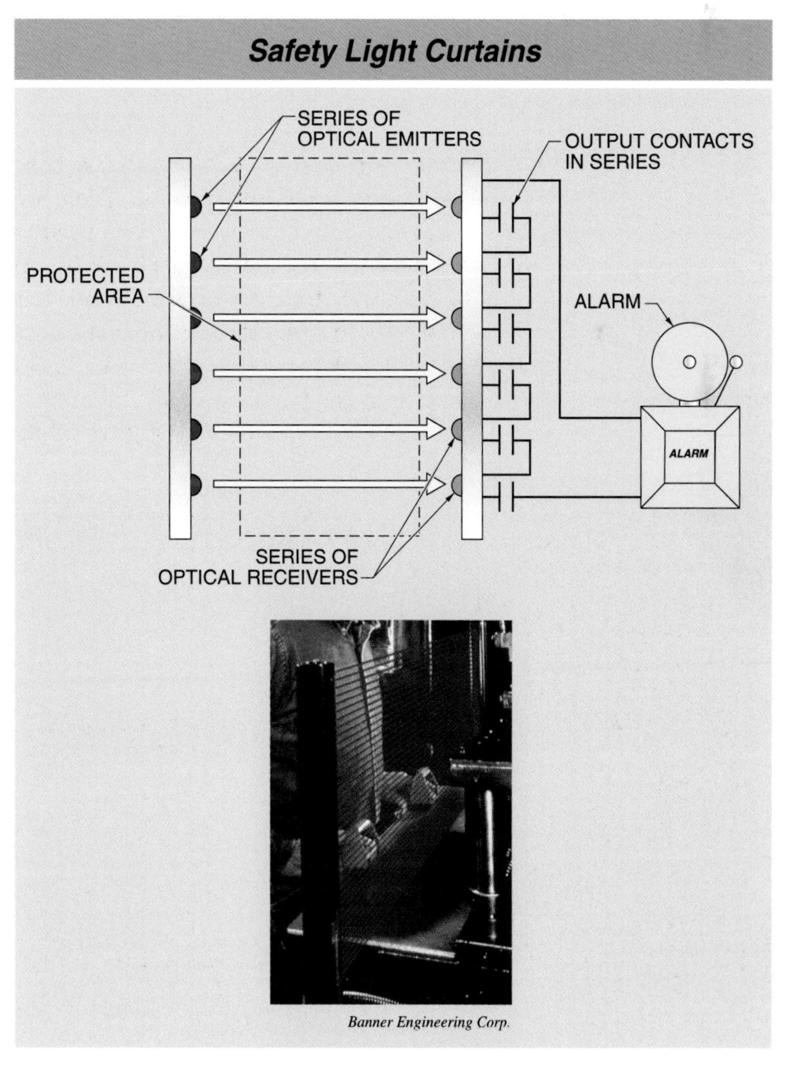

Figure 28-9. A safety light curtain is used to monitor an area and shut down a piece of equipment if an object enters the protected area.

KEY TERMS

- *sensing range:* The distance from the end of a proximity sensor to where an object can be detected.
- *actuation direction:* The direction an object moves relative to a proximity sensor.
- *rotary speed sensing:* The use of sensors to measure the speed of a rotating object.
- *web:* A continuous length of material that is fed through a process.
- *safety light curtain:* A series of closely spaced light sources mounted on a rail and used as a safety device to shut down equipment if an operator reaches into a protected space.

REVIEW QUESTIONS

1. List and describe several factors that affect sensor installation.
2. Define rotary speed sensing and list several applications.
3. How is frequency used for rotary speed sensing?
4. What are the methods used for web guiding?
5. How does a safety light curtain operate?

SECTION 8

TRANSMISSION AND COMMUNICATION

TABLE OF CONTENTS

SECTION OBJECTIVES

Chapter 29

- Define transmission and important terms associated with transmission.
- Identify the methods of electric transmission, and describe current transmission systems.
- Describe voltage, pulse, frequency, and tone transmission systems.
- Describe the use of formulas for converting between measurements and transmission values.

Chapter 30

- List the numbering systems and codes that are important for communications.
- Describe how to convert numbers between different bases.

Chapter 31

- Define digital communications and describe the main types of network configurations, addressing, and protocols.
- List and describe the types of circuits used in digital communications wiring formats.
- Identify common cable and wiring formats.

Chapter 32

- Define fieldbus, and describe the network classifications for fieldbus systems.
- Identify the major fieldbus systems.

SECTION

8 TRANSMISSION AND COMMUNICATION

Chapter 33

- Define wireless transmission and list its advantages and disadvantages.
- Describe the relationship between communication speed and distance.
- List and describe the different types of antennas.
- Compare the different types of spread spectrum transmission methods, wireless standards, and security standards.
- Describe the different industrial requirements and applications.

Chapter 34

- Explain the importance of loop impedance in a current transmission system.
- Define ground loop and identify ways that a ground loop can be avoided.
- Describe electromagnetic interference (EMI).
- List the common connectors.
- List the types of transmitters and describe a smart transmitter.

INTRODUCTION

A lack of standards led to early systems operating with analog DC voltage signals ranging up to 10 V or DC current signals ranging up to 50 mA. By the end of the decade, manufacturers had settled on the 4 mA to 20 mA signal.

By the early 1980s, digital controls were starting to appear in significant numbers. The development of the HART® protocol helped the transition to digital systems by combining digital and analog signals on the same wires. The introduction of digital standards like FOUNDATION Fieldbus, Profibus, and DeviceNet™ has helped to standardize digital communications.

ADDITIONAL ACTIVITIES

1. After completing Section 8, take the Quick Quiz® included with the Digital Resources.
2. After completing each chapter, answer the questions and complete the activities in the *Instrumentation and Process Control Workbook.*
3. Review the Flash Cards for Section 8 included with the Digital Resources.
4. Review the following related Media Clips for Section 8 included with the Digital Resources:
   - Fiber-Optic Signal
   - Grounding
   - Ground Loops
   - Isolated Devices
5. After completing each chapter, refer to the Digital Resources for the Review Questions in PDF format.

ATPeResources.com/QuickLinks • Access Code: 467690

SECTION 8 TRANSMISSION AND COMMUNICATION

Transmission Signals

Chapter 29

OBJECTIVES

- Define transmission and important terms associated with transmission.
- Identify the methods of electric transmission, and describe current transmission systems.
- Describe voltage, pulse, frequency, and tone transmission systems.
- Describe the use of formulas for converting between measurements and transmission values.

TRANSMISSION SIGNALS

A number of transmission signals have been developed and used as technology has evolved. *Transmission* is a standardized method of conveying information from one device to another. Radio and television signals that are broadcast over the airwaves are examples of transmission. A *transmission signal* is the data sent from one device to another by a specific method. A specific radio station or TV channel is an example of a transmission signal.

Each new development in the basic technology of transmission has initially produced a group of competitive transmission signal ranges, which usually are consolidated into one standard. This consolidation has already happened in pneumatic and electric transmission and it is now occurring in the fields of digital and wireless communications.

Data Types

In industrial process instrumentation, transmitted information is data. Data can be transmitted in many formats and over many different types of media. In addition, the data may be analog or digital in form.

Analog Data. *Analog data* is a continuous range of values from a minimum to a maximum that can be related to a transmission signal range. Any desired range of analog data can be conveyed by a transmission signal. An example of analog data is a simple pressure gauge. The pressure gauge dial has a range of values from a minimum to a maximum and the pointer indicates the value of pressure being measured. Intermediate analog values have an equivalent proportional transmission signal value.

A burner controller receives measurements from instruments and transmits instructions to valves and other control devices.

Although an analog transmission signal is proportional to the measurement range, there is no information conveyed by the analog transmission signal that identifies the measurement range or the units. The range and units information that the receiver displays must be matched to the specific source by human actions. For example, when a temperature transmitter with a measuring range of 200°F to 1000°F is connected to a recorder by a transmission signal, the recorder is only aware of the transmission signal that it receives and from that it puts the pen to an equivalent position on a chart. It is the responsibility of a person to select the correct chart to put on the recorder so that the recorder pen indicates the correct value between 200°F and 1000°F. **See Figure 29-1.**

Engineering Units and Transmission Signals

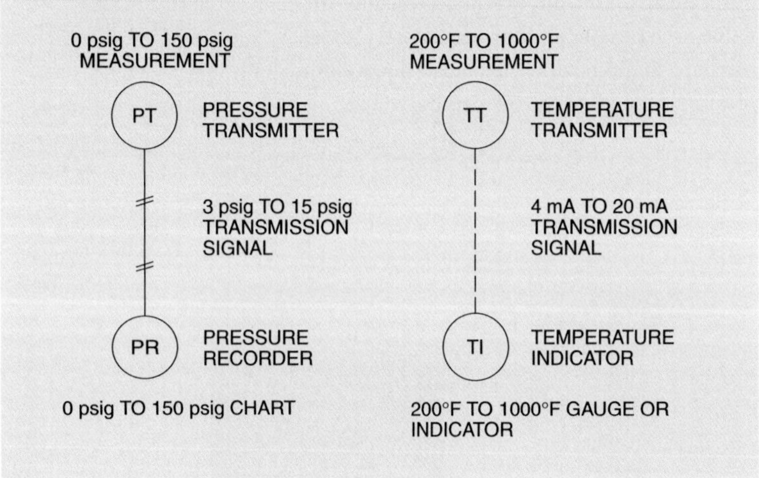

Figure 29-1. Different engineering measurement units can be sent with different types of transmission signals.

Digital Data. *Digital data* is a series of discontinuous ON/OFF signals that is transmitted electrically. In addition to the actual measurement values, digital transmission can contain extra information about sensor and transmitter configuration, data range, network addressing, calibration information, alarm state, and many other things. For example, a transmitter that is enabled to communicate digitally can convey information to a controller about the state of the transmitter and the process in addition to the actual process data.

ELECTRIC TRANSMISSION SYSTEMS

Electric transmission was developed to eliminate the problems and distance limitations of pneumatic transmission and to allow the use of a higher level of technology. Electric transmission did not become popular until solid-state circuitry became commonplace. Typical methods of electric transmission are current, voltage, pulse, frequency, and tone transmissions.

Current Transmission

Current transmission is an electric transmission method in which a transmitting unit regulates the current in a transmission loop. Current transmission is normally not affected by distance and the associated voltage losses. A current signal is also virtually immune to noise pickup. However, there are situations where current transmissions can still have problems. For example, if the transmitting distance is exceptionally long, the voltage losses along the transmission wires plus the receiver voltage drop can exceed the voltage capacity of the power supply. When this happens, the transmission signal cannot satisfy the required range.

Current transmission in the United States has been standardized on the range of 4 mA to 20 mA. In Europe, a standard of 0 mA to 20 mA may also be used. The advantage of the 4 mA to 20 mA is that the receiver instruments can distinguish between a measured value of 0 mA and a transmitter or network that is not functioning. Receiver instruments need to convert the 4 mA to 20 mA transmission signal into a voltage signal before it can be used. This is done by passing the transmission current through a precision resistor. This resistor may be internal or external to the receiver instrument. The voltage drop is calculated as follows:

$$E = I \times R$$

where

E = voltage (in V)

I = current (in A)

R = resistance (in Ω)

For example, a typical receiver has a 250 ohm (Ω) resistor. **See Figure 29-2.** The current is a minimum of 4 mA (0.004 A) and a maximum of 20 mA (0.020 A). What are the minimum and maximum voltages developed across the resistor?

$E = I \times R$

$E = 0.004 \times 250$

$E = \mathbf{1\ VDC}$

and

$E = I \times R$

$E = 0.020 \times 250$

$E = \mathbf{5\ VDC}$

This shows that a 1 VDC to 5 VDC signal is developed across a typical resistor due to the 4 mA to 20 mA current flow. There are voltage drops in the circuit, but the current remains constant in all portions of the loop.

A typical current transmission loop includes a power supply, a transmitter, and a receiver or receivers. The receivers may be indicators, recorders, or other devices in a control circuit. A loop is typically powered by a 24 VDC power supply that is electrically isolated from ground and AC power. A portion of the loop voltage is used to power 2 wire transmitter circuits. The transmitter regulates the 4 mA to 20 mA signal in proportion to the measured variable. The current passes through the resistor at the receiver. The voltage drop at the receiver is proportional to the measured variable. The circuit arrangement can take many different forms as long as the power supply terminal polarities are connected to the same polarity on the first device and after that the wires connect to the opposite polarities. **See Figure 29-3.**

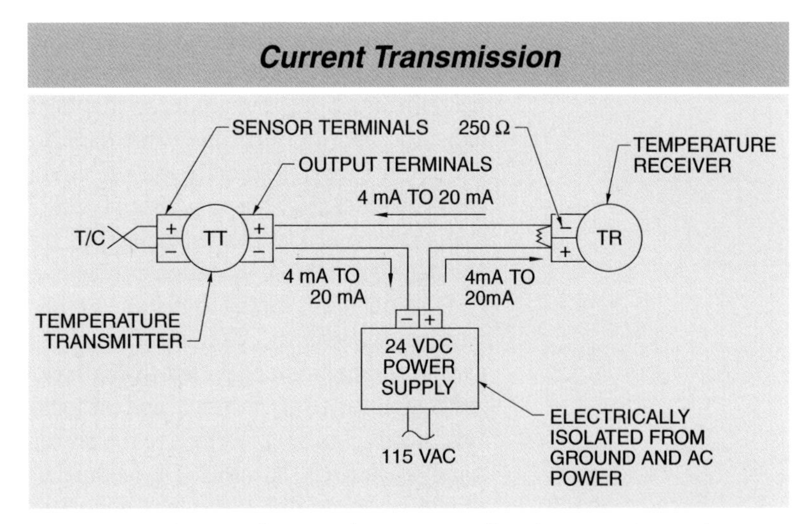

Figure 29-2. A transmitter regulates current flow in proportion to the measurement range. The power supply for the loop and the receiver that converts the current to a voltage signal are also in the loop.

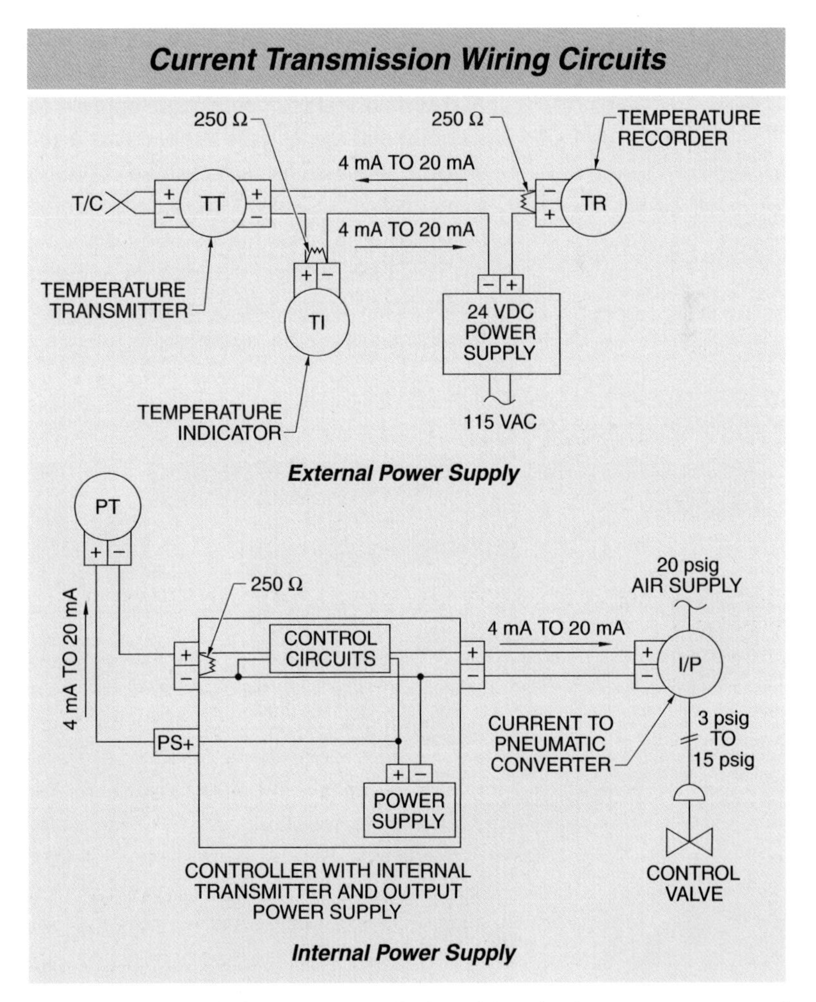

Figure 29-3. A typical current transmission wiring circuit can use an external power supply or a power supply in the controller.

TECH FACT

Pneumatic instrumentation can be used in any area where the National Electric Code® requires hazardous area classifications. Pneumatics can be the lowest total-cost option for Class 1, Division 1 service.

For example, current passes from the power supply to the receiver, the receiver to the transmitter, the transmitter to the local indicator, and the local indicator back to the power supply. Typically, the wiring is arranged as follows: the power supply positive (+) terminal is connected to the receiver positive terminal (+), the receiver negative (–) terminal is connected to the transmitter positive (+) terminal, the transmitter negative (–) terminal is connected to the local indicator positive (+) terminal, and the local indicator negative (–) terminal to the power supply negative (–) terminal. The technician should be aware that some manufacturers may have different polarity standards.

Voltage Transmission

Voltage transmission is an electric transmission method in which a transmitting unit regulates the voltage in a transmission loop. A specified voltage potential is applied to a circuit and the voltage drop across a resistor at the receiver is measured. Voltage transmission is a very convenient form for transferring information between devices, but is subject to errors caused by voltage losses in the connecting wires and voltage variations from other sources.

Data signals are transmitted wherever needed throughout a production operation.

Typical sources of voltage variations in a control circuit are leakage of AC power from the power supply of the transmitter or receiver or the signal wires acting as an antenna in the presence of nearby AC fields. Both types of variation can be prevented with suitable isolation in the transmitter and receiver and by proper field wiring.

The voltage loss along the transmission line is dependent on the current flow through the wires. The resistance of the receiving device primarily determines the current flow. The resistance of receiving devices is generally referred to as input impedance. Devices with low input impedance usually cause higher voltage drops in the connecting wires and thus higher measurement errors.

For example, if a circuit has a voltage source of 3 VDC, two 5 Ω connecting wires, and a 250 Ω input impedance, the total loop resistance is 260 Ω (250 + 5 + 5 = 260) and the error caused by the voltage drop can be calculated from the voltage and resistance values. **See Figure 29-4.** The equations used to calculate current and voltage are as follows:

$$I = \frac{E}{R}$$

and

$$E = I \times R$$

where

E = voltage (in V)

I = current (in A)

R = resistance (in Ω)

The current flow is as follows:

$$I = \frac{E}{R}$$

$$I = \frac{3}{260}$$

$$I = \mathbf{0.0115\ A}\text{, or } \mathbf{11.5\ mA}$$

The voltage drop through the wires is as follows:

$$E = I \times R$$

$$E = 0.0115 \times 10$$

$$E = \mathbf{0.115\ V}$$

A voltage drop of 0.115 V in a 3 V circuit is a 3.83% error (0.115 ÷ 3 = 0.0383). This is why devices used in a voltage-loop circuit are designed with an input impedance of 1 megohm or greater, which reduces the current flow to nearly zero and minimizes any potential voltage loss due to the signal wire resistances.

Voltage signals can be used for the transmission of information if the distances are short, with low wiring resistance, and with measuring instruments that have high input impedance (more than 1 megohm). Common voltage signals are 1 VDC to 5 VDC; 0 VDC to 10 VDC; and 0 VDC to 5 VDC.

Pulse Transmission

Pulse transmission is an electric transmission method consisting of a rapid change in voltage from a low value to a high value and then back to the low value. Pulse transmissions can be calibrated to represent an incremental quantity of material. A summation of the pulses represents the total quantity over a period of time.

Various flowmeters use pulse outputs because pulses are inherent in the transmitter measurement and they provide greater accuracy. Turbine, vortex shedding, paddle wheel, Coriolis mass, and positive-displacement flowmeters can all provide pulse outputs. Each pulse is calibrated to represent an incremental flow quantity. Each pulse is also a constant amplitude and duration. The time between pulses is variable. **See Figure 29-5.**

For example, the blade of a turbine flowmeter produces a pulse as it passes by an electrical pickup coil. This transmission method has the advantage that for many instruments, no external power or electronic circuitry is required to generate the signal. Some caution is required in that the voltage amplitude of the pulse can increase with the frequency of the pulse. Additional circuitry can regulate the peak voltage to avoid damage to receivers.

Frequency Transmission

Frequency transmission is an electrical transmission method in which the frequency of a signal is proportional to the measured value. The frequency transmitter uses standard voltage or current signals as inputs and converts them into a range of frequencies. Typical frequency ranges are 9 Hz to 15 Hz and 18 Hz to 30 Hz. The frequency signals are carried over conventional analog telephone wires. The receiver senses the frequencies and reconverts them to standard transmission signals. Digital transmission of information over telephone wires has mostly replaced frequency transmission.

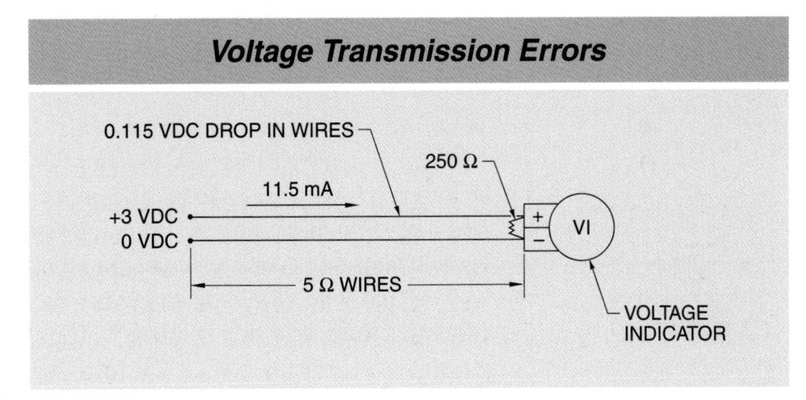

Figure 29-4. Voltage transmission errors are determined by the measurement device's input impedance and the resistance of the connecting wires.

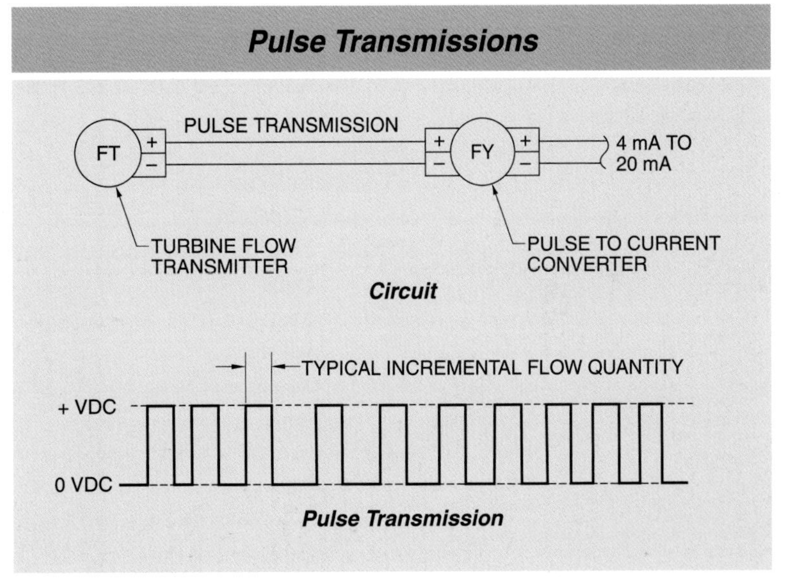

Figure 29-5. Pulse transmission is an accurate form of conveying measurement information and is commonly used with transmitters that normally generate pulse signals.

Tone Transmission

Tone transmission is an electrical transmission method consisting of a pure audible tone where the duration of the tone is proportional to the measurement value. The tone signal can be carried over digital phone wires or by microwaves. For example, telephone modems use tones to dial digital phone numbers and to send tone pulses over phone wires. Tone transmission is relatively slow when compared to modern digital communications and is now rarely used.

MEASUREMENT AND TRANSMISSION RELATIONSHIPS

When working with transmission signals, whether during calibration or troubleshooting, it is very useful to be able to convert a specific measurement value into its equivalent transmission value or the other way around. The following equations and examples show how this is done.

Measurement to Transmission Conversions

Converting a measurement value to its equivalent transmission value is useful for determining if a transmitter is functioning correctly. The formula is as follows:

$$TV = \frac{M - MRL}{MRU - MRL} \times (TRU - TRL) + TRL$$

where

TV = transmission value

M = measurement

MRL = measurement range lower value

MRU = measurement range upper value

TRU = transmission range upper value

TRL = transmission range lower value

For example, what is the transmission value with a measurement range of 0°F to 150°F, a transmission signal of 4 mA to 20 mA, and a measurement of 0°F? **See Figure 29-6.**

$$TV = \frac{M - MRL}{MRU - MRL} \times (TRU - TRL) + TRL$$

$$TV = \frac{80 - 0}{150 - 0} \times (20 - 4) + 4$$

$$TV = \frac{80}{150} \times 16 + 4$$

$$TV = 0.533 \times 16 + 4$$

$$TV = 8.53 + 4$$

$$TV = \mathbf{12.53\ mA}$$

Transmission to Measurement Conversions

It is also easy to determine a measurement value from a transmission value. This is useful for checking receiver instruments by introducing a signal in place of the signal from the transmitter. The formula is as follows:

$$M = \frac{TV - TRL}{TRU - TRL} \times (MRU - MRL) + MRL$$

where

M = measurement

TV = transmission value

MRL = measurement range lower value

MRU = measurement range upper value

TRL = transmission range lower value

TRU = transmission range upper value

For example, what is the measurement when the measurement range is 50°F to 250°F, the transmission range is 4 mA to 20 mA, and the transmission signal is 10.23 mA?

$$M = \frac{TV - TRL}{TRU - TRL} \times (MRU - MRL) + MRL$$

$$M = \frac{10.23 - 4}{20 - 4} \times (250 - 50) + 50$$

$$M = \frac{6.23}{16} \times 200 + 50$$

$$M = 0.389 \times 200 + 50$$

$$M = 77.8 + 50$$

$$M = \mathbf{127.8°F}$$

TECH FACT

Sampling is used to convert analog transmission signals to digital signals.

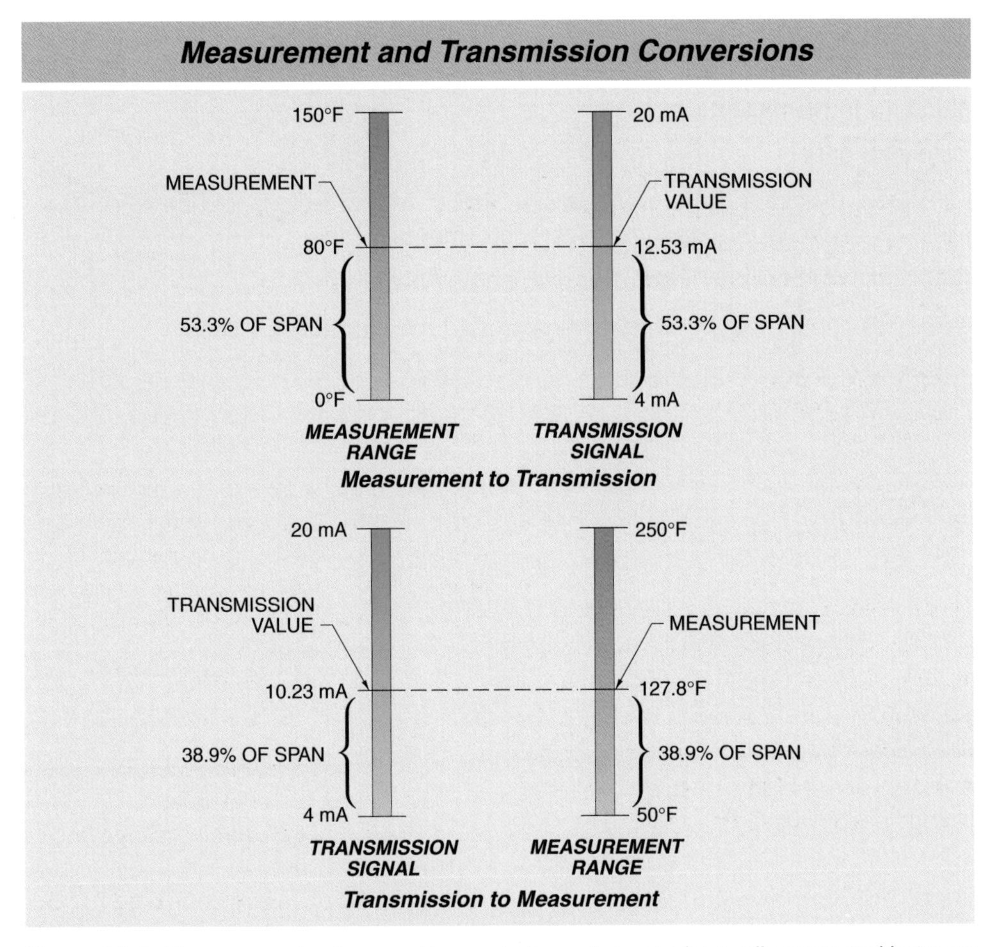

Figure 29-6. Measurement values and transmission values can be easily converted between forms because they both represent the same fraction of span.

KEY TERMS

- *transmission:* A standardized method of conveying information from one device to another.
- *transmission signal:* The data sent from one device to another by a specific method.
- *analog data:* A continuous range of values from a minimum to a maximum that can be related to a transmission signal range.
- *digital data:* A series of discontinuous ON/OFF signals that is transmitted electrically.
- *current transmission:* An electric transmission system in which a transmitting unit regulates the current in a transmission loop.
- *voltage transmission:* An electric transmission method in which a transmitting unit regulates the voltage in a transmission loop.
- *pulse transmission:* An electric transmission method consisting of a rapid change in voltage from a low value to a high value and then back to the low value.
- *frequency transmission:* An electrical transmission method in which the frequency of a signal is proportional to the measured value.
- *tone transmission:* An electrical transmission method consisting of a pure audible tone where the duration of the tone is proportional to the measurement value.

REVIEW QUESTIONS

1. Compare transmission and transmission signals.
2. Compare analog and digital data.
3. Define current transmission, describe the current range used in transmission, and explain how the current is converted to a voltage for use in receiver instruments.
4. What is pulse transmission, and how can it be used with some flowmeters?
5. Why may it be necessary to convert between transmission and measurement values?

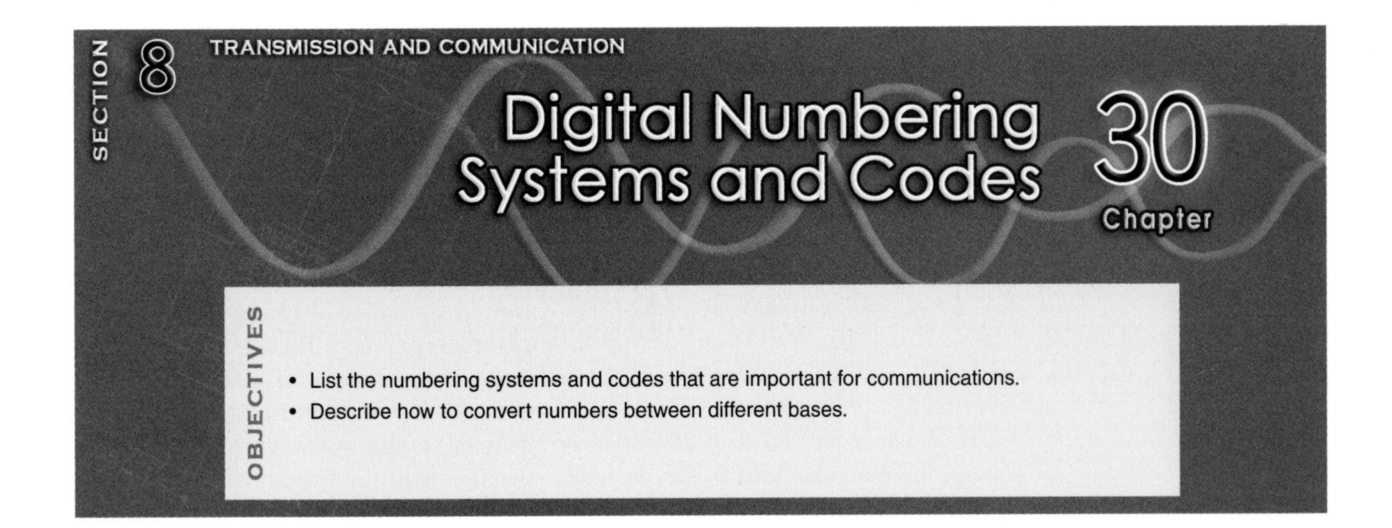

NUMBERING SYSTEMS

There are many numbering systems that can be used to describe the same quantities. Every numbering system has a base, or radix, and uses a set of symbols to represent quantities. The base of a numbering system determines the total number of unique symbols used by that system, and the largest-valued symbol always has a value of one less than the base.

For example, in the common decimal system the base is 10 and the largest-valued symbol is 9, which is one less than the base. A subscript is used to distinguish the base used when there is the possibility of confusion between different numbering systems. Common numbering systems and codes include decimal, binary, octal, hexadecimal, binary coded decimal systems, and ASCII.

Decimal Numbers

A *decimal number* is a number given in a base of 10. The symbols used in this system are the familiar digits 0, 1, 2, 3, 4, 5, 6, 7, 8, and 9. For numbers larger than 9, a place value, or weight, is assigned to each position that a number greater than 9 holds. The weighted value of each position can be expressed as the base (10 in this case) raised to the power of *n*, the position.

For example, the number 10 to the power of 2, or 10^2, is 10×10, or 100, and 10 to the power of 3, or 10^3, is $10 \times 10 \times 10$, or 1000. For the decimal system, then, the position weights from right to left are 1, 10, 100, 1000, etc. **See Figure 30-1.** This method for computing the value of a number is known as the sum-of-weights position method.

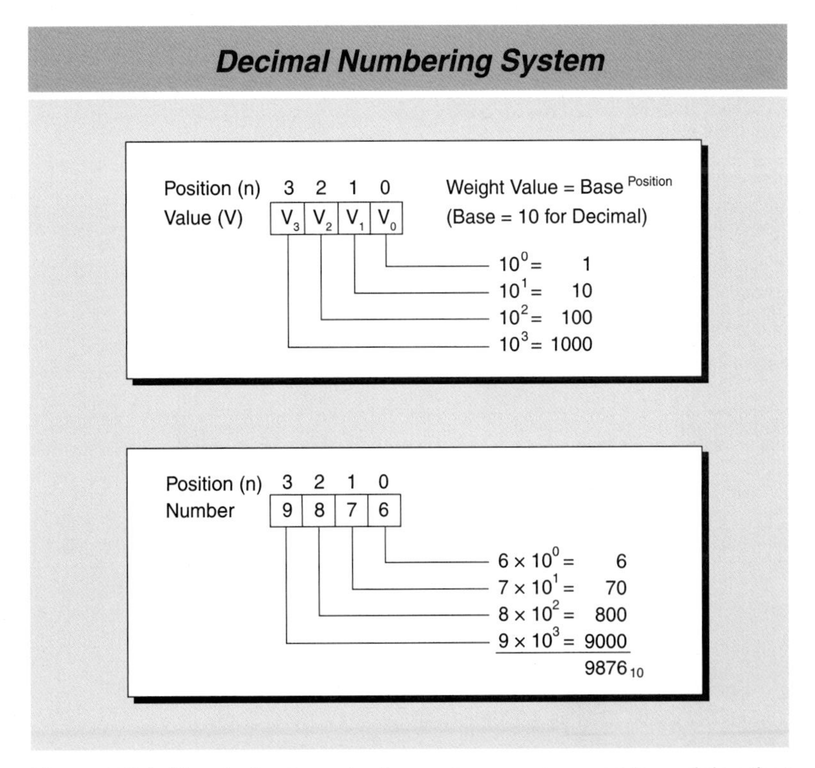

Figure 30-1. The decimal numbering system can be used to explain other numbering systems.

Multiplying each digit by the weighted value of its position and then summing the results gives the value of the equivalent decimal number. Remember that any number raised to the zero power equals 1. For example, the number 9876 can be expressed as:

$$9876_{10} = 9 \times 10^3 + 8 \times 10^2 + 7 \times 10^1 + 6 \times 10^0$$

$$9876_{10} = 9 \times 1000 + 8 \times 100 + 7 \times 10 + 6 \times 1$$

$$9876_{10} = 9000 + 800 + 70 + 6$$

$$9876_{10} = \mathbf{9876}$$

This method is used with all bases to convert the numbers back to the decimal numbering system.

DIGITAL NUMBERING AND CODES

A *digital signal* is a group of low-level DC voltage pulses that can be used to convey information. Digital signals are primarily used for communications, which are becoming more important in instrumentation systems. Before communications can be discussed, it is necessary to first understand the basics of digital numbering systems. A *digital numbering system* is a method of coding information in terms of two-position ON/OFF voltage signals.

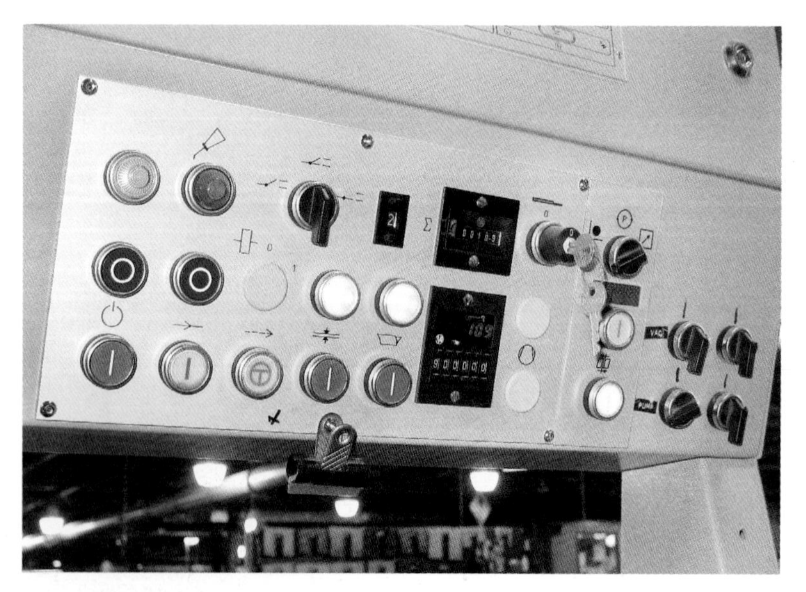

Information about the binary ON/OFF status of switches must be transmitted to a controller.

Binary Numbers

A *binary number* is a number given in a base of 2. The symbols used in this system are 0 and 1. A *bit* is a binary digit consisting of a 0 or a 1. A *nibble* is a group of four bits and is the minimum number of bits needed to represent a single decimal number. A *byte* is a group of eight bits. A *word* is a group of bits handled together by a computer system. The number of bits in a word can be different on different computer systems. **See Figure 30-2.** Just as in the decimal system, the weighted value of each position is expressed as the base raised to the power of *n*, the position. For the binary numbering system, the weighted values of each position, from right to left, are 1, 2, 4, 8, 16, 32, 64, etc.

The decimal equivalent of a binary number can be calculated by multiplying each bit by the position weighting. For example, what is the decimal equivalent of the binary number 10110110_2? The number can be expressed as:

$$10110110_2 = 1 \times 2^7 + 0 \times 2^6 + 1 \times 2^5 + 1 \times 2^4 + 0 \times 2^3 + 1 \times 2^2 + 1 \times 2^1 + 0 \times 2^0$$

$$10110110_2 = 1 \times 128 + 0 \times 64 + 1 \times 32 + 1 \times 16 + 0 \times 8 + 1 \times 4 + 1 \times 2 + 0 \times 1$$

$$10110110_2 = 128 + 0 + 32 + 16 + 0 + 4 + 2 + 0$$

$$10110110_2 = \mathbf{182_{10}}$$

An alternative way to explain this procedure is to say that the decimal equivalent of a binary number can be determined by using the decimal equivalent for the binary number in each position. As shown above, the decimal equivalent numbers start at the right and are doubled for each position to the left. The equivalent decimal numbers, which are associated with a binary 1, are added together to obtain the total decimal number as follows:

$$10110110_2 = 128 + 0 + 32 + 16 + 0 + 4 + 2 + 0$$

A binary number can be either signed or unsigned. A "signed" binary number can be positive (no sign) or negative (–). The sign uses the left-most position of a binary number. The use of a signed binary number reduces the size of the largest decimal number represented. A 16-bit unsigned binary number is always positive and can represent a maximum decimal number of 65,535. A 16-bit signed binary number can represent a decimal range of –32,767 to +32,767.

Binary numbers are very useful because they can be used to represent physical situations that have only two alternatives, such as ON/OFF, open/closed, high/low, or true/false. In digital electronics there can only be two states: voltage or no voltage, current or no current, a magnetic field or no magnetic field, and light or no light. These are called binary states. Usually, 0 is called a false state and 1 is called true state (Visual Basic® programming also uses –1 as true).

Binary Numbering System

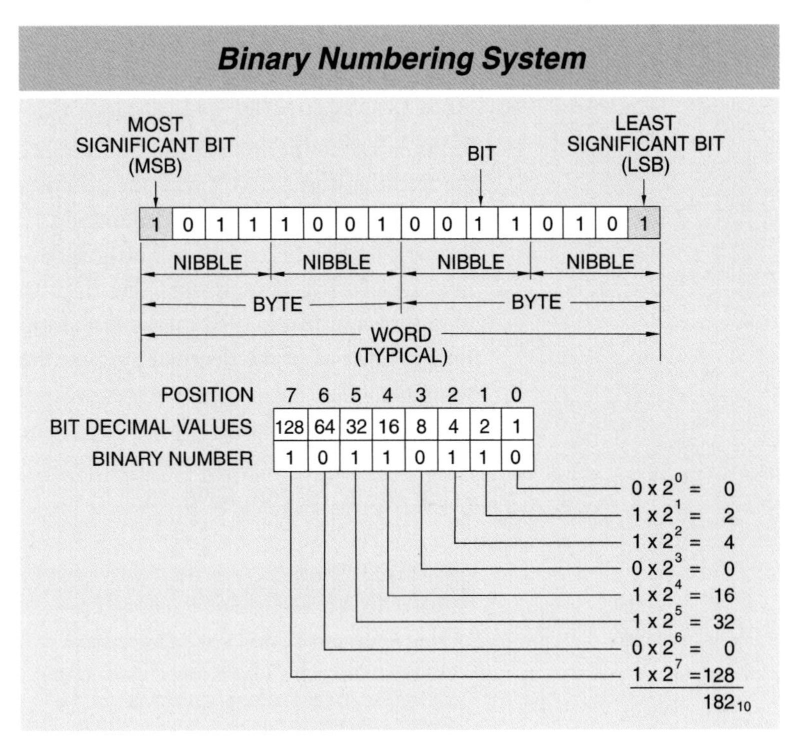

Figure 30-2. The binary numbering system is a base 2 numbering system made up of bits, nibbles, bytes, and words.

TECH FACT

The sexagesimal numbering system has a base of 60. Sexagesimal numbers are used when measuring minutes and seconds in time and when measuring angles.

Octal Numbers

An *octal number* is a number given in a base of 8. The symbols used in this system are the familiar digits 0, 1, 2, 3, 4, 5, 6, and 7. The octal numbering system can be used to represent binary numbers using fewer digits. **See Figure 30-3.** The octal numbering system uses one digit to represent three bits in a binary system. Just as in the decimal system, the weighted value of each position is expressed as the base raised to the power of n.

The octal numbering system is rarely used anymore. It was used extensively in older computer systems and some programming languages. Octal numbering has been replaced by the hexadecimal numbering system in most applications.

Decimal-Binary-Octal Comparison

Decimal	*Binary*	*Octal*
0	0	0
1	1	1
2	10	2
3	11	3
4	100	4
5	101	5
6	110	6
7	111	7
8	1000	10
9	1001	11
10	1010	12
11	1011	13
12	1100	14
13	1101	15
14	1110	16
15	1111	17

BINARY NUMBER	1	1 1 0	0 0 1	1 1 1	1 0 1	0 1 1
3-BIT GROUPS	1	1 1 0	0 0 1	1 1 1	1 0 1	0 1 1
OCTAL DIGITS	1	6	1	7	5	3

Figure 30-3. Octal numbers are a way of presenting binary numbers with fewer digits because each three-bit group of binary numbers is equivalent to a single octal digit.

Hexadecimal (Hex) Numbers

A *hexadecimal (hex) number* is a number given in a base of 16. The symbols used in the hexadecimal numbering system are the familiar digits 0 to 9 with the addition of A, B, C, D, E, and F for the numbers 10 through 15. **See Figure 30-4.** The hexadecimal numbering system uses one digit to represent four bits in the binary numbering system. Just as in the decimal system, the weighted value of each position is expressed as the base raised to the power of *n*. The weighted values of each position, from right to left, are 1, 16, 256, 4096, etc.

Many flowmeters generate pulses representing a quantity of material. The accuracy of the flow measurement depends on the number of pulses assigned to a given mass or volume of flow. The number of pulses per second is the frequency of the meter.

It is very easy to convert between binary and hexadecimal numbers. Each four-digit grouping of binary numbers is directly converted to its hex equivalent. For example, what is the hex equivalent of the 16-digit binary number 1111000110100110_2? Starting from the right, break up the binary number into groups of four. Convert each four-digit grouping to hex digits. The result is $F1A6_{hex}$. This shows that a four-digit hex number can represent a 16-digit binary number.

The decimal equivalent of a hexadecimal number can be calculated by multiplying each hex digit by the position weighting. **See Figure 30-5.** For example, what is the decimal equivalent of the hex number $F1A6_{hex}$? The number can be expressed as:

$$F1A6_{hex} = 15 \times 16^3 + 1 \times 16^2 + 10 \times 16^1 + 6 \times 16^0$$

$$F1A6_{hex} = 15 \times 4096 + 1 \times 256 + 10 \times 16 + 6 \times 1$$

$$F1A6_{hex} = 61{,}440 + 256 + 160 + 6$$

$$F1A6_{hex} = \mathbf{61{,}862}$$

Decimal-Binary-Hexadecimal Comparison

Decimal	Binary	Hexadecimal
0	0	0
1	1	1
2	10	2
3	11	3
4	100	4
5	101	5
6	110	6
7	111	7
8	1000	8
9	1001	9
10	1010	A
11	1011	B
12	1100	C
13	1101	D
14	1110	E
15	1111	F

BINARY NUMBER	1	1	1	1	0	0	0	1	1	0	1	0	0	1	1	0
4-BIT GROUPS	1 1 1 1				0 0 0 1				1 0 1 0				0 1 1 0			
HEX DIGITS	F				1				A				6			

Figure 30-4. Each hexadecimal digit can be used to represent four bits.

Binary Coded Decimal (BCD)

A *binary coded decimal (BCD) system* is a numbering system that uses groups of four bits to represent a group of decimal numbers. The binary coded decimal system was introduced as a convenient way for humans to handle numbers that must be input to digital machines and to interpret numbers that are output from machines. Four bits in BCD represent each digit in the decimal system. The need for accuracy usually requires the use of 16 bits, so that a four-digit number can be represented. The largest decimal number that can be represented by 16 bits in BCD format is 9999. **See Figure 30-6.**

ASCII

An *alphanumeric code* is a number code that uses a combination of letters, symbols, and decimal numbers to process information for computers and printers. The most common alphanumeric code is the American Standard Code for Information Interchange (ASCII). *American Standard Code for Information Interchange (ASCII)* is a seven-bit or eight-bit alphanumeric code used to represent the basic alphabet plus numbers and special symbols. The seven-bit ASCII code includes all the codes necessary to communicate with peripherals and interfaces. The eight-bit ASCII code includes one more bit for error checking.

Decimal Conversions

It is often necessary to convert decimal numbers to other bases. To convert a decimal number to its equivalent in any base, a series of divisions is performed. The conversion process starts by dividing the decimal number by the base. If there is a remainder, it is placed in the least significant digit (LSD) right-most position of the new base number. If there is no remainder, a 0 is placed in the LSD position. The result of the division is then brought down, and the process is repeated until the final result of the successive divisions is 0. This methodology is a bit cumbersome, but is the easiest conversion method to understand and use.

For example, what is the binary equivalent of the decimal number 35? **See Figure 30-7.** Divide the 35 by 2, getting 17 with a remainder of 1. The 1 is the LSD. Next, divide the 17 by 2, getting 8 with a remainder of 1. This 1 is the next LSD. Next, divide the 8 by 2, getting 4 with a remainder of 0. This 0 is the next LSD. Next, divide the 4 by 2, getting 2 with a remainder of 0. This 0 is the next LSD. Next, divide the 2 by 2, getting 1 with a remainder of 0. This 0 is the next LSD. Last, divide the 1 by 2, getting 0 with a remainder of 1. This is the final digit. The number 35_{10} is equivalent to 100011_2.

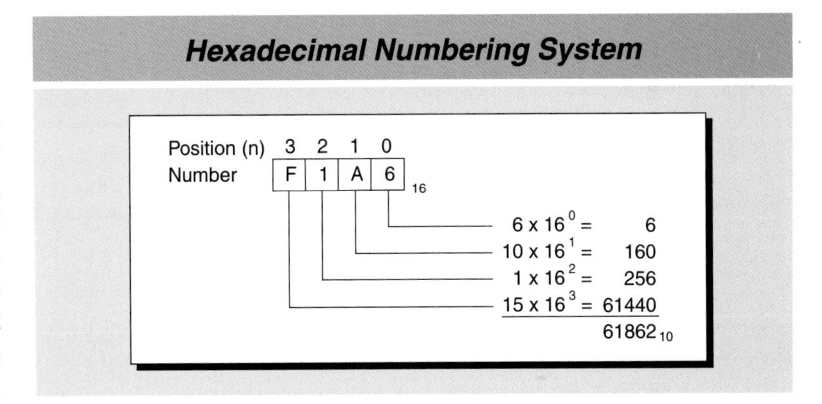

Figure 30-5. Hexadecimal numbers can be converted to decimal numbers by using the sum-of-weights position method.

Decimal-Binary-BCD Comparison

Decimal	Binary	BCD
0	0	0000
1	1	0001
2	10	0010
3	11	0011
4	100	0100
5	101	0101
6	110	0110
7	111	0111
8	1000	1000
9	1001	1001

Figure 30-6. Binary coded decimal (BCD) numbers are based on groups of four bits that are used to represent a number from 0 to 9. Multiple groups of four bits are combined to represent larger numbers.

Decimal Conversion

Division	Remainder
35 ÷ 2 = 17	1
17 ÷ 2 = 8	1
8 ÷ 2 = 4	0
4 ÷ 2 = 2	0
2 ÷ 2 = 1	0
1 ÷ 2 = 0	1

$35_{10} = 100011_2$

Division	Remainder
1355 ÷ 16 = 84	11
84 ÷ 16 = 5	4
5 ÷ 16 = 0	5

$1355_{10} = 54B_{hex}$

Figure 30-7. Decimal numbers can be converted to any other base through a series of divisions.

Digital communication relies on binary data. The voltage alternates between high and low. Each high-voltage level represents one and each low-voltage level represents zero. Other bits are added to communicate beginning and end.

The method used is the same to convert decimal numbers to other bases, except that the decimal number is divided by that base instead of by 2. For example, what is the hex equivalent of 1355_{10}? Divide the 1355 by 16, getting 84 with a remainder of 11. Remember that decimal 11 is equivalent to hex B. Therefore, B_{hex} is the LSD. Divide 84 by 16, getting 5 with a remainder of 4. This is the next LSD. Next divide 5 by 16, getting 0 with a remainder of 5. This is the last digit. The number 1355_{10} is equivalent to $54B_{hex}$.

KEY TERMS

- *bit:* A binary digit consisting of a 0 or a 1.
- *decimal number:* A number given in a base of 10.
- *digital signal:* A group of low-level DC voltage pulses that can be used to convey information.
- *digital numbering system:* A method of coding information in terms of two-position ON/OFF voltage signals.
- *binary number:* A number given in a base of 2.
- *nibble:* A group of four bits and is the minimum number of bits needed to represent a single decimal number.
- *byte:* A group of eight bits.
- *word:* A group of bits handled together by a computer system.
- *octal number:* A number given in a base of 8.
- *hexadecimal (hex) number:* A number given in a base of 16.
- *binary coded decimal (BCD) system:* A numbering system that uses groups of four bits to represent a group of decimal numbers.
- *alphanumeric code:* A number code that uses a combination of letters, symbols, and decimal numbers to process information for computers and printers.
- *American Standard Code for Information Interchange (ASCII):* A seven-bit or eight-bit alphanumeric code used to represent the basic alphabet plus numbers and special symbols.

REVIEW QUESTIONS

1. What are the numbering systems and codes that are important for communication?
2. Describe the different symbols used and the place value for each position in decimal, binary, and hex numbering.
3. How is a binary number converted to a decimal number?
4. What is the binary coded decimal (BCD) system, and why is it used?
5. What is the American Standard Code for Information Interchange (ASCII)?

Digital Communications

Chapter 31

OBJECTIVES

- Define digital communications and describe the main types of network configurations, addressing, and protocols.
- List and describe the types of circuits used in digital communications wiring formats.
- Identify common cable and wiring formats.

NETWORK PRINCIPLES

Digital communications is a method of sending and receiving binary information, in the form of low-level DC voltage pulses, between multiple devices using a common wiring format and procedure. Digital signals are used to communicate over input/output (I/O) bus networks. **See Figure 31-1.**

A *bus* is a data transmission path in which digital signals are dropped off or picked up at each point where a device is attached to the wires. An *input/output (I/O) bus network* is a communications system that allows distributed control systems (DCSs), PLCs, or other controllers to communicate with I/O devices and each other. The method of communication is similar to the way that local area networks let supervisory PLCs and controllers communicate with individual PLCs and controllers. An I/O bus network configuration decentralizes control, yielding larger and faster control systems.

Network communications is a very complex, dynamic subject and the technology continues to change rapidly. Communications is an integral part of modern instrumentation systems. The use of modular systems for DCSs and PLCs necessitated the development of fast, high-capacity, secure, reliable communications between the separate modules. Each DCS and PLC manufacturer developed its own communications method, and most of them are proprietary. Manufacturers of DCSs tend to call these communication links highways while PLC manufacturers call them networks, but they basically perform the same function.

The growth of digital technology required that communications be extended to foreign devices and be open for anyone to use. A foreign device is manufactured by someone other than the developer of the communications system. These conventional communications methods are still widely used to interface with foreign devices.

Bus networks interface with both discrete and analog devices, but all information is transmitted digitally. Digital communications allows more than one field device to be connected to a wire due to addressing capabilities and the device's ability to recognize data. These digital signals are less susceptible than other types of signals to signal degradation caused by electromagnetic interference (EMI) and radio frequencies generated by analog electronic equipment in the process environment.

TECH-FACT

Instrument air is typically provided with a dewpoint of −40°F. Desiccant air dryers use chemical beads, called desiccant, to adsorb water vapor from compressed air before it enters the air system.

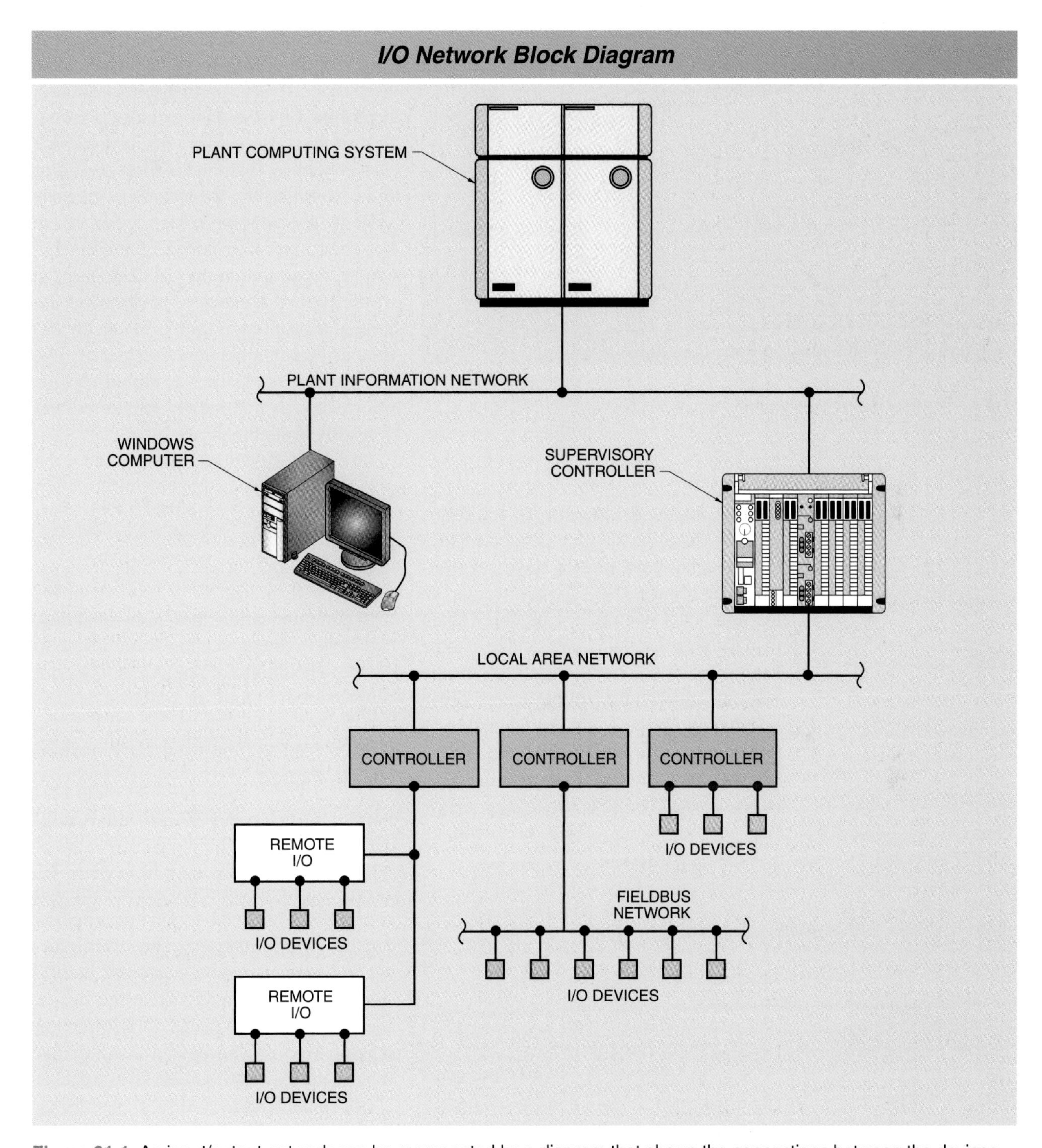

Figure 31-1. An input/output network can be represented by a diagram that shows the connections between the devices.

Network Configurations

Networks can be assembled in a number of different configurations. These configurations are sometimes called topologies. A network topology can be in the form of a bus, ring, or star system. Networks can be made from a combination of all three configurations. The different topologies are designed to serve different network requirements.

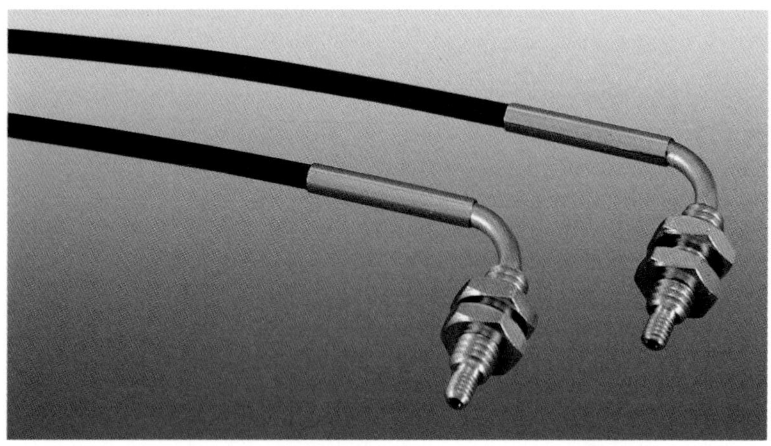

Pepperl + Fuchs, Inc.

Optical fiber cables can be used to bend light signals around an elbow.

Bus Networks. A *bus network* is a topology in which devices are connected to a common bus in a master-slave relationship. **See Figure 31-2.** The connections to the individual devices can be daisy chain or multidrop format. Each device is able to receive all information packets on the bus but ignores everything except those information packets that are addressed to that device. Information reception is very fast and many devices can receive the same information at the same time.

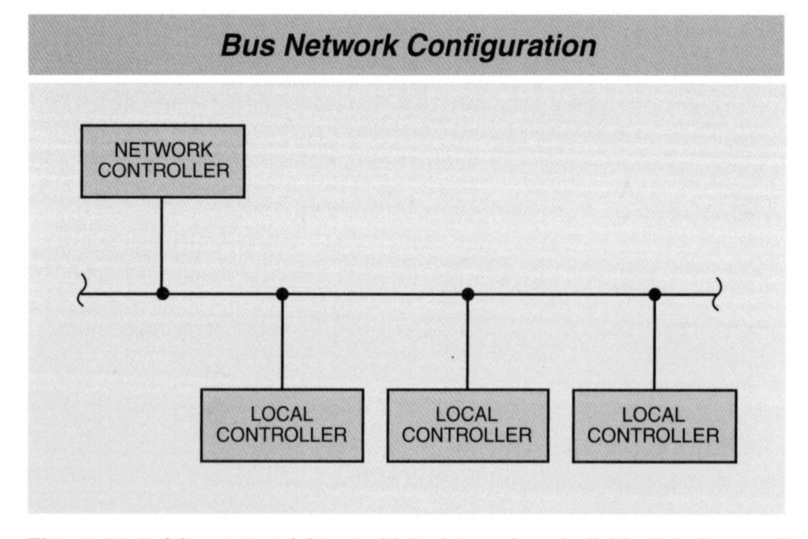

Figure 31-2. A bus network has multiple drops where individual devices and controllers are connected.

Information transmission from a device can only be done at specific times. Permission to send normal information is determined by the possession of a token. The token is passed from device to device and is held by that device for a period of time determined by the designers of the bus system. The communication speed of the bus determines the number of devices that can be supported on the bus. The purpose of the limited number of devices and the limited transmission time is to ensure that the bus does not become overloaded. The bus loading and other communications information is continually monitored and is readily available for analysis.

In critical situations, a device can use a priority interrupt to transmit information on the bus even though the device does not have possession of the token. The bus communications format is ideally designed for use with industrial systems. It offers high communication rates, high reliability, and direct communication from device to device. The failure of one device does not cause system failure. In addition, devices can easily be added to the bus. This is the predominant system used for industrial process control.

Ring Networks. A *ring network* is a topology in which devices are connected in series. A ring network permits each device to pass information on to the next device in series. **See Figure 31-3.** This information consists of information packets being forwarded on to other devices around the ring and any new information that the device needs to send. Each packet of information is addressed to one device on the ring. When a device on the ring receives a packet of information addressed to it, that device processes the information and does not pass it on to the next device on the ring.

Since each device only has a fixed time to communicate, the total amount of information being passed from one device to another is limited. If several devices on the ring need to transmit information, the network passes on the information in the order in which it was received. A ring network configuration is seldom used for industrial process systems because it is too slow, and a failure of any device can cause a system failure.

Star Networks. A *star network* is a topology in which every peripheral device is connected to a master controller with a pair of wires. **See Figure 31-4.** In this format, there is no direct communication between the peripheral devices. The master processor passes information received from one peripheral device on to another peripheral device. The master processor controls all communications. A failure of the master processor can cause a system failure or shutdown, but a failure of a peripheral device does not. A tree topology combines star networks on a bus. A star network configuration is rarely used for industrial systems because of the time it takes to get information from one peripheral to another.

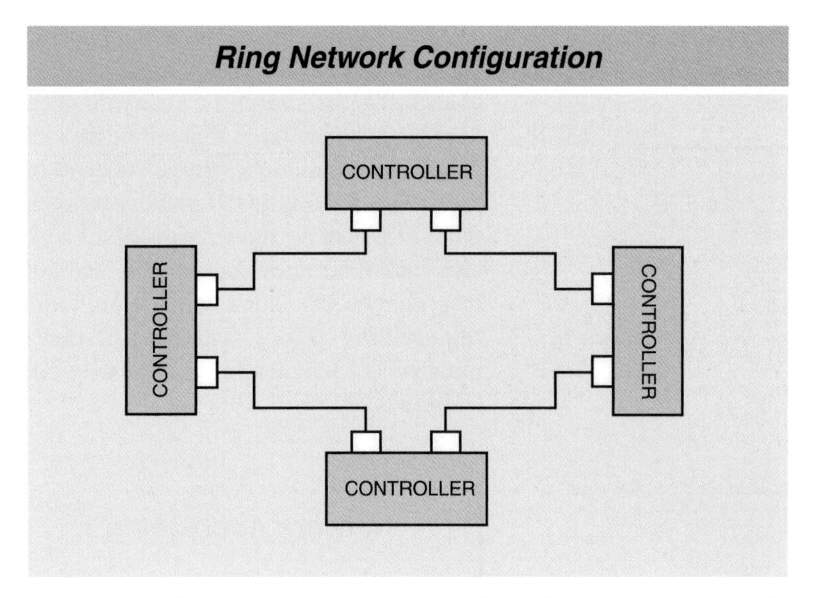

Figure 31-3. A ring network configuration has all devices connected in series.

Network Addressing

A *network address* is a unique number assigned to each device on a network. Addressing of the devices in a bus network occurs during the configuration, or programming, of the devices in the system. Some systems automatically assign addresses based on a scan of the devices connected to the network. Depending on the controller, this addressing can be done either directly on the bus network via a PC and a gateway, through a PC connected directly to the bus network interface, or through the controller's RS-232 port. Some bus network devices have switches that can be used to define that device's address, while others have a predefined address associated with each node drop.

TECH-FACT

A physical topology is described by the wiring that links the nodes of a network. The physical topology determines how the electrical signals are sent through the network. A logical topology is described by the flow of information through a network.

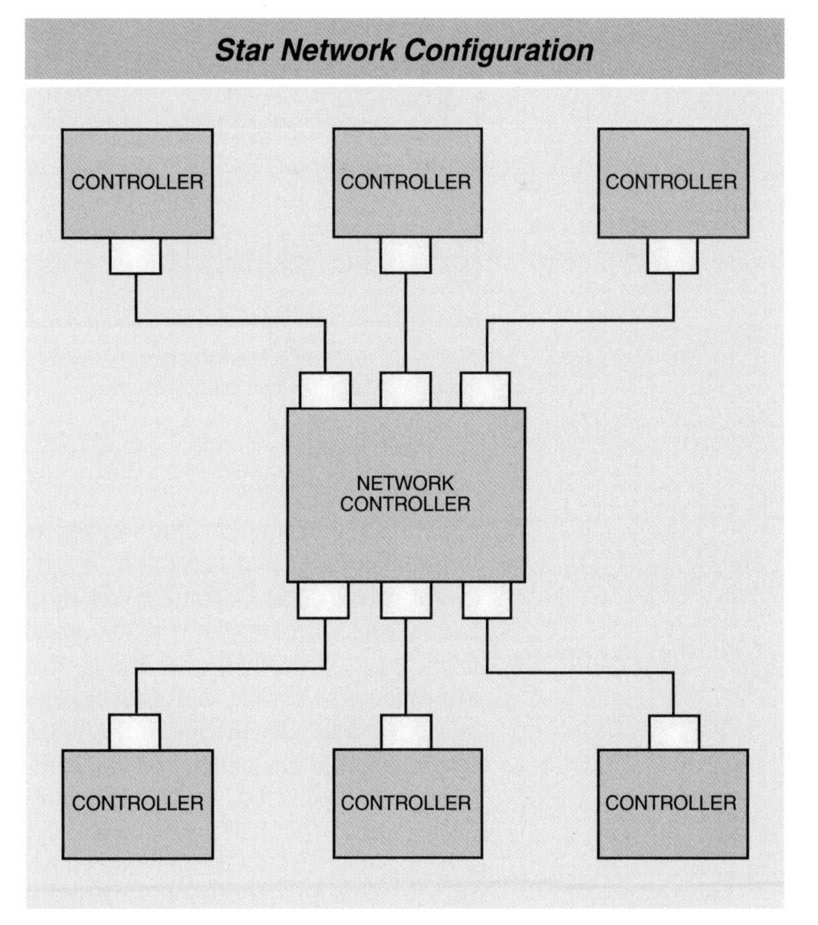

Figure 31-4. A star network configuration has all devices individually connected to a controller.

Many devices have a set of dual in-line package (DIP) switches where the position of each switch represents a binary number. Summing the switches that are in the ON position determines the network address of the device. **See Figure 31-5.** For example, the switches in the figure in positions 1, 4, and 7 are ON, representing 1, 8, and 64, respectively. The other switches are OFF, representing zeroes. The sum of these numbers is 73, indicating that the network address is 73.

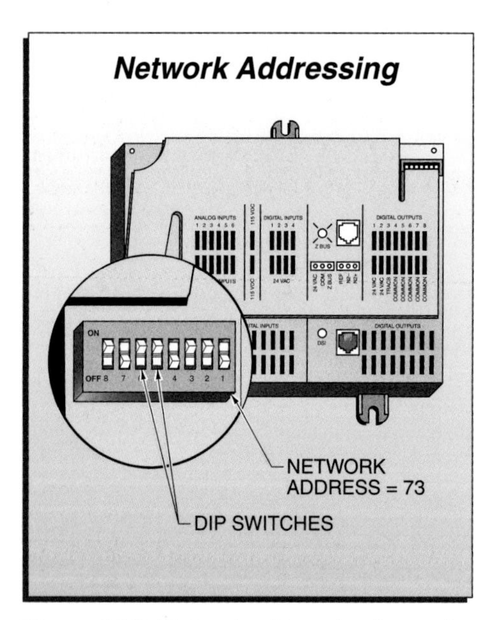

Figure 31-5. Network addressing is used to logically locate devices on a network.

Network Protocols

Networks follow a protocol to implement the transmission and reception of data over a network line that might consist of coaxial cable, twisted-pair wire, or optical fiber cable. A *protocol* is a set of rules that determines the format and transmission method of data transmission. In 1979, the International Organization for Standardization (ISO) published the Open Systems Interconnection (OSI) reference model, also known as ISO IS 7498, to provide guidelines for network protocols. Each protocol is intended to be open so that any device can be connected.

The OSI model divides the functions that protocols must perform into seven hierarchical layers. **See Figure 31-6.** Each layer interfaces only with its adjacent layers. The OSI model further subdivides the second layer into two sublayers, 2A and 2B, called medium access control (MAC) and logical link control (LLC), respectively. The physical layer (layer 1) and the MAC sublayer (layer 2A) are usually implemented with hardware, while the remaining layers are implemented with software. The physical layer is used to translate messages into signals on the control wiring and signals back to messages. The hardware components of layers 1 and 2A are generally referred to as modems (or transceivers) and drivers (or controllers).

In actual practice, a network requires only layers 1, 2, and 7 of the protocol model to operate. In fact, many device bus networks use only these three layers. The other layers are added as more services are required, such as error-free delivery, routing, session control, and data conversion. Most modern networks contain all or most of the OSI layers to allow connection to other networks and devices.

Specialized technical organizations, as opposed to standards committees such as the ISO, have made the largest efforts toward the standardization of network protocols. The ISO, however, will accept and validate a network standard as long as it complies with the protocol architecture defined by the OSI model.

The data package consists of a defined number of bits that begin with a start bit and end with a parity and a stop bit. A parity bit is used to set the number of ones in a bit string to either an even or odd number, depending on the protocol. The actual message is the bit string in the middle of the string. If the total number of digits received does not match the required string length or if the parity bit is wrong, the receiver requests that the data be sent again.

TECH FACT

The OSI Model is a standard description of the various layers of data communication commonly used in computer networks. The purpose is to create a framework to be used as the basis for defining standard communication protocols to divide the problem of computer-to-computer communication into several smaller pieces.

DIGITAL COMMUNICATIONS WIRING FORMATS

Digital communications can use either a parallel or series wiring format. The choice is dependent on the quantity of the information that needs to be transferred, the required speed, and the distance. There are different standards associated with both parallel and series wiring formats. These standards ensure that all equipment manufactured for a particular format is compatible with all other equipment.

Transistor-Transistor Logic (TTL) Circuits

Digital devices are typically connected from one transistor circuit to another using a method called transistor-transistor logic (TTL). *Transistor-transistor logic (TTL)* is the connection method that allows separate solid-state devices to be wired together to pass information. All digital communications incorporate transistor-transistor logic as the physical method of communication from one device to another. For example, the RS-232 communications standard uses TTL circuits inside every communications device. This is usually transparent to the user because it is part of each communications standard.

It is sometimes necessary to interconnect two digital devices that are not directly compatible. For example, the BCD output from a weight indicator may need to be sent to the discrete inputs of a PLC. If the PLC TTL inputs and outputs are not compatible with the weigh system inputs and outputs, a TTL conversion board must be inserted as an interface between the two systems.

OSI Reference Model

Layer	Layer Name	Function
Layer 7	Application	The level seen by users; the user interface
Layer 6	Presentation	Control functions requested by the user; data is restructured from the other standard formats; code and data conversion
Layer 5	Session	System-to-system connection; log-in and log-off controlled here; establishes connections and disconnections
Layer 4	Transport	Provides reliable data transfer between end devices; network connections for a given transmission are established by protocol
Layer 3	Network	Outgoing messages are divided into packets; incoming packets are assembled into messages for higher levels, establishing connections between equipment on the network
Layer 2	Data link	Outgoing messages are assembled into frame and acknowledgments; error detection or error correction is performed
Layer 1	Physical	Parameters, such as signal voltage-swing bit duration, and electrical connections, are established

Figure 31-6. The OSI reference model is used to define how the pieces of a network work together.

Modern building automation systems communicate over a digital network.

Source and Sink Circuits

Source and sink circuits are often used to electrically isolate the transmitter from the receiver. A *source circuit* is a circuit in a transmitting device that provides a positive voltage signal that can be detected in a receiver circuit. The positive voltage creates a current through a light emitting diode (LED) in a receiver. The LED is one-half of an optical device that electrically isolates the transmitter from the receiver. A common wire is needed to complete the current flow. **See Figure 31-7.** The transmitting device supplies the power.

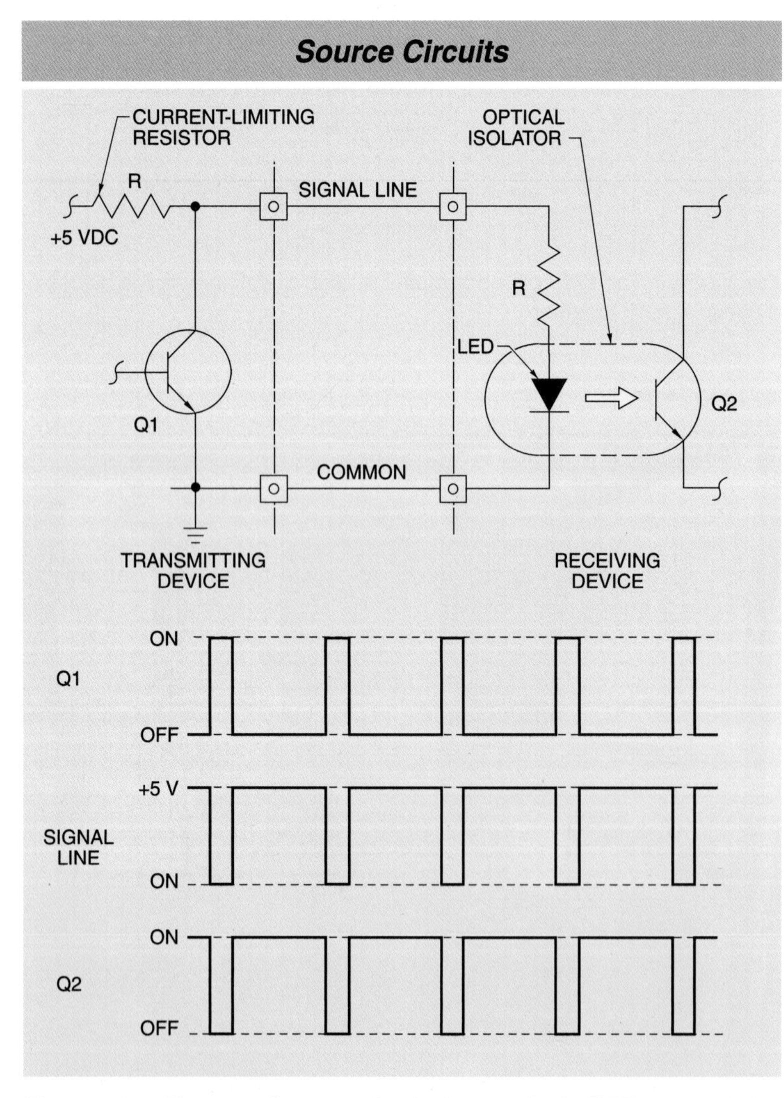

Figure 31-7. The use of source circuits is a method of TTL transmission where the source of power is in the transmitting unit.

When transistor Q1 is turned on (conducting), it pulls down the voltage in the signal line by shorting it to ground. This turns off the receiver's LED portion of transistor Q2. With no light from the LED, the transistor portion of Q2 turns off (stops conducting). When transistor Q1 is turned off, the voltage in the signal line rises to about +5 VDC, allowing the LED at transistor Q2 to emit and the transistor to turn on (conduct). The transistor Q1 may be turned on by any discrete input signal. This discrete signal could be a command signal instructing another digital device to take some action, or a reply that a message has been received, or any other digital signal.

A *sink circuit* is a circuit in a transmitting device that takes the positive voltage generated by a receiving device and shunts it to ground, lowering the voltage at the receiver circuit. The shunting element is an optically isolated transistor that provides electrical isolation between the two devices. A common wire is needed to complete the current flow. **See Figure 31-8.** The receiving device supplies power.

> **TECH-FACT**
>
> *Digital communication is subject to electromagnetic interference from crosstalk at the ends of the cable, impulse noise from lightning and switching surges, and RF noise from strong radio signals. Noise from these sources can be reduced by proper wire termination, grounding, shielding, and surge protectors.*

When the LED in Q3 in the transmitting device is conducting and emitting light, the transistor in Q3 is turned on (conducting) and pulls down the voltage in the signal line by shorting it to the receiver's ground. The low voltage in the signal line lowers the voltage at the base of Q4, turning it off. When the LED at Q3 is off, the transistor at Q3 turns off and the signal line voltage rises to about +5 VDC. This turns on Q4.

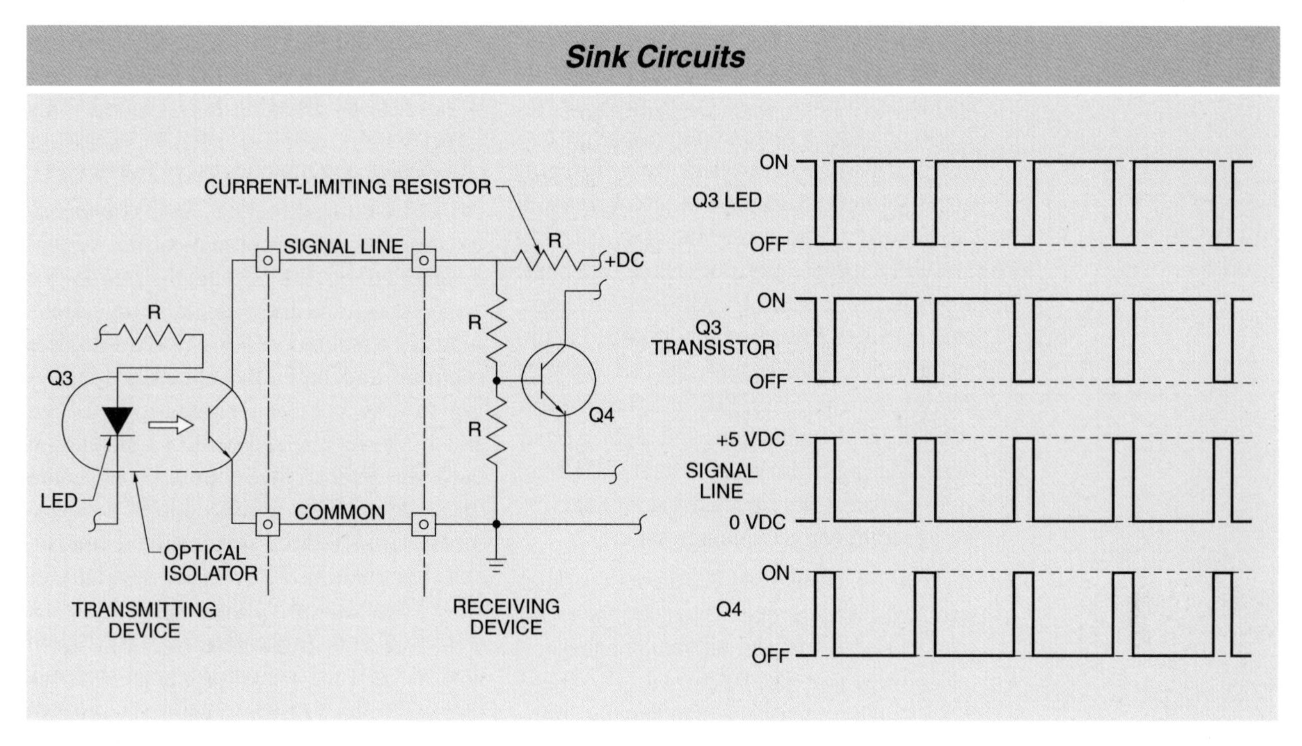

Figure 31-8. The use of sink circuits is a method of TTL transmission where the source of power is in the receiving unit.

Parallel Circuits

Parallel wiring is a method of communications wiring where multiple bits of information are transferred at the same time on multiple wires. Usually 8 to 16 bits of data are conveyed over 25 lines. The non-data lines are used for transmission protocol disciplines and handshaking. This discipline includes instructions that control the transfer of information such as "ready to receive data," "data received," "transmission error," etc. **See Figure 31-9.**

Parallel transmission is faster than serial transmission but can only be used for short distances. As distances become greater, the bits on the lines become less synchronized. Communications inside computers frequently use parallel formats because of the greater speed and the short distances. There are many different types of parallel formats, but the most common are IEEE 488, IEEE 1284, and Centronics®.

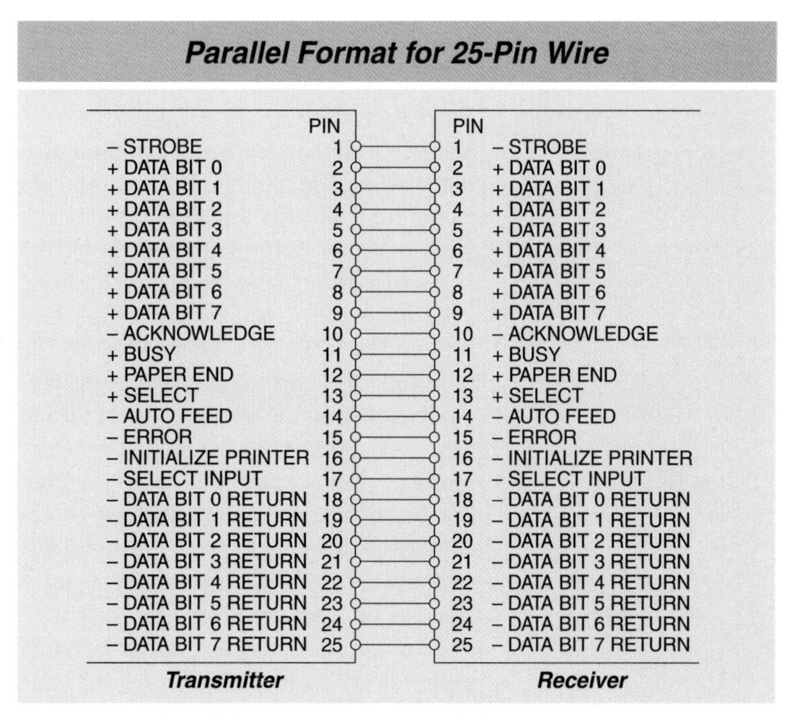

Parallel Format for 25-Pin Wire

Transmitter	PIN	PIN	Receiver
– STROBE	1	1	– STROBE
+ DATA BIT 0	2	2	+ DATA BIT 0
+ DATA BIT 1	3	3	+ DATA BIT 1
+ DATA BIT 2	4	4	+ DATA BIT 2
+ DATA BIT 3	5	5	+ DATA BIT 3
+ DATA BIT 4	6	6	+ DATA BIT 4
+ DATA BIT 5	7	7	+ DATA BIT 5
+ DATA BIT 6	8	8	+ DATA BIT 6
+ DATA BIT 7	9	9	+ DATA BIT 7
– ACKNOWLEDGE	10	10	– ACKNOWLEDGE
+ BUSY	11	11	+ BUSY
+ PAPER END	12	12	+ PAPER END
+ SELECT	13	13	+ SELECT
– AUTO FEED	14	14	– AUTO FEED
– ERROR	15	15	– ERROR
– INITIALIZE PRINTER	16	16	– INITIALIZE PRINTER
– SELECT INPUT	17	17	– SELECT INPUT
– DATA BIT 0 RETURN	18	18	– DATA BIT 0 RETURN
– DATA BIT 1 RETURN	19	19	– DATA BIT 1 RETURN
– DATA BIT 2 RETURN	20	20	– DATA BIT 2 RETURN
– DATA BIT 3 RETURN	21	21	– DATA BIT 3 RETURN
– DATA BIT 4 RETURN	22	22	– DATA BIT 4 RETURN
– DATA BIT 5 RETURN	23	23	– DATA BIT 5 RETURN
– DATA BIT 6 RETURN	24	24	– DATA BIT 6 RETURN
– DATA BIT 7 RETURN	25	25	– DATA BIT 7 RETURN

Figure 31-9. Parallel communication typically uses 25 parallel wires to transfer information from one device to another. It is fast but generally only good for short distances.

IEEE 488. IEEE 488 is a popular parallel interface protocol that was developed to connect and control programmable instruments. It also provides a standard interface between instruments and controllers from different manufacturers. IEEE 488 is also known as the general purpose interface bus (GPIB). The GPIB specification does not say anything about the instruments. The GPIB specification only specifies a function that is added to the instrument and the signals that pass through that instrument. The GPIB interface system consists of 16 signal lines and eight ground lines. The signal lines use a total of 16 bits, eight for data, three for handshaking, and five for general interface management.

IEEE 1284. IEEE 1284 is a modern parallel interface protocol that covers an enhanced parallel port (EPP) format and an enhanced capabilities port (ECP) format. The IEEE 1284 standard defines a standard for high-speed data communications between a PC or a controller and an external device or instrument. The standard is fully backward-compatible with older standards like Centronics. The EPP and ECP formats use hardware on computer chips to assist in data transfer to speed up the communications process.

Series Circuits

Series wiring is a method of communications wiring where bits of information are transferred one after another over a pair or pairs of wires. Information is transferred as a string of bits with leading and trailing handshaking bits and a final error bit. Two common factors used to establish a specific series interface method are the specific pin-out terminals in the wiring connectors and the protocol that is used. Common bit parity protocol settings are none, even, or odd. More sophisticated error-checking protocols are used for high-speed transmissions. Echoing back of the transmission signal is one way of checking for errors.

Simplex transmission, which is rarely used, allows communication in only one direction. Half-duplex transmission allows communication in one direction at a time. Full duplex transmission is when transmission and reception take place simultaneously. **See Figure 31-10.** Series transmission is very robust and can be sent over long distances. It only requires a pair of wires for simplex or half-duplex and three wires for full duplex.

RS-232 Communication. *RS-232 communication* is a serial communications system developed by the Electronics Industry Association and the Telecommunications Industry Association. RS-232 is a complete standard and includes standards specifying pin wiring configurations, voltage and signal levels, and control information between devices that communicate using the standard. This is a full duplex arrangement that includes wires for data, control, and signal timing. The original standard for RS-232 used a low voltage representing a 1 and a high voltage representing a 0. Modern devices typically include integrated circuits that invert the signals to match the modern convention of a high voltage representing a 1. **See Figure 31-11.**

The RS-232 standard was originally developed in the 1960s to connect computers to modems. The standard has been updated periodically since the 1960s. The most common version of the standard is RS-232C, which is also known as EIA-232. The most recent version is RS-232E. Any use other than computer-to-modem and modem-to-modem communications is not fully supported by the standard. Many companies market software to allow other devices to communicate through a computer serial port with the RS-232 standard. Software and devices from one company may not work properly with software and devices from another company.

RS-232 connectors were originally either 9-pin or 25-pin D-type connectors. Some computer connectors are now circular plugs and sockets. RS-232 is still in use, along with RS-422 and RS-423, which improve the speed and noise immunity. *RS-422 communication* is an improved version of RS-232 that permits data speeds up to 100 kBd and distances up to 4000 ft. RS-422 uses a differential driver and a four-conductor cable. *RS-423 communication* is a single-ended open communications variant of RS-422 that uses two data lines.

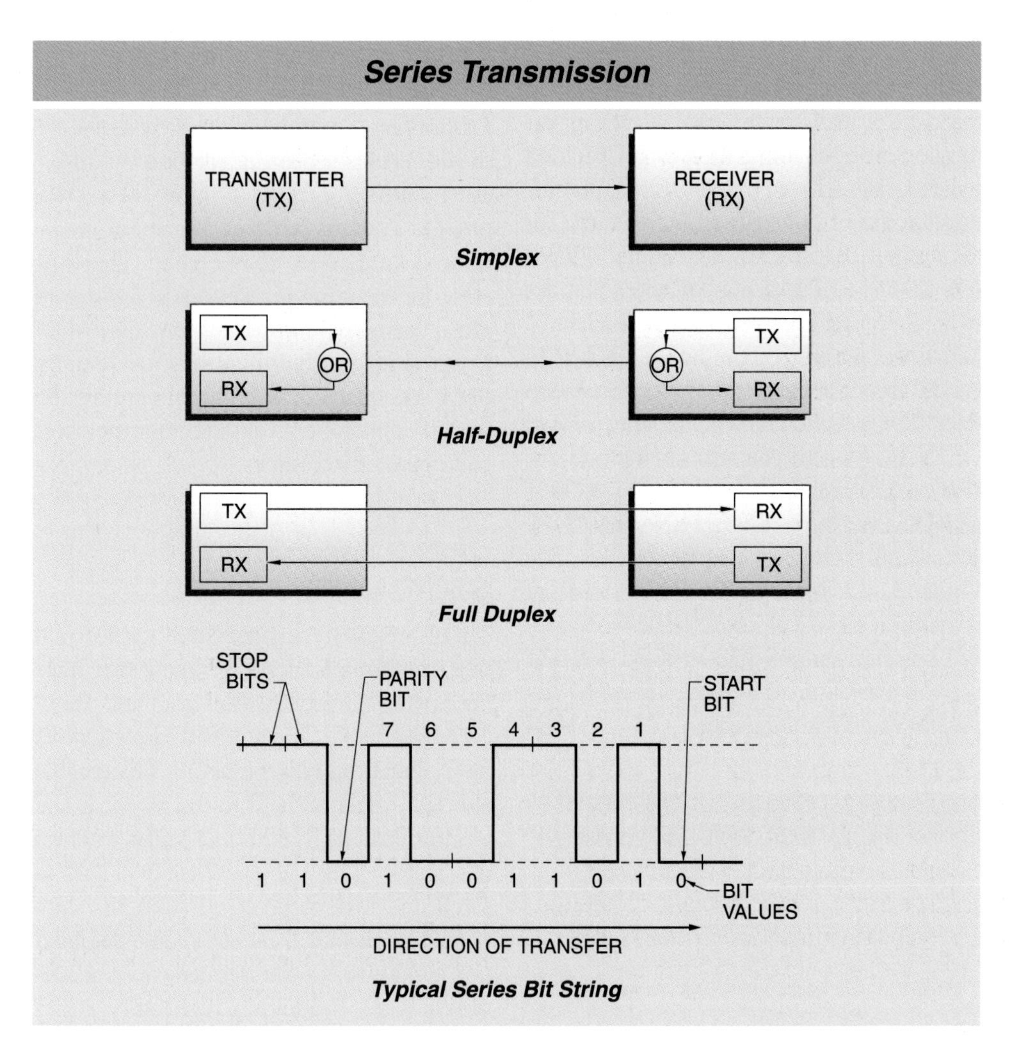

Figure 31-10. Series communication uses a minimum of two wires to transfer information from one device to another. It is slower than parallel transmission but more robust.

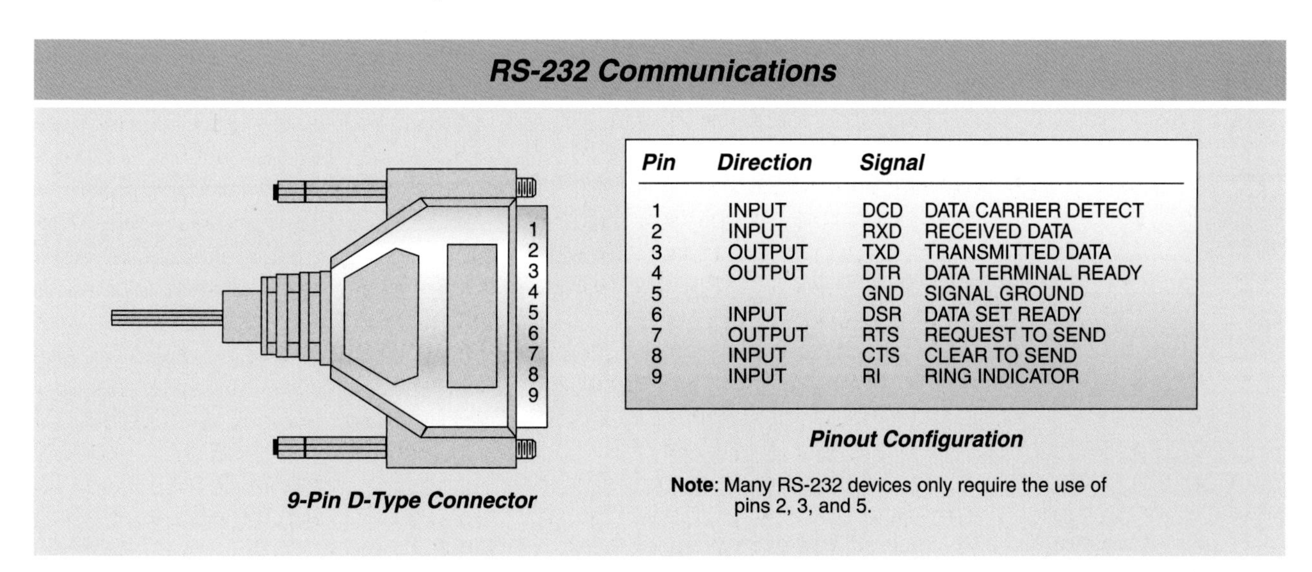

Pin	Direction	Signal	
1	INPUT	DCD	DATA CARRIER DETECT
2	INPUT	RXD	RECEIVED DATA
3	OUTPUT	TXD	TRANSMITTED DATA
4	OUTPUT	DTR	DATA TERMINAL READY
5		GND	SIGNAL GROUND
6	INPUT	DSR	DATA SET READY
7	OUTPUT	RTS	REQUEST TO SEND
8	INPUT	CTS	CLEAR TO SEND
9	INPUT	RI	RING INDICATOR

Figure 31-11. RS-232 is a full duplex form of serial communication that is commonly used for instrument connections to computers. There are many protocols that can be used with this wiring arrangement.

RS-485 Communication. *RS-485 communication* is an open communications system that uses a twin duplex coaxial wiring arrangement consisting of two unshielded twisted pairs with an overall shield, permitting the use of a multidrop architecture for the various devices. An *unshielded twisted pair (UTP) wire* is a pair of wires that are twisted around each other with no electromagnetic shielding. The data is carried in the difference between the two transmission lines. The reception data is the same format.

The two twisted pair wires of the twin duplex coaxial cable are the transmission (TX+ and TX–) and the receiving (RX+ and RX–). The overall shield provides the common. **See Figure 31-12.** There are adapter devices that provide an electrical interface between RS-485 systems and RS-232 systems. RS-485 networks are limited to a maximum of 255 possible network addresses.

> **TECH FACT**
>
> *Receivers designed for voltage loops should not be used in 4 mA to 20 mA current loops. The very high impedance of voltage-loop receivers reduces the current to a value below that required for current signaling.*

Proprietary Communications Systems

Proprietary communications systems are those that have been developed by different manufacturers to support their DCS or PLC systems without desiring to make their systems open to everyone. The software for the communications is purchased along with the control system hardware. Proprietary communications are usually the high-speed networks connecting the various controllers, data storage devices, and operator interfaces.

CABLE AND WIRING STANDARDS

There are many off-the-shelf cables and other communications products that have been proven in an office environment. However, these cables and products rarely survive in an industrial environment. Considerable attention must be paid to the physical and electromagnetic hazards that cables will face in a factory. Oils can cause cable jackets to swell and lose mechanical strength. High temperatures can also cause cable degradation and low temperatures can cause brittleness in wires. In addition, high temperatures cause significant signal attenuation.

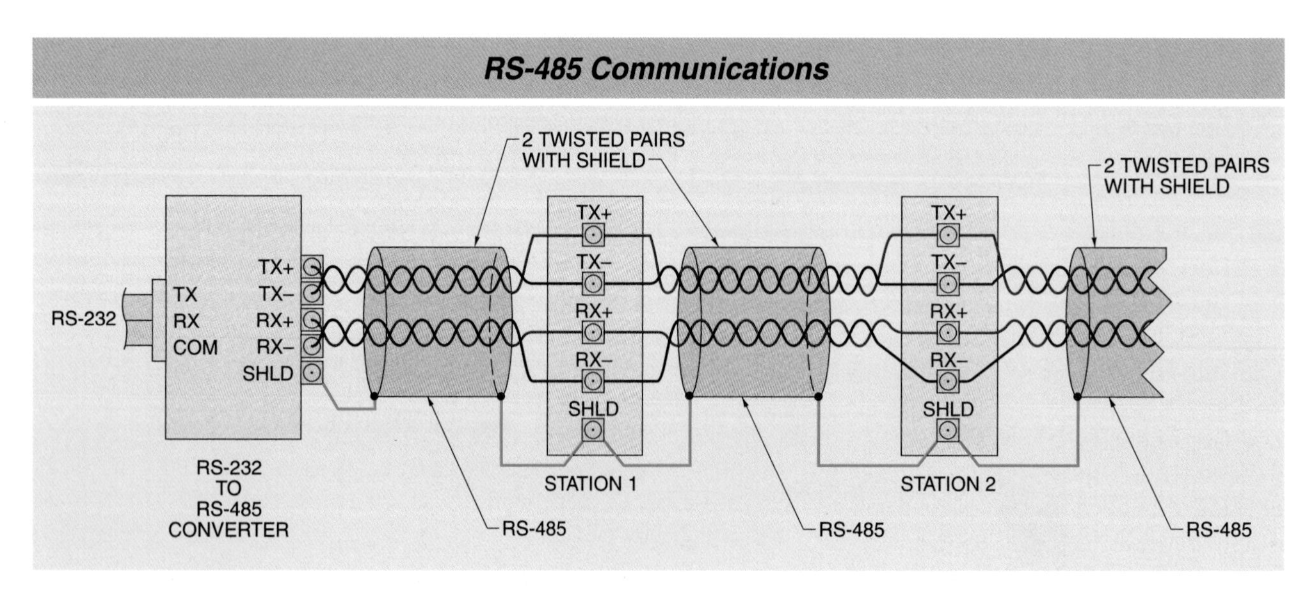

Figure 31-12. RS-485 communication uses a twin duplex cable.

Connecting Media

Category (CAT) rating numbers representing media quality are used to rate connecting media. Network wiring specifications list the required CAT rating for the connecting media. The following are connecting media CAT ratings:

- CAT 1—Older wire and cable used for telephone systems. Category 1 wire and cable is normally limited to voice transmission only.
- CAT 2—Indoor wire and cable systems used for voice applications and data transmission up to 1 MBd.
- CAT 3—Indoor wire and cable systems used for voice applications and data transmission up to 16 MBd. Data networks use a minimum of CAT 3 wire or cable.
- CAT 4—Indoor wire and cable systems used for voice applications and data transmission up to 20 MBd.
- CAT 5—Indoor wire and cable systems used for voice and data transmission up to 100 MBd. Category 5 cable is one of the most commonly used cables in networks.
- CAT 5e—Indoor wire and cable systems used for voice and data transmission up to 100 MBd. Category 5e wire and cable allow long runs because wire is packaged tightly with greater electrical balance between wire pairs.
- CAT 6—Indoor wire and cable systems used for voice and data transmission up to 250 MBd. CAT 6 is the fastest communication medium available with unshielded twisted-pair wiring.
- CAT 7—Indoor wire and cable systems used for voice and data transmission starting at 600 MBd. Each pair of wires is shielded with foil and the entire cable is shielded with copper braid screen.

Communications cables are classified as XbaseY where X is the communication speed in megabits per second, base is baseband, and Y is an indicator of the type of media. *Baseband* is a communication method where digital signals are sent without frequency shifting. Y may be the category or an abbreviation of the media. Common values for X are 5 and 10. Common values for Y are 2, 5, and T. For example, 10base5 is a cable that supports baseband communications at 10 MBd over a category 5 cable. A tester can be used to determine cable type and functionality.

Cleaver-Brooks

Modern controllers can be placed near the process being controlled.

Optical Fiber

Optical fiber communications is the newest and fastest type of digital medium for transferring information. In optical fiber communications, digital electronic communications are converted to a solid-state laser light-transmitted signal and sent through a glass or plastic fiber to a receiver where it is reconverted to electronic communications. Optical fiber strands are constructed of two layers that are designed to reflect the light down the fiber with small losses. **See Figure 31-13.**

Optical fiber communications can handle time-critical data faster and in greater volume than any other method. It can be used with any communications protocol. Optical fiber is inherently immune to EMI and can carry much more information than copper cable.

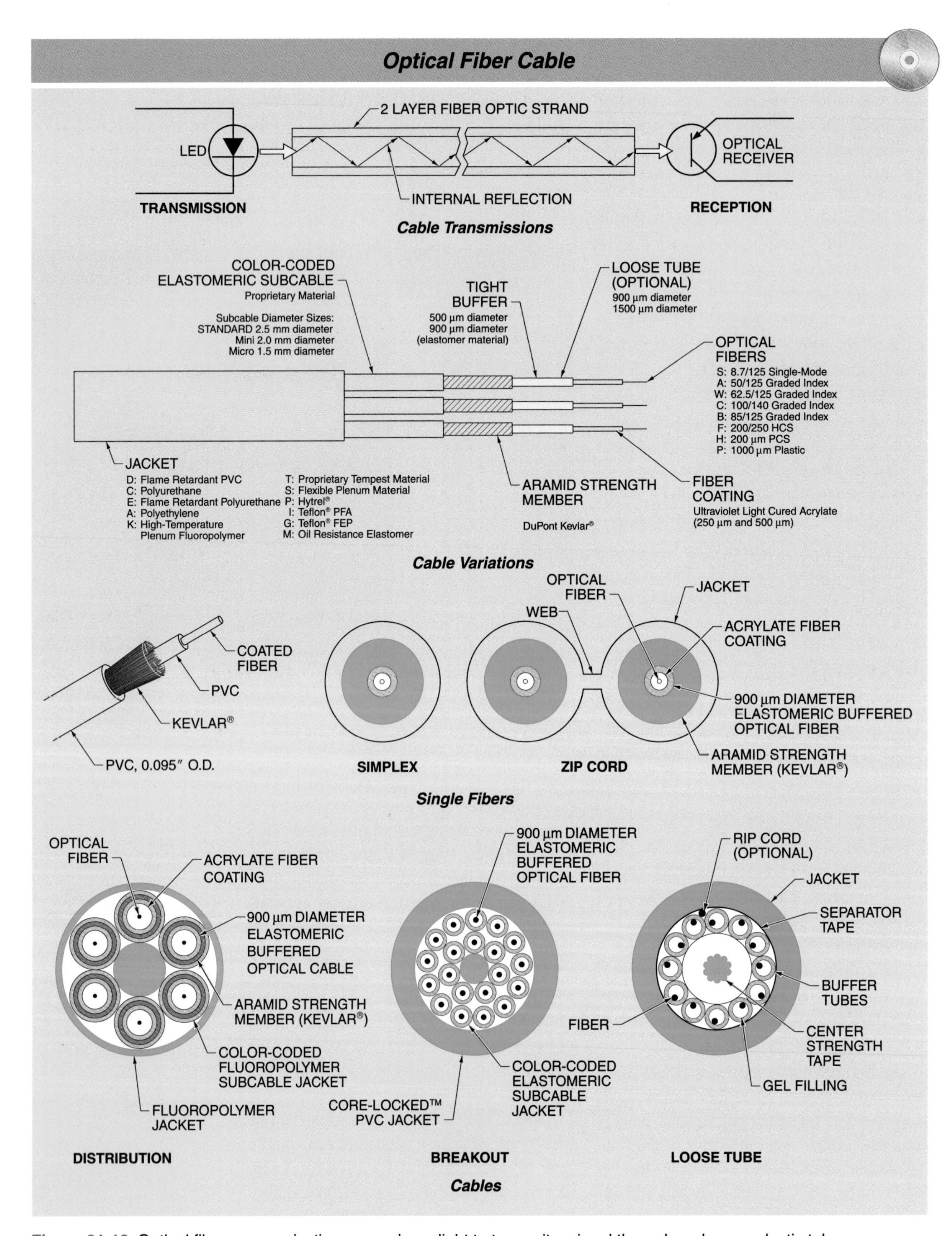

Figure 31-13. Optical fiber communication uses a laser light to transmit a signal through a glass or plastic tube.

KEY TERMS

- *digital communications:* A method of sending and receiving binary information, in the form of low-level DC voltage pulses, between multiple devices using a common wiring format and procedure.
- *bus:* A data transmission path in which digital signals are dropped off or picked up at each point where a device is attached to the wires.
- *input/output (I/O) bus network:* A communications system that allows distributed control systems (DCSs), PLCs, or other controllers to communicate with I/O devices and each other.
- *bus network:* A topology in which devices are connected to a common bus in a master-slave relationship.
- *ring network:* A topology in which devices are connected in series.
- *star network:* A topology in which every peripheral device is connected to a master controller with a pair of wires.
- *network address:* A unique number assigned to each device on a network.
- *protocol:* A set of rules that determines the format and transmission method of data transmission.
- *transistor-transistor logic (TTL):* The connection method that allows separate solid-state devices to be wired together to pass information.
- *source circuit:* A circuit in a transmitting device that provides a positive voltage signal that can be detected in a receiver circuit.
- *sink circuit:* A circuit in a transmitting device that takes the positive voltage generated by a receiving device and shunts it to ground, lowering the voltage at the receiver circuit.
- *parallel wiring:* A method of communications wiring where multiple bits of information are transferred at the same time on multiple wires.
- *series wiring:* A method of communications wiring where bits of information are transferred one after another over a pair or pairs of wires.
- *RS-232 communication:* A serial communications system developed by the Electronics Industry Association and the Telecommunications Industry Association.
- *RS-422 communication:* An improved version of RS-232 that permits data speeds up to 100 kBd and distances up to 4000 ft.
- *RS-423 communication:* A single-ended open communications variant of RS-422 that uses two data lines.
- *RS-485 communication:* An open communications system that uses a twin duplex coaxial wiring arrangement consisting of two unshielded twisted pairs with an overall shield, permitting the use of a multidrop architecture for the various devices.
- *unshielded twisted pair (UTP) wire:* A pair of wires that are twisted around each other with no electromagnetic shielding.
- *baseband:* A communication method where digital signals are sent without frequency shifting.

REVIEW QUESTIONS

1. Define digital communications.
2. Compare the three main types of network configurations.
3. What is the OSI model, and how is the physical layer implemented?
4. Compare source and sink circuits.
5. What is the difference between parallel and serial communication?

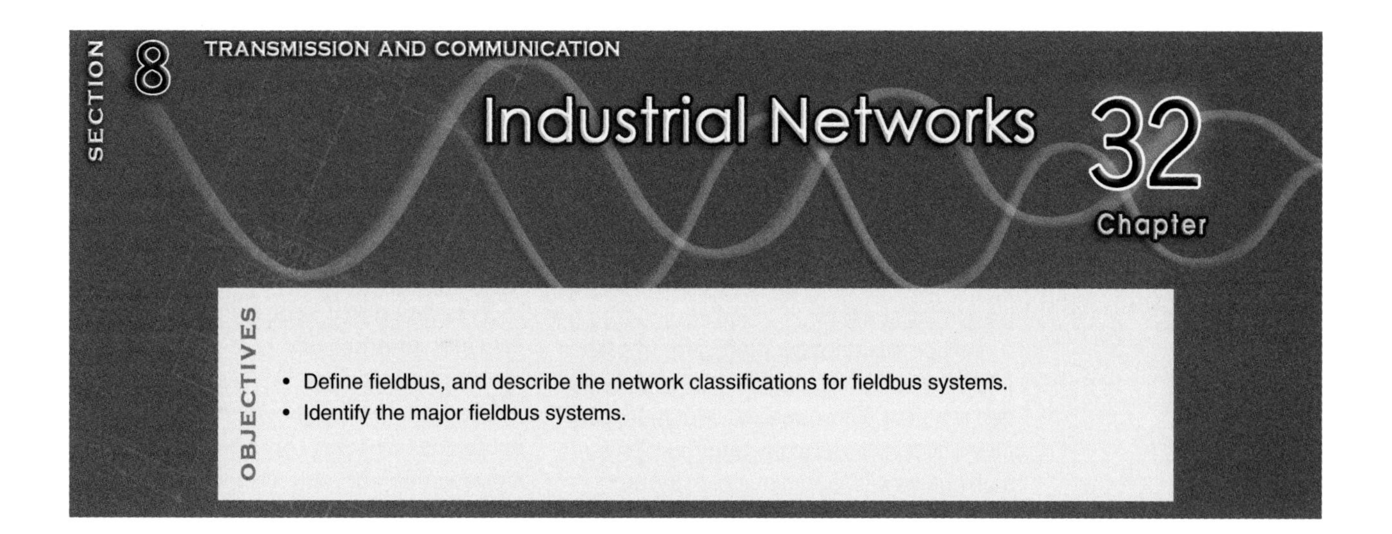

FIELDBUS SYSTEMS

A *fieldbus* is an open, digital, serial, two-way communication network that connects with high-level information devices. These information devices include smart process control valves and transmitters, flowmeters, and other complex field devices that are typically used with control equipment in process control applications. **See Figure 32-1.**

Some fieldbus networks can handle relatively large amounts of data (several hundred bytes), consisting of information about the process, as well as information about the field devices themselves. The majority of devices used in fieldbus networks are analog. The larger information packets required for process bus networks reduce network communication speed. Therefore, fieldbus networks are most applicable to the control of analog I/O devices, which do not require fast response times.

Fieldbus networks can transmit large amounts of information to a control system, thus greatly enhancing the operation of a plant or process. For example, a fieldbus-compatible flowmeter can provide information about the flow rate of the process as well as diagnostic information about the meter itself. Implementation of this type of system without a fieldbus network may be too costly and cumbersome because of the number of wire runs necessary to transmit this type of process data.

A *4-wire fieldbus network* is a fieldbus network that uses two wires for communication and two separate wires to power field instruments. A *2-wire fieldbus network* is a fieldbus network that uses communication wires to furnish power to field instruments. This means that a 2-wire fieldbus network does not require power to be available at the instrument site.

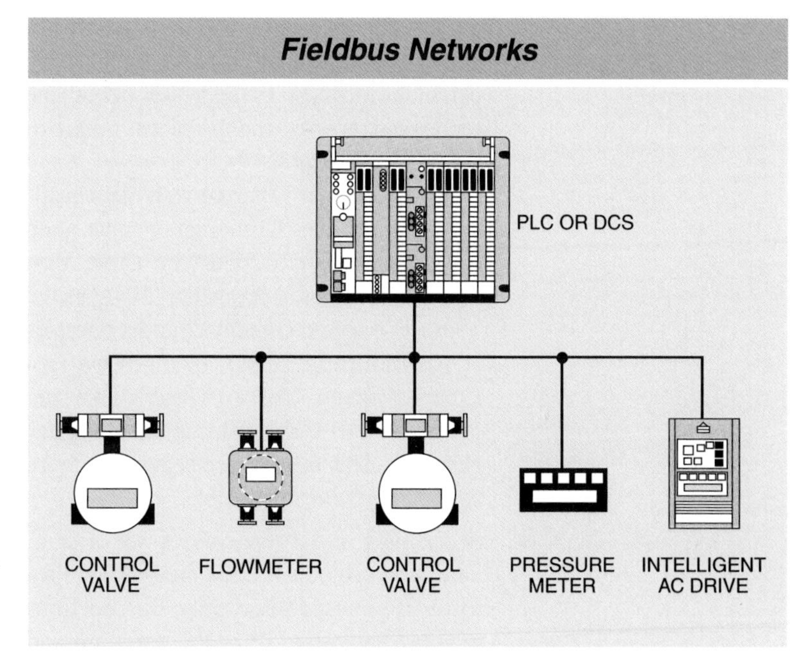

Figure 32-1. A fieldbus network connects many field devices together in a plant network.

Network Classifications

Industrial networks can be broadly classified as sensor bus (bit), device bus (byte), and fieldbus (message) networks. Sensor bus networks typically transfer only a few bits of information at a time at very high speeds. They are used with simple binary devices such as switches, pushbuttons, and proximity sensors.

Device bus networks typically transfer a few bytes to a few hundred bytes of information at a time. They are used with more complex devices such as temperature and pressure transmitters and variable-speed drives.

Fieldbus networks typically have real-time bidirectional transfer of many bytes of information at a time between smart sensors and a controller. They are used to interconnect intelligent systems such as controllers, PLCs, operator interfaces, and fieldbus instruments.

Ethernet

Ethernet is a high-speed interference-immune local area network (LAN) covered by IEEE 802.3 standards. Ethernet can be used with either twisted-pair or optical fiber cables. **See Figure 32-2.** It breaks the data into smaller packets that are sent separately. A *packet* is a unit of data sent across a network that includes part of the message, the addresses of both the sender and receiver, and the packet's place in the entire message.

There are four Ethernet data rates available. The 10 Base-T Ethernet operates at up to 10 Mbps over twisted-wire cable. Fast Ethernet operates at 100 Mbps, 10 times the 10 Base-T speed. Gigabit Ethernet operates at 1000 Mbps (1 Gbps), 10 times the Fast Ethernet speed. The 10 Gigibit Ethernet operates at 10 Gbps, 10 times the Gigabit Ethernet. The faster speeds require more twisted-pair wires in a cable.

Standards and Protocols. A set of standards and protocols called the Internet Protocol (IP) is used to describe the functions of most networks. IP is the common and primary network (OSI layer 3) protocol in the suite. It defines the addresses by which the network can transmit the packets from source to destination. IP contains the lower level specifications such as TCP, but also specifications for e-mail, terminal emulation, and file transfer. The advantage of IP is that the packets are routed the best way to their destinations. They can be moved around bottlenecks in the system.

Transmission Control Protocol (TCP) provides in-order delivery of the packets between two devices. TCP establishes communications between devices between which the packets are sent. This protocol maintains a connection until the receipt of a packet is verified. This can use a lot of overhead and may not be suited for real-time applications. In that case, the User Datagram Protocol (UDP) can be used. UDP is often used for real-time communications such as voice and I/O traffic. It does not guarantee delivery of the packets or the order of the packet, but reduces transmission time.

Standard Ethernet uses the Carrier Sense Multiple Access with Collision Detection (CSMA/CD) Protocol. This means that messages are not prioritized and the network is non-deterministic. Whenever two or more devices communicate at the same time, the data collision corrupts the message and the message is lost. The devices wait a random amount of time before sending again. The CSMA/CD Protocol makes it very difficult for Ethernet communications to manage time-sensitive tasks.

Physical Layers. The twisted pair wires are molded into plastic connectors that resemble oversize phone plug connectors. For example, a cat 5 unshielded twisted pair consists of four twisted pairs of copper wire, typically terminated by an RJ-45 connector. Most hubs use RJ-45 unshielded twisted-pair connectors. Cables with RJ-45 connectors can be purchased commercially in different lengths or fabricated locally. The two common wiring standards for Ethernet connectors, T568A and T568B, should not be mixed.

Since baseband communication uses digital switching instead of frequency modulation, there is only one data channel on a cable.

Commercially available 100 Base-T cables include patch cables and crossover cables. A patch cable is used for connections from a node to a hub. Patch cables use four wires connected straight through from connector to connector. A crossover cable is used for connections from node to node. **See Figure 32-3.**

Patch cables and crossover cables contain send and receive wires. In a crossover cable, the send wires from each connector pair are wired to the receive wires of the other connector pair. A crossover cable is used when a computer must be directly connected to a controller to monitor system performance or collect troubleshooting information. Patch cables and crossover cables are color-coded or labeled for easy identification. The two cables are not interchangeable and could cause a communication failure if misapplied.

Industrial Ethernet. Industrial Ethernet use is becoming more common for use as industrial communication highways. It provides a common high-speed communication highway that is independent of any one control manufacturer's digital network. Industrial Ethernet can connect to any number of open or proprietary protocols such as Modbus, Sinec H1, Profibus, CANopen, DeviceNet, or FOUNDATION Fieldbus. This allows more interoperability with different systems being used in a control network.

At present, an automation system usually consists of a primary control system with hardware, communications, and many independent control package systems. A control package system is a stand-alone unit with its own control system and field instruments and is designed to perform specific, limited tasks. A package system may be used from boilers, water treating plants, compressors, waste treatment plants, evaporator systems, burners, incinerators, etc.

TECH FACT

Baseband networks transmit only one message at a time. Broadband networks transmit more than one message at a time by modulating the frequency of the signals.

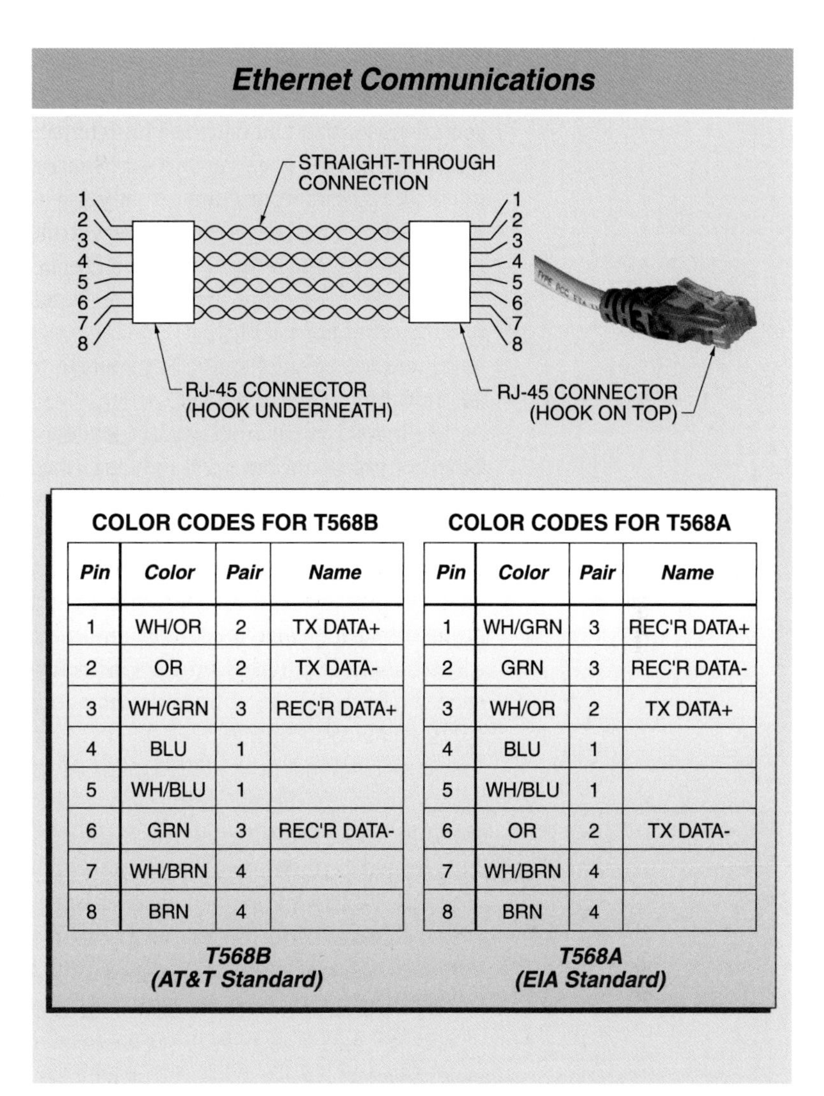

COLOR CODES FOR T568B

Pin	Color	Pair	Name
1	WH/OR	2	TX DATA+
2	OR	2	TX DATA-
3	WH/GRN	3	REC'R DATA+
4	BLU	1	
5	WH/BLU	1	
6	GRN	3	REC'R DATA-
7	WH/BRN	4	
8	BRN	4	

T568B (AT&T Standard)

COLOR CODES FOR T568A

Pin	Color	Pair	Name
1	WH/GRN	3	REC'R DATA+
2	GRN	3	REC'R DATA-
3	WH/OR	2	TX DATA+
4	BLU	1	
5	WH/BLU	1	
6	OR	2	TX DATA-
7	WH/BRN	4	
8	BRN	4	

T568A (EIA Standard)

Figure 32-2. Ethernet communications typically use an 8-wire RJ-45 connector. A straight-through connection is for hub-to-hub wiring.

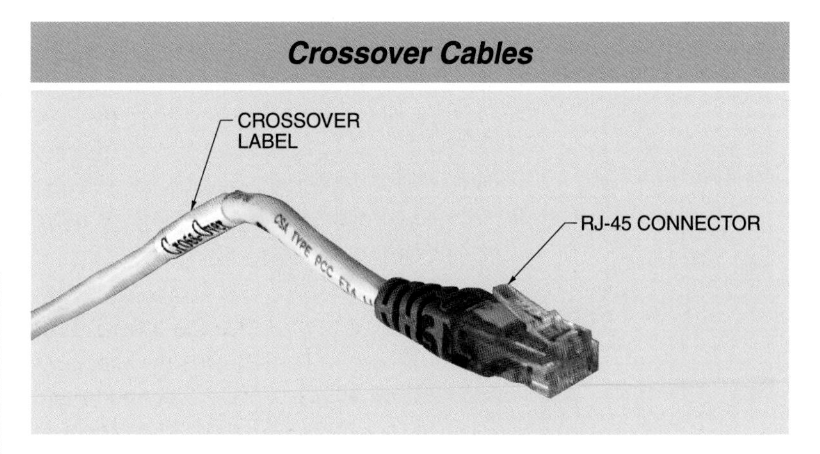

Figure 32-3. A crossover cable is used to connect nodes within a network.

Each of these package systems must maintain continuous communication with the main control system and a human machine interface (HMI). In the past, it was always a problem to establish a common communication gateway between the individual package units with the main highway and control system. The Industrial Ethernet has now become the choice because all of the established data highways have interfaces available to communicate to the Industrial Ethernet.

The Industrial Ethernet uses the standard Ethernet protocols but with industrialized hardware designed to handle the more stringent environmental conditions encountered in industrial settings. Some changes have been made to the standard Ethernet protocols to increase speed and minimize collisions. This is based on the OSI reference model but is modified for industrial applications. **See figure 32-4.** Packets are sent and received in specified periods of time and makes the network support predictable. This is essential for real-time, deterministic traffic. Industrial Ethernet uses full-duplex standards and other methods so that collisions do not unacceptably influence transmission times. In addition, industrial networks use network switches to separate a large system into logical subnetworks divided by addresses, protocols, or applications.

<table>
<tr><th colspan="6">Industrial Ethernet</th></tr>
<tr><td>Device Profiles</td><td>Transmitters</td><td>Valves</td><td>Drives</td><td>Actuators</td><td rowspan="4">Fieldbus-Specific</td></tr>
<tr><td rowspan="3">Application</td><td colspan="4">Application Object Library</td></tr>
<tr><td colspan="4">Data Management Services</td></tr>
<tr><td colspan="4">Message Routing, Connection Mangement</td></tr>
<tr><td>Layer 4 (Transport)</td><td colspan="2">TCP</td><td colspan="2">UDP</td><td rowspan="4">Quality of Service Parameters</td></tr>
<tr><td>Layer 3 (Network)</td><td colspan="4">IP</td></tr>
<tr><td>Layer 2 (Data Link)</td><td colspan="4">Ethernet MAC/LLC</td></tr>
<tr><td>Layer 1 (Physical)</td><td colspan="4">Physical Layer</td></tr>
</table>

Figure 32-4. Industrial networking uses User Datagram Protocol (UDP) to address real-time response concerns.

Switches. Ethernet switches have made it possible for several users to send information over the same network at the same time without slowing each other down. There are no hubs in a fully switched network, so each Ethernet network has a dedicated segment for every node. Because of the duplex architecture, each switch can send and receive information at the same time, effectively doubling the apparent speed of the network. Ethernet switches usually work at Layer 2 of the OSI reference model.

Switches provide a number of advantages over hubs and other LAN devices. Switches have very low latency, which is the time it takes a packet to move between a source and a target. The Industrial Ethernet uses the same basic technologies that have become common in the use of conventional Ethernet networks.

Intelligent managed switches provide additional performance, management, diagnostics, and security capabilities over unmanaged switches. Intelligent switches minimize packet loss during congested communication. This makes it possible to prioritize critical traffic. Intelligent switches can feature both broadcast and multicast. This makes it possible to dynamically configure the interfaces so that traffic is only forwarded to ports associated with requested data. In addition, intelligent switches allow traffic analyzers to monitor any port in a network. They can also provide diagnostic analysis to identify problems, even potential problems.

FIELDBUS IMPLEMENTATIONS

Fieldbus networks are gradually replacing the commonly used analog networks, which are based on the 4 mA to 20 mA standard for analog devices. Fieldbus provides greater accuracy and repeatability in process applications, as well as adding bidirectional communication between the field devices and the controller. A problem with fieldbus systems is that they add another level of digital technology that has to be supported by specially trained service personnel. It is difficult for medium-size and small plants to economically obtain this needed expertise. This limits the use of fieldbus systems to large plants until the technology can be more widely assimilated.

There are very complex technical issues that confront the use of fieldbus systems when communications are digital and the controllers are embedded in the transmitters or control valve positioners. For example, spare parts and sufficiently detailed device records including firmware revision numbers are required and configurations can be complex.

Every field device is a special unit. It is no longer a simple matter of maintaining a few models of a given transmitter with compatible ranges and materials of construction. For each instrument, transmitter, controller, and auxiliary device, technicians now must maintain complete records that include the firmware revision number, which digital boards are installed, and the needed configuration. Swapping out bad boards from otherwise good transmitters causes a serious record-keeping problem.

The two most commonly used digital process bus network protocols are FOUNDATION, sponsored by the Fieldbus Foundation, and Profibus, sponsored by the Profibus Trade Organization. Other common process bus networks include MODBUS®, HART®, CANbus, DeviceNet™, SDS, ControlNET™, ASI, BACnet® and LonWorks. Many of these process bus protocols come in several versions. Most protocols have versions that operate on top of Ethernet.

FOUNDATION

FOUNDATION is a process bus network protocol that uses a 2-wire, digital, serial communications system to connect field equipment, such as intelligent sensors and actuators, with controllers, such as PLCs. **See Figure 32-5.** This process bus network offers the following advantages:

- standard physical wiring interface
- bus-powered devices on a single pair of wires
- intrinsic safety options
- reduced wiring due to multidrop devices
- reduced control room space requirements
- compatibility among all fieldbus equipment
- digital communications reliability
- distributed control

The three types of devices on a FOUNDATION fieldbus segment are a link active scheduler (LAS), a bridge, and a basic device. On startup, the system passes a token to determine what devices are on the network and the parameters of each device. Multiple variables from field devices can be obtained in addition to the primary measurement. Three types of messages carried on the bus are client/server, report distribution, and publisher/subscriber. The cable consists of one blue (data) and one brown (power) 18 AWG twisted pair, a green/yellow 18 AWG ground, and an overall shield with a bare drain wire.

The primary difference between FOUNDATION and other fieldbus systems is that the software contains function blocks that allow the development of a field control system (FCS) without the use of a PLC or DCS. Other fieldbus systems have to rely on a PLC or DCS to implement an FCS. FOUNDATION can also be used with PLCs or DCSs, as with other fieldbus systems.

FOUNDATION fieldbus has two common protocols. The H1 protocol operates at a relatively slow 31.25 kb/s and is optimized for traditional analog process control applications. The HSE protocol operates at 10 Mb/s or 100 Mb/s. It is designed for high-speed control applications and plant information networks. HSE uses low-cost commercial off-the-shelf (COTS) Ethernet equipment. A typical FOUNDATION network has H1 connections between devices and HSE connections between computers. Combining H1 and HSE allows for good integration of control systems with supervisory applications.

Fieldbus network are digital communication techniques optimized for use in the process-oriented industries.

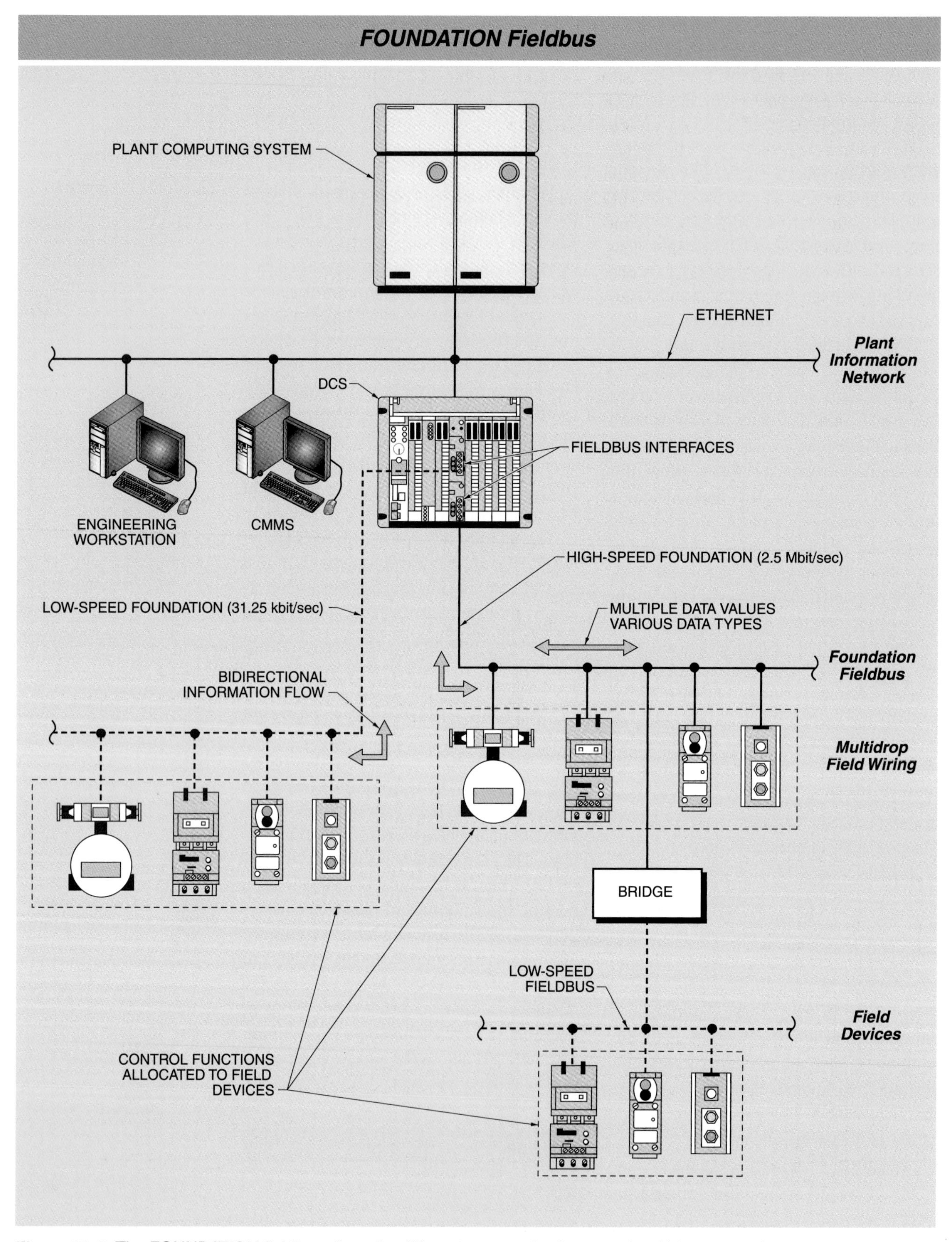

Figure 32-5. The FOUNDATION fieldbus allows for different communication speeds within a network.

Profibus

Profibus is a process bus network capable of communicating information between a master controller (or host) and an intelligent slave process field device as well as from one host to another. The master devices control the bus network and determine the data communication on the bus. The slave devices are typically I/O devices, control valves, drives, and transmitters. Slave devices are called passive stations. Profibus networks require a separate controller or computer to implement control strategies.

A Profibus connection is half-duplex over a shielded twisted-pair cable. Data rates of 12 Mb/s are available up to 100 meters. Slower data rates of 94 kb/s are available up to 1200 meters. Profibus can be used with a number of transmission methods. **See Figure 32-6.** RS-485 is used by Profibus-FMS and Profibus-DP for universal applications in manufacturing and process automation. The IEC 1158-2 transmission method is used by the PA protocol.

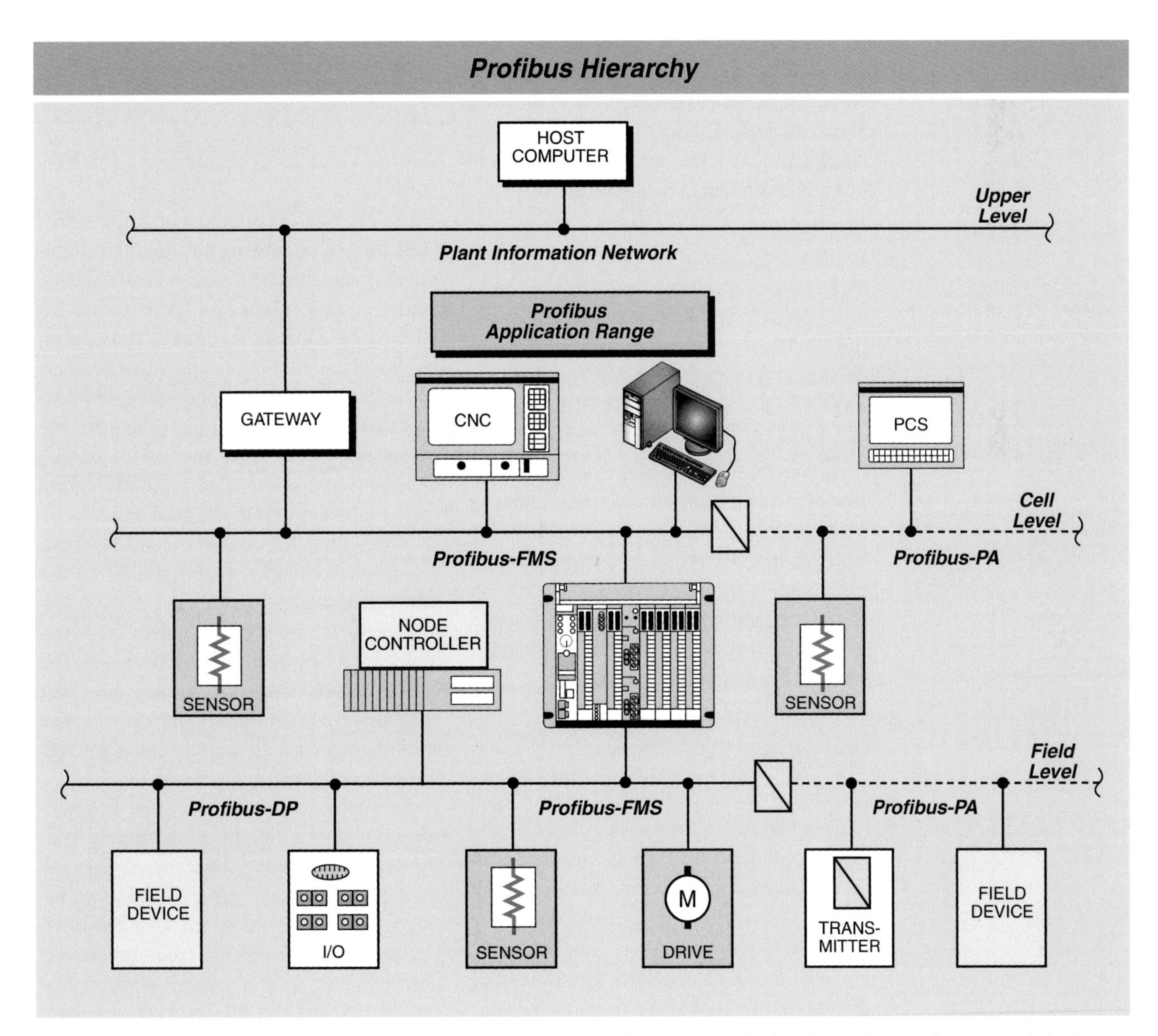

Figure 32-6. The Profibus fieldbus has several levels of communications standards, depending on the types of devices to be connected.

The IEC 1158-2 transmission method is optimized for use in the chemical and petrochemical industries. It is intrinsically safe and allows the field devices to be powered over the bus. A current of 10 mA is sent through the loop to provide power. Modulation is ±9 mA above or below the main current. The four types of Profibus networks are Profibus-FMS, Profibus-DP, and Profibus-PA.

Profibus-FMS. *Profibus-FMS* is a version of Profibus used for communicating between the upper level, the cell level, and the field device level of the Profibus hierarchy. Cell level control occurs at individual (or cell) areas, which exercise the actual control during production. The controllers at the cell level must communicate with other supervisory systems. The Profibus-FMS uses the fieldbus message specification to execute its communication tasks between hierarchical levels.

Profibus-DP. *Profibus-DP* is a 4-wire performance-optimized version of the Profibus network used primarily for factory automation applications. It is designed to handle time-critical communications between devices in factory automation systems. The Profibus-DP can replace 24 VDC, HART, and 4 mA to 20 mA wiring interfaces.

Profibus-PA. *Profibus-PA* is a 2-wire version of the Profibus network used primarily for process automation. The Profibus-PA network has bus-powered stations and intrinsic safety. It includes device description and function block capabilities, field device interoperability, and can be used to manage smart instrumentation devices.

MODBUS

MODBUS is a messaging structure with master-slave communication between intelligent devices and is independent of the physical interconnecting method. MODBUS works equally well using RS-232, RS-485, optical fiber, radio, cellular, etc. The message contains the address of the slave, the command (e.g., "read register" or "write register"), the data, and a checksum.

MODBUS has two different transmission modes, ASCII and RTU. ASCII transmission mode is 10 bits long with each eight-bit byte in a message representing two ASCII characters. RTU transmission is 11 bits long with each eight-bit byte representing two hexadecimal characters. **See Figure 32-7.** MODBUS also uses two sets of wires, a separate pair for power and a twisted pair for communication. MODBUS-TCP is a version of MODBUS used on top of Ethernet.

TECH-FACT

Computers use a binary numbering system because it is easy to measure the presence or absence of a voltage.

HART

HART is a hybrid communications protocol consisting of newer two-way digital communication and the traditional 4 mA to 20 mA analog signal. HART is an acronym for Highway Addressable Remote Transducer. HART is called a hybrid protocol because it combines digital and analog communication on the same wire. It can communicate a single variable using a 4 mA to 20 mA analog signal while also communicating added information on a digital signal.

The digital information is carried by a low-level modulation superimposed on the standard 4 mA to 20 mA current loop. The bit stream is organized into eight-bit bytes, which are further grouped into a message. The data bytes are formed into packets to which are added communications protocol bits. The resulting structure is one start bit, eight data bits, one odd parity bit, and one stop bit.

The protocol is a master-slave arrangement with token passing and a timer to limit the transaction time. This means that the field device does not generally transmit information until asked by the host controller. For example, if a HART communications device detects a fault condition in the process or in itself, the device cannot transmit this information to the host until the host specifically requests that information.

CANbus

CANbus is a device bus network based on the widely used CAN electronic chip technology that is used inside automobiles to control internal components such as brakes and other systems. A CANbus network is an open-protocol system featuring variable-length messages (up to eight bytes), nondestructive arbitration, and advanced error management. A 4-wire cable plus shield—two wires for power, two for signal transmission, and a shield used as a fifth wire—provides the communication link with field devices. **See Figure 32-8.** This communication can be either master-slave or peer to peer. The data transmission rate depends on the length of the trunk cable.

A CANbus network defines both the media access control method and the physical signaling of the network, while providing error detection. The media access control function determines when each device on the bus is enabled. A CANbus scanner or an I/O processor provides the interface between a controller and a CANbus network. The scanner converts the serial data from the CANbus network to a form usable by the controller processor. **See Figure 32-9.** Two common types of CANbus networks are DeviceNet and the Smart Distributed System (SDS).

DeviceNet. A DeviceNet byte-wide network can support 64 nodes and a maximum of 2048 field I/O devices. In a DeviceNet network, the controller connects to the field devices over a 2-wire network in a trunk line configuration with either single drops off the trunk or branched drops through multiport interfaces at the device locations. DeviceNet protocol uses a carrier sense multiple access/bitwise arbitration (CSMA/BA) bus arbitration method that ensures that the highest priority message always gets transmitted. Input/output data have the highest priority. DeviceNet is typically used to communicate between controlled devices such as motors and stop/start stations.

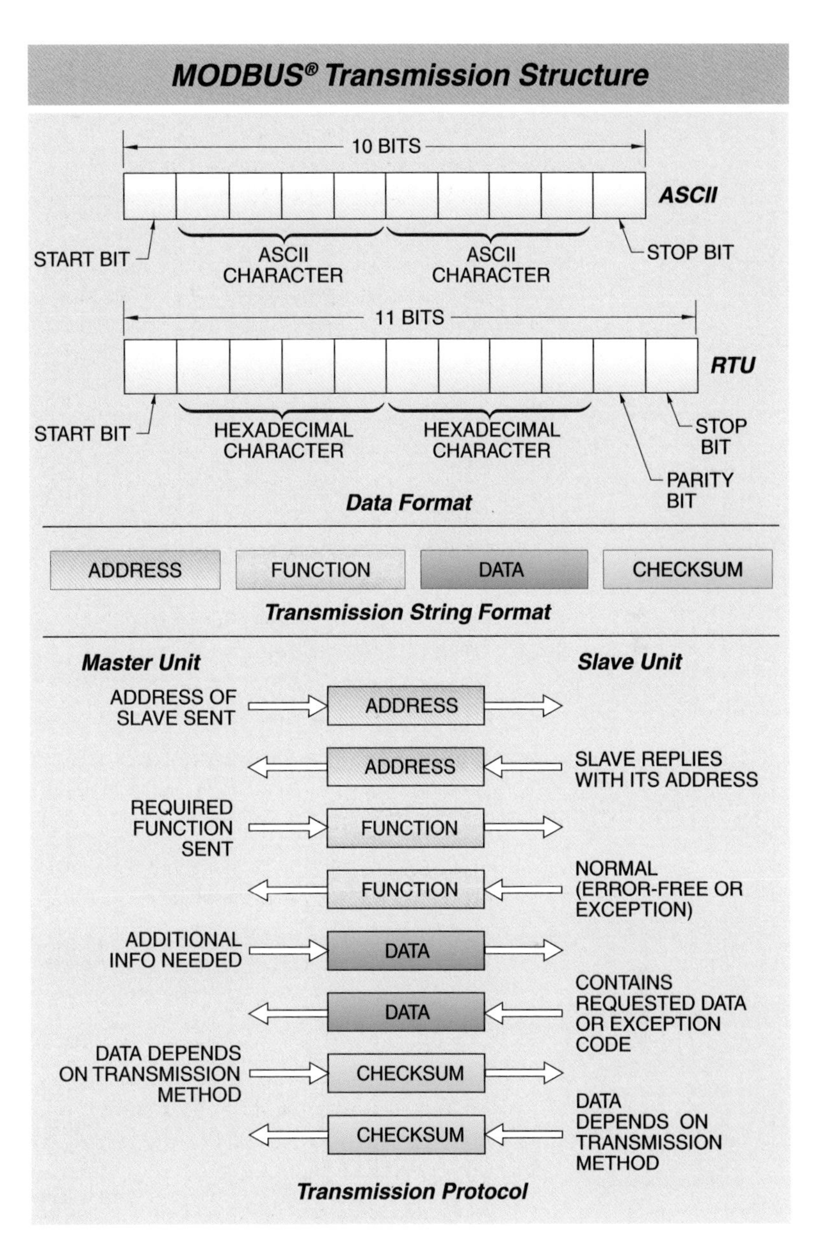

Figure 32-7. The MODBUS® fieldbus is a master-slave network with either ASCII or RTU communications.

TECH-FACT

Networks can be categorized in many overlapping ways. Common categories are message capacity, transmission rate, types of nodes, and topology. They can also be classified as sensor bus, device bus, and fieldbus networks.

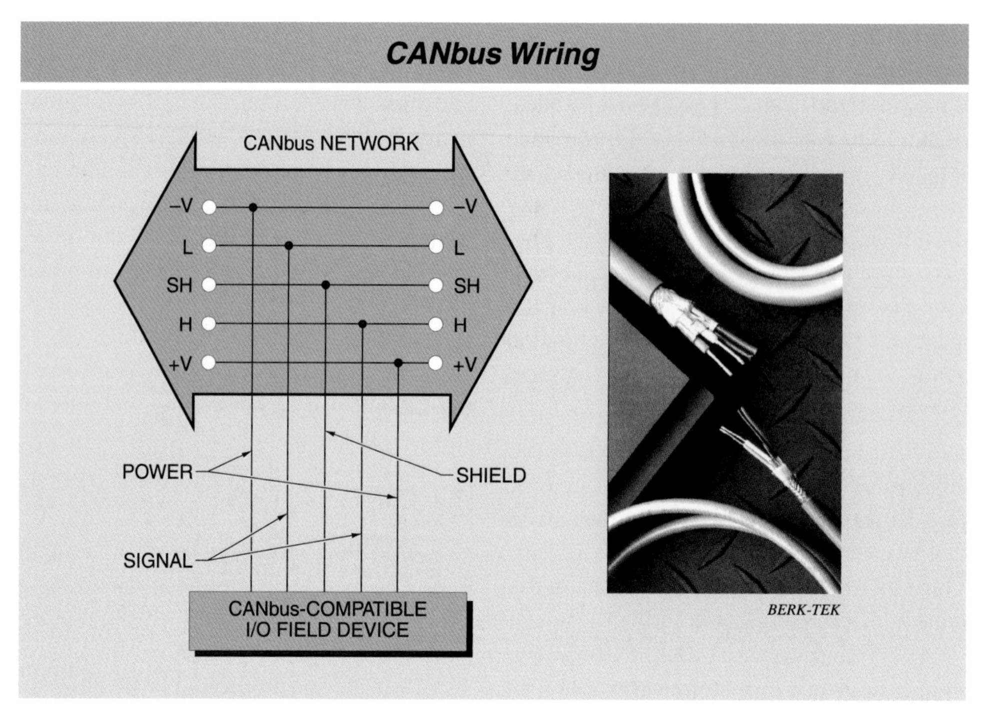

Figure 32-8. The CANbus network uses four wires for communications and a shield for grounding.

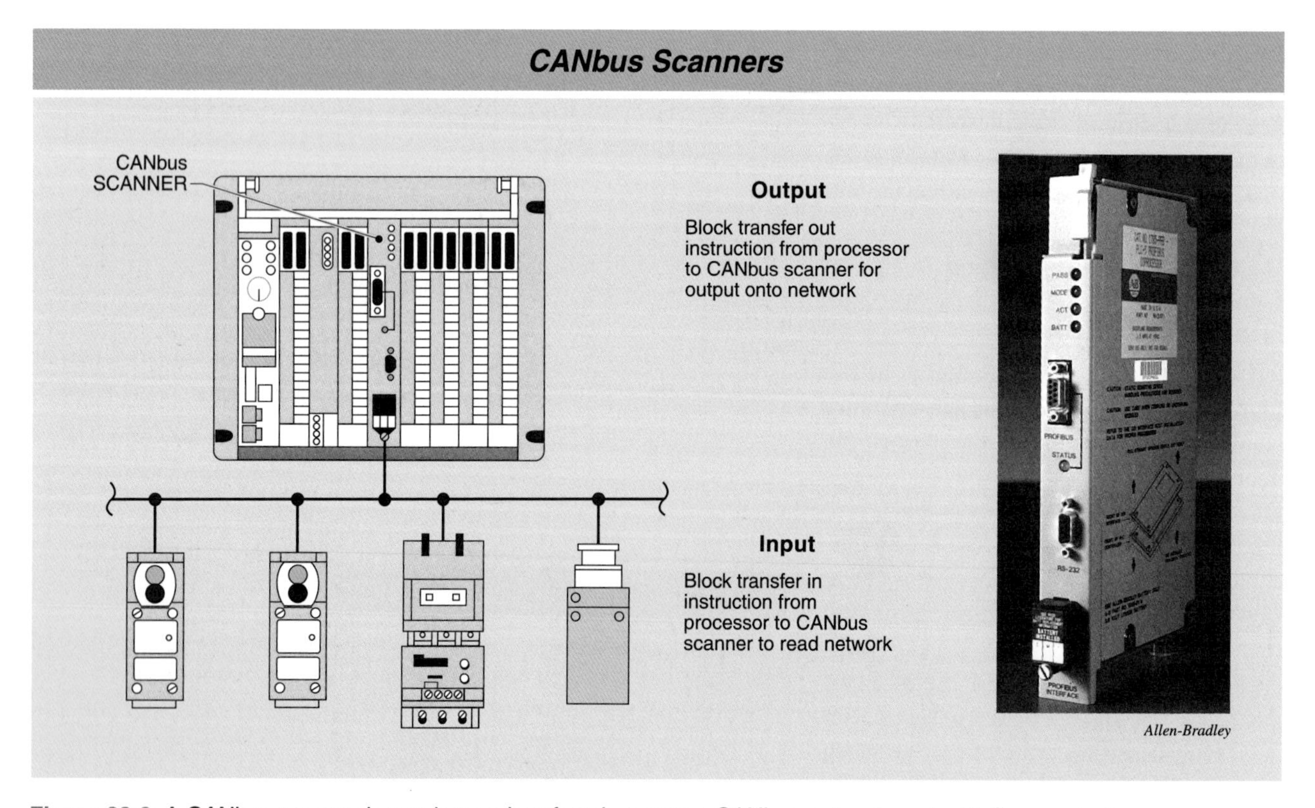

Figure 32-9. A CANbus scanner is used as an interface between a CANbus network and a PLC or controller.

Smart Distributed System (SDS). Like DeviceNet, the Smart Distributed System (SDS) network can also support 64 nodes. When using a 4-to-1 multiport I/O interface module, an SDS network can connect up to 126 nonintelligent I/O devices in any combination of inputs and outputs. This multiport interface to nonintelligent field devices has a slave CAN chip inside the interface, which provides status information about the nodes connected to the interface. **See Figure 32-10.**

Because an SDS network can transmit many bytes of information in the form of variable-length messages, it can also support many intelligent devices that can translate several bytes of information from the network into 16 or 32 bits of ON/OFF information. For example, a smart solenoid valve manifold can have up to 16 connections, thereby receiving 16 bits (two bytes) of data from the network and controlling the status of 16 valve outputs. However, this device uses only one address of the 126 possible addresses.

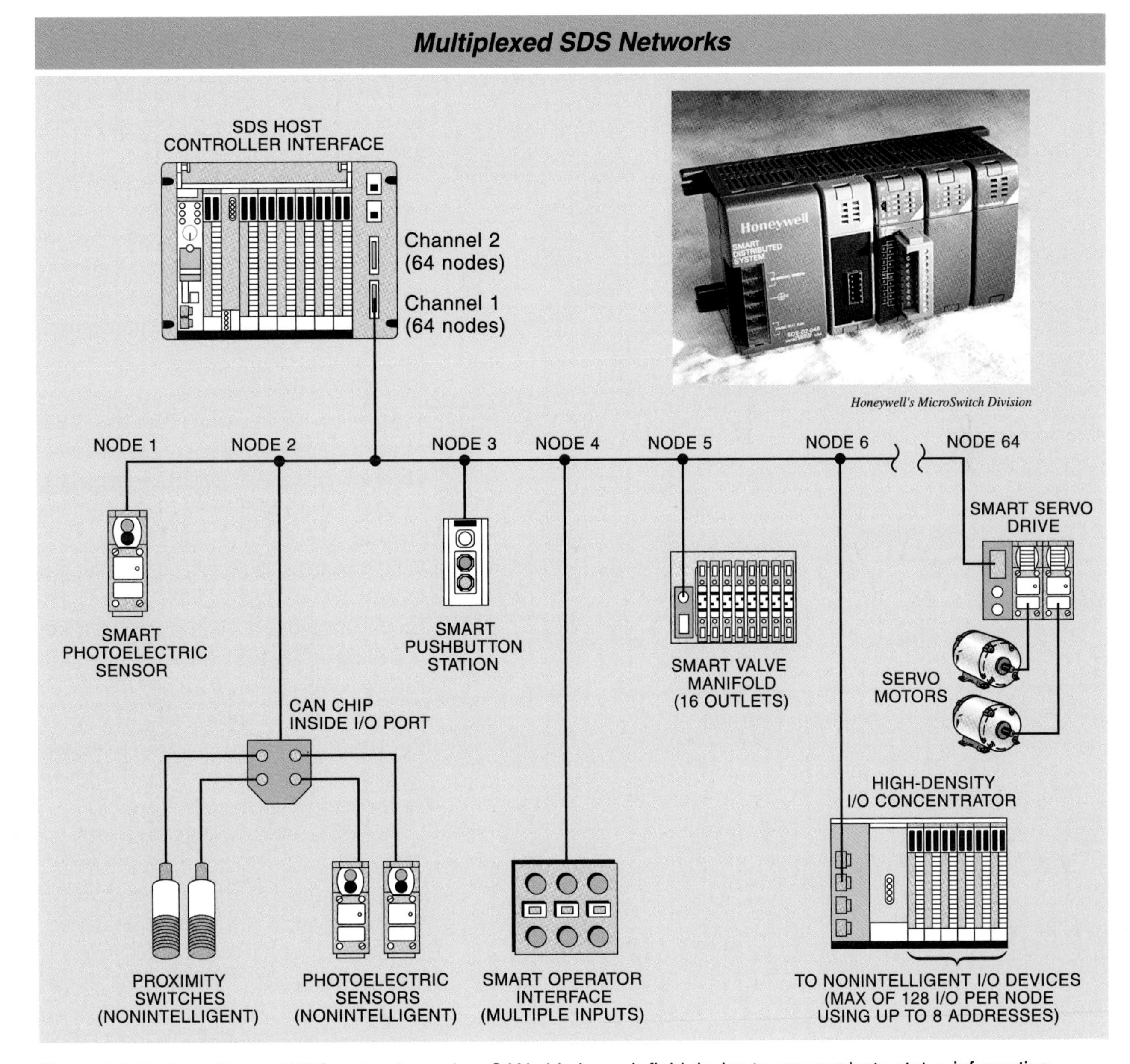

Figure 32-10. A multiplexed SDS network needs a CAN chip in each field device to communicate status information.

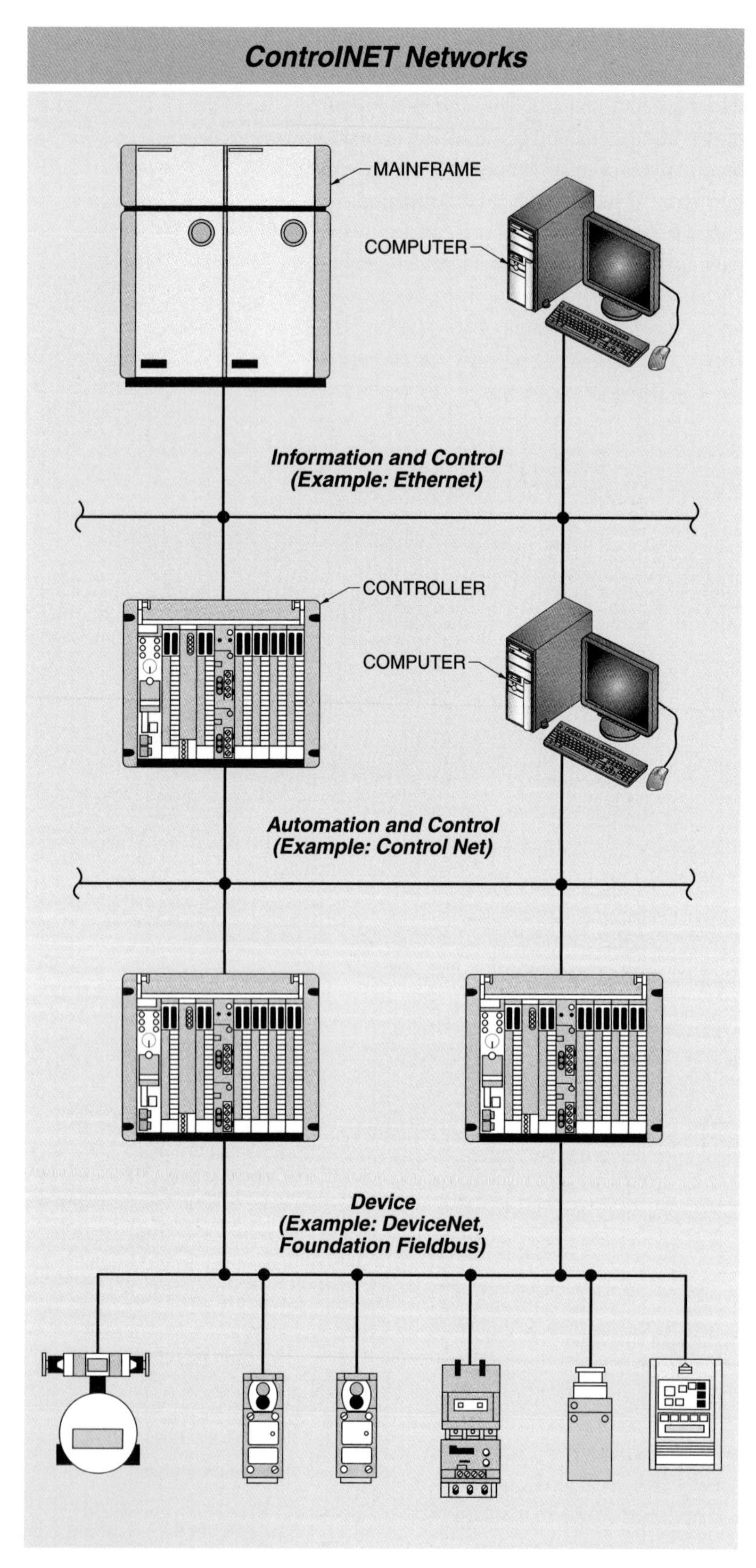

Figure 32-11. A ControlNET network provides real-time high-speed transmission of time-critical data.

ControlNET

ControlNET is a real-time device network providing for high-speed transport of time-critical I/O data, messaging, peer-to-peer communications, and the uploading and downloading of programs. Transmission speed is 5 Mbits/sec. The protocol is a producer/consumer communications model that uses no addresses. All devices on the network can listen and respond if the communication involves an individual device. Multiple controllers can direct the communications. The communication modes can be master-slave, multimaster, or peer-to-peer. Network access is controlled by a concurrent time domain multiple access (CTDMA) time slice algorithm that ensures that data is delivered in a reliable and predictable manner. **See Figure 32-11.**

Communication can be by a coaxial R6/U cable or optical fiber system. Up to 48 nodes can be handled without a repeater and 96 nodes with a repeater. Power is not carried by the network and must be provided externally for the node devices and the field equipment.

AS Interface (ASI)

AS Interface (ASI) is a sensor bus network designed to connect digital sensors and actuators where low cost and simplicity are important. The ASI protocols allow no more than 248 I/O field devices in 31 field-located slaves with four inputs and four outputs each. An ASI network requires a 24 VDC power supply connected through a 2-wire, unshielded, untwisted cable. The bus cable wires carry both data and power to all stations. An ASI network is very fast, with an update time of 5 msec or less. Three bits are used for station addresses, and each station on the bus is limited to four bits for sensing and four bits for output.

The ASI network protocol is based on the ASI protocol chip. Therefore, the I/O devices connected to this type of network must contain this chip. Typical ASI-compatible devices include proximity switches, limit switches, photoelectric sensors, and standard off-the-shelf field devices. However, in an application using an off-the-shelf device, the ASI chip is located in the node. **See Figure 32-12.**

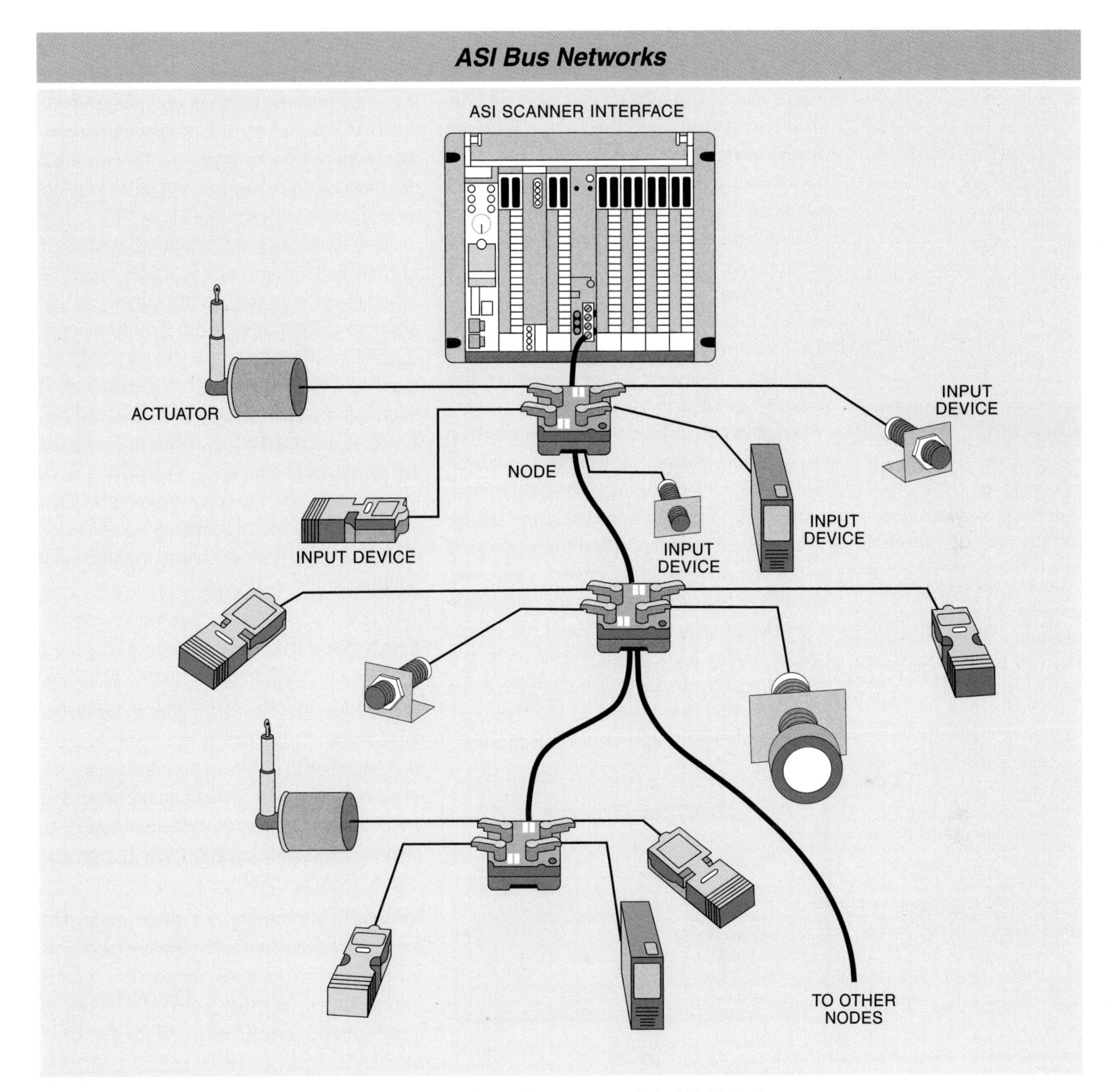

Figure 32-12. An ASI network is a low-cost network used to connect digital field devices.

BACnet

The American Society of Heating, Refrigerating and Air-Conditioning Engineers (ASHRAE®) has formed a committee to standardize building automation system communications for the HVAC industry. The ASHRAE committee has developed a standard specifically designed for building automation system communications, the building automation and control network (BACnet).

BACnet is a communications protocol that uses standard connecting media and enables controllers to be interoperable. *Interoperability* is the ability of devices produced by different manufacturers to communicate and share information. Common benefits of interoperability are the ability to do tasks in a common format regardless of the manufacturer. For example, the owner should be able to view data from all devices, change setpoints, and set up and modify schedules.

The controllers in a BACnet system are interoperable but not interchangeable. This means that one controller cannot be substituted for another controller from a different manufacturer. Interchangeability is normally impossible due to different electrical and mounting characteristics of the controllers.

BACnet is commonly used with intranet networks. Intranets are the most commonly used commercial communications systems in building automation systems today. BACnet determines the points and objects that have a common look and contain the same information.

An *object* is a collection of information that can be accessed over a network in a standardized way. All information in a BACnet control system uses objects as data structures. Analog input devices require specific information such as object name and type, present value, and status flags. For example, a temperature sensor may have an object name of Zone 101, object type of Analog Input, present value of 76.4, a status flag of Normal, a high limit of 80, and a low limit of 60. **See Figure 32-13.** There can be many other properties included in an object.

TECH-FACT

Bandwidth is an imprecise measure of communication speed. Bandwidth may be reported as information transfer in bits/sec, or baud, or as the switching frequency in Hz.

BACnet Objects

Variable	Value
Object_Name	Zone 101
Object_Type	Analog Input
Present_Value	76.4
Status_Flag	Normal
High_Limit	80
Low_Limit	60

Figure 32-13. A BACnet object is used to define a variable.

Controller Conformance Classes. The BACnet standard is also grouped into controller conformance classes. The conformance classes organize controllers according to the type of work expected to be performed. Functions are then assigned to each conformance class. Conformance statements recognize that not all controllers require all of the same functions.

For example, a variable air volume terminal box controller requires different points than a supervisory controller that shares data. The HVAC industry has developed conformance classes that show the degree to which a particular manufacturer's equipment supports the BACnet standard. Conformance statements are public documents that are meant to be shared with customers and other vendors as needed. **See Figure 32-14.**

The six classes are identified as Class 1 to Class 6. This does not mean that one class is higher or better than another. It means that different classes support different functions.

BACnet was originally known as ASHRAE standard 135-1995. Now it has been approved as an international standard as ISO 16484-5. The BACnet standard is subject to change, is always under review, and is viewed as a work in progress. Common applications of BACnet are in HVAC control systems, security systems, fire detection and alarm systems, and building utility interfaces.

LonWorks

A *local operating network (LON)* is a network of intelligent devices sharing information using LonTalk®. *LonTalk* is the open protocol standard used in LonWorks control networks. The LonTalk protocol was first introduced in 1990, and in October of 1999, the American National Standards Institute (ANSI) approved it as ANSI/CEA 709.1.

Network Variables. A typical node that implements the LonTalk protocol performs local control functions and shares control information with other LonWorks nodes in a peer-to-peer way. Control values are structured by individual nodes as network variables and transmitted onto the network in message packets. **See Figure 32-15.**

LonWorks nodes share information by sending and receiving control data as network variables. A *network variable* is a basic unit of shared control information that conforms to a certain data type. Many network variables represent simple numerical values and their units, such as a temperature in degrees Celsius (°C) or pressure in pounds per square inch (psi). Other network variables can represent text strings, node status, alarm data, enumerations, or structured setpoints.

BACnet Protocol Implementation

BACNET PROTOCOL IMPLEMENTATION CONFORMANCE STATEMENT

Vendor Name: ACME
Product Name: A40
Product Model Numbers: MS-A40 1010-0, MS-A40 1310-0, FA-A40 1010-0, FA-A40 1310-0

PRODUCT DESCRIPTION

Acme A40 Supervisory Controller is designed to manage a small building or campus of buildings. The A40 efficiently supervises the networking of Application Specific Controllers (ASCs) and provides facility management features including weekly scheduling, optimal start, alarm management, and trending.

Facility personnel can review the system status and modify control parameters for the A40 Supervisory Controller and its associated ASCs using a VT100 Terminal or an X3 Workstation.

With the addition of a network card, multiple A40s can communicate over an Ethernet peer-to-peer network, providing increased functionality for systems that are more complex.

BACnet Conformance Class Supported

Class 1 ☐	Class 3 ☒	Class 5 ☐
Class 2 ☐	Class 4 ☐	Class 6 ☐

BACnet Functional Groups Supported

Clock ☒	Files ☐
HHWS ☐	Reinitialize ☒

BACnet Standard Application Services Supported

Application Service	Initiates Requests	Executes Requests	Application Service	Initiates Requests	Executes Requests
CreateObject	☐	☒	UnconfirmedPrivate Transfer	☒	☒
DeleteObject	☐	☒	ReinitializeDevice	☐	☒
ReadProperty	☐	☒	UTCTimesSynchronization	☒	☒
ReadPropertyMultiple	☒	☒	Who-Has	☒	☒
WriteProperty	☐	☒	I-Have	☒	☒
WritePropertyMultiple	☒	☒	Who-Is	☒	☒
ConfirmedPrivateTransfer	☒	☒	I-Am	☒	☒

Figure 32-14. A BACnet protocol implementation is used to define the conformance class of a device.

LonWorks Technology

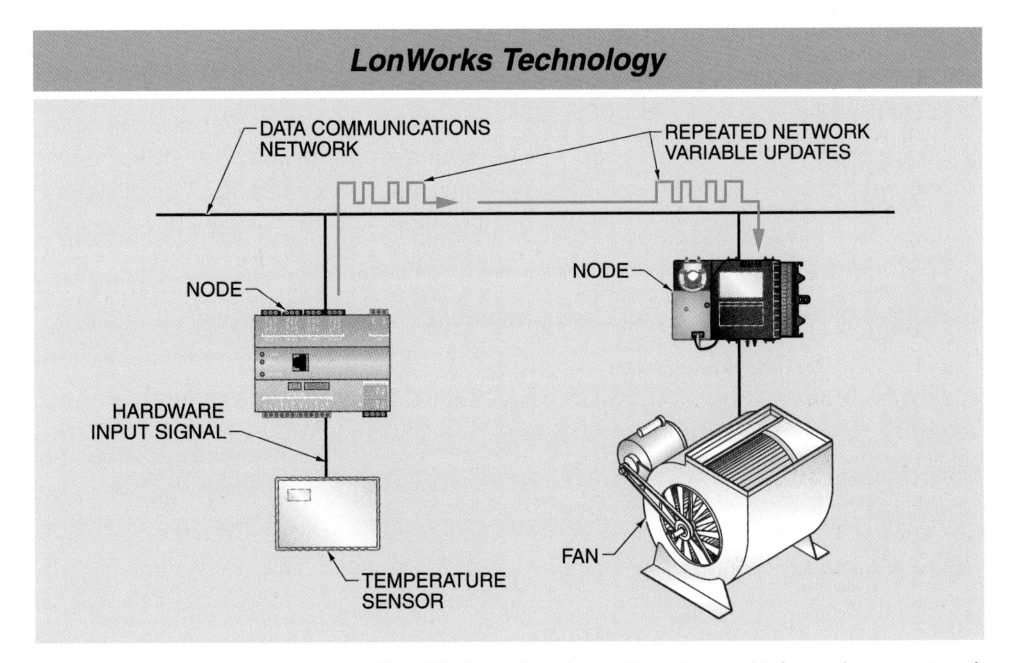

Figure 32-15. The basic concept of LonWorks is the structuring of control information as network variables, which are shared with other nodes on a communications network.

KEY TERMS

- *fieldbus:* An open, digital, serial, two-way communication network that connects with high-level information devices.
- *4-wire fieldbus network:* A fieldbus network that uses two wires for communication and two separate wires to power field instruments.
- *2-wire fieldbus network:* A fieldbus network that uses communication wires to furnish power to field instruments.
- *Ethernet:* A high-speed interference-immune local area network (LAN) covered by IEEE 802.3 standards.
- *packet:* A unit of data sent across a network that includes part of the message, the addresses of both the sender and receiver, and the packet's place in the entire message.
- *FOUNDATION:* A process bus network protocol that uses a 2-wire, digital, serial communications system to connect field equipment, such as intelligent sensors and actuators, with controllers, such as PLCs.
- *Profibus:* A process bus network capable of communicating information between a master controller (or host) and an intelligent slave process field device as well as from one host to another.
- *Profibus-FMS:* A version of Profibus used for communicating between the upper level, the cell level, and the field device level of the Profibus hierarchy.
- *Profibus-DP:* A 4-wire performance-optimized version of the Profibus network used primarily for factory automation applications.
- *Profibus-PA:* A 2-wire version of the Profibus network used primarily for process automation.
- *MODBUS®:* A messaging structure with master-slave communication between intelligent devices and is independent of the physical interconnecting method.
- *HART®:* A hybrid communications protocol consisting of newer two-way digital communication and the traditional 4 mA to 20 mA analog signal.
- *CANbus:* A device bus network based on the widely used CAN electronic chip technology that is used inside automobiles to control internal components such as brakes and other systems.
- *ControlNET™:* A real-time device network providing for high-speed transport of time-critical I/O data, messaging, peer-to-peer communications, and the uploading and downloading of programs.
- *AS Interface (ASI):* A sensor bus network designed to connect digital sensors and actuators where low cost and simplicity are important.
- *BACnet®:* A communications protocol that uses standard connecting media and enables controllers to be interoperable.
- *interoperability:* The ability of devices produced by different manufacturers to communicate and share information.
- *object:* A collection of information that can be accessed over a network in a standardized way.
- *local operating network (LON):* A network of intelligent devices sharing information using LonTalk®.
- *LonTalk®:* The open protocol standard used in LonWorks control networks.
- *network variable:* A basic unit of shared control information that conforms to a certain data type.

REVIEW QUESTIONS

1. What is the difference between 2-wire and 4-wire fieldbus networks, and why is it significant?
2. Why does the CSMA/CD protocol make it difficult to use the standard Ethernet format in industrial communications?
3. What is the difference between H1 and HSE connections in a FOUNDATION network and, why can both be used?
4. Define Profibus-FMS, Profibus-DP, and Profibus-PA.
5. Why is HART considered a hybrid protocol?

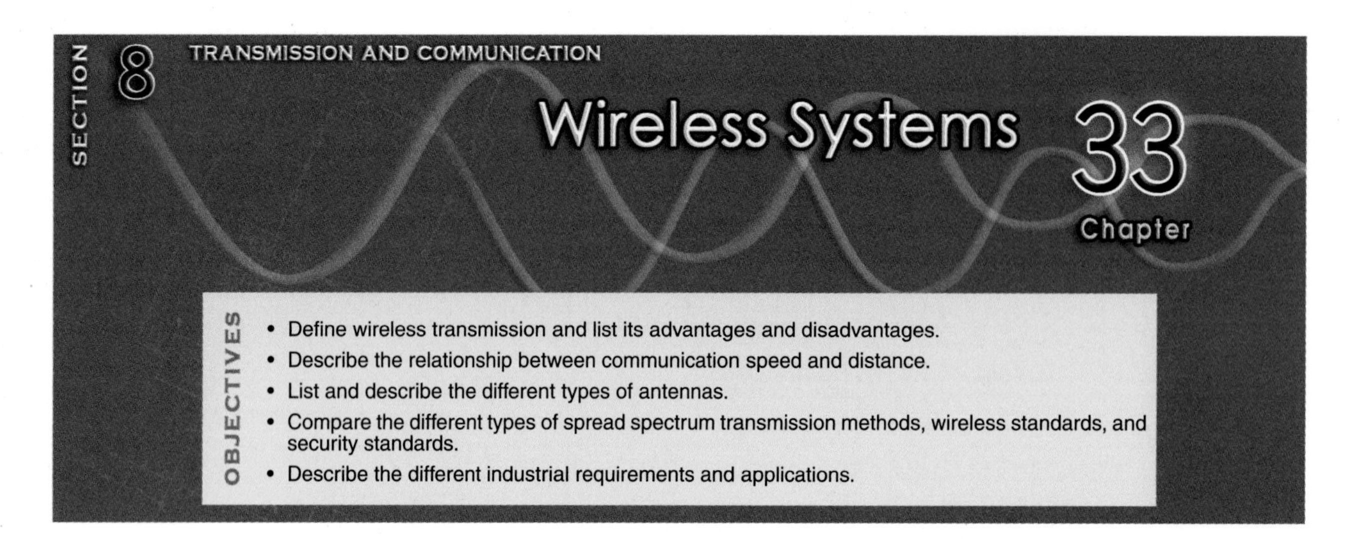

WIRELESS TRANSMISSION

The advent of cellular phone technology and wireless Internet access with data communications has opened the door to industrial wireless transmission. *Wireless transmission* is the method of using radio frequencies to transfer information from one device to another. **See Figure 33-1.** Wireless transmission can also include infrared transmission, but this is rarely used because of its line-of-sight restrictions.

A major advantage of wireless transmission is that an instrument can be easily placed in a remote area without having to run network wires to the instrument. In addition, an instrument can be physically relocated to a different part of a process without the need to rewire the system. For example, a building automation system controller and temperature sensor can be moved from one office to another while maintaining communication with the control system.

A disadvantage of wireless transmission is that remote devices require power even though they do not require wires for communication. The instruments need power at the point of installation, or they must be low-power devices that can operate on batteries. In addition, electromagnetic interference from other radio frequency devices, such as variable-speed drives, fluorescent lights, local controllers, and computer monitors, can interfere with wireless signals. Wireless repeaters can be used to overcome problems due to weak wireless signals. A wireless repeater simply receives a signal from a transmitter and retransmits the signal to another receiver. The proper placement of wireless repeaters is essential.

Wireless radio technology has been used in the form of fixed frequency radio in homes and cars for decades. Radio transmission involves obtaining a license from the government that established an area, a bandwidth, and power limit. The license is intended to prevent other users from interfering with the licensed territory. The problem with this for industrial data collection is the cost of the license and the waiting time. In addition, commercial radio transmits from a central location. Industrial applications are usually just the opposite with a signal coming to a central location from a variety of field locations.

Frequency and Broadcast Power

In 1987, the Federal Communications Commission (FCC) allocated industrial, scientific, and medical (ISM) spread spectrum bands. **See Figure 33-2.** These bands are 900 MHz (902 MHz to 928 MHz), 2.4 GHz (2.4 GHz to 2.4835 GHz), and 5.8 GHz (5.15 GHz to 5.25 GHz, 5.25 GHz to 5.35 GHz, and 5.725 GHz to 5.850 GHz). The 2.4 GHz band is generally accepted globally as an unlicensed band with limited transmission power of 100 mW. In the United States, this limit is set at 1 W. The 900 MHz and 5.8 GHz bands are not accepted globally.

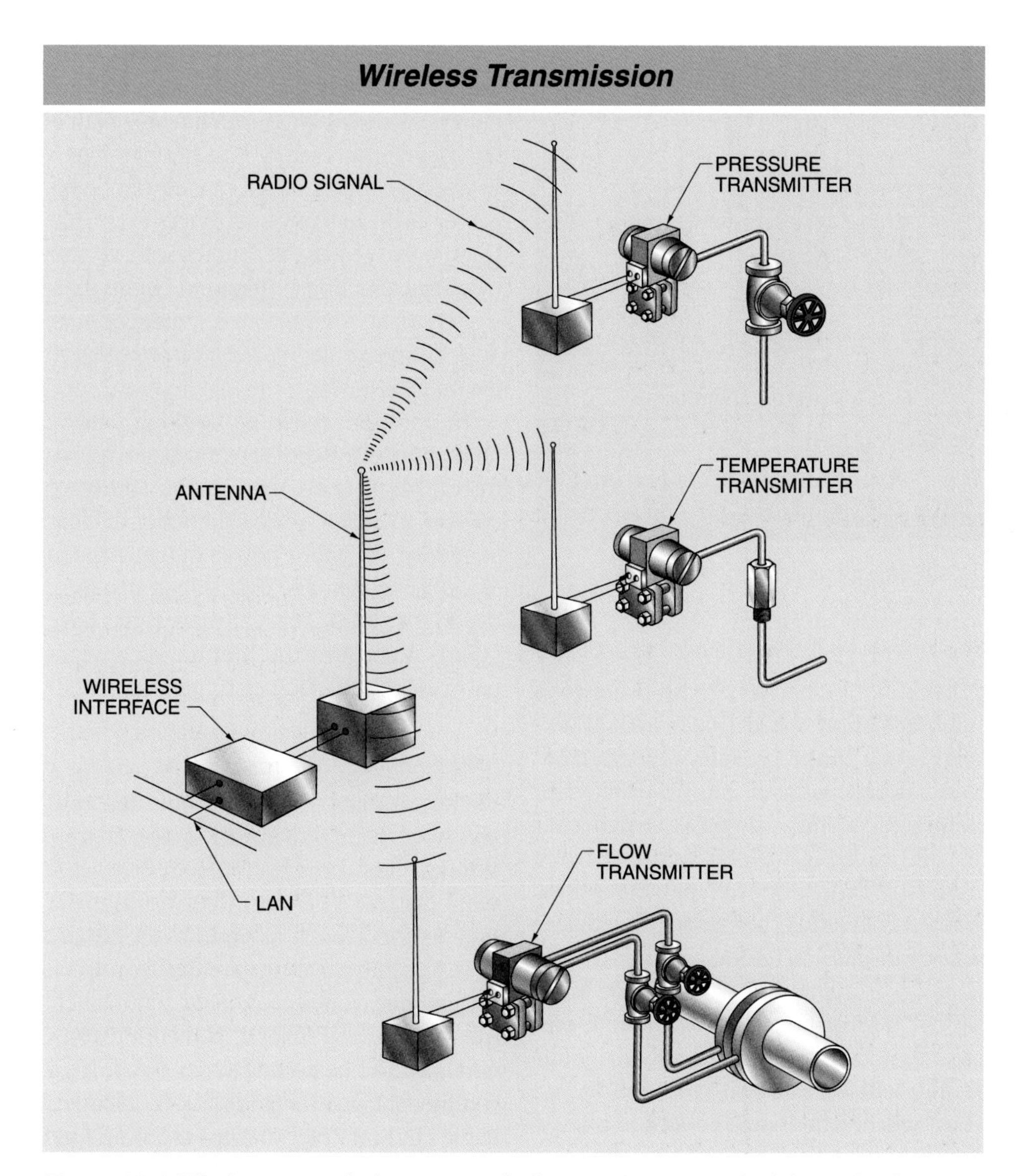

Figure 33-1. Wireless transmission uses radio frequencies to transfer information from one device to another.

ISM Frequency Bands

Frequency Band	Advantages	Disadvantages	Comment
900 MHz (902 MHz to 928 MHz)	More robust with less interference; travels farther through obstacles	Narrow bandwidth and slow speeds	Not available globally
2.4 GHz (2.4 GHz to 2.4835 GHz)	Higher bandwidth allows large data transfers	Attentuates quickly preventing transmission through objects	Accepted globally; lower cost components because of econmomics of scale
5.8 GHz (5.15 GHz to 5.25 GHz) (5.25 GHz to 5.35 GHz) (5.725 GHz to 5.850 GHz)	Highest bandwidth allows largest data transfers; least congested ISM band	Attenuates very quickly; may require low-loss cables and high-gain antennas	Not available globally

Figure 33-2. The three ISM frequency bands are 900 MHz, 2.4 GHz, and 5.8 GHz.

Wireless communications can be used in large plants to reduce the need for wiring.

There is a clear relationship between communication speed (bit rate) and distance. At a lower bit rate, the available power is concentrated into fewer bits and the power per bit is high. At a higher bit rate, the available power is spread out over more bits and the power per bit is low. When there is more power per bit, there is better transmission through barriers. If 1 W of power is applied to a transmitter sending out information at a slow speed, that signal will travel farther than a transmitter sending out information at a high speed. Thus, where data has to be moved a large distance, a small number of bits will have a better chance of making it to the receiver than a large number of bits.

Power Sources

A problem with wireless transmitters is the power source for the wireless devices. The power limitations and large size of NiCad batteries are too demanding for wireless transmission. Newer battery compositions have allowed for smaller and longer-lasting power sources. The signaling methods have also improved, allowing devices to sleep (power down to a minimally operating state) when not sending and wake (power up to a normal operating state) upon receiving a certain signal.

There are energy-harvesting technologies that allow devices to operate without batteries or with small rechargeable batteries. *Energy harvesting* is a strategy that a device uses for operation whereby it obtains power from its surrounding environment. Energy-harvesting technologies can use mechanical actions, vibrations, light, thermal gradients, or other energy sources in its area to operate its electronics or recharge the batteries.

In addition, mesh networking can be used to reduce the distance a signal has to travel. *Mesh networking* is the ability of devices to relay signals for other devices to reduce the radio power output needed to deliver a message over a long distance. **See Figure 33-3.** These advancements have greatly improved the lifetime of devices between battery charges or replacements.

Interference

There are many concerns about the reliability of wireless sensors and actuators. Wireless networks are vulnerable to disruptions from a variety of electromagnetic sources. Accidental disruptions are caused by stray radio waves in the same frequency, perhaps due to new equipment or transients from people and objects. Network disruption may also be caused by poorly isolated equipment power supplies or by changes in the environment, such as installing foil-based wallpaper and moving large metallic equipment or furniture. Awareness of the potential causes of wireless network interference contributes to improving the success rate of these implementations.

TECH FACT

Wireless communication within a plant can generate signals that can unexpectedly cause problems. Without proper shielding, a wireless signal can activate or deactivate other nearby equipment. An unplanned startup or shutdown can cause an unexpected risk of injury to personnel and damage to equipment.

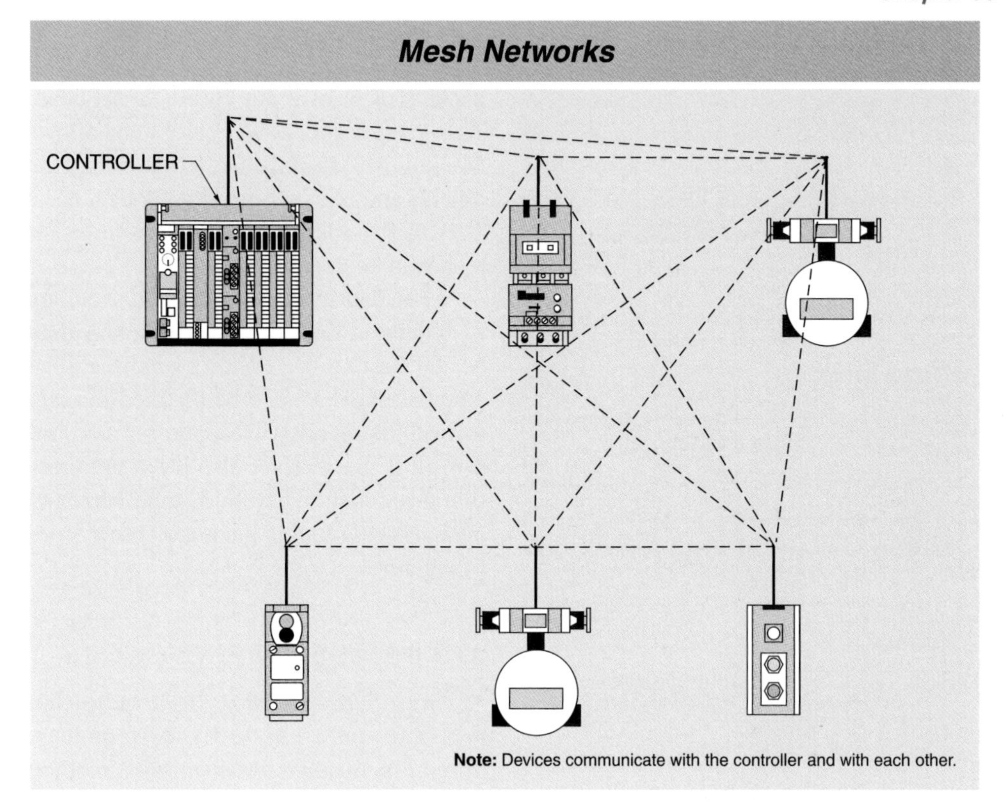

Figure 33-3. Mesh networks relay information from one device to another to reduce the power needed to send messages over long distances.

ANTENNAS AND WAVELENGTH

An *antenna* is a device that sends or receives radio signals through the air. A *wavelength* is the peak-to-peak length of an electromagnetic wave. A wavelength can be calculated from the frequency as follows:

$$\lambda = \frac{300}{f}$$

where

λ = wavelength (in m)

f = frequency (in MHz)

For example, with a radio frequency (f) of 900 MHz, the wavelength is calculated as follows:

$$\lambda = \frac{300}{f}$$

$$\lambda = \frac{300}{900}$$

$$\lambda = \mathbf{0.33\ m\ (33\ cm)}$$

The free end of an antenna is a node where the signal voltage is zero. An antenna that is 33 cm long matches the wavelength. Half-wave and quarter-wave antennas are also natural complements of a 33 cm wave with high emission efficiency. A quarter-wavelength antenna is just over 8 cm, or a little over 3″ long.

Antenna Types

An *omnidirectional antenna* is an antenna constructed of a single wire or bar that radiates energy equally in all directions. **See Figure 33-4.** A *ground plane antenna* is a standard omnidirectional antenna with a horizontal rod of the same length attached to the bottom of the antenna.

A *dipole antenna* is an antenna constructed of two straight quarter-length rods in line with each other where the two ends nearest each other are connected to the transmitter. The orientation can be horizontal or vertical. Dipoles emit their energy in more of a two-dimensional semicircular or doughnut pattern. The main signal strength is in two directions perpendicular to the antenna. Because of the two-dimensional pattern, the transmitting and receiving antennas must be orientated in the same direction.

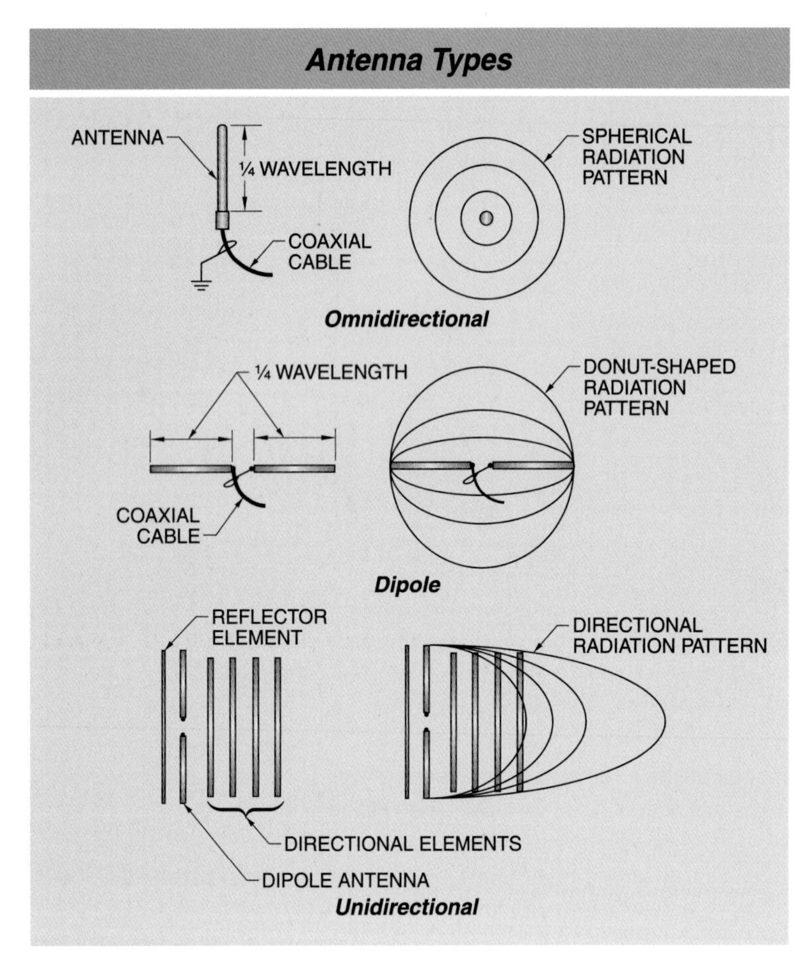

Figure 33-4. The three most common types of antennas are omnidirectional, dipole, and unidirectional.

A *unidirectional antenna,* or *Yagi antenna,* is a dipole antenna with additional elements placed in front of it at specific distances to focus transmission energy in one direction. With this type of antenna, the transmission antenna and the receiving antenna need to be aligned.

Antenna Gain

Antenna gain is a measure of how well an antenna focuses transmission energy. The transmission power is limited to 1 W. With an omnidirectional antenna, the 1 W is spread spherically in all directions so that only a small fraction of the transmission power is directed toward the receiver. One type of omnidirectional antenna uses measurement control regulation-radio analog digital (MCR-RAD) to transmit small packets of only about 16 bits at 96 baud. This gives higher power per bit and allows MCR-RAD signals to better penetrate objects and transmit over longer distances. This allows inexpensive quarter-wave antennas to be used.

When data must be transmitted over long distances, a unidirectional antenna may be the best choice. With a unidirectional antenna, the 1 W can be directed directly toward the receiver. Reception is also improved if the receiver also has a unidirectional antenna. Therefore, unidirectional antennas have higher gains than other types of antennas.

TRANSMISSION PROTOCOLS

A *single channel radio* is a radio that transmits on a single fixed frequency. *Spread spectrum* is a transmission method that uses multiple frequencies to transmit data. Spread spectrum technology was originally developed by the military for secure communication. The data to be transmitted is broken up into packets that are sent out over a continually changing series of frequencies. The three methods of signal spreading allowed by the FCC include frequency hopping spread spectrum (FHSS), direct sequence spread spectrum (DSSS), and orthogonal frequency division multiplexing (OFDM).

Frequency Hopping Spread Spectrum (FHSS)

Frequency hopping spread spectrum (FHSS) is a transmission method that takes packets of information and then transmits the packets across a frequency band by selecting pseudorandom frequencies within the band. **See Figure 33-5.** Both the transmitter and receiver know the pseudorandom frequencies. Each frequency contains a separate piece of information and is transmitted at high power. This gives the transmission the ability to overcome many sources of noise that may arise. At the receiver, the packets are checked for errors and decoded back into the original packet.

Although interference can knock out a packet of information, the rest of the packets generally reach the receiver. Therefore, FHSS technology tolerates interference. In addition, constantly and randomly changing the frequency makes it impossible for someone to tap into the signal. The robustness of the data transmission makes FHSS technology very attractive for the high EMI/RFI world of industrial applications.

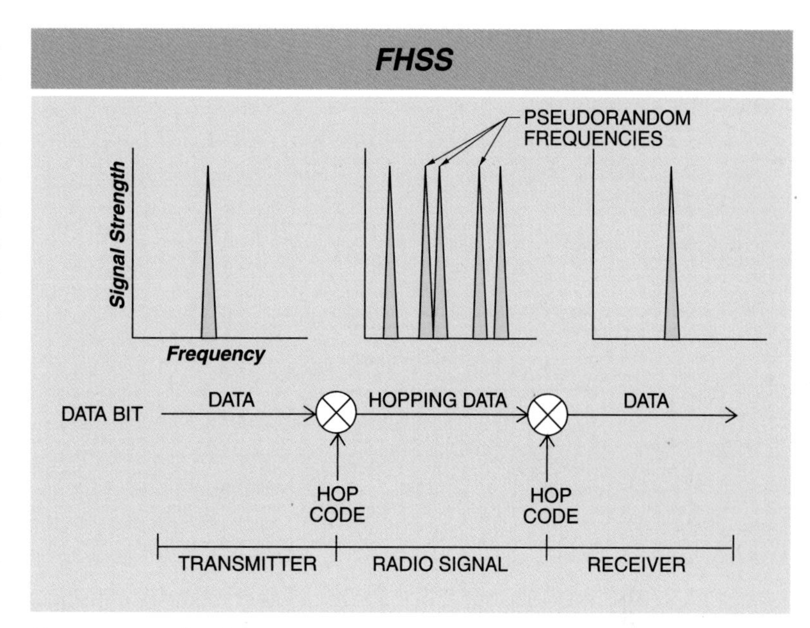

Figure 33-5. The frequency hopping spread spectrum (FHSS) transmits packets of information across a frequency band by selecting pseudorandom frequencies within the band.

Direct Sequence Spread Spectrum (DSSS)

Direct sequence spread spectrum (DSSS) is a transmission method that divides data into packets and sends the packets over a wide portion of a frequency band. **See Figure 33-6.** The divided packets are grouped together with a bit sequence called a chipping code. This takes the packets and further divides them into several segments, breaking one piece of information into several pieces. The packets are then transmitted at the same time across a series of frequencies called a channel.

With DSSS, the information can be transmitted with much lower power consumption than with the FHSS system. DSSS radios can lose data if there is excessive interference or there is other equipment on the same bandwidth. If one or more bits are damaged due to interference during transmission, the data will be restored but at a lowered transmission speed.

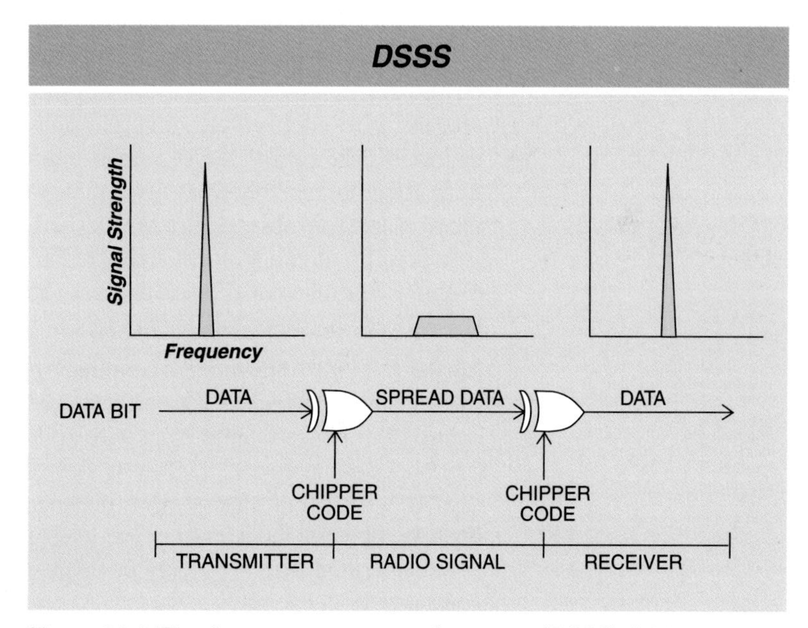

Figure 33-6. The direct sequence spread spectrum (DSSS) divides data into packets and sends the packets over a wide portion of the frequency band.

Orthogonal Frequency Division Multiplexing (OFDM)

Orthogonal frequency division multiplexing (OFDM) is a spread-spectrum transmission method that divides data packets and then further divides them before they are transmitted across a series of frequencies. The smaller pieces of data can be transmitted at greater speed than with FHSS or DSSS. **See Figure 33-7.**

A 2-wire transmitter must be able to operate with less than 4 mA.

Wireless Standards

There are several standards that are used with wireless communications. Different standards have different communications range and speed, security protocols, and operating frequencies.

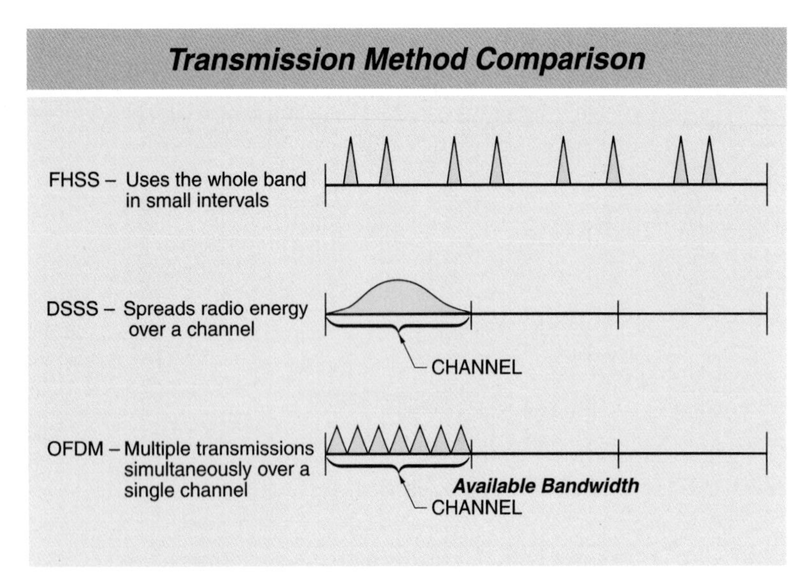

Figure 33-7. FHSS, DSSS, and OFDM have different methods of using a spread spectrum to send data packets.

Bluetooth®. *Bluetooth®* is a wireless communications protocol designed to link instruments and devices that are in close proximity to each other, such as appliances in a home or devices on a plant floor. Bluetooth is somewhat susceptible to interference from other radio transmissions. However, Bluetooth is an extremely redundant protocol that makes it suitable as an ad hoc network between capable devices.

ZigBee®. *ZigBee®* is a wireless communications protocol designed for monitoring and control for distances up to 75 m with up to 250 nodes per network. ZigBee is slower than Bluetooth but allows more distance between devices and allows more devices to be connected with extremely low power consumption.

IEEE 802.11x. *IEEE 802.11x* is a set of wireless communications protocols for high-speed communication. There are several different versions of IEEE 802.11x standards that address different needs. IEEE 802.11a/b/g describes the variations of the wireless Ethernet standard. **See Figure 33-8.** Some of the devices based on this standard are called Wi-Fi, W-LAN, and wireless Ethernet.

IEEE 802.11a. The IEEE 802.11a standard describes operation in the 5 MHz band and uses OFDM to obtain transmission rates of up to 54 Mbps. One advantage is that it operates at the same date rates as IEEE 802.11g but escapes the sometimes-crowded 2.4 GHz channels.

IEEE 802.11b. The IEEE 802.11b standard uses the 2.4 GHz band with DSSS modulation. The transmission rate is limited to 11Mbps.

IEEE 802.11g. The IEEE 802.11g standard uses the OFDM technology of 802.11a in a globally accepted 2.4 GHz band. Transmission rates can be as high as 54 Mbps.

ISA-100.11a. The standard 100.11a is a recently approved consensus standard that addresses wireless manufacturing and control systems. The standard is intended to provide reliable and secure operation in noncritical monitoring, alarming, and control applications. The standard defines the protocol suite, system management, gateway, and security specifications for processes that can tolerate a latency of about 100 ms. The standard includes specifications for coexistence with other wireless standards.

SECURITY AND ENCRYPTION

A major concern for wireless networks is the possibility of intentional network disruptions. A person can intercept wireless signals and either attempt to jam network communications or gain control of nodes or systems. It may be possible for someone even outside the facility to infiltrate the network. With a wireless access point, a physical connection to the network is not needed.

A wireless network is considered breached or hacked when an undesired client device gains access to the network. This access happens when the key has been found or cracked. Depending on the systems involved, a hacker could cause serious equipment damage. Encryption, authentication, and other security measures are used to minimize the potential for security breaches.

IEEE 802.11 a/b/g			
	802.11a	**802.11b**	**802.11g**
Frequency	5.8 GHz	2.4 GHz	2.4 GHz
Speed	54 Mbps	11 Mbps	54 Mbps
Compatibility	802.11a	802.11b	Backwards to 802.11b
Usage	Least	Most	Replacing 802.11b
Pros	Unaffected by 2.4 GHz traffic; channels have no overlap	Large user base; long distance with low cost	Higher speed than 802.11b
Cons	Shortest distance	Low network speed	Moderate distance; low energy per bit

Figure 33-8. The IEEE 802.11x standards communicate over different frequencies and at different data rates.

When public standard technology such as those based on standard IEEE 802.11 is used, security concerns can be greater since the standard is public knowledge. In the public 802.11, there are several different levels of security or encryption. WEP was the first level of security used.

WEP

WEP, or *wired equivalency privacy,* is a security standard used for wireless Ethernet networks. This level of encryption is very susceptible to wireless attack. Software tools can be found online that allow unauthorized access to any WEP encrypted network. WEP cannot be secure and should not be used for wireless networking. **See Figure 33-9.**

WPA

WPA, or *Wi-Fi protected access,* is a security standard based on the same type of encryption that is used in WEP with advancements such as authentication and dynamic keys. The authentication process was added to help restrict unauthorized access. There are two levels of authentication.

Overall, WPA is a good form of security against many types of attack, although WPA security is still vulnerable to more sophisticated wireless attackers. The IEEE committees recognized this problem and developed a new IEEE.11i security standard known as WPA2.

WPA2

WPA2 is a security standard that uses advanced encryption standard (AES) encryption. AES has been used by the government to protect all of their wireless communications. Many security experts consider AES to be unbreakable with today's tools. Future software and hardware tools may allow WPA2 to be broken.

WPA2 allows for pre-authentication. Pre-authentication allows a client device the ability to connect with one access point while becoming authenticated with another. This gives the client the ability to roam seamlessly from access point to access point without losing a wireless connection. Overall, WPA2 is currently the best security that can be used to protect wireless transmitted information. Products claiming WPA and WPA2 security must be tested and certified by the Wi-Fi Alliance.

Encryption Comparison		
Encryption Method	**Description**	**Security**
WEP	Weak key; can be hacked with low security knowledge	Poor
WPA	Same encryption as WEP with added authentication; can be hacked with high security knowledge	Good
WPA2	Highest level of security; considered unhackable with today's tools	Best

Figure 33-9. Different encryption methods have different levels of security.

A wireless network can bring remote signals into a control room.

PSK and 802.1x

PSK, or *pre-shared key,* is a security standard with a lower-level authentication for small office/home office (SOHO) networks that do not require the higher-level security of 802.1x. *IEEE 802.1x* is a security standard with a higher-level authentication security than PSK. IEEE 802.1x was developed by the IEEE to provide secure communication between the client device and authentication server such as remote authentication dial-in user service.

Dynamic Keys

Dynamic keys are a large factor in the security of a wireless network. A *dynamic key* is the component of a wireless system that encrypts the information being sent using a continuously changing key. Temporal key integrity protocol (TKIP) is used to accomplish this.

INDUSTRIAL REQUIREMENTS AND APPLICATIONS

The limitations on data throughput, performance, and RF behavior means that wireless Ethernet based on IEEE 802.11 technology is not yet capable of handling critical process control applications. However, other wireless applications are readily implemented. They can be as simple as tank farm inventory data to more complex tasks like handling information from mobile data sources such as laptops, bar code readers, etc.

Mobile Computing

A common application for mobile computing is to support workforce data management applications. A hot spot is created for mobile clients using an IEEE 802.11 product as an access point. For example, mobile clients can be the laptops or pocket PCs of utility or wastewater maintenance personnel. These workers can receive electronic work orders, service problems, or request parts or assistance.

Video Surveillance

Plant security, perimeter monitoring, and video surveillance have become very important subjects. **See Figure 33-10.** An IP camera usually includes embedded setup and control software making it easy to access from any PC. Alarms can trigger the control system to record a video.

Network Bridging and Remote Access

It is often necessary to connect two separate Ethernet networks. For example, a remote local wastewater treatment plant with a pumping station may need to be connected with the main treatment facility. A wireless network called a network bridge is used to connect the two independent Ethernet networks.

Cellular Communications

Cellular communications are becoming increasingly popular in process applications where traditional wired and wireless communications are not possible. Many cellular networks do not permit machine-to-machine (M2M), or continuous always-on, connections without specific modem approval. To be acceptable, the modem must first comply with the PCS Type Certification Review Board (PTCRB), which verifies that the modem will not cause harm to the cellular network. After approval, specific carriers can approve the modem.

Cellular communications can be an efficient means of accessing data that is not accessible by traditional means. It is important that cellular modems be approved to ensure that they will operate successfully over the network. In the United States, carriers typically do not allow the use of voice networks like global system for mobile communications (GSM) for M2M data communication applications. This is because the modems could monopolize network bandwidths. It is also a disadvantage for the modem user because of billing arrangements that charge by the minute.

Data networks, such as general packet radio service (GPRS) communications, are designed to provide better network architecture for data communications. Data networks offer an always-on connection and provide billing based on the amount of data sent. The three types of GPRS communications are dynamic IP, virtual static IP, and virtual static IP with VPN.

Dynamic IP. If an application requires that a remote device periodically report in with status updates or reports by exception, a dynamic IP service can be used. **See Figure 33-11.** In this type of application, a host may only respond to client requests and may not initiate communication to the client. The client is assigned a dynamic IP address, which may change each time it contacts the host.

Firewalls may assign the same IP address to many different client modems at the same time by using a process known as port mapping. Therefore, the client modem will have to identify itself when it contacts its host. In operation, the remote device requests a temporary IP address. The remote device uses that temporary IP address and communicates with the host. The host replies to the device request but cannot initiate a request.

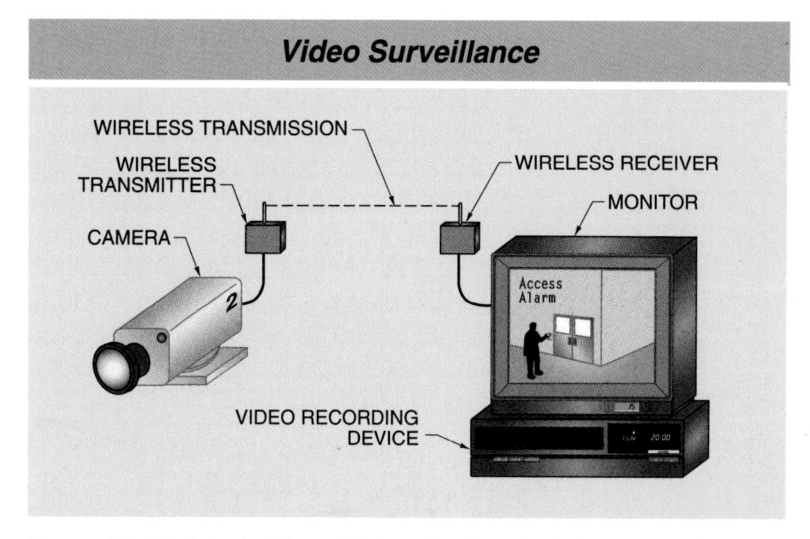

Figure 33-10. A typical industrial application of wireless transmission is video surveillance.

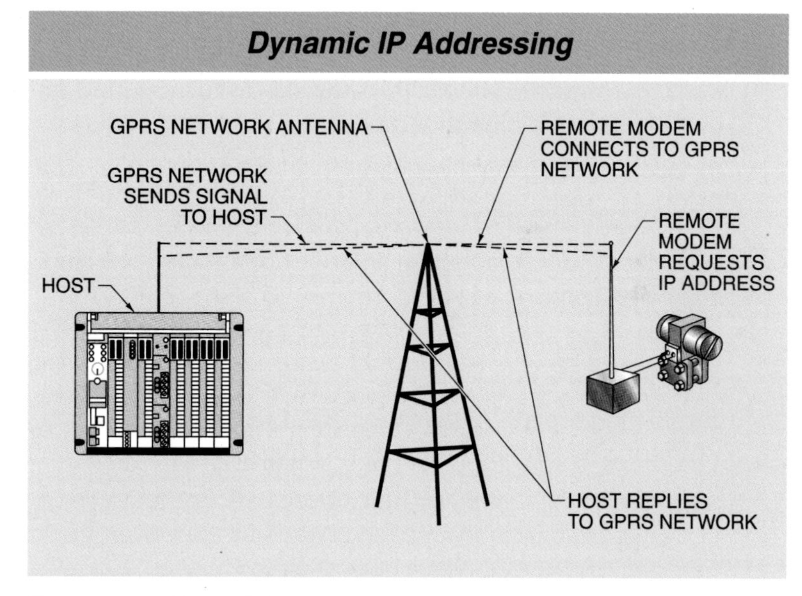

Figure 33-11. Dynamic IP service can be used when a remote device needs to periodically report in with status updates or reports by exception.

Virtual Static IP (VSIP). If a dynamic IP is not desired, a virtual static IP (VSIP) connection can be set up to the service provider that bypasses the firewall. This way the remote devices and the host are protected from the Internet. This approach requires specialized hardware and networking knowledge.

VSIP is the most robust approach for private connections. With VSIP, smaller deployments can be assigned fixed Internet addresses that are protected behind a firewall. The firewall is designed so that only a designated host can access these devices. The devices are assigned dynamic addresses, but these addresses are tracked so that they can be mapped to the fixed addresses. This service is very flexible and allows the modem to act as both a server and a client device.

VSIP with Virtual Private Network (VPN). The VSIP with VPN is similar to VSIP service with the addition of a VPN connection. The VPN setup is established with the IT department and the service provider once the setup is completed. The IT department manages the users. There can be multiple users with access to the modem and remote device while maintaining one static IP address as the host. It also allows users to be mobile and have access to modem data from anywhere in the world.

KEY TERMS

- *wireless transmission:* The method of using radio frequencies to transfer information from one device to another.
- *energy harvesting:* A strategy that a device uses for operation whereby it obtains power from its surrounding environment.
- *mesh networking:* The ability of devices to relay signals for other devices to reduce the radio power output needed to deliver a message over a long distance.
- *antenna:* A device that sends or receives radio signals through the air.
- *wavelength:* The peak-to-peak length of an electromagnetic wave.
- *omnidirectional antenna:* An antenna constructed of a single wire or bar that radiates energy equally in all directions.
- *ground plane antenna:* A standard omnidirectional antenna with a horizontal rod of the same length attached to the bottom of the antenna.
- *dipole antenna:* An antenna constructed of two straight quarter-length rods in line with each other where the two ends nearest each other are connected to the transmitter.
- *unidirectional antenna,* or *Yagi antenna:* A dipole antenna with additional elements placed in front of it at specific distances to focus transmission energy in one direction.
- *antenna gain:* A measure of how well an antenna focuses transmission energy.
- *single channel radio:* A radio that transmits on a single fixed frequency.
- *spread spectrum:* A transmission method that uses multiple frequencies to transmit data.
- *frequency hopping spread spectrum (FHSS):* A transmission method that takes packets of information and then transmits the packets across a frequency band by selecting pseudorandom frequencies within the band.
- *direct sequence spread spectrum (DSSS):* A transmission method that divides data into packets and sends the packets over a wide portion of a frequency band.
- *orthogonal frequency division multiplexing (OFDM):* A spread-spectrum transmission method that divides data packets and then further divides them before they are transmitted across a series of frequencies.

KEY TERMS *(continued)*

- *Bluetooth®:* A wireless communications protocol designed to link instruments and devices that are in close proximity to each other, such as appliances in a home or devices on a plant floor.
- *ZigBee®:* A wireless communications protocol designed for monitoring and control for distances up to 75 m with up to 250 nodes per network.
- *IEEE 802.11x:* A set of wireless communications protocols for high-speed communication.
- *WEP,* or *wired equivalency privacy:* A security standard used for wireless Ethernet networks.
- *WPA,* or *Wi-Fi protected access:* A security standard based on the same type of encryption that is used in WEP with advancements such as authentication and dynamic keys.
- *WPA2:* A security standard that uses advanced encryption standard (AES) encryption.
- *PSK,* or *pre-shared key:* A security standard with a lower-level authentication for small office/home office (SOHO) networks that do not require the higher-level security of 802.1x.
- *IEEE 802.1x:* A security standard with a higher-level authentication security than PSK.
- *dynamic key:* The component of a wireless system that encrypts the information being sent using a continuously changing key.

REVIEW QUESTIONS

1. What are the advantages and disadvantages of wireless transmission?
2. What is the relationship between communication speed and distance?
3. Define and compare frequency hopping spread spectrum (FHSS) and direct sequence spread spectrum (DSSS).
4. Define the security standards WPA and WPA2, and explain how they are different.
5. Why should general packet radio service (GPRS) be used for cellular communication instead of the more common global system for mobile communications (GSM)?

Practical Transmission and Communication

Chapter 34

OBJECTIVES

- Explain the importance of loop impedance in a current transmission system.
- Define ground loop and identify ways that a ground loop can be avoided.
- Describe electromagnetic interference (EMI).
- List the common connectors.
- List the types of transmitters and describe a smart transmitter.

ALLOWABLE LOOP IMPEDANCE

Allowable loop impedance is the resistance limit that is the summation of all of the receiver impedances and connecting wiring resistance in the current transmission loop. This limit excludes the impedance of the sending unit. A loop with too much impedance is not able to generate enough current to deliver the 20 mA that is necessary for signaling. The allowable loop impedance, excluding the transmitter, is based on the use of a 24 VDC power supply and is usually 650 Ω to 1000 Ω, but varies for different transmitters. In a DC circuit, impedance is the same as resistance. Impedance is measured in ohms and is usually used when referring to measurement inputs.

Calculating Impedance

The Ohm's law equation to calculate resistance or impedance is as follows:

$$R = \frac{E}{I}$$

where

R = resistance or impedance (in Ω)

E = voltage (in V)

I = current (in A)

The equation to determine the maximum allowable total loop impedance of a circuit powered by a 24 VDC power supply with a 20 mA (0.020 A) current requirement is applied as follows:

$$R = \frac{E}{I}$$

$$R = \frac{24}{0.020}$$

$$R = \mathbf{1200\ \Omega}$$

This means that the total loop impedance is 1200 Ω for a 24 VDC circuit, including the transmitter. The sum of the transmitter impedance, the receiver impedances, and any impedance from the wiring itself must be less than or equal to 1200 Ω. If the circuit impedance is higher than 1200 Ω, the maximum current is less than the 20 mA required for signaling.

For example, if a 2-wire transmitter with a 24 VDC power supply has an internal impedance of 550 Ω, the balance of the loop must be limited to less than 650 Ω to keep the total loop impedance below the limit of 1200 Ω. This means that the transmitter has an allowable loop impedance limit of 650 Ω.

If each receiver has an impedance of 250 Ω, how many receivers can be used in the circuit? Since each receiver has an impedance of 250 Ω, two receivers have an impedance of 500 Ω, and three receivers have an impedance of 750 Ω. The total loop impedance of a circuit with a transmitter and two receivers is 1050 Ω (550 + 250 + 250 = 1050). This circuit allows the proper amount of current flow to allow a 20 mA signal to the receivers. The total loop impedance of a circuit with a transmitter and three receivers is 1300 Ω (550 + 250 + 250 + 250 = 1300). This circuit has a total loop impedance of more than 1200 Ω and does not allow a 20 mA signal to travel to the receivers. **See Figure 34-1.**

Calculating Current

The actual maximum current for a circuit with a total resistance of 1300 Ω can be determined from Ohm's law.

$$I = \frac{E}{R}$$

where

I = current (in A)

E = voltage (in V)

R = resistance (in Ω)

$$I = \frac{E}{R}$$

$$I = \frac{24}{1300}$$

$$I = \mathbf{0.0185\ A}$$

This means that the circuit with three receivers will not be able to generate a current signal greater than 0.0185 A (18.5 mA). A receiver calibrated for 20 mA for a maximum signal will report erroneous values.

Current Repeaters

When a loop cannot generate enough loop current, a current repeater should be used or the retransmission option of a process monitor or controller for one or more of the allowed receivers. **See Figure 34-2.** A current repeater makes it possible to add additional receivers by retransmitting the current output on a second loop. The current repeater acts as a transmitter by using the 4 mA to 20 mA current output of the primary transmitter to produce a duplicate isolated current signal.

TECH FACT

A common air dryer configuration has two desiccant towers with one on-line, drying the instrument air, and the other off-line, being regenerated.

Current Transmission Loop Impedance

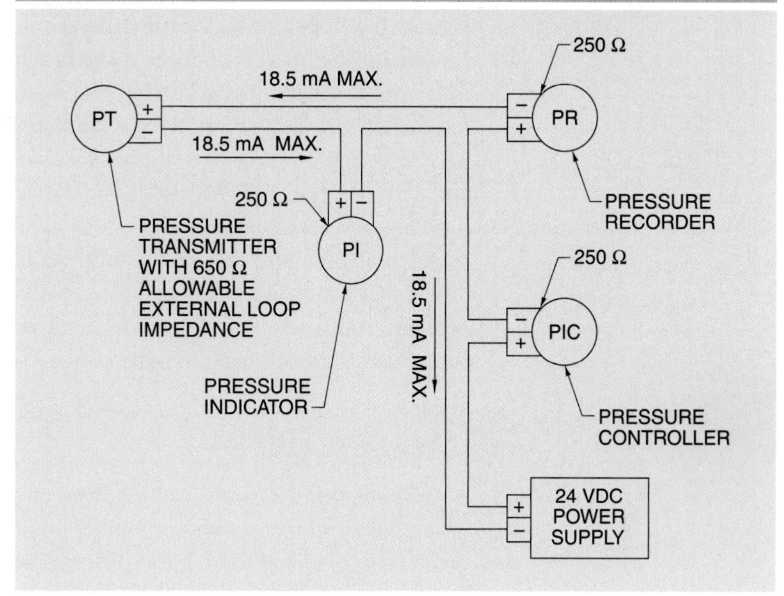

Figure 34-1. The total current in a circuit is less than 20 mA when the allowable loop impedance is exceeded.

Current Transmission Loop with Current Repeater

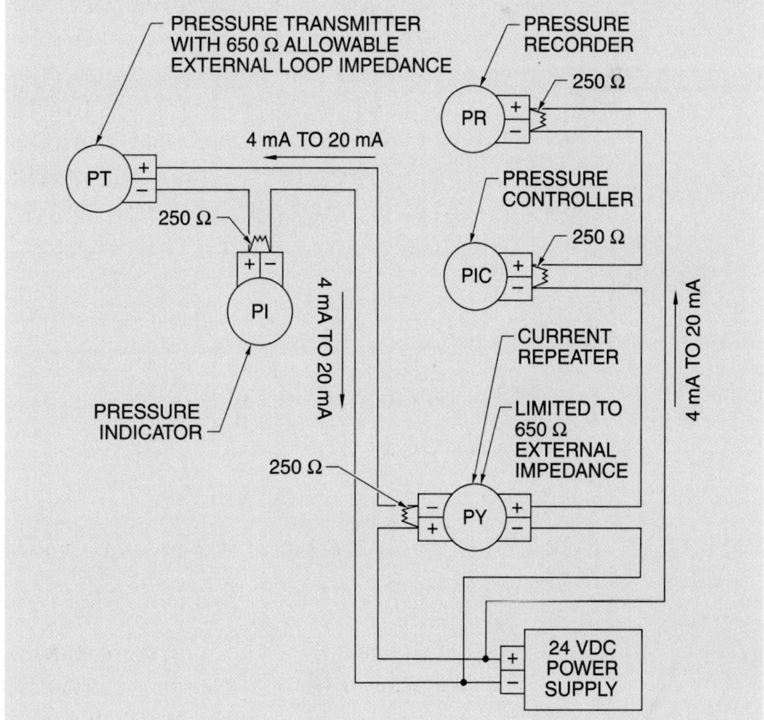

Figure 34-2. A current repeater allows more receivers to be used than would be allowed in a simple current loop circuit.

The use of a current repeater reduces the external loop impedance of the primary loop so that the impedance is below the allowable loop impedance of the transmitter. This creates an additional current loop with the current repeater acting as the transmitter for the second loop. If two of the receivers are within a few feet of each other, they may be able to share a 1 VDC to 5 VDC signal and thus stay below the allowable impedance limit. The retransmission option available with some process monitors or controllers serves the same purpose.

GROUND LOOPS

A *ground loop* is the current flow from one grounded point to a second grounded point in the same powered loop due to differences in the actual ground potential. The resulting current flowing through a ground loop can create false signals to the receivers. Thus the receivers in those loops do not see the same current that is produced by the transmitters. **See Figure 34-3.**

Field devices communicate with a network through transmitters.

The cost to wire an industrial operation typically ranges from $40 to $2000 per foot. Wireless communication eliminates these wiring costs. However, there still may be costs associated with running power to the field devices.

A common arrangement is to have a number of loops powered by a single power supply. In order to prevent ground loops, there should be only one ground point for all the loops. Ground loop problems can be inadvertently introduced into transmission loops and then become very difficult to identify and eliminate. It is much better to be aware of the potential problem and prevent it by only using isolated devices.

Isolated Devices

Transmitters, primary elements, receivers, power supplies, and other devices are available as isolated devices. An *isolated device* is an electric component that provides protection against ground loops because it has no electrical connection from the current transmission loop signal to any other electrical circuits or grounds.

ELECTROMAGNETIC INTERFERENCE (EMI)

Cables used in industrial environments are often subject to electromagnetic interference (EMI). A possible solution to EMI problems is to use optical fiber cable. However, optical fiber cable is generally not as robust as copper wire cable. In addition, optical fiber is more expensive than copper cable. A properly balanced unshielded twisted pair provides significant immunity to EMI. A *balanced line* is a signal transmission cable in which two conductors carry the signal and create a balanced circuit alternating from positive to negative. The paired wires balance out each other's signals and help to nullify the effect of EMI from nearby wires or electromagnetic fields.

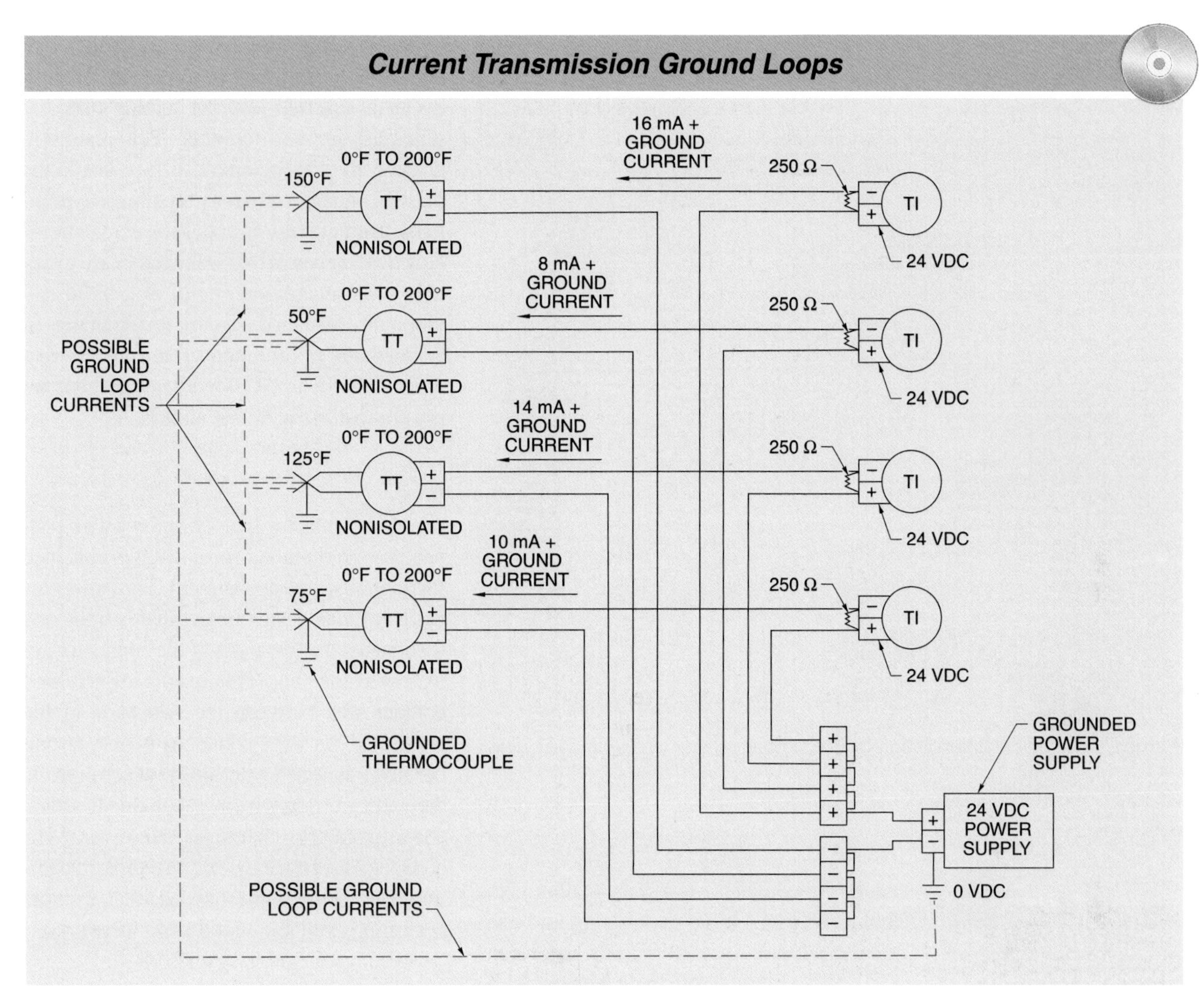

Figure 34-3. Current transmission ground loops are the leakage currents from nonisolated devices by routes outside of the designed wiring paths. Ground loops generate errors in the transmission signals.

Another possible solution to EMI problems is to use shielded cable. **See Figure 34-4.** This cable includes an aluminum foil shield that can provide significant immunity to EMI when properly grounded. The key is proper grounding, which is very difficult to do. Improper grounding can lead to ground loop currents or to significant noise in the line. Routing of cables through conduit for protection is a significant improvement. The conduit provides grounding and shielding as well as protection from chemicals and oils in the environment. However, the use of unshielded cable in conduit can affect the electrical performance of the cable.

CONNECTORS

All cables must be connected to devices at each end. Connectors can be a weak spot in network wiring. Connectors are subject to miswiring, improper terminations and shield connections, and mechanical damage from rough handling. There are several different standards emerging for connector wiring, but there is no overriding standard for industrial networks, even within a single type of connector. The most common types of cable connectors are the RJ-45 and the M12 connectors. Other types of connectors are standardized for different industrial networks.

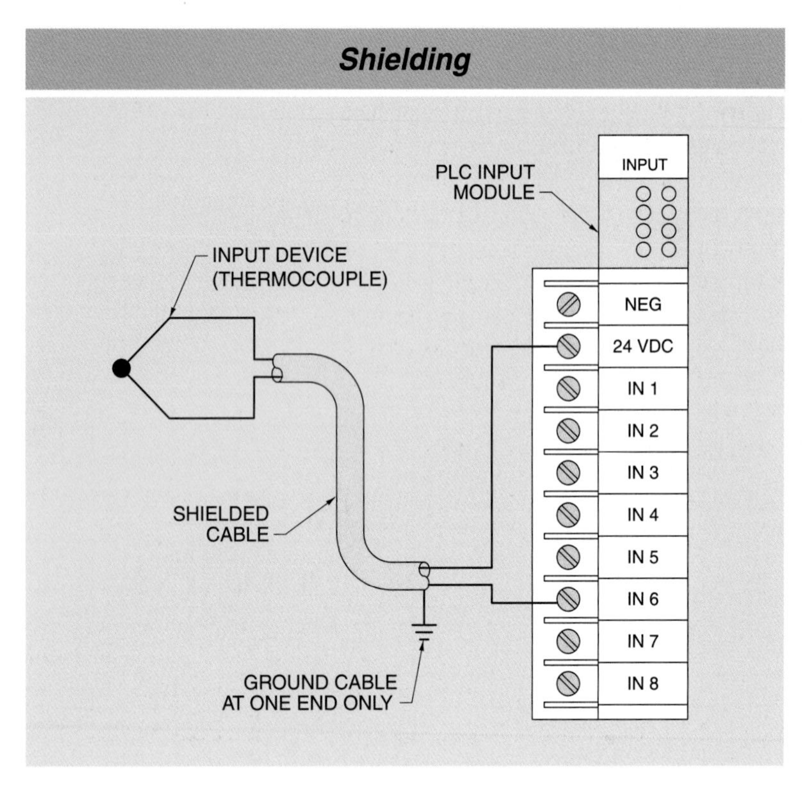

Figure 34-4. A possible solution to EMI problems is to use shielded cable.

RJ-45

An *RJ-45 connector* is a snap-in connector that looks like a large phone plug and contains eight pins instead of four as in standard phone connectors. **See Figure 34-5.** RJ-45 connectors are widely used in office environments and where the connection is inside a protected enclosure.

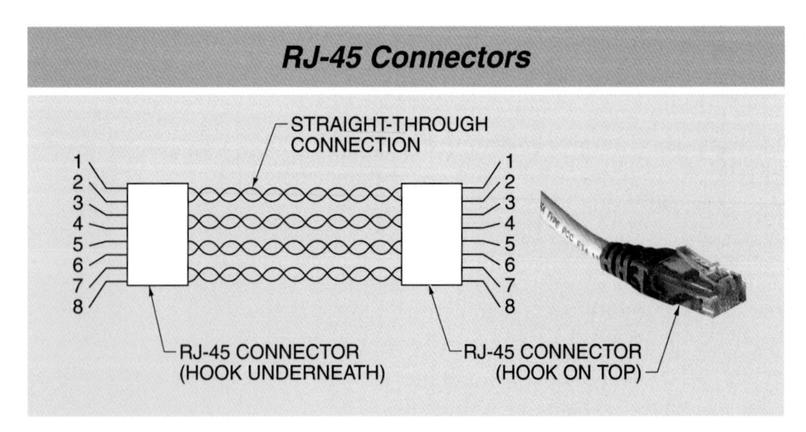

Figure 34-5. An RJ-45 connector is a snap-in connector with eight pins instead of four as in similar phone connectors.

RJ-45 connectors have exposed wires in both ends of the connector. When the connecter is inserted into the mating part, the wires are pressed together. This results in a relatively small contact surface inside the plastic connector that may be subject to problems from corrosion and poor connections. However, newer RJ-45 connectors are available with sealed watertight, dust-tight, and vibration-resistant designs to resist damage in an industrial environment. There are different types of hardened RJ-45 connectors that are not compatible with one another.

M12

An *M12 connector* is a 4-wire or 8-wire connector with threaded metal fittings designed for an industrial environment. The connector has a circular metal contact that is tightened together as the two parts of the connector are tightened together. This results in increased contact area between the two parts of the connector. An M12 connector is very strong because it is sealed with epoxy, which protects the internal parts from mechanical shock and exposure to chemicals. However, M12 connectors are not expected to be able to communicate at the highest speeds that are now becoming available in industrial networks.

TRANSMITTERS

Transmitters come in several wiring configurations. A *2-wire transmitter* is a transmitter that uses the 24 VDC power from a current transmission loop to power the transmitter. A *3-wire transmitter* is a transmitter that uses a DC power supply, with the negative lead of the power supply and the negative lead for the transmitter being the same wire that is used to establish a zero level for the circuit. The positive wire from the power supply is the source of power for the transmitter and the positive wire from the transmitter sends a signal to the controller. The power supply is separate. A *4-wire transmitter* is a transmitter that uses two wires to provide power to the transmitter and two wires for the current output. The separate power supply may be DC or AC.

Smart Transmitters

A *smart transmitter* is a microprocessor-based signal transmitter that combines digital and analog signals so that it can handle multiple inputs and outputs, communicate and change configuration details, and signal alarms and error conditions. **See Figure 34-6.** Smart transmitters offer greater ease and accuracy in the calibration procedure and provide the ability to have their performance examined remotely. In addition, the transmitter can be instructed to generate any specified output signal. It is also possible, if the system wiring and digital communications method are compatible, to transfer measured information to the control system digitally. A smart transmitter allows the user to configure the transmitter as follows:

- Set the lower and upper measurement range, alarm points, transmission output, and signal processing.
- Activate and configure special features such as averaging time, peak hold, or directions to close alarm contacts.
- Recalibrate the input and output of the transmitter.

A smart transmitter consists of three sections that perform different functions. **See Figure 34-7.** First, the sensor circuit input section converts the analog sensor signal to a digital signal through an A/D converter. The input may be any type of signal from an instrument. Smart transmitters generally do not require calibration as often as the solid state transmitters they replaced. In many cases, the word "certification" should be used instead of the word "calibration." If a smart transmitter does not meet the manufacturer's specifications, it may be necessary to replace it.

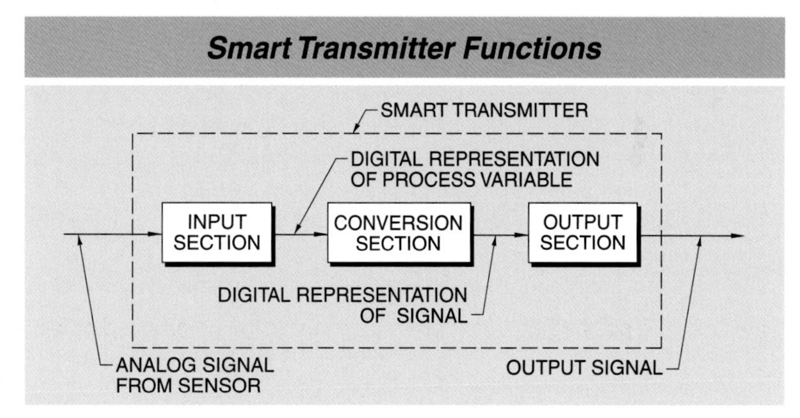

Figure 34-7. Smart transmitters consist of three sections that work together to convert an analog sensor signal to a linear output to a controller.

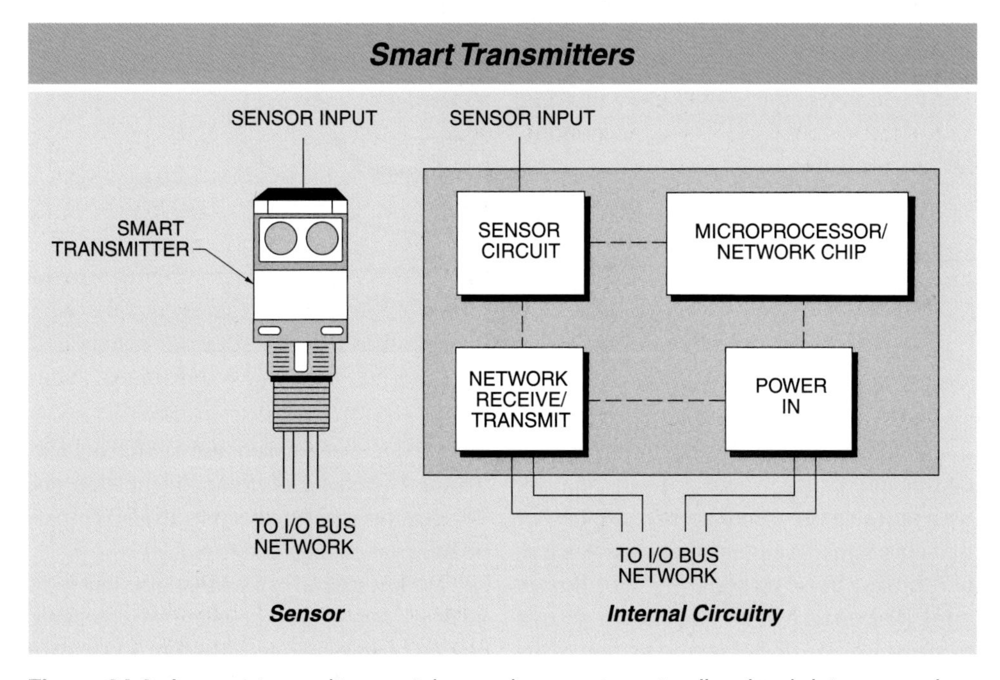

Figure 34-6. A smart transmitter contains a microprocessor to allow local data conversions and communication with a controller.

HART® Transmitter Calibration

Configuration and calibration of transmitters is usually done with a hand-held smart transmitter communicator. Although the majority of smart transmitters use the HART® communications protocol, there are a few other legacy systems that use different digital communications methods. Thus a hand-held communicator used for one manufacturer's smart transmitters may not work with all smart transmitters.

A common misconception is that the accuracy and stability of HART instruments eliminates the need for calibration or that recalibration can be accomplished by re-ranging field instruments. All smart transmitters need regular calibration. Regular calibration is also prudent because performance checks often uncover other system problems like solidified or congealed pressure lines.

A typical calibration adjustment for conventional transmitters involves setting the zero and span values by making a multiple point test between the minimum and maximum values. A smart transmitter calibration requires more than that because the input section and the output section require separate trim adjustment. If the transmitter zero and span are the only changes made during a calibration, the internal digital signals will no longer agree with the external analog signals.

For example, if a transmitter has drifted 1% so that the minimum value now corresponds to a 4.16 mA output, adjusting the zero and span will correct the output to 4.00 mA. However, the internal digital value of 4.16 mA was not changed with the zero and span adjustment. A HART transmitter combines the digital and analog signals and this can cause problems if the digital and analog signals do not agree.

A HART transmitter requires a sensor trim adjustment on the input section. A digital loop calibrator is used to set a range of applied analog inputs and a communicator is used to read the outputs from the input section. Some tools combine the calibrator and communicator in one device. If the error is out of range, follow the manufacturer's instructions to adjust the sensor trim. This is not the same as adjusting the transmitter zero and span.

The output section also needs to be calibrated. Use a communicator to send fixed values to the output section and use the calibrator to measure the resulting output current. Follow the manufacturer's instructions to adjust the output trim. This may also be called a 4 mA to 20 mA trim, current loop trim, or D/A trim.

A common HART transmitter parameter called damping can have an adverse effect on tests and adjustments. Damping adds a delay between the time the instrument receives a signal at the input section and the time that signal is detectable at the output section. This delay may be more than the time the calibrator waits between setting the input and reading the output. Therefore, the damping constant should be set to zero during a calibration and reset to its original value after the calibration.

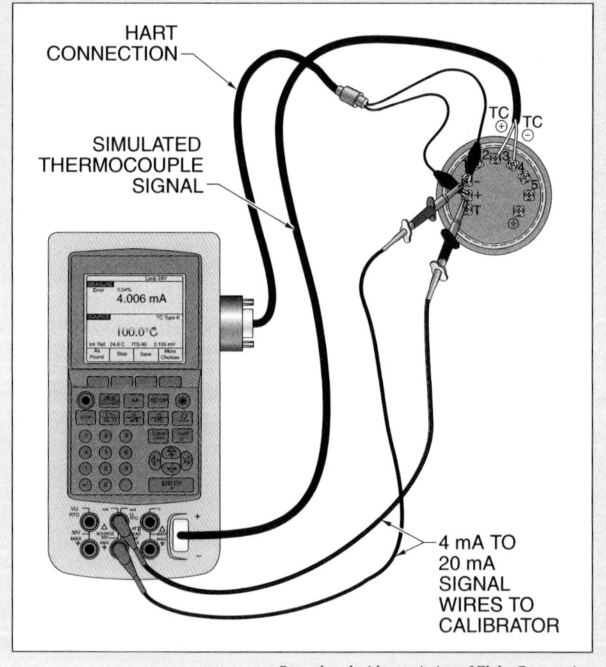

Reproduced with permission of Fluke Corporation.

Next, the digital signal goes through a data conversion section where a microprocessor uses an equation or lookup table to convert the input signal to engineering units, such as temperature in °F, pressure in psi, or flow in gpm. The mathematical conversion section then changes the digital representation of the process value to a digital representation of a 4 mA to 20 mA signal. This may be a linear conversion, but pressure measurements may require a square root extraction function, and temperature measurements from an RTD may require a nonlinear conversion.

The last section of a smart transmitter is the network output section where the calculated digital representation of the 4 mA to 20 mA signal is loaded into a D/A converter to produce the actual analog electrical signal.

KEY TERMS

- *allowable loop impedance:* The resistance limit that is the summation of all of the receiver impedances and connecting wiring resistance in the current transmission loop.
- *ground loop:* The current flow from one grounded point to a second grounded point in the same powered loop due to differences in the actual ground potential.
- *isolated device:* An electric component that provides protection against ground loops because it has no electrical connection from the current transmission loop signal to any other electrical circuits or grounds.
- *balanced line:* A signal transmission cable in which two conductors carry the signal and create a balanced circuit alternating from positive to negative.
- *RJ-45 connector:* A snap-in connector that looks like a large phone plug and contains eight pins instead of four as in standard phone connectors.
- *M12 connector:* A 4-wire or 8-wire connector with threaded metal fittings designed for an industrial environment.
- *2-wire transmitter:* A transmitter that uses the 24 VDC power from a current transmission loop to power the transmitter.
- *3-wire transmitter:* A transmitter that uses a DC power supply, with the negative lead of the power supply and the negative lead for the transmitter being the same wire that is used to establish a zero level for the circuit.
- *4-wire transmitter:* A transmitter that uses two wires to provide power to the transmitter and two wires for the current output.
- *smart transmitter:* A microprocessor-based signal transmitter that combines digital and analog signals so that it can handle multiple inputs and outputs, communicate and change configuration details, and signal alarms and error conditions.

REVIEW QUESTIONS

1. What is allowable loop impedance, and why is it important?
2. What is the cause of a ground loop, and what are ways to prevent it?
3. What are two ways to avoid problems with electromagnetic interference (EMI)?
4. Define two common types of connectors.
5. Describe three common wiring arrangements for transmitters.

SECTION 9

AUTOMATIC CONTROL

TABLE OF CONTENTS

SECTION OBJECTIVES

Chapter 35

- Define automatic control and identify common terms associated with it.
- Explain process dynamics and define the terms associated with it.
- Identify the functions of controllers and define these functions.

Chapter 36

- Define control strategy and identify common control strategies.
- List and define the common types of advanced control strategies.

Chapter 37

- Describe the purpose of controller tuning and tuning coefficients.
- Identify the different tuning performance standards.
- Explain the methods of tuning controllers.

Chapter 38

- List and define the different types of digital controllers and control systems.
- Define electric controller and describe the types of electric controllers.
- Describe the common operator interfaces.
- Describe the various configuration formats.

SECTION

9 AUTOMATIC CONTROL

INTRODUCTION

Simple feedback controllers have been used since antiquity. In 270 BC, the Greek inventor Ktesibios invented a float regulator for a water clock. The regulator kept a water tank at a constant level to ensure constant flow out of a spigot into another tank. This device was developed throughout the Middle Ages until it was finally abandoned in favor of mechanical clocks.

In about 1624, Cornelius Drebbel of Holland invented an automatic regulator that he used to keep an oven at a constant temperature. This idea was developed over the years and in the 1770s the first industrial temperature regulators were developed.

In the late 1600s, the pressure regulator was developed for use as a safety valve. In the mid-1700s, the float regulator was first used in boilers. By the end of the century, float regulators were in common use in boilers all over the world. By the early 1800s, the pressure regulator and the float regulator were combined into one device by Boulton and Watt for use in their steam engines.

In 1868, James Clerk Maxwell was the first person to supply a rigorous mathematical analysis of a feedback control system when he studied Watt's flyball governor. These ideas were developed throughout the 20th century into our modern version of control theory.

ADDITIONAL ACTIVITIES

1. After completing Section 9, take the Quick Quiz® included with the Digital Resources.
2. After completing each chapter, answer the questions and complete the activities in the *Instrumentation and Process Control Workbook*.
3. Review the Flash Cards for Section 9 included with the Digital Resources.
4. Review the following related Media Clips for Section 9 included with the Digital Resources:
   - Air Handling Unit Controller
   - Alarm Panel
   - Controller I/O Wiring
   - Controller Power Supplies
   - Controller Programming
   - Ladder Diagrams
   - Operator Interface
   - Receiver Controllers
   - Stand-Alone Controller
5. After completing each chapter, refer to the Digital Resources for the Review Questions in PDF format.

ATPeResources.com/QuickLinks • Access Code: 467690

SECTION 9 AUTOMATIC CONTROL

Automatic Control and Process Dynamics

Chapter 35

OBJECTIVES

- Define automatic control and identify common terms associated with it.
- Explain process dynamics and define the terms associated with it.
- Identify the functions of controllers and define these functions.

AUTOMATIC CONTROL

Automatic control is the equipment and techniques used to automatically regulate a process to maintain a desired outcome. Automatic controls are designed to handle dynamic situations where there are unplanned changes. The three components of an automatic control system are the process variable, the control variable, and the controller.

A *process variable (PV)* is the dependent variable that is to be controlled in a control system. A *control variable (CV),* or *manipulated variable,* is the independent variable in a process control system that is used to adjust the dependent variable, the process variable. A *controller* is a device that compares a process measurement to a setpoint and changes the control variable (CV) to bring the process variable (PV) back to the setpoint. A process may use chemical, thermal, or physical methods to convert an input to an output.

An automatic control system adds two elements to the three basic components of automatic control. A *primary element* is a sensing device that detects the condition of the process variable. Thermocouples, pressure diaphragms, and flowmeters are examples of primary elements. The primary element may be combined with a device that converts a process measurement, such as a voltage or movement of a diaphragm, into a scaled value, such as temperature or pressure. Indicators, recorders, and transmitters are examples of devices that scale a process variable. A *final element* is a device that receives a control signal and regulates the amount of material or energy in a process. Common types of final elements are control valves, variable-speed drives, and dampers.

An automotive cruise control is an example of automatic control. The automobile is accelerated to a set speed and then the cruise control is engaged to maintain that desired speed. The process is the engine and all the other components that affect the speed. The PV is the set speed. The CV is the fuel flow to the engine. The controller is the cruise control system. The primary element is the speedometer system that measures the set speed. The final element is the throttle plate that controls the fuel flow. **See Figure 35-1.**

TECH-FACT

Duty cycling is a common control strategy in building automation systems. Commercial buildings with a large number of electric baseboard heaters or small exhaust fans often use duty cycling as a control strategy. Because a duty cycle alternates the areas it heats, loss of comfort may occur when the load is OFF.

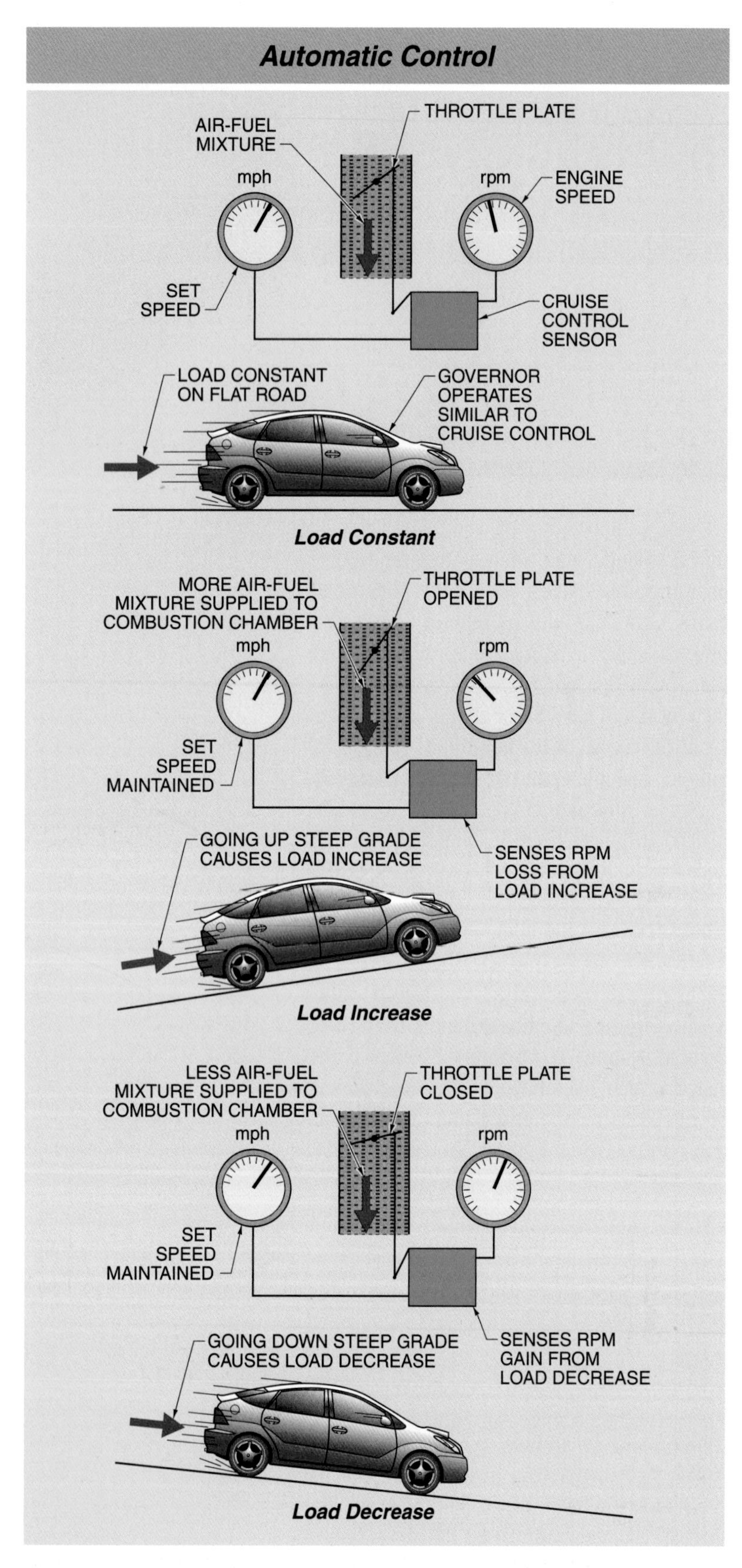

Figure 35-1. An automobile cruise control is an example of automatic control.

A home heating system is another example of automatic control. A thermostat is adjusted to a predetermined setpoint to maintain a comfortable living environment. The PV is the indoor air temperature. The CV is the energy flow to the furnace, which can be natural gas, coal, oil, propane, or electricity. The process is the house and the heating system. The controller is the thermostat. The primary element is the temperature-measuring device in the thermostat, often a bimetallic element. The final element is the valve that controls the fuel flow to the furnace.

Self-Regulation

Some automatic processes are self-regulating. For example, a tank of liquid that discharges from the bottom displays self-regulation as it is filling or emptying. **See Figure 35-2.** The flow out of the tank is steady when there is a steady flow into the tank. If the flow into the tank increases, the hydrostatic pressure increases and the flow out of the tank increases. As the flow into the tank returns to its normal rate, the flow out of the tank returns to its original rate. The level in the tank varies with changes in flow.

Another self-regulating process is the temperature of a pure boiling liquid. The temperature of a pure boiling liquid at constant pressure does not change no matter how much additional heat is applied. Once a pure liquid reaches the temperature of its boiling point, any additional energy that is applied to the liquid is used to convert liquid to vapor.

An example of a process that is not self-regulating is a chemical reaction. **See Figure 35-3.** An *exothermic reaction* is a chemical reaction that generates heat during the reaction and increases the temperature. The heat released by the reaction increases the temperature and the rate of reaction. Heat energy must be removed to prevent a runaway reaction and explosion. An *endothermic reaction* is a chemical reaction that consumes heat, and therefore more heat energy must be added to sustain the reaction. The energy used by the reaction decreases the temperature and the rate of reaction. Heat energy, such as from steam or hot water, must be added to keep the reaction going.

PROCESS DYNAMICS

The degree of difficulty of controlling a process depends on the characteristics of the process variable, the selected control variable, and the process itself. Control systems must be designed to work with the process dynamics to produce the best control. *Process dynamics* are the attributes of a process that describe how a process responds to load changes imposed upon it. These attributes are gain, dead time, and lags. Most processes possess all three process attributes. In addition, measurement signal processing may affect the process dynamics.

Load Changes

A *load change* is a change in process operating conditions that changes the process variable (PV) and must be compensated for by a change in the control variable (CV). In most processes, a load change is a change in the amount of material being handled, but it can also be changes in temperature or pressure of process feed streams or energy sources.

For example, a heat exchanger is designed to transfer heat from one fluid to another. There are two separate sections in a heat exchanger. One section contains the process fluid. The other section has another fluid, such as steam or chilled refrigerant. A change in the inlet temperature or flow rate of the process fluid through the heat exchanger requires a change in the steam or refrigerant flow rate in order to maintain the outlet temperature of the process fluid. **See Figure 35-4.** The only load changes that are a concern to a control system are those that can cause a change in the PV.

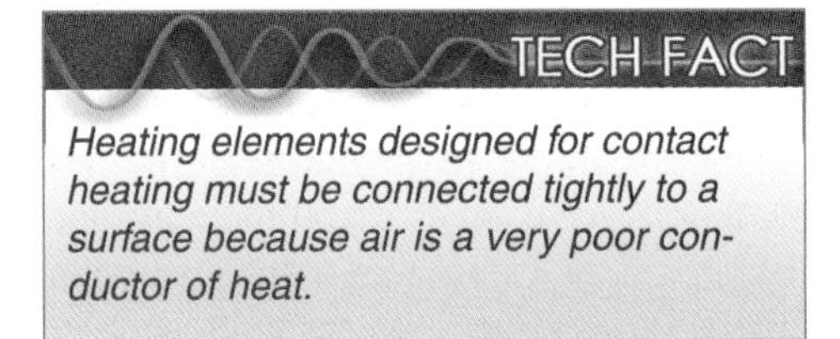

Heating elements designed for contact heating must be connected tightly to a surface because air is a very poor conductor of heat.

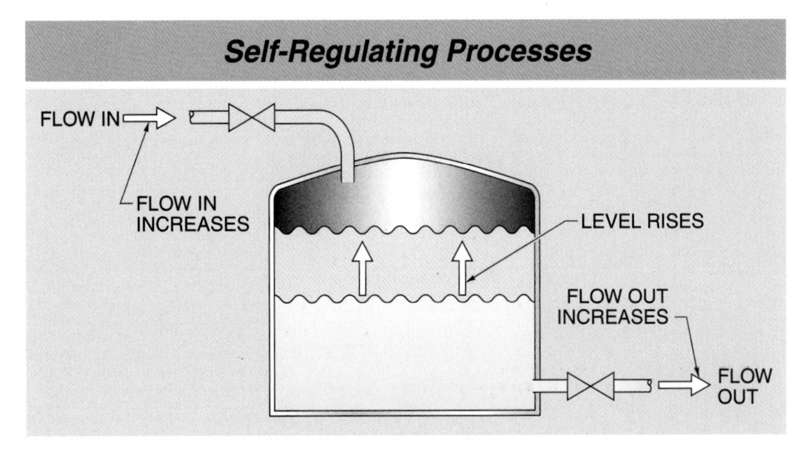

Figure 35-2. The flow in and out of a tank is a self-regulating process. The outlet flow changes when the inlet flow changes.

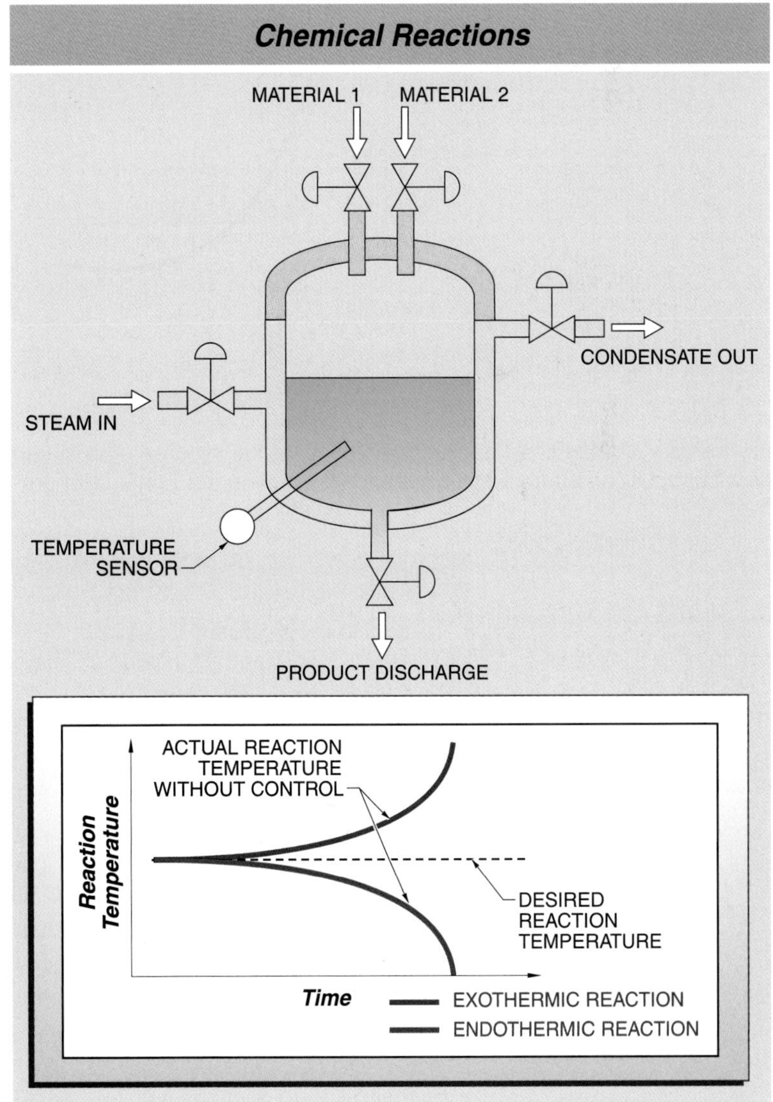

Figure 35-3. Exothermic and endothermic reactions in a process are not self-regulating.

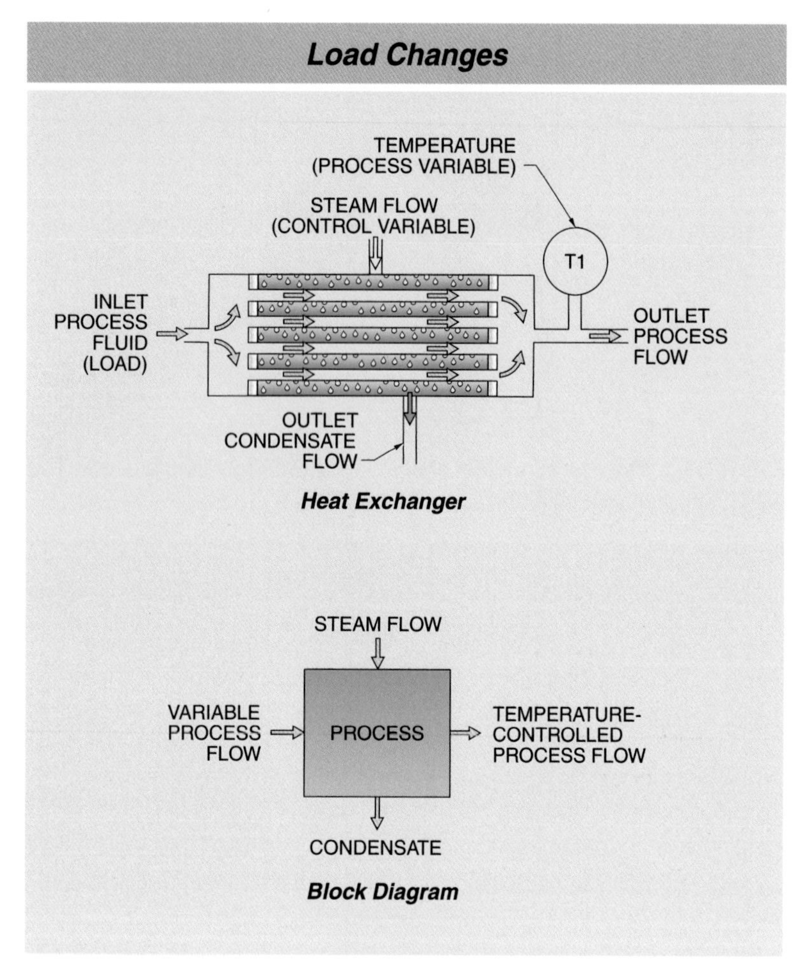

Figure 35-4. A load change causes an upset to the process. The steam flow rate must be changed to compensate for changes in the inlet process fluid.

The calculation-programming block in controller software can be used to enable the controller to use the correct values if a nonlinear sensor is used.

Process Gain

Gain is a ratio of the change in output to the change in input of a process. Gain can be measured using any unit. An example of gain is heating a process fluid with steam in a heat exchanger. The process fluid can be any fluid handled in the manufacturing process. The gain is the change in temperature of a process fluid through a heat exchanger due to a change in the steam flow rate through the other side of the heat exchanger. **See Figure 35-5.**

A *process curve,* or *process reaction curve,* is a plot of the process variable (PV) against the control variable (CV). For example, a process curve for a heat exchanger can be drawn by plotting the process fluid temperature (PV) against steam flow rate (CV) for a constant process fluid flow. The gain of the process at any combination of steam flow rate and process fluid temperature is equal to the slope of the curve at that point. The slope is determined by taking the difference of two temperatures on the curve divided by the difference of the two associated steam flow rates. In this case, the gain is as follows:

$$K = \frac{\text{change in output}}{\text{change in input}}$$

$$K = \frac{T_2 - T_1}{F_2 - F_1}$$

where

K = gain

T_1 = process fluid temperature at the low steam flow (in °F)

T_2 = process fluid temperature at the high steam flow (in °F)

F_1 = steam flow rate at low steam flow (in lb/hr)

F_2 = steam flow rate at high steam flow (in lb/hr)

For example, a heat exchanger is used to raise the temperature of a process fluid from 95°F (T_1) to 105°F (T_2) by increasing the flow of steam from 380 lb/hr (F_1) to 500 lb/hr (F_2). The gain is calculated as follows:

$$K = \frac{T_2 - T_1}{F_2 - F_1}$$

$$K = \frac{105 - 95}{500 - 380}$$

$$K = \frac{10}{120}$$

K = **0.083 F/(lb/hr)**

This means that for every 1 lb/hr change in steam flow, there is a corresponding 0.083°F change in temperature of the process fluid. It takes a 100 lb/hr change in steam flow to cause an 8.3°F change in temperature of the process fluid.

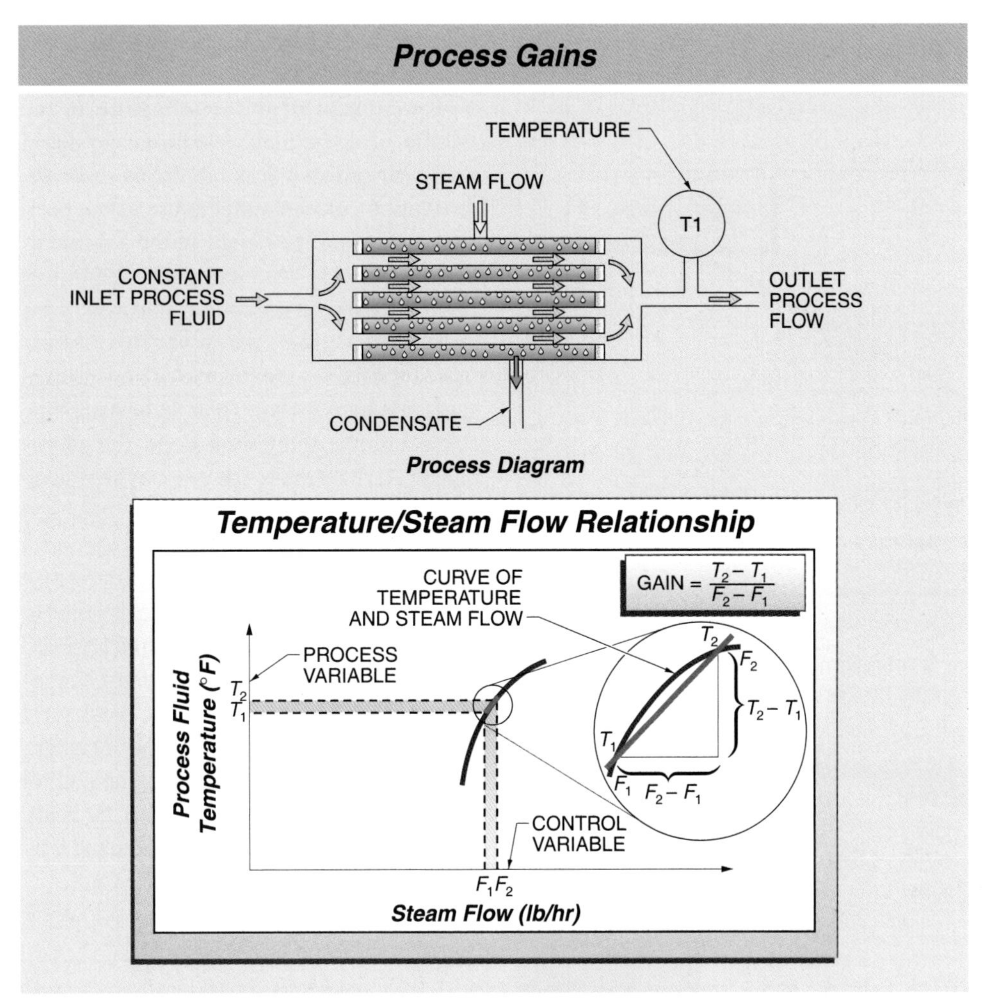

Figure 35-5. The gain of a process at a specific operating point can be determined by the slope of the process curve at that operating point.

Linear Process Gain. A *linear process* is a process where the gain at any value of the process variable (PV) is the same as the gain at any other value. The process curve for a linear process is a straight line. An example of a linear process is a heat exchanger where a change in the steam flow rate linearly changes the process fluid temperature. **See Figure 35-6.** Most processes are approximately linear, but there are some that are very nonlinear.

Nonlinear Process Gain. A *nonlinear process* is a process where the gain changes at different points on a process curve. In a nonlinear process, a plot of the PV against the CV is a curved line. An example of a nonlinear process is pH control. There is a nonlinear relationship between the pH value (the hydrogen ion concentration) and the amount of reagent required to change the pH of a solution. At the part of the graph with the steep slope, the gain is very high. This means that a small change in the CV, the amount of reagent added, causes a large change in the PV, the pH. At the parts of the graph with a shallow slope, it takes a large change in the CV to cause an equivalent change in the pH.

TECH FACT

The first computer ever used for a control system was the DIGITAC developed by Hughes Aircraft Company for the Air Force. This computer was first used in 1954 to control an airplane in flight.

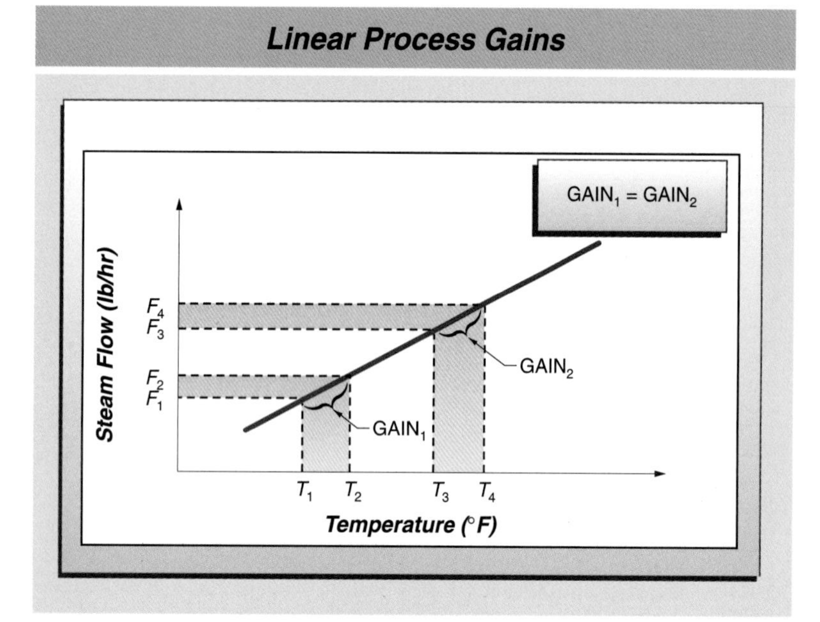

Figure 35-6. The process curve is a straight line for a linear process. This means that the gains at all the points on the curve are equal.

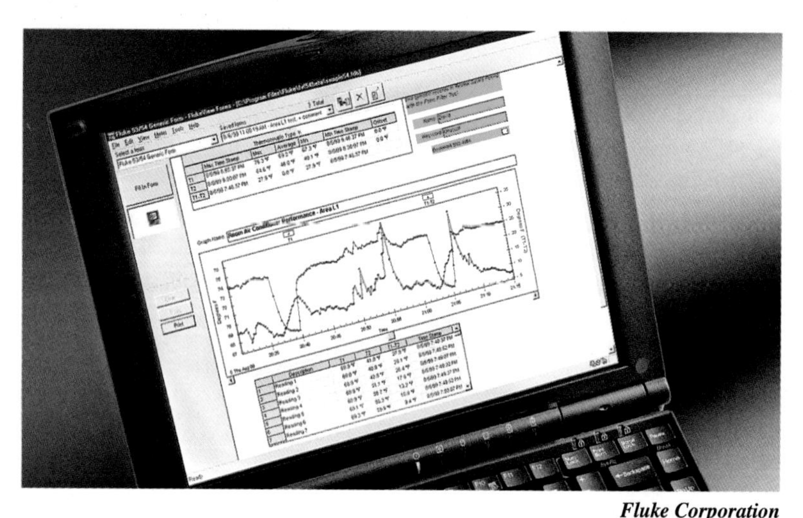

Fluke Corporation

A laptop computer can be used to monitor the operation of a control system.

For example, the control of boiler water pH is critical to the safe operation of a boiler. If the pH is too low, corrosion of the metal surfaces begins. The amount of reagent needed to change the pH by one unit depends on the starting pH. At the point on the curve where the gain is very high, the amount of reagent needed to change the pH by one unit is very small. At the point on the curve where the gain is lower, the amount of reagent needed is larger. **See Figure 35-7.** The operator needs to understand this relationship in order to manage the chemical additions used to control pH.

The control of the temperature in the middle of a distillation column is also a nonlinear process. It is typical to show the distillation column temperature as the horizontal axis and the height of the column as the vertical axis. An increase in steam flow raises the curve in the column and a decrease in steam flow lowers the curve. This orientation of the axis is the reverse of how normal gain curves are shown. Thus, at the part of the curve with the shallowest slope, the gain is higher. At the parts of the graph with a steep slope, the gain is lower.

Nonlinear processes are difficult to control because the gain of the process is different for every value of the PV. A controller is tuned to provide steady control at the desired setpoint. A disturbance to the process that changes the measured value can result in the control system either going into oscillation or becoming very sluggish. In either case, the control system may not be able to recover by itself. This is why it is necessary to understand how process gain can affect a control system.

Lags

A *lag* is a delay in the response of a process that represents the time it takes for a process to respond completely when there is a change in the inputs to the process. A lag is caused by the capacitance and resistance of a process. *Capacitance* is the ability of a process to store material or energy. Capacitance is present in processes with storage tanks, surge tanks, or piping systems with a large volume. *Resistance* is an opposition to the potential that moves material or energy in or out of a process. Resistance is commonly seen as the resistance to flow through ducts, pipes, and fittings.

Potential is a driving force that causes material or energy to move through a process. Potential may be fluid pressure, temperature difference, or electrical voltage. Fluid flows from high pressure to low pressure. Heat flows from high temperature to low temperature. Current flows from high voltage to low voltage.

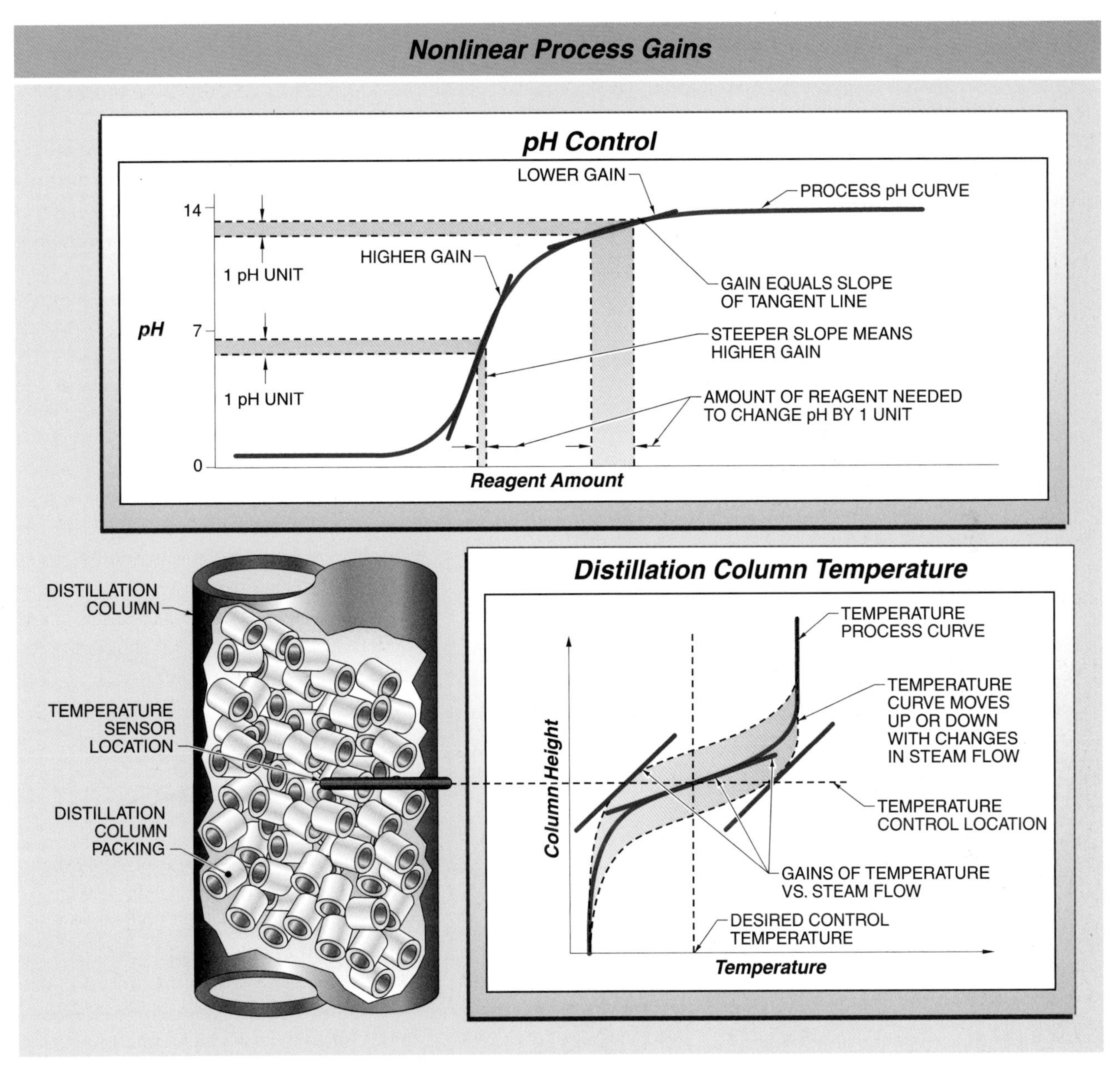

Figure 35-7. A nonlinear process has different gains at different points on a process curve.

The combination of a single capacitance and resistance results in the formation of a lag with a single time constant. A lag in a process with a single time constant has the same properties as a lag produced by a combination of resistance and capacitance elements in an electrical circuit. A measure of the capacity of a process is its response to a step change in the setpoint or CV. A *step change* is a sudden change in an input variable in a process that is managed by a controller. Typical step changes are a setpoint change or a load change.

TECH FACT

System response should be verified before implementing PID control algorithms. Some systems, such as steam heat, have quick changes in pressure and heat while others, such as hot water heat in a fan coil, change slowly.

Dimensionless Gain

Dimensionless gain is a gain expressed as a number without measurement units. The dimensionless gain of a process is calculated from the slope of the process curve adjusted for the ranges of the process variable and the control variable. Dimensionless gain is calculated by dividing the output and input values by their respective ranges as follows:

$$K = \frac{(T_2 - T_1)/(TR)}{(F_2 - F_1)/(FR)}$$

where

K = gain

T_1 = process fluid temperature at the low steam flow (in °F)

T_2 = process fluid temperature at the high steam flow (in °F)

F_1 = steam flow rate at low steam flow (in lb/hr)

F_2 = steam flow rate at high steam flow (in lb/hr)

TR = temperature range (in °F)

FR = steam flow range (in lb/hr)

The input to a process is the control variable. For this example, the input is the steam flow through a heat exchanger. The range of the control variable is the difference between the maximum and minimum steam flows that can be obtained by the control system.

The output of a process is the process variable. For this example, the output is the temperature of the process liquid exiting the heat exchanger. The range of the process variable is the difference between the maximum and minimum temperatures obtained at the maximum and minimum steam flows.

For example, for T_1 = 95°F, T_2 = 105°F, F_1 = 500 lb/hr, and F_2 = 380 lb/hr, with a temperature range of 0°F to 200°F and a steam flow range of 0 lb/hr to 1000 lb/hr, the calculation of the dimensionless gain is as follows:

$$K = \frac{(T_2 - T_1)/(TR)}{(F_2 - F_1)/(FR)}$$

$$K = \frac{(105 - 95)/(200 - 0)}{(500 - 380)/(1000 - 0)}$$

$$K = \frac{10/200}{120/1000}$$

$$K = \frac{0.05}{0.12}$$

K = **0.42**

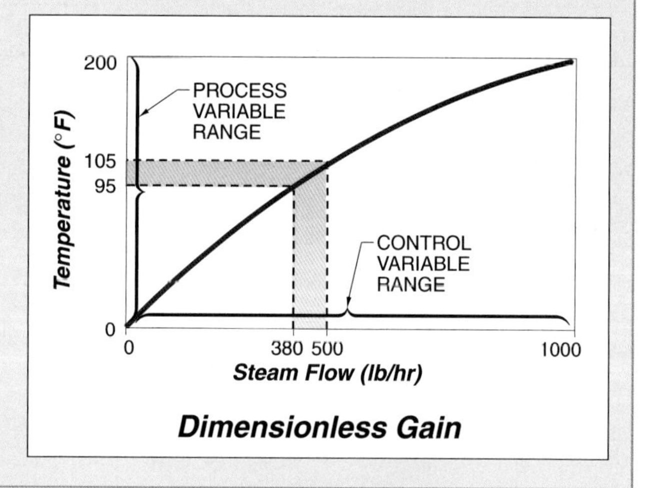

Dimensionless Gain

For a process with a single time constant, the time constant can be determined from the plotted response of the process to a step change. A *time constant* (τ) is the time required for a process to change by 63.2% of its total change when an input to the process is changed. This is also called the first-order time constant. As the response continues, it comes closer and closer to the final value caused by the step change. When the time is equal to five time constants, the response is 99.3% of the final output change. This means that it takes a time equal to one time constant for a process output to change by 63.2% of the total amount it will change in response to a step change in input. **See Figure 35-8.**

For a water tank with a drain valve at the bottom and an inlet water source that can vary in flow, the outlet flow is related to the level in the tank. As the level in the tank increases, the flow out of the tank also increases. A step increase in the inlet flow increases the level until the outlet flow equals the inlet flow. A plot of the level over time is a single-time-constant response curve. Flow out of a tank is proportional to the square root of height. Therefore, if the water level in the tank is doubled, the total outflow increases by $\sqrt{2}$, or by a factor of 1.414.

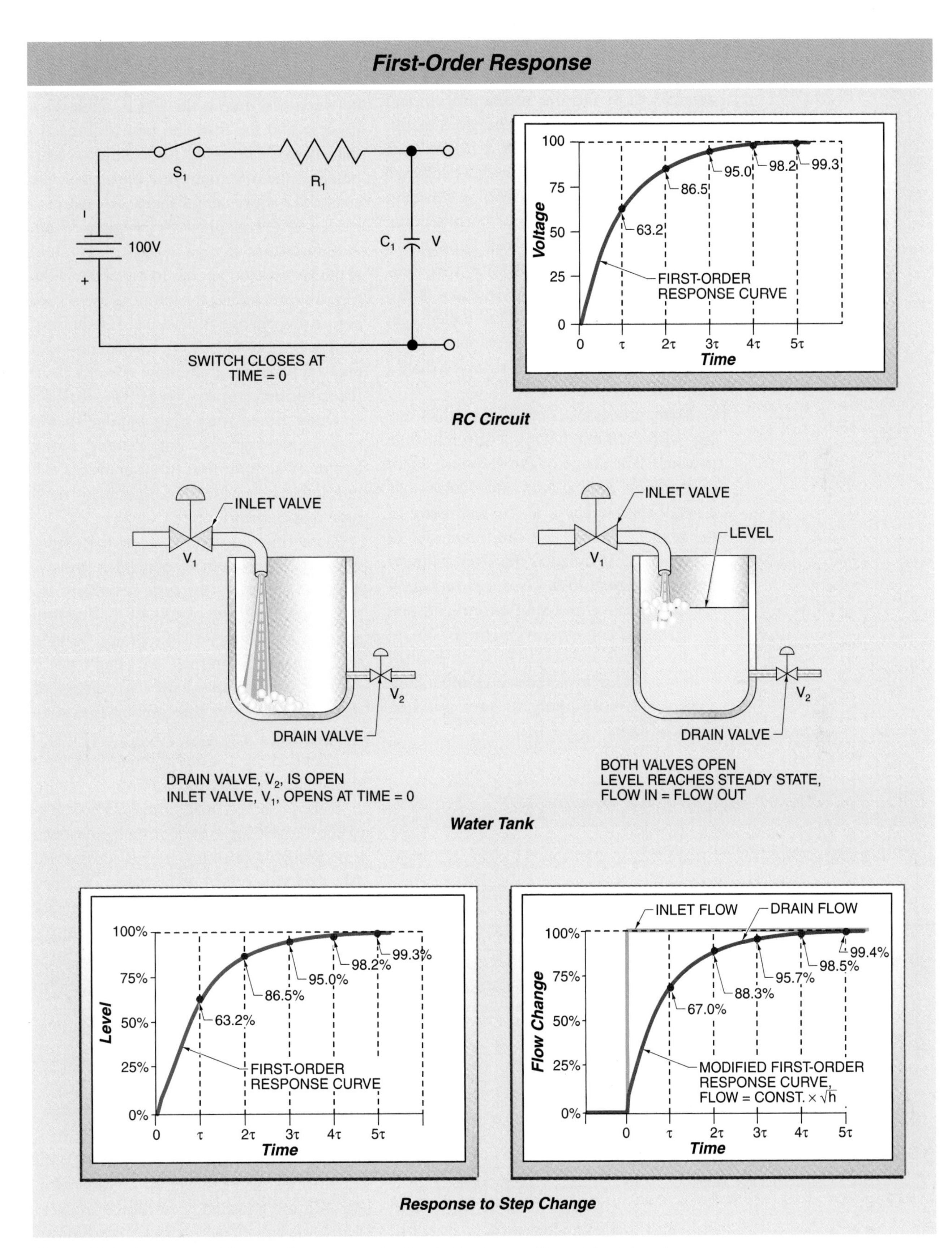

Figure 35-8. A process with a single time constant is similar to an RC electrical circuit.

Since the time constant of a process is a measure of the capacity of a system, it is determined by the size of the process and the rate of material and energy transfer. Therefore, a process with a larger time constant responds more slowly to a change than a process with a smaller time constant. For example, a thick-wire thermocouple has more capacity to absorb heat than a thin-wire thermocouple. A thick-wire thermocouple has a larger time constant than a thin-wire thermocouple. This means that a thick-wire thermocouple responds more slowly to a change in temperature than a thin-wire thermocouple.

Many processes contain more than one lag, with each one having a different time constant. The shape of the response curve of a process having two capacitances has a small reverse curve at the beginning of the response curve and the remainder of the response is dragged out over a longer time. **See Figure 35-9.** The simple equation for the response of a single-time-constant process does not apply when there is more than one time constant. Process control systems that are involved with temperature control are more likely to have multiple time constants.

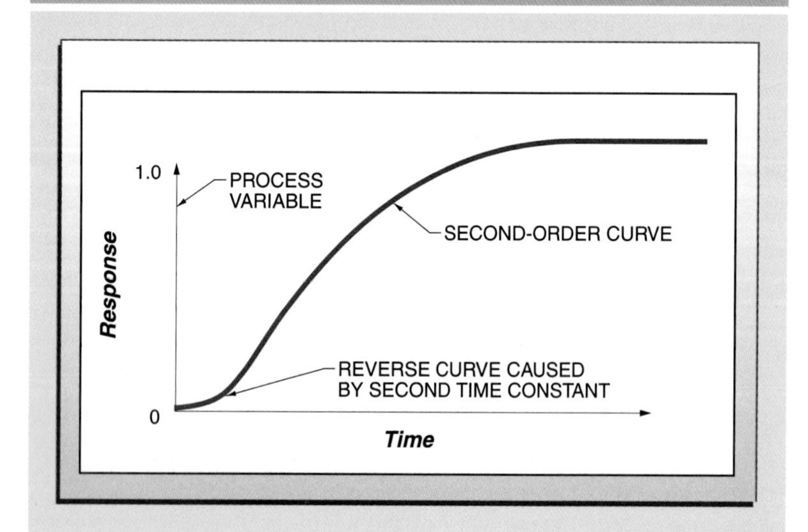

Figure 35-9. A process with more than one time constant has a reverse curve at the beginning of the response.

Dead Time

Dead time is the period of time that occurs between the time a change is made to a process and the time the first response to that change is detected. An example is temperature measurement in a pipe when the sensor is a significant distance downstream from a heat exchanger. **See Figure 35-10.** A temperature change in the process fluid at the heat exchanger has to travel down the pipe until it reaches the temperature sensor before any change is detected. If a thermocouple is mounted 100 ft downstream of a heat exchanger and the flow velocity is 20 ft/sec, the dead time is 5 sec ($100 \div 20 = 5$).

Dead times may also appear in the measurement process. For example, a long length of sample line to an analyzer can introduce a considerable amount of dead time to a control loop.

Dead time adversely affects the ability of a control system to control a process because it extends the time between control actions and the results of that control action. Everything possible should be done to eliminate or minimize the dead time of a process. Sometimes a process consists of many small lags that can combine to act like a dead time. These have the same detrimental effect on the control of the process as a single, larger dead time.

If the observed dead time is 25% or less of the largest lag time constant, the dead time probably does not have a detrimental effect on the control performance. A dead time greater than 25% of the largest lag time constant starts degrading the controllability of the system. **See Figure 35-11.** Sensors, transmitters, and control valves also have gains and lag times, although the lag times are usually small compared to the process lags.

Measurement Signal Conditioning

Transmitters and controllers are increasingly available with signal conditioning capabilities. Typical signal conditioning capabilities are analog-to-digital conversions, signal amplification, integration, square root extraction, and scaling. Some types of signal conditioning can have an impact on the total loop gain and lags. These features can cause serious control problems if used improperly. Common signal conditioning capabilities that can cause changes in loop gain and lags are transmitter dynamic gain, measurement signal filtering, and some types of measurement conditioning, such as peak picking and peak hold.

Transmitter Dynamic Gain. *Transmitter dynamic gain* is the amount of output change from a transmitter for a specific input change. The choice of the measurement range affects the transmitter dynamic gain. A measurement range that is selected to be much wider than necessary for the normal operating range reduces the milliamp output change for a change in temperature measurement.

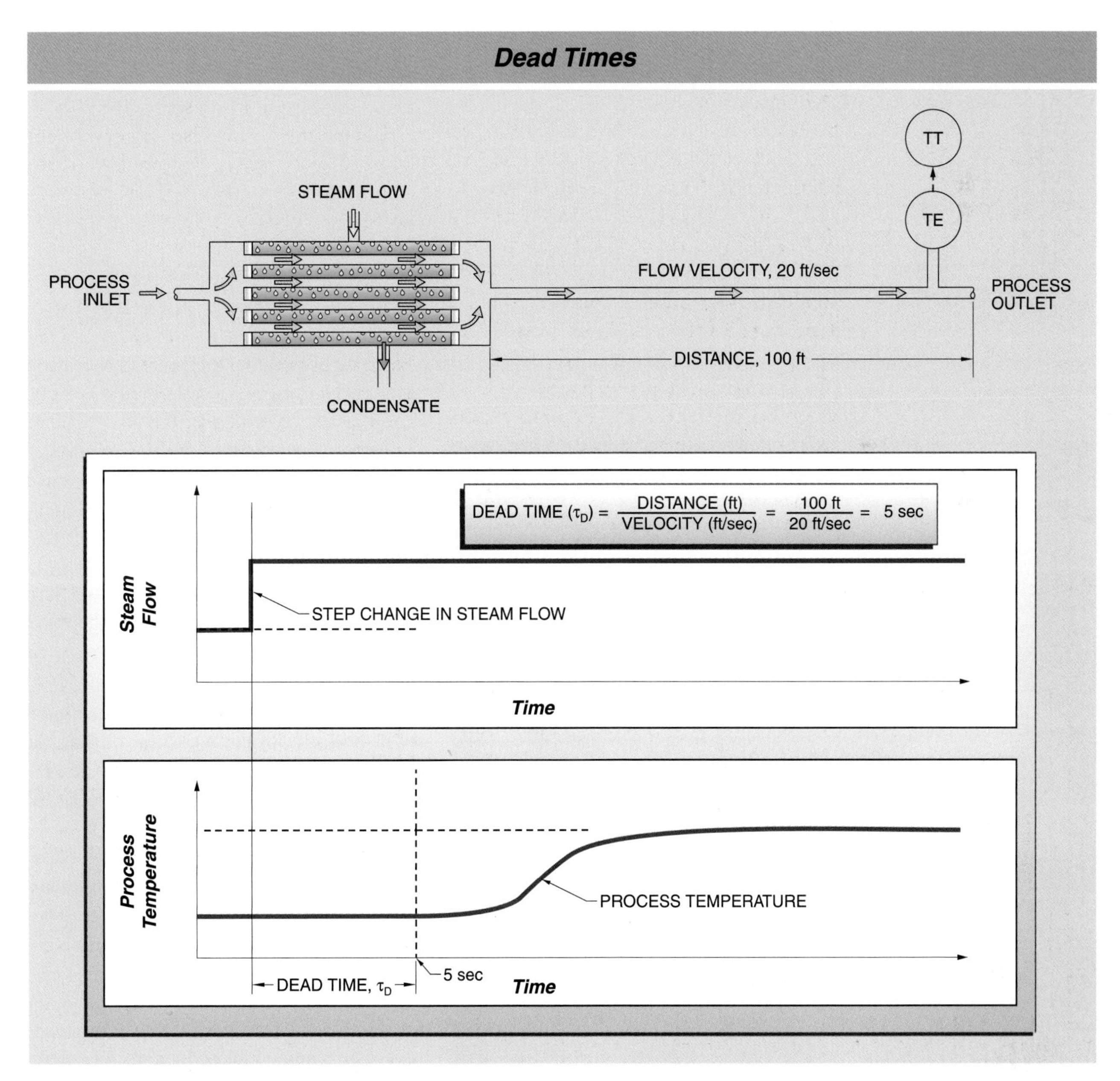

Figure 35-10. Dead time is the period of time between when a change occurs and when it is first detected. Multiple small time constants act as dead time.

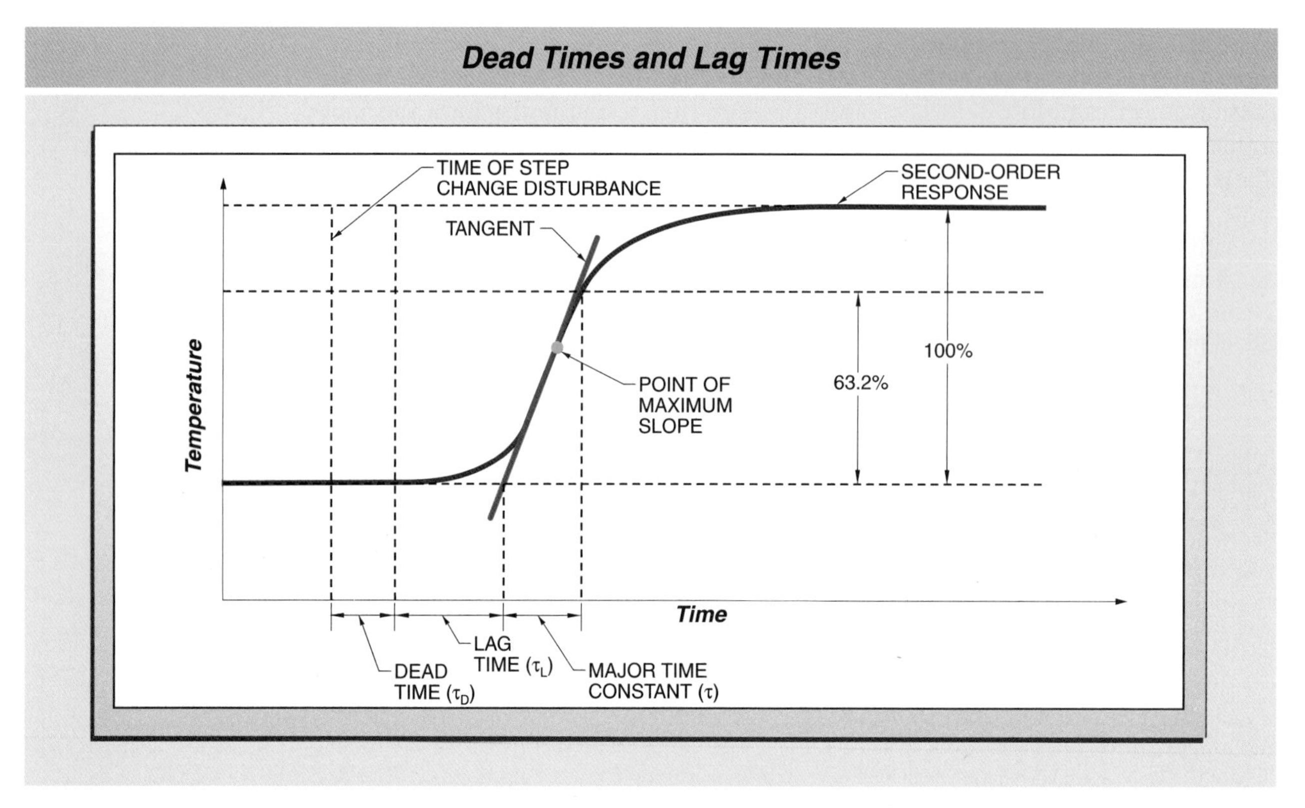

Figure 35-11. If the dead time is larger than 25% of the largest time constant, it can be detrimental to the controllability of the process.

Many instruments are used to measure and record the operating conditions of a boiler.

An example of transmitter dynamic gain can be seen with a thermocouple instrument that has a nominal measurement range of 0°F to 1600°F. **See Figure 35-12.** This means that the 4 mA to 20 mA output range corresponds to the full 0°F to 1600°F input temperature range. The lowest temperature, 0°F, corresponds to 4 mA, the lowest current output. The highest temperature, 1600°F, corresponds to 20 mA, the highest current output. A 100°F change in temperature from 250°F to 350°F results in a 1.00 mA change in output from 6.5 mA to 7.5 mA.

If normal process temperatures are in the range of 250°F to 410°F, a measurement range is selected to match that range. The lowest temperature, 250°F, corresponds to 4 mA and 410°F corresponds to 20 mA. This means that the same 100°F change in temperature results in a 10.00 mA change in output, a much higher dynamic gain than in the first case. This greatly improves the resolution of the temperature measurement.

In addition, the accuracy of the instrument is increased, since accuracy is based on the full-scale measurement range.

Measurement Signal Filtering. Transmitters and controllers frequently contain the ability to filter, damp, or average the measurement. This is often done to smooth out high-frequency noisy measurements to improve the control of the process. However, if too much damping is selected, it can be detrimental to the control of the process. In addition to transmitter signal filtering, the process itself can introduce damping. For example, pulsation dampers, needle valves, or long lengths of tubing between the process and the sensor can inadvertently introduce damping.

Measurement Conditioning. Measurement conditioning, such as peak picking or peak hold, is used on some infrared pyrometer transmitters. Peak picking causes the output of the transmitter to hold the highest sensed temperature for a period of time until the next high temperature is sensed. This introduces a special type of dead time to the control loop.

Peak picking is used to measure the temperature of discrete parts being heated by a furnace. **See Figure 35-13.** The temperature of each piece is held in memory until the next piece passes the IR detector. This solves the problem of the furnace burners being changed because the temperature decreases between discrete pieces.

CONTROL FUNCTIONS

Controllers are made up of various functions, such as adjustable setpoints, setpoint tracking, manual output, and bumpless transfer, which are designed to work together to provide the desired control actions. Not all controllers provide the same functions. Some controllers are relatively simple and do not offer all of the functions. Other controllers are very complex, with numerous options available to the user. Most of these functions are also available in pneumatic or electric controllers, but may take different forms.

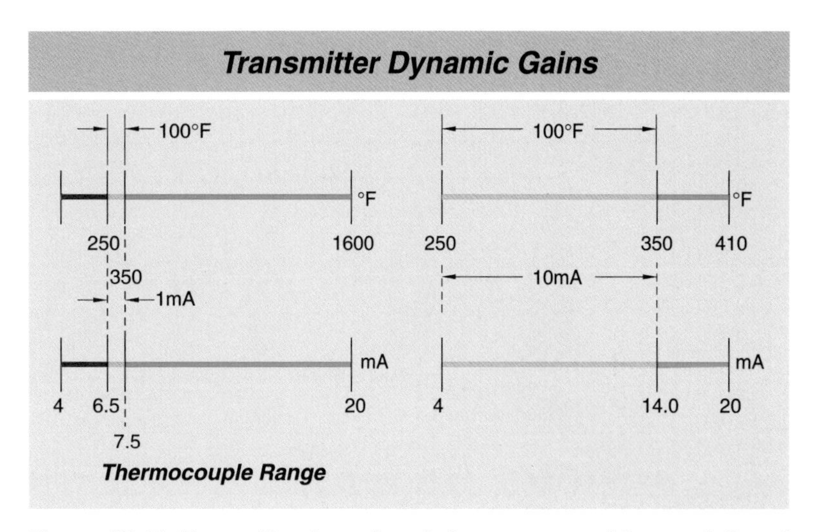

Figure 35-12. Transmitter dynamic gain is a measure of the resolution of a transmitter.

Setpoints

A *setpoint (SP)* is the desired value at which a process should be controlled and is used by a controller for comparison with the process variable (PV). The setpoint can be manually entered into the controller or automatically entered by a remote system. An example of a setpoint is the temperature setting of a room thermostat. The engineering units of the setpoint are the same as for the measurement. All controllers must have a setpoint to be able to provide a control function.

Error is the difference between a process variable and a setpoint. The use of the term "error" does not imply that there is a mistake or inaccuracy in the measurement. It simply means the difference between the PV and the SP. Controllers use the error to calculate the output to a final element. *Offset* is a steady-state error that is a permanent part of a system. Offset has occasionally been used instead of error to describe the difference between the PV and the SP.

TECH-FACT

James Prescott Joule (1818-1889), an English physicist, was the first to determine the relationship between heat energy and mechanical energy and established the mechanical theory of heat.

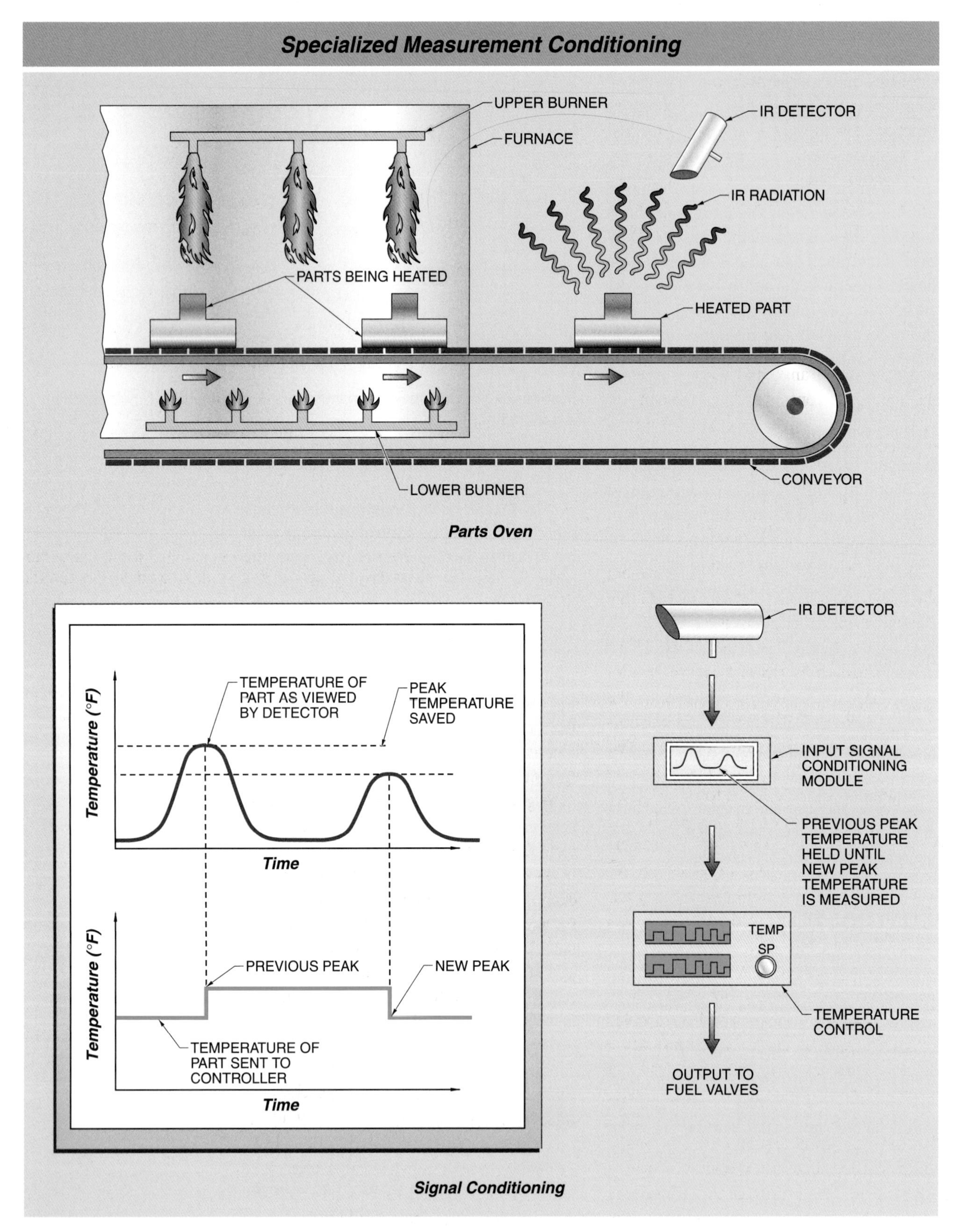

Figure 35-13. Signal conditioning modifies a measured value in order to simplify control.

Feedback

Feedback is a control design used where a controller is connected to a process in an arrangement such that any change in the process is measured and used to adjust action by the controller. **See Figure 35-14.** In this arrangement, the controller continuously monitors the results of its actions. The vast majority of all control systems use feedback.

Control Loops

A *control loop* is a control system in which information is transferred from a primary element to the controller, from the controller to the final element, and from the final element to the process. A *closed loop* is a control system that provides feedback to the controller on the state of the process variable due to changes made by the final control element.

For example, a conveyor fills a hopper above a grinder with process material that must be properly sized. **See Figure 35-15.** The hopper has a level controller. If the hopper starts to fill faster than the grinder takes away the material, the level in the hopper rises. As the level rises, the controller detects the increase in level (PV). The level controller adjusts its output signal to a variable speed drive (CV) to slow down the conveyor feeding the hopper. If the hopper level drops, the level controller adjusts its output signal to speed up the conveyor. The controller continuously adjusts the speed of the conveyor to maintain the level at the desired point.

Another example of a closed loop is a thermostat controlling a hot water room heater. The thermostat controller receives the measurement of the air temperature (PV) and makes adjustments to the hot water valve to adjust the flow of hot water (CV) through the radiator. Changes in the flow of hot water cause changes in the room temperature that are sensed by the thermostat. The controller continuously adjusts the water valve to maintain the room temperature at the desired point.

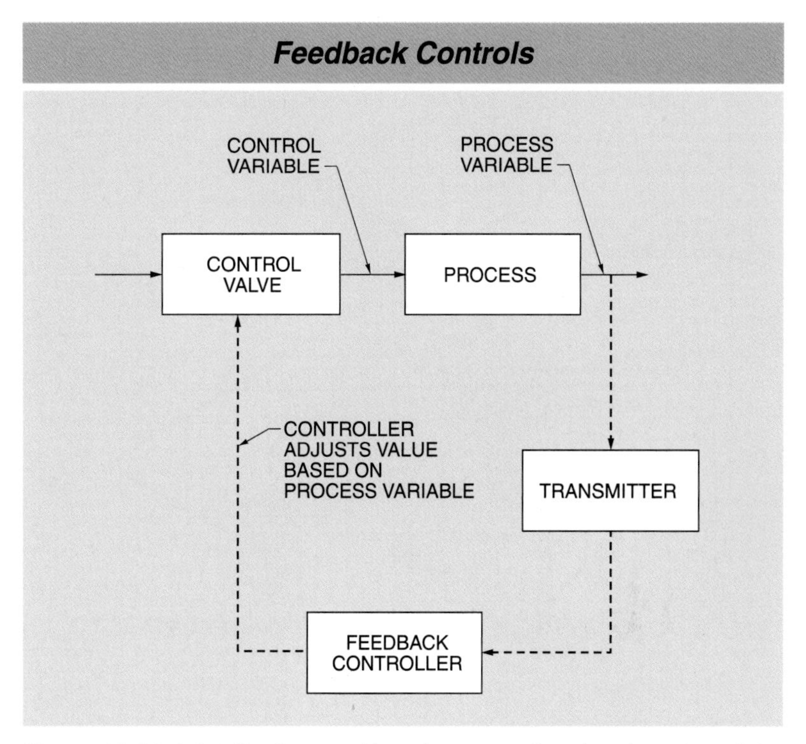

Figure 35-14. A feedback control loop is arranged so that the value of the process variable is sensed by the controller and used to adjust the control variable.

An *open loop* is a control system that sends a control signal to a final element but does not verify the results of that control. For example, a conveyor fills a hopper above a grinder. There is no feedback on the status of the grinder or the hopper because there is no control loop and the speed is manually controlled or not controlled at all. An open loop may also exist because the transfer of information is interrupted at some point in the loop. The usual point of this interruption is at the output of the controller. When a controller is in the manual mode, the normal output of the controller is blocked and a manually adjusted output signal is used to adjust the CV.

An *open loop process response curve* is a graph of the results of a step change in the manually adjusted output signal of an open loop controller that results in a change in the process measurement. The resulting curve provides the information to determine the process attributes of gain, lag, and dead time.

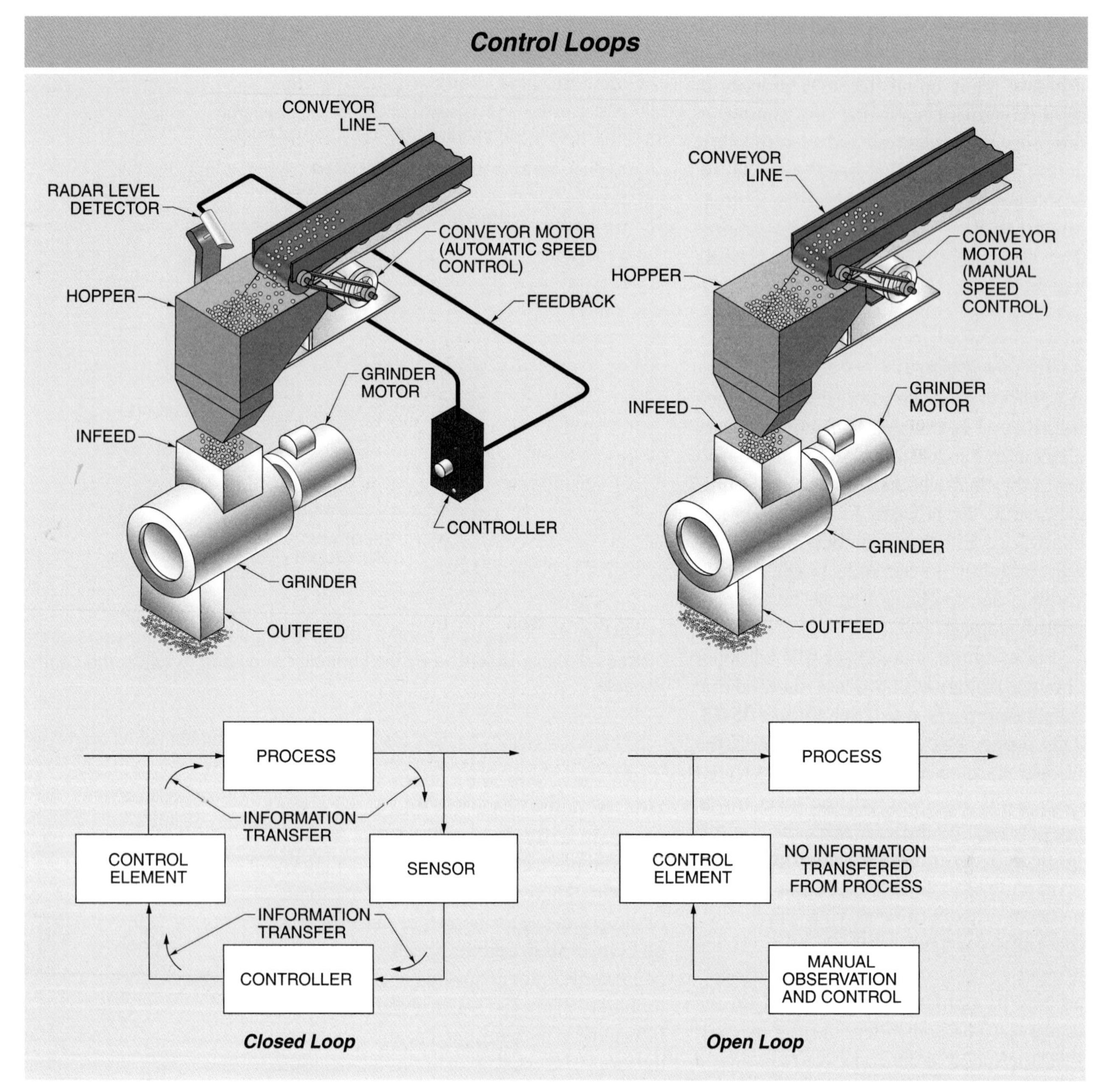

Figure 35-15. A closed loop is a loop in which a controller automatically controls a process, and an open loop is a loop in which there is no automatic control.

Control Actions

A controller can be configured to use either direct or reverse action. *Direct action* is a form of control action where the controller output increases with an increase in the measurement of the process variable (PV). *Reverse action* is a form of control action where the controller output decreases with an increase in the measurement of the process variable (PV). **See Figure 35-16.**

The control action for a digital controller is selected during configuration. An electric controller has a switch to change the action. Selecting the correct controller action is absolutely essential for the control system to work properly. The proper controller action depends on the type of process and the final element failure action. A change in the measurement of the PV must result in an output change to the control valve that brings the PV back to the setpoint. For example, valves may be designed to either open or close when the controller signal increases. Selecting the wrong controller action results in the controller output winding up to either 0% or 100% and not changing from that position.

Some single-loop controllers have a toggle switch that can be used to select the desired action of the controller. The control switch on a home thermostat connected to a furnace and air conditioner is an example of this type of switch. During the winter, when the switch is set to heat, the system operates in a reverse-acting mode. During the summer, when the switch is set to cool, the system operates in a direct-acting mode. The closed loop system remains the same, except for the behavior of the controller. The process behavior, which changes from winter to summer, necessitates the switch from reverse to direct action.

Direct Action. A water chiller, where water is chilled by a refrigerant, has a fail-closed vapor return control valve so that an air supply or control signal failure closes the valve. Opening the valve releases the evaporating vapor and lowers the pressure of the refrigerant and its equilibrium temperature. This lowers the chilled water outlet temperature. A drop in chilled water outlet temperature requires an increase in the equilibrium vapor pressure of the refrigerant to increase the outlet water temperature. An increase in the equilibrium vapor pressure of the refrigerant is accomplished by closing the control valve. Therefore, the controller has to use direct action.

A backpressure control system with a fail-closed control valve also has to use direct action. An increase in pressure requires the control valve to open, which is accomplished with an increase in output. Thus the controller has to use direct action.

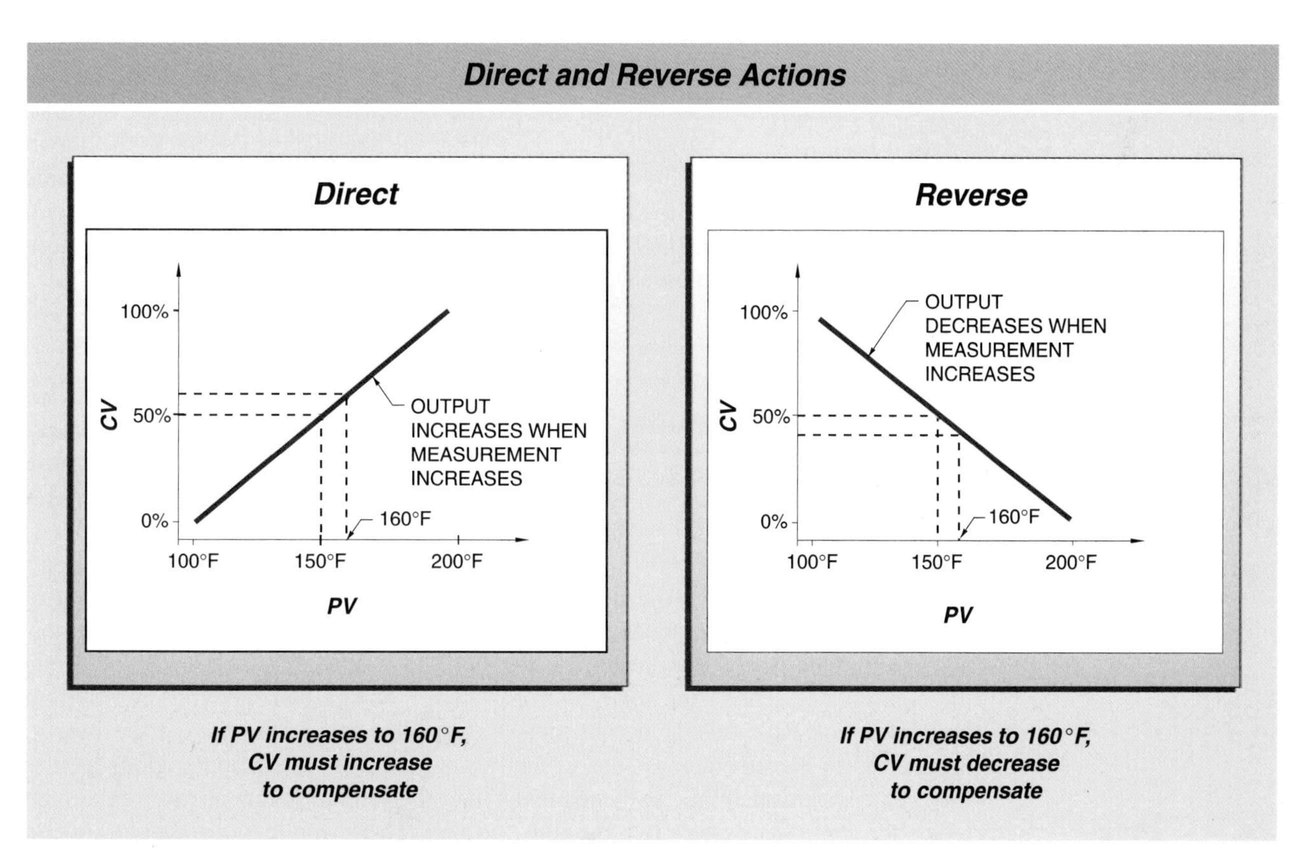

Figure 35-16. Controllers can be configured for direct or reverse action.

A clock gauge is an instrument used for measuring level. The hands represent feet and inches of depth.

Reverse Action. Most processes require reverse-acting controllers. Many valves are configured to be fail-closed for safety reasons. For processes that require an increase in the CV and a fail-closed valve, reverse action is required.

For example, a heating process that uses a fail-closed steam control valve needs to use reverse action. An increase in the process temperature above the setpoint must close the heating control valve to reduce the amount of heating. This requires a decrease in the output. Thus the controller must use reverse action. **See Figure 35-17.**

A cooling process that uses a fail-open control valve to throttle water to a heat exchanger also has to be reverse-acting. An increase in temperature requires an increase in cooling water, which is obtained by a decreased output. Thus the controller must use reverse-action.

Automatic and Manual Control

Controllers usually have the ability to operate in automatic or manual mode. In automatic mode, an output developed by the controller is sent to the final control element. In manual mode, an output value determined by the operator is sent to the final control element. While in the manual mode, the controller output from the automatic actions is blocked by an open circuit.

Older controllers experienced a change, or bump, in the output when switching from automatic to manual or back again unless a manual balancing procedure was followed. This bump was caused by the manual and automatic output being at different values. Modern controllers usually have bumpless transfer techniques incorporated into their design.

Bumpless transfer is a controller function that eliminates any sudden change in output value when the controller is switched from automatic to manual mode or back again. This is accomplished by the use of two memory and tracking functions. **See Figure 35-18.**

When in automatic mode, the controller output is fed back to the manual output memory module. Thus the manual value tracks the output of the controller. Since the manual value is always the same as the controller output, there is no bump during the transfer from automatic to manual.

When in the manual mode, the output from the controller functions is overridden by the controller output (the manual value). The manual output is fed back to the automatic output memory module. Thus the automatic value tracks the output of the controller. Since the automatic value is always the same as the controller output, there is no bump during the transfer from manual to automatic.

Setpoint Tracking

Modern controllers offer the choice of whether or not to use the setpoint tracking technique. *Setpoint tracking* is the technique of storing the PV in the setpoint memory module while the controller is in manual. The purpose of the setpoint tracking is to make the setpoint equal to the measurement so that when the controller is switched back to automatic mode there is no deviation and thus no resulting controller actions to change the output. Some control system applications require that the setpoint value always remain the same. These applications should not use the setpoint tracking feature.

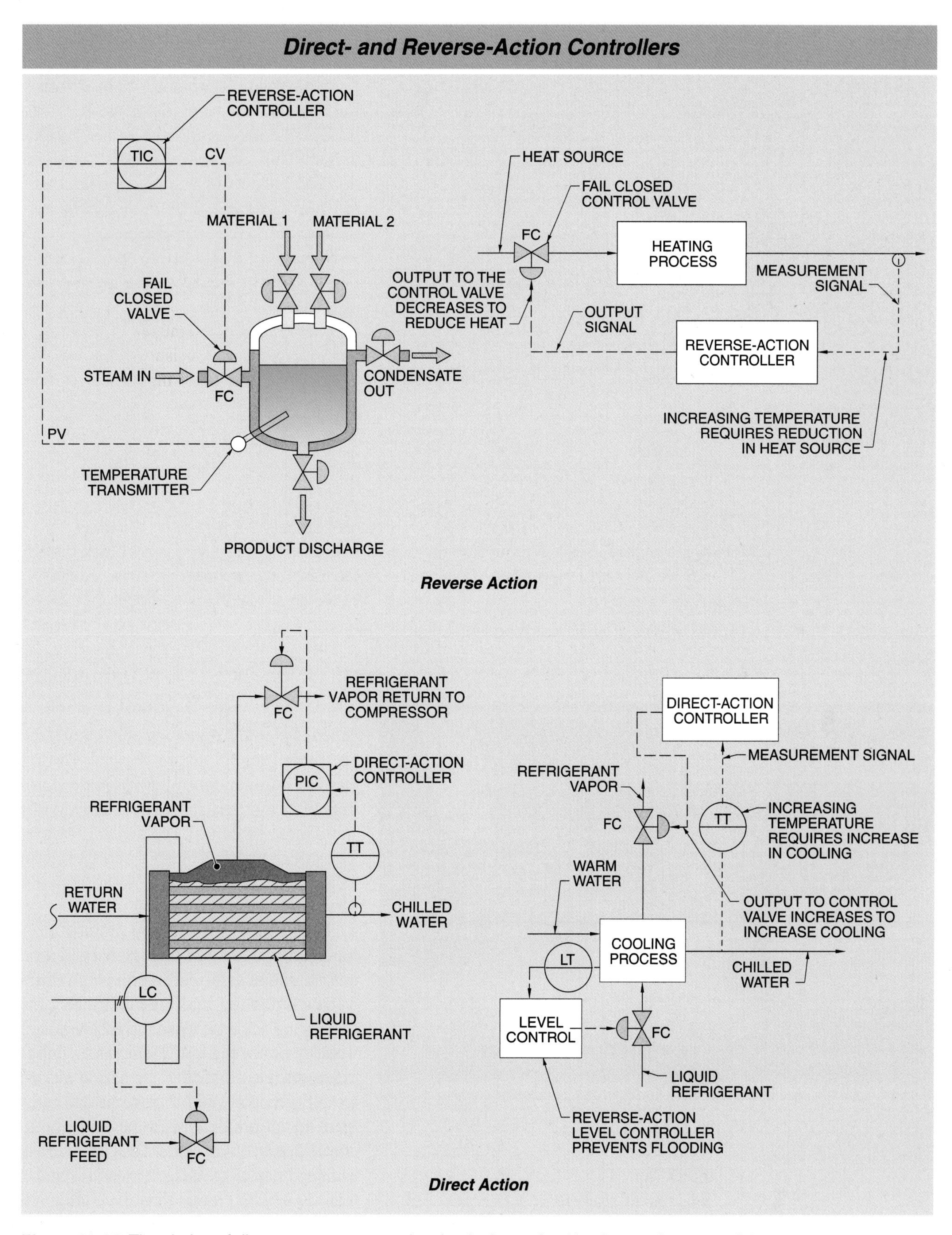

Figure 35-17. The choice of direct or reverse control action is determined by the requirements of the process.

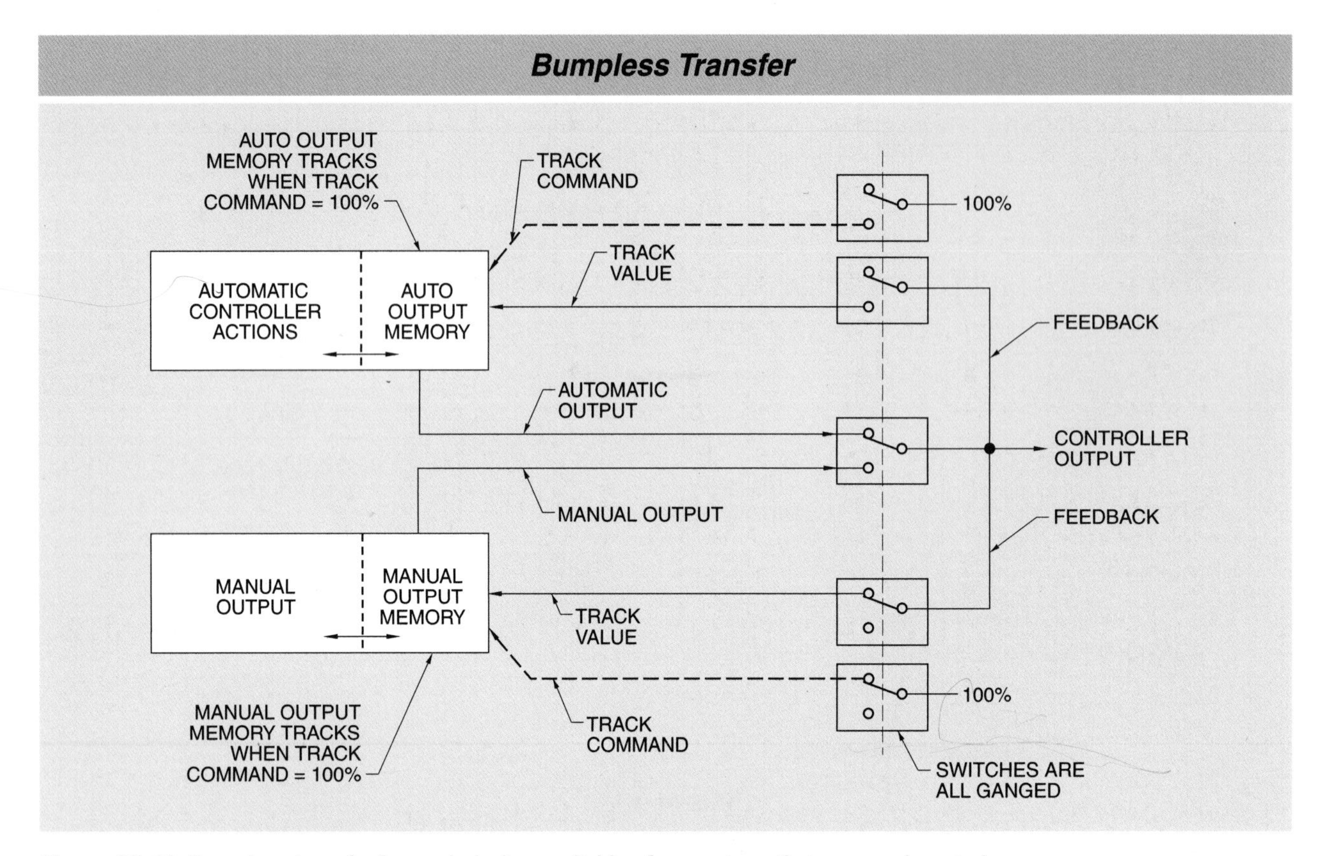

Figure 35-18. Bumpless transfer is needed when switching from automatic to manual control.

A control system may use a control valve to control the operation of a heat exchanger.

Alarms

An *alarm function* is the process of monitoring the state of a condition or a variable and comparing it to a preselected condition or value. Appropriate action is taken to alert the operating personnel if the monitored conditions are satisfied. The upper level digital controllers provide alarm functions that are used to actuate relays or solid-state switches. The switch contacts can either open or close when the alarm is actuated. Alarm switches can be high, low, or deviation.

High alarms monitor the process measurement and the alarm is actuated when the measurement value exceeds the high alarm value configured into the controller. Low alarms are actuated when the measurement value is below the low alarm value configured into the controller. Deviation alarms usually compare the PV to the SP and compare the absolute value of the deviation to the deviation alarm value. Thus the alarm is actuated when the measurement deviation, high or low, is greater than the configured value. The advantage of the deviation alarm is that it is functional at any setpoint.

KEY TERMS

- *automatic control:* The equipment and techniques used to automatically regulate a process to maintain a desired outcome.
- *process variable (PV):* The dependent variable that is to be controlled in a control system.
- *control variable (CV),* or *manipulated variable:* The independent variable in a process control system that is used to adjust the dependent variable, the process variable.
- *controller:* A device that compares a process measurement to a setpoint and changes the control variable (CV) to bring the process variable (PV) back to the setpoint.
- *primary element:* A sensing device that detects the condition of the process variable.
- *final element:* A device that receives a control signal and regulates the amount of material or energy in a process.
- *exothermic reaction:* A chemical reaction that generates heat during the reaction and increases the temperature.
- *endothermic reaction:* A chemical reaction that consumes heat, and therefore more heat energy must be added to sustain the reaction.
- *process dynamics:* The attributes of a process that describe how a process responds to load changes imposed upon it.
- *load change:* A change in process operating conditions that changes the process variable (PV) and must be compensated for by a change in the control variable (CV).
- *gain:* A ratio of the change in output to the change in input of a process.
- *process curve,* or *process reaction curve:* A plot of the process variable (PV) against the control variable (CV).
- *linear process:* A process where the gain at any value of the process variable (PV) is the same as the gain at any other value.
- *nonlinear process:* A process where the gain changes at different points on a process curve.
- *lag:* A delay in the response of a process that represents the time it takes for a process to respond completely when there is a change in the inputs to the process.
- *capacitance:* The ability of a process to store material or energy.
- *resistance:* An opposition to the potential that moves material or energy in or out of a process.
- *potential:* A driving force that causes material or energy to move through a process.
- *step change:* A sudden change in an input variable in a process that is managed by a controller.
- *time constant (τ):* The time required for a process to change by 63.2% of its total change when an input to the process is changed.
- *dead time:* The period of time that occurs between the time a change is made to a process and the time the first response to that change is detected.
- *transmitter dynamic gain:* The amount of output change from a transmitter for a specific input change.
- *setpoint (SP):* The desired value at which a process should be controlled and is used by a controller for comparison with the process variable (PV).
- *error:* The difference between a process variable and a setpoint.

KEY TERMS *(continued)*

- *offset:* A steady-state error that is a permanent part of a system.
- *feedback:* A control design used where a controller is connected to a process in an arrangement such that any change in the process is measured and used to adjust action by the controller.
- *control loop:* A control system in which information is transferred from a primary element to the controller, from the controller to the final element, and from the final element to the process.
- *closed loop:* A control system that provides feedback to the controller on the state of the process variable due to changes made by the final control element.
- *open loop:* A control system that sends a control signal to a final element but does not verify the results of that control.
- *open loop process response curve:* A graph of the results of a step change in the manually adjusted output signal of an open loop controller that results in a change in the process measurement.
- *direct action:* A form of control action where the controller output increases with an increase in the measurement of the process variable (PV).
- *reverse action:* A form of control action where the controller output decreases with an increase in the measurement of the process variable (PV).
- *bumpless transfer:* A controller function that eliminates any sudden change in output value when the controller is switched from automatic to manual mode or back again.
- *setpoint tracking:* The technique of storing a PV in the setpoint memory module while the controller is in manual.
- *alarm function:* The process of monitoring the state of a condition or a variable and comparing it to a preselected condition or value.

REVIEW QUESTIONS

1. What are the three components of automatic control in an automatic control system?
2. Define gain and explain the difference between the gain in linear and nonlinear processes.
3. Why do lag and dead time make it more difficult to control a process?
4. What is the difference between a closed loop and an open loop in a control system?
5. Define direct action and reverse action, and explain why it is important to know the correct type of action to control a process.

CONTROL STRATEGIES

A *control strategy* is a method of selecting the response of a controller to produce the desired outputs that maintain a process at a setpoint. The choice of control strategy is determined by the dynamics of the process to be controlled. Common control strategies are ON/OFF control, proportional control, and feedforward control.

ON/OFF Control

ON/OFF, or *two-position, control* is a method of changing the output of a controller that provides only an ON or OFF signal to the final element of the process. The two states are arranged on either side of the setpoint. An ON/OFF controller is typically used for systems that have a large storage capacity and can tolerate some deviations from the setpoint.

For example, a room thermostat is usually an ON/OFF controller. Air temperature deviations from the setpoint can be tolerated and the amount of energy stored in the air is usually large enough that the system responds slowly when the furnace is OFF. Common examples of ON/OFF devices are air conditioning compressors, electric heating stages, gas valves, refrigeration compressors, and constant-speed fans.

ON/OFF controllers usually have discrete switches as the output element. The switches are often Form C single-pole double-throw contacts. This arrangement allows either the normally open (NO) or the normally closed (NC) contacts to be used for the output. The selection of the correct output contacts is necessary for the controller to properly control the process.

For example, ON/OFF control is used in a heating process where the output switch turns on a heater when the temperature is below the setpoint. Thus, the switch contacts to be used for this application are the NC contacts so that the heater runs constantly except when the temperature is above the setpoint. An air conditioning application uses the NO contacts to turn on the air conditioner circuit to cool off a room so that the air conditioner is always off except when the temperature is above the setpoint. **See Figure 36-1.**

TECH FACT

Transmitters are selected so that the variable they sense falls near the middle of their range. HVAC mixed air applications have temperatures near 55° F and use transmitters with a range of 0° F to 100° F.

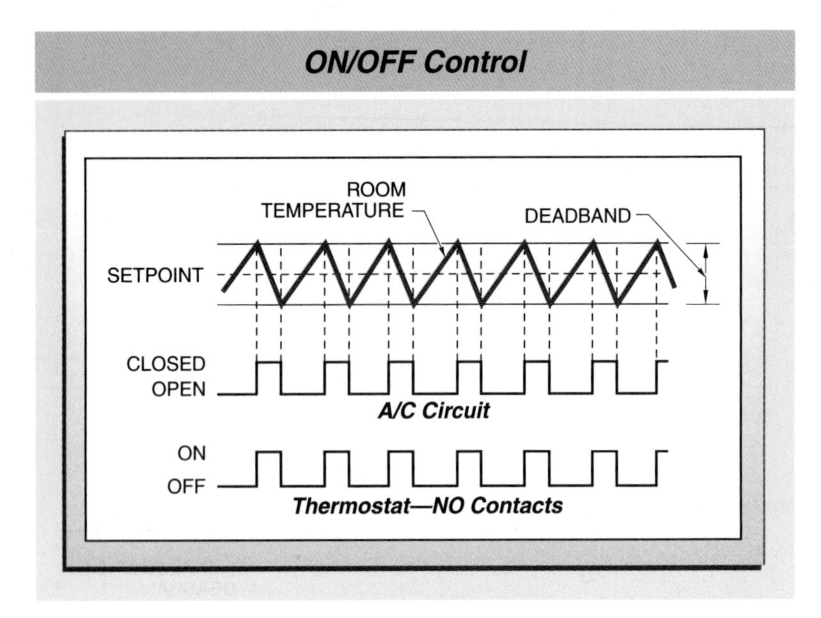

Figure 36-1. ON/OFF control is commonly used in HVAC control systems.

ON/OFF controllers usually have a narrow fixed deadband. A *deadband* is the range of values where a change in measurement value does not result in a change in controller output. Deadband is usually intentional to eliminate the chattering of the contacts from frequent ON/OFF switching. The performance of ON/OFF controllers depends on the magnitude of the lags and dead times of the process. *Dead time* is the period of time that occurs between the time a change is made to a process and the time the first response to that change is detected.

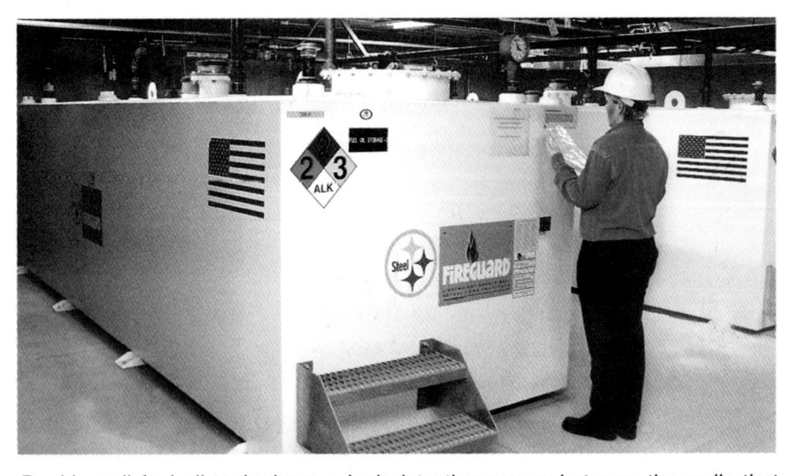

Double-wall fuel oil tanks have a leak-detecting sensor between the walls that eliminates the need for containment dikes.

Large lags and/or dead times result in overshoot and undershoot of the setpoint and a long cycle time. *Overshoot* is the change of a process variable (PV) that exceeds the upper deadband value when there is a disturbance to a system. *Undershoot* is the change of a process variable (PV) that goes below the lower deadband value when there is a disturbance to the system. Processes with very small lags and dead times maintain the measurement much closer to the setpoint, but have very short cycles. **See Figure 36-2.**

In some applications, such as in HVAC control, a heat anticipator is used. A *heat anticipator* is a small electric heater that is part of a thermostat. The anticipator warms up a bimetallic element and turns heating equipment ON before the controller normally activates. Some ON/OFF controllers have an adjustable dead band to limit output cycling.

A more sophisticated ON/OFF controller is a time proportional ON/OFF controller. A *time proportional ON/OFF controller* is an ON/OFF controller that has a predetermined output period during which the output contact is held closed (or power is ON) for a variable portion of the output period. During the remainder of the period the contact is open. **See Figure 36-3.**

The portion of the output ON time varies with the difference between the measurement and the setpoint and length of time that the measurement has been away from the setpoint. The output fixed period should be selected so that the ON cycle time is suited to the characteristics of the process. Processes with a large time constant usually have longer fixed periods. Processes with a small time constant must have short fixed periods.

Proportional Control

Proportional (P) control is a method of changing the output of a controller by an amount proportional to an error. The amount of output change is determined by the proportional gain. Proportional control is typically used for processes that are subject to large load changes and operating throughput changes.

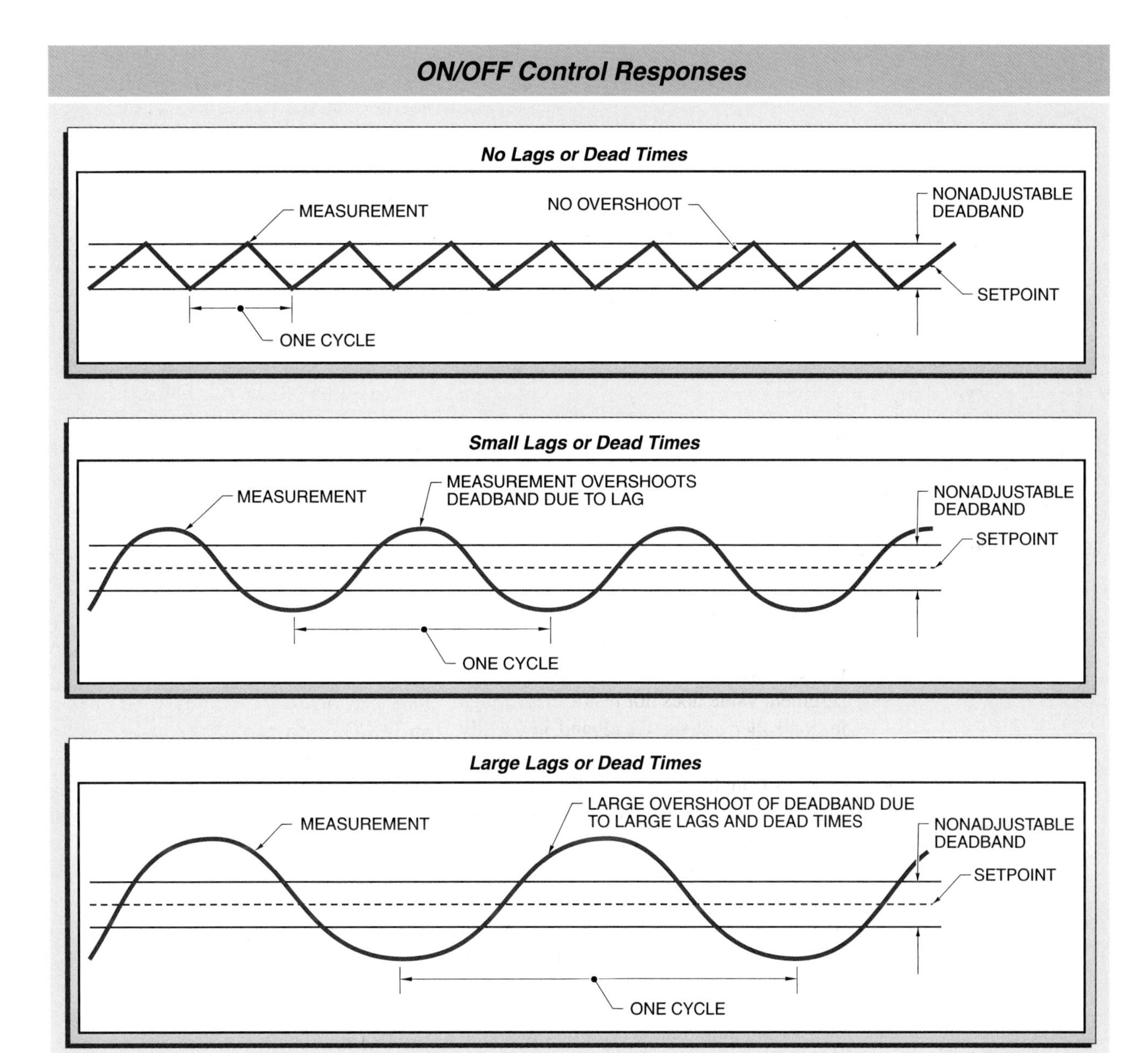

Figure 36-2. ON/OFF control works best for processes with little or no lag or dead time.

A proportional controller is designed to have an analog output that has a continuous range of possible output values between minimum and maximum limits. The output range is the difference between the minimum and maximum limits. Digital and electric controllers typically have 4 mA to 20 mA outputs. Pneumatic controllers have 3 psig to 15 psig outputs. The analog output varies within these ranges according to the action of the control function on the error.

The percent change in output for a percent change in input represents the proportional gain. The proportional gain is designed to be adjustable to best match the needs of the process, the transmitter, and the control valve. It is important to understand the proportional relationship between the controller input and its output.

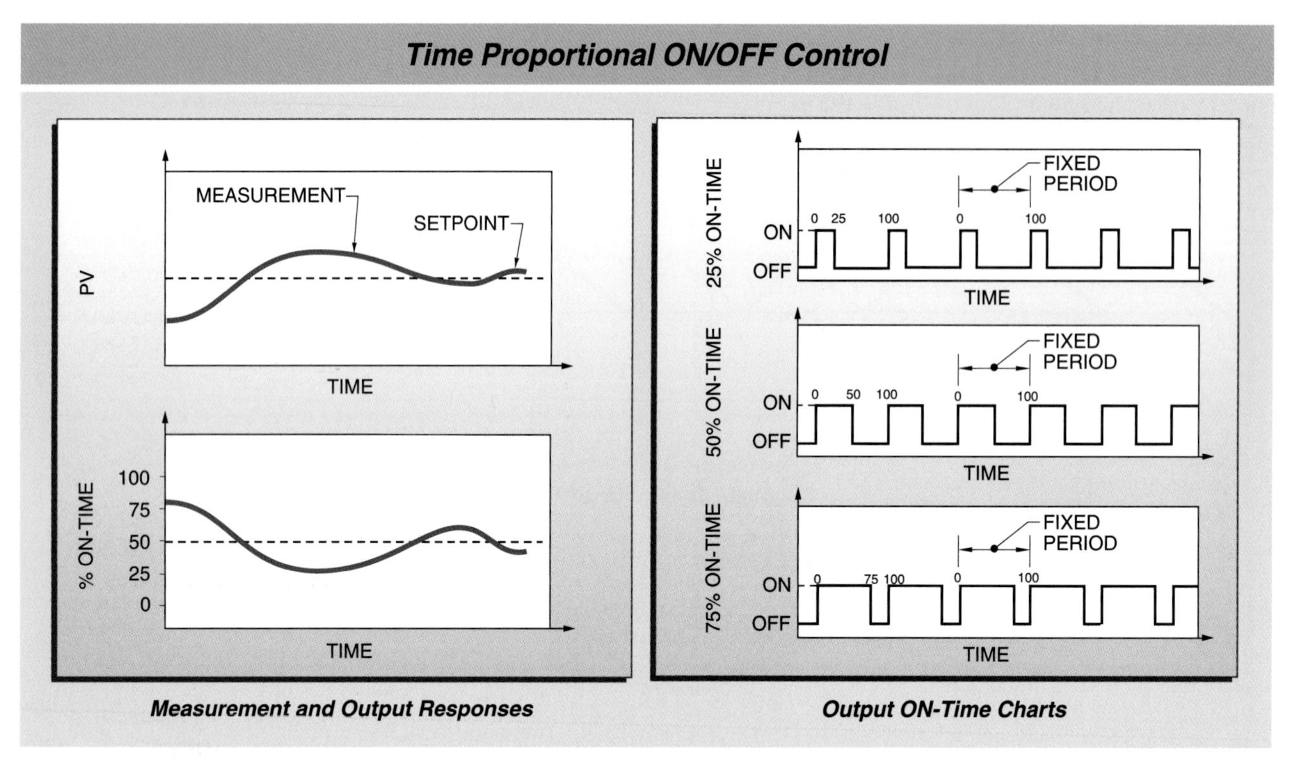

Figure 36-3. A smoother control can be obtained by using a time proportional ON/OFF controller. This controller uses a fixed cycle time with a variable ON time.

Gain in this context is commonly used to describe two different functions. *Controller gain* is the gain, or sensitivity, of a controller itself. *Proportional gain* is the gain, or sensitivity, of a proportional term only. The terms may have the same numeric value, but have a very different influence on the control system. Changing a proportional gain only changes the influence of the proportional term in a controller calculation. Changing a controller gain changes the influence of all terms. These terms are often confused in common usage. The user needs to be very careful about using these terms.

An example of a proportional control system is a tank used to provide a solvent to a number of users. **See Figure 36-4.** The flow going to the users is the process load. The tank is filled from a source that is controlled by a float-actuated control valve. The gain of the control arrangement, which is similar to the way the process gain is calculated, is the change in flow output divided by the change in level input, and can be calculated as follows:

$$K = \frac{(F_2 - F_1)/(FR)}{(L_2 - L_1)/(LR)}$$

where
F_1 = flow in
F_2 = flow out
FR = flow range
L_1 = initial level
L_2 = final level
LR = level range

For example, the float, linkage, and pivot to the control valve are arranged so that at a level of 35% the valve is 100% open, and at a level of 65% the valve is fully closed. Depending on how much solvent is being used, the level in the tank changes until the inflow equals the outflow. Each different load requires a different valve position to make the inflow equal the outflow. Each different valve position has an equivalent float position.

It should be noted that there is no fixed level control point. The level in the tank must change to make the inflow equal the outflow. The gain for this control arrangement is calculated as follows:

$$K = \frac{(F_2 - F_1)/(FR)}{(L_2 - L_1)/(LR)}$$

$$K = \frac{(100\% - 0\%)/(100\%)}{(65\% - 35\%)/(100\%)}$$

$$K = \frac{(100\%)/(100\%)}{(30\%)/(100\%)}$$

$$K = \frac{100}{30}$$

$$K = \mathbf{3.33}$$

If the same tank and controls are used, except that the pivot is now located closer to the control valve, the effect on the control system can be compared. The level in the tank now has to be at 10% for the valve to be 100% open, and the valve does not fully shut until the level is at 90%. The gain of this control arrangement can also be calculated as follows:

$$K = \frac{(F_2 - F_1)/(FR)}{(L_2 - L_1)/(LR)}$$

$$K = \frac{(100\% - 0\%)/(100\%)}{(90\% - 10\%)/(100\%)}$$

$$K = \frac{100}{80}$$

$$K = \mathbf{1.25}$$

This shows that a control system with a larger gain requires less input change to generate the same output change. In other words, the controller acts as a multiplier on the input change. The gain is the multiplication factor.

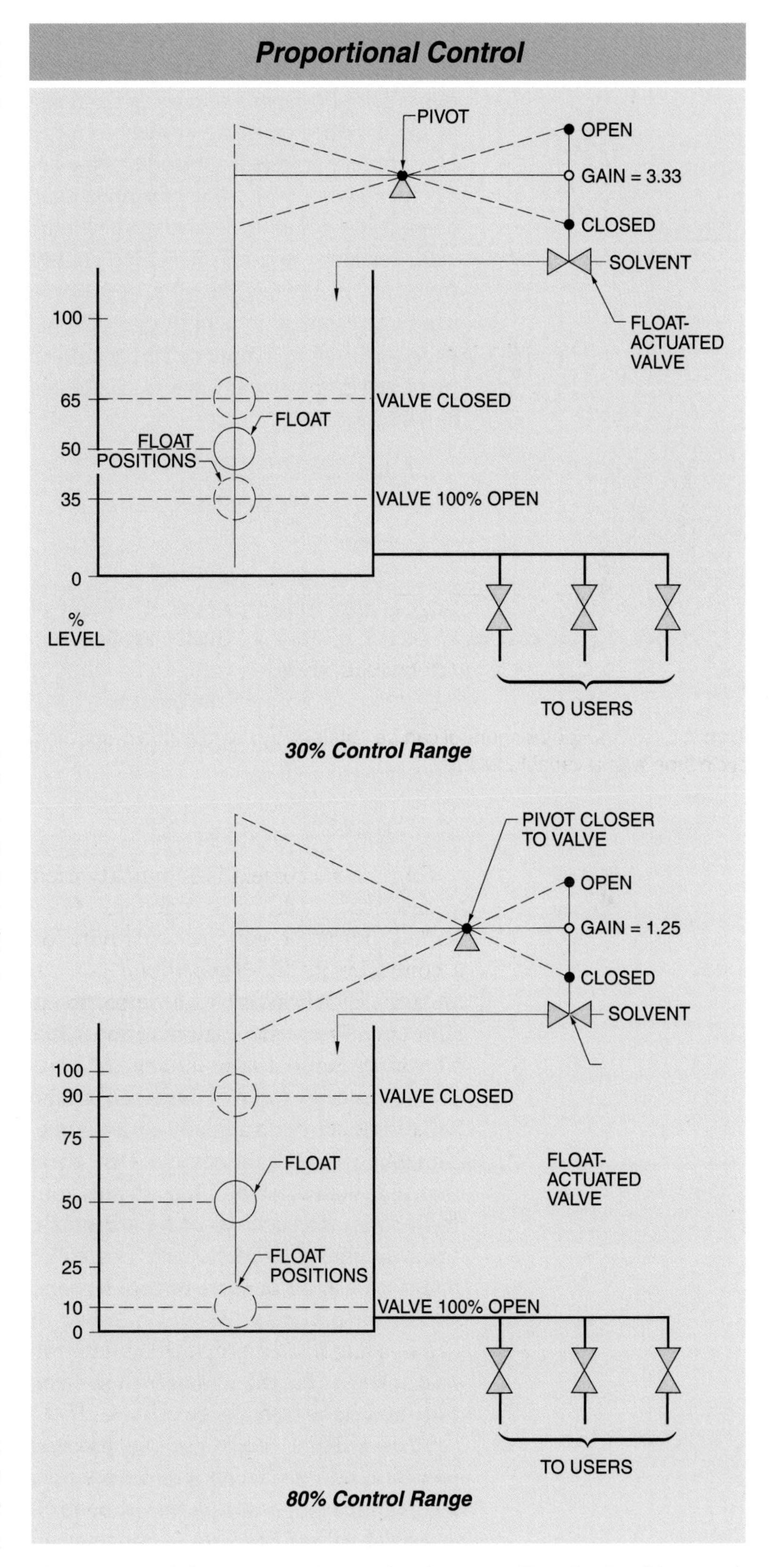

Figure 36-4. A float-actuated control valve controlling the level in a tank demonstrates the principles of proportional control.

Proportional Band. Some controllers use proportional band (PB) instead of proportional gain. *Proportional band* is the range of input values that corresponds to a full range of output from a controller, stated as a percentage. A high value of proportional band corresponds to a low proportional gain, resulting in a less sensitive control function. A low value of proportional band corresponds to a high proportional gain, resulting in a more sensitive control function. Proportional band is calculated as follows:

$$PB = \frac{control\ range}{process\ range}$$

where
PB = proportional band
control range = range of values of the controller
process range = range of values of the process

Proportional band and proportional gain are related as follows:

$$K = \frac{100\%}{PB} \text{ or } PB = \frac{100\%}{K}$$

where
K = gain
PB = proportional band

In the HVAC industry, the term "throttling range" is often used instead of PB. *Throttling range* is the number of units of the process variable that causes the actuator to move through its entire range. Throttling range is the same as PB except that it is stated in the units of the controlled variable instead of a percentage. For example, if a temperature controller opens a fresh air damper to cool a room, the throttling range is the number of degrees that the temperature in the room must change for the damper to go from fully closed to fully open.

For example, what are the PB and proportional gain when a process has a temperature range that spans from 60°F to 180°F and the controller only needs to control over the range from 90°F to 150°F? **See Figure 36-5.**

$$PB = \frac{\text{control range}}{\text{process range}}$$

$$PB = \frac{150 - 90}{180 - 60}$$

$$PB = \frac{60}{120}$$

$$PB = \mathbf{0.50} \text{ or } \mathbf{50\%}$$

$$K = \frac{100\%}{PB}$$

$$K = \frac{100\%}{50\%}$$

$$K = \mathbf{2}$$

In this example, 50% of the input is scaled to 100% of the output. This is analogous to placing a pivot point on a lever arm connecting input and output.

Output Bias. *Output bias,* or *manual reset,* is a controller function that positions a final element in a central position when the process variable is at setpoint. A controller that only has a proportional action must also have the ability to include an output bias. It is necessary to have the final element at a position that allows the final element to move in either direction.

The output bias is incorporated into the controller during the configuration or when a pneumatic or electronic controller is built. The equation relating the controller output to the control action, error, proportional gain, and bias is as follows:

$$output = CA \times \frac{error}{MR} \times K \times OR + bias$$

where
CA = control action (direct = +1; reverse = −1)
error = deviation between PV and SP
MR = measurement range
K = proportional gain
OR = output range
bias = output bias

TECH-FACT

Possible sources of leaks in pneumatic actuator systems include the diaphragm, air line connections, and the input line itself.

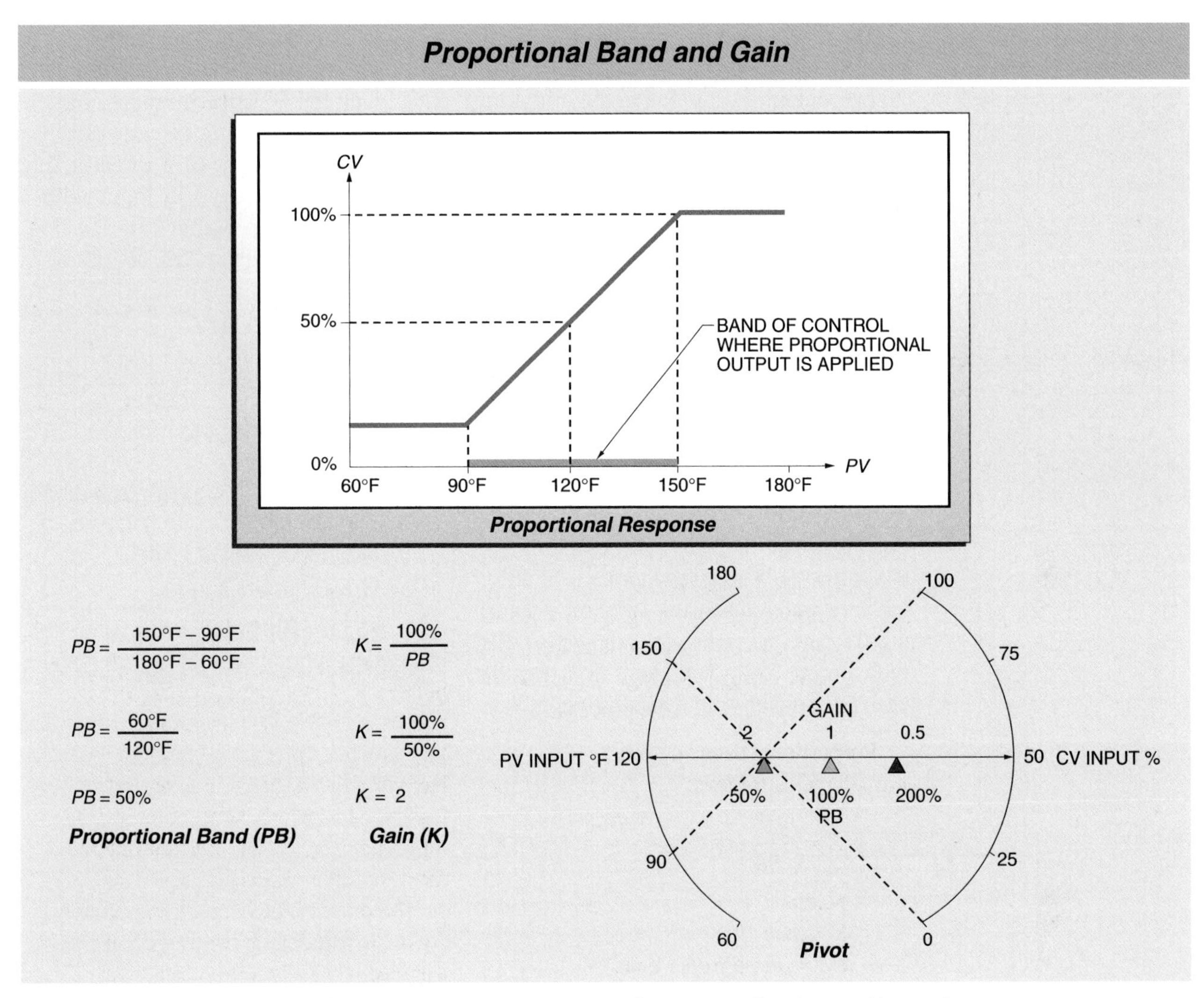

Figure 36-5. Proportional band is the range of measurements where proportional control is used.

If the PV is at the setpoint, the error is zero. Then if the bias is zero, the controller output is zero. For example, in a process where the liquid level in a tank is being controlled at a setpoint of 10 ft, the actual level is 10 ft, and the bias is 0, the controller output is as follows:

$$output = CA \times \frac{error}{MR} \times K \times OR + bias$$

$$output = CA \times \frac{(10-10)}{MR} \times K \times OR + 0$$

$$output = CA \times \frac{(0)}{MR} \times K \times OR + 0$$

$$output = \mathbf{0}$$

For this example, the value of the proportional gain, output range, measurement range, and controller action do not matter because the PV is equal to the SP. The error is equal to zero and this is multiplied by all the other factors. Without bias, the controller output goes to zero as the PV approaches the SP. This means that as the controller output goes to zero, the controller completely closes or completely opens the control valve. The tank either starts draining or starts filling, depending on whether the control valve is opened or closed. This is not acceptable controller action. Therefore a bias is used.

The most common output bias value is 50%. The midpoint of the standard 4 mA to 20 mA output is 12 mA and the midpoint of the standard pneumatic 3 psig to 15 psig output is 9 psig. Therefore, a proportional-only electric controller has a 12 mA output when the measurement is at the setpoint. A proportional-only pneumatic controller has a 9 psig output when the measurement is at the setpoint. The outputs are set to these values when the measurement is at the setpoint no matter where the setpoint is on the measurement scale.

Offset. The functioning of a controller's proportional action is clearer if a simple representation of a controller is discussed. **See Figure 36-6.** Each scale has a range of 0% to 100%. When the setpoint is at 50%, the left scale shows the setpoint pointer at 50% and the right scale shows the PV indicator at 75%, 50%, and 25%. Beside each PV pointer is an output value.

With a direct-acting controller and a proportional gain of 2.0, the output is 50% when the PV measurement is at the setpoint of 50%. Using the equation for calculating the output of a direct-acting controller for a change in input with a range of 100%, the output can be calculated when the PV is at 75% as follows:

$$output = CA \times \frac{error}{MR} \times K \times OR + bias$$

$$output = (+1) \times \frac{(75\% - 50\%)}{100\%} \times 2.0 \times 100\% + 50\%$$

$$output = \frac{(25\%)}{100\%} \times 2.0 \times 100\% + 50\%$$

$$output = 25\% \times 2.0 + 50\%$$

$$output = 50\% + 50\%$$

$$output = \mathbf{100\%}$$

Similarly, when the measurement is at 25%, the output is at 0%.

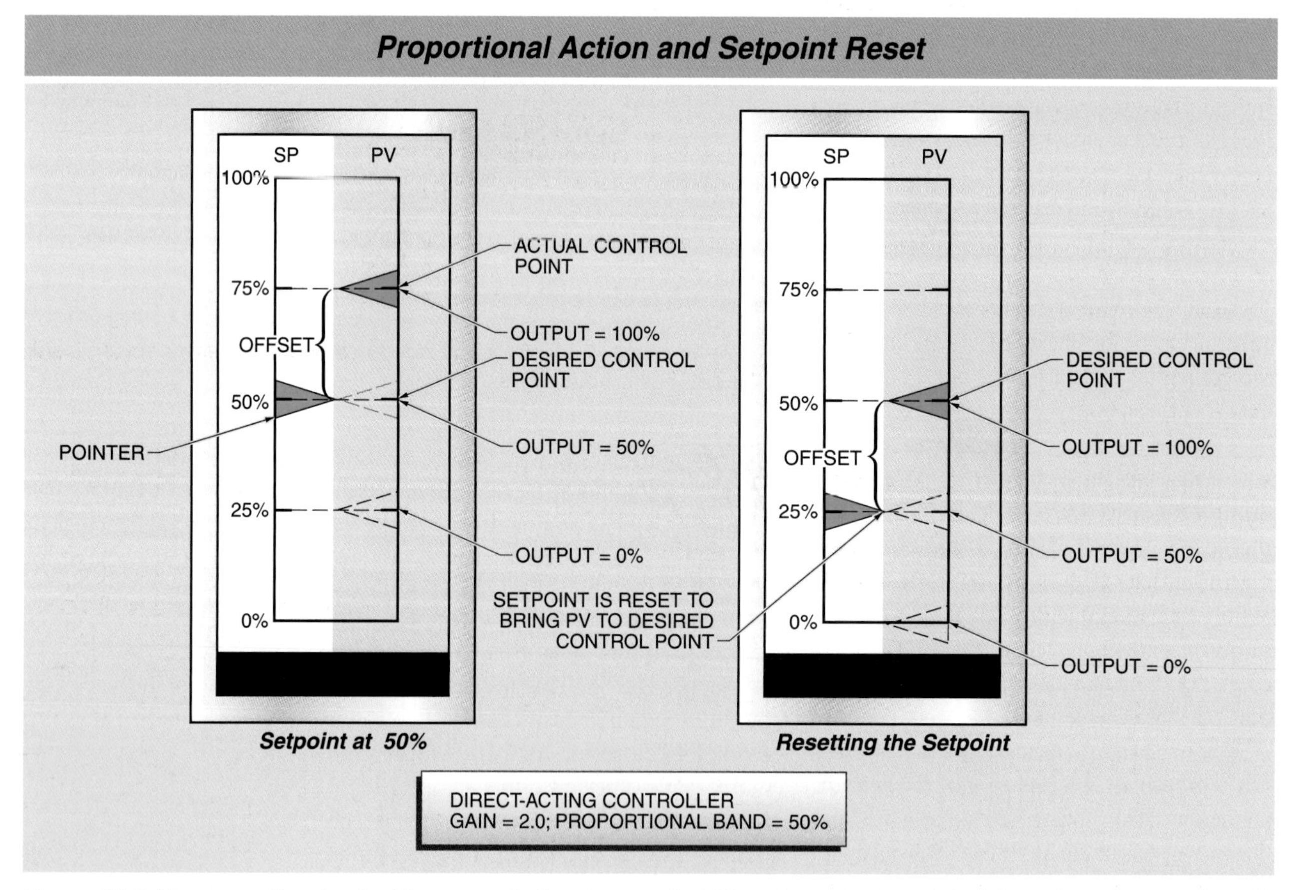

Figure 36-6. The proportional output from a controller changes for different measurement deviations from the setpoint with a controller gain of 2.0.

This shows that with a controller PB of 50% (gain of 2.0), it only takes a 50% change in the PV measurement to obtain a 100% change in output. This also shows that it requires an offset of the PV from the setpoint to obtain an output other than 50%. *Offset* is a permanent difference in measurement between a process variable (PV) and a setpoint (SP) as a result of proportional control action.

This offset is not a transient condition and is necessary to bring the control system to a new steady-state condition after a load change to the process. This means that during most control situations, the control system is not controlling at the desired setpoint. This is usually not a satisfactory condition when a process is being controlled. One solution is to reset the setpoint value to bring the measurement back to the desired PV. If the setpoint is at 50%, but the PV has stabilized at 75%, moving the setpoint down to 25% brings the measurement down to 50%, the desired control point.

A second example uses the same direct-acting controller and the same setpoint and measurement scales, but the setpoint pointer is now at 75% and the gain is 1.0. **See Figure 36-7.** Measurement indicators are shown at 100%, 75%, 50%, and 25%. The associated outputs are shown at each measurement position. Using the previous equation for the calculation of the outputs, at a PV of 25%, the controller output is 0%, and at a PV of 100%, the controller output is 75%.

This shows that when the PV is at the setpoint, the output is still 50%, even though the setpoint is now at 75% on the measurement scale. This also shows that the measurement offset is greater, as compared to the first example, to develop the same output. Also note that the output cannot reach 100% because the PV cannot be offset that far from the setpoint.

The two examples show that a smaller gain has a greater offset, and the larger gain has a smaller offset. This seems to indicate that the way to minimize the offset is to make the controller gain large. The problem with this approach is that the appropriate controller gain is not an independent variable, but is determined by the interaction of all the other gains in the control loop. A controller gain that is too large causes the control loop to go into oscillation. A control loop that is stable usually has a total loop gain of about 0.5.

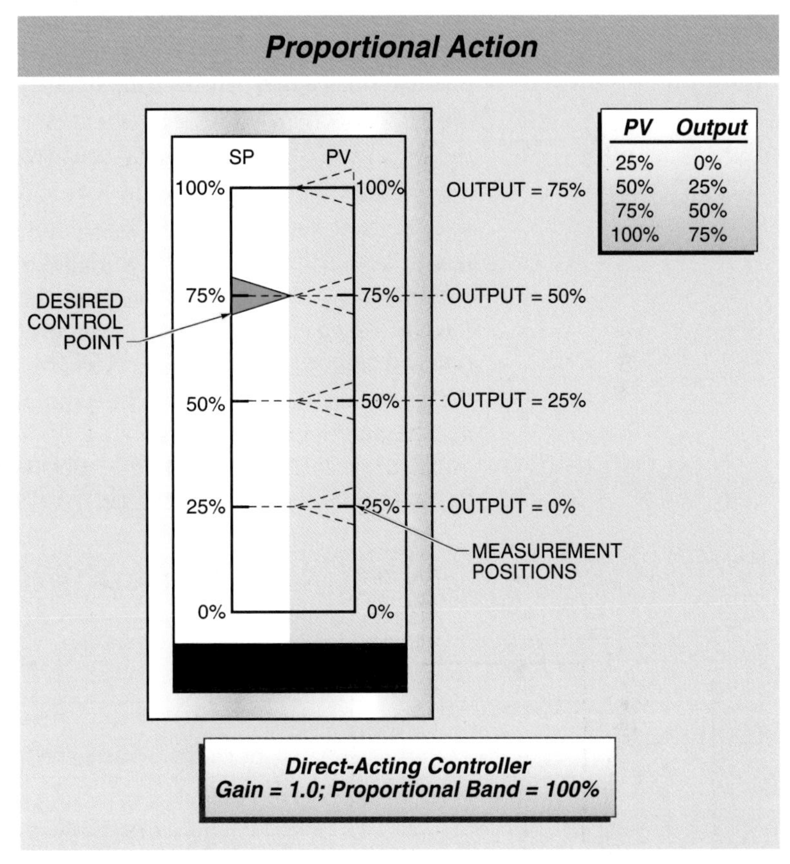

Figure 36-7. A gain of 1.0 means that not all possible outputs are attainable if the setpoint is not at 50%.

A deaerator is used to remove dissolved and entrained gases from boiler water. The controls for the deaerator are linked to the main boiler controls to ensure safety.

The function of the controller gain or proportional band is to compensate for all the other gains in the loop so that the product of all the gains results in a stable system. The selection of the controller gain is one portion of the tuning of a controller. This also means that proportional-only control can work with processes with large capacity and small dead times where changes are slow and small deviations from the setpoint can be tolerated.

Open Loop Response Diagram. Another way of showing the relationship between measurement deviations from a setpoint, gain, control action, and controller output is by the use of a controller open loop response diagram. An *open loop response diagram* is a curve that shows the controller response to a given measurement without the controller actually being connected to the process. **See Figure 36-8.** These diagrams are not the same as the process curves that are used to examine the attributes of a process. These open loop controller response diagrams apply only to the controller.

The controller is evaluated on a test bench where the output of the controller is measured in response to a simulated signal of the measurement. The controller is not connected to a process. Therefore it has no feedback for any of its output changes.

A step change deviation of the measurement from the setpoint results in a proportionate change in the output. When the measurement returns to the setpoint, the output returns to its original value.

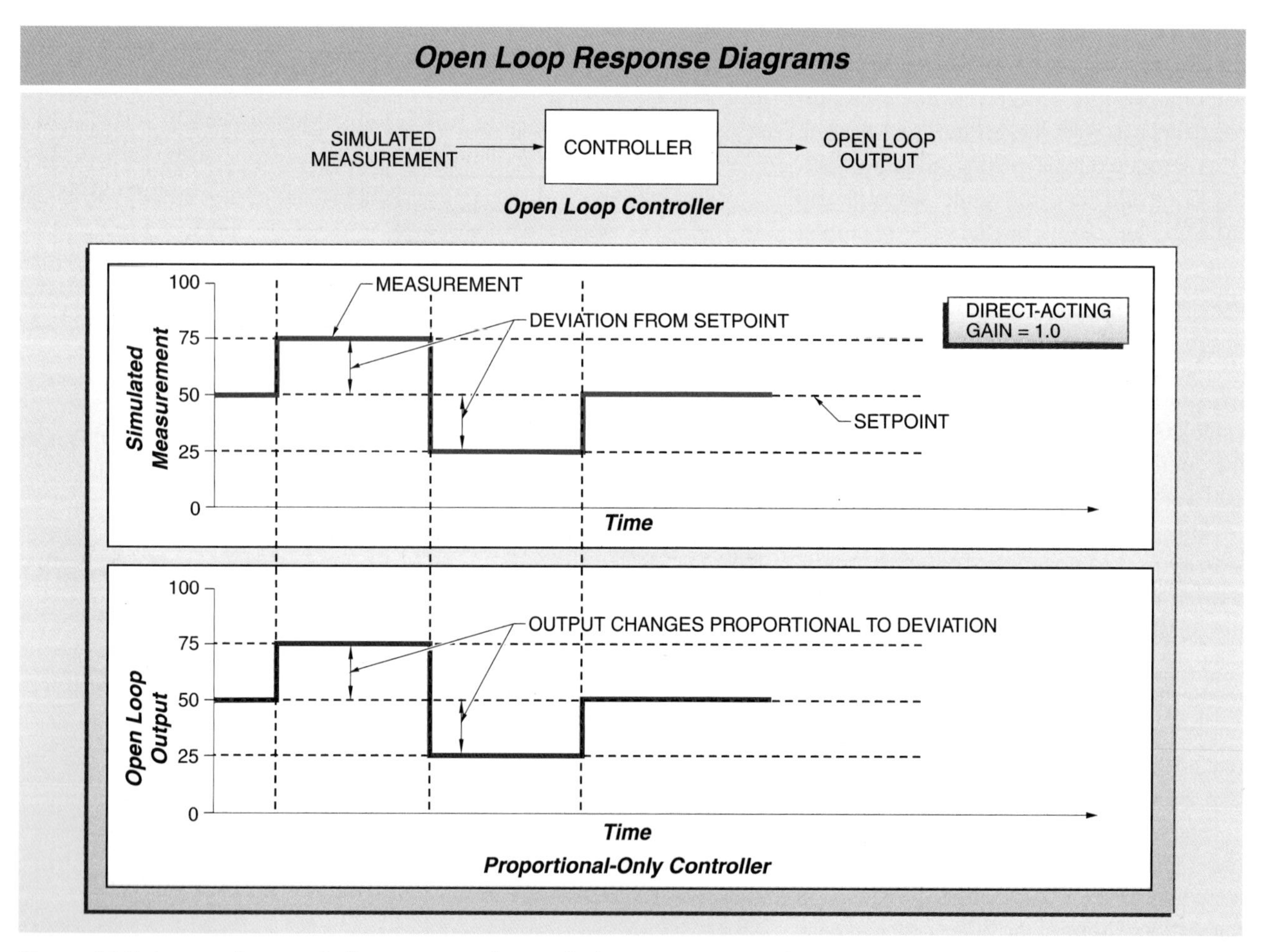

Figure 36-8. An open loop controller response diagram is the measurement of an output over time for simulated inputs. This is an effective way of showing how controller actions function.

Integral (I) Control

Integral (I) control is a method of changing the output of a controller by an amount proportional to an error and the duration of that error. *Automatic reset,* or *reset action,* is the integral controller mode, named so because in older proportional-only controllers the operator had to manually reset the setpoint. The mathematical function of integration is the summation of the error over a period of time. The sum of the errors continues to increase as long as the measurement is not at the setpoint. As the sum of the errors continues to increase, the controller output increases until the process variable is brought back to the setpoint. **See Figure 36-9.**

Integral controllers require the user to enter a value of integral time, reset rate, or integral gain as part of the tuning process. *Integral time (T_I), or reset time,* is the time it takes for a controller to change a control variable (CV) by 1% for a 1% change in the difference between the control variable and the setpoint (SP). This typically has units of minutes per repeat. A *reset rate* is the reciprocal of integral time. This typically has units of repeats per minute. An *integral gain (K_I)* is the integral time multiplied by the controller gain, if present. Some new controllers, based on the IEC 1131 standard, specify that controllers use seconds instead of minutes. Anyone working with controllers should become comfortable with the different units.

TECH-FACT

When averaging a number of sensors, ensure that the sensors are in the correct location and are not being affected by other environmental factors. A common problem is when a sensor is damaged and giving a false reading.

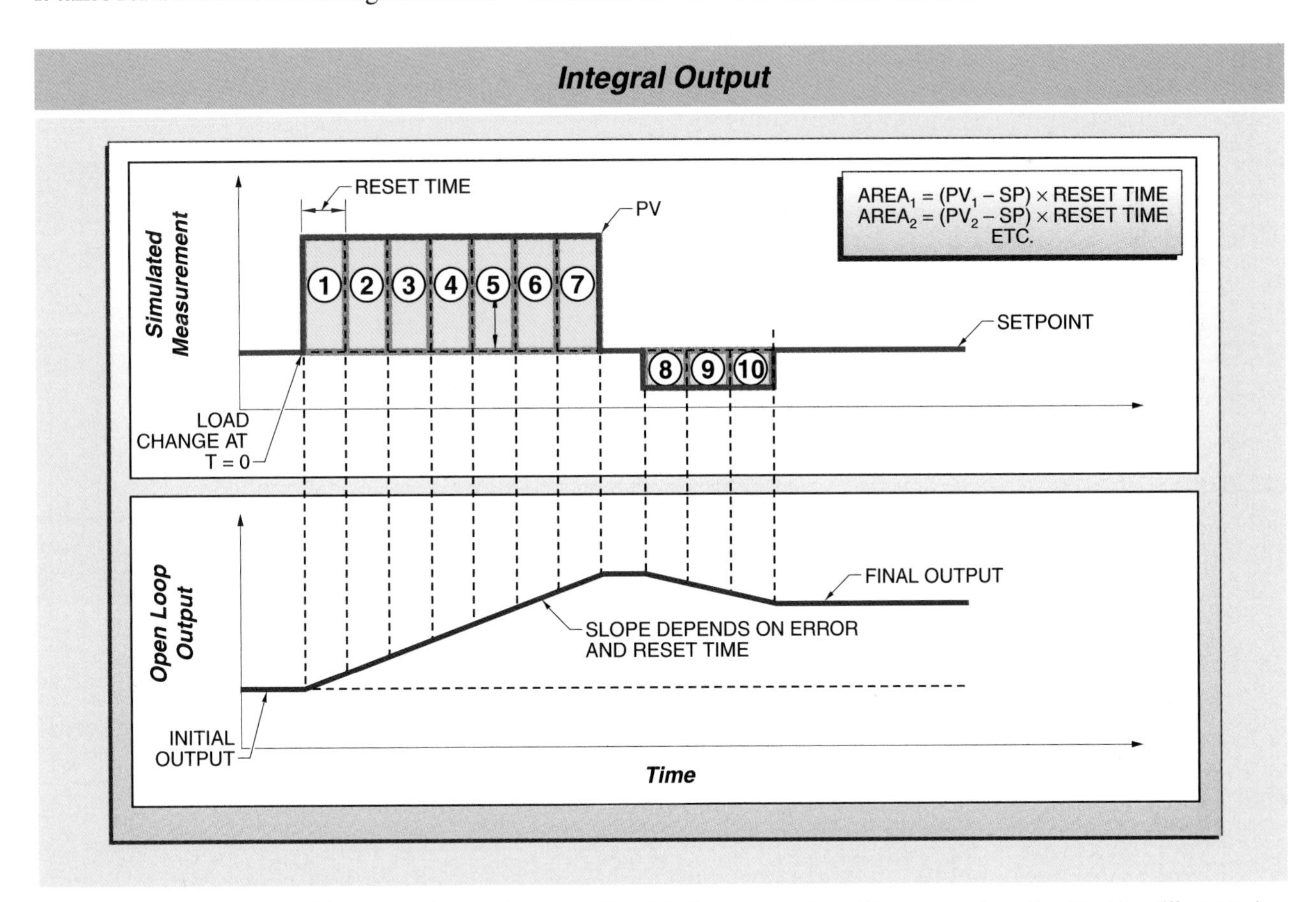

Figure 36-9. The output of the integral action is proportional to the summation of the areas described by the difference between the measurement and the setpoint for small increments of time, over a period of time.

A proportional-only controller has a manually determined output bias and a fixed output at the setpoint. When a controller has an integral control action, the output does not have a preset output bias and there is no fixed output value when the measurement is at the setpoint. An integral controller includes automatic calculation of the output bias. In other words, the integral controller automatically resets the output bias every time it calculates the error. The output bias at any given time is the sum of the errors up to that time multiplied by the integral gain.

For integral control, the open loop controller response shows a step change in measurement, which then maintains a constant deviation for a period of time before it returns to the setpoint. **See Figure 36-10.** The output, due to the integral action by itself, gradually increases until the measurement returns to the setpoint. At that point, the output stops changing.

Integral-only control is a rarely used controller action. There are some specialized control applications where no proportional action is desired because the response is too rapid. The action that is desired is the slow output change developed by integral action.

An example where integral-only control is suitable is the automatic correction of a flow ratio system where the ratio setting is adjusted by an analysis of the resulting product. The basic flow ratio control can adjust for any rapid disturbances to the ratio and an analysis control of the product using the integral-only controller function can handle small long-term corrections. **See Figure 36-11.**

Integral action is nearly always combined with a proportional action, forming a proportional-integral (PI) controller. *Proportional-integral (PI) control* is proportional control combined with integral control. The addition of integral control to a controller typically decreases the stability of most controlled processes. This slight disadvantage is more than offset by the ability of integral control to bring the measurement back to the setpoint.

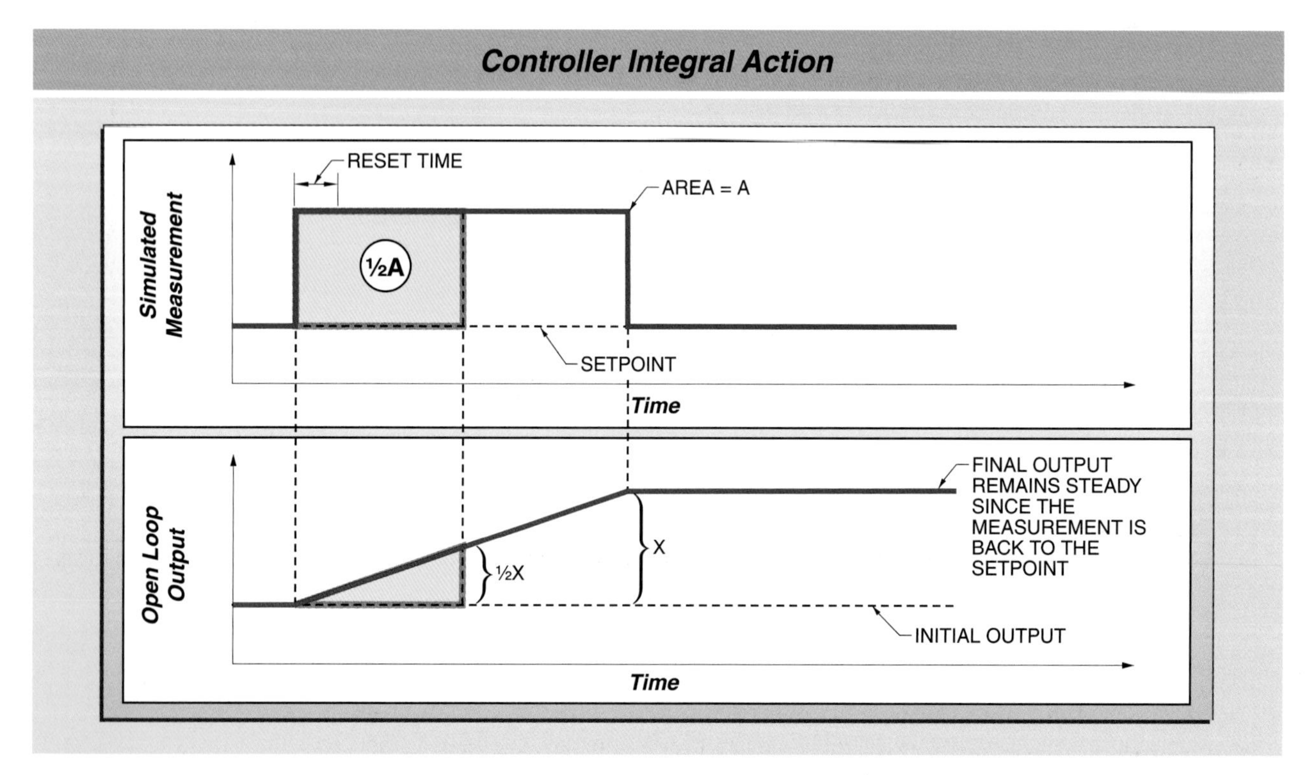

Figure 36-10. An open loop controller response for an integral-only control action shows the relationship between the measurement deviation area and the output.

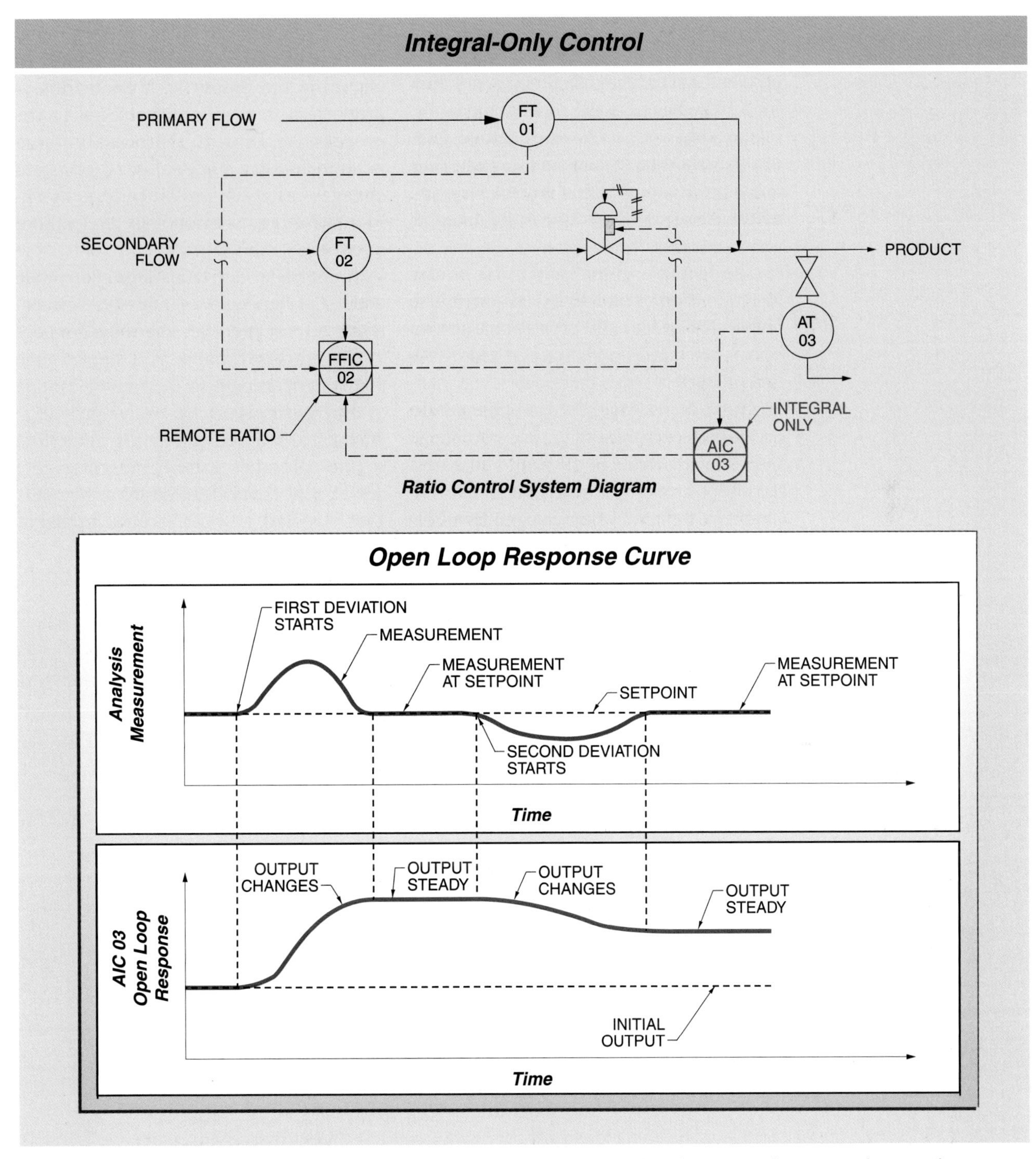

Figure 36-11. The open loop response from an integral controller shows that the output changes as long as the process measurement is away from the setpoint.

Almost all controllers should contain the integral function. This is especially true for liquid flow control loops, which tend to be very fast and very noisy. In this case, integral control can provide a damping or filtering action. The integral action, when combined with a proportional control action, solves the offset problem experienced in proportional-only controllers.

The output due to the integral action is related to the proportional gain. If the proportional and the integral outputs are plotted for a step change in the measurement, the relationship can be shown. It can be seen that there is a step change in the output due to the proportional action and then a gradually increasing output due to the integral action. **See Figure 36-12.**

The increase in the output due to the integral action eventually equals the initial output change due to the proportional action. This is the repeat in the integral term. The time it takes to reach that point is the integral time. Increasing the value of the min/R decreases the response of the integral action. Increasing the value of the R/min increases the response of the integral action. Thus it is important to know which integral term the controller uses before making any changes.

Derivative (D) Control

Derivative (D) control is a method of changing the output of a controller in proportion to the rate of change of the process variable. If a process volume is large and has a very slow response to changes in the controller output signal, PI control may continuously undershoot or overshoot the setpoint.

A controller derivative term, derivative time (T_D), has units of minutes. The derivative term modifies the measurement of the process variable as a step change that is proportional to the rate of change of the measurement. Derivative gain (K_D) is equal to the derivative time multiplied by the controller gain, if present. Some new digital controllers, based on the IEC 1131 standard, now use seconds instead of minutes for the time.

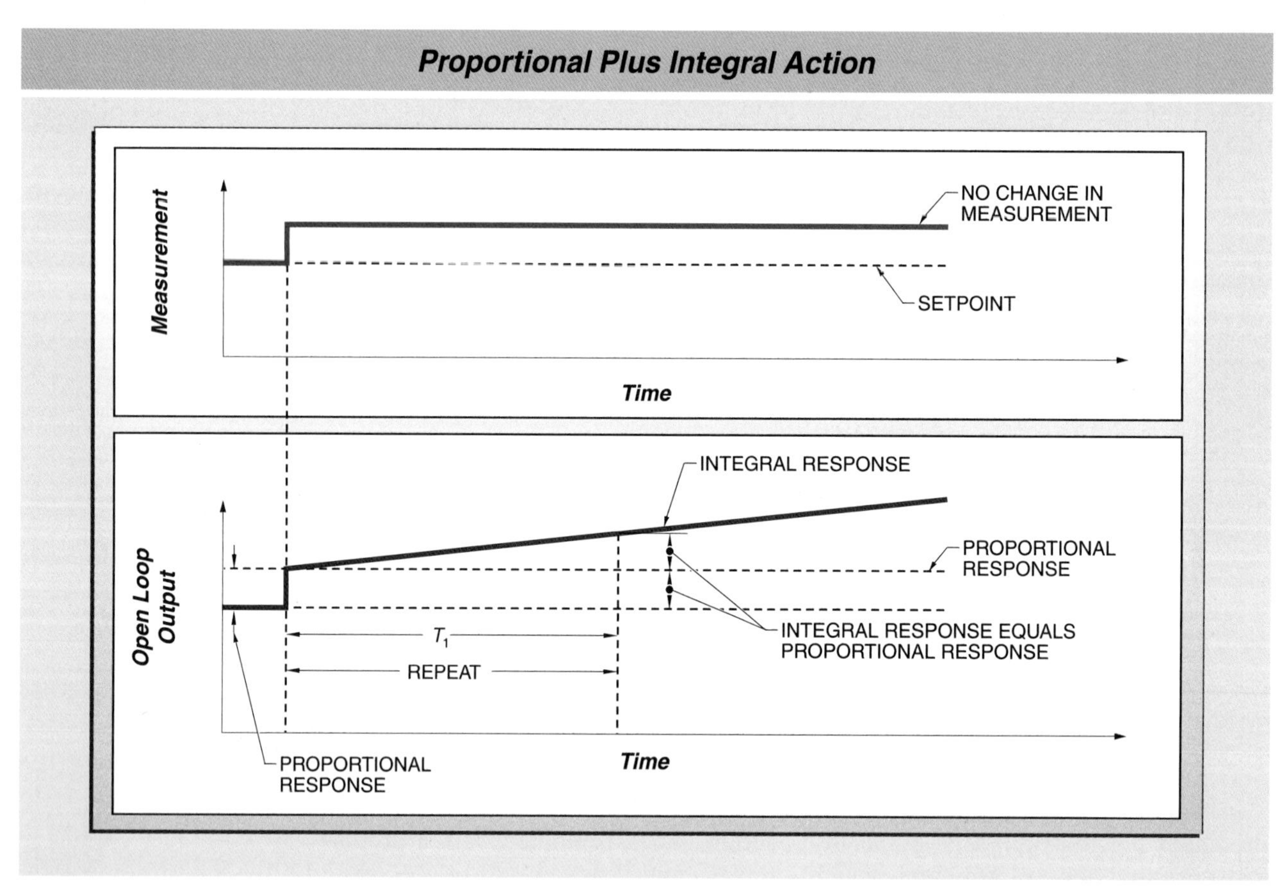

Figure 36-12. The open loop controller response from a proportional plus integral control action shows how the integral time is defined.

An open loop controller response for a ramped measurement change can be used to demonstrate the derivative action when the measurement change is ramped. A step change causes a spike in the output. **See Figure 36-13.** The derivative action generates a step increase in response to a ramped measurement change. In the controller, the step increase is then added to the actual measurement value. This makes the modified measurement lead the actual measurement. The modified measurement value is then used for comparison to the setpoint and other controller actions. The result to the output is also a step change.

When derivative action was part of the controller feedback system in older style controllers, the derivative time and the proportional band were related. The step change in the output due to the derivative action is the same amount of change that would have been developed by the proportional action had it started responding to the ramped measurement at a time equal to the derivative time before the change happened. **See Figure 36-14.** In modern controllers, the derivative function has its own proportional gain independent of the controller gain adjustment. The derivative time is not directly related to the old definition.

Older pneumatic and electric miniature and large-case controllers had the derivative action actuated by the error. This is an acceptable action when the error is caused by changes in the process. However, this causes problems when the setpoint is changed. A setpoint change causes a large error that results in a large derivative action. This is not the desired controller action. This means that the setpoint must be changed slowly on older controllers so that the derivative action does not cause large disturbances to the output. Modern controllers have a derivative action that responds only to the changes in the PV.

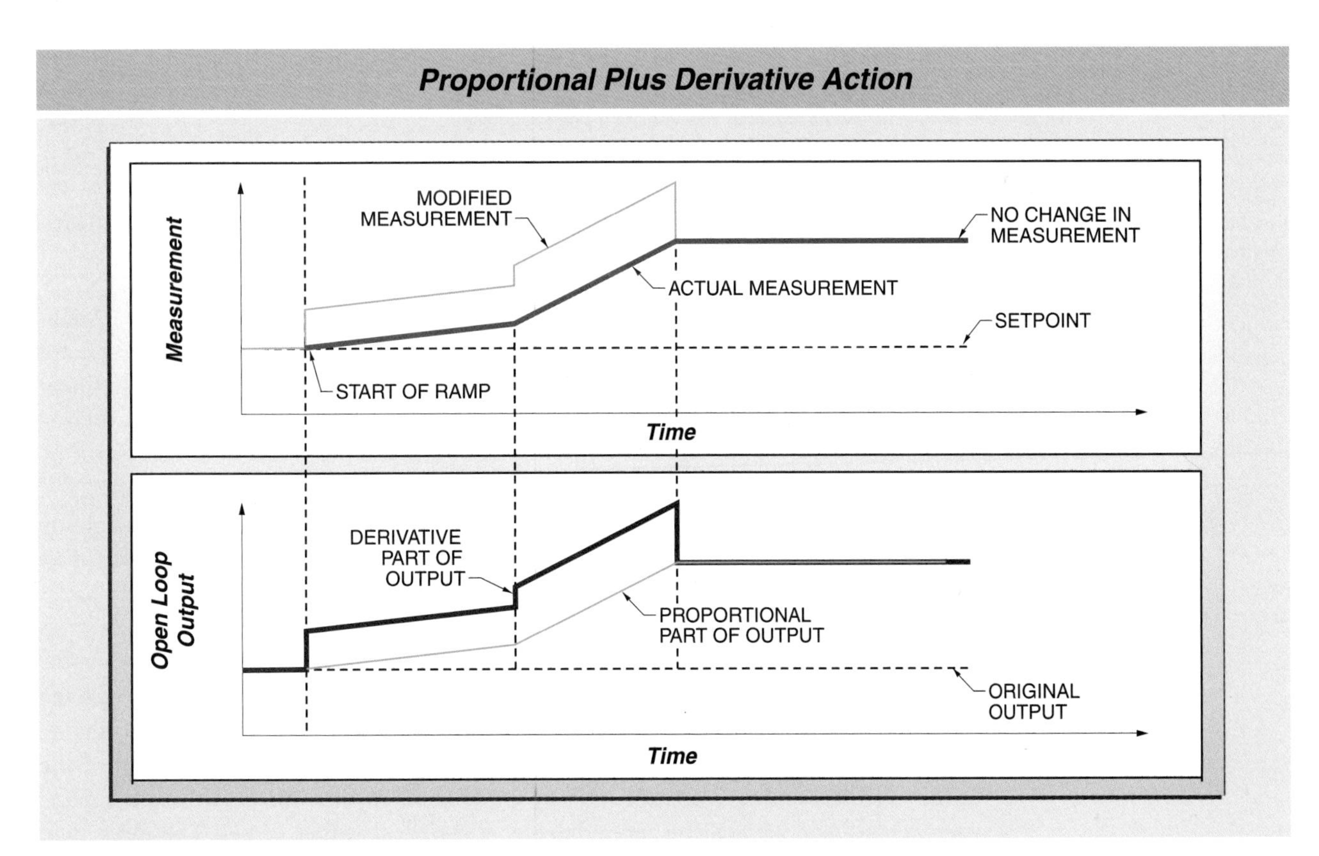

Figure 36-13. An open loop controller response for a proportional plus derivative control action shows how changes in the measurement affect the output.

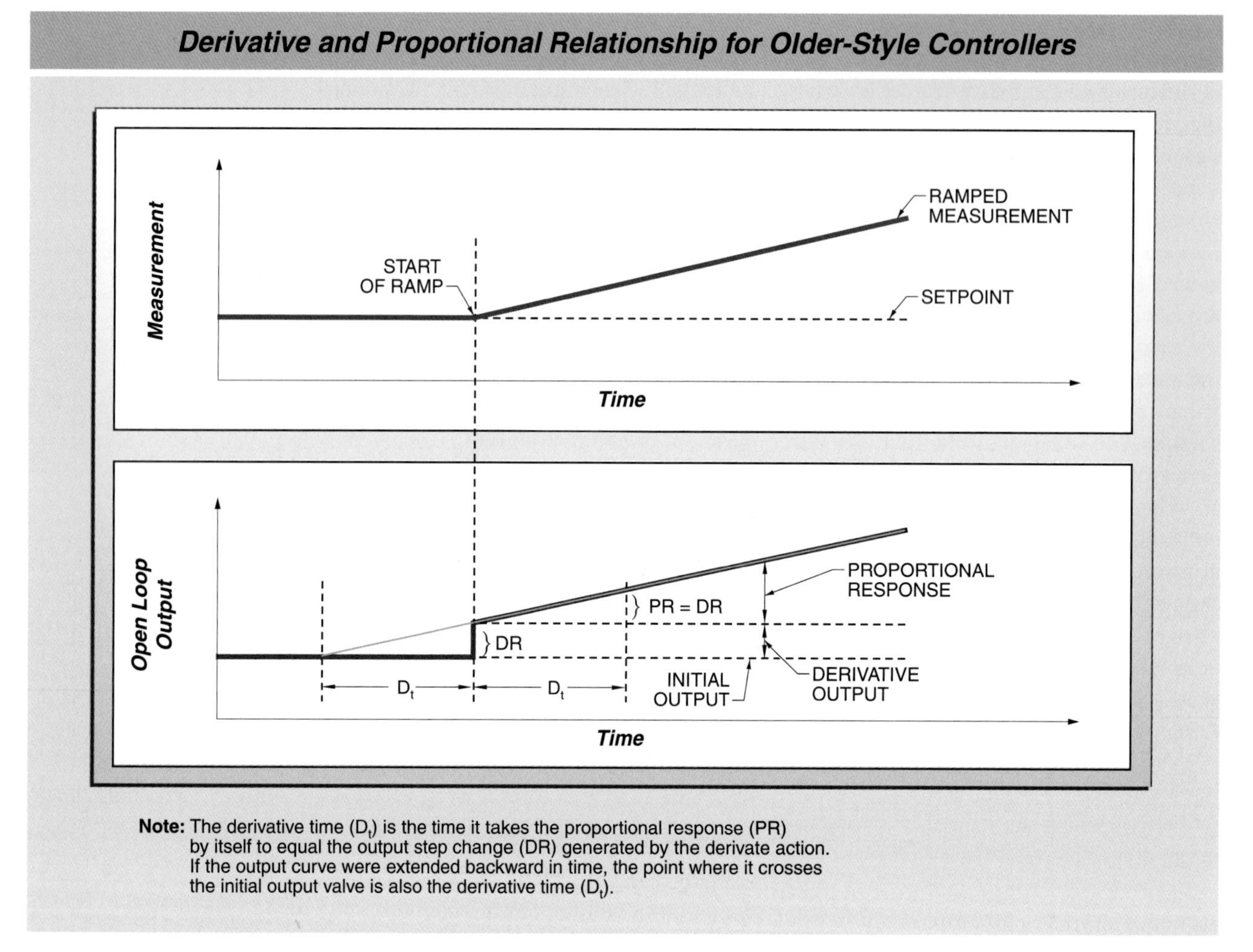

Figure 36-14. An open loop controller response for an older-style proportional derivative control action shows how the derivative time is defined.

Derivative action is used when the process has large time constant lags and there is a need to minimize measurement deviations. *Proportional-integral-derivative (PID) control* is proportional control combined with both integral control and derivative control. Derivative action should never be used with any process that has any noise in the measurement, since this generates spikes in the output. Derivative action should not be used for flow control because flow control is very noisy.

Adaptive Gain Control

Adaptive gain control is a control strategy where a nonlinear gain can be applied to a nonlinear process. The error is multiplied by a variable that is separate from, but similar to, the proportional gain. Adaptive gain can be used to compensate for the variable gain effects of a nonlinear process. Adaptive gain acts like a revised proportional gain where the gain is set by a variable function. Any nonlinear process, such as pH control or distillation column temperature, is a good application for an adaptive gain controller.

In a nonlinear control system such as pH control, the gain of the process over the measured range can be obtained from the slope at various points of the process curve. **See Figure 36-15.** The reciprocal of the pH gains results in a set of inverse gains at different values of pH. The individual inverse gains are divided by the largest inverse gain value to obtain a set of unitized inverse gains (gains that vary from 0 to 1.0) for different values of pH. A characterized curve with the pH as the input produces an output that is the inverse gain of the process at any measured pH.

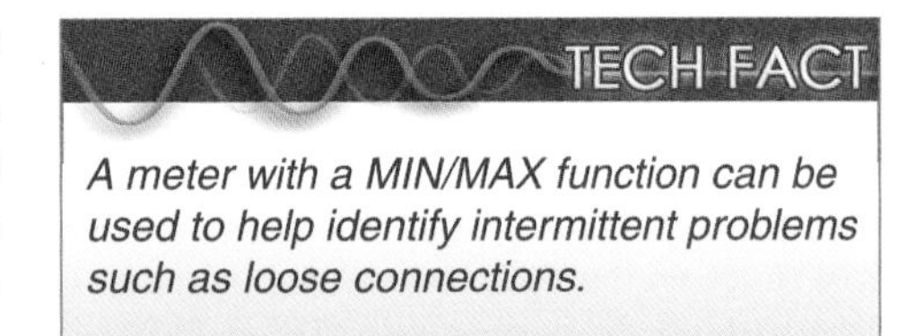

A meter with a MIN/MAX function can be used to help identify intermittent problems such as loose connections.

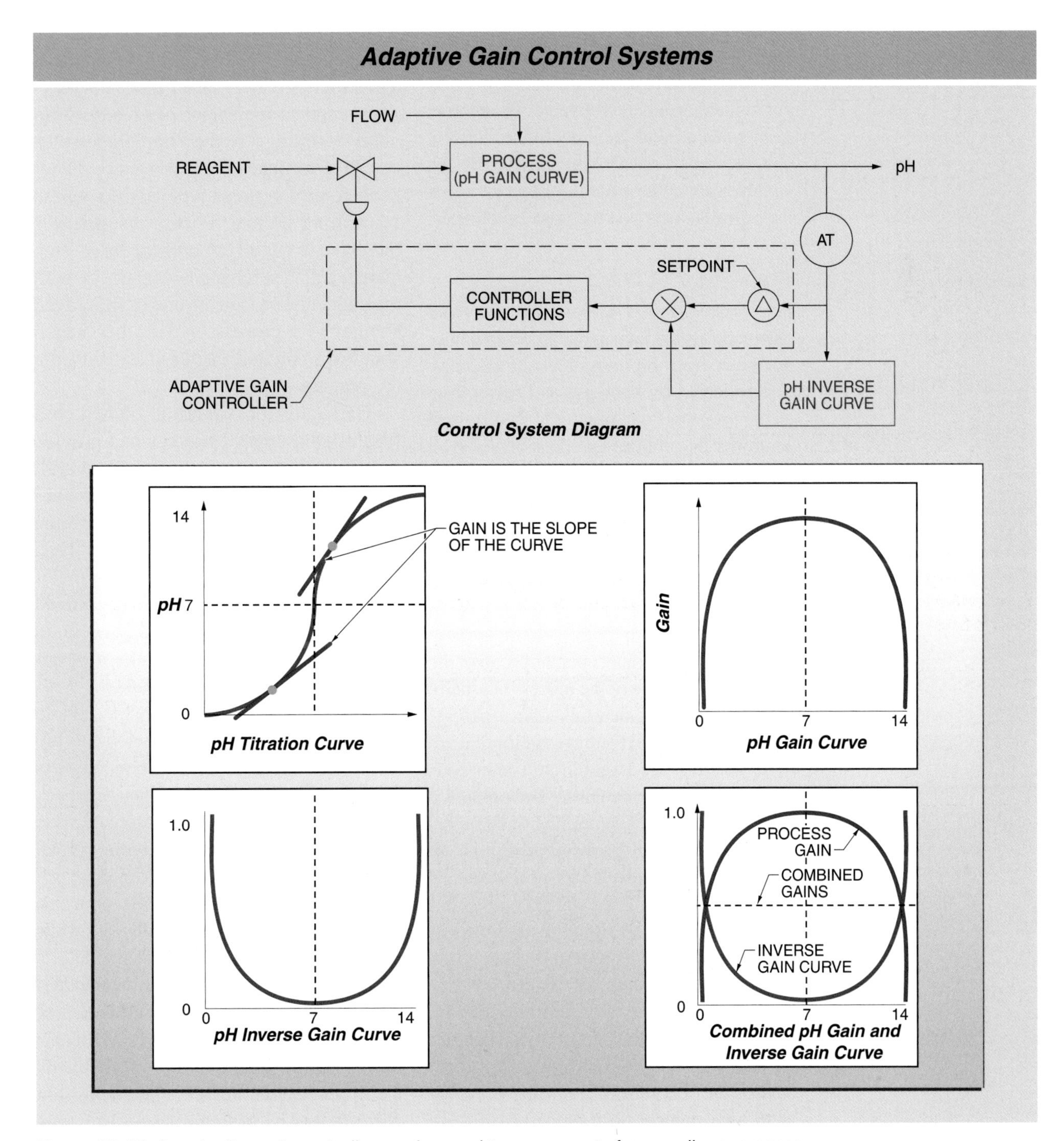

Figure 36-15. An adaptive gain controller can be used to compensate for a nonlinear process.

The pH is measured and the results transmitted to the adaptive gain controller. The controller compares the pH to the setpoint and calculates the error. The pH is also transmitted to a function that uses the inverse gain curve to determine an inverse gain. The calculated error is multiplied by the inverse gain. The resulting modified error is sent on to the controller functions. The controller uses the modified error in its normal proportional function calculation to determine the proper output to the reagent control valve. The inverse gain, when used as the input to the adaptive gain controller, exactly compensates for the variable gain of the process. The result is that the controller acts as if the process is linear.

Feedforward Control

Feedforward control is a control strategy that only controls the inputs to a process without feedback from the output of the process. Theoretically, by knowing and controlling all the properties of a process, a feedforward controller can produce a product satisfying all requirements. Feedforward control systems have an advantage over feedback control systems in that they are designed to compensate for any disturbances before they affect the product. Feedforward control systems also eliminate the lag of a feedback control loop. In practice, feedforward control systems are very difficult to design and operate because the depth of knowledge of the process is usually insufficient to implement this control strategy.

The feedforward control systems that are in service usually incorporate some feedback signals to compensate for unknown process reactions. The burning of some fuels with air is a well-known process and can be confidently controlled by a feedforward process. However, even with a combustion process, a feedback signal in the form of a flue gas analysis is required when a low percentage of excess air is desired. For example, a feedforward controller can be used to position an induced draft fan damper based on an analysis of the incoming combustion air. A feedback controller provides trim to the damper position based on stack gas analysis.

If frequent load changes occur in a process and a feedback controller cannot manage the changes, a feedforward system can be added to regulate a product stream before it enters a process. A feedforward controller is used to control the effects of the load change while a feedback controller is used to control other effects on the process. For example, a feedback system can be used to provide control of the output of a continuous reactor. **See Figure 36-16.**

A practical application of a feedforward control strategy is the supplement of a standard feedback control system. Feedback control systems typically do well at controlling normal process disturbances but have difficulty handling large step changes. These changes can occur with the sudden shutdown of one or more feed sources to a process or with the sudden shutdown of a unit that derives feed from the process.

These are the situations in which a feedforward control strategy can provide significant improvement in the overall control performance. The feedforward action can force all the appropriate final control elements, such as control valves, to predetermined positions when a major disturbance is sensed. The predetermined positions are those that would be required for steady-state operation at the new operating conditions.

The feedback controllers are placed into tracking mode while the final elements are forced into the new positions. The controllers are then placed back into normal feedback control when the forcing is complete. This eliminates the time it would take the normal feedback controls to arrive at the same positions for the final control elements. It also eliminates normal process oscillations while a process is brought to a new setpoint.

This type of control depends on predetermining the appropriate final control-element positions for the various types of disturbances. In the past, the only way to obtain the appropriate control-element position changes was by trial and error. This took a long time because large disturbances would not be intentionally introduced into an operating process. Also, the required information had to be extracted from the available process records.

An alternative method to trial and error is the creation of a real-time dynamic simulation of the appropriate control variables. The whole process does not need to be included in the simulation. In most processes, there may be only one or two key variables (pressure, temperature, flow, etc.) that need to be considered in the simulation. This makes the programming of the simulation much easier.

Existing DCS and PLC systems typically have sufficient computing power to support the mathematical equations used in the simulation. The simulation would be run as a subroutine in the normal control system. The simulation needs to run at a scan rate faster than the fastest process scan. This is to ensure that the simulation program is run completely for every controller scan. The normal feedback control system can then be programmed to use either the simulation or the actual process as the source of process information and the destination of the control signals.

Obviously, the simulation cannot be run at the same time as the actual process, so a selection is made to use either the simulation or the actual process. When the simulation is used, any number of disturbances can be introduced to test proposed feedforward control actions and evaluate their performance. The finalized feedforward control strategies can remain as a portion of the total control system.

Since the simulation acts like the actual process, the feedback controllers have to be tuned to provide the appropriate responses for normal operation before any feedforward testing is started. Determining the proper controller tuning settings before the actual process is started is another major benefit of the simulation. The simulation can be used to help train the operators if it is available in sufficient time before the process is due to be started up.

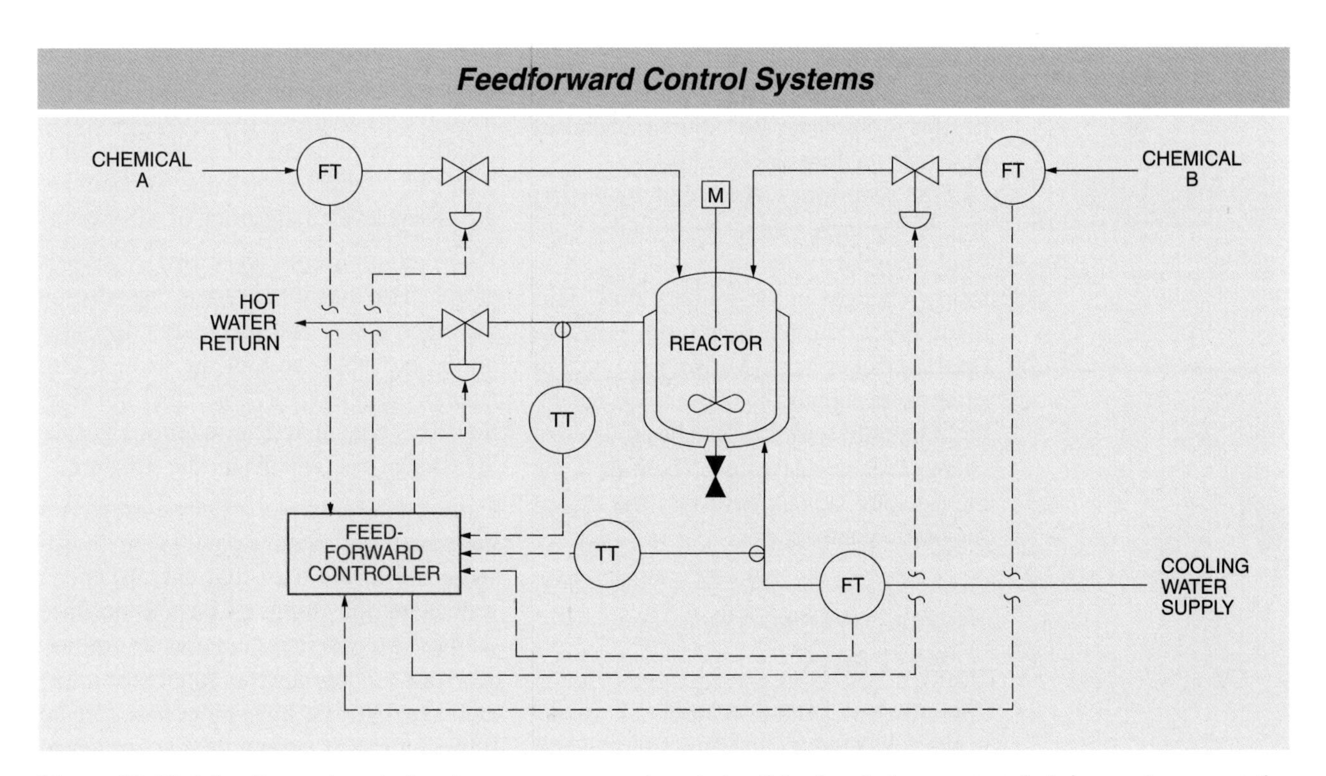

Figure 36-16. A feedforward control system measures and controls all the inputs to a process but does not measure the output.

Programming errors can be a cause of system problems. Many HVAC applications have a time schedule that should be checked to determine if the HVAC load is active or inactive.

Cascade Control

Cascade control is a control strategy where a primary controller, which controls the ultimate measurement, adjusts the setpoint of a secondary controller. The secondary variable affects the primary variable. The primary objective in cascade control is to divide a control process into two portions, where a secondary control loop is formed around a major disturbance. This leaves only minor disturbances to be controlled by the primary controller.

The proper selection of the primary and secondary controllers is very important. The response of the secondary controller must be about four times faster than the primary controller. The primary and secondary controllers are very interconnected, and there are many interlock signals built into the modern digital cascade controllers.

Ratio Control

Ratio control is a control strategy used to control a secondary flow to the predetermined faction, or flow ratio, of a primary flow. It is primarily used for flow applications. The flow measurement of an independent flow, called the primary flow, can be used to automatically adjust a secondary flow using a flow ratio controller.

The secondary flow can be either smaller or larger than the primary flow. The secondary flow then is adjusted to follow changes in the primary flow. This maintains a constant flow ratio. The flow ratio controller works with either linear or square root signals. The flow ratio controller has a ratio scale that can be adjusted to change the secondary flow. There are some interlocking signals built into the digital flow ratio controllers.

ADVANCED CONTROL STRATEGIES

The advanced control strategies discussed in the following section are at the highest level of digital control strategies. These are advanced control strategies that are difficult to implement but are very valuable when they can be used. The most successful applications of these strategies incorporate multiple strategies in a hybrid manner. Typical types of advanced control strategies are artificial intelligence, fuzzy logic, and neural networks.

Artificial Intelligence (AI)

Artificial intelligence (AI) is a form of computer programming that simulates some of the thinking processes of the human brain. It has long been recognized that the human brain is the most complex and adaptable reasoning device known. Artificial intelligence is an attempt to identify and formulate procedures that simulate some of the human reasoning processes so that they can be used by a computer for control.

Diagnostic Systems. A *diagnostic system* is a logic-based system that monitors various operational conditions to detect faults. The type of fault is identified and a message is conveyed to facility personnel. No attempt is made to correct the fault. Diagnostic systems have been installed in all automobiles for many years. This type of system is what the mechanic uses when a diagnostic analyzer is connected to the automobile to identify the problem. Diagnostic systems have only a small knowledge base and are the lowest level of the group of AI systems.

Knowledge Systems. A *knowledge system* is an enhanced diagnostic system that not only identifies a fault but also makes decisions on the probable cause of the fault. It also attempts to check for false input information. This type of system requires a greater knowledge base and diagnostic structure.

Expert Systems. An *expert system* is a program that contains a database of known facts and a rule base that uses "if-then" statements for groups of facts to produce outputs. An expert system uses the knowledge of a human expert to supply the information to build the knowledge base and the rules. An expert system uses an inference engine to compare input conditions to all the listed rules to attempt to find a match.

If a match is found, the output is determined from the rules. The output can be another input, which results in a further search of the rules. This procedure is continued until there is a conclusion.

An expert system differs from a standard "if-then" program in that if a change is needed due to additional relationships, the program does not have to be rewritten, as is the case for a conventional program. Only the knowledge base and a few rules need to be changed. The inference engine is separate from the knowledge base and the rules. The control approach of an expert system is called data-driven as compared to procedural or fixed sequence approaches. Expert systems are the most widely used of the AI systems.

Fuzzy Logic

Fuzzy logic is a system that uses mathematical or computational reasoning based on undefined, or fuzzy, sets of data derived from analog inputs to establish single or multiple defined outputs. The algorithms use no discrete values. They use statements such as "The room is cool." The three parts of a fuzzy logic control system are input fuzzification, fuzzy rule processing, and fuzzy output calculations to establish an output. **See Figure 36-17.** Fuzzy logic can be applied to any control system, but is specifically suited for applications that require human intuition and experience.

Fuzzy Input Calculations. An input can be linked to a fuzzy set, which contains membership functions and labels. The actual input value can be converted to a graded value on one or more membership functions. For example, a fuzzy logic control system for a cooling application can have five membership sets ranging from "cold" through "too cool," "normal," "too warm," to "hot." **See Figure 36-18.**

The horizontal axis is the input condition (temperature) and the vertical axis is the output (air-conditioner motor speed). A single input can trigger more than one output condition. For example, if the temperature is 137.5°F, then the temperature is part of two input curves, 50% "too cool" and 50% "normal." Therefore, the input triggers two outputs. The "too cool" input condition signals a "less speed" output and the "normal" input triggers a "normal speed" output. Since the fuzzy logic controller can only have one output, it completes the fuzzy output calculations to determine the actual final output value.

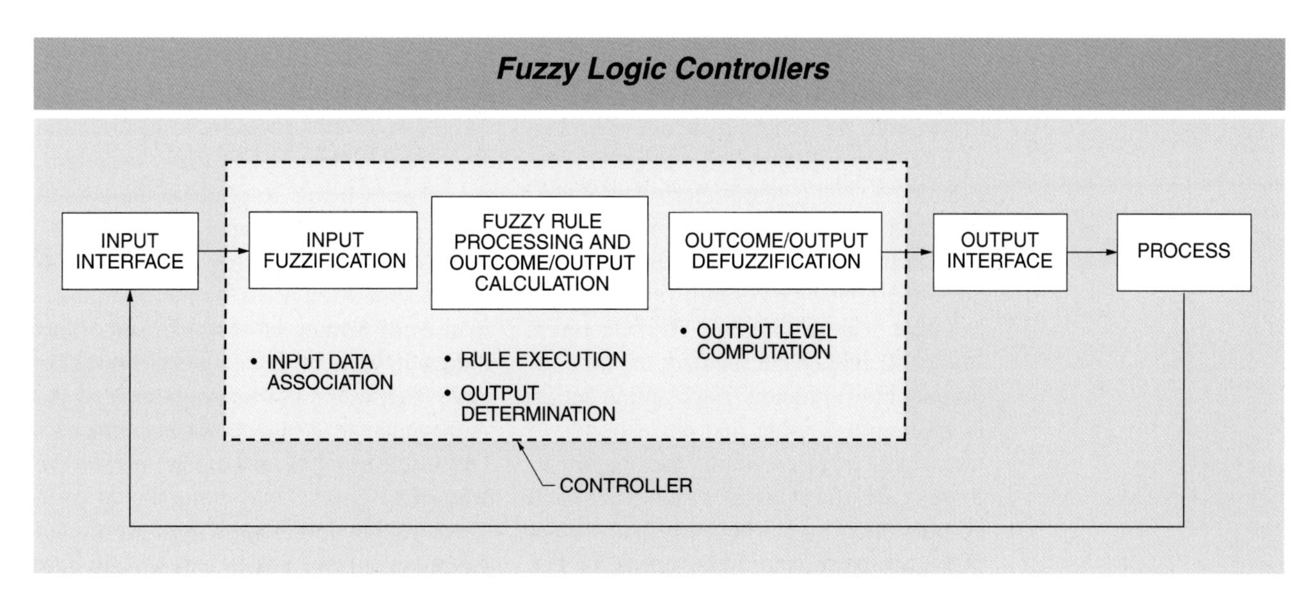

Figure 36-17. A fuzzy logic control system block diagram consists of input fuzzification, rule processing, and output defuzzification.

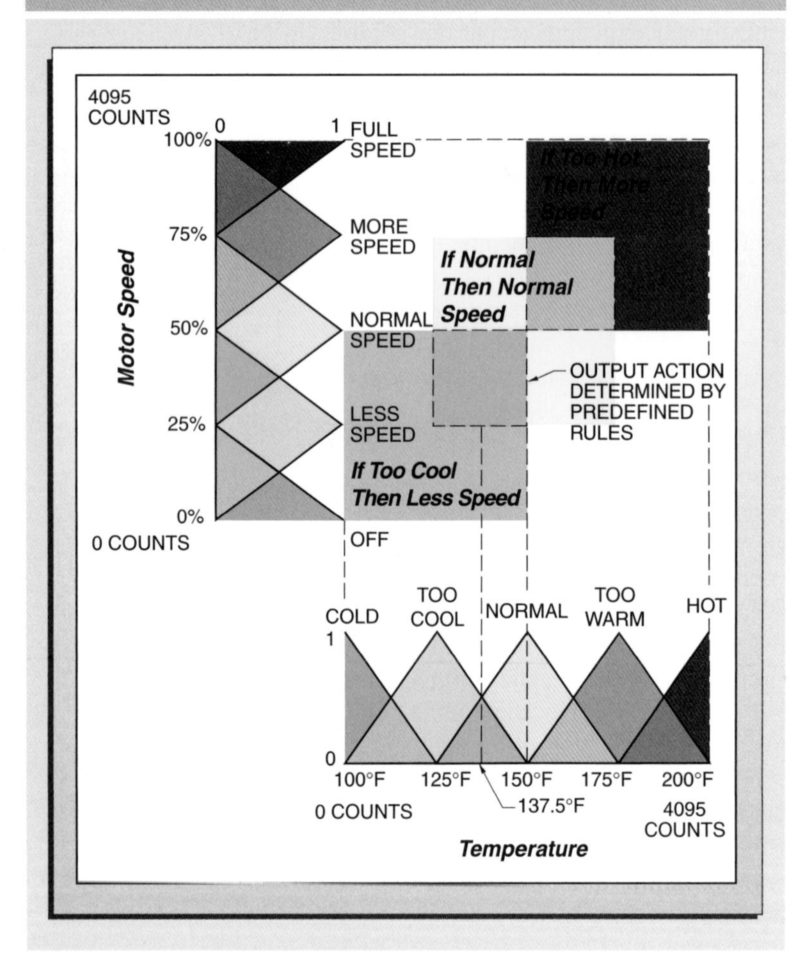

Figure 36-18. Fuzzy logic places measurements into membership functions and then determines the proper output for that membership function.

Fuzzy Output Calculations. Once an input value is assigned to one or more membership functions, the output can be calculated. The programmer of the fuzzy logic system is required to define rules to determine the output action for each possible input membership function. For example, Rule 1 might be, "If the room is 'too cool,' reduce the speed of the air conditioner blower motor." Each output action is applied at a value that corresponds to the scores from the input memberships. In this case, the controller multiplies each output action by 0.5 because the room is 50% "too cool" and 50% "normal." The resulting control action is between the actions required for the two memberships.

Neural Networks

A *neural network* is a computer program that simulates the parallel inputs and linking of information of the human brain to arrive at an educated output determination. The term "neural network" comes from the way neurons in the brain link to each other to form an associated network. There is no memory and no program in a neural network.

A neural network controller consists of a large number of identical processing elements, with biased links to a similar number of receiver sites. Each of the biases has a weighted value so that not all the receiver sites see the same value applied to a single input. Some neural networks contain two layers of interlinked sites. Each receiver obtains weighted values from multiple inputs. These inputs are combined to form a receiver site value. **See Figure 36-19.**

When analog signals are applied to a number of inputs, an output pattern is obtained. This pattern consists of the values in the individual receivers. At this point, the output pattern obtained does not mean anything. The neural network has to be trained to associate the output pattern to a specific output action. The two ways of training are supervised and unsupervised training.

Supervised training consists of inputting the correct output pattern for a specific set of input values. The neural network compares the correct pattern with its generated pattern and readjusts bias factors between the inputs and the receivers according to a correction algorithm. With time and many examples, the network learns to produce the correct output pattern for any input arrangement.

Unsupervised training involves techniques that strengthen or inhibit the responses of certain network elements. This is done in ways other than what is used for supervised training. The magnitude of the applied change is called a "learning factor." The best learning factors are in the range of 0.2 to 0.3.

Neural network control systems are acceptable if all that is desired is a result. It is not a good system to use if it is important to know how the neural system arrived at its result. The best neural network applications are when humans know how to do the task, there are no known formula solutions but lots of examples, and there are examples with values in one set associating objects in another set. An example of a neural network system application is machine recognition of printed or handwritten words.

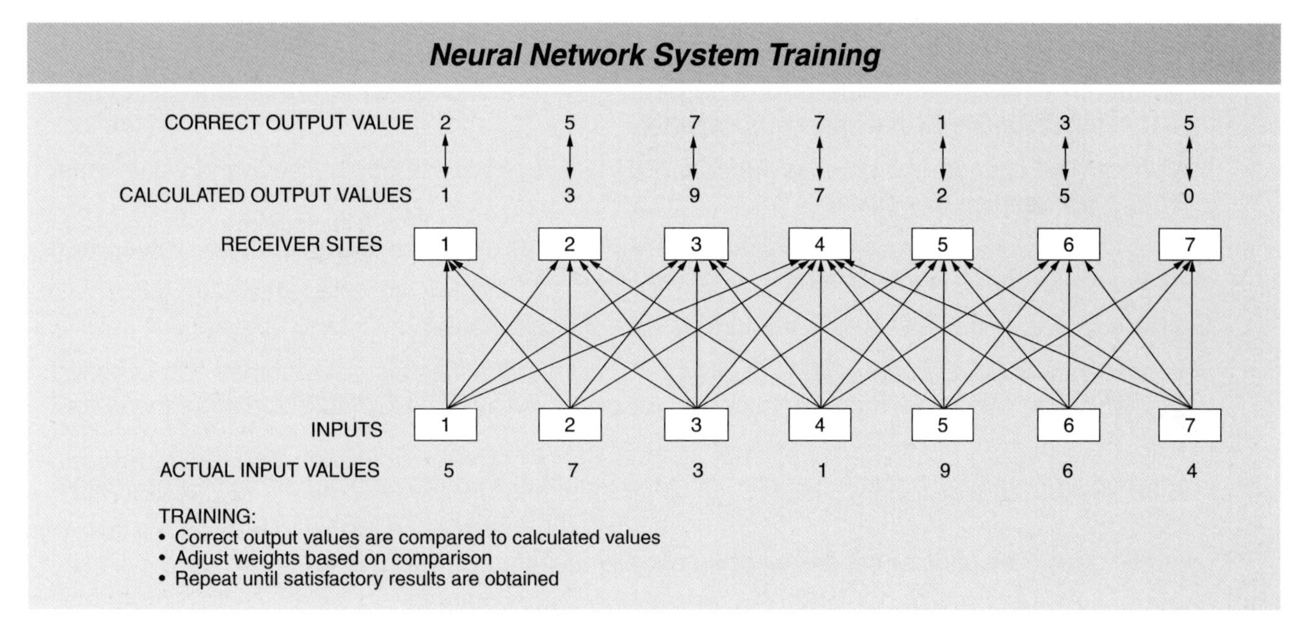

Figure 36-19. A neural network uses a number of identical simple inputs, each of which is connected to multiple receivers with weighted values.

KEY TERMS

- *ON/OFF,* or *two-position, control:* A method of changing the output of a controller that provides only an ON or OFF signal to the final element of the process.
- *deadband:* The range of values where a change in measurement value does not result in a change in controller output.
- *dead time:* The period of time that occurs between the time a change is made to a process and the time the first response to that change is detected.
- *overshoot:* The change of a process variable (PV) that exceeds the upper deadband value when there is a disturbance to a system.
- *undershoot:* The change of a process variable (PV) that goes below the lower deadband value when there is a disturbance to the system.
- *heat anticipator:* A small electric heater that is part of a thermostat.
- *time proportional ON/OFF controller:* An ON/OFF controller that has a predetermined output period during which the output contact is held closed (or power is ON) for a variable portion of the output period.
- *proportional (P) control:* A method of changing the output of a controller by an amount proportional to an error.
- *controller gain:* The gain, or sensitivity, of a controller itself.
- *proportional gain:* The gain, or sensitivity, of a proportional term only.
- *proportional band:* The range of input values that corresponds to a full range of output from a controller, stated as a percentage.
- *throttling range:* The number of units of the process variable that causes the actuator to move through its entire range.
- *output bias,* or *manual reset:* A controller function that positions a final element in a central position when the process variable is at setpoint.
- *offset:* A permanent difference in measurement between a process variable (PV) and a setpoint (SP) as a result of proportional control action.
- *open loop response diagram:* A curve that shows the controller response to a given measurement without the controller actually being connected to the process.
- *integral (I) control:* A method of changing the output of a controller by an amount proportional to an error and the duration of that error.
- *automatic reset,* or *reset action:* The integral controller mode, named so because in older proportional-only controllers the operator had to manually reset the setpoint.
- *integral time (T_I),* or *reset time:* The time it takes for a controller to change a control variable (CV) by 1% for a 1% change in the difference between the control variable and the setpoint (SP).
- *reset rate:* The reciprocal of integral time.
- *integral gain (K_I):* The integral time multiplied by the controller gain, if present.
- *proportional-integral (PI) control:* Proportional control combined with integral control.
- *derivative (D) control:* A method of changing the output of a controller in proportion to the rate of change of the process variable.

KEY TERMS *(continued)*

- *proportional-integral-derivative (PID) control:* Proportional control combined with both integral control and derivative control.
- *adaptive gain control:* A control strategy where a nonlinear gain can be applied to a nonlinear process.
- *feedforward control:* A control strategy that only controls the inputs to a process without feedback from the output of the process.
- *cascade control:* A control strategy where a primary controller, which controls the ultimate measurement, adjusts the setpoint of a secondary controller.
- *ratio control:* A control strategy used to control a secondary flow to the predetermined fraction, or flow ratio, of a primary flow.
- *artificial intelligence (AI):* A form of computer programming that simulates some of the thinking processes of the human brain.
- *diagnostic system:* A logic-based system that monitors various operational conditions to detect faults.
- *knowledge system:* An enhanced diagnostic system that not only identifies a fault but also makes decisions on the probable cause of the fault.
- *expert system:* A program that contains a database of known facts and a rule base that uses "if-then" statements for groups of facts to produce outputs.
- *fuzzy logic:* A system that uses mathematical or computational reasoning based on undefined, or fuzzy, sets of data derived from analog inputs to establish single or multiple defined outputs.
- *neural network:* A computer program that simulates the parallel inputs and linking of information of the human brain to arrive at an educated output determination.

REVIEW QUESTIONS

1. What is the difference between ON/OFF control and proportional control?
2. How are proportional band and proportional gain related?
3. Define output bias and explain why it is necessary.
4. Define proportional-integral (PI) control and list some advantages and disadvantages of PI control.
5. What are the typical types of advanced control strategies?

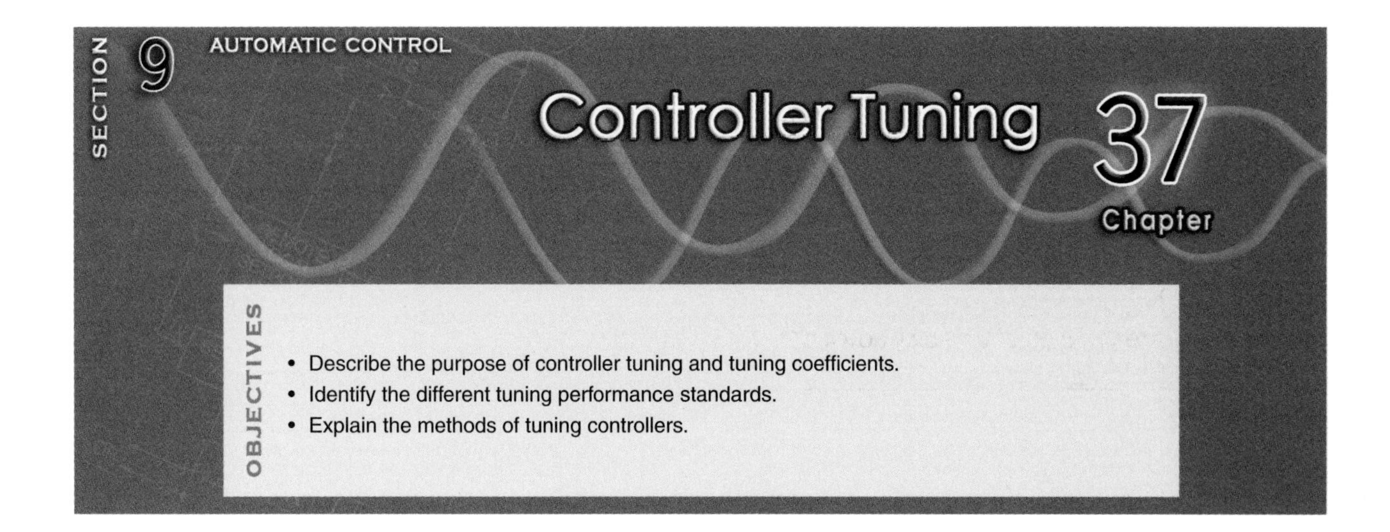

TUNING COEFFICIENTS

Every process has different characteristics and the controllers must be optimized for that process. *Controller tuning* is the process of determining the tuning coefficients for the PID controller proportional gain, integral time, and derivative time to obtain a desired controller response to process disturbances. Common concerns with controller tuning are selecting the proper tuning coefficients, developing appropriate performance standards, using self-tuning controllers, and choosing a manual tuning method.

Controllers from different manufacturers use different terms for the tuning coefficients. Common terms for proportional control are controller gain (K_C), proportional gain (K_P), and proportional band (PB). The terms K_C and K_P are often mixed in common usage. Often only the K is used without any subscripts. Common terms for integral control are reset time (T_I), reset rate (R_I), and reset gain (K_I). Common terms for derivative control are derivative time (T_D) and derivative gain (K_D). These coefficients are all used when tuning controllers for PID control. However, not all combinations are used. **See Figure 37-1.**

This variation in terminology can be very confusing. Fortunately, any loop that can be tuned with one set of terms works just as well with another set of terms. The terms are related as follows:

$$K_P = K_C = \frac{100}{PB}$$

$$T_I = \frac{1}{R_I} = \frac{K_C}{K_I}$$

$$T_D = \frac{K_D}{K_C}$$

Before digital controllers were developed, proportional band, reset rate or time, and derivative time were almost universally used. Since then, some manufacturers kept the terms that were used with pneumatic or electric analog controllers. Other manufacturers changed the coefficients to individual gains for each function. Still other manufacturers switched to a single controller gain that applied to all the controller PID functions.

Common PID Tuning Coefficients

Proportional	Integral (Reset)	Derivative
PB	R_I	T_D
PB	T_I	T_D
K_C	R_I	T_D
K_C	T_I	T_D
K_P	K_I	K_D

Figure 37-1. Different manufacturers use different tuning coefficients. Each row represents a combination that may be used together.

TUNING PERFORMANCE STANDARDS

Performance standards are a way of measuring the effectiveness of controller tuning and controller action. There are a number of very different desired performance standards, depending on the process being controlled. The performance standards can often be defined by the closed loop response of a second-order system. Common controller performance standards are decay ratio, overshoot, and dead time. **See Figure 37-2.**

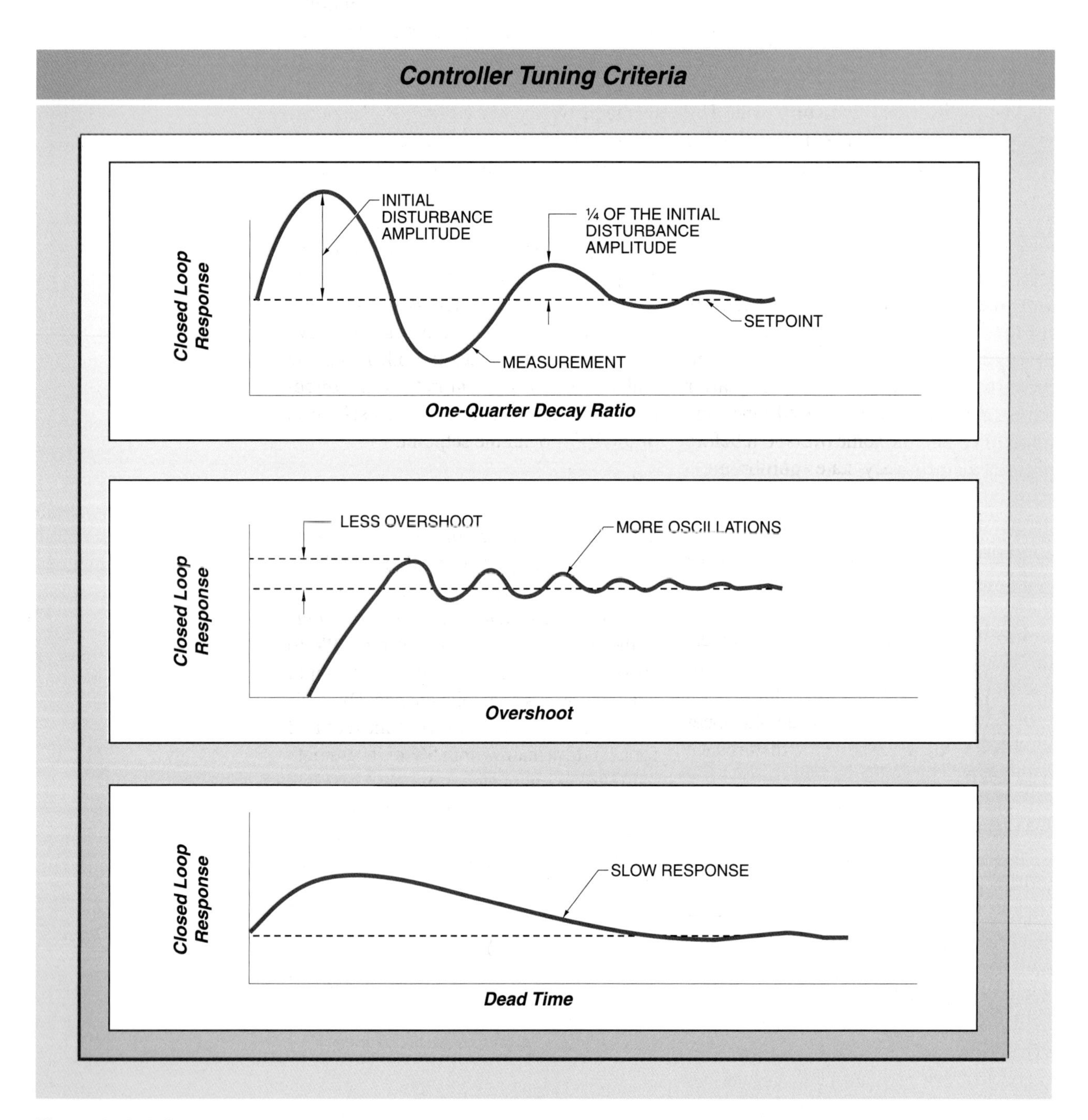

Figure 37-2. Different controller tuning criteria are used depending on the needs of different processes.

Decay Ratio

A very common controller performance standard is the decay ratio. The *decay ratio* is a measure of how quickly an overshoot decays from one oscillation to the next as the controller brings the process to the setpoint. A common decay ratio used as a performance standard is the one-quarter decay ratio. A *one-quarter decay ratio* is a response where the amount of overshoot decays to one-fourth of the previous amplitude of the overshoot every whole cycle after being upset by a disturbance. The one-quarter decay ratio is a compromise between rapid initial response and a quick return to the setpoint.

One reason the one-quarter decay ratio controller performance standard is so popular is that it is very easy to judge when it is obtained. Another reason is that the response is very close to the optimum in which the amplitude and time away from the setpoint is minimized. One-quarter decay ratio tuning can be used when the process can tolerate some overshoot before reaching a new steady-state equilibrium.

Overshoot

There are times when the one-quarter decay ratio is not acceptable. For example, the rate of reaction of a chemical process increases with increases in temperature. An exothermic reaction gives off heat as a result of the reaction. *Overshoot* is the change of a process variable (PV) that exceeds the upper deadband value when there is a disturbance to a system.

If the controller allows the reactor to get too hot because of overshoot, the reaction rate can increase beyond the ability of the reactor cooling system to remove the excess heat. The temperature control of chemical reactors generally must be tuned to have very little temperature overshoot from disturbances, such as changes in setpoint. Excessive overshoot is not desirable since it can start a runaway reaction. The process may be allowed to oscillate around the new setpoint for longer periods of time than with the one-quarter decay ratio response. **See Figure 37-3.**

Dead Time

Dead time is the period of time that occurs between the time a change is made to a process and the time the first response to that change is detected. Processes with a lot of dead time must be tuned to respond slowly and minimize overshoot. Therefore, the proportional gain is small and the integral time (min/R) is large. Under these controller settings, the response to a disturbance does not have any oscillation.

A controller must not change the control element, typically a valve, faster than the dead time allows the process to respond. The process needs time to respond to changes in the control valve or else the valve will be driven to its limits before the process responds. The valve remains at its extreme position until the process responds and the measurement moves across the setpoint. The controller then changes the valve quickly until it reaches the other extreme and crosses the setpoint in the other direction. This results in continuous oscillation of the PV around the setpoint.

Other Standards

There are other performance standards used in industry. These other standards may be part of a specifications package for controller performance standards. *Rise time* is the length of time required for a PV to cross the ultimate value after a step input change, such as a setpoint change. *Dynamic response time* is the length of time required for a PV to remain within 5% of its ultimate value following a step input change. *Gain margin* is the factor by which controller gain may be increased before instability occurs, and therefore is a measure of relative stability. **See Figure 37-4.**

TECH FACT

A boiler operating control is a boiler pressure control that senses steam pressure and automatically signals the burner to initiate the burner startup and shutdown sequence as needed to meet steam demand.

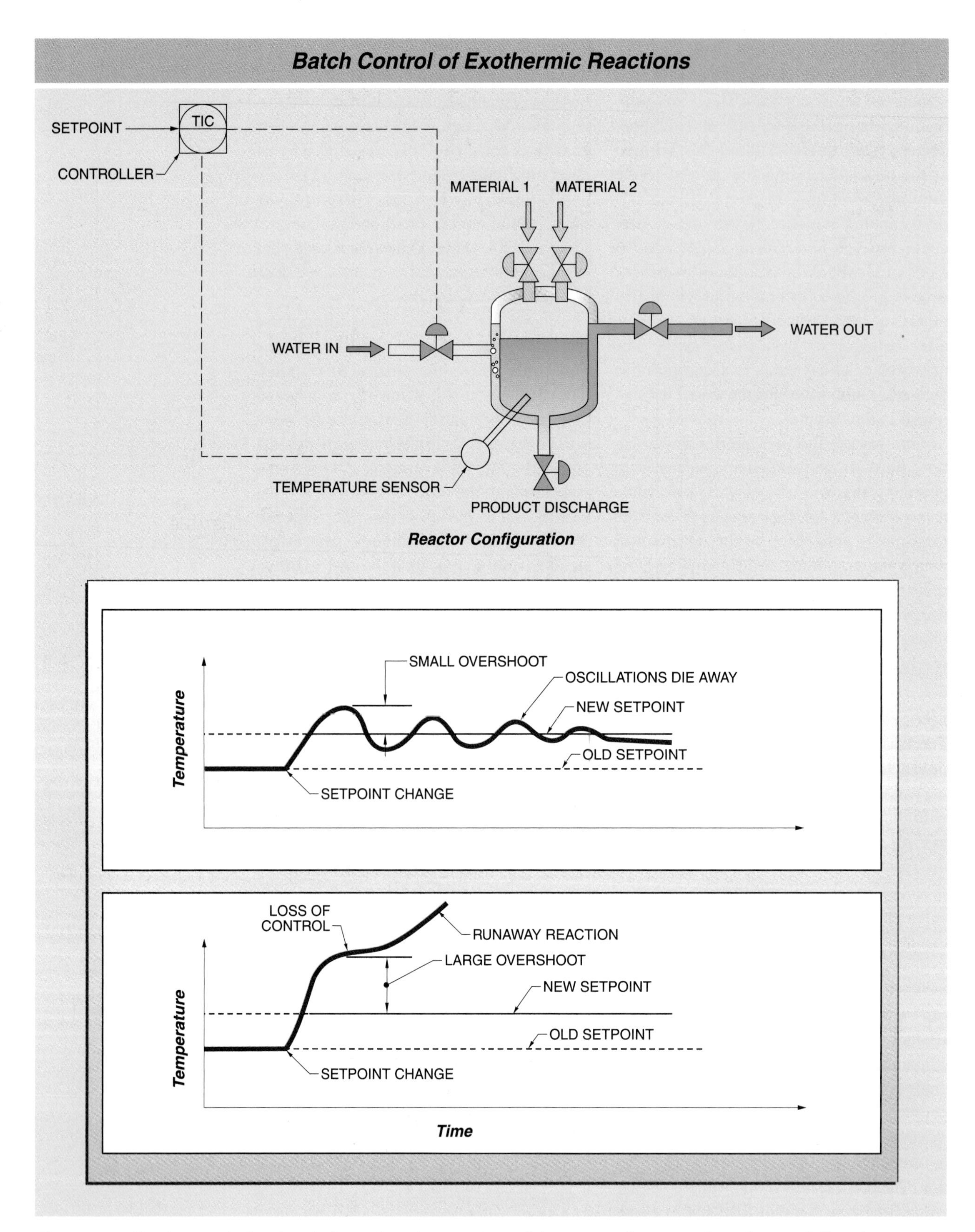

Figure 37-3. Batch control of a chemical reaction often requires minimum overshoot to reduce the possibility of a runaway chemical reaction.

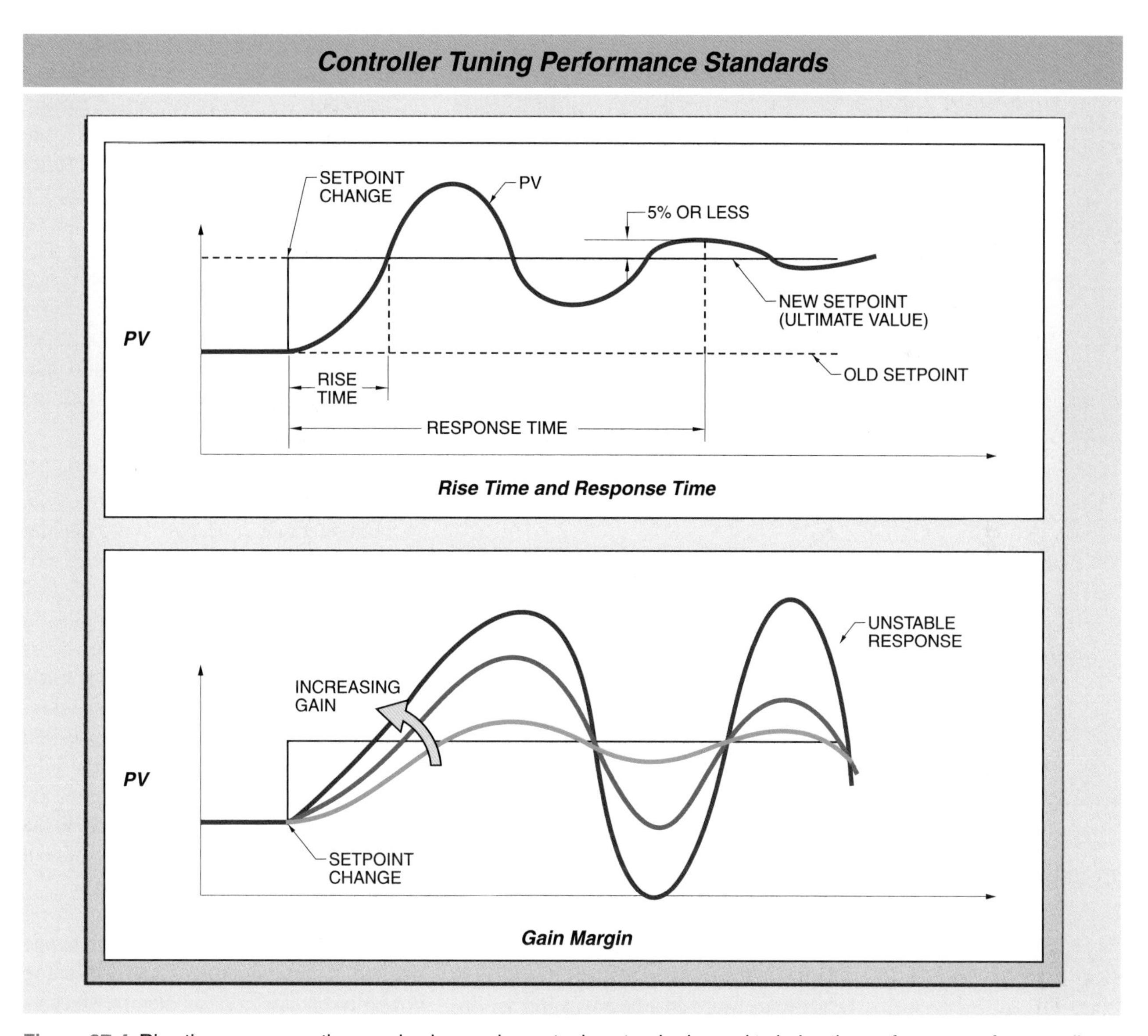

Figure 37-4. Rise time, response time, and gain margin are tuning standards used to judge the performance of a controller.

TUNING METHODS

A *self-tuning controller* is a controller that has built-in algorithms or pattern recognition techniques that periodically test the process and make changes to the controller tuning settings while the process is operating. Many stand-alone microprocessor-based controllers have built-in self-tuning programs. A *stand-alone controller* is a controller that has its power supplies, input signal processing, controller functions, output signals, and displays contained in the same case. The self-tuning methods differ between manufacturers, but all do a reasonable job for common control loops. However, there are many occasions when manual tuning is needed to obtain acceptable control.

Although the majority of digital controllers have some sort of built-in automatic tuning algorithm, they usually take a period of time before the proper tuning is achieved. During this period, the controller could be badly out of control. It is far better to be able to make a relatively quick manual tuning and then let the automatic tuning algorithm complete the tuning.

Pump controls are a common part of automatic control loops.

There are many controllers in service that do not have automatic tuning algorithms. In addition, a particular process may need to be tuned to a different set of standards than what the automatic tuning controllers achieve. As more experience is obtained using manual tuning techniques, the tuning procedure becomes easier and quicker. The goal is to have the tuning parameters set close to the optimum settings when the controller is initially installed.

Pretuning Checks

There are three checks that must be made to a control loop prior to placing it in operation. First, ensure that the controller action, direct or reverse, is selected correctly. Begin by assuming that the PV (measurement) increases, then mentally follow the actions taken by each succeeding portion of the control loop, ending back at the measurement. The result should be the opposite of the original action. If the result is not the opposite of the initial action, the controller action needs to be reversed.

Second, determine which controller functions are needed for the process. As a general rule, PI controllers satisfy most control loop requirements. The derivative function is needed for the temperature control of exothermic processes and in other processes that contain a large capacitance. The derivative function is usually not applied to flow control, analysis control, level control, or any noisy measurement.

Third, determine the initial tuning values that are to be used for the chosen controller functions. The initial controller settings must be safe. This means that the measurement must not go into oscillation when it is put into automatic control. The safest way is to select low proportional gains (high proportional bands) and larger integral settings (min/R). If the controller oscillates, set the proportional gain to one-half of the original value. Repeat until the process is stable.

Setpoint Step Change Method

A common tuning procedure is the setpoint step change method. The *setpoint step change method* is a manual tuning method that consists of making small changes in the setpoint and observing the responses. A series of controller tuning changes is made until the controller response is satisfactory. The objective is to obtain the desired controller response. It is important to pick a time when the process operation is steady so that there are no other disturbances to the process while the response to the intentional disturbance is being completed.

Use the following procedure to tune a controller with the setpoint step change method. It must be remembered that tuning controllers is not a casual exercise since an operating process is involved.

1. The controller must be in auto mode and the measurement should be steady. Note the controller output value so that if the control loop starts to oscillate it can be switched to manual and the output adjusted to its original value.
2. Start by adjusting the proportional gain. Make a small change in the controller setpoint and watch the control system response. If an automatic recording device is not available, write down the measurement and output values frequently until the measurement is no longer moving. **See Figure 37-5.** Compare the control system response to the curves shown and make the

proportional gain tuning adjustments shown next to the curve.

3. Repeat step 2 until the controller response is satisfactory. Each setpoint change should be in the direction opposite from that of the previous change so that the process remains near its original setpoint.

4. At this point, the integral time can be adjusted. Make a small change in the controller setpoint and watch the control system response. Compare the control system response to the curves shown and make the integral time tuning adjustments shown next to the curve. **See Figure 37-6.**

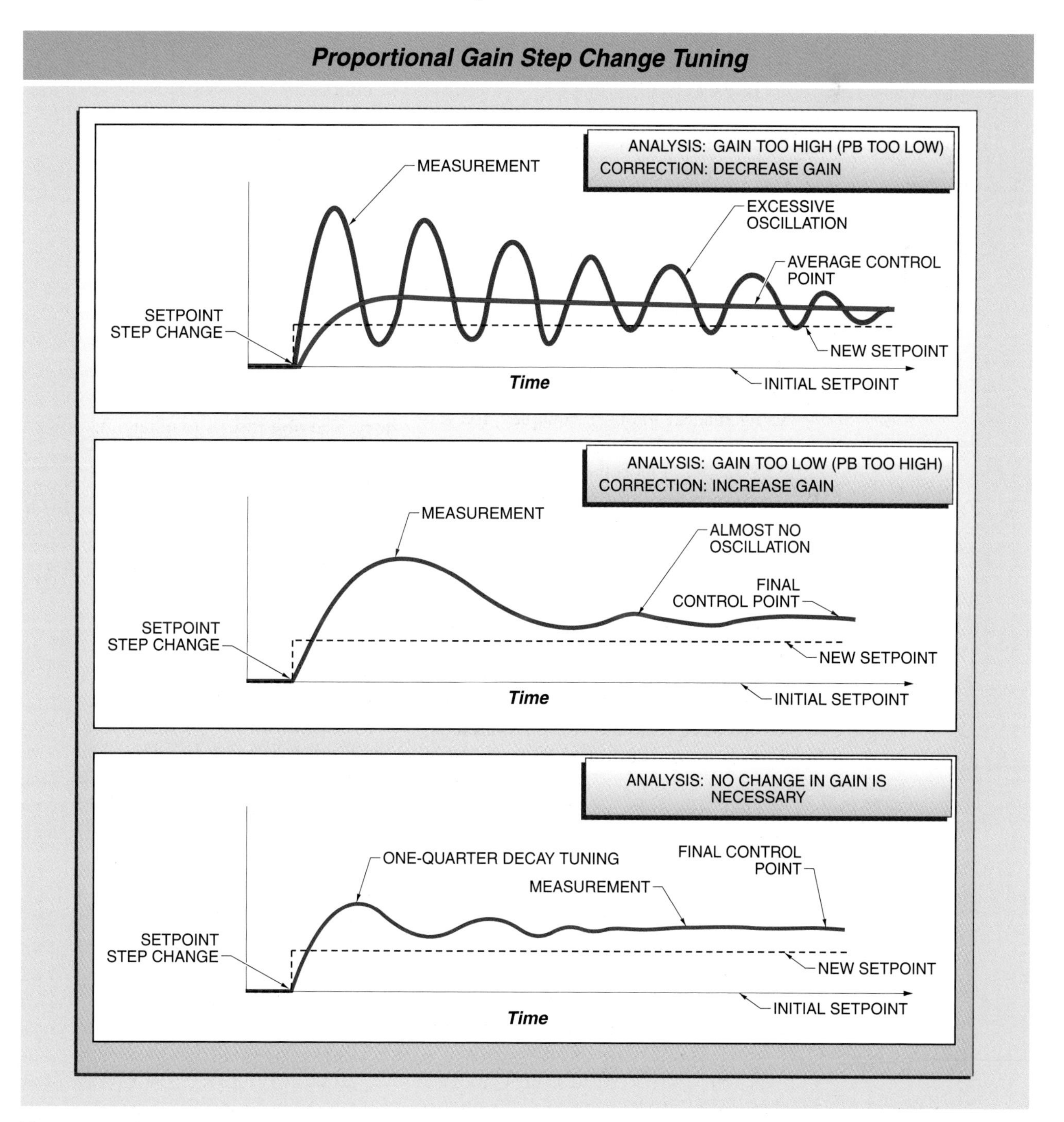

Figure 37-5. Closed loop controller responses for a proportional control action can be used to make tuning changes.

5. Repeat step 4 until the controller response is satisfactory. Adding integral action may require an increase in the proportional gain.
6. If a derivative function is needed in the controller, set it to a value in minutes no larger than the integral time. If derivative is added at this time it may be necessary to go back to step 1 and reduce the proportional gain.

The setpoint step change method takes time to accomplish, but it is a safe method that has a low impact on the process operation.

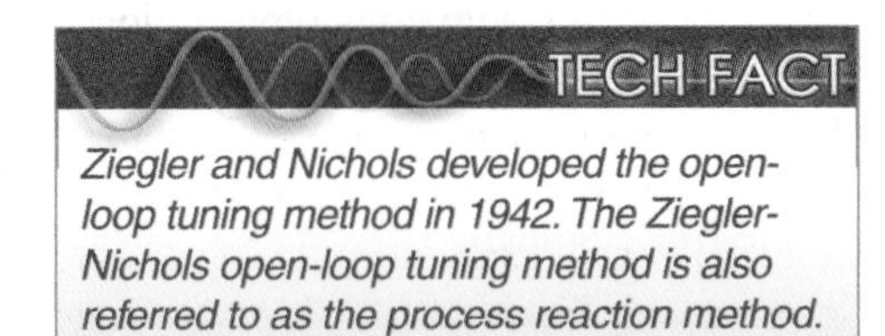

Ziegler and Nichols developed the open-loop tuning method in 1942. The Ziegler-Nichols open-loop tuning method is also referred to as the process reaction method.

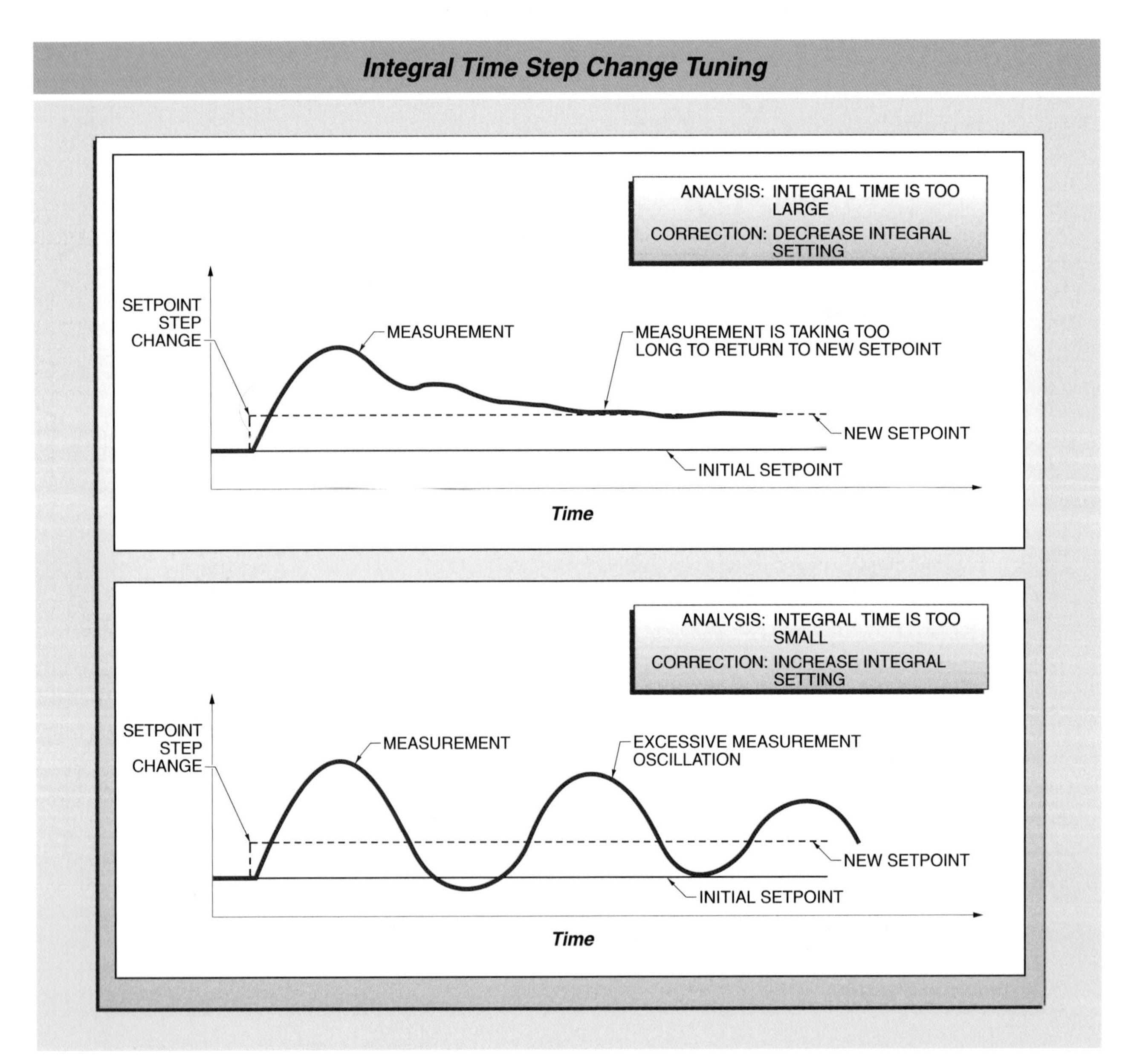

Figure 37-6. Closed loop controller responses for an integral control action can be used to make tuning changes.

Tuning Map Method

The *tuning map method* is a procedure for controller tuning that compares process curves to one of numerous typical closed loop response curves. The position of the matching response curve on a map of various curves provides the directions to be taken for changes in controller tuning settings. **See Figure 37-7.** The advantage of this method is that all the controller parameters can be adjusted at the same time. The disadvantage of this method is that it can be very difficult to match the present controller response to the correct response on the tuning maps. Selecting the wrong response on the map can lead to making the wrong tuning changes.

For example, if a PID process response curve looks like the curve labeled A in the figure and curve B is the desired response, the adjustments required can be determined from the tuning map. The proportional band needs to be decreased because point A is in the right column and point B is in the middle column. The derivative time must be increased because point A is in the bottom row and point B is in the middle row. The reset time must be increased because point A is in the front set of curves and point B is in the middle set of curves. The tuning map gives the direction of change for the tuning constants but it does not give the magnitude of the required changes.

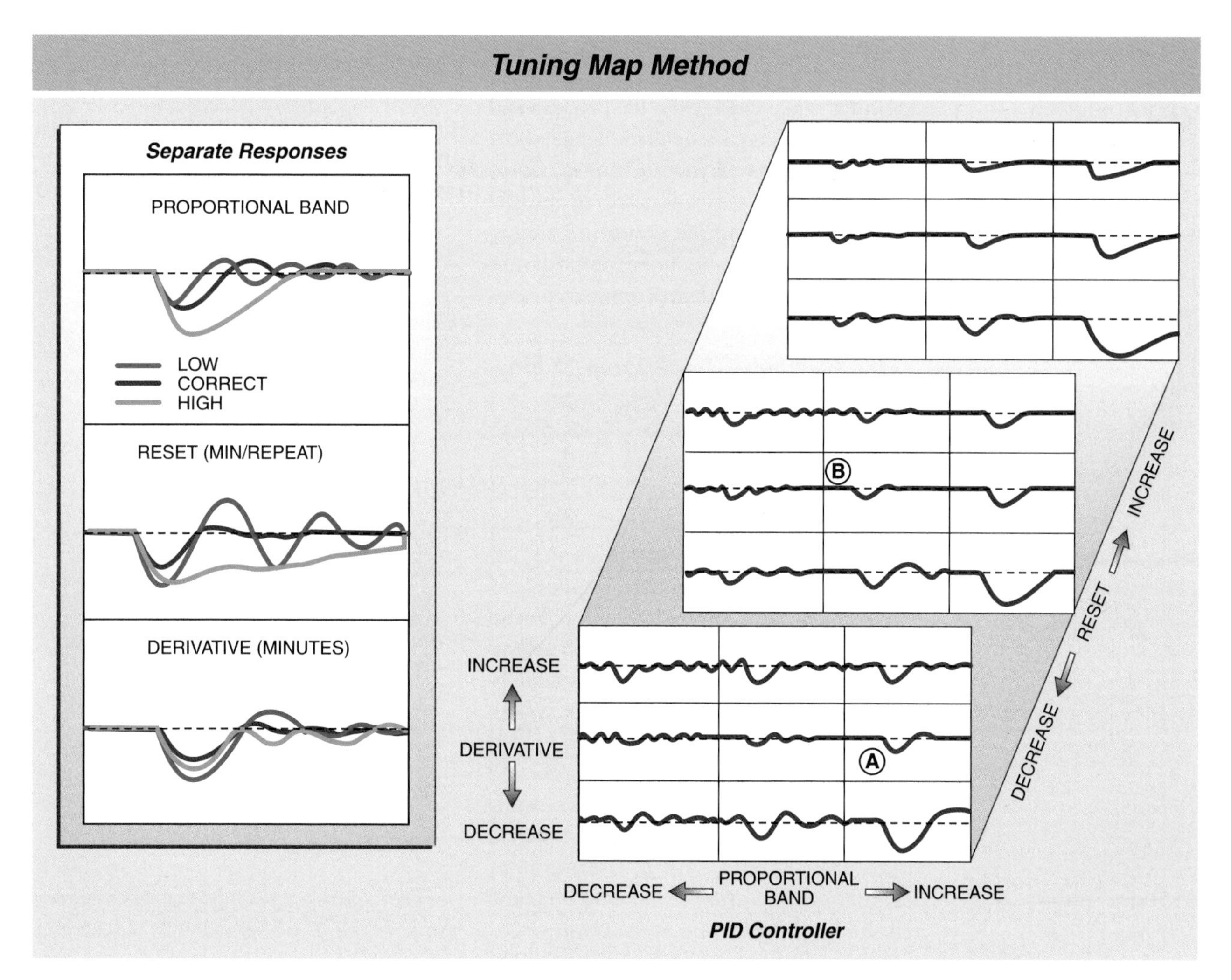

Figure 37-7. The tuning map method is used to recommend the direction of changes to tuning coefficients.

TECH-FACT

Caution must be used in programming control loops. Control functions will execute in the order specified, even if that order is incorrect. Scheduling the analog output first, the PID next, and the analog input last adds a large and needless delay to the overall loop because the loop cycles three times before a change in output occurs.

Ziegler-Nichols Tuning Methods

There are several other methods for the manual tuning of controllers that have been published in many sources. The most common of these other tuning methods are the Ziegler-Nichols open-loop tuning method, or process reaction curve method, and the Ziegler-Nichols closed-loop tuning method, or ultimate period method. Both methods make changes to the process and measure process parameters related to how the system responds to the changes. These process parameters are used in simple equations to determine controller tuning constants. Ziegler-Nichols methods are empirical and intended to achieve a one-quarter decay ratio.

Ziegler-Nichols Closed Loop Tuning. *Ziegler-Nichols closed loop tuning* is a method of tuning a controller by increasing the gain until the system cycles at the point of instability. This method works well, but forcing an active process into oscillation may be dangerous. Anyone using this method must be sure to understand the implications of forcing the process into oscillation. To use this method, remove the integral and derivative functions from the controller. Gradually increase the controller gain until the system starts to oscillate. **See Figure 37-8.**

The setpoint may need to be adjusted to create a disturbance to start the oscillation. An *ultimate gain (K_u)* is the closed loop proportional gain at the point of oscillation. An *ultimate period (T_u)* is the closed loop cycle time at the point of oscillation. The values of the ultimate gain and the ultimate period are used with the Ziegler-Nichols equations to determine the PID controller gain, integral time, and derivative time as follows:

$$K = 0.6 \times K_u$$

$$T_I = \frac{T_u}{2}$$

$$T_D = \frac{T_u}{8}$$

where

K = proportional gain

K_u = ultimate gain

T_I = integral time

T_u = ultimate time

T_D = derivative time

For example, a PID controller is used to control the temperature of a process. The process is at steady state with a temperature of 150°F and a gain of 2. Increase the gain to 3 and the setpoint to 155°F. At this point, the process begins to oscillate and the setpoint is returned to 150°F. The oscillation quickly begins to damp out. Increase the gain to 4 and observe. In this case, the oscillation begins to increase in magnitude. This means that the gain is too high, so the gain is reduced to 3.5. At this point, the oscillation reaches steady state. Measure the period of the oscillation. In this case, the period is 10 minutes. The PID tuning constants are calculated as follows:

$$K = 0.6 \times K_u$$

$$K = 0.6 \times 3.5$$

$$K = \mathbf{2.10}$$

$$T_I = \frac{T_u}{2}$$

$$T_I = \frac{10}{2}$$

$$T_I = \mathbf{5\ min}$$

$$T_D = \frac{T_u}{8}$$

$$T_D = \frac{10}{8}$$

$$T_D = \mathbf{1.25\ min}$$

The controller settings for this situation are K = 2.10, T_I = 5 min, and T_D = 1.25 min. This achieves a response close to a one-quarter decay response.

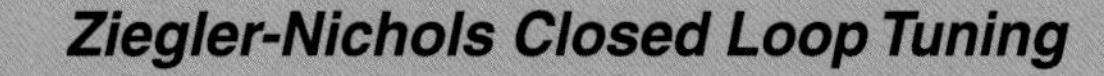

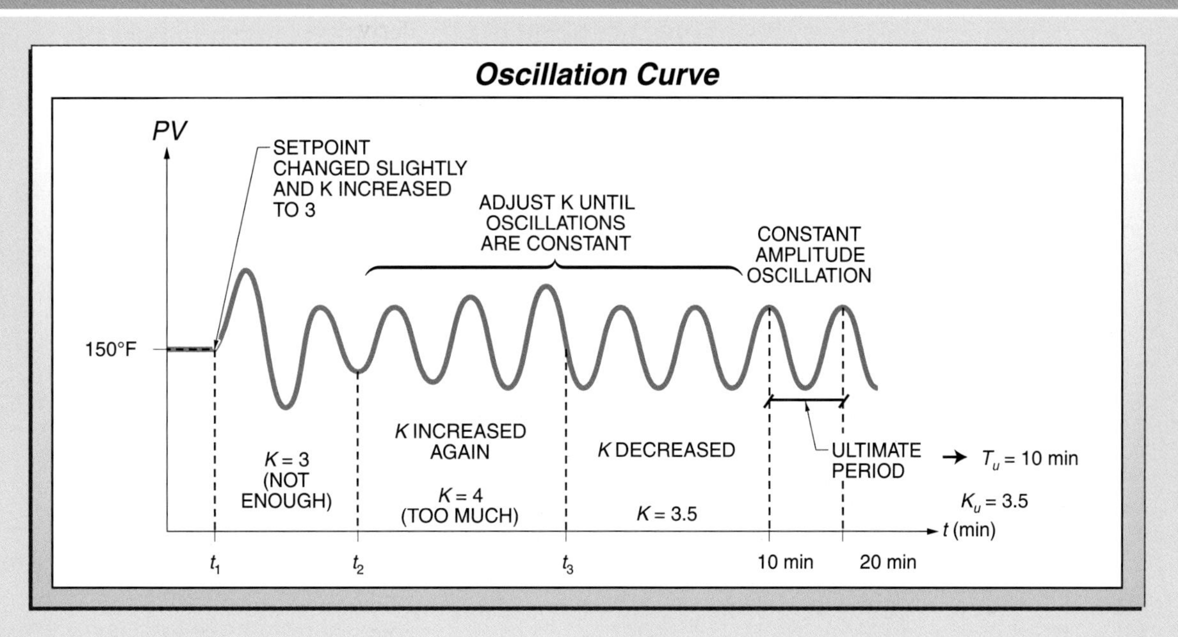

Type of Controller	Loop Tuning Constant	Tuning Equation
Proportional (P)	K	$K = (0.5)(K_u)$
Proportional-Integral (PI)	K	$K = (0.45)(K_u)$
	T_I	$T_I = \frac{T_u}{1.2}$
Proportional-Integral-Derivative (PID)	K	$K = (0.6)(K_u)$
	T_I	$T_I = \frac{T_u}{2}$
	T_D	$T_D = \frac{T_u}{8}$

Equations

Figure 37-8. Ziegler-Nichols closed loop tuning sets a process into oscillation and then measures the period and gain.

Ziegler-Nichols Open Loop Tuning. *Ziegler-Nichols open loop tuning* is a method of tuning a controller based on open loop response to a step input. Graph the process reaction curve after making a step change in the process. **See Figure 37-9.** The process reaction curve shows a slow initial response away from steady state, an increasing response, and a slow final response toward a new steady state. Draw a tangent at the inflection point of the curve where the slope is at its maximum. The rate of change (R) is the slope of the tangent line. The lag time (L) is the time from the step change to the time when the tangent crosses the original steady-state value. The values L and R are used to determine the controller settings as follows:

$$K = \frac{1.2 \times \Delta CV}{L \times R}$$

$$T_I = 2 \times L$$

$$T_D = 0.5 \times L$$

where
K = proportional gain
ΔCV = change in CV
L = lag time
R = slope of tangent
T_I = integral time
T_D = derivative time

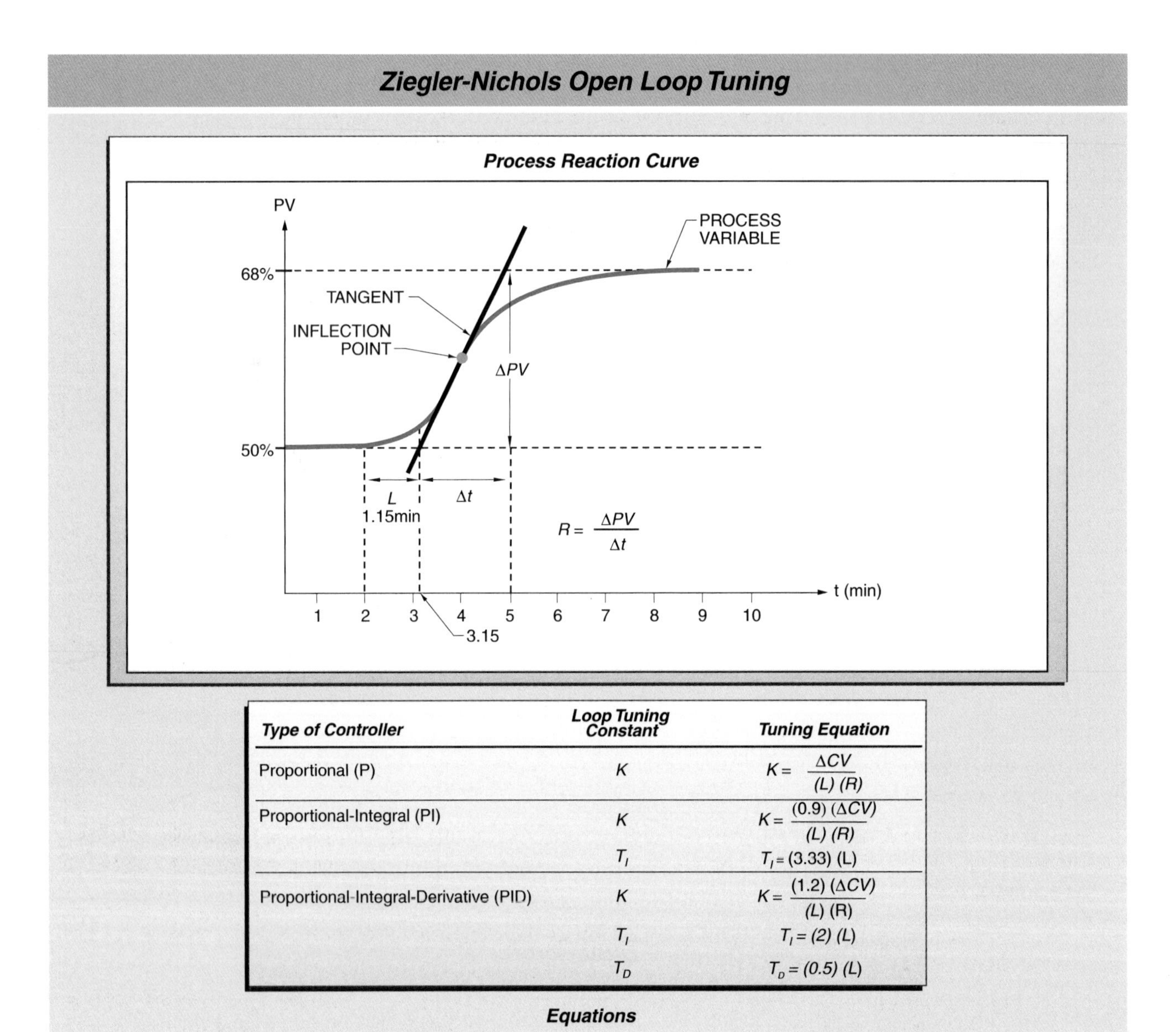

Type of Controller	Loop Tuning Constant	Tuning Equation
Proportional (P)	K	$K = \frac{\Delta CV}{(L)(R)}$
Proportional-Integral (PI)	K	$K = \frac{(0.9)(\Delta CV)}{(L)(R)}$
	T_I	$T_I = (3.33)(L)$
Proportional-Integral-Derivative (PID)	K	$K = \frac{(1.2)(\Delta CV)}{(L)(R)}$
	T_I	$T_I = (2)(L)$
	T_D	$T_D = (0.5)(L)$

Figure 37-9. Ziegler-Nichols open loop tuning measures the response of a process to a step change in input.

For example, the Ziegler-Nichols open-loop tuning method was used to obtain a process response curve. The controller output was increased from 50% to 61%, causing the PV to increase from 50% to 68%. The tangent line intersects the original steady-state line at 3.15 minutes. The change was made at 2 minutes. Therefore, the lag time is 1.15 minutes (3.15 – 2 = 1.15). The slope of the tangent line is the rise over the run. The rise is the change in the PV, or 18% (68% – 50% = 18%).

The Ziegler-Nichols open-loop tuning method is also known as the reaction curve tuning method. The setpoint is changed slightly and the process is monitored to determine the time delay, the amount of change, and the rate of change in the process. These measured values are used to calculate the gain, reset, and derivative tuning constants. This method only works with a stable process that does not go into oscillation with a change in setpoint.

The run is the amount of time for the tangent line to run from the original steady state to the new steady state. The tangent intersects the original steady state at 3.15 minutes and it intersects the new steady state at 5 minutes. Therefore, the run is 1.85 minutes (5 – 3.15 = 1.85). The slope is 18% ÷ 1.85 min, or 9.73 %/min. The PID tuning constants, using the Ziegler-Nichols open loop equations, are calculated as follows:

$$K = \frac{1.2 \times \Delta CV}{L \times R}$$

$$K = \frac{1.2 \times 11}{1.15 \times 9.73}$$

$$K = \mathbf{1.18}$$

$$T_I = 2 \times L$$

$$T_I = 2 \times 1.15$$

$$T_I = \mathbf{2.30\ min}$$

$$T_D = 0.5 \times L$$

$$T_D = 0.5 \times 1.15$$

$$T_D = \mathbf{0.575\ min}$$

An advantage of this method is that it is safer than the closed loop method because it does not force the process into oscillation. A disadvantage is that it is sometimes difficult to identify all the key parameters with enough accuracy to establish the correct tuning values.

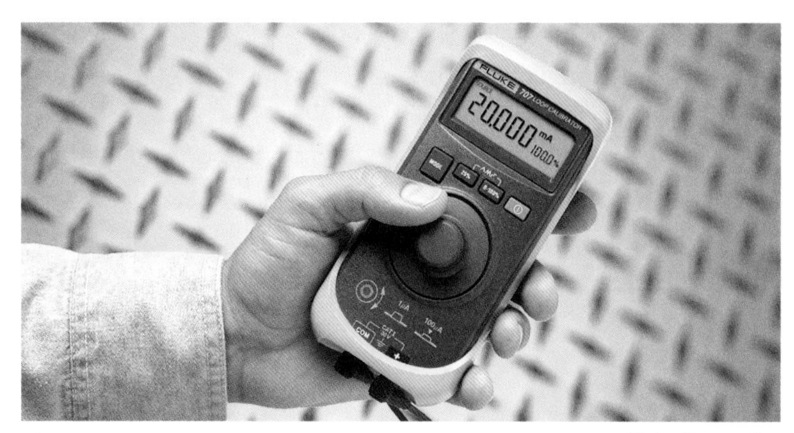

Fluke Corporation

A loop calibrator is used to calibrate instruments, transmitters, and controllers in a control loop.

KEY TERMS

- *controller tuning:* The process of determining the tuning coefficients for the PID controller proportional gain, integral time, and derivative time to obtain a desired controller response to process disturbances.
- *decay ratio:* A measure of how quickly an overshoot decays from one oscillation to the next as the controller brings the process to the setpoint.
- *one-quarter decay ratio:* A response where the amount of overshoot decays to one-fourth of the previous amplitude of the overshoot every whole cycle after being upset by a disturbance.
- *overshoot:* The change of a process variable (PV) that exceeds the upper deadband value when there is a disturbance to a system.
- *dead time:* The period of time that occurs between the time a change is made to a process and the time the first response to that change is detected.
- *rise time:* The length of time required for a PV to cross the ultimate value after a step input change, such as a setpoint change.
- *dynamic response time:* The length of time required for a PV to remain within 5% of its ultimate value following a step input change.
- *gain margin:* The factor by which controller gain may be increased before instability occurs, and therefore is a measure of relative stability.
- *self-tuning controller:* A controller that has built-in algorithms or pattern recognition techniques that periodically test the process and make changes to the controller tuning settings while the process is operating.

KEY TERMS (continued)

- *stand-alone controller:* A controller that has its power supplies, input signal processing, controller functions, output signals, and displays contained in the same case.
- *setpoint step change method:* A manual tuning method that consists of making small changes in the setpoint and observing the responses.
- *tuning map method:* A procedure for controller tuning that compares process curves to one of numerous typical closed loop response curves.
- *Ziegler-Nichols closed loop tuning:* A method of tuning a controller by increasing the gain until the system cycles at the point of instability.
- *ultimate gain (K_u):* The closed loop proportional gain at the point of oscillation.
- *ultimate period (T_u):* The closed loop cycle time at the point of oscillation.
- *Ziegler-Nichols open loop tuning:* A method of tuning a controller based on open loop response to a step input.

REVIEW QUESTIONS

1. List and describe the most common controller performance standards.
2. What are the three common pretuning checks that must be made before tuning a controller?
3. What is the setpoint step change method of tuning, and what are the advantages and disadvantages of this method?
4. What is the tuning map method, and what are the advantages and disadvantages of this method?
5. Define the two types of Ziegler-Nichols tuning methods and list advantages and disadvantages of each.

SECTION 9 AUTOMATIC CONTROL

Digital and Electric Controllers

Chapter 38

OBJECTIVES

- List and define the different types of digital controllers and control systems.
- Define electric controller and describe the types of electric controllers.
- Describe the common operator interfaces.
- Describe the various configuration formats.

DIGITAL CONTROLLERS

A *digital controller* is a controller that uses microprocessor technology and special programming to perform the controller functions. Instead of mechanical linkages, pneumatic pressures with bellows or diaphragms, or electronic circuits with operational amplifiers, a digital controller uses mathematical equations. Analog inputs are converted to digital numbers that are processed by the controller equations and then converted back to analog outputs.

Digital controllers have provided a whole new approach to the use of control systems. In the past, the type of controller that was desired had to be exactly spelled out in the instrument specification sheet. That process is now considerably simplified and has been replaced with the configuration of the controller after it has been delivered. Digital controllers offer lower costs, smaller size, greater accuracy, self-tuning, and intercommunication between controllers and other systems.

Common usage often describes controllers as small or large. It is difficult to place a label like this on any type of digital system, since either type of digital system can be used for either small or large installations. The term "small" generally refers to those digital devices that are more economical for use with small process systems. The smaller systems also are more limited in the number of I/O points that can be handled without a reduction in scan time.

The term "large" generally refers to those digital systems that are the most economical choice for use with large processes. Large distributed control systems and programmable logic controllers offer large data storage capabilities and a tremendously increased capability for creative control strategies. In the past, distributed control systems and programmable logic control systems were quite distinct from each other, but as digital technology has progressed, the systems have been growing closer to each other in configuration and programming software, appearance, and architecture.

A burner control system uses inputs from many instruments to control the operation of a boiler.

Stand-Alone Digital Controllers

A *stand-alone digital controller* is a general type of microprocessor-based controller with all required operating components enclosed in one housing. Many stand-alone controllers are single-loop controllers, but many also include multiple loops. The selection of the desired controller functions, the input type and measurement range, controller actions, alarms, and other functions is done during configuration. **See Figure 38-1.**

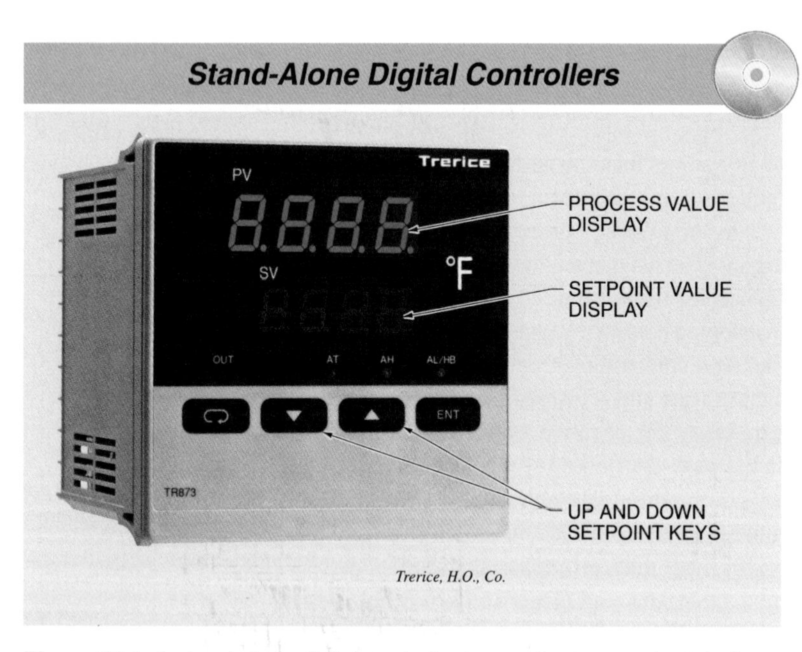

Figure 38-1. A stand-alone digital controller has a simple operator interface.

Configuration is the selection of preprogrammed software packages embedded in a controller representing available features that can be chosen. Each digital controller manufacturer develops and supplies their own configuration method for their controllers. The configuration choices offer a wide variety of some very powerful functions.

Stand-alone digital controllers typically mount in control panels. Digital controllers are available in sizes from 1/32″ DIN (approximately 24 mm × 45 mm) up to 1/4″ DIN (approximately 92 mm × 92 mm). DIN is an international standard for sizing instrument cases and is based on metric dimensions.

Stand-alone digital controllers usually have only one controller function and one output, but may have two or more inputs. The inputs usually accept any type of signal, but may not provide DC power for a transmitter. There are other stand-alone digital controllers that can contain multiple control functions and multiple inputs and outputs. Most controllers contain a self-tuning function to adjust the controller settings automatically. The method that is used to accomplish this varies and it may take some time, but the results are generally fairly good.

The display on the front of the instrument typically has digital values of the PV, the setpoint, and the output. The larger and the more expensive controllers also have bar graph displays of these three variables. There are also up and down pushbuttons for the setpoint adjustment. The controllers may or may not have a manual function with auto/manual switching. Stand-alone digital controllers generally have no provision for data collection or trend displays.

Direct Computer Control Systems

A *direct computer control system* is a control system that uses a computer as the controller. The development of more robust and secure personal computer software, which has a true interruptible operating system strategy, has led to a greater acceptance of this arrangement for process control. There are manufacturers that offer standard rack-mounted I/O hardware that can communicate directly with a PC when the proper I/O scanning software is installed. **See Figure 38-2.**

TECH-FACT

A boiler pressure control starts a burner, controls the firing rate, or shuts down the burner based on the steam pressure. A high limit control is a boiler pressure control that measures steam pressure and shuts down a burner if the pressure control fails.

There are also separate control and display software systems that allow the user to develop the desired control strategies. These system arrangements can offer significant cost savings, since the PC hardware is less expensive than a normal DCS controller. As the software and hardware become more reliable, the use of this digital control system arrangement continues to increase.

Distributed Control Systems (DCSs)

A *distributed control system (DCS)* is a control system where the individual functions that make up a control system are distributed among a number of physical pieces of equipment that are connected by a high-speed digital communication network. **See Figure 38-3.** DCS systems, since they are designed to control slow-changing chemical and petrochemical processes, work very well with scan speeds of about 0.5 seconds. *Scan speed* is the time that it takes to access all the I/O points and go through the configuration program. The programming of the DCS can use a number of configuration and programming methods that are different for each manufacturer.

The distributed units that house the various functions are usually rack-mounted in cabinets. The main units consist of dual 24 VDC power supplies, analog input modules, discrete input modules, analog output modules, discrete output modules, and controller modules. The controller modules, which contain the configurations for multiple control loops and other functions, can handle any number of controllers and are only limited by the memory size. The controller modules are specially designed and built by different DCS manufacturers. The continual advances in digital technology impose a large burden on the manufacturers to redesign and improve their equipment while still being backward compatible with earlier systems.

Information from the input modules is made available to the high-speed communication network to be used by any device or program in the system. In addition, a number of digital signals such as Ethernet, RS-232, MODBUS, and so on, can be imported from special controllers like PLC and PC systems. All the inputs, outputs, and internal information variables are given tag names, which are then used throughout the system to identify the pieces of information.

A digital storage device called a historian can be added to the communication network to provide long-term data records. A historian stores data in a format that allows easy retrieval for reports or trend analysis.

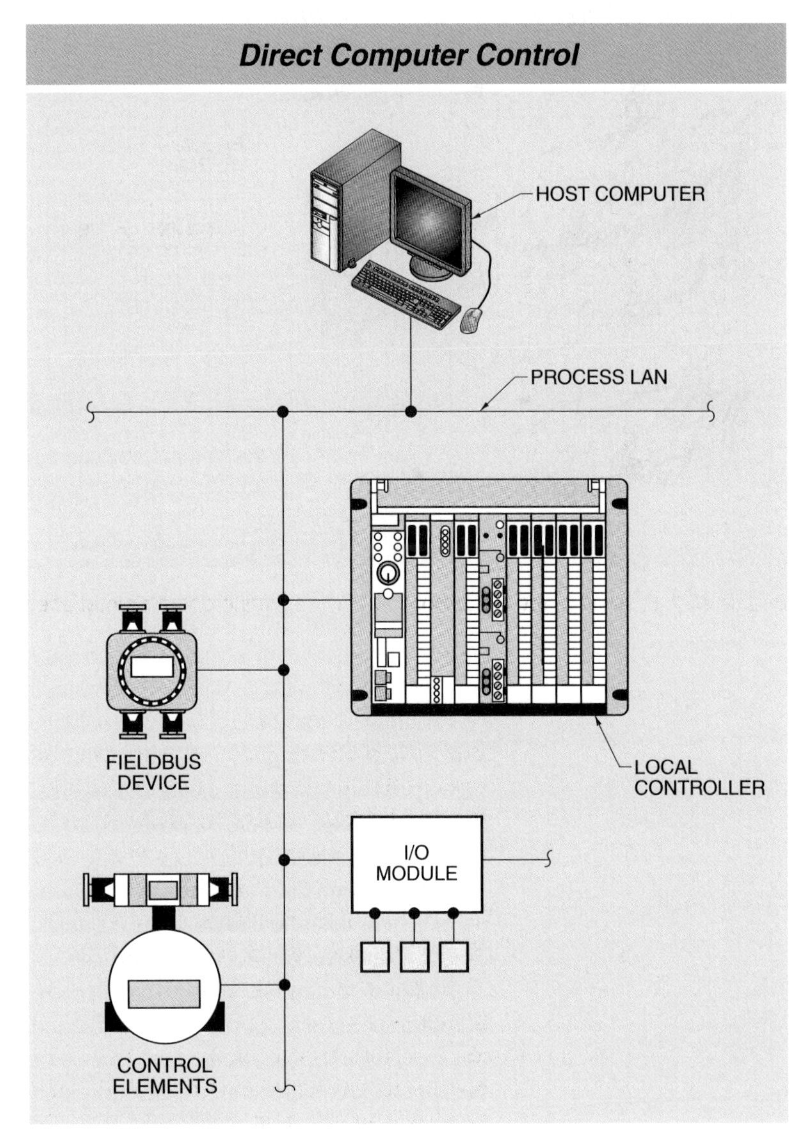

Figure 38-2. A direct computer control system uses a PC to monitor and control a process.

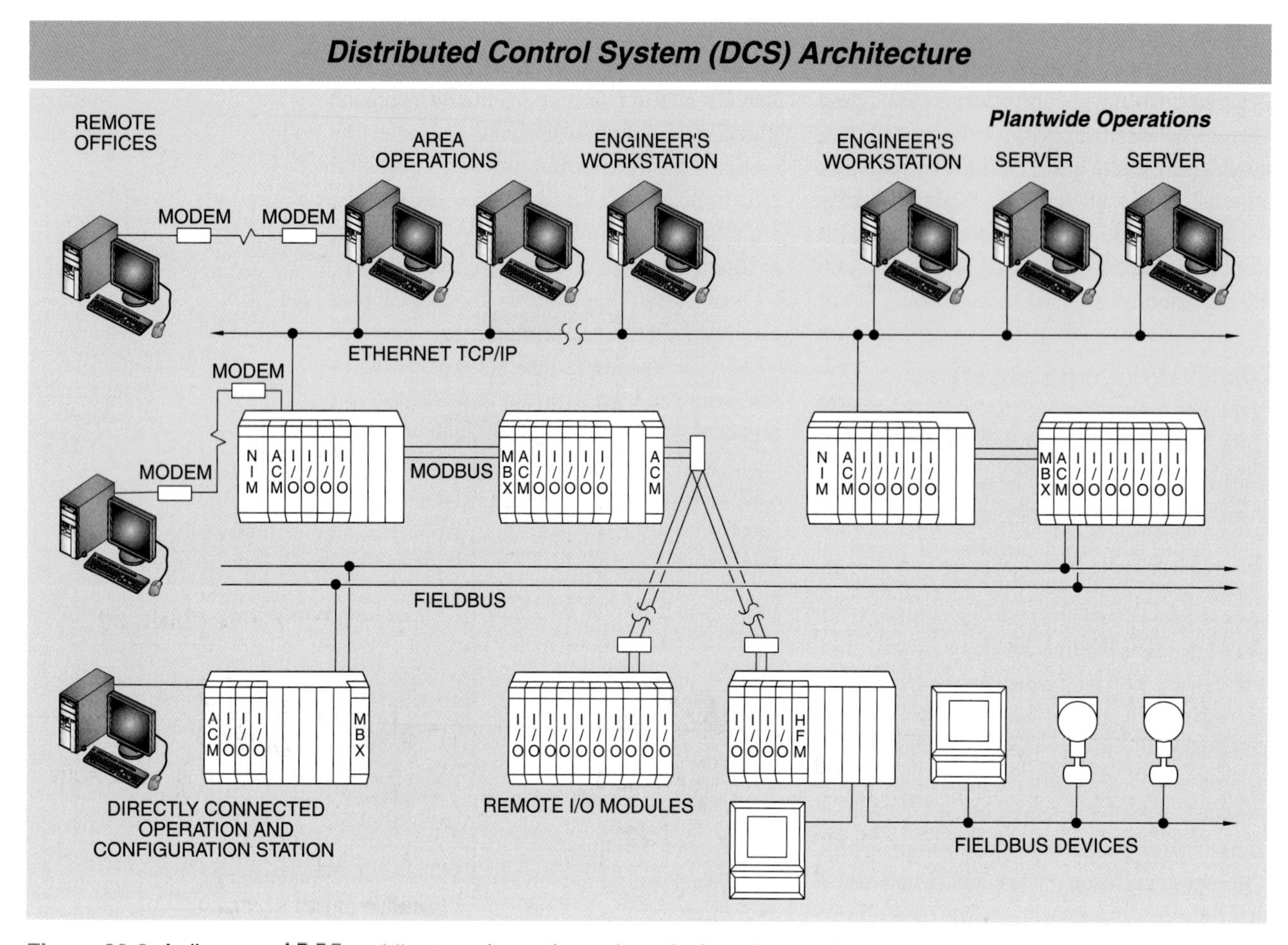

Figure 38-3. A diagram of DCS architecture shows the various devices that can be supported.

Programmable Logic Controllers (PLCs)

A *programmable logic controller (PLC)* is a control system with an architecture very similar to that of a DCS, with self-contained power supplies, distributed inputs and outputs, and a controller module, all connected on high-speed digital communication networks. **See Figure 38-4.** A PLC is designed to be more rugged than a DCS, since PLCs were originally designed for mounting on the production floor in discrete manufacturing areas. The PLC discrete manufacturing applications also needed to have very high scan speeds. The scan speeds of PLCs are now variable and can be chosen to match the requirements of the application. In general, a PLC has much faster scan speeds than a DCS.

Most PLCs are programmed using a ladder logic format, but some of the newer large systems can use other programming methods. Each manufacturer offers a different form of ladder logic programming software that does not work on equipment from other manufacturers.

Many varieties of discrete input and output modules are available, such as isolated and nonisolated I/O, powered and nonpowered, transistor-transistor logic (TTL), relays, and high-speed counters. Analog inputs and outputs are also available in assorted AC and DC voltages, isolated and nonisolated I/O, and 4 mA to 20 mA signals. **See Figure 38-5.** In addition, a number of digital communications signals can be imported. Long-term data storage is accomplished using a separate PC system and special interconnection software. The data can then be transferred to standard office or database programs.

Recent years have seen an increase in the number of new small PLCs designed for small applications. Some have fixed built-in inputs and outputs while others support removable I/O modules. They are considerably simpler than large PLC systems and usually use ladder logic or a simplified instruction set for programming. Typically, there is no data storage capability available in these systems. However, they can pass information to a conventional PC where it can be stored. Operator interface devices provide access to the process for the operating personnel.

ELECTRIC CONTROLLERS

Electric controllers were developed to overcome the shortcomings of pneumatic controllers. Electric controllers were quicker in response, could control over long distances, and did not need the additional expense of a dry air system. The two common categories of electric controllers are null-balance and analog output controllers. Null-balance electric controllers were the first ones to be developed. Analog controllers were developed when transistors became more common.

Null-Balance

A *null-balance controller* is an electric controller that generates an output signal to a final control element based on a measurement from a process and requires a proportional feedback signal representing the position of the final element. **See Figure 38-6.** The measurement can be any electrical signal such as a thermocouple voltage, the voltage across a resistance element, a pH or ORP voltage, or any similar signal.

The controllers use input amplifiers that are designed to handle the different types of input signals. The controller output can be either an ON/OFF signal, or a proportional voltage or current, which goes to a motor-driven final control element. The final element position is monitored by a feedback potentiometer whose signal is sent back to the controller. The feedback signal is compared to the desired output signal, and when the two signals are in a null-balance condition, the output signal is terminated. The feedback signal can also be used in special circuits to provide proportional, reset, and derivative action to the output signal.

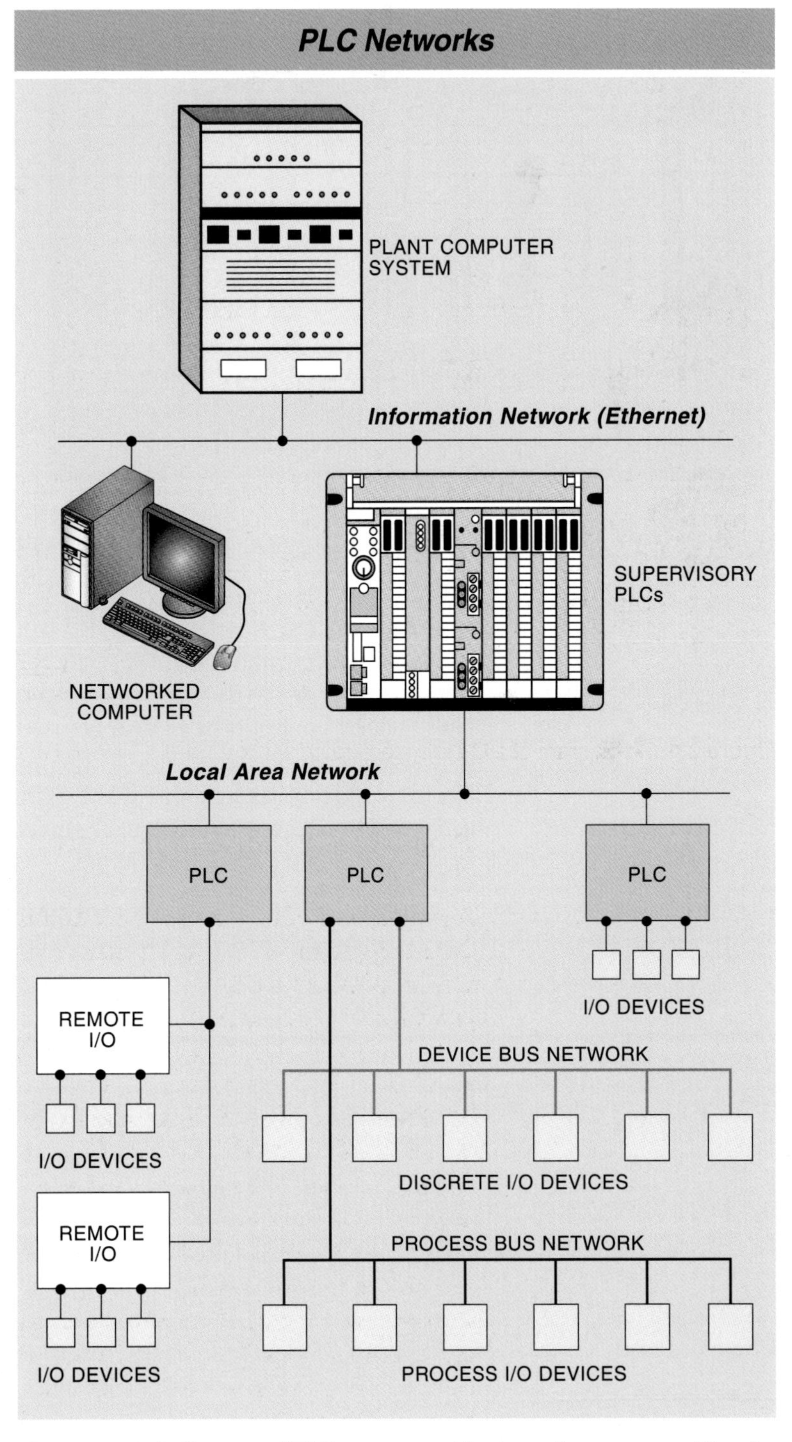

Figure 38-4. A diagram of PLC system architecture shows many of the devices that can be supported by the various communication methods.

Programmable Logic Controllers

Figure 38-5. A modular PLC system contains the devices required to perform desired functions.

A null-balance electric controller is a common control method when the final control element is driven by an electric motor. Null-balance controllers can also be used with silicon controlled rectifiers (SCRs) for electrical heating applications if the SCR can provide an appropriate feedback signal that is equivalent to the power output.

Analog Output

An *analog output controller* is an electric controller that has 4 mA to 20 mA DC electric current as the output. Analog output controllers contain adjustable setpoints, process measurements, derivative action on the measurement signal, direct and reverse controller action, PID controller actions, and auto/manual output selection. The controller output is an electric signal, but the controller circuitry is electronic, rather than electric, since this type of controller was not developed until after the development of transistorized circuits. These controller functions are provided by circuit boards or modules that provide the various functions using operational amplifiers, resistors, and capacitors in hard-wired circuits.

Electric controllers provided improved control performance over pneumatic controllers, but their dominance was short-lived. The rapid development of digital technology soon brought a new form of controller that is dramatically better than the analog output electric controller.

PID loop settings are not always transferable between controllers because of differences in timing that are influenced by time delays.

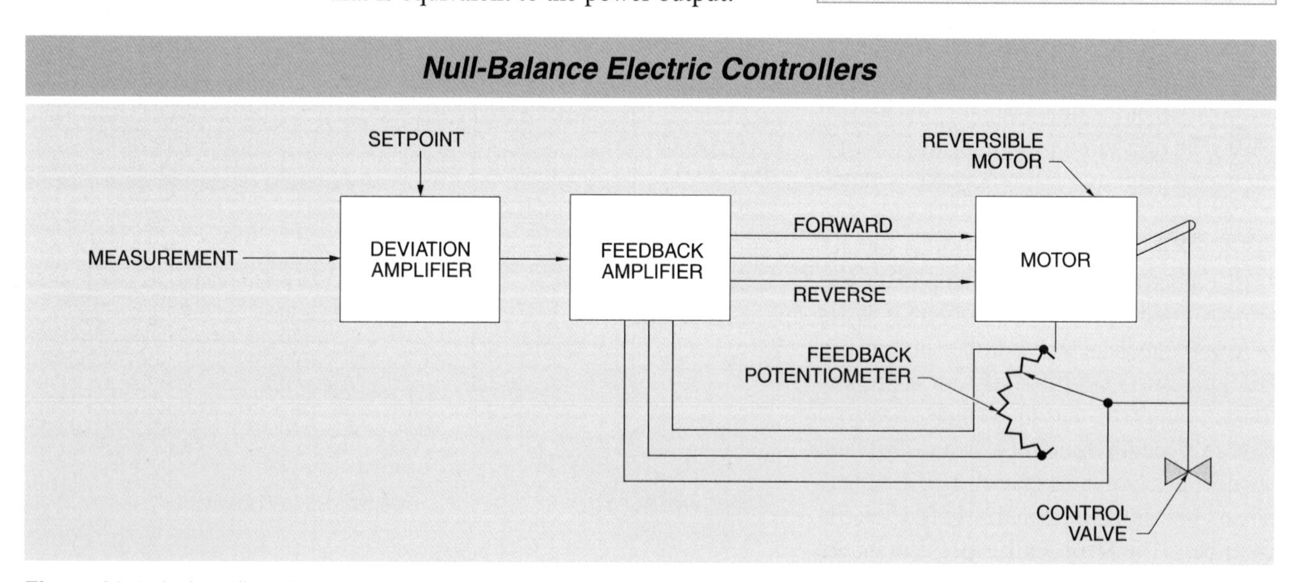

Figure 38-6. A simplified diagram of a null-balance type of electric controller shows the position of the control-valve feedback potentiometer.

OPERATOR INTERFACES

An *operator interface* is a view into a digital control system through which an operator can observe and control a process. The operator interface can also be called a "human machine interface" (HMI) or "man machine interface" (MMI). At a minimum, an operator interface consists of a keyboard and a display screen.

Various screen pages provide views of different parts of the process with static or dynamic pictures, controller faceplates, measured variables, pop-up windows, pushbuttons, and areas to insert keypad values. A number of pages are devoted to alarm displays and alarm summary. In addition, if the digital control system supports data collection, the display can be configured to provide trend records and historical data displays.

Stand-Alone Display Systems

A *stand-alone display system* is a display system designed for use with smaller digital control systems and has the electronics and display packaged together in a panel-mounted enclosure. **See Figure 38-7.** The display screen typically is a liquid crystal display (LCD), often in color, and can be a touch screen. A *touch screen* is a display system designed so that touching specific spots on the screen produces an action. The keyboard is usually of a special design so that it can also activate specific spots on the screen. Stand-alone display systems usually do not have the ability to display trends or to collect historical data.

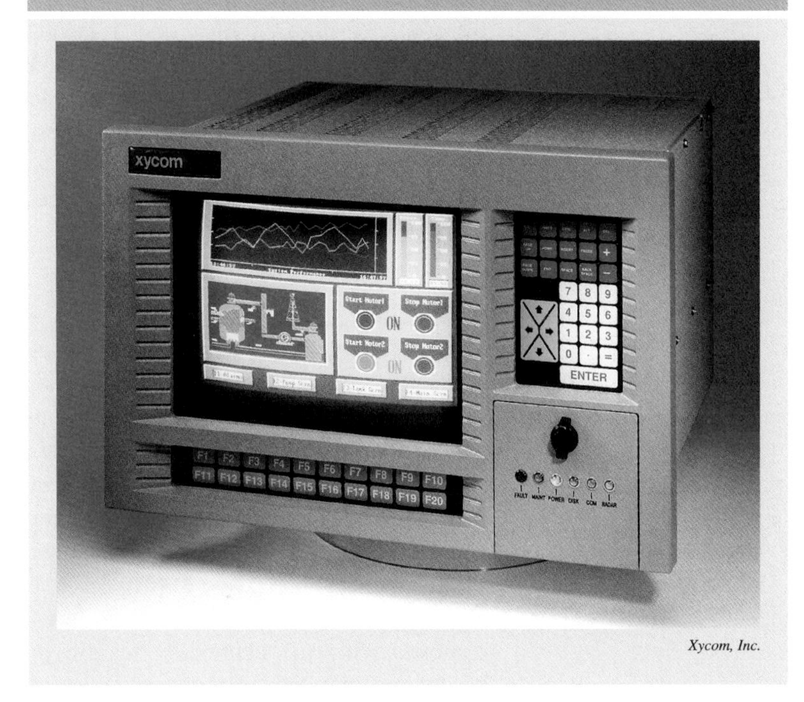

Figure 38-7. Stand-alone display systems are used to show the status of a process.

PC-Based Display Systems

A *PC-based display system* is a display system that uses standard personal computer hardware and special software to display full graphics, control functions, alarms, trends, etc. The special software must be compatible with the digital control system and have a communication method supported by both systems. These are usually third-party systems. Industrial computers are hardened to survive in a factory environment. **See Figure 38-8.**

Figure 38-8. An industrial hardened PC display system is used in a factory environment.

These software systems have started to replace the older custom-designed systems offered by the digital control system manufacturers. The advantage of the PC-based display system is that it is easy to upgrade the hardware as PC technology advances. PC-based display systems are usually capable of displaying trend and historical process data.

Digital Recorders

A *digital recorder* is a device that uses local memory to record process data, which is then displayed on digital display screens that can have the appearance of the old-style strip chart recorders. The screens can be for a ¼ DIN-sized case or for larger cases, which are similar in size to the large electronic strip charts. **See Figure 38-9.** A digital recorder typically has 6 to 24 input channels, with each channel being able to accept any input type. A number of custom digital displays can be designed, with any group of inputs displayed on each chart. A display record can be scrolled backward to review older data without disturbing the present recording of data.

Digital Recorders

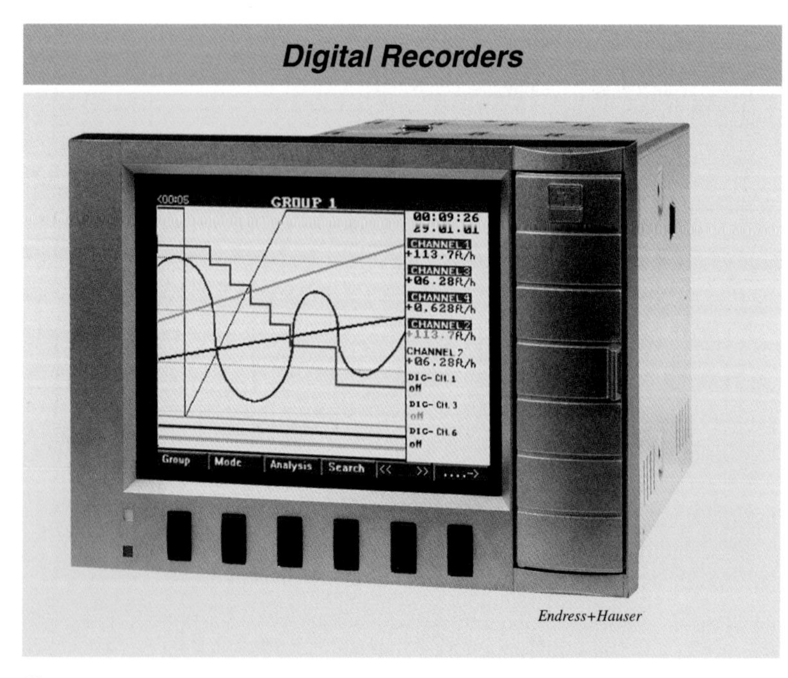

Figure 38-9. A digital recorder shows trends of process data.

The recorders have limited memory. When the memory is full, the recorder overwrites the oldest data. If the recorded data is important and is needed for a long-term history, it can easily be transferred to permanent storage. Most manufacturers provide software that can be run on a PC to allow the user to re-examine older recorded data and print out hard copies.

CONFIGURATION FORMATS

A *configuration format* is one of the various programming methods that have been developed to provide a simplified method for instructing the various digital control systems on how to control a process. Configuration is the process of selecting the functions used to program a controller. The choices may include the number and type of inputs, input engineering units, type of control functions, controller tuning, math functions, and the number and types of outputs. The separate lines of products of each manufacturer have different options, thus requiring different configuration procedures.

A stand-alone digital controller normally offers many choices of preprogrammed functions for control loops. The functions also offer numerous designs that allow the controller to perform its desired function.

Some configuration formats are more commonly referred to as programming languages, since they do not rely as much on packets of preprogrammed functions. Programming languages include ladder logic, structured text, and sequential function charts. While there has been some standardization in configuration formats for large DCSs, this standardization has not been well implemented in small systems or in stand-alone digital controllers.

Some sort of configuration storage procedure, either electronic or on paper, must be instituted to save the configuration and to record any changes. Digital controllers can fail and completely lose their memory. If a copy of the latest configuration is not available, the plant personnel may find it hard to reconstruct the proper configuration.

Pick and Choose

Pick and choose is a configuration format where the configuration is selected from a list of available functions. A pick-and-choose format is most commonly used with stand-alone digital controllers. There are many different types of configuration procedures. They all follow the same basic procedure of providing a simplified method of entering the configuration and moving through it step-by-step and making choices at each step. Sometimes sections of the configuration can be skipped if there is no interest in those sections. Once the configuration is completed, it is stored in memory chips powered by an onboard battery.

The various manufacturers often provide a paper configuration form that can be used to record the configuration entered. **See Figure 38-10.** Some manufacturers offer configuration software that can be run on a PC to develop the configuration and then download it to the controller.

Function Blocks

Function block configuration is a configuration method that uses a library of functions provided by the manufacturer. A *function block* is a stand-alone procedure designed to perform a specific task and can be linked with other function blocks to form complex control strategies. Function block configuration is used by medium and large DCS systems and by some of the stand-alone digital controllers. **See Figure 38-11.**

TECH-FACT

A function block receives information about a control system from inputs, performs some logic or algorithmic processing task, and generates one or more outputs. Inputs and outputs are either physical signals or network variables that are shared between function blocks.

Pick-and-Choose Configuration Forms

Controller ID:	Document Number:
	Serial Number:
Location:	Revision:

Data Specifications		Remarks	Standard Value	Scale Values	
				0	100
Analog Inputs	AI 1				
	AI 2				
	AI 3				
	AI 4				
Analog Outputs	OUT 1				
	OUT 2				
	OUT 3				
	OUT 4				
Digital Inputs	DI 1				
	DI 2				
	DI 3				
	DI 4				

Controller Mode			
Controller Type			
Controller Gain			
Controller Reset			
Controller Deriv. Time			
Data Setup	Input 1	Input 2	Input 3
Input Type			
Unit Specification			
Range Max.			
Range Min.			
Scale Max.			
Scale Min.			

Figure 38-10. A pick-and-choose configuration form is used to select and record controller settings.

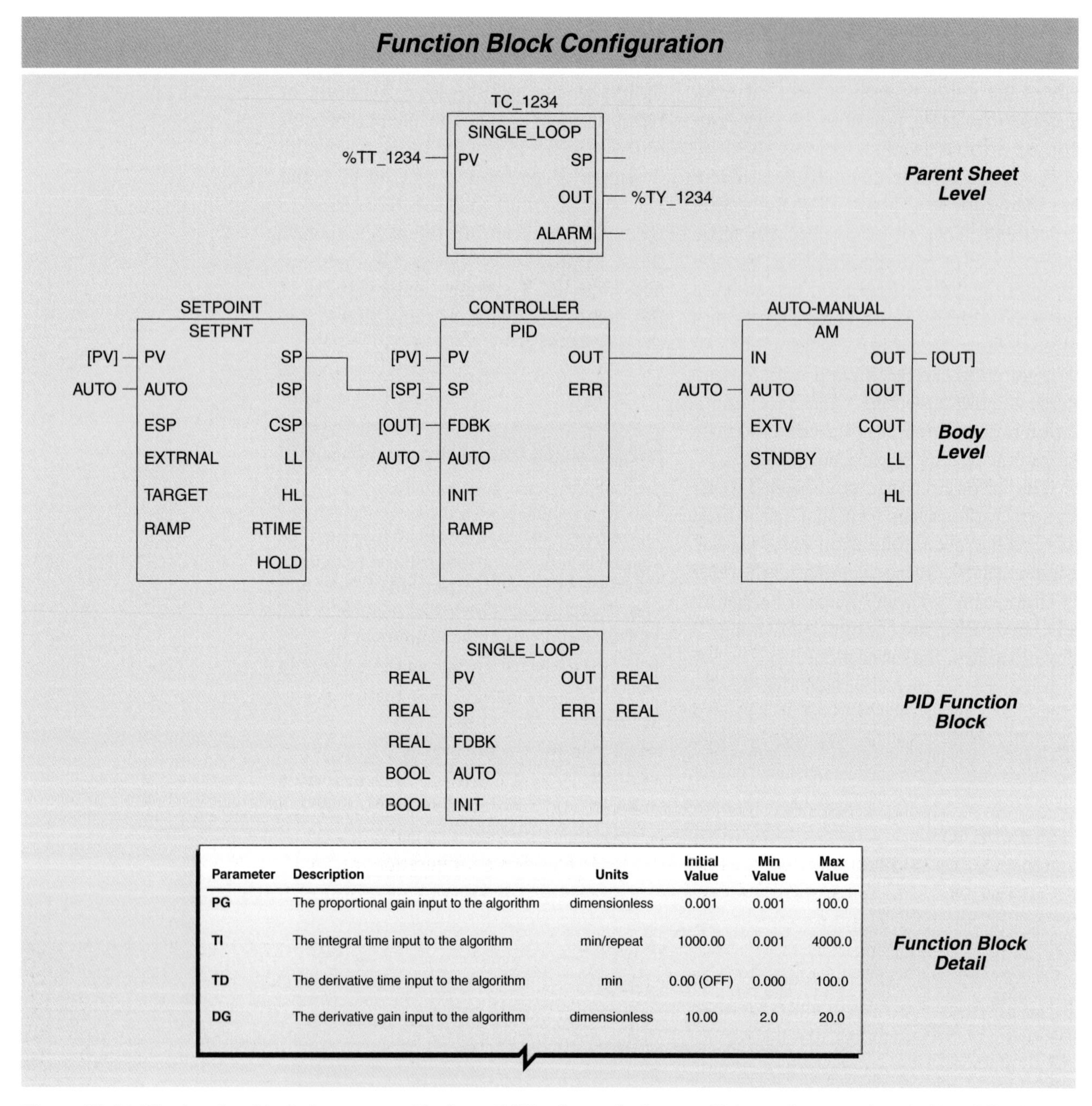

Parameter	Description	Units	Initial Value	Min Value	Max Value
PG	The proportional gain input to the algorithm	dimensionless	0.001	0.001	100.0
TI	The integral time input to the algorithm	min/repeat	1000.00	0.001	4000.0
TD	The derivative time input to the algorithm	min	0.00 (OFF)	0.000	100.0
DG	The derivative gain input to the algorithm	dimensionless	10.00	2.0	20.0

Figure 38-11. The function block diagram used for large DCSs shows the layers of information associated with a PID block.

> **TECH-FACT**
>
> *A function block configuration property is an adjustable value that affects function block or network variable behavior. Configuration property values affect how a node manages its own application and how network-variable message traffic is sent. Examples include alarm limits, setpoints, default values, time delays, and network-variable update intervals.*

An input function block requires the selection of the engineering measurement range and units, whether or not a square root extraction is to be done, and how much filtering is desired. Another function block is for the type of controller function desired, either proportional (P), proportional-integral (PI), proportional-integral-derivative (PID), adaptive gain PID, and so on. Each controller block also has a number of other properties that need to be selected, such as direct or reverse action, output tracking, tuning constants, output bias, start-up values, and so on.

The configuration blocks are often presented as a hierarchical set of pages or sheets. For example, a parent sheet level contains the overall control configuration including inputs, outputs, and alarms. The body level sheets present the different function blocks and how the outputs from one block are connected to the inputs of other blocks. The actual function block is configured at the lowest level.

Each block has to be given a unique address to distinguish it from every other block so that information can be directed to a particular block. Function block configuration is ideally suited to analog control systems. Many manufacturers provide a preconfigured set of function blocks connected to represent a typical type of controller. The factory configuration preselects the required blocks and their interconnection, but not the data needed for each block. That type of information is dependent on the specific application.

Ladder Logic

Ladder logic is a configuration method that consists of two vertical rails, the left one being the source and the right one being the end, and the sequential rungs of logic between the two. Ladder logic was originally used to represent electrical logic circuits. Because the ladder logic format was in common use prior to the emergence of the PLC, it was adapted as the configuration method for this type of controller.

The ladder logic language has evolved to provide mathematical functions, trip points, timing functions, and PID controllers. The most recent versions of this programming method include access to a huge choice of additional I/O status information that must be considered when developing a program.

The ladder logic configuration method is ideally suited to discrete logic. Discrete logic deals with ON/OFF elements instead of analog systems. A group of logic elements arranged across the page from left to right is a rung of logic. **See Figure 38-12.** Logic elements can be linked to analog elements to turn those analog functions ON or OFF. Very complex control systems can be developed with ladder logic programming.

A disadvantage of ladder logic formats is that they are more difficult to understand and to troubleshoot than many other configuration methods. For this reason, considerable effort should be devoted to extensive organization of the program and the associated remarks. The understanding of complex programs also benefits from a supplemental document such as a flowchart.

Structured Text

Structured text is a type of configuration that is very similar to Microsoft® Visual Basic® or older structured programming languages. Structured text uses terms and formats as in Visual Basic programs. **See Figure 38-13.** Running structured text programs uses a lot of controller time and can slow down a control system if used too much. It does provide an efficient method of performing complex mathematical calculations when called up by other configuration methods.

Pumps are often placed in banks. This allows a wide range of flows while still allowing efficient operation.

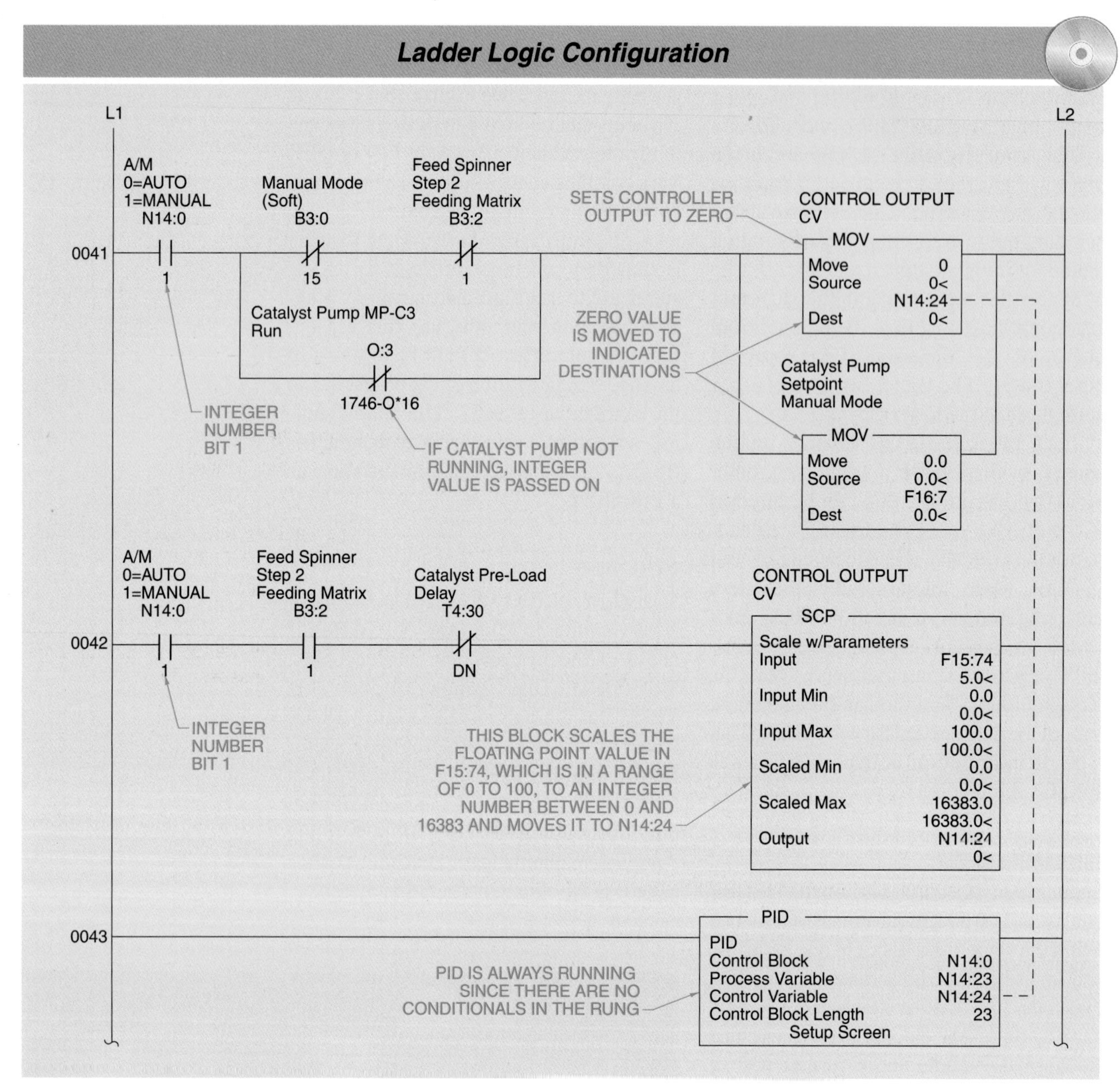

Figure 38-12. Ladder logic configuration is commonly used for programmable logic controllers and distributed control systems. Comments are added to explain the logic of the rungs.

Sequential Function Charts

A *sequential function chart* is a type of configuration format consisting of a series of conditional statements, parallel paths, and action blocks, which begins with a Start command and ends with an existing scan. **See Figure 38-14.** This style of configuration is used to control batch processes with sequential steps, such as charging; pre-heating; setting of various temperature values during the batch; cooling; and dumping. The sequential function chart starts and stops other configurations as needed to implement the batch program. Any individual command, or step, can have multiple actions take place.

For example, a simple process to fill a tank with water, heat the water to a predetermined temperature, and then wait a predetermined time before draining the tank can be accomplished with a simple five-step function chart. The first step is highlighted with a double outline. This is the Start step where the tank is already empty. When the tank is empty, the function chart allows the controller to move on to the next step where the discharge valve is closed and the pump turned ON.

When the tank is full, the function chart allows the controller to move on to the next step where the pump is shut OFF and the heater is turned ON. When the water reaches the setpoint, the function chart allows the controller to move on to the next step where the heater is turned OFF and a timer is started. When the timer has completed its cycle, the function chart allows the controller to move on to the final step where the discharge valve is opened. When the tank is empty, the function chart allows the controller to move back to the Start step.

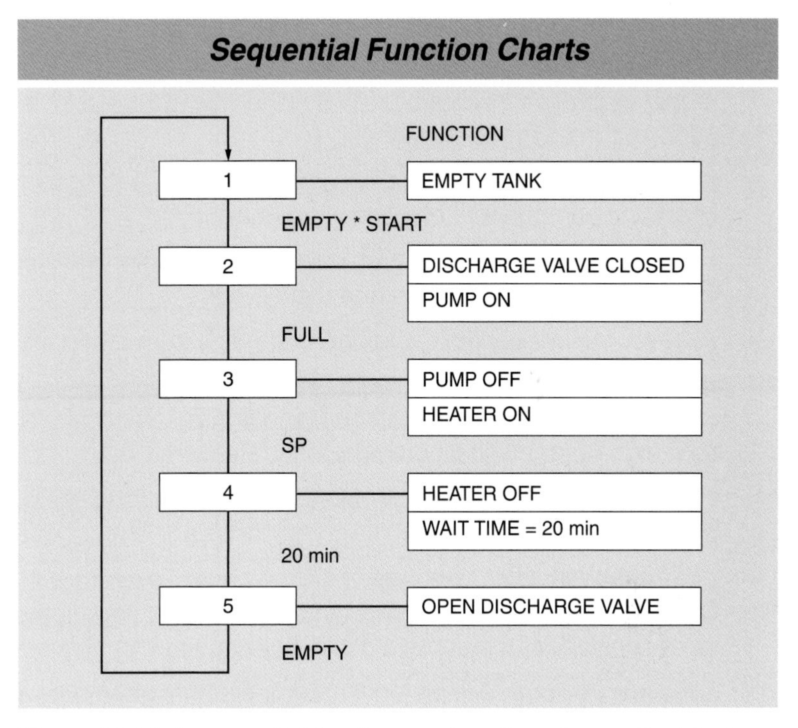

Figure 38-14. A sequential function chart configuration method for large DCS systems can be used to control batch processes.

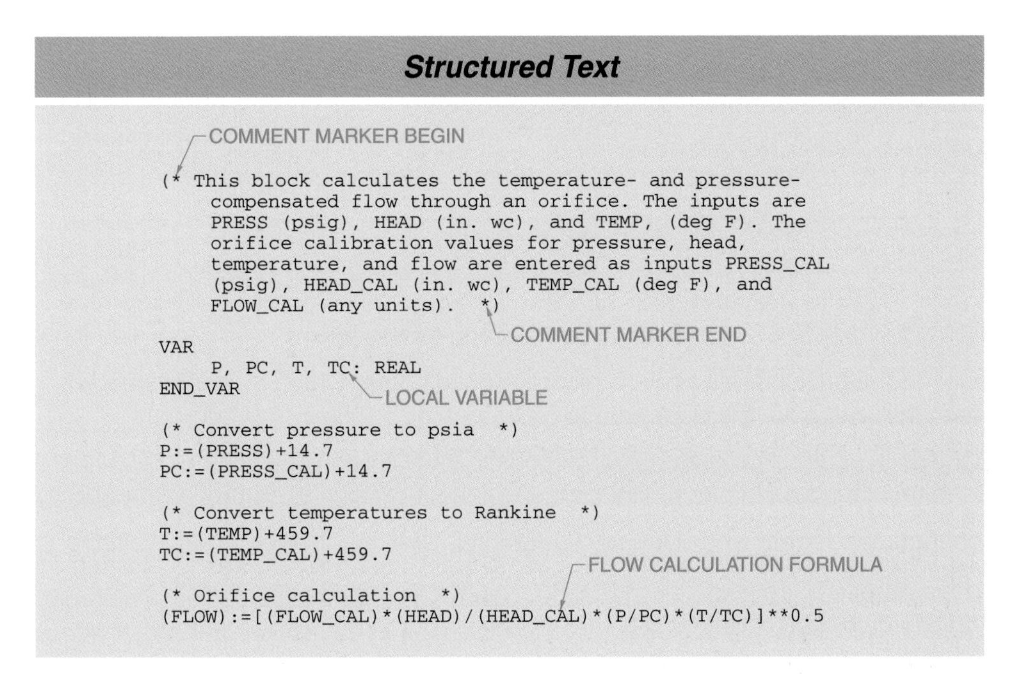

Figure 38-13. Structured text configuration is a method used to program large DCS systems.

KEY TERMS

- *digital controller:* A controller that uses microprocessor technology and special programming to perform the controller functions.
- *stand-alone digital controller:* A general type of microprocessor-based controller with all required operating components enclosed in one housing.
- *configuration:* The selection of preprogrammed software packages embedded in a controller representing available features that can be chosen.
- *direct computer control system:* A control system that uses a computer as the controller.
- *distributed control system (DCS):* A control system where the individual functions that make up a control system are distributed among a number of physical pieces of equipment that are connected by a high-speed digital communication network.
- *scan speed:* The time that it takes to access all the I/O points and go through the configuration program.
- *programmable logic controller (PLC):* A control system with an architecture very similar to that of a DCS, with self-contained power supplies, distributed inputs and outputs, and a controller module, all connected on high-speed digital communication networks.
- *null-balance controller:* An electric controller that generates an output signal to a final control element based on a measurement from a process and requires a proportional feedback signal representing the position of the final element.
- *analog output controller:* An electric controller that has 4 mA to 20 mA DC electric current as the output.
- *operator interface:* A view into a digital control system through which an operator can observe and control a process.
- *stand-alone display system:* A display system designed for use with smaller digital control systems and has the electronics and display packaged together in a panel-mounted enclosure.
- *touch screen:* A display system designed so that touching specific spots on the screen produces an action.
- *PC-based display system:* A display system that uses standard personal computer hardware and special software to display full graphics, control functions, alarms, trends, etc.
- *digital recorder:* A device that uses local memory to record process data, which is then displayed on digital display screens that can have the appearance of the old-style strip chart recorders.
- *configuration format:* One of the various programming methods that have been developed to provide a simplified method for instructing the various digital control systems on how to control a process.
- *pick and choose:* A configuration format where the configuration is selected from a list of available functions.
- *function block configuration:* A configuration method that uses a library of functions provided by the manufacturer.
- *function block:* A stand-alone procedure designed to perform a specific task and can be linked with other function blocks to form complex control strategies.
- *ladder logic:* A configuration method that consists of two vertical rails, the left one being the source and the right one being the end, and the sequential rungs of logic between the two.

KEY TERMS *(continued)*

- *structured text:* A type of configuration that is very similar to Microsoft® Visual Basic® or older structured programming languages.
- *sequential function chart:* A type of configuration format consisting of a series of conditional statements, parallel paths, and action blocks, which begins with a Start command and ends with an existing scan.

REVIEW QUESTIONS

1. Define the common types of digital controllers.
2. Define the common types of electric controllers.
3. What is an operator interface, and what are the common types of operator interfaces?
4. What is a configuration format, and what are some of the common decisions that need to be made during setup?
5. Define the common types of configuration formats.

SECTION

FINAL ELEMENTS 10

TABLE OF CONTENTS

SECTION OBJECTIVES

Chapter 39

- Define throttling control valve and describe its function.
- Explain the use of ON/OFF control actions and describe the different types available.

Chapter 40

- Define regulator and describe the different types of regulators.
- Define damper and describe the different types of dampers.

Chapter 41

- List the different types of actuators and describe their operation.
- Explain the use of positioners and describe their operation.

Chapter 42

- Define variable-speed drive and explain its use.
- List and describe the different types of electric power controllers.

SECTION

10 FINAL ELEMENTS

INTRODUCTION

A final element is a device that receives a control signal and regulates the amount of material or energy in a process. The most common type of final element is the control valve. The first valves were sluice gates developed by the Egyptians that were used to control the flow of irrigation water. The Romans developed advanced plumbing systems that used plug valves or stopcocks to control water flow and check valves to prevent backflow. The diaphragm valve was also developed in ancient Greek and Roman times when it was used to control the water and temperature of the hot baths. A leather diaphragm was manually closed over a weir to create a simple control valve.

In the late 18th century, James Watt placed a butterfly valve in the steam line to a steam engine. A governor was connected to the butterfly valve. As the engine speed increased, the governor caused a sliding collar to operate a lever that closed the butterfly valve and reduced the amount of steam to the cylinder.

In the early 1900s, an engineer named Saunders was charged with a project to cut costly power losses from the valves used to supply air and water in underground mines. Saunders revived the idea of the control valves used in the Roman baths. He used this concept to develop the first modern diaphragm valve.

ADDITIONAL ACTIVITIES

1. After completing Section 10, take the Quick Quiz® included with the Digital Resources.
2. After completing each chapter, answer the questions and complete the activities in the *Instrumentation and Process Control Workbook.*
3. Review the Flash Cards for Section 10 included with the Digital Resources.
4. Review the following related Media Clips for Section 10 included with the Digital Resources:
   - Air Compressors
   - Barometric Dampers
   - Damper Actuators
   - Dampers
   - Pressure Regulators
   - Sliding Stem Valves
   - Valve Actuators
5. After completing each chapter, refer to the Digital Resources for the Review Questions in PDF format.

ATPeResources.com/QuickLinks • Access Code: 467690

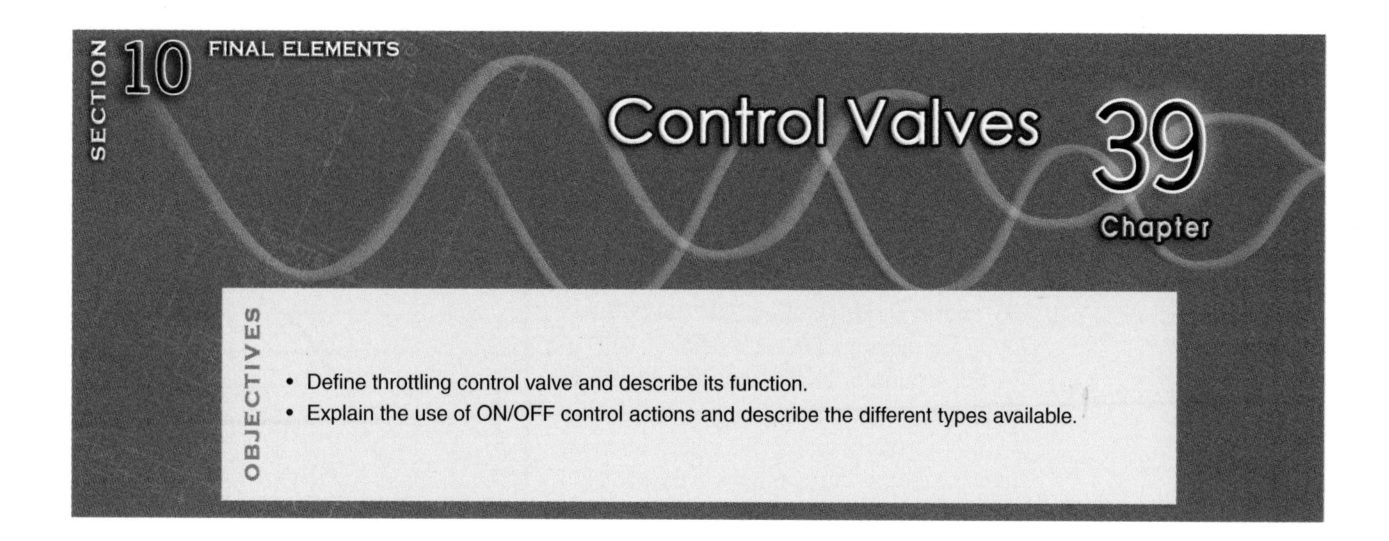

CONTROL VALVES

A process control system is only as good as the components that actually regulate the flow of material or energy into or out of a process. A *final element* is a device that receives a control signal and regulates the amount of material or energy in a process. Control valves are the most common type of final element. Regulators are self-operating control valves. Other types of final elements include variable-speed drives, dampers, and electric power controllers.

A *throttling control valve* is a valve and actuator assembly that is able to modulate fluid flow at any position between fully open and fully closed in response to signals from a controller. The capability of throttling control valves to precisely adjust the flow of the controlled fluid is vital for accurate control. **See Figure 39-1.** The majority of the control valves described in this chapter are designed strictly for throttling service. A few of the valves, such as ball, plug, and diaphragm valves, were originally designed primarily as ON/OFF valves, but were modified to provide throttling service. Butterfly valves were originally designed for ON/OFF service, but they perform well in throttling services without any modification.

A *way* is a path through a valve from an inlet port to an outlet port. The number of ways required depends on the application. A control valve is placed in different positions to stop, start, or change the direction of fluid flow. A *position* is a location within a valve where the internal parts are placed to direct fluid through the valve. A control valve normally has two or three positions.

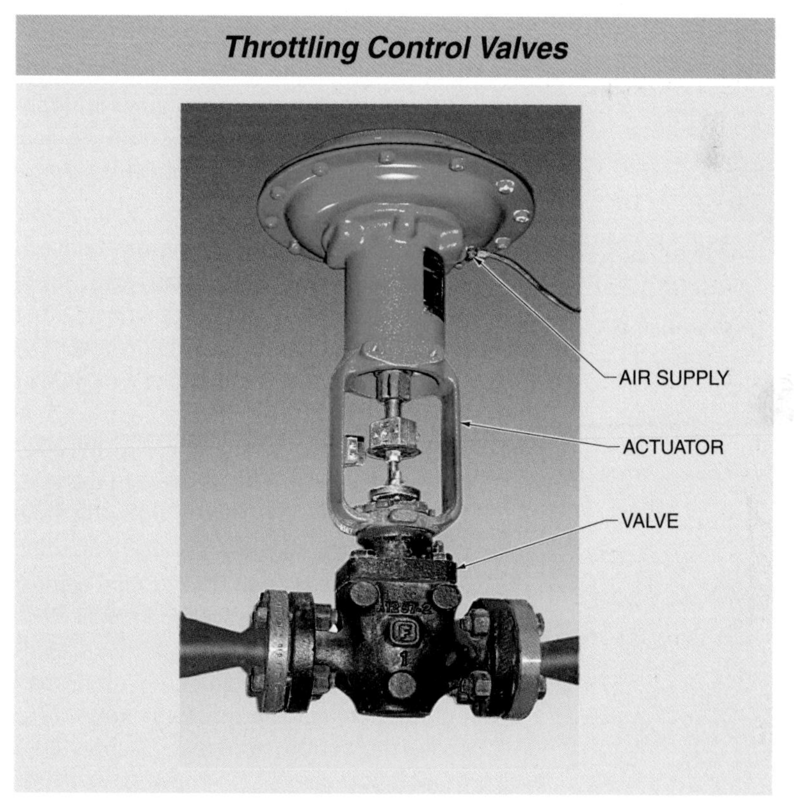

Figure 39-1. Control valves are used to regulate the flow of materials and energy into and out of a process. Throttling control valves are a very common type of final element.

A *normally closed (NC) valve* is a valve that does not allow pressurized fluid to flow out of the valve in the spring-actuated (de-energized) position. A *normally open (NO) valve* is a valve that allows pressurized fluid to flow out of the valve in the spring-actuated (de-energized) position. The terms NC and NO are used in reference to electricity and to fluid handling. The terms have opposite meanings in each field.

For example, NC in the fluid handling field means that fluid does not flow in the valve's normal position. NC in the electrical field means that current flows in the switch's normal position. Care must be taken not to confuse the meaning of NC and NO when working on fluid handling equipment that is electrically controlled.

Control Valve Properties

The physical properties of control valves are related to the fluid properties that affect flow and differential pressure measurement. Control valve capacities are affected by differential pressure, fluid density, viscosity, temperature, vapor pressure, and velocity. The control valve sizing equations incorporate factors that address all the fluid physical properties that affect control valve sizing.

Valve Flow Characteristics. A *valve flow characteristic* is the relationship between valve flow capacity (in percent) and control valve open travel (in percent), with all other factors that affect flow held constant. The flow through a control valve at a particular instant is determined by the effective valve opening that is itself determined by the position of the disc or plug relative to the seat. Different valve types have different flow characteristics that are usually described graphically as percent flow plotted against percent opening (or lift). **See Figure 39-2.**

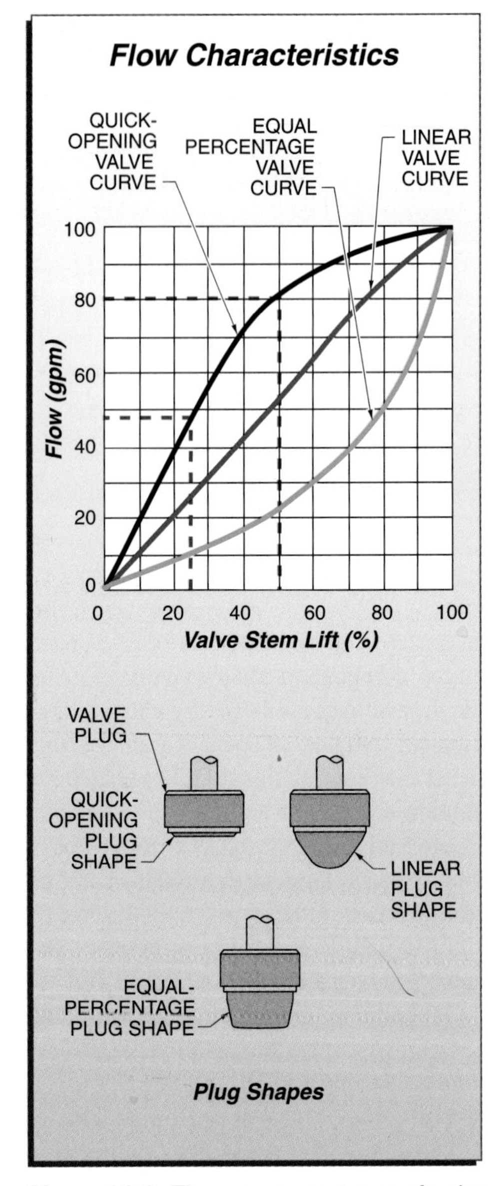

Figure 39-2. Three common types of valve flow characteristics are quick-opening, linear, and equal-percentage.

A *quick-opening valve* is a valve that allows flow to increase quickly with a small initial opening. The quick-opening curve shows that at 50% open there is about 80% of total flow, and at 25% open there is just under 50% of total flow. This limits throttling capability and makes the valve best suited to ON/OFF operation.

A *linear valve* is a valve that allows flow to increase at the same rate as an opening. The linear valve curve shows that at 50% open there is about 50% of total flow, and at 25% open there is 25% of total flow. The throttling action for this characteristic is good but the rangeability is somewhat limited at about 10:1. *Rangeability* is the ratio of the maximum flow to the minimum flow at the desired measurement accuracy.

TECH-FACT

Packing consolidation is the shortening of a packing stack as voids are eliminated in, between, and around the packing rings and leaking increases. Consolidation occurs when the packing wears or when an uneven stress distribution in the packing exists.

An *equal-percentage valve* is a valve that allows the flow rate percentage to change by an amount equal to the change in the opening percentage. For example, if the valve opening increases by 10% over the previous valve opening, the flow rate increases by 10% over the previous flow rate. This improves the control at the smaller openings and expands the rangeability of the valve to as much as 50:1. Globe, split body, cage, ball, plug, and eccentric valves all are available with characterized internals for linear or equal-percentage flow.

Piping Resistance Effects. Valve flow characteristics are determined at constant conditions. This means that the amount of flow for a given valve opening is determined at a constant pressure drop across the valve. However, installed valves rarely have a constant pressure drop because the frictional pressure drop in pipes and other fittings changes with the flow rate.

When sizing control valves, it is common practice to have about one-third of the system frictional pressure drop occur through the control valve. This means that two-thirds of the system frictional pressure drop is from piping items. The pressure drop through pipe and fittings varies when the flow through them changes. One of the purposes of the control valve is to compensate for the change in pressure drop in the piping system. The system frictional pressure drop does not include the static head needed for a change in elevation, since this pressure drop is constant for all flow rates.

It is common practice to size a control valve at only slightly more than the maximum design requirements. This is done to provide excess capacity to allow for control and to cover inaccurate sizing information.

For example, a linear control valve is typically sized with a pressure drop of one-third of the total frictional loss. A pump is usually used to pressurize the system. A pump curve is a graph of the pressure head plotted against the liquid flow delivered by the pump. The system curve is a graph of the pressure head plotted against the liquid flow through the system. The system curve takes into account any friction losses due to piping resistance.

The piping resistance includes the effects of elbows, tees, reducers, expanders, and pipe lengths without the presence of the control valve. At any given flow, the pressure drop across the valve is the value of the pump curve minus the value of the pump curve minus the value of the system curve at that flow rate. In other words, the inlet pressure to the valve at each flow rate is given by the pump curve. The outlet pressure at the same flow rate is given by the system curve. Therefore, the pressure drop across the control valve will be different at different flow rates.

Manual valves are often used for starting up and shutting down a process.

The varying pressure drop, the rangeability of the control valve, and the amount a valve is oversized all contribute to the flow characteristics through the valve. These factors are all used to calculate the installed characteristics of a control valve from the inherent characteristics of the valve. As a result of these factors, an equal-percentage valve behaves more like a linear valve and a linear valve behaves more like a quick opening valve. **See Figure 39-3.**

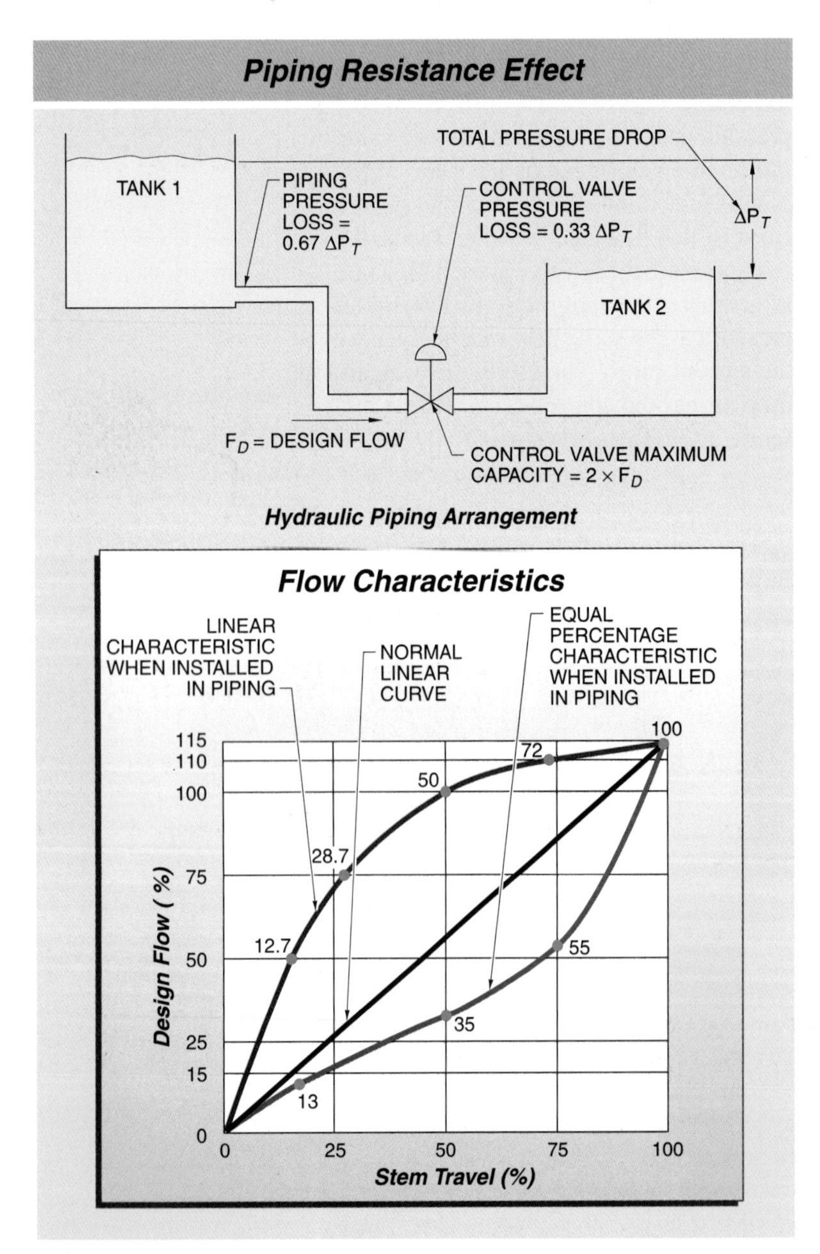

Figure 39-3. The valve flow characteristics are modified by piping resistance as the flow changes.

Flashing and Cavitation. Control valves for liquid service may be subject to phenomena known as flashing and cavitation. When liquids flow through a restriction, there is an increase in flow velocity to allow the liquid to pass though the smaller opening. When the flow velocity increases, the pressure decreases. *Flashing* is a process in which a portion of a liquid converts to a vapor as it passes through a control valve because the pressure has fallen below the vapor pressure of the liquid.

Flashing can occur when the upstream pressure is too low or when the inlet fluid temperature is too high and some of the liquid flashes to its vapor form. **See Figure 39-4.** As flashing occurs, the fluid stream becomes two-phase with liquid and vapor. The presence of the vapor increases the volume that has to pass through the valve at the same mass flow rate. As the volume increases, the total amount of liquid that can flow through the valve decreases and a limited flow condition exists.

Flashing causes wear patterns that resemble a sanded and polished surface. The wear patterns occur at the point of lowest pressure and highest velocity. If flashing is a possibility, there are special valve sizing calculations that result in the selection of a different valve to handle the increased volume. When severe flashing conditions are experienced, a straight-through flow valve should be used.

After fluids have passed through a restriction, the velocity returns to the previous value and some of the pressure drop is recovered and is applied to the flowing fluid. *Cavitation* is a process in which vapor bubbles in a flowing liquid collapse inside a control valve as the pressure begins to increase. The collapsing bubbles result in implosions causing considerable noise inside the valve. The implosions result in tearing away of some of the valve internal surface. Therefore, the damaged metal surface will show a spiky roughness. This is very different than the smooth surface created by erosion. **See Figure 39-5.**

The exact location of the cavitation damage can be anywhere downstream of the trim. The location of erosion damage is predictable and is located at the point of highest velocity in the valve. Cavitation is violent and noisy while erosion is often quiet and smooth.

Cavitation is usually due to a flashing condition developed by high velocity and excessive pressure drop inside the valve. It can also be caused by dissolved gases in the liquid that form bubbles when the pressure is reduced. Cavitation is very noisy and sounds like gravel flowing through a valve. It causes damage to the valve trim assembly in the form of an eroded or porous surface. It can sometimes be avoided by using more than one valve to handle the pressure drop. More often, cavitation is handled by using special valve designs. Both flashing and cavitation can cause erosion, calling for the use of valves made of hardened materials. Pumps are also subject to damage from cavitation.

Critical Pressure Ratio. Gas flow through a valve that is at a fixed position depends on the pressure ratio. An increase in pressure drop (decrease in pressure ratio) through the valve causes an increase in flow until the critical pressure ratio is reached. The *critical pressure ratio* is a ratio of downstream pressure to upstream pressure where the gas velocity out of a valve is at sonic velocity, and further decreases in downstream pressure no longer increase the flow.

Sonic velocity is the speed of sound in a gas. When the upstream pressure increases above the design conditions, the gas velocity remains constant but the total mass flowing through the valve increases because of an increase in density of the gas. *Choked flow* is a condition where the actual pressure ratio is less than the critical pressure ratio and flow is restricted. Absolute pressure must be used when calculating pressure ratios. The critical pressure ratio for most gases is about 0.5. For example, the critical pressure ratio for air is 0.527. The pressure ratio is calculated as follows:

$$PR = \frac{P_2}{P_1}$$

where

PR = pressure ratio

P_2 = downstream pressure (in psia)

P_1 = upstream pressure (in psia)

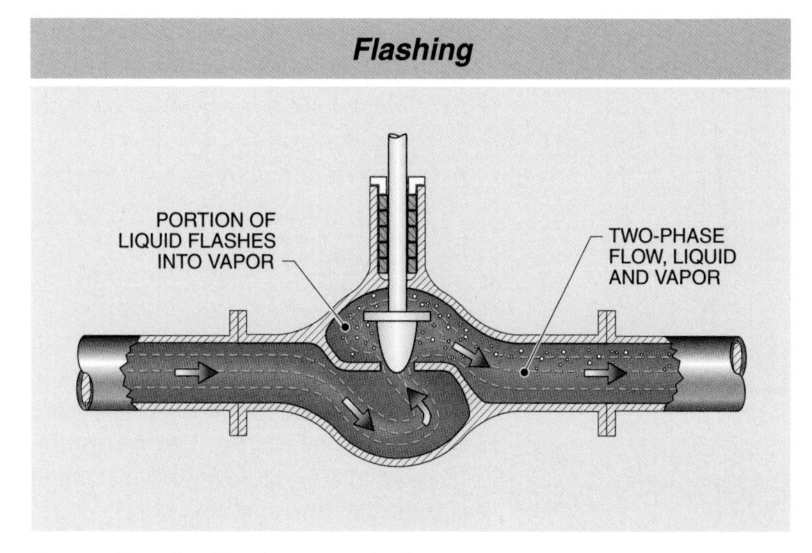

Figure 39-4. Flashing in a control valve occurs when low pressure causes a liquid to flash into vapor.

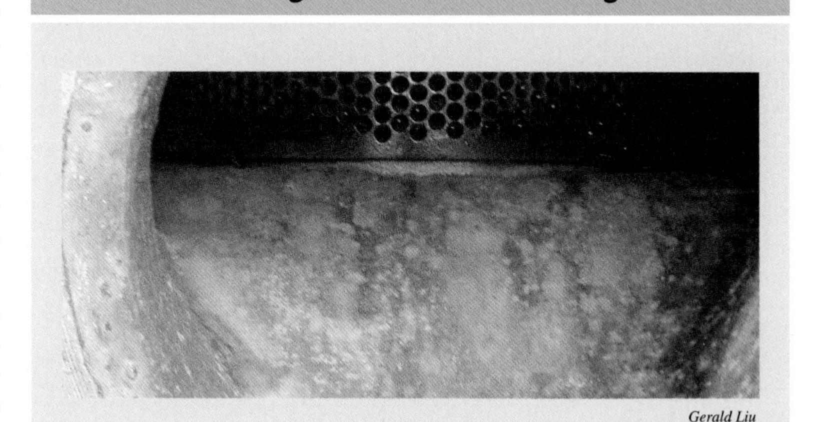

Figure 39-5. Flashing and cavitation cause wear and damage to internal valve parts.

After the critical pressure ratio is reached, the flow does not increase any further, as long as the upstream pressure and the valve position are constant. **See Figure 39-6.** For example, if the upstream air pressure is 100 psig and downstream air pressure is 40 psig, the pressure ratio can be calculated to determine whether a choked flow condition exists. The pressure ratio calculation uses the absolute pressure by adding 14.7 to the gauge pressure. The downstream absolute pressure is divided by the upstream absolute pressure as follows:

$$PR = \frac{P_2}{P_1}$$

$$PR = \frac{40 + 14.7}{100 + 14.7}$$

$$PR = \frac{54.7}{114.7}$$

$$PR = \mathbf{0.477}$$

For this example, the calculated pressure ratio is less than the critical pressure ratio. This means that the airflow is at a maximum and a choked flow condition exists.

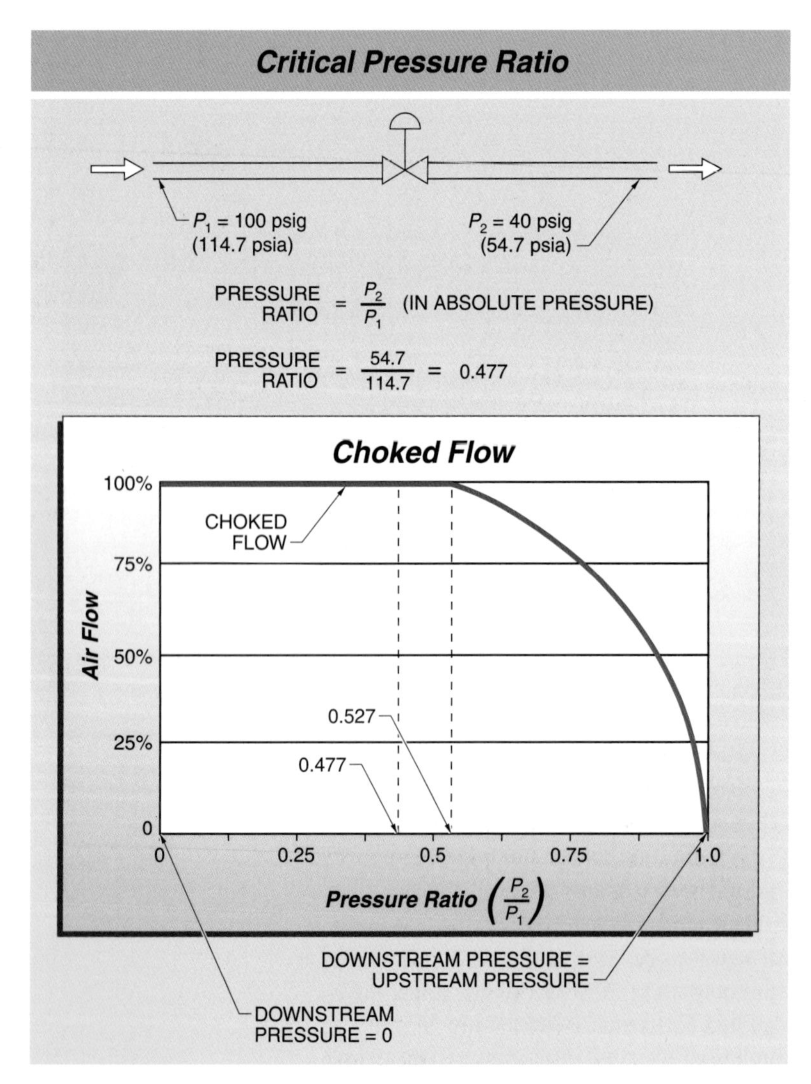

Figure 39-6. Choked flow occurs when the downstream absolute pressure is approximately half the upstream absolute pressure.

Airflow tests have determined that the internal configuration of a control valve affects its flow capacity when the gas or vapor is near its critical pressure drop. To compensate, the rated flow capacity of a valve must be corrected by applying factors that describe the internal configuration of the valve (obtained from the manufacturer's sizing tables) and the effect of piping reducers next to the valve.

Valve Noise. Flashing, cavitation, and sonic velocities all generate considerable amounts of control valve noise. Valve manufacturers provide equations for calculating the amount of noise that is generated. OSHA has definitive limits on the noise level allowed in a plant. Depending on the amount of noise that is produced, there are various solutions. Most control valve manufacturers offer special control valve plug and seat designs that distribute the pressure drop, which reduces the generated noise. There are also mufflers that can be installed directly downstream of the control valve to quiet the noise. As a last resort, there are specially designed control valves for reducing the generation of noise. These are very expensive but they do a superior job.

Valve Sizing. A valve for throttling control is essentially a variable orifice where the flow rate depends on the resistance to flow and the difference in pressure between the inlet and the outlet. Therefore, the principles and variables that apply to orifice sizing also apply to control valve sizing. Determining the optimum size of a control valve for an application is very important. An excessively large valve results in poor control from too much cycling while a valve that is too small can become a hazard if it cannot pass the required flow.

When the differential pressure is expected to change during service, the lowest differential pressure estimate should be the one used for sizing the valve. When the flow and differential pressure conditions are expected to change during service, two sizing calculations should be done. The first calculation is done at the maximum flow and minimum pressure drop, and the other at the minimum flow and maximum pressure drop. However, flow calculations and valve sizing are very complex and only a simplified method is presented here.

Performing a control valve calculation is accomplished by inserting a set of operating conditions into the equations. The most important part of control valve sizing is selecting the most appropriate operating conditions. The vendor should be responsible for final sizing. Control valve manufacturers run extensive tests on their control valves under controlled conditions to determine the exact valve coefficients at various openings. The basic sizing equation for liquid flow is as follows:

$$C_V = Q \times \sqrt{\frac{\rho}{\Delta P}}$$

where

C_V = valve coefficient

Q = rate of flow (in gpm)

ρ = specific gravity of the flowing fluid

ΔP = pressure drop across valve (in psi)

When the objective is to determine the valve size, the required valve coefficient is calculated. To perform this calculation, the desired flow rate, pressure drop, and specific gravity are entered into the formula. For example, a valve for water service is to be sized. The desired flow rate is 20 gpm and the desired maximum pressure drop is 10 psi. Water has a specific gravity of 1.0. The valve coefficient is calculated as follows:

$$C_V = Q \times \sqrt{\frac{\rho}{\Delta P}}$$

$$C_V = 20 \times \sqrt{\frac{1.0}{10}}$$

$$C_V = 20 \times 0.316$$

$$C_V = \mathbf{6.32}$$

Sliding Stem Valves

A *sliding stem control valve* is a throttling valve that has a stem (shaft) attached to a plug or disc at one end and an actuator at the other end. The valve is opened and closed by the stem movement moving the plug relative to the port inside the valve body. A *valve stem* is a valve component that consists of a metal shaft that transmits the force of the actuator to the valve plug. The stem passes through a section called a bonnet that fastens to the valve body. **See Figure 39-7.**

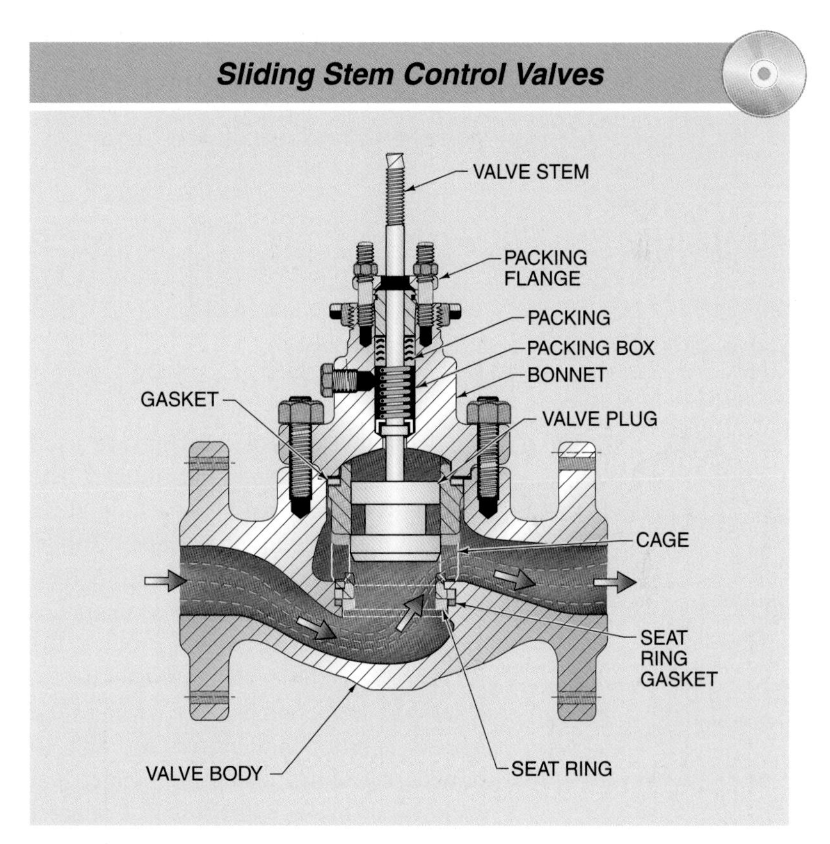

Figure 39-7. A sliding stem control valve uses the linear motion of the stem to open or close the valve.

A *valve bonnet* is a packing enclosure that is bolted or threaded to the top of the valve body. The bonnet contains the valve packing and provides lubrication and leak protection. A *valve body* is a casting or forging with an enclosed port and integral threaded or flanged inlet and outlet openings. A *valve plug* is a machined disc or shaped piece that regulates the flow of a material by changing the size of the valve opening.

Control valves are the most common type of final element.

Different types of packing are used for different fluids and temperatures. The packing is compressed by a packing gland and may be made up of O-rings, V-rings, braided PTFE material, or braided carbon fibers. For very high- or low-temperature service, an extended bonnet protects the packing from the extreme temperatures. If leakage to the atmosphere can create a hazard, some manufacturers are able to provide a bellows inside the bonnet that encloses and moves with the stem. Common bellows materials are 316 SS, Hastelloy® C, Monel®, Nickel 200, and PTFE. The valve shafts used for bellows-sealed valves are designed to prevent shaft rotation, which can damage the bellows.

Globe Valves. A *globe valve* is a throttling valve where fluid flow enters horizontally, makes a turn through the plug and seat, and then makes another turn to exit the valve. The body of a globe valve is a single casting with one or two integral ports plus an opening for the bonnet. Globe valves are available in sizes from less than 1″ up to 24″ and are made of various materials including iron, steel, stainless steel, and bronze.

A *single-port globe valve* is a globe valve that consists of a single valve plug and seat ring through which a fluid flows. Single-port globe valves are typically available in sizes as small as ¼″ threaded NPT and as large as 16″ flanged. **See Figure 39-8.** Single-port valves are used in all services because they offer a wide range of materials of construction. They are available with different flow characteristics and can close tightly. A single-port globe valve seal is rated by its leakage classification as follows:

- Class I—no testing limits for either water or air
- Class II—0.5% of valve capacity for either water or air
- Class III—0.1% of valve capacity for either water or air
- Class IV—0.01% of valve capacity for either water or air
- Class V—0.0005 ml per min water/in. of valve seat size/psi differential
- Class VI—0.0005 ml per min air/in. of valve seat size/psi differential

Control valve manufacturers list the leakage classifications of their valves in their catalogs. Most manufacturers offer different trim packages so that different leakage classifications can be chosen from the standard valve design.

As the size of the single port globe valve increases, the force applied to the valve plug and stem by the process fluid pressure can become very large. This requires very large and expensive actuators. The forces can be greatly reduced by using a double-port globe valve or the more modern balanced plug cage valve.

A *double-port globe valve* is a globe valve that consists of two plugs and seat rings through which a fluid flows. A double-port globe valve reduces the force on the valve plug, but manufacturing and assembly tolerances make it practically impossible to close both seats tightly. **See Figure 39-9.** The upper seat is larger than the lower one so that the one-piece stem and plugs can be removed from the body.

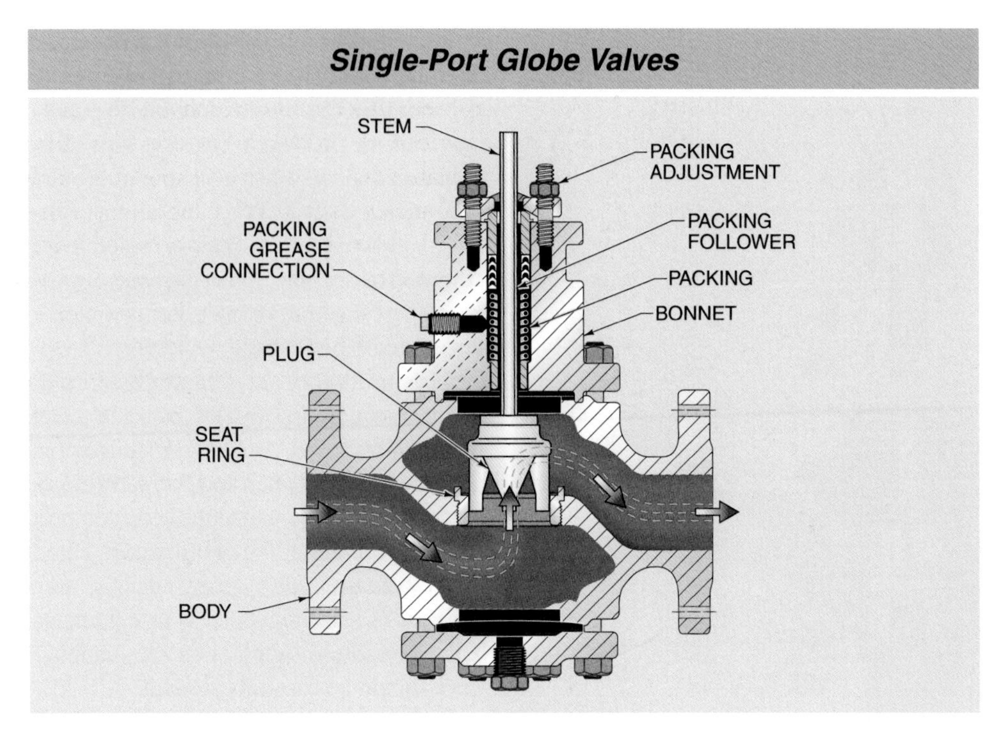

Figure 39-8. A single-port globe valve has one path through the valve.

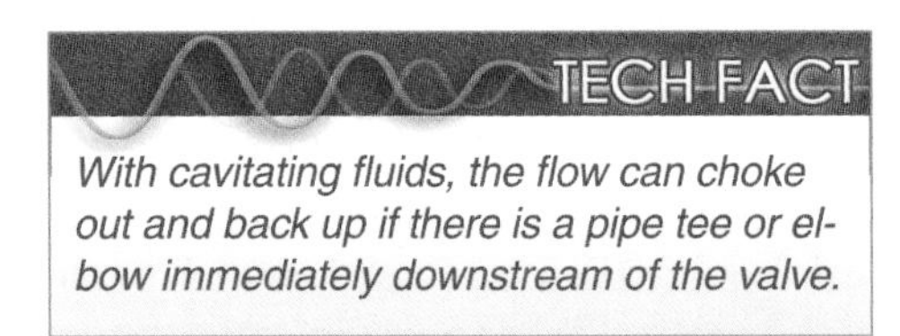

In a double-port valve, there is a division of the flow through the two ports. The division of flow reduces the imbalance of the forces acting upon the plugs. The reduced force imbalance makes it possible to use a smaller actuator than for a single-port valve. The seats in a double-port valve are screwed into the body. This makes it very difficult to remove them during servicing. The more modern cage-style globe valve with a balanced plug offers the same advantages as the double-port valve, with a design that is less expensive and easier to maintain.

A *three-way globe valve* is a globe valve that consists of three pipe connections and is used for mixing, blending, or flow division or diversion applications. **See Figure 39-10.** Three-way valves are not throttling control valves since the total flow through the ports is constant for any valve plug position.

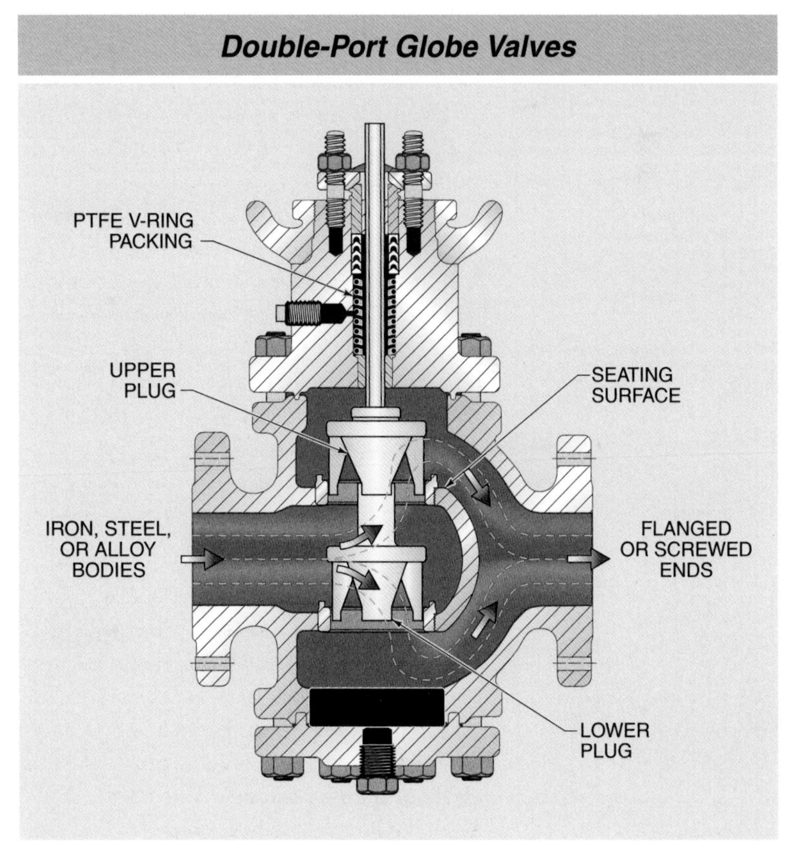

Figure 39-9. A double-port globe valve has two paths through the valve to reduce forces on the valve plug.

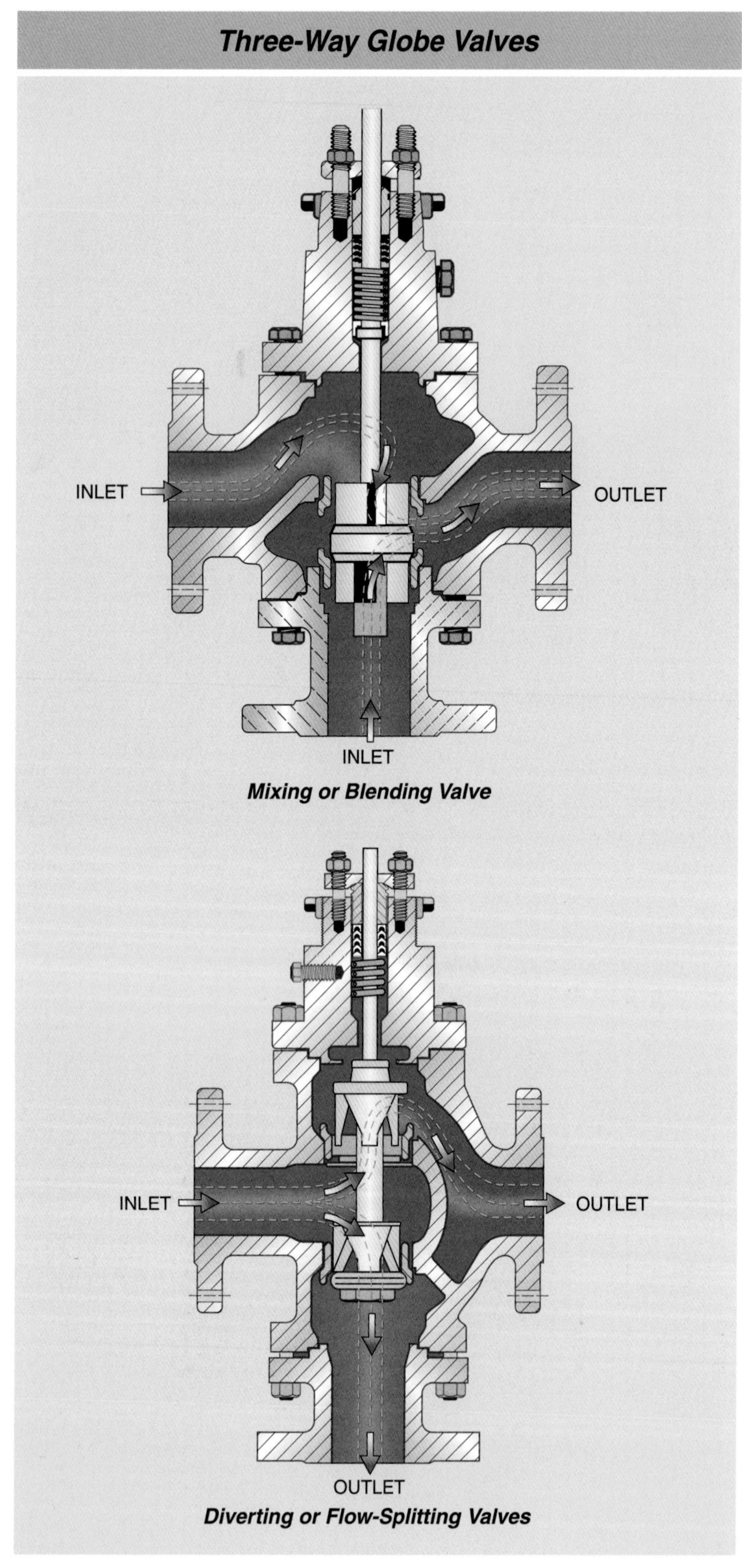

Figure 39-10. A three-way globe valve is used for mixing, blending, diverting, or flow splitting.

A three-way mixing valve has two inlets and one outlet. The mixing valve common port handles the mixed streams. Typically, one inlet stream is at the opposite side of the common and the other inlet stream is at the bottom of the valve. When the control valve plug is down, the side inlet is routed to the common outlet port. When the control valve plug is up, the bottom inlet port is routed to the common outlet port.

The position of the actuator determines which inlet port is normally open and which is normally closed. At an intermediate position, the normally open and normally closed ports are partially open to the common port, allowing mixing of the two streams. Three-way valves are made by many manufacturers and may use different porting arrangements. The information supplied by the manufacturer should be carefully studied.

A three-way mixing valve is commonly used to provide tempered water in a heating system. Hot water is mixed with cooler water to provide a tempered water source. The hot water source is connected to one of the inlet ports and ambient water is connected to the other inlet port. The position of the mixing valve plug determines the proportion of the two streams. A downstream temperature control system measures the temperature and adjusts the valve plug position to maintain the desired tempered water temperature. Depending on the use of the tempered water, the control valve actuator can be selected to fail to 100% of the ambient water (most likely) or to 100% of the hot water.

A *diverting,* or *bypass, valve* is a three-way valve that has one inlet and two outlets. Diverting valves have a common inlet port, a side outlet port, and a bottom outlet port. The choice of which exit port is normally open depends on the choice of the action of the actuator. A common application is to use a diverting valve to send a water stream through or around a heater. The normally open port is typically selected for the bypass stream since this is usually the safest failure position. A partially open valve passes only a portion of the water through the heater. Diverting valves are often used in applications that require ON/OFF flow. Diverting valves cannot be used in mixing applications. As with mixing valves, different manufacturers may have different configurations.

A *cage globe valve* is a modified single-port globe valve with a removable cylindrical cage holding a seat ring in place. The cage side openings are provided in specific shapes to provide different flow characteristics for the cage valve. This allows the plug to be a simple cylindrical shape that closes off portions of the cage openings. The cage and seat ring assembly design change to the globe valve was developed to eliminate the difficulty with replacement of threaded seats in older style globe valves. In a cage valve, the bonnet, stem, plug, cage, and seat ring can be removed and replaced together. **See Figure 39-11.**

Some styles of cage valves provide a method for equalizing the pressure drop across the plug. Equalizing the pressure reduces the imbalance in force acting on the plug and makes it possible to use a smaller actuator. The balanced plug design allows the use of a smaller actuator, but at the expense of not being able to provide a tight shutoff. There are also cage designs that minimize turbulence and noise.

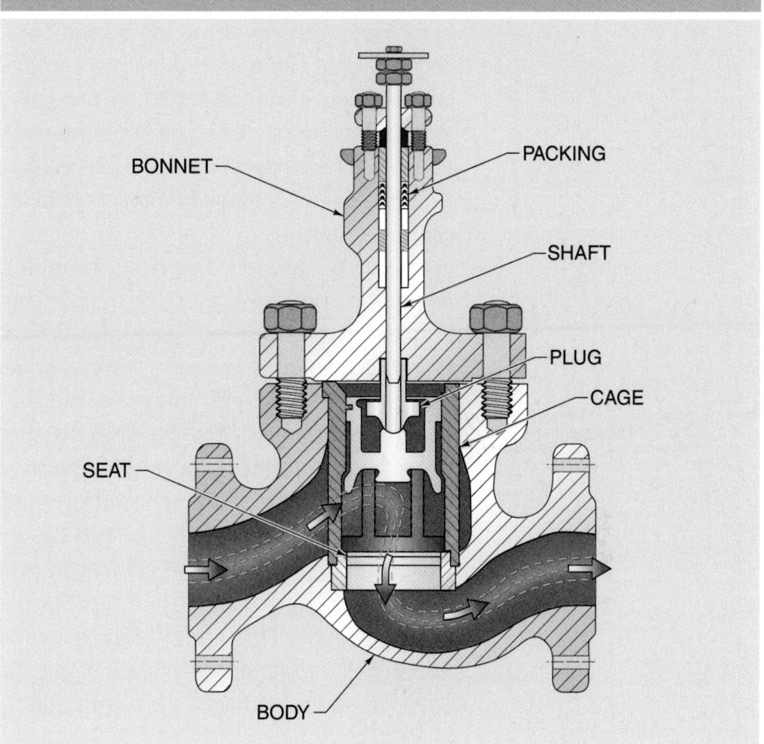

Figure 39-11. A cage globe valve has a cage that creates the desired flow characteristics.

Split Body Valves. A *split body valve* is a valve that consists of a two-piece body, with the lower half of the body being the inlet and the upper half of the body being the outlet, and a single-port assembly sandwiched between them. Split body valves can be assembled as either a straight-through or an angle valve. This makes them useful as pipe elbows in tight places. **See Figure 39-12.**

The seat assembly can be replaced by simply separating and reconnecting the two body halves. Common valve plugs are available in V-port or parabolic shapes. This is a versatile and simple design available in sizes from ½″ to 24″ pipe sizes. A construction design using separable flanges makes a split body valve better suited than other types of valves to the use of exotic materials for use in severely corrosive environments.

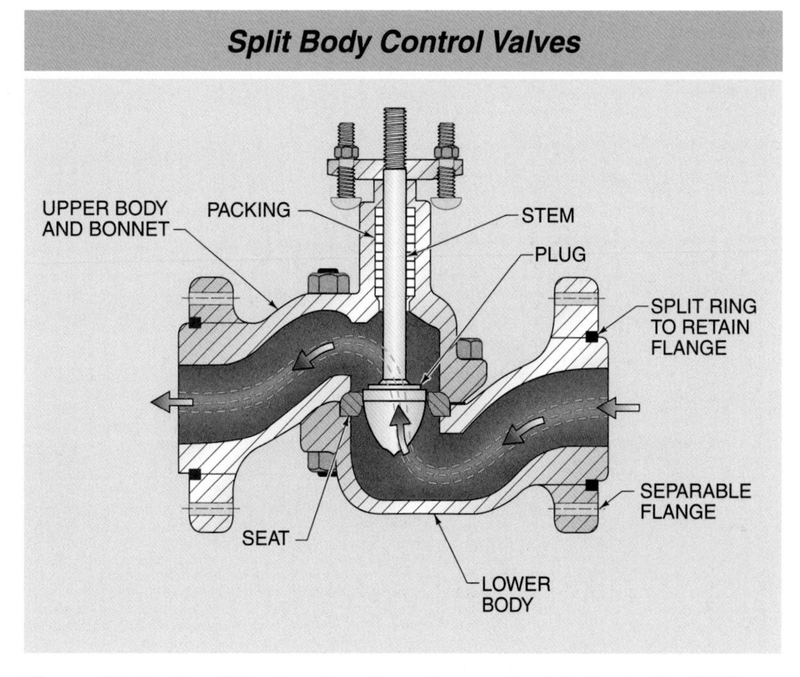

Figure 39-12. A split body valve allows for easy installation in the field.

Diaphragm Valves. A *diaphragm valve* is a throttling control valve consisting of a one-piece body incorporating an internal weir and a flexible diaphragm with a molded elastomer backing attached to the valve stem. Diaphragm valves are sometimes referred to as Saunders Patent valves because a mining engineer named Saunders held the original patent.

When fully open, the flow through a diaphragm valve is almost straight except for passing over the weir. **See Figure 39-13.** When closed, the stem and compressor act against the diaphragm and press the diaphragm against the top of the weir. Because the open flow area is narrow, diaphragm valves have a limited throttling control range. There is a variation of the standard design where the diaphragm follower closes the outside edges of the diaphragm before the center is compressed. This greatly improves the throttling control range by concentrating the open area in the center of the diaphragm.

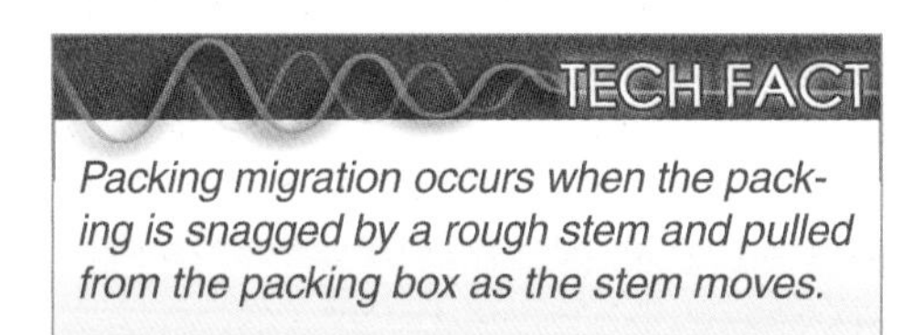

The principle advantages of a diaphragm valve are tight shutoff and corrosion resistance. Diaphragm valves are available in sizes from ½″ to 8″ in a wide selection of bodies and diaphragms. A modified diaphragm valve eliminates the weir and has a movable diaphragm and plug that presses against a bottom diaphragm in a pinching action that is better for slurries and liquids with suspended solids. This form of diaphragm valve is not suited to control and should only be used for ON/OFF services.

Rotary Shaft Valves

A *rotary shaft valve* is a throttling control valve used to change the flow of materials by means of the movement of a rotating wafer, contoured disc, ball, or plug. A shaft attached to the valve internals passes through a seal or packing to the actuator linkage. Throttling service works best with the shaft connected with a rack and pinion or crank arm to a spring and diaphragm or sliding stem piston actuator. Rotary valves have stem seals instead of bonnets. The lubrication and sealing design of the rotating shaft varies depending on the sealing requirements.

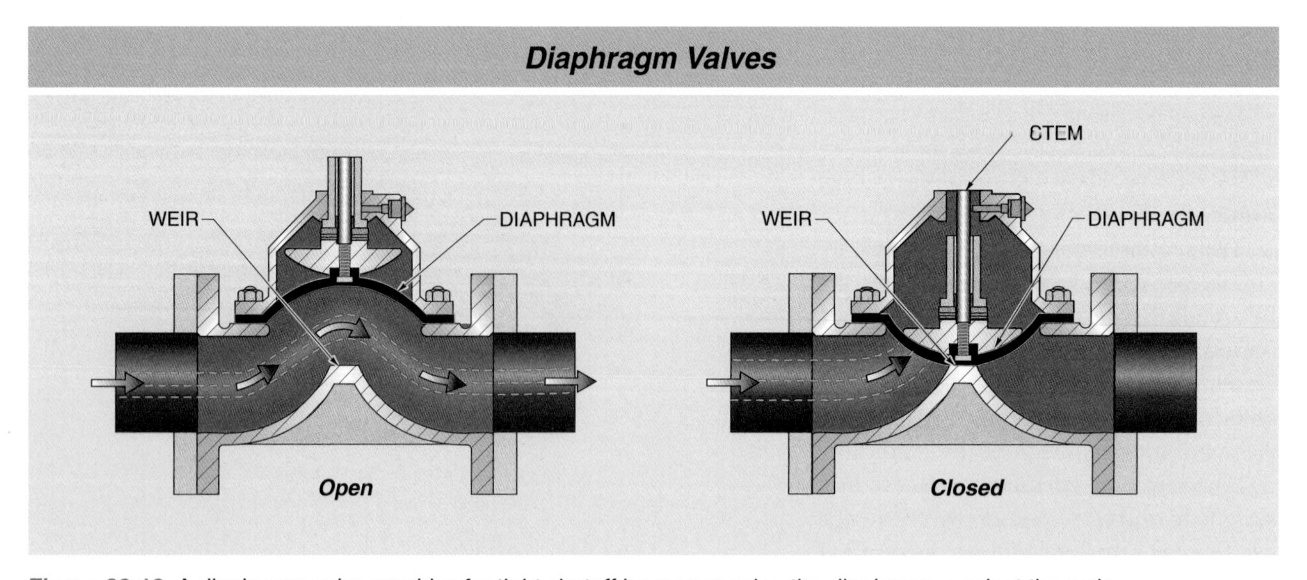

Figure 39-13. A diaphragm valve provides for tight shutoff by compressing the diaphragm against the weir.

Butterfly Valves. A *butterfly valve* is a valve with a disc that is rotated perpendicular to the valve body. **See Figure 39-14.** Standard butterfly valves are made in diameters from 2″ up to 24″. Larger sizes are available as special orders. Butterfly valves can be made of many materials, with unlined and lined bodies, and with metal or coated discs. Some butterfly valve designs have eccentrically positioned discs to provide for tighter shutoff.

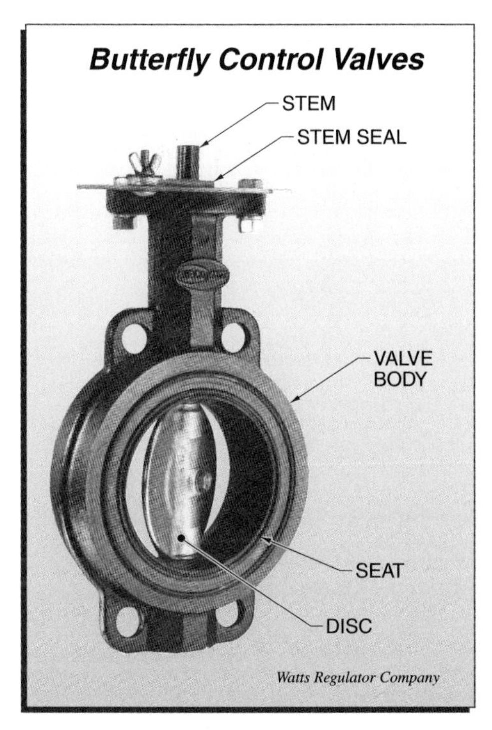

Figure 39-14. A butterfly valve has a disc that rotates to modulate flow.

Butterfly valves are usually supplied with wafer-style bodies. Wafer-style bodies are those without their own flanges and are inserted between flange pairs and secured by through-bolts. A full-bodied valve has cast flanges that are attached independently to the two adjacent piping flanges. This allows one of the piping flanges to be removed without the butterfly valve falling out of the piping.

Butterfly valves have a flow characteristic close to an equal percentage curve, but they are susceptible to instability over portions of their operating ranges due to dynamic forces acting upon the disc. A conventional butterfly valve has an effective operational range for throttling services of 20° to 60° open. There are special butterfly designs that can handle throttling services for a range of 20° to 90° open. Butterfly valves should not be allowed to throttle at less than 20° open.

Butterfly valves are relatively inexpensive compared with other valves. They are the only cost-effective solution for very large diameter piping. However, the differential pressure across a butterfly valve can create torque forces on the disc that must be opposed by the actuator. This can limit the pressure differential range of the valve.

Ball Valves. A *ball valve* is a throttling valve consisting of a straight-through valve body enclosing a ball with a hole through the center. The ball is rotated by a shaft attached to the top of the ball that connects it to an actuator. **See Figure 39-15.** There are seat rings that include seals in contact with the ball at the inlet and outlet ports of the body. Most ball valves have a hole in the ball that is smaller than the valve size. A full port ball valve is one with a hole matching the inlet and outlet diameter of the valve body. Replaceable seat rings are available with reduced ports specially shaped for specific flow characteristics.

Figure 39-15. A ball valve works best for ON/OFF service.

Ball valves are available unlined or with linings in many combinations of materials. Ball valves are available in sizes from ½″ to 8″. Standard ball valves can be used for throttling over a limited range, but they are better for ON/OFF applications. Ball valves are especially useful for slurries and liquids with suspended solids due to the straight-through design. A ball valve generally has higher capacity and lower cost than a similar size globe valve.

A modification of the ball valve is called a V-ball and has a notched partial ball. The V-ball provides greater throttling rangeability and an equal-percentage flow characteristic, both of which are much better than a standard ball valve. The V-ball valve is an excellent throttling valve. **See Figure 39-16.** It is also designed to provide tight shutoff and can handle slurry services.

V-Ball Control Valves

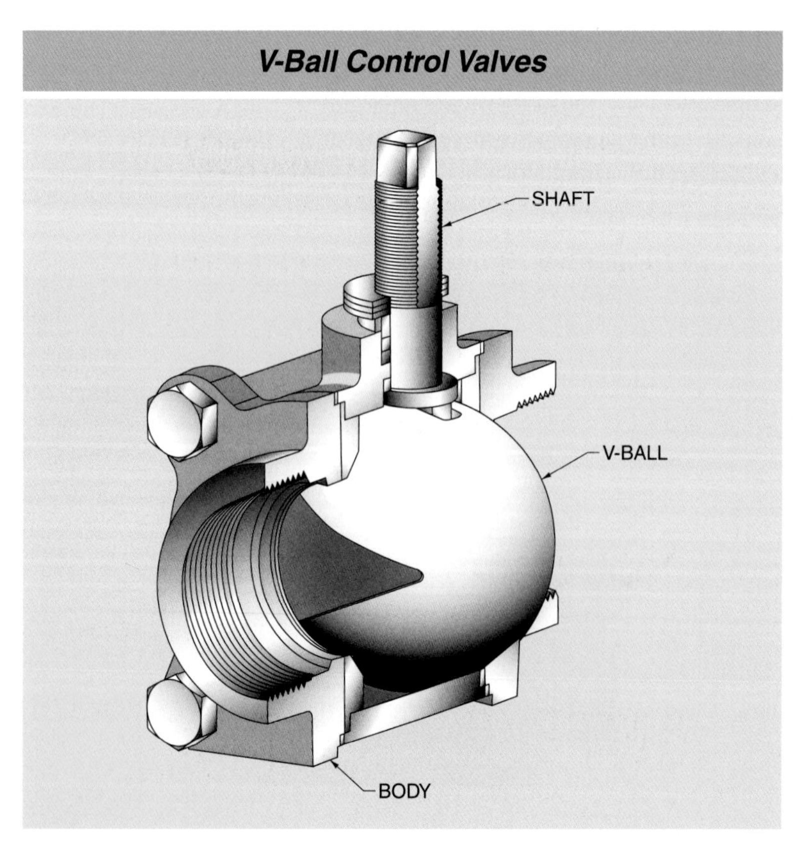

Figure 39-16. A V-ball valve is a modified ball valve that allows accurate throttling.

Plug Valves. A *plug valve* is a modified ball valve consisting of a tapered cylinder with a slot through it. Plug valves offer very tight shutoff at the expense of higher operating torque. **See Figure 39-17.** Conventional plug valves do not have a good flow characteristic for throttling services. A plug valve is primarily an ON/OFF valve. Plug valves are available with a selection of slot shapes that alter the normal relationship between flow and opening to provide better throttling control. Some plug valves have inserted sleeves that match the opening. Like butterfly and ball valves, plug valves are available with lined bodies. The sizes range from ½″ to 24″ for unlined valves and from ½″ to 12″ for lined valves.

Plug Valves

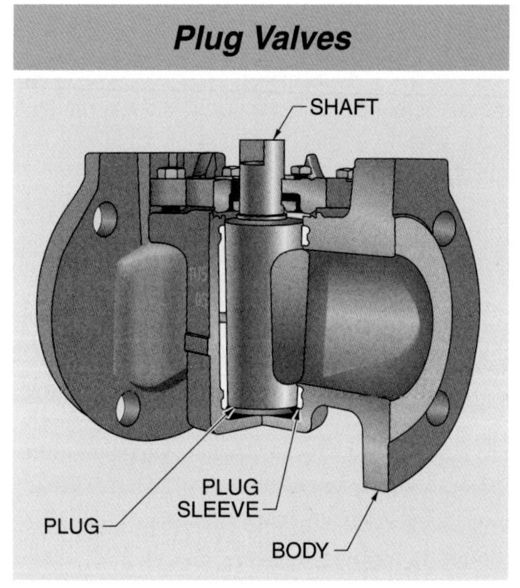

Figure 39-17. A plug valve provides improved sealing compared to many other types of valves.

Eccentric Cam Valves. An *eccentric cam valve* is a throttling control valve with a rotating shaft with an attached convex disc, at a right angle to the shaft, that closes against a seat ring for tight shutoff and extended seat life. **See Figure 39-18.** Eccentric cam control valves were developed as an alternative to globe valves to provide lighter weight bodies, the equivalent of cage trim, and easy change-out of internal parts. The flow path is straight through, which reduces the pressure drop from inlet to outlet. Reverse direction flow decreases valve capacity. There are many choices of materials of construction, and connection sizes are available from ½″ to 8″.

Eccentric Cam Control Valve

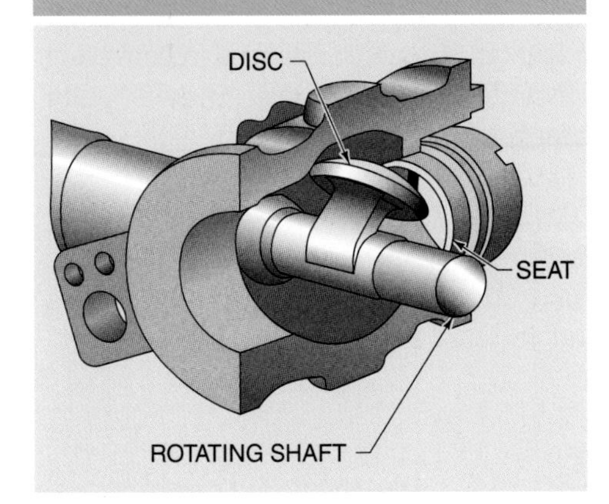

Figure 39-18. An eccentric cam control valve has a disc that rotates out of the way to allow flow.

ON/OFF CONTROL ACTIONS

An *ON/OFF control valve* is a valve used to start and stop the flow of materials in the pipelines that control chemicals and energy sources for automatic sequencing, batching, or safety operations. Most ON/OFF control valves are two-way or three-way valves. Many ON/OFF control valves are simple solenoid valves. Solenoid valves can be two-way, three-way, or four-way and have two, three, four, or five ports with two or three positions. The Joint Industry Conference (JIC) has established standard drawing symbols for solenoid valves. **See Figure 39-19.** Common ON/OFF actions use direct-acting valves, spring-return and double-acting actuators, and limit switches.

Direct-Acting Valves (Solenoids)

Direct-acting valves are the simplest form of ON/OFF valves. In a direct-acting valve, an electrical solenoid is directly attached to the moving parts controlling the fluid flow through the valve.

Solenoid Valves

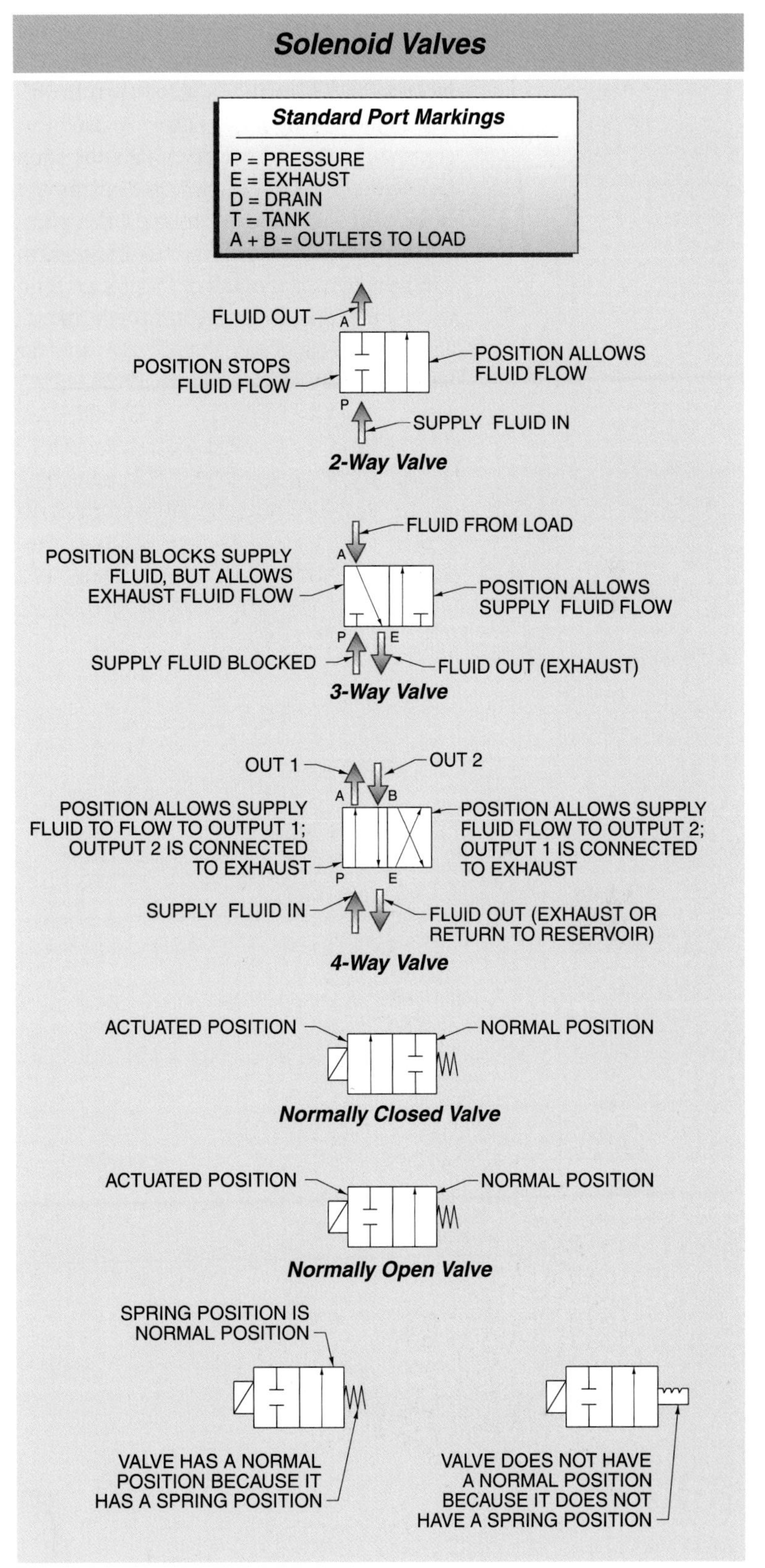

Figure 39-19. Solenoids can be used as valve actuators. There are many designs to allow different types of control actions.

For example, a two-way solenoid valve can be used to directly control the flow of a process fluid. These types of valves are available in NO or NC forms. The sizes are usually smaller than 1″ because solenoids often lack the power to close the valve against the process pressure. There are some larger pilot-operated solenoid valves that are designed to use the upstream pressure of the process fluid to close the valve. A special form of the direct-acting valve is the safety shutoff valve used for burner fuel valves. **See Figure 39-20.**

A gas pressure regulator and gas flow control valve are used to control the fuel flow to a burner. The safety shutoff valve is held open as long as the burner system is operating properly. Unsafe operation trips the safety valve, allowing its spring to close the valve. Solenoids are usually controlled by an electrical output from a logic or sequenced program. In this application, solenoids are used to convert an ON/OFF electrical signal to an ON/OFF supply of fuel.

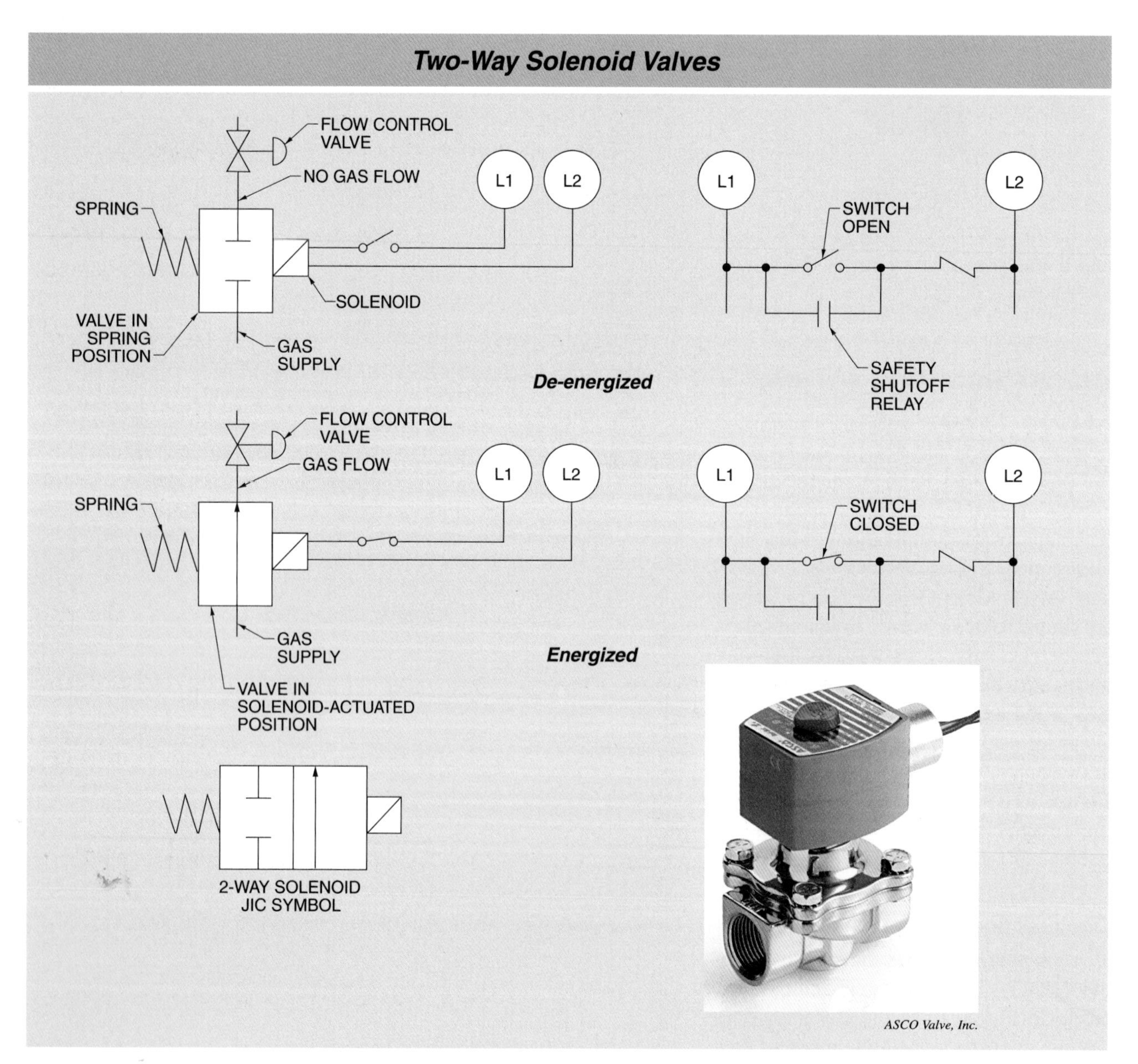

Figure 39-20. Two-way solenoid valves allow simple ON/OFF operation.

KEY TERMS

- *final element:* A device that receives a control signal and regulates the amount of material or energy in a process.
- *throttling control valve:* A valve and actuator assembly that is able to modulate fluid flow at any position between fully open and fully closed in response to signals from a controller.
- *way:* A path through a valve from an inlet port to an outlet port.
- *position:* A location within a valve where the internal parts are placed to direct fluid through the valve.
- *normally closed (NC) valve:* A valve that does not allow pressurized fluid to flow out of the valve in the spring-actuated (de-energized) position.
- *normally open (NO) valve:* A valve that allows pressurized fluid to flow out of the valve in the spring-actuated (de-energized) position.
- *valve flow characteristic:* The relationship between valve flow capacity (in percent) and control valve open travel (in percent), with all other factors that affect flow held constant.
- *quick-opening valve:* A valve that allows flow to increase quickly with a small initial opening.
- *linear valve:* A valve that allows flow to increase at the same rate as an opening.
- *rangeability:* The ratio of the maximum flow to the minimum flow at the desired measurement accuracy.
- *equal-percentage valve:* A valve that allows the flow rate percentage to change by an amount equal to the change in the opening percentage.
- *flashing:* A process in which a portion of a liquid converts to a vapor as it passes through a control valve because the pressure has fallen below the vapor pressure of the liquid.
- *cavitation:* A process in which vapor bubbles in a flowing liquid collapse inside a control valve as the pressure begins to increase.
- *critical pressure ratio:* The ratio of downstream pressure to upstream pressure where the gas velocity out of a valve is at sonic velocity, and further decreases in downstream pressure no longer increase the flow.
- *sonic velocity:* The speed of sound in a gas.
- *choked flow:* A condition where the actual pressure ratio is less than the critical pressure ratio and flow is restricted.
- *sliding stem control valve:* A throttling valve that has a stem (shaft) attached to a plug or disc at one end and an actuator at the other end.
- *valve stem:* A valve component that consists of a metal shaft that transmits the force of the actuator to the valve plug.
- *valve bonnet:* A packing enclosure that is bolted or threaded to the top of the valve body.
- *valve body:* A casting or forging with an enclosed port and integral threaded or flanged inlet and outlet openings.
- *valve plug:* A machined disc or shaped piece that regulates the flow of a material by changing the size of the valve opening.
- *globe valve:* A throttling valve where the fluid flow enters horizontally, makes a turn through the plug and seat, and then makes another turn to exit the valve.

KEY TERMS (continued)

- *single-port globe valve:* A globe valve that consists of a single valve plug and seat ring through which a fluid flows.
- *double-port globe valve:* A globe valve that consists of two plugs and seat rings through which a fluid flows.
- *three-way globe valve:* A globe valve that consists of three pipe connections and is used for mixing, blending, or flow division or diversion applications.
- *diverting,* or *bypass, valve:* A three-way valve that has one inlet and two outlets.
- *cage globe valve:* A modified single-port globe valve with a removable cylindrical cage holding a seat ring in place.
- *split body valve:* A valve that consists of a two-piece body, with the lower half of the body being the inlet and the upper half of the body being the outlet, and a single-port assembly sandwiched between them.
- *diaphragm valve:* A throttling control valve consisting of a one-piece body incorporating an internal weir and a flexible diaphragm with a molded elastomer backing attached to the valve stem.
- *rotary shaft valve:* A throttling control valve used to change the flow of materials by means of the movement of a rotating wafer, contoured disc, ball, or plug.
- *butterfly valve:* A valve with a disc that is rotated perpendicular to the valve body.
- *ball valve:* A throttling valve consisting of a straight-through valve body enclosing a ball with a hole through the center.
- *plug valve:* A modified ball valve consisting of a tapered cylinder with a slot through it.
- *eccentric cam valve:* A throttling control valve with a rotating shaft with an attached convex disc, at a right angle to the shaft, that closes against a seat ring for tight shutoff and extended seat life.
- *ON/OFF control valve:* A valve used to start and stop the flow of materials in the pipelines that control chemicals and energy sources for automatic sequencing, batching, or safety operations.

REVIEW QUESTIONS

1. What is the difference between normally open and normally closed valves and how are the terms NO and NC used differently with valves and with electricity?
2. Define valve flow characteristic and explain the difference between quick-opening, linear, and equal-percentage valves.
3. What is the difference between flashing and cavitation?
4. What is the difference between sliding stem and rotary shaft valves?
5. Define ON/OFF control valve.

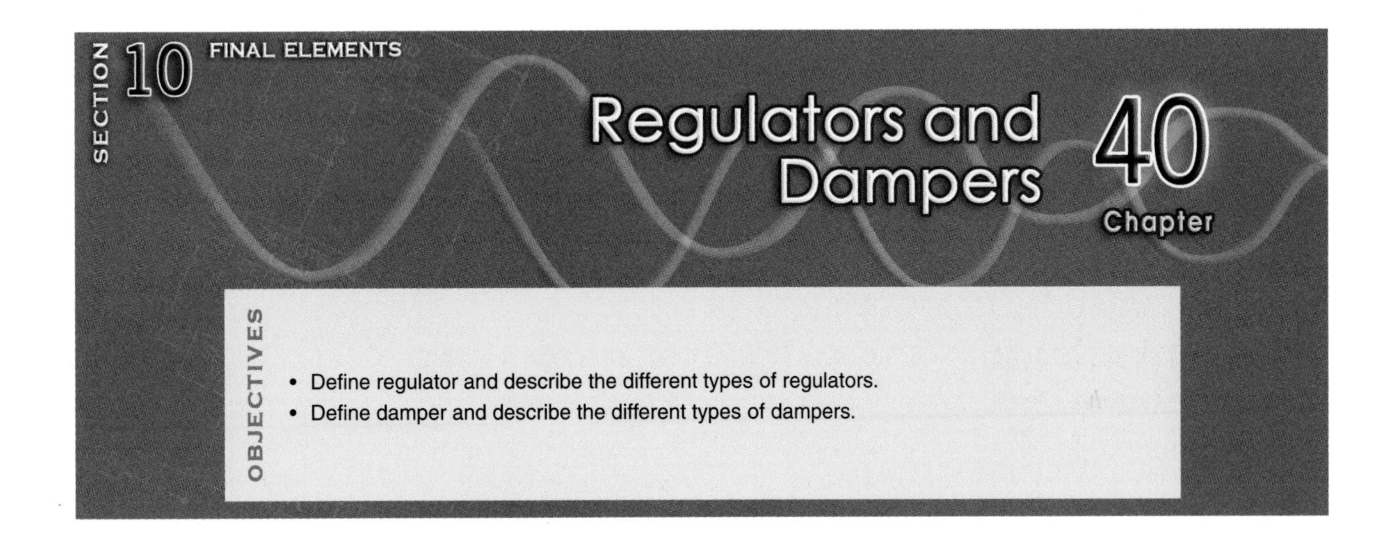

REGULATORS

A *regulator* is a self-operating control valve for pressure and temperature control. Regulators require no other source of energy than the process itself. The manufacturers of regulators provide flow capacity and pressure drop tables for sizing the regulators. The main types of regulators are pressure regulators, used to control downstream pressure; backpressure regulators, used to control upstream pressure; and temperature self-operating regulators, used to automatically control temperature.

Pressure Regulators

A *pressure regulator* is an adjustable valve that is designed to automatically control the pressure downstream of the regulator. The three basic components of most regulators are a loading mechanism, a primary or sensing element, and a final or control element. The loading mechanism determines the setpoint. This is typically a spring, but may also be a balancing air pressure. The primary element senses the force placed on the loading mechanism. The primary element is typically a diaphragm. Primary elements transmit the force to the final element. The control element is a valve that actually accomplishes the pressure drop.

Regulators are commonly used to control water pressure in a distribution system that must be held to a value that ensures supply to all sectors of the system. For example, the pressure of makeup water to a hot water heating system is commonly controlled to about 12 psig to 18 psig with a pressure regulator. Likewise, fuel gas pressure must be regulated before entering a burner. The main types of pressure regulators are spring-loaded, air-loaded, pilot-operated, and differential pressure regulators.

Spring-Loaded Pressure Regulators. A *spring-loaded pressure regulator* is a regulator consisting of a throttling element such as a valve plug connected to a pressure-sensing diaphragm that is opposed by a spring and contained in a single housing. **See Figure 40-1.** The diaphragm serves as the separator between the spring chamber and the valve body in addition to being the primary element. The spring pressure against the diaphragm is adjusted with a manual setting knob or screw. The controlled downstream pressure is applied to the underside of a diaphragm through an internal path within the valve body.

TECH FACT

A normally closed pneumatic valve has no flow when the valve is de-energized. A normally closed electric switch does have current flow when the switch is de-energized. It is important to keep this distinction clear.

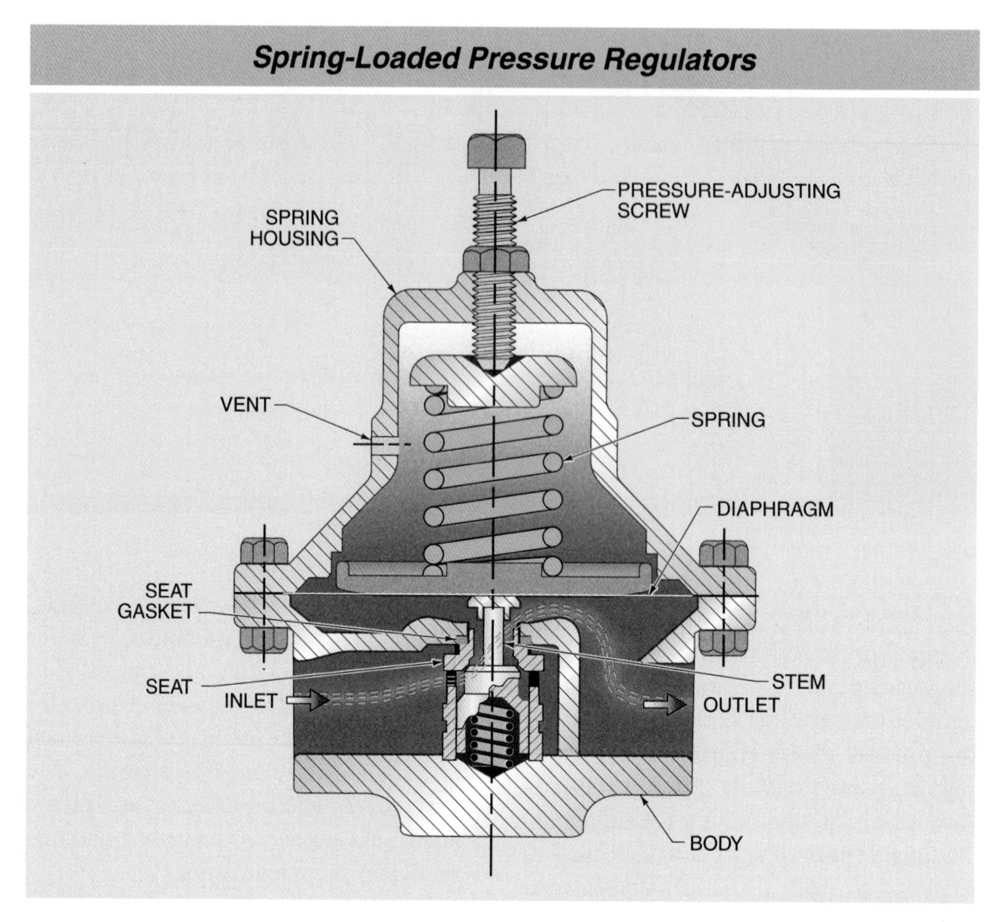

Figure 40-1. A spring-loaded pressure regulator balances spring pressure against downstream pressure to control the downstream pressure to a desired value.

The metal or elastomer diaphragm is attached at its center to the valve stem below the diaphragm and to the spring above the diaphragm. At the other end of the stem is the valve plug that opens or closes a valve port. If the controlled pressure rises above the pressure set by the spring adjustment, the diaphragm compresses the spring, causing the valve plug to move toward the seat. This action decreases the flow through the regulator and reduces the controlled downstream pressure.

Some regulators are designed for external pressure sensing. In addition, some types of pressure-reducing regulators are designed to automatically vent excess pressure to maintain more precise pressure control. The venting capacity is only a small portion of the regulator capacity and is not designed to provide protection in the event of a regulator failure.

Self-contained pressure regulators are subject to built-in operational characteristics called droop and lockup. **See Figure 40-2.** *Droop* is a drop in pressure below a set value when there is high flow demand. *Lockup* is an increase in pressure above a set value when there is low flow demand.

An adjusting screw is turned to compress or release the regulator spring to set the downstream pressure. If the flow demand through the regulator increases after the pressure is set, the compression of the spring changes to handle the additional flow and results in a downstream pressure that is less than the original set pressure. Reductions in flow result in lockup as the compression of the spring changes to handle the reduced flow. Droop is expressed as the percentage of the set pressure.

For example, an application requires 40 psig process gas pressure but the building supply is 100 psig. A pressure regulator is installed to reduce the line pressure to the desired value. When there is no demand for the process gas, no flow is required and the regulator valve is closed. As the demand for the process gas begins, the regulator valve starts to open. If the demand is high enough, the regulator valve opens completely.

The opening and closing of the valve is controlled by the amount of force on the spring caused by the backpressure from downstream of the regulator. When the flow demand decreases, the downstream pressure increases, and the increasing pressure increases the force on the diaphragm and starts to close the valve plug. When the flow demand increases, the reverse action occurs and the valve plug opens. This changes the force on the spring and the setpoint changes, resulting in droop. Droop varies with the change in flow demand and is usually small but may be as much as 30%. Manufacturers provide information on the droop at different flows.

Simple regulators are available in sizes from ¼″ to 2″ in many combinations of materials for body, spring and spring chamber, diaphragm, plug, and seat. Regulators are suitable for liquid or gas service.

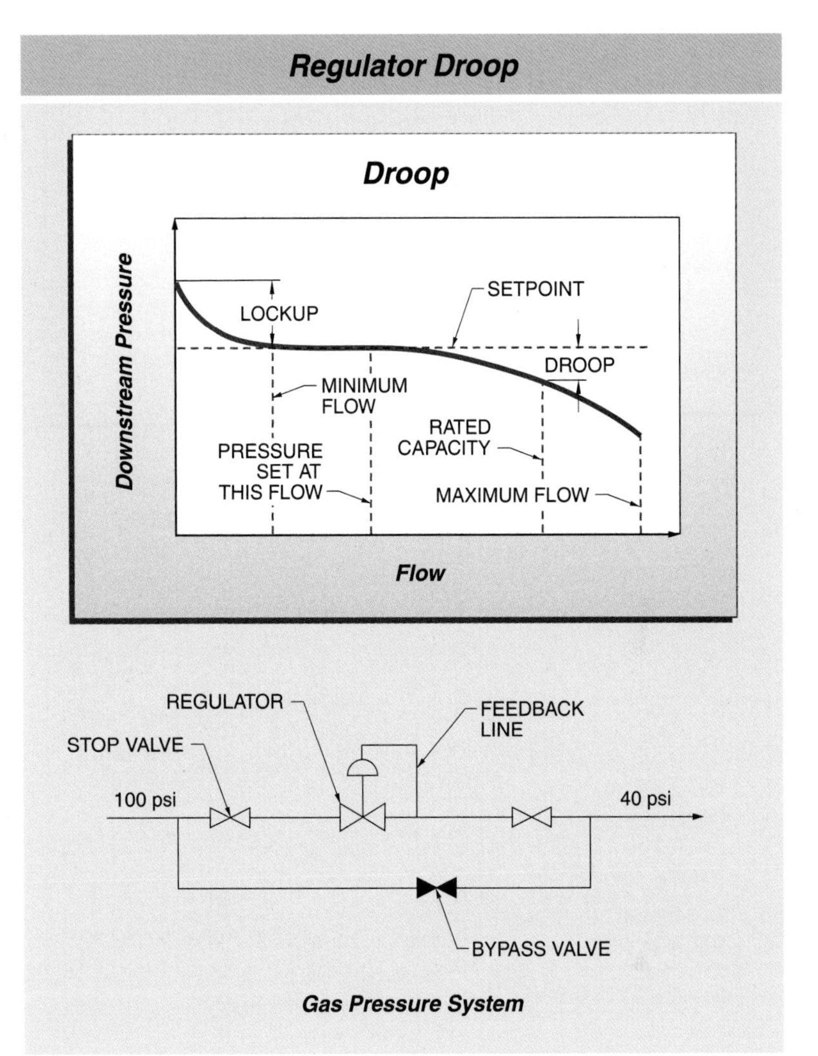

Figure 40-2. Regulator droop occurs when the flow demand increases.

Air-Loaded Pressure Regulators. An *air-loaded pressure regulator* is a regulator that uses air pressure instead of the force of a spring to oppose downstream pressure. The air must constantly flow to the regulator and be allowed to bleed from the spring chamber to allow the set pressure to be changed. An air-loaded regulator provides the capability to develop more complex control arrangements. An air-loaded pressure regulator can also be used when the regulator is inaccessible and needs to be adjusted remotely. A remote air regulator must have an internal overpressure vent so that the set pressure can be adjusted in both directions.

A zero regulator is a special form of air-loaded pressure regulator that is used to regulate fuel gas flow in proportion to combustion airflow for burners. The combustion air pressure is used as a set pressure for the regulated gas pressure. Maintaining equal gas and combustion air pressures ensures consistent air-fuel ratios for any firing rate. **See Figure 40-3.** This is based on equal pressure drops through the combustion air and natural gas piping. A zero regulator, along with a small blanketing pressure regulator, is also commonly used to control the pressure of a gas blanket in the vapor space of a pressurized liquid storage vessel.

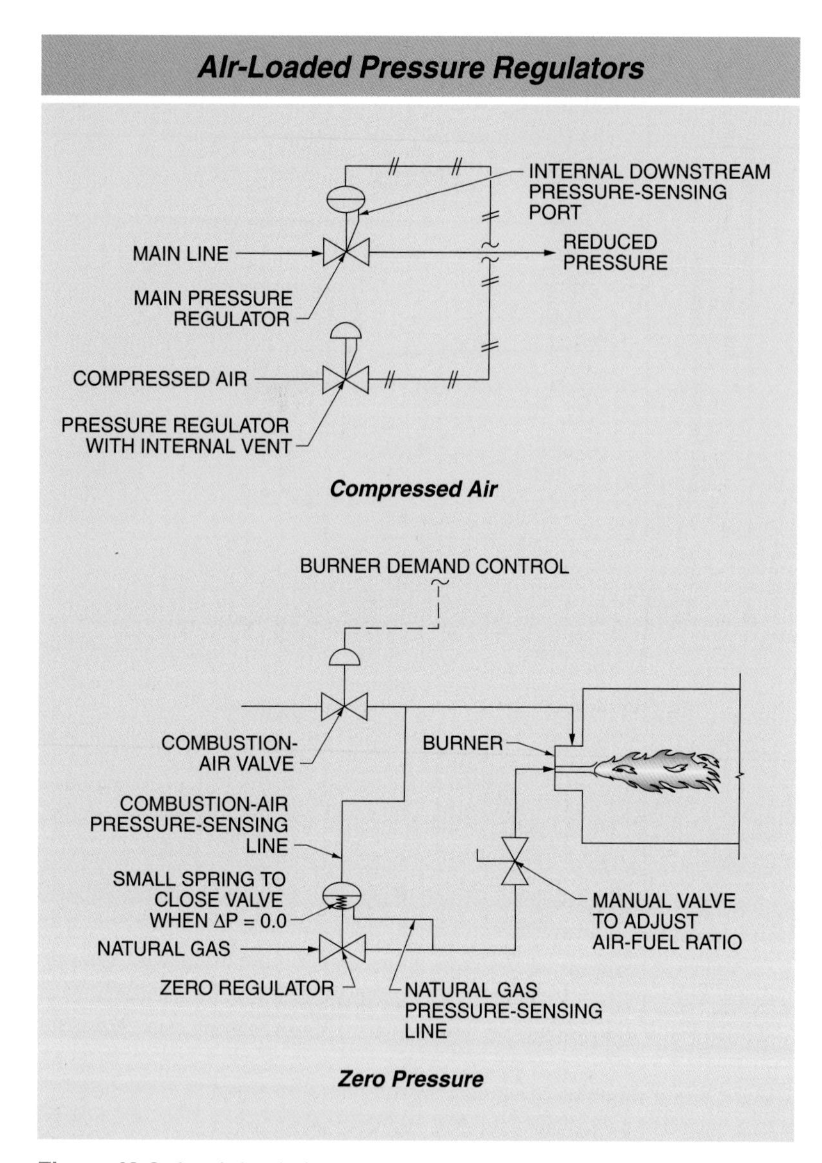

Figure 40-3. An air-loaded pressure regulator can be used to maintain two flow streams in the same ratio.

Pilot-Operated Pressure Regulators. A *pilot-operated pressure regulator* is a regulator that uses upstream fluid as a pressure source to power the diaphragm of a larger valve. **See Figure 40-4.** Pilot regulators can be either external or internal to the main valve. The purpose of the pilot regulator is to provide more accurate pressure control by nearly eliminating droop. Pilot regulators are typically used with pressure-reducing valves up to 4″ in diameter. Larger sizes, up to 24″, are available.

The upstream pressurized fluid enters the inlet of the main valve and moves through the pilot supply line to the pilot regulator. This reduces the pressure to the value set by the pilot valve range spring adjustment. *Loading pressure* is the pressure above a main diaphragm. The force created by the loading pressure overcomes the force of the valve spring and opens the valve plug.

As the valve plug opens, the fluid starts to flow through the main regulator valve. The fluid fills the area below the main diaphragm at the pressure of the downstream fluid. The downstream pressure opposes the movement of the diaphragm and maintains the valve plug in the correct position to maintain the desired downstream pressure. The downstream pressure is also transferred to the pilot regulator where it opposes the spring pressure and tends to close the pilot valve plug. The pilot regulator adjusts the loading pressure to compensate.

Differential Pressure Regulators. A *differential pressure regulator* is a regulator that controls the pressure difference between the outlet pressure of a regulator and a fluid loading pressure supplied from an external source. The output pressure applies a force through a pitot tube against the underside of a diaphragm while the loading pressure applies its force against the top side of the same diaphragm. A spring adds a fixed force to that created by the loading pressure. **See Figure 40-5.** The output pressure will always be higher than the loading pressure by the amount of pressure added by the spring. This ensures that the flow through the differential pressure regulator is constant regardless of inlet or outlet pressure variations.

A typical application of a differential pressure regulator is to maintain a constant differential pressure on a compressor shaft seal relative to a process pressure. A differential pressure regulator can also be used with an orifice to provide a constant-flow device by maintaining a constant differential pressure across the orifice. This is a common accessory for purge rotameters where the purge rotameter and flow-adjusting valve take the place of the orifice.

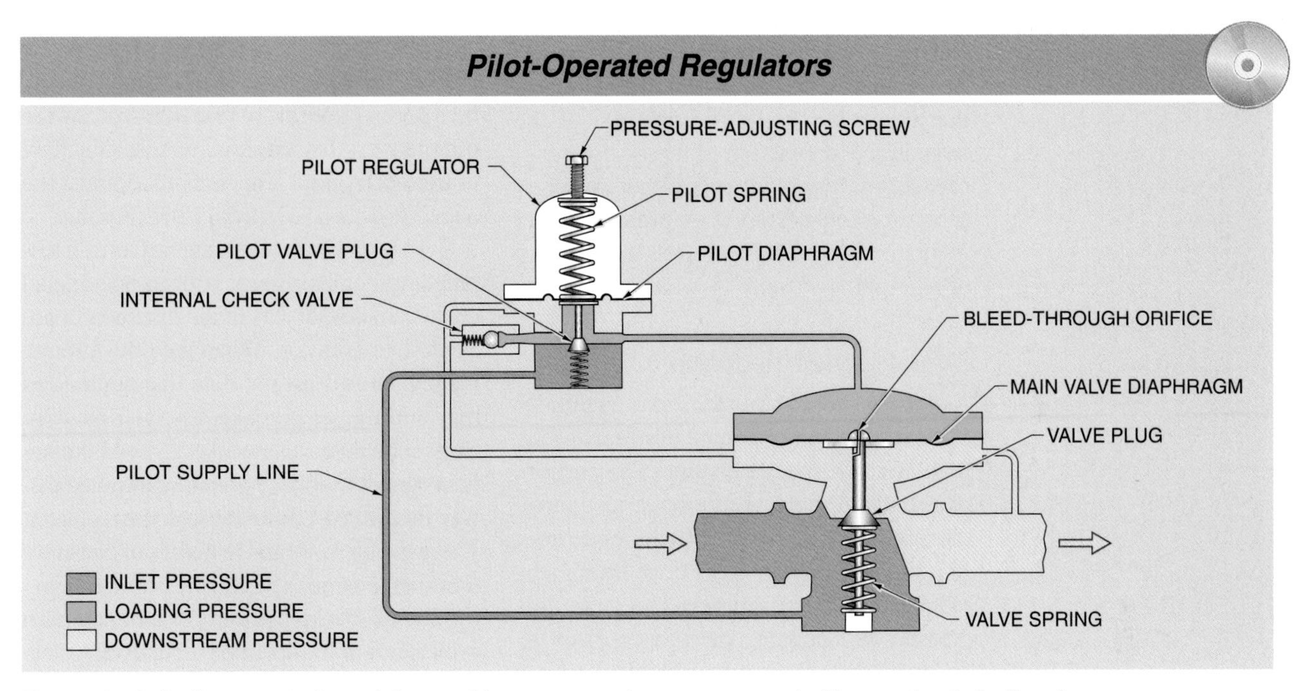

Figure 40-4. A pilot-operated regulator provides more precise pressure control by nearly eliminating droop.

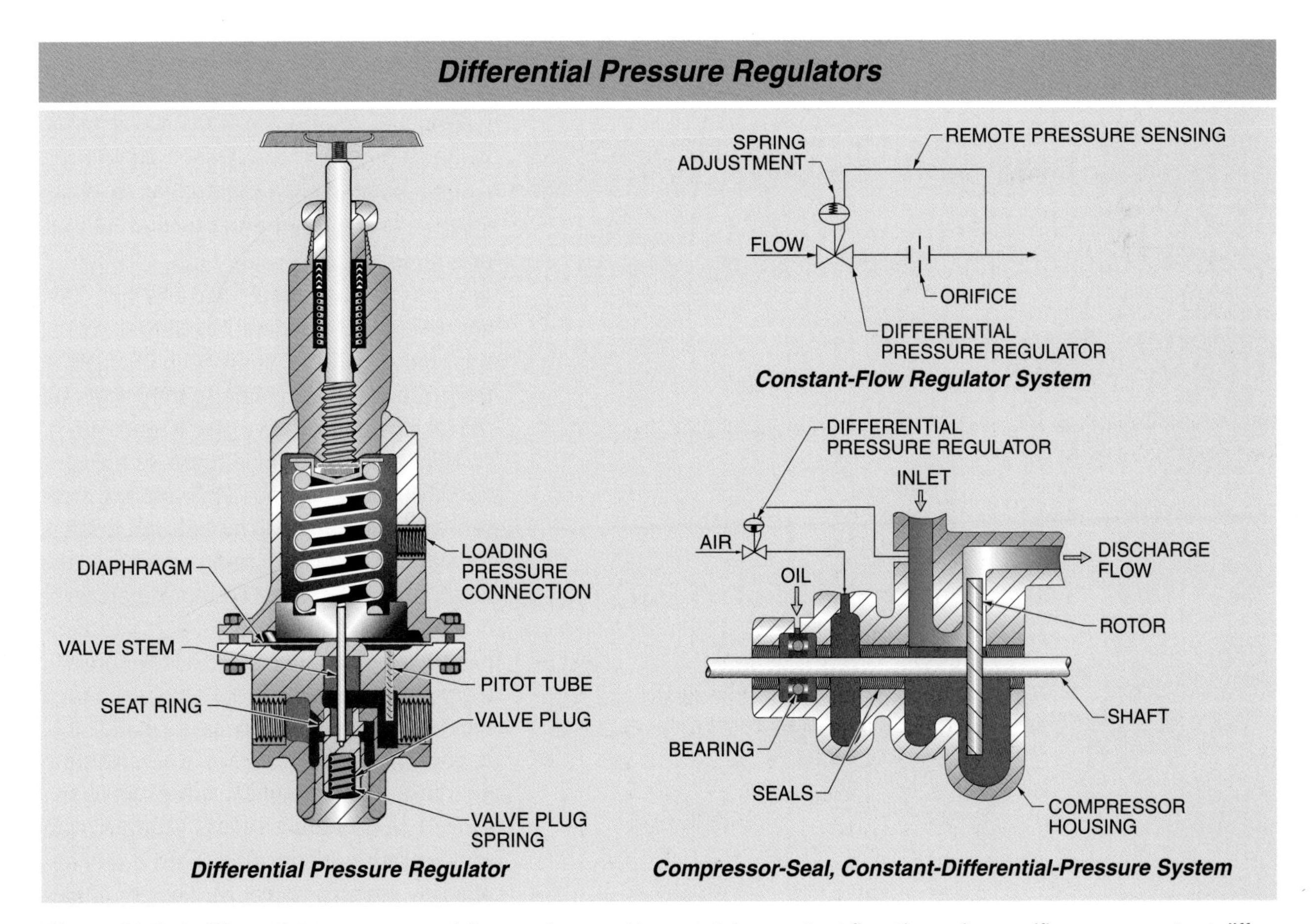

Figure 40-5. A differential pressure regulator can be used to maintain constant flow through an orifice or a constant differential pressure across an oil seal.

Pressure-Relief Regulators

A *pressure-relief regulator* is a regulator that limits the pressure upstream of the regulator. The function of a pressure-relief regulator is to provide relief if an overpressure situation develops. A pressure-relief regulator may exhaust to the atmosphere or to a piping system.

Backpressure Regulators

A *backpressure regulator* is a regulator that maintains the pressure upstream of the regulator to a specified value. The purpose of a backpressure regulator is to control the pressure in a vessel or pipeline to a setpoint by modulating the flow through the regulator. Backpressure regulators resemble pressure-reducing regulators except that the upstream pressure acts against the underside of the diaphragm, and the plug and seat relationship is reversed. **See Figure 40-6.**

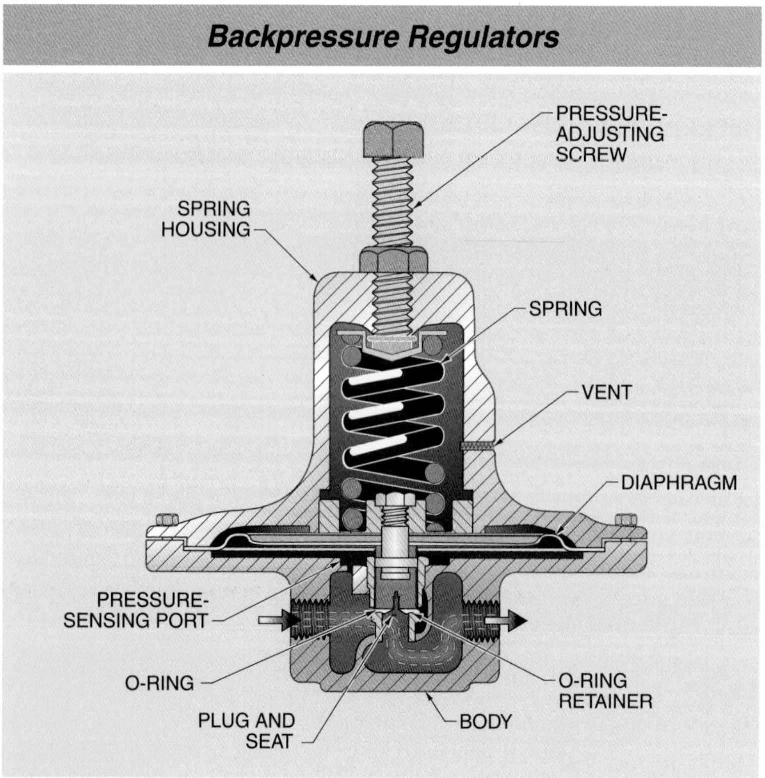

Backpressure Regulators

Figure 40-6. A backpressure regulator is used to maintain a constant upstream pressure.

When the upstream pressure rises, the diaphragm compresses the spring, causing the plug to move away from the seat. As the plug moves, the opening increases the flow to the outlet port. This acts to exhaust the added flow and lower the inlet pressure.

Like a pressure-reducing valve, a backpressure regulator has a built-in operational characteristic that acts in the direction opposite to that of droop. When the flow through the regulator is greater than that required to maintain the set pressure, the inlet pressure rises, sometimes as much as 25%. Manufacturers provide tables or graphs of the overpressure for different flows through the regulator. The available sizes and materials of construction are the same as for the pressure-reducing regulators. Backpressure regulators are also available as pilot-operated regulators.

Temperature Self-Operating Regulators

A *temperature self-operating regulator* is a regulator that is used to automatically control a process to a defined temperature. A temperature self-operating regulator is a combination of a thermal filled system and a valve. The thermal filled system consists of a temperature-sensing bulb, a length of flexible capillary tubing, and a bellows that are assembled as a unit and filled with a fluid that changes volume with changes in temperature. The remaining components are the spring and the valve. **See Figure 40-7.**

Heat acting on the bulb causes the contained fluid to expand. This applies pressure to the bellows. The bellows applies force to the adjusting spring and moves the connected valve plug. The spring pressure is adjusted to modify the operating temperature setting. Heating service requires a temperature regulator that closes the plug with an increase in temperature at the bulb. A cooling service requires a temperature regulator that opens the plug when the sensed temperature rises. Temperature self-operating regulators are generally applied to systems that have large volumes to be heated or cooled where the sensed temperature changes slowly.

DAMPERS

A *damper* is an adjustable blade or set of blades used to control the flow of air. **See Figure 40-8.** Dampers are usually not provided with actuators. It is the responsibility of the system designer to determine the force and travel requirements. The designer is required to design the actuator mounting location and to design the interconnecting linkages. Dampers are commonly classified as parallel-, opposed-, and round-blade dampers.

Parallel-Blade Dampers

A *parallel-blade damper* is a damper in which adjacent blades are parallel and move in the same direction with one another. Parallel-blade dampers are the most common dampers used in HVAC systems. Parallel-blade dampers are less expensive than other damper designs and provide better air mixing when used to blend outside air and return air. Parallel-blade dampers are often used in systems that require a damper to be either fully open or fully closed and require less maintenance than opposed-blade dampers. A disadvantage of parallel-blade dampers is that they do not provide the same precise air control that opposed-blade dampers provide. Many basic air handling units and packaged exhaust fans contain parallel-blade dampers.

Opposed-Blade Dampers

An *opposed-blade damper* is a damper in which adjacent blades are parallel and move in opposite directions from one another. Opposed-blade dampers are more expensive than parallel-blade dampers, but opposed-blade dampers provide better flow characteristics than parallel-blade dampers. An opposed-blade damper linkage is more complicated than a parallel-blade damper linkage because duplicate drive arms are required to drive adjacent blades in opposite directions, requiring increased maintenance. Opposed-blade dampers are used in applications that require two air streams to mix in order to prevent cold spots and possible freeze-up.

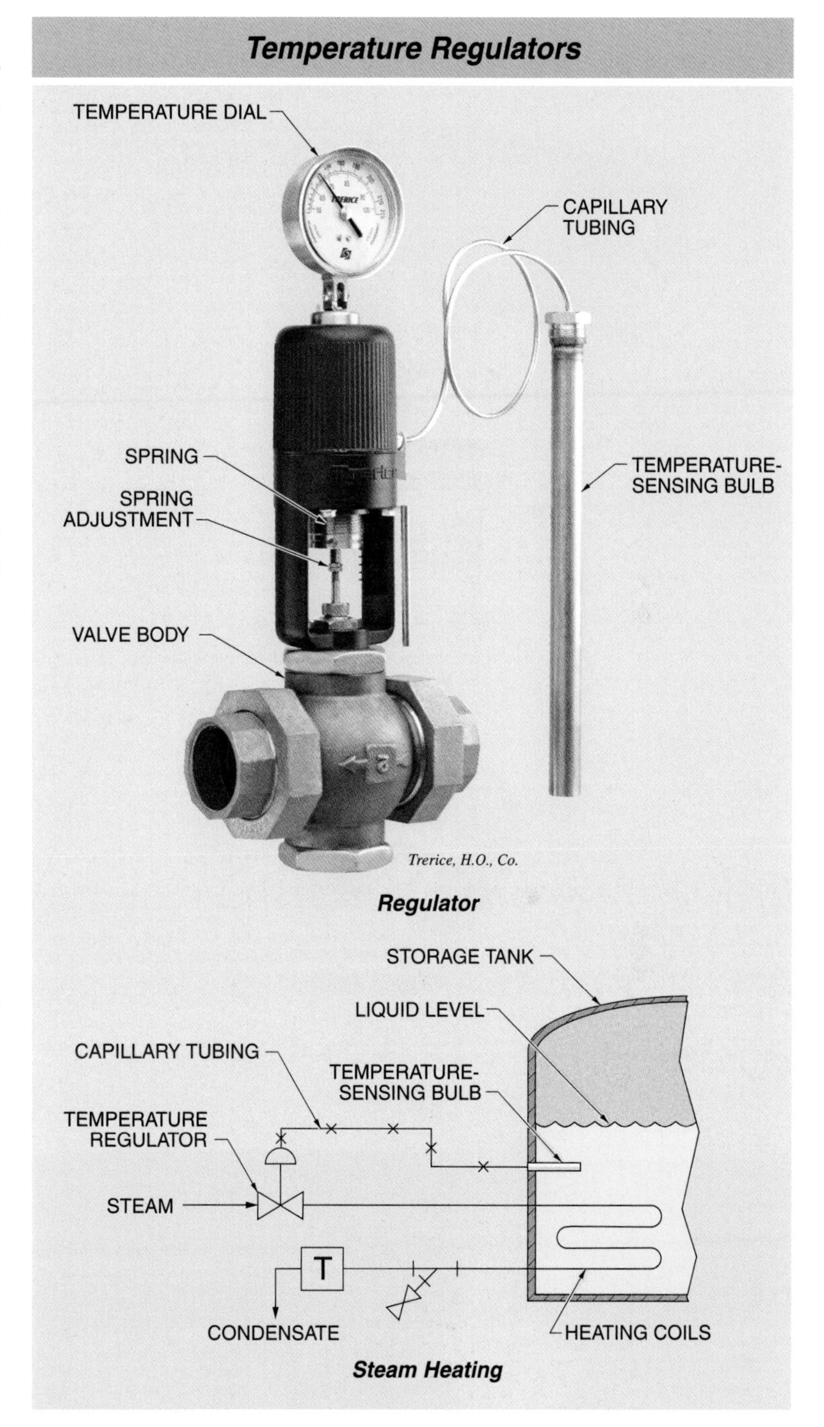

Figure 40-7. A temperature regulator combines a temperature-sensing bulb with a control valve.

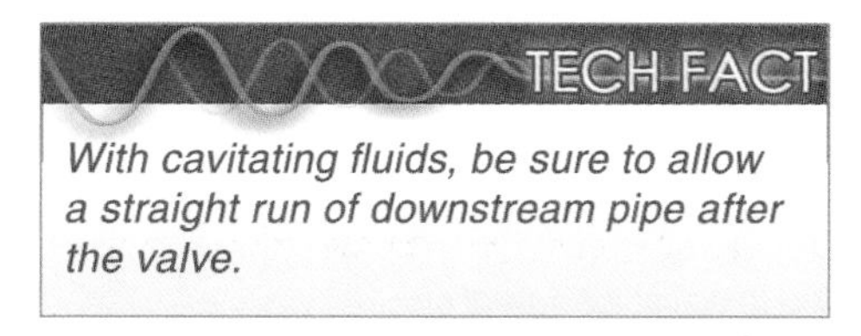

With cavitating fluids, be sure to allow a straight run of downstream pipe after the valve.

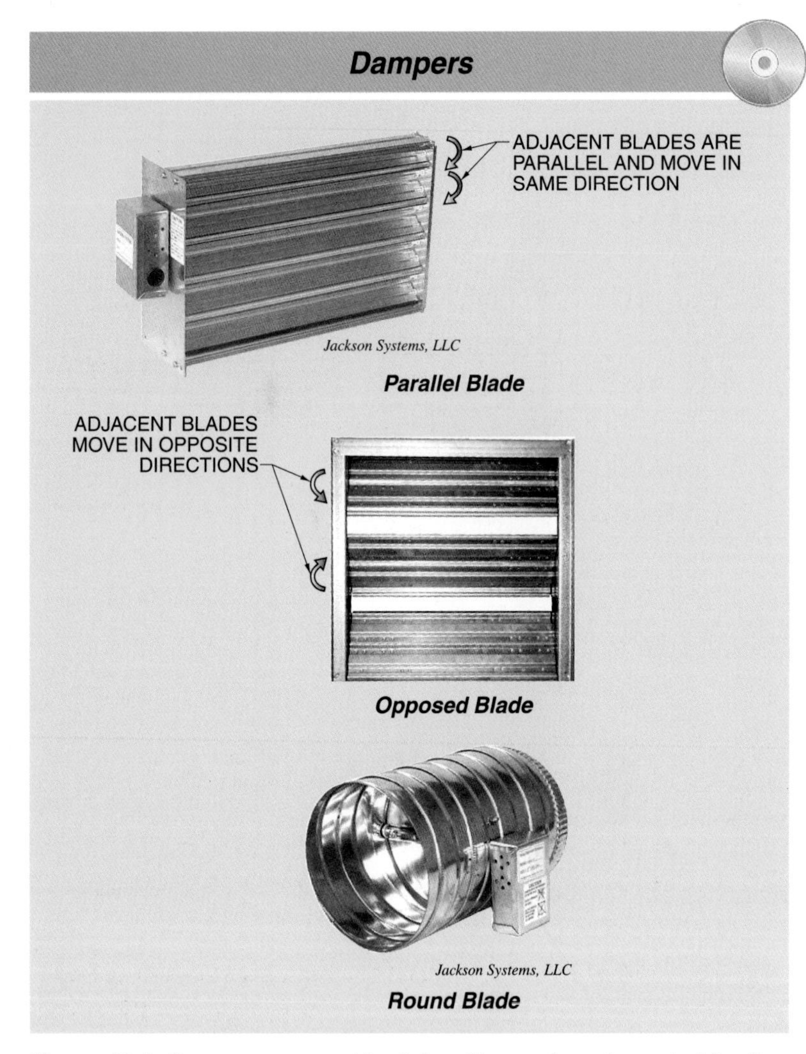

Figure 40-8. Dampers are used in air handling systems to control the flow of air.

Dampers can be manually controlled to adjust the flow of air.

TECH FACT

The term "hunting" refers to an undesirable oscillation of significant magnitude that continues after the external cause disappears. Hunting is sometimes called "cycling" or "limit cycle" and is evidence of operation near the stability limit. In control valves, hunting appears as an oscillation in the loading pressure to the actuator. It may be necessary to retune the control system or recalibrate the valve and actuator.

Round-Blade Dampers

A *round-blade damper* is a damper that has a circular blade designed to fit into round ductwork. Round-blade dampers use small-diameter diaphragm actuators and are simple in operation. Round-blade dampers have a nonlinear flow characteristic but are widely used in small variable air volume terminal boxes.

Care should be taken with round-blade dampers to avoid strapping the duct too tightly at the terminal box. If the duct gets bent out of round, the damper may bind. This binding may prevent the damper from working properly.

Damper Construction

Dampers are normally constructed from welded steel that is treated to resist corrosion and rust. Dampers also contain blade seals made of neoprene that seal when the blades contact each other. **See Figure 40-9.** Damper blade seals can become brittle and crack when exposed to temperature variations and humidity extremes. Cracked blade seals lead to excessive air leakage and poor temperature control.

The actuator moves the damper crank arm. The drive blade transmits the force to the other blades through a set of linkages. The linkage from the actuator must be connected to the drive blade only. Blades should be locked with a bolt or key instead of a setscrew, which could vibrate loose. If the damper blades are mounted vertically instead of horizontally, thrust bearings should be used on the bottom to support the weight of the blades.

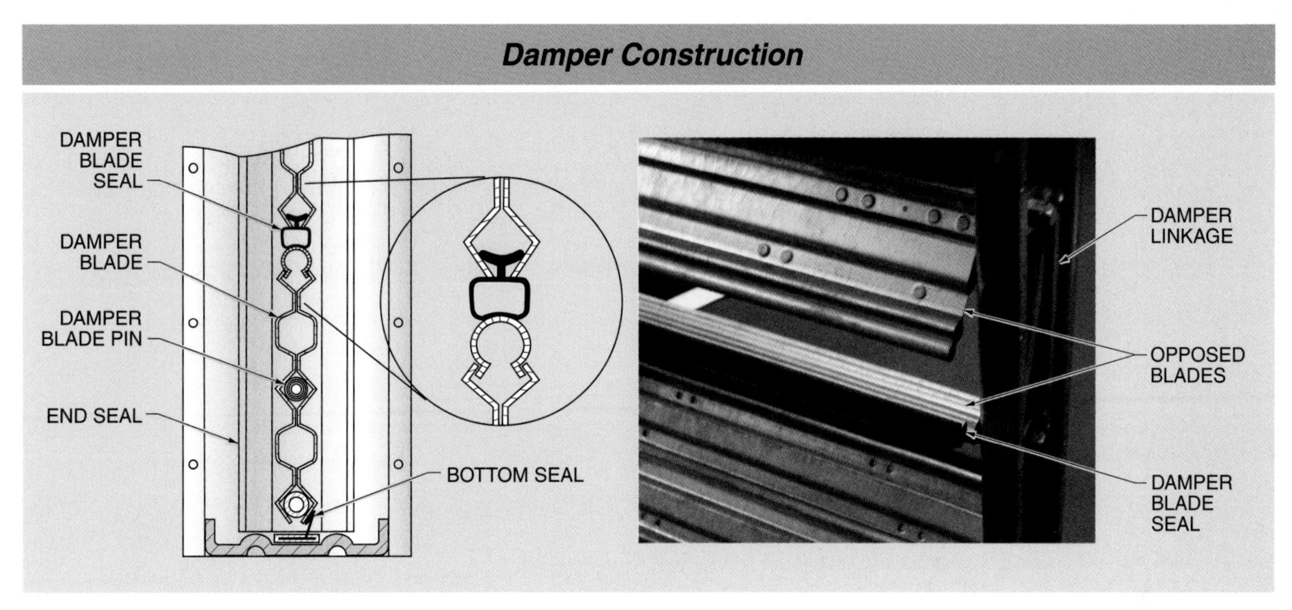

Figure 40-9. Damper seals provide a snug fit to minimize leakage.

KEY TERMS

- *regulator:* A self-operating control valve for pressure and temperature control.
- *pressure regulator:* An adjustable valve that is designed to automatically control the pressure downstream of the regulator.
- *spring-loaded pressure regulator:* A regulator consisting of a throttling element such as a valve plug connected to a pressure-sensing diaphragm that is opposed by a spring and contained in a single housing.
- *droop:* A drop in pressure below a set value when there is high flow demand.
- *lockup:* An increase in pressure above a set value when there is low flow demand.
- *air-loaded pressure regulator:* A regulator that uses air pressure instead of the force of a spring to oppose downstream pressure.
- *pilot-operated pressure regulator:* A regulator that uses upstream fluid as a pressure source to power the diaphragm of a larger valve.
- *loading pressure:* The pressure above a main diaphragm.
- *differential pressure regulator:* A regulator that controls the pressure difference between the outlet pressure of a regulator and a fluid loading pressure supplied from an external source.
- *pressure-relief regulator:* A regulator that limits the pressure upstream of the regulator.
- *backpressure regulator:* A regulator that maintains the pressure upstream of the regulator to a specified value.
- *temperature self-operating regulator:* A regulator that is used to automatically control a process to a defined temperature.
- *damper:* An adjustable blade or set of blades used to control the flow of air.

KEY TERMS *(continued)*

- *parallel-blade damper:* A damper in which adjacent blades are parallel and move in the same direction with one another.
- *opposed-blade damper:* A damper in which adjacent blades are parallel and move in opposite directions from one another.
- *round-blade damper:* A damper that has a circular blade designed to fit into round ductwork.

REVIEW QUESTIONS

1. What is a regulator, and what are the common types of regulators?
2. List and define the common types of pressure regulators.
3. What is the difference between droop and lockup, and how do they occur?
4. What is a backpressure regulator, and how may it be used?
5. List and define common types of dampers.

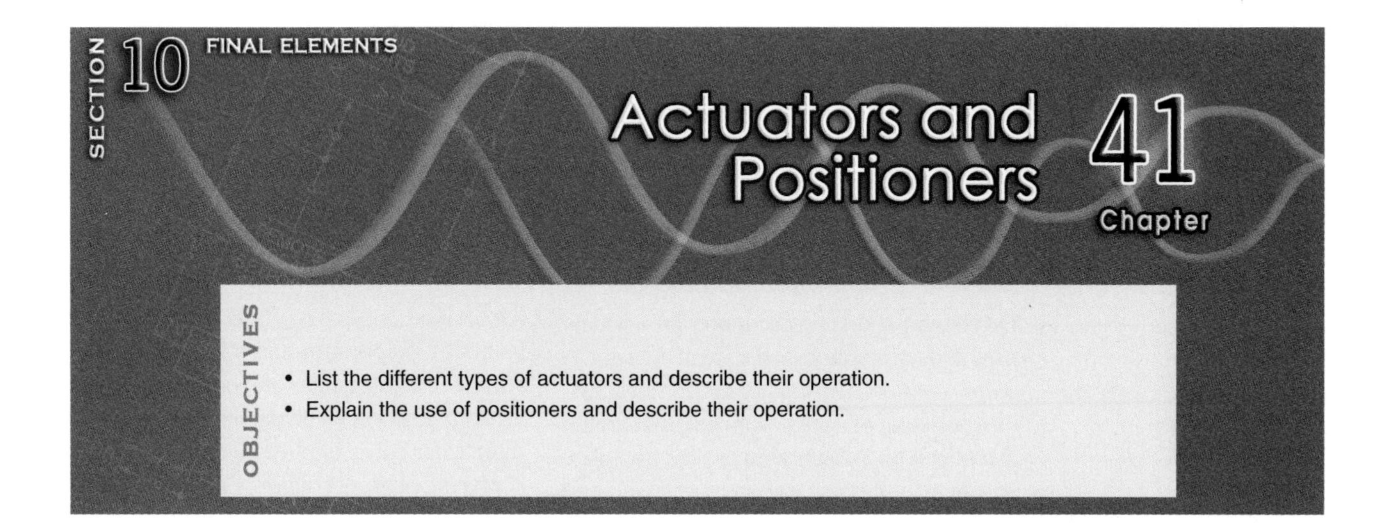

ACTUATORS

An *actuator* is a device that provides the power and motion to manipulate the moving parts of a valve or damper used to control fluid flow through a final element. The requirements of an actuator are speed, power, precision, and resolution. An actuator must respond quickly to a change in control signal and have enough power to overcome the process pressure and mechanical friction of the moving parts. Positioning needs to be repeatable for the same control signal. In addition, the actuator must be able to work with minimal maintenance for long periods of time.

Valve actuators are used for direct mounting on sliding stem valves, rotary valves, and dampers. The actuator strokes are designed to match the control valve body stem travel requirements and the actuator yokes are designed to match the valve-mounting boss. **See Figure 41-1.** For this reason, valve bodies and the associated actuator must be purchased from the same manufacturer.

Larger valves require longer actuator strokes. Actuators designed for rotary valves and for dampers have longer strokes to allow a longer crank arm to reduce the forces on the connecting linkage. Actuators for the largest dampers are floor mounted using a spring and diaphragm or a spring-return piston to provide the power. These large actuators require positioners for positive positioning.

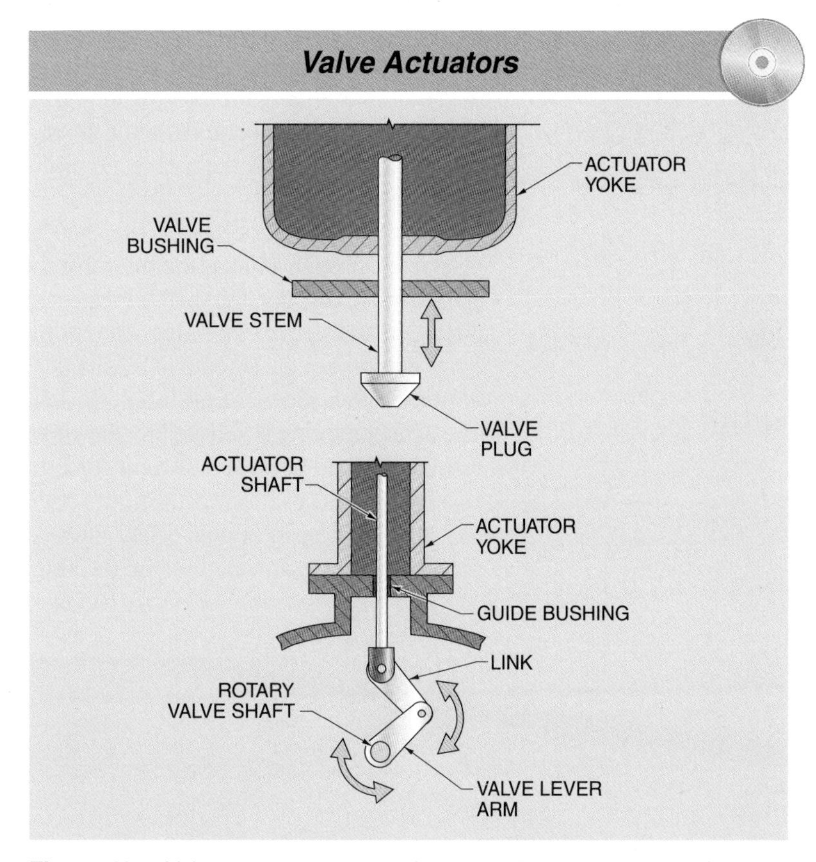

Figure 41-1. Valve actuators are used to move the internal parts of a valve to modulate flow.

Most actuators are supplied with the control valve body and have a nameplate that indicates the model number and spring range of the actuator, the valve model, body size, port size and flow characteristic, pressure, temperature, and C_v rating. If the original spring has been replaced by a spring with a different spring range, or the nameplate is obscured, the spring range should be tested. In addition, some manufacturers have color-coded springs that indicate the spring range.

Diaphragm-and-Spring Actuators

Diaphragm-and-spring actuators are a widely used and long-established style of actuator. A *diaphragm-and-spring actuator* is an actuator consisting of a large-diameter diaphragm chamber with a diaphragm backed by a plate attached to the actuator stem and opposed by the actuator spring. **See Figure 41-2.** The spring and shaft assembly converts the air pressure change at the diaphragm into mechanical movement. The majority of diaphragms are flat with a corrugation ring to allow easier movement, but some are designed to roll back on themselves to allow large travels.

In a rotating valve actuator, the shaft is connected through a linkage to the valve. In a linear valve actuator, the shaft is connected directly to the stem of the valve. The spring range determines the amount of mechanical movement for a given change in air pressure. The actuator spring is selected by the valve manufacturer to provide sufficient force to close the valve (fail closed action) or to open the valve (fail open action) when there is pressure from the process fluid on the plug.

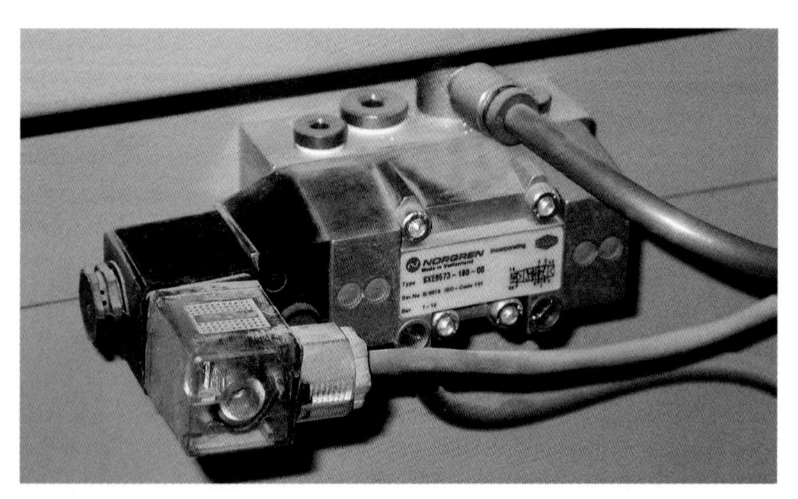

A solenoid pilot-actuated air valve is used to switch the flow of air.

A diaphragm-and-spring actuator is usually installed in a vertical position, but it can also be used upside down or in a horizontal position (with special support). A diaphragm-and-spring actuator can satisfy the speed, power, and precision requirements if a diaphragm of proper size is selected and the control valve stem is sufficiently lubricated.

Direct and Reverse Action. Two common forms of diaphragm-and-spring actuators are direct-acting and reverse-acting. A *direct-acting actuator* is an actuator that extends the shaft when air is applied to the diaphragm. A *reverse-acting actuator* is an actuator that retracts the shaft when air is applied to the diaphragm. **See Figure 41-3.** Actuator actions are chosen to match valve designs and to provide the desired fail-safe positions.

In a control system requiring a valve to control the amount of heat supplied to a process, the most common failsafe mode is to remove the heat during power failure or loss of signal. Therefore, the valve must close automatically during a failure. This means that the plug moves in the direction that causes it to come in contact with the seat. When power and a control signal are available, the plug moves away from its seat and allows flow. This can only be achieved by having an actuator that causes the stem to move away from the seat on an increasing signal. This is a reverse-acting actuator.

A cooling process requires the opposite action. The valve must open automatically during a failure. This can only be achieved by having an actuator that causes the steam to move toward the seat on an increasing signal. This is a direct-acting actuator.

Actuator Pressure and Force. Pneumatic pressure acting upon the diaphragm provides the power required to actuate a valve. The pneumatic pressure range is selected by the manufacturer to satisfy the forces needed to work with the spring to obtain a full valve stroke. This pressure range is usually 3 psig to 15 psig, but may be different, depending on the process conditions. The diaphragms used in these actuators are quite large in comparison to the valve port. Smaller actuators have an area of about 50 sq in.

Diaphragm-and-Spring Actuators

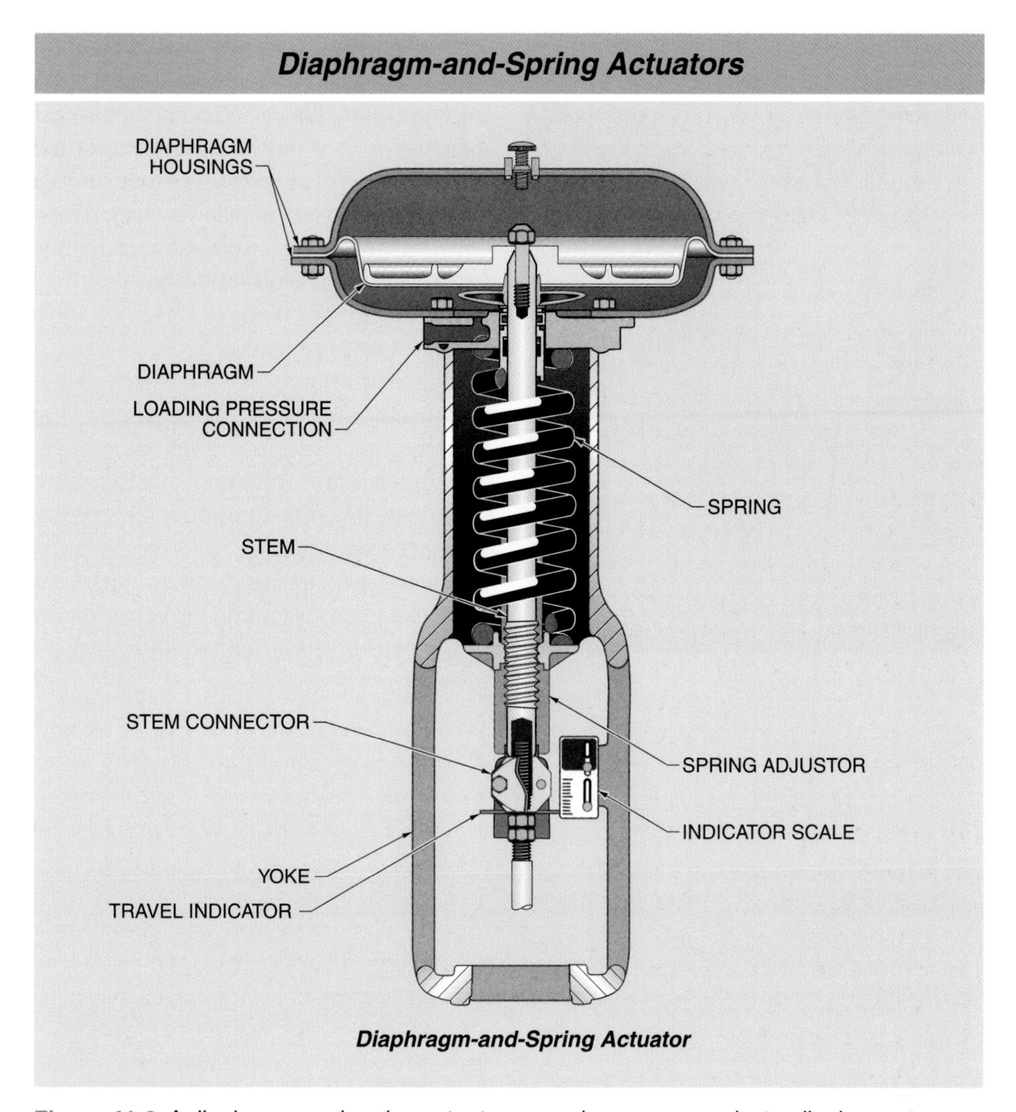

Figure 41-2. A diaphragm-and-spring actuator uses air pressure against a diaphragm to compress a spring and move a valve stem.

Direct- and Reverse-Acting Actuators

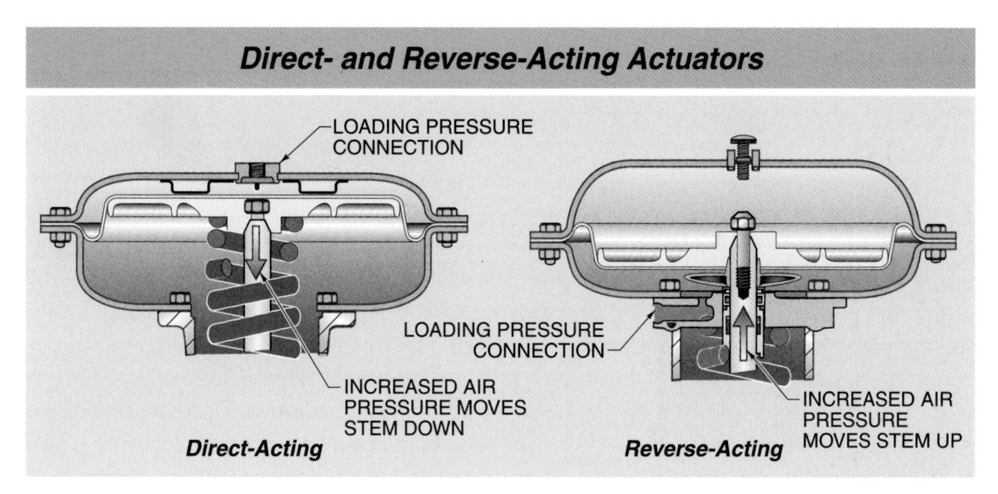

Figure 41-3. Diaphragm-and-spring actuators can be configured for direct or reverse action.

Damper actuators can be mounted on a bracket near ductwork.

Smaller actuators are used for valves with ports of up to about 1 sq in. Applying a pressure of 15 psig to the diaphragm provides a force of 750 lb. The normal valve travel for a valve this size is ¾″. Therefore, the spring must be very strong because it opposes 750 lb at the end of the ¾″ travel. Even with only 3 psig applied to the diaphragm there is 150 lb of force being generated.

The diaphragms are sandwiched between two heavy backing plates that transfer the force to the actuator stem. The diaphragm actuators usually have sufficient power to minimize the need to adjust the spring range. The combination of spring and control pressure range ensures that upon air failure, the valve returns to the safe position, either fully open or fully closed.

Calibration

Valves, the actuators that move them, and the electronic circuits that control them are all subject to the effects of aging soon after they are installed. The valve seat wears not only from the repeated seating of the valve, but also from the fluid flow through it. Depending on the application, a valve can be stroked tens of thousands of times per year. This can cause screws to reposition, springs to weaken, and mechanical linkages to loosen. This results in valves that do not fully open or close, close prematurely, or operate erratically and cause improper regulation of the fluid under their control. This is known as calibration drift. To keep a system operating properly, a good preventive maintenance program that requires periodic checks of valve positioners is required.

A current source, such as a loop calibrator, is required to calibrate an electric positioner. First, disconnect the current signal wires from the positioner and connect the calibrator source current. Adjust the source current to 4 mA and allow the valve to stabilize. Watch the valve stem while adjusting the source current to a value less than 4 mA, such as 3.9 mA. There should be no movement of the valve stem. Next, adjust the current to a value just above 4 mA. The valve stem should start to move. If necessary, adjust the zero on the positioner to ensure that the valve is fully open or fully closed at 4 mA and only begins to move when the current exceeds that minimum value.

Next, adjust the source current to 20 mA. The valve should move to its other extreme. This is the span position check. Allow the valve to stabilize. Adjust the current to a value just above 20 mA, such as 20.1 mA. There should be no movement of the valve stem. Next, adjust the current to a value just below 20 mA, such as 19.9 mA. The valve should begin to move. If necessary, adjust the span on the positioner to ensure that the valve is fully open or fully closed at 20 mA and only begins to move when the current is less than that maximum value.

If the valve stem has a position marker or reference, linearity can be checked at this time. Set the source current to 12 mA. The valve should move to the 50% lift position. Many valve positioners have an interaction between the zero and span adjustments. Therefore, it is best to ensure proper valve positioner adjustment by repeating the test of the fully closed and fully open positions until no further adjustment is necessary.

Pneumatic Sliding Stem Piston Actuators

A *pneumatic sliding stem piston actuator* is an actuator where the movement of a piston replaces the action of the diaphragm and a balancing air pressure replaces the function of the spring. A pressure regulator may be used to provide the balancing air pressure. **See Figure 41-4.** There must be individually controlled pressures on each side of the piston for the piston to move to any position. A piston actuator requires a positioner for throttling service. The biggest advantage of a piston actuator is that higher air pressures, up to 100 psig, can be used. This provides greater power to the valve.

A sliding stem piston actuator can be used with nearly any valve style. However, as with the diaphragm-and-spring actuator, the actuator should be supplied by the manufacturer of the valve body for the two to match requirements. When a sliding stem piston actuator is used with rotary valves, the linear motion is converted to a rotary action by a crank arm or a rack and pinion. Because a piston actuator usually has no spring, it has no failsafe position. There are solutions to this problem that require auxiliary air tanks and switching valves to take over if the normal air supply fails.

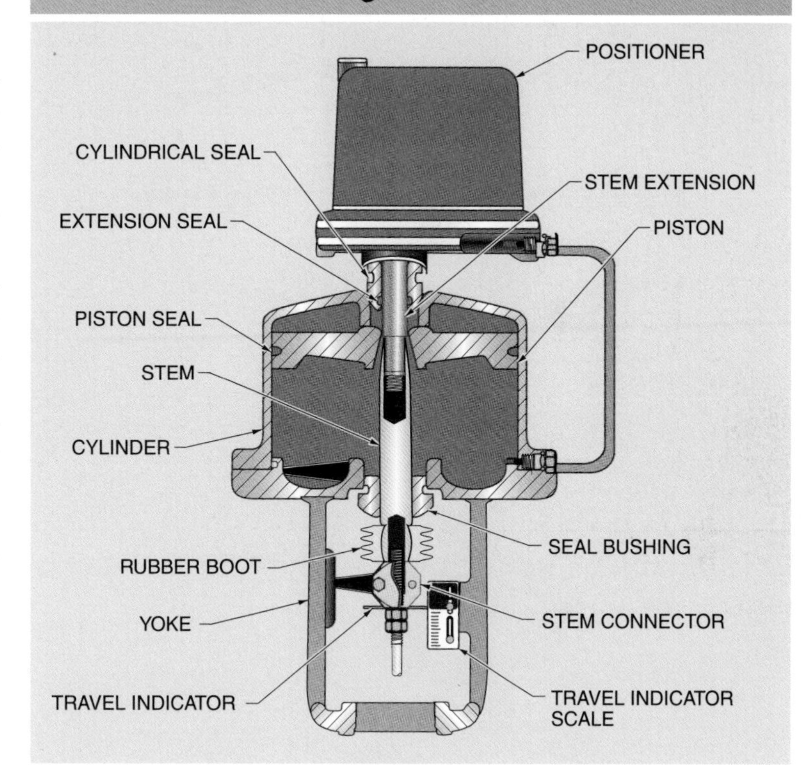

Figure 41-4. A pneumatic sliding stem piston actuator uses a piston with balancing air pressures to move a valve stem. High air pressures can be used to overcome large forces on the valve plug.

Pneumatic Rotary Piston Actuators

A *pneumatic rotary piston actuator* is an actuator that can be attached directly to most rotary valves and is primarily designed for ON/OFF action. The actuator can be a spring-return actuator where the spring action provides a valve failure position. Air is applied to the piston to oppose the spring and move the valve. **See Figure 41-5.** A rotary piston actuator can also be a double-acting type where the valve remains in its last position on air failure. In this type, air is usually applied to one side of the piston or the other to move the valve.

Unlike many other types of actuators, rotary actuators are quite versatile in having kits that allow the actuator to be mounted on almost any plug, ball, or butterfly valve. Each valve manufacturer has at least one design of spring-return or double-acting rotary piston actuator that they support and sell. While it is usually advantageous to purchase the actuator from the manufacturer of the valve body, this is not necessary.

The addition of a rotary positioner allows a valve to be set to any degree of opening. Positioners are usually used with the spring return of a piston actuator so that there is a defined failure position.

A variation of the double-acting piston actuator is the vane actuator where the moving piston is attached to one side of the actuator shaft. The vane is designed to rotate 90° when air is applied to either side of the actuator. Vane actuators can be obtained with spring-return accessories and with rotary positioners.

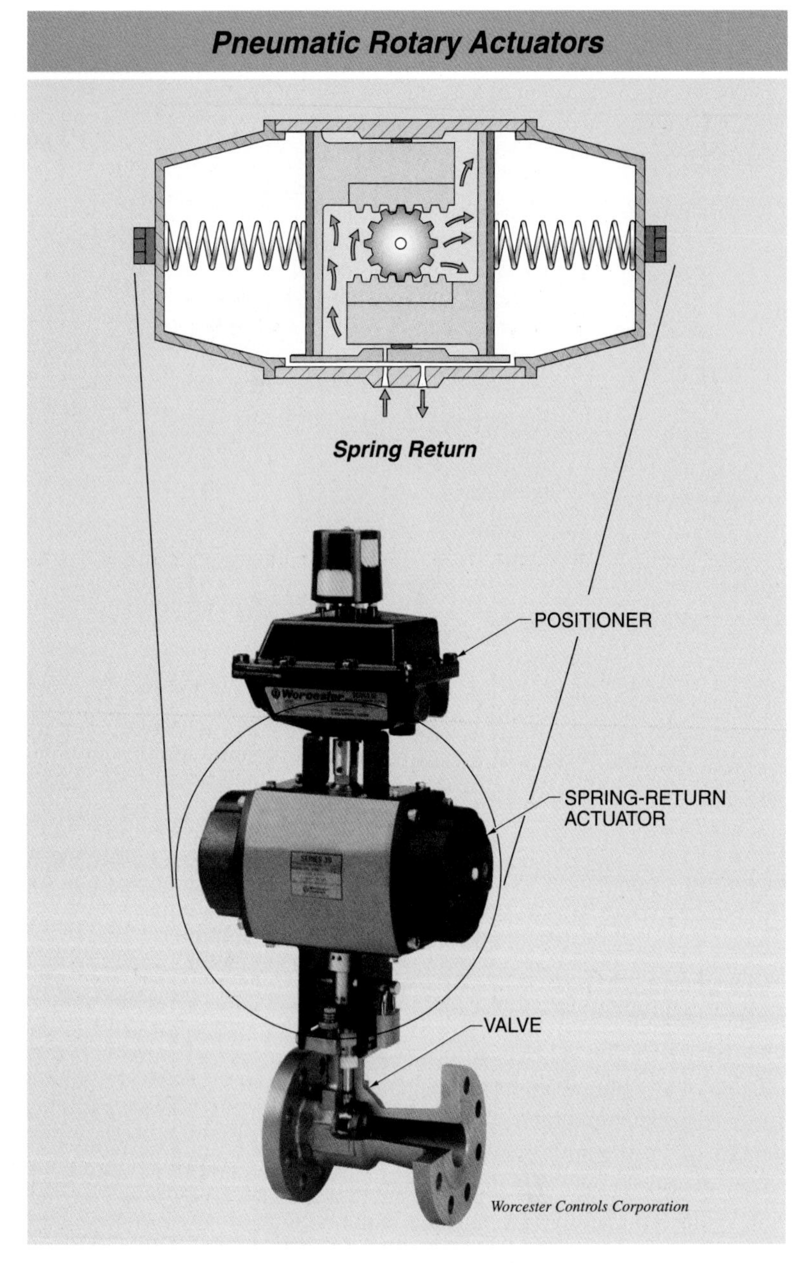

Figure 41-5. Pneumatic rotary piston actuators are primarily designed for ON/OFF action.

It is important for rotary piston actuators to receive a small amount of oil lubrication with the actuating air. Normal plant compressed air may contain sufficient carryover oil from the air compressors that nothing special needs to be done for the piston actuators. Newer plants are increasingly using oil-less air compressors and dryers for the instrument air supplies. In modern plants, it is likely that these piston actuators are installed with instrument air. In these cases it is necessary to provide a self-contained lubricator in the air supply to each actuator. Failure to do this results in an extremely short life for the actuator.

Electric Actuators

An *electric actuator* is an actuator consisting of an electric motor that is connected through gearing to the valve stem or shaft. Electric actuators are most commonly used for ON/OFF services. **See Figure 41-6.** Throttling service requires a positioner to compare the location of the valve stem or shaft against the control signal. If a difference is detected, the motor is turned ON to move the valve in the correct direction to eliminate the difference. All electric actuators, whether ON/OFF or throttling, have built-in travel limit switches that prevent the electric motor from trying to power the valve past the fully open or fully closed position. Except in small sizes, electric actuators are more expensive than an equivalent pneumatic actuator.

TECH FACT

The key to troubleshooting a valve is knowing the fail-safe mode. When a valve is not in its fail-safe position, a control signal is positioning the valve in that position.

Damper Actuators

Damper actuators come in a wide variety of sizes and forms. The sizes range from small, simple actuators used in HVAC systems up to large floor-mounted units that can provide 4700 lb-ft of torque. Actuators can be powered by rolling diaphragms, vane-type actuators, double-acting piston actuators, and electric drives. **See Figure 41-7.** The one common feature of all damper actuators is that each installation has to be individually engineered and designed. The reason for this is that different manufacturers make dampers and damper actuators, and each mounting is different from the others.

The actuator must be selected to provide the appropriate power and stroke to match the damper.

In addition, all the linkages need to be designed. The linkages and lever arms have to be selected so that the actuation of the damper is linear over its stroke. Except for the smallest sizes, all damper actuators use positioners to ensure accurate positioning of the damper. The most complex applications use specially cut positioner feedback cams to change the damper flow characteristic.

Spring-Return Solenoid Actuators

A *spring-return actuator* is a piston or diaphragm actuator with an internal spring to force the actuator shaft to one end of its travel. Air is connected to the actuator through a three-way solenoid valve so that when pressure is applied, the air compresses the spring and moves the shaft to its opposite position. **See Figure 41-8.** A *three-way solenoid valve* is a solenoid that shuts off the air supply and vents air from the actuator or cylinder when the solenoid is de-energized. When the solenoid is energized, the supply air is routed to the actuator or diaphragm.

Throttling valves with diaphragm-and-spring actuators can also be used to perform ON/OFF functions with the addition of a three-way solenoid valve in the air line to the actuator or cylinder. In this way, a throttling control valve can be equipped to provide an overriding emergency shutdown action. A three-way solenoid valve can be used with direct pneumatically controlled valves, a valve with an I/P converter, or a valve with a positioner. The only consideration is that the solenoid valve must be able to directly vent the diaphragm. This same solenoid valve arrangement can be used with spring-return rotary piston actuators.

TECH FACT

Butterfly valves may be the only economical choice for modulating flow in very large pipes.

Rotary Electric Actuators

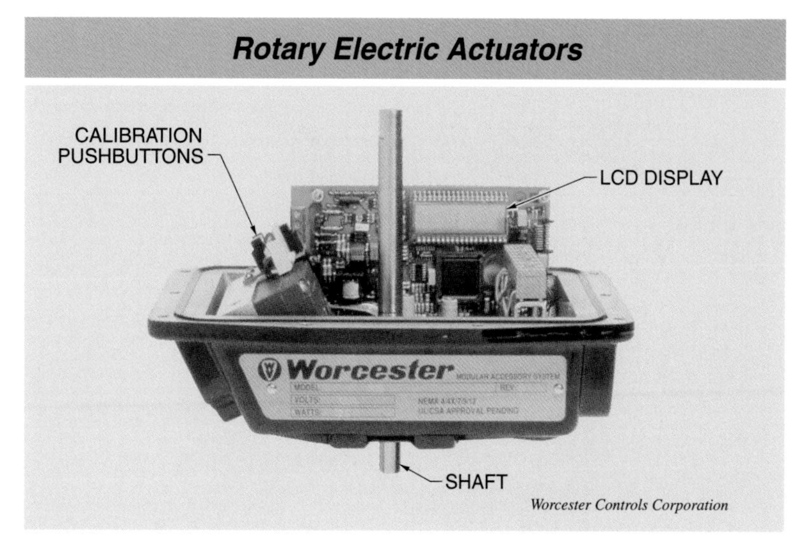

Figure 41-6. An electric actuator uses a motor and gears to rotate a valve shaft.

Damper Actuators

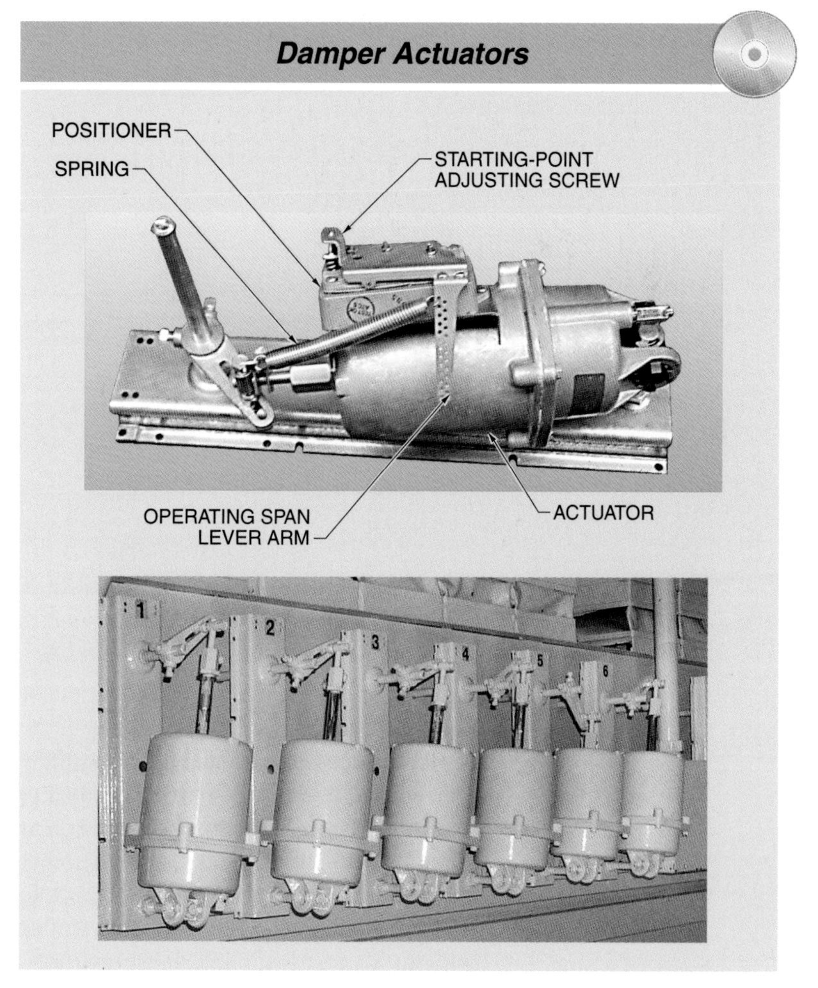

Figure 41-7. Damper actuators provide greater movement than valve positioners in order to modulate the position of dampers.

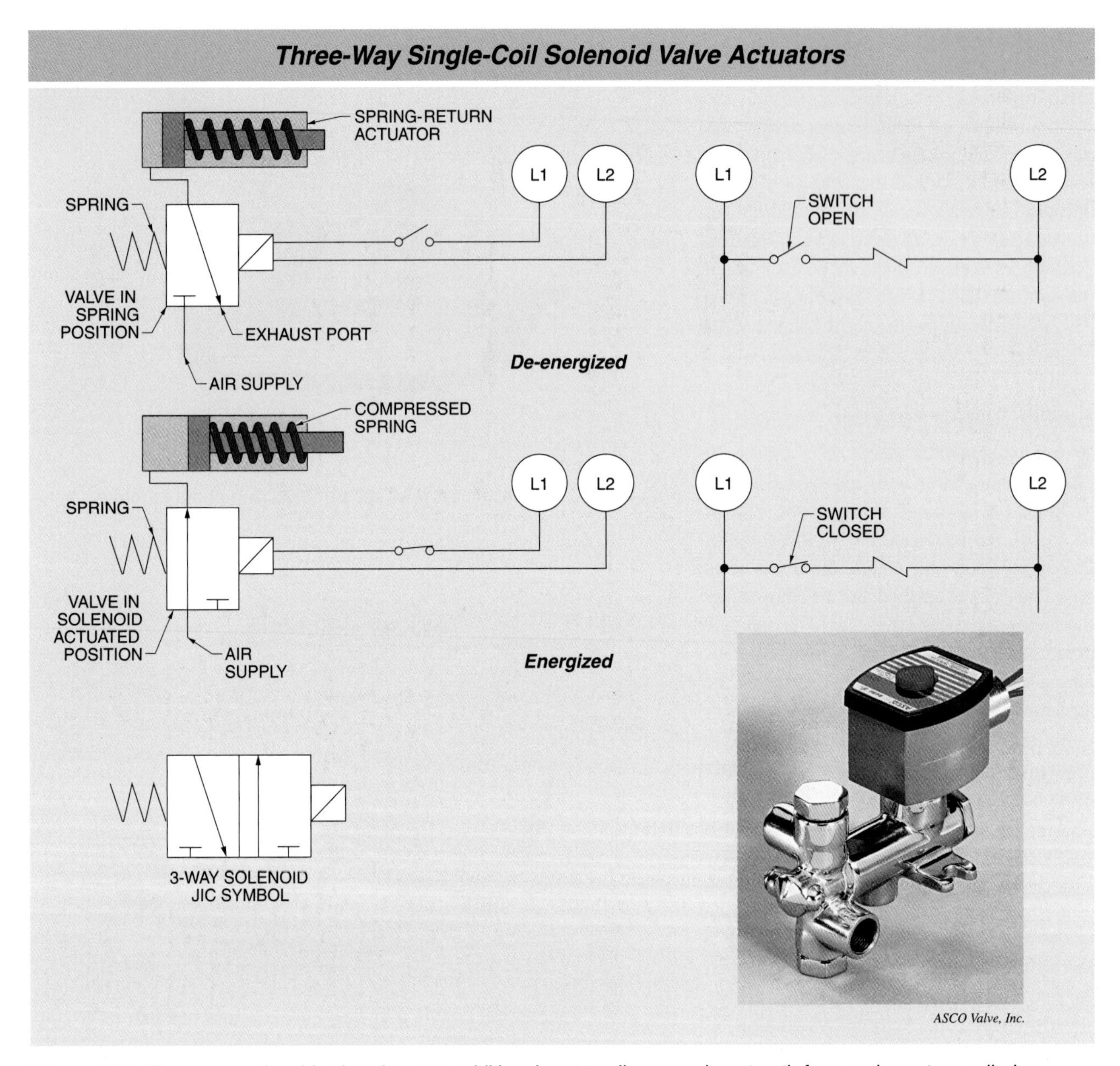

Figure 41-8. Three-way solenoid valves have an additional port to allow an exhaust path for a spring-return cylinder.

Double-Acting Solenoid Actuators

A *double-acting actuator* is a piston actuator that has an air connection at each end of the piston and does not contain a spring to allow for a known failure position. Air applied to one end of the cylinder moves the piston to the other end. Switching the air to the other end of the cylinder moves the piston back to its original position. The piston cannot move without air being applied to the cylinder. Double-acting rotary piston actuators require the use of four-way solenoid valves.

A *four-way solenoid valve* is a single- or dual-coil solenoid that has an air supply, a vent, and two cylinder ports. A single-coil solenoid has an internal spring to return the internal parts to one position. The solenoid valve connects the air supply to one end of the cylinder and vents the other end of the cylinder when the solenoid coil is de-energized. When the coil is energized, the air supply and vent are switched to the opposite ends of the cylinder. **See Figure 41-9.**

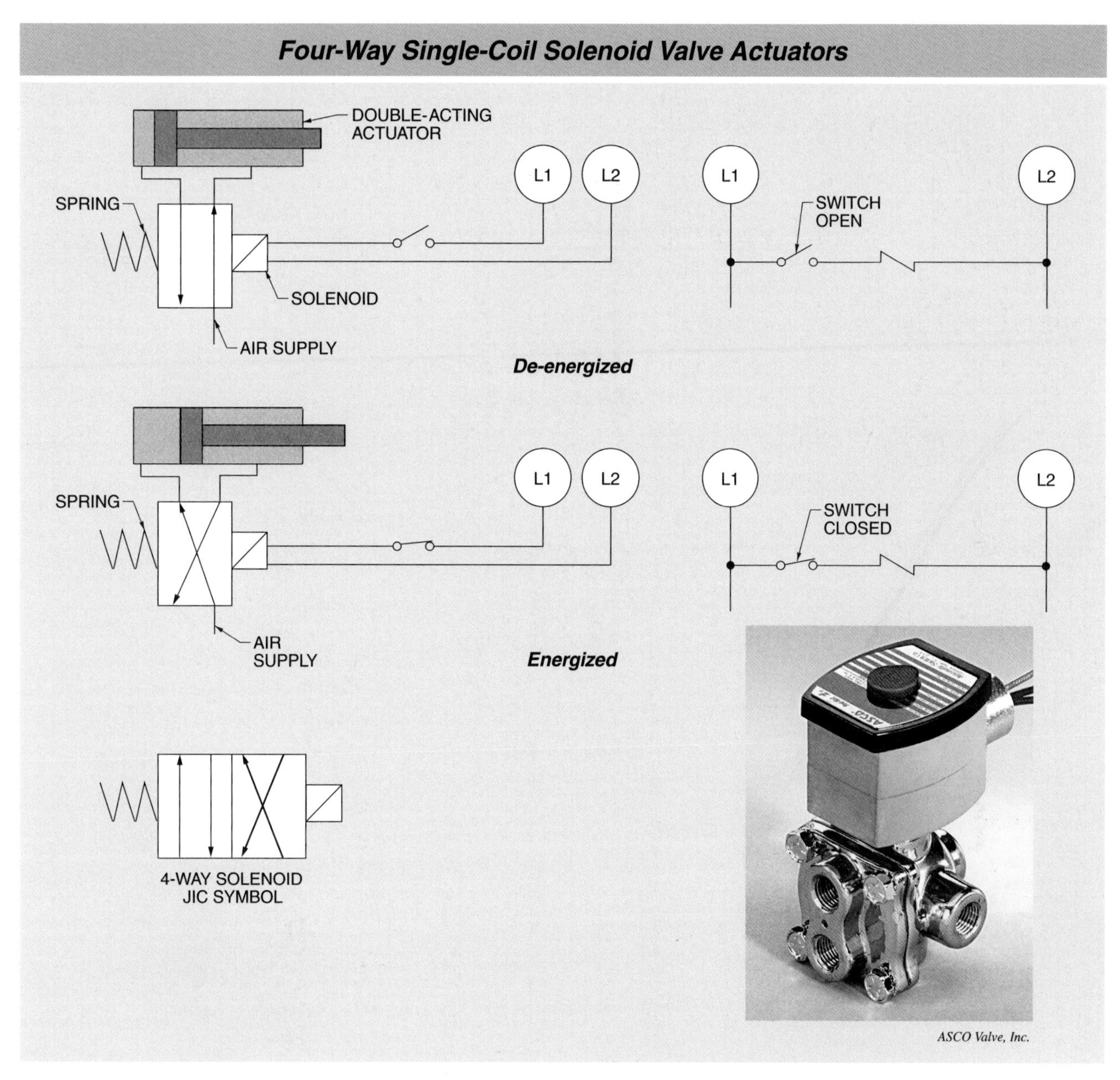

Figure 41-9. Four-way single-coil solenoid valves allow a fail-safe position for a double-acting actuator.

This assembly provides some safety action since the spring places the valve in its normal position and moves the valve to a predetermined position with a loss of electrical power to the coil, as long as air pressure is available. There is no safety action with the loss of air supply since there is no air pressure under these conditions to change the piston position.

A *four-way dual-coil solenoid valve* is a valve connected to a solenoid actuator in the same way as a single-coil solenoid valve except that there is no spring in the solenoid. **See Figure 41-10.** Dual-coil solenoids require one coil to be momentarily energized to move the valve in one direction. Momentarily powering the other solenoid coil moves the actuator in the opposite direction. This arrangement provides a fail-last position action. This is because the valve remains in its last position on either electrical or air failure. In actual practice the coils can be either momentarily or continuously energized. The use of momentary actuation places less stress on the coils.

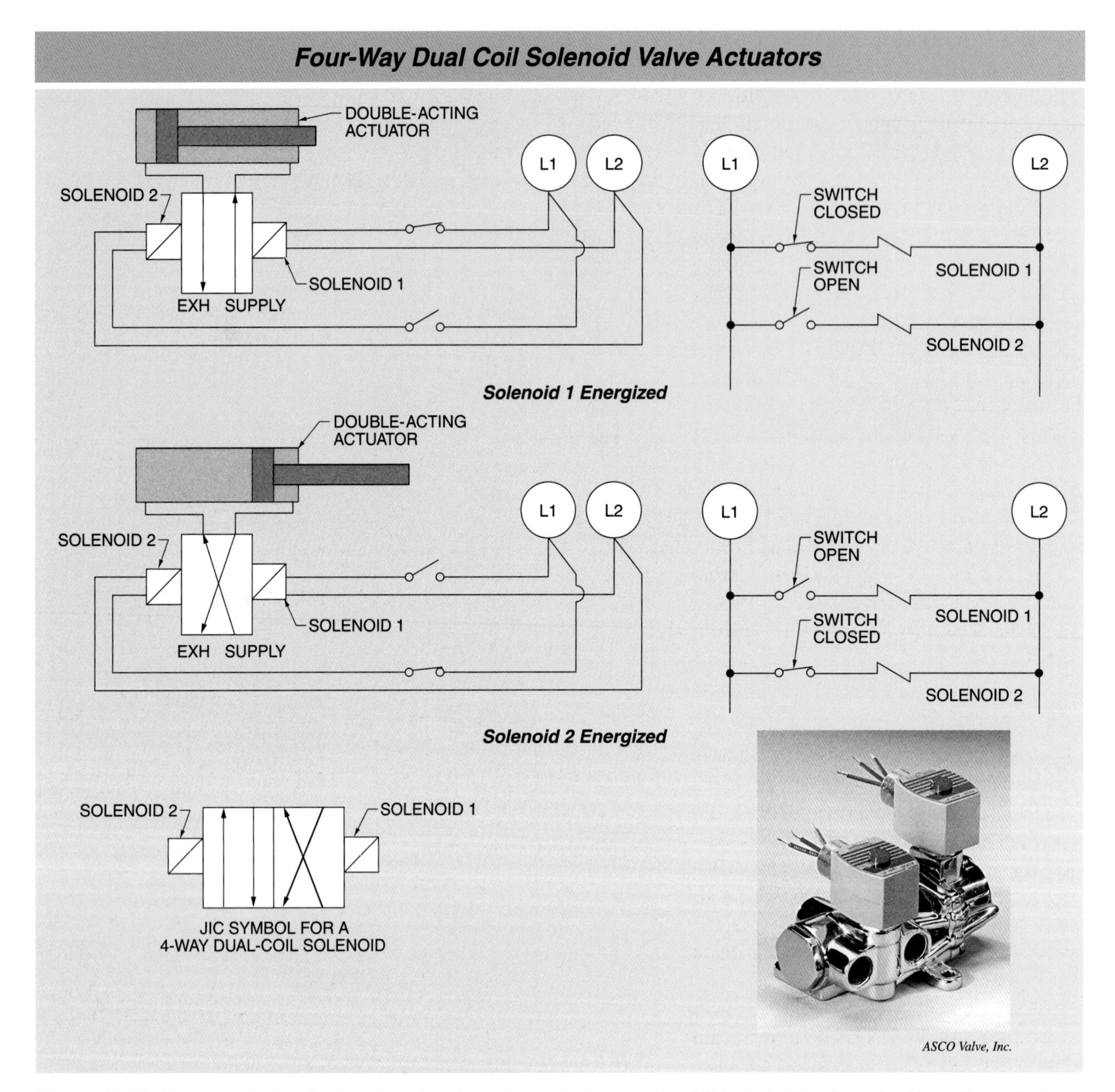

Figure 41-10. Four-way dual-coil solenoid valves have two coils to provide a fail-last position for a double-acting actuator.

Current-to-Pneumatic (I/P) Transducers

A *current-to-pneumatic (I/P) transducer* is a device that can convert the electronic controller output signal, usually 4 mA to 20 mA, into the standard pneumatic output, 3 psig to 15 psig (or similar metric range). **See Figure 41-11.** As the current changes, the magnetic field around the coils changes proportionally. This causes the balance beam to rotate slightly, covering or uncovering the nozzle. The balance beam acts as the flapper in a typical pneumatic system. This rotation of the balance beam changes the pressure in the upper section of the pneumatic relay and changes the output signal. The output signal is fed back through a bellows or zero spring to the balance beam to counterbalance the forces on the balance beam due to the magnetic forces.

An I/P transducer can be used to provide pneumatic power to diaphragm-and-spring actuators or to provide an input signal to a pneumatic positioner. Most transducers are connected directly to controlled devices such as valve actuators, damper actuators, and valves. The transducer output can often be adjusted by a controller to match the actual spring range of the actuator.

For special situations, an I/P transducer can be calibrated to reverse the output signal so that a 4 mA to 20 mA input produces a 15 psig to 3 psig output. Some transducers can also be configured for split range operation for full outputs with either 4 mA to 12 mA or 12 mA to 20 mA. I/P transducers can be purchased already mounted on a valve, or purchased separately and mounted independently from the valve. Cases are available in general-purpose, weatherproof, and explosion-proof construction.

Air Supply

The key to reliable pneumatic operation is the use of clean, dry air. Desiccant dryers are the most common method used to dry instrument air. **See Figure 41-12.** A desiccant dryer uses a material such as silica gel to adsorb water from compressed air during an adsorption cycle. The collected water is then purged out of the desiccant during a regeneration cycle.

Dew points of –40°F are common for instrument air as it leaves a desiccant dryer. Oil-less air compressors are used for instrument air systems because the oil contained in the air from a standard air compressor contaminates the air dryer desiccant. Air dryer desiccants have a greater affinity for oil than water, and once they have adsorbed the oil they cannot adsorb water.

It is best to maintain a separate air compressor and distribution system for the instrument air supply system to prevent possible contamination by process fluids. Plant compressed air is sometimes used to purge process vessels or piping. With systems that combine plant air and instrument air, there have been many occasions where process fluids have been allowed to flow back into the plant air system. When process fluids get into the instrument air system, instrument damage may follow.

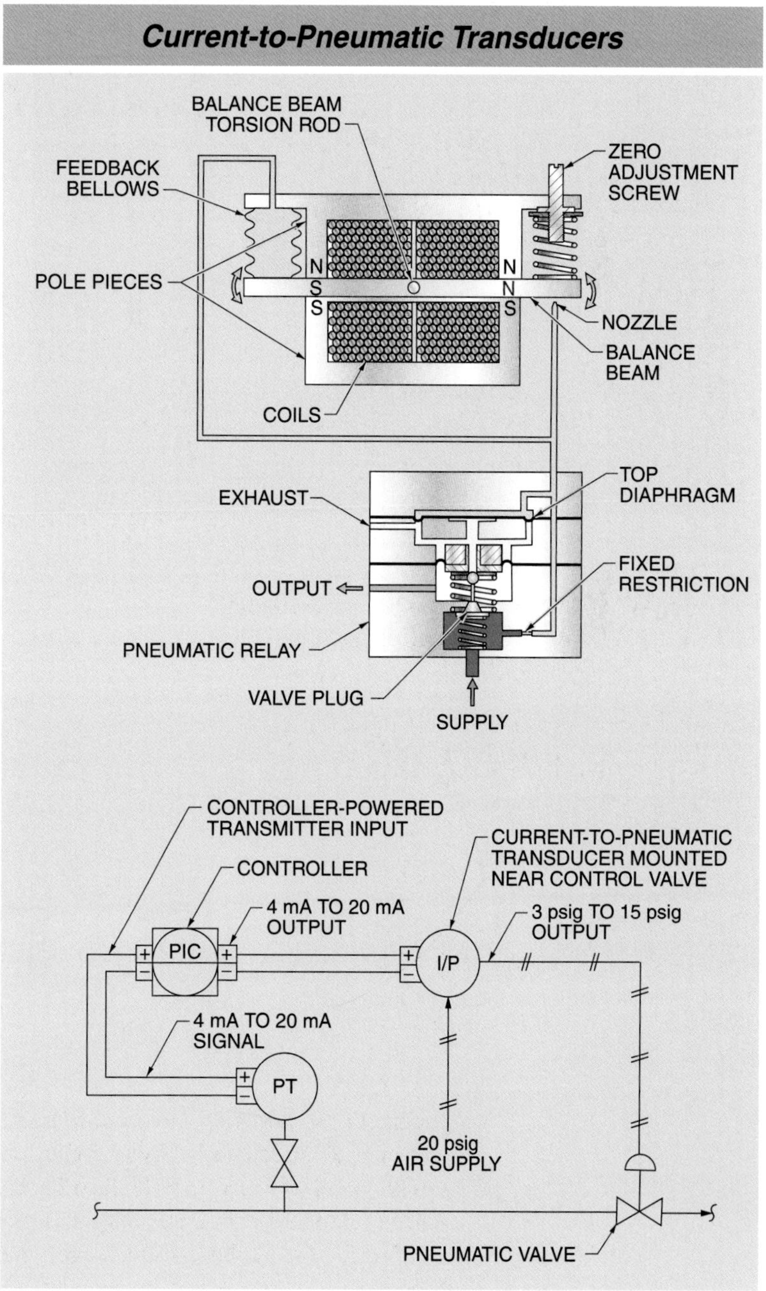

Current-to-Pneumatic Transducers

Figure 41-11. A current-to-pneumatic transducer converts a 4 mA to 20 mA control signal into an air pressure signal used to actuate a control valve.

Figure 41-12. Desiccant dryers are the most common method used to dry instrument air.

Safety valves do not use actuators. Safety valves open when the pressure against the disc exceeds the spring pressure.

TECH FACT

While normally among the easiest devices to set up, positioners are often the most confusing of all auxiliary devices. Positioners have a start point and span adjustment, both of which must be calibrated properly. If a control valve is actuated and fails to respond, or goes to full open or full closed and stays there, try reversing the controller or the positioner action to get it to operate properly.

Flappers and Nozzles. A flapper and nozzle arrangement is the basic element in all pneumatic devices and is the element that initiates the pneumatic signals. **See Figure 41-13.** Clean air regulated to 20 psig is fed through a restriction to a nozzle. A *flapper* is a flat piece of metal installed at right angles to a nozzle tip that is held in position a few thousandths of an inch away from the nozzle tip by a linkage to a measurement device such as the diaphragm of a pressure gauge.

A tubing branch between the restriction and the nozzle carries a pressure signal to a pneumatic relay. The nozzle diameter is larger than the restriction diameter so when the flapper is moved away from the nozzle the pressure drops to zero. The flapper and nozzle combination is used to generate a backpressure signal of 0.75 psig to 1.5 psig to the pneumatic relay.

Relays. A *pneumatic relay* is a pneumatic amplifier used to take a signal from a flapper and nozzle and boost the signal to a standard range required for a process. The relay has an air supply, typically 20 psig, that enters the supply chamber. The opening between the supply chamber and the output signal chamber has an air supply valve plug that limits the airflow between chambers. The size of the opening is controlled by a relay vent plug that is connected to a diaphragm that moves in response to the air from the flapper and nozzle. A feedback bellows using the pneumatic transmission signal is connected to the measurement device linkages to balance the forces and stabilize the pneumatic output. **See Figure 41-14.**

POSITIONERS

A *positioner* is a device used to ensure positive position of a valve or damper actuator. All positioners have a feedback link from the valve shaft, providing accurate valve position information. Positioners can work with linear or rotary actuation and can be used to increase the actuation power.

Linear Actuator Positioners

A positioner compares the input signal from the controller to the actual valve position and applies a pneumatic output to the actuator to ensure that the valve moves to the proper position. If the valve experiences resistance to movement, the positioner continues to apply more pressure until the valve moves to the desired position. For example, a controller sends a signal to a linear pneumatic positioner. **See Figure 41-15.** If the signal is pneumatic, it goes directly to the bellows. If the signal is 4 mA to 20 mA, it goes through an I/P transducer and then to the bellows.

The change in air pressure in the bellows causes the end of the bellows to move. The movement of the bellows causes the beam to pivot on the input axis. As the beam pivots, the spacing between the flapper and nozzle changes. This changes the nozzle back pressure to the relay, which in turn changes the pressure being sent to the actuator diaphragm.

As the pressure at the actuator diaphragm changes, the valve stem moves, either up or down. There is a linkage that transfers the valve stem movement to the cam. The movement of the cam against a roller on the back of the beam causes the beam to pivot on the feedback axis. The bellows is mounted on the feedback axis, so movement of the cam has no effect on the bellows. The movement of the beam changes the spacing between the flapper and nozzle. Eventually, the forces balance and the valve stem is at the desired position.

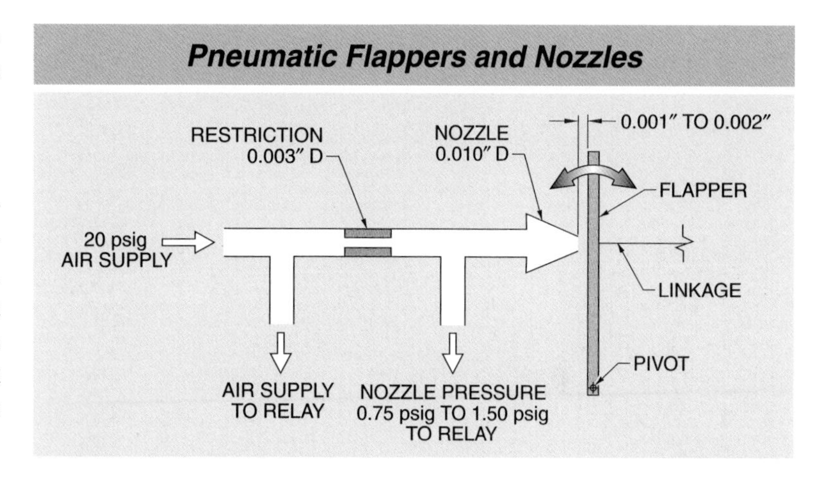

Figure 41-13. A flapper and nozzle arrangement is the basic element in all pneumatic devices and is the element that initiates the pneumatic signals.

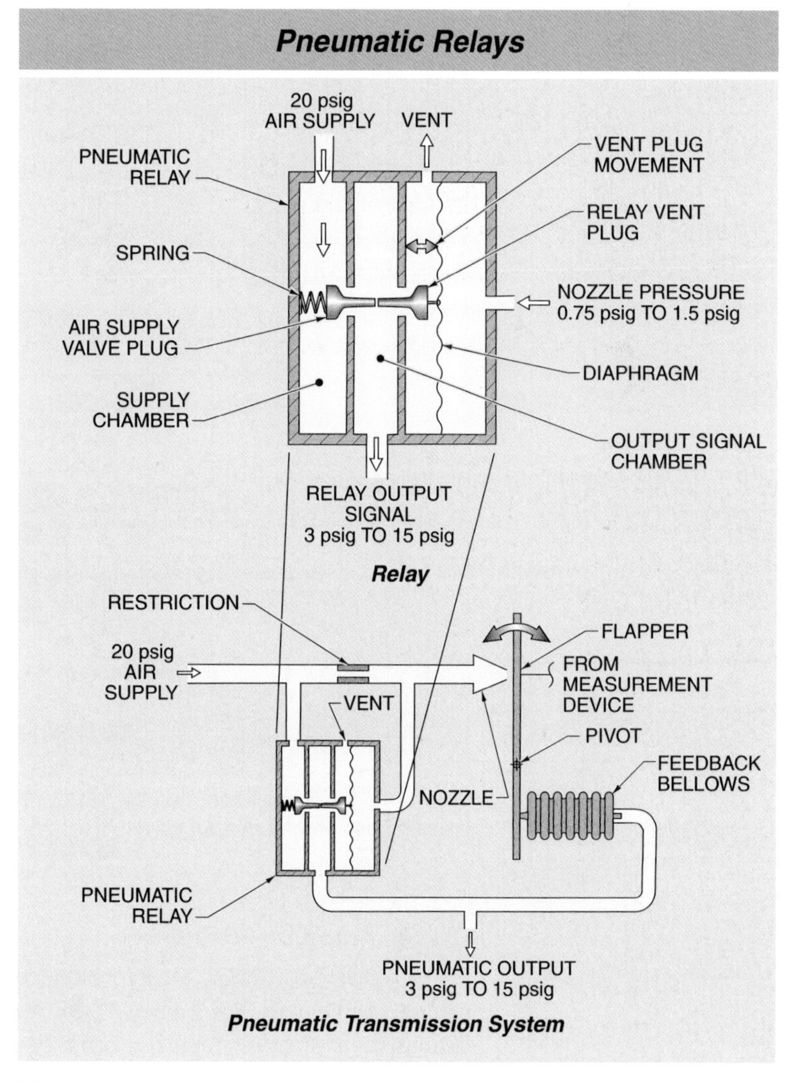

Figure 41-14. A pneumatic relay takes a signal from the flapper and nozzle and boosts it to a standard range required for the process.

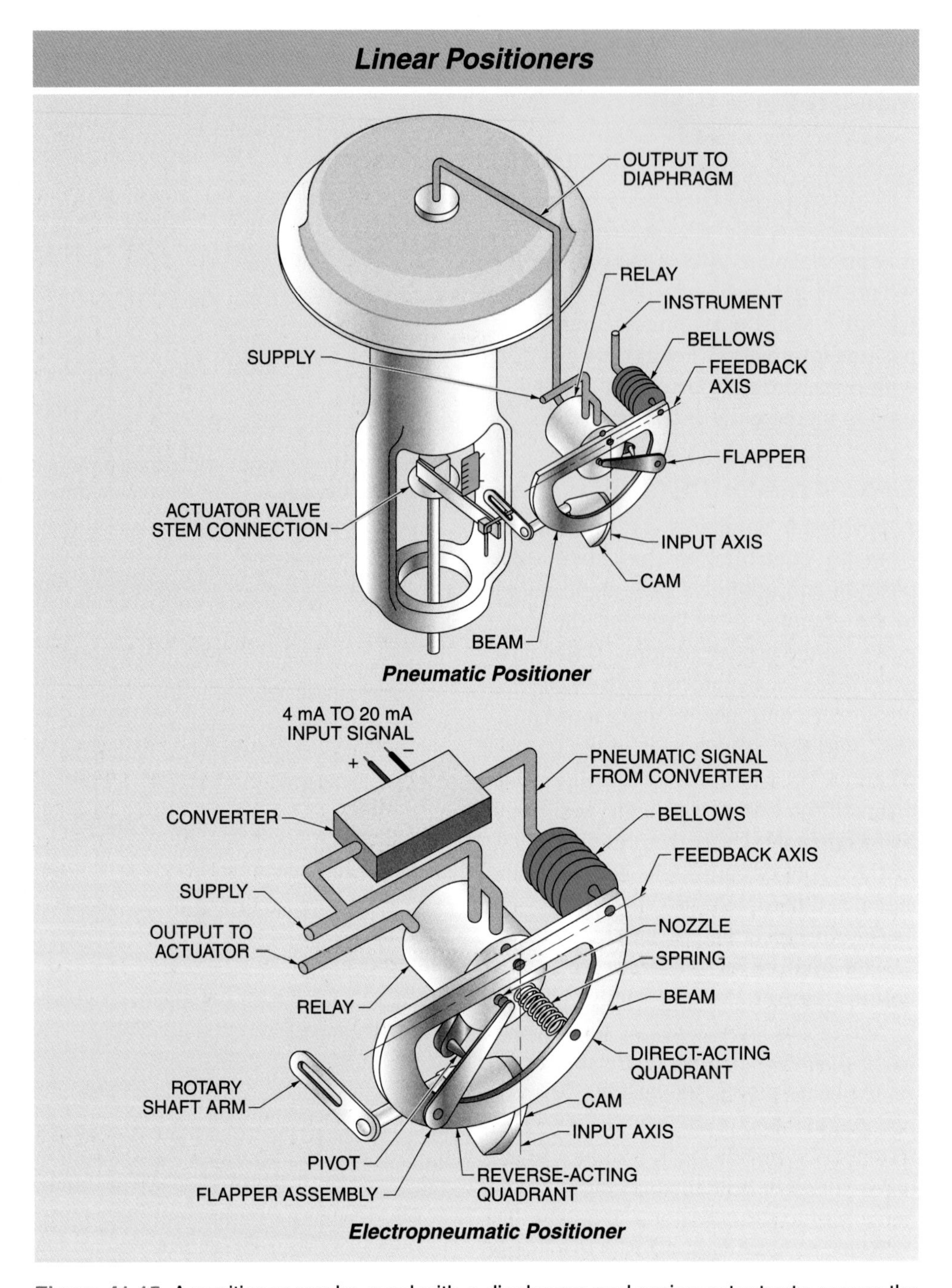

Figure 41-15. A positioner can be used with a diaphragm-and-spring actuator to ensure the precise position of a control valve.

Rotary Actuator Positioners

For a rotary positioner, a force balance method can be used. **See Figure 41-16.** An increased air pressure from the controller moves the diaphragm downward and compresses the feedback spring. The balance arm moves the cylinder in the pilot valve and furnishes supply air to the actuator. The air to the actuator rotates the valve stem, moving the positioner spindle. The spindle and cam rotate, causing the lower arm to pivot upward and compressing the feedback spring. This motion continues until the two forces are equal and the positioner is at equilibrium.

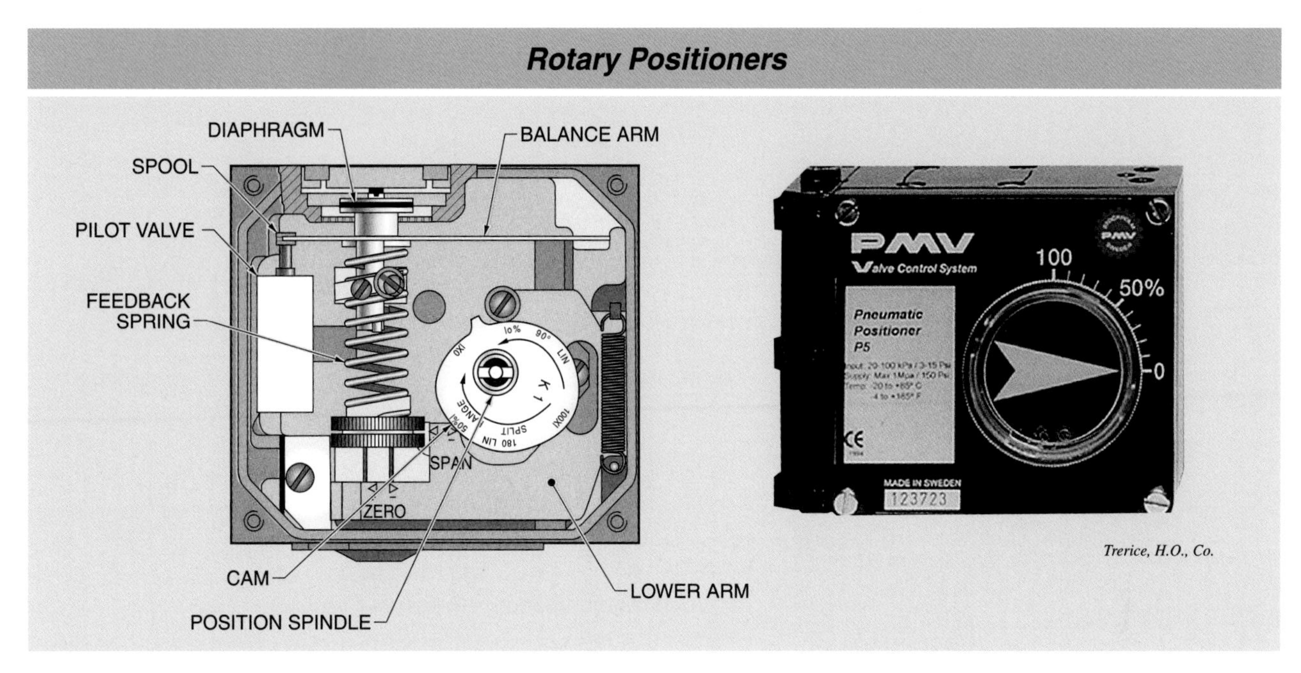

Figure 41-16. A pneumatic rotary positioner can be used to ensure the accurate position of a rotary valve by using the location of the positioner spindle to compress a spring.

Actuation Power

Positioners are used to provide increased actuation power. Most positioners used with diaphragm-and-spring actuators use a 3 psig to 15 psig pneumatic output, but if needed, a 6 psig to 30 psig output is available as long as the actuator is designed for that pressure. The higher pressures allow the use of a smaller actuator.

Positioners can also be used to increase the control-air flow capacity to an actuator. For example, the output pressure of a controller is piped to the inlet port of a positioner. The positioner is attached to a separate regulated air supply and can deliver a greater flow of air to the actuator than the controller. The positioner ensures that the actuators extend smoothly and that any forces opposing their opening are overcome. **See Figure 41-17.**

Split Range Operation

Valve positioners can also be used for split range operation. *Split range operation* is a control configuration where a single control signal, 4 mA to 20 mA or 3 psig to 15 psig, can be directed to two throttling control valves or dampers equipped with electropneumatic positioners. One of these valves uses the 4 mA to 12 mA portion of the control signal and expands it to a 3 psig to 15 psig signal to the valve diaphragm. The other valve positioner does the same for the 12 mA to 20 mA portion of the control signal. This same method is used for 3 psig to 15 psig pneumatic control systems and pneumatic positioners. This arrangement can be used for a heating and cooling process or a sequential actuator operation. **See Figure 41-18.**

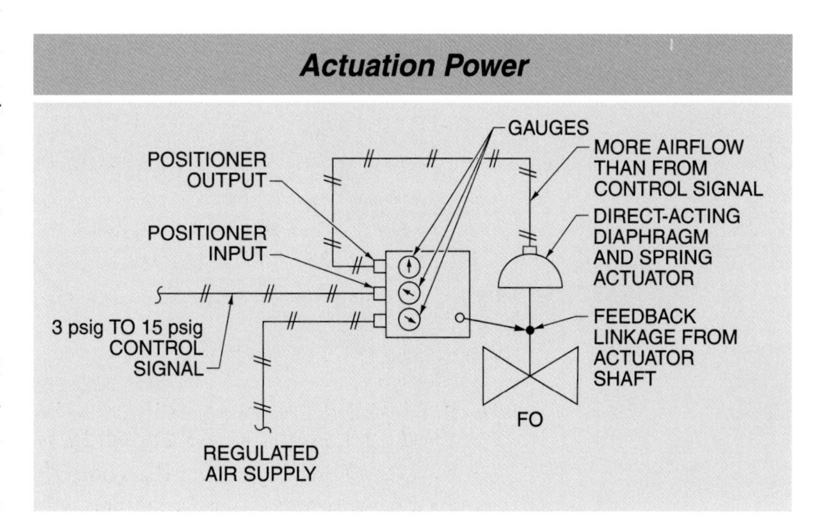

Figure 41-17. Positioners can be used to increase the pressure of the air sent to an actuator.

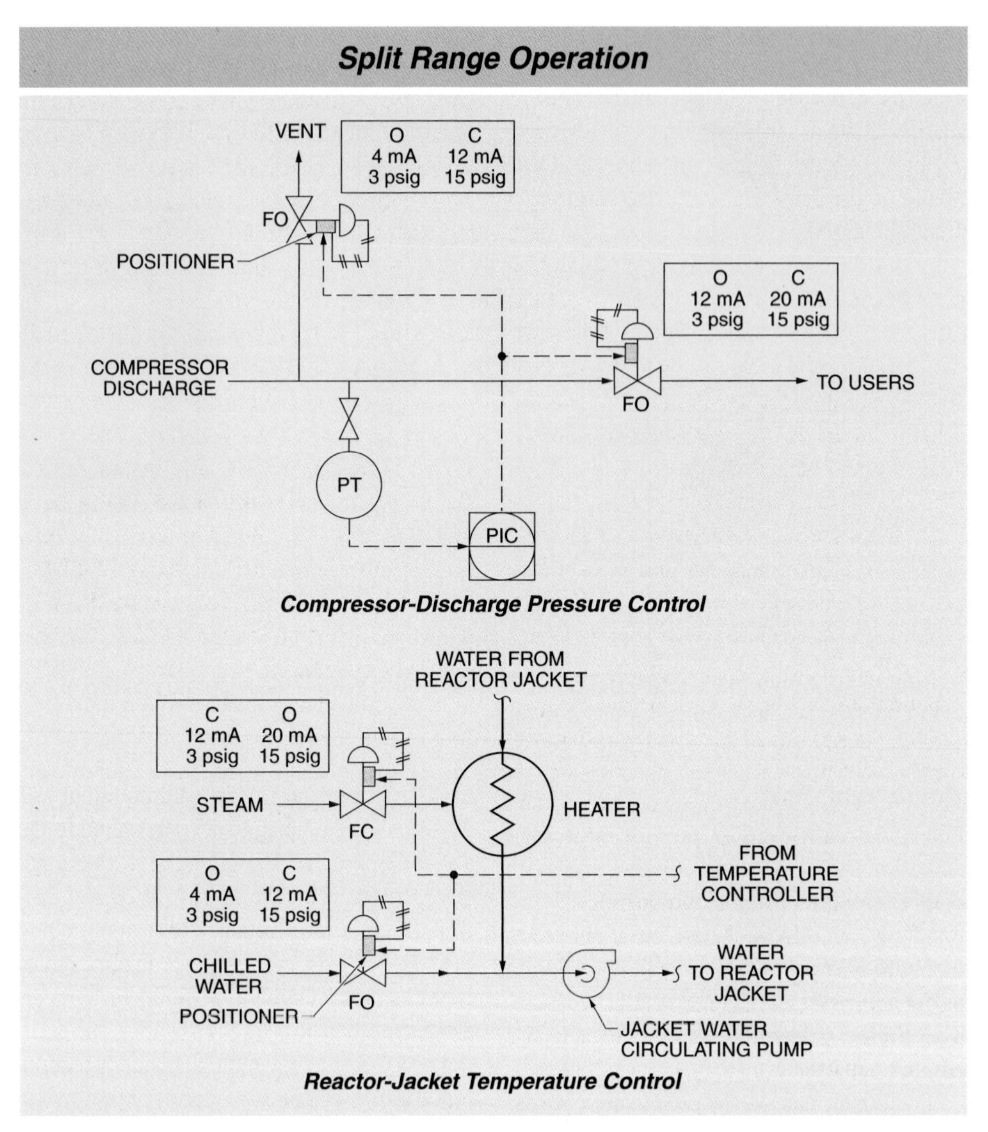

Figure 41-18. Each positioner in a split-range control system uses part of the control signal.

Solenoid-actuated valves are commonly used to stop and start the flow of fluids in pipes.

Compressor-Discharge Pressure Control. A compressor-discharge pressure control system may use sequential split range valve operation. Upon increasing pressure, the valve to the users opens first. As the pressure continues to increase, the controller then opens the vent valve enough to control the pressure. If the users start taking more gas than is available, the vent valve closes and then the user valve is throttled to limit how much gas the users can take. The failsafe condition is that both valves open to prevent damage to the compressor.

Reactor-Jacket Temperature Control. A chemical reactor-jacket temperature control system can use two valves—one for chilled water and the other for steam—that are used in heating or cooling a heat transfer fluid. At the beginning of a batch, the chemicals must be heated to initiate the reaction. The controller calls for heat. As the control signal current increases from 4 mA to 12 mA, the chilled water valve starts to close until the signal reaches the midpoint of 12 mA. At this point, both valves are closed and the heat transfer fluid is circulated without being heated or cooled.

The split ranging is designed so that the two valves are not open at the same time, and with a 12 mA control signal both valves are closed. As the control signal increases further, the steam valve starts to open. As the reactor temperature approaches its setpoint, the control signal starts decreasing the steam. As the reaction starts in the reactor, the reactor temperature increases and the cooling water valve is opened. The fail-safe condition is that the steam valve closes and the cooling valve opens to prevent the reactor from overheating.

The control transfers smoothly from one valve to the other. Split ranging is a very useful tool for the control of more complex control loops. It is usually not advisable to split the control signal into more than three parts.

Reverse Action

A positioner can make it possible to reverse the controller output signal to the valve. The controller output action, direct or reverse, is determined by what is necessary for proper feedback control, but the desired valve failure position on loss of air supply may require the reverse action. An example is a controller signal of 4 mA to 20 mA and a positioner output of 15 psig to 3 psig. This does not change the valve failure position on loss of air supply, which is determined by the valve design. This arrangement does reduce the safety of the loop since a loss of the control signal, while the air supply is maintained, causes the valve to go to a potentially unsafe position.

Flow Characteristic Change

If a positioner includes feedback through a cam, the positioner can be altered from its normal linear action to some other type of action by changing the shape of the cam. The normal equal-percentage characteristic of a butterfly valve can be made to act like a linear valve by using a quick-opening cam in the positioner.

Positioners for reciprocating-stem piston actuators use higher pressures than other positioners. Like other pneumatic devices, air flows through a restriction and some type of flapper and nozzle arrangement. Balancing air pressures act on a piston to move a shaft. **See Figure 41-19.** A cam is used to change the movement of the feedback lever. Depending on the shape of the cam, this can change the flow to any desired characteristic.

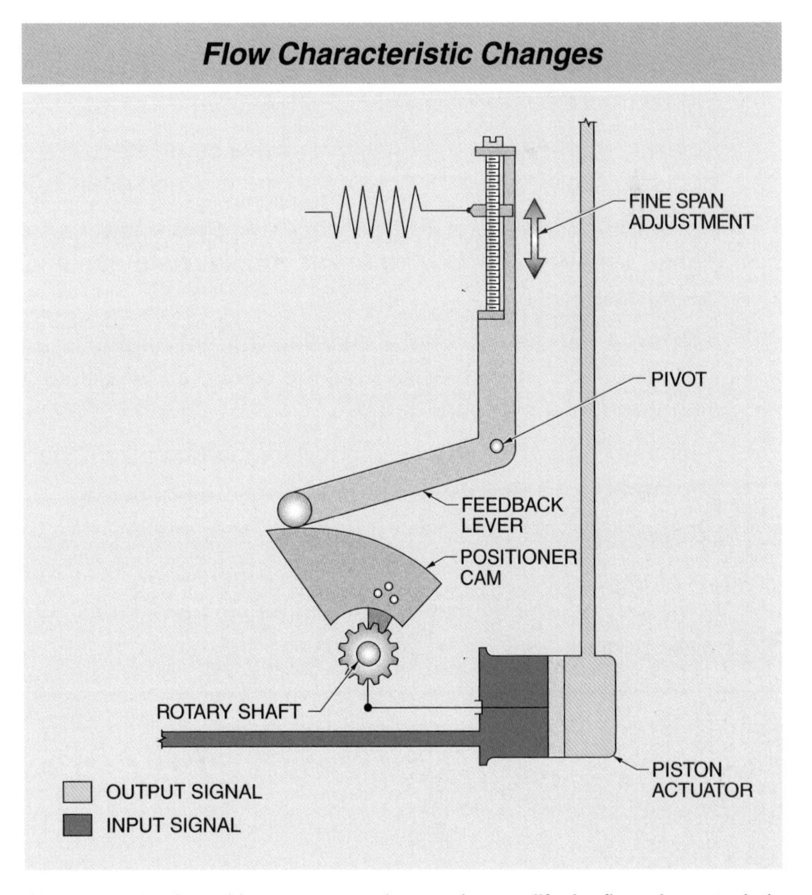

Figure 41-19. A positioner cam can be used to modify the flow characteristic of a control valve.

KEY TERMS

- *actuator:* A device that provides the power and motion to manipulate the moving parts of a valve or damper used to control fluid flow through a final element.
- *diaphragm-and-spring actuator:* An actuator consisting of a large-diameter diaphragm chamber with a diaphragm backed by a plate attached to the actuator stem and opposed by the actuator spring.
- *direct-acting actuator:* An actuator that extends the shaft when air is applied to the diaphragm.
- *reverse-acting actuator:* An actuator that retracts the shaft when air is applied to the diaphragm.
- *pneumatic sliding stem piston actuator:* An actuator where the movement of a piston replaces the action of the diaphragm and a balancing air pressure replaces the function of the spring.
- *pneumatic rotary piston actuator:* An actuator that can be attached directly to most rotary valves and is primarily designed for ON/OFF action.
- *electric actuator:* An actuator consisting of an electric motor that is connected through gearing to the valve stem or shaft.
- *spring-return actuator:* A piston or diaphragm actuator with an internal spring to force the actuator shaft to one end of its travel.
- *three-way solenoid valve:* A solenoid that shuts off the air supply and vents air from the actuator or cylinder when the solenoid is de-energized.
- *double-acting actuator:* A piston actuator that has an air connection at each end of the piston and does not contain a spring to allow for a known failure position.
- *four-way solenoid valve:* A single- or dual-coil solenoid that has an air supply, a vent, and two cylinder ports.
- *four-way dual-coil solenoid valve:* A valve connected to a solenoid actuator in the same way as a single-coil solenoid valve except that there is no spring in the solenoid.
- *current-to-pneumatic (I/P) transducer:* A device that can convert the electronic controller output signal, usually 4 mA to 20 mA, into the standard pneumatic output, 3 psig to 15 psig (or similar metric range).
- *flapper:* A flat piece of metal installed at right angles to a nozzle tip that is held in position a few thousandths of an inch away from the nozzle tip by a linkage to a measurement device such as the diaphragm of a pressure gauge.
- *pneumatic relay:* A pneumatic amplifier used to take a signal from a flapper and nozzle and boost the signal to a standard range required for a process.
- *positioner:* A device used to ensure positive position of a valve or damper actuator.
- *split range operation:* A control configuration where a single control signal, 4 mA to 20 mA or 3 psig to 15 psig, can be directed to two throttling control valves or dampers equipped with electropneumatic positioners.

REVIEW QUESTIONS

1. What is an actuator, and what are some of the requirements for an actuator?
2. What is the difference between direct-acting and reverse-acting actuators, and why would one be chosen instead of the other in a process?
3. Define sliding stem and rotary piston actuators.
4. What is the difference between diaphragm-and-spring and electric actuators?
5. Define positioner and explain how a positioner operates.

Variable-Speed Drives and Electric Power Controllers

Chapter 42

OBJECTIVES

- Define variable-speed drive and explain its use.
- List and describe the different types of electric power controllers.

VARIABLE-SPEED DRIVES

A *variable-speed drive* is a device that varies the speed of an electric motor. Many final elements like blowers, fans, compressors, and pumps are powered by electric motors. The ability to vary the speed of the motor allows the final element to vary the amount of material moved. An AC variable-speed drive changes the frequency of the voltage applied across the motor contacts to change the motor speed. **See Figure 42-1.**

Industrial AC motors are usually powered by three-phase synchronous electricity, where the frequency of each phase is 60 Hz. This means that the motor's rotating magnetic field in the stationary windings drags the armature at almost the same speed as the rotating magnetic field. Standard motor speeds for a squirrel-cage induction motor are about 1750 rpm or 3500 rpm for rotating magnetic fields of 1800 rpm and 3600 rpm.

Changing the frequency to something less than 60 Hz reduces the motor running speed. The motor must be designed to work with variable AC frequencies. The development of inexpensive solid-state variable-speed drives has increased the use of these devices, replacing conventional control valves in flow applications. Variable-speed drives are attractive alternatives to control valves since they save energy and money in most applications.

There are also variable-speed DC motors available. A variable-speed DC drive changes the magnitude of the voltage across the motor contacts to change the motor speed. A common type of DC motor is the direct-drive DC torque motor. Torque motors are usually used for high-torque positioning systems and in slow-speed control systems.

Blowers, Fans, and Compressors

Blowers, fans, and compressors are ideal pieces of equipment for use with variable-speed drives. There is no clear distinction between a fan and a blower, but a fan generally refers to a device that generates a low pressure used to move air when used in free air or against little or no resistance. A blower is used to generate a higher pressure used to move air through a duct or other restriction. In both cases, changes in speed result in proportional changes in airflow. Blowers and fans are commonly available in centrifugal and axial configurations. Compressors are used to pressurize to a gas. A typical example is to provide compressed air for a process. Compressors are commonly available in screw and reciprocating configurations.

A valve design that equalizes the pressure drop across the valve allows the use of a smaller actuator.

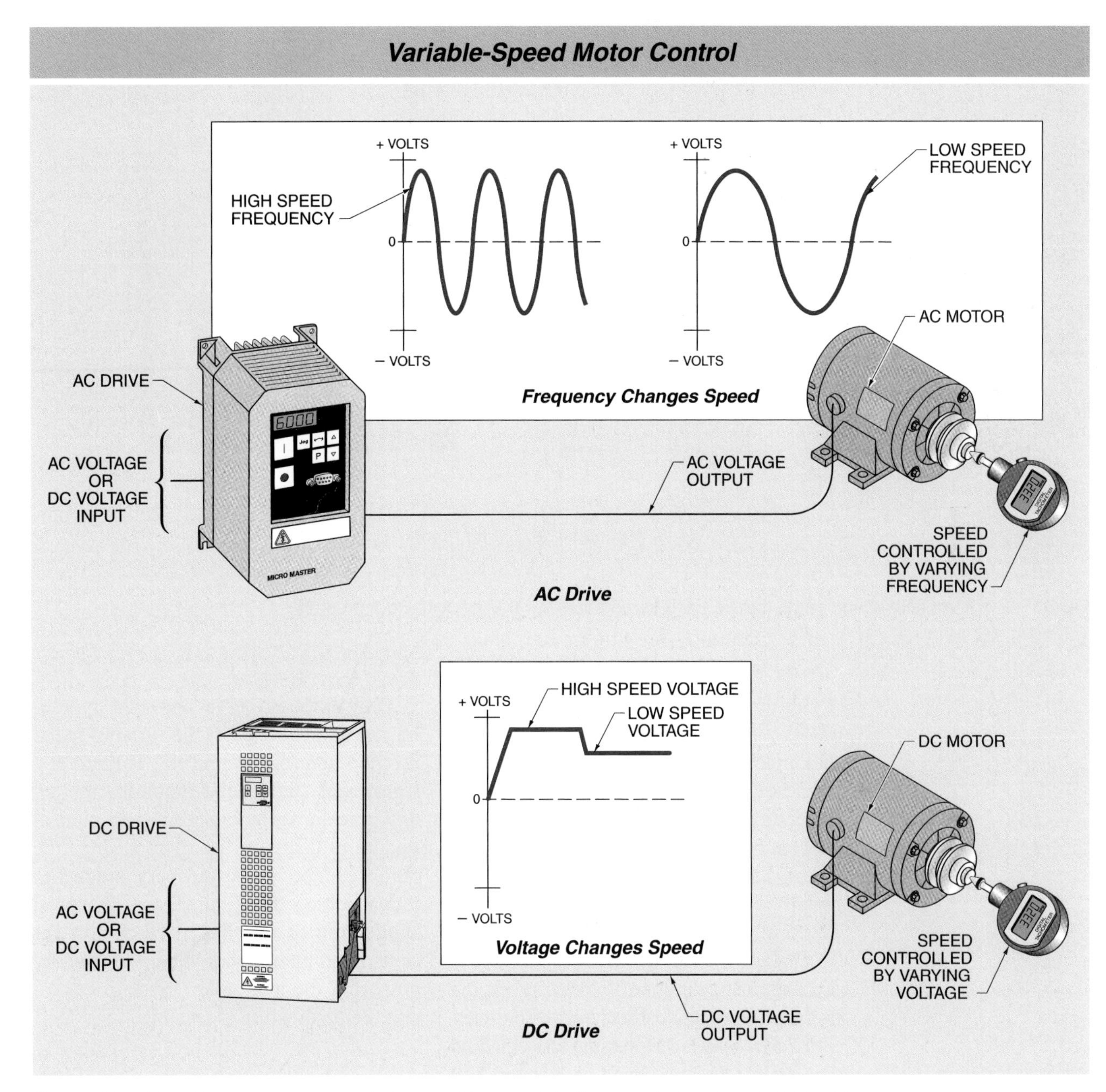

Figure 42-1. Variable-speed motor control is used to modulate the speed of motors that drive blowers, compressors, pumps and other devices.

Pumps

A *centrifugal pump* is a pump used to provide a constant discharge pressure at any flow rate for a given operating speed. Head and flow are both related to the speed of rotation of the pump impeller. Higher pump speeds increase the head and the flow. A variable-speed-drive pump is commonly used as a circulating pump used to move water through a boiler or piping system. **See Figure 42-2.** A *positive-displacement pump* is a pump used to provide a constant flow rate at any discharge pressure for a given operating speed. This is an ideal application for variable-speed drives. The combination of variable speed and variable stroke that is available with some positive-displacement pumps can provide exceptional rangeability.

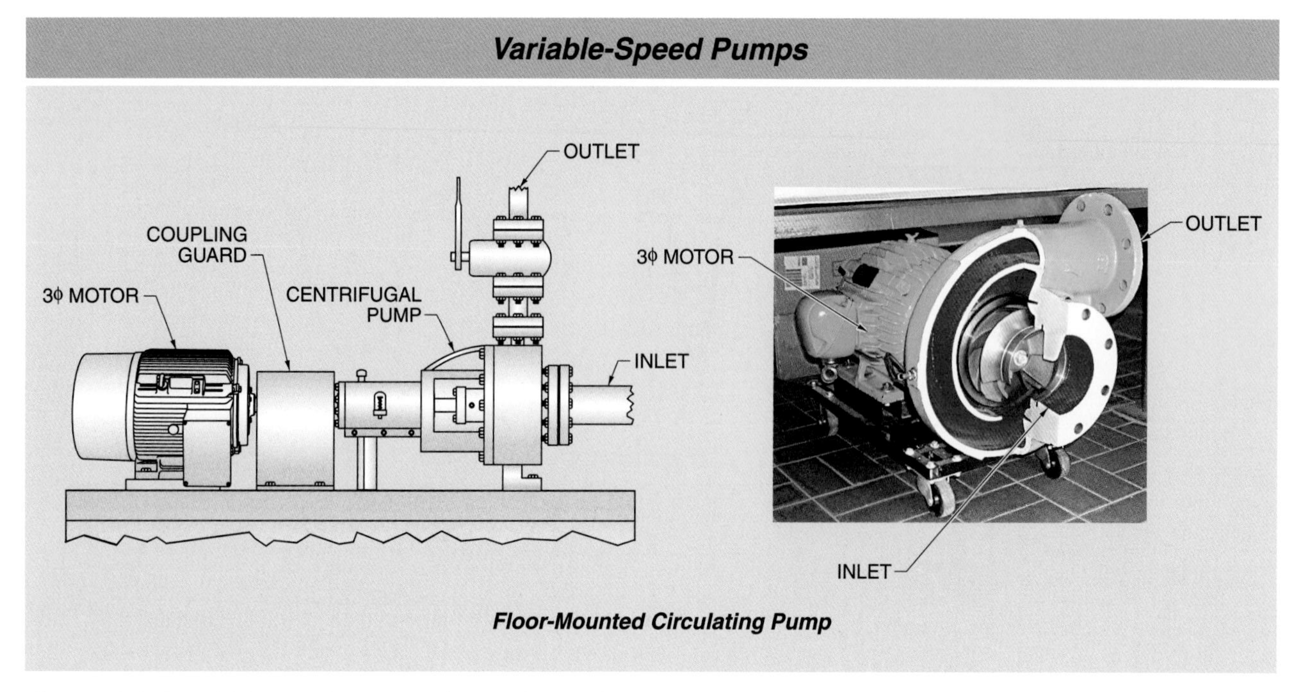

Figure 42-2. A variable-speed pump can be used to modulate the flow of water to a process.

Pump and Hydraulic Curves. A *pump curve* is a plot of the pump discharge pressure, or head, plotted against flow for various pump rotational speeds. The discharge pressure is proportional to the rotational speed squared and the flow is proportional to the rotational speed. Variable speed control of centrifugal pumps is possible if the process application requires only a small static head.

Static head is the pressure of a fluid due to a change in the elevation of a discharge piping system and remains constant for all flow rates. *Frictional head* is the pressure loss of a fluid due to flow through piping or ductwork and varies with the flow squared. Pump manufacturers provide pump curves for different types, sizes, and speeds of their pumps. Applications that have high static heads should be evaluated by the preparation of a combined pump and system hydraulic curve.

The hydraulic curve of the pump and piping system can be superimposed on top of the pump curve. A *hydraulic curve* is a flow curve that takes into account the static and frictional heads of the process. **See Figure 42-3.** A hydraulic curve describes the minimum requirements of the process.

Only the pump curves above the hydraulic curve can be used by the variable-speed drive to control the process. If there is a significant static head required for low flow rates, a variable-speed-drive control system probably will not work since the pump cannot be slowed very much before reaching that limit. Every proposed application of variable speed control for centrifugal pumps should be evaluated by preparing the hydraulic curves to ensure that the application is feasible.

ELECTRIC POWER CONTROLLERS

There are many applications that require a controller to modulate the power to an electric heating element. However, process controllers generally cannot handle the current load required to provide power to electric heating elements. In this case, the final element is the power controller. Common types of power controllers are electromechanical relays, solid-state relays (SSR), and silicon-controlled rectifiers (SCR).

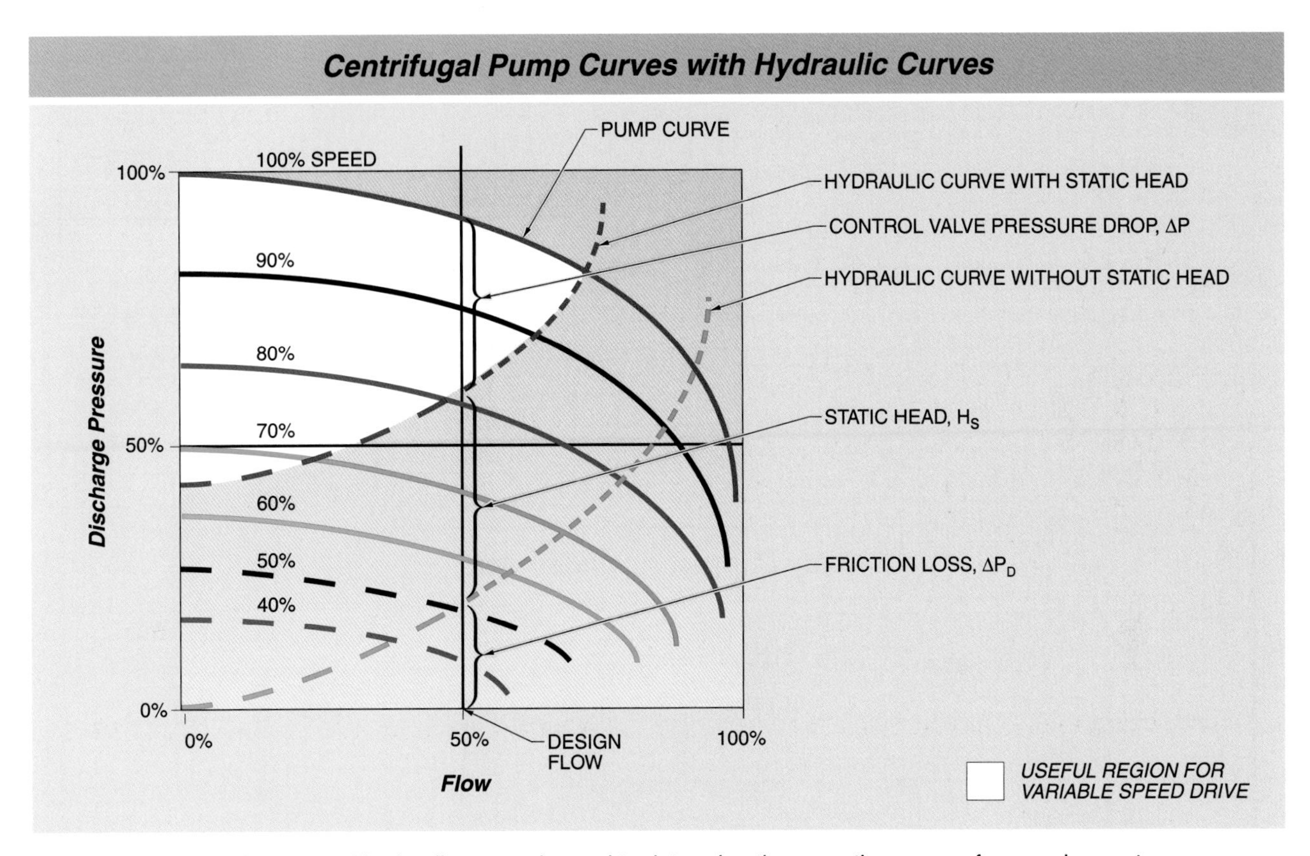

Figure 42-3. A set of pump and hydraulic curves is used to determine the operating range of a pumping system.

Electromechanical Relays

An *electromechanical relay* is a device that controls one electrical circuit by opening and closing contacts in another circuit through the action of magnetic force. In an electromechanical relay circuit, a coil opens and closes contacts by a magnetic force that is developed each time the coil is energized. Electromechanical relays are ON/OFF devices. The best way to use electromechanical relays to regulate the power to an electric heating element is by using ON/OFF or time proportional control. Relay contacts are described by their number of breaks, poles, and throws. **See Figure 42-4.**

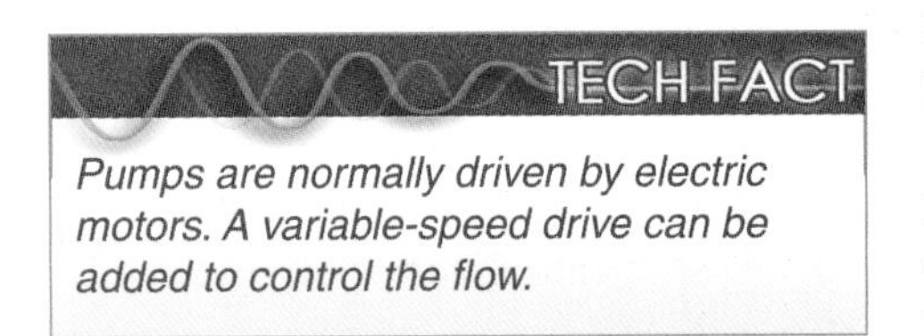

Pumps are normally driven by electric motors. A variable-speed drive can be added to control the flow.

A *break* is a place on a contact that opens or closes an electrical circuit. A single-break (SB) contact breaks an electrical circuit in one place. A double-break contact breaks the electrical circuit in two places. All contacts are single-break or double-break. Single-break contacts are normally used when switching low-power devices such as alarms and lights. Double-break contacts are used when switching high-power devices such as solenoids and heaters.

A *pole* is a completely isolated circuit that a relay can switch. A single-pole contact can carry current through only one circuit at a time. A double-pole contact can carry current through two circuits simultaneously. Relays are available with 1 to 12 poles.

A *throw* is the number of closed contact positions per pole in a relay. A single-throw contact can control only one circuit. A double-throw contact can control two circuits.

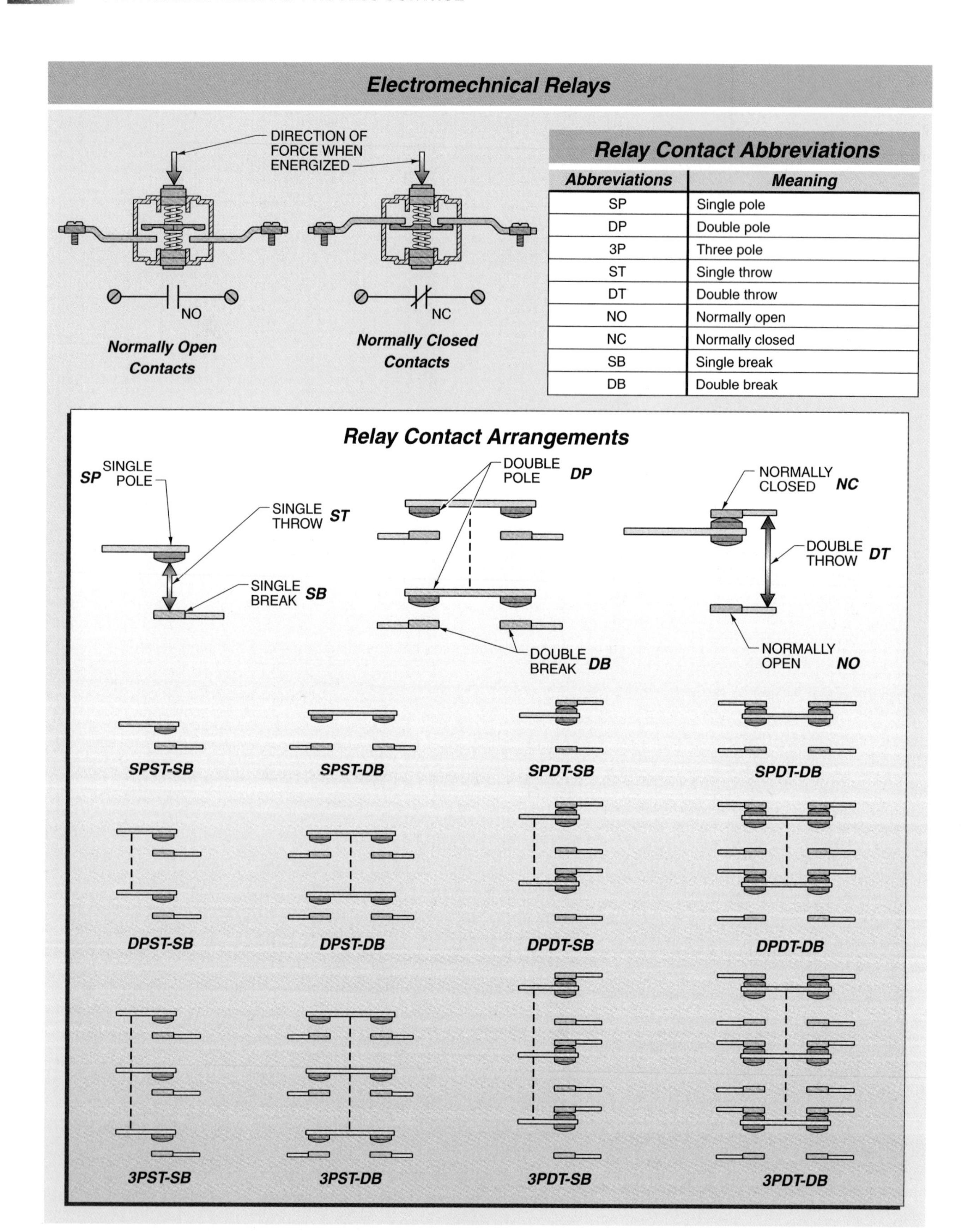

Relay Contact Abbreviations

Abbreviations	Meaning
SP	Single pole
DP	Double pole
3P	Three pole
ST	Single throw
DT	Double throw
NO	Normally open
NC	Normally closed
SB	Single break
DB	Double break

Figure 42-4. Electromechanical relays come in many configurations.

Solid-State Relays

A *solid-state relay* is a semiconductor switching device that uses a low-current DC input to switch an AC circuit. Solid-state relays typically are able to switch one circuit at a time. Solid-state relays can switch moderately high currents. Solid-state relays can last for billions of cycles because they have no moving parts. Because solid-state relays have no exposed electronic parts, their circuits are relatively unaffected by outside environments such as dusty conditions or volatile gases. However, high temperatures affect the electronic circuits inside the relay and special care must be taken to prevent exposure to excessive heat.

There are several methods used when switching solid-state relays. The most common type of switching is zero switching. A *zero switching relay* is a solid-state relay where a load becomes energized when a control input voltage is applied and AC load voltage crosses zero. **See Figure 42-5.** This means that there may be a small lag between application of a control voltage to the relay and when the load is actually powered. The load is de-energized when the control voltage is discontinued and the load current crosses zero. Zero switching relays are typically used for ON/OFF control circuits for resistive loads such as heating elements in ovens and extruders for forming plastic.

Another method of switching a solid-state relay is analog switching. An *analog switching relay* is a solid-state relay that has a continuous range of output voltages within the relay's rated range. An analog switching relay has a built-in synchronizing circuit that controls the amount of output voltage as a function of the input voltage. This allows for proportional control and a ramp-up function of the load. **See Figure 42-6.** Analog switching relays are used for closed loop applications such as temperature control with feedback from a temperature sensor to the controller.

Output Circuits. An output circuit of a solid-state relay is the part of the relay that actually conducts current when activated. The output circuit performs the same function as the mechanical contacts of an electromechanical relay.

The temperature of a steam trap can be used to troubleshoot steam system problems.

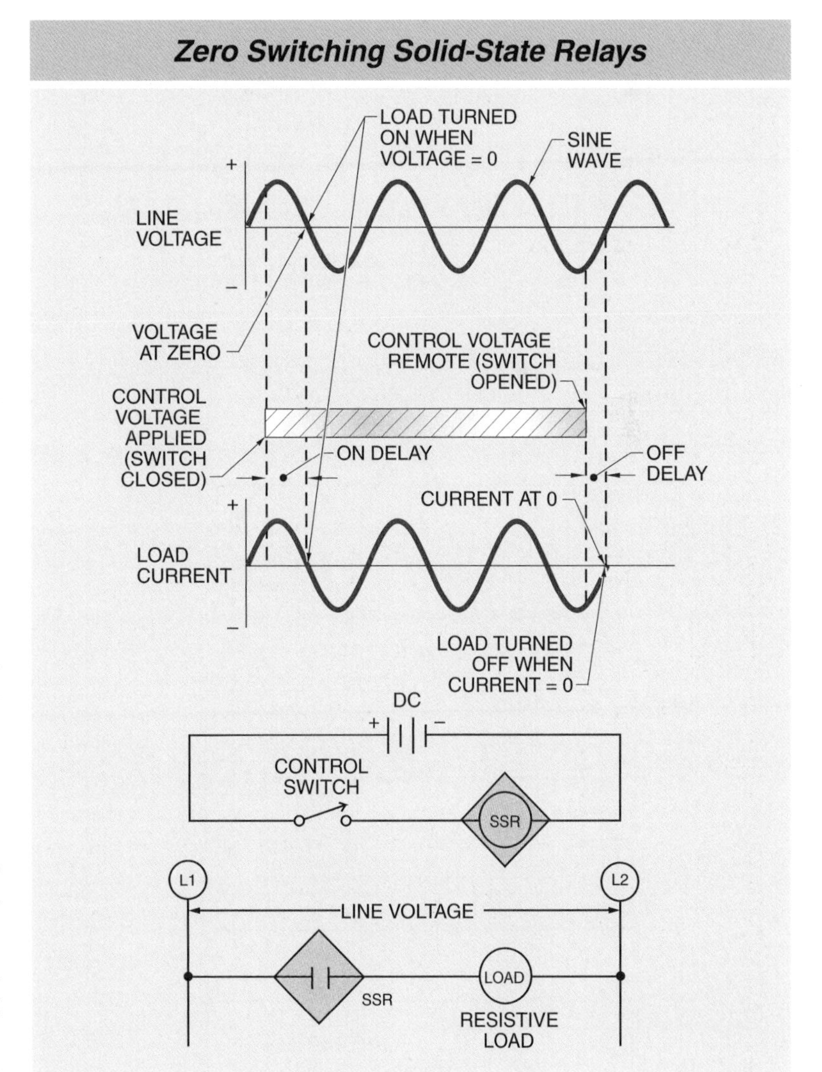

Figure 42-5. Zero switching relays are a very common type of solid-state relay that actuates when the voltage becomes zero.

A *transistor* is a solid-state switching device that is used to switch very-low-current DC loads. When a high-current DC or AC load must be switched, a transistor is connected to a second interface. The second interface is normally a mechanical or solid-state relay. Solid-state relays are the most common choice when using transistors.

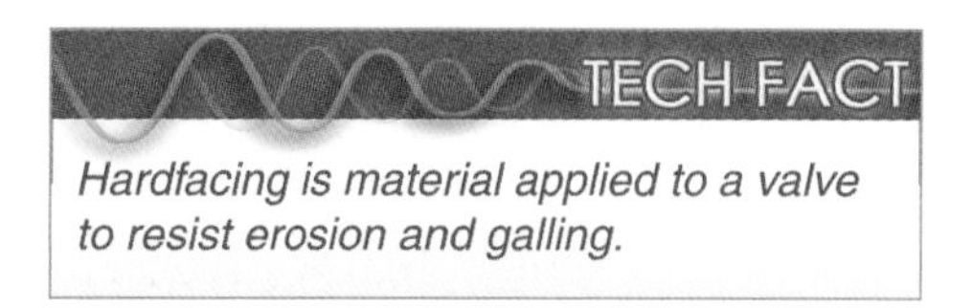

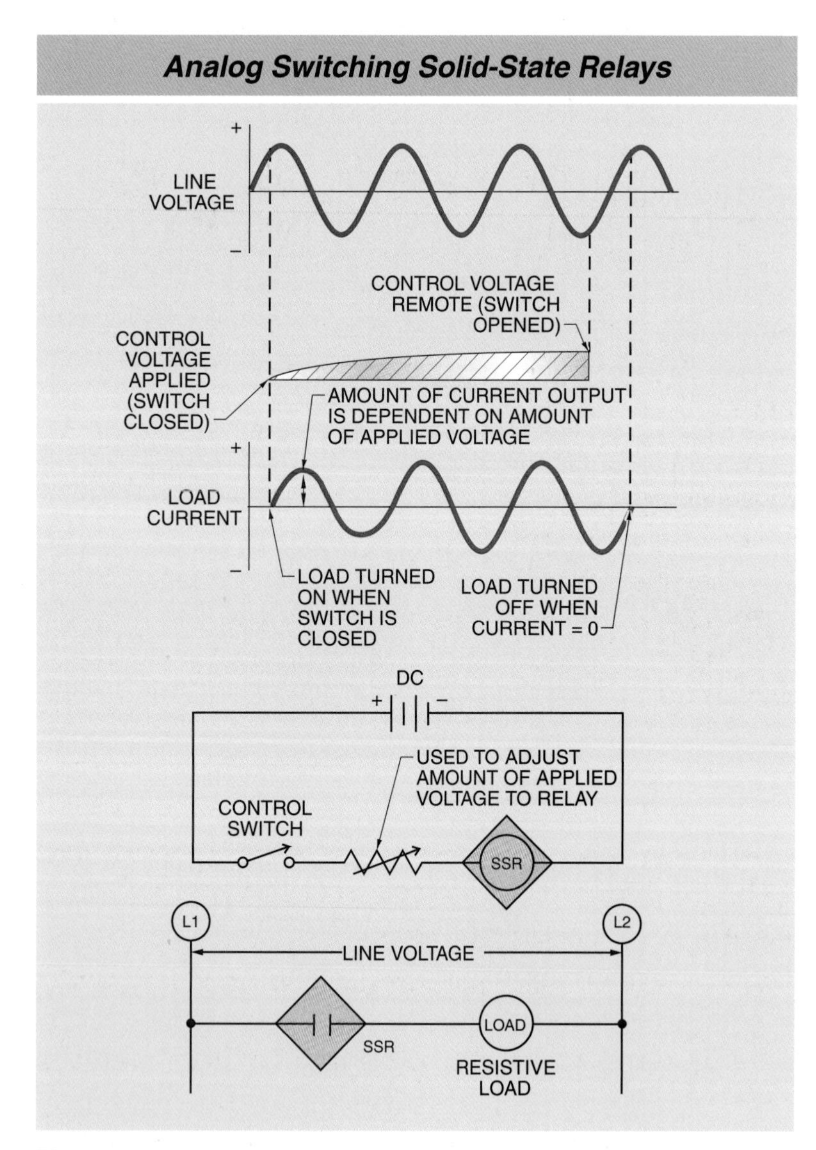

Figure 42-6. Analog switching relays are solid-state relays that are able to modulate the current being switched.

A *thyristor* is a solid-state switching device that switches current ON by a quick pulse of control current to its gate. Thyristors are often used to switch high-current loads such as motor controls and heating elements. The most common types of thyristors are silicon-controlled rectifiers and triacs.

A *silicon-controlled rectifier (SCR)* is a solid-state power controller that provides proportional current to a heating element in response to an analog control signal. **See Figure 42-7.** An SCR is able to use a controller signal (4 mA to 20 mA DC or 1 VDC to 5 VDC) to regulate current as high as 1200 A. An SCR conducts only in one direction so it is ideal for controlling DC loads.

An SCR may be used to control AC loads when two SCRs are connected in parallel and in opposite directions. This arrangement allows for better cooling in the thyristor circuit. An SCR continues to conduct as long as the current in the load circuit is higher than the holding current value of the SCR. The current must be broken by some other means in order to de-energize an SCR. As solid-state devices, SCRs have no moving parts and provide long operating life. Also, like all other solid-state devices, they are sensitive to heat.

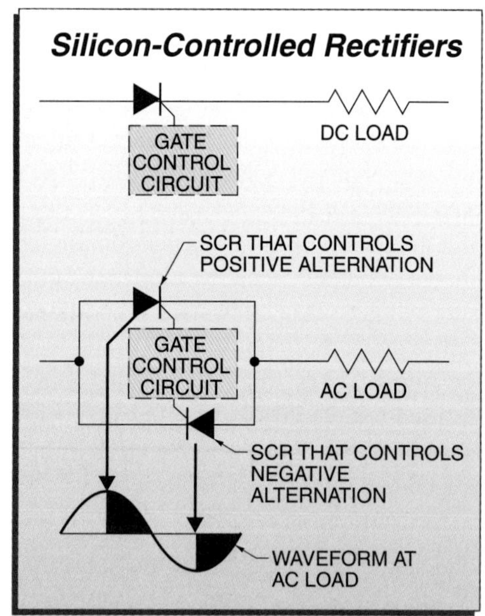

Figure 42-7. A silicon-controlled rectifier is a one-direction thyristor that switches by allowing current to flow when a control signal switches a gate to a conducting state.

A *triac* is a thyristor that is triggered into conduction in either direction by a small current to its gate. Like an SCR, a triac stops conducting when the voltage drops below the holding current value. Since the triac is used in AC circuits, a continuous trigger current is supplied to the gate. Schematically, a triac is drawn as two SCRs in parallel but in opposite directions, with one gate. **See Figure 42-8.** Both SCRs and triacs can be used to vary the amount of power to a load in addition to acting as switches. This feature allows them to control the speed of motors, the amount of heat delivered from heating elements, and the brightness of incandescent lamps.

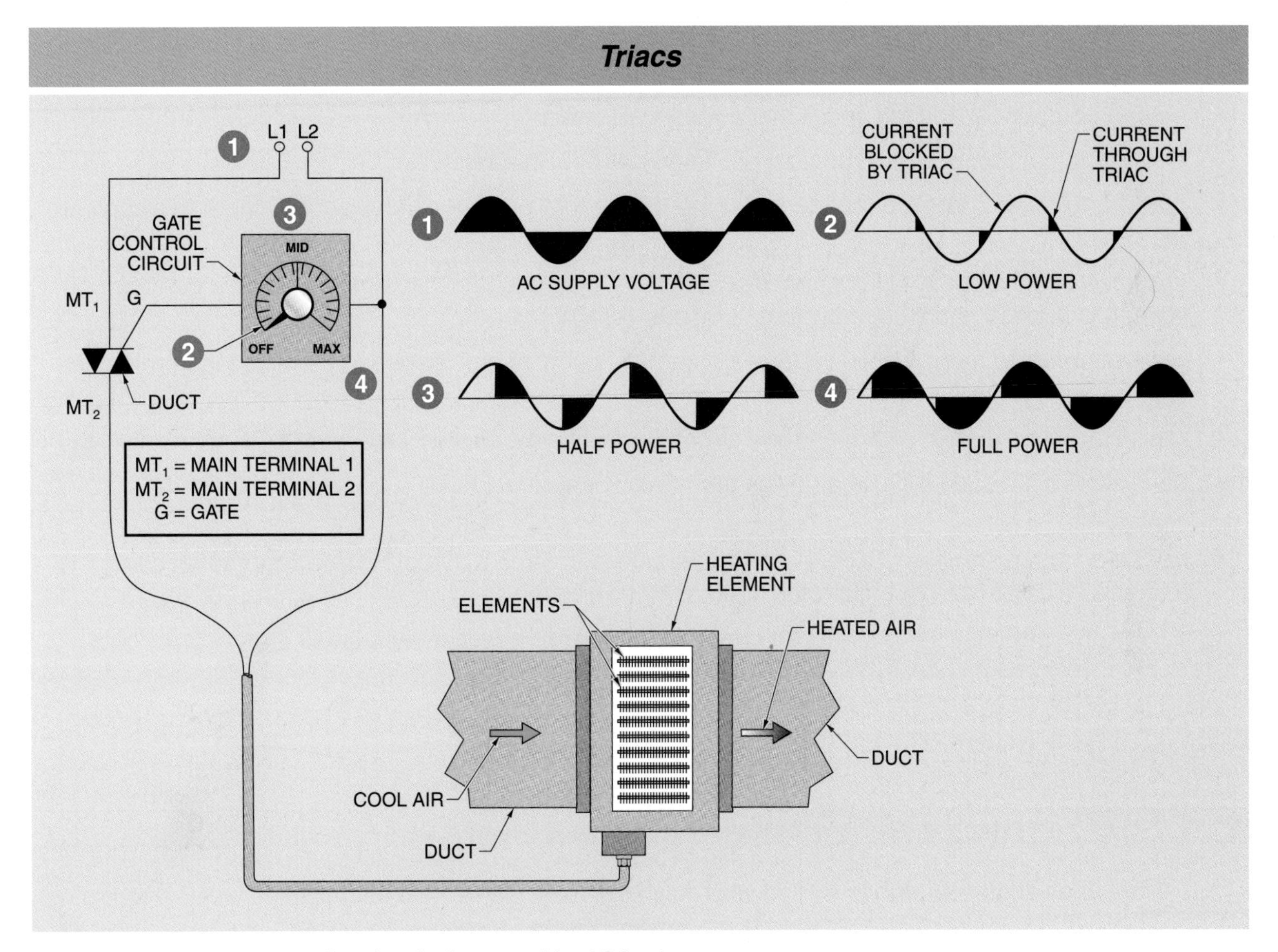

Figure 42-8. A triac is a two-direction thyristor used for AC loads.

KEY TERMS

- *variable-speed drive:* A device that varies the speed of an electric motor.
- *centrifugal pump:* A pump used to provide a constant discharge pressure at any flow rate for a given operating speed.
- *positive-displacement pump:* A pump used to provide a constant flow rate at any discharge pressure for a given operating speed.
- *pump curve:* A plot of the pump discharge pressure, or head, plotted against flow for various pump rotational speeds.

KEY TERMS (continued)

- *static head:* The pressure of a fluid due to a change in the elevation of a discharge piping system and remains constant for all flow rates.
- *frictional head:* The pressure loss of a fluid due to flow through piping or ductwork and varies with the flow squared.
- *hydraulic curve:* A flow curve that takes into account the static and frictional heads of the process.
- *electromechanical relay:* A device that controls one electrical circuit by opening and closing contacts in another circuit through the action of magnetic force.
- *break:* A place on a contact that opens or closes an electrical circuit.
- *pole:* A completely isolated circuit that a relay can switch.
- *throw:* The number of closed contact positions per pole in a relay.
- *solid-state relay:* A semiconductor switching device that uses a low-current DC input to switch an AC circuit.
- *zero switching relay:* A solid-state relay where a load becomes energized when a control input voltage is applied and AC load voltage crosses zero.
- *analog switching relay:* A solid-state relay that has a continuous range of output voltages within the relay's rated range.
- *transistor:* A solid-state switching device that is used to switch very-low-current DC loads.
- *thyristor:* A solid-state switching device that switches current ON by a quick pulse of control current to its gate.
- *silicon-controlled rectifier (SCR):* A solid-state power controller that provides proportional current to a heating element in response to an analog control signal.
- *triac:* A thyristor that is triggered into conduction in either direction by a small current to its gate.

REVIEW QUESTIONS

1. What is a variable speed drive, and why may it be used with a final element?
2. Define centrifugal pump and positive-displacement pump.
3. What is the difference between static head and frictional head?
4. How does an electromechanical relay operate?
5. What is a solid-state relay, and what are two common ways to switch a solid-state relay?

SECTION 11

SAFETY SYSTEMS

TABLE OF CONTENTS

SECTION OBJECTIVES

Chapter 43

- Explain the use of personal protective equipment (PPE).
- List the types of valves used in safety systems and describe their use.
- Explain the function and use of rupture discs.
- Describe the operation of burner control systems.
- Explain the function of alarm systems.
- Describe the use of hazardous atmosphere detectors.

Chapter 44

- Describe the role of the National Electric Code® in establishing hazardous location classifications.
- Describe electrical protections used with safety systems.

Chapter 45

- Define safety instrumented system (SIS) and describe its importance.
- Describe the factors used to evaluate a process.
- List the technology options available to use in an SIS.

SECTION

11 SAFETY SYSTEMS

INTRODUCTION

The earliest efforts to improve safety in manufacturing used the "learn from experience" approach. As accidents and incidents occurred, causes were identified and procedures were changed to try to keep accidents from happening. For example, the National Fire Protection Association (NFPA) was formed to safeguard people and property against electrical hazards. OSHA keeps records of accidents and targets industries for improvement when accident rates are higher than in other industries. The EPA keeps records in spill/accident databases and targets improvements in operations that have the most incidents. This approach has been fairly effective in preventing incidents.

A series of major industrial incidents in the 1970s and 1980s led to new requirements on the design and operation of manufacturing facilities that handle hazardous chemicals. A new standard, ANSI/ISA-S84.01-1996 (ISA S84), was approved as a nationally recognized consensus standard for procedures to evaluate the use of hazardous chemicals. OSHA and the EPA do not require the use of this standard, but both agencies require companies to use good engineering practice in the design of safety systems.

Organizations such as Underwriters Laboratories, Inc. (UL), FM Global, and the Canadian Standards Association (CSA®) are the three main approval agencies in North America for products used in hazardous (classified) locations. Other organizations such as the National Fire Protection Association (NFPA), the International Society of Automation (ISA), the National Electrical Manufacturers Association (NEMA®), and the International Electrotechnical Commission (IEC) are involved in the development of standards for hazardous areas.

ADDITIONAL ACTIVITIES

1. After completing Section 11, take the Quick Quiz® included with the Digital Resources.
2. After completing each chapter, answer the questions and complete the activities in the *Instrumentation and Process Control Workbook.*
3. Review the Flash Cards for Section 11 included with the Digital Resources.
4. Review the following related Media Clips for Section 11 included with the Digital Resources:
   - Electrical Enclosures
   - Lockout
5. After completing each chapter, refer to the Digital Resources for the Review Questions in PDF format.

ATPeResources.com/QuickLinks • Access Code: 467690

SECTION 11 SAFETY SYSTEMS

Safety Devices and Equipment

Chapter 43

OBJECTIVES

- Explain the use of personal protective equipment (PPE).
- List the types of valves used in safety systems and describe their use.
- Explain the function and use of rupture discs.
- Describe the operation of burner control systems.
- Explain the function of alarm systems.
- Describe the use of hazardous atmosphere detectors.

SAFETY EQUIPMENT

A *safety system* is a system that consists of an individual device or an assembly of devices that form a system designed to protect personnel from injury and production equipment from damage. Essentially, any instrument can be classified as a safety device if it is part of a safety system. However, several types of instruments and other devices are used specifically as individual safety devices. Common types of individual safety devices are safety and safety relief valves, rupture discs, burner control systems, flame detectors, fuel safety shutoff valves, alarm systems, and hazardous atmosphere detectors. Safety devices and systems are designed to have significantly higher reliability than standard instrumentation systems.

Personal Protective Equipment

A key part of a safety system is safe workers. Manufacturing and chemical plants can be very dangerous environments. Some of the simplest and most important individual safety devices are the personal protective equipment that people wear while working. *Personal protective equipment (PPE)* is any clothing or device worn by a worker to prevent injury. **See Figure 43-1.** All PPE must meet the requirements specified in OSHA 29 CFR 1900-1999, applicable ANSI and MSHA standards, and other safety mandates. A list of typical PPE that is required is as follows:

- hard hats or helmets
- steel-toed safety shoes
- safety glasses with side shields
- hearing protection
- chemical goggles
- face shields
- chemical or dust masks
- special outer clothing
- long-sleeved shirts
- special chemical boots
- appropriate gloves for the work being done

Not all of this equipment is needed in every plant or at all times. Each plant has its own requirements. Some of this equipment is only necessary in the most hazardous areas of a plant. These areas should be clearly identified with warning signs. Visitors to a plant are normally provided with the required safety equipment before entering the plant. An employee whose job it is to keep visitors safe is usually assigned to guide visitors.

Employees are trained to recognize and avoid hazardous situations in their work area. In many plants, visitors are required to watch a safety video before entering the plant. Contractors regularly working at a plant generally have greater freedom and access to the plant and usually do not have guides.

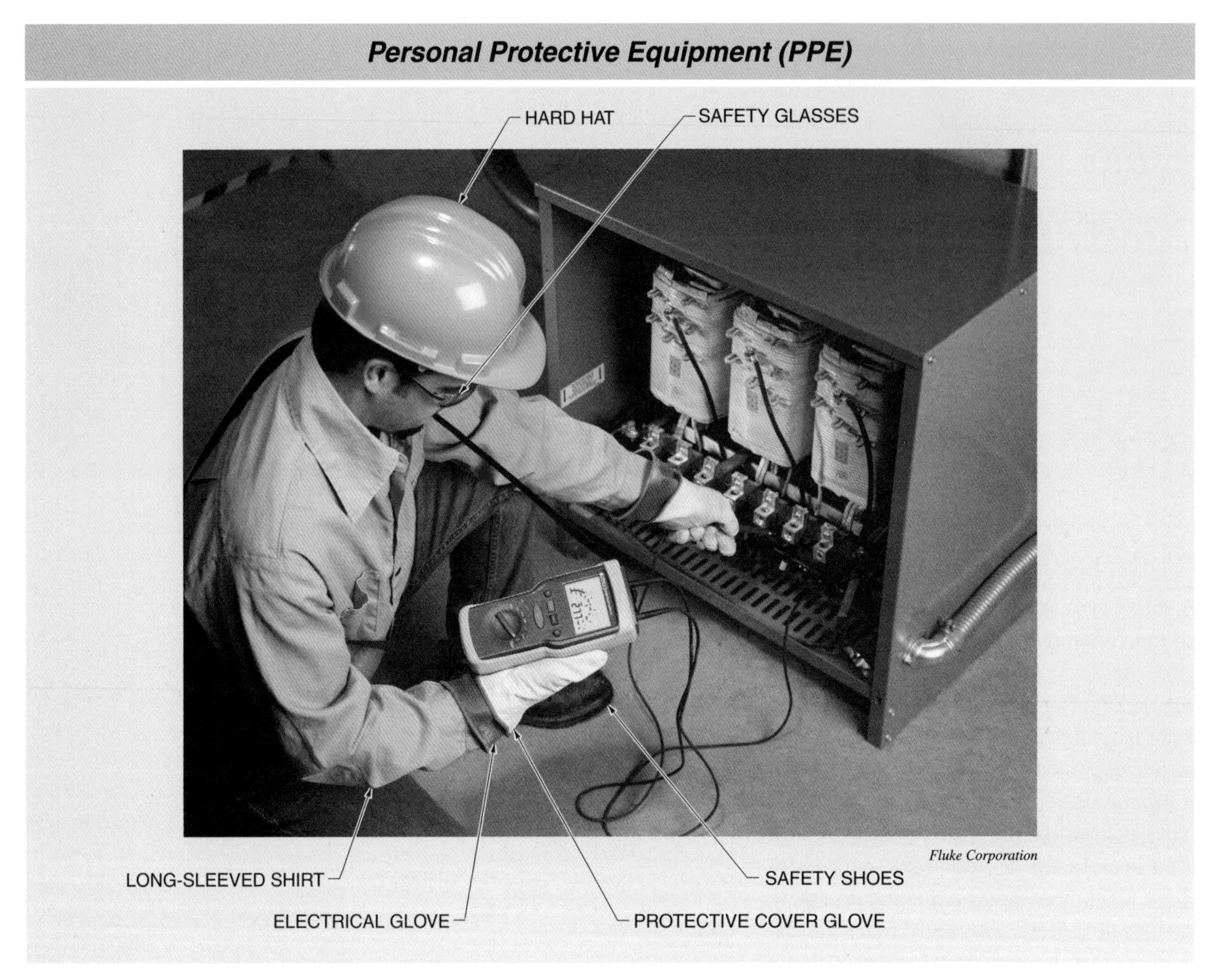

Figure 43-1. Personal protective equipment (PPE) is used to protect individuals from workplace hazards.

Overalarming causes frequent alarms that are eventually ignored. As a result, critical situations may not be handled properly.

Electrical Personal Protection

NFPA 70E®: Standard for Electrical Safety in the Workplace is a voluntary standard for electrical safety-related work practices. While OSHA does not enforce NFPA 70E, it does use the standard to determine whether an employer took sufficient steps to protect workers.

Electrical shock occurs when a worker makes contact with an energized component and the worker also has contact with electrical ground. An arc flash is an extremely high temperature discharge. It typically occurs when a conductive object makes contact with an energized component and ground. An arc blast vaporizes any metal and produces pressure waves that blast the plasma ball of vaporized metal in all directions.

For electrical workers, PPE may include arc-rated face shields, fire-resistant clothing, UV-rated safety glasses or goggles, rubber gloves with leather protectors, and other types of equipment. **See Figure 43-2.** Electrical workers may also be required to use some types of PPE that are not arc-rated such as respirators.

VALVES

Valves have several uses in safety systems. Some valves are used to relieve pressure in vessels and piping. Other valves are used to shut off the flow of a material in the case of an emergency. Common types of valves used to relieve pressure in safety systems include safety valves, relief valves, and safety relief valves. Common types of shutoff valves include excess flow valves and fuel-safety shutoff valves.

Safety Valves

A *safety valve* is a gas- or vapor-service valve that opens very quickly when the inlet pressure exceeds the spring setpoint pressure. **See Figure 43-3.** These valves are used in applications where the valve must open quickly to reduce dangerous conditions. The quick-opening action, or pop, is caused by the design of the valve disc. The force applied to the valve disc is the product of the pressure and the surface area of the disc. The disc has a special design that provides two surfaces.

A *huddling chamber* is a recessed area in a safety valve disc that increases the surface area and the total force applied. The smaller surface area is always exposed to the pressure. When the pressure is high enough, the force due to the pressure on the smaller surface overcomes the force of the spring and the disc is lifted off the seat. As the disc rises, the huddling chamber is exposed to the pressure because the exiting vapors are restricted by the special design of the disc. The lifting force increases because the surface area is greatly increased. The sudden increase in force causes the valve to pop open.

While the safety valve is open, the total force on the valve disc is applied over the larger surface area and the contents of the pressurized boiler or vessel escape out the valve. As the pressure decreases due to the outflow through the safety valve, the force on the disc decreases and the spring force overcomes the force on the disc and closes the valve. As the valve closes, the force is applied only to the smaller disc area again and the valve snaps completely closed.

Electrical Personal Protective Equipment

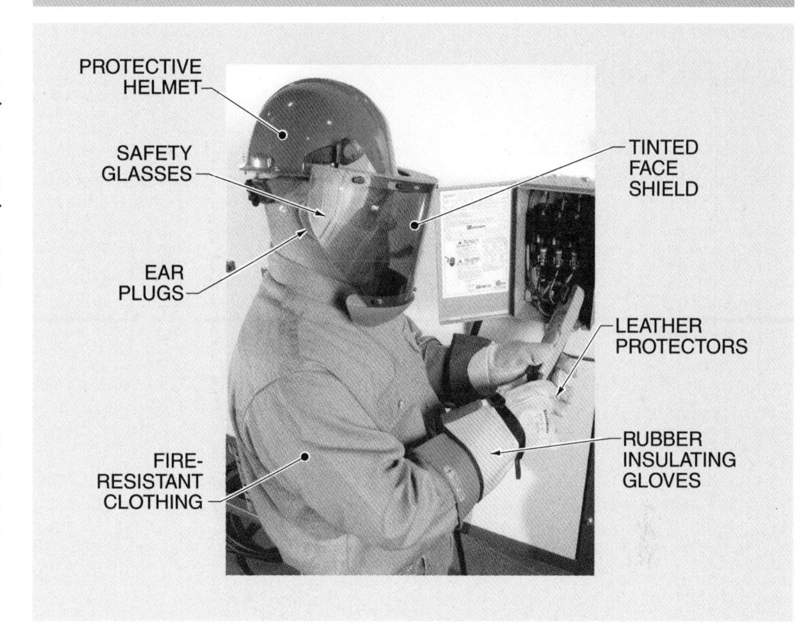

Figure 43-2. Electrical personal protective equipment protects workers from electrical hazards.

The total pressure is reduced below the original pressure setpoint because of the contents that were removed. *Blowdown,* or *blowback,* is the amount of pressure drop below the pressure setpoint of a safety valve. For boilers, blowdown is typically 2 psi to 8 psi. A larger popping effect increases the blowdown.

Safety valves are used for fired vessels such as boilers and follow the *ASME Boiler and Pressure Vessel Code, Section I.* Safety valves are designed to have an overpressure of 3%. *Overpressure* is the amount of pressure above the setpoint of a safety valve necessary to develop the full relieving capacity and is expressed as a percentage of the set pressure.

The *safety valve capacity* is the amount of steam, in lb/hr, that a safety valve is capable of venting at the rated pressure. The rated pressure is the set pressure plus the overpressure. The safety valve capacity is listed on the data plate attached to the safety valve. The safety valve capacity must equal or exceed the boiler steam pounds per hour rating.

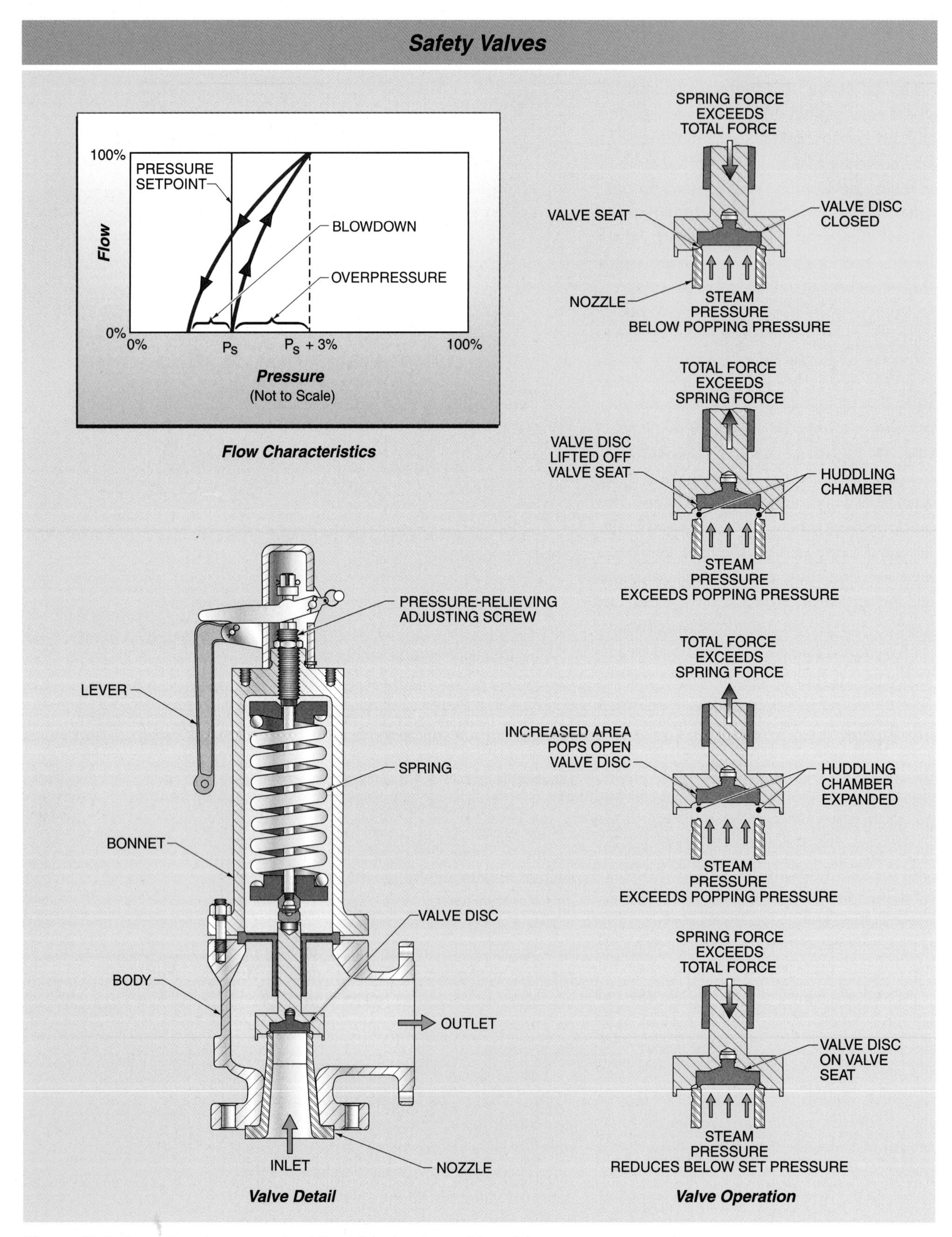

Figure 43-3. A safety valve uses a huddling chamber to enable quick snap-open operation.

Safety valves have no adjustment available to change the pop action of the valve. The majority of safety valve applications actually use safety relief valves. This is done because the design is suitable for either safety or relief applications. Valves used for safety applications are usually provided with a testing lever arm to help lift the disc from the seat at pressures lower than the set pressure.

Rupture Pin Safety Valves. A *rupture pin safety valve* is a safety valve with no spring; it is held closed by a pin or a thin rod. When the forces applied against the disc by the process pressure exceed the strength of the pin, the pin buckles and opens the valve. **See Figure 43-4.** Since the valve cannot reseat once the pin buckles, this type of valve acts like a rupture disc. The advantage of the rupture pin safety valve is that it has a more precise pressure relief point than a standard rupture disc.

A variation of the rupture pin safety valve is the rupture pin emergency shutdown valve, which is held in a normally open position by the pin. When the differential pressure across the disc seal exceeds the strength of the pin, the pin buckles and the valve closes. Once the valve closes it can only be reopened by disassembly and replacement of the pin.

Relief Valves

A *relief valve* is a valve that opens in proportion to the pressure above a setpoint. These valves are used for unfired vessel applications, protection against overpressure downstream of pressure regulators, protection against overpressure due to fire, or thermal expansion protection. A relief valve cannot be changed to a safety valve. Most relief valves are small in size and are used mostly for thermal expansion protection. Safety relief valves handle the large majority of relief valve applications.

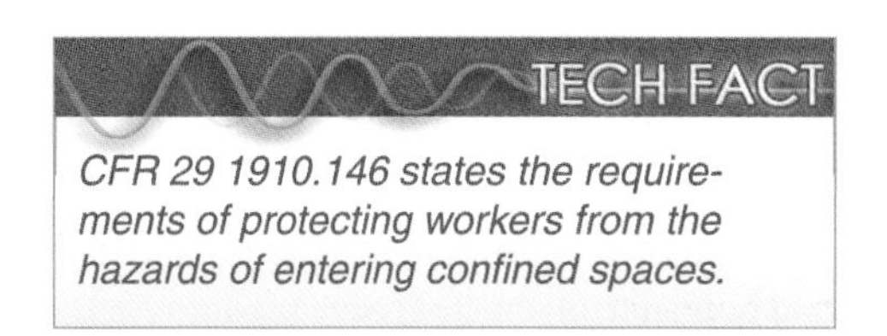

CFR 29 1910.146 states the requirements of protecting workers from the hazards of entering confined spaces.

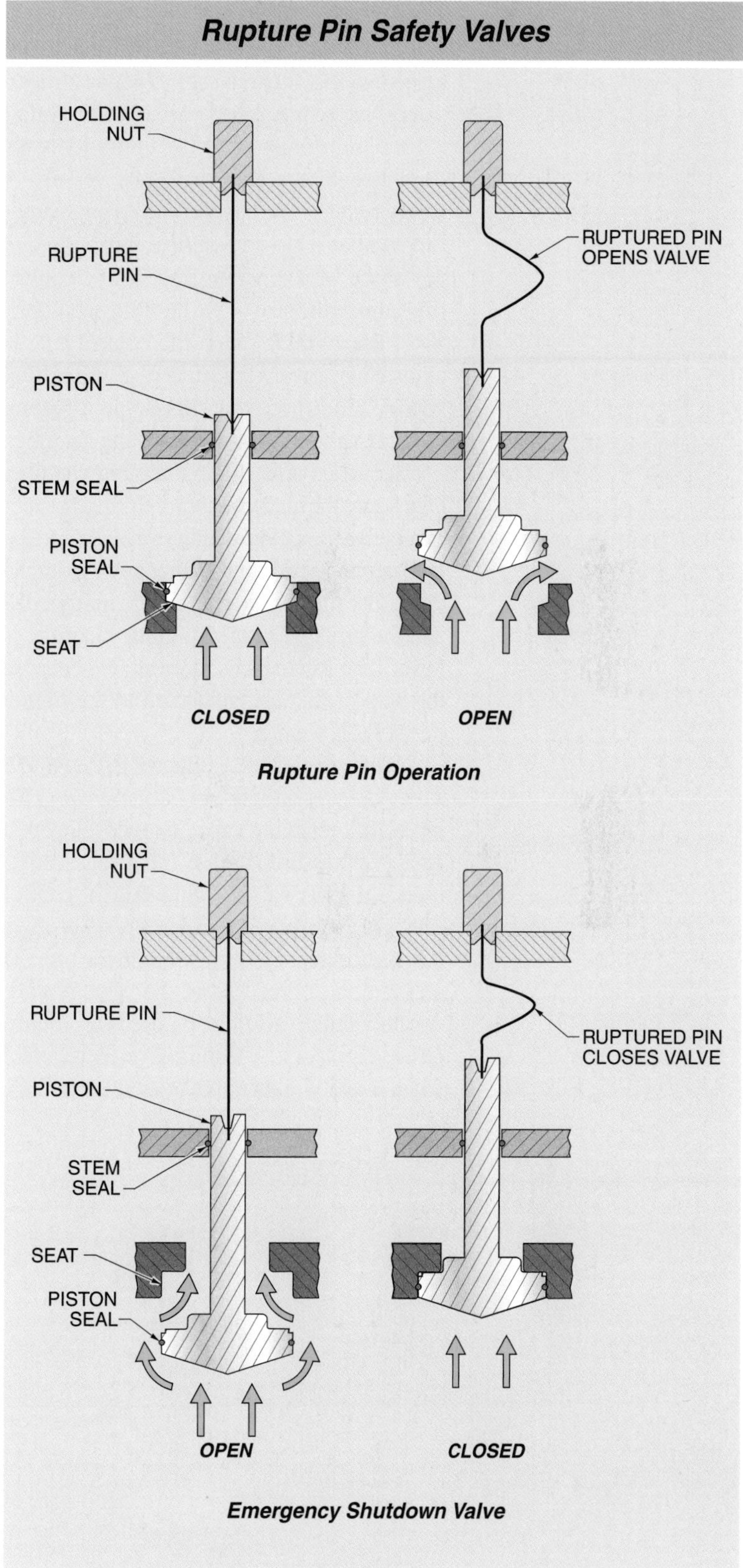

Figure 43-4. A rupture pin safety valve can be used to open a valve at a set pressure or as an emergency shutdown valve.

Safety Relief Valves

A *safety relief valve* is a valve that is designed so that it can be set to act as either a safety valve or a relief valve. Safety relief valves are usually used for unfired vessels such as compressed air receiver vessels. A safety relief valve has a control ring, sometimes called a blowdown ring, that is used to adjust the valve opening speed. The control ring adjusts the size of the huddling chamber. **See Figure 43-5.** When the control ring is adjusted up toward the valve disc, the exiting vapors are restricted at this point, allowing pressure to build up in the huddling chamber.

The added area of the huddling chamber exposed to the higher pressure makes the safety relief valve act like a safety valve. When the control ring is adjusted farther away from the valve disc, the exiting vapors are unimpeded and the pressure in the huddling chamber is lower. The valve opens more slowly because the area of the disc subjected to the higher pressures is smaller.

A safety relief valve adheres to the *ASME Boiler and Pressure Vessel Code, Section VIII,* where full relieving capacity is developed at 10% overpressure. A safety relief valve can be used as a safety valve by adjusting the control ring. The more pop action a safety relief valve has, the further below the setpoint the process pressure must drop before the valve reseats. A safety relief valve can be used for liquids, gases, or vapors. A testing lever is required when a safety relief valve is used for boilers.

Lockout/tagout procedures must be followed to ensure the safety of personnel working on equipment.

Hazardous Material Safety Relief Valves. A *hazardous material safety relief valve* is a special valve design that has the standard safety valve internal parts protected by an internal diaphragm seal. A calibrated breaking pin supports the diaphragm. When the pressure reaches the set value, the pin breaks and the diaphragm ruptures. The pressure then opens the valve. The valve reseats when the pressure drops below the set value. Hazardous material safety relief valves are approved for use on tank cars and tank trucks where standard safety valves are not approved for these services.

Valve Construction

Safety and safety relief valves may have an adjusting screw in the bonnet that has a limited ability to change the preset factory pressure. The manufacture and testing of these valves must conform to codes established by ASME International. Typically, small valves have threaded connections while larger valves have flanged or welded connections. The small ½″ and ¾″ valves combine the inlet nozzle and seat with the lower body housing. These valves are available in a limited selection of materials and in most cases are used for liquid thermal expansion services.

Safety relief valves 1″ and larger have a separate inlet nozzle screwed into the lower body housing. The nozzle base is flared out to provide the gasket surface for the inlet connection. The other end of the nozzle forms the valve seat. Nozzles are available in a wide selection of materials. The valve disc is made of the same materials as the nozzle and can be supplied with an O-ring to provide a tight closure after being opened. The outlet connection is larger than the inlet, is part of the upper body casting, and faces to the side. The bonnet, which contains the spring, can be either open or closed. A closed bonnet provides more protection for the spring assembly.

A balanced disc is required in applications where exiting vapors discharge to a fixed downstream pressure or develop a backpressure due to long discharge lines. The upper area of the disc is maintained at atmospheric pressure by a bellows that separates the disc and bonnet from the discharge piping system.

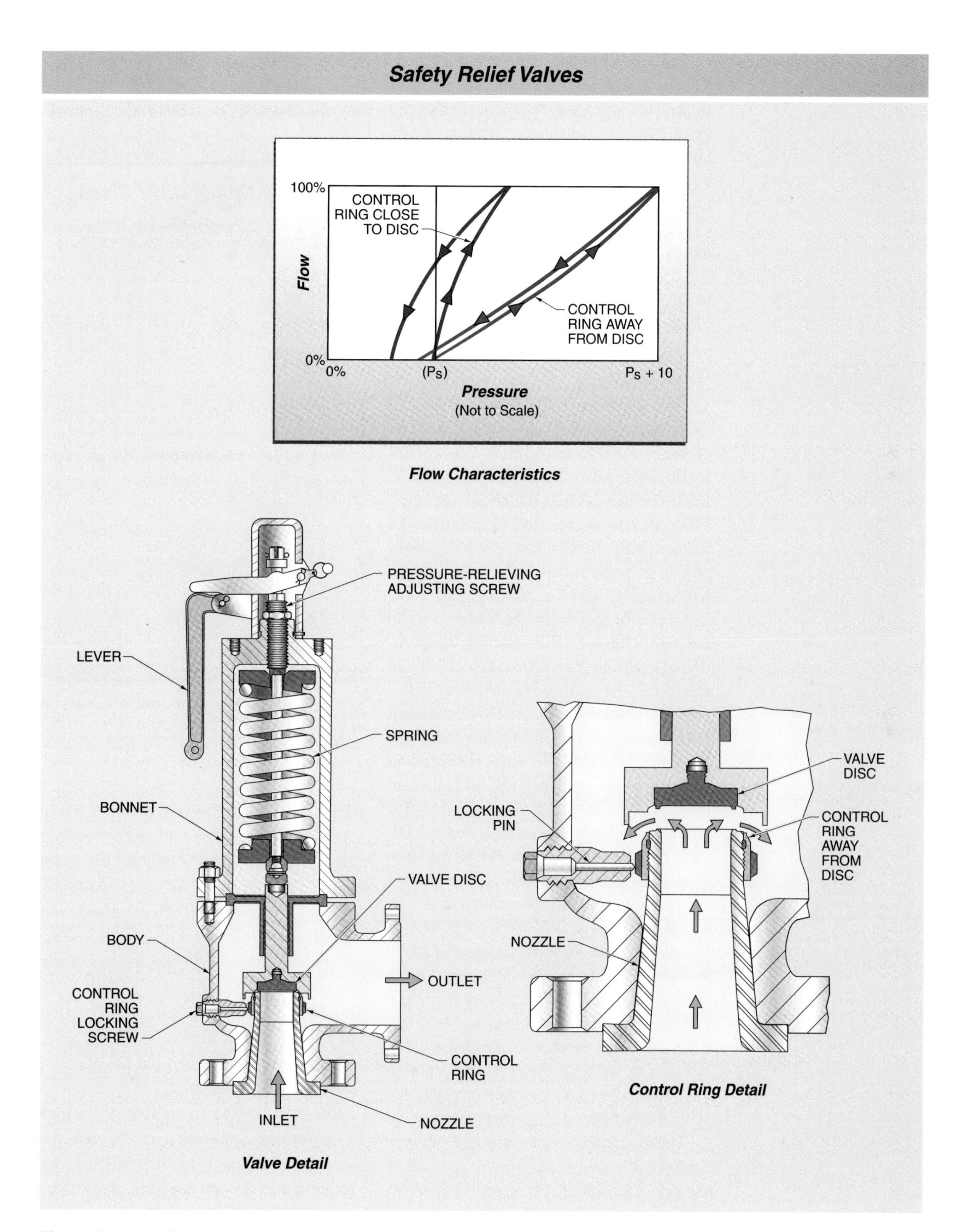

Figure 43-5. A safety relief valve uses a control ring to adjust the opening speed of the valve.

The inlet piping to the safety and relief valves should be as straight and as short as possible and be of the same size as the valve inlet. The discharge piping can be horizontal or make a 90° bend upward. In all cases the discharge should be directed toward a safe area away from personnel. The low point of any vertical discharge piping must have a small drain to remove collected moisture. The installation must be strong enough to withstand mechanical forces on the valve when it is discharging.

Valve Sizing

The sizing equations for safety, relief, and safety relief valves are available from any of the manufacturers of these valves. The equations are also contained in the ASME Code for valves used to relieve pressure. These equations are used to calculate the required relieving device nozzle area based on the set pressure, allowed overpressure, temperature, specific gravity, and other factors associated with the material to be relieved. This is a relatively simple operation of inserting the required values into the equation. The main difficulty is to determine the amount of material to be relieved.

The manufacturers of valves used to relieve pressure have agreed on a letter coding system to represent the various nozzle areas that are available. This letter code along with the valve inlet and outlet piping sizes identify a specific valve size. **See Figure 43-6.** The piping sizes vary somewhat depending on the type of safety relief valve and the manufacturer.

The relieving device set pressure is based on the weakest item in the system to be protected. The set pressure for a vessel is the design pressure. The ASME International certified pressure limit is one and one-half times the design pressure for pressure vessels. A relieving device used to protect a piping system must be set at a pressure lower than the lowest-rated piece of equipment in the system.

Thermal expansion relieving devices are set at the design pressure of the side of the exchanger being protected. A valve that relieves pressure must be sized to handle the flow from the worst possible relieving condition. Four common worst-case conditions are fire, runaway chemical reaction, blocked discharge, and thermal expansion.

Safety Relief Valve Sizes

Nozzle Area*	Safety Relief Valve Sizes (Inlet†, Nozzle Code, Outlet†)		
0.110	1	D	2
0.196	1	E	2
0.307	1½	F	2
0.503	1½	G	2½
0.785	1½	H	3
1.287	2	J	3
1.838	3	K	4
2.853	3	L	4
3.60	4	M	6
4.34	4	N	6
6.38	4	P	6
11.05	6	Q	8
16.0	6	R	8
26.0	8	T	10

* in sq. in.
† in in.

Figure 43-6. Safety relief valves are made in standard sizes with standard codes to describe the nozzle.

Fire. A fire condition is a situation where a pressurized vessel is in a fully involved external fire that transfers heat into the vessel contents and converts the contained liquid into vapor. The amount of heat transferred into atmospheric or low-pressure vessels can be calculated using equations in the American Petroleum Institute (API) 2000 standard, which takes into consideration the surface area of the vessel, the effective area subjected to the fire, and the effect of insulation on the vessel.

The API 2000 also contains equations to convert the fire energy, in Btu, into the equivalent flow rate of air. API 520 provides similar equations for use with pressurized vessels. The Compressed Gas Association (CGA) pamphlet S-1 part 3 contains equations that can be used to determine the flow rate of equivalent free air that needs to be released due to a fire condition.

Runaway Chemical Reaction. Chemical reactions in pressure vessels are usually exothermic reactions. An *exothermic reaction* is a chemical reaction that generates heat during the reaction and increases the temperature. As the chemicals warm up, the rate of reaction increases even more. This generates more heat. A runaway reaction can occur due to a loss of cooling or due to a reaction rate exceeding the cooling capacity. This excess heat raises the temperature of the vessel contents, converting the liquid to vapor. This vapor must be released from the vessel to prevent an overpressure situation.

Blocked Discharge. A blocked discharge situation occurs when all exits from a vessel or piping system are blocked and inlet flows that exceed the vessel or piping pressure limits must be relieved. A common application is the pressure protection of downstream piping and equipment due to an upstream pressure-reducing valve failure. The relieving device must be able to handle the fully open capacity of the pressure-reducing valve.

Thermal Expansion. Thermal expansion relief is for liquid-filled vessels or piping systems that can be sealed off and can absorb heat from either a heat source or from ambient temperature. A typical example is a process heat exchanger that uses water as a cooling medium. If the inlet and outlet of the water side of the exchanger are closed, heat transferred from the process fluid raises the temperature of the water and causes a severe pressure increase in the water that can easily damage the heat exchanger.

A small relief valve is needed to relieve the excess pressure. The flow capacity is not a factor since it only requires the relief of a small quantity of water to reduce the pressure. However, if the source of heat has a temperature high enough to cause the cool fluid to vaporize at the relieving pressure, then the relief valve must be sized to relieve the vapor.

Excess Flow Valves

An *excess flow valve* is a device that shuts off the flow if the flow exceeds a specific flow rate. Excess flow valves are usually used for liquid services, but they can also be used for gases and vapors. A common application is a gas station pump hose. Gas station pump hoses are all provided with excess flow valves to shut off the flow of gasoline if a hose and nozzle are damaged and start to spill.

Tank cars and tank trucks are equipped with excess flow valves on the liquid and vapor lines to contain the material in the tank if a hose is broken during unloading. The valve consists of only a ball in a housing. When the differential pressure across the ball becomes too high, the ball seals the outlet line. These valves do not release until the differential pressure across the ball becomes zero allowing the ball to drop.

Fuel Safety Shutoff Valves

A *fuel safety shutoff valve* is a special spring-actuated valve used to stop the fuel flow to a burner system. **See Figure 43-7.** Fuel safety shutoff valves are designed to be very reliable and to have no leakage. The internal valve trim is a sliding plate over a circular port.

Fuel Safety Shutoff Valves

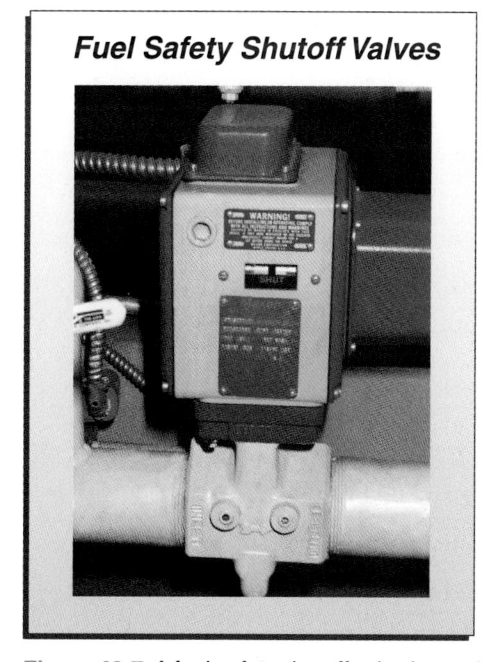

Figure 43-7. A fuel safety shutoff valve is used to stop the fuel flow to a burner.

Fuel safety shutoff valves are opened by either a manual cocking lever or by an electrical actuator. Many styles of electrical limit switches are available that can be used to confirm the valve position for the burner control system. The fuel safety shutoff valve can only be opened if the safety system energizes the latching solenoid. A failure of any one of the burner system permissive conditions de-energizes the latching solenoid, allowing the spring to close the valve.

The NFPA requires that fuel shutoff valve systems be installed in a double block-and-bleed configuration. A *double block-and-bleed fuel shutoff* is an arrangement with two fail-closed safety shutoff valves in series with a fail-open vent valve in a tee connection between the two. This valve arrangement is also called a double block-and-vent configuration. **See Figure 43-8.** The valves are opened and closed together in a programmed sequence. The sequence is established using the open and closed limit switches in the individual shutoff valves of the system.

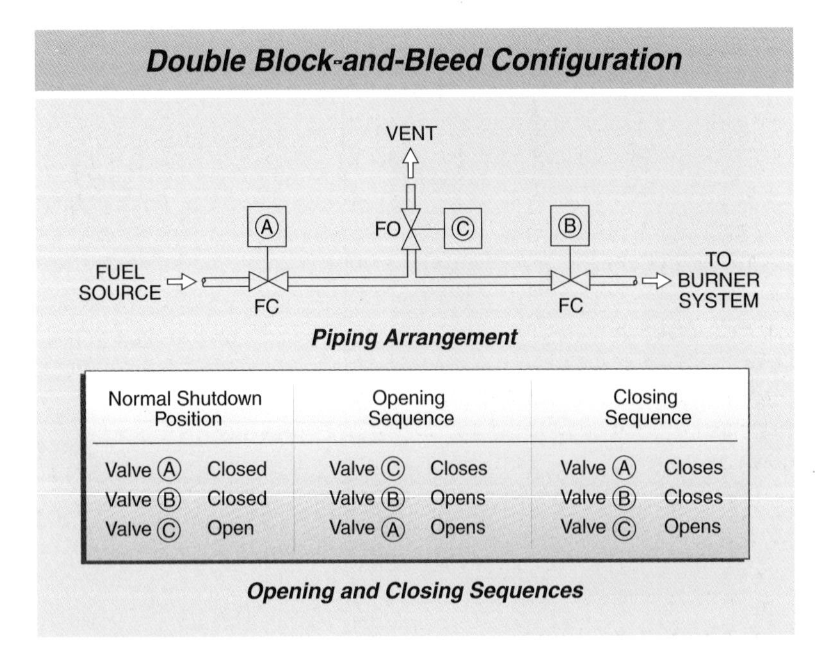

Normal Shutdown Position		Opening Sequence		Closing Sequence	
Valve Ⓐ	Closed	Valve Ⓒ	Closes	Valve Ⓐ	Closes
Valve Ⓑ	Closed	Valve Ⓑ	Opens	Valve Ⓑ	Closes
Valve Ⓒ	Open	Valve Ⓐ	Opens	Valve Ⓒ	Opens

Figure 43-8. A double block-and-bleed configuration is a safe way of protecting a fuel system from fuel accumulation in the lines.

The opening sequence is to close the vent valve, open the downstream shutoffvalve, and then open the upstream shutoffvalve. The closing sequence is to close the upstream shutoff valve, close the downstream shutoff valve, and then open the vent valve.

Conservation Vents

A *conservation vent* is a pressure relieving device used to protect low-pressure vessels from damage caused by changes in pressure. Trapped vapors need to be displaced when low-pressure storage vessels have liquids pumped into them, and air has to replace the liquid volume being removed when liquids are pumped out of the vessel. Pumps have more than enough discharge pressure to rupture the vessel or generate a negative pressure capable of collapsing the vessel if vents are not provided.

Conservation vents can be selected to relieve pressure, vacuum, or both. A conservation vent usually vents directly to the atmosphere or allows air to enter the vessel for vacuum applications. This is called an end-of-line style. In some installations, the conservation vent is piped to a vapor collection system designed to process the vessel's vapors before being discharged to the atmosphere. For those installations, the conservation vent is built with a flanged outlet connection. This is called a pipe-away style. The materials of construction of the conservation vents can be steel, aluminum, or various types of plastics and fiberglass.

The vessels for which the conservation vents are used have very low maximum allowable pressures. Therefore, the fluids contained in these vessels must have low vapor pressures. The conservation vent set pressure has to be set higher than the vapor pressure of the fluid. The set pressures are determined by weighted pallets. A *pallet* is a flat plate that covers the vapor outlet or air inlet ports of a conservation vent. The set pressure is the weight of the pallet divided by the port area. The pressures are usually expressed as ounces per square inch (osi). The pallets have a resilient surface where it seals against the ports to prevent leakage.

The size of the conservation vent is determined by the volume of vapors that have to be vented and the developed pressure over the set pressure. The developed pressure is the increase in pressure necessary to lift the pallet to relieve the necessary volume of vapor. The combination of the set pressure and the developed pressure has to be less than the maximum allowable pressure of the vessel.

Manufacturers can calculate the relieving capacity when they know the size, set pressure, and the maximum allowable vessel pressure. Alternatively, manufacturers can calculate the size when they know the relieving capacity, the set pressure, and the maximum allowable vessel pressure. There are families of curves showing the relieving capacity compared to the developed pressure at different set pressures for different sizes contained in the manufacture's catalogs.

The manufacturer's calculations and charts reflect actual test results. However, at low developed pressures and flows, the test results of flow capacities are inconsistent. Therefore, manufacturers generally do not provide flow data for certain ranges.

Flame Arresters

A *flame arrester* is a device used to prevent an external source of fire from propagating into a low-pressure vessel containing flammable vapor. Design details vary by manufacturer, but basically they work to reduce the combustion temperature below the vapor ignition point before the flame reaches the vessel interior. The temperature is reduced by transferring the heat of the burning vapors to the metal of the flame arrester. This is done by routing the vapors through narrow metal passages that have a large surface area in contact with the vapor.

Flame arresters can be used alone in vent piping or as a vessel's atmospheric vent in conjunction with conservation vents. The size of the flame arrester is based on the vapor flow and the allowable pressure drop. The length of the flame arrester element is determined by the vapor composition and the associated flame front velocity. *Flame front velocity* is the speed of the vapor burning edge traveling into a combustible mixture. Flame arrestors must be metallic, but they may be made of a variety of materials. A shortcoming of the flame arrester design is that it has to be at ambient temperature to function properly. If it becomes too hot it cannot remove sufficient heat to extinguish the burning vapor. Detonation arresters are similar in design to flame arrestors except that they are designed for high flame front velocities.

RUPTURE DISCS

A *rupture disc* is a safety device that breaks to open a discharge device and is used to prevent damage in pipelines and pressure vessels due to excessive pressure. A rupture disc is constructed of an impervious material supported in a structure mounted between a pair of flanges and works by bursting at a preset pressure. Rupture discs are fast, single-action devices that must be replaced after each action.

Rupture discs are often used in conjunction with a safety valve. The rupture disc can be mounted before or after a safety valve to protect the safety valve from corrosive process fluids. When a rupture disc is used with a safety valve, its rupture setting should be at the same value as the safety valve. There should also be a telltale to monitor the pressure in the space between the rupture disc and the safety valve. A buildup of pressure in the enclosed space due to a pinhole leak can prevent the disc from rupturing on a sudden increase in process pressure.

A rupture disc can also fail to burst because the normal operating pressure trapped between the rupture disc and the safety valve inlet opposes the excessive process pressure, reducing the applied differential. A telltale can be a restricted vent, a pressure switch connected to an annunciator, or a whistle. The restricted vent or whistle relieves the pressure buildup but may not be noticed. The pressure switch provides notification, but does not prevent the buildup of pressure.

Disc Types

There are several types of rupture discs. The choice of disc depends on the application. Common types of rupture discs include conventional metal, reverse buckling, composite, and graphite composite rupture discs.

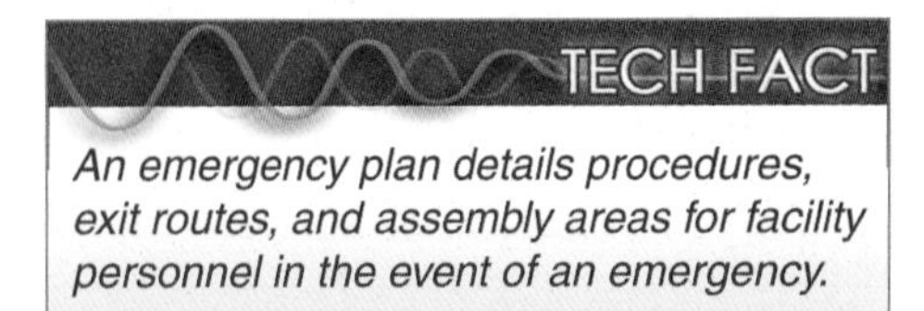

An emergency plan details procedures, exit routes, and assembly areas for facility personnel in the event of an emergency.

Conventional Metal Rupture Discs. A *conventional metal rupture disc* is a dished, thin metal sheet that is placed between a pair of holders with the process pressure applied to the concave (hollow) side of the disc. **See Figure 43-9.** The assembly is mounted between standard pipe flanges. The relieving pressure is determined by the metal thickness and tensile strength. A wide variety of materials are available for the discs. Frequent process pressures greater than 80% of the set pressure applied to the disc can cause premature failure.

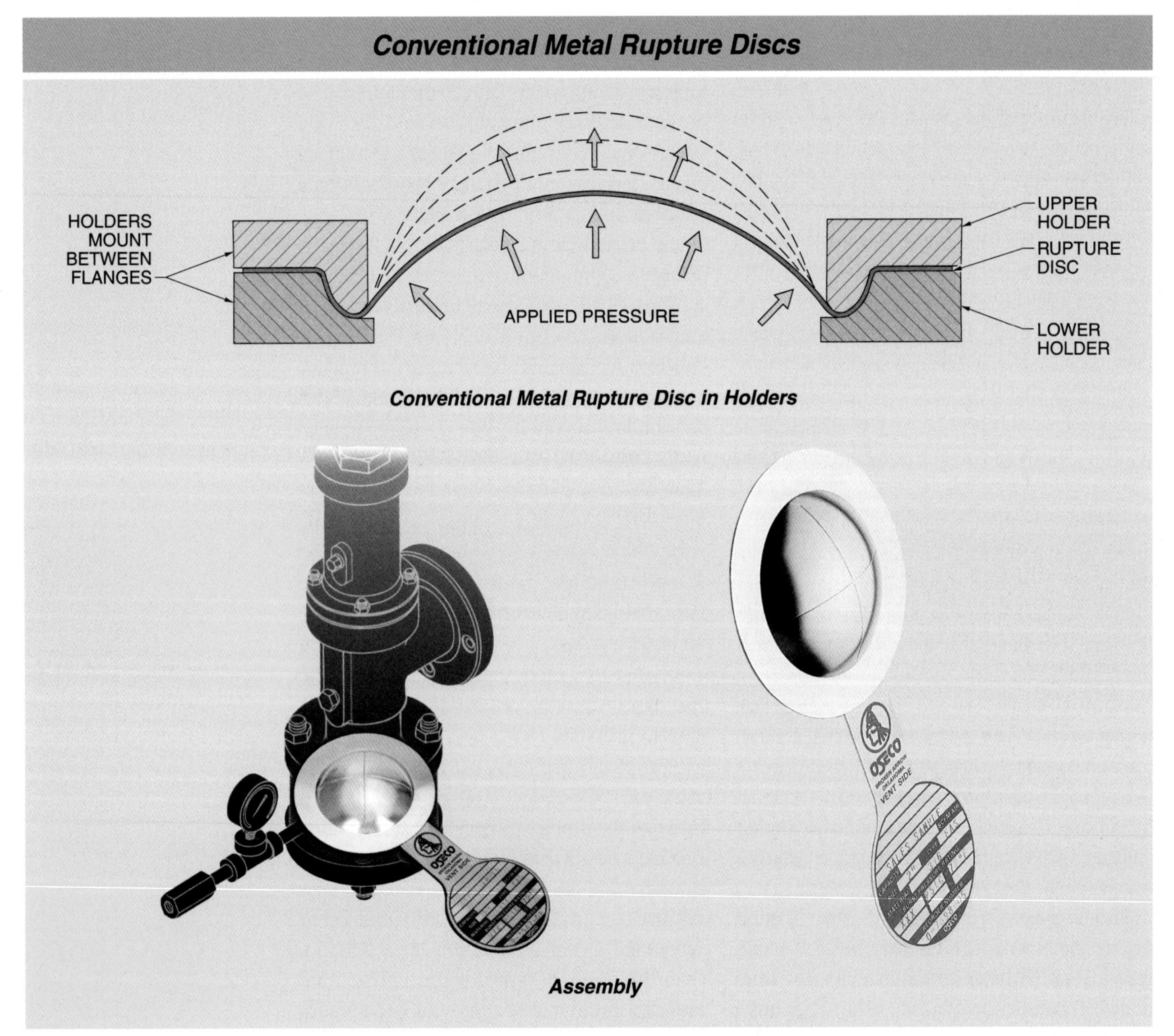

Figure 43-9. The use of a conventional metal rupture disc is a relatively inexpensive way to relieve pressure in a vessel.

This type of disc tends to release broken pieces downstream and cannot be used under a safety relief valve. If vacuum conditions can be present in the process, a vacuum support should be used with the disc to prevent its failure under vacuum conditions. This is needed because the disc can fail at a lower pressure applied to the convex side.

Reverse Buckling Rupture Discs. A *reverse buckling rupture disc* is a dished, thin metal sheet that is placed between a pair of holders with the process pressure applied to the convex side and with the downstream holder having a set of pointed knife blades. **See Figure 43-10.** When the disc starts to buckle from the pressure, it is pushed against the blades, which then cut the disc cleanly. Bulged rupture discs are weaker when the pressure is applied to the convex side, thus the disc material can be made a little thicker and the rupture point can be determined more precisely. In addition, the disc does not need a vacuum support because the convex side of the disc faces the pressure.

A reverse buckling disc can be subjected to pressures up to 90% of the relieving pressure without damaging the disc. Reverse buckling discs can be installed under a safety valve to protect the valve from corrosion attack. Reverse buckling discs do not release loose pieces that can become trapped within the safety valve. However, great care must be taken when using reverse-buckling rupture discs. The knife blades can become dull from corrosion and fail to open at the design pressure. Regular maintenance checks are required.

Composite Rupture Discs. A *composite rupture disc* is a dished, thin metal sheet that is perforated to provide weak points and combined with a thin polymer sheet to prevent leakage. **See Figure 43-11.** Lower rupture pressures can be obtained with a composite disc than with a conventional rupture disc, since the metal is weakened by the perforations. The perforations also control the breakage pattern to minimize loose pieces. A vacuum support is also available for these discs if they are used in vacuum service.

Graphite Composite Rupture Discs. A *graphite composite rupture disc* is a rupture disc composed of graphite powder in a polymer cement and is used where there are corrosive conditions. Additional corrosion protection can be provided by a TFE coating applied to the process side. **See Figure 43-12.** Graphite rupture discs are also available to handle both pressure and vacuum relief in the same device. When graphite discs rupture, they release chunks of material downstream.

Rupture Disc Sizing and Selection

The process of determining the required capacity of a rupture disc follows the same guidelines described for valves that relieve pressure. In fact, the same calculations are used to determine the vapor release rate. The equations to determine the required size of the rupture disc are available from the various manufacturers of rupture discs. The results of these equations are influenced by the same factors described for valves.

Part of the selection process of rupture discs is determining whether the process fluid will be a gas or a liquid. Rupture discs designed for the gas phase only cannot be used for the liquid phase. For example, many reverse buckling rupture discs are applied in gas conditions only. They depend on the energy stored in the compressed gas to reverse the disc and cause it to rupture. For liquids, this type of energy is not available and the rupture disc does not operate properly.

Operating Ratio. An *operating ratio* is the ratio of process operating pressure to rupture disc burst pressure. Rupture discs are available in a variety of predetermined operating ratios. The higher the operating ratio, the more expensive the rupture disc is. A rupture disc design with a higher operating ratio than needed for a process is a waste of money. Furthermore, the service life is reduced if a rupture disc is placed in a working condition where the designed operating ratio is higher than the needed operating ratio. Therefore, any rupture disc that is not designed for the needed operating ratio should not be used in that application.

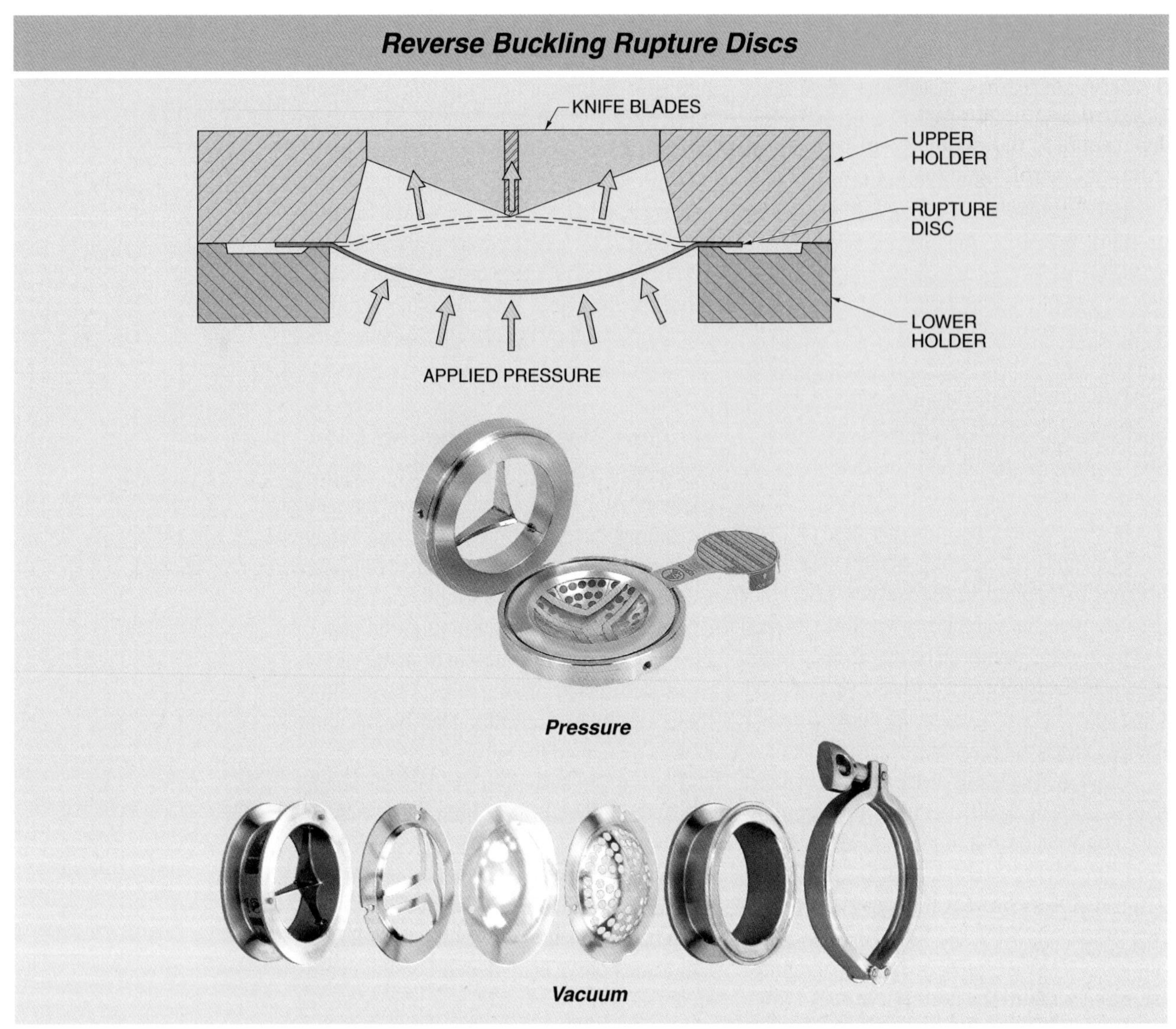

Figure 43-10. A reverse buckling disc puts the disc metal under compression loading for a more precise rupture point.

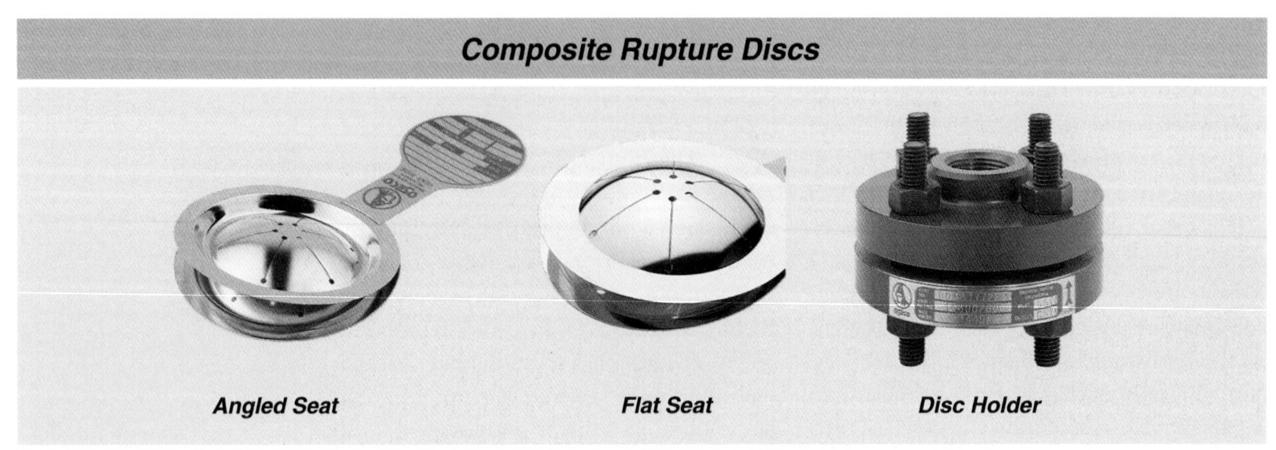

Figure 43-11. A composite rupture disc prevents debris and metal shards from blowing downstream from the disc assembly.

BURNER CONTROL SYSTEMS

A *burner control system* is a specially designed solid-state electronics package that provides the sequencing and safety logic for controlling industrial burners. **See Figure 43-13.** A burner control system is designed to cover various levels of complexity and accept the signals from various types of flame detectors. A burner control system sequences and monitors safe startup, normal operation, and shutdown of the burner system. Burner control systems do not control the burning rate or duration. The burning rate and duration are controlled by other instrument systems. Burner control systems can be used with programmable logic controller (PLC) or safety PLC systems.

Graphite Composite Rupture Discs

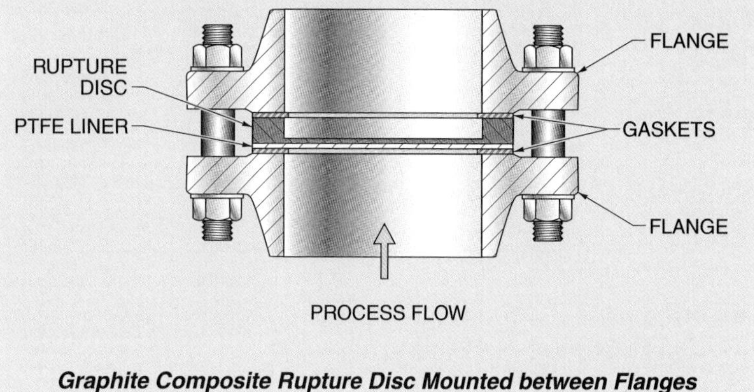

Graphite Composite Rupture Disc Mounted between Flanges

Figure 43-12. A graphite composite rupture disc is very resistant to corrosive chemicals.

TECH FACT

Boiler fittings are components directly attached to a boiler that are required for the operation of the boiler.

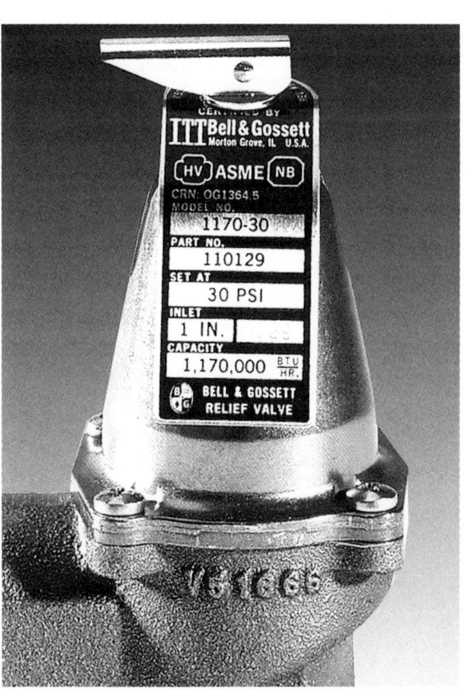

Bell & Gossett

Relief valves are manufactured and tested according to ASME Code requirements.

Flame Detectors

A *flame detector* is a device that can provide a usable signal to a burner control system by detecting the presence of a flame. The purpose of a flame detector is to prevent the accumulation of a combustible mixture in the burner chamber and eliminate the possibility of an explosion. The flame detector is used to ensure that a flame is present so that the burner control system can allow the ignition or continued burning of combustible fuels. Flame detectors are designed to detect an actual flame and not just a hot surface. If no flame is detected, the burner control system prevents the main gas valve from opening or shuts down the main gas valve and pilot.

The most common industrial flame detectors are those based on the emission of ultraviolet (UV) radiation from flames. **See Figure 43-14.** The combustion of carbon-based fuels emits enough UV radiation to be detected. It is necessary for the detector to actually "see" the flame, so the unit needs to be mounted close to and in direct line of sight of the flame. Sometimes the unit needs extra cooling, such as a water jacket, to protect the device from the heat.

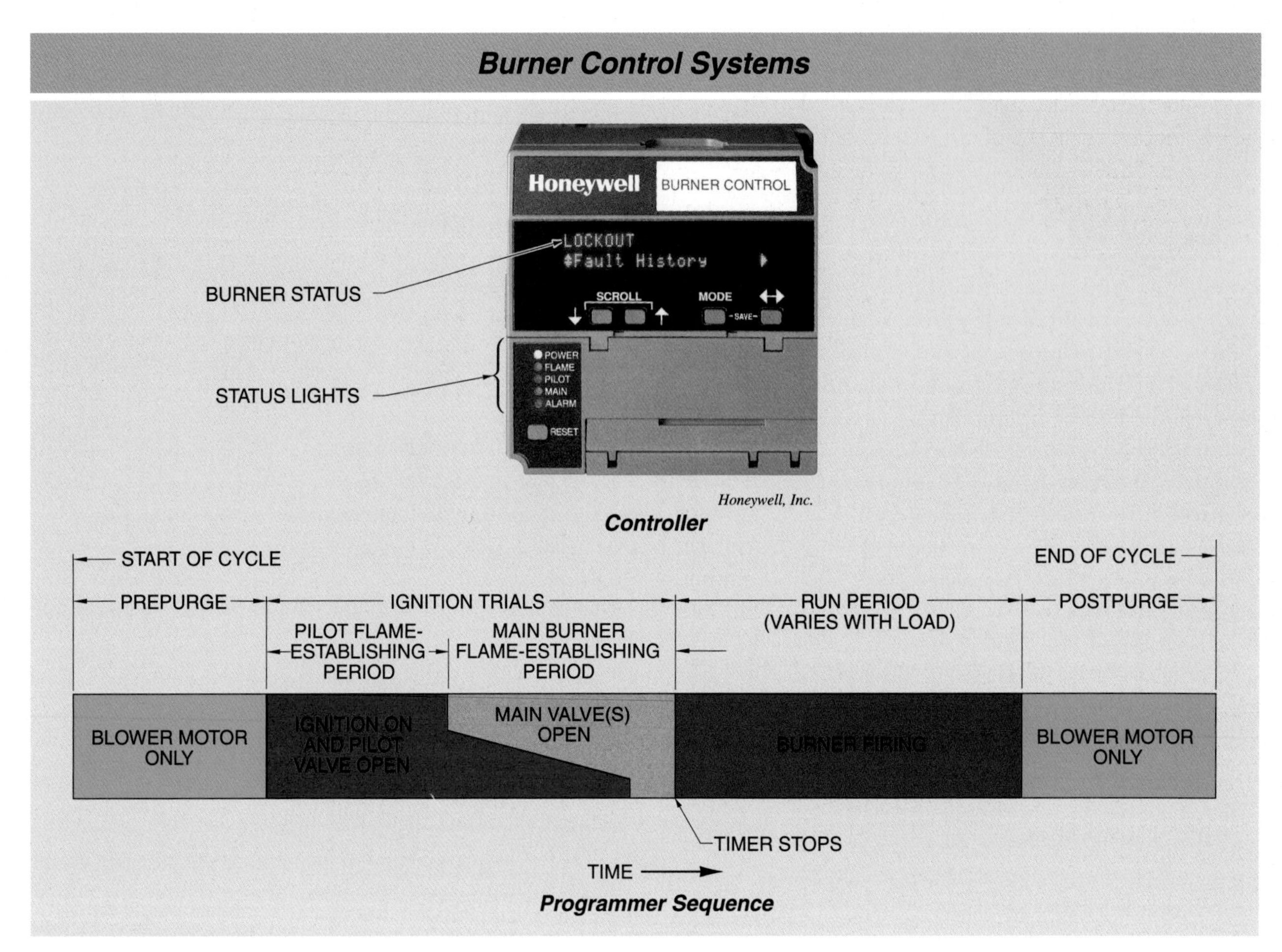

Figure 43-13. A burner control system is used to sequence the startup, operation, and shutdown of a boiler.

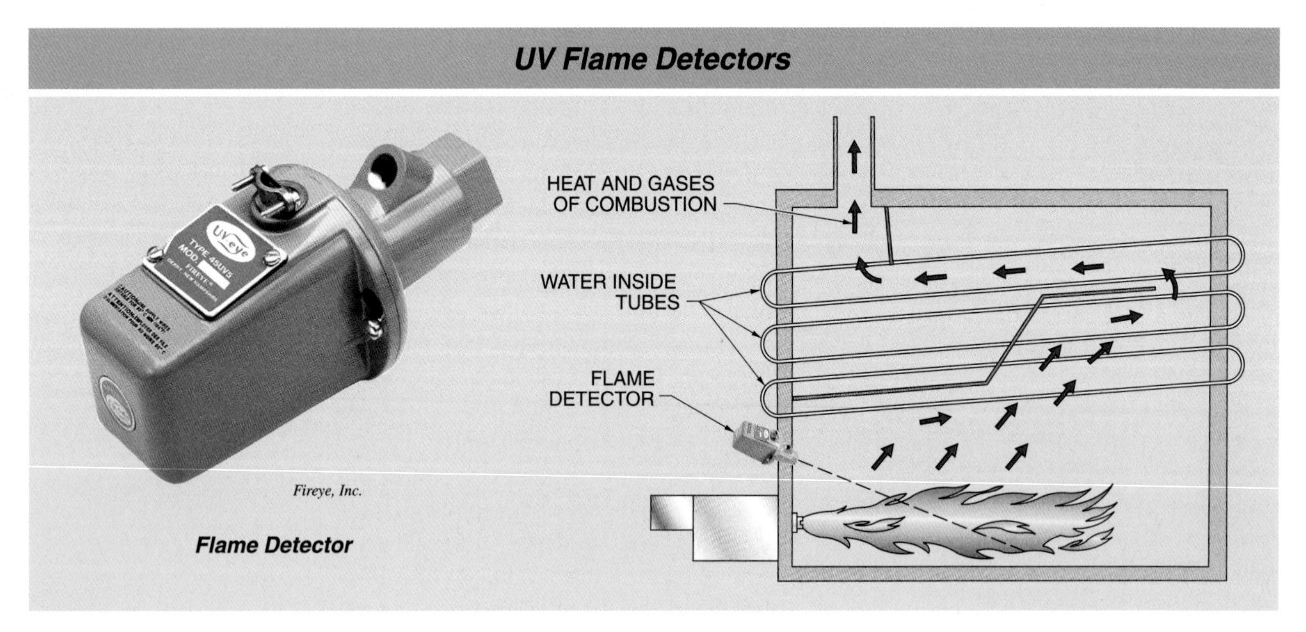

Figure 43-14. A UV flame detector is a safety device that allows a combustion operation to continue only when a flame is detected.

There are other sources of UV radiation that can also be detected by a flame detector. Care must be taken to have the detector aimed at a flame and not at any of these other sources. Some of these other sources of UV include hot refractory materials above 2300°F, bright incandescent light, lasers, and high-voltage coronas. In addition, smoke, dirt, dust, and oil mist can block the radiation and attenuate the signal, causing false alarms. There are other optical flame detectors based on visible light and infrared (IR) radiation. In addition, there are flame rods that must be in direct contact with the flame. **See Figure 43-15.**

Flame detectors for commercial applications are generally simpler than process industry flame detectors. The most common flame detector for commercial applications like home furnaces, hot water tanks, clothes dryers, and other similar burners is a thermocouple. Thermocouples are usually positioned to be heated by the pilot flame. Other types of flame sensors that are used are filled system and expanding metal rod temperature switches.

ALARM SYSTEMS

An *alarm system* is a device that provides a visual and audible signal to the operator that something in the process is not normal. The alarm input is usually a discrete ON/OFF signal that can originate from a monitoring device such as a pressure, temperature, or limit switch. The alarm contact actions can be either "close to alarm" or "open to alarm." *Open to alarm* is an alarm that provides a fail-safe alarm signal and should always be used with safety systems. *Close to alarm* is an alarm that can be used for normal nonsafety alarms. This arrangement minimizes nuisance alarms, but may fail to provide an alarm signal if there is a broken wire or communications failure.

Chemical plant security is a mandatory part of the International Council of Chemical Associations (ICCA) Responsible Care® doctrine for all member companies.

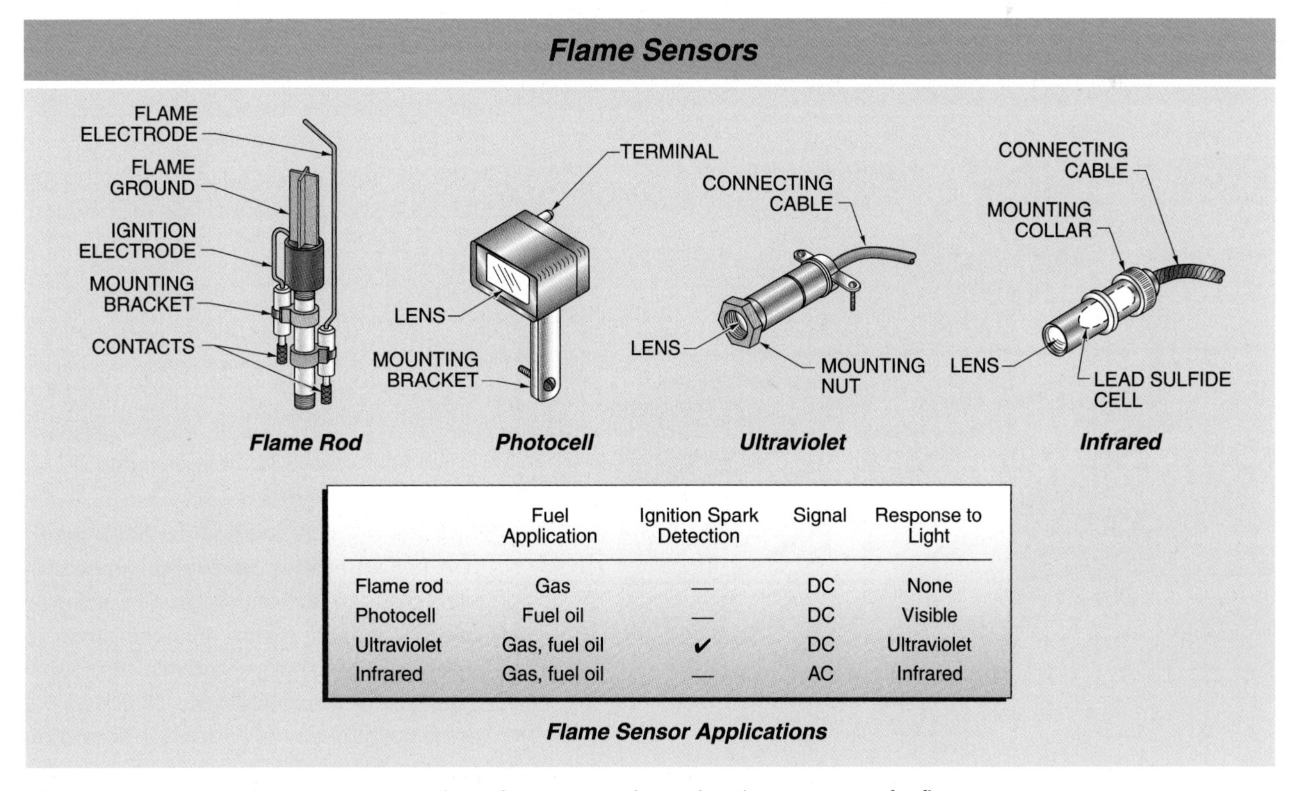

	Fuel Application	Ignition Spark Detection	Signal	Response to Light
Flame rod	Gas	—	DC	None
Photocell	Fuel oil	—	DC	Visible
Ultraviolet	Gas, fuel oil	✔	DC	Ultraviolet
Infrared	Gas, fuel oil	—	AC	Infrared

Figure 43-15. Flame detectors use a variety of sensors to determine the presence of a flame.

Annunciators

An *annunciator* is a solid-state or relay-based system for monitoring and alarming plant process operations. Annunciators are stand-alone devices that can work with any control system. An annunciator has back-lighted windows marked with a description of the type of alarm and provides an audible signal when there is an alarm condition. **See Figure 43-16.** An alarm state typically results in a flashing light behind the appropriate window and an audible signal. Pressing a SILENCE pushbutton shuts off the audible signal and switches the flashing light to steady. The steady light remains lit until the alarm condition clears.

When using annunciators, the term "normal" means the normal operational state of the process. While an annunciator alarm may be in a normal state, the alarm system electrical contacts may be either normally open (NO) or normally closed (NC).

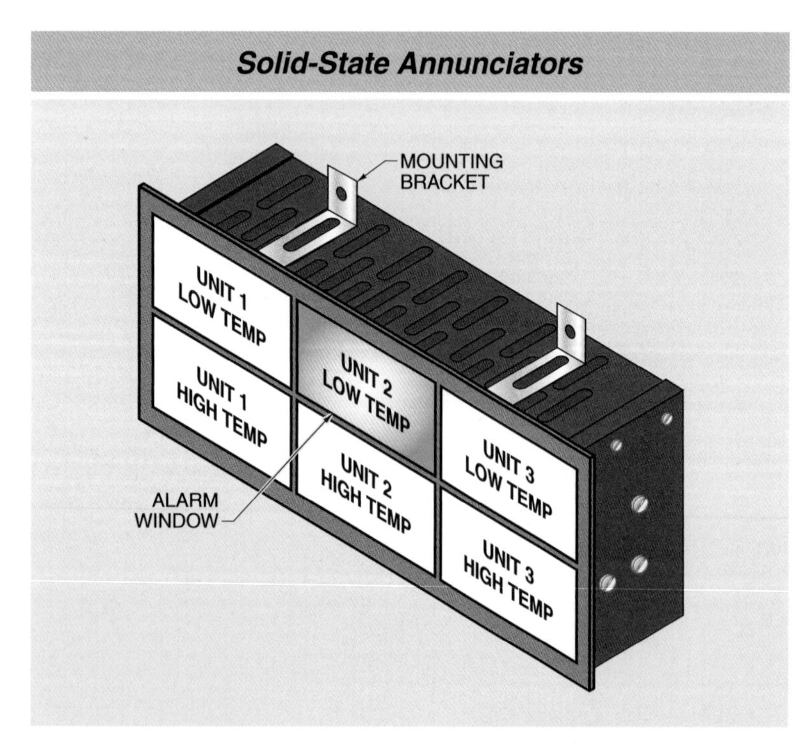

Figure 43-16. Annunciators combine audible and visual alarms into one system.

Standards have been developed that provide a description for all the light and audible signal actions used for alarms and subsequent alarms. **See Figure 43-17.** Many of these standards are for the operational sequences. For example, sequence M-1-2 has an audible alarm and a flashing light upon an alarm condition. After the operator acknowledges the alarm, the audible alarm is silenced, but the flashing light continues. After the operator resets the flash, the audible alarm remains silenced and the flashing light changes to a steady light. When the process returns to normal, the light remains on until the operator resets to normal.

Digital Alarm Systems

A *digital alarm system* is an alarm that provides the same functions as an annunciator except that the functions are accomplished by programming a distributed control system (DCS) or PLC. In addition to the normal visual and audible signals, digital alarm systems have several very significant advantages.

Digital alarm systems record the time and date of all alarm actions as well as the time and date the alarm is cleared. The digital alarm system maintains this record until the memory is filled, at which point the oldest alarm records are overwritten. If an accident occurs, the whole alarm history is recorded and is available to help determine the sequence of events. Digital alarm systems can have individual alarms assigned to one of many levels of alarm severity classifications.

First-Out Alarms

A *first-out alarm* is a feature that is available in annunciators and digital alarm systems that can identify the first of multiple closely tripped alarms. It is common in safety systems that a shutdown signal also causes several other alarms to trip as the process shuts down. Since it is important to know what the initiating factor was for the shutdown, a method called first-out was developed. A group of alarms are linked together so that only the first one tripped flashes, while the others light but do not flash. There are other special sequences that utilize two different speeds of flashing and reset pushbuttons. Any annunciator manufacturer catalog provides detailed information on all the available alarm sequences.

HAZARDOUS ATMOSPHERE DETECTORS

Atmospheres can become hazardous due to the accidental release of combustible gases or toxic chemicals such as chlorine, hydrogen sulfide, or carbon monoxide. It is the purpose of hazardous atmosphere detectors to detect small traces of these gases when they are at concentrations too low to ignite or at concentrations well below harmful levels. Early detection of hazardous conditions can allow plant operators to minimize releases and provide greater safety.

Combustible Gas Detectors

A *combustible gas detector* is a hazardous atmosphere detector used to measure low concentrations of combustible gases and vapors in the atmosphere. Combustible gas detectors are usually placed in and around locations where combustible gases or vapors can be present during abnormal operations. The type of combustible gas expected to be present is determined by the selection of location and elevation. It is vital to detect the presence of a combustible atmosphere before it can spread into an area where it could be ignited.

The *lower explosive limit (LEL)* is the lowest concentration of a combustible gas or vapor in air that can be ignited. The *upper explosive limit (UEL)* is the highest concentration of a combustible gas or vapor in air that can be ignited. The lower flammable limit (LFL) and upper flammable limit (UFL) are sometimes used in place of the LEL and UEL. Different combustible gases or vapors have different LELs and UELs. **See Figure 43-18.**

Typical Annunciator Operational Sequences

M-1-2

Basic Flashing, Separate Flasher Reset, and Manual Reset	Test or Alert	Acknowledge	Flash Reset	Return to Normal	Reset to Normal
Visual	Flashing	Flashing	Steady ON	Steady ON	OFF
Audible	ON	OFF	OFF	OFF	OFF

F3M-3

Tri-Flash, First Out Manual Reset	Test or First Alert	Subsequent Alert	Acknowledge	First Out Reset	Return to Normal	Reset to Normal
First Out Visual	Intermittent Fast Flashing	Intermittent Fast Flashing	Slow Flashing	Steady ON	Steady ON	OFF
Subsequent Visual	OFF	Fast Flashing	Steady ON	Steady ON	Steady ON	OFF
Audible	ON	ON	OFF	OFF	OFF	OFF

Figure 43-17. Annunciator operational sequences are the states of the lights and audible alarms during the various steps of an alarm sequence.

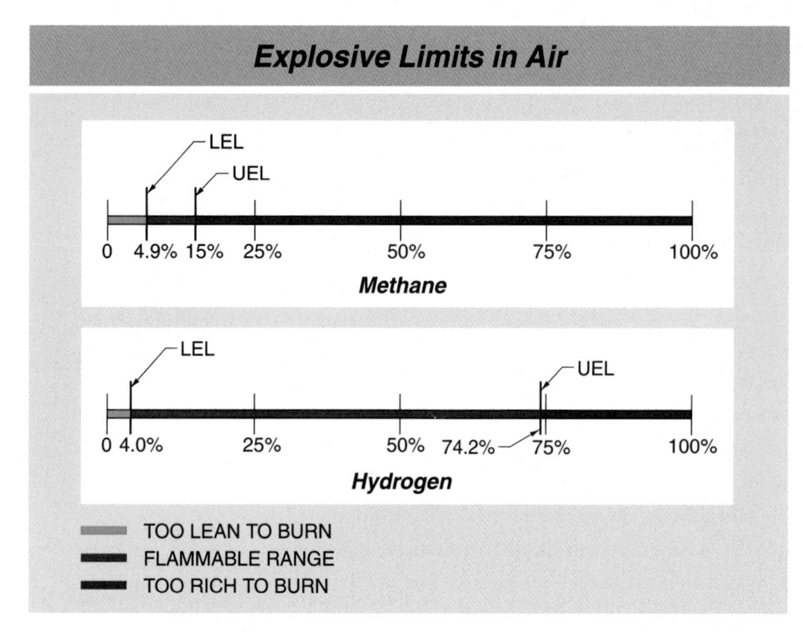

Figure 43-18. Upper and lower explosive limits in air vary considerably for different chemicals.

Catalytic Combustible Sensors. The sensors used in combustible gas detectors need to be able to respond to any combustible material. A *catalytic combustible sensor* is a common type of combustible gas sensor that uses a pair of resistors encased in ceramic beads, with the sensor element coated with a catalyst and the reference element coated with impervious glass. **See Figure 43-19.**

In the presence of air, the catalytic material on the sensor element promotes the reaction of any combustible material at temperatures low enough that the sensor cannot initiate an explosion. The reaction heats the sensor element to a temperature greater than the reference element. The difference in temperature is a measure of the concentration of combustibles in the air.

Catalytic sensors only work in the presence of sufficient oxygen to support combustion and are not very accurate for very low levels of combustible gases. Certain materials that may be present, such as lead or silicone, can damage the sensor catalyst.

A catalytic combustible gas sensor usually provides an analog signal expressed as a percentage of the LEL of methane (natural gas). Other sensor designs provide only a discrete ON/OFF signal when there are combustible gases or vapors present above a predetermined concentration. Exposure to high concentrations of a combustible gas or vapor can permanently disable a sensor.

Semiconductor MOS Sensors. Another common type of combustible gas detector is a sensor consisting of a semiconductor silicon material with a strip of metal oxide semiconductor (MOS) coating. Each end of the MOS coating has an electrode attached and the resistance of the coating is measured. The resistance of the MOS coating changes when the sensor is heated in the presence of combustible materials. **See Figure 43-20.** The semiconductor sensors are very sensitive and are primarily used for low concentrations of combustible gases. Some sensors have heaters with a high demand for power and require large batteries or a wired power source.

Infrared Sensors. A less common type of combustible gas detector is the IR gas detector. It has an infrared sensor that uses an IR radiation source to send a beam through a stream of gas to a mirror. The mirror reflects the beam back to a detector that measures the attenuation at a particular wavelength. The amount of attenuation is proportional to the concentration of gases in the gas stream. An IR gas detector only works with gases that absorb IR radiation. These detectors work in low oxygen levels but are relatively expensive.

Toxic Gas Detectors

A *toxic gas detector* is a hazardous atmosphere detector used to measure toxic gases or vapors that are not combustible but are harmful to people. This category includes chemicals such as chlorine, phosgene, ammonia, carbon monoxide, and many other gases. The sensor must be selected and installed to be responsive to a single chemical. Toxic gas detectors are used in and around areas in a plant where there is a likelihood of an inadvertent release of a toxic gas or vapor to the atmosphere. Early detection is imperative in order to determine the size of the release and the direction of drift so that corrective actions can be taken.

A *permissible exposure limit (PEL)* is a regulatory limit on the amount of allowed workplace exposure to a hazardous chemical. The Occupational Safety and Health Administration (OSHA) establishes a PEL based on an 8-hour time-weighted average exposure. Each different chemical is evaluated for toxicity and a PEL is established.

Other government agencies also regulate exposure to chemicals. For example, carbon monoxide is a poisonous byproduct of combustion processes. The OSHA PEL for this chemical is 50 parts per million (ppm) of air averaged over an 8-hour shift. The National Institute for Occupational Safety and Health (NIOSH) has recommended that the PEL for carbon monoxide should be 35 ppm with a maximum exposure of 200 ppm.

There are several designs of toxic gas detectors. A simple and relatively inexpensive toxic gas analyzer uses a sensing cell with an inert dispersal barrier or flame arrestor with an internal chemical sensitive to the desired hazardous gas or vapor. The toxic gas or vapor can migrate through the barrier and activate the cell. **See Figure 43-21.** The cell is set for a specific concentration range of the toxic material and the output is proportional to the concentration. This design can be permanently disabled by exposure to high concentrations of the toxic material and then the cell must be replaced.

Another design uses a chemical reagent that is placed in contact with a sample gas or vapor in a detector cell. A pump is used to carry an atmospheric sample into the analyzer. The reagent is specific for the selected hazardous gas or vapor and reacts with the gas if it is present. The cell generates an electrical signal dependent on the concentration of the toxic gas or vapor. The reagent needs to be replaced periodically to ensure the proper operation of the analyzer.

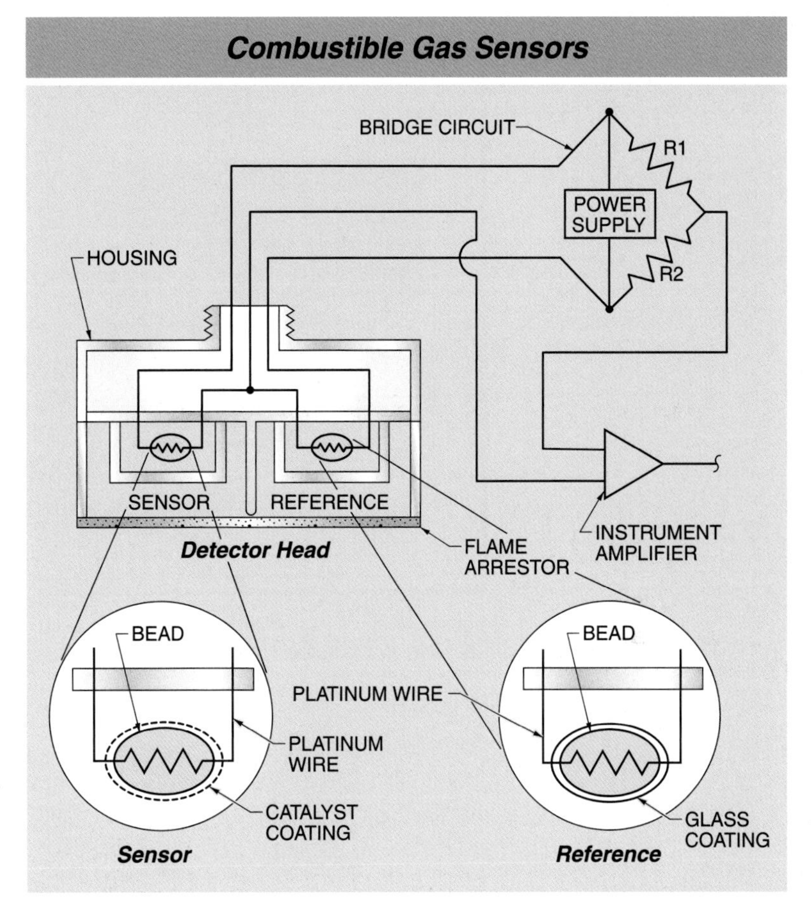

Figure 43-19. A combustible gas sensor detects the presence of a flammable gas before the concentration becomes high enough to become a danger.

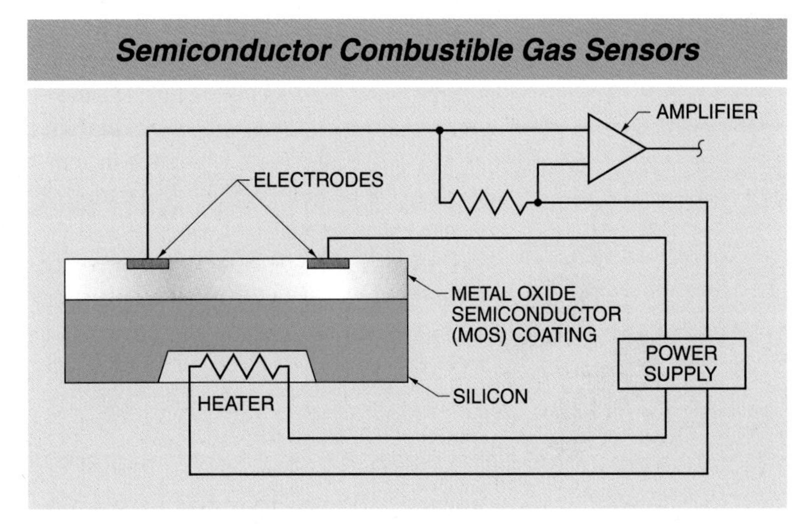

Figure 43-20. A semiconductor combustible gas sensor measures changes in the resistance of the MOS coating to detect combustible gases.

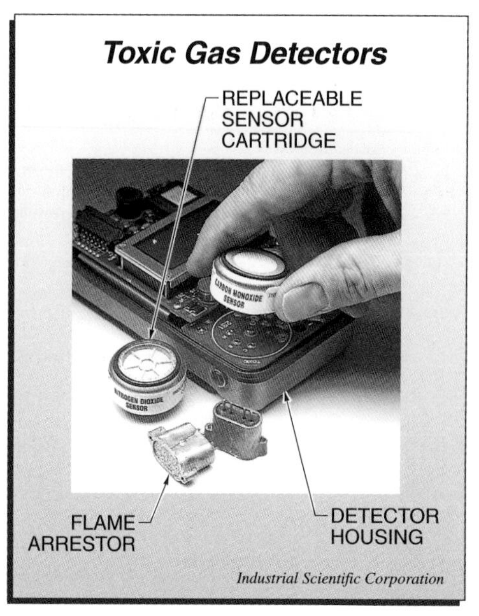

Figure 43-21. A toxic gas detector uses replaceable sensor cartridges to detect different gases.

Fire suppression systems are an important safety system for facilities that use flammable materials.

KEY TERMS

- *safety system:* A system that consists of an individual device or an assembly of devices that form a system designed to protect personnel from injury and production equipment from damage.
- *personal protective equipment (PPE):* Any clothing or device worn by a worker to prevent injury.
- *safety valve:* A gas- or vapor-service valve that opens very quickly when the inlet pressure exceeds the spring setpoint pressure.
- *huddling chamber:* A recessed area in a safety valve disc that increases the surface area and the total force applied.
- *blowdown,* or *blowback:* The amount of pressure drop below the pressure setpoint of a safety valve.
- *overpressure:* The amount of pressure above the setpoint of a safety valve necessary to develop the full relieving capacity and is expressed as a percentage of the set pressure.
- *safety valve capacity:* The amount of steam, in lb/hr, that a safety valve is capable of venting at the rated pressure.
- *rupture pin safety valve:* A safety valve with no spring; it is held closed by a pin or a thin rod.
- *relief valve:* A valve that opens in proportion to the pressure above a setpoint.
- *safety relief valve:* A valve that is designed so that it can be set to act as either a safety valve or a relief valve.

KEY TERMS (continued)

- *hazardous material safety relief valve:* A special valve design that has the standard safety valve internal parts protected by an internal diaphragm seal.
- *exothermic reaction:* A chemical reaction that generates heat during the reaction and increases the temperature.
- *excess flow valve:* A device that shuts off the flow if the flow exceeds a specific flow rate.
- *fuel safety shutoff valve:* A special spring-actuated valve used to stop the fuel flow to a burner system.
- *double block-and-bleed fuel shutoff:* An arrangement with two fail-closed safety shutoff valves in series with a fail-open vent valve in a tee connection between the two.
- *conservation vent:* A pressure relieving device used to protect low-pressure vessels from damage caused by changes in pressure.
- *pallet:* A flat plate that covers the vapor outlet or air inlet ports of a conservation vent.
- *flame arrester:* A device used to prevent an external source of fire from propagating into a low-pressure vessel containing flammable vapor.
- *flame front velocity:* the speed of the vapor burning edge traveling into a combustible mixture.
- *rupture disc:* A safety device that breaks to open a discharge device and is used to prevent damage in pipelines and pressure vessels due to excessive pressure.
- *conventional metal rupture disc:* A dished thin metal sheet, placed between a pair of holders, with the process pressure applied to the concave (hollow) side of the disc.
- *reverse buckling rupture disc:* A dished thin metal sheet placed between a pair of holders, with the process pressure applied to the convex side and with the downstream holder having a set of pointed knife blades.
- *composite rupture disc:* A dished thin metal sheet, perforated to provide weak points and combined with a thin polymer sheet to prevent leakage.
- *graphite composite rupture disc:* A rupture disc composed of graphite powder in a polymer cement and is used where there are corrosive conditions.
- *operating ratio:* The ratio of process operating pressure to rupture disc burst pressure.
- *burner control system:* A specially designed solid-state electronics package that provides the sequencing and safety logic for controlling industrial burners.
- *flame detector:* A device that can provide a usable signal to a burner control system by detecting the presence of a flame.
- *alarm system:* A device that provides a visual and audible signal to the operator that something in the process is not normal.
- *open to alarm:* An alarm that provides a fail-safe alarm signal and should always be used with safety systems.
- *close to alarm:* An alarm that can be used for normal nonsafety alarms.
- *annunciator:* A solid-state or relay-based system for monitoring and alarming plant process operations.
- *digital alarm system:* An alarm that provides the same functions as an annunciator except that the functions are accomplished by programming a distributed control system (DCS) or PLC.
- *first-out alarm:* A feature that is available in annunciators and digital alarm systems that can identify the first of multiple closely tripped alarms.

KEY TERMS *(continued)*

- *combustible gas detector:* A hazardous atmosphere detector used to measure low concentrations of combustible gases and vapors in the atmosphere.
- *lower explosive limit (LEL):* The lowest concentration of a combustible gas or vapor in air that can be ignited.
- *upper explosive limit (UEL):* The highest concentration of a combustible gas or vapor in air that can be ignited.
- *catalytic combustible sensor:* A common type of combustible gas sensor that uses a pair of resistors encased in ceramic beads, with the sensor element coated with a catalyst and the reference element coated with impervious glass.
- *toxic gas detector:* A hazardous atmosphere detector used to measure toxic gases or vapors that are not combustible but are harmful to people.
- *permissible exposure limit (PEL):* A regulatory limit on the amount of allowed workplace exposure to a hazardous chemical.

REVIEW QUESTIONS

1. How are safety valves, relief valves, and safety relief valves defined, and where might each be used?
2. What are several worst-case conditions that must be considered when sizing pressure-relief devices?
3. What is a rupture disc, and what are several common types of rupture discs?
4. What is a burner control system, and how does it work with a flame detector?
5. Define lower explosive limit (LEL) and upper explosive limit (UEL).

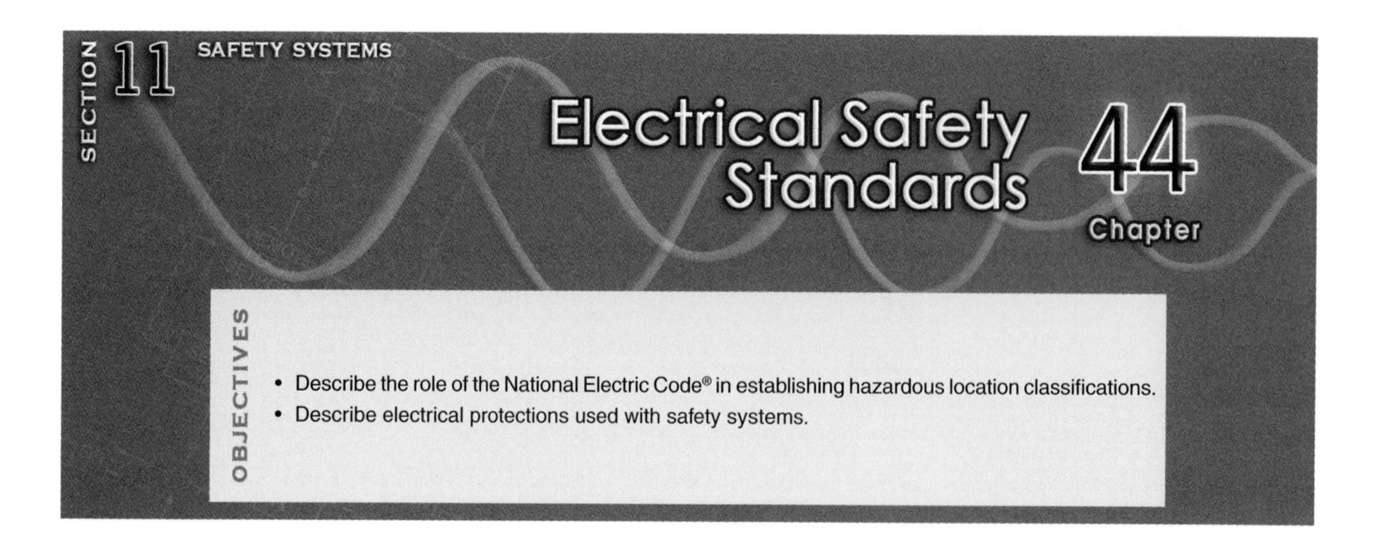

NATIONAL ELECTRICAL CODE® (NEC®)

Many processes handle combustible liquids, gases, vapors, or dusts that can be present in the atmosphere during portions of the operation and can be ignited by electrical energy. To prevent the ignition of combustible material by electrical energy, a number of standards have been developed for the design and installation of electrical systems.

The National Electrical Code® (NEC®) is sponsored by the NFPA. The primary function of the NEC® is to safeguard people and property against electrical hazards. The NEC® is not intended to be used as an instruction manual nor a design specification. It contains the rules and necessary provisions for electrical systems, but leaves the design and layout to others. The NEC® provides the definitions of the various hazardous classifications and the electrical hardware and the appropriate installations required for those classifications.

Hazardous Location Classifications

The classification system evaluates the various types of hazardous substances and groups them according to their type and potential for hazard. Electrical classifications for hazardous locations are identified by class, division, and group. **See Figure 44-1.** It is the responsibility of knowledgeable personnel at each manufacturing facility to specify the appropriate hazardous zone, location, size, and classifications. Each area has specific wiring requirements.

Classes. A *class* is a term used to define the general nature of a hazardous gas or vapor in the surrounding atmosphere. A *Class I location* is a hazardous location in which sufficient quantities of flammable gases and vapors are present in the air to cause an explosion or ignite hazardous materials. A *Class II location* is a hazardous location in which a sufficient quantity of combustible dust is present in the air to cause an explosion or ignite hazardous materials. A *Class III location* is a hazardous location in which easily ignitable fibers, or flyings, are present in the air but not in a sufficient quantity to cause an explosion or ignite the hazardous materials.

Divisions. A *division* is a classification assigned to each class based upon the likelihood of the presence of a hazardous substance in the atmosphere. Each of the three classes contains two divisions. A *Division 1 location* is a hazardous location in which the hazardous substance is normally present in the air in sufficient quantities to cause an explosion or ignite hazardous materials. The Division 1 location classification also includes areas where service or maintenance operations could cause ignitable concentrations of flammable gases or vapors to occur frequently.

Hazardous Location Classifications	
Classes	***Likelihood that a flammable or combustible concentration is present***
I	Sufficient quantities of flammable gases and vapors present in air to cause an explosion or ignite hazardous materials
II	Sufficient quantities of combustible dust are present in air to cause an explosion or ignite hazardous materials
III	Easily ignitable fibers or flyings are present in air, but not in a sufficient quantity to cause an explosion or ignite hazardous materials
Divisions	***Location containing hazardous substances***
1	Hazardous location in which hazardous substance is normally present in air in sufficient quantities to cause an explosion or ignite hazardous materials
2	Hazardous location in which hazardous substance is not normally present in air in sufficient quantities to cause an explosion or ignite hazardous materials
Groups	***Atmosphere containing flammable gases or vapors or combustible dust***
	Class I
A	Acetylene
B	Hydrogen, fuel, and combustible process gases containing more than 30% hydrogen by volume or gases of equivalent hazard such as butadiene, ethylene oxide, propylene oxide, and acrolein
C	Ethyl ether and ethylene or gases of equivalent hazard
D	Acetone, ammonia, benzene, butane, cyclopropane, ethanol, gasoline, hexane, methane, methanol, naphtha, natural gas, propane, or gases of equivalent hazards
	Class II
E	Combustible metal dusts, including aluminum, magnesium, and their commercial alloys or other combustible dust whose particle size, abrasiveness, and conductivity present similar hazards in connection with electrical equipment
F	Carbonaceous dusts, coal black, charcoal, coal, or coke dusts that have more than 8% total entrapped volatiles or dusts that have been sensitized by other material so they present an explosion hazard
G	Flour dust, grain, wood, plastic, and chemicals
	Class III
None	

Figure 44-1. The NEC® divides hazardous locations into classes, divisions, and groups based on the chemicals handled in that area.

A *Division 2 location* is a hazardous location in which the hazardous substance is not normally present in the air in sufficient quantities to cause an explosion or ignite hazardous materials. The Division 2 location classification is often applied to areas near Division 1 locations where hazardous substances can migrate or move into the location. Division 2 locations are also those areas that only have flammable vapors or gases during abnormal conditions.

Groups. A *group* is a term used to classify an atmosphere containing particular flammable gases or vapors or combustible dust. Groups A to D are associated with Class I. *Group A* is the term used for an atmosphere containing acetylene. *Group B* is the term used for an atmosphere containing hydrogen, fuel, and combustible process gases containing 30% hydrogen by volume, or gases or vapors of equivalent hazard such as butadiene, ethylene oxide, propylene oxide, and acrolein.

Enclosures can contain electrical components in addition to electrical wiring.

Group C is the term used for an atmosphere containing ethyl ether, ethylene, or gases or vapors of equivalent hazard. *Group D* is the term used for an atmosphere containing acetone, ammonia, benzene, butane, cyclopropane, ethanol, gasoline, hexane, methane, methanol, naphtha, natural gas, propane, or gases or vapors of equivalent hazard.

Groups E to G are associated with Class II. *Group E* is the term used for an atmosphere containing metal dusts such as aluminum or magnesium dust. *Group F* is the term used for an atmosphere containing carbonaceous dusts such as carbon black, coal, or coke dust. *Group G* is the term used for an atmosphere containing dust not included in E and F. This includes dusts from foodstuffs such as flour or grain, or from plastics and chemicals.

There is no group designation associated with Class III. This Class typically includes wood fibers such as excelsior, textile fibers such as cotton or sisal, and synthetic fibers such as rayon.

ELECTRICAL PROTECTION

There are four types of electrical protection commonly used for instruments in North America. These types of electrical protection are used to protect personnel and property from the dangers of explosive gases. *Nonincendive protection* is a type of protection in which electrical equipment is incapable, under normal conditions, of causing the ignition of a specified flammable gas or vapor in an air mixture due to arcing or thermal effect. This includes glass-encased mercury switches, reed switches, or fixed wiring.

Dust ignition-proof protection is a type of protection that excludes ignitable amounts of dust or amounts that can affect performance or rating. When electrical equipment is installed and protected in accordance with code requirements, it does not allow arcs, sparks, or heat otherwise generated or liberated inside the enclosure to cause ignition of exterior accumulations or atmospheric suspensions of a specified dust.

Explosionproof protection is a type of protection that uses an enclosure that is capable of withstanding an explosion of a gas or vapor within it and preventing the ignition of an explosive gas or vapor that may surround it. The enclosure operates at an external temperature such that a surrounding explosive gas or vapor cannot be ignited.

Intrinsically safe protection is a type of protection in which electrical equipment, under normal or abnormal conditions, is incapable of releasing sufficient electrical or thermal energy to cause the ignition of a specific hazardous atmospheric mixture in its most easily ignitable concentration. Intrinsically safe equipment contains electrical components that could be a source of ignition, but the electrical energy is limited to safe levels and the piece of equipment has been tested to prove it is intrinsically safe.

Electrical Enclosures

An *enclosure* is a case or housing for equipment or other apparatus that provides protection for controllers, motor drives, or other devices. **See Figure 44-2.** The National Electrical Manufacturers Association (NEMA) classifies enclosures according to their purpose and intended resistance to the ingress of moisture and hazardous materials. In the United States, equipment is tested to the NEMA 250 standard. Some enclosures are designed to include cooling units to remove heat generated by the enclosure contents. The most common types of enclosures are discussed below. There are other, specialized NEMA classifications that are not discussed here.

The total cross-sectional area of the wires in a conduit seal used in classified areas must not exceed 25% of the area of the rigid metal conduit unless otherwise approved.

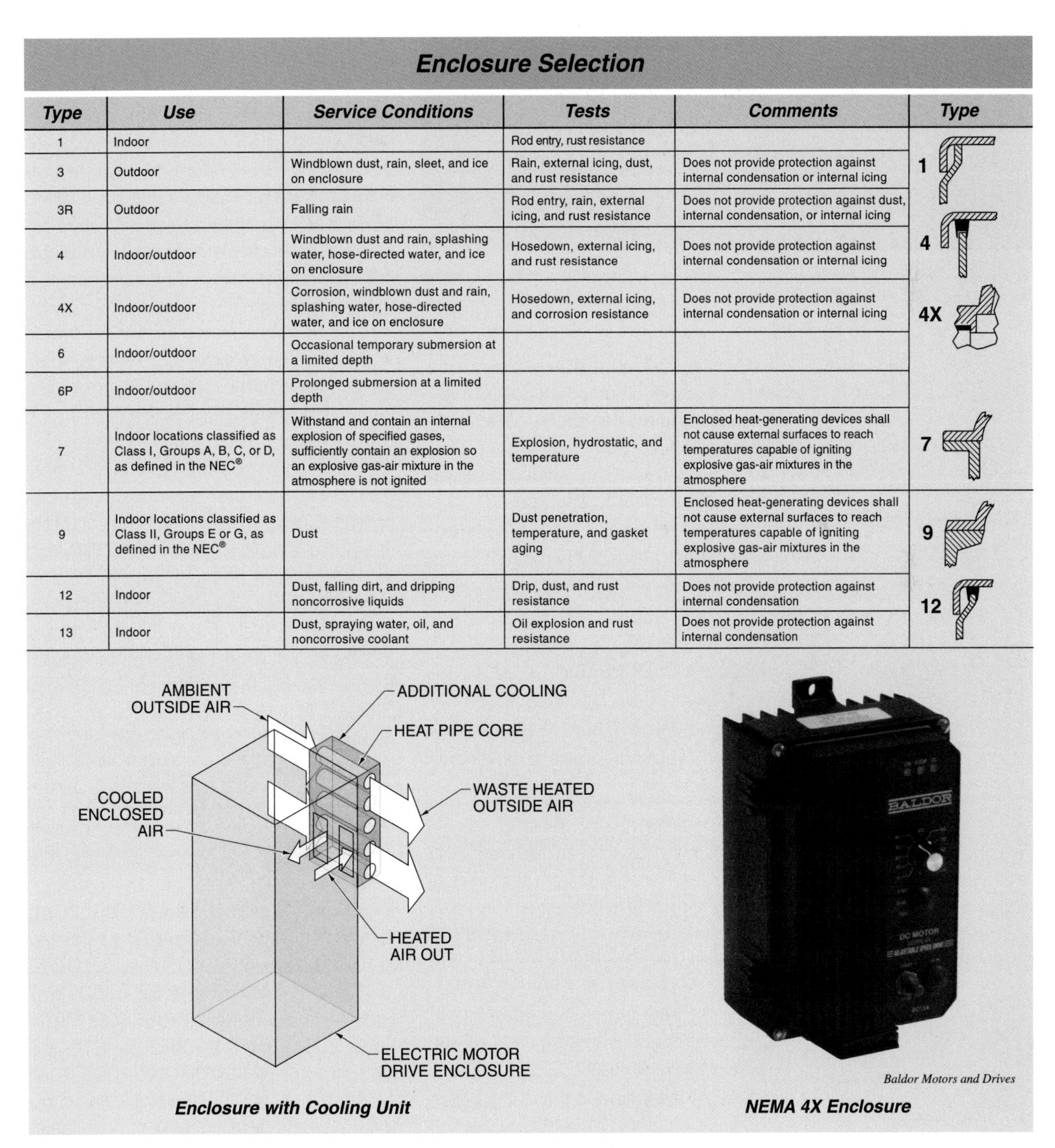

Enclosure Selection

Type	Use	Service Conditions	Tests	Comments
1	Indoor		Rod entry, rust resistance	
3	Outdoor	Windblown dust, rain, sleet, and ice on enclosure	Rain, external icing, dust, and rust resistance	Does not provide protection against internal condensation or internal icing
3R	Outdoor	Falling rain	Rod entry, rain, external icing, and rust resistance	Does not provide protection against dust, internal condensation, or internal icing
4	Indoor/outdoor	Windblown dust and rain, splashing water, hose-directed water, and ice on enclosure	Hosedown, external icing, and rust resistance	Does not provide protection against internal condensation or internal icing
4X	Indoor/outdoor	Corrosion, windblown dust and rain, splashing water, hose-directed water, and ice on enclosure	Hosedown, external icing, and corrosion resistance	Does not provide protection against internal condensation or internal icing
6	Indoor/outdoor	Occasional temporary submersion at a limited depth		
6P	Indoor/outdoor	Prolonged submersion at a limited depth		
7	Indoor locations classified as Class I, Groups A, B, C, or D, as defined in the NEC®	Withstand and contain an internal explosion of specified gases, sufficiently contain an explosion so an explosive gas-air mixture in the atmosphere is not ignited	Explosion, hydrostatic, and temperature	Enclosed heat-generating devices shall not cause external surfaces to reach temperatures capable of igniting explosive gas-air mixtures in the atmosphere
9	Indoor locations classified as Class II, Groups E or G, as defined in the NEC®	Dust	Dust penetration, temperature, and gasket aging	Enclosed heat-generating devices shall not cause external surfaces to reach temperatures capable of igniting explosive gas-air mixtures in the atmosphere
12	Indoor	Dust, falling dirt, and dripping noncorrosive liquids	Drip, dust, and rust resistance	Does not provide protection against internal condensation
13	Indoor	Dust, spraying water, oil, and noncorrosive coolant	Oil explosion and rust resistance	Does not provide protection against internal condensation

Enclosure with Cooling Unit *NEMA 4X Enclosure*

Figure 44-2. NEMA classifies electrical enclosures according to intended use.

The choice of enclosure for electrical equipment depends on the environment in which the enclosure will be located.

Enclosures for Nonhazardous Locations. The most common enclosure types for nonhazardous locations include the following:

- Type 1 enclosure—For indoor use with no unusual service conditions
- Type 3 enclosure—For outdoor use; sealed tight to windblown dust and rain and is sleet- and ice-resistant
- Type 3R enclosure—The same as a Type 3 enclosure with the addition of protection against falling rain
- Type 4 enclosure—For indoor and outdoor use; sealed tight to windblown dust and rain, splashing water, hose-directed water, and ice
- Type 4X enclosure—The same as a Type 4 enclosure with the addition of 316 SS, fiberglass, or other plastic construction for corrosion resistance
- Type 6 enclosure—For indoor and outdoor use; intended for occasional or temporary submersion to a limited depth
- Type 12 enclosure—For indoor use; sealed to dust, falling dirt, and dripping noncorrosive fluids
- Type 13 enclosure—For indoor use; sealed tight to dust, spraying water, oil, and noncorrosive coolant

Purged Enclosures. A *purged enclosure* is a nonhazardous enclosure that is pressurized and purged with air to allow it to be used in hazardous areas. Some enclosures cannot be made explosionproof. This may be because of the required size or because of non-explosionproof instruments that must be mounted in the face of the cabinet. However, such an enclosure must be located in a hazardous area. **See Figure 44-3.**

The NEC® allows an enclosure to be rated one classification higher if it is purged with air. Thus an enclosure and exposed instruments that are normally classified as General Purpose can be rated for Class I, Division 2 areas when purged with air. Class I, Division 2 enclosures and instruments can be rated for Class I, Division 1 areas if they are purged.

Purged cabinets are usually provided with a fast purge flow that is used when the cabinet is initially closed to ensure that there are no combustible vapors remaining in the cabinet. The cabinets are provided with low-pressure relief valves to handle the excess flow and limit the maximum pressure. The purged cabinets must be kept at a positive pressure, about 0.2″ WC, and monitored by a pressure switch. Loss of pressure initiates an alarm and can terminate power to the cabinet. Purged cabinets in Class I, Division 2 areas can be opened without having the power shut off. Purged cabinets in Class I, Division 1 areas are required to have the power shut off if the cabinet is opened.

Enclosures are available as predesigned and assembled purging system packages to satisfy Class I, Division 1 or Division 2 areas.

Enclosures for Hazardous (Classified) Locations. Type 7 enclosures are for use in Class I, Division 1 or 2, Group A, B, C, or D environments for indoor use. Type 7 enclosures are designed to contain an internal explosion of specified gases. Type 9 enclosures are for use in Class II, Division 1 or 2, Group E, F, or G environments for indoor use. Type 9 enclosures are sealed to keep out dust.

Enclosure Temperature Codes. A *temperature code* is a designation that specifies the maximum surface skin temperature obtained by an enclosure during testing by approval agencies. **See Figure 44-4.** Any equipment that does not exceed 185°F (65°C) is not required to be marked with the temperature code. The allowed surface temperature limit is determined by the degree of difficulty in igniting a flammable mixture and the amount of heat capacity in the enclosure that prevents a temperature rise and a resulting internal explosion.

Hazardous Area Wiring

The wiring requirements in hazardous areas are all defined in the NEC®. If there are ever any differences between the NEC® and other standards, the NEC® is the final authority. A sampling of a few of the major requirements is covered in the following sections. This does not take the place of the NEC®.

Conduit Seals. Conduit seals must be installed properly to ensure that the hazardous substance does not migrate through the raceway system. Several factors are involved in selecting the proper seal fitting for the application. Seal fittings can be designed for either vertical or horizontal mounting in the conduit system and in male and female configurations. In some cases, the seal fittings may be mounted in any position.

Conduit seals must be installed in accordance with the manufacturer's instructions and the requirements of the NEC®. These instructions may contain additional installation and location specifications. For example, the manufacturer's instructions may give the preferred side of a classified area boundary where the seal shall be installed. Always read the manufacturer's instructions when installing conduit seals. In addition, the sealing compound used to construct the seal also contains a set of installation instructions and directions for making the seal. Always read and follow these instructions to make a good conduit seal.

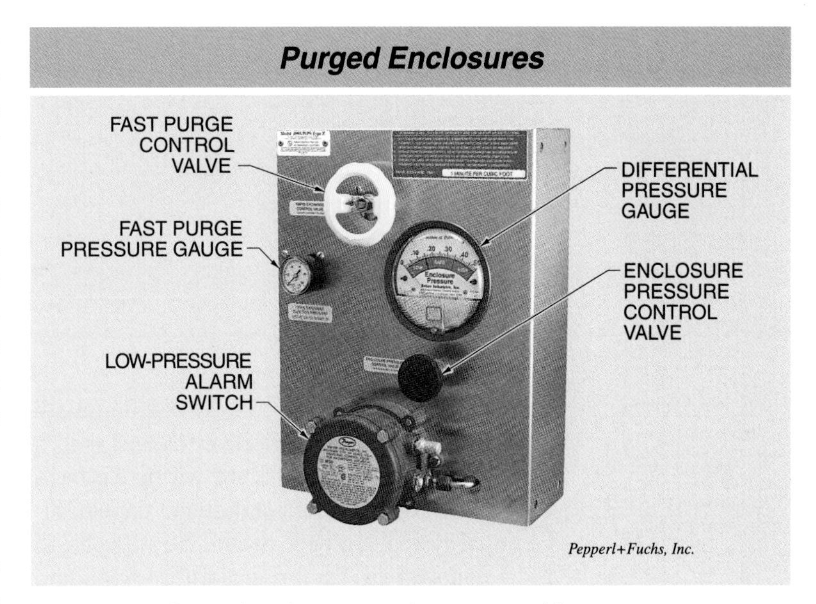

Figure 44-3. Purged enclosures are kept at a positive pressure to prevent flammable gases from entering the enclosure.

Enclosure Surface Temperature Codes

Class I	
Maximum Surface Temperature (°F)	**Identification Number**
842	T1
572	T2
536	T2A
500	T2B
446	T2C
419	T2D
392	T3
356	T3A
329	T3B
320	T3C
275	T4
248	T4A
212	T5
185	T6

Class II

Skin temperature of devices not subject to overloading shall not exceed 329°F. Motor transformers and other equipment subject to overloading shall not have skin temperature over 248°F.

Figure 44-4. Enclosure temperature codes are used to determine which devices are allowed in classified areas.

Design Responsibility

Designers and installers of electrical systems in hazardous locations have a responsibility to ensure that the installation of the electrical system does not jeopardize the safety of personnel or the public at large. Often, the determination of the classification for a particular area can be a very complex issue, with many factors involved. The owners of the process or the building or structure where the classified area is located, their engineering support staffs, and the insurance underwriters are responsible for establishing the electrical classification for these areas. Consulting engineers or contractors are responsible for questioning area classifications, but the plant owners have the final authority to establish the classification.

An *EYS seal* is a conduit coupling with an angled side connection closed with a screwed plug. After all the wiring has been pulled through the conduit and the electrical system has been tested, the side plug is removed and the fitting is filled with liquid sealing material. This material quickly hardens, providing a gas-proof seal inside the conduit and around the wires. Once the seal is poured, the wiring cannot be changed without cutting out the EYS seal, repairing the conduit, and pulling new wires. It is a common practice to pull extra wires in conduits that have seals so that they are available if needed.

Class I, Division 1, All Groups. All enclosures, even those only containing terminal strips, must be of explosionproof design. All fittings must be explosionproof. Flexible conduit can be used only if it is an explosionproof design. Conduit can become a passageway for explosive gases to move from one area to another. Therefore, seals must be used in every conduit within 18″ of any instrument or junction box, and all conduits must be sealed at the boundary between hazardous and nonhazardous areas. **See Figure 44-5.**

Class I, Division 2, All Groups. Rigid threaded conduit or approved flexible conduit is required for all individual wires, but approved cables can be used between termination boxes and between hazardous and nonhazardous areas with appropriate cable seals. Enclosures for instruments must be of explosionproof design, but those enclosures containing terminal strips or other nonincendive devices can be general-purpose.

An EYS seal must be used in every conduit within 18″ of any instrument or explosionproof enclosure. All fittings between the instrument and the EYS seal are to be explosionproof. Flexible conduit, if used, can be general-purpose with provision for a continuous ground. All conduits must be sealed at the boundary between hazardous and nonhazardous areas. **See Figure 44-6.**

Intrinsically Safe Systems

An *intrinsically safe system* is a system designed to limit electrical and thermal energy to a level that cannot ignite combustible atmospheres and to avoid some of the high installation costs associated with explosionproof installations. For a system to be classified as being intrinsically safe, the instruments, barriers, and wiring must all satisfy the intrinsic requirements. The NEC® contains all the requirements for an intrinsically safe installation.

For example, all intrinsically safe conduit, wiring, or cables must be entirely separate from any nonintrinsic wiring. The conduit and wiring in the hazardous area can be general-purpose, but the wiring at the boundary between hazardous and nonhazardous areas must be sealed. During the testing or maintenance of intrinsically safe instruments, care must be exercised that any test instruments in combination with the instrument circuits do not cause a spark when the case is open.

Always use proper lockout/tagout procedures when working on circuits.

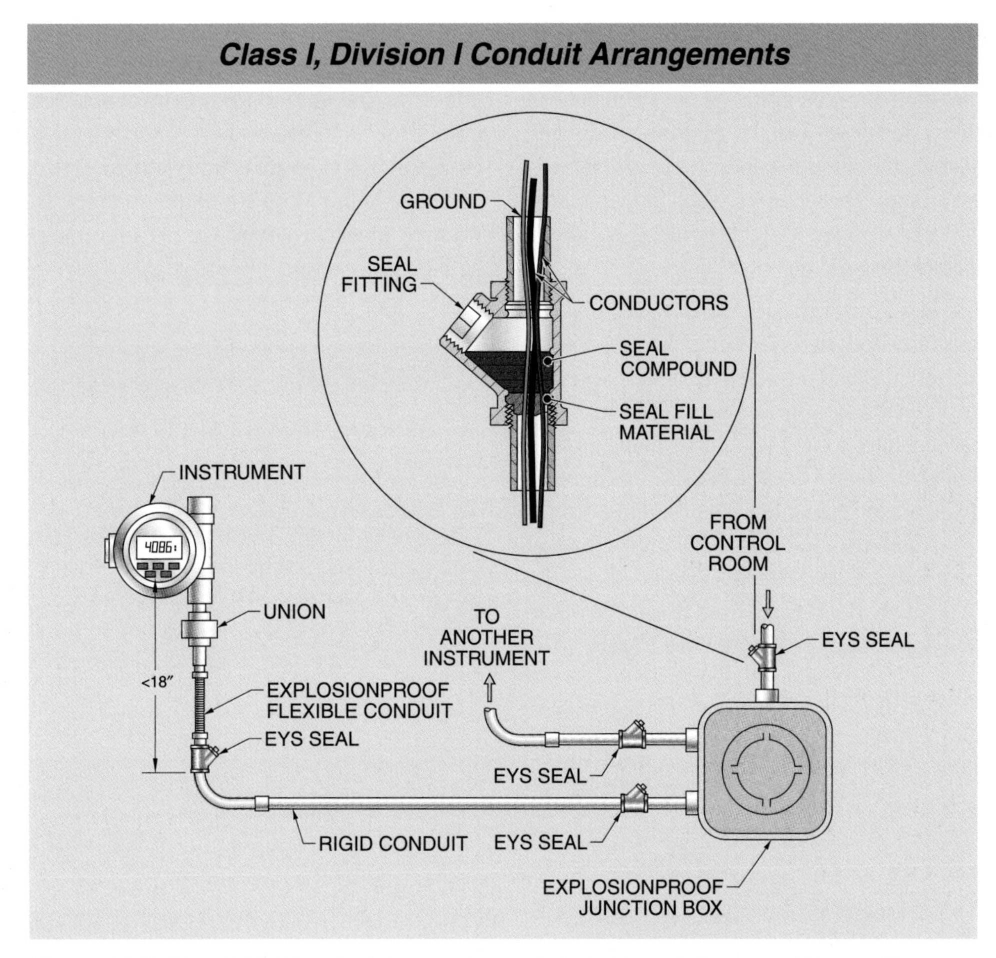

Figure 44-5. Class I, Division 1 wiring requirements include seals to prevent flammable gases from flowing through conduit.

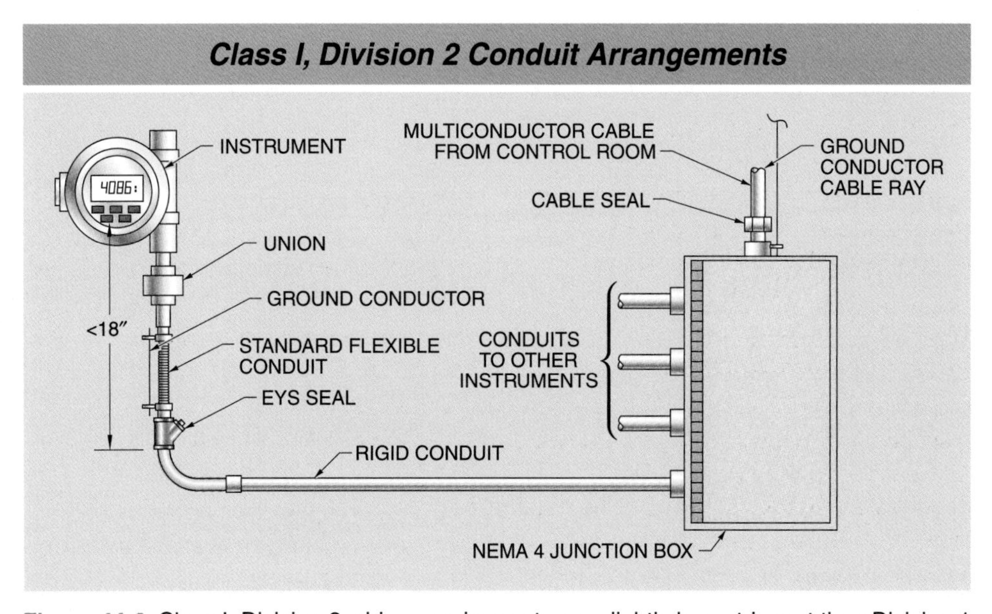

Figure 44-6. Class I, Division 2 wiring requirements are slightly less stringent than Division 1 requirements.

Intrinsically Safe Instruments. An *intrinsically safe instrument* is an instrument that has restricted voltage levels and current flows designed into the instrument as well as a very limited number of electrical energy storage components, such as capacitors and transformers, which could generate an electrical spark.

Intrinsic Barriers. An *intrinsic barrier* is a specially designed electronic circuit containing resistors and diodes that is used to prevent any electrical ignition energy from being carried into a hazardous area along power, signal, or control wiring. Intrinsic barriers are available to protect low-level signals such as from thermocouples and RTDs and high-level systems such as 115 VAC power or signals. In addition, barriers are used to send power to transmitters and solenoids. **See Figure 44-7.**

It is extremely important that the correct barriers are used for the application. Selecting the incorrect barrier does not protect the system and may interfere with the signal. Barriers are usually rail-mounted in a junction box located in the nonhazardous area close to where the wiring enters the hazardous area.

> **TECH FACT**
>
> *Process hazard analysis (PHA) is a procedure in which a process is evaluated for potential hazards, potential consequences are identified, and an assessment is made of whether present safeguards are adequate. PHA is a major requirement of OSHA's process safety management (PSM) standard in 29 CFR 1910.119.*

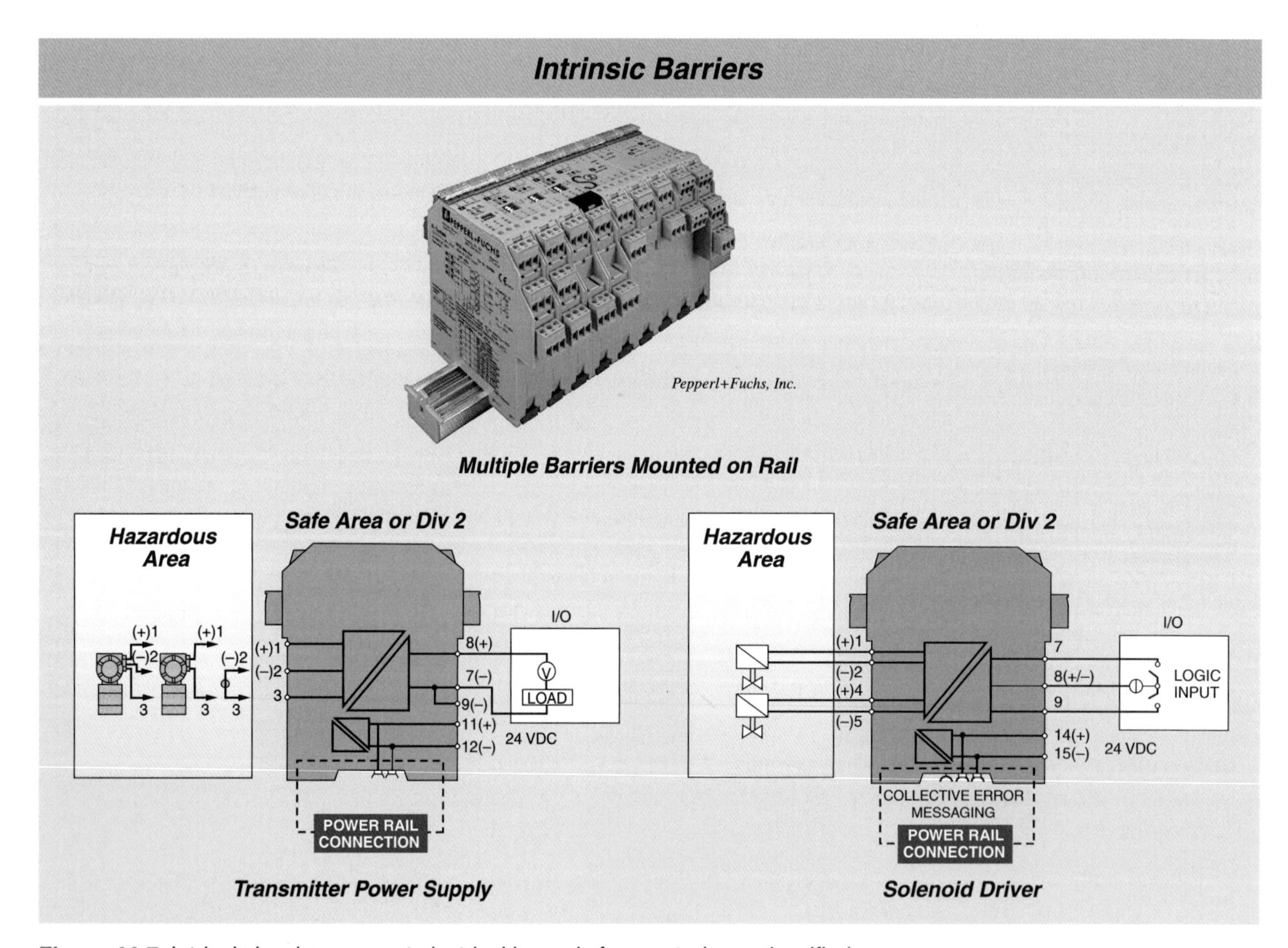

Figure 44-7. Intrinsic barriers prevent electrical hazards from entering a classified area.

KEY TERMS

- *class:* A term used to define the general nature of a hazardous gas or vapor in the surrounding atmosphere.
- *Class I location:* A hazardous location in which sufficient quantities of flammable gases and vapors are present in the air to cause an explosion or ignite hazardous materials.
- *Class II location:* A hazardous location in which a sufficient quantity of combustible dust is present in the air to cause an explosion or ignite hazardous materials.
- *Class III location:* A hazardous location in which easily ignitable fibers, or flyings, are present in the air but not in a sufficient quantity to cause an explosion or ignite hazardous materials.
- *division:* A classification assigned to each class based upon the likelihood of the presence of a hazardous substance in the atmosphere.
- *Division 1 location:* A hazardous location in which a hazardous substance is normally present in the air in sufficient quantities to cause an explosion or ignite hazardous materials.
- *Division 2 location:* A hazardous location in which the hazardous substance is not normally present in the air in sufficient quantities to cause an explosion or ignite hazardous materials.
- *group:* A term used to classify an atmosphere containing particular flammable gases or vapors or combustible dust.
- *Group A:* The term used for an atmosphere containing acetylene.
- *Group B:* The term used for an atmosphere containing hydrogen, fuel, and combustible process gases containing 30% hydrogen by volume, or gases or vapors of equivalent hazard such as butadiene, ethylene oxide, propylene oxide, and acrolein.
- *Group C:* The term used for an atmosphere containing ethyl ether, ethylene, or gases or vapors of equivalent hazard.
- *Group D:* The term used for an atmosphere containing acetone, ammonia, benzene, butane, cyclopropane, ethanol, gasoline, hexane, methane, methanol, naphtha, natural gas, propane, or gases or vapors of equivalent hazard.
- *Group E:* The term used for an atmosphere containing metal dusts such as aluminum or magnesium dust.
- *Group F:* The term used for an atmosphere containing carbonaceous dusts such as carbon black, coal, or coke dust.
- *Group G:* The term used for an atmosphere containing dust not included in E and F.
- *nonincendive protection:* A type of protection in which electrical equipment is incapable, under normal conditions, of causing the ignition of a specified flammable gas or vapor in an air mixture due to arcing or thermal effect.
- *dust ignition-proof protection:* A type of protection that excludes ignitable amounts of dust or amounts that can affect performance or rating.
- *explosionproof protection:* A type of protection that uses an enclosure that is capable of withstanding an explosion of a gas or vapor within it and preventing the ignition of an explosive gas or vapor that may surround it.
- *intrinsically safe protection:* A type of protection in which electrical equipment, under normal or abnormal conditions, is incapable of releasing sufficient electrical or thermal energy to cause the ignition of a specific hazardous atmospheric mixture in its most easily ignitable concentration.

KEY TERMS *(continued)*

- *enclosure:* A case or housing for equipment or other apparatus that provides protection for controllers, motor drives, or other devices.
- *purged enclosure:* A nonhazardous enclosure that is pressurized and purged with air to allow it to be used in hazardous areas.
- *temperature code:* A designation that specifies the maximum surface skin temperature obtained by an enclosure during testing by approval agencies.
- *EYS seal:* A conduit coupling with an angled side connection closed with a screwed plug.
- *intrinsically safe system:* A system designed to limit electrical and thermal energy to a level that cannot ignite combustible atmospheres and to avoid some of the high installation costs associated with explosionproof installations.
- *intrinsically safe instrument:* An instrument that has restricted voltage levels and current flows designed into the instrument as well as a very limited number of electrical energy storage components, such as capacitors and transformers, which could generate an electrical spark.
- *intrinsic barrier:* A specially designed electronic circuit containing resistors and diodes that is used to prevent any electrical ignition energy from being carried into a hazardous area along power, signal, or control wiring.

REVIEW QUESTIONS

1. Define class, division, and group.
2. Define the four types of electrical protections.
3. What is an enclosure, and what organization classifies enclosure types in the United States?
4. What is the purpose of installing conduit seals?
5. What is an intrinsically safe system, and where can the requirements for intrinsically safe systems be found?

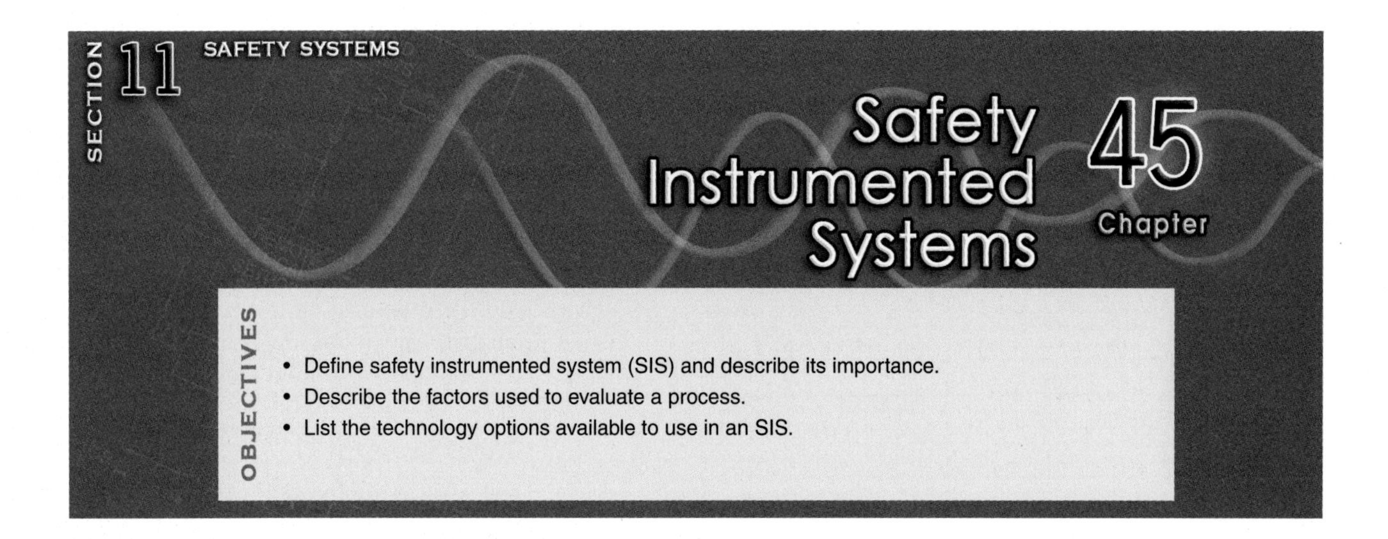

EVALUATION OF PROCESSES

A *safety instrumented system (SIS)* is a system consisting of sensors, logic solvers, and final control elements that bring a process to a safe state when normal operating conditions are violated. Safety instrumented systems have been in existence for many years. The use of an SIS to determine the required degree of safety is very complex, but the evaluation needs to consider risk, which is comprised of event severity and event likelihood, and the integrity level needed to protect against that risk. Standards have been developed for the safe automation of chemical processes because of the potentially serious impact that an accident can have on people, the environment, and equipment.

The American Institute of Chemical Engineers (AIChE), through the Center for Chemical Process Safety (CCPS), has issued an SIS document that is commonly used in the chemical industry called the *Guidelines for Safe Automation of Chemical Processes*. Another widely accepted SIS document is ISA S84, *Application of Safety Instrumented Systems for the Process Industries*. The Occupational Safety and Health Administration has publicly identified the ISA S84 standard as a "good engineering practice" when following OSHA regulations on management of highly hazardous chemicals.

A European standard that is becoming increasing well known is IEC 61511, *Functional Safety: Safety Instrumented Systems for the Process Industry Sector*. This is a process industry sector implementation of an earlier, more general, standard known as IEC 61508. IEC 61511 gives standards for the specification, design, installation, operation, and maintenance of an SIS.

Risks

Any process involving a hazardous material has an inherent level of risk. **See Figure 45-1.** *Risk* is a measure of the probability and severity of adverse effects of a process failure. Risk is used to evaluate how often a failure can happen and the consequences if a failure does happen. One common theme in all safety systems is to have multiple levels of protection, which increase the integrity level and reduce the residual risk.

> **TECH FACT**
>
> *Excessive alarms contributed to a 1994 fire at a refinery in Milford Haven, UK. Two operators had to respond to 275 alarms within 10 minutes without knowing which emergency was the most vital. Excessive alarms also contribute to operator complacency when operators are not able to distinguish between routine and emergency alarms.*

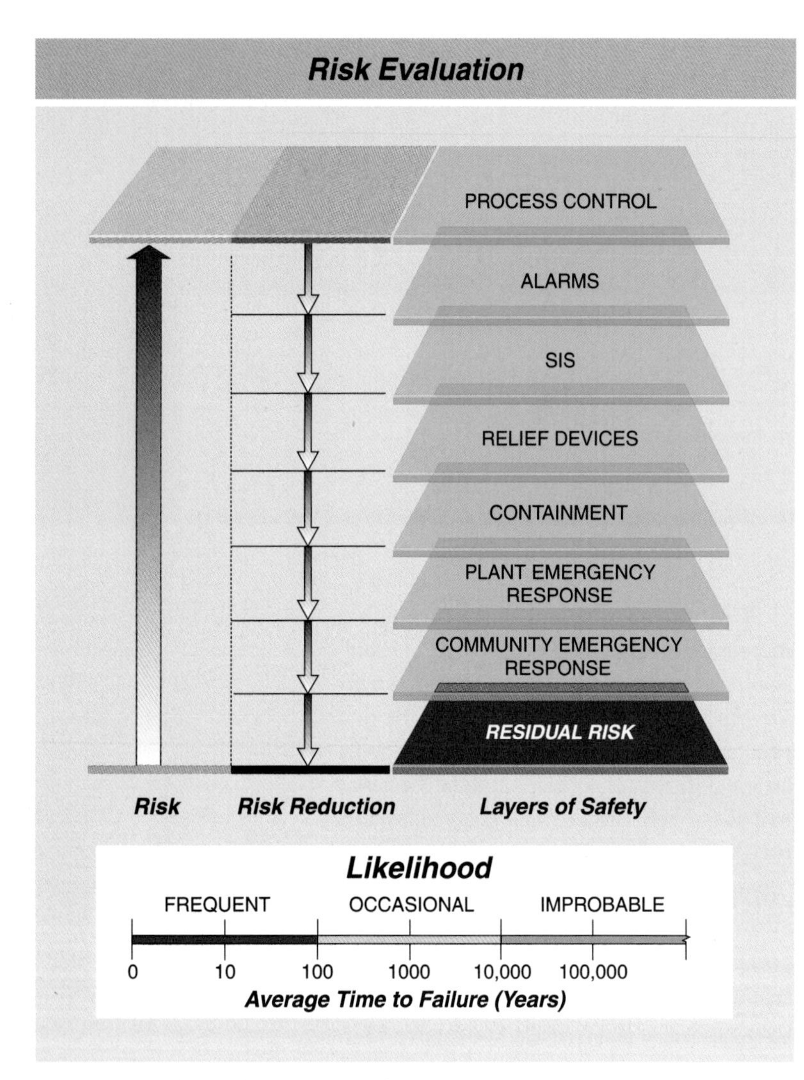

Figure 45-1. A combination of safety layers reduces the inherent process risk to a small residual risk level.

Processes can be evaluated to determine the level of protection that is needed to obtain the required degree of safety. For example, if the hazardous material is always contained within vessels and piping, that physical separation is the first level of protection. Alarm systems are an additional level of protection. Operating procedures and training can be an additional level of protection. A pressure-relieving device is a level of protection. A conventional control system is a level of protection. Typical levels of protection are as follows:

- process controls
- alarms and operator intervention
- safety instrumented systems
- physical protection (relief devices)
- physical containment (dikes)
- plant emergency response
- community emergency response

Two factors that must be taken into account in evaluating risk are the severity and likelihood of an event. The required integrity levels are determined after the severity and likelihood are determined.

Event Severity

The severity of an event can be classified as minor, serious, or severe. A minor incident is one where the impact is initially limited to local areas of the event with potential for broader consequences if corrective action is not taken. A serious incident is one that could cause any serious injury or fatality onsite or offsite or property damage of $1 million offsite or $5 million onsite. A severe incident is one that is five or more times worse than a serious accident.

Event Likelihood

The likelihood of an event can be classified as improbable, occasional, or frequent. An improbable likelihood means that a failure or series of failures has a very low probability of occurrence within the expected lifetime of the plant (less than 1 in 10,000 per year). This is equivalent to saying that the expected average time to a failure event is more than 10,000 years.

An occasional likelihood means that a failure or series of failures has a low probability of occurrence within the expected lifetime of the plant (1 in 100 to 1 in 10,000 per year). This is equivalent to saying that the expected average time to a failure event is between 100 and 10,000 years. A frequent likelihood means that a failure or series of failures can reasonably be expected to occur within the expected lifetime of the plant (more than 1 in 100 per year). This is equivalent to saying that the expected average time to a failure event is less than 100 years.

TECH FACT

Wood or fiberglass ladders must be used when working on or around electric wiring and circuits. A metal ladder can conduct electricity and become part of the circuit.

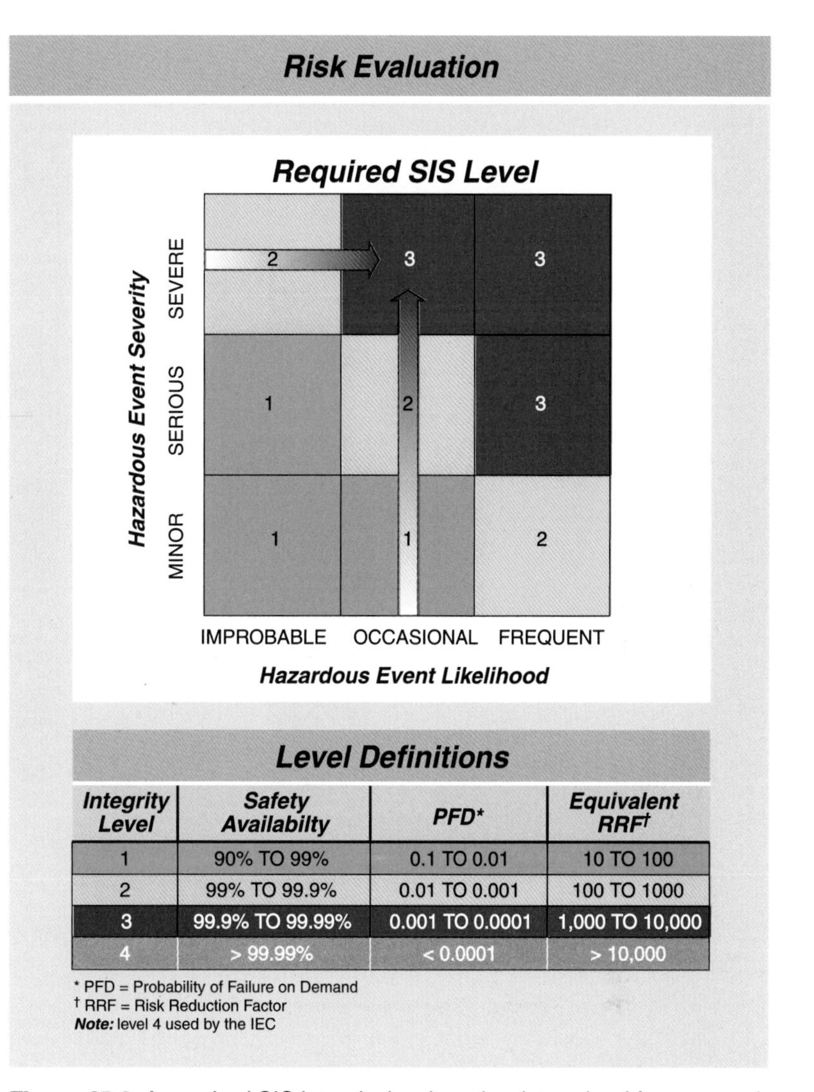

Integrity Level	Safety Availabilty	PFD*	Equivalent RRF†
1	90% TO 99%	0.1 TO 0.01	10 TO 100
2	99% TO 99.9%	0.01 TO 0.001	100 TO 1000
3	99.9% TO 99.99%	0.001 TO 0.0001	1,000 TO 10,000
4	> 99.99%	< 0.0001	> 10,000

* PFD = Probability of Failure on Demand
† RRF = Risk Reduction Factor
Note: level 4 used by the IEC

Figure 45-2. A required SIS integrity level can be determined from an evaluation of event severity and event likelihood.

Integrity Levels

A determination of the required SIS integrity level is obtained from a chart of hazardous event severity and event likelihood. **See Figure 45-2.** For example, a process that has a hazardous event severity rating of "severe" and an event likelihood rating of "occasional" has a required SIS integrity level of 3. The IEC, ISA, and AIChE define integrity levels in terms of safety availability, probability of failure on demand (PFD), and risk reduction factor (RRF). The RRF is equal to 1/PFD. The levels of protection and the SIS integrity level determine the type of technology required.

An alternate qualitative risk ranking is also available. This risk ranking determines the SIS level required based on the consequences, frequency and exposure, possibility of avoidance, and probability of occurrence. **See Figure 45-3.** For example, a failure event with a consequence of "several deaths" (Cc), frequency of "rare to frequent" (Fa), possibility of avoidance of "almost impossible" (Pb), and a probability of occurrence of "slight" (W2) requires an SIS integrity level of 2.

TECHNOLOGY OPTIONS

Once it has been determined that an SIS is needed, the next step is to decide what type of technology is to be used. In the past, the only choice was a hardwired system using relays and timers, lights, and pushbuttons. These systems, while being very reliable, were difficult and expensive to build and very difficult to change when used for large systems. The development of the PLC made it much easier to handle large SIS applications. It soon became apparent that one PLC was insufficient to provide the required level of safety because the logic was in the programming.

Programming always has a question of reliability. Therefore, two independent PLCs with multiple inputs and outputs were commonly used. Using two independent PLCs meant that various polling methods needed to be used to determine which input and output signals were valid and what action to take. The complexity of the design and the costs soon became unacceptable. As a result, the safety PLC was developed. A safety PLC contains redundant logic and self-checking systems, and has the ability to handle multiple inputs and outputs at a lower cost.

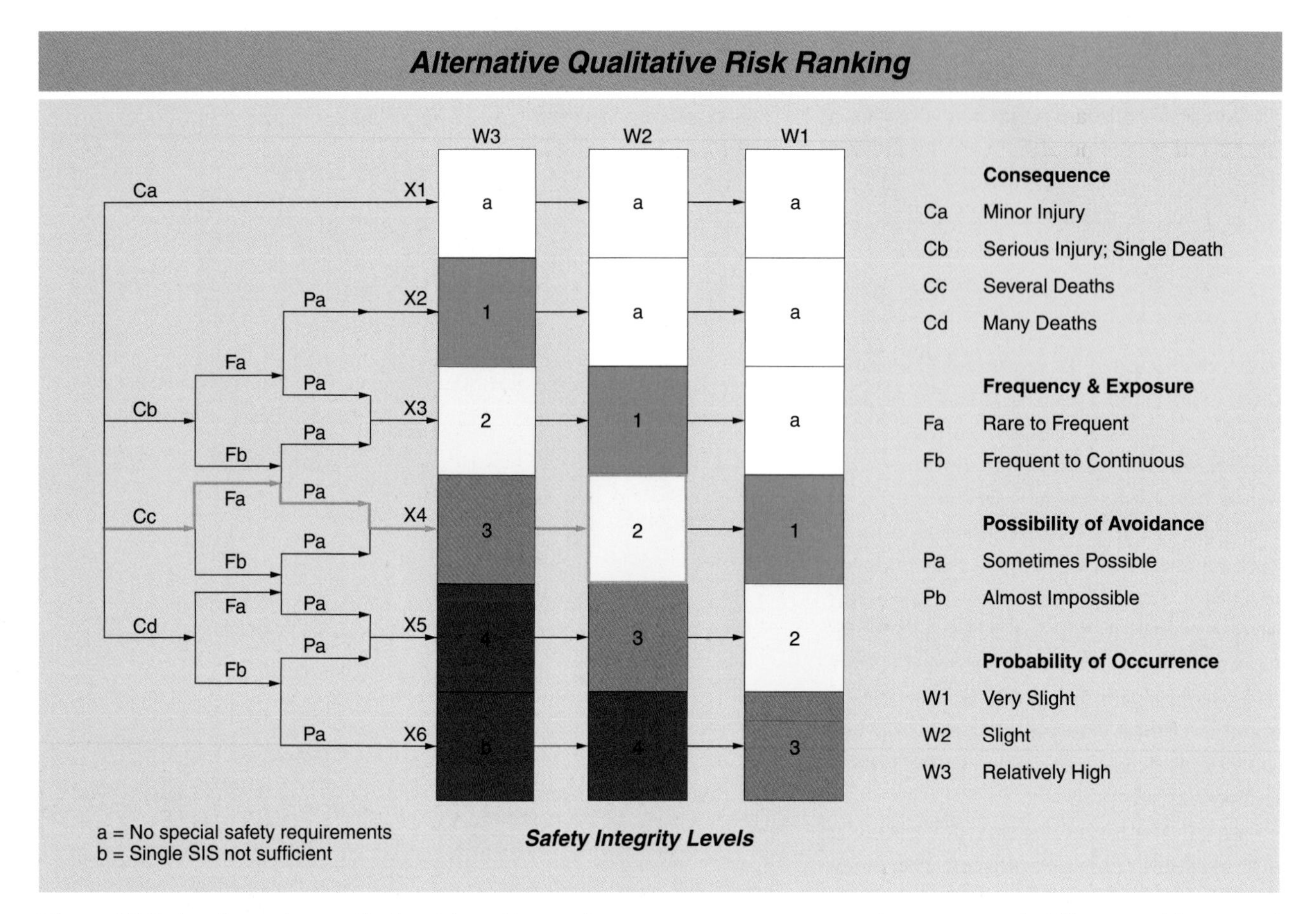

Figure 45-3. An alternative qualitative risk ranking evaluates event consequences, frequency, probability of avoidance, and probability of occurrence to determine a required SIS integrity level.

Hardwired Systems

A *hardwired system* is an SIS that uses discrete logic components interconnected with physical paths for logic signals and contains no programmable memory devices. These systems are considered safe because the safety logic is nonvolatile. Hardwired systems experience an increasing number of nuisance alarms as the systems get older and individual components start to fail. This is because the probability of one component failing in a complex system is very high. In addition, the costs of making changes to a hardwired system are very high. Hardwired systems can be constructed of pneumatic devices, electric relays, and solid-state devices in any combination.

Programmable Electronic Systems

A *programmable electronic system* is a solid-state electronic device where the safety logic is programmed into the device and stored in electronically programmed read-only memory (EPROM) chips or random-access memory (RAM). Solid-state devices that use EPROMs cannot lose their memory, but are difficult to change and are designed for special safety applications. Other advantages of programmable electronic systems are the ability to add calculation capabilities and serial communications.

Disadvantages of programmable electronic systems are a higher initial cost and dependence on software. Most programmable electronic systems use RAM memory and even though board-mounted, long-life batteries maintain the memory, there are significantly greater chances of failure, thus the need for redundancy and polling strategies. *Polling* is the process of comparing redundant safety signals to determine the proper response. Some of the PLC arrangements that have been used to provide the necessary reliability include microprocessors, single PLCs, dual PLCs, and triplicated PLCs.

A dual PLC can be programmed to perform the following polling strategies:

- 1oo2 (1 out of 2) polling: Any one trip signal of the redundant pair can shut down the system.
- 2oo2 (2 out of 2) polling: Both redundant tripping signals are needed to shut down the system.
- 1oo2D (1 out of 2 with diagnosis) polling: Any one trip signal of the redundant pair can shut down the system with proof that it is valid.

A triple modular redundancy (TMR) PLC system consists of three independent systems that receive the same inputs, but from separate sensing devices, and operate using the same programs. The output action that is taken is based on the majority consensus (2oo3) from the three PLCs.

Conventional PLC

A conventional PLC is a significant improvement over hardwired systems when working with large safety systems because a PLC can be modified as needed. The problem with a conventional PLC is that it must have external hardwired circuits to provide input, output, and logic solver diagnostics as well as output protection in order to be classified as an SIS.

Safety PLC

A *safety PLC* is a highly reliable PLC that includes fail-safe designs, built-in self-diagnostics, and a fault-tolerant architecture. Safety PLCs are also designed to meet third-party approval criteria because of the special hardware and software. Safety PLC architectures can be simple, low-cost, safe single systems (1oo1), or they can be more complex, with dual (1oo2D) or TMR system architectures, to satisfy the SIS integrity level needed. **See Figure 45-4.**

Safety PLC System Testing. Whenever a safety system is established, it needs to be tested to ensure that it is functioning properly. The concern is usually not about the safety logic after the initial installation commissioning is successfully completed, but about the individual sensing instruments, such as pressure switches, temperature switches, and limit switches. Testing can be a very difficult task because it is not easy to simulate all the process conditions and then selectively fail each one.

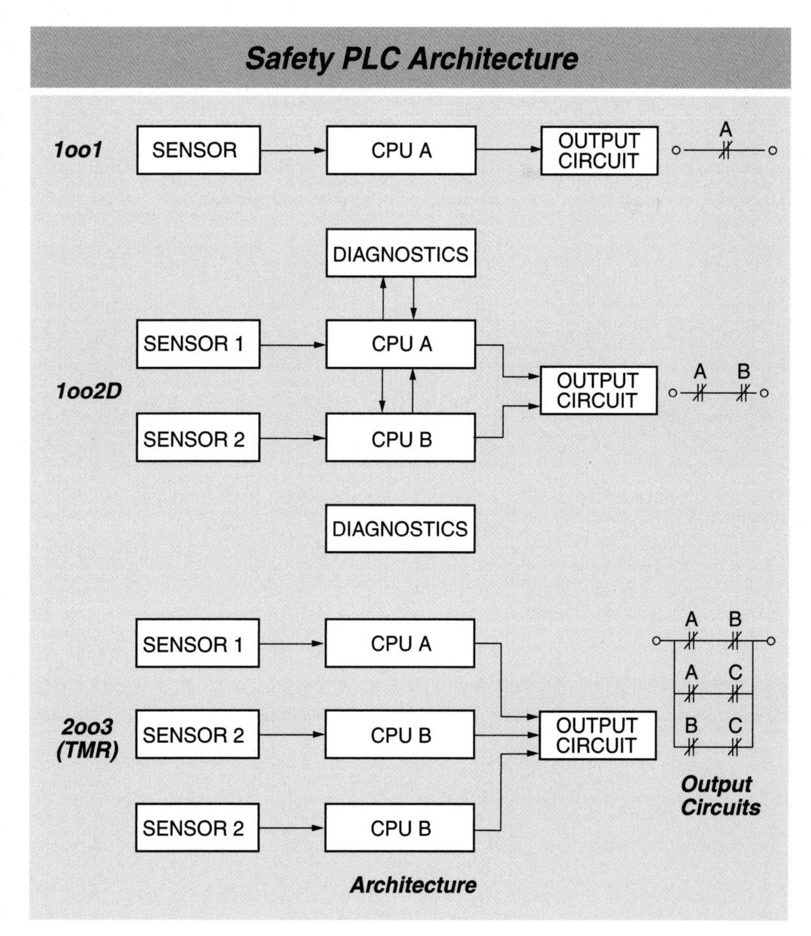

Figure 45-4. The choice of safety PLC architecture depends on the application.

One method is to intentionally shut down the normally running process by manually failing a sensor and observing the safety action, restarting the process, and repeating the shutdown with a different sensor. Most plant management personnel do not approve of this procedure because of the time involved, loss of production, and the chance of damage to the process equipment.

Deionizers are used as part of a water treatment system. Water must be treated to remove contaminants that can cause safety problems in a boiler.

Usually a compromise testing procedure is selected that is conducted during a process shutdown. This method simulates the process conditions for those sensors for which this can be done easily, such as pressure switches, limit switches, and flame detectors. The remaining devices only have their electrical signals simulated. In this manner, any single failure or combination of input failures can be simulated.

Prior to the testing, all the instruments should be calibrated. It is often necessary to modify some of the sensor process connections to facilitate the process simulation. If there is any serious concern about the reliability of any particular sensor, redundant sensors should be installed in such a way that either one could shut down the process.

A testing program must be clearly thought out and have testing procedures prepared prior to the testing. Full documentation of the testing procedure and the results of the testing is necessary because it is the first document examined if there is an actual failure of the system at a later date.

System Monitoring

System monitoring is the process of providing notification to an operator that there has been a failure in the process controlled by an SIS. Monitoring may be done in a portion of the safety system or in the total system. Monitoring is usually done with an alarm system. The alarm system can be a part of the digital SIS or it can be a separate, conventional annunciator.

KEY TERMS

- *safety instrumented system (SIS):* A system consisting of sensors, logic solvers, and final control elements that bring a process to a safe state when normal operating conditions are violated.
- *risk:* A measure of the probability and severity of adverse effects of a process failure.
- *hardwired system:* An SIS that uses discrete logic components interconnected with physical paths for logic signals and contains no programmable memory devices.

KEY TERMS *(continued)*

- *programmable electronic system:* A solid-state electronic device where the safety logic is programmed into the device and stored in electronically programmed read-only memory (EPROM) chips or random-access memory (RAM).
- *polling:* The process of comparing redundant safety signals to determine the proper response.
- *safety PLC:* A highly reliable PLC that includes fail-safe designs, built-in self-diagnostics, and a fault-tolerant architecture.
- *system monitoring:* The process of providing notification to an operator that there has been a failure in the process controlled by an SIS.

REVIEW QUESTIONS

1. Define safety instrumented system (SIS) and list several organizations and standards that define processes for safe automation of chemical processes.
2. What is the approach taken to reduce risk?
3. How may the severity and likelihood of an event be classified?
4. What are the advantages and disadvantages of a hardwired system?
5. What are the common safety PLC architectures?

SECTION 12

INSTRUMENTATION AND CONTROL APPLICATIONS

TABLE OF CONTENTS

SECTION OBJECTIVES

Chapter 46

- Compare the continuous and batch processes.
- Describe common uses for split range control valves.
- Explain the use of high and low selectors with controllers.
- Explain the use of limit controls.
- Explain how cascade control works and describe its common applications.

Chapter 47

- List examples of temperature control.
- Describe jacketed reactors.

Chapter 48

- Describe smooth transfer of pressure control.
- Explain the different uses of level control.

Chapter 49

- Describe the common applications of flow ratio control.
- Explain the use and operation of lead-lag air-fuel ratio control.

SECTION

12 INSTRUMENTATION AND CONTROL APPLICATIONS

Chapter 50

- Describe a common use for analysis control.
- Describe the common multivariable applications.
- Explain the control of conveyor systems.

INTRODUCTION

The expansion of the chemical industry after World War II led to the need for complete solutions to manufacturing problems on a much larger scale than ever before. Relatively simple piecework assembly operations were being replaced by large automated assembly plants and continuous processing operations like refineries and chemical plants. These integrated manufacturing operations needed integrated control systems that combined many types of instruments and controllers into systems that controlled an entire process.

Typically, large boilers require pressure, level, and steam flow measurements to adjust the makeup water. Conductivity measurements are made on the returned condensate. Ratio controllers maintain the air-fuel mixture, and stack gas analysis is used to measure the oxygen content to fine-tune the air-fuel mixture.

Complex cascade control systems are used to improve the control of processes. Multiple variables are measured with one controller determining how another controller adjusts the inputs to a process. A primary controller maintains a chemical reaction at the specified temperature setpoint by cascading control to a split range valve system that controls the reactor-jacket chilled water temperature.

ADDITIONAL ACTIVITIES

1. After completing Section 12, take the Quick Quiz® included with the Digital Resources.
2. After completing each chapter, answer the questions and complete the activities in the *Instrumentation and Process Control Workbook.*
3. Review the Flash Cards for Section 12 included with the Digital Resources.
4. Review the following related Media Clips for Section 12 included with the Digital Resources:
   - Combustion
   - Heat Exchangers
   - Oxygen Trim Monitor
5. After completing each chapter, refer to the Digital Resources for the Review Questions in PDF format.

ATPeResources.com/QuickLinks • Access Code: 467690

SECTION 12 INSTRUMENTATION AND CONTROL APPLICATIONS

General Control Techniques

Chapter 46

OBJECTIVES

- Compare the continuous and batch processes.
- Describe common uses for split range control valves.
- Explain the use of high and low selectors with controllers.
- Explain the use of limit controls.
- Explain how cascade control works and describe its common applications.

INSTRUMENT SYMBOLS

Individual instruments are put together with controllers, transmission or communications systems, and final control elements to form control loops. The selection of the instruments for a control loop must take into consideration the characteristics of the process being controlled. Knowledge of the processing equipment and how it functions, the properties of the material being processed, and the range of control of both the measurement and the throughput is necessary to successfully control a process.

There are several instrumentation standards and control techniques that are common to all measurement areas. All control systems use diagrams with standard instrument symbols. Instrument symbols are used in this chapter to represent the instruments and the interconnections between them. Instrumentation symbols are widely used in industry to describe the function of the controls. The International Society of Automation (ISA) has developed symbols and notation commonly used on instrumentation diagrams.

Modern digital instrumentation systems have the ability to implement very complex control strategies that are difficult to show using standard symbols. In these cases, a different symbol form needs to be used to show the internal workings of a controller. Symbols originally developed and issued by the Scientific Apparatus Makers Association (SAMA), and now issued by the Measurement, Control and Automation Association (MCAA), can be used for these applications and are used in this section for the descriptions of cascade and ratio controllers. Common general control techniques include various split range control valve arrangements, high and low selectors, high and low limits, and cascade controllers.

NDC Infrared Engineering

Modern industrial processes require tight control of variables as a product passes through a machine.

TYPES OF PROCESSES

The two common types of processes are continuous and batch processes. Although the individual instruments and controls that are used for continuous and batch processes are essentially the same, continuous and batch processes require different control strategies.

Continuous Processes

A *continuous process* is a type of processing operation in which continuous raw material is received and the resulting product is continually fed to the next operation. Storage volumes (surge capacity) are usually provided at strategic locations in the sequence of operations so that a problem in one segment of the manufacturing operation does not immediately affect other segments of the operation.

Multiple parallel operations can be installed so that individual units can be shut down without shutting down the entire operation. A common example of this is an electrical power station where there are multiple boilers and generators. Any one boiler and generator set can be shut down without the total production being shut down. It is common for continuous operations to run for months or longer without being shut down.

For example, a continuous process can be used to make a brine solution from solid salt and water. **See Figure 46-1.** Water is pumped into a saturator. The water flow controller (FC) adjusts the flow rate of the water into the saturator by adjusting the control valve. Solid salt is added to the saturator and the resulting raw brine is transferred to a pump tank. The brine is pumped from the pump tank to the chemical treatment tank. A level controller (LC) maintains the level in the pump tank by adjusting a control valve after the centrifugal pump.

Control systems can be integrated with phone, e-mail, and pager systems to notify the proper personnel of alarms.

Chemicals are added to the chemical treatment tank. The chemical flow controllers (FC) set the flow rate of the chemicals into the chemical treatment tank. The treated brine is transferred to the clarifier where contaminants are precipitated out. The sludge is drained out of the bottom of the clarifier. The clarified brine is transferred to a pump tank. The brine is pumped from the pump tank to the filters. A level controller (LC) maintains the level in the pump tank by adjusting a control valve after the pump.

Batch Processes

A *batch process* is a type of processing operation in which multiple feeds are received and the resulting product is fed all at once to the next operation. In reality, most batch process plants also have numerous continuous operations for handling byproducts of the reaction and to purify the product. Batch processes are usually much more complex than continuous processes since they usually involve automated charging, transferring, advancing or repeating processing steps, and emergency operations.

For example, a batch process can be used to make various chemical products. **See Figure 46-2.** Totalizing flow controllers (FQIC) add chemicals sequentially. Different ratios of the chemicals result in different final products. The temperature of the water that circulates through the jacket controls the temperature inside the reactor. The batch reactor is heated to operating temperature to start the reaction by setting the batch reactor temperature controller to its desired setpoint. This instructs the circulated water temperature controller to open the steam valve to the heater in order to heat the circulating water.

When the temperature inside the reactor reaches the setpoint, the batch reactor temperature controller sends a signal to the circulated water return controller to switch from heating with steam to cooling with chilled water. The circulated water controller modulates the chilled water valve to maintain the reactor temperature. After the reaction is complete, the finished product is pumped to storage and the process is repeated.

Continuous Processes

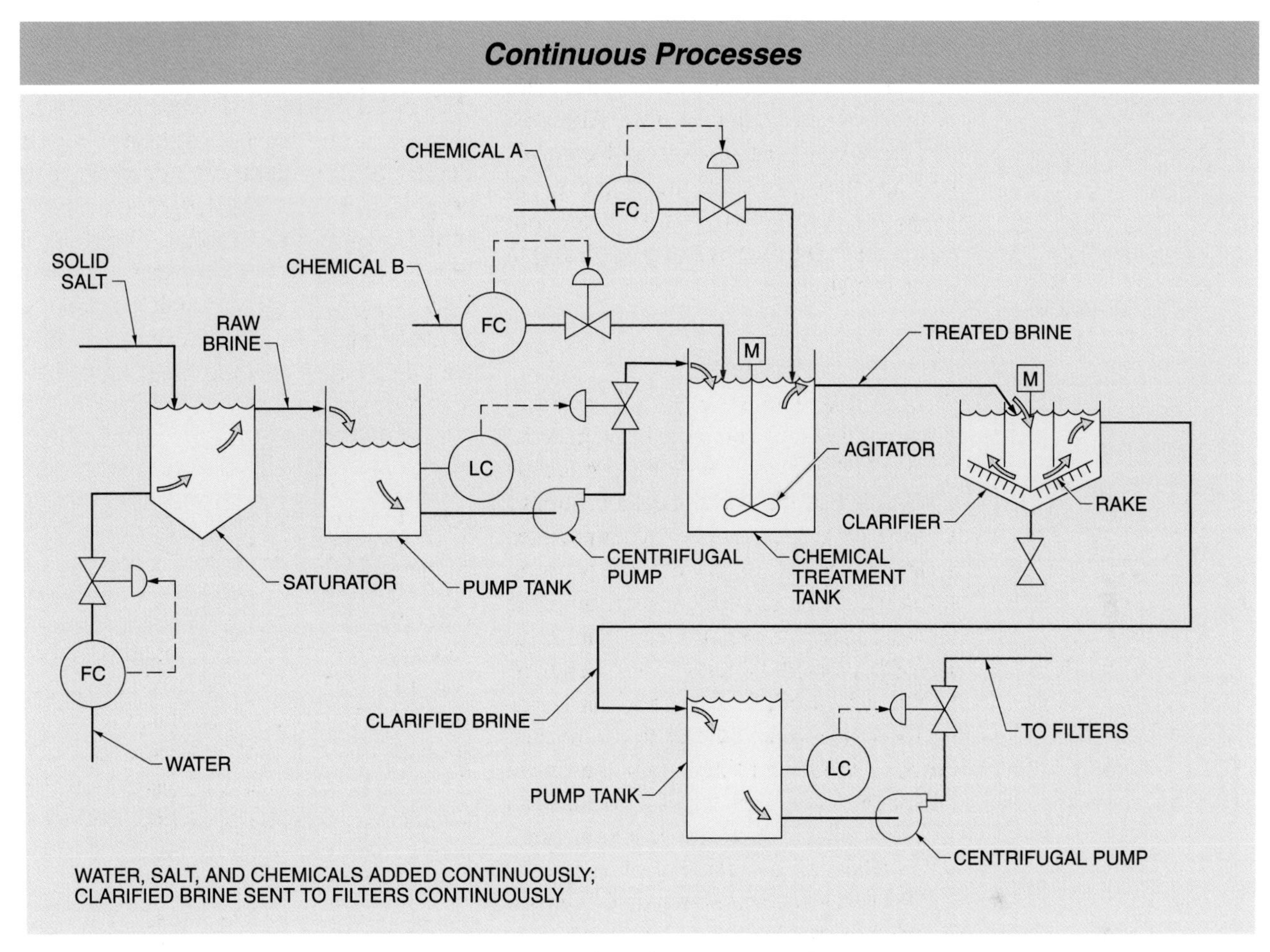

Figure 46-1. A continuous process has regular material flows from one part of the process to another.

SPLIT RANGE CONTROLS

Split range control is an application where two or more control valves operate using the same controller output signal range, with each valve using a portion of the controller output signal. The control signals can be split using current-to-pneumatic (I/P) transducers or valve positioners. The preferred method is to use valve positioners to do the split ranging. When the control signal is split, the output from the positioner is its normal full range.

For example, a positioner can use half of the normal 4 mA to 20 mA signal and have the full 3 psig to 15 psig output. This means that the positioner can be calibrated to use 4 mA to 12 mA of the control signal and still allow full valve movement.

Fluke Corporation

Power quality issues can cause problems in a control loop.

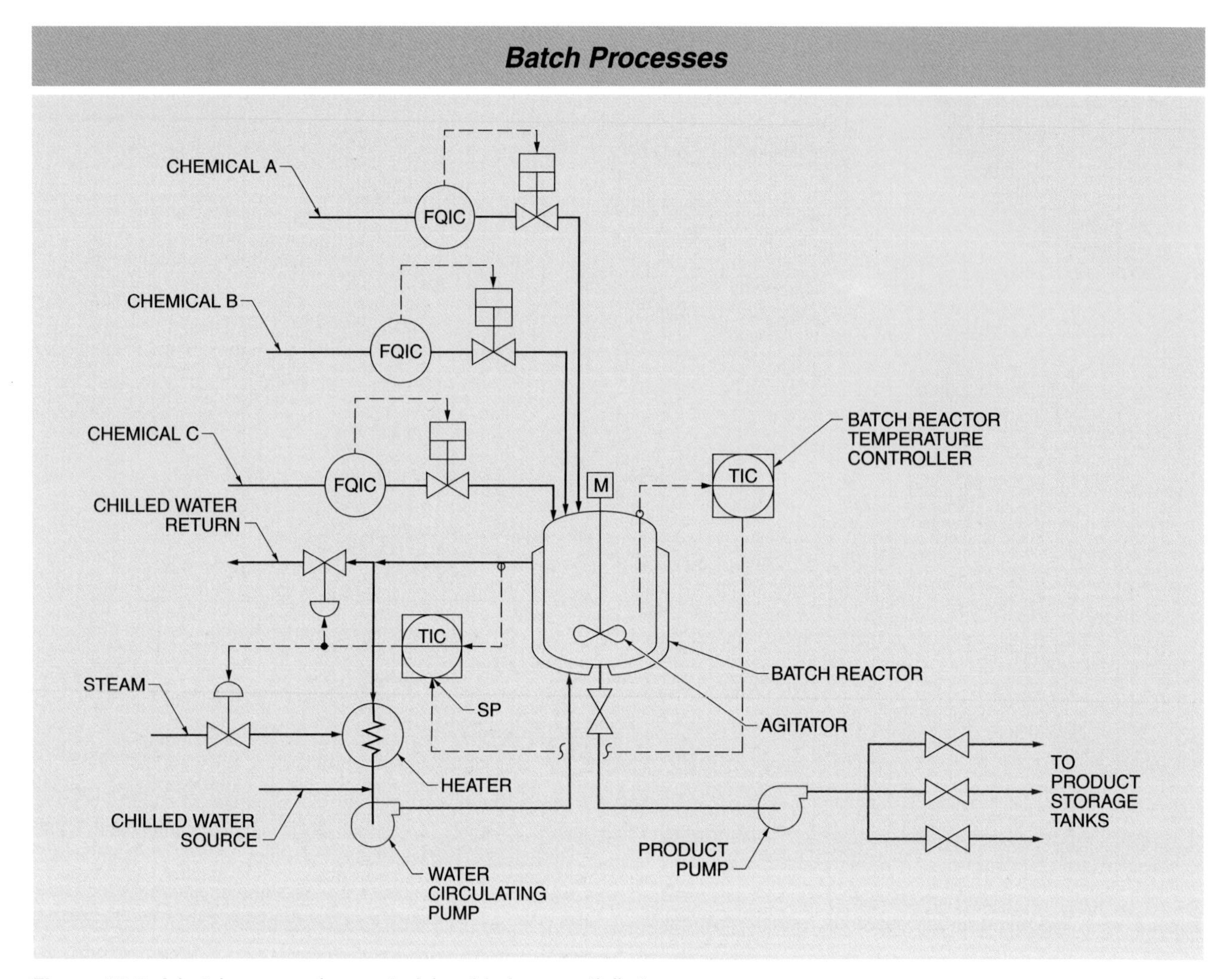

Figure 46-2. A batch process has materials added sequentially to a process.

The use of split range control valves is a very powerful technique that can be used to implement many different control strategies. Split ranging is accomplished at the expense of some loss of precision of adjustment of the control valves. The controller output range is fixed and accuracy is lost when the output is divided between two or more control valves.

TECH-FACT

Data trending information can be used to track equipment-energy use, efficiency, and the need for equipment maintenance.

Heating and Cooling Services

Heating and cooling services are among the most common uses of split range control valves. The heating and cooling control valves each use one-half of the controller output signal. Heating control valves are usually selected to have a fail closed action, and cooling control valves are usually selected to have a fail open action. Heating and cooling control valves should not both be open at the same time because this wastes energy. For example, a process fluid flows first through a shell-and-tube cooler and then through a shell-and-tube steam heater. **See Figure 46-3.**

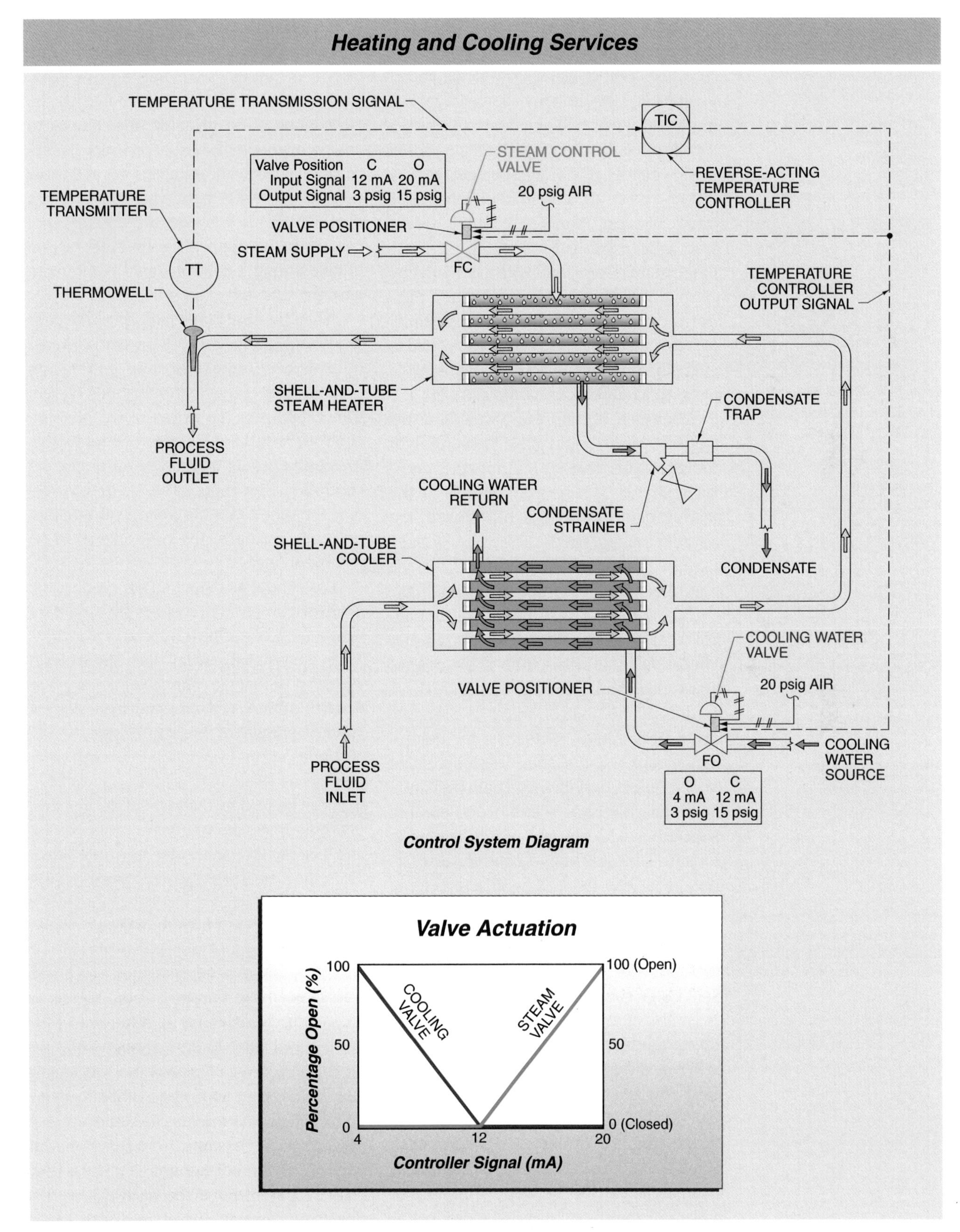

Figure 46-3. Heat exchangers are used to heat or cool a process fluid.

Highly alkaline liquids can etch the inside of sight glasses and glass tube rotameters. Etched glass is hard to see through and can cause erroneous readings.

The safest failure action is with the cooling water fully ON and the steam shut OFF. Thus, the cooling water flow to the cooler is controlled by an equal-percentage, fail open control valve. Steam flow to the heater is controlled by an equal-percentage, fail closed control valve. Equal-percentage flow characteristics are a good choice for these valves because equal-percentage valves have a large control range. The process temperature is measured with a temperature sensor, such as a thermocouple, in a thermowell in an elbow located downstream of both exchangers. The sensor is connected to a temperature transmitter (TT) that sends a signal to a temperature controller (TIC).

When there is a temperature increase, the controller needs to increase the cooling or decrease the heating. Since the cooling control valve is fail open and the heating control valve is fail closed, the controller output needs to decrease with an increase in process temperature. Thus, the controller action has to be reverse-acting.

If the 4 mA to 12 mA portion of the control signal is assigned to the cooling water control valve, it opens fully if the control signal should fail. When the 12 mA to 20 mA portion of the control signal is assigned to the steam valve, it closes if the control signal drops below 12 mA. This assignment of split ranges provides a smooth transition from full cooling at 4 mA to no cooling or heating at 12 mA and to full heating at 20 mA. This split range control valve arrangement can be used with any heating and cooling system.

Priority Usage Selections

Split range control valves can also be used to establish the order in which control valves are operated and thus the priority between different operations. One of the most common applications is the division of a fluid flow to a number of different users by the specified priority assigned to each of the users. This can be the distribution of a supplemental fuel between a number of boilers or the distribution of a product to several processing operations.

For example, hydrogen gas is a by-product produced by a process and is distributed to two different users. **See Figure 46-4.** Hydrogen gas not used is vented to the atmosphere. The primary objective is to always maintain the header pressure. This is accomplished by throttling the gas to the two users to ensure that they do not take more gas than is available or by venting gas that is not needed. In this application, the safest action to take during a control signal failure is to have the vent valve fail fully open.

When the vent valve fails open, the consequences are less severe on the upstream hydrogen compression and generation systems than if the vent valve fails closed. This establishes the action of the pressure controller and all of the other control valves. Increasing header pressure requires that the controller allow more gas to go to the users or to the vent. Since the control valve failure positions are open, the controller must be reverse-acting.

The controller output signal needs to be divided between three control valves, with the vent valve being at the lower end of the range. Thus, the vent valve is fail open (air to close) and the positioner is calibrated for a range of 4 mA to 9.3 mA, with an output to the control valve diaphragm of 3 psig to 15 psig.

Under normal operation, the vent valve only opens when there is more gas available than can be held by the users. If the gas supply diminishes, the vent valve is closed first and then the lowest priority user is throttled. Thus, the lowest priority user control valve is fail open and the positioner is calibrated for a range of 9.3 mA to 14.7 mA, with an output to the valve diaphragm of 3 psig to 15 psig.

A continued shortage of gas supply results in the lower priority user control valve being fully closed and the higher priority user control valve being throttled. Thus the higher priority user control valve is fail open and is calibrated for a range of 14.7 mA to 20 mA, with an output to the control valve diaphragm of 3 psig to 15 psig. A linear control valve flow characteristic is the best choice for all three valves, since a change in flow for a change in control signal is the same anywhere in the throttling range.

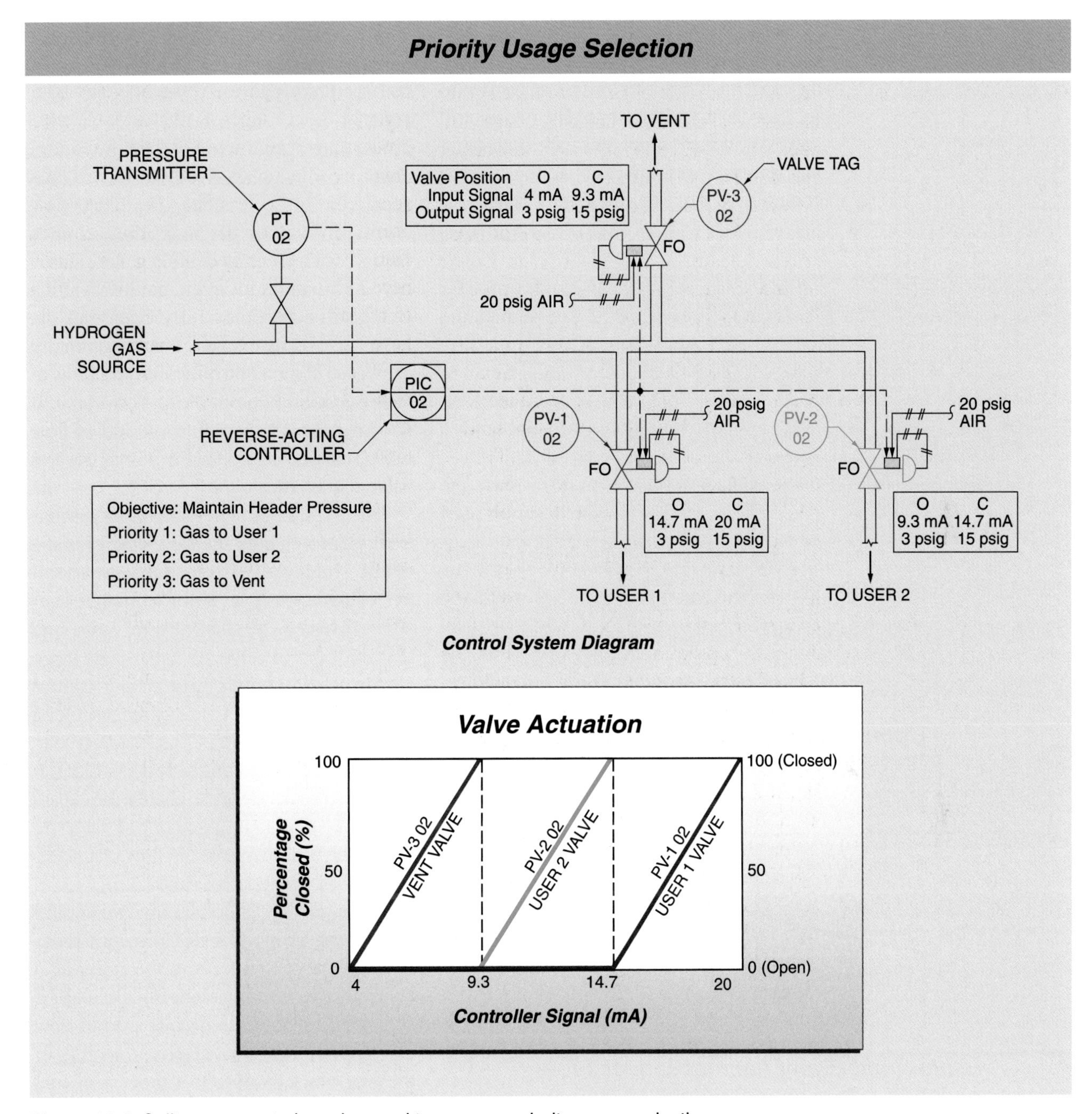

Figure 46-4. Split range control can be used to manage priority usage selection.

Throttling Range Increase

A less common, but still used, split-ranging technique is to increase the flow control range of a control valve. Occasionally, there is a need for a control system to handle a greater range of flow than it is possible to obtain from a single control valve. Two control valves, calibrated to operate sequentially, can be used to handle a greater range of flow. This greater range of control is accomplished at the expense of precision of control. The controller output range is 4 mA to 20 mA, and accuracy is lost when it is divided between two control valves.

For example, two fail closed linear flow control valves, with one valve being five times the capacity of the other valve, can be used to increase the throttling range. The two control valves are piped in parallel with each other and the positioners are calibrated to split the control range equally. **See Figure 46-5.**

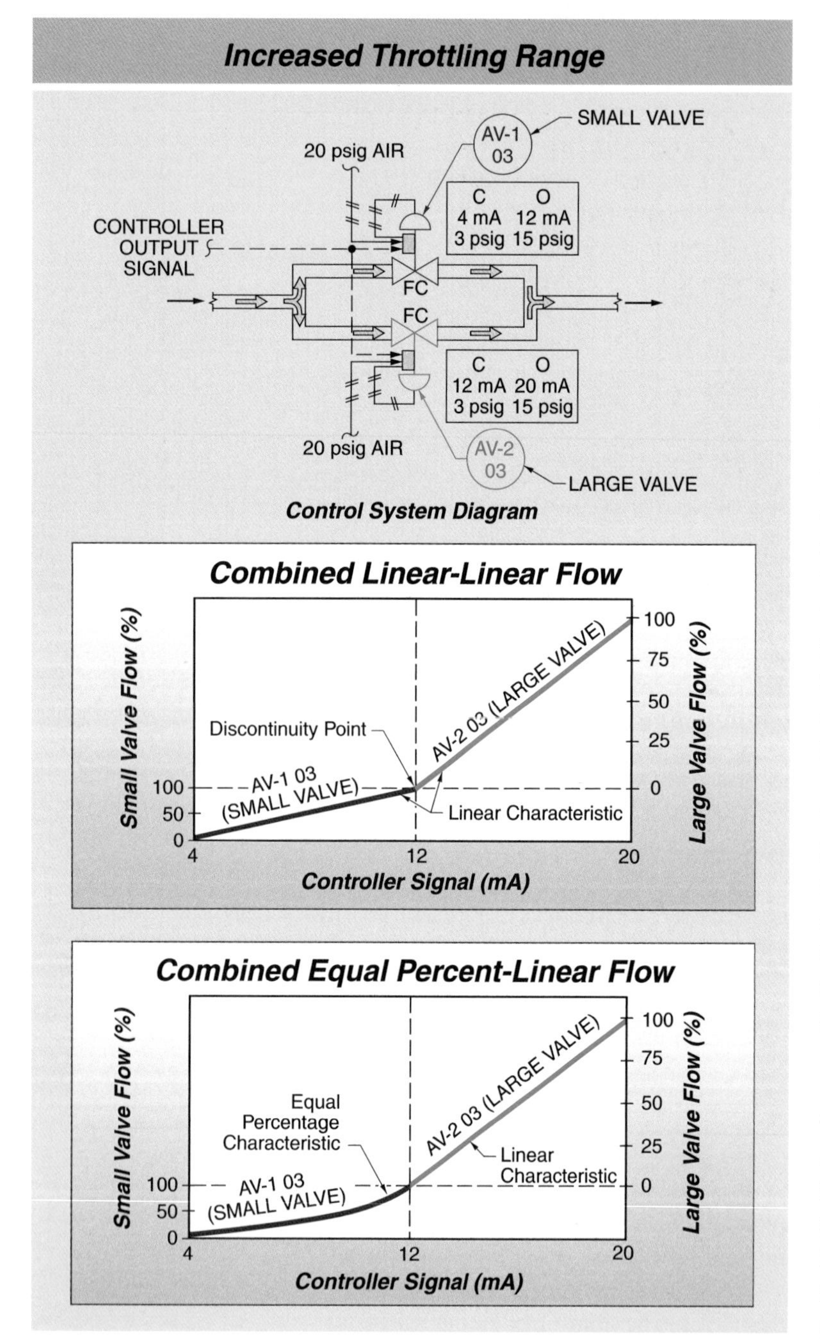

Figure 46-5. An increased throttling range can be accomplished with a split range control system.

The small control valve operates over a signal input range of 4 mA to 12 mA and the large control valve operates over a signal input range of 12 mA to 20 mA. This arrangement increases the flow control range by a factor of five over the original small control valve alone. The linear flow characteristics of the individual control valves, when combined into one flow curve, have a discontinuity at the point where the small valve becomes fully open and the large valve starts to open. The discontinuity can cause some controllability problems at the discontinuity point of the flow curve. It is best to have a smooth transition of flow when transferring control from one control valve to the other.

One possible way to remove this discontinuity is to change the flow characteristic of the small valve from linear to equal-percentage. This provides a smoother transition between the small control valve and the large control valve. An equal-percentage characteristic control valve has a control range of at least 25:1. In conjunction with the large valve, which is 5 times larger than the small valve, the control range is 125:1. It might also be necessary to change the capacity of the small and large control valves to obtain the best merge of the flow characteristics of the two control valves.

In the process of combining two control valves, the proportions of the split ranging, the valve sizing, and the individual control valve flow characteristics can all be adjusted to obtain the desired overall flow curve. Obtaining a satisfactory merging of the two control valves is a process of trial and error.

HIGH AND LOW SELECTORS

A *selector* is a device that compares two or three input signals and passes the highest or the lowest signal to the output of the device. High or low selectors are available in pneumatic, electric, or digital systems, and are very useful in implementing control strategies. Not all digital controllers have high and low selector devices available in their configurations.

Only the most powerful stand-alone digital controllers, DCS systems, and PLC systems have the internal programming to fully implement all control strategies.

High Selectors

A high selector can be used to select the highest of a group of temperature measurements to be used as the controller input. This is done if it is dangerous to the process for any measurement to be too high. The highest temperature is the one used for control and all the others are at a lower value. **See Figure 46-6.**

For example, a chemical reactor may have several thermocouples inserted through the wall to measure the internal temperature. Each temperature transmitter sends a signal to a temperature indicator as well as to the high selector (TY 04). The highest temperature is sent to the temperature controller (TIC 04), which then sends a signal to a cooling water control valve (TV 04). The transfer from one measurement to another is smooth and causes no upset to the control system.

Low Selectors

A low selector can be used to select the lowest of several level measurements as the input to a level controller to ensure that none of the levels are lower than the controller setpoint. **See Figure 46-7.** For example, three storage tanks are piped to a single pump. A level transmitter on each tank sends a signal to the low selector (LY 08). The lowest signal is sent to the level controller (LIC 08), which then sends a signal to the control valve (LV 08). This prevents any tank from being pumped dry. This control system is needed if the tanks should not be allowed to equalize or if the piping resistance between the tanks is too high for the levels to equalize by themselves.

> **TECH FACT**
>
> *The ASME Boiler and Pressure Vessel Code has been the accepted standard for many years for the construction of pressure vessels. The ASME code is legally enforceable in many jurisdictions.*

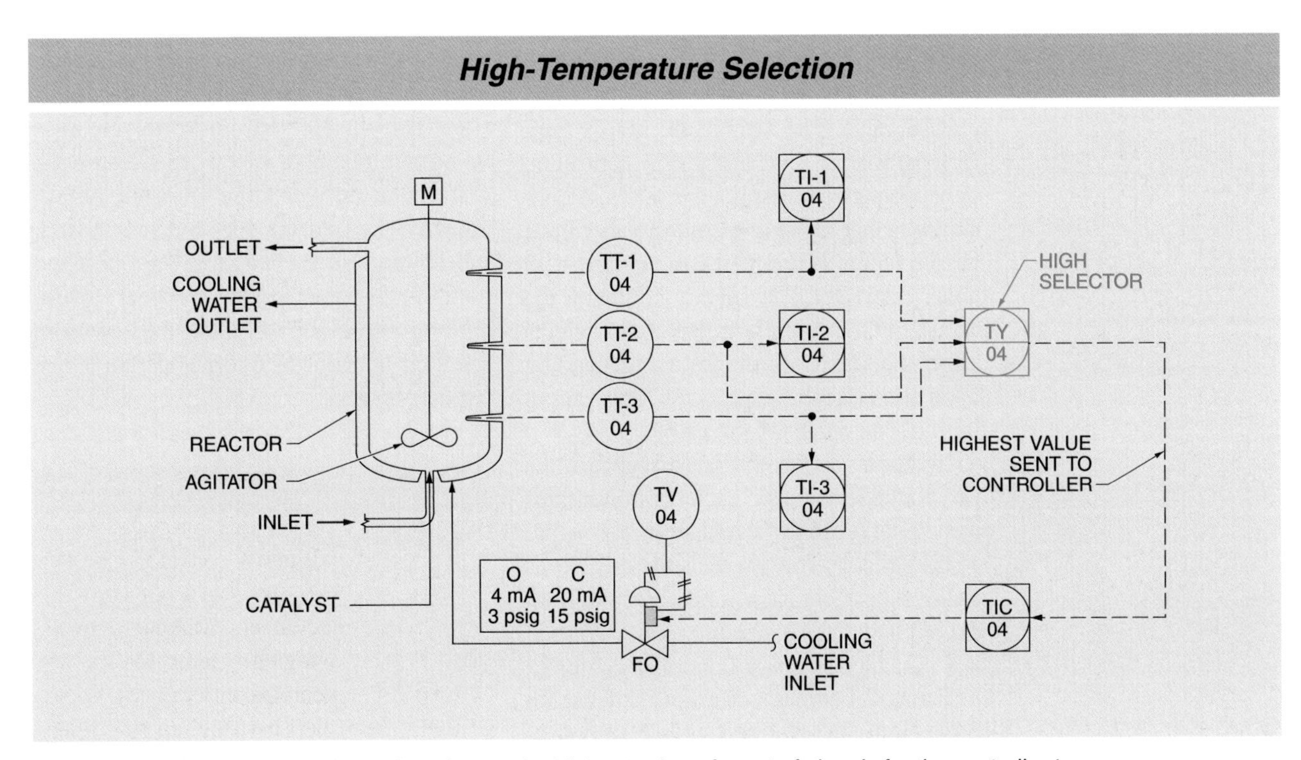

Figure 46-6. A high selector is used to choose the highest value of a set of signals for the controller to use.

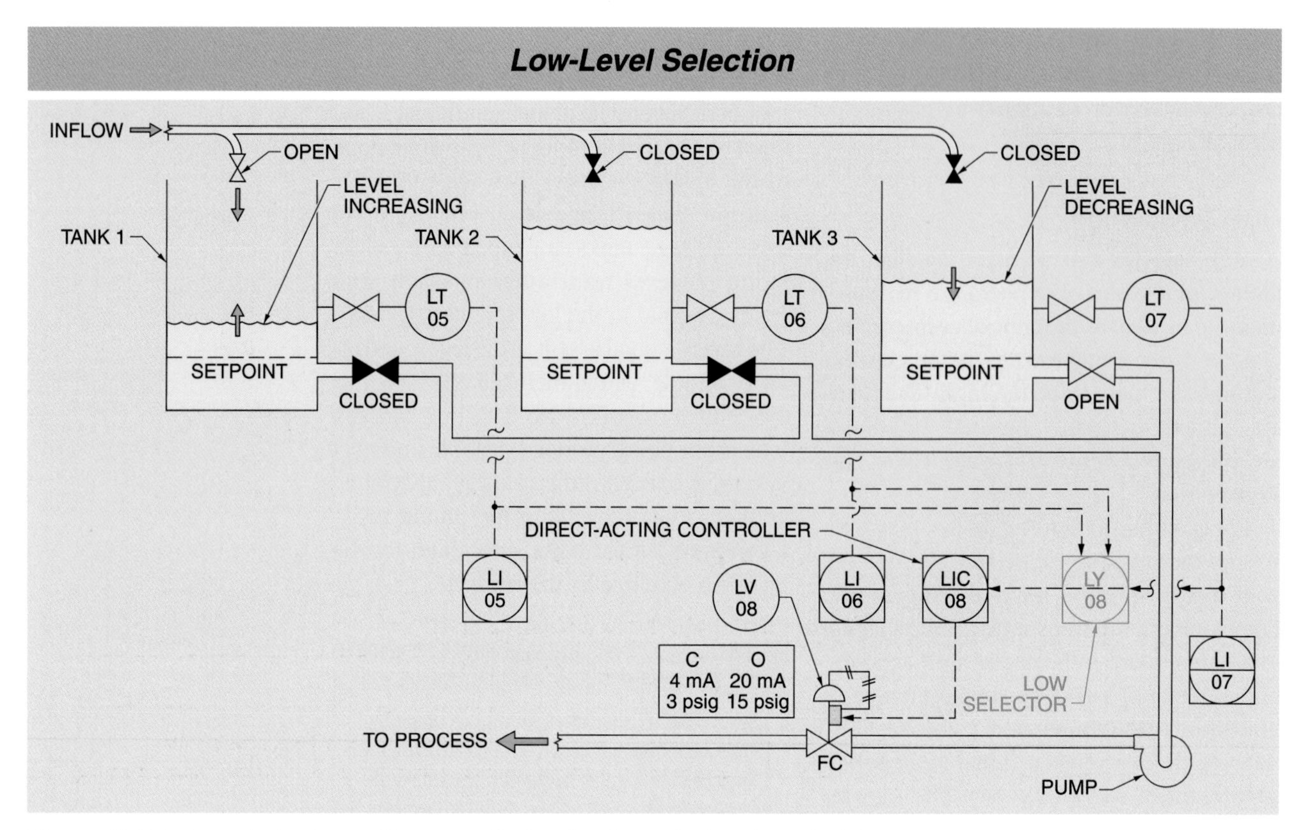

Figure 46-7. A low selector is used to ensure that no level goes below the setpoint.

Storage Tank Level and Flow Control

A common application of high and low selectors is the selection of one of two controller outputs for the control of a single control valve. The signal selection is typically between a level and a flow controller, or between two pressure controllers. A constraint is that the single control valve must have a failure action that can be used with either controller. The selector compares the output of the two controllers and selects either the highest or the lowest, depending upon the selector being used. This output limit can be either 0% or 100%.

To ensure a smooth transfer between the two controller outputs, each controller must receive its feedback signal from the output of the selector. This prevents the other controller, whose output was not selected, from experiencing reset windup. *Reset windup* is a condition where a controller continues to change its output, because of a deviation between the setpoint and the measurement, until the output reaches its limit.

The transfer between the two controllers is usually clean with no back-and-forth switching when the output from the selector is sent to each controller feedback connection. The unselected controller changes its output in such a way as to increase the difference between the selected output signals. The transfer does not switch back unless there is a change in the operating conditions. For example, a process can use a flow control loop to control the flow from a storage tank during normal operation. **See Figure 46-8.**

TECH FACT

The western water irrigation industry uses the term miner's inch. A miner's inch is the amount of water that discharges through a square inch of opening under a certain head of pressure. The miner's inch varies from 8.977 gpm to 11.690 gpm, depending on the state where the measurement is being made.

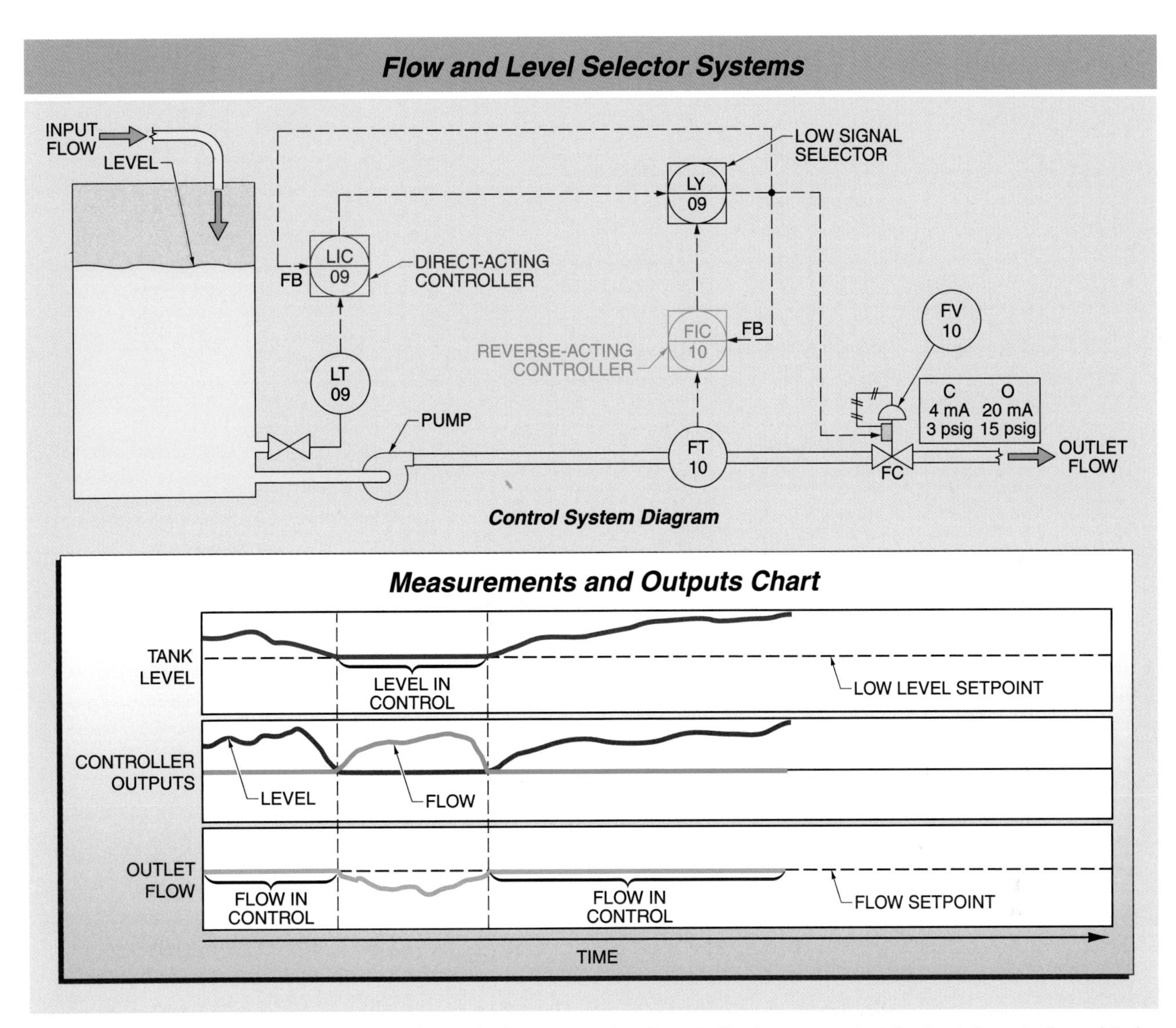

Figure 46-8. A flow control system out of a tank also uses a level controller to ensure that the tank is not allowed to be pumped dry.

If the input flow into the storage tank is less than the normal controlled outlet flow, the level in the tank decreases. At a predetermined low setpoint, the level controller signal drops below the flow controller signal, is selected by the low signal selector, and takes over control of the control valve (FV 10). The control valve is throttled to maintain the tank level at its low setpoint to avoid draining the tank. With an increase in inlet flow to the storage tank, the level controller increases the flow through the control valve until it reaches the setpoint of the flow controller (FIC 10). At that point the flow controller output becomes lower than the level controller output, is selected by the low selector, and takes over control of the control valve.

LIMIT CONTROLS

The purpose of a limit device is to put limits on a signal. **See Figure 46-9.** A high limit device compares an input signal to an internally stored value. The input signal is allowed to pass through the limit device unchanged as long as the signal is below the high limit value.

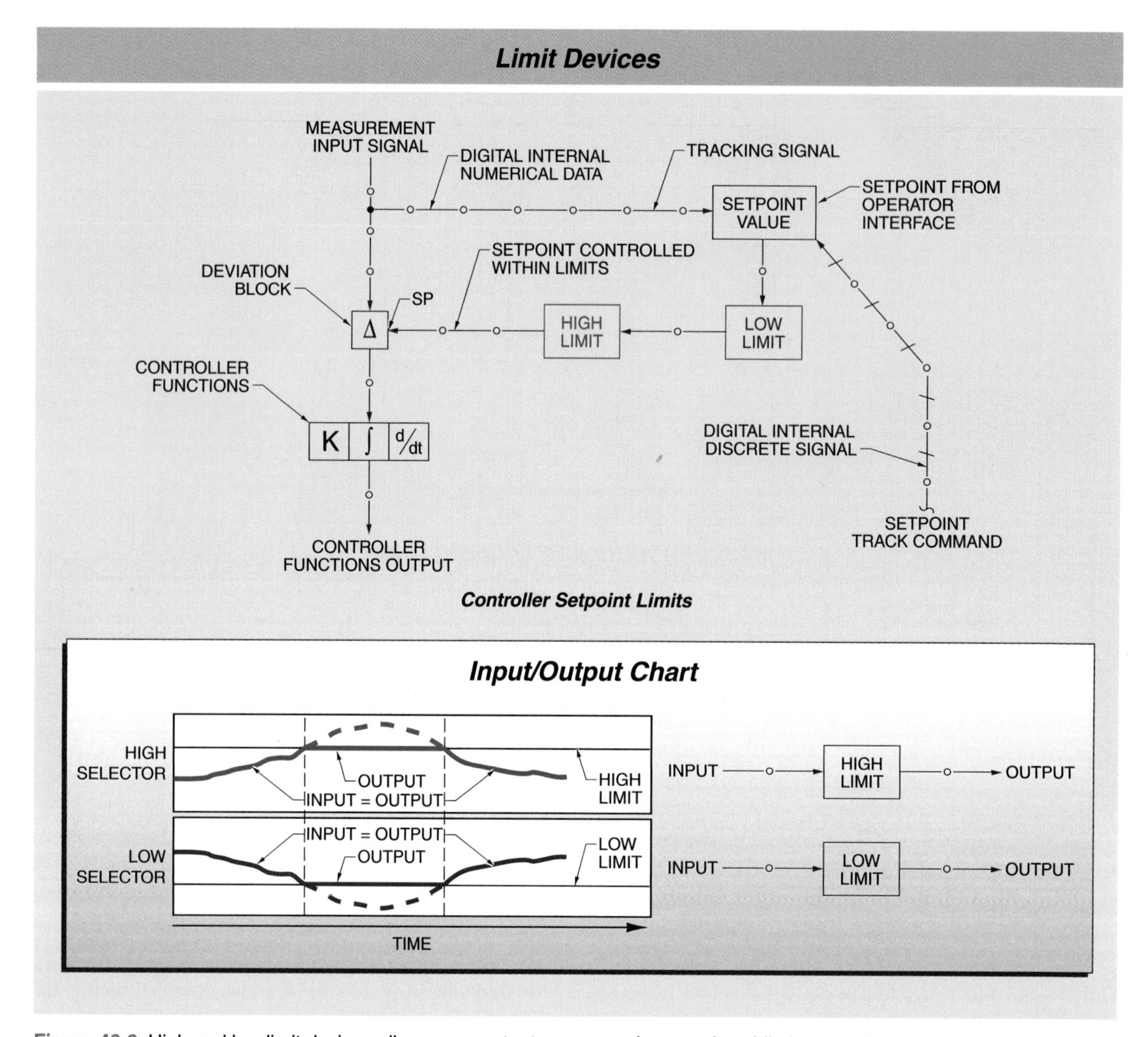

Figure 46-9. High and low limit devices allow an operator to program the setpoint while keeping it within predetermined limits.

An input signal greater than the high limit value is blocked and the high limit value becomes the output. A low limit works in a similar way. An input signal less than the low limit value is blocked and the low limit value becomes the output. A high and low limit device has both high and low limits. Limit devices are available in pneumatic, electric, or digital systems. As with signal selectors, limit devices may not be available for the more complex control strategies in some digital controllers.

A common application is in limiting controller setpoint values so that they cannot be set higher or lower than predetermined values. The operator setpoint is sent through a high limit device and a low limit device to control the setpoint to a value between the limits. Then, the controller compares the setpoint to the input signal and uses the controller function to calculate an output. Limits are used only infrequently with controller output signals and are never used on measurement inputs to controllers.

Gap Action Control

Gap action control is a control strategy that operates only when a measurement value is outside the high and low limits of a predefined range, or gap. When the measurement is between the high and low gap values, the controller output holds its last output prior to the measurement entering the gap. This means that the controller setpoint and the measurement value are the same. When the measurement and setpoint values are the same, there is no deviation and the controller makes no changes in its output. **See Figure 46-10.**

A gap action controller is a standard proportional plus integral (PI) controller except that the setpoint is obtained from the measurement value after it has gone through a high and low limit. The high and low limits are the high and low gap values. When the measurement signal goes above the gap upper limit (upper set-point), the measurement signal is blocked at the high limit device and the limit device output is fixed at the upper setpoint value. The controller now has a deviation and starts changing the output to bring the measurement back to the setpoint.

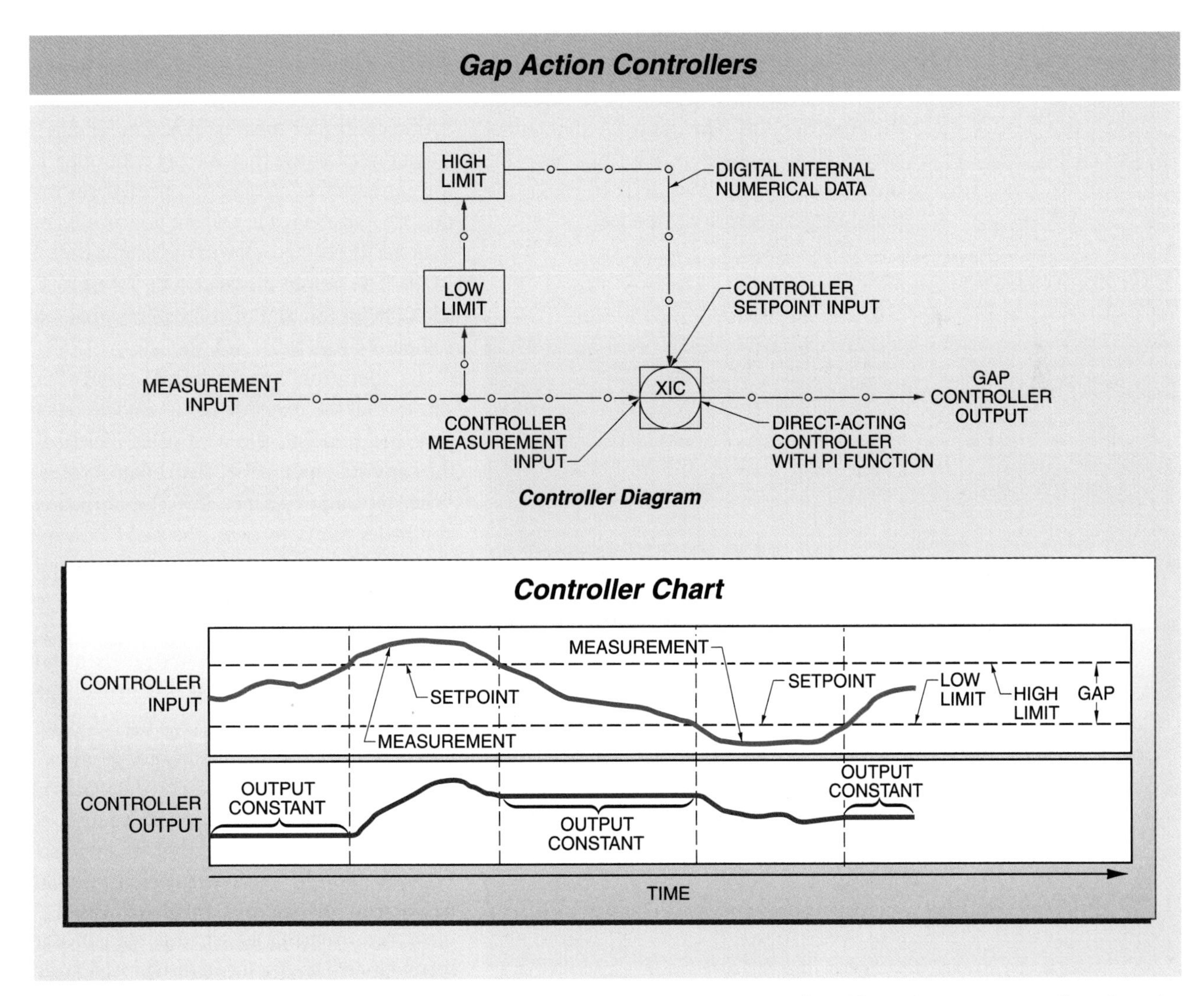

Figure 46-10. A gap action controller allows a process to drift within the gap limits while acting as a controller when the process is outside the gap limits.

Once the measurement is again within the gap, the controller output stops changing. A similar action occurs when the measurement goes below the lower setpoint. The controller again acts to bring the measurement back into the gap.

Precision Throttling. A gap action controller is used for those applications where control action is only desired above or below predetermined values. Precision throttling is an application where a gap action controller is a good solution to a control problem. There are a number of processes that require a precise throttling adjustment where precision is not available at that valve opening.

For example, a pH neutralization application takes a large quantity of reagent to get close to the neutral point but only a very small amount of reagent to maintain the control point. The control valve, at the required open position, may not have the precise adjustment required to make the small reagent additions needed.

Cleaver-Brooks

Temperature control of heat-treatment processes is critical in the manufacture of pressure vessels.

To implement this strategy, a gap action controller is used with a large fail closed reagent control valve to maintain the pH at a value near the setpoint and a small control valve to provide precise trim. **See Figure 46-11.** The pH controller (AIC 20) output goes to the input of the gap action controller and is also connected to a small fail closed, linear throttling control valve (AV 20) that regulates reagent to maintain the desired pH control point. The high and low limits of the gap action controller are set at 75% and 25%, respectively.

The output of the gap action controller goes to the large reagent valve. The large reagent control valve (ZV 21) should be 20 to 30 times larger than the small reagent valve and should have an equal-percentage flow characteristic. The pH controller tuning should follow normal practices. The gap action controller must be tuned to be much slower in response than the pH controller. If the gap action controller output changes too quickly, the change in reagent flow is greater than the throttling range of the small control valve. The output modules (AY 20 and ZY 21) convert the digital controller signals to standard 4 mA to 20 mA signals.

In operation, when the pH controller senses that the measurement value is lower than the setpoint, the controller increases the current output to the small control valve. When the output reaches 75%, the gap action controller starts to open the large control valve. The large valve continues to open until the control signal to the small control valve goes below 75%. At that point, the large control valve remains at a fixed position and trimming of the reagent flow is done with the small control valve to maintain the pH at the control point.

CASCADE CONTROL

Cascade control is a control strategy where a primary controller, which controls the ultimate measurement, adjusts the setpoint of a secondary controller. The secondary controller controls a less important measurement that influences control of the primary variable.

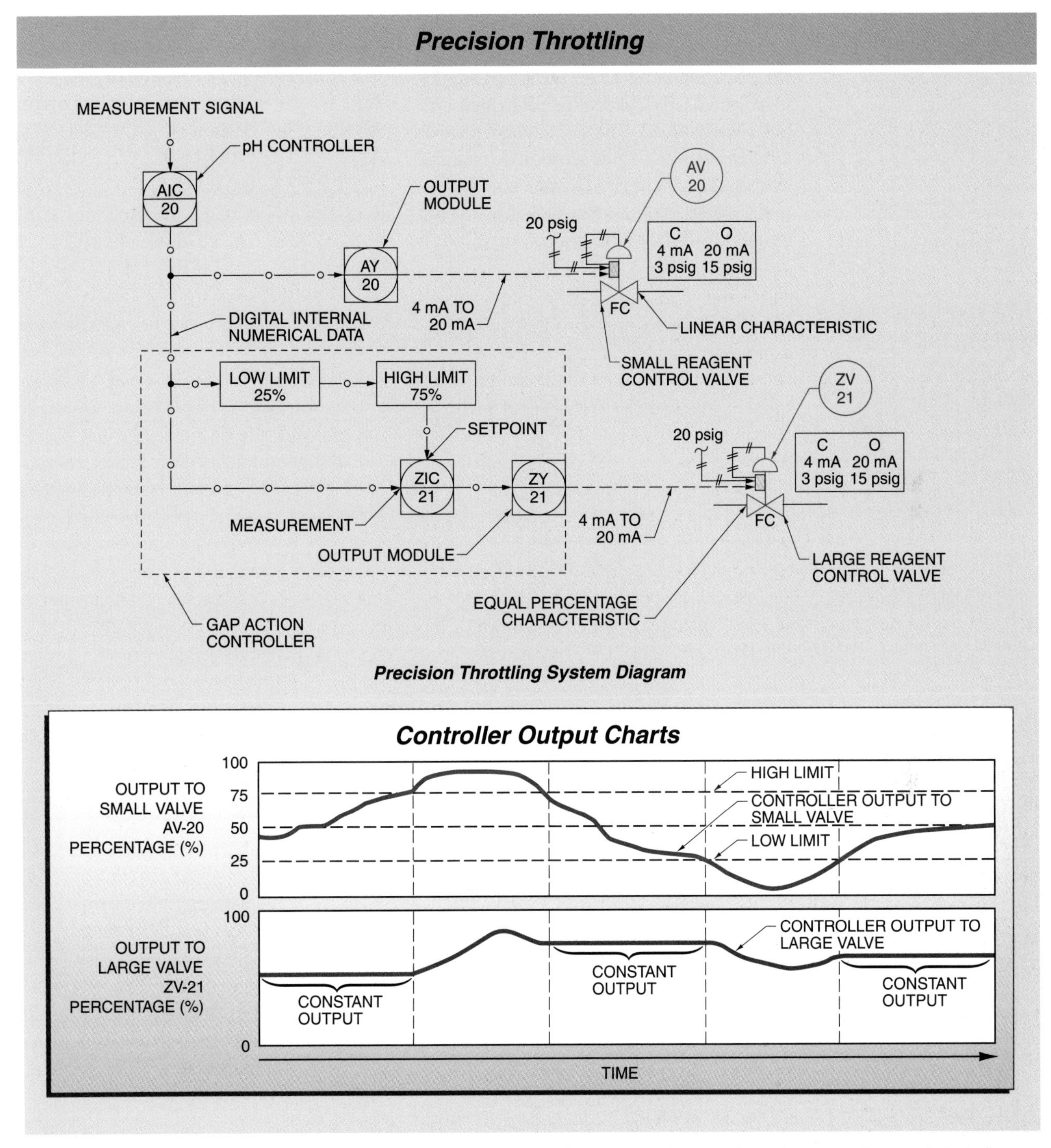

Figure 46-11. Precision throttling of valves is needed for control of processes that need precise control at a point where one valve is not sufficient.

There are a few key requirements that are necessary for a cascade control system to function properly. First, changes in the secondary controlled measurement must directly affect the primary measurement. Second, the response speed of the secondary controller must be at least four times faster than the primary controller. This is to ensure that the secondary controller has time to completely respond to setpoint changes from the primary controller before new changes are made.

For example, the temperature control of a steam-heated reactor is a good application for cascade control when the steam supply is subject to large changes in steam pressure. **See Figure 46-12.** The steam control valve (TV 25) can be directly throttled from the temperature controller (TIC 25), but changes in the steam pressure cause changes in the steam flow through the control valve.

The dimensions of an orifice plate must satisfy a number of conditions. The orifice plate thickness must be less than 2.5% of the pipe inside diameter and less than 12.5% of the orifice bore. If the plate must be thicker, the downstream side of the orifice bore is beveled to reduce the effective thickness.

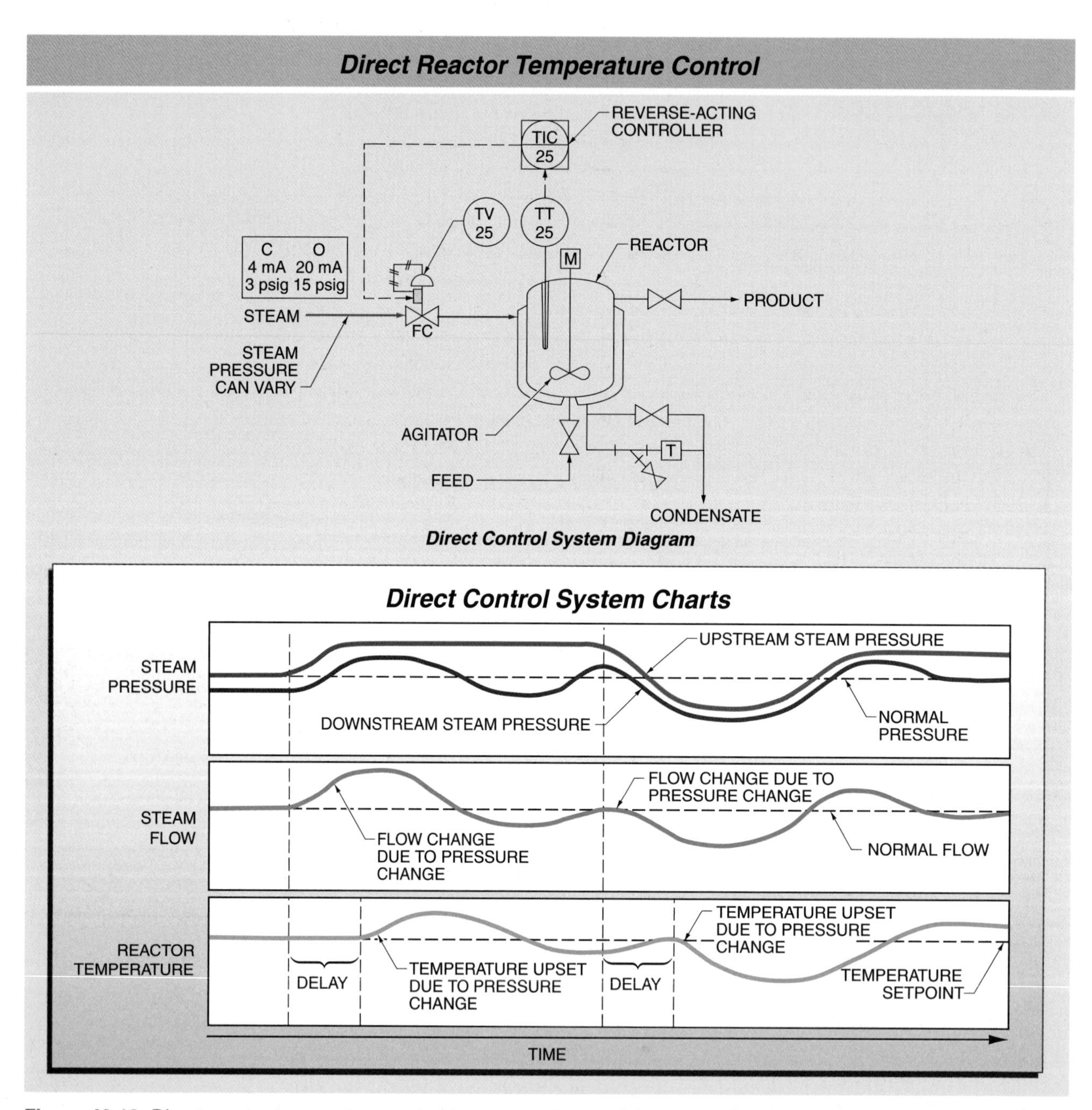

Figure 46-12. Direct reactor temperature control is not very successful at controlling temperature upsets due to varying steam pressure.

A reactor and its contents have a large mass and this mass acts as a lag to the control system because it takes time to change the temperature of a large mass. Therefore, the temperature sensor may not sense a change in reactor temperature resulting from the changes in steam pressure caused by changes in steam flow for a considerable amount of time. The result is very poor control with wide swings of the temperature above and below the setpoint as the changes in steam pressure changes the flow and affects the reactor temperature.

A solution to the problem of poor control is to install a pressure transmitter (PT 26) downstream of the steam control valve (PV 26) and connect the transmitter to a pressure controller (PIC 26). **See Figure 46-13.** The pressure controller, instead of the temperature controller, is used to adjust the steam valve. The temperature controller (TIC 25) is then used to set the pressure controller setpoint.

This is the arrangement of a standard cascade controller. The pressure controller (secondary controller) immediately corrects for changes in steam pressure before the change in steam pressure can significantly affect the heat input rate to the reactor. Changes in the temperature in the reactor cause the temperature controller (primary controller) to adjust the setpoint of the pressure controller (secondary controller).

Internal Functions

Cascade control is frequently offered in many modern stand-alone digital controllers but usually with restricted operational choices. Digital cascade controllers that are used in DCS and PLC systems are usually fully functional, with auto/manual operations in both primary and secondary controllers. The operating modes include fully automatic operation, local setpoint operation of the secondary controller, and manual operation of either controller. The preconfigured cascade controllers available in some digital controllers simplify the communications requirements by eliminating some of the operating modes.

The internal operation of a cascade controller is explained with the use of special symbols and connecting lines based on existing MCAA symbols. **See Figure 46-14.** The digital symbol key is color-coded to help explain each operation.

The internal functions of fully operational primary and secondary cascaded controllers and the associated communications are complex. This complexity is hidden inside modern digital cascade controllers and becomes important during manual operation. This internal digital complexity replaces the complicated manual balancing operations that were necessary when individual pneumatic or electronic controllers were used to form a cascade controller during auto/manual switching.

Connecting two digital controllers together in a cascade operation while still allowing manual control is more difficult than it seems. For stand-alone digital controllers, manual operation usually means that the secondary controller is placed in manual while the primary controller remains in automatic. For a DCS or PLC, manual operation usually means that both controllers are placed in manual. The best way to understand the difference is to ensure an understanding of the internal functions of the cascade controller.

Exhaust stack oxygen analysis is used to fine-tune a combustion process.

Cascade Reactor Temperature Control

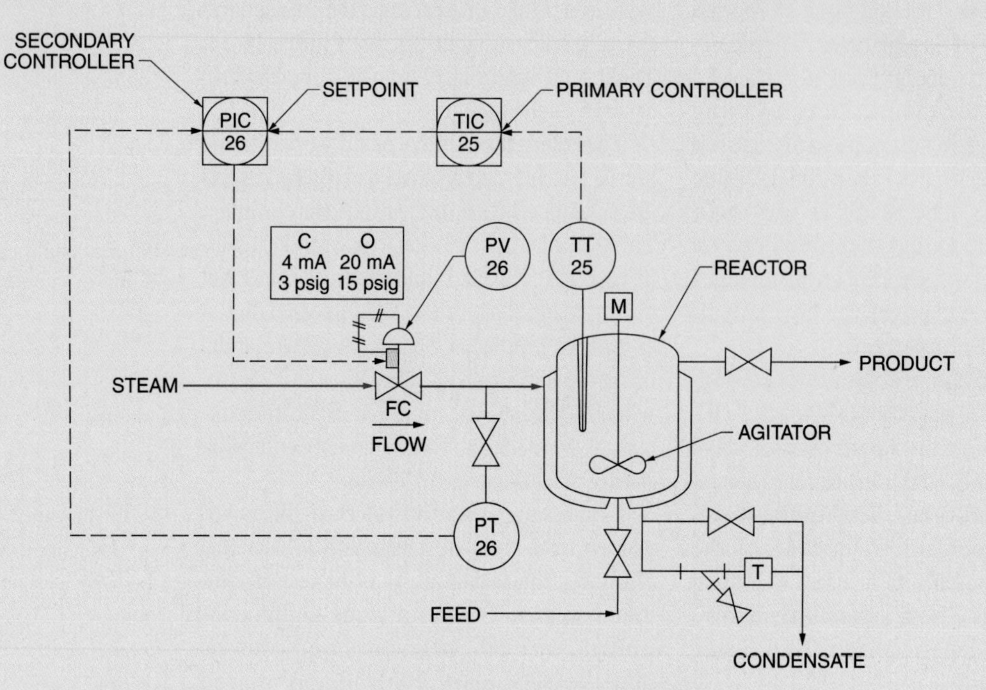

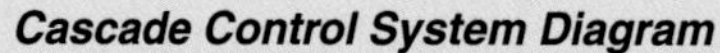
Cascade Control System Diagram

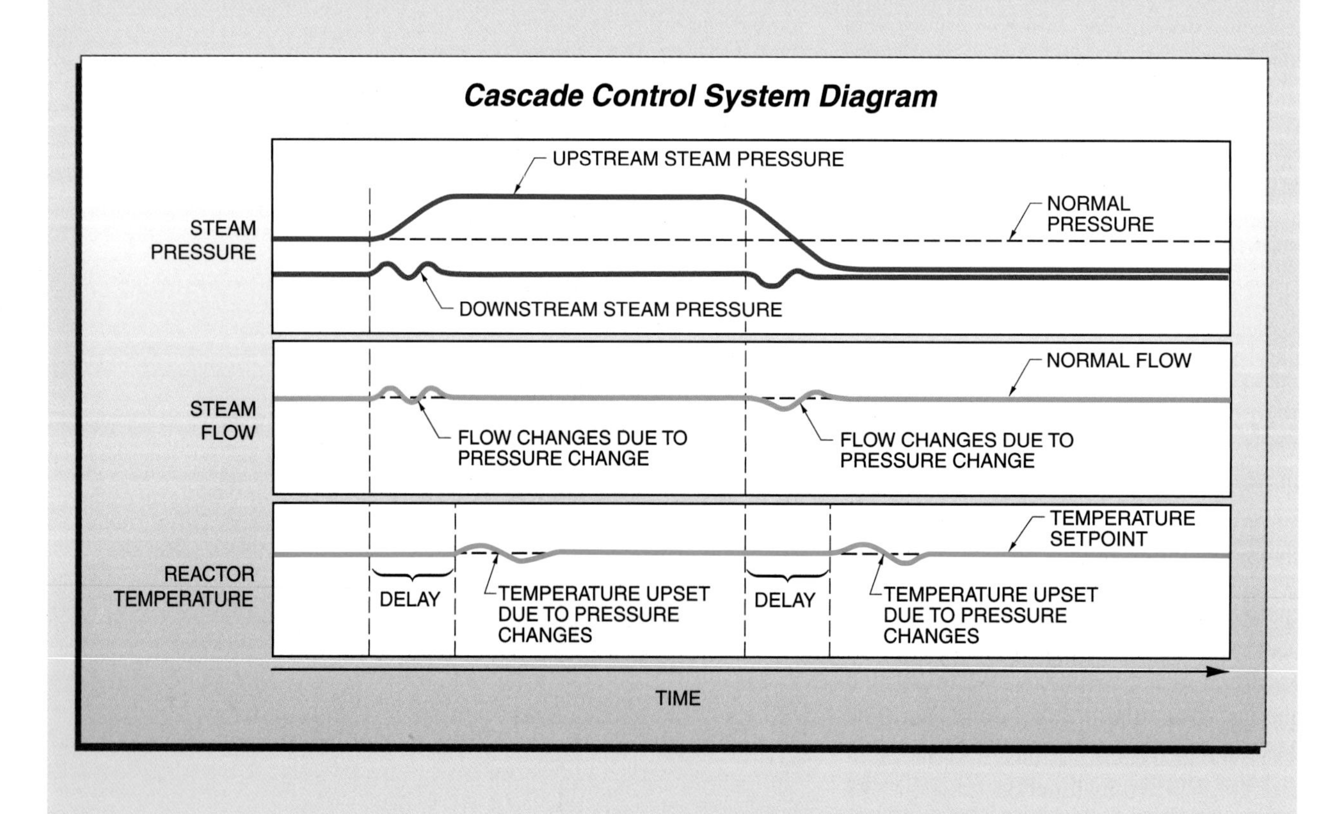

Figure 46-13. Cascade reactor temperature control is very successful at reducing temperature upsets.

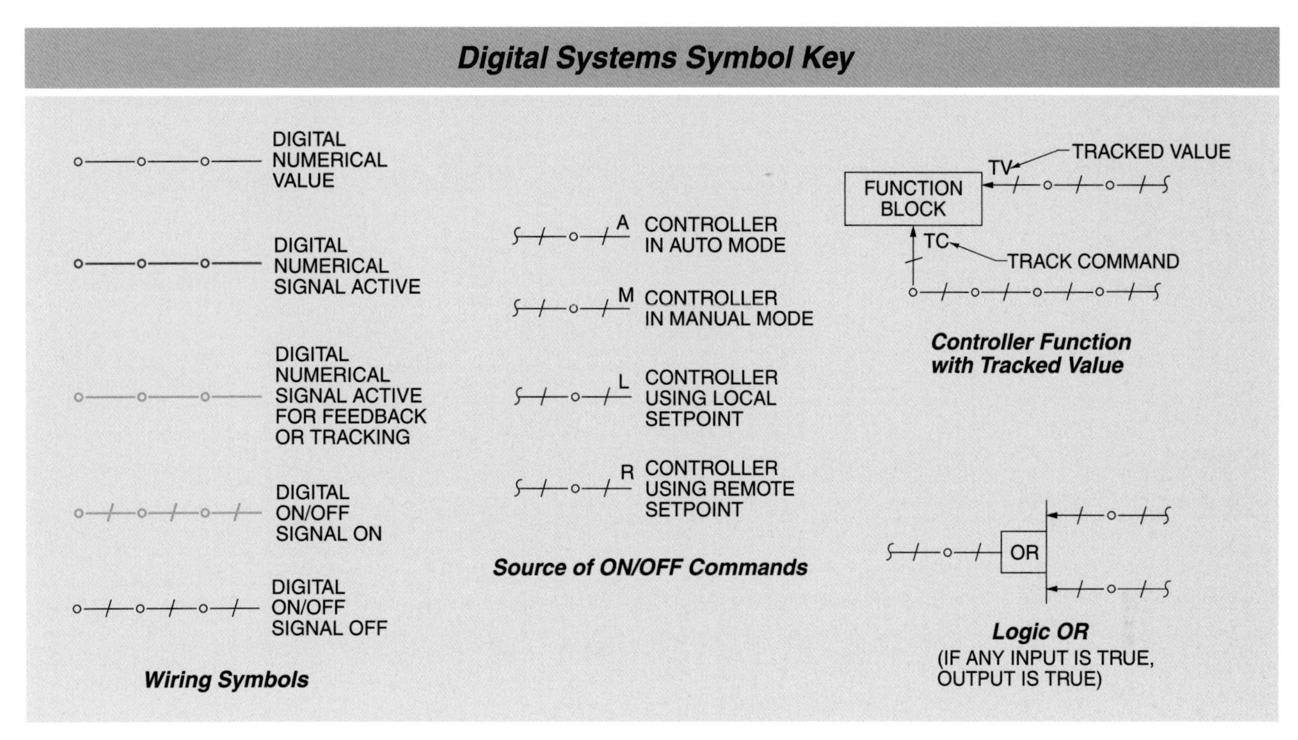

Figure 46-14. Common symbols are used to describe the communications within digital systems.

The basic arrangement of a cascade controller consists of a primary controller that sends its output to a secondary controller. Both controllers have setpoint tracking and bumpless transfer between auto and manual modes. The secondary controller receives the primary controller output signal at a remote/local (R/L) setpoint switch. This is also called an internal/external setpoint switch. **See Figure 46-15.**

The R/L setpoint switch is controlled from the face of the secondary controller or from the operator station. The R/L setpoint switch determines if the setpoint is generated locally (internally), as in a standard controller, or is obtained remotely (externally) from the output of the primary controller. Regardless of the source of the setpoint value, it must be sent back to the primary controller external feedback. The status of all of the auto/manual and the remote/local functions must be communicated to the various functions to keep both controllers synchronized. This is the major source of the communication complexity.

One additional complexity experienced in digital cascade controllers is the need to change the engineering units of a signal when it goes from one controller to another. Digital controllers use engineering units in their internal calculations. The output from the primary controller is 0% to 100% and must be converted to the secondary controller setpoint engineering units before the signal can be used in the R/L setpoint switch.

The setpoint engineering units have to be the same as the measurement engineering units, such as °F, gpm, or psig. The feedback signal must be converted back to 0% to 100% before the primary controller can use it. Pneumatic and electric cascade controllers do not internally use engineering units and thus do not need scaling.

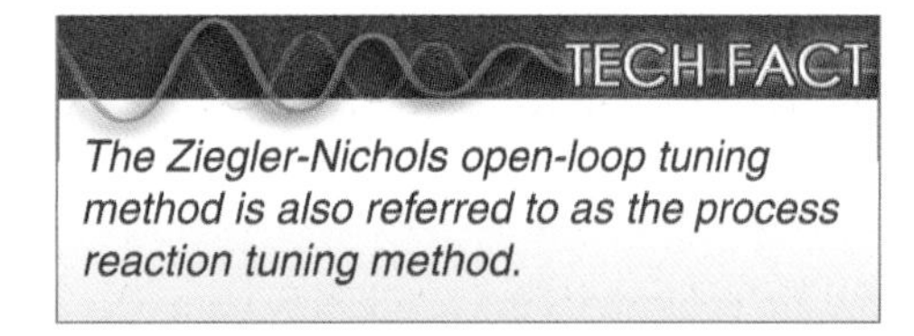

The Ziegler-Nichols open-loop tuning method is also referred to as the process reaction tuning method.

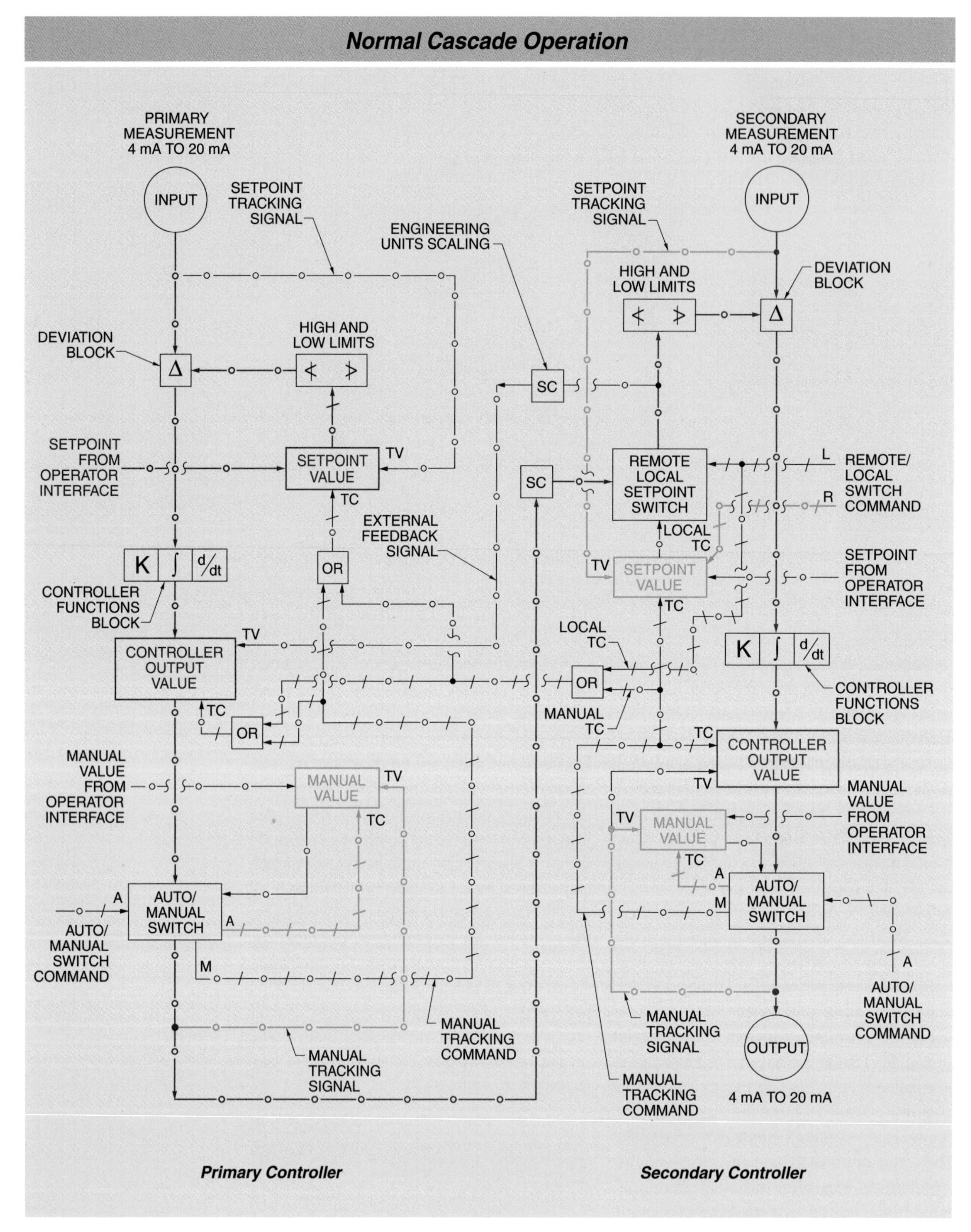

Figure 46-15. A cascade controller in normal operation has an operator-determined setpoint for the primary controller. The primary controller then calculates the setpoint for the secondary controller.

Normal Cascade Operation

In normal cascade operation, the operator sets the primary controller setpoint from the controller face or operator station. The active control signal path is shown in red. The setpoint is compared to the primary measurement in the deviation block and the difference is sent to the controller functions block. The primary controller output resulting from the controller functions operating on the difference is sent to the auto/manual (A/M) switch station. Since the A/M switch station is in automatic, the station passes the controller output on to the units scaling block.

The manual value block receives a tracking command from the A/M switch station, as shown by the green TC line. This means the manual value block tracks (follows) the controller output, as shown on the orange TV line. The manual value block tracks the controller output so that if the A/M station switch is changed to manual, the controller output stays the same and allows a bumpless transfer of control.

The output from the primary controller is scaled and passed on to the R/L setpoint switch in the secondary controller, as shown by the red line. The R/L setpoint switch is in the remote position, as shown by the green TC line to the setpoint value block. The secondary controller setpoint value block tracks the secondary controller measurement value, as shown by the orange TV line. This is done so that if the setpoint is switched to the local position, there is no change in the setpoint value.

The remote setpoint from the primary controller output passes through the R/L setpoint switch and is compared to the secondary controller measurement signal in the deviation block. The difference is passed on to the secondary controller functions block. The resulting secondary controller output is passed through the A/M station, since it is in automatic, to the cascade controller output. The manual value block receives a tracking command from the A/M switch station, as shown by the green TC line. This means the manual value block tracks the controller output to allow a bumpless transfer of control if switched to manual.

Secondary Controller Local Setpoint

When the secondary controller is switched to a local setpoint, the R/L setpoint switch is in the local position, as shown by the green line. **See Figure 46-16.** In the local position, the secondary controller setpoint value block is active, as shown in red. It is adjusted at an operator interface. The remote setpoint signal, which is the primary controller output, is blocked at the R/L switch. The local setpoint signal is sent from the setpoint value block to the deviation block where it is compared to the secondary measurement value, as shown by the red lines. The deviation is sent to the secondary controller functions block where a controller output value is calculated. The output value passes through the A/M switch, since the switch is in automatic, to become the secondary controller output.

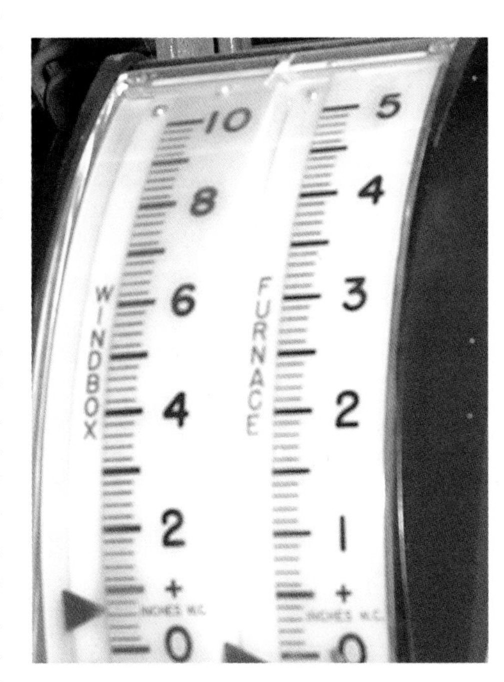

Draft pressure gauges are used to monitor the chimney draft for a combustion process.

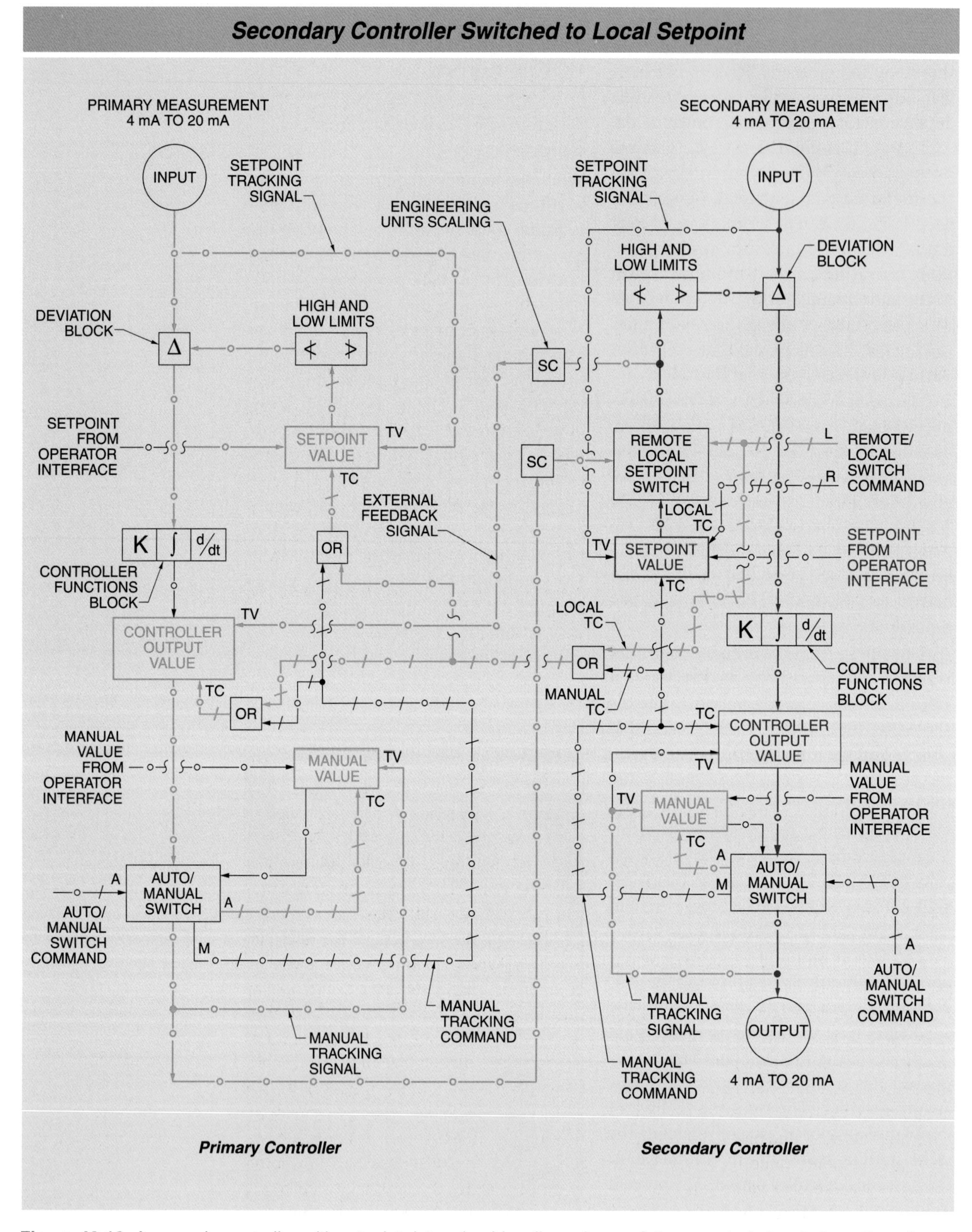

Figure 46-16. A cascade controller with setpoint determined locally no longer follows control signals from the primary controller.

The primary controller output value block receives a tracking command, as shown by the green TC line. This means that the primary controller is forced to track the feedback signal from the outlet of the secondary controller R/L switch, as shown by the orange TV line going to the primary controller outlet value block. In addition, the primary controller setpoint value block receives the same tracking command and is forced to track the primary measurement value, as shown by the orange TV line. The primary and secondary controller A/M switch stations are both in auto and are tracking their respective output values.

This is important during maintenance and calibration procedures. The secondary controller can be switched to local setpoint to allow the primary measurement sensor to be calibrated or replaced. However, the primary controller and sensor must be returned to normal operation and allowed to reach equilibrium before switching back to normal cascade control operations. If this is not done, the primary controller setpoint value block tracks an erroneous value and there may be control problems when switching back.

Primary Controller Manual Operation

When the primary controller is in manual operation, the output of the primary controller is determined by the operator-adjusted manual value. **See Figure 46-17.** The secondary controller is still operating in automatic mode, with the setpoint determined by the manual output of the primary controller, as shown by the red line. Because the primary controller A/M station is in manual, as shown by the green line, the primary controller output value tracks the secondary controller feedback value, as shown by the orange line to the primary controller output value block. The primary controller setpoint value block tracks the primary measurement value. The secondary controller A/M switch station is still in automatic, so the manual value still tracks the cascade controller output.

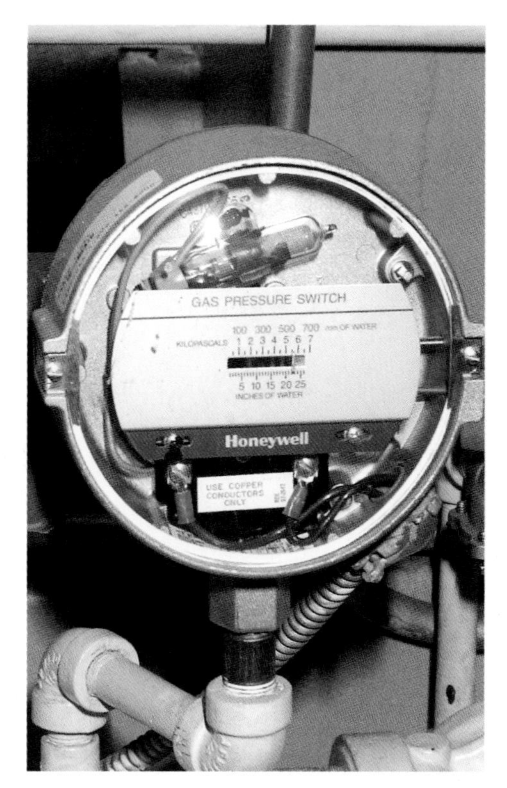

A gas pressure switch is used to shut off the flow of gas.

Secondary Controller Manual Operation

When the secondary controller is in manual operation, the secondary controller A/M switch station is placed in manual, as shown by the green line to the A/M switch station. **See Figure 46-18.** The operator has direct control of the controller output from the operator interface.

The secondary controller output value block is forced to track the secondary controller manual output, as shown by the orange line. The secondary controller setpoint value block tracks the secondary measurement value and the primary controller setpoint value block tracks the primary measurement value, as shown by the orange lines to those blocks. Since the primary controller A/M switch station is in automatic, the manual value block tracks the primary controller output, as shown by the orange line to the manual value block.

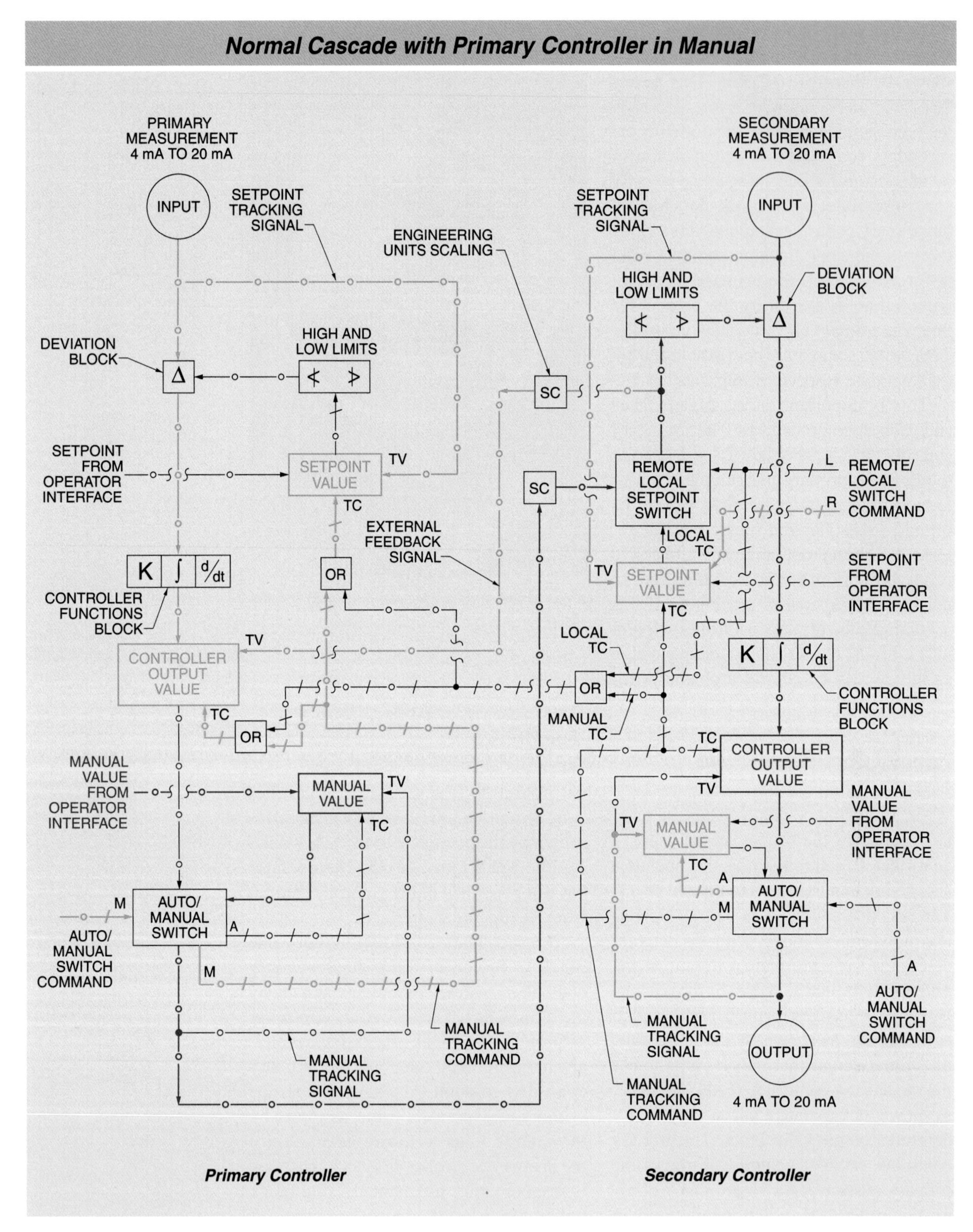

Figure 46-17. A cascade controller with the primary controller in manual no longer uses the primary measurement to control the secondary controller.

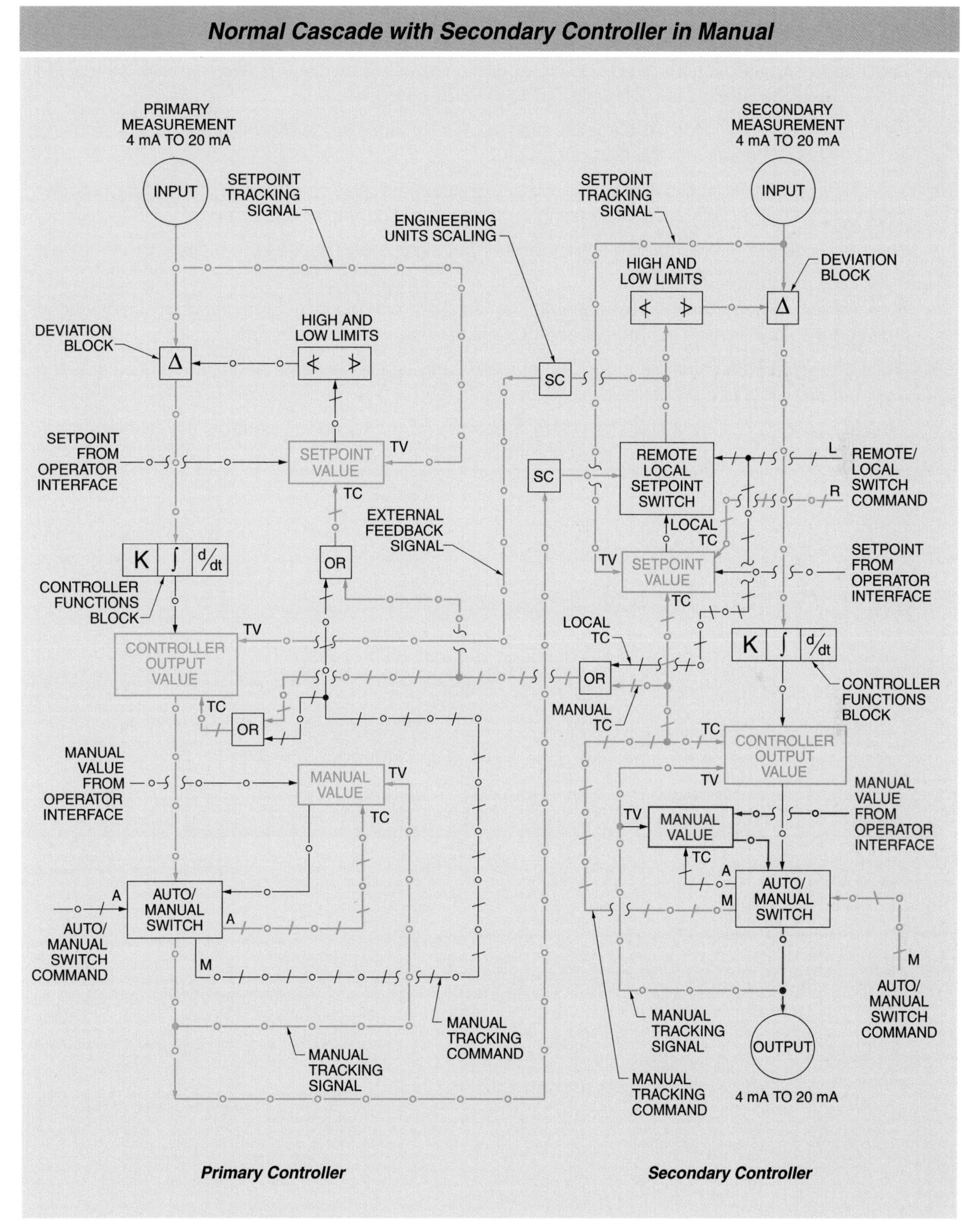

Figure 46-18. A cascade controller with the secondary controller in manual operation implements several tracking functions to enable bumpless transfer of control when switching back to automatic operation.

KEY TERMS

- *continuous process:* A type of processing operation in which continuous raw material is received and the resulting product is continually fed to the next operation.
- *batch process:* A type of processing operation in which multiple feeds are received and the resulting product is fed all at once to the next operation.
- *split range control:* An application where two or more control valves operate using the same controller output signal range, with each valve using a portion of the controller output signal.
- *selector:* A device that compares two or three input signals and passes the highest or the lowest signal to the output of the device.
- *reset windup:* A condition where a controller continues to change its output, because of a deviation between the setpoint and the measurement, until the output reaches its limit.
- *gap action control:* A control strategy that operates only when a measurement value is outside the high and low limits of a predefined range, or gap.
- *cascade control:* A control strategy where a primary controller, which controls the ultimate measurement, adjusts the setpoint of a secondary controller.

REVIEW QUESTIONS

1. What is the difference between a continuous and a batch process?
2. What are several applications of split range control?
3. Define selector, and give an example of why a high selector may be used for a process.
4. What is gap action control, and how does a gap action controller operate?
5. What is cascade control, and what are some of the requirements necessary for a cascade control system to function properly?

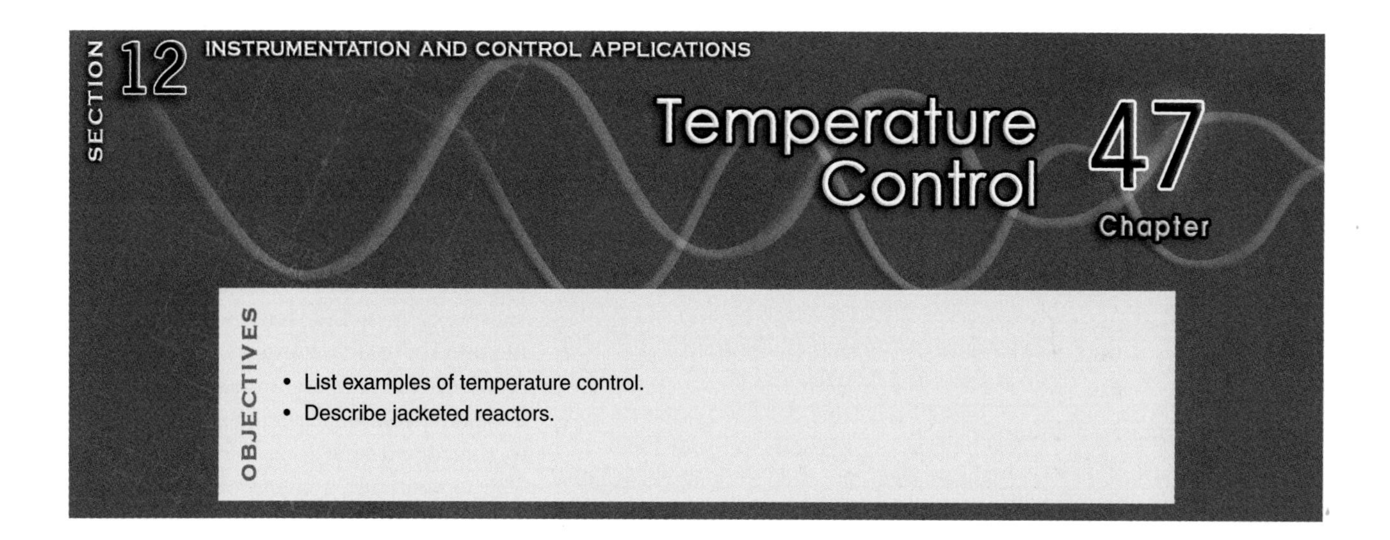

Temperature Control

- List examples of temperature control.
- Describe jacketed reactors.

HEAT EXCHANGERS

Temperature control is the most common control application. The applications can range from high temperatures to extremely low temperatures. At the high-temperature end, burning fuels are used for steam generation, to provide heat to hot oil heat transfer fluid, to promote the cracking of heavy organic compounds, or to directly heat objects in furnaces. At the low-temperature end are refrigeration systems, building air conditioning systems, and cryogenics, such as the liquefaction of gases. In between these extremes is a vast array of heat-exchanging applications using cooling water, steam, or other heat transfer fluids. **See Figure 47-1.**

A very common way of transferring heat, either for cooling or for heating services, is with the use of a heat exchanger. A *heat exchanger* is any piece of equipment that transfers heat from one material to another. There are many designs of heat exchangers, but the common element is a relatively thin layer of material separating the cooling or heating fluid from the process fluid. The barrier material is typically a conductive metal that allows heat to flow from the higher temperature fluid to the lower temperature fluid.

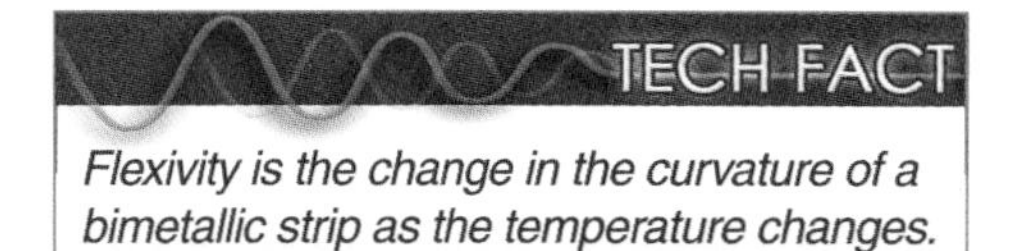

Flexivity is the change in the curvature of a bimetallic strip as the temperature changes.

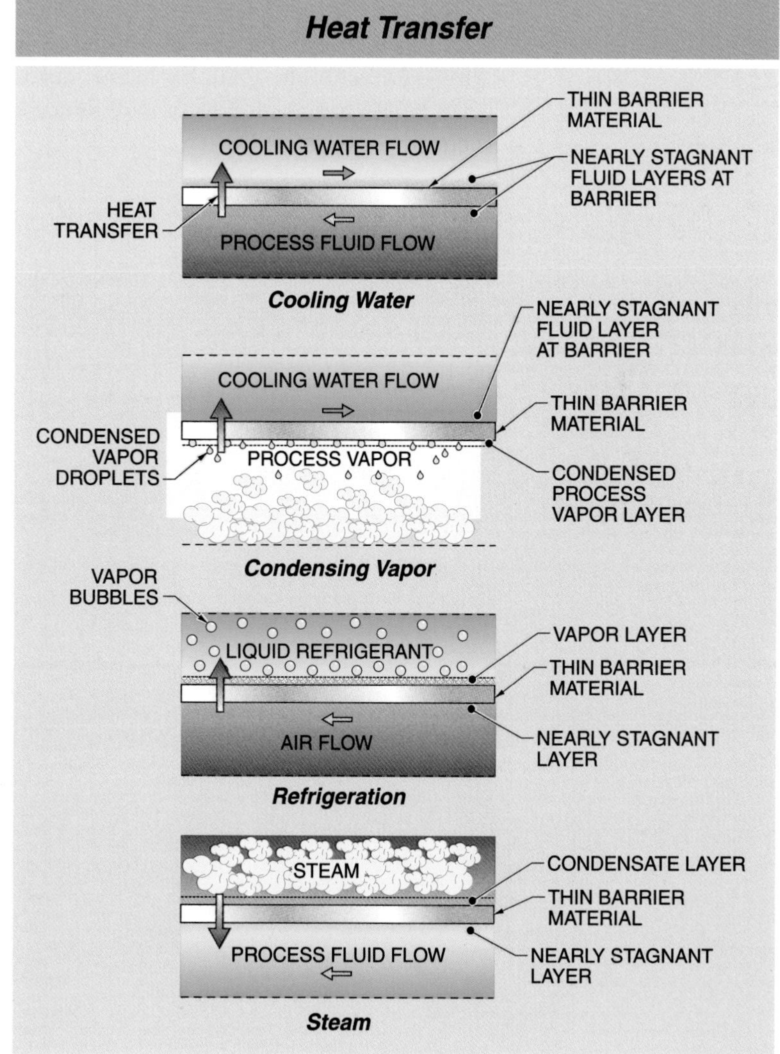

Figure 47-1. Heat transfer across a barrier can include liquids, gases, or condensing vapors on either side of the barrier.

The process fluid can be a vapor or a liquid and the heating or cooling fluid can also be a vapor or a liquid. The amount of heat transferred depends on the temperature difference between the two fluids, the area of the exchanger where heat is transferred, and the heat transfer coefficient. There are different heat transfer coefficient values for every different combination of process heating and cooling fluids and for different designs of heat exchangers.

One of the most common types of heat exchangers is a shell-and-tube heat exchanger. Shell-and-tube exchangers usually consist of a cylindrical shell containing the heating or cooling fluid. The process fluid is usually contained in tubes that run lengthwise through the shell. The ends of the tubes are sealed in tube sheets that close off each end of the shell. Exchanger heads are bolted to the shell to provide process piping connections.

A plate and frame heat exchanger uses thin metal sheets instead of tubes to separate the hot and cold flows.

Cooling Service

The most common cooling service is with water on one side of the exchanger and the hot process fluid—a gas or a liquid—on the other side. **See Figure 47-2.** The temperature of the outlet process fluid is measured and controlled to a desired value. The temperature is measured with a temperature-sensing element inserted into a thermowell mounted in the outlet piping. The thermowell allows the temperature-sensing element to be removed for servicing without having to shut down and drain the process fluid. The temperature-sensing element can be directly connected to a controller or to a transmitter (TT 04). The normal transmission mode is a 4 mA to 20 mA signal.

The cooling water enters the bottom side of the exchanger at the end closest to the process outlet and exits from the top of the exchanger at the end closest to the process inlet. This is done to ensure that the exchanger is always filled with cooling water and to use the cooling water efficiently. The control valve (TV 04) is usually in the inlet cooling water piping so that the lowest water pressure is in the heat exchanger. If there are leaks between the process side and the cooling water side of the heat exchanger, it is usually better for the process fluid to go into the cooling water.

The control valve is sized to handle about twice the maximum design flow, and an equal-percentage control valve characteristic is chosen to provide a large turndown. Cooling services usually do not require a control valve that has a tight shutoff, but it normally should have a fail open action. If temperature control is an important variable, a positioner should be used with the control valve; otherwise an I/P transducer can be used.

Any stand-alone microprocessor controller, DCS, or PLC is a suitable choice to control the temperature. A proportional-plus-integral (PI) controlled function should be selected and the action should be reverse. Heat exchangers have small thermal mass; therefore, they have very small lags and respond quickly to changes.

Cooling Service Heat Exchangers

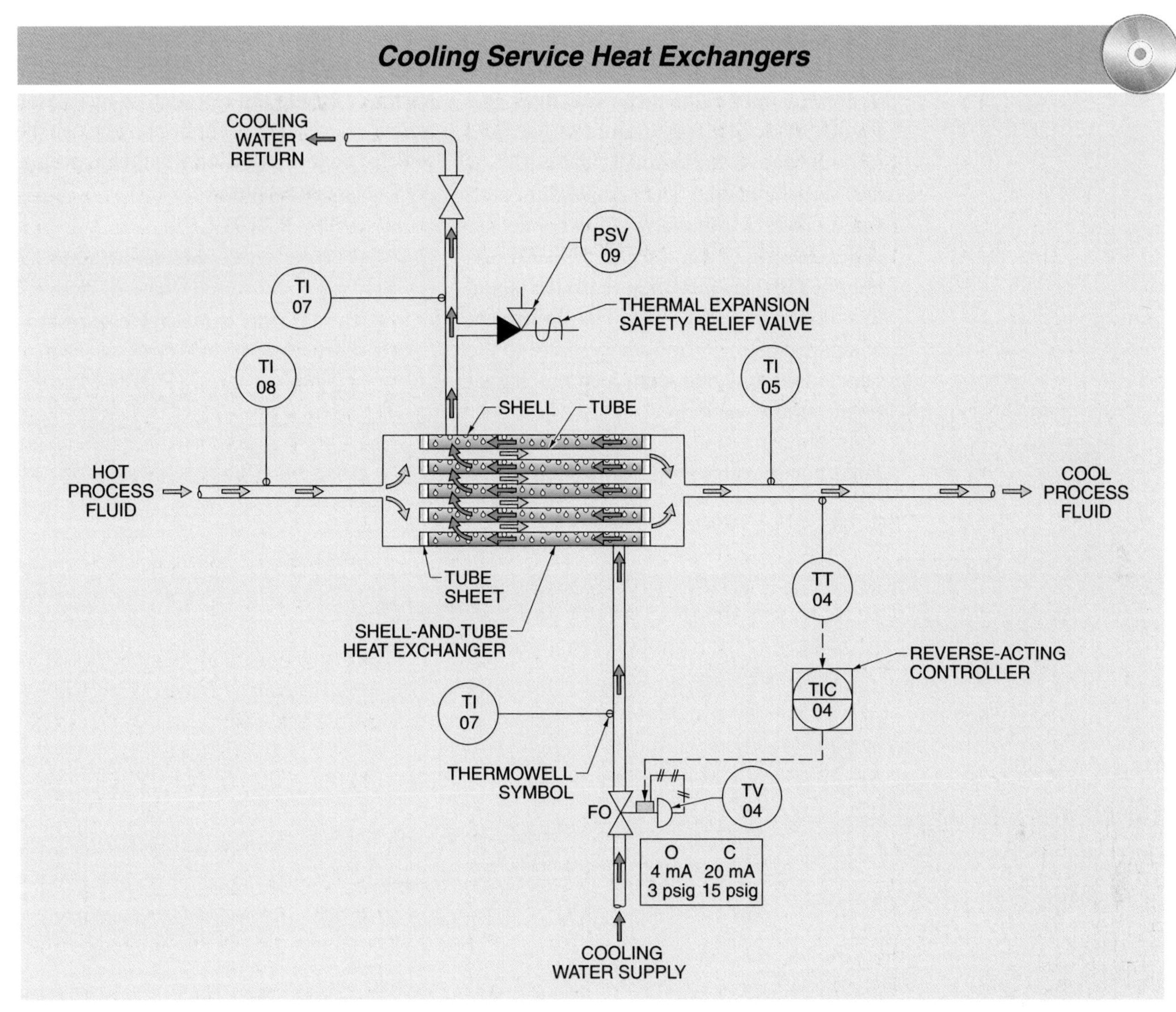

Figure 47-2. A cooling service heat exchanger often uses cooling water to lower the temperature of a process fluid.

If the cooling water in the exchanger can be closed off by valves, the trapped water must be protected from excessive pressures produced by thermal expansion of the cooling water. The protection is provided by the use of a thermal expansion safety relief valve (PSV 09) located within the trapped water piping. Typically, a ½″ or ¾″ thermal expansion safety relief valve is used, with the relief pressure set at the design pressure of the water side of the exchanger.

If the process fluid temperature is high enough to boil the cooling water, the thermal safety relief needs to be replaced with a standard relief valve sized to handle the steam that can be generated. The diagram shows a number of local temperature indicators (TI) that are used to troubleshoot the system if problems occur.

Steam-Controlled Heat Exchangers

Many heating services use steam as a heating fluid. A large quantity of heat is released when steam is condensed to water (condensate). The quantity of heat contained in steam vapor and condensate is listed in steam tables as enthalpy. It must also be remembered that steam at a temperature of 212°F has a pressure of 0.0 psig (14.7 psia).

A steam-controlled heat exchanger has the steam throttled with a control valve (TV 01) on the inlet to the heat exchanger. **See Figure 47-3.** The temperature of the fluid to be heated is measured at the outlet of the heat exchanger. The temperature sensor, located in a thermowell, is connected either directly to a controller or to a transmitter (TT 01), which then sends the signal to a controller (TIC 01). The controller compares the measured temperature to the selected setpoint and sends a control signal to the control valve.

In almost all heating applications, the steam control valve should have a fail closed action and, if possible, should close tightly. It is safer for most process applications for the source of heat to fail closed. The fail closed action and the heating service dictate that the selected controller action should be reverse. A PI controller should be used.

TECH FACT

On March 20, 1905, a boiler explosion at the R.B. Grover shoe factory, in Brockton, Massachusetts, killed 58 people and injured 117 others. This disaster led to the first rules and regulations for the safe construction of boilers.

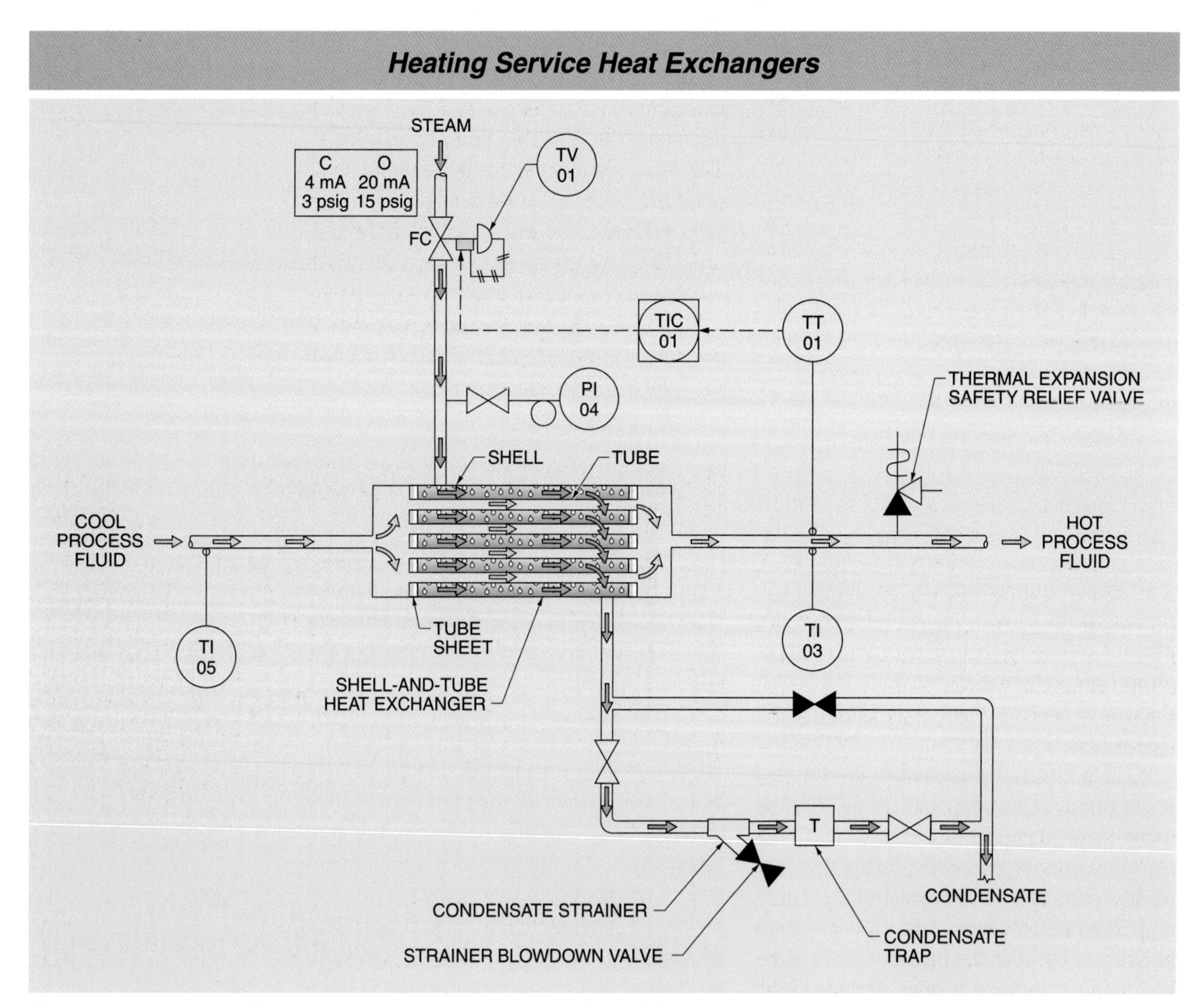

Figure 47-3. A heating service heat exchanger often uses steam to raise the temperature of a process fluid.

If the process fluid is a liquid and can be trapped within the heat exchanger by closed valves, the heat exchanger must be protected from overpressure due to thermal expansion. The usual protection device is a thermal expansion pressure relief valve or a rupture disc. If the process fluid can boil due to the heat of the steam, the relieving device must be sized to handle the process vapor flow.

The steam vapor is converted to condensate in the process of transferring heat to the process fluid. The condensate must then be removed from the heat exchanger to allow more steam to enter. Thus the steam is connected to the top of the heat exchanger and the condensate is removed from the bottom of the exchanger.

A condensate trap is used to ensure that the steam pressure is maintained in the exchanger. A condensate trap keeps the steam in the system and allows the condensate to exit. Removing condensate from the exchanger requires a small positive steam pressure in the exchanger. If the steam does not have enough pressure to push out the condensate, the condensate collects in the exchanger. A buildup of condensate shields the heat exchange surface from contact with the steam. This situation causes serious control problems that cannot be corrected by instrumentation changes.

Condensate-Level-Controlled Heat Exchangers

Some heat exchangers are used in a vertical orientation and are used for heating distillation columns or strippers. **See Figure 47-4.** In these applications, the purpose of the heat exchanger is to convert some of the process fluid to vapor so that the vapor can be carried up the column. In most cases, the conventional method of throttling the inlet steam is the best choice.

In some special cases where the process fluid boils at a low temperature, there can be problems with condensate removal from the exchanger. Some process fluids vaporize very readily and only a very low steam temperature is appropriate. However, if the required steam pressure is too low, the pressure may not be high enough to force condensate out of the bottom of the exchanger.

A different control solution can be implemented by placing the exchanger in a vertical orientation. If the lowest pressure steam at 0.0 psig has too much heat transfer, reducing the effective surface area of the exchanger can reduce the heat transfer. Using a fail closed equal-percentage or linear-characteristic throttling valve (TV 15) in the condensate exit piping allows the condensate to back up in the exchanger and reduces the effective area for heat transfer.

The steam is supplied through a low-pressure regulator (PCV 11). If too much heat is being transferred, the condensate is throttled to allow its level to rise in the exchanger shell. This reduces the effective surface area through which the steam can transfer its heat because the hot steam is only in contact with part of the heat transfer area.

A temperature sensor in a thermowell is connected to a transmitter (TT 15), which then sends the signal to a controller (TIC 15). The controller compares the measured temperature to the setpoint and generates an output. The control action needs to be reverse and a PI control function is the best choice. In some cases, a cascade control arrangement is used, with the temperature controller output determining the setpoint of a condensate level controller.

The sodium in salt pellets is used in water softener to replace the calcium, iron, and magnesium ions present in typical hard water.

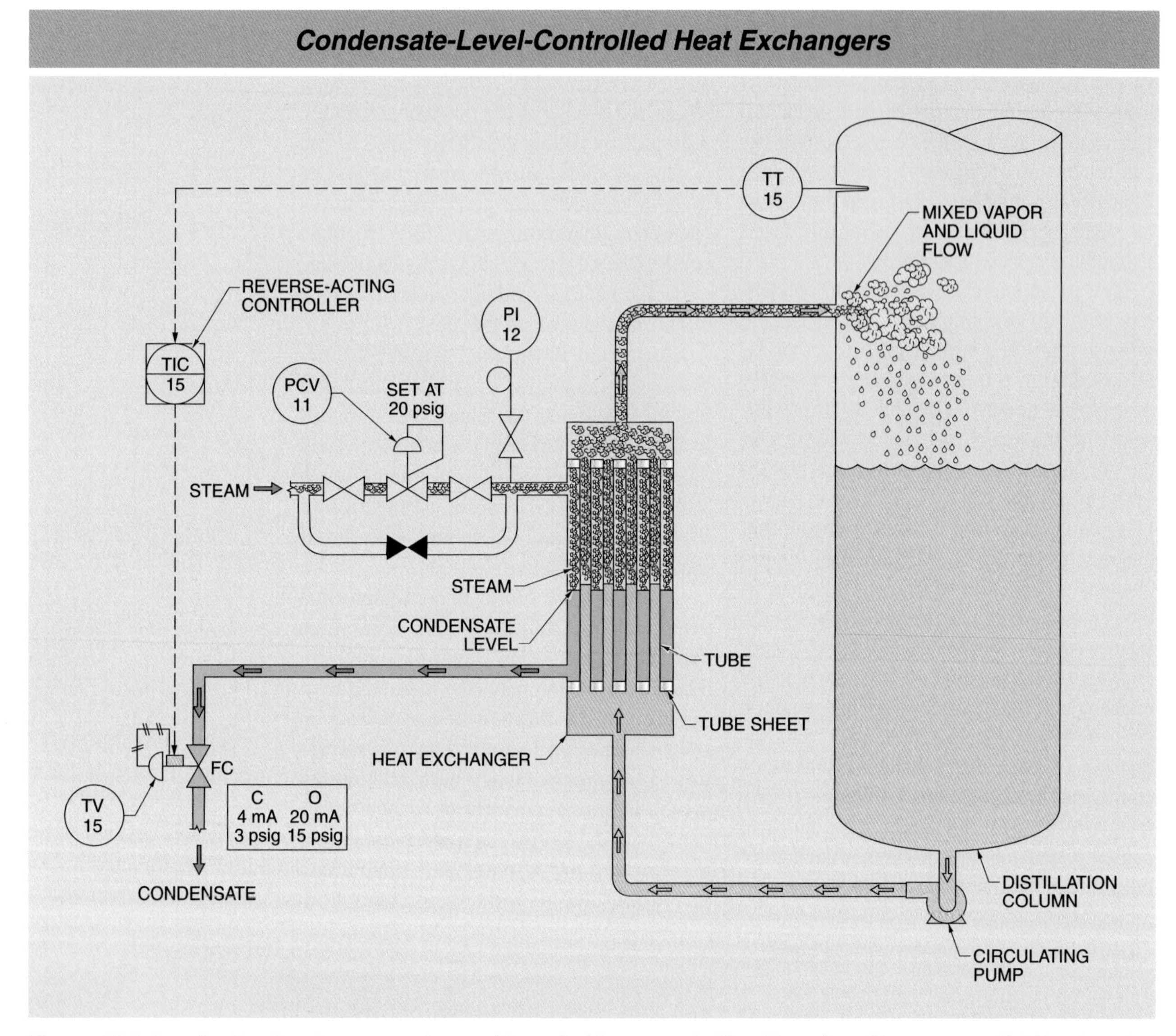

Figure 47-4. A vertical heat exchanger can be used to control the amount of heat transferred to a process fluid by changing the level of condensate in the exchanger.

JACKETED REACTORS

A *jacketed reactor* is a vertical vessel used in batch processing whose sides and bottom are covered with a steel shell spaced about an inch from the outside of the reactor and used to contain either a cooling or heating fluid. **See Figure 47-5.** A jacketed reactor works like a heat exchanger to transfer heat from one fluid to another. Jacketed reactors need to be stirred with an agitator to help transfer heat between the reactor walls and the process material. Sometimes baffles are installed in the reactor to increase the turbulence and the heat transfer. A jacketed reactor usually has baffles in the jacket to improve the heat transfer from the water.

In steam heating service, the reactor jacket is subject to the same condensate removal problems experienced with heat exchangers. Jacketed reactors contain a large thermal mass in the reactor and thus are slow to change temperature. This must be taken into consideration when designing a control system.

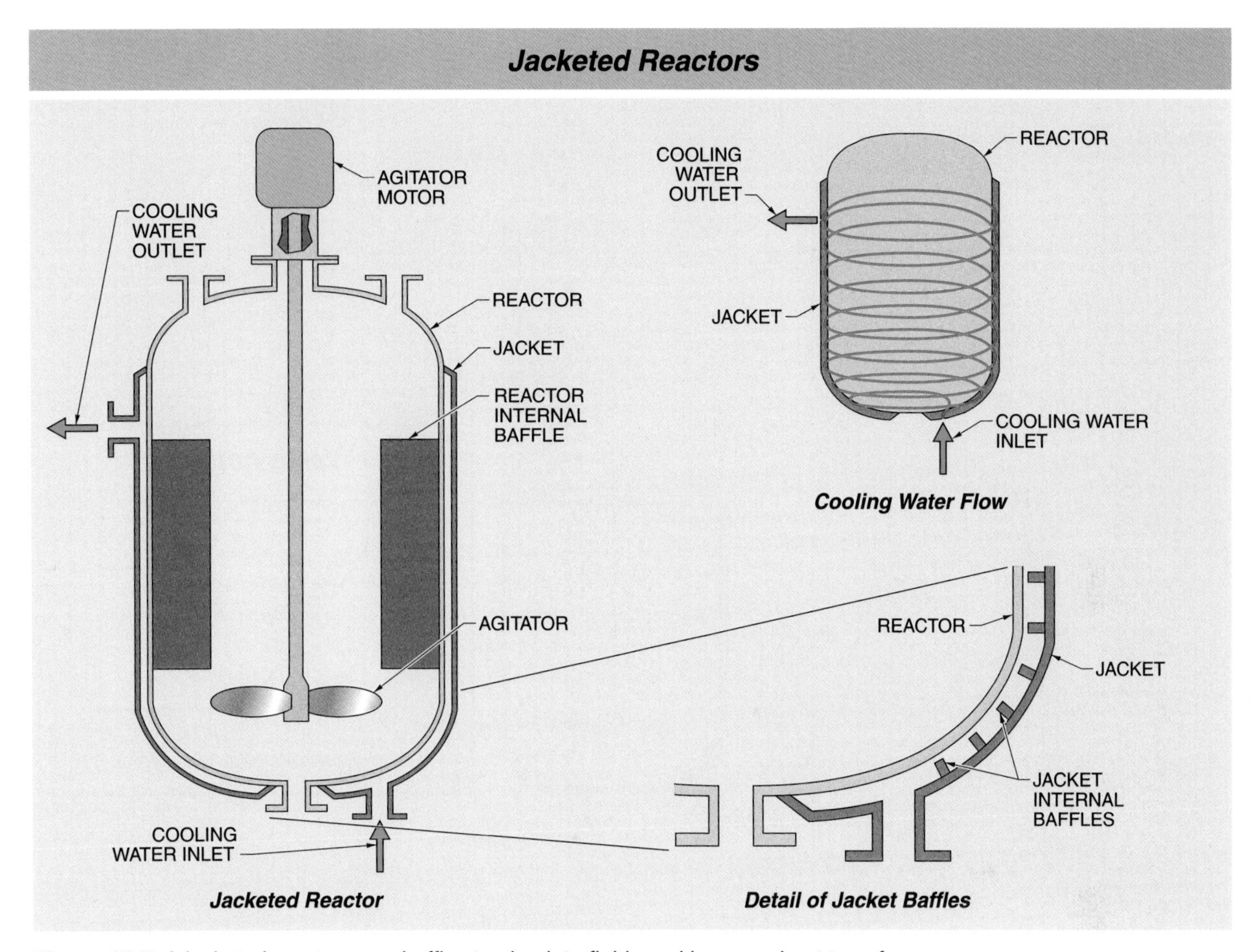

Figure 47-5. A jacketed reactor uses baffles to circulate fluids and increase heat transfer.

Circulated Water Temperature Control

A *circulated water system* is a system that provides both heating and cooling for batch reactors. For example, the contents of a reactor typically need to be heated to reach the reaction temperature. **See Figure 47-6.** After the reaction begins, the cooling system needs to remove the heat generated by the reaction in order to maintain the reactor temperature at the setpoint. Applying steam to the jacket and then switching to cooling water is a cumbersome method and does not provide good temperature control.

In a circulated water system, water is pumped into the bottom of the reactor jacket where baffles are used to direct the water around the jacket to the exit at the top. The water from the top of the jacket is returned to the water circulating pump after passing through a steam heated exchanger. A chilled water supply is also connected to the pump suction. A branch line from the jacket exit carries hot return water back to the water chiller system or the drain.

Two split range control valves with positioners are typically used to control the amount of heat that is being added to or removed from the reactor. A normally closed equal-percentage valve (TV-2 02) controls the steam to the heat exchanger and a normally open equal-percentage valve (TV-1 02) controls the chilled water return flow out of the jacket. The control valve in the chilled water return line is called the chilled water control valve.

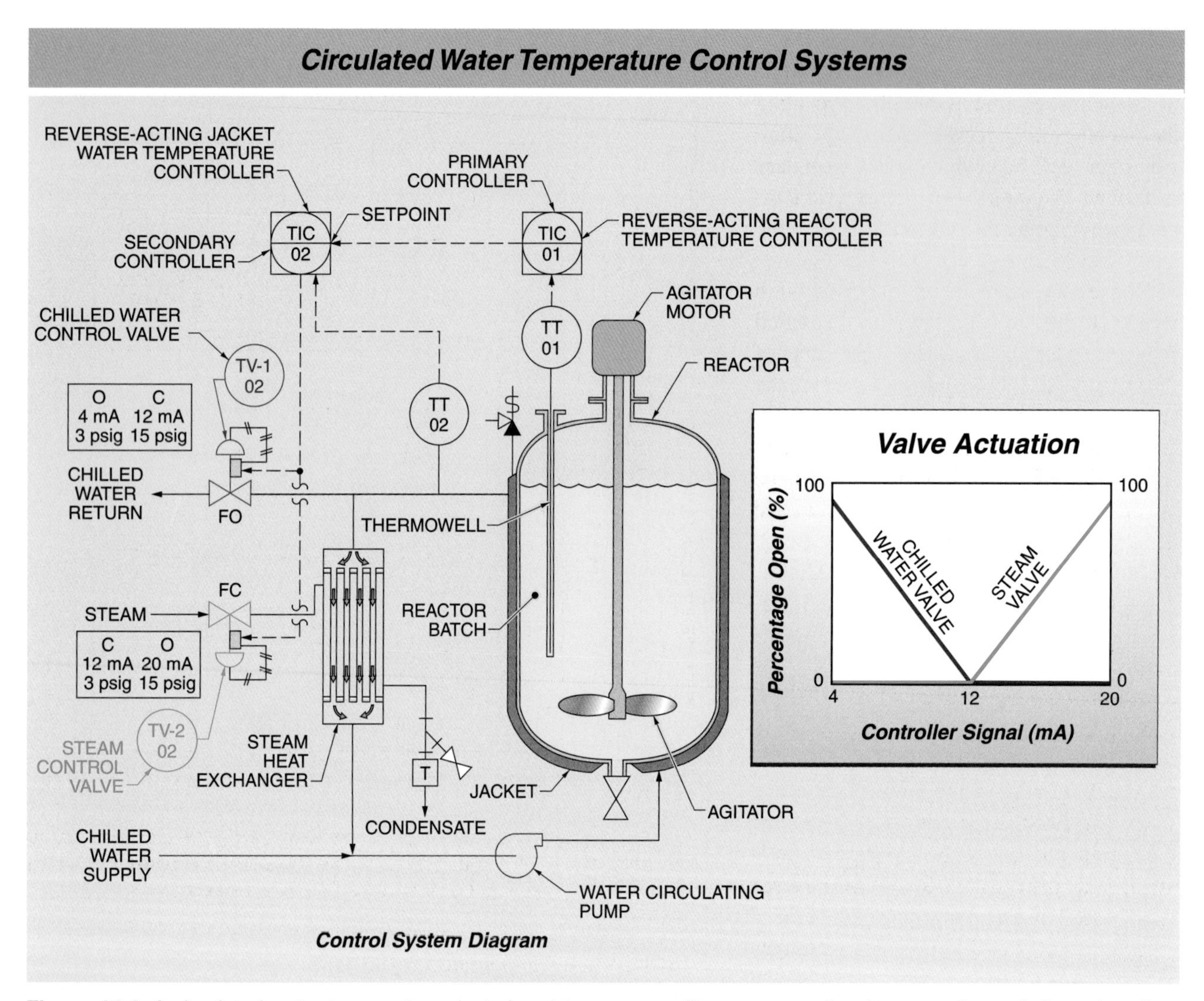

Figure 47-6. A circulated water temperature control system uses a split range operation to use water and steam to adjust the temperature of the circulated water.

Heat exchangers can be used to recover heat energy from wastewater.

When the control valve opens and lets water leave the loop, more chilled water immediately enters the loop to take its place. The chilled water control valve is fully open at a controller signal of 4 mA and fully closed at 12 mA and above. The steam control valve is fully closed at 12 mA and below and fully open at 20 mA. The split range configuration ensures that the heating and cooling valves are not open at the same time. The location of the chilled water control valve maintains the circulated water loop at the chilled water supply pressure, which prevents the water from boiling during the heating cycle.

Temperature control is a cascade strategy, with the reactor temperature controller (TIC 01), which is the primary controller, providing the setpoint for the jacket water temperature controller (TIC 02), which is the secondary controller. The output from the reactor temperature controller is the setpoint value of the jacket water temperature controller.

The reactor temperature is measured with a sensor located in a long thermowell inserted through the top of the reactor. The output of the sensor is then transmitted (TT 01) to the reactor temperature controller. The reactor temperature controller needs to be reverse-acting with PID control functions. The large reactor mass requires that the controller have a high gain, a large integral time (min/repeat), and an equally large derivative time (min).

A temperature sensor in a thermowell is located in the jacket water outlet piping and a signal is transmitted (TT 02) to the jacket water temperature controller. The jacket water temperature controller needs to be reverse-acting with a PI control function. The circulated water loop temperature is much more responsive than the reactor temperature. Therefore, the quick response time requires that the controller have a small gain with an integral time (min/repeat) of only a few minutes.

The jacket water temperature controller measures a change in the return water temperature caused by a change in the batch reaction rate before the reactor temperature controller detects any temperature change. Thus the jacket water temperature controller starts to adjust the heating or cooling rate before the reactor temperature controller calls for any changes. This is one of the advantages of a cascade control system.

In operation, the reactor is charged and the reactor temperature controller setpoint is set to the desired batch temperature. **See Figure 47-7.** Since the reactor temperature is lower than the setpoint, the reactor temperature controller output is high. This sets the jacket water temperature controller setpoint to a high temperature. The steam control valve is opened fully and hot water circulates through the water loop. This raises the reactor temperature at the fastest possible rate.

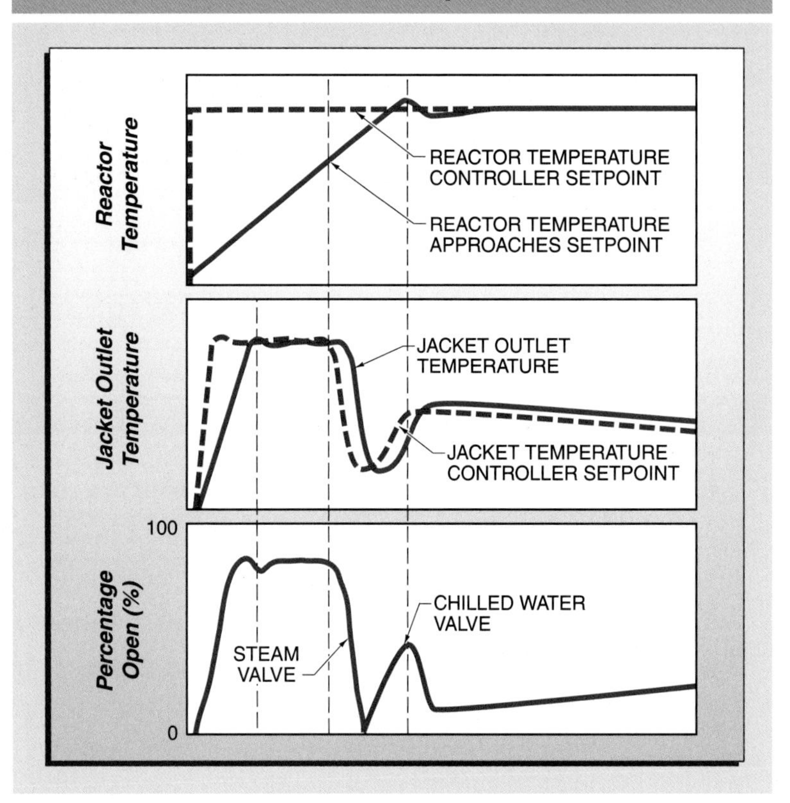

Figure 47-7. The steam valve is opened to heat up the reactor contents as quickly as possible. After the reactor reaches its setpoint, the reaction starts generating heat that must be removed by the chilled water.

As the reactor temperature approaches the setpoint, the derivative function starts lowering the output and the steam valve starts to close. The reactor temperature controller calls for cooling before the reactor temperature reaches the setpoint. With proper tuning of the controller, the reactor batch temperature ends its heating cycle at the setpoint with little or no overshoot.

At this point of the batch cycle, the reactor temperature controller is calling for very little cooling. Once the reactor temperature is at the operating temperature, the batch materials start reacting at faster rates and generating heat that must be removed. The reactor temperature controller continues to lower the jacket water controller setpoint as the batch reaction continues. The jacket water temperature controller allows more and more chilled water into the circulation loop to maintain the required jacket temperature.

KEY TERMS

- *heat exchanger:* Any piece of equipment that transfers heat from one material to another.
- *jacketed reactor:* A vertical vessel used in batch processing whose sides and bottom are covered with a steel shell spaced about an inch from the outside of the reactor, and which is used to contain either a cooling or heating fluid.
- *circulated water system:* A system that provides both heating and cooling for batch reactors.

REVIEW QUESTIONS

1. What is the design of typical heat exchangers?
2. What is the typical flow of fluids through a heat exchanger used for cooling service?
3. What is the typical fail action of the steam valve used for a steam controlled heat exchanger?
4. Define jacketed reactor and explain how heat transfer in a jacketed reactor is improved.
5. How can a circulated water system be used to provide both heating and cooling to a reactor?

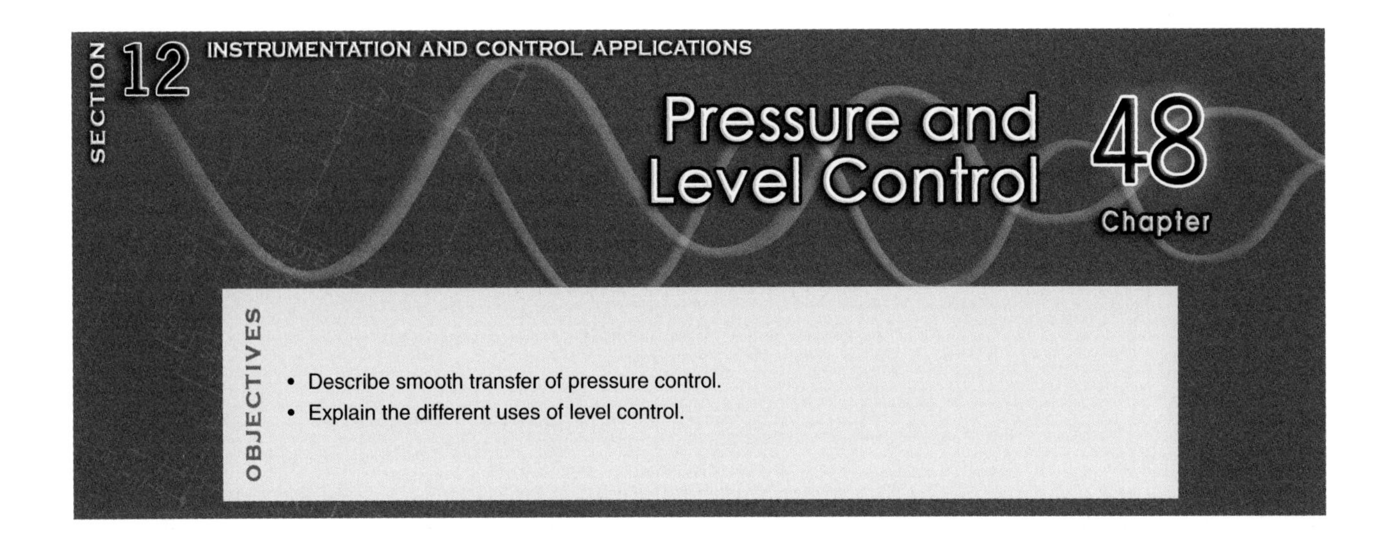

Pressure and Level Control

OBJECTIVES

- Describe smooth transfer of pressure control.
- Explain the different uses of level control.

SMOOTH TRANSFER OF PRESSURE CONTROL

A pressure control system is a common type of control application. Most pressure control applications are simple pressure-reducing or backpressure control systems. A more complex pressure control application was covered in the previous discussion on split ranging control valves. The control logic in that application can be used with many pressure control applications with only minor adjustments.

The transfer of pressure control from one process operation to another in a smooth and controlled manner can be very difficult at times, especially if the process is sensitive to disturbances. For example, a process generates a gas at a pressure of a few inches of water. **See Figure 48-1.** At startup, the gas is vented through a backpressure control system. Once the process is stabilized and at normal operating conditions, it is usually desirable to transfer the gas flow from the vent to a compressor system. The gas needs to be at a higher pressure before it can be used in other processes.

Very narrow high and low operating pressure limits are required by the gas generation process. Exceeding these limits shuts down the compressor system and possibly even the gas generation system. Using manual valves and separate pressure controllers to transfer the gas from the vent system to the compressor is difficult and can easily disrupt the pressure control.

The control arrangement consists of a single reverse-acting pressure controller (PIC 01), with a PI control mode, which obtains a signal from a pressure transmitter (PT 01). The pressure transmitter is connected to the process piping near where the gas is being generated. The control valve used for the vent (PV-1 01) has a fail open action with a linear flow characteristic and a direct-acting positioner.

The second control valve (PV-2 01) is located in the suction piping to the compressor, and it has a fail-closed action with a linear flow characteristic and a reverse-acting positioner. The fail-closed action of this control valve is selected for safety. The fail-closed action is not compatible with the reverse action of the controller, so the positioner action must be reversed to make the control valve move in the correct direction.

TECH FACT

Manufacturers of pressure vessels, in compliance with the ASME code, must have an agreement with a qualified inspection agency to certify that each vessel is manufactured according to the code. Modifications to a vessel after purchase may invalidate the certification.

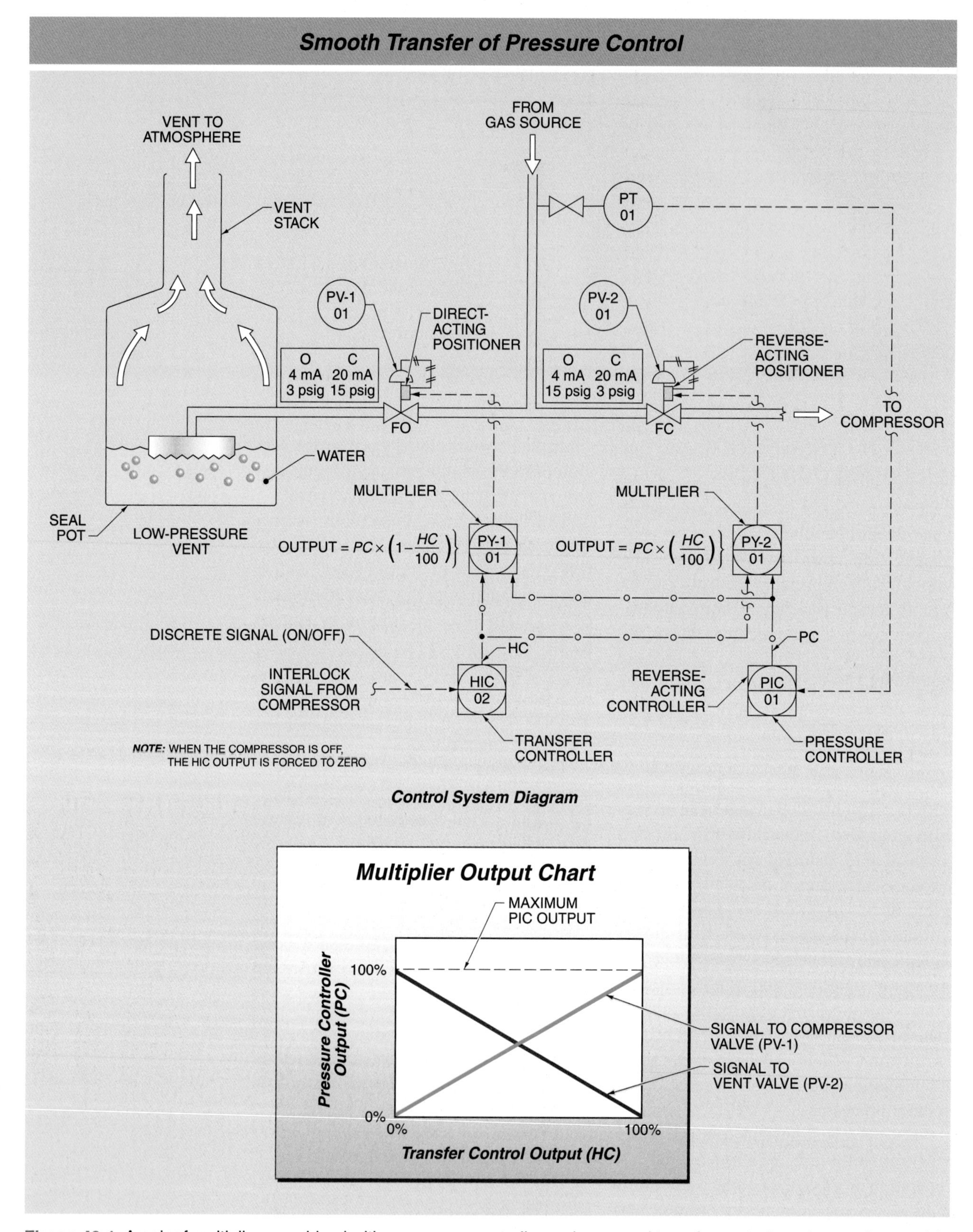

Figure 48-1. A pair of multipliers combined with a pressure controller and a manual transfer control can be used to provide a smooth transfer of pressure control between two control valves.

Both control valves need to be able to have a tight shutoff. Globe-style control valves are the ideal choice for this application. If globe valves cannot be obtained in the proper sizes, V-ball valves can be used, with the normal equal-percentage flow characteristics being linearized. A quick-opening cam in the positioner can be used to accomplish this linearization. The linear flow characteristics are necessary so that during the transfer of control, any change of flow in one valve is equaled by an opposite change of flow in the other valve.

Each control valve receives a control signal from its own signal multiplier module. Each of the multipliers receives control signals from the pressure controller and from the output of a manual loading station transfer controller (HIC 02). For the manual transfer controller signal, 0% represents all the gas going to the vent and 100% represents all the gas going to the compressor. The manual regulator provides a method to smoothly change the control from a venting operation to a compression operation without causing sudden changes to the process. The multiplier (PY-1 01) output used with the vent valve (PV-1 01) is calculated as follows:

$$output = PC \times \left(1 - \frac{HC}{100\%}\right)$$

where

output = multiplier (0% to 100%)

PC = pressure controller (0% to 100%)

HC = manual transfer (0% to 100%)

For example, if the pressure controller output, *PC*, is 72% and the manual transfer controller output, *HC*, is set at 40%, then the multiplier output is calculated as follows:

$$output = PC \times \left(1 - \frac{HC}{100\%}\right)$$

$$output = 72\% \times \left(1 - \frac{40\%}{100\%}\right)$$

$$output = 72\% \times (1 - 0.4)$$

$$output = 72\% \times 0.6$$

$$output = \mathbf{43.2\%}$$

The multiplier (PY-2 01) output used with the compressor suction valve (PV-2 01) is calculated as follows:

$$output = PC \times \left(\frac{HC}{100\%}\right)$$

where

output = multiplier (0% to 100%)

PC = pressure controller (0% to 100%)

HC = manual transfer (0% to 100%)

For example, if the controller output, *PC*, is 72% and the manual transfer controller, *HC*, is set at 40%, then the multiplier output is calculated as follows:

$$output = PC \times \left(\frac{HC}{100\%}\right)$$

$$output = 72\% \times \left(\frac{40\%}{100\%}\right)$$

$$output = 72\% \times 0.4$$

$$output = \mathbf{28.8\%}$$

Some multipliers only work properly when the inputs and outputs are scaled to be from 0 to 1. Scaling converts actual engineering unit ranges into ranges of 0 to 1. A range of 0% to 100% inputs, when divided by 100, results in scaled inputs of 0 to 1.

Boiler water analysis is used to determine when to pump treatment chemicals out of storage tanks.

Scaled ranges of 0 to 1 can be multiplied, divided, squared, or have the square root calculated with no errors in the mathematical operations. This is not the case if percentages are used in mathematical operations. A reverse scaling factor then has to be applied to the results of the mathematical operations to obtain the equivalent in engineering units. In this example, the multiplier output is scaled to obtain an output of 0% to 100%.

If the outputs of the two multipliers are plotted, with the pressure controller output signal on the vertical axis and the transfer control output signal on the horizontal axis, the functions of the two multipliers are easier to understand. When the manual transfer controller output is 0%, the entire output signal from the pressure controller is directed to the vent valve (PV-1 01), which then opens.

The valve (PV-2 01) to the compressors is fully closed. When the manual transfer controller output is 50%, half of the pressure controller output goes to the vent valve and the other half goes to the compressor valve. When the transfer controller output is 100%, all of the pressure controller output goes to the compressor valve and the vent valve is fully closed.

During the whole transfer process, the pressure controller is always in full control and can make changes in both control valves. The use of linear flow characteristics for the two control valves minimizes the changes in total flow capacity of the control valves when going through the transfer operation. To ensure safety, the transfer controller, HIC 02, is interlocked with the compressor system. If the compressor system shuts down, the manual transfer controller output is set to a 0% value. This action switches the pressure control to the vent valve and closes the valve to the compressor. With this interlock, the compressor must be running before the transfer can be started.

LEVEL CONTROL

Level control is another common type of control application. Most level applications are simple single control loop systems where the level controller is connected to a single final control element and regulates the flow into or out of a vessel. The following level control applications do not require the use of a control valve and are commonly used for smaller boiler systems.

Pump-Up Level Control

Pump-up level control is a control arrangement used to start a pump when the level of liquid drops below a setpoint and stop the pump when the liquid level reaches the desired level. This operation controls the level of the water between the two control points. Pump-up level control is a very common method of supplying feedwater to small boilers. **See Figure 48-2.** This is often called boiler water level control or feedwater control in boiler operations.

In a boiler, the high and low levels are only a few inches apart. Conductive-probe level sensors or float-switch level sensors are commonly used to send a signal to a controller. Safety is provided by a high level alarm (LSH 02), a low level alarm (LSL 03), and a low low level shutdown (LSLL 04).

Many applications require methods of measuring and handling continuous webs of material.

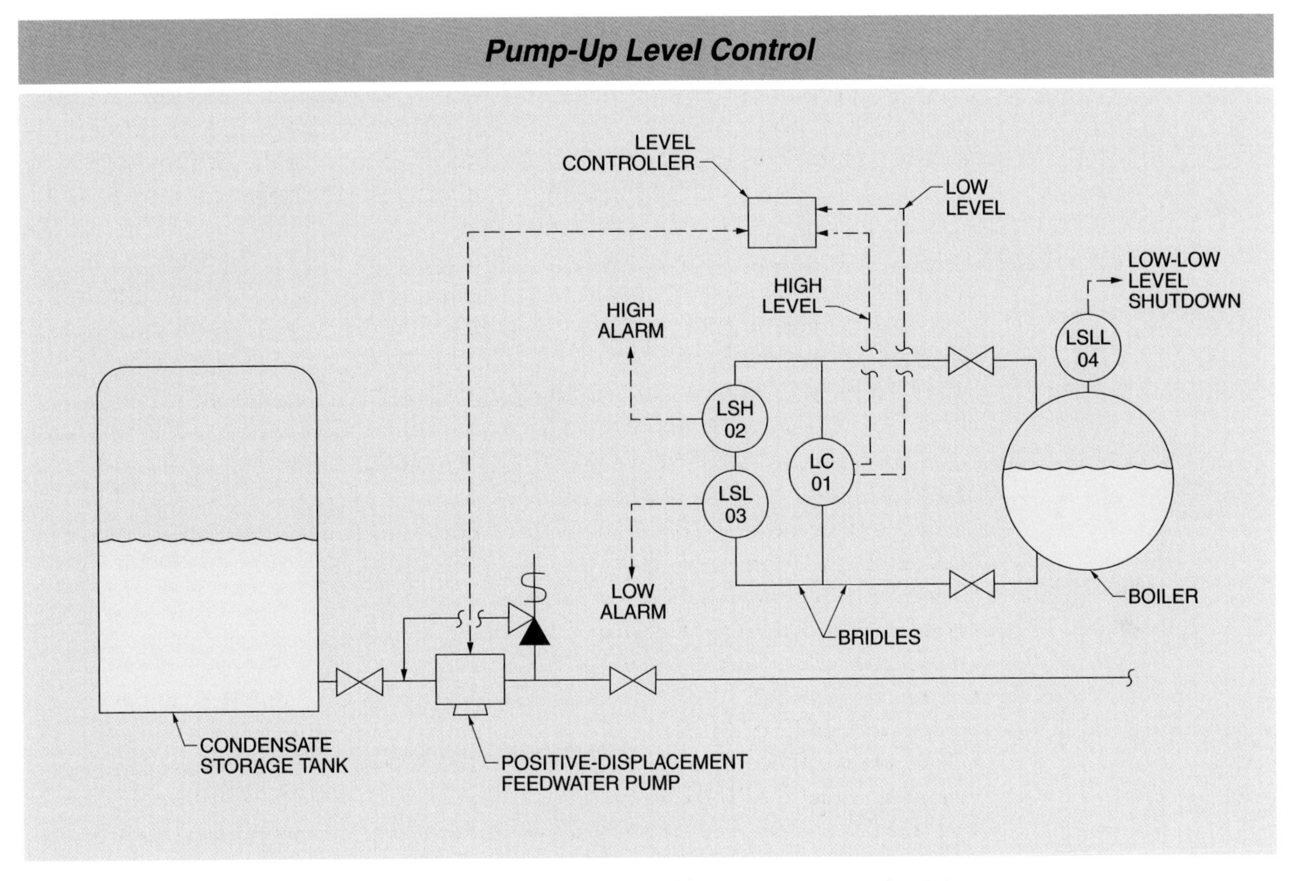

Figure 48-2. Pump-up level control is used to keep a liquid level between two control points.

Conductive-Probe Level Sensors. Conductive-probe level sensors consist of two electrodes that pass through a sealed fitting into a stand leg outside of the boiler. **See Figure 48-3.** A stand leg is a section of pipe large enough in diameter and length for the electrodes. The stand leg is piped to the top and bottom of the boiler. This piping arrangement is called a bridle.

A bridle is commonly used in the petrochemical industry where it minimizes the number of connections to a pressure vessel. The two electrodes are at different lengths, with the tips of the electrodes representing the high and low level points for the pump control. The common for the two electrical circuits is the metal of the boiler or the stand leg piping. If necessary, a third electrode longer than the other two electrodes can be the circuit common.

The NEC® uses Fine Print Notes throughout the code to explain information that is contained in the code. Fine Print Notes do not contain any mandatory provisions but do provide supplemental material.

Float-Switch Level Sensors. A boiler feedwater pump can be controlled with a pair of float-switch level sensors. One switch is positioned at the high-level point (LSH 05) and the other at the low level point (LSL 06). Another low low level float switch (LSLL 07) is included for safety and shuts down the boiler if tripped. The float switches are mounted to a stand leg piped to the top and bottom of the boiler. Each float switch actuates a normally open and a normally closed Form C electrical switch.

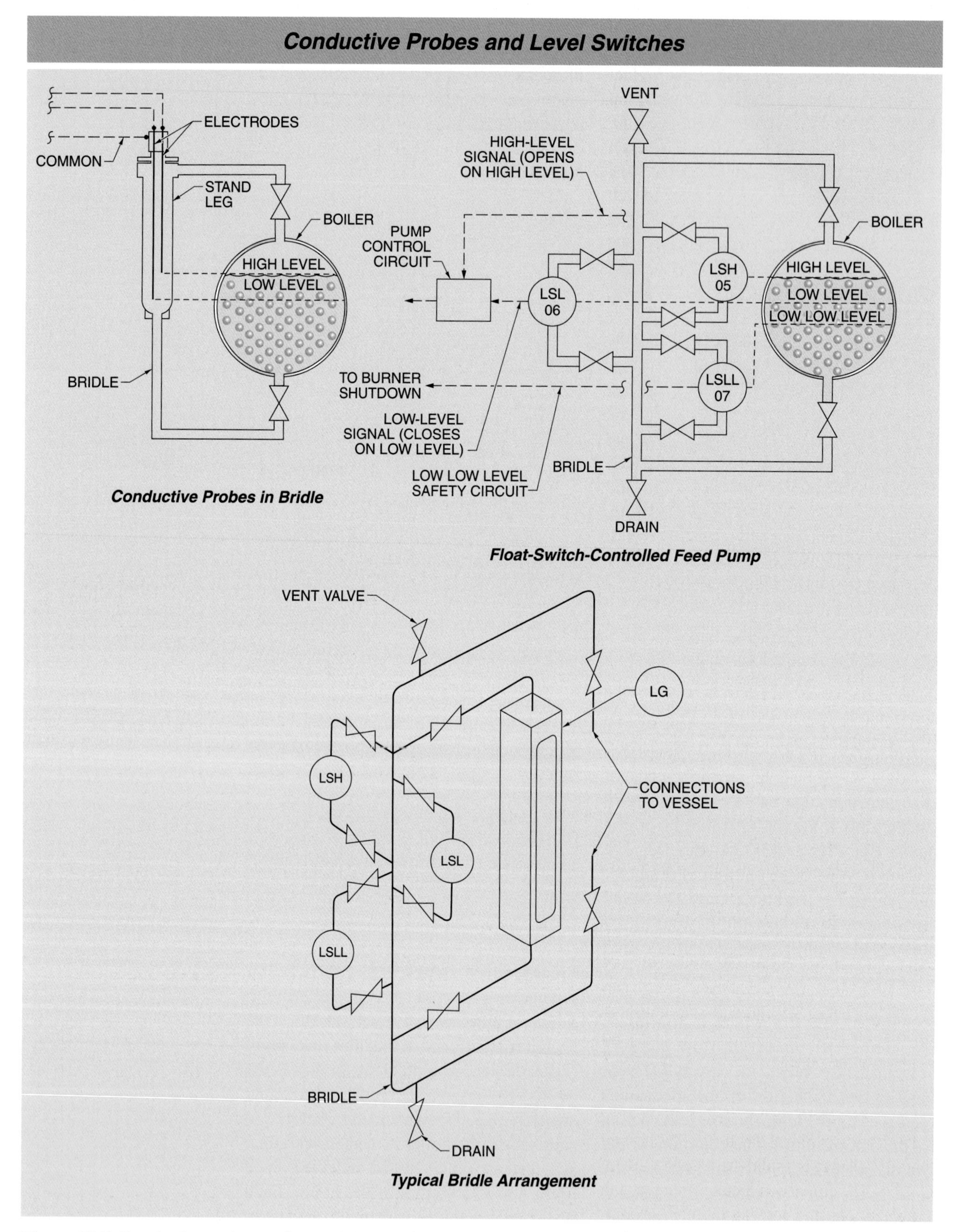

Figure 48-3. Conductive probes or float switches can be used to control the water level in a boiler.

Pump-Up Control Circuits. The electrical control circuit for a boiler feedwater pump is the same whether the level sensor is a conductive probe or a float switch. The control circuit is very similar to a motor start/stop circuit. **See Figure 48-4.** The low level NC contact closes when the water drops below the low level point. This energizes relay R1, which closes the NO holding contact (R1-1). The holding contact is parallel with the low level NC contact. Relay R1 also closes a second NO contact (R1-2) which provides power to the feedwater pump.

When the level starts rising, the low level NC switch opens. Relay 1 remains energized through the NO holding contact (R1-1). When the water level reaches the high-level point, the high-level NC contact opens, interrupting the power to relay R1. This opens the relay contacts (R1-1 and R1-2) and shuts off the feedwater pump. The chart shows the boiler level changes, the high- and low-level contact action, and the pump ON and OFF operation.

Pump-Down Level Control

Pump-down level control is a control arrangement used to control the transfer of collected steam condensate to a common condensate storage tank. The operation is based on maintaining the level of water in the condensate tank. The two levels are a high level, where the feedwater pump is started, and a low level, where the feedwater pump is stopped. The high probe is typically set at about 75% of the full tank level and the low probe is typically set at about 25%. Conductive-probe level sensors are commonly used to send a signal to a controller. **See Figure 48-5.** If the vessel is pressurized or if its elevation is higher than the destination of the liquid, a solenoid valve or an electrically actuated valve can be used in place of the pump.

Pump-Down Control Circuits. There is a separate electrical control relay for each of the high and low level electrodes. It is best if this control relay is a solid-state device instead of a direct electrical conductance type, so that the electrodes and sensing circuits are only subjected to low voltages. The high and low level circuits each energize their own output relays when the condensate water level is in contact with the level electrode and the circuit common. The relays have a normally open and a normally closed Form C electrical contact, which can be used in the pump control circuit.

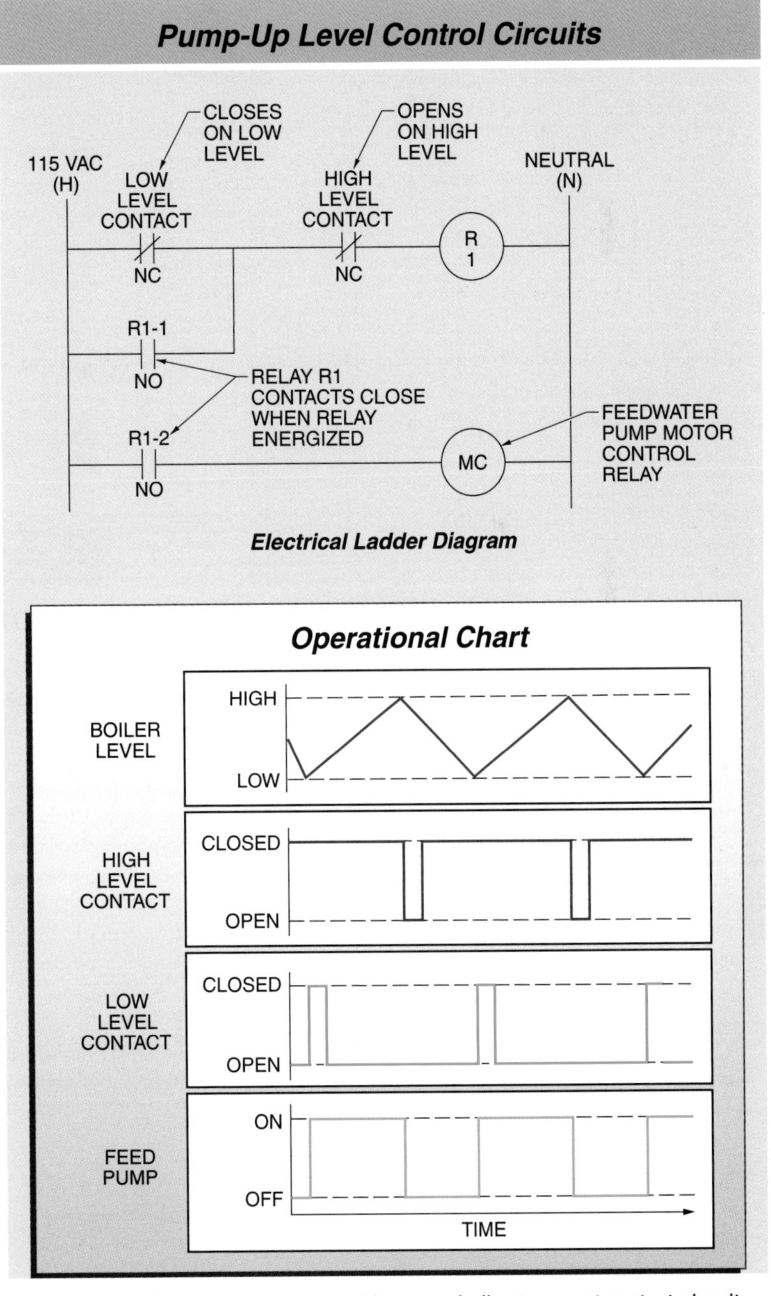

Figure 48-4. Pump-up level control is very similar to a motor start circuit.

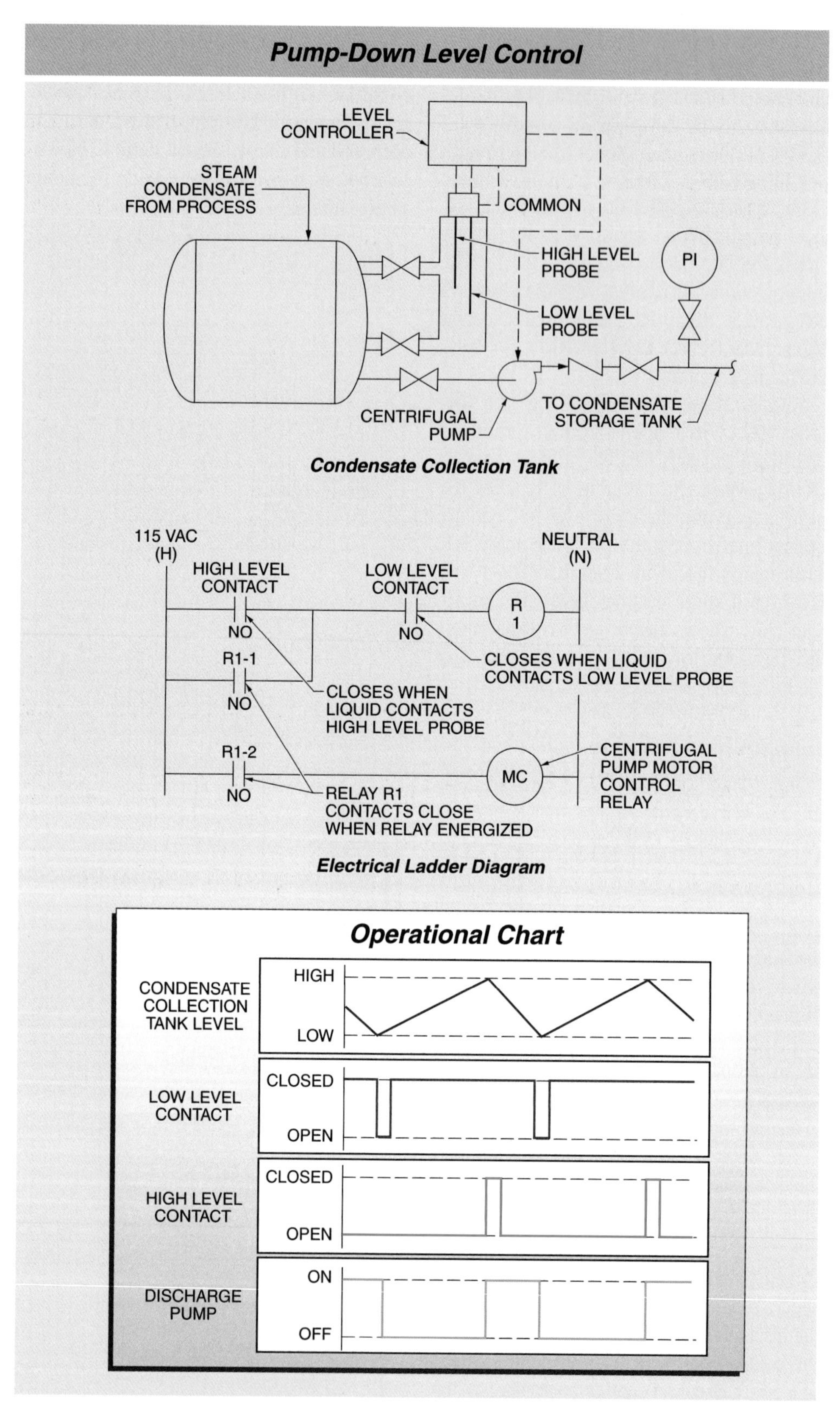

Figure 48-5. Pump-down level control is used to partially pump out a tank while maintaining the level within predetermined limits.

The control circuit description starts with an empty vessel. As the condensate starts to collect in the vessel, it rises until it contacts the low level probe, which closes the low level contact. When the level rises to the high-level probe, the high-level contact is closed. This completes the circuit and energizes relay R1. When relay R1 is energized, its contacts close, latching in the level circuits and energizing the motor control relay. The level drops as the condensate is pumped out until the level drops to the low level probe. When the level breaks contact with the low level probe, the low level contact opens, dropping out relay R1 and shutting off the condensate pump.

KEY TERMS

- *pump-up level control:* A control arrangement used to start a pump when the level of liquid drops below a setpoint and stop the pump when the liquid level reaches the desired level.
- *pump-down level control:* A control arrangement used to control the transfer of collected steam condensate to a common condensate storage tank.

REVIEW QUESTIONS

1. What is the fail action of a positioner and control valve on the suction piping to a compressor in the smooth transfer of pressure control application?
2. How do signals get from controllers to the control valves in the smooth transfer of pressure control application?
3. What is the purpose of pump-up level control?
4. How are conductive-probe level sensors and bridles used to measure level?
5. What is the purpose of pump-down level control?

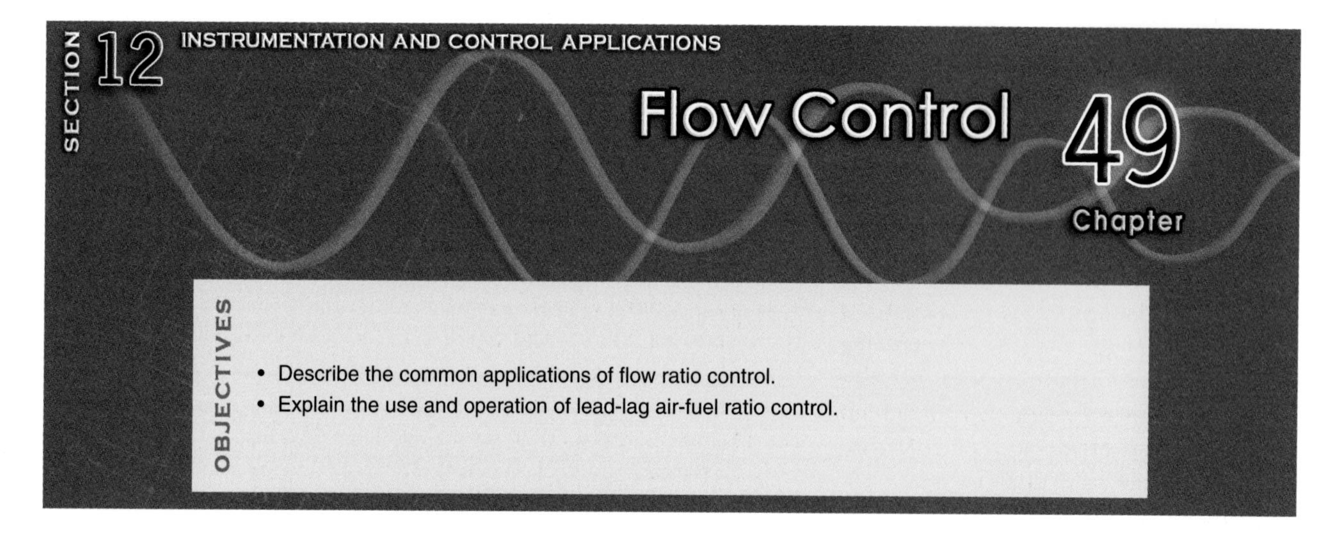

RATIO CONTROLLERS

Flow control loops are used to set flow rates in continuous processes or in feeding the proper quantities of chemicals to batch reactors. A very common type of flow control is automatic control of the ratio of one flow to another. *Ratio control* is a control strategy used to control a secondary flow to the predetermined fraction, or flow ratio, of a primary flow. The secondary flow measurement is compared to the setpoint and the secondary flow controller modulates a control valve to maintain the desired flow rate. **See Figure 49-1.**

Flow ratio controllers are available as pneumatic, electric, or digital controllers and can accept either linear or square root flow signals. Since pneumatic and electric ratio controllers do not use engineering units in their internal operations, the flow ranges selected for the primary and secondary flow transmitters are used to determine the ratio of the two flows.

> **TECH FACT**
> *The lift of a safety valve is the distance the valve disc lifts from the seat when the valve is fully open.*

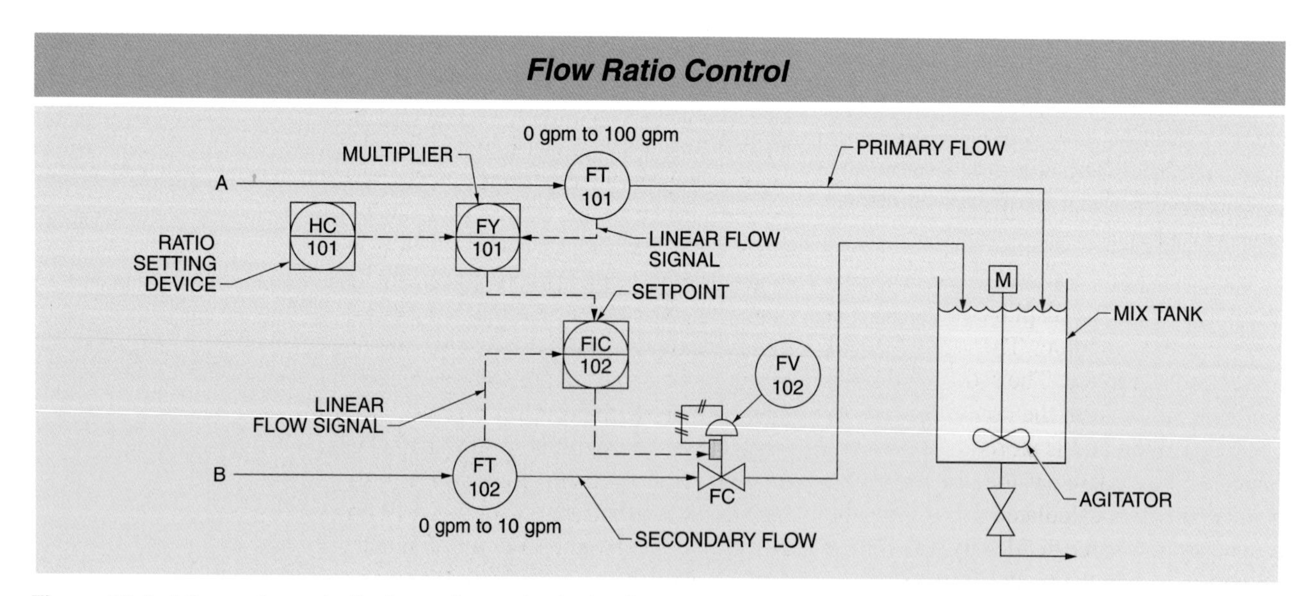

Figure 49-1. A flow ratio controller is used to maintain the flow rates of two streams at a constant ratio.

For example, if the primary flow is 81 gpm and the secondary flow is 7.5 gpm, the basic flow ratio is 0.0926 (7.5 ÷ 81 = 0.0926). This is the ratio used in a digital controller. If the ratio controller is pneumatic or electric instead of digital, the ratio setting is determined by dividing the actual flow ratio by the flow range ratio. If the flow transmitter (FT 101) for the larger flow is sized for a flow range of 0 gpm to 100 gpm and the flow transmitter (FT 102) for the smaller flow is sized for a flow range of 0 gpm to 10 gpm, the flow range ratio is 0.1 because the smaller range is 0.1 times the size of the larger range. In this case, the flow ratio is 0.926 (0.0926 ÷ 0.1 = 0.926).

The setpoint for the secondary controller is the primary flow multiplied by the ratio factor. For digital controllers, the ratio factor is 0.0926. For pneumatic and electric ratio controllers, the ratio factor is 0.926.

Ratio Controller Functions

Digital flow ratio controllers, which are available from different manufacturers, vary in their internal complexity. The complexity depends on the features offered. Not all controllers offer all functions. In most cases, what is offered is not fully described by the manufacturer. A description of a flow ratio controller with all of the functions is presented here in order to help understand the internal operation. A secondary flow ratio controller is similar to a secondary cascade controller. The flow ratio controller contains the flow ratio value and the multiplier module. Both the primary and the secondary flows can be displayed on the ratio controller.

Ratio Controller Normal Flow. The primary flow signal is displayed and goes to the multiplier module. The active signal route is shown in red. The ratio value is set by the operator from the controller face or operator station and is stored in a memory location. This memory location can also be loaded with a calculated ratio value, the secondary flow value divided by the primary flow value, when the track command is active. The ratio value in memory is sent to the multiplier module. The ratio value is multiplied by the primary flow. The output of the multiplier module is sent to the remote input of the remote/local (R/L) setpoint switch as the setpoint for the secondary controller. **See Figure 49-2.**

The R/L setpoint switch is used to determine if the secondary controller is to function as a ratio controller or as a simple flow controller. In the remote position, the setpoint value tracks the secondary flow measurement, as shown by the orange lines. This allows the controller to be switched to the local setpoint mode of operation with no change in the setpoint value. The output from the R/L selector switch is used as a setpoint for the secondary flow controller. The secondary flow is compared to the setpoint in the deviation block. The difference is sent to the controller functions block.

The difference can be either positive or negative depending on the action of the controller and the actual secondary flow. The difference from the deviation block is acted on by the controller functions and the result is sent to the automatic/manual (A/M) switch. The A/M switch is in automatic, which allows the control signal to go to the flow ratio controller output. The manual value is instructed by the track command (TC), as shown in green, to track the output signal, as shown in orange. This provides a bumpless transfer when switching from automatic to manual.

Ratio Controller Local Setpoint. Switching the normal flow ratio operation to local setpoint mode generates a few changes. **See Figure 49-3.** With the R/L setpoint switch in the local position, the R/L setpoint switch sends a track command (TC), as shown in green, through the R/L setpoint switch to the ratio value block. Thus the ratio value block tracks the output of the divide module. The output of the divide module is the secondary flow divided by the primary flow. The R/L setpoint switch uses the operator-generated flow setpoint value and compares it to the secondary flow measurement in the deviation block. The other operations of the flow controller work as described in the normal flow ratio operation.

TECH FACT

Before installing a safety instrumented system (SIS), an evaluation should be made to determine if adequate non-SIS layers of protection are present.

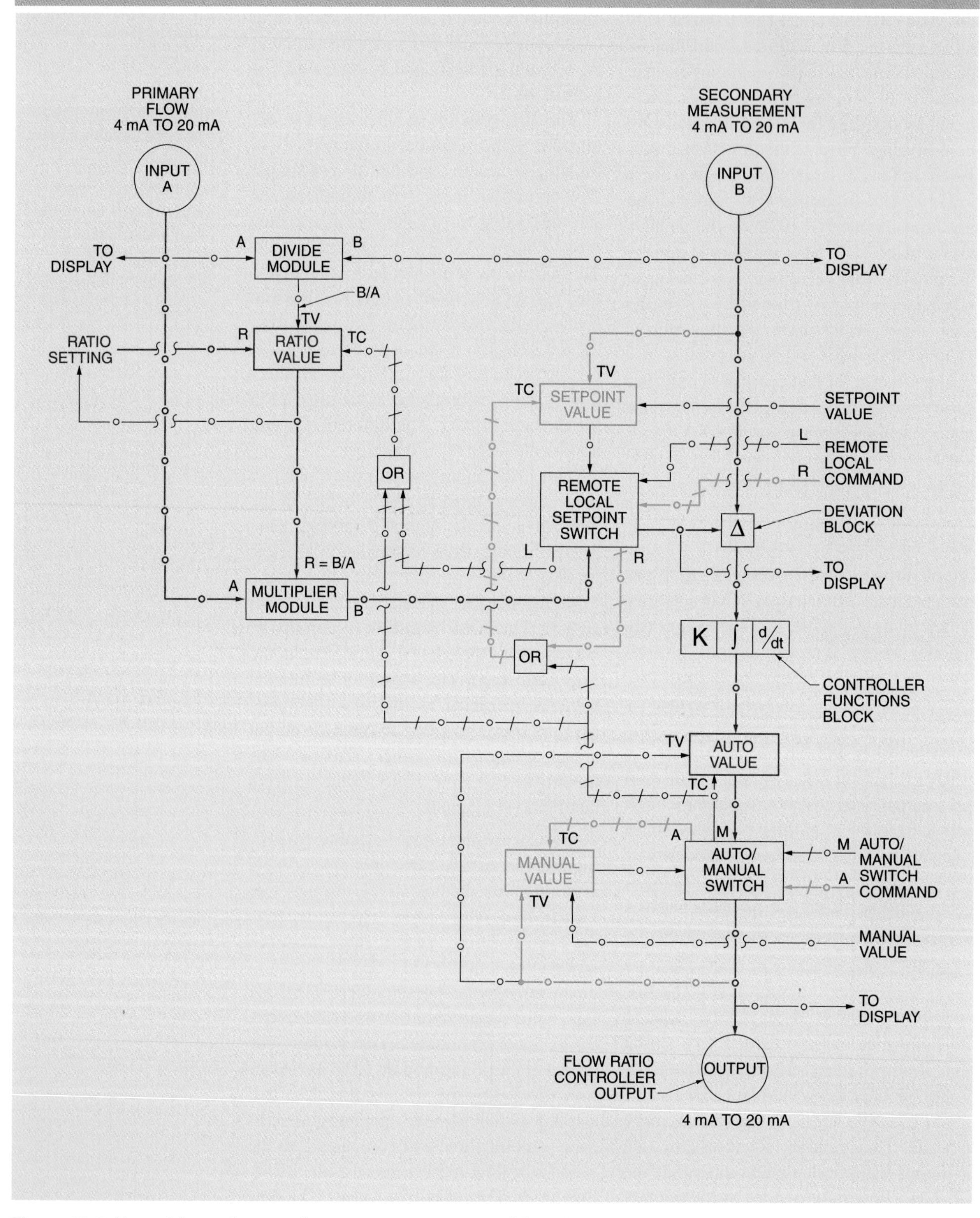

Figure 49-2. Normal flow ratio operation uses a measurement of the primary flow and a ratio value to calculate the setpoint for the secondary flow controller.

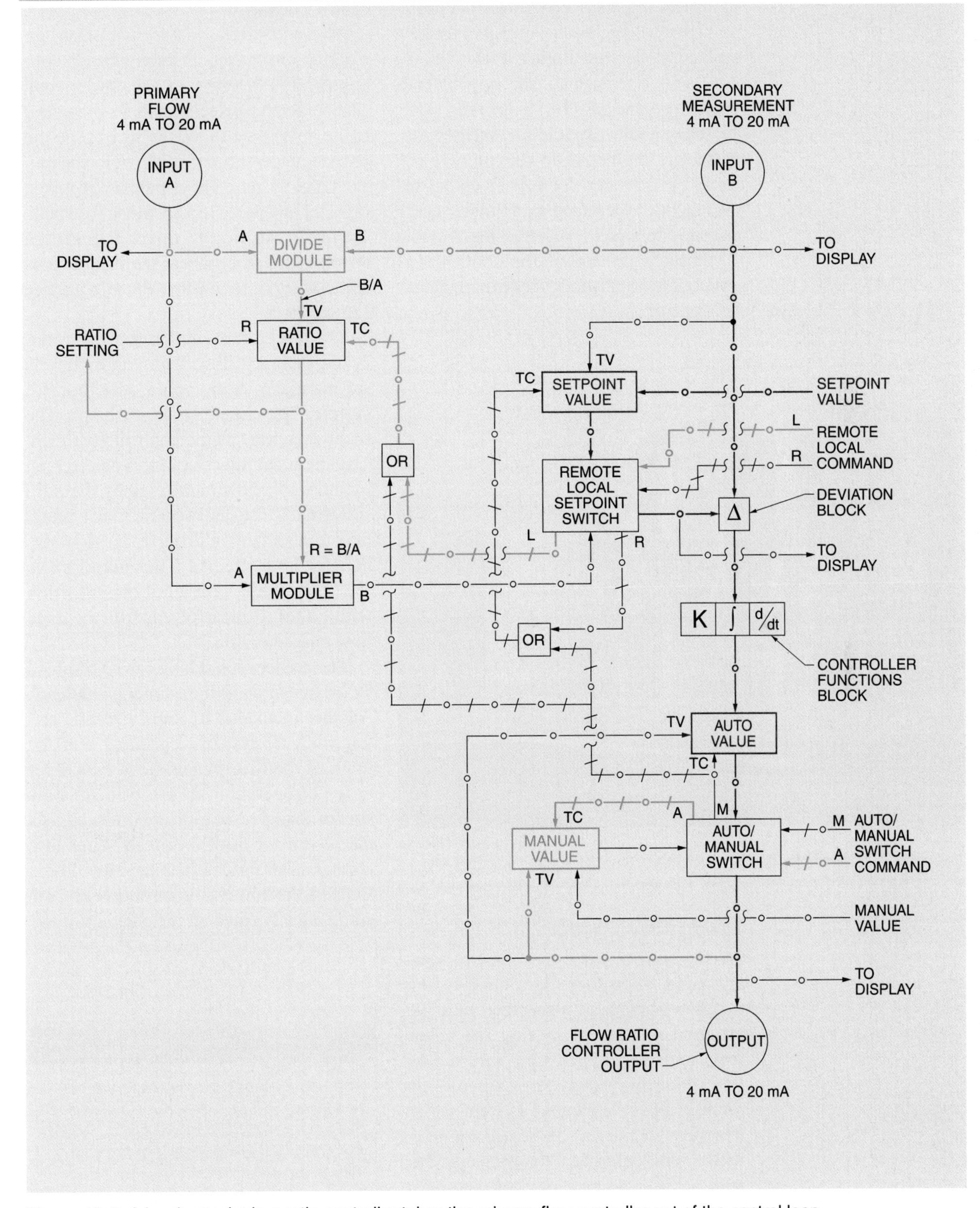

Figure 49-3. A local setpoint in a ratio controller takes the primary flow controller out of the control loop.

Ratio Controller Manual Operation. Switching the normal flow ratio operation to manual setpoint mode allows the operator to manually adjust the output of the flow ratio controller. **See Figure 49-4.** With the A/M switch in manual, the switch sends tracking commands (TC) to the ratio value, the setpoint value, and the auto value blocks, as shown in green. The auto value block tracks the flow ratio controller's output and the other two track their appropriate values, as shown in orange. This tracking provides a bumpless transfer if the controller is switched to automatic or back to normal flow ratio operation.

Cleaver-Brooks

Conductivity measurements can be used when purifying water for industrial use.

Mixing Ratio Control

Mixing ratio control is a control strategy used when there is a need to mix two flow streams in a specified ratio. An efficient method of mixing the chemicals and water is in-line with a flow ratio control system. For example, the dilution of ion-exchange system water-treatment chemicals can be accomplished with a ratio controller. Ion-exchange systems are used to treat raw water to make it suitable for use as boiler feedwater. The ion-exchange beds require regeneration with diluted solutions of acid and caustic chemicals.

The acid and caustic chemicals are usually purchased in commercial concentrations because of the lower cost. The acid and caustic chemicals then need to be mixed with water to obtain the proper concentrations required by the ion-exchange system. The two flows are mixed together and piped to a chemical feed tank. Typically, an in-line mixer is installed downstream from where the two streams come together to improve the mixing. **See Figure 49-5.**

The largest flow, in this case the water, is the primary flow. The water flow is set manually or by some other control system. The chemical flow, which is the secondary flow, is the controlled stream. The chemical flow setpoint is determined by the water flow multiplied by the ratio set value. It is usually better to use linear flow transmitters or linearize the flow signals for both streams. The control valve (FV 21) should have a fail closed action with either a linear or equal-percentage flow characteristic.

The proper ratio setting can be obtained by a calculation of the two required flows. Another method is to adjust the ratio setting while the system is operating, until the flows of the two streams are at the proper ratio. In operation, as the primary flow stream changes, the ratio controller adjusts the secondary flow stream to maintain a constant flow ratio. When the water flow is stopped, the flow ratio controller (FFIC 21) automatically shuts off the chemical flow.

The effectiveness of independent protection layers (IPLs) is described in terms of the probability that they will fail to perform their required functions upon demand. This is called probability of failure on demand (PFD).

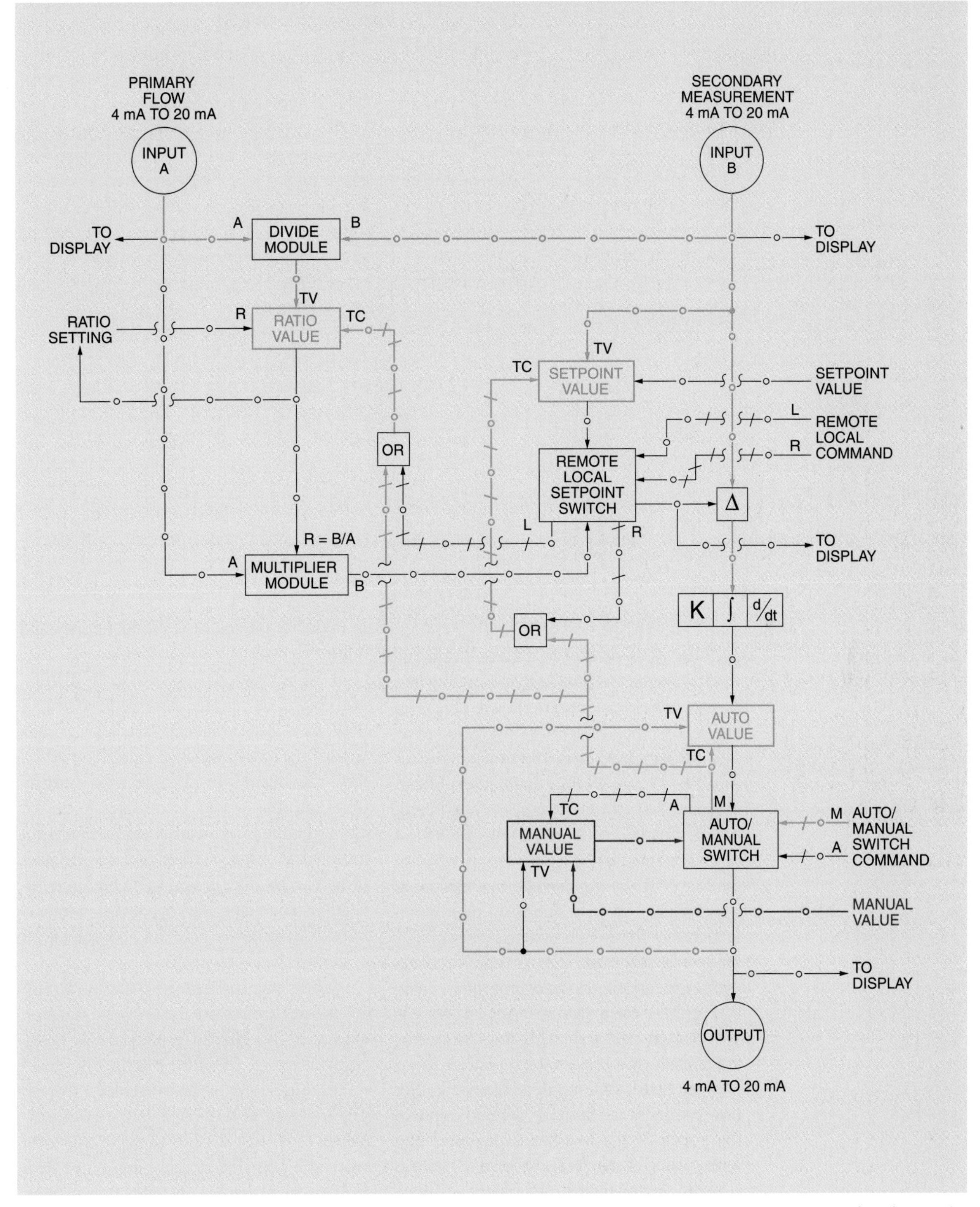

Figure 49-4. A manual flow ratio operation implements several tracking operations to enable bumpless transfer of control.

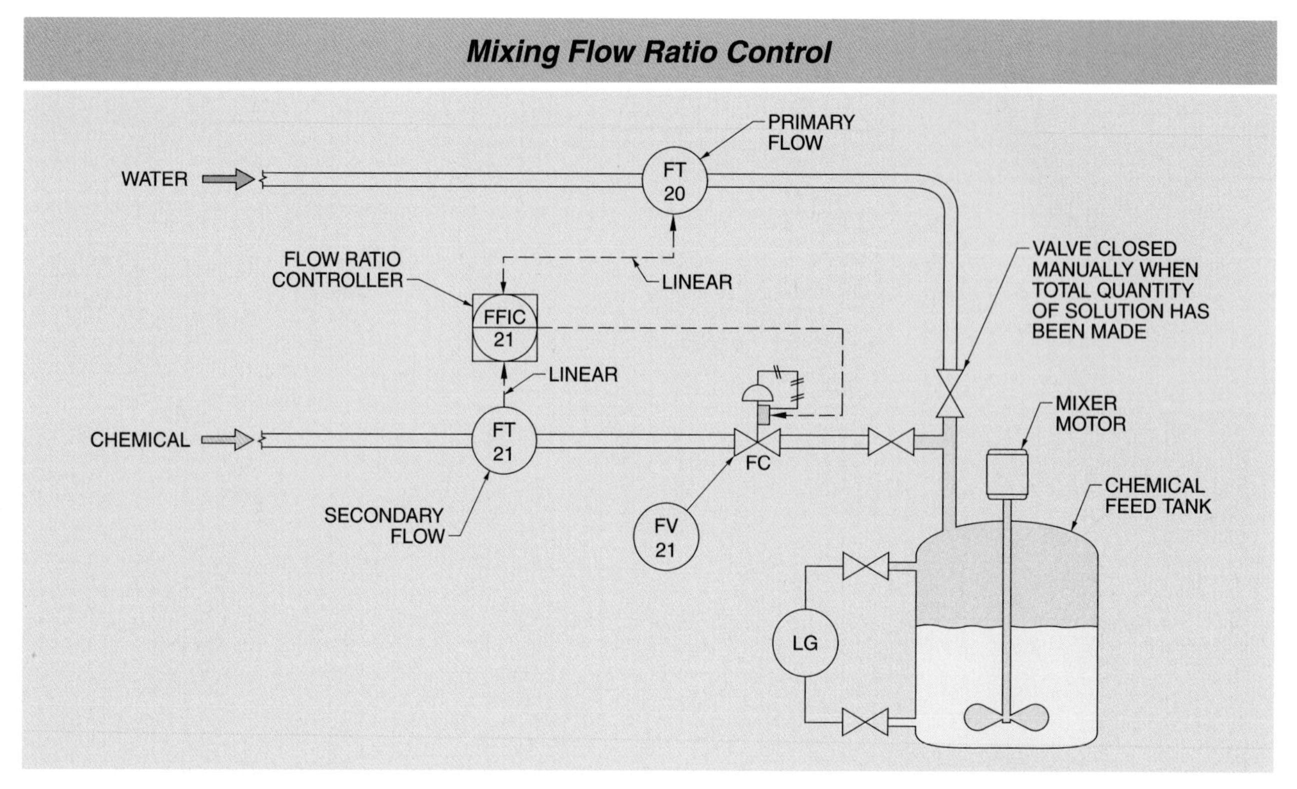

Figure 49-5. A common application of ratio control is to add mixing water with a chemical to get a predetermined concentration.

Splitting Ratio Control

Splitting ratio control is a control strategy used when there is a need to split a flow into two separate constant-ratio flows. This is an efficient method of splitting overheads flow in a distillation column into product and reflux. **See Figure 49-6.** A distillation column is used to separate liquids that have different boiling points. The material with the lowest boiling point ends up as overheads vapor at the top of the column while the remaining material ends up at the bottom of the column as bottoms liquid. The overheads vapor is removed and sent through a condensing heat exchanger, and then to an overheads receiver.

From the overheads receiver, the flow is split into product and reflux flows. *Reflux* is the portion of a condensed overhead vapor pumped back to the top of a column. Reflux flow back into a column is done to improve the purity of the overheads product. The *reflux ratio* is the reflux flow back into a column divided by the total overheads flow. A constant reflux ratio is desired in order to maintain the required product purity and minimize cost.

If the raw material feed rate to the distillation column is variable, the overheads vapor flow rate is also variable. When the overheads vapor flow rate is variable, a constant reflux flow rate results in changes in the reflux ratio and thus results in changes in purity. The remaining overheads material that is not returned as reflux is the overheads product flow. The product is sent to a column overheads storage tank. The product flow rate is throttled in order to maintain a constant level in the overheads receiver.

A differential pressure level transmitter (LT 05) is used to measure the level and transmit the signal. Increases in feed to the distillation column result in more overheads material accumulating in the overheads receiver. The increase in level signals the level controller (LIC 05) to start opening the level valve (LV 05). The increase in product flow requires an increase in the reflux flow in order to maintain the reflux ratio.

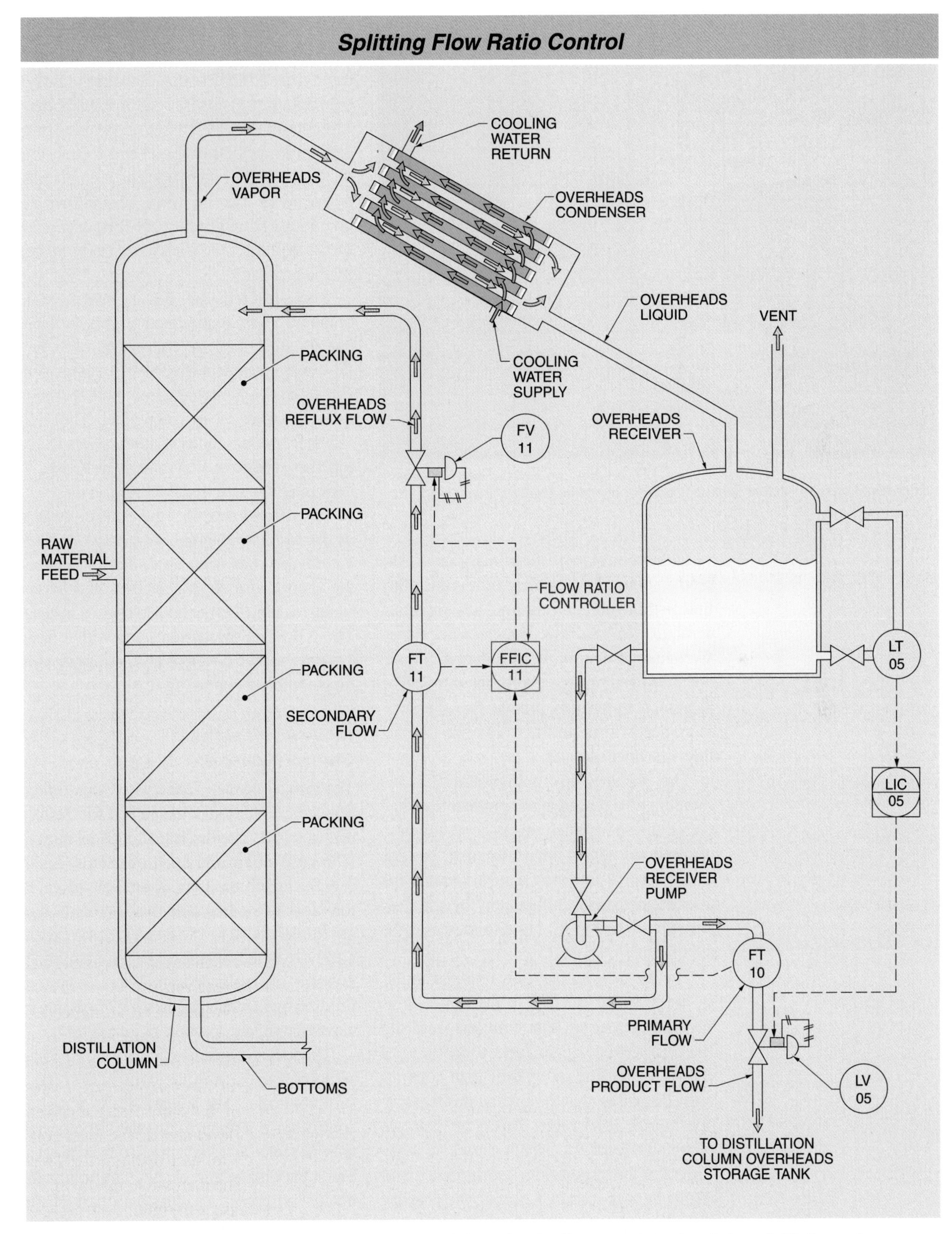

Figure 49-6. A flow ratio controller can be used to control the ratio of reflux to total overheads in a distillation column.

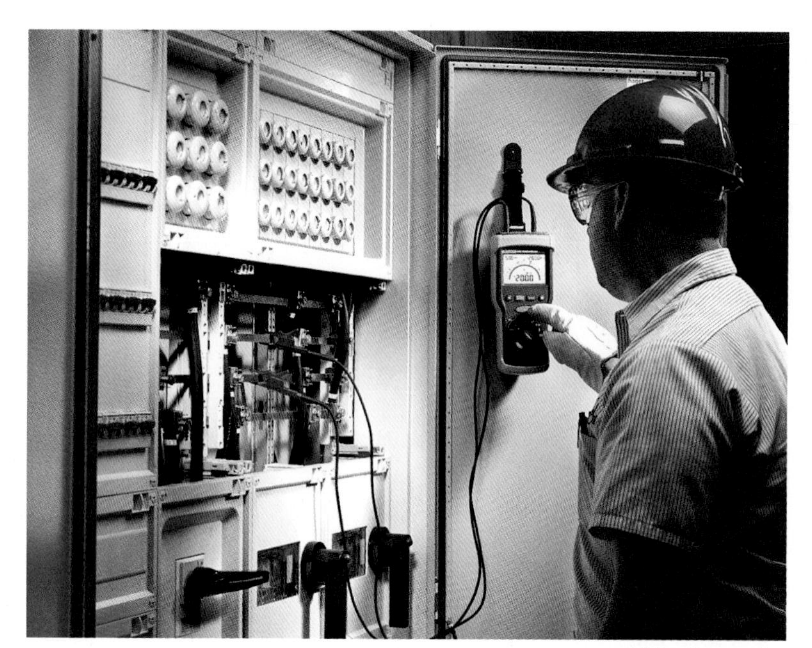

Fluke Corporation

Technicians are often called upon to troubleshoot instrumentation loops.

In the control loop, the primary flow is the product flow. The secondary flow is the reflux flow to the distillation column. It is important to minimize reflux flow changes, since excessive changes disrupt the distillation column operation and change overheads purity. The location of the primary product flow measurement needs to be in the overheads product flow line after the split.

The primary flow transmitter (FT 10) should not be placed in the product flow line before the split. When this is done, the ratio of the flows equals the reflux ratio, but this causes control problems. The problem is that a small increase in reflux flow by the flow ratio controller (FFIC 11) also increases the total flow.

The increase in total flow requires another, but smaller, increase in the reflux flow. These increases in flow continue until the level changes enough to start closing the level control valve. The reflux flow then continuously decreases until the level controller starts opening the level valve. This is an example of positive feedback, which causes an overcorrection in response to a disturbance and results in poor control. Locating the primary flow measurement in the overheads product flow eliminates any positive feedback.

AIR-FUEL RATIO CONTROL

Medium to large boilers use a more sophisticated fuel and combustion-air control strategy than the strategy that is used for small boilers. The control strategy is based on the basic principle of flow control of fuel and combustion air. Many large boiler systems are designed to use more than one fuel. This allows the operation to be able to use the least expensive fuel at any time, or to use a waste gas produced by the process.

Each different fuel has its own specific combustion-air requirements. This problem can be simplified by converting each fuel flow to its equivalent Btu value. An equivalent Btu value of any fuel requires about the same quantity of air as any other fuel. The individual Btu values for each flow are then added together. The total Btu value is then used to determine the required combustion air.

In order to simplify the control system in the example that follows, only one fuel, natural gas, has been used. This eliminates the need to use equivalent Btus and allows standard cubic feet per hour to be used instead. The following discussion starts with a basic system and then develops this in steps into the final complete control system.

Basic Air-Fuel Ratio Control Systems

The control system consists of two linked flow control loops for the air and for the natural gas fuel. The theoretical air-fuel ratio is 9.56 cu ft of air to 1 cu ft of natural gas to exactly match the correct amount of air to the fuel. An air-fuel ratio that can be used in gas burners is 10 to 1 to provide some excess airflow over the theoretical ratio minimum. In practice, a simple air-to-fuel control system needs to use a much greater air-to-fuel ratio to ensure complete combustion of the fuel.

The fuel and airflow control loops receive setpoint signals from the boiler steam pressure controller. **See Figure 49-7.** A digital control system requires that the setpoint signals be scaled to match the two flow ranges. Since pneumatic and electric controllers do not use engineering units as a part of the transmission signal, scaling is not needed for these types of controllers.

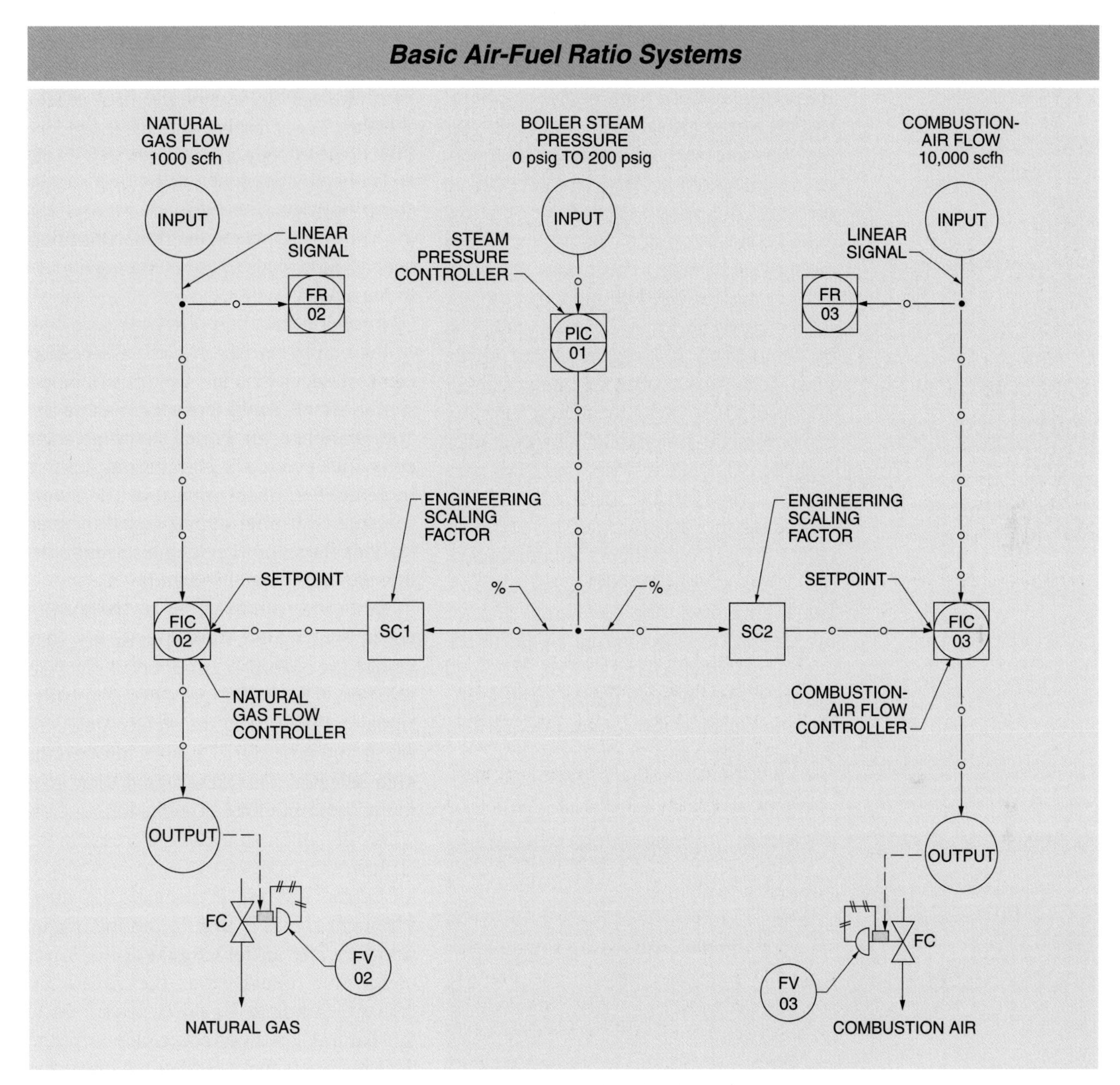

Figure 49-7. A basic air-fuel ratio control system controls air and fuel flows at a predetermined ratio.

The scaling is done by selection of the measurement ranges. In this simplified control arrangement, an increase in the output from the boiler steam pressure controller (PIC 01) raises the setpoints to the fuel flow controller (FIC 02) and the combustion-air flow controller by an equal percentage. Since the selected air and fuel flow ranges establish the operating air-to-fuel ratio, this ratio is maintained for any burning rate.

All burner systems need to operate with an excess of combustion air. If the excess combustion air is very small, any change in burning rate (load change) can temporarily reduce the excess combustion air to zero or less. This is not a safe condition because it can result in pockets of explosive combustible gases. A lead-lag air-fuel ratio control system, which solves this problem, can be added to the basic air-fuel ratio system.

Lead-Lag Air-Fuel Ratio Control Systems

A *lead-lag air-fuel control system* is a control system where increases in combustion-air flow lead increases in fuel flow and decreases in combustion-air flow lag decreases in fuel flow. A linking system solves the low excess combustion air situation experienced during load changes. This means that extra air is always provided during load changes. A lead-lag air-fuel ratio control system can be obtained by linking the output of the steam pressure controller, the combustion-air and fuel flow measurements, and the combustion-air flow and fuel flow controller setpoints. The linking is accomplished with the use of high and low signal selectors.

The output of the steam pressure controller (PIC 01) goes to the inputs of both the low (PY-1 01) and high (PY-2 01) signal selectors. **See Figure 49-8.** The low signal selector output supplies the setpoint signal to the natural gas flow controller (FIC 02), and the high signal selector output supplies the setpoint signal to the combustion-air flow controller (FIC 03).

Combustion processes need multiple controls that integrate safety and efficiency.

In addition, the low signal selector receives an input from the combustion-air flow measurement and the high signal selector receives an input from the fuel flow measurement. If the control strategy is being developed entirely in a digital control system, the two flow signals and the setpoint signals to the flow controllers need to be scaled before being introduced to the selectors.

In operation, the boiler pressure controller output signal, the airflow measurement signal, and the fuel flow measurement signal are all equal at a steady-state operation. When the boiler steam pressure controller needs a higher burning rate to increase the steam pressure, the steam pressure controller output signal increases. This increased signal goes to both the low and high signal selectors.

Since the boiler pressure controller signal is now higher than the steady-state value, the signal is stopped at the low selector. The boiler pressure controller signal is higher than the steady-state value, so the new signal passes through the high selector. This new signal then goes to the setpoint of the combustion-air flow controller and results in an increase in the airflow.

As the airflow increases, the measurement signal is sent to the low signal selector. The airflow signal is the lower of the two inputs, so as the airflow increases, this flow signal is passed on to the natural gas flow controller setpoint. This increases the fuel flow, but only after the airflow has already increased. Thus the airflow increase leads the natural gas flow increase when there is a demand for increased burning rate.

A signal for a decrease in burning demand from the boiler pressure controller is stopped by the high signal selector but is passed through the low selector to the fuel flow controller setpoint. This results in the natural gas flow being reduced before the combustion-air flow is reduced. Thus, the airflow decrease lags the natural gas flow when there is a need for a decreased burning rate.

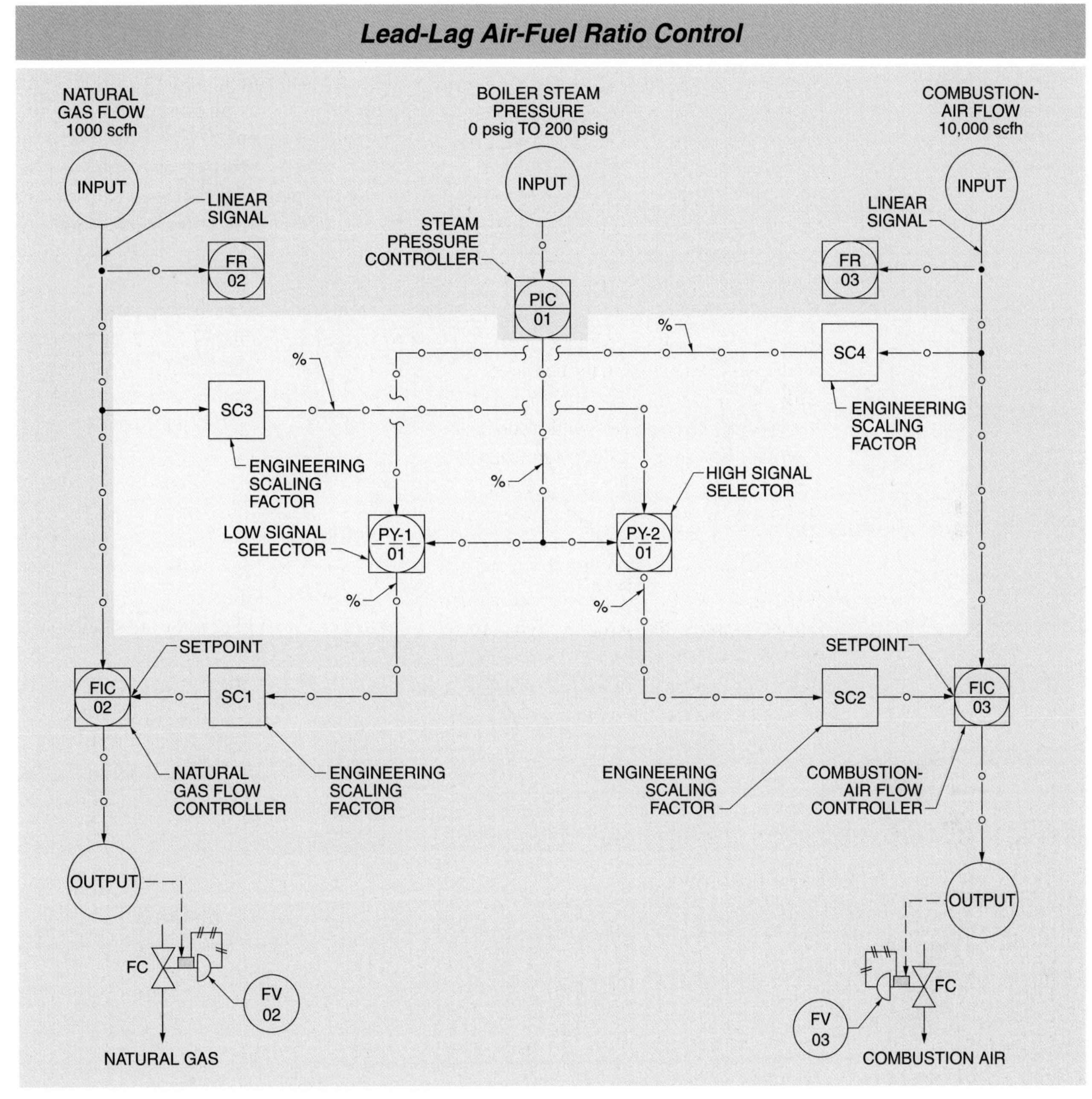

Figure 49-8. A lead-lag air-fuel control system improves safety by ensuring that the air-fuel mixture always maintains excess air during process upsets or load changes.

Oxygen Trim Control

Oxygen trim control is a control strategy used to adjust the air-fuel ratio of a burner system. **See Figure 49-9.** The air-fuel ratio can be adjusted according to the results from a flue gas oxygen analyzer. The oxygen analyzer output goes to an oxygen controller (AIC 05). The oxygen controller is programmed to maintain a specified percentage of excess oxygen. The oxygen controller needs to be tuned to have a slow response so that there are no sudden changes in the controller output. The controller output goes to a high and low limit module (AY-1 05) in order to limit the effect on the lead-lag control system. The signal then goes to the summing modules (AY-2 05 and AY-3 05).

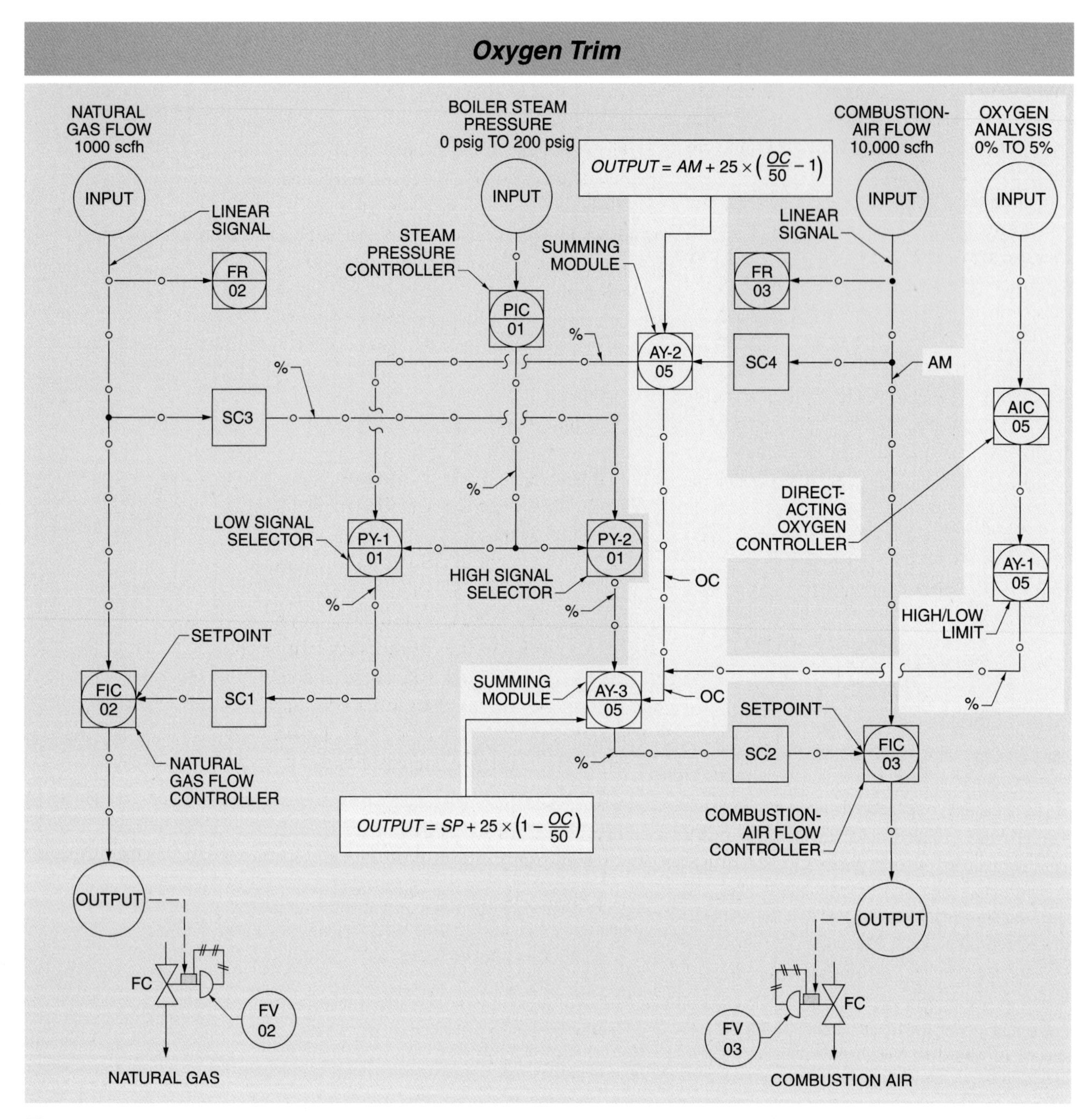

Figure 49-9. An oxygen analyzer is used to adjust an air-fuel ratio to obtain the desired oxygen level in the flue gas.

These modules modify the signals going to the combustion-air and natural gas flow controllers. The summing modules need to be placed where the input and output signals are a percentage signal rather than in engineering units. Any change in the signal to the combustion-air flow controller setpoint must be balanced with an exactly equal inverse change in the airflow measurement signal that goes to the natural gas flow controller.

This is best explained by observing that the summing module (AY-3 05) is used to change the combustion-air flow setpoint. This setpoint change is based on changes to the natural gas flow and changes in the oxygen analysis. Changes in the natural gas flow are passed through to the combustion-air flow controller to maintain the air-fuel ratio. Changes in the oxygen analysis are used to adjust the air-fuel ratio by changing the airflow without changing the fuel flow.

When the airflow setpoint increases, the combustion-air valve opens slightly and increases the airflow. The increased airflow signal is then sent through the other summing module (AY-2 05) to the low signal selector (PY-1 01). The summing module (AY-2 05) reduces the airflow measurement signal going to the low signal selector (PY-1 01) by the exact amount that the airflow was increased. The result is that the airflow is increased, but the natural gas flow remains the same. Thus the air-fuel ratio is increased. The summing module (AY-3 05) modification of the airflow signal being sent to the airflow controller is calculated as follows:

$$output = SP + 25 \times \left(1 - \frac{OC}{50}\right)$$

where
$output$ = summing module (0% to 100%)
SP = setpoint (0% to 100%)
OC = oxygen controller (0% to 100%)

For example, if the excess oxygen in the flue gas is less than desired, the burner operation is too rich. The oxygen controller output needs to act to increase the air-fuel ratio. If the oxygen controller output is 40% and the setpoint is 70%, the modified setpoint signal to the combustion-air flow controller is calculated as follows:

$$output = SP + 25 \times \left(1 - \frac{OC}{50}\right)$$

$$output = 70 + 25 \times \left(1 - \frac{40}{50}\right)$$

$$output = 70 + 25 \times (1 - 0.8)$$

$$output = 70 + 25 \times 0.2$$

$$output = 70 + 5$$

$$output = \mathbf{75\%}$$

This shows that the summing module increases the combustion-air controller setpoint from 70% to 75%. This increases the relative amount of air in the air-fuel mixture and therefore increases the air-fuel ratio.

The other summing module (AY-2 05) is used to modify the signal to the natural gas flow controller. The summing module modification of the airflow signal being sent to the low signal selector is calculated as follows:

$$output = AM + 25 \times \left(\frac{OC}{50} - 1\right)$$

where
$output$ = summing module (0% to 100%)
AM = airflow (0% to 100%)
OC = oxygen controller (0% to 100%)

While the first summing module is acting to increase the amount of airflow relative to the natural gas flow, the same signal is sent to the other summing module (AY-2 05). As the oxygen flow measurement increases from its original 70% to its new 75%, the modified setpoint signal to the low signal selector is calculated as follows:

$$output = AM + 25 \times \left(\frac{OC}{50} - 1\right)$$

$$output = 75 + 25 \times \left(\frac{40}{50} - 1\right)$$

$$output = 75 + 25 \times (0.8 - 1)$$

$$output = 75 + 25 \times (-0.2)$$

$$output = 75 + (-5)$$

$$output = \mathbf{70\%}$$

The net result of the oxygen trim is to increase the air-fuel ratio by increasing the combustion-air flow controller setpoint to 75% and leaving the natural gas flow at 70%. In actual practice, at the same time that the airflow controller setpoint is being increased, the natural gas flow is being decreased. Not until the new airflow has been measured does the natural gas flow return to its original value. This is why changes due to oxygen measurements must be done slowly.

TECH FACT

A high limit or low limit may be used on a sensor signal when using multiple measurements of the same object. This strategy eliminates errors due to failed sensors.

In operation, a 50% signal from the oxygen flue gas controller provides no modification to the lead-lag control system. A zero output signal from the oxygen trim controller adds 25% to the normal combustion-air flow setpoint signal and increases the combustion-air flow. **See Figure 49-10.** Thus, a loss of the oxygen trim controller output increases the combustion-air flow. This is designed into the system as a safety measure. The oxygen trim control strategy automatically adjusts the air-fuel ratio to maintain the desired excess air. Alternatively, the oxygen controller can be put into manual and the output adjusted to change the excess air.

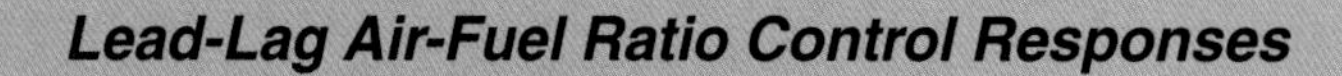

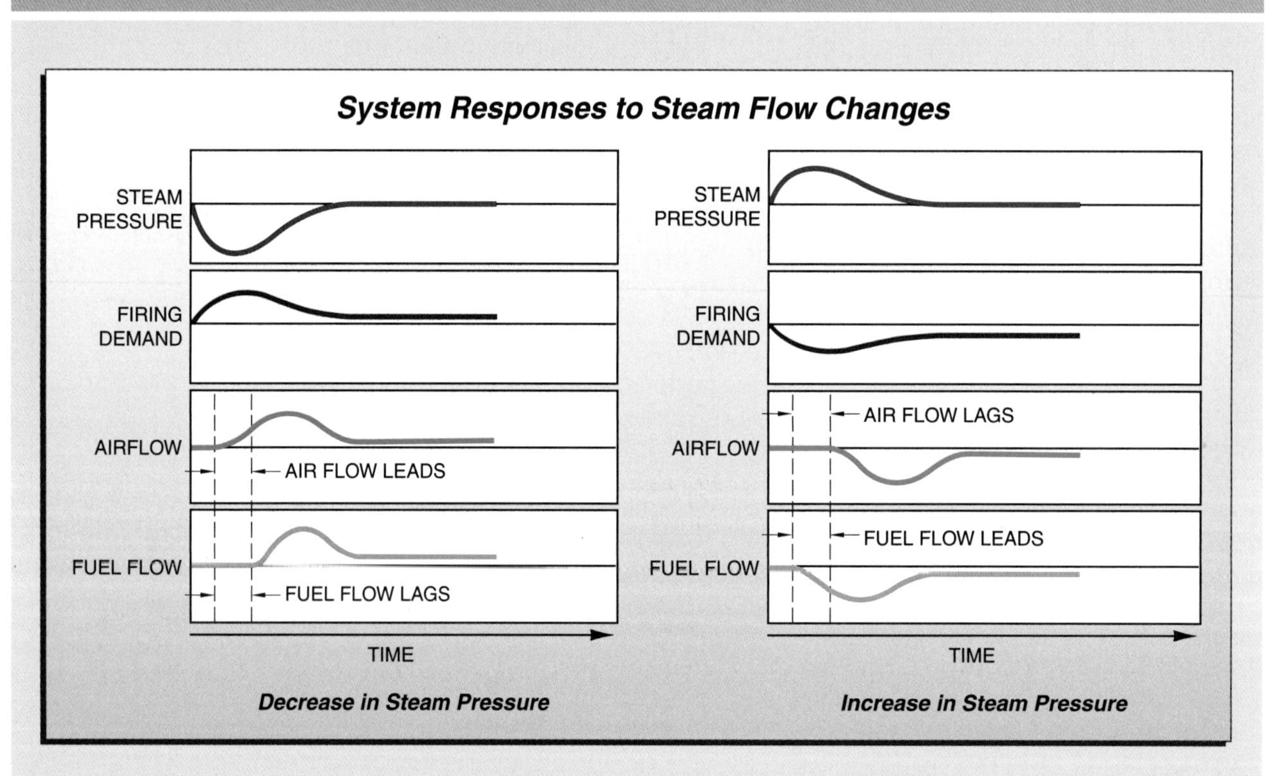

Oxygen Controller Output %	AY-3 Output Signal to Airflow Controller %	Increase in Airflow %	AY-2 Output Signal to Low Selector %	Output from Low Selector %
0	+25	+25	−25	0.0
25	+12.5	+12.5	−12.5	0.0
50	0	0	0	0.0
75	−12.5	−12.5	+12.5	0.0
100	−25	−25	+25	0.0

Oxygen Trim Addition Module Signal Values

Figure 49-10. The lead-lag air-fuel system allows an airflow change of up to 25% above or below the nominal flow rate.

KEY TERMS

- *ratio control:* A control strategy used to control a secondary flow to the predetermined fraction, or flow ratio, of a primary flow.
- *mixing ratio control:* A control strategy used when there is a need to mix two flow streams in a specified ratio.
- *splitting ratio control:* A control strategy used when there is a need to split a flow into two separate constant-ratio flows.
- *reflux:* The portion of a condensed overhead vapor pumped back to the top of a column.
- *reflux ratio:* The reflux flow back into a column divided by the total overheads flow.
- *lead-lag air-fuel control system:* A control system where increases in combustion-air flow lead increases in fuel flow and decreases in combustion-air flow lag decreases in fuel flow.
- *oxygen trim control:* A control strategy used to adjust the air-fuel ratio of a burner system.

REVIEW QUESTIONS

1. Define ratio control.
2. How is the remote/local (R/L) switch used for normal-flow ratio control operation?
3. Define and compare mixing and splitting ratio control.
4. Why is it important for burner systems to operate with excess combustion air?
5. What is lead-lag air-fuel control, and why is it used?

Analysis and Multivariable Control

Chapter 50

OBJECTIVES

- Describe a common use for analysis control.
- Describe the common multivariable applications.
- Explain the control of conveyor systems.

ANALYSIS CONTROL

Analysis control loops are seldom used because of the complexity and lack of reliability of many analyzers. Simpler analyzers are more likely to be used in an automatic control loop. Monitoring the conductivity of recovered condensate is one of the more common control applications.

Condensate Conductivity Monitoring

Condensate is the water formed when steam condenses. Condensate is often collected for reuse in boilers. This is done to lower the costs associated with treating raw water and to reduce the discharge to water treatment systems. Returned condensate requires little or no treatment and is preferable for use in boilers as long as the quality is acceptable. Condensate quality can be measured with a conductivity instrument. A low conductivity measurement represents good quality condensate. Condensate with conductivity greater than about 5 μS is usually not used for boiler feedwater.

Condensate conductivity is usually measured while the condensate is being piped from the condensate collection system to the boiler feedwater system. **See Figure 50-1.** The piping is split into two branches. One branch goes to the boiler feedwater storage tank and the other to the condensate waste tank. A normally closed automated ON/OFF valve (AV-1 01) is installed in the line to the boiler feedwater system and a normally open automated ON/OFF valve (AV-2 01) is installed in the line to the condensate waste tank. The condensate conductivity is measured upstream of the junction of the piping branches.

The condensate conductivity instrument (AE 01) has a setpoint that controls an ON/OFF output switch (ASH 01). The output switch controls electrical power to the two ON/OFF valves. When the condensate quality is poor, the conductivity is above the setpoint and the output switch is open. This de-energizes the two ON/OFF valves and diverts the condensate to the waste tank. When the condensate conductivity is below the setpoint, the output switch is closed. This energizes the two ON/OFF valves and directs the condensate to the boiler feedwater system.

In operation, contaminated condensate produces a high conductivity, which is measured, and the control action diverts the condensate to the waste tank. When the condensate conductivity returns to a value below the setpoint, the condensate is routed back to the boiler feedwater system.

TECH FACT

The NEC® allows Type MI cable to be used in Class 1, Division 1 locations. When Type MI cable is listed as an acceptable wiring method, special explosionproof fittings are required.

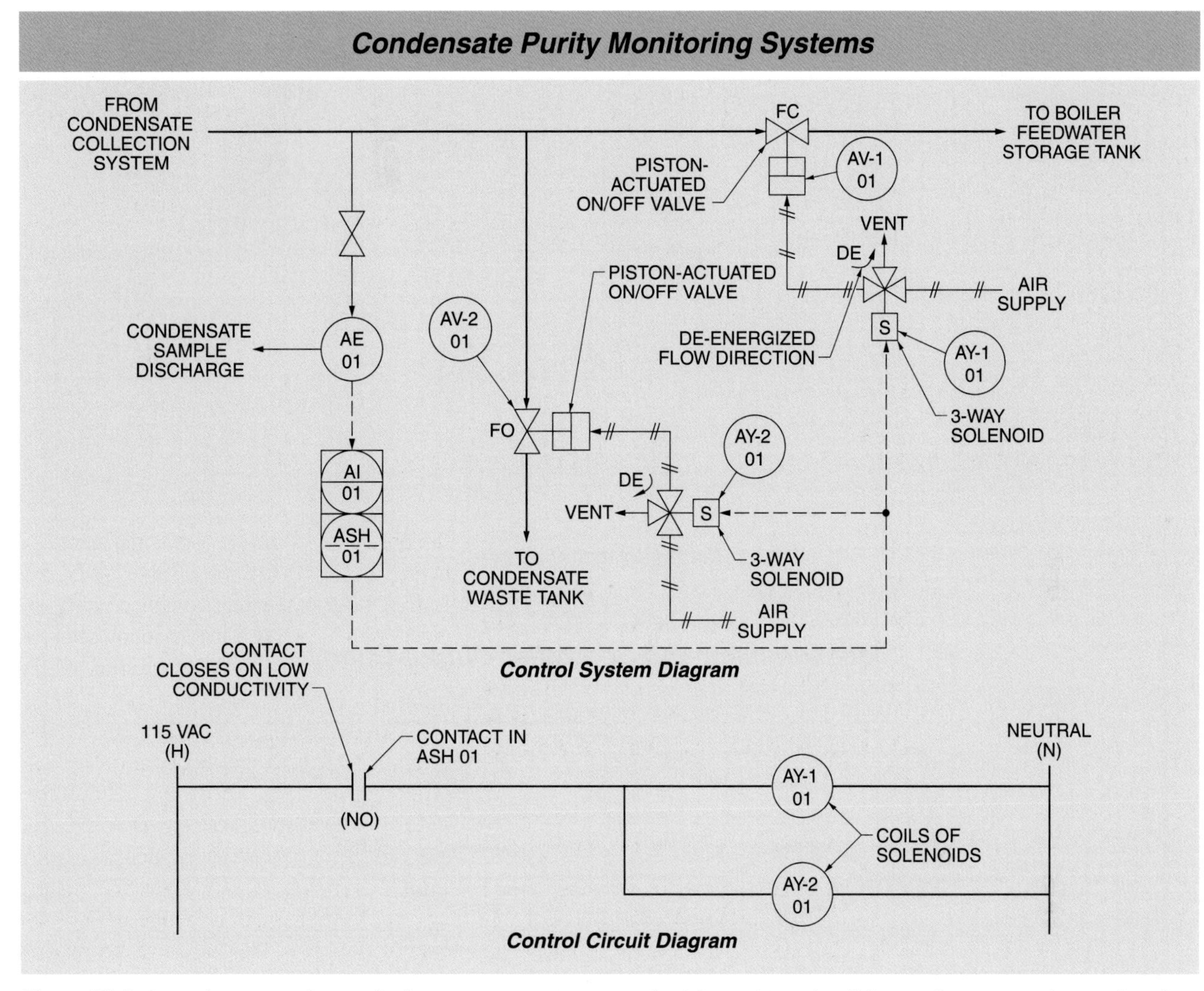

Figure 50-1. A condensate purity monitoring system measures conductivity to determine if the condensate can be used again.

MULTIVARIABLE CONTROL

A multivariable application is one that includes at least two different measurements combined into a single control system. Common applications are combinations of level and flow control.

Flow High and Low Level Selector System

There are some processes where the fluid transfer between two operations is normally handled by flow control, but if the feed tank level becomes too low or too high, the level controller takes over control from the flow controller. The level controller then adjusts the control valve to prevent the feed tank level from exceeding the low or high setpoints. This control strategy maintains a constant flow during normal conditions but prevents the tank from overflowing or from being pumped dry during abnormal conditions.

A control system that maintains constant flow, except when a tank level is at a level limit, consists of a standard flow control loop and a gap action level controller connected to a normally closed control valve through low and high signal selectors. **See Figure 50-2.** The flow transmitter (FT 51) is an inline meter such as a magnetic flowmeter, a vortex shedding meter, or a transmitting rotameter. The transmitter signal is sent to the flow controller (FIC 51).

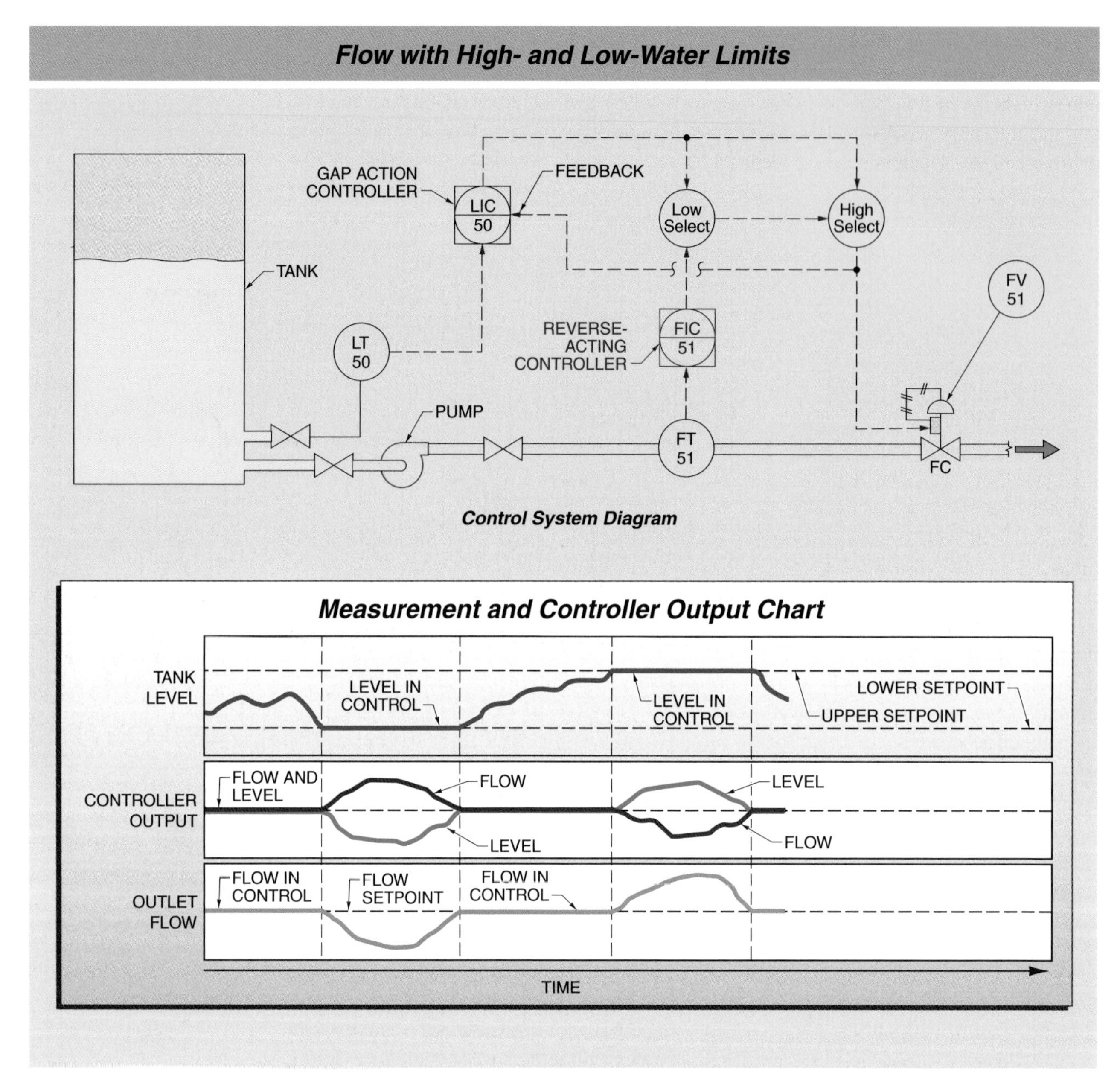

Figure 50-2. A flow controller maintains constant flow out of a tank until the level goes above or below the gap action controller setpoints.

The reverse-acting flow controller has proportional and integral functions. The flow controller output goes to a low selector. The output of the low selector goes to the high selector. The output of the high selector goes to the control valve (FV 51). A signal from the output of the high selector is connected to the feedback of the flow controller and the gap-action level controller.

The level transmitter (LT 50) is a flush-mounted d/p cell, but any suitable level measurement method can be used. The transmitter output goes to a gap action level controller (LIC 50). The gap action, proportional and integral, direct-acting controller has its control action locked at a constant output value when the level measurement is between the low and high setpoints. When the measurement is outside of the gap, the gap action level controller functions as a normal controller. The output of the gap action controller goes to the low and high selectors. A signal from the output of the high selector goes to the feedback of the gap action controller.

Under normal conditions, the flow controller modulates the control valve to maintain a flow equal to the flow setpoint. Under these conditions, the level measurement is in the gap and the gap action level controller has no active control action. The gap action level controller output tracks the signal to the control valve.

Under abnormal conditions, when the fluid flow into the tank becomes less than the controlled outlet flow, the level in the tank decreases until the level is at the low setpoint of the gap action controller. At this point, the gap action level controller starts working and lowers its output signal. The gap action controller setpoint is sent to both the low and high selectors. The gap action level controller output is lower than the output from the flow controller. The low signal selector then selects the gap action controller output and sends it to the high selector.

The two inputs to the high selector are now equal. Therefore, the gap action controller output is passed on to the control valve. The control valve is throttled to reduce the flow and maintain the level at the low setpoint. The flow controller output increases because the actual flow is below the controller setpoint. The flow controller output will wind up when the level controller has control of the valve. However, the normal tuning of a flow controller is very quick so that the output will unwind very quickly when the flow controller needs to retake control of the valve. The level controller needs the feedback from the output to the control valve to allow a smooth transfer between it and the flow controller.

If the inlet flow to the feed tank increases so that the level starts to rise again, the gap action level controller opens the control valve to increase the flow. When the level rises above the low level setpoint, control is transferred back to the flow controller. If the feed tank inlet flow continues to increase, the level eventually reaches the high level setpoint. At this point, the gap action controller starts working and increases its output. The low selector still selects the flow controller output since it is the lower of the two inputs, but the high selector selects the gap action controller output and passes it on to the control valve.

A condensate sampling system is a convenient place to collect a condensate sample for off-line analysis.

Boiler Drum Level and Feedwater Flow Control

Small boilers are often designed as firetube boilers. A *firetube boiler* is a boiler in which heat and gases of combustion pass through tubes surrounded by water. The level is measured in the boiler itself. Large boilers are often designed as watertube boilers. A *watertube boiler* is a boiler in which water passes through tubes surrounded by gases of combustion.

Watertube boilers are connected to a steam drum at the top of the boiler. The steam drum is where the steam separates from the water. It is also the location where the boiler water level is measured and controlled. The steam drum of a watertube boiler is fairly small relative to the size of a firetube boiler. This difference in size makes it more difficult to measure the level in a watertube boiler.

TECH FACT

The NEC® uses a dual system of measurement. Metric units are shown using the International System (SI) of units. The U.S. customary units are shown using standard inch-pound units. The SI units appear first and the U.S. customary units follow and are shown in parentheses.

A sudden increase in steam demand temporarily lowers the steam pressure and causes the water level in the steam drum to swell. Swell is caused by bubbles of steam suddenly forming in the water because of the drop in pressure. Swell displaces the water into the steam drum and results in erroneous high level readings that cause the level controller to reduce the feedwater flow. This is the wrong action, since the increased steam demand and the lowered steam pressure result in an increase in the firing rate. An increase in the firing rate requires more feedwater.

As the feedwater flow decreases, the pressure continues to drop and there is a further increase in the quantity of steam bubbles. As the increased steam demand continues, the water level quickly falls as steam is removed. The result is a greater boiler water deficiency that results from the increase in steam demand by itself.

The solution to this problem is to combine the boiler drum level controller output with the steam flow measurement. **See Figure 50-3.** The level controller (LIC 71) needs to be reverse-acting with proportional and integral functions. The level controller output and the steam flow signals are combined in a summation module (FY 75). The output of the summation module is calculated as follows:

$$output = 0.75 \times SF + 0.25 \times \left(\frac{LC}{100}\right) \times SFR$$

where

$output$ = output of summation module (in lb/hr)

SF = steam flow (in lb/hr)

LC = level controller (0% to 100%)

SFR = steam flow range (in lb/hr)

For example, if the steam flow range is 100,000 lb/hr, the steam flow demand increases from 50,000 lb/hr to 60,000 lb/hr, and the level controller output is 50%, the summation module output is calculated as follows:

$$output = 0.75 \times SF + 0.25 \times \left(\frac{LC}{100}\right) \times SFR$$

$$output = 0.75 \times 60{,}000 + 0.25 \times \left(\frac{50}{100}\right) \times 100{,}000$$

$$output = 45{,}000 + 0.25 \times 0.5 \times 100{,}000$$

$$output = 45{,}000 + 12{,}500$$

$$output = \mathbf{57{,}500\ lb/hr}$$

This shows that the water flow setpoint is changed to 57,500 lb/hr based on a combination of the steam flow measurement out of the boiler and the level measurement. As the higher steam demand continues, the water level in the boiler decreases and the controller output increases. This eventually brings the water flow control setpoint and the steam flow rate to the same value.

Pneumatic or electric systems do not use engineering units, so the inputs need to be scaled so that 0% to 100% values are used in the summation module. The SFR factor is not needed if the signals are being scaled.

The factors are selected to give the steam flow a greater influence than the level controller output in changing the feedwater flow setpoint. This is done because the boiler feedwater is added using a feedwater flow control loop. The feedwater flow controller range, in lb/hr, is the same as the steam flow range. Thus a 10% increase in steam flow results in a 7.5% increase in feedwater flow. The remainder of the signal to the feedwater flow controller comes from the level controller output to minimize the swell effect. The level controller output is needed to compensate for errors in the flow metering and to ensure that the level is always correct. This system works very well for the control of boiler steam drum level.

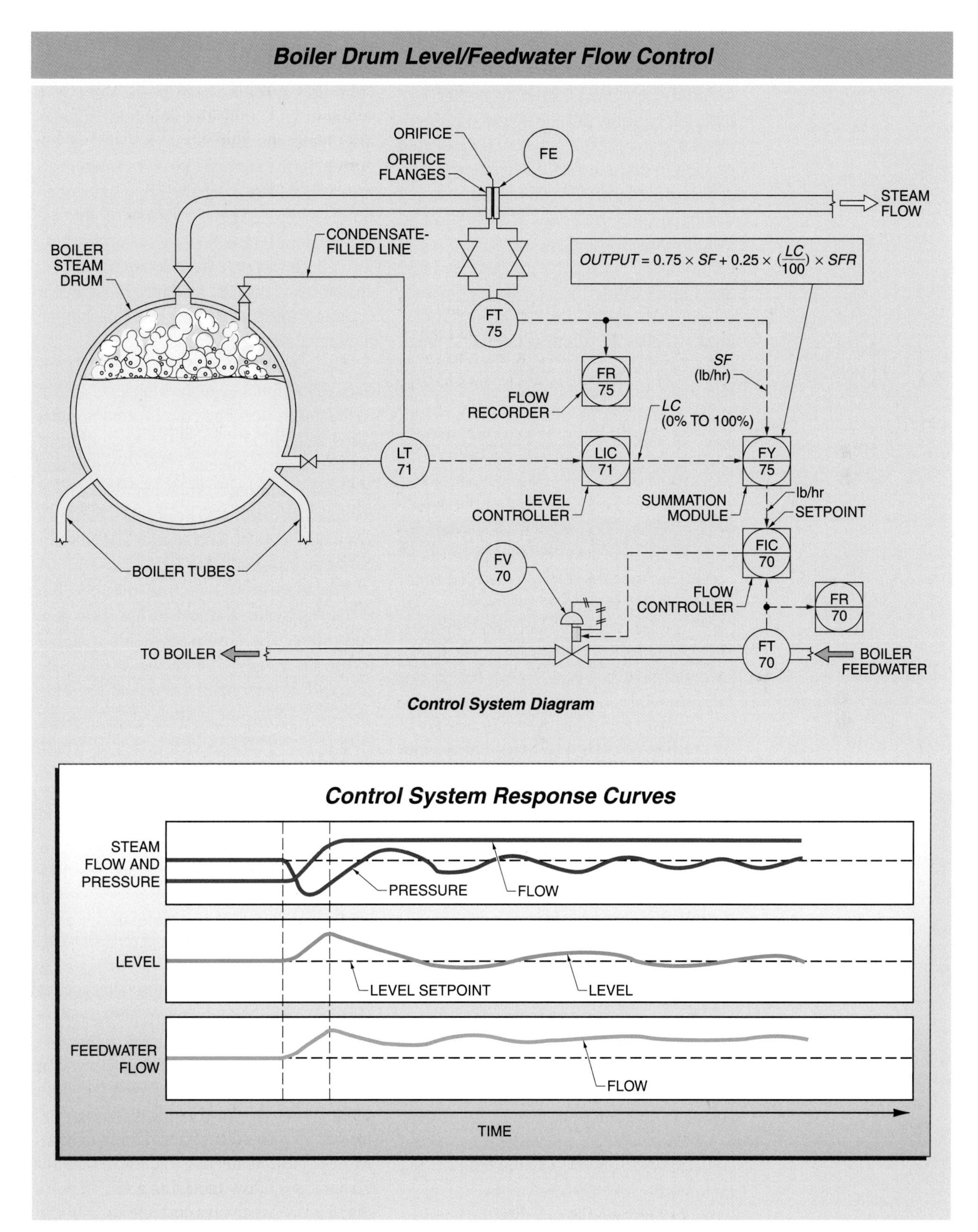

Figure 50-3. Boiler feedwater flow is determined by measuring the steam flow and the level change in the boiler steam drum.

Conveyor Systems

Conveyors are commonly used to move dry materials from one location to another. Dry materials typically include flowable bulk solids like grain, gravel, powdered or pelleted plastics, or discrete manufactured parts. The conveyor system used in this example moves dry bulk solids material from a storage tank, through a slide gate valve, to a screw conveyor and bucket elevator, to a feed bin. **See Figure 50-4.** The conveyor combines a level measurement with speed sensors, limit switches, an alarm system, and many types of pushbuttons. There are electrical interlocks between the different pieces of equipment to ensure that each downstream piece of the conveyor is operating before upstream pieces are allowed to operate.

TECH-FACT

HVAC technicians often work with health department officials to prevent mold growth. Mold growth may be caused by poor airflow, excessive heat and humidity conditions, or interior water leaks.

Normal Operational Sequences. The conveyor start sequence is initiated by a signal from the low level sensor to the controller. The conveyor equipment starts sequentially from the end of the conveyor back to the beginning.

The low level switch in the feed bin (LSL 01) sends a signal to the bucket elevator controller. The controller sends a start signal to the bucket elevator motor. The motor starts and the bucket elevator starts moving. The speed switch (SSL 03) sends a signal back to the controller to confirm that the bucket elevator is moving.

The bucket elevator controller sends a signal to the screw conveyor controller. The screw conveyor controller sends a start signal to the screw conveyor motor. The motor starts and the screw conveyor starts moving. The speed switch (SSL 04) sends a signal back to the controller to confirm that the screw conveyor is moving.

The screw conveyor controller sends a signal to the slide gate controller. The controller sends a signal to the three-way solenoid (KY 05). The solenoid actuates and opens the slide gate (KV 05) at the bottom of the storage tank. The open position is confirmed by the slide gate open limit switch (ZSH 05). At this point, the bulk material starts flowing out of the tank, through the screw conveyor, up the bucket elevator, and into the feed bin. The transfer continues until the feed bin level reaches the high level point.

The conveyor shutdown sequence is initiated by a signal from the high level sensor to the controller. The conveyor equipment stops sequentially from the beginning of the conveyor to the end. The conveyor and elevator could have problems restarting if loaded with material. The normal shutdown sequence empties the conveyor and elevator before being shut down.

The high level switch in the feed bin (LSH 02) sends a signal to the slide gate controller. The controller sends a signal to the three-way solenoid (KY 05). The solenoid de-energizes and closes the slide gate (KV 05) at the bottom of the storage tank. The closed position is confirmed by the slide gate closed limit switch (ZSL 05). At this point, the bulk material stops flowing out of the tank. The screw conveyor and bucket elevator keep running.

When the storage tank slide gate valve is closed, a signal is sent to a delay timer (TDO 09), which is connected to the running permissive of the screw conveyor. This timer allows the material in the screw conveyor to be discharged to the bucket elevator before the screw conveyor is shut down.

When the screw conveyor is stopped, a delay timer (TDO 08) is started which is connected to the bucket elevator controller. This timer allows the material in the elevator to be carried into the feed bin before the elevator is shut down. At this point, all parts of the conveyor are empty, the feed bin is full, and the slide gate on the bottom of the storage tank is closed.

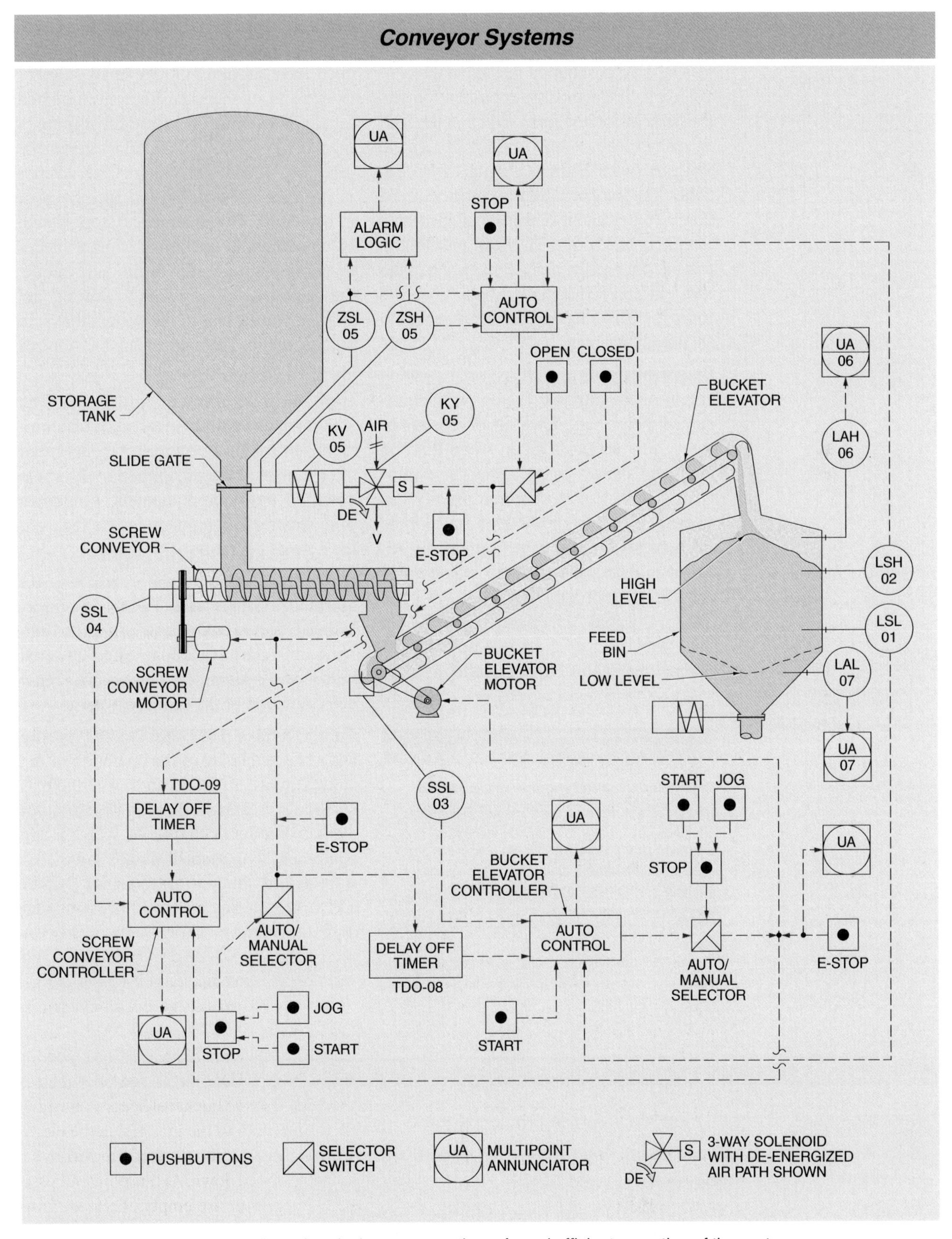

Figure 50-4. Conveyor systems have interlocks to ensure the safe and efficient operation of the system.

Manual Initiation of Auto Sequences. The automatic fill sequence can be started by a manual-start pushbutton located in the control room. The manual-start pushbutton acts the same as the low level switch in the feed bin, and the entire conveyor starts up as if the low level switch had sent a signal to the controller. The automatic filling sequence can be stopped by a manual-stop pushbutton located in the control room. The stop pushbutton acts the same as the high-level switch in the feed bin and sequentially shuts down the conveyor. This empties all the equipment in the normal manner.

Emergency Shutdown (E-Stop). Actuation of any of the multiple E-stop pushbuttons immediately shuts down all of the equipment. This action is taken when there is imminent danger to personnel or equipment. The E-stop pushbutton usually has a detent so that the pushbutton stays in the actuated position until it is manually returned to its normal position. This ensures that the system cannot restart until the E-stop is cleared. Since actuating the E-stop shuts down all the equipment, some of the equipment remains full of material.

TECH FACT

Many temperature measurement errors are caused by unintended thermocouple junctions. A thermocouple junction is formed any time wires of two different metals are connected together. Care must be taken to ensure that the proper extension wires are used.

Manual Operation. The bucket elevator, the screw conveyor, and the slide gate valve each have an automatic/manual selector switch and start/stop pushbuttons mounted locally. Switching the selector from auto to manual bypasses all the interlocks from the other equipment. The motor starter latches in the manual start, but the speed sensor is bypassed. The emergency stop system is maintained in operation. If other pieces of equipment are still in the auto mode, they automatically respond to starting and stopping of the part of the system that is in manual mode. This may mean that some of the equipment remains full of material.

Individual Equipment Shutdown. The shutdown of any individual piece of equipment due to a low operating rotational speed or a motor overload immediately shuts down all upstream equipment. Shutdown of downstream equipment is delayed as in the normal operation.

Alarming. All abnormal conditions are alarmed. Alarms associated with interlocked pieces of equipment are grouped into "first-out" alarm functions where the first item to trip is identified. This allows a rapid determination of the shutdown cause.

Status Display. The control room typically has a status display of the operation of the conveyor system. This can be as simple as some colored lights indicating which equipment is running or shut down, whether any equipment is in manual mode, and the E-stop status. If the control system is a DCS or a PLC, the display can be a fully animated digital display.

KEY TERMS

- *condensate:* The water formed when steam condenses.
- *firetube boiler:* A boiler in which heat and gases of combustion pass through tubes surrounded by water.
- *watertube boiler:* A boiler in which water passes through tubes surrounded by gases of combustion.

REVIEW QUESTIONS

1. Define condensate, and explain why it is used.
2. How are low and high signal selectors used with normal flow control to maintain the level in a feedwater tank within lower and upper limits?
3. Define firetube and watertube boiler, and describe where the water level is measured.
4. How does swell give erroneous level readings in a boiler?
5. What are some of the measurements that need to be made to operate a conveyor as described in this chapter?

REVIEW QUESTION ANSWERS

SECTION 1
INTRODUCTION TO INSTRUMENTATION

Chapter 1
Instrumentation Overview

1. Process control instrumentation is the technology of using instruments to measure and control manufacturing, conversion, or treating processes to create the desired physical, electrical, and chemical properties of materials. Process control instrumentation measures, controls, and interacts with computer, electrical, hydraulic, and mechanical systems.
3. Some four-year engineering schools and two-year technical colleges provide training in process control instrumentation as part of another program. A few technical colleges have complete programs in instrumentation as well as individual courses within other programs. The large majority of technicians working with instrumentation and control systems have learned on the job. Many process control instrument manufacturers provide training for their products. There are many trade union training programs that help technicians gain and develop the skills needed to be successful. A new trend in training is using computer-based process simulators.
5. Technical societies are organizations composed of groups of engineers and technical personnel united by professional interest. Private organizations are organizations that develop standards from an accumulation of knowledge and experience with materials, methods, and practices. Trade associations are organizations that represent producers and distributors of specific products.

Chapter 2
Fundamentals of Process Control

1. A process variable is the dependent variable that is to be controlled in a control system. A control variable, or manipulated variable, is the independent variable in a process control system that is used to adjust the dependent variable, the process variable. A setpoint is the desired value at which the process should be controlled and is used by the controller as a reference for comparison with the process variable.
3. Reproducibility is the closeness of agreement among repeated measured values when approached from both directions. Response time is the time it takes an element to respond to a change in the value of the measured variable or to produce a 100% change in the output signal due to a 100% change in the input signal. Fidelity is the ability of an element to follow a change in the value of an input. Dynamic error is the difference between a changing value and the momentary instrument reading or the controller action. Hysteresis is a property of physical systems that do not react immediately to the forces applied to them or do not return completely to their original state. Stability is the ability of a measurement to exhibit only natural, random variation where there are no known identifiable external effects causing the variation. Linearity is the closeness to which multiple measurements approximate a straight line on a graph. Nonlinearity is the degree to which multiple measurements do not approximate a straight line on a graph.
5. ON/OFF control is where a controller activates or deactivates the final element depending on whether the measured variable is above or below the setpoint. Proportional control is a control strategy that uses the difference between the setpoint and the process variable to determine a control output that is sent to a final element. A time proportional ON/OFF controller is a control strategy that has a predetermined output period during which the output contact is held closed (or power is ON) for a variable portion of the output period.

Chapter 3
Piping and Instrumentation Diagrams

1. A piping and instrumentation diagram (P&ID) is a schematic diagram of the relationship between instruments, controllers, piping, and system equipment. A P&ID typically includes the instrumentation and designations; all valves and their designations; mechanical equipment; flow direction; control inputs and outputs and interlocks; and details of the pipe connections, sample lines, and other fittings.
3. Letters are combined to show multiple pieces of information. The first letter identifies the function of an instrument and the variable that is being measured. Succeeding letters are used to modify the functional description.
5. A P&ID cannot show the complete control strategy implemented in distributed control systems, programmable logic controllers, computer control, and more complex control systems. The most complex systems use additional types of documents to describe how the system is to function.

SECTION 2
TEMPERATURE MEASUREMENT

Chapter 4
Temperature, Heat, and Energy

1. Temperature is the degree or intensity of heat measured on a definite scale. Temperature is an indirect measurement of the heat energy contained in molecules.

3. Conduction is heat transfer that occurs when molecules in a material are heated and the heat is passed from molecule to molecule through the material. When heat is transferred through conduction, there is no flow of material. Convection is heat transfer by the movement of gas or liquid from one place to another caused by a pressure difference. The two types of convection are natural convection and forced convection. Radiation is heat transfer by electromagnetic waves emitted by a higher-temperature object and absorbed by a lower-temperature object. When heat is transferred by radiation, there is no flow of material.

5. Heat capacity is the amount of energy needed to change the temperature of a material by a certain amount. Specific heat is the ratio of the heat capacity of a liquid to the heat capacity of water at the same temperature.

Chapter 5
Thermal Expansion Thermometers

1. Materials usually expand when heated and contract when cooled. The coefficient of linear expansion is the amount a unit length of a material lengthens or contracts with temperature changes. The coefficient of volumetric expansion is the amount a unit volume of a material expands or contracts with temperature changes. The expansion of materials is usually approximately linear with changes in temperature. This allows the expansion to be used to indicate temperature.

3. When one material has a greater coefficient of thermal expansion than another material, the difference in expansion can be used as a measure of temperature by direct reading or by connection to a mechanical linkage. A bimetallic thermometer is a thermal expansion thermometer that uses a strip consisting of two metal alloys with different coefficients of thermal expansion that are fused together and formed into a single strip, and a pointer or indicating mechanism calibrated for temperature reading.

5. The difference in height between the bulb and the pressure spring can introduce error, especially in a partially filled vapor-pressure system. Since this system is not filled under pressure, as are the liquid- and gas-filled systems, any column of fluid can create a pressure that causes an erroneous reading. Liquid-filled systems may be filled under a little more pressure than a vapor-pressure system, but it is not so high that elevation differences between the bulb and the indicator can be completely ignored. If there is 50′ or more of difference, the head of fluid affects the measurement.

Chapter 6
Electrical Thermometers

1. A thermocouple is an electrical thermometer consisting of two dissimilar metal wires joined at one end and a voltmeter to measure the voltage at the other end of the two wires. When the temperature changes at the hot junction, a measurable voltage is generated across the cold junction.

3. A Seebeck voltage cannot be measured directly because a voltmeter is connected to a thermocouple and the copper leads to the voltmeter create new thermocouple junctions that generate Seebeck voltages that oppose the thermocouple voltage. Therefore, it is necessary to determine the temperatures at the junctions before the thermocouple voltage can be measured. A simple way to determine the temperatures at the new junctions is to hermetically seal those junctions and immerse them in ice water to form a cold junction. Cold junction compensation measures the temperature at the voltmeter junctions and calculates the equivalent reference voltage.

5. A difference thermocouple is a pair of thermocouples connected together to measure a temperature difference between two objects. A thermopile is an electrical thermometer consisting of several thermocouples connected in series to provide a higher voltage output. An averaging thermocouple is an electrical thermometer consisting of a set of parallel-connected thermocouples that is commonly used to measure an average temperature of an object or area. A thermocouple pyrometer consists of a plain electrical meter with a measurement range of 20 mV to 50 mV, a thermocouple, and a balancing resistor. A null-current thermocouple consists of a circuit and a voltage generator that can be adjusted to exactly balance the voltage output of a thermocouple.

Chapter 7
Infrared Radiation Thermometers

1. An IR thermometer measures the infrared radiation (IR) emitted by an object to determine its temperature. As the temperature increases, the amount of emitted infrared radiation from the surface increases dramatically and the wavelength of peak emittance becomes shorter.

3. Emissivity is the ability of a body to emit radiation and is the ratio of the relative emissive power of any radiating surface to the emissive power of a blackbody radiator at the same temperature. Reflectivity is the ability of an object to reflect radiation. Bodies with a high emissivity reflect very little. Transmissivity is the ability of objects like glass to allow infrared radiation to pass through. Background temperature compensation is a process by which some IR thermometers allow the temperature of a reflected source to be measured or specified.

5. A disappearing filament pyrometer is a high-temperature thermometer that has an electrically heated, calibrated tungsten filament contained within a telescope tube. The current to the filament is manually adjusted until the apparent brightness matches that of the target source. When the brightness of the filament matches the brightness of the hot source, the filament disappears from view. Measuring circuitry in the pyrometer converts the filament current value to a temperature reading on a temperature indicator.

Chapter 8
Practical Temperature Measurement and Calibration

1. All temperature-measuring instruments take time to respond to changes in temperature. For a first order system, a time constant can be used to develop a model that describes the process. A time constant is the time required for a process to change by 63.2% of its total change when an input to the process is changed.

3. A ground loop is current flow from one grounded point to a second grounded point in the same powered loop due to differences in the actual ground potential. The combination of the isolator, transmitter, and power supply allows the use of grounded thermocouples without any danger of ground loop electrical currents between thermocouples since there is only one ground point in the sensing portion of the circuit.
5. A dry well calibrator is a temperature-controlled well or box where a thermometer can be inserted and the output compared to the known dry well temperature. A microbath is a small tank containing a stirred liquid used to calibrate thermometers. A blackbody calibrator is a device used to calibrate infrared thermometers.

SECTION 3
PRESSURE MEASUREMENT

Chapter 9
Pressure and Force

1. Pressure is force divided by the area over which that force is applied. Pressure increases as force increases or as area decreases.
3. Pneumatic pressure is the pressure of air or another gas. This may be in the form of compressed air that is used to do work or it may be in the form of compressed air that is used to send a signal in a pneumatic control system. Hydraulic pressure is the pressure of a confined hydraulic liquid that has been subjected to the action of a pump. The pressurized hydraulic fluid can be used to move objects or do other work.
5. The four common pressure scales are absolute, gauge, vacuum, and differential pressure. Absolute pressure is pressure measured with a perfect vacuum as the zero point of the scale. Gauge pressure is pressure measured with atmospheric pressure as the zero point of the scale. Vacuum pressure is pressure less than atmospheric pressure measured with atmospheric pressure as the zero point of the scale. Differential pressure is the difference in pressure between two measurement points in a process.

Chapter 10
Mechanical Pressure Instruments

1. A manometer is a device for measuring pressure with a liquid-filled tube. In a manometer, a fluid under pressure is allowed to push against a liquid in a tube. The movement of the liquid is proportional to the pressure.
3. The purpose of the angled tube is to lengthen the scale for easier reading. For example, when the angle is 30°, 1 vertical inch becomes 2″ on the inclined scale. The scale is then stretched proportionally to enable an accurate reading. This type of manometer is used for low-pressure applications because it is difficult to accurately read low pressures in a vertical tube.
5. A diaphragm is a mechanical pressure sensor consisting of a thin, flexible disc that flexes in response to a change in pressure. A capsule is a mechanical pressure sensor consisting of two convoluted metal diaphragms with their outer edges welded, brazed, or soldered together to provide an empty chamber between them. A pressure spring is a mechanical pressure sensor consisting of a hollow tube formed into a helical, spiral, or C shape. A bellows is a mechanical pressure sensor consisting of a one-piece, collapsible, seamless metallic unit with deep folds formed from thin-wall tubing with an enclosed spring to provide stability, or with an assembled unit of welded sections. A double-ended piston is a mechanical pressure sensor consisting of a differential pressure gauge with a piston that admits pressurized fluid at each end.

Chapter 11
Electrical Pressure Instruments

1. In order to measure a change in resistance, it must occur in a circuit. A bridge circuit is commonly used, with the strain gauge resistance occupying one or more arms of the bridge. When the balance of the bridge is upset due to a change in the resistance of one arm (the strain gauge), the imbalance can be detected and amplified.
3. A capacitance pressure transducer is a diaphragm pressure sensor with a capacitor as the electrical element. In a typical industrial capacitive pressure sensor, part of the diaphragm is one plate and the mounting surface is the other. When pressure distorts the diaphragm and alters the distance between the plates, the capacitance of the sensor changes. The change in the capacitance of the sensor causes a variation in the impedance that varies with the applied pressure.

   A differential pressure transmitter sends the output of a d/p cell to another location where the signal is used for recording, indicating, or control. Many electronic differential pressure cells consist of a diaphragm surrounded by silicone oil where the process pressure is applied to one side of the diaphragm. Another process pressure or atmospheric pressure is applied to the other side of the diaphragm. The differential capacitance between the center diaphragm and the plates is converted to a digital or to a 4 mA to 20 mA analog signal.
5. A piezoelectric pressure transducer is a diaphragm pressure sensor combined with a crystalline material that is sensitive to mechanical stress in the form of pressure. As the crystal is compressed, a small electric potential is developed across the crystal.

Chapter 12
Practical Pressure Measurement and Calibration

1. A dry leg is an impulse line that is filled with a noncondensing gas. A dry leg can be used when process vapors are noncorrosive and nonplugging, and when their condensation rates, at normal operating temperatures, are very low. A wet leg is an impulse line filled with fluid that is compatible with the pressure-measuring device.
3. Adding snubbers (pulsation dampers) to inlet lines limits pulsations and surges. A porous filter, a ball check valve, or a variable orifice may serve as a snubber. Some snubbers have a moving piston within the body of the device that also cleans out any scale and sediment.
5. For many years, the calibration of pressure-measuring instruments has been performed using manometers, deadweight testers, and digital electronic test gauges.

SECTION 4
LEVEL MEASUREMENT

Chapter 13
Mechanical Level Instruments

1. Point level measurement is a method of level measurement where the only concern is whether the amount of material is within the desired limits. Continuous level measurement is a method of tracking the change of level over a range of values.

3. There are many applications where it is more convenient to measure the pressure at the bottom of a tank than to measure the actual location of the top of the liquid. For example, a tank may be sealed to prevent the escape of volatile or toxic fluids.

5. A paddle wheel switch is a point level measuring device consisting of a drive motor and a rotating paddle wheel mounted inside a tank. As the level in the tank rises and touches the paddle wheel, the torque required to turn the paddle wheel increases. The increased torque activates a switch that can be used to stop or start equipment or signal an alarm.

Chapter 14
Electrical Level Instruments

1. A capacitance probe is a part of a level measuring instrument and consists of a metal rod inserted into a tank or vessel, with a high-frequency alternating voltage applied to it and a means to measure the current that flows between the rod and a second conductor. The amount of capacitance depends on the dielectric constant, the surface area of the conductors, and the distance between the conductors. As the level in the vessel rises, the capacitance increases because the material in the vessel replaces the air or vapor between the conductors. The granular solid or liquid material has a higher dielectric constant than air, and the changed capacitance is measured and used to determine the level.

3. A conductivity probe is a point level measuring system consisting of a circuit of two or more probes or electrodes, or an electrode and the vessel wall where the material in the vessel completes the circuit as the level rises in the vessel. Conductivity level probes can only be used with conductive liquids. The length of the center element establishes the level control point.

   An inductive probe is a point level measuring instrument consisting of a sealed probe containing a coil, an electrical source that generates an alternating magnetic field, and circuitry to detect changes in inductance. As the level changes, the magnetic field of the probe interacts with the conductive material. The interaction is detected by measuring the inductive reactance.

5. A thermal dispersion sensor is a point level measuring instrument consisting of two probes that extend from the detector into the vessel, with one of the probe tips being heated. The detector monitors the difference in temperature between the heated probe tip and the unheated probe. When the liquid covers the probe tips, the temperature of the heated probe drops because the heat is removed by the liquid. The decreased differential temperature is detected and a switch is activated.

Chapter 15
Ultrasonic, Radar, and Laser Level Instruments

1. Two crystals of an ultrasonic sensor are enclosed in a probe but are separated by a small integral air gap. When the gap is exposed to air or vapor, the ultrasonic signal is not able to pass through the gap in sufficient strength to be received. When the level rises and material fills the gap, the ultrasonic signal from the transmitter passes through the gap because liquids and granular solids carry sound waves more efficiently than air or vapor.

3. A frequency modulated continuous wave (FMCW) radar is a level measuring sensor consisting of an oscillator that emits a continuous microwave signal that repeatedly varies its frequency between a minimum and maximum value, a receiver that detects the signal, and electronics that measure the frequency difference between the signal and the echo. The reflected echo signal has a different frequency than the emitted signal that is being generated at that instant. These differences vary directly with the distance between the emitter and the surface of the material in the vessel.

5. A laser level instrument is a level measuring instrument consisting of a laser beam generator, a timer, and a detector mounted at the top of a vessel. A crystal-emitted pulsing laser beam with a wavelength of about 900 nm is directed at the surface of the process material. The laser beam is reflected back to the emitter where a very accurate timing device measures the out-and-back interval. The travel time varies with the level.

Chapter 16
Nuclear Level Instruments and Weigh Systems

1. Nuclear level sensors are used for process materials that are extremely hot, corrosive, toxic, or under very high pressure, and so are not suitable for intrusive level detectors.

3. A load cell is a device used to weigh large items and typically consists of either piston-cylinder devices that produce a hydraulic output pressure or strain gauge assemblies that provide electrical output proportional to the applied load.

5. A hydraulic load cell transfers the pressure acting on the cell from the weight of the vessel and its contents to a piston. The piston pressurizes the system hydraulic fluid in a diaphragm chamber. This change in pressure varies with the load acting on the load cell. The pressure measurement can be converted to a level measurement when the density of the material and the configuration of the tank are known.

Chapter 17
Practical Level Measurement and Calibration

1. Funnel flow is the flow of a bulk solid where the material empties out of the bottom of a silo and the main material flow is down the center of the silo, with stagnant areas at the sides and bottom of the silo. Mass flow is the flow of a bulk solid where all material in a silo flows down toward the bottom at the same rate.

3. Newer digital indicating transmitters have the ability to provide calibration for the load cells. It is also possible to use a load cell simulator to set the range of the weigh

cell transmitter. A simple calibration method uses physical weights piled onto the vessel a little at a time, checking the indicated weight against the actual applied weight. Another calibration method uses known amounts of liquid added to the vessel as the calibration weight. The final method uses a portable load cell calibration system, a readout instrument, and hydraulic jacks.

5. A transmitter can be protected from a corrosive process fluid by connecting capillary tubing, with diaphragm seals, that is filled with liquid of constant specific gravity.

SECTION 5 FLOW MEASUREMENT

Chapter 18 Fluid Flow

1. Flow rate is the quantity of fluid passing a point at a particular moment. Total flow is the quantity of fluid that passes a point during a specific time interval. For example, the flow rate of pumping a fluid may be given in gallons per hour and the total flow is the total gallons pumped.
3. Absolute viscosity is the resistance to flow of a fluid and has units of centipoise (cp). Kinematic viscosity is the ratio of absolute viscosity to fluid density and has units of centistokes (cS).
5. Boyle's law is a gas law that states that the absolute pressure of a given quantity of gas varies inversely with its volume provided the temperature remains constant. Charles' law is a gas law that states that the volume of a given quantity of gas varies directly with its absolute temperature provided the pressure remains constant. Gay-Lussac's law is a gas law that states that the absolute pressure of a given quantity of a gas varies directly with its absolute temperature provided the volume remains constant. The three gas laws can be combined into one equation that shows the relationship between volume, pressure, and temperature.

Chapter 19 Differential Pressure Flowmeters

1. The most common primary element is the orifice plate. Other primary elements include flow nozzles, venturi tubes, low-loss flow tubes, and pitot tubes.
3. The Bernoulli equation is an equation stating that the sum of the pressure heads of an enclosed flowing fluid is the same at any two locations.
5. Flow measurement is only accurate as long as the flowing conditions remain the same as when the system was designed. Changes in pressures and temperatures are common in gas and vapor flow measurements. Flowing conditions that differ from the original flowmeter design calculation can result in significant flow measurement errors.

Chapter 20 Mechanical Flowmeters

1. A differential pressure flowmeter maintains a constant flow area and measures the differential pressure. A variable-area flowmeter maintains a constant differential pressure and allows the area to change with flow rate.
3. A nutating disc meter is a positive-displacement flowmeter for liquids where the liquid flows through the chambers, causing a disk to rotate and wobble (nutate). A rotating-impeller meter is a positive-displacement flowmeter where the fluid flows into chambers defined by the shape of the impellers. A sliding-vane meter is a positive-displacement flowmeter where the fluid fills a chamber formed by sliding vanes mounted on a common hub rotated by the fluid. An oscillating-piston meter is a positive-displacement flowmeter for liquids in which the fluid fills one piston chamber while the other piston chamber is emptied.
5. A weir is an open-channel flow measurement device consisting of a flat plate that has a notch cut into the top edge and is placed vertically in a flow channel. A Parshall flume is a special form of open-channel flow element that has a horizontal configuration similar to a venturi tube, with converging inlet walls, a parallel throat, and diverging outlet walls.

Chapter 21 Magnetic, Ultrasonic, and Mass Flowmeters

1. A magnetic meter, or magmeter, is a flowmeter consisting of a stainless steel tube lined with nonconductive material, with two electrical coils mounted on the tube like a saddle. As a flowing conductive liquid moves within the nonconductive tube and passes through the magnetic field of the coils, a voltage is induced and detected by the electrodes. The voltage depends on the strength of the magnetic field, the distance between the electrodes, and the conductivity and velocity of the liquid.
3. A Coriolis meter is a mass flowmeter consisting of specially formed tubing that is oscillated at a right angle to the flowing mass of fluid. The flow passes through two tubes of equal length and shape. The two sections of tubing are made to oscillate at their natural frequency. The fluid accelerates as it is vibrated and causes the tubing to twist back and forth while the tube oscillates. Two detectors develop a sine wave current due to the opposite oscillations of the two sections of tubing. The tubes twist in opposite directions, resulting in the sine waves being out of phase. The degree of phase shift varies with the mass flow through the meter.
5. A belt weighing system is a mass flow measuring system consisting of a specially constructed belt conveyor and a section that is supported by electronic weigh cells. Solids are carried onto the weighing section. The weight of solids on the measured section divided by the length of the measured section multiplied by the conveyor speed results in the flow in pounds per unit time (flow = weight ÷ length × speed).

Chapter 22 Practical Flow Measurement

1. For orifice plates, the three common variations are flange taps, vena contracta taps, and pipe taps. The names are based on the location of the taps. Flange taps are located in the orifice plate flanges 1″ upstream and downstream of the orifice plate faces. Vena contracta taps are located 1 pipe diameter upstream and about ½ pipe diameter downstream of the orifice plate faces. Pipe taps are located 2½ pipe diameters upstream and 8 pipe diameters downstream of the orifice plate faces.

3. A blocking valve is a valve used at the differential measuring instrument to provide a convenient location to isolate the instrument from the impulse, equalizing, or venting lines and to provide a way to equalize the high- and low-pressure sides of the differential instrument. Equalizing the instrument is necessary so that the instrument can be periodically calibrated and zeroed.

5. A flow switch is a device used to monitor flowing streams and to provide a discrete electrical or pneumatic output action at a predetermined flow rate. Common types of flow switches include differential pressure, blade, thermal, and rotameter switches.

SECTION 6
ANALYZERS

Chapter 23
Gas Analyzers

1. A representative sample is a sample from a process in which the composition of the sample is the same as in the process piping. If the sample is not representative of the process, the analyzer results are not accurate.

3. A radiant-energy absorption analyzer is a gas analyzer that uses the principle that different gases absorb different, very specific, wavelengths of electromagnetic radiation in the infrared (IR) or ultraviolet (UV) regions of the electromagnetic spectrum. A nondispersive infrared (NDIR) analyzer is a radiant-energy absorption analyzer consisting of an IR electromagnetic radiation source, an IR detector, and two IR absorption chambers. An ultraviolet (UV) analyzer is a radiant-energy absorption analyzer consisting of a UV electromagnetic radiation source, a sample cell, and a detector that measures the absorption of UV radiation by specific molecules.

5. An opacity analyzer is a gas analyzer consisting of a collimated (focused beam) light source and an analyzer to measure the received light intensity. The received light is measured by a silicon photodiode and converted to an output signal. An increase in particulate matter absorbs and scatters the transmitted light, reducing the light received at the detector. Thus, the decrease in light intensity is proportional to the amount of particulate matter.

Chapter 24
Humidity and Solids Moisture Analyzers

1. Absolute humidity is the ratio of the mass of water vapor to the mass of dry air. It is typically expressed in pounds of water per pound of dry air, grains of water per cubic foot of dry air (there are 7000 grains to a pound), or grams of water per cubic centimeter of dry air. Relative humidity (rh) is the ratio of the actual amount of water vapor in the air to the maximum amount of water vapor possible at the same temperature. Relative humidity is equivalent to the ratio of the actual water vapor pressure in the air to the maximum water vapor pressure at the same temperature.

3. A hygrometer is a humidity analyzer that measures the physical or electrical changes that occur in various materials as they absorb or release moisture. Some hygrometers employ human hair, animal membrane, or other materials that lengthen or stretch when they absorb water. Another type of electrical hygrometer uses a capacitance probe as its primary element. A thermohygrometer is a combination of a hygrometer, pressure sensor, and temperature-sensing instrument with digital processing to calculate relative humidity, absolute humidity, dry bulb temperature, dewpoint reading, and other properties for local display or transmission. The three actual measurements of temperature, pressure, and relative humidity allow the digital calculation of the other values.

5. Two wavelengths of infrared radiation are selected, a measuring wavelength and a reference wavelength. The amount of energy reflected by the process material is compared at each wavelength through two optical filters. One of the filters transmits light at the measuring wavelength, and the other at the reference wavelength. The changes due to factors other than moisture are cancelled by using the ratio of the two reflected signals. Both the reference and measuring wavelengths are subject to extraneous factors, but the moisture additionally absorbs only the measuring wavelength. The analyzer responds only to the changes in moisture content because the wavelengths selected are unique to the presence or absence of water.

Chapter 25
Liquid Analyzers

1. A hydrometer is a liquid density analyzer with a sealed float consisting of a hollow, tubular glass cylinder with the upper portion much smaller in diameter, a scale on the small-diameter portion, and weights at the lower end to make it float upright, with the upper portion partially above the surface of the liquid. The position of the hydrometer in the liquid depends on the density of the liquid. A less dense liquid causes the hydrometer to be positioned lower in the liquid because a greater volume of the liquid has to be displaced to equal the weight of the hydrometer.

3. A Newtonian liquid is a liquid whose viscosity does not change with applied force. A non-Newtonian liquid is a liquid whose viscosity changes (usually decreases) when force is applied. The force may be from gravity causing the fluid to flow or it may be the force from mixing, pumping, or agitation.

5. The refractive index is the amount of bending of a light beam as it moves between fluids with different refractive index values. A refractive index analyzer is a liquid analyzer consisting of a light source directed into a prism that has a flat surface in contact with the liquid to be measured.

Chapter 26
Electrochemical and Composition Analyzers

1. A conductivity analyzer is an electrochemical analyzer that measures the electrical conductivity of liquids and consists of two electrodes immersed in a solution. The electrical conductivity is proportional to the concentration of ions in a solution.

3. An oxidation-reduction potential (ORP) analyzer consists of a metal measuring electrode and a standard reference electrode that measure the voltage produced by an electrochemical reaction between the metals of the electrodes and the chemicals in solution. The voltage is proportional to the chemical concentration. Oxidation means a molecule or element loses electrons. Reduction means electrons are added to a molecule. Electrons move from one electrode to the other. This movement produces the ORP voltage.

5. A near infrared (NIR) liquid analyzer uses infrared radiation to measure the organic molecules in a sample. The analyzer design is similar to the analyzer used to measure moisture in solids except that NIR light normally passes through the sample. To measure the moisture content of a product, two wavelengths of IR radiation are used in the analyzer. One wavelength is used for measurement and the other for reference.

   A Fourier transform infrared (FTIR) analyzer uses a Michelson interferometer to examine a sample with a broad spectrum of infrared radiation. An FTIR analyzer passes a beam containing many different frequencies of light and measures how much of that beam is absorbed or reflected by the sample. Next, the beam is modified to use a different combination of frequencies to make a second measurement. This process is repeated many times to create an interferogram.

SECTION 7 POSITION MEASUREMENT

Chapter 27 Mechanical and Proximity Switches

1. A mechanical switch is a switch that requires physical contact with an object to actuate a switch mechanism. The switch-actuating mechanism can have many forms. One of the simplest forms is a simple button type that requires a direct pressure against the button. Mechanical switches are also designed with a pivoting arm that actuates the internal switch mechanism.

3. An inductance proximity sensor is a proximity sensor that consists of a sensor coil, an oscillator, a trigger circuit, and an output switching circuit. The oscillator generates a radio frequency (RF) signal that is emitted from the coil. Eddy currents are generated in an electrically conductive object as it enters the RF field. These eddy currents draw energy from the oscillator. The trigger circuit detects when the oscillation stops and sends a signal to the output switching circuit that changes the state of the output.

   A capacitance proximity sensor is a proximity sensor that consists of a sensor plate, an oscillator, a trigger circuit, and an output switching circuit. As the target object nears the sensor, the circuit detects the increasing capacitance between the object and the sensor plate. The circuit then starts oscillating and increases the amplitude of the oscillation as the object gets closer to the sensor plate. The trigger circuit detects when the oscillation begins and sends a signal to the output switching circuit that changes the state of the output.

5. A photoelectric sensor is a proximity sensor that uses visible light and infrared radiation sources to detect target objects. A diffused-mode photoelectric sensor is a photoelectric sensor that directs its source against a target object and detects a reflection from the target object. A retro-reflective mode photoelectric sensor is a photoelectric sensor that uses a focused beam directed across the path of a target object and reflected back to the sensor. A through-beam mode photoelectric sensor is a photoelectric sensor that uses a beam aimed directly at a target object with a separate receiver to sense the beam.

Chapter 28 Practical Position Measurement

1. A sensing range is the distance from the end of a proximity sensor to where an object can be detected. An actuation direction is the direction an object moves relative to a proximity sensor. Depending on the style of the sensor, the mounting method can cause problems with the ability of the sensor to detect the target object.

3. Speed sensors measure how often an object moves past the sensor. The sensor detects a target on the rotating object and determines the rate at which the target moves past the sensor. The rate is expressed as a frequency.

5. A safety light curtain is a series of closely spaced light sources mounted on a rail and used as a safety device to shut down equipment if an operator reaches into a protected space. Optical sensors used as receivers are also mounted on another rail using the same spacing as the light sources. The two rails are mounted directly opposite each other so that an area of space has a curtain of light. An object that enters the curtain breaks some of the light beams and generates an output. Light curtains can be used to protect operators from hazardous areas and moving parts.

SECTION 8 TRANSMISSION AND COMMUNICATION

Chapter 29 Transmission Signals

1. Transmission is a standardized method of conveying information from one device to another. Radio and television signals that are broadcast over the airwaves are examples of transmission. A transmission signal is the data sent from one device to another by a specific method. A specific radio station or TV channel is an example of a transmission signal.

3. Current transmission is an electric transmission method in which a transmitting unit regulates the current in a transmission loop. Current transmission in the United States has been standardized on the range of 4 mA to 20 mA. Receiver instruments need to convert the 4 mA to 20 mA transmission signal into a voltage signal before it can be used. This is done by passing the transmission current through a precision resistor. This resistor may be internal or external to the receiver instrument.

5. Converting a measurement value to its equivalent transmission value is useful for determining if a transmitter is functioning correctly. Converting a transmission value to its equivalent measurement value is useful for checking receiver instruments by introducing a signal in place of the signal from the transmitter.

Chapter 30 Digital Numbering Systems and Codes

1. Common numbering systems and codes include decimal, binary, octal, hexadecimal, binary coded decimal systems, and ASCII.

3. The decimal equivalent of a binary number can be calculated by multiplying each bit by the position weighting. The decimal equivalent of a binary number can be determined by using the decimal equivalent for the binary number in each position. As shown above, the decimal equivalent numbers start at the right and are doubled for each position to the left. The equivalent decimal numbers, which are associated with a binary 1, are added together to obtain the total decimal number.

5. The most common alphanumeric code is the American Standard Code for Information Interchange (ASCII). American Standard Code for Information Interchange (ASCII) is a seven-bit or eight-bit alphanumeric code used to represent the basic alphabet plus numbers and special symbols. The seven-bit ASCII code includes all the codes necessary to communicate with peripherals and interfaces. The eight-bit ASCII code includes one more bit for error checking.

Chapter 31
Digital Communications

1. Digital communications is a method of sending and receiving binary information, in the form of low-level DC voltage pulses, between multiple devices using a common wiring format and procedure. Digital signals are used to communicate over input/output (I/O) bus networks.

3. The OSI model divides the functions that protocols must perform into seven hierarchical layers. Each layer interfaces only with its adjacent layers. The OSI model further subdivides the second layer into two sublayers, 2A and 2B, called medium access control (MAC) and logical link control (LLC), respectively. The physical layer (layer 1) and the MAC sublayer (layer 2A) are usually implemented with hardware, while the remaining layers are implemented with software. The physical layer is used to translate messages into signals on the control wiring and signals back to messages. The hardware components of layers 1 and 2A are generally referred to as modems (or transceivers) and drivers (or controllers).

5. Parallel wiring is a method of communications wiring where multiple bits of information are transferred at the same time on multiple wires. Usually 8 to 16 bits of data are conveyed over 25 lines. The non-data lines are used for transmission protocol disciplines and handshaking. Parallel transmission is faster than serial transmission but can only be used for short distances. As distances become greater, the bits on the lines become less synchronized.

   Series wiring is a method of communications wiring where bits of information are transferred one after another over a pair or pairs of wires. Information is transferred as a string of bits with leading and trailing handshaking bits and a final error bit. Series transmission is very robust and can be sent over long distances. It only requires a pair of wires for simplex or half-duplex and three wires for full duplex.

Chapter 32
Industrial Networks

1. A 4-wire fieldbus network is a fieldbus network that uses two wires for communication and two separate wires to power field instruments. A 2-wire fieldbus network is a fieldbus network that uses the communication wires to furnish power to field instruments. This means that a 2-wire fieldbus network does not require power to be available at the instrument site.

3. The H1 protocol operates at a relatively slow 31.25 kb/s and is optimized for traditional analog process control applications. The HSE protocol operates at 10 Mb/s or 100 Mb/s. It is designed for high-speed control applications and plant information networks. HSE uses low-cost commercial off-the-shelf (COTS) Ethernet equipment. A typical FOUNDATION network has H1 connections between devices and HSE connections between computers. Combining H1 and HSE allows for good integration of control systems with supervisory applications.

5. HART is a hybrid communications protocol consisting of newer two-way digital communication and the traditional 4 mA to 20 mA analog signal. HART is an acronym for Highway Addressable Remote Transducer. HART is called a hybrid protocol because it combines digital and analog communication on the same wire. It can communicate a single variable using a 4 mA to 20 mA analog signal while also communicating added information on a digital signal.

Chapter 33
Wireless Systems

1. A major advantage of wireless transmission is that an instrument can be easily placed in a remote area without having to run wires to the instrument. In addition, an instrument can be physically relocated to a different part of a process without the need to rewire the system.

   A disadvantage of wireless transmission is that remove devices require power even though they do not require wires for communication. In addition, electromagnetic interference from other radio frequency devices, such as variable-speed drives, fluorescent lights, local controllers, and computer monitors, can interfere with wireless signals.

3. Frequency hopping spread spectrum (FHSS) is a transmission method that takes packets of information and then transmits the packets across a frequency band by selecting pseudorandom frequencies within the band. Each frequency contains a separate piece of information and is transmitted at high power. This gives the transmission the ability to overcome many sources of noise that may arise.

   Direct sequence spread spectrum (DSSS) is a transmission method that divides data into packets and sends the packets over a wide portion of the frequency band. With DSSS, the information can be transmitted with much lower power consumption than with the FHSS system.

5. In the United States, carriers typically do not allow the use of voice networks like global system for mobile communications (GSM) for M2M data communication applications. This is because the modems could monopolize network bandwidths. It is also a disadvantage for the modem user because of billing arrangements that charge by the minute.

   Data networks, such general packet radio service (GPRS) communications, are designed to provide better network architecture for data communications. Data networks offer an always-on connection and provide billing based on the amount of data sent.

Chapter 34
Practical Transmission and Communication

1. Allowable loop impedance is the resistance limit that is the summation of all of the receiver impedances and connecting wiring resistance in the current transmission loop. A loop with too much impedance is not able to generate enough current to deliver the 20 mA that is necessary for signaling.

3. A balanced line is a signal transmission cable in which two conductors carry the signal and create a balanced circuit alternating from positive to negative. The paired wires balance out each other's signals and help to nullify the effect of EMI from nearby wires or electromagnetic fields. Another possible solution to EMI problems is to use shielded cable. This cable includes an aluminum foil shield that can provide significant immunity to EMI when properly grounded.
5. A 2-wire transmitter is a transmitter that uses the 24 VDC power from the current transmission loop to power the transmitter. A 3-wire transmitter is a transmitter that uses a DC power supply, with the negative lead of the power supply and the negative lead for the transmitter being the same wire that is used to establish a zero level for the circuit. The positive wire from the power supply is the source of power for the transmitter and the positive wire from the transmitter sends a signal to the controller. The power supply is separate. A 4-wire transmitter is a transmitter that uses two wires to provide power to the transmitter and two wires for the current output. The separate power supply may be DC or AC.

SECTION 9
AUTOMATIC CONTROL

Chapter 35
Automatic Control and Process Dynamics

1. Automatic control is the equipment and techniques used to automatically regulate a process to maintain a desired outcome. A process variable (PV) is the dependent variable that is to be controlled in a control system. A control variable (CV), or manipulated variable, is the independent variable in a process control system that is used to adjust the dependent variable, the process variable. A controller is a device that compares a process measurement to a setpoint and changes the control variable (CV) to bring the process variable (PV) back to the setpoint.
3. A lag is a delay in the response of a process that represents the time it takes for a process to respond completely when there is a change in the inputs to the process. A process with a larger time constant responds more slowly to a change than a process with a smaller time constant. Dead time is the period of time that occurs between the time a change is made to a process and the time the first response to that change is detected. Dead time adversely affects the ability of a control system to control a process because it extends the time between control actions and the results of that control action.
5. Direct action is a form of control action where the controller output increases with an increase in the measurement of the process variable (PV). Reverse action is a form of control action where the controller output decreases with an increase in the measurement of the process variable (PV). Selecting the correct controller action is absolutely essential for the control system to work properly. The proper controller action depends on the type of process and the final element failure action. A change in the measurement of the PV must result in an output change to the control valve that brings the PV back to the setpoint.

Chapter 36
Control Strategies

1. ON/OFF, or two-position control, is a method of changing the output of a controller that provides only an ON or OFF signal to the final element of the process. The two states are arranged on either side of the setpoint. Proportional (P) control is a method of changing the output of a controller by an amount proportional to an error. The amount of output change is determined by the proportional gain.
3. Output bias, or manual reset, is a controller function that positions a final element in a central position when the process variable is at setpoint. A controller that only has a proportional action must also have the ability to include an output bias. It is necessary to have the final element at a position that allows the final element to move in either direction.
5. Typical types of advanced control strategies are artificial intelligence, fuzzy logic, and neural networks.

Chapter 37
Controller Tuning

1. Common controller performance standards are decay ratio, overshoot, and dead time. The decay ratio is a measure of how quickly an overshoot decays from one oscillation to the next as the controller brings the process to the setpoint. Overshoot is the change of a process variable (PV) that exceeds the upper deadband value when there is a disturbance to a system. Dead time is the period of time that occurs between the time a change is made to a process and the time the first response to that change is detected.
3. The setpoint step change method is a manual tuning method that consists of making small changes in the setpoint and observing the responses. The setpoint step change method takes time to accomplish, but it is a safe method that has a low impact on the process operation.
5. Ziegler-Nichols closed loop tuning is a method of tuning a controller by increasing the gain until the system cycles at the point of instability. This method works well, but forcing an active process into oscillation may be dangerous.

   Ziegler-Nichols open loop tuning is a method of tuning a controller based on open loop response to a step input. An advantage of this method is that it is safer than the closed loop method because it does not force the process into oscillation. A disadvantage is that it is sometimes difficult to identify all the key parameters with enough accuracy to establish the correct tuning values.

Chapter 38
Digital and Electric Controllers

1. A stand-alone digital controller is a general type of microprocessor-based controller with all required operating components enclosed in one housing. A direct computer control system is a control system that uses a computer as the controller. A distributed control system (DCS) is a control system where the individual functions that make up a control system are distributed among a number of physical pieces of equipment that are connected by a high-speed digital communication network. A programmable logic controller (PLC) is a control system with an architecture very similar to that of a DCS, with self-contained power supplies, distributed inputs and outputs, and a controller module, all connected on high-speed digital communication networks.

3. An operator interface is a view into a digital control system through which an operator can observe and control a process. A stand-alone display system is a display system designed for use with smaller digital control systems and has the electronics and display packaged together in a panel-mounted enclosure. A PC-based display system is a display system that uses standard personal computer hardware and special software to display full graphics, control functions, alarms, trends, etc. A digital recorder is a device that uses local memory to record process data, which is then displayed on digital display screens that can have the appearance of the old-style strip chart recorders.
5. Pick and choose is a configuration format where the configuration is selected from a list of available functions. Function block configuration is a configuration method that uses a library of functions provided by the manufacturer. Ladder logic is a configuration method that consists of two vertical rails, the left one being the source and the right one being the end, and the sequential rungs of logic between the two. Structured text is a type of configuration that is very similar to Microsoft® Visual Basic® or older structured programming languages. A sequential function chart is a type of configuration format consisting of a series of conditional statements, parallel paths, and action blocks, which begins with a Start command and ends with an existing scan.

SECTION 10 FINAL ELEMENTS

Chapter 39 Control Valves

1. A normally closed (NC) valve is a valve that does not allow pressurized fluid to flow out of the valve in the spring-actuated (de-energized) position. A normally open (NO) valve is a valve that allows pressurized fluid to flow out of the valve in the spring-actuated (de-energized) position. The terms NC and NO are used in reference to electricity and to fluid handling. The terms have opposite meanings in each field. For example, NC in the fluid handling field means that fluid does not flow in the valve's normal position. NC in the electrical field means that current flows in the switch's normal position.
3. Flashing is a process in which a portion of a liquid converts to a vapor as it passes through a control valve because the pressure has fallen below the vapor pressure of the liquid. Flashing can occur when the upstream pressure is too low or when the inlet fluid temperature is too high and some of the liquid flashes to its vapor form. Cavitation is a process in which vapor bubbles in a flowing liquid collapse inside a control valve as the pressure begins to increase. The collapsing bubbles result in implosions causing considerable noise inside the valve.
5. An ON/OFF control valve is a valve used to start and stop the flow of materials in the pipelines that control chemicals and energy sources for automatic sequencing, batching, or safety operations.

Chapter 40 Regulators and Dampers

1. A regulator is a self-operating control valve for pressure and temperature control. The main types of regulators are pressure regulators, used to control downstream pressure; backpressure regulators, used to control upstream pressure; and temperature self-operating regulators, used to automatically control temperature.
3. Droop is a drop in pressure below a set value when there is high flow demand. Lockup is an increase in pressure above a set value when there is low flow demand. An adjusting screw is turned to compress or release the regulator spring to set the downstream pressure. If the flow demand through the regulator increases after the pressure is set, the compression of the spring changes to handle the additional flow and results in a downstream pressure that is less than the original set pressure. Reductions in flow result in lockup as the compression of the spring changes to handle the reduced flow.
5. Dampers are commonly classified as parallel-, opposed-, and round-blade dampers. A parallel-blade damper is a damper in which adjacent blades are parallel and move in the same direction with one another. An opposed-blade damper is a damper in which adjacent blades are parallel and move in opposite directions from one another. A round-blade damper is a damper that has a circular blade designed to fit into round ductwork.

Chapter 41 Actuators and Positioners

1. An actuator is a device that provides the power and motion to manipulate the moving parts of a valve or damper used to control fluid flow through a final element. The requirements of an actuator are speed, power, precision, and resolution. An actuator must respond quickly to a change in control signal and have enough power to overcome the process pressure and mechanical friction of the moving parts. Positioning needs to be repeatable for the same control signal. In addition, the actuator must be able to work with minimal maintenance for long periods of time.
3. A pneumatic sliding stem piston actuator is an actuator where the movement of a piston replaces the action of the diaphragm and a balancing air pressure replaces the function of the spring. A pneumatic rotary piston actuator is an actuator that can be attached directly to most rotary valves and is primarily designed for ON/OFF action.
5. A positioner is a device used to ensure positive position of a valve or damper actuator. A positioner compares the input signal from the controller to the actual valve position and applies a pneumatic output to the actuator to ensure that the valve moves to the proper position. If the valve experiences resistance to movement, the positioner continues to apply more pressure until the valve moves to the desired position.

Chapter 42 Variable-Speed Drives and Electric Power Controllers

1. A variable-speed drive is a device that varies the speed of an electric motor. Many final elements like blowers, fans, compressors, and pumps are powered by electric motors. The ability to vary the speed of the motor allows the final element to vary the amount of material moved.
3. Static head is the pressure of a fluid due to a change in the elevation of a discharge piping system and remains constant for all flow rates. Frictional head is the pressure loss of a fluid due to flow through the piping or ductwork and varies with the flow squared.

5. A solid-state relay is a semiconductor switching device that uses a low-current DC input to switch an AC circuit. A zero switching relay is a solid-state relay where a load becomes energized when a control input voltage is applied and AC load voltage crosses zero. An analog switching relay is a solid-state relay that has a continuous range of output voltages within the relay's rated range.

SECTION 11
SAFETY SYSTEMS

Chapter 43
Safety Devices and Equipment

1. A safety valve is a gas- or vapor-service valve that opens very quickly when the inlet pressure exceeds the spring setpoint pressure. These valves are used in applications where the valve must open quickly to reduce dangerous conditions. A relief valve is a valve that opens in proportion to the pressure above the setpoint. These valves are used for unfired vessel applications. A safety relief valve is a valve that is designed so that it can be set to act as either a safety valve or a relief valve. These valves are usually used for unfired vessels such as compressed air receiver vessels.
3. A rupture disc is a safety device that breaks to open a discharge device and is used to prevent damage in pipelines and pressure vessels due to excessive pressure. Common types of rupture discs include conventional metal, reverse buckling, composite, and graphite composite rupture discs.
5. The lower explosive limit (LEL) is the lowest concentration of a combustible gas or vapor in air that can be ignited. The upper explosive limit (UEL) is the highest concentration of a combustible gas or vapor in air that can be ignited. The lower flammable limit (LFL) and upper flammable limit (UFL) are sometimes used in place of the LEL and UEL.

Chapter 44
Electrical Safety Standards

1. Electrical classifications for hazardous locations are identified by class, division, and group. A class is a term used to define the general nature of a hazardous gas or vapor in the surrounding atmosphere. A division is a classification assigned to each class based upon the likelihood of the presence of a hazardous substance in the atmosphere. A group is a term used to classify an atmosphere containing particular flammable gases or vapors or combustible dust.
3. An enclosure is a case or housing for equipment or other apparatus that provides protection for controllers, motor drives, or other devices. The National Electrical Manufacturers Association (NEMA) classifies enclosures according to their purpose and intended resistance to the ingress of moisture and hazardous materials.
5. An intrinsically safe system is a system designed to limit the electrical and thermal energy to a level that cannot ignite combustible atmospheres and to avoid some of the high installation costs associated with explosionproof installations. The NEC® contains all the requirements for an intrinsically safe installation.

Chapter 45
Safety Instrumented Systems

1. A safety instrumented system (SIS) is a system consisting of sensors, logic solvers, and final control elements that bring a process to a safe state when normal operating conditions are violated. The American Institute of Chemical Engineers (AIChE), through the Center for Chemical Process Safety (CCPS), has issued an SIS document that is commonly used in the chemical industry called the *Guidelines for Safe Automation of Chemical Processes.* Another widely accepted SIS document is ISA S84, *Application of Safety Instrumented Systems for the Process Industries.* A European standard that is becoming increasing well known is IEC 61511, *Functional Safety: Safety Instrumented Systems for the Process Industry Sector.*
3. The severity of an event can be classified as minor, serious, or severe. The likelihood of an event can be classified as improbable, occasional, or frequent.
5. A safety PLC is a highly reliable PLC that includes fail-safe designs, built-in self-diagnostics, and a fault-tolerant architecture. Safety PLCs are also designed to meet third-party approval criteria because of the special hardware and software. Safety PLC architectures can be simple, low-cost, safe single systems (1oo1), or they can be more complex, with dual (1oo2D) or TMR system architectures, to satisfy the SIS integrity level needed.

SECTION 12
INSTRUMENTATION AND CONTROL APPLICATIONS

Chapter 46
General Control Techniques

1. A continuous process is a type of processing operation in which continuous raw material is received and the resulting product is continually fed to the next operation. Storage volumes (surge capacity) are usually provided at strategic locations in the sequence of operations so that a problem in one segment of the manufacturing operation does not immediately affect other segments of the operation.

   A batch process is a type of processing operation in which multiple feeds are received and the resulting product is fed all at once to the next operation. In reality, most batch process plants also have numerous continuous operations for handling byproducts of the reaction and to purify the product. Batch processes are usually much more complex than continuous processes since they usually involve automated charging, transferring, advancing or repeating processing steps, and emergency operations.
3. A selector is a device that compares two or three input signals and passes the highest or the lowest signal to the output of the device. A high selector can be used to select the highest of a group of temperature measurements to be used as the controller input. This is done if it is dangerous to the process for any measurement to be too high.

5. Cascade control is a control strategy where a primary controller, which controls the ultimate measurement, adjusts the setpoint of a secondary controller. There are a few key requirements that are necessary for a cascade control system to function properly. First, changes in the secondary controlled measurement must directly affect the primary measurement. Second, the response speed of the secondary controller must be at least four times faster than the primary controller. This is to ensure that the secondary controller has time to completely respond to setpoint changes from the primary controller before new changes are made.

Chapter 47
Temperature Control

1. There are many designs of heat exchangers, but the common element is a relatively thin layer of material separating the cooling or heating fluid from the process fluid. The barrier material is typically a conductive metal that allows heat to flow from the higher temperature fluid to the lower temperature fluid.

   Shell-and-tube exchangers usually consist of a cylindrical shell containing the heating or cooling fluid. The process fluid is usually contained in tubes that run lengthwise through the shell. The ends of the tubes are sealed in tube sheets that close off each end of the shell. Exchanger heads are bolted to the shell to provide process piping connections.

3. In almost all heating applications, the steam control valve should have a fail-closed action and if possible, should close tightly. It is safer for most process applications for the source of heat to fail-closed. The fail-closed action and the heating service dictate that the selected controller action should be reverse.

5. The contents of a reactor typically need to be heated to reach the reaction temperature. After the reaction begins, the cooling system needs to remove the heat generated by the reaction in order to maintain the reactor temperature at the setpoint. In a circulated water system, water is pumped into the bottom of the reactor jacket where baffles are used to direct the water around the jacket to the exit at the top. The water from the top of the jacket is returned to the water circulating pump after passing through a steam heated exchanger. A chilled water supply is also connected to the pump suction. A branch line from the jacket exit carries hot return water back to the water chiller system or the drain.

Chapter 48
Pressure and Level Control

1. The second control valve (PV-2 01) is located in the suction piping to the compressor, and it has a fail-closed action with a linear flow characteristic and a reverse-acting positioner. The fail-closed action of this control valve is selected for safety. The fail-closed action is not compatible with the reverse action of the controller, so the positioner action must be reversed to make the control valve move in the correct direction.

3. Pump-up level control is a control arrangement used to start a pump when the level of liquid drops below a setpoint and stop the pump when the liquid level reaches the desired level. This operation controls the level of the water between the two control points.

5. Pump-down level control is a control arrangement used to control the transfer of collected steam condensate to a common condensate storage tank. The operation is based on maintaining the level of water in the condensate tank. The two levels are a high level, where the feedwater pump is started, and a low level, where the feedwater pump is stopped. The high probe is typically set at about 75% of the full tank level and the low probe is typically set at about 25%.

Chapter 49
Flow Control

1. Ratio control is a control strategy used to control a secondary flow to the predetermined fraction, or flow ratio, of a primary flow.

3. Mixing ratio control is a control strategy used when there is a need to mix two flow streams in a specified ratio. An efficient method of mixing water treatment chemicals and water is in-line with a flow ratio control system. Splitting ratio control is a control strategy used when there is a need to split a flow into two separate constant-ratio flows. This is an efficient method of splitting overheads flow in a distillation column into product and reflux.

5. A lead-lag air-fuel control system is a control system where increases in combustion-air flow lead increases in fuel flow and decreases in combustion-air flow lag decreases in fuel flow. A linking system solves the low excess combustion air situation experienced during load changes. This means that extra air is always provided during load changes.

Chapter 50
Analysis and Multivariable Control

1. Condensate is the water formed when steam condenses. Condensate is often collected for reuse in boilers. This is done to lower the costs associated with treating raw water and to reduce the discharge to water treatment systems. Returned condensate requires little or no treatment and is preferable for use in boilers as long as the quality is acceptable.

3. A firetube boiler is a boiler in which heat and gases of combustion pass through tubes surrounded by water. The level is measured in the boiler itself. Large boilers are often designed as watertube boilers. A watertube boiler is a boiler in which water passes through tubes surrounded by gases of combustion. Watertube boilers are connected to a steam drum at the top of the boiler. The steam drum is where the steam separates from the water. It is also the location where the boiler water level is measured and controlled.

5. The conveyor combines a level measurement with speed sensors, limit switches, an alarm system, and many types of pushbuttons. There are electrical interlocks between the different pieces of equipment to ensure that each downstream piece of the conveyor is operating before upstream pieces are allowed to operate.

APPENDIX

Table of Contents

The following are in the Instrumentation and Process Control Resources included with the Digital Resources.

Conversion Tables
- Flow Rate Conversion Table
- Pressure Equivalents

Properties of Substances
- Periodic Table of the Elements
- Physical Properties of Liquids
- Physical Properties of Solids
- Physical Properties of Vapors
- Psychrometric Chart

RTD Resistance Tables
- 10-Ohm Copper – 0.00427
- 100-Ohm Platinum – 0.00385
- 100-Ohm Platinum – 0.00390
- 100-Ohm Platinum – 0.00392
- 120-Ohm Nickel – 0.00672
- 507.5-Ohm Nickel-Iron – 0.00520
- 604-Ohm Nickel-Iron – 0.00518

Steam Tables
- Absolute Pressure
- Gauge Pressure
- Superheated Steam

Supplemental Topics
- Pneumatic Transmission and Control
- Principles of Electricity

Test Tool Connections
- AC Current Measurement
- AC Voltage Measurement
- DC Current Measurement
- DC Voltage Measurement
- Diode Testing Measurement
- Frequency Measurement
- Resistance Measurement

Thermocouple Voltage Tables
- Type B
- Type E
- Type J
- Type K
- Type N
- Type R
- Type S
- Type T

Instrument Tag Identification					
	First Letter		Second Letter		
	Measured or Initiating Variable	Modifier	Readout or Passive Function	Output Function	Modifier
A	Analysis		Alarm		
B	Burner Flame		User's Choice	User's Choice	User's Choice
C	Conductivity (Electrical)			Control	
D	Density (Mass) or Specific Gravity	Differential			
E	Voltage (EMF)		Primary Element		
F	Flow Rate	Ratio (Fraction)			
G	Gaging (Dimensional)		Glass		
H	Hand (Manually Initiated)				High
I	Current (Electrical)		Indicate		
J	Power	Scan			
K	Time or Time Schedule			Control Station	
L	Level		Light (Pilot)		Low
M	Moisture or Humidity				Middle or Intermediate
N	User's Choice		User's Choice	User's Choice	User's Choice
O	User's Choice		Orifice (Restriction)		
P	Pressure or Vacuum		Point (Test Connection)		
Q	Quantity or Event	Integrate or Totalize			
R	Radioactivity, radiation		Record or Print		
S	Speed or Frequency	Safety		Switch	
T	Temperature			Transmit	
U	Multivariable		Multifunction	Multifunction	Multifunction
V	Viscosity, Vibration			Valve, Damper, or Louver	
W	Weight or Force		Well		
X	Unclassified		Unclassified	Unclassified	Unclassified
Y	Event or State			Relay or Compute	
Z	Position			Drive, Actuate, or Unclassified Final Control Element	

Selected Instrumentation Symbols

General Instrument Symbols — Balloons

Instrument for single measured variable * with any number of functions.

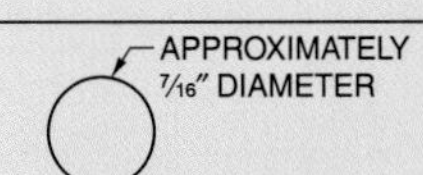

LOCALLY MOUNTED

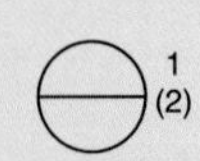

MOUNTED ON BOARD 1 (OR BOARD 2). BOARD 2 MAY ALTERNATIVELY BE DESIGNATED BY A DOUBLE HORIZONTAL LINE INSTEAD OF A SINGLE LINE, WITH THE DESIGNATION OUTSIDE THE BALLOON OMITTED.

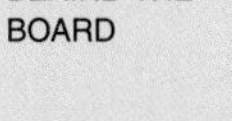

MOUNTED BEHIND THE BOARD

6TE
2584-23

LOCALLY MOUNTED INSTRUMENT WITH LONG TAG NUMBER. (6 IS OPTIONAL AND IS PLANT NUMBER.) ALTERNATIVELY, A CLOSED CIRCLE MAY BE ENLARGED.

Instrument for two measured variables.* Optionally, single-variable instrument with more than one function. Additional tangent balloons my be added as required.

LOCALLY MOUNTED

MOUNTED ON MAIN BOARD

Control Valve Body Symbols

GLOBE, GATE, OR OTHER IN-LINE TYPE NOT OTHERWISE IDENTIFIED

ANGLE

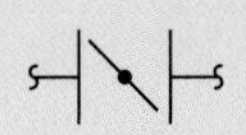

BUTTERFLY, DAMPER, OR LOUVER

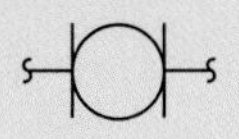

ROTARY PLUG OR BALL

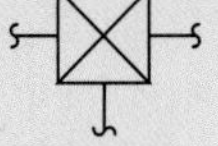

THREE-WAY

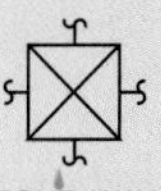

FOUR-WAY

Actuator Symbols

Diaphragm, spring-opposed

WITHOUT POSITIONER OR OTHER PILOT

PREFERRED FOR DIAPHRAGM THAT IS ASSEMBLED WITH PILOT SO THAT ASSEMBLY IS ACTUATED BY ONE CONTROLLED INPUT (SHOWN TYPICALLY WITH ELECTRIC INPUT TO ASSEMBLY)

Diaphragm, spring-opposed, with positioner and overriding pilot valve that pressurizes diaphragm when actuated.

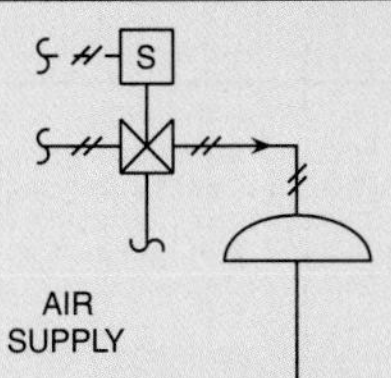

PREFERRED ALTERNATIVE

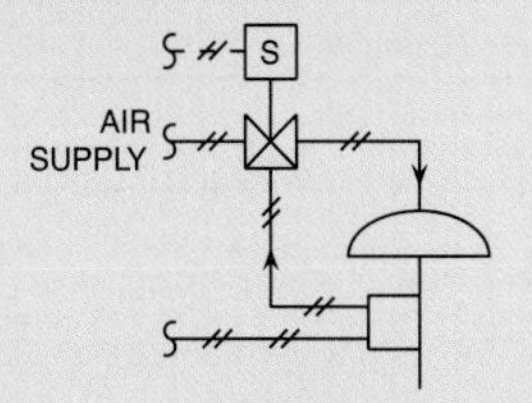

OPTIONAL ALTERNATIVE

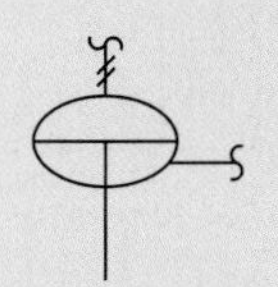

DIAPHRAGM, PRESSURE-BALANCED

Cylinder, without positioner or other pilot

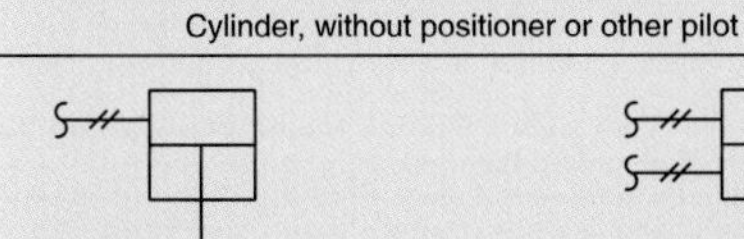

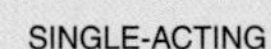

SINGLE-ACTING

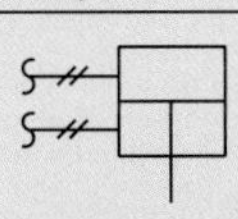

DOUBLE-ACTING

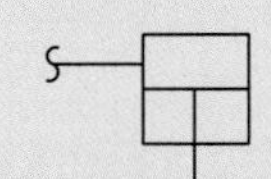

PREFERRED FOR ANY CYLINDER THAT IS ASSEMBLED WITH PILOT SO THAT ASSEMBLY IS ACTUATED BY ONE CONTROLLED INPUT

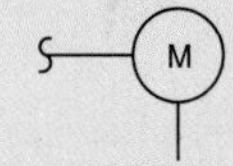

ROTARY MOTOR (SHOWN TYPICALLY WITH ELECTRIC SIGNAL)

Symbols for Self-Actuated Regulators, Valves, and Other Devices

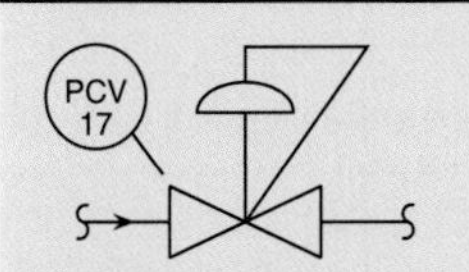

PRESSURE-REDUCING REGULATOR, SELF-CONTAINED

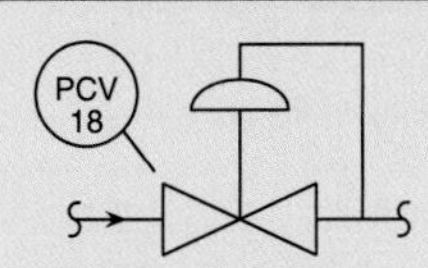

PRESSURE-REDUCING REGULATOR WITH EXTERNAL PRESSURE TAP

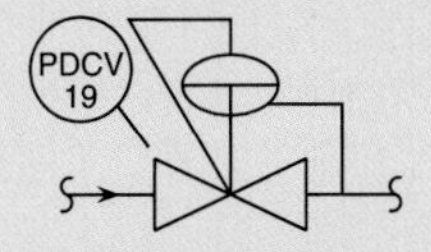

DIFFERENTIAL-PRESSURE-REDUCING REGULATOR WITH INTERNAL AND EXTERNAL PRESSURE TAPS

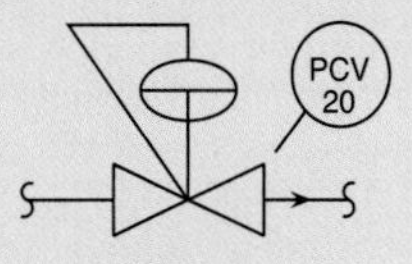

BACKPRESSURE REGULATOR, SELF-CONTAINED

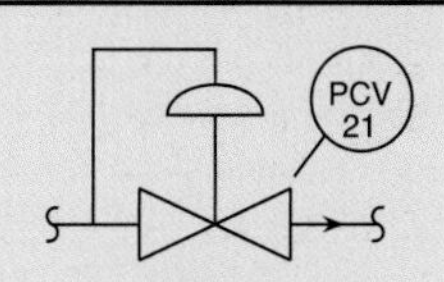

BACKPRESSURE REGULATOR WITH EXTERNAL PRESSURE TAP

Symbols for Actuator Action in Event of Actuator Power Failure

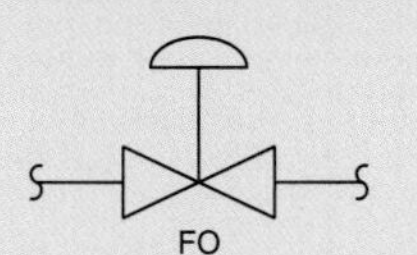

TWO-WAY VALVE, FAIL OPEN

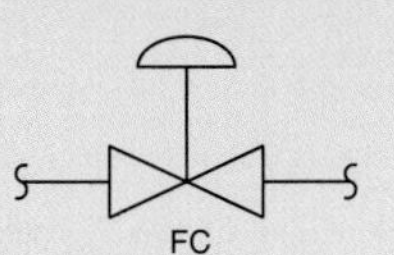

TWO-WAY VALVE, FAIL CLOSED

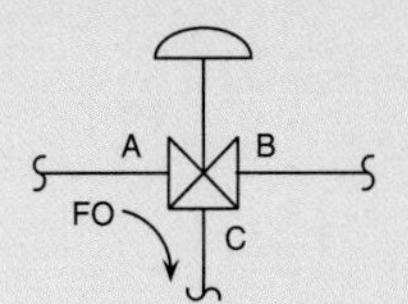

THREE-WAY VALVE, FAIL OPEN TO PATH A-C

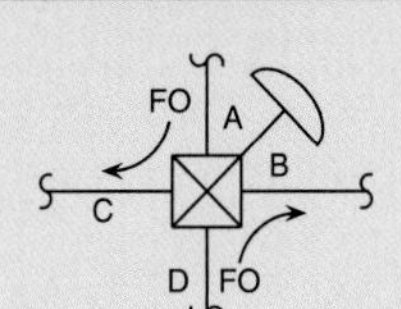

FOUR-WAY VALVE, FAIL OPEN TO PATHS A-C AND D-B

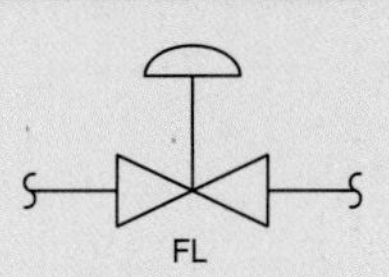

ANY VALVE, FAIL LOCKED (POSITION DOES NOT CHANGE)

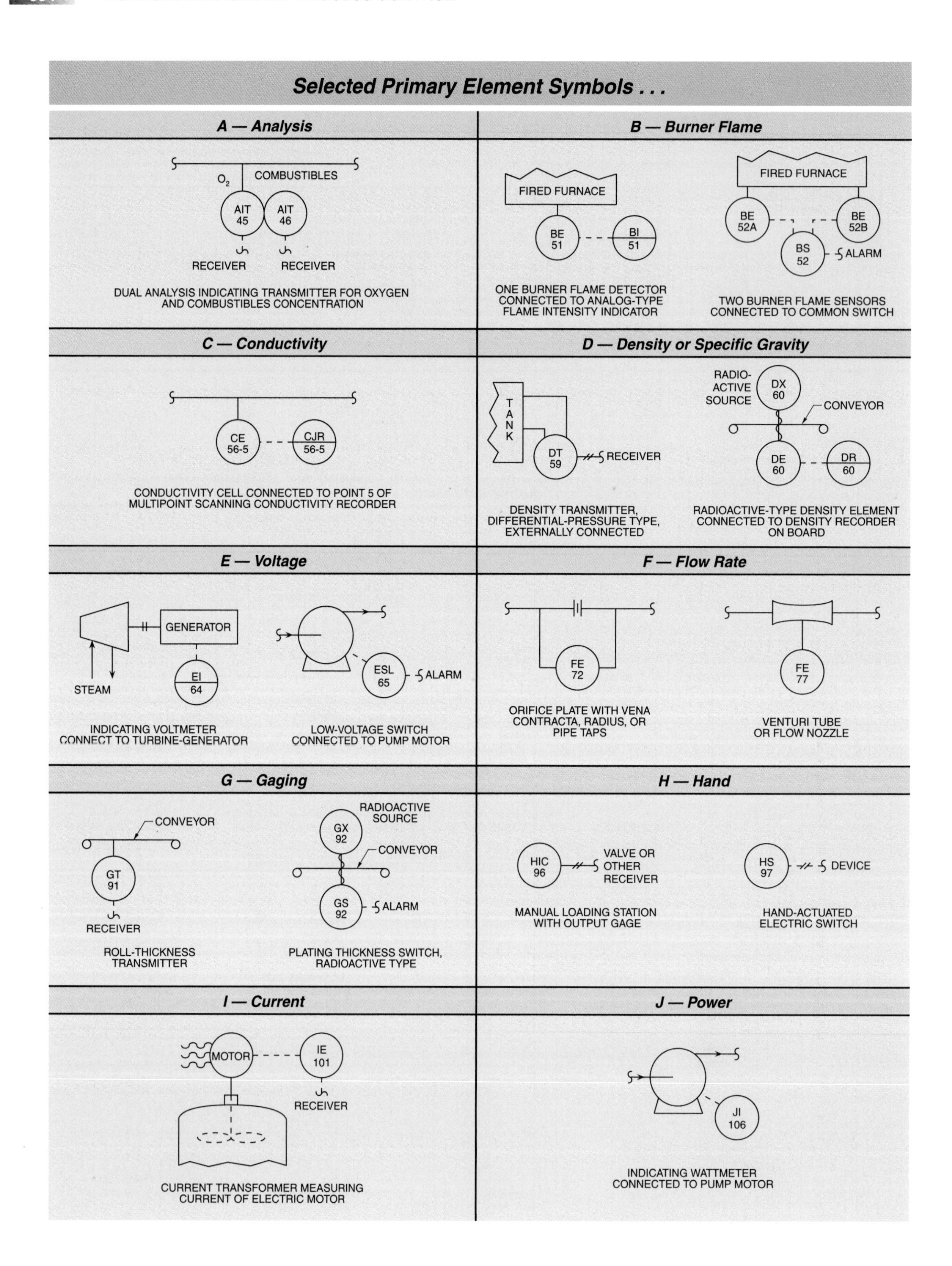
Selected Primary Element Symbols . . .
A — Analysis
O2
COMBUSTIBLES
AIT 45
AIT 46
RECEIVER
RECEIVER
DUAL ANALYSIS INDICATING TRANSMITTER FOR OXYGEN AND COMBUSTIBLES CONCENTRATION
B — Burner Flame
FIRED FURNACE
BE 51
BI 51
ONE BURNER FLAME DETECTOR CONNECTED TO ANALOG-TYPE FLAME INTENSITY INDICATOR
FIRED FURNACE
BE 52A
BE 52B
BS 52
ALARM
TWO BURNER FLAME SENSORS CONNECTED TO COMMON SWITCH
C — Conductivity
CE 56-5
CJR 56-5
CONDUCTIVITY CELL CONNECTED TO POINT 5 OF MULTIPOINT SCANNING CONDUCTIVITY RECORDER
D — Density or Specific Gravity
TANK
DT 59
RECEIVER
DENSITY TRANSMITTER, DIFFERENTIAL-PRESSURE TYPE, EXTERNALLY CONNECTED
RADIO-ACTIVE SOURCE
DX 60
CONVEYOR
DE 60
DR 60
RADIOACTIVE-TYPE DENSITY ELEMENT CONNECTED TO DENSITY RECORDER ON BOARD
E — Voltage
GENERATOR
STEAM
EI 64
INDICATING VOLTMETER CONNECT TO TURBINE-GENERATOR
ESL 65
ALARM
LOW-VOLTAGE SWITCH CONNECTED TO PUMP MOTOR
F — Flow Rate
FE 72
ORIFICE PLATE WITH VENA CONTRACTA, RADIUS, OR PIPE TAPS
FE 77
VENTURI TUBE OR FLOW NOZZLE
G — Gaging
CONVEYOR
GT 91
RECEIVER
ROLL-THICKNESS TRANSMITTER
RADIOACTIVE SOURCE
GX 92
CONVEYOR
GS 92
ALARM
PLATING THICKNESS SWITCH, RADIOACTIVE TYPE
H — Hand
HIC 96
VALVE OR OTHER RECEIVER
MANUAL LOADING STATION WITH OUTPUT GAGE
HS 97
DEVICE
HAND-ACTUATED ELECTRIC SWITCH
I — Current
MOTOR
IE 101
RECEIVER
CURRENT TRANSFORMER MEASURING CURRENT OF ELECTRIC MOTOR
J — Power
JI 106
INDICATING WATTMETER CONNECTED TO PUMP MOTOR

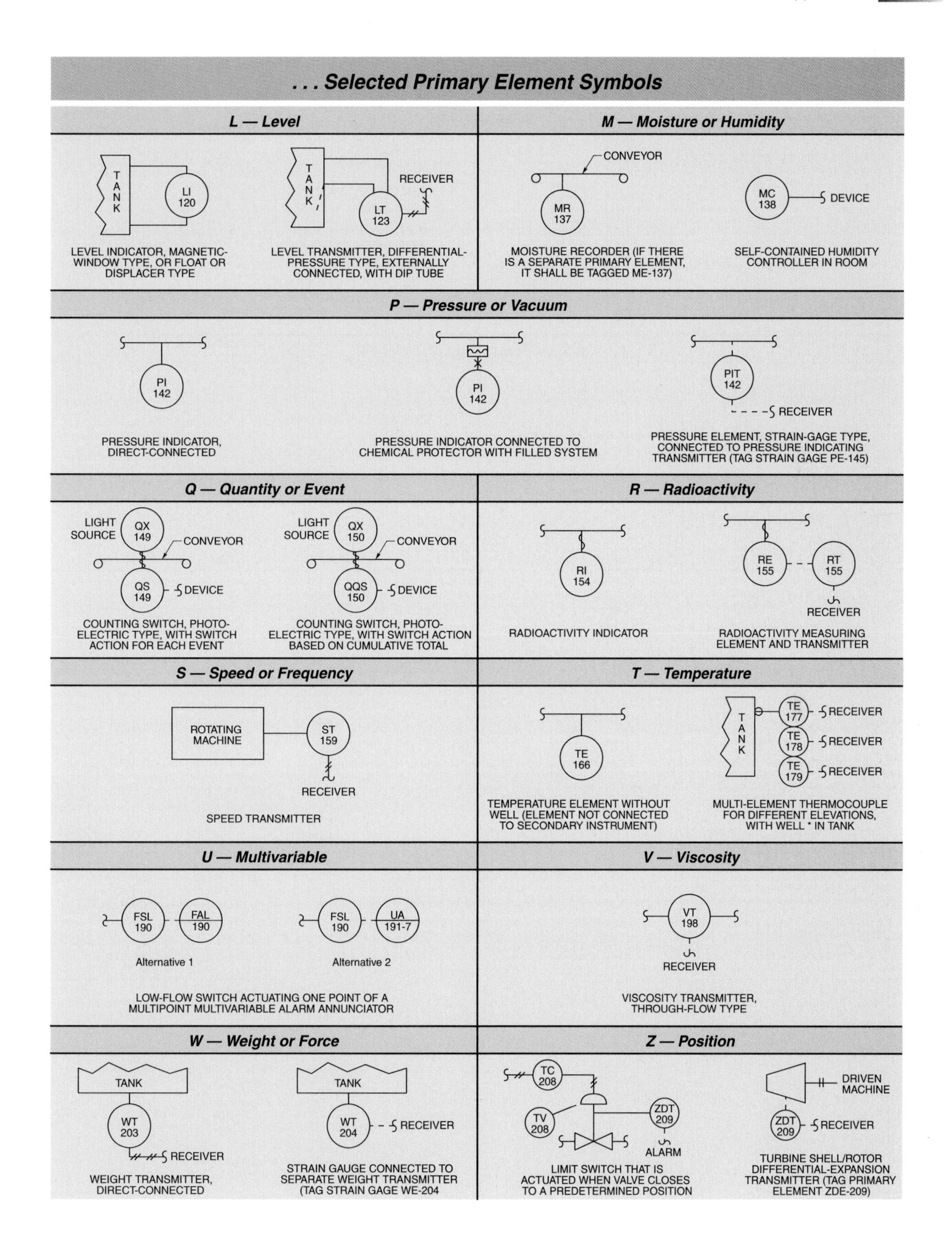
. . . Selected Primary Element Symbols
L — Level
TANK
LI 120
LEVEL INDICATOR, MAGNETIC-WINDOW TYPE, OR FLOAT OR DISPLACER TYPE
RECEIVER
LT 123
LEVEL TRANSMITTER, DIFFERENTIAL-PRESSURE TYPE, EXTERNALLY CONNECTED, WITH DIP TUBE
M — Moisture or Humidity
CONVEYOR
MR 137
MOISTURE RECORDER (IF THERE IS A SEPARATE PRIMARY ELEMENT, IT SHALL BE TAGGED ME-137)
MC 138
DEVICE
SELF-CONTAINED HUMIDITY CONTROLLER IN ROOM
P — Pressure or Vacuum
PI 142
PRESSURE INDICATOR, DIRECT-CONNECTED
PRESSURE INDICATOR CONNECTED TO CHEMICAL PROTECTOR WITH FILLED SYSTEM
PIT 142
RECEIVER
PRESSURE ELEMENT, STRAIN-GAGE TYPE, CONNECTED TO PRESSURE INDICATING TRANSMITTER (TAG STRAIN GAGE PE-145)
Q — Quantity or Event
LIGHT SOURCE
QX 149
CONVEYOR
QS 149
DEVICE
COUNTING SWITCH, PHOTO-ELECTRIC TYPE, WITH SWITCH ACTION FOR EACH EVENT
QX 150
QQS 150
COUNTING SWITCH, PHOTO-ELECTRIC TYPE, WITH SWITCH ACTION BASED ON CUMULATIVE TOTAL
R — Radioactivity
RI 154
RADIOACTIVITY INDICATOR
RE 155
RT 155
RECEIVER
RADIOACTIVITY MEASURING ELEMENT AND TRANSMITTER
S — Speed or Frequency
ROTATING MACHINE
ST 159
RECEIVER
SPEED TRANSMITTER
T — Temperature
TE 166
TEMPERATURE ELEMENT WITHOUT WELL (ELEMENT NOT CONNECTED TO SECONDARY INSTRUMENT)
TE 177
TE 178
TE 179
RECEIVER
MULTI-ELEMENT THERMOCOUPLE FOR DIFFERENT ELEVATIONS, WITH WELL * IN TANK
U — Multivariable
FSL 190
FAL 190
Alternative 1
UA 191-7
Alternative 2
LOW-FLOW SWITCH ACTUATING ONE POINT OF A MULTIPOINT MULTIVARIABLE ALARM ANNUNCIATOR
V — Viscosity
VT 198
RECEIVER
VISCOSITY TRANSMITTER, THROUGH-FLOW TYPE
W — Weight or Force
TANK
WT 203
RECEIVER
WEIGHT TRANSMITTER, DIRECT-CONNECTED
WT 204
STRAIN GAUGE CONNECTED TO SEPARATE WEIGHT TRANSMITTER (TAG STRAIN GAGE WE-204
Z — Position
TC 208
TV 208
ZDT 209
ALARM
LIMIT SWITCH THAT IS ACTUATED WHEN VALVE CLOSES TO A PREDETERMINED POSITION
DRIVEN MACHINE
TURBINE SHELL/ROTOR DIFFERENTIAL-EXPANSION TRANSMITTER (TAG PRIMARY ELEMENT ZDE-209)

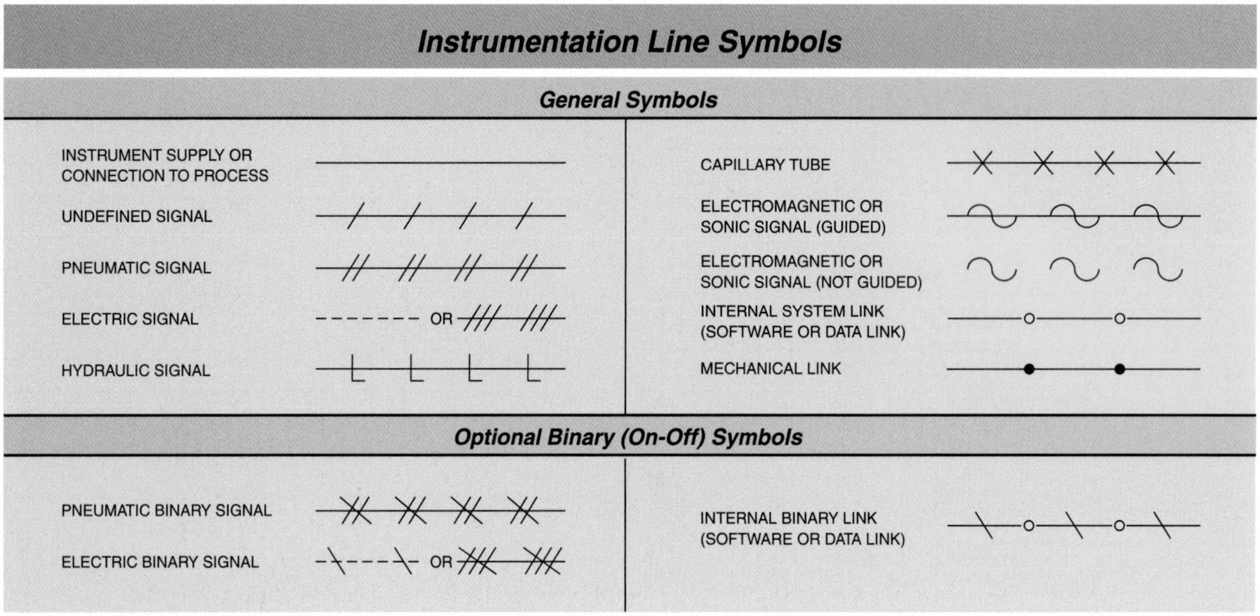

Instrumentation Line Symbols

General Symbols

INSTRUMENT SUPPLY OR CONNECTION TO PROCESS	CAPILLARY TUBE
UNDEFINED SIGNAL	ELECTROMAGNETIC OR SONIC SIGNAL (GUIDED)
PNEUMATIC SIGNAL	ELECTROMAGNETIC OR SONIC SIGNAL (NOT GUIDED)
ELECTRIC SIGNAL — OR	INTERNAL SYSTEM LINK (SOFTWARE OR DATA LINK)
HYDRAULIC SIGNAL	MECHANICAL LINK

Optional Binary (On-Off) Symbols

PNEUMATIC BINARY SIGNAL	INTERNAL BINARY LINK (SOFTWARE OR DATA LINK)
ELECTRIC BINARY SIGNAL — OR	

Periodic Table of the Elements

PERIODIC TABLE OF THE ELEMENTS

Key (example): 92 U Uranium 238.029 -21-9-2 — Atomic Number, Symbol, Name, Atomic Weight, Electron Configuration

	I	II											III	IV	V	VI	VII	0
	1 H Hydrogen 1.0079 1																	2 He Helium 4.00260 2
	3 Li Lithium 6.941 2-1	4 Be Beryllium 9.01218 2-2											5 B Boron 10.81 2-3	6 C Carbon 12.011 2-4	7 N Nitrogen 14.0067 2-5	8 O Oxygen 15.9994 2-6	9 F Fluorine 18.99840 2-7	10 Ne Neon 20.179 2-8
	11 Na Sodium 22.98977 2-8-1	12 Mg Magnesium 24.305 2-8-2											13 Al Aluminum 26.98154 2-8-3	14 Si Silicon 28.086 2-8-4	15 P Phosphorus 30.97376 2-8-5	16 S Sulfur 32.06 2-8-6	17 Cl Clorine 35.453 2-8-7	18 Ar Argon 39.948 2-8-8
2	19 K Potassium 39.098 -8-8-1	20 Ca Calcium 40.08 -8-8-2	21 Sc Scandium 44.9559 -8-9-2	22 Ti Titanium 47.90 -8-10-2	23 V Vanadium 50.9414 -8-11-2	24 Cr Chromium 51.996 -8-13-1	25 Mn Manganese 54.9380 -8-13-2	26 Fe Iron 55.847 -8-14-2	27 Co Cobalt 58.9332 -8-15-2	28 Ni Nickel 58.70 -8-16-2	29 Cu Copper 63.546 -8-18-1	30 Zn Zinc 65.38 -8-18-2	31 Ga Gallium 69.72 -8-18-3	32 Ge Germanium 72.59 -8-18-4	33 As Arsenic 74.9216 -8-18-5	34 Se Selenium 78.96 -8-18-6	35 Br Bromine 79.904 -8-18-7	36 Kr Krypton 83.80 -8-18-8
2 8	37 Rb Rubidium 85.4678 -18-8-1	38 Sr Strontium 87.62 -18-8-2	39 Y Yttrium 88.9059 -18-9-2	40 Zr Zirconium 91.22 -18-10-2	41 Nb Niobium 92.9064 -18-12-1	42 Mo Molybdenum 95.94 -18-13-1	43 Tc Technetium 97 -18-13-2	44 Ru Ruthenium 101.07 -18-15-1	45 Rh Rhodium 102.9055 -18-16-1	46 Pd Palladium 106.4 -18-18-0	47 Ag Silver 107.868 -18-18-1	48 Cd Cadmium 112.40 -18-18-2	49 In Indium 114.82 -18-18-3	50 Sn Tin 118.69 -18-18-4	51 Sb Antimony 121.75 -18-18-5	52 Te Tellurium 127.60 -18-18-6	53 I Iodine 126.9045 -18-18-7	54 Xe Xenon 131.30 -18-18-8
2 8 18	55 Cs Cesium 132.9054 -18-8-1	56 Ba Barium 137.34 -18-8-2	57–71	72 Hf Hafnium 178.49 -32-10-2	73 Ta Tantalum 180.9479 -32-11-2	74 W Tungsten 183.85 -32-12-2	75 Re Rhenium 186.207 -32-13-2	76 Os Osmium 190.2 -32-14-2	77 Ir Iridium 192.22 -32-15-2	78 Pt Platinum 195.09 -32-17-1	79 Au Gold 196.9665 -32-18-1	80 Hg Mercury 200.59 -32-18-2	81 Tl Thallium 204.37 -32-18-3	82 Pb Lead 207.2 -32-18-4	83 Bi Bismuth 208.9804 -32-18-5	84 Po Polonium 209 -32-18-6	85 At Astatine 210 -32-18-7	86 Rn Radon 222 -32-18-8
2 8 18 32	87 Fr Francium 223 -18-8-1	88 Ra Radium 226.0254 -18-8-2	89–103	104 Rf Rutherfordium 267 -32-10-2	105 Db Dubnium 268 -32-11-2	106 Sg Seaborgium 271 -32-12-2	107 Bh Bohrium 270 -32-13-2	108 Hs Hassium 277 -32-14-2	109 Mt Meitnerium 276 -32-15-2	110 Ds Darmstadtium 281 -32-17-1	111 Rg Roentgenium 280 -32-18-1	112 Cn Copernicium 285 -32-18-2	113 Uut (Temporary name) 284 -32-18-3	114 Uuq (Temporary name) 289 -32-18-4	115 Uup (Temporary name) 288 -32-18-5	116 Uuh (Temporary name) 293 -32-18-6	117 Uus (Temporary name)	118 Uuo (Temporary name) 294 -32-18-8

57 La Lanthanum 138.9055 -18-9-2	58 Ce Cerium 140.12 -19-9-2	59 Pr Praseodymium 140.9077 -21-8-2	60 Nd Neodymium 144.24 -22-8-2	61 Pm Promethium 145 -23-8-2	62 Sm Samarium 150.4 -24-8-2	63 Eu Europium 151.96 -25-8-2	64 Gd Gadolinium 157.25 -25-9-2	65 Tb Terbium 158.9254 -26-9-2	66 Dy Dysprosium 162.50 -28-8-2	67 Ho Holmium 164.9304 -29-8-2	68 Er Erbium 167.26 -30-8-2	69 Tm Thulium 168.9342 -31-8-2	70 Yb Ytterbium 173.04 -32-8-2	71 Lu Lutetium 174.97 -32-9-2
89 Ac Actinium 227 -18-9-2	90 Th Thorium 323.0381 -18-10-2	91 Pa Protactinium 231.0359 -20-9-2	92 U Uranium 238.029 -21-9-2	93 Np Neptunium 237.0482 -22-9-2	94 Pu Plutonium 244 -24-8-2	95 Am Americium 243 -25-8-2	96 Cm Curium 247 -25-9-2	97 Bk Berkelium 247 -27-8-2	98 Cf Californium 251 -28-8-2	99 Es Einsteinium 254 -29-8-2	100 Fm Fermium 257 -30-8-2	101 Md Mendelevium 258 -31-8-2	102 No Nobelium 255 -32-8-2	103 Lr Lawrencium 260 -32-9-2

Hazardous Locations

Hazardous Location – A location where there is an increased risk of fire or explosion due to the presence of flammable gases, vapors, liquids, combustible dusts, or easily-ignitable fibers or flyings.

Location – A position or site.

Flammable – Capable of being easily ignited and of burning quickly.

Gas – A fluid (such as air) that has no independent shape or volume but tends to expand indefinitely.

Vapor – A substance in the gaseous state as distinguished from the solid or liquid state.

Liquid – A fluid (such as water) that has no independent shape but has a definite volume. A liquid does not expand indefinitely and is only slightly compressible.

Combustible – Capable of burning.

Ignitable – Capable of being set on fire.

Fiber – A thread or piece of material.

Flyings – Small particles of material.

Dust – Fine particles of matter.

Classes	Likelihood that a flammable or combustible concentration is present
I	Sufficient quantities of flammable gases and vapors present in air to cause an explosion or ignite hazardous materials
II	Sufficient quantities of combustible dust are present in air to cause an explosion or ignite hazardous materials
III	Easily ignitable fibers or flyings are present in air, but not in a sufficient quantity to cause an explosion of ignite hazardous materials.

Divisions	Location containing hazardous substances
1	Hazardous location in which hazardous substance is normally present in air in sufficient quantities to cause an explosion or ignite hazardous materials
2	Hazardous location in which hazardous substance is not normally present in air in sufficient quantities to cause an explosion or ignite hazardous materials

Groups	Atmosphere containing flammable gases or vapors or combustible dust	
Class I	Class II	Class III
A B C D	E F G	none

DIVISION I EXAMPLES

Class I:

- Spray booth interiors
- Areas adjacent to spraying or painting operations using volatile flammable solvents
- Open tanks or vats of volatile flammable liquids
- Drying or evaporation rooms for flammable vents
- Areas where fats and oil extraction equipment using flammable solvents are operated
- Cleaning and dyeing plant rooms that use flammable liquids that do not contain adequate ventilation
- Refrigeration or freezer interiors that store flammable materials
- All other locations where sufficient ignitable quantities of flammable gases or vapors are likely to occur during routine operations

Class II:

- Grain and grain products
- Pulverized sugar and cocoa
- Dried egg and milk powders
- Pulverized spices
- Starch and pastes
- Potato and wood flour
- Oil meal from beans and seeds
- Dried hay
- Any other organic material that may produce combustible dusts during their use or handling

Class III:

- Portions of rayon, cotton, or other textile mills
- Manufacturing and processing plants for combustible fibers, cotton gins, and cotton seed mills
- Flax processing plants
- Clothing manufacturing plants
- Woodworking plants
- Other establishments involving similar hazardous processes or conditions

GLOSSARY

A

absolute humidity: The ratio of the mass of water vapor to the mass of dry air. Also known as humidity ratio.

absolute pressure: Pressure measured with a perfect vacuum as the zero point of the scale.

absolute viscosity: The resistance to flow of a fluid; has units of centipoise (cP).

absolute zero: The lowest temperature possible, where there is no molecular movement and the energy is at a minimum.

absolute zero pressure: A perfect vacuum.

accuracy: The degree to which an observed value matches the actual value of a measurement over a specified range.

actuation direction: The direction an object moves relative to a proximity sensor.

actuator: A device that provides the power and motion to manipulate the moving parts of a valve or damper used to control fluid flow through a final element.

adaptive gain control: A control strategy where a nonlinear gain can be applied to a nonlinear process.

admittance: The ability of a circuit to conduct alternating current; the reciprocal of impedance.

air-loaded pressure regulator: A regulator that uses air pressure instead of the force of a spring to oppose downstream pressure.

alarm function: The process of monitoring the state of a condition or a variable and comparing it to a preselected condition or value.

alarm system: A device that provides a visual and audible signal to the operator that something in the process is not normal.

allowable loop impedance: The resistance limit that is the summation of all of the receiver impedances and connecting wiring resistance in the current transmission loop.

alphanumeric code: A number code that uses a combination of letters, symbols, and decimal numbers to process information for computers and printers.

American Standard Code for Information Interchange (ASCII): A seven-bit or eight-bit alphanumeric code used to represent the basic alphabet plus numbers and special symbols.

analog data: A continuous range of values from a minimum to a maximum that can be related to a transmission signal range.

analog output controller: An electric controller that has 4 mA to 20 mA DC electric current as the output.

analog switching relay: A solid-state relay that has a continuous range of output voltages within the relay's rated range.

analysis: The process of measuring the physical, chemical, or electrical properties of chemical compounds so that the composition and quantities of the components can be determined.

analyzer: An instrument used to provide an analysis of a sample from a process.

analyzer calibration: The process of substituting known sample compositions for the normal process sample so that the analyzer can be adjusted to read the correct values.

analyzer sampling system: A system of piping, valves, and other equipment that is used to extract a sample from a process stream, condition it if necessary, and convey it to an analyzer.

aneroid barometer: An instrument consisting of a mechanical pressure sensor with a linkage to a pointer that is used to measure atmospheric pressure without the use of a liquid manometer.

annunciator: A solid-state or relay-based system for monitoring and alarming plant process operations.

antenna: A device that sends or receives radio signals through the air.

antenna gain: A measure of how well an antenna focuses transmission energy.

area: The number of unit squares equal to the surface of an object.

armored gauge glass: A gauge glass that uses flat glasses enclosed in metal bodies and covers to protect against breakage when used in boilers or high-pressure vessels, or for other safety requirements.

artificial intelligence (AI): A form of computer programming that simulates some of the thinking processes of the human brain.

AS Interface (ASI): A sensor bus network designed to connect digital sensors and actuators where low cost and simplicity are important.

atmospheric pressure: The pressure due to the weight of the atmosphere above the point where it is measured.

automatic control: The equipment and techniques used to automatically regulate a process to maintain a desired outcome.

automatic reset: The integral controller mode, named so because in older proportional-only controllers the operator had to manually reset the setpoint. Also known as reset action.

averaging pitot tube: A pitot tube consisting of a tube with several impact openings inserted through the wall of the pipe or duct and extending across the entire flow profile.

averaging thermocouple: An electrical thermometer consisting of a set of parallel-connected thermocouples that is commonly used to measure an average temperature of an object or area.

B

background temperature compensation: A process by which some IR thermometers allow the temperature of a reflected source to be measured or specified.

backpressure regulator: A regulator that maintains the pressure upstream of the regulator to a specified value.

BACnet: A communications protocol that uses standard connecting media and enables controllers to be interoperable.

balanced line: A signal transmission cable in which two conductors carry the signal and create a balanced circuit alternating from positive to negative.

balloon: A circular symbol used to identify the function and loop number of an instrument or device on a P&ID. Also known as a bubble.

ball valve: A throttling valve consisting of a straight-through valve body enclosing a ball with a hole through the center.

bandwidth: The range of frequencies from the minimum to the maximum value in an FMCW radar level sensor.

barometer: A manometer used to measure atmospheric pressure.

barometric pressure: A pressure reading made with a barometer.

baseband: A communication method where digital signals are sent without frequency shifting.

batch process: A type of processing operation in which multiple feeds are received and the resulting product is fed all at once to the next operation.

beam-breaking photometric sensor: A point level measuring instrument consisting of a light source and a detector that indicates a level of the contents of a vessel when the beam is broken.

bellows: A mechanical pressure sensor consisting of a one-piece, collapsible, seamless metallic unit with deep folds formed from thin-wall tubing with an enclosed spring to provide stability, or with an assembled unit of welded sections.

belt weighing system: A mass flow measuring system consisting of a specially constructed belt conveyor and a section that is supported by electronic weigh cells.

Bernoulli equation: An equation stating that the sum of the pressure heads of an enclosed flowing fluid is the same at any two locations.

beta ratio (d/D): The ratio between the diameter of the orifice plate (d) and the internal diameter of the pipe (D).

bias: A systematic error or offset introduced into a measurement system.

bimetallic element: A bimetallic strip that is usually wound into a spiral, helix, or coil and allows movement for a given change in temperature.

bimetallic thermometer: A thermal expansion thermometer that uses a strip consisting of two metal alloys with different coefficients of thermal expansion that are fused together and formed into a single strip and a pointer or indicating mechanism calibrated for temperature reading.

binary coded decimal (BCD) system: A numbering system that uses groups of four bits to represent a group of decimal numbers.

binary number: A number given in a base of 2.

bit: A binary digit consisting of a 0 or a 1.

blackbody: An ideal body that completely absorbs all radiant energy of any wavelength falling on it and reflects none of this energy from the surface.

blackbody calibrator: A device used to calibrate infrared thermometers.

blade switch: A flow switch consisting of a thin, flexible blade inserted into a pipeline.

blocking valve: A valve used at the differential measuring instrument to provide a convenient location to isolate the instrument from the impulse, equalizing, or venting lines and to provide a way to equalize the high- and low-pressure sides of the differential instrument.

blowback: *See* blowdown (def. 2).

blowdown: **1.** The process of discharging water and undesirable accumulated material. **2.** The amount of pressure drop below the pressure setpoint of a safety valve. Also known as blowback.

Bluetooth®: A wireless communications protocol designed to link instruments and devices that are in close proximity to each other, such as appliances in a home or devices on a plant floor.

Boyle's law: A gas law that states that the absolute pressure of a given quantity of gas varies inversely with its volume provided the temperature remains constant.

break: A place on a contact that opens or closes an electrical circuit.

bridging: A condition arising in a silo when material has built up over the feeder, blocking all flow out of the silo.

British thermal unit (Btu): The amount of energy necessary to change the temperature of 1 lb of water by 1°F from 59°F to 60°F.

bubble: *See* balloon.

bubbler: A level measuring instrument consisting of a tube extending to the bottom of a vessel; a pressure gauge, single-leg manometer, transmitter, or recorder; a purge flowmeter or sight feed bubble; and a pressure regulator.

buffered solution: A solution of an acid or a base and another chemical compound in water where one of the parts of the compound is more reactive with the acid or base than the other.

bulb: A cylinder larger in diameter than a capillary tube that contains the vast majority of the fluid in a thermometer.

bulk solid: A granular solid, such as gravel, sand, sugar, grain, cement, or other solid material that can be made to flow.

bumpless transfer: A controller function that eliminates any sudden change in output value when the controller is switched from automatic to manual mode or back again.

burner control system: A specially designed solid-state electronics package that provides the sequencing and safety logic for controlling industrial burners.

bus: A data transmission path in which digital signals are dropped off or picked up at each point where a device is attached to the wires.

bus network: A topology in which devices are connected to a common bus in a master-slave relationship.

butterfly valve: A valve with a disc that is rotated perpendicular to the valve body.

bypass meter: A combination of a rotameter with an orifice plate used to measure flow rates through large pipes.

bypass valve: *See* diverting valve.

byte: A group of eight bits.

C

cable and weight system: An intermittent full-range level measuring assembly consisting of a manual or remotely operated switch, a relay and a servomotor, a plumb bob for a weight, and a cable.

cage globe valve: A modified single-port globe valve with a removable cylindrical cage holding a seat ring in place.

calorie (cal): The amount of energy necessary to change the temperature of 1 g of water by 1°C from 14.5°C to 15.5°C.

CANbus: A device bus network based on the widely used CAN electronic chip technology that is used inside automobiles to control internal components such as brakes and other systems.

capacitance: **1.** The ability of an electrical device to store charge as the result of the separation of charge. **2.** The ability of a process to store material or energy.

capacitance pressure transducer: A diaphragm pressure sensor with a capacitor as the electrical element.

capacitance probe: A part of a level measuring instrument that consists of a metal rod inserted into a tank or vessel, with a high-frequency alternating voltage applied to it and a means to measure the current that flows between the rod and a second conductor.

capacitance proximity sensor: A proximity sensor that consists of a sensor plate, an oscillator, a trigger circuit, and an output switching circuit.

capacitor: An electrical device that stores electrical energy by means of an electrostatic field.

capillary tube: A small diameter tube with an interior hole 10 thousandths of an inch (0.010″) in diameter and designed to contain very little liquid.

capsule: A mechanical pressure sensor consisting of two convoluted metal diaphragms with their outer edges welded, brazed, or soldered together to provide an empty chamber between them.

cascade control: A control strategy where a primary controller, which controls the ultimate measurement, adjusts the setpoint of a secondary controller.

catalytic combustible sensor: A common type of combustible gas sensor that uses a pair of resistors encased in ceramic beads, with the sensor element coated with a catalyst and the reference element coated with impervious glass.

categorical scale: A measurement scale that uses unique identifiers such as numbers or labels to represent a variable. Also known as a nominal scale.

cavitation: A process in which vapor bubbles in a flowing liquid collapse inside a control valve as the pressure begins to increase.

centrifugal pump: A pump used to provide a constant discharge pressure at any flow rate for a given operating speed.

Charles' law: A gas law that states that the volume of a given quantity of gas varies directly with its absolute temperature provided the pressure remains constant.

choked flow: A condition where the actual pressure ratio is less than the critical pressure ratio and flow is restricted.

chromatograph analyzer: An instrument consisting of a small stainless steel tube packed with a porous inert material such as silica gel or alumina, an injection valve assembly, and a detector.

chromatographic column: A stainless steel tube or length of capillary tubing of a chromatography instrument after the tube is filled with packing and ready to use.

chromatography: The process of separating components of a sample transported by an inert carrier stream through a variety of media.

circulated water system: A system that provides both heating and cooling for batch reactors.

class: A term used to define the general nature of a hazardous gas or vapor in the surrounding atmosphere.

Class I location: A hazardous location in which sufficient quantities of flammable gases and vapors are present in the air to cause an explosion or ignite hazardous materials.

Class II location: A hazardous location in which a sufficient quantity of combustible dust is present in the air to cause an explosion or ignite hazardous materials.

Class III location: A hazardous location in which easily ignitable fibers, or flyings, are present in the air but not in a sufficient quantity to cause an explosion or ignite hazardous materials.

clear-tube rotameter: A rotameter consisting of a clear tube to allow visual observation of the flow rate.

closed loop: A control system that provides feedback to the controller on the state of the process variable due to changes made by the final control element.

close to alarm: An alarm that can be used for normal non-safety alarms.

coefficient of discharge: The ratio of the actual velocity to the theoretical velocity of flow through an orifice.

coefficient of linear expansion: The amount a unit length of a material lengthens or contracts with temperature changes.

coefficient of volumetric expansion: The amount a unit volume of a material expands or contracts with temperature changes.

cold junction: The end of a thermocouple used to provide a reference point. Also known as the reference junction.

cold junction compensation: The process of using automatic compensation to calculate temperatures when the reference junction is not at the ice point and is often achieved by measuring the temperature of the cold junction with a thermistor.

combustible gas detector: A hazardous atmosphere detector used to measure low concentrations of combustible gases and vapors in the atmosphere.

composite rupture disc: A dished thin metal sheet that is perforated to provide weak points and combined with a thin polymer sheet to prevent leakage.

composition analyzer: An analyzer used to measure the quantity of multiple components in a single sample from a process.

compressible fluid: A fluid where the volume and density change when subjected to a change in pressure.

condensate: The water formed when steam condenses.

conduction: Heat transfer that occurs when molecules in a material are heated and the heat is passed from molecule to molecule through the material.

conductivity analyzer: An electrochemical analyzer that measures the electrical conductivity of liquids and consists of two electrodes immersed in a solution.

conductivity cell constant: The ratio of the size of the actual electrodes to those of the standard conductivity cell.

conductivity probe: A point level measuring system consisting of a circuit of two or more probes or electrodes, or an electrode and the vessel wall where the material in the vessel completes the circuit as the level rises in the vessel.

configuration: The selection of preprogrammed software packages embedded in a controller representing available features that can be chosen.

configuration format: One of the various programming methods that have been developed to provide a simplified method for instructing the various digital control systems on how to control a process.

conservation vent: A pressure relieving device used to protect low-pressure vessels from damage caused by changes in pressure.

continuous hydrometer: A liquid density analyzer consisting of a float that is completely submerged in a liquid and has chains attached to the bottom to change the effective float weight to match changes in density of the liquid.

continuous level measurement: A method of tracking the change of level over a range of values.

continuous process: A type of processing operation in which continuous raw material is received and the resulting product is continually fed to the next operation.

control element: A device that compares a process measurement to a setpoint and changes the control variable (CV) to bring the process variable (PV) back to the setpoint. Also known as a controller.

controller: *See* control element.

controller gain: The gain, or sensitivity, of a controller itself.

controller tuning: The process of determining the tuning coefficients for the PID controller proportional gain, integral time, and derivative time to obtain a desired controller response to process disturbances.

control loop: A control system in which information is transferred from a primary element to the controller, from the controller to the final element, and from the final element to the process.

ControlNET: A real-time device network providing for high-speed transport of time-critical I/O data, messaging, peer-to-peer communications, and the uploading and downloading of programs.

control variable (CV): The independent variable in a process control system that is used to adjust the dependent variable, the process variable. Also known as a manipulated variable.

convection: Heat transfer by the movement of gas or liquid from one place to another caused by a pressure difference.

conventional metal rupture disc: A dished thin metal sheet, placed between a pair of holders, with the process pressure applied to the concave (hollow) side of the disc.

Coriolis meter: A mass flowmeter consisting of specially formed tubing that is oscillated at a right angle to the flowing mass of fluid.

critical angle: The one angle at which there is no refraction and all light is reflected.

critical pressure ratio: A ratio of downstream pressure to upstream pressure where the gas velocity out of a valve is at sonic velocity, and further decreases in downstream pressure no longer increase the flow.

current-to-pneumatic (I/P) transducer: A device that can convert the electronic controller output signal, usually 4 mA to 20 mA, into the standard pneumatic output, 3 psig to 15 psig (or similar metric range).

current transmission: An electric transmission system in which a transmitting unit regulates the current in a transmission loop.

cutting in the manometer: The procedure for operating the manometer three-way manifold valves to connect a manometer to a process without blowing out the manometer fluid.

cutting out the manometer: The procedure for operating the manometer three-way manifold valves to disconnect the manometer from the process.

D

damper: An adjustable blade or set of blades used to control the flow of air.

deadband: The range of values where a change in measurement value does not result in a change in controller output.

dead time: The period of time that occurs between the time a change is made to a process and the time the first response to that change is detected.

deadweight tester: A hydraulic pressure-calibrating device that includes a manually operated screw press, a weight platform supported by a piston, a set of weights, and a fitting to connect the tester to a gauge.

dead zone: A condition where there is no response to a change because the change is less than the sensitivity of an element.

decalibration: The process of unintentionally altering the physical makeup of a thermocouple wire so that it no longer conforms to the limits of the voltage-temperature curves.

decay ratio: A measure of how quickly an overshoot decays from one oscillation to the next as the controller brings the process to the setpoint.

decimal number: A number given in a base of 10.

density: Mass per unit volume.

derivative (D) control: A method of changing the output of a controller in proportion to the rate of change of the process variable.

dew cell: A fiberglass cylindrical core impregnated with lithium chloride and wrapped with a winding of gold wire around the core that serves as a heating element.

dewpoint: The temperature to which air must be cooled for the air to be saturated with water and any further cooling results in water condensing out.

diagnostic system: A logic-based system that monitors various operational conditions to detect faults.

diaphragm: A mechanical pressure sensor consisting of a thin, flexible disc that flexes in response to a change in pressure.

diaphragm-and-spring actuator: An actuator consisting of a large-diameter diaphragm chamber with a diaphragm backed by a plate attached to the actuator stem and opposed by the actuator spring.

diaphragm seal: A metal or elastomeric diaphragm clamped between two metal housings.

diaphragm valve: A throttling control valve consisting of a one-piece body incorporating an internal weir and a flexible diaphragm with a molded elastomer backing attached to the valve stem.

dielectric: The insulating material between the conductors of a capacitor.

dielectric constant: The ratio of the insulating ability of a material to the insulating ability of a vacuum.

difference thermocouple: A pair of thermocouples connected together to measure a temperature difference between two objects.

differential pressure: The difference in pressure between two measurement points in a process.

differential pressure regulator: A regulator that controls the pressure difference between the outlet pressure of a regulator and a fluid loading pressure supplied from an external source.

differential pressure switch: A flow switch consisting of a pair of pressure-sensing elements and an adjustable spring that can be set at a specific value to operate an output switch.

diffused mode photoelectric sensor: A photoelectric sensor that directs its source against a target object and detects a reflection from the target object.

digital alarm system: An alarm that provides the same functions as an annunciator except that the functions are accomplished by programming a distributed control system (DCS) or PLC.

digital communications: A method of sending and receiving binary information, in the form of low-level DC voltage pulses, between multiple devices using a common wiring format and procedure.

digital controller: A controller that uses microprocessor technology and special programming to perform the controller functions.

digital data: A series of discontinuous ON/OFF signals that is transmitted electrically.

digital numbering system: A method of coding information in terms of two-position ON/OFF voltage signals.

digital recorder: A device that uses local memory to record process data, which is then displayed on digital display screens that can have the appearance of the old-style strip chart recorders.

digital signal: A group of low-level DC voltage pulses that can be used to convey information.

dipole antenna: An antenna constructed of two straight quarter-length rods in line with each other where the two ends nearest each other are connected to the transmitter.

direct-acting actuator: An actuator that extends the shaft when air is applied to the diaphragm.

direct action: A form of control action where the controller output increases with an increase in the measurement of the process variable (PV).

direct computer control system: A control system that uses a computer as the controller.

direct sequence spread spectrum (DSSS): A transmission method that divides data into packets and sends the packets over a wide portion of a frequency band.

disappearing filament pyrometer: A high-temperature thermometer that has an electrically heated, calibrated tungsten filament contained within a telescope tube.

displacement hydrometer: A liquid density analyzer consisting of a fixed-volume and fixed-weight cylinder heavier than the fluid and supported by a spring or a lever connected to a torque tube that acts as a spring.

displacer: A liquid level measuring instrument consisting of a buoyant cylindrical object, heavier than the liquid, that is immersed in the liquid and connected to a spring or torsion device that measures the buoyancy of the cylinder.

distributed control system (DCS): A control system where the individual functions that make up a control system are distributed among a number of physical pieces of equipment that are connected by a high-speed digital communication network.

diverting valve: A three-way valve that has one inlet and two outlets. Also known as a bypass valve.

division: A classification assigned to each class based upon the likelihood of the presence of a hazardous substance in the atmosphere.

Division 1 location: A hazardous location in which a hazardous substance is normally present in the air in sufficient quantities to cause an explosion or ignite hazardous materials.

Division 2 location: A hazardous location in which the hazardous substance is not normally present in the air in sufficient quantities to cause an explosion or ignite hazardous materials.

Doppler ultrasonic meter: A flowmeter that transmits an ultrasonic pulse diagonally across a flow stream, which reflects off turbulence, bubbles, or suspended particles and is detected by a receiving crystal.

double-acting actuator: A piston actuator that has an air connection at each end of the piston and does not contain a spring to allow for a known failure position.

double block-and-bleed fuel shutoff: An arrangement with two fail-closed safety shutoff valves in series with a fail-open vent valve in a tee connection between the two.

double-ended piston: A mechanical pressure sensor consisting of a differential pressure gauge with a piston that admits pressurized fluid at each end.

double-port globe valve: A globe valve that consists of two plugs and seat rings through which a fluid flows.

drift: A gradual change in a variable over time when the process conditions are constant.

droop: A drop in pressure below a set value when there is high flow demand.

dry bulb temperature: The ambient air temperature measured by a thermometer that is freely exposed to the air but shielded from other heating or cooling effects.

dry leg: An impulse line that is filled with a noncondensing gas.

dry well calibrator: A temperature-controlled well or box where a thermometer can be inserted and the output compared to the known dry well temperature.

dust ignition-proof protection: A type of protection that excludes ignitable amounts of dust or amounts that can affect performance or rating.

dynamic error: The difference between a changing value and the momentary instrument reading or the controller action.

dynamic key: The component of a wireless system that encrypts the information being sent using a continuously changing key.

dynamic response time: The length of time required for a PV to remain within 5% of its ultimate value following a step input change.

E

eccentric cam valve: A throttling control valve with a rotating shaft with an attached convex disc, at a right angle to the shaft, that closes against a seat ring for tight shutoff and extended seat life.

elastic deformation element: A device consisting of metal, rubber, or plastic components such as diaphragms, capsules, springs, or bellows that flex, expand, or contract in proportion to the pressure applied within them or against them.

electric actuator: An actuator consisting of an electric motor that is connected through gearing to the valve stem or shaft.

electrical impedance moisture analyzer: An analyzer consisting of two conductive plates, separated by a nonconductive rib, that are pressed against the material to be measured.

electrical proximity sensor: A proximity sensor that uses inductance and capacitance properties to detect the presence of an object.

electrical transducer: A device that converts input energy into output electrical energy.

electrochemical oxygen analyzer: An analyzer that measures an electric current generated by the reaction of oxygen with an electrolytic reagent.

electromechanical relay: A device that controls one electrical circuit by opening and closing contacts in another circuit through the action of magnetic force.

emissivity: The ability of a body to emit radiation and is the ratio of the relative emissive power of any radiating surface to the emissive power of a blackbody radiator at the same temperature.

enclosure: A case or housing for equipment or other apparatus that provides protection for controllers, motor drives, or other devices.

endothermic reaction: A chemical reaction that consumes heat, and therefore more heat energy must be added to sustain the reaction.

energy harvesting: A strategy that a device uses for operation whereby it obtains power from its surrounding environment.

equal-percentage valve: A valve that allows the flow rate percentage to change by an amount equal to the change in the opening percentage.

error: The difference between a process variable and a setpoint.

Ethernet: A high-speed interference-immune local area network (LAN) covered by IEEE 802.3 standards.

excess flow valve: A device that shuts off the flow if the flow exceeds a specific flow rate.

exothermic reaction: A chemical reaction that generates heat during the reaction and increases the temperature.

expert system: A program that contains a database of known facts and a rule base that uses "if-then" statements for groups of facts to produce outputs.

explosionproof protection: A type of protection that uses an enclosure that is capable of withstanding an explosion of a gas or vapor within it and preventing the ignition of an explosive gas or vapor that may surround it.

EYS seal: A conduit coupling with an angled side connection closed with a screwed plug.

F

falling piston viscosity analyzer: A viscosity analyzer consisting of a precision tube and a piston with a timed fall through a constant-temperature liquid.

feedback: A control design used where a controller is connected to a process in an arrangement such that any change in the process is measured and used to adjust action by the controller.

feedforward control: A control strategy that only controls the inputs to a process without feedback from the output of the process.

fidelity: The ability of an element to follow a change in the value of an input.

fieldbus: An open, digital, serial, two-way communication network that connects with high-level information devices.

final element: A device that receives a control signal and regulates the amount of material or energy in a process.

firetube boiler: A boiler in which heat and gases of combustion pass through tubes surrounded by water.

first-out alarm: A feature that is available in annunciators and digital alarm systems that can identify the first of multiple closely tripped alarms.

flame arrester: A device used to prevent an external source of fire from propagating into a low-pressure vessel containing flammable vapor.

flame detector: A device that can provide a usable signal to a burner control system by detecting the presence of a flame.

flame front velocity: The speed of the vapor burning edge traveling into a combustible mixture.

flapper: A flat piece of metal installed at right angles to a nozzle tip that is held in position a few thousandths of an inch away from the nozzle tip by a linkage to a measurement device such as the diaphragm of a pressure gauge.

flashing: A process in which a portion of a liquid converts to a vapor as it passes through a control valve because the pressure has fallen below the vapor pressure of the liquid.

float: A point level measuring instrument consisting of a hollow ball that floats on top of a liquid in a tank and is attached to the instrument.

flowing condition: The pressure and temperature of the gas or vapor at the point of measurement.

flow nozzle: A primary flow element consisting of a restriction shaped like a curved funnel that allows a little more flow than an orifice plate and reduces the straight run pipe requirements.

flow rate: The quantity of fluid passing a point at a particular moment.

flow switch: A device used to monitor flowing streams and to provide a discrete electrical or pneumatic output action at a predetermined flow rate.

fluid: Any material that flows and takes the shape of its container.

force: Anything that changes or tends to change the state of rest or motion of a body.

forced convection: The movement of a gas or liquid due to a pressure difference caused by the mechanical action of a fan or pump.

Form C electrical contact: A single-pole double-throw switch that allows both NO and NC contacts.

FOUNDATION: A process bus network protocol that uses a 2-wire, digital, serial communications system to connect field equipment, such as intelligent sensors and actuators, with controllers, such as PLCs.

Fourier transform infrared (FTIR) analyzer: An infrared analyzer that uses a Michelson interferometer to examine a sample with a broad spectrum of infrared radiation.

four-way dual-coil solenoid valve: A valve connected to a solenoid actuator in the same way as a single-coil solenoid valve except that there is no spring in the solenoid.

four-way solenoid valve: A single- or dual-coil solenoid that has an air supply, a vent, and two cylinder ports.

4-wire fieldbus network: A fieldbus network that uses two wires for communication and two separate wires to power field instruments.

4-wire transmitter: A transmitter that uses two wires to provide power to the transmitter and two wires for the current output.

frequency hopping spread spectrum (FHSS): A transmission method that takes packets of information and then transmits the packets across a frequency band by selecting pseudorandom frequencies within the band.

frequency modulated continuous wave (FMCW) radar: A level measuring sensor consisting of an oscillator that emits a continuous microwave signal that repeatedly varies its frequency between a minimum and maximum value, a receiver that detects the signal, and electronics that measure the frequency difference between the signal and the echo.

frequency transmission: An electrical transmission method in which the frequency of a signal is proportional to the measured value.

frictional head: The pressure loss of a fluid due to flow through piping or ductwork; varies with the flow squared.

fuel safety shutoff valve: A special spring-actuated valve used to stop the fuel flow to a burner system.

function block: A stand-alone procedure designed to perform a specific task and can be linked with other function blocks to form complex control strategies.

function block configuration: A configuration method that uses a library of functions provided by the manufacturer.

funnel flow: The flow of a bulk solid where the material empties out of the bottom of a silo and the main material flow is down the center of the silo, with stagnant areas at the sides and bottom of the silo.

fuzzy logic: A system that uses mathematical or computational reasoning based on undefined, or fuzzy, sets of data derived from analog inputs to establish single or multiple defined outputs.

G

gain: A ratio of the change in output to the change in input of a process.

gain margin: The factor by which controller gain may be increased before instability occurs, and therefore is a measure of relative stability.

gap action control: A control strategy that operates only when a measurement value is outside the high and low limits of a predefined range, or gap.

gas analyzer: An instrument that measures the concentration of an individual gas in a gaseous sample.

gas-filled pressure-spring thermometer: A pressure-spring thermometer that measures the increase in pressure of a confined gas (kept at constant volume) due to a temperature increase.

gauge glass: A continuous level measuring instrument that consists of a glass tube connected above and below the liquid level in a tank and that allows the liquid level to be observed visually.

gauge pressure: Pressure measured with atmospheric pressure as the zero point of the scale.

Gay-Lussac's law: A gas law that states that the absolute pressure of a given quantity of a gas varies directly with its absolute temperature provided the volume remains constant.

globe valve: A throttling valve where the fluid flow enters horizontally, makes a turn through the plug and seat, and then makes another turn to exit the valve.

graphite composite rupture disc: A rupture disc composed of graphite powder in a polymer cement and is used where there are corrosive conditions.

gravimetric moisture analysis: A method of measuring moisture levels in solids by determining the loss of weight from evaporation when the sample is dried.

graybody: A body that emits and reflects a portion of all wavelengths of radiation equally.

ground loop: Current flow from one grounded point to a second grounded point in the same powered loop due to differences in the actual ground potential.

ground plane antenna: A standard omnidirectional antenna with a horizontal rod of the same length attached to the bottom of the antenna.

group: A term used to classify an atmosphere containing particular flammable gases or vapors or combustible dust.

Group A: The term used for an atmosphere containing acetylene.

Group B: The term used for an atmosphere containing hydrogen, fuel, and combustible process gases containing 30% hydrogen by volume, or gases or vapors of equivalent hazard such as butadiene, ethylene oxide, propylene oxide, and acrolein.

Group C: The term used for an atmosphere containing ethyl ether, ethylene, or gases or vapors of equivalent hazard.

Group D: The term used for an atmosphere containing acetone, ammonia, benzene, butane, cyclopropane, ethanol, gasoline, hexane, methane, methanol, naphtha, natural gas, propane, or gases or vapors of equivalent hazard.

Group E: The term used for an atmosphere containing metal dusts such as aluminum or magnesium dust.

Group F: The term used for an atmosphere containing carbonaceous dusts such as carbon black, coal, or coke dust.

Group G: The term used for an atmosphere containing dust not included in E and F.

guided wave radar: A level measuring detector consisting of a cable or rod as the wave carrier extending from the emitter down to the bottom of the vessel and electronics to measure the transit time.

H

hardwired system: An SIS that uses discrete logic components interconnected with physical paths for logic signals and contains no programmable memory devices.

HART: A hybrid communications protocol consisting of newer two-way digital communication and the traditional 4 mA to 20 mA analog signal.

hazardous material safety relief valve: A special valve design that has the standard safety valve internal parts protected by an internal diaphragm seal.

head: The actual height of a column of liquid.

heat: A form of energy.

heat anticipator: A small electric heater that is part of a thermostat.

heat capacity: The amount of energy needed to change the temperature of a material by a certain amount.

heat exchanger: Any piece of equipment that transfers heat from one material to another.

heat sink: A heat conductor used to remove heat from sensitive electronic parts.

heat transfer: The movement of thermal energy from one place to another.

hexadecimal (hex) number: A number given in a base of 16.

high-pressure liquid chromatography (HPLC): A type of liquid chromatography that uses high pressure to force the liquid sample through a column at a faster rate than the liquid would normally travel.

hot junction: The joined end of the thermocouple that is exposed to the process where the temperature measurement is desired. Also known as the measuring junction.

huddling chamber: A recessed area in a safety valve disc that increases the surface area and the total force applied.

humidity: The amount of water vapor in a given volume of air or other gases.

humidity analyzer: An instrument that measures the amount of humidity in air.

humidity ratio: *See* absolute humidity.

hydraulic curve: A flow curve that takes into account the static and frictional heads of the process.

hydraulic pressure: The pressure of a confined hydraulic liquid that has been subjected to the action of a pump.

hydrometer: A liquid density analyzer with a sealed float consisting of a hollow, tubular glass cylinder with the upper portion much smaller in diameter, a scale on the small-diameter portion, and weights at the lower end to make it float upright, with the upper portion partially above the surface of the liquid.

hydrostatic pressure: The pressure due to the head of a liquid column.

hygrometer: A humidity analyzer that measures the physical or electrical changes that occur in various materials as they absorb or release moisture.

hygroscopic material: A material that readily absorbs and retains moisture with increases in humidity and releases moisture with decreases in humidity.

hysteresis: A property of physical systems that do not react immediately to the forces applied to them or do not return completely to their original state.

I

IEEE 802.11x: A set of wireless communications protocols for high-speed communication.

IEEE 802.1x: A security standard with a higher-level authentication security than PSK.

impulse line: The tubing or piping connection that connects the flowmeter taps to any of the differential pressure instruments.

inclined-tube manometer: A manometer with a reservoir serving as one end and the measuring column at an angle to the horizontal to reduce the vertical height.

incompressible fluid: A fluid where there is very little change in volume when subjected to a change in pressure.

inductance: The property of an electric circuit that opposes a changing current flow.

inductance pressure transducer: A diaphragm or bellows pressure sensor with electrical coils and a movable ferrite core as the electrical element.

inductance proximity sensor: A proximity sensor that consists of a sensor coil, an oscillator, a trigger circuit, and an output switching circuit.

inductive probe: A point level measuring instrument consisting of a sealed probe containing a coil, an electrical source that generates an alternating magnetic field, and circuitry to detect changes in inductance.

input/output (I/O) bus network: A communications system that allows distributed control systems (DCSs), PLCs, or other controllers to communicate with I/O devices and each other.

instrument loop: A control system in which one or more instruments are connected together to perform a task.

integral (I) control: A method of changing the output of a controller by an amount proportional to an error and the duration of that error.

integral gain (K_I): The integral time multiplied by the controller gain, if present.

integral time (T_I): The time it takes for a controller to change a control variable (CV) by 1% for a 1% change in the difference between the control variable and the setpoint (SP). Also known as reset time.

integrator: A calculating device that totalizes the amount of flow during a specified time period.

interferogram: The spectrum developed by an FTIR analyzer and is similar to the interference pattern developed by a hologram.

interoperability: The ability of devices produced by different manufacturers to communicate and share information.

interval scale: A measurement scale where there are defined intervals between the values on the scale but the zero value is arbitrary.

intrinsically safe instrument: An instrument that has restricted voltage levels and current flows designed into the instrument as well as a very limited number of electrical energy storage components, such as capacitors and transformers, which could generate an electrical spark.

intrinsically safe protection: A type of protection in which electrical equipment, under normal or abnormal conditions, is incapable of releasing sufficient electrical or thermal energy to cause the ignition of a specific hazardous atmospheric mixture in its most easily ignitable concentration.

intrinsically safe system: A system designed to limit electrical and thermal energy to a level that cannot ignite combustible atmospheres and to avoid some of the high installation costs associated with explosionproof installations.

intrinsic barrier: A specially designed electronic circuit containing resistors and diodes that is used to prevent any electrical ignition energy from being carried into a hazardous area along power, signal, or control wiring.

IR thermometer: A thermometer that measures the infrared radiation (IR) emitted by an object to determine its temperature.

isolated device: An electric component that provides protection against ground loops because it has no electrical connection from the current transmission loop signal to any other electrical circuits or grounds.

J

jacketed reactor: A vertical vessel used in batch processing whose sides and bottom are covered with a steel shell spaced about an inch from the outside of the reactor and used to contain either a cooling or heating fluid.

K

kinematic viscosity: The ratio of absolute viscosity to fluid density and has units of centistokes (cS).

knowledge system: An enhanced diagnostic system that not only identifies a fault but also makes decisions on the probable cause of the fault.

L

ladder logic: A configuration method that consists of two vertical rails, the left one being the source and the right one being the end, and the sequential rungs of logic between the two.

lag: A delay in the response of a process that represents the time it takes for a process to respond completely when there is a change in the inputs to the process.

laminar flow: Smooth fluid flow that has a flow profile that is parabolic in shape with no mixing between the streamlines.

laser level instrument: A level measuring instrument consisting of a laser beam generator, a timer, and a detector mounted at the top of a vessel.

law of intermediate metals: A law stating that the use of a third metal in a thermocouple circuit does not affect the voltage, as long as the temperature of the three metals at the point of junction is the same.

law of intermediate temperatures: A law stating that in a thermocouple circuit, if a voltage is developed between two temperatures T_1 and T_2, and another voltage is developed between temperatures T_2 and T_3, the thermocouple circuit generates a voltage that is the sum of those two voltages when operating between temperatures T_1 and T_3.

lead-lag air-fuel control system: A control system where increases in combustion-air flow lead increases in fuel flow and decreases in combustion-air flow lag decreases in fuel flow.

linearity: The closeness to which multiple measurements approximate a straight line on a graph.

linear process: A process where the gain at any value of the process variable (PV) is the same as the gain at any other value.

linear valve: A valve that allows flow to increase at the same rate as an opening.

linear-voltage differential transformer (LVDT): An inductance transducer consisting of two coils wound on a single nonconductive tube.

line scanner IR thermometer: A thermometer that uses a rotating mirror with a single, very fast response detector or a linear array of IR detectors to measure successive areas with a single stationary device.

liquid analyzer: An instrument that measures the properties of a liquid.

liquid density analyzer: An analyzer that measures the density of a liquid by measuring related variables such as buoyancy, the pressure developed by a column of liquid, or the natural frequency of a vibrating mass of liquid; or by nuclear radiation absorption.

liquid-filled pressure-spring thermometer: A pressure-spring thermometer that is filled with a liquid under pressure.

liquid-in-glass thermometer: A thermal expansion thermometer consisting of a sealed, narrow-bore glass tube with a bulb at the bottom filled with a liquid.

load cell: A device used to weigh large items and typically consists of either piston-cylinder devices that produce hydraulic output pressure or strain gauge assemblies that provide electrical output proportional to the applied load.

load change: A change in process operating conditions that changes the process variable (PV) and must be compensated for by a change in the control variable (CV).

loading pressure: The pressure above a main diaphragm.

local operating network (LON): A network of intelligent devices sharing information using LonTalk®.

lockup: An increase in pressure above a set value when there is low flow demand.

LonTalk®: The open protocol standard used in LonWorks control networks.

lower explosive limit (LEL): The lowest concentration of a combustible gas or vapor in air that can be ignited.

low-loss flow tube: A primary flow element consisting of an aerodynamic internal cross section with the low-pressure connection at the throat.

low water fuel cutoff: A boiler fitting that shuts the burner OFF in the event of a low water condition.

M

M12 connector: A 4-wire or 8-wire connector with threaded metal fittings designed for an industrial environment.

magmeter: *See* magnetic meter.

magnetically coupled level gauge: A gauge that consists of a stainless steel float containing a magnet riding in a stainless steel tube where the level indicator consists of horizontally pivoted magnetized vanes painted yellow or white on one side and black on the other in a housing bolted to the level tube.

magnetic meter: A flowmeter consisting of a stainless steel tube lined with nonconductive material, with two electrical coils mounted on the tube like a saddle. Also known as a magmeter.

magnetostrictive sensor: The part of a continuous level measuring system consisting of an electronics module, a waveguide, and a float containing a magnet that is free to move up and down a pipe that is inserted into a vessel from the top.

manipulated variable: *See* control variable (CV).

manometer: A device for measuring pressure with a liquid-filled tube.

manual reset: *See* output bias.

mass flow: The flow of a bulk solid where all material in a silo flows down toward the bottom at the same rate.

mass flowmeter: A flowmeter that measures the actual quantity of mass of a flowing fluid.

measuring element: A device that establishes a scaled value for the measured process variable.

measuring junction: *See* hot junction.

mechanical switch: A switch that requires physical contact with an object to actuate a switch mechanism.

mesh networking: The ability of devices to relay signals for other devices to reduce the radio power output needed to deliver a message over a long distance.

metal-tube tapered rotameter: A rotameter consisting of a tapered metal tube and a rod-guided float.

metering-cone meter: A flowmeter consisting of a straight tube and a tapered cone, instead of a tapered tube, with an indicator that moves up and down the cone with changes in flow.

microbath: A small tank containing a stirred liquid used to calibrate thermometers.

microwave: The band of electromagnetic radiation between infrared and VHF broadcast frequencies, covering the range of approximately 3 mm to 3 m wavelengths.

microwave moisture analyzer: A solids moisture analyzer consisting of a transmitter that directs a microwave beam onto the material whose moisture is to be measured and a receiver.

mixing ratio control: A control strategy used when there is a need to mix two flow streams in a specified ratio.

MODBUS®: A messaging structure with master-slave communication between intelligent devices and is independent of the physical interconnecting method.

N

natural convection: The unaided movement of a gas or liquid caused by a pressure difference due to a difference in density within the gas or liquid.

near infrared (NIR) liquid analyzer: An analyzer that uses infrared radiation to measure the organic molecules in a sample.

near infrared (NIR) moisture analyzer: A solids moisture analyzer that measures the reflectance of the process material and calculates moisture content.

negative gauge pressure: Gauge pressure that is less than atmospheric pressure.

network address: A unique number assigned to each device on a network.

network variable: A basic unit of shared control information that conforms to a certain data type.

neural network: A computer program that simulates the parallel inputs and linking of information of the human brain to arrive at an educated output determination.

Newtonian liquid: A liquid whose viscosity does not change with applied force.

nibble: A group of four bits and is the minimum number of bits needed to represent a single decimal number.

nominal scale: *See* categorical scale.

nondispersive infrared (NDIR) analyzer: A radiant-energy absorption analyzer consisting of an IR electromagnetic radiation source, an IR detector, and two IR absorption chambers.

non-graybody: A body that emits and reflects radiation to a varying degree depending on the wavelength of the infrared radiation.

nonincendive protection: A type of protection in which electrical equipment is incapable, under normal conditions, of causing the ignition of a specified flammable gas or vapor in an air mixture due to arcing or thermal effect.

nonlinearity: The degree to which multiple measurements do not approximate a straight line on a graph.

nonlinear process: A process where the gain changes at different points on a process curve.

non-Newtonian liquid: A liquid whose viscosity changes (usually decreases) when force is applied.

nonradiometric thermal imager: An imager in which a surface-temperature image is generated but the actual temperature at a specific position is unknown.

normally closed (NC) valve: A valve that does not allow pressurized fluid to flow out of the valve in the spring-actuated (de-energized) position.

normally open (NO) valve: A valve that allows pressurized fluid to flow out of the valve in the spring-actuated (de-energized) position.

nuclear level detector: A level measuring system consisting of a radioactive source that directs radiation through a vessel to a detector, such as a Geiger counter, on the other side.

nuclear radiation density analyzer: A liquid density analyzer consisting of a suitable radioactive isotope source producing gamma rays directed through a chamber containing the liquid to be measured.

nuclear solids moisture analyzer: An analyzer that measures the amount of moisture in solids by measuring the speed of neutrons that strike the object.

null-balance controller: An electric controller that generates an output signal to a final control element based on a measurement from a process and requires a proportional feedback signal representing the position of the final element.

null-current thermocouple: A circuit and a voltage generator that can be adjusted to exactly balance the voltage output of a thermocouple.

nutating disc meter: A positive-displacement flowmeter for liquids where the liquid flows through the chambers, causing a disk to rotate and wobble (nutate).

O

object: A collection of information that can be accessed over a network in a standardized way.

octal number: A number given in a base of 8.

offset: **1.** A steady-state error that is a permanent part of a system. **2.** A permanent difference in measurement between a process variable (PV) and a setpoint (SP) as a result of proportional control action.

omnidirectional antenna: An antenna constructed of a single wire or bar that radiates energy equally in all directions.

one-color IR thermometer: A thermometer that measures infrared radiation using one IR detector. Also known as a single-color IR thermometer.

one-quarter decay ratio: A response where the amount of overshoot decays to one-fourth of the previous amplitude of the overshoot every whole cycle after being upset by a disturbance.

on-line analyzer: An instrument that is located in the process area and obtains frequent or continuous samples from the process.

ON/OFF control: A method of changing the output of a controller that provides only an ON or OFF signal to the final element of the process. Also known as two-position control.

ON/OFF control valve: A valve used to start and stop the flow of materials in the pipelines that control chemicals and energy sources for automatic sequencing, batching, or safety operations.

opacity analyzer: A gas analyzer consisting of a collimated (focused beam) light source and an analyzer to measure the received light intensity.

open loop: A control system that sends a control signal to a final element but does not verify the results of that control.

open loop process response curve: A graph of the results of a step change in the manually adjusted output signal of an open loop controller that results in a change in the process measurement.

open loop response diagram: A curve that shows the controller response to a given measurement without the controller actually being connected to the process.

open to alarm: An alarm that provides a fail-safe alarm signal and should always be used with safety systems.

operating ratio: The ratio of process operating pressure to rupture disc burst pressure.

operator interface: A view into a digital control system through which an operator can observe and control a process.

opposed-blade damper: A damper in which adjacent blades are parallel and move in opposite directions from one another.

optical liquid-level sensor: A liquid point level measuring instrument where a light source and a light detector, shielded from each other, are mounted in a housing and the light source is directed against the inside of a glass or plastic cone-shaped prism.

ordinal scale: A measurement scale that establishes rank by "more or less" or "larger or smaller."

orifice plate: A primary flow element consisting of a thin circular metal plate with a sharp-edged round hole in it and a tab that protrudes from the flanges.

orthogonal frequency division multiplexing (OFDM): A spread-spectrum transmission method that divides data packets and then further divides them before they are transmitted across a series of frequencies.

oscillating-piston meter: A positive-displacement flowmeter for liquids in which the fluid fills one piston chamber while the other piston chamber is emptied.

output bias: A controller function that positions a final element in a central position when the process variable is at setpoint. Also known as manual reset.

overpressure: The amount of pressure above the setpoint of a safety valve necessary to develop the full relieving capacity and is expressed as a percentage of the set pressure.

overranging: Subjecting a mechanical sensor to excessive pressure beyond the design limits of the instrument.

overshoot: The change of a process variable (PV) that exceeds the upper deadband value when there is a disturbance to a system.

oxidation-reduction potential (ORP) analyzer: An electrochemical analyzer consisting of a metal measuring electrode and a standard reference electrode that measure the voltage produced by an electrochemical reaction between the metals of the electrodes and the chemicals in solution.

oxygen trim control: A control strategy used to adjust the air-fuel ratio of a burner system.

P

packet: A unit of data sent across a network that includes part of the message, the addresses of both the sender and receiver, and the packet's place in the entire message.

paddle wheel meter: A flowmeter consisting of a number of paddles mounted on a shaft fastened in a housing, which can be inserted into a straight section of pipe.

paddle wheel switch: A point level measuring device consisting of a drive motor and a rotating paddle wheel mounted inside a tank.

pallet: A flat plate that covers the vapor outlet or air inlet ports of a conservation vent.

parallel-blade damper: A damper in which adjacent blades are parallel and move in the same direction with one another.

parallel wiring: A method of communications wiring where multiple bits of information are transferred at the same time on multiple wires.

paramagnetic oxygen analyzer: An analyzer consisting of two diamagnetic spheres filled with nitrogen that are connected with a bar to form a "dumbbell" shaped assembly.

Parshall flume: A special form of open-channel flow element that has a horizontal configuration similar to a venturi tube, with converging inlet walls, a parallel throat, and diverging outlet walls.

Pascal's law: A law stating that the pressure applied to a confined static fluid is transmitted with equal intensity throughout the fluid.

PC-based display system: A display system that uses standard personal computer hardware and special software to display full graphics, control functions, alarms, trends, etc.

Peltier effect: A thermoelectric effect where heating and cooling occurs at the junctions of two dissimilar conductive materials when a current flows through the junctions.

percent solids: The volume of solids suspended in a slurry divided by the total volume of the slurry.

permissible exposure limit (PEL): A regulatory limit on the amount of allowed workplace exposure to a hazardous chemical.

personal protective equipment (PPE): Any clothing or device worn by a worker to prevent injury.

pH: The measurement of the acidity or alkalinity of a solution caused by the dissociation of chemical compounds in water.

pH analyzer: An electrochemical analyzer consisting of a cell that generates an electric potential when immersed in a sample.

photoelectric sensor: A proximity sensor that uses visible light and infrared radiation sources to detect target objects.

pick and choose: A configuration format where the configuration is selected from a list of available functions.

piezoelectric pressure transducer: A diaphragm pressure sensor combined with a crystalline material that is sensitive to mechanical stress in the form of pressure.

pilot-operated pressure regulator: A regulator that uses upstream fluid as a pressure source to power the diaphragm of a larger valve.

piping and instrumentation diagram (P&ID): A schematic diagram of the relationship between instruments, controllers, piping, and system equipment.

pitot tube: A flow element consisting of a small bent tube with a nozzle opening facing into the flow.

plug valve: A modified ball valve consisting of a tapered cylinder with a slot through it.

pneumatic pressure: The pressure of air or another gas.

pneumatic relay: A pneumatic amplifier used to take a signal from a flapper and nozzle and boost the signal to a standard range required for a process.

pneumatic rotary piston actuator: An actuator that can be attached directly to most rotary valves and is primarily designed for ON/OFF action.

pneumatic sliding stem piston actuator: An actuator where the movement of a piston replaces the action of the diaphragm and a balancing air pressure replaces the function of the spring.

point level measurement: A method of level measurement where the only concern is whether the amount of material is within the desired limits.

pole: A completely isolated circuit that a relay can switch.

polling: The process of comparing redundant safety signals to determine the proper response.

position: A location within a valve where the internal parts are placed to direct fluid through the valve.

positioner: A device used to ensure positive position of a valve or damper actuator.

positive-displacement flowmeter: A flowmeter that admits fluid into a chamber of known volume and then discharges it.

positive-displacement pump: A pump used to provide a constant flow rate at any discharge pressure for a given operating speed.

potential: A driving force that causes material or energy to move through a process.

precision: The closeness to which elements provide agreement among measured values.

pre-shared key: *See* PSK.

pressure: Force divided by the area over which that force is applied.

pressure drop: A pressure decrease that occurs due to friction or obstructions as an enclosed fluid flows from one point in a process to another.

pressure regulator: An adjustable valve that is designed to automatically control the pressure downstream of the regulator.

pressure-relief regulator: A regulator that limits the pressure upstream of the regulator.

pressure spring: A mechanical pressure sensor consisting of a hollow tube formed into a helical, spiral, or C shape.

pressure-spring thermometer: A thermal expansion thermometer consisting of a filled, hollow spring attached to a capillary tube and bulb where the fluid in the bulb expands or contracts with temperature changes.

pressure switch: A pressure-sensing device that provides a discrete output (contact make or break) when applied pressure reaches a preset level within the switch.

pressure transmitter: A pressure transducer with a power supply and a device that conditions and converts the transducer output into a standard analog or digital output.

primary element: A sensing device that detects the condition of the process variable.

primary flow element: A pipeline restriction that causes a pressure drop used to measure flow.

process control: A system that combines measuring materials and controlling instruments into an arrangement capable of automatic action.

process curve: A plot of the process variable (PV) against the control variable (CV). Also known as a process reaction curve.

process dynamics: The attributes of a process that describe how a process responds to load changes imposed upon it.

process reaction curve: *See* process curve.

process variable (PV): The dependent variable that is to be controlled in a control system.

Profibus: A process bus network capable of communicating information between a master controller (or host) and an intelligent slave process field device as well as from one host to another.

Profibus-DP: A 4-wire performance-optimized version of the Profibus network used primarily for factory automation applications.

Profibus-FMS: A version of Profibus used for communicating between the upper level, the cell level, and the field device level of the Profibus hierarchy.

Profibus-PA: A 2-wire version of the Profibus network used primarily for process automation.

programmable electronic system: A solid-state electronic device where the safety logic is programmed into the device and stored in electronically programmed read-only memory (EPROM) chips or random-access memory (RAM).

programmable logic controller (PLC): A control system with an architecture very similar to that of a DCS, with self-contained power supplies, distributed inputs and outputs, and a controller module, all connected on high-speed digital communication networks.

proportional (P) control: A method of changing the output of a controller by an amount proportional to an error.

proportional band: The range of input values that corresponds to a full range of output from a controller, stated as a percentage.

proportional gain: The gain, or sensitivity, of a proportional term only.

proportional-integral (PI) control: Proportional control combined with integral control.

proportional-integral-derivative (PID) control: Proportional control combined with both integral control and derivative control.

protocol: A set of rules that determines the format and transmission method of data transmission.

proximity sensor: A sensor that detects the presence of an object without requiring contact with the object.

pseudo density: The density of a slurry, determined by the total weight of a slurry including the solids, divided by the volume of the slurry.

PSK: A security standard with a lower-level authentication for small office/home office (SOHO) networks that do not require the higher-level security of 802.1x. Also known as pre-shared key.

psychrometer: A humidity analyzer that uses two thermometers with the bulb of one thermometer kept moist (wet bulb) and the other bulb kept dry (dry bulb).

psychrometric chart: A graph that graphically combines the properties of moist air at standard atmospheric pressure (14.7 psi).

pulsed radar level sensor: A level measuring sensor consisting of a radar generator that directs an intermittent pulse with a constant frequency toward the surface of the material in a vessel.

pulse transmission: An electric transmission method consisting of a rapid change in voltage from a low value to a high value and then back to the low value.

pump curve: A plot of the pump discharge pressure, or head, plotted against flow for various pump rotational speeds.

pump-down level control: A control arrangement used to control the transfer of collected steam condensate to a common condensate storage tank.

pump-up level control: A control arrangement used to start a pump when the level of liquid drops below a setpoint and stop the pump when the liquid level reaches the desired level.

purged enclosure: A nonhazardous enclosure that is pressurized and purged with air to allow it to be used in hazardous areas.

purge meter: A small metal or plastic rotameter with an adjustable valve at the inlet or outlet of the meter to control the flow rate of the purge fluid.

pyrometer: An instrument used to measure the temperature of an object that is hot enough to emit visible light.

Q

quick-opening valve: A valve that allows flow to increase quickly with a small initial opening.

R

radiant-energy absorption analyzer: A gas analyzer that uses the principle that different gases absorb different, very specific, wavelengths of electromagnetic radiation in the infrared (IR) or ultraviolet (UV) regions of the electromagnetic spectrum.

radiation: Heat transfer by electromagnetic waves emitted by a higher-temperature object and absorbed by a lower-temperature object.

radiometric thermal imager: An imager in which the temperature measurement at all positions in the image is known.

range: The boundary of the values that identify the minimum and maximum limits of an element.

rangeability: The ratio of the maximum flow to the minimum measurable flow at the desired measurement accuracy. Also known as turndown ratio.

ratholing: A condition arising in a silo when material in the center has flowed out the feeder at the bottom, leaving large areas of stagnant material on the sides.

ratio control: A control strategy used to control a secondary flow to the predetermined fraction, or flow ratio, of a primary flow.

ratio IR thermometer: *See* two-color IR thermometer.

ratio scale: A measurement scale where there are defined intervals between the values on the scale and the zero value corresponds to none, or zero, of the object.

recording psychrometer: A psychrometer consisting of wet and dry bulb thermometers connected to a recorder.

reference junction: *See* cold junction.

reflectance: The ability of a material to reflect light or radiant energy.

reflectivity: The ability of an object to reflect radiation.

reflex gauge glass: A flat gauge glass with a special vertical sawtooth surface that acts as a prism to improve readability.

reflux: The portion of a condensed overhead vapor pumped back to the top of a column.

reflux ratio: The reflux flow back into a column divided by the total overheads flow.

refractive index: The amount of bending of a light beam as it moves between fluids with different refractive index values.

refractive index analyzer: A liquid analyzer consisting of a light source directed into a prism that has a flat surface in contact with the liquid to be measured.

regulator: A self-operating control valve for pressure and temperature control.

relative humidity (rh): The ratio of the actual amount of water vapor in the air to the maximum amount of water vapor possible at the same temperature.

relief valve: A valve that opens in proportion to the pressure above a setpoint.

reluctance: The property of an electric circuit that opposes a magnetic flux.

reluctance pressure transducer: A diaphragm pressure sensor with a metal diaphragm mounted between two stainless steel blocks.

repeatability: The degree to which an element provides the same result with successive occurrences of the same condition.

representative sample: A sample from a process in which the composition of the sample is the same as in the process piping.

reproducibility: The closeness of agreement among repeated measured values when approached from both directions.

reset action: *See* automatic reset.

reset rate: The reciprocal of integral time.

reset time: *See* integral time (T_I).

reset windup: A condition where a controller continues to change its output, because of a deviation between the setpoint and the measurement, until the output reaches its limit.

resistance: An opposition to the potential that moves material or energy in or out of a process.

resistance bridge: A circuit used to precisely measure an unknown resistance and consists of the unknown electrical resistance, several known resistances, and a voltage meter.

resistance pressure transducer: A diaphragm pressure sensor with a strain gauge as the electrical output element.

resistance temperature detector (RTD): An electrical thermometer consisting of a high-precision resistor with resistance that varies with temperature, a voltage or current source, and a measuring circuit.

response time: The time it takes an element to respond to a change in the value of the measured variable or to produce a 100% change in the output signal due to a 100% change in the input signal.

retro-reflective mode photoelectric sensor: A photoelectric sensor that uses a focused beam directed across the path of a target object and reflected back to the sensor.

reverse-acting actuator: An actuator that retracts the shaft when air is applied to the diaphragm.

reverse action: A form of control action where the controller output decreases with an increase in the measurement of the process variable (PV).

reverse buckling rupture disc: A dished thin metal sheet placed between a pair of holders with the process pressure applied to the convex side and with the downstream holder having a set of pointed knife blades.

Reynolds number: The ratio between the inertial forces moving a fluid and viscous forces resisting that movement.

rheology: The science of the deformation and flow of matter.

rheometer: A viscosity analyzer consisting of a heated, constant-temperature cylinder where a polymer is melted and forced through an orifice by a piston moving at a constant rate.

ring network: A topology in which devices are connected in series.

rise time: The length of time required for a PV to cross the ultimate value after a step input change, such as a setpoint change.

risk: A measure of the probability and severity of adverse effects of a process failure.

RJ-45 connector: A snap-in connector that looks like a large phone plug and contains eight pins instead of four as in standard phone connectors.

rotameter: A variable-area flowmeter consisting of a tapered tube and a float with a fixed diameter.

rotameter switch: A flow switch consisting of a shaped float, a fixed orifice, and a magnet embedded in the float that trips a magnetic sensing switch outside the tube.

rotary shaft valve: A throttling control valve used to change the flow of materials by means of the movement of a rotating wafer, contoured disc, ball, or plug.

rotary speed sensing: The use of sensors to measure the speed of a rotating object.

rotating-impeller meter: A positive-displacement flowmeter where the fluid flows into chambers defined by the shape of the impellers.

rotating spindle viscosity analyzer: A viscosity analyzer consisting of a rotating spindle in a container of the sample liquid at a controlled temperature.

round-blade damper: A damper that has a circular blade designed to fit into round ductwork.

RS-232 communication: A serial communications system developed by the Electronics Industry Association and the Telecommunications Industry Association.

RS-422 communication: An improved version of RS-232 that permits data speeds up to 100 kBd and distances up to 4000 ft.

RS-423 communication: A single-ended open communications variant of RS-422 that uses two data lines.

RS-485 communication: An open communications system that uses a twin duplex coaxial wiring arrangement consisting of two unshielded twisted pairs with an overall shield, permitting the use of a multidrop architecture for the various devices.

rupture disc: A safety device that breaks to open a discharge device and is used to prevent damage in pipelines and pressure vessels due to excessive pressure.

rupture pin safety valve: A safety valve with no spring; it is held closed by a pin or a thin rod.

S

safety instrumented system (SIS): A system consisting of sensors, logic solvers, and final control elements that bring a process to a safe state when normal operating conditions are violated.

safety light curtain: A series of closely spaced light sources mounted on a rail and used as a safety device to shut down equipment if an operator reaches into a protected space.

safety PLC: A highly reliable PLC that includes fail-safe designs, built-in self-diagnostics, and a fault-tolerant architecture.

safety relief valve: A valve that is designed so that it can be set to act as either a safety valve or a relief valve.

safety system: A system that consists of an individual device or an assembly of devices that form a system designed to protect personnel from injury and production equipment from damage.

safety valve: A gas- or vapor-service valve that opens very quickly when the inlet pressure exceeds the spring setpoint pressure.

safety valve capacity: The amount of steam, in lb/hr, that a safety valve is capable of venting at the rated pressure.

sample transportation lag: The time that it takes for a sample to get from the process through the final analysis.

saturated air: A mixture of water and air where the relative humidity is 100%.

Saybolt universal viscometer: A laboratory apparatus consisting of a temperature-controlled vessel with an orifice in the bottom for measuring the kinematic viscosity of oils and other viscous liquids.

scan speed: The time that it takes to access all the I/O points and go through the configuration program.

scintillation counter: A device that detects and measures nuclear radiation as it strikes a sensitive material, known as a phosphor, producing tiny flashes of visible light.

seal pot: A surge tank that may be installed in a wet leg to prevent volume changes from forcing the fluid into the process as well as protecting the sensing element from high temperatures as with steam applications.

Seebeck effect: A thermoelectric effect where continuous current is generated in a circuit where the junctions of two dissimilar conductive materials are kept at different temperatures.

selector: A device that compares two or three input signals and passes the highest or the lowest signal to the output of the device.

self-tuning controller: A controller that has built-in algorithms or pattern recognition techniques that periodically test the process and make changes to the controller tuning settings while the process is operating.

sensing range: The distance from the end of a proximity sensor to where an object can be detected.

sensitivity: The smallest change of a value a primary element can detect, the smallest change in input that can cause a control element to change its output, or the smallest change a final element can produce.

sensor style: The physical housing into which a sensor element is placed.

sequential function chart: A type of configuration format consisting of a series of conditional statements, parallel paths, and action blocks, which begins with a Start command and ends with an existing scan.

series wiring: A method of communications wiring where bits of information are transferred one after another over a pair or pairs of wires.

setpoint (SP): The desired value at which a process should be controlled; used by a controller for comparison with the process variable (PV).

setpoint step change method: A manual tuning method that consists of making small changes in the setpoint and observing the responses.

setpoint tracking: The technique of storing a PV in the setpoint memory module while the controller is in manual.

shaped-float and orifice meter: A flowmeter consisting of an orifice as part of the float assembly that acts as a guide.

shunt impedance: An unintended circuit caused by damaged thermocouple insulation.

siemens: The modern unit of electrical conductivity and the reciprocal of resistance.

silicon-controlled rectifier (SCR): A solid-state power controller that provides proportional current to a heating element in response to an analog control signal.

single channel radio: A radio that transmits on a single fixed frequency.

single-color IR thermometer: *See* one-color IR thermometer.

single-port globe valve: A globe valve that consists of a single valve plug and seat ring through which a fluid flows.

sink circuit: A circuit in a transmitting device that takes the positive voltage generated by a receiving device and shunts it to ground, lowering the voltage at the receiver circuit.

sliding stem control valve: A throttling valve that has a stem (shaft) attached to a plug or disc at one end and an actuator at the other end.

sliding-vane meter: A positive-displacement flowmeter where the fluid fills a chamber formed by sliding vanes mounted on a common hub rotated by the fluid.

sling psychrometer: A psychrometer consisting of two glass thermometers attached to an assembly that permits the two thermometers to be rotated through the air.

slurry: A liquid that contains suspended solids that are heavier than the liquid.

slurry density analyzer: An analyzer that is used to measure the density of a slurry.

smart transmitter: A microprocessor-based signal transmitter that combines digital and analog signals so that it can handle multiple inputs and outputs, communicate and change configuration details, and signal alarms and error conditions.

solids moisture analyzer: An instrument used to measure the amount of moisture in a solid.

solid-state relay: A semiconductor switching device that uses a low-current DC input to switch an AC circuit.

sonic velocity: The speed of sound in a gas.

source circuit: A circuit in a transmitting device that provides a positive voltage signal that can be detected in a receiver circuit.

span: The difference between the highest and lowest numbers in a range.

specific gravity: The ratio of the density of a fluid to the density of a reference fluid.

specific heat: The ratio of the heat capacity of a liquid to the heat capacity of water at the same temperature.

specific humidity: The ratio of the mass of water vapor to the mass of dry air plus moisture.

spectral response: The range of infrared wavelengths measured by an IR thermometer.

split body valve: A valve that consists of a two-piece body, with the lower half of the body being the inlet and the upper half of the body being the outlet, and a single-port assembly sandwiched between them.

split range control: An application where two or more control valves operate using the same controller output signal range, with each valve using a portion of the controller output signal.

split range operation: A control configuration where a single control signal, 4 mA to 20 mA or 3 psig to 15 psig, can be directed to two throttling control valves or dampers equipped with electropneumatic positioners.

splitting ratio control: A control strategy used when there is a need to split a flow into two separate constant-ratio flows.

spread spectrum: A transmission method that uses multiple frequencies to transmit data.

spring-loaded pressure regulator: A regulator consisting of a throttling element such as a valve plug connected to a pressure-sensing diaphragm that is opposed by a spring and contained in a single housing.

spring-return actuator: A piston or diaphragm actuator with an internal spring to force the actuator shaft to one end of its travel.

stability: The ability of a measurement to exhibit only natural, random variation where there are no known identifiable external effects causing the variation.

stand-alone controller: A controller that has its power supplies, input signal processing, controller functions, output signals, and displays contained in the same case.

stand-alone digital controller: A general type of microprocessor-based controller with all required operating components enclosed in one housing.

stand-alone display system: A display system designed for use with smaller digital control systems and has the electronics and display packaged together in a panel-mounted enclosure.

standard condition: An accepted set of temperature and pressure conditions used as a basis for measurement.

star network: A topology in which every peripheral device is connected to a master controller with a pair of wires.

static head: The pressure of a fluid due to a change in the elevation of a discharge piping system and remains constant for all flow rates.

step change: A sudden change in an input variable in a process that is managed by a controller.

strain foil gauge: A strain gauge that has the wire grid impressed on nonmetallic foil and then the assembly is mechanically bonded to the metal diaphragm.

strain gauge: A transducer that measures the deformation, or strain, of a rigid body as a result of the force applied to the body.

structured text: A type of configuration that is very similar to Microsoft® Visual Basic® or older structured programming languages.

sweeptime: The constant time for an FMCW emitter to vary the frequency from the lowest frequency to the highest.

system monitoring: The process of providing notification to an operator that there has been a failure in the process controlled by an SIS.

T

tap: A pressure connection.

tape float: A continuous level measuring instrument consisting of a floating object connected by a chain, rope, or wire to a counterweight, which is the level pointer.

tare weight: The weight of the vessel, piping, and equipment that is supported by the load cell(s).

temperature: The degree or intensity of heat measured on a definite scale.

temperature code: A designation that specifies the maximum surface skin temperature obtained by an enclosure during testing by approval agencies.

temperature self-operating regulator: A regulator that is used to automatically control a process to a defined temperature.

thermal conductivity analyzer: A gas analyzer that measures the concentration of a single gas in a sample by comparing its ability to conduct heat to that of a reference gas.

thermal dispersion sensor: A point level measuring instrument consisting of two probes that extend from the detector into the vessel, with one of the probe tips being heated.

thermal equilibrium: The state where objects are at the same temperature and there is no heat transfer between them.

thermal imager: An infrared device that uses a two-dimensional array of IR detectors to generate an image showing the temperature of an object.

thermal mass meter: A mass flowmeter consisting of two RTD temperature probes and a heating element that measure the heat loss to the fluid mass.

thermal switch: A flow switch consisting of a heated temperature sensor.

thermistor: A temperature-sensitive resistor consisting of solid-state semiconductors made from sintered metal oxides and lead wires, hermetically sealed in glass.

thermocouple: An electrical thermometer consisting of two dissimilar metal wires joined at one end and a voltmeter to measure the voltage at the other end of the two wires.

thermocouple aging: The process by which thermocouples gradually change their voltage-temperature curve due to extended time in extreme environments.

thermocouple break protection: A circuit where an electronic device sends a low-level current across the thermocouple.

thermocouple junction: The point where the two dissimilar wires are joined.

thermocouple pyrometer: An electrical thermometer consisting of a plain electrical meter with a measurement range of 20 mV to 50 mV, a thermocouple, and a balancing resistor.

thermohygrometer: A combination of a hygrometer, pressure sensor, and temperature-sensing instrument with digital processing to calculate relative humidity, absolute humidity, dry bulb temperature, dewpoint reading, and other properties for local display or transmission.

thermometer: An instrument that is used to indicate temperature.

thermoparamagnetic oxygen analyzer: An analyzer consisting of a sensing head that uses magnetic fields to generate a "magnetic wind" that carries the oxygen-containing gas sample through a sample cell and across a pair of thermistors.

thermopile: An electrical thermometer consisting of several thermocouples connected in series to provide a higher voltage output.

thermowell: A closed tube used to protect a temperature instrument from process conditions and to allow instrument maintenance to be performed without draining the process fluid.

thin-film strain gauge: A strain gauge that has the wire grid sputter-deposited on the diaphragm surface.

Thomson effect: A thermoelectric effect where heat is generated or absorbed when an electric current passes through a conductor in which there is a temperature gradient.

three-way globe valve: A globe valve that consists of three pipe connections and is used for mixing, blending, or flow division or diversion applications.

three-way solenoid valve: A solenoid that shuts off the air supply and vents air from the actuator or cylinder when the solenoid is de-energized.

3-wire transmitter: A transmitter that uses a DC power supply, with the negative lead of the power supply and the negative lead for the transmitter being the same wire that is used to establish a zero level for the circuit.

throttling control valve: A valve and actuator assembly that is able to modulate fluid flow at any position between fully open and fully closed in response to signals from a controller.

throttling range: The number of units of the process variable that causes the actuator to move through its entire range.

through-beam mode photoelectric sensor: A photoelectric sensor that uses a beam aimed directly at a target object with a separate receiver to sense the beam.

throw: The number of closed contact positions per pole in a relay.

thyristor: A solid-state switching device that switches current ON by a quick pulse of control current to its gate.

time constant (τ): The time required for a process to change by 63.2% of its total change when an input to the process is changed.

time proportional ON/OFF controller: An ON/OFF controller that has a predetermined output period during which the output contact is held closed (or power is ON) for a variable portion of the output period.

titration analyzer: A liquid analyzer consisting of an instrument where a measured quantity of a process sample is mixed with precise quantities of reagents and then quantitatively measured with a pH meter or by a colorimetric or other type of detector.

titration curve: A graph that shows the quantities of reagent required to change the pH of a solution.

tone transmission: An electrical transmission method consisting of a pure audible tone where the duration of the tone is proportional to the measurement value.

total flow: The quantity of fluid that passes a point during a specific time interval.

touch screen: A display system designed so that touching specific spots on the screen produces an action.

toxic gas detector: A hazardous atmosphere detector used to measure toxic gases or vapors that are not combustible but are harmful to people.

transducer: A device that converts one form of energy to another, such as converting pressure to voltage.

transistor: A solid-state switching device that is used to switch very-low-current DC loads.

transistor-transistor logic (TTL): The connection method that allows separate solid-state devices to be wired together to pass information.

transit time: The time it takes for a transmitted ultrasonic signal to travel from the ultrasonic level transmitter to the surface of the material to be measured and back to the receiver.

transit time ultrasonic meter: A flowmeter consisting of two sets of transmitting and receiving crystals, one set aimed diagonally upstream and the other aimed diagonally downstream.

transmission: A standardized method of conveying information from one device to another.

transmission signal: The data sent from one device to another by a specific method.

transmissivity: The ability of objects to allow radiation to pass through.

transmitter dynamic gain: The amount of output change from a transmitter for a specific input change.

triac: A thyristor that is triggered into conduction in either direction by a small current to its gate.

triple point: The condition where all three phases of a substance—gas, liquid, and solid—can coexist in equilibrium.

try cock: A valve located on a water column used to determine the boiler water level if the gauge glass is not functional.

tuning fork level detector: A point level measuring instrument consisting of a vibrating tuning fork that resonates at a particular sound frequency and the circuitry to measure that frequency.

tuning map method: A procedure for controller tuning that compares process curves to one of numerous typical closed loop response curves.

turbidity analyzer: A liquid analyzer that measures the amount of suspended solids in a liquid by the measurement of light scattering from the suspended particles.

turbine meter: A flowmeter consisting of turbine blades mounted on a wheel that measures the velocity of a liquid stream by counting the pulses produced by the blades as they pass an electromagnetic pickup.

turbulent flow: Fluid flow in which the flow profile is a flattened parabola, the streamlines are not present, and the fluid is freely intermixing.

turndown ratio: *See* rangeability.

two-color IR thermometer: A thermometer that has two IR detectors that measure the infrared radiation of two different wavelengths. Also known as a ratio IR thermometer.

two-position control: *See* ON/OFF control.

2-wire fieldbus network: A fieldbus network that uses communication wires to furnish power to field instruments.

2-wire transmitter: A transmitter that uses the 24 VDC power from a current transmission loop to power the transmitter.

U

ultimate gain (K_u): The closed loop proportional gain at the point of oscillation.

ultimate period (T_u): The closed loop cycle time at the point of oscillation.

ultrasonic flowmeter: A flowmeter that uses the principles of sound transmission in liquids to measure flow.

ultrasonic proximity sensor: A proximity sensor that uses a pulse of sound waves to detect the presence of an object.

ultrasonic sensor: A level measuring instrument that uses ultrasonic sounds to measure level.

ultraviolet (UV) analyzer: A radiant-energy absorption analyzer consisting of a UV electromagnetic radiation source, a sample cell, and a detector that measures the absorption of UV radiation by specific molecules.

unbuffered solution: A solution of a strong acid or strong base without any other chemicals that react with the acid or base.

undershoot: The change of a process variable (PV) that goes below the lower deadband value when there is a disturbance to the system.

unidirectional antenna: A dipole antenna with additional elements placed in front of it at specific distances to focus transmission energy in one direction. Also known as a Yagi antenna.

unshielded twisted pair (UTP) wire: A pair of wires that are twisted around each other with no electromagnetic shielding.

upper explosive limit (UEL): The highest concentration of a combustible gas or vapor in air that can be ignited.

U-tube manometer: A clear tube bent into the shape of an elongated letter U.

V

vacuum pressure: Pressure less than atmospheric pressure measured with atmospheric pressure as the zero point of the scale.

valve body: A casting or forging with an enclosed port and integral threaded or flanged inlet and outlet openings.

valve bonnet: A packing enclosure that is bolted or threaded to the top of the valve body.

valve flow characteristic: The relationship between valve flow capacity (in percent) and control valve open travel (in percent), with all other factors that affect flow held constant.

valve plug: A machined disc or shaped piece that regulates the flow of a material by changing the size of the valve opening.

valve stem: A valve component that consists of a metal shaft that transmits the force of the actuator to the valve plug.

vapor-pressure pressure-spring thermometer: A pressure-spring thermometer that uses the change in vapor pressure due to temperature change of an organic liquid to determine the temperature.

variable: A value measured by an instrument.

variable-area flowmeter: A meter that maintains a constant differential pressure and allows the flow area to change with flow rate.

variable-speed drive: A device that varies the speed of an electric motor.

vena contracta: The point of lowest pressure and the highest velocity downstream from a primary flow element.

venturi tube: A primary flow element consisting of a fabricated pipe section with a converging inlet section, a straight throat, and a diverging outlet section.

vibrating U-tube (Coriolis) density analyzer: A liquid density analyzer consisting of a U-tube that is fixed at the open ends and filled with liquid.

viscosity: Resistance to flow.

viscosity analyzer: A liquid analyzer that measures a liquid's resistance to flow at specific conditions.

voltage transmission: An electric transmission method in which a transmitting unit regulates the voltage in a transmission loop.

vortex shedding meter: An electrical flowmeter consisting of a pipe section with a symmetrical vertical bluff body (a partial dam) across the flowing stream.

W

water column: A boiler fitting that reduces the turbulence of boiler water to provide an accurate water level in the gauge glass.

watertube boiler: A boiler in which water passes through tubes surrounded by gases of combustion.

wavelength: The peak-to-peak length of an electromagnetic wave.

way: A path through a valve from an inlet port to an outlet port.

web: A continuous length of material that is fed through a process.

weir: An open-channel flow measurement device consisting of a flat plate that has a notch cut into the top edge and is placed vertically in a flow channel.

well drop: The ratio of the area of a well-type manometer tube to the area of a well.

well-type manometer: A manometer with a vertical glass tube connected to a metal well, with the measuring liquid in the well at the same level as the zero point on the tube scale.

WEP: A security standard used for wireless Ethernet networks. Also known as wired equivalency privacy.

wet bulb temperature: The lowest temperature that can be obtained through the cooling effect of water evaporating into the atmosphere.

wet leg: An impulse line filled with fluid that is compatible with the pressure-measuring device.

Wi-Fi protected access: *See* WPA.

wired equivalency privacy: *See* WEP.

wireless transmission: The method of using radio frequencies to transfer information from one device to another.

word: A group of bits handled together by a computer system.

WPA: A security standard based on the same type of encryption that is used in WEP with advancements such as authentication and dynamic keys. Also known as Wi-Fi protected access.

WPA2: A security standard that uses advanced encryption standard (AES) encryption.

Y

Yagi antenna: *See* unidirectional antenna.

Z

zero power sensing: A thermistor circuit with the current kept low enough that power is dissipated without causing the thermistor to self-heat enough to cause erroneous readings.

zero switching relay: A solid-state relay where a load becomes energized when a control input voltage is applied and AC load voltage crosses zero.

Ziegler-Nichols closed loop tuning: A method of tuning a controller by increasing the gain until the system cycles at the point of instability.

Ziegler-Nichols open loop tuning: A method of tuning a controller based on open loop response to a step input.

ZigBee®: A wireless communications protocol designed for monitoring and control for distances up to 75 m with up to 250 nodes per network.

zirconium oxide oxygen analyzer: An analyzer that measures an electric current generated when the analyzer is subjected to different oxygen concentrations on opposite sides of an electrode.

INDEX

Page numbers in italic refer to figures.

A

B

E

F

G

H

I

J

K

L

M

N

Q

R

S

T

U

V

W

Y

Z